ARCTIC OCEAN
Svalbard
Franz Josef Land
Cape Zhelaniya
Kara Sea
Novaya Zemlya
Norwegian Sea
North Cape
Barents Sea
Lapland
Taymyr Peninsula
Laptev Sea
New Siberian Is.
East Siberian Sea
SIBERIA
Central Siberian Plateau
VERKHOYANSK RANGE
KOLYMA RANGE
Scandinavia
Lake Ladoga
URAL MTS.
West Siberian Plain
Ob R.
Yenisey R.
Lena R.
North Sea
Volga R.
Northern European Plain
EURASIA
ASIA
Sea of Okhotsk
Kamchatka Peninsula
Sakhalin
Cape Lopatka
EUROPE
ALPS
Danube R.
Mt. Elbrus 18,510 ft (5,642 m)
Caspian Depression
Aral Sea
Lake Baikal
YABLONOVY RANGE
Amur R.
ALTAY MTS.
Kuril Is.
Kuril Tr.
Emperor Seamounts
Corsica
Balkan Peninsula
Black Sea
CAUCASUS MTS.
Qizilqum
Caspian Sea
Gobi
TIAN SHAN
Taklimakan Desert
Huang (Yellow) R.
Sea of Japan (East Sea)
Hokkaido
Japan Trench
Northwest Pacific Basin
Sardinia
Anatolia
ZAGROS MTS.
Garagum
HINDU KUSH
HIMALAYAS
Korea
Honshu
Shatsky Rise
Sicily
Mediterranean Sea
Tigris R.
Euphrates R.
Mt. Everest 29,035 ft (8,850 m)
East China Sea
Kyushu
Qattara Depression
Indus R.
Ganges R.
Chang Jiang (Yangtze R.)
Ryukyu Trench
PACIFIC OCEAN
SAHARA
Persian Gulf
Great Indian Desert
Tropic of Cancer
Ahaggar
Tibesti
Nile R.
Red Sea
Arabian Peninsula
WESTERN GHATS
Taiwan
Mid-Pacific Mountains
Air
SAHEL
Mekong R.
Hainan
Bay of Bengal
South China Sea
Mariana Is.
Kyushu-Palau Ridge
Niger R.
Lake Chad
Marra Mts.
Arabian Sea
Indochina Peninsula
Philippine Trench
Marshall Is.
AFRICA
Gulf of Aden
Cape Gwardafuy
Philippine Islands
Guam
Mariana Trench
Great Rift Valley
Somali Pen.
Cape Comorin
Sri Lanka (Ceylon)
Palau
MICRONESIA
Ethiopian Highlands
Malay Peninsula
Caroline Islands
Central Pacific Basin
Bioko
Somali Basin
Maldive Is.
Gilbert Is.
Congo R.
Mid-Indian Basin
Equator
Congo Basin
MITUMBA MTS.
Lake Victoria
Borneo
Sulawesi (Celebes)
Tomé
Kilimanjaro 19,340 ft (5,895 m)
Seychelles
Mid-Indian Ridge
Ninetyeast Ridge
Sumatra
MELANESIA
New Guinea
Tuvalu
L. Tanganyika
INDIAN OCEAN
Mascarene Plateau
Java
INDONESIA
Timor
Solomon Is.
Java Trench
Katanga Plateau
Comoro Is.
Cape York
Angola Plain
Lake Nyasa
Cocos Is.
Coral Sea
Vanuatu
Zambezi R.
Fiji Is.
Madagascar
GREAT DIVIDING RANGE
New Hebrides Tr.
Namib Desert
Mauritius
Reunion
Great Sandy Desert
New Caledonia
Kalahari Desert
Mozambique Channel
Tropic of Capricorn
Western Plateau
AUSTRALIA
Walvis Ridge
Great Victoria Desert
Simpson Desert
Cape Plain
DRAKENSBERG
Madagascar Basin
Broken Ridge
Perth Basin
Mt. Kosciusko 7,310 ft (2,228 m)
Cape of Good Hope
Great Australian Bight
Tasman Sea
Agulhas Plateau
Southwest Indian Ridge
Tasman Plain
North Is.
Crozet Basin
Crozet Is.
South Australian Basin
South Is.
Agulhas Basin
Southeast Indian Ridge
Tasmania
Prince Edward Is.
Kerguelen Is.
Kerguelen Plateau
Campbell Plateau
Atlantic-Indian Ridge
Macquarie Ridge
South Indian Basin
Enderby Plain
Weddell Plain
SOUTHERN OCEAN
Antarctic Circle
ANTARCTICA

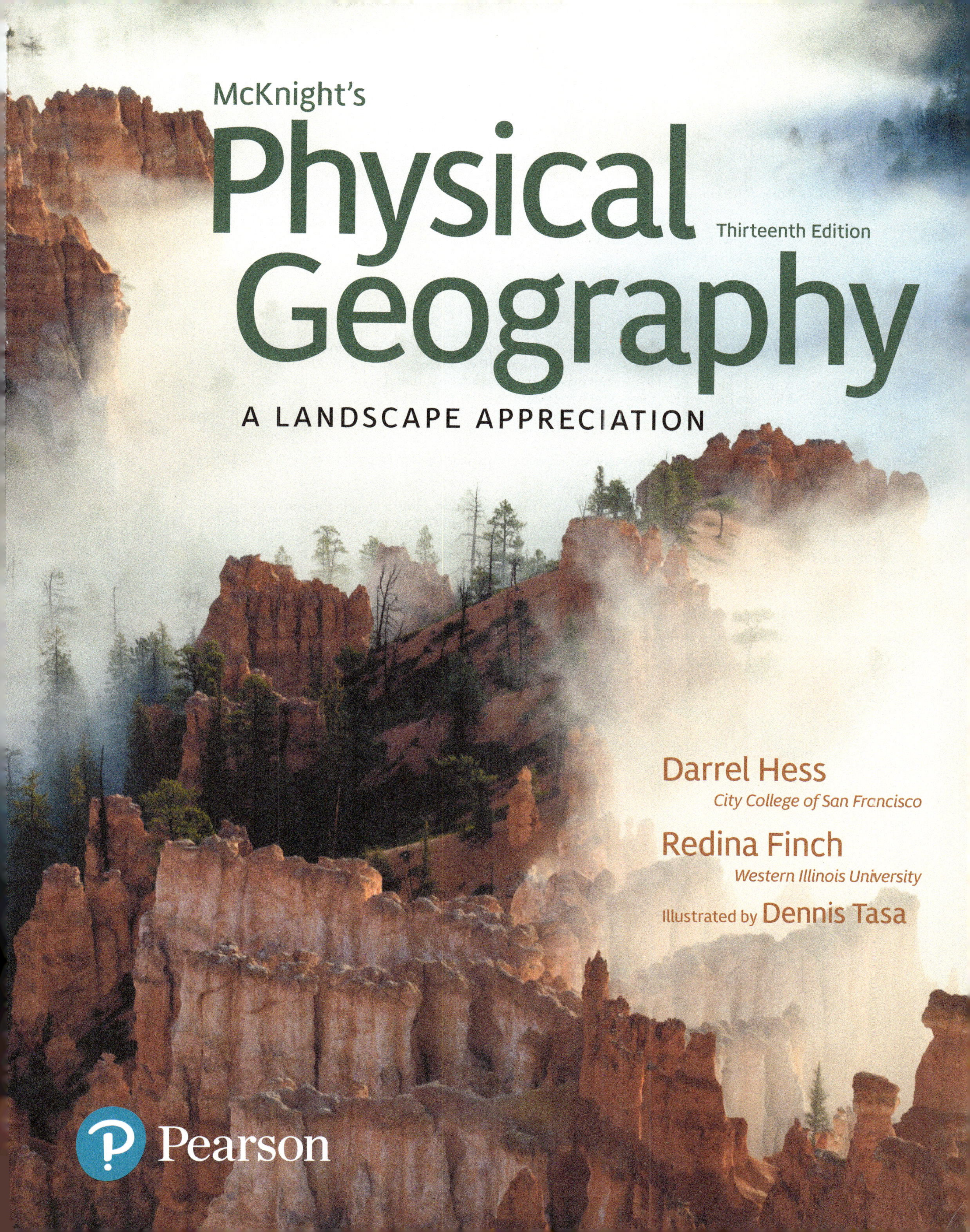

McKnight's

Physical Geography

Thirteenth Edition

A LANDSCAPE APPRECIATION

Darrel Hess
City College of San Francisco

Redina Finch
Western Illinois University

Illustrated by Dennis Tasa

Pearson

Content Development: Mary Hill, Dennis Tasa, Ginnie Simione Jutson, Matthew Walker, International Mapping
Content Management: Jeanne Zalesky, Terry Haugen, Chelsea Noack
Content Production: Michael Early, Titas Basu, Tod Regan, Ziki Dekel, Mireille Pfeffer, Katie Ostler, SPi Global
Product Management: Michael Gillespie, Aileen Pogran
Product Marketing: Candice Madden, Rosemary Morton
Rights and Permissions: Ben Ferrini, Jeamweal Bonilla, Beth Brenzel, SPi Global

Cover Image by Seth K. Hughes/Getty Images

4 2022

Library of Congress Cataloging-in-Publication Data

Library of Congress Control Number: 2020950587

Pearson

Rental
ISBN-10: 0-13-582714-0
ISBN-13: 978-0-135-82714-7

www.pearsonhighered.com

Brief Contents

Geoscience Animations

Covering the most difficult-to-visualize topics in physical geography, students can access the Geoscience Animations by clicking on the related Pearson links in the book or through the **Mastering Geography™** Study Area. Teachers can assign these media with assessments in **Mastering Geography™**.

1 Introduction to Earth
Solar System Formation
Earth-Sun Relations

2 Portraying Earth
Map Projections

3 Introduction to the Atmosphere
Coriolis Effect
Ozone Depletion

4 Insolation and Temperature
Atmospheric Energy Balance
Gulf Stream
Global Warming

5 Atmospheric Pressure and Wind
Development of Wind Patterns
Coriolis Effect
Cyclones and Anticyclones
Global Atmospheric Circulation
The Jet Stream and Rossby Waves
Seasonal Pressure and Precipitation Patterns
El Niño

6 Atmospheric Moisture
Hydrologic Cycle
Water Phase Changes
Adiabatic Processes and Atmospheric Stability
Seasonal Pressure and Precipitation Patterns

7 Atmospheric Disturbances
Cold Fronts
Warm Fronts
Midlatitude Cyclones
Hurricanes
Hurricane Hot Towers
Tornadoes

8 Climate and Climate Change
Seasonal Pressure and Precipitation Patterns
End of the Last Ice Age
Orbital Variations and Climate Change

9 The Hydrosphere
Hydrologic Cycle
The Carbonate Buffering System
Tides
Tidal Cycle
Ocean Circulation Patterns—Subtropical Gyres
Ocean Circulation Patterns—Global Conveyor-Belt Circulation
North Atlantic Deep Water Circulation
Arctic Sea Ice Decline
Groundwater Cone of Depression
The Water Table

10 Cycles and Patterns in the Biosphere
Net Primary Productivity
Biological Productivity in Midlatitude Oceans

13 Introduction to Landform Study
Metamorphic Rock Formation
Isostasy

14 The Internal Processes
Seafloor Spreading
Paleomagnetism
Convection and Plate Tectonics
Plate Boundaries
Divergent Boundaries
Breakup of Pangaea
Subduction Zones
Collision of India with Eurasia
Transform Faults and Boundaries
Hot Spot Volcano Tracks
Terrane Formation
Volcanoes
Formation of Crater Lake
The Eruption of Mount St. Helens
Igneous Features
Folding
Faulting
Seismic Waves
Seismographs

15 Weathering and Mass Wasting
Mechanical Weathering
Mass Wasting
The Eruption of Mount St. Helens

16 Fluvial Processes
Stream Sediment Movement
Oxbow Lake Formation
Floods and Natural Levee Formation
Stream Terrace Formation

18 The Topography of Arid Lands
Wind Transportation of Sediment
Desert Sand Dunes

19 Glacial Modification of Terrain
Arctic Sea Ice Minimum
End of the Last Ice Age
Isostasy
Flow of Ice within a Glacier
Glacial Processes
Orbital Variations and Climate Change

20 Coastal Processes and Terrain
Wave Motion
Wave Refraction
Tsunami
Tides
Coastal Sediment Transport
Movement of a Barrier Island
Coastal Stabilization Structures
Seamounts & Coral Reefs

Videos

Students can access videos providing engaging visualizations and real-world examples of physical geography concepts by clicking on the related Pearson links in the book or through the **Mastering Geography™** Study Area. Teachers can assign these media with assessments in **Mastering Geography™**.

1 Introduction to Earth

Mobile Field Trip: Introduction to Physical Geography

2 Portraying Earth

Studying Fires Using Multiple Satellite Sensors

3 Introduction to the Atmosphere

NPS: Science in the Park: Denali
Coriolis Effect Merry Go Round
Ozone Hole
NASA: Exploring Earth

4 Insolation and Temperature

Seasonal Radiation Patterns
Ocean Circulation Patterns—Subtropical Gyres
Seasonal Changes in Temperature
NPS: Outside Science: Isle Royale

5 Atmospheric Pressure and Wind

NPS: Point Reyes Wilderness
El Niño
El Niño Triggers Disease
La Niña
Mobile Field Trip: El Niño

6 Atmospheric Moisture

Hydrological Cycle
Mobile Field Trip: Clouds: Earth's Dynamic Atmosphere

7 Atmospheric Disturbances

NPS: Lightening Safety at Grand Canyon
NPS: Outside Science: Keeping an Eye on the Everglades
NPS: Climate Change: Everglades
2005 Hurricane Season
NASA: Hurricane Michael as Seen from Space

8 Climate and Climate Change

Mobile Field Trip: Changing Arctic
18,000 Years of Pine Pollen
Temperature and Agriculture

9 The Hydrosphere

Hydrological Cycle
NPS: Hidden Waters—Grand Canyon in Depth
Mobile Field Trip: Moving Water Across California
Mobile Field Trip: Mammoth Cave

10 Cycles and Patterns in the Biosphere

Global Carbon Uptake by Plants
Mobile Field Trip: Forest Fires

11 Terrestrial Flora and Fauna

Mobile Field Trip: Cloud Forest
Climate, Crops, and Bees
NPS: Managing Invasive Species
NPS: Outside Science: Hunting Lionfish in Biscayne National Park

12 Soils

Mobile Field Trip: The Critical Zone
NPS: Inside Canyonlands—Biological Soil Crust
Maps of Soil Moisture
California Drought

13 Introduction to Landform Study

Mobile Field Trip: Yosemite
Mobile Field Trip: Oil Sands
Black Smokers

14 The Internal Processes

Mobile Field Trip: San Andreas Fault
Mobile Field Trip: Kīlauea Volcano
Project Condor: Cinder Cones and Lava Flows
Project Condor: Monoclines
Project Condor: Anticlines and Synclines
Project Condor: Faulting versus Joints

15 Weathering and Mass Wasting

Project Condor: Joints and Faults
NPS: Geology of Arches
NPS: Rock Fall
Mobile Field Trip: Landslide!

16 Fluvial Processes

Mobile Field Trip: Streams of the Great Smoky Mountains
Project Condor: Meandering Rivers
Mobile Field Trip: Mississippi Delta
Project Condor: River Terraces and Base Level

17 Karst and Hydrothermal Processes

Mobile Field Trip: Mammoth Cave
NPS: Yellowstone: Thermofiles
NPS: Yellowstone: Geysers

18 The Topography of Arid Lands

NPS: Death Valley Explorer (E1)
Project Condor: Alluvial Fans
Mobile Field Trip: Desert Geomorphology
NPS: Death Valley Explorer (E2)

19 Glacial Modification of Terrain

Mobile Field Trip: The Glaciers of Alaska
NASA: 16 Years of Ice Loss
Mobile Field Trip: Climate Change in the Arctic
NASA: Operation IceBridge
NPS: Glacial Change at Kenai Fjords National Park
Mobile Field Trip: Yosemite
NPS: Experience Your Yosemite

20 Coastal Processes and Terrain

Mobile Field Trip: Gulf Coast
Summertime/Wintertime Beach Conditions
Movement of Sand in Beach Compartment
NPS: Padre Island National Seashore
Mobile Field Trip: Cape Cod: Sculpted by Ice & Storm

Contents

6 Atmospheric Moisture 143

7 Atmospheric Disturbances 179

8 Climate and Climate Change 211

9 The Hydrosphere 257

12 Soils 349

13 Introduction to Landform Study 379

19 Glacial Modification of Terrain 545

20 Coastal Processes and Terrain 575

Preface

McKnight's Physical Geography: A Landscape Appreciation presents the concepts of physical geography in a clear, readable way to help students comprehend Earth's physical landscape. The time-proven approach to physical geography first presented by Tom McKnight more than 35 years ago and carried through 12 editions, now has a new focus in the 13th edition—that of this country's National Parks. The parks, monuments, seashores, and recreational areas that comprise our National Park system serve as a source of inspiration for anyone interested in experiencing natural landscapes. They also provide the prefect introduction to topics that range from Earth's physical features and the processes that form them, to the opportunities they provide for scientific research, and offer a lens through which to view climate change

FEATURES OF 13TH EDITION

Users of earlier editions will see that the overall sequence of chapters and most topics remains the same, with material added and updated in several key areas. Changes to the new edition include the following:

- NEW As introduced in Chapter 1, each chapter includes at least one feature that explores a National Park. These include a ***Featured National Park*** that focuses on one particular aspect or ***Snapshots From Our National Parks*** that presents a series of images that relate to a particular topic.
- NEW In addition to the images incorporated into the National Park features, ***Our National Parks*** banners highlight more than 90 photographs that appear as numbered figures in the text, each accompanied by a locator map of the park. Also, highlighted with ***Science in the Field*** banners, photos highlight scientists and researchers at work, many in National Parks.
- NEW Each chapter now opens with the image of a National Park that is the subject of ***Seeing Geographically in Our National Parks***. This feature tests students ability to analyze and interpret what they see before they begin the chapter and then, at the end of the chapter, asks them to reassess their initial assumptions given what they have learned.

All the features originally introduced in the 12th edition, including many authored by outside contributors, have been retained and updated. These include the following:

- ***Global Environmental Change*** features written by expert contributors that present brief case studies on natural and human-caused environmental change, exploring important contemporary events and implications for the future. Topics include *Aerosol Plumes Circling the Globe* and *Changes in the South Asian Monsoon*.
- ***Energy for the 21st Century*** features cover *Transitioning from Fossil Fuels; Solar Energy; Wind Power; Strategies for Reducing Greenhouse Gas Emissions; Biofuels; Unconventional Hydrocarbons and the Fracking Revolution; Hydropower; Geothermal Energy;* and *Tidal Power*.
- ***Focus*** features focus on science and technology. Topics include *Multiyear Atmospheric and Oceanic Cycles; Soil Differences—They're All About Scale; Measuring Earth's Surface Temperature by Satellite; GOES Weather Satellites; Conveyor Belt Model of Midlatitude Cyclones; Weather Radar; and Earthquake Prediction.*
- ***People & the Environment*** features include *The Oso Landslide*; *The Great Pacific Garbage Patch*; and *The Future of the Mississippi River Delta*.

The pedagogical features that were a hallmark of the 12th edition have been retained and updated. These include the following:

- *Mobile Field Trip Videos* have students accompany acclaimed photographer and pilot Michael Collier in the air and on the ground to explore iconic landscapes of North America and beyond. "All 22 videos are" available within *Mastering Geography*.
- The entire art program was created by the geoscience illustrator Dennis Tasa and includes more than 800 diagrams and maps, done in Dennis's distinctive, accessible style.
- Each chapter includes a refined learning path, beginning with a series of **Key Questions** to help students prioritize key issues and concepts.
- Throughout each chapter, **Learning Check** questions periodically confirm student understanding of the material.
- The end-of-chapter **Learning Review** includes *Key Terms and Concepts*, *Study Questions*, and *Exercises* plus a capstone activity called *Environmental Analysis* that sends students online to use a variety of interactive science resources and data sets to perform data analysis and critical thinking tasks.

The media assets included in the 13th edition include:

- NEW More than 20 *NPS Videos* produced by the National Park Service are included with many of the National Park features and linked by short Urls. Also available within Mastering Geography.
- *Project Condor Quadcopter Videos*, take students out into the field through narrated quadcopter footage, exploring the physical processes that have helped shape North American landscapes.
- The book is supported by Mastering Geography™, the most widely used and effective online homework, tutorial, and assessment system for the sciences. Assignable media and activities include Geoscience Animations, Videos, Mobile Field Trip Videos, Project Condor Quadcopter Videos, Encounter Physical Geography Google Earth™ Explorations, GIS-inspired MapMaster™

interactive maps, coaching activities on the toughest topics in physical geography, end-of-chapter questions and exercises, reading quizzes, and Test Bank questions.

Chapter updates for the 13th edition include the following:

- In Chapter 1, the introduction of the new national parks theme in a two-page feature *National Parks: Our Geographical Laboratories*.
- In Chapter 3, the chapter was reorganized, placing *Human-Caused Atmospheric Change* at the end of the chapter; data and satellite imagery has been updated.
- In Chapter 4, the material on the greenhouse effect has been updated to reflect current terminology; the figure illustrating adiabatic cooling and warming has been clarified; data and satellite imagery has been updated.
- In Chapter 6, the material on latent heat in the atmosphere has been clarified and now includes an illustration; data and satellite imagery has been updated.
- Chapter 7 includes clarifications in the Midlatitude Cyclones section and discussion and illustrations of some of the latest storms, including Hurricanes Harvey (2017), Michael (2018), and Dorian (2019).
- Chapter 8, Climate and Climate Change, has been thoroughly updated and revised with the latest data and applications.
- Chapter 9, the material on the cryosphere has been updated, along with other data and satellite.
- Chapter 10, image of the 2019 Camp Fire has been included.
- Chapter 11, the material and data on the tropical rainforest has been updated.
- Chapter 13, the Geologic Time Scale has been updated and a discussion of absolute and relative dating has been included in the NPS feature.
- Chapter 14, the new NPS feature includes a timeline of the ongoing eruptions of Kīlauea, including the eruption in 2018.
- Chapter 19, maps and data have been updated on glaciers and ice shelves.

TO THE STUDENT

Welcome to *McKnight's Physical Geography: A Landscape Appreciation*. Take a minute to skim through this book to see some of the features that will help you learn the material in your physical geography course:

- You'll notice that the book includes many diagrams, maps, and photographs. Physical geography is a visual discipline, so studying the figures and their captions is just as important as reading through the text itself.
- Many photographs have "locator maps" to help you learn the locations of the many places we mention in the book.
- A reference map of physical features of the world is found inside the front cover of the book, and a reference map of the countries of the world is found inside the back cover.
- *Science in the Field* photo features highlight the real-world people and professions in geography and science today.
- Each chapter begins with a quick overview of the material, as well as a series of questions—think about these questions as you study the material in that chapter.
- Look at the photograph from a National Park that begins each chapter. The *Seeing Geographically in Our National Parks* questions for this photograph will get you thinking about the material in the chapter and about the kinds of things that geographers can learn by looking at a landscape.
- As you read through each chapter, you'll come across short *Learning Check* questions. These quick questions are designed to check your understanding of key information in the text section you've just read. Answers to the Learning Check questions are found in the back of the book.
- Each chapter concludes with a *Learning Review*. Begin with the *Key Terms and Concepts* questions—these will check your understanding of basic factual information and key terms (which are printed in bold type throughout the text). Then, answer the *Study Questions*—these will confirm your understanding of major concepts presented in the chapter. Finally, you can try the *Exercises*—for these problems you'll interpret maps or diagrams and use basic math to reinforce your understanding of the material you've studied.
- *Environmental Analysis* activities at the end of each chapter will direct you to interactive science resources and data sets for broader data analysis and critical thinking.
- Finish the chapter by answering the *Seeing Geographically* questions at the end of the Learning Review. To answer these questions, you'll put to use things you've learned in the chapter. As you progress through the book, you begin to recognize how much more you can "see" in a landscape after studying physical geography.
- The alphabetical glossary at the end of the book provides definitions for all of the key terms.
- All chapters include URLs that direct you to *Mobile Field Trips*, *Project Condor* Quadcopter Videos, online animations, and other videos. The animations and videos help explain important concepts in physical geography and also provide real-world case studies of physical geography in action. The animations and videos can also be accessed through the Student Study Area in MasteringGeography, and can also be assigned for credit by teachers.

ACKNOWLEDGMENTS

I first want to formally welcome Redina Finch as coauthor of *McKnight's Physical Geography*. Her involvement with the textbook began as a major contributor to the previous edition. Now as coauthor, she took primary responsibility for the revisions of our chapters on atmospheric science—although her mark is found throughout the book. It is a much better textbook because of her contributions and thoughtful critiques.

Once again Dennis Tasa has done a wonderful job on the illustrations. We've now worked together on four editions, and he continues to impress me with his ability to take rough ideas and turn them into effective and beautiful illustrations.

I am delighted that we are again including the *Mobile Field Trip* videos developed by photographer, writer, pilot, and educator, Michael Collier—many of his videos fit in perfectly with this new edition's focus on our National Parks.

A special nod of appreciation goes to my previous editor, Christian Botting, who first developed the idea of a National Parks emphasis for this new edition of the textbook.

Finally, I offer my deepest gratitude to Development Editor Mary Hill. Her sound advice and critical eye for detail all through the writing and production process has improved this book enormously.

As with previous editions, this book was a collaborative effort incorporating contributions of many scholars who wrote short boxed essays, problem sets, and activities for the book. Many thanks to all who offered their ideas for improvement for this edition, including:

Mary Bates, *College of the Canyons*
John Conley, *Santa Ana College*
Matthew Connolly, *University of Central Arkansas*
Robert Earle, *Sacramento City College*
Tracy Edwards, *Frostburg State University*
Abdullah Khan, *Los Angeles Mission College*
Heidi Lannon, *Santa Fe College*
Jennifer Lewis, *Madison Area Technical College*
Armando Mendoza, *Cypress College*
Angela Milakovic, *Bismarck State College*
Scott Phillips, *Clovis Community College*
Jon Rossiter, *Irvine Valley College*
Tiffany Seeley, *East Los Angeles College*
Ray Sumner, *Long Beach City College*
Paul Vincent, *Valdosta State University*
David Walker, *College of Southern Maryland*
Thomas Wikle, *Oklahoma State University*
Lin Wu, *Cal State Polytech University Pomona*

Over the years, scores of colleagues, students, and friends have helped me and the founding author of this book, Tom McKnight, update and improve this textbook. Their assistance has been gratefully acknowledged previously. Here we acknowledge those who have provided assistance in recent years by acting as reviewers of the text and animations that accompany it, or by providing helpful critiques and suggestions:

Victoria Alapo, *Metropolitan Community College*
Jason Allard, *Valdosta State University*
Casey Allen, *Weber State University*
Sergei Andronikov, *Austin Peay State University*
Sandra Arlinghaus, *University of Michigan*
Christopher Atkinson, *University of North Dakota*
Robert Bailis, *Stockholm Environment Institute*
Greg Bierly, *Indiana State University*
Mark Binkley, *Mississippi State University*
Peter Blanken, *University of Colorado*
Margaret Boorstein, *Long Island University*
James Brey, *University of Wisconsin Fox Valley*
David Butler, *Texas State University*
Karl Byrand, *University of Wisconsin*
Sean Cannon, *Brigham Young University–Idaho*
Wing Cheung, *Palomar College*
Jongnam Choi, *Western Illinois University*
Keith Clarke, *University of California–Santa Barbara*
Glen Conner, *Western Kentucky University*
Carlos E. Cordova, *Oklahoma State University*
Richard A. Crooker, *Kutztown University of Pennsylvania*
Kristine L. DeLong, *Louisiana State University*
Mike DeVivo, *Grand Rapids Community College*
Bryan Dorsey, *Weber State University*
Don W. Duckson, Jr., *Frostburg State University*
Robert A. Dull, *University of Texas at Austin*
Ted Eckmann, *University of Portland*
Tracy Edwards, *Frostburg State University*
Steve Emerick, *Glendale Community College*
Purba Fernandez, *De Anza College*
Jason Finley, *Los Angeles Pierce College*
Lynda Folts, *Richland College*
Doug Foster, *Clackamas Community College*
Matthew Fry, *University of North Texas*
Basil Gomez, *Indiana State University*
Jerry Green, *Miami University–Oxford*
Michael Grossman, *Southern Illinois University–Edwardsville*
Christopher Groves, *Western Kentucky University*
Andrew J. Grundstein, *University of Georgia*
Perry J. Hardin, *Brigham Young University*
Ann Harris, *Eastern Kentucky University*
Miriam Helen Hill, *Jacksonville State University*
Barbara Holzman, *San Francisco State University*
Robert M. Hordon, *Rutgers University*
Matt Huber, *Syracuse University*
Paul Hudson, *University of Texas*
Catherine Jain, *Palomar College*
Steven Jennings, *University of Colorado at Colorado Springs*
Ryan Jensen, *Brigham Young University*
Dorleen B. Jenson, *Salt Lake Community College*
Kris Jones, *Saddleback College*
Ryan Kelly, *Lexington Community College*
Joseph Kerski, *ESRI*
John Keyantash, *California State University–Dominguez Hills*
Rob Kremer, *Metropolitan State College of Denver*
Kara Kuvakas, *Hartnell College*
Steve LaDochy, *California State University*

Colin Long, *University of Wisconsin–Oshkosh*
Ryan Longman, *University of Hawaii at Manoa*
Kerry Lyste, *Everett Community College*
Michael Madsen, *Brigham Young University–Idaho*
Michael E. Mann, *Pennsylvania State University*
Kenneth Martis, *West Virginia University*
Martin Mitchell, *Minnesota State University–Mankato*
William Monfredo, *University of Oklahoma*
Mandy Munro-Stasiuk, *Kent State University*
Paul O'Farrell, *Middle Tennessee State University*
Thomas Orf, *Las Positas College*
Michael C. Pease, *Central Washington University*
Natalie Peyronnin, *Mississippi River Delta Restoration*
Stephen Podewell, *Western Michigan University*
Nick Polizzi, *Cypress College*
Jennifer Rahn, *Samford University*
Robert Rohli, *Louisiana State University*
Diana Sammataro, *DianaBrand Honey Bee Research Services*
Anne Saxe, *Saddleback College*
Randall Schaetzl, *Michigan State University*
Jeffrey Schaffer, *Napa Valley College*
John H. Scheufler, *Mesa College*
Christopher J. Seeger, *Iowa State University*
Bradley A. Shellito, *Youngstown State University*
Terry Shirley, *University of North Carolina–Charlotte*
Jorge Sifuentes, *Cuesta College*
Robert A. Sirk, *Austin Peay State University*
Valerie Sloan, *University of Colorado at Boulder*
Dale Splinter, *University of Wisconsin–Whitewater*
Stephen Stadler, *Oklahoma State University*
Herschel Stern, *Mira Costa College*
Pat Stevenson, *Natural Resources Department, Stillaguamish Tribe*
Paul Sutton, *University of South Australia*
Jane Thorngren, *San Diego State University*
Christi Townsend, *San Diego State University*
Scott Walker, *Northwest Vista College*
Timothy Warner, *West Virginia University*
Nancy Lee Wilkinson, *San Francisco State University*
Shawn Willsey, *College of Southern Idaho*
Donald Wuebbles, *University of Illinois at Urbana Champaign*
Kyungsoo Yoo, *University of Minnesota*
Kenneth Zweibel, *George Washington University*

I would also like to thank Jess Porter of the University of Arkansas at Little Rock, Stephen O'Connell of the University of Central Arkansas, Jason Allard of Valdosta State University, Richard Crooker of Kutztown University, Chris Sutton of Western Illinois University, and Andrew Mercer of Mississippi State University for their contributions to *Mastering Geography* and other supporting material.

Many of my colleagues at City College of San Francisco offered valuable suggestions on sections of the previous and current editions of the book: Ian Duncan, Carlos Jennings, Dack Lee, Chris Lewis, Joyce Lucas-Clark, Robert Manlove, Kathryn Pinna, Todd Rigg-Carriero, Kirstie Stramler, Carole Toebe, and Katryn Wiese. I also extend my appreciation to my many students over the years—their curiosity, thoughtful questions, and cheerful acceptance of my enthusiasm for geography have helped me as a teacher and as a textbook author.

Textbooks of this scope cannot be created without a production team that is as dedicated to quality as the authors. First of all, my special thanks go to both the former Pearson Acquisitions Editors Christian Botting and Thomas Hoff as well as Director of Product Michael Gillespie, Content Manager Josh Frost, and Director of Content Jeanne Zalesky; all of whom facilitated a steady transition of content creation during a tumultuous time. My thanks and admiration go to SPi Content Producer Brett Coker and Associate Content Producer Titas Basu whom kept me on track throughout the entire production process. Many thanks also to Development Editors Mary Hill, David Chelton, and Sonia Divitorrio, and Product Manager Aileen Pogran, SPi Global Project Manager Katie Ostler, Photo Researcher Jeff Puda, International Mapping Senior Project Manager Kevin Lear, Content Analyst Chelsea Noack, Marketing Coordinator Rosemary Morton, Senior Product Marketing Manager Candice Madden, Media Producer Ziki Dekel, Photo Researchers Grace Annalyn Subito and Jeamweal Bonilla, Permissions Manager Grace Subito Annalyn, and Copy Editor Sheryl Nelson.

Finally, I wish to express my appreciation for my wife, Nora. Her help, understanding, and support have once again seen me through the long hours and many months of work that went into this book.

Darrel Hess
Earth Sciences Department
City College of San Francisco
50 Phelan Avenue
San Francisco, CA 94112
dhess@ccsf.edu

I would like to thank my husband, Owen, for his support and patience as I worked some very late hours to complete this book.

Redina Finch
Earth, Atmospheric and Geographic Information Sciences Department
Western Illinois University
1 University Circle
Macomb, IL 61455
RL-Finch@wiu.edu

Digital & Print Resources

Mastering Geography*™ *with Pearson eText. The Mastering platform is the most widely used and effective online homework, tutorial, and assessment system for the sciences. It delivers self-paced tutorials that provide individualized coaching, focus on course objectives, and are responsive to each student's progress. The Mastering system helps teachers maximize class time with customizable, easy-to-assign, and automatically graded assessments that motivate students to learn outside of class and arrive prepared for lecture. Mastering Geography™ offers:

- Assignable activities that include GIS-inspired MapMaster™ interactive map activities, Encounter Google Earth™ Explorations, video activities, Geoscience Animation activities, *Mobile Field Trip* video activities, *Project Condor* Quadcopter video activities, map projections activities, GeoTutor coaching activities on the toughest topics in geography, Dynamic Study Modules that provide each student with a customized learning experience, end-of-chapter questions and exercises, reading quizzes, Test Bank questions, and more.
- A student Study Area with GIS-inspired MapMaster™ interactive maps, videos, Geoscience Animations, *Mobile Field Trip* videos, *Project Condor* Quadcopter videos, web links, glossary flashcards, *In the News* readings, chapter quizzes, PDF downloads of outline maps, an optional Pearson eText, and more.

Pearson eText gives students access to the text whenever and wherever they can access the Internet. Features of Pearson eText include:

- Now available on smartphones and tablets
- Seamlessly integrated videos and other rich media
- Fully accessible (screen-reader ready)
- Configurable reading settings, including resizable type and night reading mode
- Instructor and student note-taking, highlighting, bookmarking, and search

www.masteringgeography.com

FOR STUDENTS

Physical Geography Laboratory Manual, 13th edition by Darrel Hess. This lab manual offers a comprehensive set of more than 45 lab exercises to accompany any physical geography class. The first half covers topics such as basic meteorological processes, the interpretation of weather maps, weather satellite images, and climate data. The second half focuses on understanding the development of landforms and the interpretation of topographic maps and aerial imagery. Many exercises have problems that use Google Earth™, and the lab manual website contains maps, images, photographs, satellite movie loops, and Google Earth™ KMZ files. The 13th edition of the lab manual includes both new and revised exercises, new maps, expanded use of Google Earth™, and is now supported by a full Mastering Geography program. www.masteringgeography.com.

Goode's World Atlas, 23rd Edition (0133864642). Goode's World Atlas has been the world's premiere educational atlas since 1923—and for good reason. It features more than 250 pages of maps, from definitive physical and political maps to important thematic maps that illustrate the spatial aspects of many important topics. The 23rd Edition includes more than 160 pages of digitally produced reference maps, as well as thematic maps on global climate change, sea-level rise, CO_2 emissions, polar ice fluctuations, deforestation, extreme weather events, infectious diseases, water resources, and energy production.

Pearson's Encounter Series provides rich, interactive explorations of geoscience concepts through Google Earth™ activities, covering a range of topics in regional, human, and physical geography. Each exploration consists of a worksheet, online quizzes whose results can be emailed to teachers, and a corresponding Google Earth™ KMZ file.

- Encounter Physical Geography by Jess C. Porter and Stephen O'Connell (0321672526)
- Encounter World Regional Geography by Jess C. Porter (0321681754)
- Encounter Human Geography by Jess C. Porter (0321682203)

FOR TEACHERS

***Instructor Resource Manual* (Download)** (0134326385). The manual includes lecture outlines and key terms, additional source materials, teaching tips, and a complete annotation of chapter review questions. Available from www.pearsonhighered.com/irc and in the Instructor Resources area of *Mastering Geography*™.

***TestGen® Test Bank* (Download)** by Steve Stadler (0134326377). TestGen® is a computerized test generator that lets you view and edit Test Bank questions, transfer questions to tests, and print tests in a variety of customized formats. This Test Bank includes around 3000 multiple-choice, true/false, and short answer/essay questions. All questions are correlated against the National Geography Standards, textbook key learning concepts, and Bloom's Taxonomy. The Test Bank is also available in Microsoft Word® and importable into Blackboard. Available in the Instructor Resources area of *Mastering Geography*™.

Instructor Resource on Mastering Geography (0134326369). The Instructor Resource provides a collection of resources to help teachers make efficient and effective use of their time. All digital resources can be found in one well-organized, easy-to-access place. The IR includes:

- All textbook images as JPEGs, PDFs, and PowerPoint™ Presentations
- Pre-authored Lecture Outline PowerPoint® Presentations, which outline the concepts of each chapter with embedded art and can be customized to fit teachers' lecture requirements
- CRS "Clicker" Questions in PowerPoint™
- The TestGen software, Test Bank questions, and answers for both Macs and PCs
- Electronic files of the Instructor Resource Manual and Test Bank

This Instructor Resource content is also available online via the Instructor Resources section of *Mastering Geography*™.

Learning Catalytics is a "bring your own device" student engagement, assessment, and classroom intelligence system. With Learning Catalytics, you can:

- Assess students in real time, using open-ended tasks to probe student understanding.
- Understand immediately where students are and adjust your lecture accordingly.
- Improve your students' critical thinking skills.
- Access rich analytics to understand student performance.
- Add your own questions to make Learning Catalytics fit your course exactly.
- Manage student interactions with intelligent grouping and timing.

Learning Catalytics is a technology that has grown out of twenty years of cutting-edge research, innovation, and implementation of interactive teaching and peer instruction. Available integrated with Mastering Geography™.

Dedication

For my students over the last 30 years at

City College of San Francisco

D.H.

To my students, past, present, and future:

You inspire me!

R.L.F.

About the Authors

Darrel Hess began teaching geography at City College of San Francisco in 1990 and served as chair of the Earth Sciences Department from 1995 to 2009. After earning his bachelor's degree in geography at the University of California, Berkeley, in 1978, he served for two years as a teacher in the Peace Corps on Jeju Island, Korea. Upon returning to the United States, he worked as a writer, photographer, and audiovisual producer. His association with Tom McKnight began as a graduate student at UCLA, where he served as one of Tom's teaching assistants. Their professional collaboration developed after Darrel graduated from UCLA with a master's degree in geography in 1990. He first wrote the *Study Guide* that accompanied the fourth edition of *Physical Geography: A Landscape Appreciation*, and then the *Laboratory Manual* that accompanied the fifth edition. Along with the 9th edition, he developed the first of four special *California Editions* of the textbook. In 1999 Tom asked Darrel to join him as coauthor of the textbook. Darrel was the 2014 recipient of the American Association of Geographers (AAG) Gilbert Grosvenor Geographic Education Honors. As did Tom, Darrel greatly enjoys the outdoor world. Darrel and his wife, Nora, are avid hikers, campers, and scuba divers.

Redina L. Finch has been teaching introductory and advanced meteorology classes for almost 20 years. She earned a bachelor's degree in physics from the Florida Institute of Technology and a PhD in atmospheric sciences from the University of Illinois, Urbana-Champaign. Redina is involved in science education research and won the Western Illinois University College of Arts and Sciences award for Teaching with Technology in 2014. She is the current editor-in-chief for the National Association of Geoscience Teachers' education magazine, *In the Trenches*. Redina regularly contributes to community service activities, including Science Olympiad, Discovering the World through Science summer camp, Girl Scouts STEM program, and others. She is the Western Illinois University representative to the University Corporation for Atmospheric Research (UCAR), which runs the National Center for Atmospheric Research (NCAR). Redina is also a coauthor of *The Atmosphere: An Introduction to Meteorology*. In her spare time, Redina likes to do almost anything outdoors. People jokingly say that every picture Redina takes has clouds in it. She admits that's probably true! Redina and her husband, Owen, are lucky to have horses in the back yard and the Mississippi River right down the road.

Dennis Tasa attended the Minneapolis College of Art and Design and was then employed at a publishing company, working in book production, design, and illustration. While illustrating a geology textbook for the publisher, a close association with the author spurred an interest in the field of geology. After forming his own business, Tasa Graphic Arts, Inc., his interest in the geosciences continued, illustrating numerous textbooks including physical geography, physical geology, earth science, meteorology, historical geology, and mineralogy. From 1993 to 2018 he expanded his company to include computer software publishing. Working with noted college professors Dennis and his staff created and published award-winning educational programs in the geosciences. Now residing in Taos, New Mexico, in addition to book illustration under DK Tasa, Inc., he and his wife, Karen, enjoy camping, hiking, and painting.

Tom L. McKnight taught geography at UCLA from 1956 to 1993. He received his bachelor's degree in geology from Southern Methodist University in 1949, his master's degree in geography from the University of Colorado in 1951, and his Ph.D. in geography and meteorology from the University of Wisconsin in 1955. During his long academic career, Tom served as chair of the UCLA Department of Geography from 1978 to 1983, and was director of the University of California Education Abroad Program in Australia from 1984 to 1985. Passionate about furthering the discipline of geography, he helped establish the UCLA/Community College Geography Alliance and generously funded awards for both undergraduate and graduate geography students. His many honors include the California Geographical Society's Outstanding Educator Award in 1988, and the honorary rank of Professor Emeritus upon his retirement from UCLA. In addition to *Physical Geography: A Landscape Appreciation*, his other college textbooks include *The Regional Geography of the United States and Canada*; *Oceania: The Geography of Australia, New Zealand, and the Pacific Islands*; and *Introduction to Geography*, with Edward F. Bergman. Tom passed away in 2004—the geographic community misses him enormously.

Seeing Geographically in Our National Parks

NASA Astronaut Terry Virts took this photograph of Florida from the International Space Station in 2015. Dry Tortugas National Park is the tiny horseshoe-shaped blue patch in the ocean west of the Florida Keys. What might explain the light color of the ocean here? From this distance above the surface, what evidence of human activity can you see?

Introduction to Earth

1

HAVE YOU EVER WONDERED how we know that human activity is changing global climate? Or why Seattle residents need to worry about earthquakes but Miami residents don't? Or why kangaroos are native to Australia but not to China? Or even why the days are longer in summer than in winter? These are the kinds of questions we answer in physical geography.

If you opened this book expecting that the study of geography was going to be memorizing names and places on maps, you'll be surprised to find that geography is much more than that. Geographers study the location and distribution of things—tangible things such as rainfall, mountains, and trees, as well as less tangible things such as language, migration, and voting patterns. In short, geographers look for and explain patterns in the physical and human landscape.

In this book you learn about fundamental processes and patterns in the natural world—the kinds of things you can see whenever you walk outside: clouds in the sky, mountains, streams and valleys, and the plants and animals that inhabit the landscape. You also learn about human interactions with the natural environment—how events such as hurricanes, earthquakes, and floods affect our lives and the world around us, as well as how human activities are increasingly altering our global environment. Many examples come from one of our greatest resources for studying and protecting the natural environment: our National Parks—three are even visible in this image of Florida taken from space: Everglades, Biscayne, and Dry Tortugas.

By the time you finish this book, we hope you'll understand—in other words, appreciate—the landscape in new ways.

As you study this chapter, think about these **Key**Questions:

- **How do geographers study the world?**
- **How do we make sense of different environments on Earth?**
- **How do we describe location on Earth?**
- **Why do the seasons change?**
- **How do global time zones work?**

Mobile Field Trip videos, created by renowned Earth Science writer, photographer, and pilot, Michael Collier, are virtual field trips that explore physical geography from the air and ground. This first Mobile Field Trip introduces you to the study of physical geography.

Mobile Field Trip
Introduction to Physical Geography
https://media.pearsoncmg.com/ph/esm/esm_mcknight_physgeo_13/media/

The Study of Geography

The word **geography** comes from the Greek words meaning "Earth description." Several thousand years ago many scholars were indeed "Earth describers," and therefore geographers, more than anything else. Nonetheless, over the centuries studies veered away from generalized Earth description toward more specialized disciplines—such as geology, meteorology, economics, and biology—so geography as a field of study was somewhat overshadowed. Over the last few hundred years, however, geography reaffirmed its place in the academic world, and today geography is an expanding and flourishing field of study.

Studying the World Geographically

Geographers study how things differ from place to place—the distributional and locational relationships of things around the world (what is sometimes called the "spatial" aspect of things). Figure 1-1 shows the kinds of "things" geographers study, divided into two groups representing the two principal branches of geography. The elements of **physical geography** are natural in origin, and for this reason physical geography is sometimes called *environmental geography*. The elements of **human geography** are those of human activity; this branch includes such subfields as *cultural geography, economic geography, political geography*, and *urban geography*. The almost unlimited possible combinations of these various elements create the physical and cultural landscapes of the world that geographers study.

All of the items shown in Figure 1-1 are familiar to us, and this familiarity highlights a basic characteristic of geography as a field of learning: geography doesn't have its own body of facts or objects that only geographers study. The focus of geology is rocks, the attention of economics is economic systems, demography examines human population, and so on. Geography, on the other hand, is much broader in scope than most other disciplines, "borrowing" its objects of study from related fields. Geographers, too, are interested in rocks and economic systems and population—especially in describing and understanding their location and distribution. We sometimes say that geography asks the fundamental question, "Why is what where, and so what?"

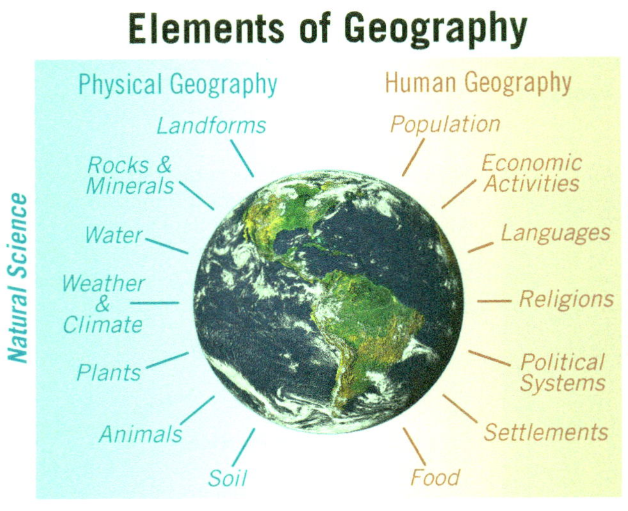

▲ Figure 1-1 The elements of geography can be grouped into two broad categories. Physical geography primarily involves the study of natural science, whereas human geography primarily entails the study of social science.

LearningCheck 1-1 **What are the differences between physical geography and human geography? (Answer on p. AK-1)**

Another basic characteristic of geography is its interest in interrelationships. We cannot understand the distribution of soils, for example, without knowing something about the rocks from which the soils were derived, the slopes on which the soils developed, and the climate and vegetation under which they developed. Similarly, it is impossible to comprehend the distribution of agriculture without an understanding of climate, topography, soil, drainage, population, economic conditions, technology, historical development, and many other factors, both physical and cultural. Because of its wide scope, geography bridges the academic gap between natural science and social science, studying all of the elements in Figure 1-1 in an intricate web of geographic interrelationships.

In this book we concentrate on the physical elements of the landscape, the processes involved in their development, their distribution, and their basic interrelationships. As we proceed from chapter to chapter, this notion of landscape development by natural processes and landscape modification by humans serves as a central focus. We pay attention to elements of human geography when they help to explain the development or patterns of the physical elements— especially the ways in which humans influence or alter the physical environment.

Global Environmental Change: Several broad geographic themes run through this book. One of these themes is *global environmental change*—both the human-caused and natural processes that are currently altering the landscapes of the world. Some of these changes can take place over a period of just a few years, whereas others require many decades or even thousands of years (Figure 1-2). We pay special attention to the accelerating impact of human activities on the global environment: in the chapters on the atmosphere we discuss such issues as human-caused climate change, ozone depletion, and acid rain, whereas in later chapters we look at issues such as rainforest removal and coastal erosion.

Rather than treat global environmental change as a separate topic, we integrate this theme throughout the book. To help with this integration, we supplement the main text with short boxed essays, such as those titled "People & the Environment" that focus on specific cases of human interaction with the natural environment, as well as boxes titled "Energy for the 21st Century" that focus on the challenge of supplementing—and perhaps eventually replacing—fossil fuels with renewable sources of energy. These essays serve to illustrate the connections between many aspects of the environment—such as the relationships between changing global temperatures, changing sea level, changing quantities of polar ice, and the changing distribution of plant and animal species—with the global economy and human society.

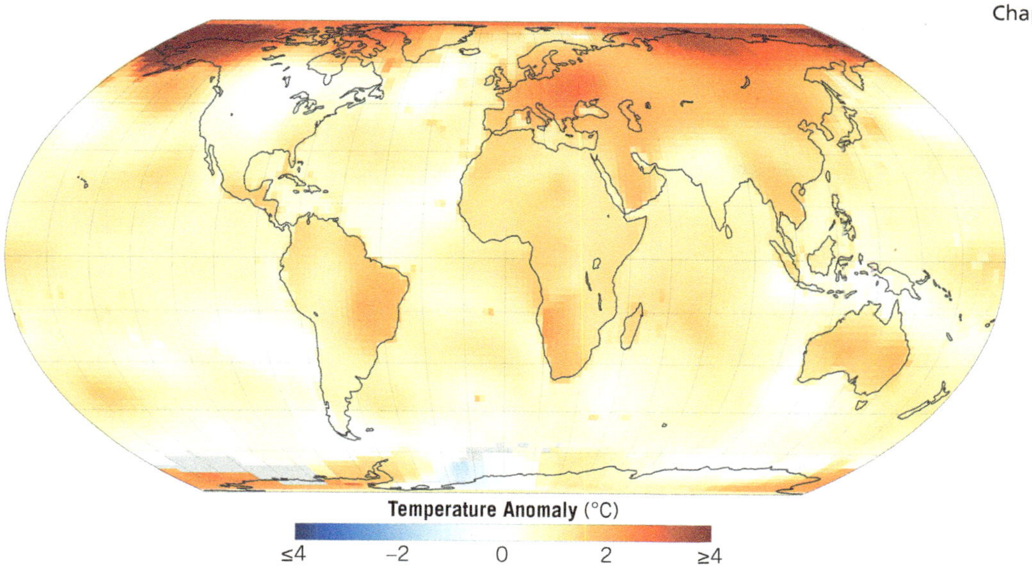

◀ **Figure 1-2** Earth's climate is changing. This image shows the difference in temperature (the temperature anomaly in °C) during the year 2019 compared with the average temperatures for the baseline period 1951 to 1980. (NASA)

Furthermore, in many chapters you'll see boxed essays titled "Global Environmental Change." These essays introduce special topics and include activities and questions that will help you understand the scope of both natural and human-caused environmental changes.

Globalization: A related but less obvious theme running through this book is *globalization*. In the broadest terms, globalization refers to the processes and consequences of an increasingly interconnected world—connections among the economies, cultures, and political systems of the world. Although globalization is most commonly associated with the cultural and economic realms, it is important to recognize the environmental components of globalization as well. For example, the loss of tropical rainforest for timber or commercial agriculture in some regions of the world is driven in part by growing demand for commodities in countries far from the tropics (**Figure 1-3**). Similarly, rapid economic growth in newly industrialized countries is contributing to the already high atmospheric greenhouse gas emissions of older industrialized countries. The economies of the world are thus interconnected in their influence on the natural environment.

Because of geography's global perspective and its interest in both the natural and human landscapes, geographers are able to offer insights into many of the world's most pressing problems—problems too complex to address from a narrower perspective. For example, the detrimental consequences of climate change cannot be addressed if we ignore the economic, social, historical, and political aspects of the issue. Similarly, global inequities of wealth and political power cannot be addressed if we ignore environmental and resource issues.

Just about everything in the world is in one way or another connected with everything else; geography helps us understand these connections.

LearningCheck 1-2 **Why are physical geographers interested in globalization?**

Geography in the National Parks

A final theme running through this book is the role of National Parks as places to study, protect, and enjoy all facets of the natural environment. Throughout the book, you'll see many topics illustrated with examples from our National Park System. Every chapter begins with an image from a Park and each chapter includes at least one special box featuring a National Park. The first box is an overview of the National Park System itself—see the box *Featured National Parks: National Parks—Our Geographical Laboratories*.

▼ **Figure 1-3** Deforestation in some parts of the tropics is influenced by consumer demand in other parts of the world. This logging operation is in Sarawak, Borneo, Malaysia.

Featured National Parks

National Parks

Our Geographical Laboratories

The American writer Wallace Stegner once said that our National Parks are "the best idea we ever had." Although we can argue that our Bill of Rights, the Emancipation Proclamation, the Nineteenth Amendment, and the Civil Rights Act of 1964 rank higher in terms of American ideals, our National Park System did indeed begin with the distinctly American idea that there are places in our country that belong to all of us, and these places should be protected and preserved for all generations to come.

History of the National Parks: The first step toward our present-day National Park System took place in 1864 when President Lincoln set aside Yosemite Valley and adjacent lands for California to protect (**Figure 1-A**). In 1872, President Grant signed the act that would lead to the creation of Yellowstone National Park—the first national park in the world (**Figure 1-B**). The 1906 Antiquities Act allowed a president—with simply a signature—to protect a specific natural, historic, or cultural feature as a National Monument. Over the decades, many National Monuments established in this way, such as Grand Canyon (**Figure 1-C**) and Death Valley, became full-fledged National Parks. In 1916, the Organic Act created the National Park Service (NPS) within the Department of the Interior, with the mission to

> *. . . conserve the scenery and the natural and historic objects and the wild life therein and to provide for the enjoyment of the same in such manner and by such means as will leave them unimpaired for the enjoyment of future generations.*
>
> From the National Park Service *Organic Act* of 1916.

▲ **Figure 1-A** Yosemite in 1899. African American "Buffalo Soldiers" of the Twenty-Fourth Infantry were the first protectors of Yosemite National Park—before the National Park Service was established, the U.S. Army was responsible for administering National Parks.

Today the National Park System has more than 400 units. The NPS oversees National Parks, National Monuments, National Seashores, National Recreation Areas, National Memorials, National Historic Parks, National Battlefields, and National Preserves, along with many other components. The differences among these designations relate to how each was established, its purpose, the types of activities permitted, and the kind of protection afforded. These sites range in size from a single house to areas twice the size of Maryland. In all, about 340,000 square kilometers (130,000 square miles) of land are administered by the National Park Service—an area greater than the entire country of Finland.

Although the mission of the NPS has not wavered over the last century, in some regards what we conserve and how we conserve our natural and historic heritage has evolved. Early on, the Parks were thought of primarily as wild places with little regard for the people originally living there—for example, Native Americans and their settlements were removed from Yosemite Valley by the 1960s. But over time, the NPS has come to acknowledge that even some of the darkest chapters of American history are as important for us to remember and preserve as are wild places; the National Park System now incorporates sites such as the Gettysburg battlefield of the Civil War, the Manzanar internment camp where Japanese Americans were imprisoned during World War II, and Little Rock Central High School where in 1957 nine African American students passed through an angry white crowd to desegregate a public school.

The Parks as Preserves: For physical geographers, the units of our National Park System offer some of the best places in the country to observe processes operating in the natural environment. In many Parks, large tracts of land have been set aside where the landscape is allowed to change with as little influence from people as possible. Parks provide some of the last sanctuaries for animals now gone from the rest of the country; there are expanses of natural grassland not yet tilled under for agriculture; coastlines not yet lined with houses; and

◀ **Figure 1-B** Yellowstone National Park. The Midway Geyser Basin with Excelsior Geyser and Grand Prismatic Spring seen from the air.

▲ **Figure 1-C** Grand Canyon National Park. Visitors at Mather Point along the South Rim.

there are Parks with truly dark skies where city dwellers can glimpse the Milky Way for perhaps the first time in their lives.

The Parks as Laboratories: In addition to being places where we can visit largely unspoiled landscapes, National Parks are also critical laboratories for the study of both natural and anthropogenic (human-caused) environmental change. Within our Parks, we can observe how forest health changes in the decades after a wildfire; how whole ecosystems are altered when a once-exterminated predator is reintroduced; and how a changing climate affects the size of glaciers, the distribution of birds, or the pattern of vegetation.

Because our National Parks include such a variety of environments—from the tropics to the Arctic, and from low deserts to high mountains—the research conducted here will only become more important in the years to come.

Featured National Parks: In this book we illustrate many aspects of physical geography with examples from our National Parks (a term we'll use rather broadly to include any unit in the National Park System). In addition to the many photographs, "Mobile Field Trips" and videos you'll see, we also provide a series of short case studies from the National Parks. These "Featured National Park" boxes encompass a wide range of topics, environments, and locations within the National Park System (**Figure 1-D**).

The examples in this book only scratch the surface of what you can see and learn about in our National Parks. To find out more about National Parks near you, visit *https://www.nps.gov/findapark*.

Questions

1. Which unit of the National Park System is closest to your home?
2. Why are National Parks important "laboratories" for studying environmental change?

▼ **Figure 1-D** Map showing the location of National Parks featured in this textbook.

Other Parks Included in This Book:

(NHP = National Historic Park, NM = National Monument, NMe = National Memorial, NP = National Park, NPr = National Preserve, NR = National River, NRA = National Recreation Area, NS = National Seashore)

Acadia NP, *ME*
Arches NP, *UT*
Big Bend NP, *TX*
Cape Hatteras NS, *NC*
Capitol Reef NP, *UT*
Congaree NP, *SC*
Craters of the Moon NM, *ID*
Devils Postpile NM, *CA*
Dry Tortugas NP, *FL*
Gates of the Arctic NP, *AK*
Gateway Arch NP, *MO*
Haleakalā NP, *HI*
Joshua Tree NP, *CA*
Katmai NP, *AK*
Kenai Fjords NP, *AK*
Lyndon B. Johnson NHP, *TX*
Mount Rushmore NMe, *SD*
National Marine Sanctuary of *American Samoa*
New River Gorge NR, *WV*
North Cascades NP, *WA*
Organ Pipe Cactus NM, *AZ*
Padre Island NS, *TX*
Rainbow Bridge NM, *UT*
Saguaro NP, *AZ*
Santa Monica Mountains NRA, *CA*
Sunset Crater NM, *AZ*
Theodore Roosevelt NP, *ND*
Voyageurs NP, *MN*
White Sands NP, *NM*
Wind Cave NP, *SD*
Wrangell–St. Elias NP, *AK*
Yukon–Charley Rivers NPr, *AK*
Zion NP, *UT*

Geography and Science

Because physical geography is concerned with processes and patterns in the natural world, knowledge in physical geography advance primarily through the study of science. It is useful for us to say a few words about science in general.

The Process of Science

Science is often described—although somewhat simplistically—as a process that follows the *scientific method*:

1. Observe phenomena that stimulate a question or problem.
2. Offer an educated guess—a *hypothesis*—about the answer.
3. Design an experiment to test the hypothesis.
4. Predict the outcome of the experiment if the hypothesis is supported or if the hypothesis is not supported.
5. Conduct the experiment and observe what actually happens.
6. Draw a conclusion or formulate a simple generalized "rule" based on the results of the experiment.

In practice, however, science doesn't always work through experimentation; in many fields of science, data collection through observation of phenomena is the basis of knowledge. In some regards science is best thought of as a process—or perhaps even as an attitude—for gaining knowledge. The scientific approach is based on observation, experimentation, logical reasoning, skepticism of unsupported conclusions, and the willingness to modify or even reject long-held ideas when new evidence contradicts them. For example, up until the 1950s most Earth scientists thought it impossible that the positions of continents could change over time. However, as we see in Chapter 14, by the late 1960s enough new evidence had been gathered to convince them that their earlier ideas were wrong—the configuration of continents has changed and continues to change.

Although the term "scientific proof" is sometimes used by the general public, strictly speaking, science does not "prove" ideas. Instead, science works by eliminating alternative explanations—eliminating explanations that aren't supported by evidence. In fact, in order for a hypothesis to be "scientific," there must be some test or possible observation that could *disprove* it. If there is no way to disprove an idea, then that idea simply cannot be supported by science.

The word "theory" is often used in everyday conversation to mean a "hunch" or conjecture. However, in science a theory represents the highest order of understanding for a body of information—a logical, well-tested explanation that encompasses a wide variety of facts and observations. Thus, the "theory of plate tectonics" presented in Chapter 14 represents an empirically supported, broadly accepted, overarching framework for understanding processes that operate within Earth.

The acceptance of scientific ideas and theories is based on a preponderance of evidence, not on "belief" and not on the pronouncements of "authorities." New observations and new evidence often cause scientists to revise their conclusions and theories or those of others. Much of this self-correcting process for refining scientific knowledge takes place through peer-reviewed journal articles. Peers—that is, fellow scientists—scrutinize a scientific report for sound reasoning, appropriate data collection, and solid evidence before it is published; reviewers need not agree with the author's conclusions, but they strive to ensure that the research meets rigorous standards of scholarship before publication.

Because new evidence may prompt scientists to change their ideas, good science tends to be somewhat cautious in its stated conclusions. For this reason, the findings of many scientific studies are prefaced by phrases such as "the evidence suggests" or "the results most likely show." In some cases, different scientists interpret the same data quite differently and so disagree in their conclusions. Frequently, studies find that "more research is needed." The kind of uncertainty sometimes inherent in science may lead the general public to question the conclusions of scientific studies—especially when presented with a simple, and perhaps comforting, nonscientific alternative. However, this very uncertainty often compels scientists to push forward in the quest for knowledge and understanding.

In this book we present the fundamentals of physical geography as supported by scientific research and evidence. In some cases, we describe how our current understanding of a phenomenon developed over time; in other cases we point out where uncertainty remains, where scientists still disagree, or where intriguing questions still remain.

LearningCheck 1-3 **Why is the term "theory" sometimes misunderstood by the general public?**

Numbers and Measurement Systems

Because so much of science is based on observation and measurable data, any thorough study of physical geography entails the use of mathematics. Although this book introduces physical geography primarily in a conceptual way without the extensive use of mathematical formulas, numbers and measurement systems are nonetheless important for us. Throughout the book, we use numbers and simple formulas to help illustrate concepts—the most obvious of which are numbers used to describe distance, size, weight, or temperature.

Two quite different measurement systems are used today. In the United States, much of the general public is most familiar with the *English System* of measurement—with measurements such as miles, pounds, and degrees Fahrenheit. However, most of the rest of the world—and the entire scientific community—uses the **International System** of measurement (abbreviated **S.I.** from the French *Système International*; also called the "metric system")—with measurements such as kilometers, kilograms, and degrees Celsius.

TABLE 1-1 Unit Conversions—Quick Approximations

	S.I. to English Units	English to S.I. Units
Distance:	1 centimeter = a little less than ½ inch	1 inch = about 2½ centimeters
	1 meter = a little more than 3 feet	1 foot = about ⅓ meters
	1 kilometer = about ⅔ mile	1 yard = about 1 meter
		1 mile = about 1½ kilometers
Volume:	1 liter = about 1 quart	1 quart = about 1 liter
		1 gallon = about 4 liters
Mass:	1 gram = about 1/30 ounce	1 ounce = about 30 grams
	1 kilogram = about 2 pounds	1 pound = about ½ kilogram
Temperature:	1°C change = 1.8°F change	1°F change = about 0.6°C change

For exact conversion formulas, see Appendix I.

This book gives measurements in both S.I. and English units. Table 1-1 provides some quick approximations of the basic equivalents in each; detailed tables of conversion formulas between English and S.I. units appear in Appendix I.

Environmental Spheres and Earth Systems

From the standpoint of physical geography, the surface of Earth is a complex interface where four principal components of the environment meet and to some degree overlap and interact (Figure 1-4). These four components are often referred to as Earth's *environmental spheres*.

Earth's Environmental Spheres

The solid, inorganic portion of Earth is sometimes called the **lithosphere**[1] (*litho* is Greek for "stone"), comprising the rocks of Earth's crust as well as the unconsolidated particles of mineral matter that overlie the solid bedrock. The lithosphere's surface is shaped into an almost infinite variety of landforms, both on the seafloors and on the surfaces of the continents and islands.

The gaseous envelope of air that surrounds Earth is the **atmosphere** (*atmos* is Greek for "air"). It contains the complex mixture of gases needed to sustain life. Most of the atmosphere is close to Earth's surface, being densest at sea level and rapidly thinning with increased altitude. It is a very dynamic sphere, kept in almost constant motion by solar energy and Earth's rotation.

The **hydrosphere** (*hydro* is Greek for "water") comprises water in all its forms. The oceans contain the vast majority of the water found on Earth and are the moisture source for most precipitation. A subcomponent of the hydrosphere is known as the **cryosphere** (*cryo* comes from the Greek word for "cold")—water frozen as snow and ice.

The **biosphere** (*bios* is Greek for "life") encompasses all the parts of Earth where living organisms can exist; in its broadest and loosest sense, the term also includes the vast variety of earthly life-forms (properly referred to as *biota*).

[1]As we will see in Chapter 13, in the context of plate tectonics and our study of landforms, the term "lithosphere" is used specifically to refer to large "plates" consisting of Earth's crustal and upper mantle rock.

▲ **Figure 1-4** Earth's physical landscape is composed of four overlapping, interacting systems called "spheres." The atmosphere is the air we breathe. The hydrosphere is the water of rivers, lakes, and oceans, the moisture in soil and air, as well as the snow and ice of the cryosphere. The biosphere is the habitat of all life, as well as the life-forms themselves. The lithosphere is the soil and bedrock that cover Earth's surface. This scene shows Wonder Lake and Denali (formerly Mt. McKinley) in Denali National Park, Alaska.

These "spheres" are not discrete entities but rather are considerably interconnected. This intermingling is apparent when we consider an ocean—a body that is clearly a major component of the hydrosphere yet may contain a vast quantity of fish and other organisms that are part of the biosphere. An even better example is soil, which is composed largely of bits of mineral material (lithosphere) but also contains life-forms (biosphere), along with air (atmosphere), soil moisture (hydrosphere), and perhaps frozen water (cryosphere) in its pore spaces.

The environmental spheres can help us broadly organize concepts for the systematic study of Earth's physical geography and are used that way in this book.

LearningCheck 1-4 **Briefly define the lithosphere, atmosphere, hydrosphere, cryosphere, and biosphere.**

Earth Systems

Earth's environmental spheres operate and interact through a complex of *Earth systems*. By "system" we mean a collection of things and processes that are connected and operate as a whole. In the human realm, for example, we talk of a global "financial system" that encompasses the exchange of money between institutions and individuals, or of a "transportation system" that involves the movement of people and commodities. In the natural world, systems entail the interconnected flows and storage of energy and matter.

Closed Systems: Effectively self-contained systems, which are therefore isolated from influences outside that system, are called *closed systems*. It is rare to find closed systems in nature. Earth as a whole is essentially a closed system with regard to matter—currently there is no significant increase or decrease in the amount of matter (the "stuff") of Earth, although relatively small but measurable amounts of meteoric debris arrives from space, and tiny amounts of gas are lost to space from the atmosphere. Energy, on the other hand, constantly enters and exits the Earth system.

Open Systems: Most Earth systems are *open systems*—both matter and energy are exchanged across the system boundary. Matter and energy that enter the system are called *inputs*, and losses from the system to its surroundings are called *outputs*. For example, as we see in Chapter 19, a glacier behaves as an open system (**Figure 1-5**). The material inputs to a glacier include water in the form of snow and ice, along with rocks and other debris picked up by the moving ice; the material outputs of a glacier include the meltwater and water vapor lost to the atmosphere, as well as the rock and debris transported and eventually deposited by the ice. The most obvious energy input into a glacial system is solar radiation, which melts the ice by warming the surrounding air and by direct absorption into the ice itself. But also at work are less obvious exchanges of energy that involve latent heat—energy stored by water during melting and evaporation, and released during freezing and condensation. (Latent heat is discussed in detail in Chapter 6.)

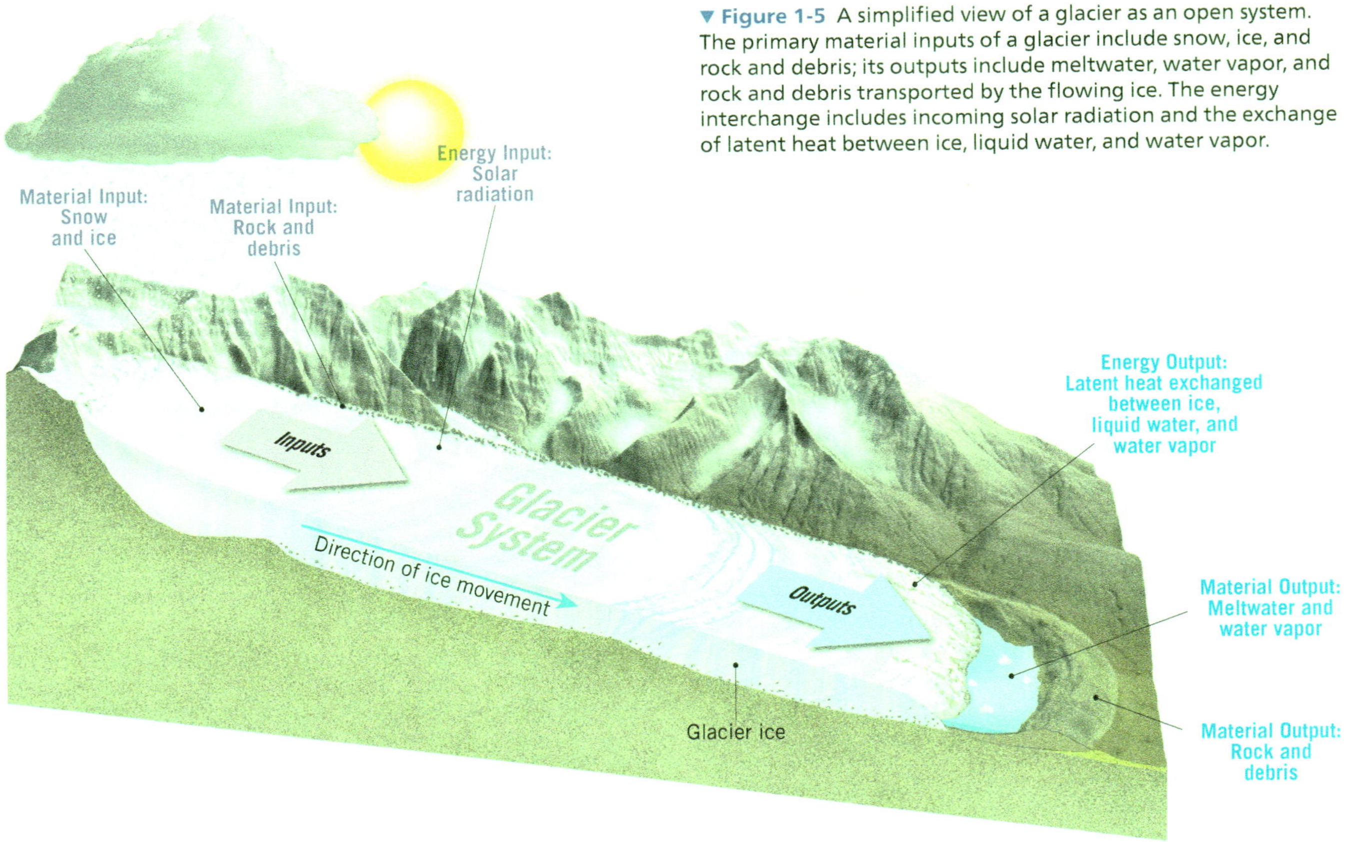

▼ **Figure 1-5** A simplified view of a glacier as an open system. The primary material inputs of a glacier include snow, ice, and rock and debris; its outputs include meltwater, water vapor, and rock and debris transported by the flowing ice. The energy interchange includes incoming solar radiation and the exchange of latent heat between ice, liquid water, and water vapor.

Equilibrium: When inputs and outputs balance over time, the conditions within a system remain the same; we describe such a system as being in *equilibrium*. For example, a glacier remains the same size over many years if its inputs of snow and ice are balanced by the loss of an equivalent amount of ice through melting. If, however, the balance between inputs and outputs changes, equilibrium is disrupted—increasing snowfall for several years, for example, can cause a glacier to grow until a new equilibrium size is reached.

Interconnected Systems: In physical geography we study the myriad of interconnections among Earth's systems and subsystems. Continuing with our example of a glacier: the system of an individual glacier is interconnected with many other Earth systems, including Earth's solar radiation budget (discussed in Chapter 4), wind and pressure patterns (discussed in Chapter 5), and the hydrologic cycle (discussed in Chapter 6). If inputs or outputs in those systems change, a glacier may also change. For example, if air temperature increases through a change in Earth's solar radiation budget, both the amount of water vapor available to precipitate as snow and the rate of melting of that snow may change, causing an adjustment in the size of the glacier.

LearningCheck 1-5 **What does it mean when we say a system is in equilibrium?**

Feedback Loops: Some systems produce outputs that "feed back" into that system, reinforcing change. As we see in Chapter 8, over the last few decades, increasing temperatures in the Arctic have reduced the amount of highly reflective summer sea ice. As the area of sea ice has diminished, the darker, less reflective ocean has absorbed more solar radiation, contributing to the temperature increase—which in turn has reduced the amount of sea ice even more, further reducing reflection and increasing absorption. Were Arctic temperatures to decrease, an expanding cover of reflective sea ice would reduce absorption of solar radiation and so reinforce a cooling trend. These are examples of *positive feedback loops*—change within a system continuing in one direction.

Conversely, *negative feedback loops* tend to inhibit a system from changing—in this case, increasing a system input tends to *decrease* further change, keeping the system in equilibrium. For example, an increase in air temperature may increase the amount of water vapor in the air; the extra water vapor may in turn condense and increase the cloud cover—which can reflect incoming solar radiation and so prevent a further temperature increase.

Although systems may resist change through negative feedback loops, a system may reach a *tipping point* or threshold. Beyond that point, the system becomes unstable and changes abruptly until it reaches a new equilibrium. For instance, as we see in Chapter 9, the increasing freshwater runoff from melting glaciers in the Arctic could one day disrupt the energy transfer of the slow deep ocean *thermohaline circulation* in the Atlantic Ocean, triggering a sudden change in climate.

The preceding examples are not intended to confuse you but rather to illustrate the great complexity of Earth's interconnected systems! Because of this complexity, in this book we often first describe one process or Earth system in isolation before we present its interconnections with other systems.

LearningCheck 1-6 **What is the difference between a positive feedback loop and a negative feedback loop?**

Earth and the Solar System

Earth is part of a larger *solar system*—an open system with which Earth interacts. Earth is an extensive rotating mass of mostly solid material that orbits the enormous ball of superheated gases we call the Sun. The geographer's concern with spatial relationships properly begins with the relative location of this "spaceship Earth" in the universe.

Animation
Solar System Formation
https://media.pearsoncmg.com/ph/esm/esm_mcknight_physgeo_13/media/

The Solar System

Earth is one of eight planets of our solar system, which also contains more than 200 natural satellites or "moons" revolving around the planets; an uncertain number of smaller *dwarf planets*, such as Pluto; scores of comets (bodies composed of frozen liquid and gases together with small pieces of rock and metallic minerals); more than 995,000 asteroids (small, rocky, and sometimes icy objects, mostly less than a few kilometers in diameter); and millions of meteoroids (most of them the size of sand grains).

The medium-massed star we call the Sun is the central body of the solar system and makes up more than 99.8 percent of its total mass. The solar system is part of the Milky Way Galaxy, which consists of at least 100 billion stars arranged in a disk-shaped spiral that is about 100,000 light-years in diameter and 10,000 light-years thick at the center. (One light-year equals the distance a beam of light travels over a period of one year—about 9.5 trillion kilometers.) The Milky Way Galaxy is only one of hundreds of billions of galaxies in the universe.

Origins: The origin of Earth, and indeed of the universe, is incompletely understood. Scientists generally accept that the universe began with a cosmic event called the *big bang*. The most widely held view is that the big bang took place about 13.7 billion years ago—similar to the age of the oldest known stars. The big bang began in a fraction of a second as an infinitely dense and infinitesimally small bundle of energy containing all of space and time started to expand in all directions at extraordinary speeds, pushing out the fabric of space and filling the universe with the energy and matter we see today.

Our solar system originated between 4.5 and 5 billion years ago when a *nebula*—a huge, cold, diffuse cloud of gas and dust—began to contract inward due to gravitational collapse, forming a hot, dense *protostar*. This hot center became our Sun, and the cold revolving disk of gas and dust around it eventually condensed and coalesced to form the planets and other bodies.

All of the planets revolve around the Sun in elliptical orbits, with the Sun located at one focus. (If we look "down" on the

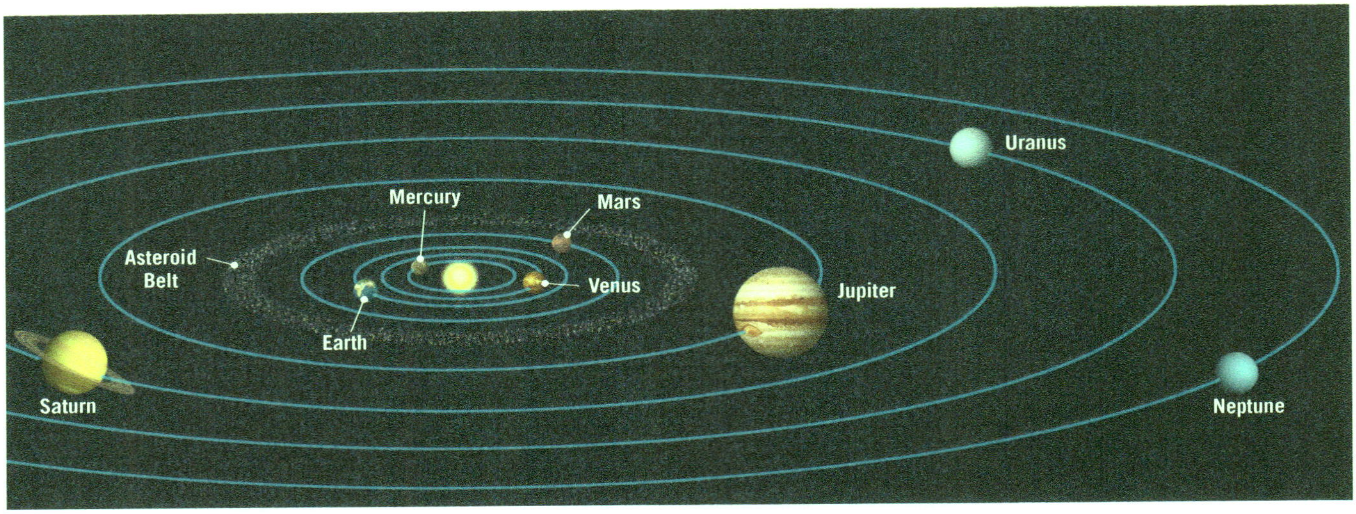

▲ **Figure 1-6** The solar system (not drawn to scale). The Sun is not exactly at the center of the solar system—the planets revolve around the Sun in elliptical orbits. The Kuiper Belt, which includes dwarf planets such as Pluto, begins beyond Neptune.

solar system from a vantage point high above the North Pole of Earth, the planets appear to orbit in a counterclockwise direction around the Sun.) All the planetary orbits are in nearly the same plane (**Figure 1-6**), perhaps revealing their relationship to the original spinning direction of the nebular disk.

The Planets: The four inner *terrestrial planets*—Mercury, Venus, Earth, and Mars—are generally smaller, denser, and less oblate (more nearly spherical) than the four outer *Jovian planets*—Jupiter, Saturn, Uranus, and Neptune. The inner planets are composed principally of mineral matter and, except for airless Mercury, have diverse but relatively shallow atmospheres. The four large Jovian planets are more massive (although they are less dense) and less perfectly spherical because they rotate more rapidly. The Jovian planets have deep atmospheres and are mostly composed of elements such as hydrogen and helium—liquid near the surface, but frozen toward the interior—as well as ices of compounds such as methane and ammonia.

It was long thought that tiny Pluto was the ninth and outermost planet in the solar system. However, astronomers have discovered other icy bodies that are similar to Pluto and orbit the Sun beyond Neptune in what is known as the *Kuiper Belt* or *trans-Neptunian region*. In June 2008, the International Astronomical Union reclassified Pluto as a special type of dwarf planet known as a *plutoid*. Several dozen yet-to-be-discovered plutoids and other dwarf planets may be in the outer reaches of our solar system.

LearningCheck 1-7 **Contrast the characteristics of the terrestrial and Jovian planets in our solar system.**

The Size and Shape of Earth

Is Earth large or small? The answer to this question depends on your frame of reference. If the frame of reference is the universe, Earth is almost infinitely small. The diameter of our planet is only about 13,000 kilometers (7900 miles), a tiny distance at the scale of the universe. For example, the Moon is 385,000 kilometers (239,000 miles) from Earth, the Sun is 150,000,000 kilometers (93,000,000 miles) away, and the nearest star is 40 trillion kilometers (25 trillion miles) distant.

The Size of Earth: In a human frame of reference, however, Earth is impressive in size. Its surface varies in elevation from the highest mountain peak, Mount Everest, at about 8848 meters (29,029 feet) above sea level, to the deepest oceanic trench, the Mariana Trench of the Pacific Ocean, at about 11,033 meters (36,198 feet) below sea level—a total difference in elevation of about 19,881 meters (65,227 feet).[2]

Although prominent on a human scale of perception, the difference between the highest and lowest places is minor on the planetary scale (**Figure 1-7**). If Earth were the size of a basketball, Mount Everest would be an imperceptible pimple no greater than 0.17 millimeter (about 7 thousandths of an inch) high. Similarly, the Mariana Trench would be a tiny crease only 0.21 millimeter (about 8 thousandths of an inch) deep—this represents a depression smaller than the thickness of a sheet of paper.

Our perception of the relative size of topographic irregularities on Earth is often distorted by maps and globes that emphasize such landforms. To portray any noticeable appearance of topographic variation, the vertical dimension on such maps is usually exaggerated 8 to 20 times—as are many diagrams used in this book. Furthermore, many diagrams illustrating features of the atmosphere also exaggerate relative sizes to convey important concepts.

[2]As more precise data become available through the use of GPS (see Chapter 2), the accepted elevations of points on Earth can change. The "height modernization" program of the National Oceanic and Atmospheric Administration in the United States is scheduled for completion by 2023; once the new vertical "datum" is established, the precise elevations of some places may change—but only on paper!

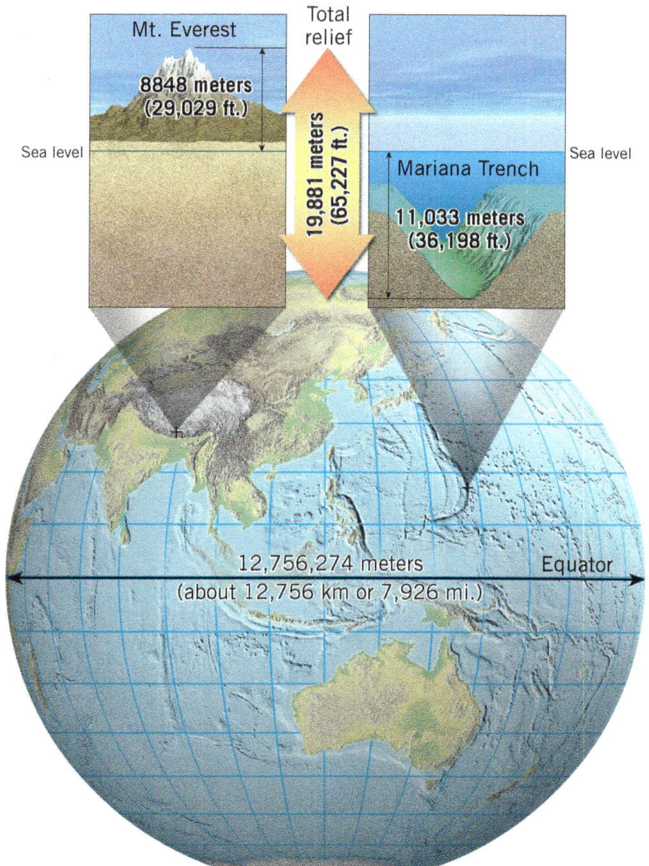

▲ **Figure 1-7** Earth is large relative to the size of its surface features. Earth's maximum relief (the difference in elevation between the highest and lowest points) is 19,881 meters (65,227 feet) or about 20 kilometers (12 miles) from the top of Mount Everest to the bottom of the Mariana Trench in the Pacific Ocean.

More than 2600 years ago, Greek scholars correctly reasoned that Earth has a spherical shape. About 2200 years ago, Eratosthenes, the director of the Greek library at Alexandria, calculated the circumference of Earth trigonometrically. He determined the angle of the noon Sun's rays at Alexandria and at the city of Syene, 800 kilometers (500 miles) away. From these angular and linear distances, he was able to estimate an Earth circumference of almost 43,000 kilometers (26,700 miles), which is reasonably close to the actual figure of 40,000 kilometers (24,900 miles).

The Shape of Earth: Earth is not quite spherical. The cross section revealed by a cut through the equator would be circular, but a similar cut from pole to pole would be an ellipse rather than a circle. Any rotating body has a tendency to bulge around its equator and flatten at the polar ends of its rotational axis. Although the rocks of Earth may seem quite rigid, they are pliable enough to allow Earth to develop a bulge around its middle. The slightly flattened polar diameter of Earth is 12,714 kilometers (7900 miles), whereas the slightly bulging equatorial diameter is 12,756 kilometers (7926 miles), a difference of only about 0.3 percent (**Figure 1-8**). Thus, our planet is properly described as an *oblate spheroid* rather than a true sphere. However, because this variation from true sphericity is exceedingly small, in most cases in this book we will treat Earth as if it were a perfect sphere.

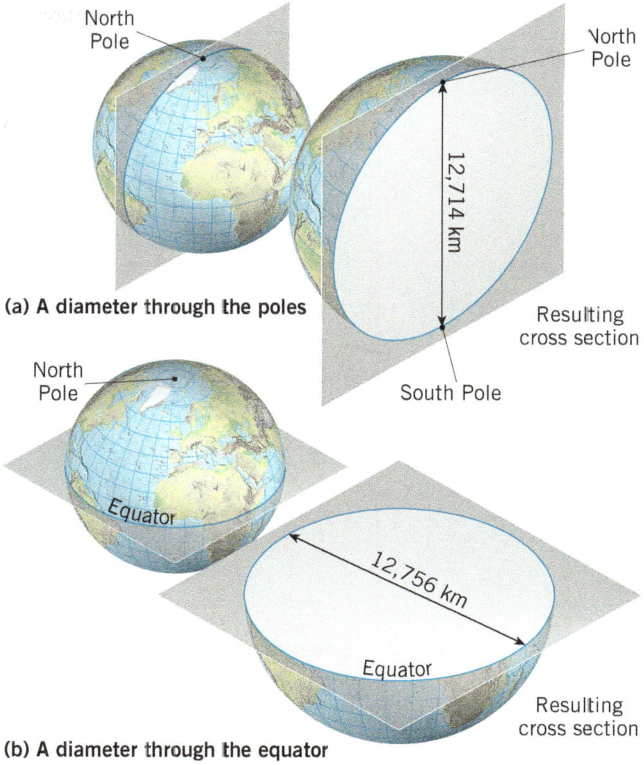

▲ **Figure 1-8** Earth is not quite a perfect sphere. Its surface flattens slightly at the North Pole and the South Pole and bulges out slightly around the equator. Thus, (a) a cross section through the poles has a diameter slightly less than the diameter of (b) a cross section through the equator.

LearningCheck 1-8 **What are Earth's highest and lowest points, and what is the approximate elevation difference between them?**

The Geographic Grid—Latitude and Longitude

Any understanding of the distribution of geographic features over Earth's surface requires some system of accurate location. The simplest technique for achieving this is a grid system consisting of two sets of lines that intersect at right angles, allowing the location of any point on the surface to be described by the appropriate intersection (**Figure 1-9**). Such a rectangular grid system has been reconfigured for Earth's spherical surface.

If our planet wasn't rotating, the problem of describing location would be more difficult than it is: imagine trying to describe the location of a particular point on a perfectly round, perfectly clean ball. Because Earth does rotate, we can use its rotation axis as a starting point to describe locations.

Earth's rotation axis is an imaginary line passing through Earth that connects the points on the surface called the **North Pole** and the **South Pole** (**Figure 1-10**). Furthermore, if we visualize an imaginary plane passing through Earth halfway between the poles and perpendicular to the axis of rotation, we have another valuable reference

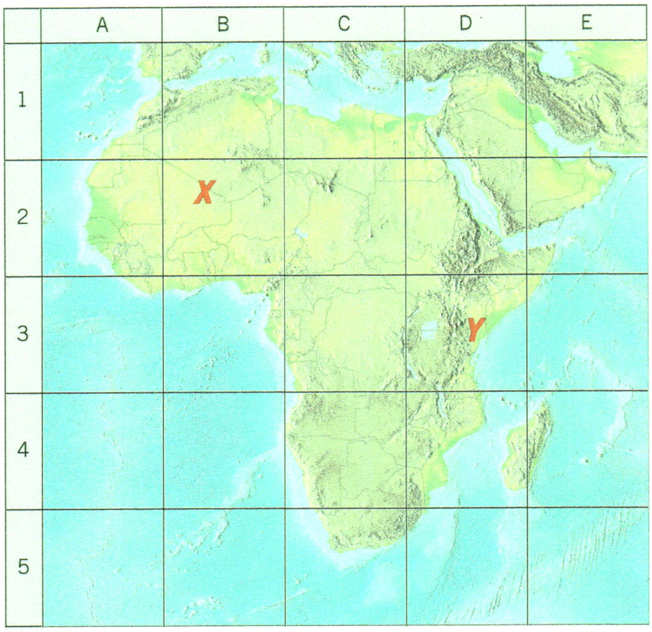

▲ **Figure 1-9** An example of a grid system. The location of point *X* can be described as 2B or as B2; the location of *Y* is 3D or D3.

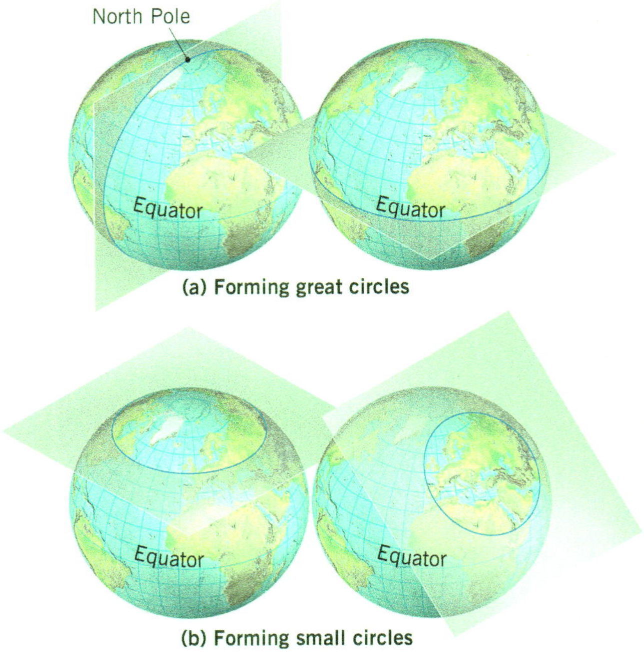

▲ **Figure 1-11** (a) A great circle results from the intersection of Earth's surface with any plane that passes through Earth's center. (b) A small circle results from the intersection of Earth's surface with any other plane.

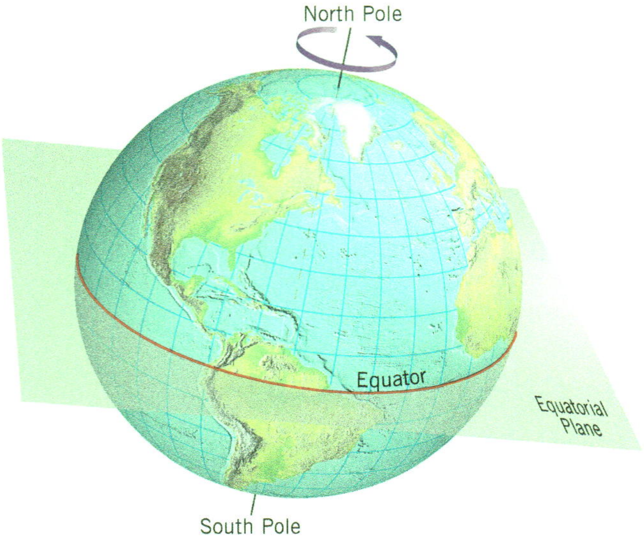

▲ **Figure 1-10** Earth spins around its rotation axis, an imaginary line that passes through the North Pole and the South Pole. An imaginary plane bisecting Earth midway between the two poles defines the equator.

feature: the *plane of the equator*. Where this plane intersects Earth's surface is the imaginary midline of Earth, called simply the **equator**. We use the North Pole, South Pole, rotational axis, and equatorial plane as natural reference features for measuring and describing locations on Earth's surface.

Great Circles: Any plane that passes through the center of a sphere bisects that sphere (divides it into equal halves) and creates what is called a **great circle** where it intersects the surface of the sphere (Figure 1-11a). The equator is a great circle. Planes passing through any other part of the sphere produce what are called *small circles* where they intersect the surface (Figure 1-11b).

Great circles have two properties of special interest:

1. A great circle is the largest circle that can be drawn on a sphere, dividing its surface into two equal halves, or *hemispheres*. As we see later in this chapter, the dividing line between the daytime and nighttime halves of Earth is a great circle.
2. A path between two points along the arc of a great circle is always the shortest route between those points. Such routes on Earth are known as *great circle routes*. (We discuss great circle routes in more detail in Chapter 2.)

The geographic grid used as the locational system for Earth is based on the principles just discussed. This locational system is closely linked with the various positions of Earth in its orbit around the Sun. Earth's grid system, called a *graticule*, consists of lines of *latitude* and *longitude*.

LearningCheck 1-9 **What is a great circle? Provide one example of a great circle.**

Latitude

Latitude is a description of location expressed as an angle north or south of the equator. As shown in Figure 1-12, we can project a line from any location on Earth's surface to the center of Earth. The angle between this line and the equatorial plane is the latitude of that location.

Latitude is expressed in *degrees*, *minutes*, and *seconds*. There are 360 degrees (°) in a circle, 60 minutes (′) in one degree, and 60 seconds (″) in one minute. With the advent of GPS navigation (discussed in Chapter 2), it is increasingly common to see latitude and longitude designated using decimal notation. For example, 38°22′47″ N can be written 38°22.78′ N (47″ is 78 percent of one minute) or even 38.3797° N (22′47″ is 37.97 percent of one degree).

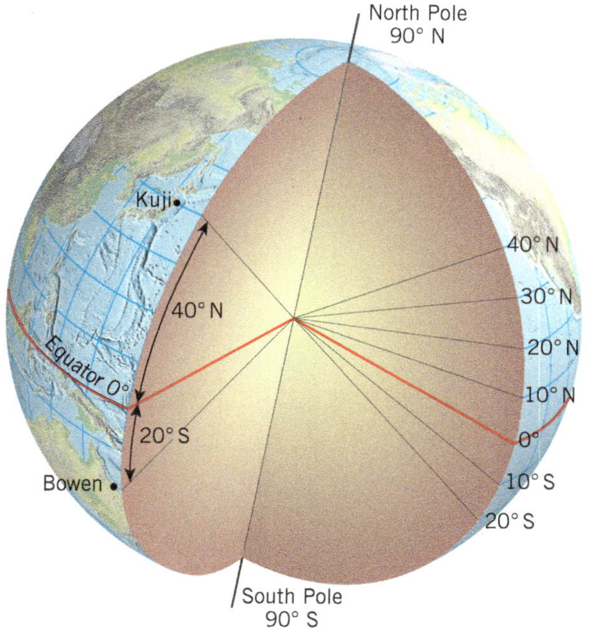

▲ **Figure 1-12** Measuring latitude. An imaginary line from Kuji, Japan, to Earth's center makes an angle of 40° with the equator. Therefore, Kuji's latitude is 40° N. An imaginary line from Bowen, Australia, to Earth's center makes an angle of 20°, giving this city a latitude of 20° S.

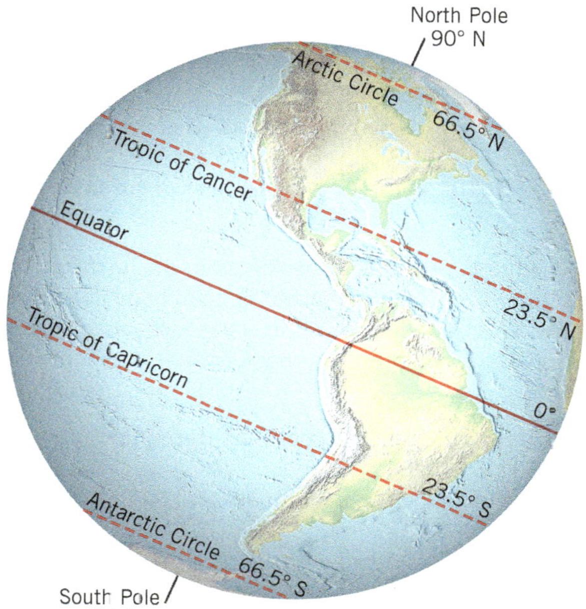

▲ **Figure 1-14** Seven important latitudes. As we will see when we discuss the seasons, these latitudes represent special locations where rays from the Sun strike Earth's surface on certain days of the year.

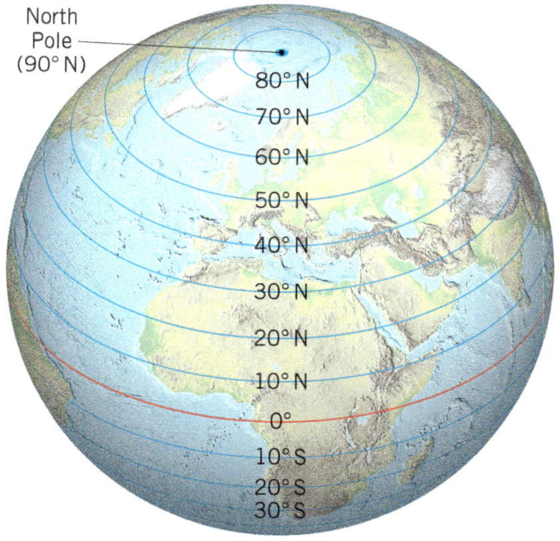

▲ **Figure 1-13** Lines of latitude indicate north–south location. They are called parallels because they are always parallel to each other.

◀ **Figure 1-15** The equator, like all other parallels of latitude, is an imaginary line. Here at Mitad del Mundo near Quito, Ecuador, its nearby location is commemorated by a monument.

Latitude varies from 0° at the equator to 90° north at the North Pole and 90° south at the South Pole. Any position north of the equator is north latitude, and any position south of the equator is south latitude.[3] The equator itself is simply assigned a latitude of 0°.

A line connecting all points of the same latitude is called a **parallel**—because it is parallel to all other lines of latitude (**Figure 1-13**). The equator is the parallel of 0° latitude, and it, alone of all parallels, constitutes a great circle. All other parallels are small circles—all aligned in true east–west directions on Earth's surface.

Although we could visualize an unlimited number of parallels, seven latitudes are of particular significance in a general study of Earth (**Figure 1-14**):

1. Equator, 0° (**Figure 1-15**)
2. Tropic of Cancer, 23.5° N
3. Tropic of Capricorn, 23.5° S
4. Arctic Circle, 66.5° N
5. Antarctic Circle, 66.5° S
6. North Pole, 90° N
7. South Pole, 90° S

[3]North latitudes are sometimes designated with positive numbers and south latitudes with negative numbers; east longitudes are designated with positive numbers and west longitudes with negative numbers.

The North Pole and South Pole are points rather than lines, but we can think of them as infinitely small parallels. The significance of these seven parallels is explained later in this chapter when we discuss the seasons.

LearningCheck 1-10 **Why are lines of latitude called parallels?**

Descriptive Zones of Latitude: Regions on Earth are sometimes described as falling within general bands or zones of latitude. The following common terms associated with latitude are used throughout this book (notice that some terms overlap):

- *Low latitude*—generally between the equator and 30° N and 30° S
- *Midlatitude*—between about 30° and 60° N and 30° and 60° S
- *High latitude*—latitudes greater than about 60° N and 60° S
- *Equatorial*—within a few degrees of the equator
- *Tropical*—within the tropics (between 23.5° N and 23.5° S)
- *Subtropical*—slightly poleward of the tropics, generally around 25°–30° N and 25°–30° S
- *Polar*—within a few degrees of the North or South Pole

Nautical Miles: Each degree of latitude on the surface of Earth covers a north–south distance of about 111 kilometers (69 miles). The distance varies slightly with latitude because of the flattening of Earth at the poles. The distance measurement of a *nautical mile*—and the description of speed known as a *knot* (one nautical mile per hour)—is defined by the distance covered by one minute of latitude (1′), the equivalent of about 1.15 statute ("ordinary") miles or about 1.85 kilometers.

Longitude

Longitude describes east–west location on Earth. Like latitude, it is an angular description of location measured in degrees, minutes, and seconds.

Longitude is represented by imaginary lines extending from pole to pole and crossing all parallels at right angles. These lines, called **meridians**, are not parallel to one another except where they cross the equator. Notice that any pair of meridians is farthest apart at the equator, becoming increasingly close together northward and southward and finally converge at the poles (Figure 1-16).

Establishing the Prime Meridian: The equator is a natural baseline from which to measure latitude, but no such natural reference line exists for longitude. Consequently, for most of recorded history, there was no accepted longitudinal baseline; each country selected its own "prime meridian" as the reference line for east–west measurement. At least 13 different prime meridians were in use in the 1880s.

In 1884, an international conference was convened in Washington, D.C., to establish global time zones and select a single prime meridian. After weeks of debate, the conference chose the meridian passing through the Royal Observatory at Greenwich, England, just east of central London, as the **prime meridian** for all longitudinal measurement (Figure 1-17). The principal argument for adopting the Greenwich meridian as the prime meridian was a practical one: more than two-thirds of the world's shipping lines already used the Greenwich meridian as a navigational base.

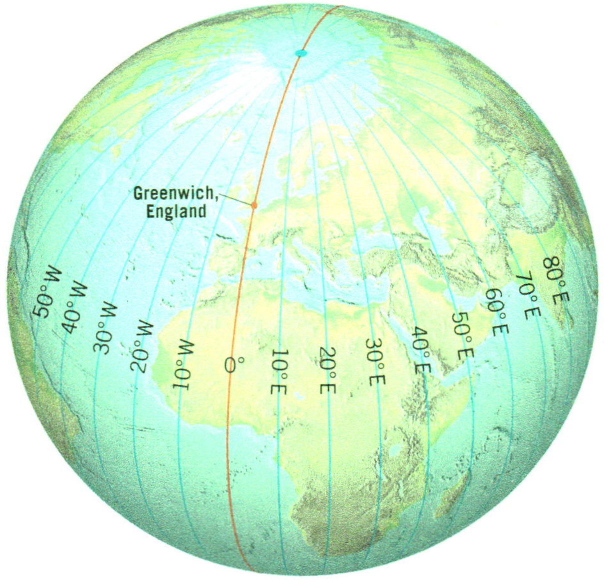

▲ Figure 1-16 Lines of longitude, or meridians, indicate east–west location and all converge at the poles.

▲ Figure 1-17 The prime meridian of the world, longitude 0°0′0″ at Greenwich, England, which is about 8 kilometers (5 miles) from the heart of London.

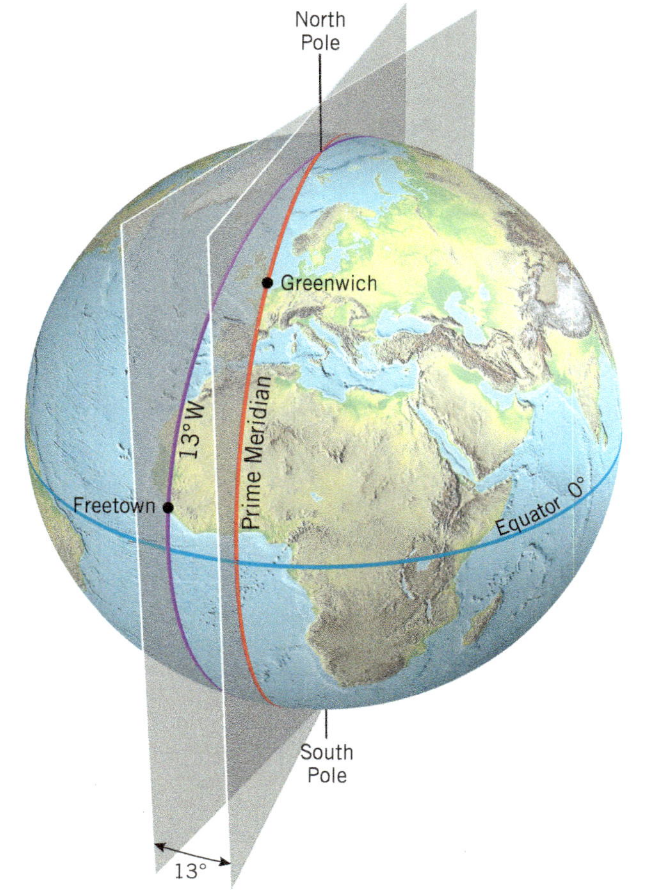

▲ **Figure 1-18** The meridians that mark longitude are defined by intersecting imaginary planes passing through the poles. Shown here are the planes for the prime meridian through Greenwich, England, and the meridian through Freetown, Sierra Leone, at 13° west longitude.

Thus, an imaginary north–south plane passing through Greenwich and through Earth's axis of rotation represents the plane of the prime meridian. The angle between this plane and a plane passed through any other point and the axis of Earth is a measure of longitude. For example, the angle between the Greenwich plane and a plane passing through the center of the city of Freetown (in the western African country of Sierra Leone) is 13 degrees, 15 minutes, and 12 seconds. Because the angle is formed west of the prime meridian, the longitude of Freetown is written 13°15′12″ W (**Figure 1-18**).

Measuring Longitude: Longitude is measured both east and west of the prime meridian to a maximum of 180° in each direction. Exactly halfway around the globe from the prime meridian, in the middle of the Pacific Ocean, is the 180° meridian (**Figure 1-19**). All places on Earth, then, have a location that is either east longitude or west longitude, except for points exactly on the prime meridian (described simply as 0° longitude) or exactly on the 180th meridian (described as 180° longitude).

The distance between any two meridians varies predictably. At the equator, the surface length of one degree of longitude is about the same as that of one degree of latitude. However, because meridians converge at the poles, the distance covered by one degree of longitude decreases poleward (**Figure 1-20**), diminishing to zero at the poles, where all meridians meet.

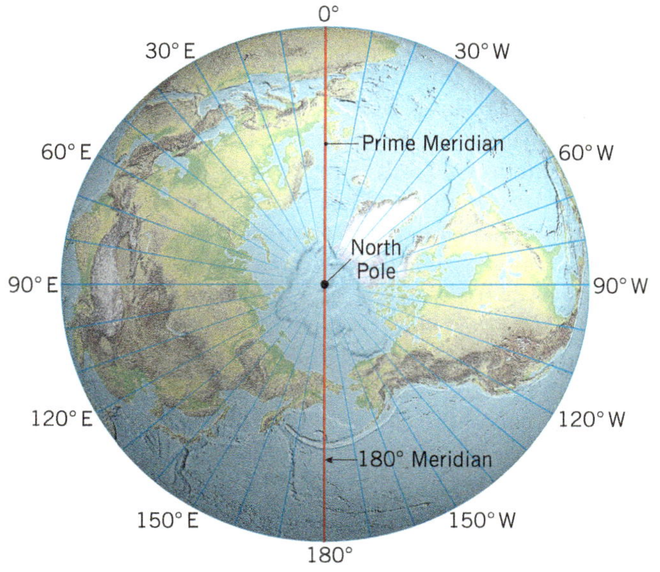

▲ **Figure 1-19** A polar view of meridians radiating from the North Pole. Think of each line as the top edge of an imaginary plane passing through both poles. All the planes are perpendicular to the plane of the page.

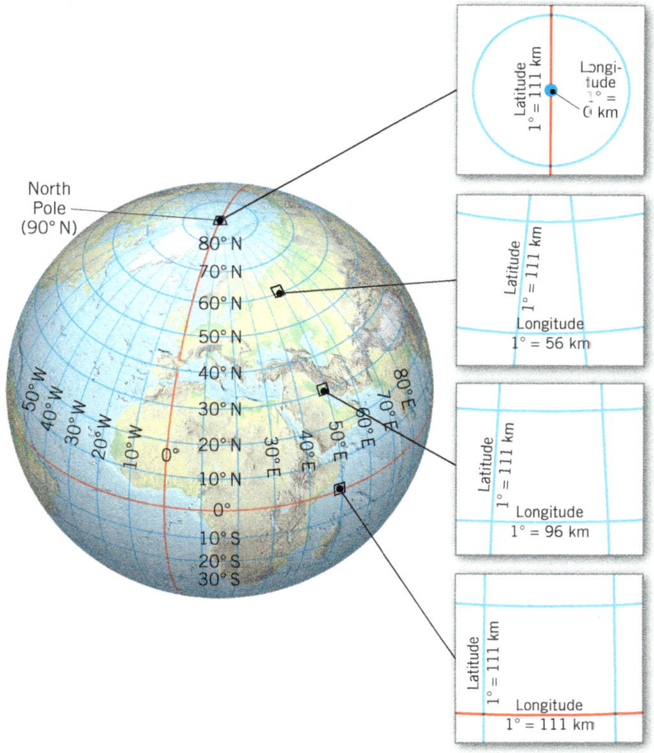

▲ **Figure 1-20** The complete grid system of latitude and longitude—the graticule. Because the meridians converge at the poles, the distance of 1° of longitude is greatest at the equator and diminishes to zero at the poles, whereas the distance of 1° of latitude varies only slightly (due to the slight flattening of Earth at the poles).

Locating Points on the Geographic Grid

The network of intersecting parallels and meridians creates a geographic grid over the entire surface of Earth (see Figure 1-20). The location of any place on Earth's surface can be described with great precision by reference to detailed latitude and longitude data. For example, at the 1964 World's Fair in New York City, a time capsule was buried. For reference purposes, the U.S. Coast and Geodetic Survey determined that the capsule was located at 40°44′34.089″ N and 73°50′43.842″ W. At some time in the future, if a hole were to be dug at the spot indicated by those coordinates, it would be within 15 centimeters (6 inches) of the capsule.

LearningCheck 1-11 **Are locations in North America described by east longitude or west longitude? Locations in China?**

Earth–Sun Relations and the Seasons

Animation **Earth-Sun Relations** https://media.pearsoncmg.com/ph/esm/esm_mcknight_physgeo_13/media/

Nearly all life on Earth depends on solar energy; therefore, the relationship between Earth and the Sun is of vital importance. Because of the constant motions of Earth, this relationship does not remain the same throughout the year. We begin with a description of Earth movements and the relationship of Earth's axis to the Sun, and then we offer an explanation of the change of seasons.

Earth Movements

Two basic Earth movements—its daily rotation on its axis and its annual revolution around the Sun—along with the inclination and "polarity" of Earth's rotation axis, combine to change Earth's orientation to the Sun. This change produces the change of seasons.

Earth's Rotation on Its Axis: Earth rotates from west to east on its axis (Figure 1-21), a complete **rotation** requiring 24 hours. (From the vantage point of looking down at the North Pole from space, Earth is rotating counterclockwise.) The Sun, the Moon, and the stars appear to rise in the east and set in the west—this is, of course, an illusion created by the steady eastward spin of Earth.

Rotation causes all parts of Earth's surface except the poles to move in a circle around Earth's axis. Although the speed of rotation varies by latitude (see Figure 1-21), it is constant at any given place on Earth; thus we experience no sense of motion. This is the same reason that we have little sense of motion on a smooth jet airplane flight at cruising speed—only when speed changes, such as during takeoff and landing, does motion become apparent.

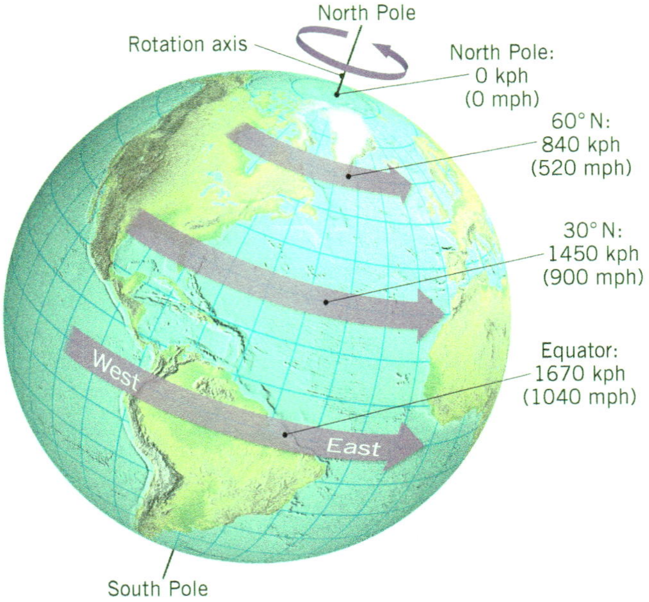

▲ **Figure 1-21** Earth rotates from west to east. Looking down at the North Pole from above, Earth appears to rotate counterclockwise. The speed of Earth's rotation is constant but varies by latitude, being greatest at the equator and effectively diminishing to zero at the poles. The speed of rotation at different latitudes is shown in kilometers per hour (kph) and miles per hour (mph).

Rotation has several important effects on the physical characteristics of Earth's surface:

1. The most obvious effect of Earth's rotation is the *diurnal* (daily) alternation of daylight and darkness, as portions of Earth's surface are turned first toward and then away from the Sun. This variation in exposure to sunlight greatly influences local temperature, humidity, and wind movements. Except for organisms that live in caves or in the deep ocean, nearly all forms of life have adapted to this sequential pattern of daylight and darkness. For example, we humans fare poorly when our *circadian* (24-hour cycle) rhythms are disrupted by long-distance, high-speed air travel, leaving us with a sense of fatigue known as "jet lag."
2. The rotation of Earth brings any point on the surface through the increasing and then decreasing gravitational pull of the Moon and the Sun. Although the land areas of Earth are too rigid to be significantly moved by these oscillating gravitational attractions, oceanic waters move onshore and then recede in a rhythmic pattern of *tides*, discussed further in Chapter 9.
3. Earth's constant rotation also causes an apparent deflection in the paths of both wind and ocean currents—to the right in the Northern Hemisphere and to the left in the Southern Hemisphere. This phenomenon is called the *Coriolis effect* and is discussed in detail in Chapter 3.

Earth's Revolution around the Sun: Another significant Earth motion is its **revolution** or orbit around the Sun. Each revolution takes 365 days, 5 hours, 48 minutes, and 46 seconds, or 365.242199 days. This is known formally as

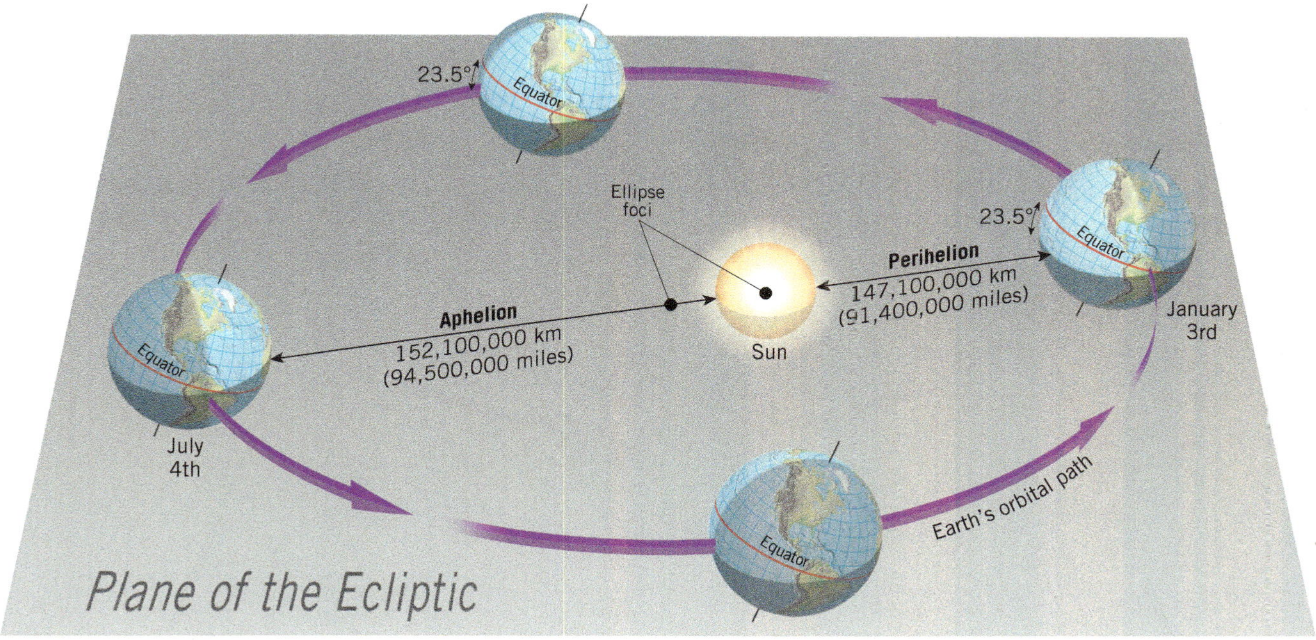

▲ **Figure 1-22** Earth reaches perihelion (its closest point to the Sun) on about January 3 and aphelion (its farthest point from the Sun) on about July 4. The plane of the ecliptic is the orbital plane of Earth. Because Earth's rotation axis is tilted, the plane of the ecliptic and the equatorial plane do not coincide. The path Earth follows in its revolution around the Sun is an ellipse with the Sun not exactly in the center. In this diagram the elliptical shape of Earth's orbit is greatly exaggerated. The Sun is actually much larger than Earth, but both are shown exaggerated in size. Notice that Earth's axis always points in the same direction throughout its orbit.

the *tropical year* and for practical purposes is usually simplified to 365.25 days.

The path followed by Earth in its journey around the Sun is not a true circle but an *ellipse* (**Figure 1-22**). Because of this elliptical orbit, the Earth–Sun distance is not constant. Rather, it varies from approximately 147,100,000 kilometers (91,400,000 miles) at the closest or **perihelion** position (*peri* is from the Greek and means "around" and *helios* means "Sun") on about January 3, to approximately 152,100,000 kilometers (94,500,000 miles) at the farthest or **aphelion** position (*ap* is from the Greek and means "away from") on about July 4. The average Earth–Sun distance is defined as one *astronomical unit* (1 AU)—149,597,871 kilometers (92,960,117 miles).

Earth is 3.3 percent closer to the Sun during the Northern Hemisphere winter than during the Northern Hemisphere summer, an indication that variations in the distance between Earth and the Sun do *not* cause the change of seasons. Instead, two additional factors in the relationship of Earth to the Sun—*inclination* and *polarity*—work together with rotation and revolution to produce the change of seasons.

LearningCheck 1-12 **Distinguish between Earth's rotation and its revolution.**

Inclination of Earth's Axis: Earth's rotation axis is not perpendicular to the imaginary plane defined by Earth's orbital path around the Sun, called the **plane of the ecliptic** (see Figure 1-22). Rather, the axis is tilted about 23.5° from the perpendicular (**Figure 1-23**) and maintains this tilt throughout the year. This tilt is referred to as the **inclination of Earth's axis**.

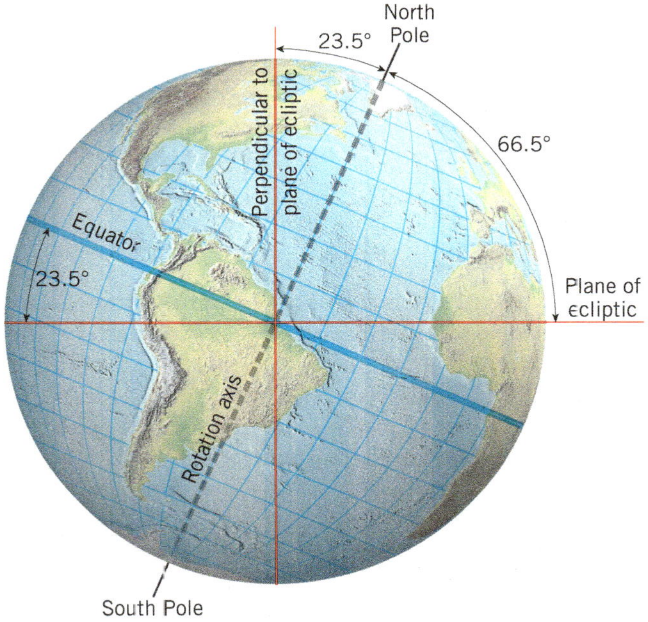

▲ **Figure 1-23** Earth's rotation axis is inclined 23.5° from a line perpendicular to the plane of the ecliptic.

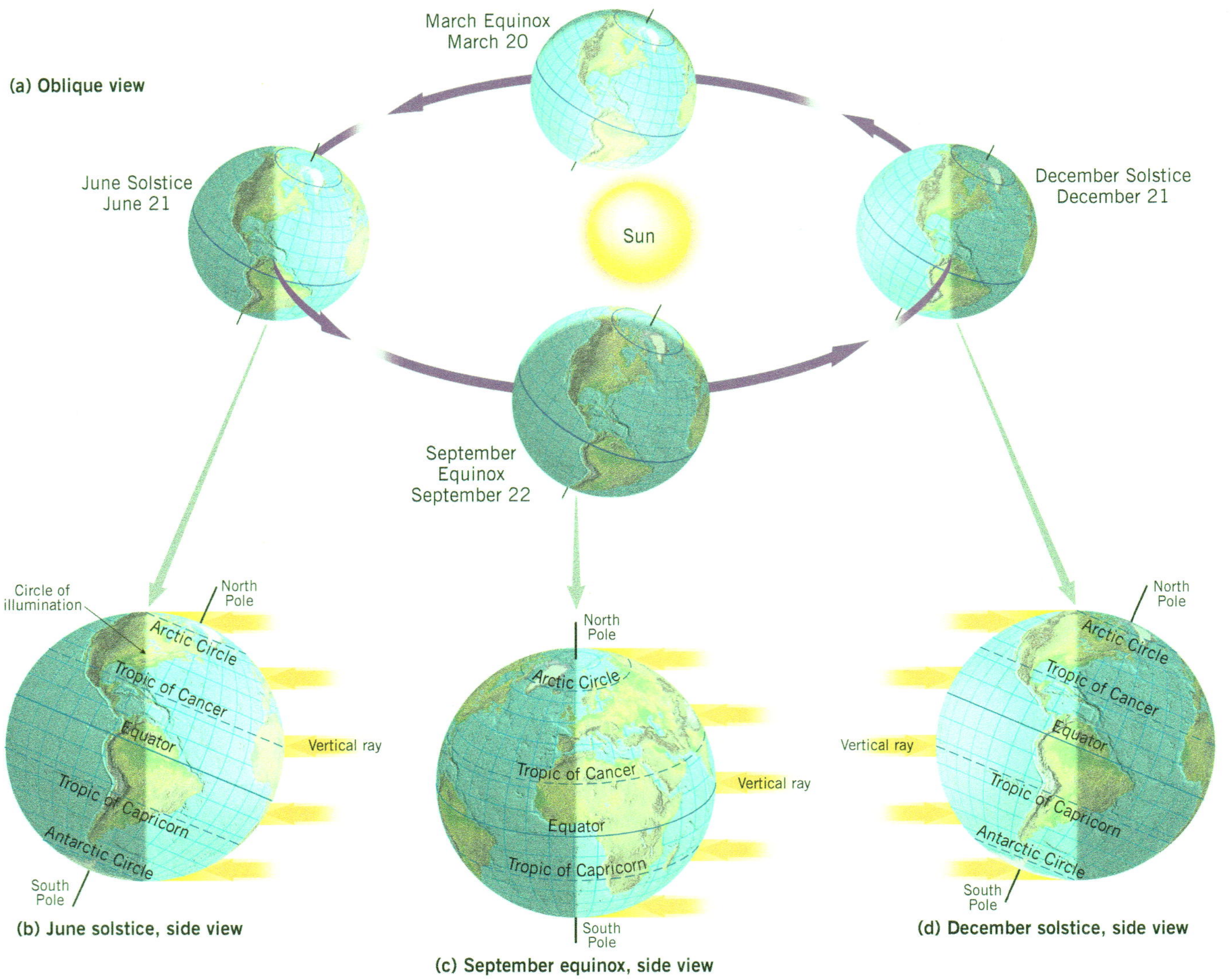

▲ **Figure 1-24** (a) The annual march of the seasons, showing Earth–Sun relations on the June solstice, September equinox, December solstice, and March equinox (the dates shown are approximate). (b) On the June solstice, the vertical rays of the noon Sun strike 23.5° N latitude. The circle of illumination is the dividing line between the daylight and nighttime halves of Earth. (c) On the March equinox and September equinox, the vertical rays of the noon Sun strike the equator. (d) On the December solstice, the vertical rays of the noon Sun strike 23.5° S latitude.

Polarity of Earth's Axis: Not only is Earth's rotation axis inclined relative to its orbital path, but also no matter where Earth is in its orbit around the Sun, the axis always points in the same direction relative to the stars—toward the North Star, Polaris (Figure 1-24). This characteristic is called the **polarity of Earth's axis** (or **parallelism**, because at any time of the year Earth's axis is parallel to its orientation at all other times).

The combined effects of rotation, revolution, inclination, and polarity result in the seasonal patterns experienced on Earth. Notice in Figure 1-24 that at one point in Earth's orbit, during the Northern Hemisphere summer, the North Pole is oriented most directly toward the Sun, whereas six months later, during the Northern Hemisphere winter, the North Pole is oriented most directly away from the Sun. This is the most fundamental feature of the annual march of the seasons.

LearningCheck 1-13 **Does the North pole lean toward the Sun throughout the year? If not, how does the North pole's orientation change during the year?**

The Annual March of the Seasons

During a year, the changing relationship of Earth to the Sun results in variations in day length and in the angle at which the Sun's rays strike the surface of Earth. These changes are most obvious in the mid- and high latitudes, but important variations take place within the tropics as well.

TABLE 1-2 Conditions on Equinoxes and Solstices

	March Equinox	June Solstice	September Equinox	December Solstice
Latitude of vertical rays of Sun	0°	23.5° N	0°	23.5° S
Day length at equator	12 hours	12 hours	12 hours	12 hours
Day length in midlatitudes of Northern Hemisphere	12 hours	Becomes longer with increasing latitude north of equator	12 hours	Day length becomes shorter with increasing latitude north of equator
Day length in midlatitudes of Southern Hemisphere	12 hours	Becomes shorter with increasing latitude south of equator	12 hours	Day length becomes longer with increasing latitude south of equator
24 hours of daylight	Nowhere	From Arctic Circle to North Pole	Nowhere	From Antarctic Circle to South Pole
24 hours of darkness	Nowhere	From Antarctic Circle to South Pole	Nowhere	From Arctic Circle to North Pole
Season in Northern Hemisphere	Spring	Summer	Autumn	Winter
Season in Southern Hemisphere	Autumn	Winter	Spring	Summer

As we discuss the annual march of the seasons, we pay special attention to three conditions:

1. The latitude receiving the vertical rays of the Sun (rays striking the surface at a right angle), referred to as the **declination of the Sun**.
2. The **solar altitude** (the height of the noon Sun above the horizon) at different latitudes.
3. The length of day (number of daylight hours) at different latitudes.

Initially, we emphasize the conditions on four special days of the year: the June solstice, the September equinox, the December solstice, and the March equinox (see Figure 1-24a and Table 1-2). As we describe the change of seasons, the significance of the "seven important parallels" discussed earlier in this chapter will become clear. We begin with the June solstice.

June Solstice: On the **June solstice**, which occurs on or about June 21 (the exact date varies slightly from year to year), Earth reaches the position in its orbit where the North Pole is oriented most directly toward the Sun. On this day, the vertical rays of the Sun strike the **Tropic of Cancer**, 23.5° north of the equator (Figure 1-24b). Were you at the Tropic of Cancer on this day, the Sun would be directly overhead in the sky at noon (in other words, the solar altitude would be 90°). The Tropic of Cancer marks the northernmost latitude reached by the vertical rays of the Sun during the year.

The dividing line between the daylight half of Earth and nighttime half of Earth is a great circle called the **circle of illumination**. On the June solstice, the circle of illumination bisects the equator (Figure 1-24b), so on this day the equator receives equal day and night—12 hours of daylight and 12 hours of darkness. However, as we move north of the equator, the portion of each parallel in daylight increases—in other words, day length increases. Conversely, day length decreases as we move south of the equator.

Notice in Figure 1-24b that on the June solstice, the circle of illumination reaches 23.5° *beyond* the North Pole to a latitude of 66.5° N. As Earth rotates, all locations north of 66.5° remain continuously in daylight and so experience 24 hours of daylight. By contrast, all points south of 66.5° S are always outside the circle of illumination and so have 24 continuous hours of darkness. These special parallels defining the equatorward limit of 24 hours of light and dark on the solstices are called the *polar circles*. The northern polar circle, at 66.5° N, is the **Arctic Circle**; the southern polar circle, at 66.5° S, is the **Antarctic Circle**.

The June solstice is called the *summer solstice* in the Northern Hemisphere and the *winter solstice* in the Southern Hemisphere. (These are commonly called the "first day of summer" and the "first day of winter" in their respective hemispheres.)

LearningCheck 1-14 **What is the latitude of the vertical rays of the Sun on the June solstice?**

September Equinox: Three months after the June solstice, on about September 22, Earth experiences the **September equinox**. Notice in Figure 1-24c that the vertical rays of the Sun strike the equator. Notice also that the circle of illumination just touches both poles, bisecting all other parallels—on this day all locations on Earth experience 12 hours of daylight and 12 hours of darkness. (The word "equinox" comes from the Latin, meaning "the time of equal days and equal nights.") At the equator—and only at the equator—every day of the year has virtually 12 hours of daylight and

12 hours of darkness; all other locations have equal day and night only on an equinox.

The September equinox is called the *autumnal equinox* in the Northern Hemisphere and the *vernal equinox* in the Southern Hemisphere. (These are commonly called the "first day of fall" and the "first day of spring" in their respective hemispheres.)

December Solstice: On the **December solstice**, which occurs on about December 21, Earth reaches the position in its orbit where the North Pole is oriented most directly away from the Sun. The vertical rays of the Sun now strike 23.5° S, the **Tropic of Capricorn** (Figure 1-24d). Once again, the circle of illumination reaches to the far side of one pole and falls short on the near side of the other pole, so areas north of the Arctic Circle are in continuous darkness, whereas areas south of the Antarctic Circle are in daylight for 24 hours.

The relationships between Earth and the Sun on the June solstice and the December solstice are very similar; the conditions in each hemisphere are simply reversed. The December solstice is called the *winter solstice* in the Northern Hemisphere and the *summer solstice* in the Southern Hemisphere (the "first day of winter" and the "first day of summer," respectively).

March Equinox: Three months after the December solstice, on approximately March 20, Earth experiences the **March equinox**. The relationships of Earth and the Sun are virtually identical on the March and September equinoxes (compare Figures 1-24a and 1-24c). The March equinox is called the *vernal equinox* in the Northern Hemisphere and the *autumnal equinox* in the Southern Hemisphere (the "first day of spring" and the "first day of fall," respectively).

LearningCheck 1-15 **How much does day length at the equator change during the year?**

Seasonal Transitions

In the preceding discussion of the solstices and equinoxes, we emphasized the conditions on just four special days of the year. It is important to understand the transitions in day length and Sun angle that take place on other days as well.

Latitude Receiving the Vertical Rays of the Sun: The vertical rays of the Sun strike Earth only between the Tropic of Cancer and the Tropic of Capricorn. After the March equinox, the vertical rays of the Sun migrate north from the equator, striking the Tropic of Cancer on the June solstice (the day the Sun is highest in the sky for all latitudes north of the Tropic of Cancer). After the June solstice, the vertical rays migrate south, striking the equator again on the September equinox and reaching the Tropic of Capricorn on the December solstice (the day the Sun is lowest in the sky in the Northern Hemisphere). Following the December solstice, the vertical rays migrate northward, reaching the equator once again on the March equinox. The changing latitude of the vertical rays of the Sun during the year is shown graphically in Figure 1-25.

Day Length: Only at the equator is day length constant throughout the year—virtually 12 hours of daylight every day of the year.

For all regions in the Northern Hemisphere up to the latitude of the Arctic Circle, following the shortest day of the year on the December solstice, the number of hours of daylight gradually increases, reaching 12 hours on the March equinox. After the equinox, day length continues to increase until the longest day of the year, on the June solstice. (During this period, day length is diminishing in the Southern Hemisphere.)

Following the longest day of the year in the Northern Hemisphere on the June solstice, the pattern is reversed: the days get shorter in the Northern Hemisphere—reaching 12 hours on the September equinox. Day length continues

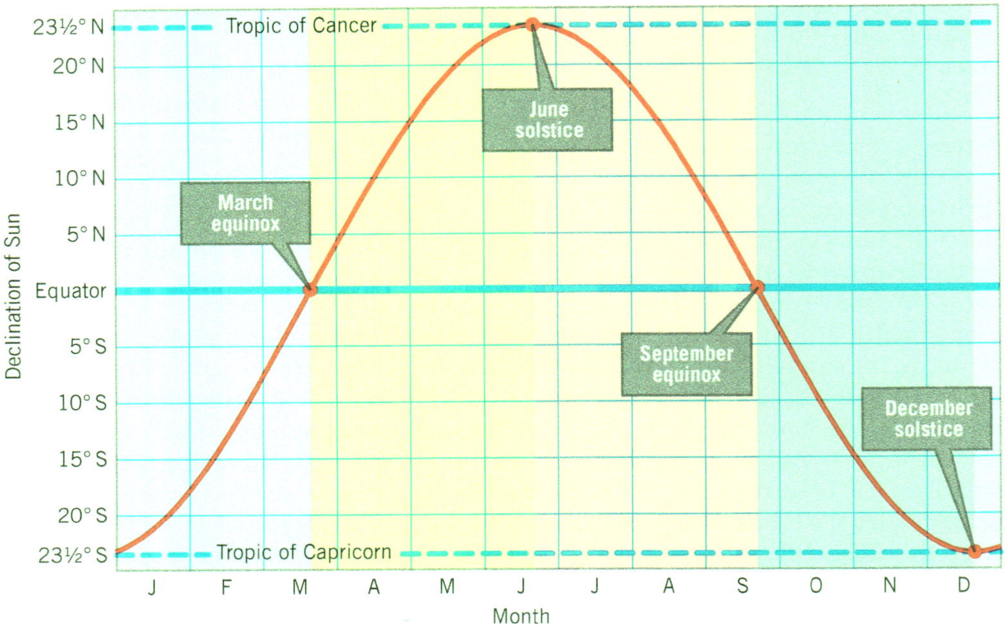

◀ **Figure 1-25** The latitude receiving the vertical rays of the noon Sun (the declination of the Sun) throughout the year. The vertical rays strike the Tropic of Cancer on the June solstice and the Tropic of Capricorn on the December solstice, crossing the equator on the equinoxes.

TABLE 1-3 Day Length and Noon Sun Angle on the June Solstice

Latitude	Day Length	Noon Sun Angle (degrees above horizon)
90° N	24 h	23.5
60° N	18 h 53 min	53.5
30° N	14 h 05 min	83.5
0°	12 h 07 min	66.5
30° S	10 h 12 min	36.5
60° S	05 h 52 min	6.5
90° S	0	0

Source: After Robert J. List, *Smithsonian Meteorological Tables*, 6th rev. ed. Washington, D.C.: Smithsonian Institution, 1963, Table 171.

to diminish until the shortest day of the year, on the December solstice. (During this period, day length is increasing in the Southern Hemisphere.)

Overall, the annual variation in day length is the least in the tropics and the greatest at high latitudes (Table 1-3).

LearningCheck 1-16 **On which days of the year do the vertical rays of the Sun strike the equator?**

Day Length in the Arctic and Antarctic: The patterns of day and night in the Arctic and Antarctic deserve special mention. For an observer exactly at the North Pole, the Sun rises on the March equinox and is above the horizon continuously for the next six months—circling the horizon higher and higher each day until the June solstice, after which it circles lower and lower until setting on the September equinox.

Week by week following the March equinox, a growing region surrounding the North Pole experiences 24 hours of daylight—until the June solstice, when the entire region from the Arctic Circle to the North Pole experiences 24 hours of daylight. Following the June solstice, the region of 24 hours of daylight diminishes week by week until the September equinox—when the Sun sets at the North Pole and remains below the horizon continuously for the next six months.

Week by week following the September equinox, the region around the North Pole experiencing 24 hours of darkness grows until the December solstice—when the entire region from the Arctic Circle to the North Pole experiences 24 hours of darkness. Following the December solstice, the region experiencing 24 hours of darkness diminishes week by week until the March equinox—when the Sun again rises at the North Pole.

In the Antarctic region of the Southern Hemisphere, these seasonal patterns are simply reversed.

Significance of Seasonal Patterns

Both day length and the angle at which the Sun's rays strike Earth determine the amount of solar energy received at any particular latitude. In general, the higher the Sun is in the sky, the more effective is the warming. Furthermore, short periods of daylight in winter and long periods of daylight in summer contribute to seasonal differences in temperature in the mid- and high-latitude regions.

Thus, the tropical latitudes are generally always warm because they have high Sun angles and consistent, near-12-hour days all year long. Conversely, the polar regions are consistently cold because they always have low Sun angles—even the 24-hour days in summer do not compensate for the low angle of incidence of sunlight. Seasonal temperature differences are large in the midlatitudes because of sizable seasonal variations in Sun angles and length of day. This topic will be explored further in Chapter 4.

LearningCheck 1-17 **For how many months of the year does the North Pole go without sunlight?**

Telling Time

To comprehend time around the world, we need an understanding of both (1) the geographic grid of latitude and longitude and (2) Earth–Sun relations.

In prehistoric times, the rising and setting of the Sun were probably the principal means of telling time. Local solar noon was determined by watching for the moment when an object cast its shortest shadows. The Romans used sundials to tell time (Figure 1-26) and gave great importance to the noon position, which they called the *meridiem*—the Sun's highest (*meri*) point of the day (*diem*). Our use of A.M. (ante meridiem: "before noon") and P.M. (post meridiem: "after noon") was derived from the Roman world.

When nearly all transportation was by foot, horse, or sailing vessel, it was difficult to compare time at different localities. Each community set its own time by correcting its local clocks to high noon at the moment of the shortest shadow.

▼ **Figure 1-26** A typical sundial. The edge of the vertical gnomon slants upward from the dial face at an angle equal to the latitude of the sundial, pointing toward the North Pole in the Northern Hemisphere and the South Pole in the Southern Hemisphere. As the Sun appears to move across the sky during the course of a day, the position of the shadow cast by the gnomon changes. The time shown on this sundial is about 2:00 P.M.

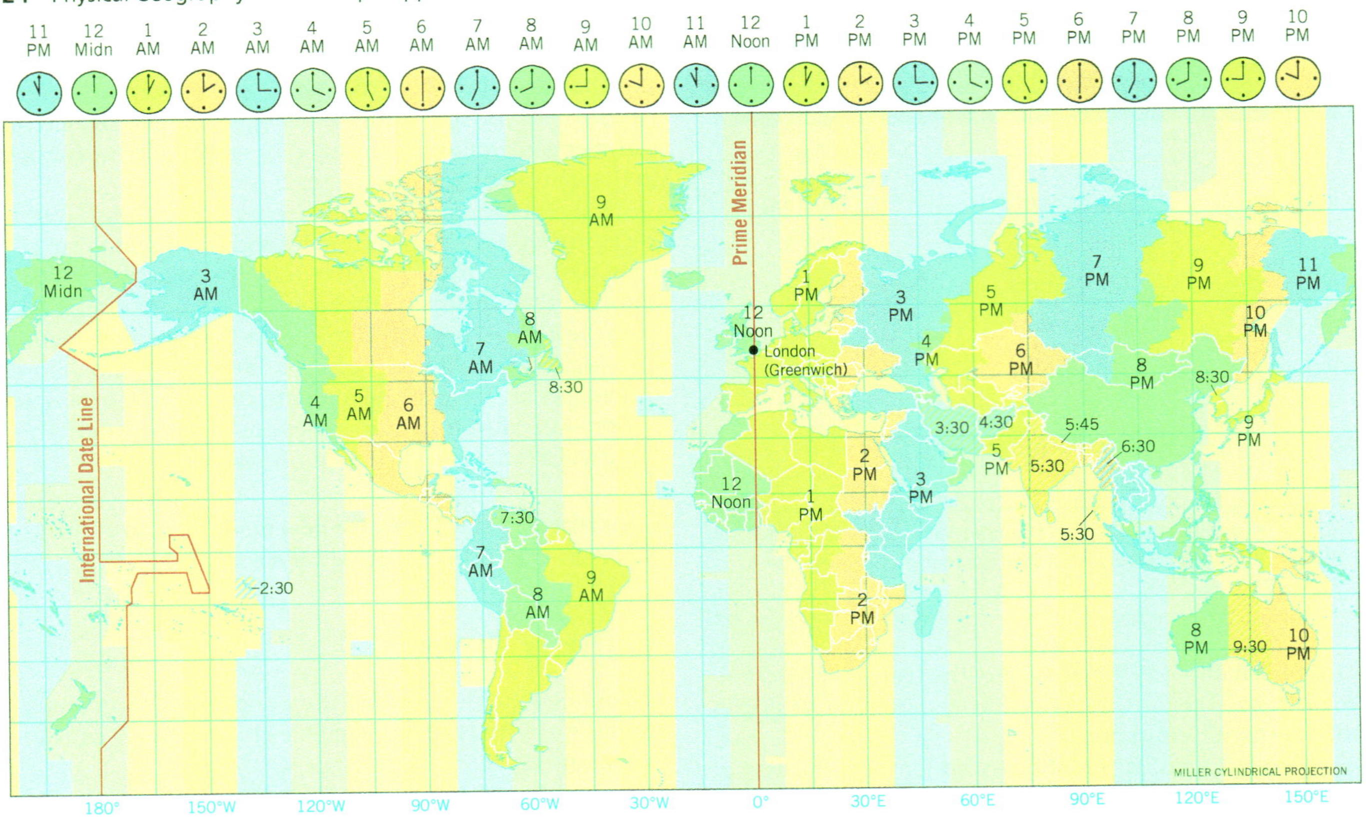

▲ Figure 1-27 The time zones of the world, each based on central meridians spaced 15° apart. Especially over land areas, these boundaries have been significantly adjusted.

Standard Time

As the telegraph and railroad began to speed communications and passengers between cities, the use of many different local times created increasing problems. Eventually, the railroads stimulated the development of a standardized time system.

At the 1884 International Prime Meridian Conference in Washington, D.C., countries established 24 central meridians, 15° of longitude apart, in order to divide the world into standard **time zones**. The mean (averaged) local solar time of the Greenwich prime meridian was chosen as the standard for the entire system. The prime meridian became the center of a time zone that extends 7.5° of longitude to the west and 7.5° to the east of the prime meridian. Similarly, the meridians that are multiples of 15° both east and west of the prime meridian were set as the *central meridians* for the other time zones (Figure 1-27). When you cross into the next time zone from west to east, the time becomes one hour later.

Although **Greenwich Mean Time (GMT)** is now referred to as **Coordinated Universal Time (UTC)**, the prime meridian is still the reference for standard time. To know the exact local time, we usually need to know only how many hours later or earlier our local time zone is compared with the time in Greenwich. Notice that a few countries, such as India, do not adhere to standard one-hour-interval time zones.

Most countries lie totally within a single time zone. However, some large countries may encompass several zones: Russia extends across 11 time zones; including Alaska and Hawai'i, the United States spreads over six (Figure 1-28a).

In international waters, time zones are exactly 15° wide. Over land areas, however, boundaries vary to coincide with political and economic boundaries. For example, the Central Standard Time Zone of the United States, centered on 90° W, extends all the way to 105° W (which is the central meridian of the Mountain Standard Time Zone) in Texas to keep most of that state within the same zone. But El Paso, Texas, is officially within the Mountain Standard Time Zone in accord with its role as a major market center for southern New Mexico, which observes Mountain Standard Time. At another extreme, China extends across four 15° zones, but the entire country, at least officially, observes the time of the 120° east meridian near Beijing.

In each time zone, the central meridian marks the location where clock time is the same as mean Sun time (i.e., the Sun reaches its highest point in the sky at 12:00 noon). On either side of that meridian, clock time does not coincide with Sun time. The deviation between the two is shown for one U.S. zone in Figure 1-28b.

LearningCheck 1-18 **What happens to the hour when you cross from one time zone to the next from west to east?**

International Date Line

In 1519, Ferdinand Magellan set out westward from Spain, sailing for East Asia with 270 men in five ships. Three years later, the remnants of his crew (18 men in one ship) successfully completed the first circumnavigation of the globe. Although a careful log had been kept, the crew found that

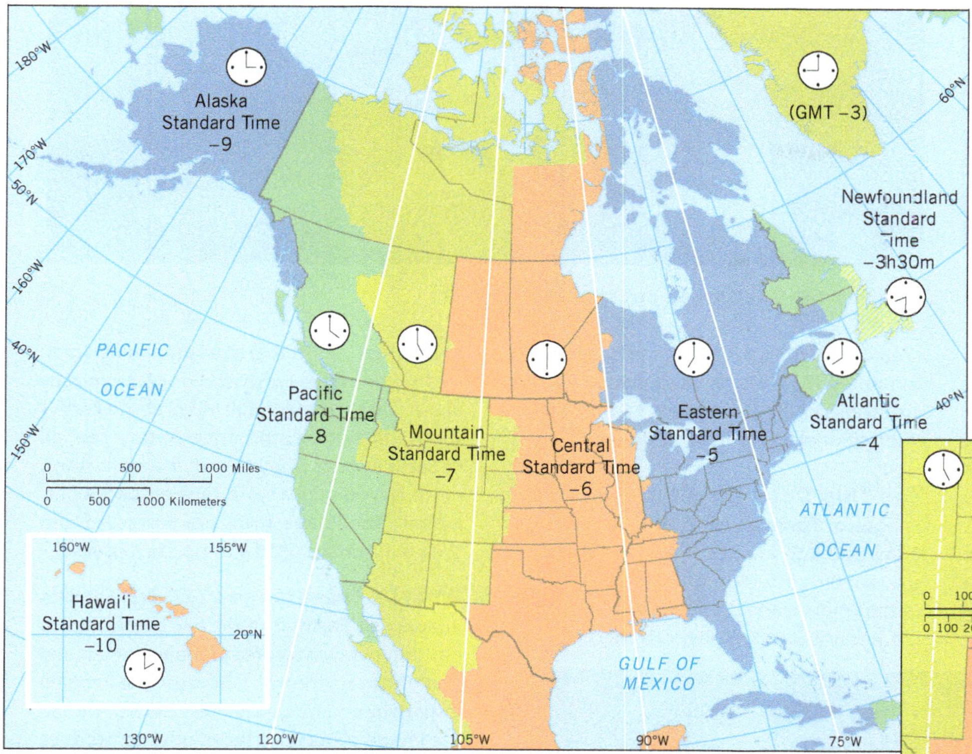

(a) North American time zones

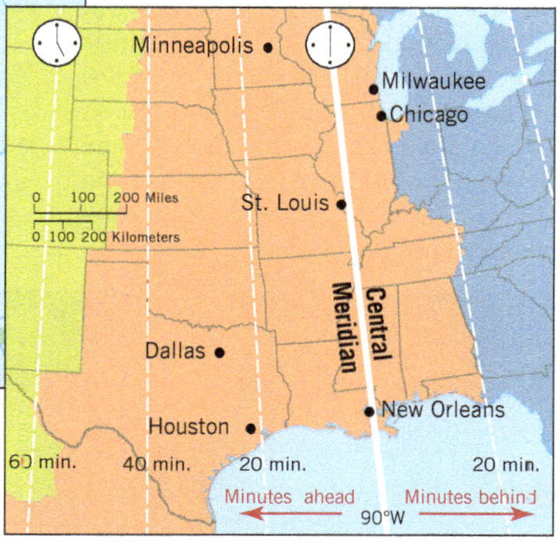

(b) Clock time versus Sun time

◀ **Figure 1-28** (a) Times zones for Canada, the United States, northern Mexico, and part of Greenland. The number in each time zone refers to the number of hours earlier than UTC (GMT). (b) Standard clock time versus Sun time. The Sun reaches its highest point in the sky at 12:00 noon in St. Louis and New Orleans because both cities lie on the central meridian. East of the central meridian, the Sun is highest in the sky a few minutes before standard time noon; to the west, local solar noon is a few minutes after. In Chicago, for instance, the Sun is highest in the sky at 11:50 A.M.; in Dallas, at 12:28 P.M.

their calendar was one day short of the correct date. This was the first human experience with time change on a global scale, the realization of which eventually led to the establishment of the **International Date Line**.

One advantage of establishing the Greenwich meridian as the prime meridian is that its opposite arc is in the Pacific Ocean. The 180th meridian, transiting the sparsely populated mid-Pacific, was chosen as the meridian at which new days begin and old days exit from the surface of Earth. The International Date Line deviates from the 180th meridian in the Bering Sea to include all of the Aleutian Islands of Alaska within the same day and again in the South Pacific to keep islands of the same group—such as Fiji and Tonga—within the same day (**Figure 1-29**). The extensive eastern displacement of the date line in the central Pacific is due to the widely scattered locations of the many islands of the country of Kiribati.

The International Date Line is in the middle of the time zone defined by the 180° meridian. Consequently, there is no time (i.e., hourly) change when you cross the International Date Line—only the calendar day changes, not the clock. When you cross the International Date Line from west to east, it becomes one day earlier (e.g., from January 2 to January 1); when you move across the line from east to west, it becomes one day later (e.g., from January 1 to January 2).

LearningCheck 1-19 **What happens to the day when you cross the International date line from west to east?**

Satellite images of Earth can help us trace the expansion of human activities taking place day and night around the globe. Commerce, transportation, industry, and many aspects of urban life are no longer constrained by darkness

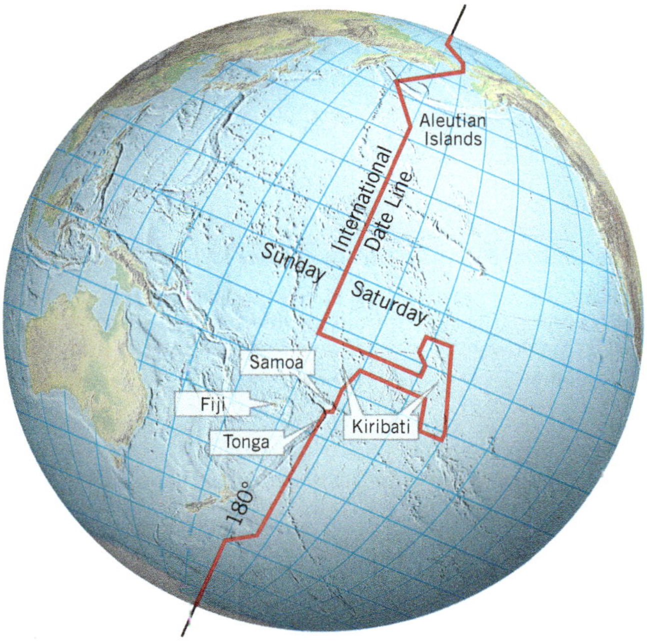

▲ **Figure 1-29** The International Date Line generally follows the 180th meridian, but it deviates around various island groups—most notably Kiribati. When you cross the International Date Line as you travel from west to east on Sunday, the day becomes Saturday; when you cross the International Date Line from east to west, Saturday becomes Sunday.

Global Environmental Change

Images of Earth at Night

▲ **Figure 1-E** Satellite composite of Earth at night.

• **Paul Sutton,** University of South Australia

Images of Earth at night appear to map the world's largest cities (**Figure 1-E**). Imagine what these images would look like if you viewed Earth from space at night 100,000 years ago. Would there be clues of human presence? It is unlikely that human-made fires would be visible from space unless they were forest and grass fires lit as a land-use practice. Only in the last 200 years or so have we used nocturnal high-elevation perspectives to monitor human presence. Balloons were used during the U.S. Civil War to count enemy campfires. Most of the images you see today are from satellites or photos taken from the International Space Station (**Figure 1-F**)

The "Big Data" Reality: Most modern images of Earth at night are composites of many satellite images. There is no point in space or time in which you could take a "selfie" of the whole planet to show Earth as a rectangle. Because Earth is spherical, only half of it is in darkness (nighttime) at any moment. Satellites that observe Earth at night are typically about 800 kilometers (500 miles) above Earth's surface in a polar orbit; they do not "see" all of Earth at once. They record images in strips, much as you might wrap tape around a basketball. The image strips are processed and then reassembled into a complete mosaic.

▲ **Figure 1-F** Image of Europe at night showing city lights and aurora borealis taken by NASA astronaut Terry Virts from the International Space Station. (A portion of the solar panels that power the International Space Station appears in the upper right.)

What Do These Images Tell Us? Satellite imagery of Earth at night has been used to develop proxy measures of many human and nonhuman phenomena. Because the imagery is spatially and temporally referenced, we can use image processing techniques to produce maps of various features around the world, such as fires, lightning strikes, lantern fishing, or gas flaring, as well as human settlements or cities. These "maps" are digital data products that we can manipulate mathematically and statistically. By doing so, we can use a map of city lights as a model of human population density. Applying a different set of mathematical parameters to such data can produce a map of urban area, economic activity, energy consumption, ecological footprint, or carbon dioxide emissions. The myriad possibilities are still being explored.

Questions

1. Where in the universe could you take a photo to produce an image of Earth like that in Figure 1-E?
2. What does it mean to say that nighttime imagery is used as a proxy measure of energy consumption?

or time differences between distant cities—see the box *Global Environmental Change: Images of Earth at Night*.

Daylight-Saving Time

To conserve energy during World War I, Germany ordered all clocks set forward by an hour. This practice allowed the citizenry to "save" an hour of daylight by shifting the daylight period into the usual evening hours, thus reducing the consumption of electricity for lighting. The United States began a similar "summer time" policy in 1918, but Hawai'i and parts of Arizona have exempted themselves from observance of **daylight-saving time** under the Uniform Time Act.

Canada, Australia, New Zealand, and most of the nations of western Europe have also adopted daylight-saving time. In the Northern Hemisphere, many nations, such as the United States, begin daylight-saving time on the second Sunday in March (in spring we "spring forward" one hour) and resume standard time on the first Sunday in November (in fall we "fall back" one hour). In the tropics, the lengths of day and night change little seasonally, and there is not much twilight. Consequently, daylight-saving time would offer little or no savings there.

After studying this chapter, you should be able to answer the following questions. Key terms from each text section are shown in **bold type.** Definitions for key terms are also found in the glossary at the back of the book.

Learning Review 1

Key Terms and Concepts

The Study of Geography (p. 4)

1. What is the study of **geography**? Contrast **physical geography** and **human geography**.

Geography and Science (p. 8)

2. If an idea or a theory cannot be disproven by some possible observation, experiment, or test, can such an idea or theory be supported by science? Explain your reasoning.
3. What is the approximate English System of measurement equivalent of one kilometer in the **International System (S.I.)**?

Environmental Spheres and Earth Systems (p. 9)

4. Briefly describe the environmental "spheres": **atmosphere**, **hydrosphere**, **cryosphere**, **biosphere**, and **lithosphere**.
5. Contrast *closed systems* and *open systems*.
6. What does it mean when a system is in *equilibrium*?
7. How does a *positive feedback loop* differ from a *negative feedback loop*?

Earth and the Solar System (p. 11)

8. In what ways do the inner (terrestrial) and outer (Jovian) planets differ from each other?
9. Compare the size of Earth to that of its surface features and atmosphere.
10. Is Earth perfectly spherical? Explain.

The Geographic Grid—Latitude and Longitude (p. 13)

11. Define the following terms: **latitude**, **longitude**, **parallel**, **meridian**, and **prime meridian**.
12. Latitude ranges from _____° to _____° north and south, whereas longitude ranges from _____° to _____° east and west.
13. State the latitude (in degrees) of the following "special" parallels: **equator**, **North Pole**, **South Pole**, **Tropic of Cancer**, **Tropic of Capricorn**, **Arctic Circle**, and **Antarctic Circle**.
14. What is a **great circle**? A *small circle*? Provide examples of both.

Earth–Sun Relations and the Seasons (p. 18)

15. Describe and explain the four factors in Earth–Sun relations associated with the change of seasons: **rotation**, **revolution** around the Sun, **inclination of Earth's axis**, and **polarity (parallelism) of Earth's axis**.
16. Does the **plane of the ecliptic** coincide with the plane of the equator? Explain.
17. On which day of the year is Earth closest to the Sun (**perihelion**)? Farthest from the Sun (**aphelion**)?
18. Provide the approximate dates of the following special days of the year: **March equinox**, **June solstice**, **September equinox**, and **December solstice**.
19. What is the **circle of illumination**?
20. What is meant by the **solar altitude**?
21. Briefly describe Earth's orientation to the Sun during summer and winter in the Northern Hemisphere.
22. Beginning with the March equinox, describe the changing **declination of the Sun** during the year.
23. In the midlatitudes of the Northern Hemisphere, on which day of the year is the Sun highest in the sky? Lowest in the sky?
24. For the equator, describe the approximate number of daylight hours on the following days: March equinox, June solstice, September equinox, and December solstice.
25. What is the longest day of the year (the day with the greatest number of daylight hours) in the midlatitudes of the Northern Hemisphere? In the Southern Hemisphere?
26. For the North Pole, describe the approximate number of daylight hours on the following days: March equinox, June solstice, September equinox, and December solstice.
27. For how many months of the year does the North Pole have no sunlight at all?

Telling Time (p. 23)

28. What happens to the hour when you cross a **time zone** boundary from west to east?
29. What is meant by **UTC (Coordinated Universal Time)** and **Greenwich Mean Time (GMT)**?
30. What happens to the day when you cross the **International Date Line** from east to west?
31. When **daylight-saving time** begins in the spring, you would adjust your clock from 2:00 A.M. to _____.

Study Questions

1. Why are physical geographers interested in globalization of the economy?
2. Why is a distance covered by 1° of longitude at the equator different from the distance covered by 1° of longitude at a latitude of 45° N?
3. What is the significance of aphelion and perihelion in Earth's seasons?
4. In terms of the change of seasons, explain the significance of the Tropic of Cancer, the Tropic of Capricorn, the Arctic Circle, and the Antarctic Circle.
5. Is the noon Sun ever directly overhead in Detroit, Michigan (42° N)? If not, on which day of the year is the noon Sun highest in the sky there? Lowest?
6. What would be the effect on the annual march of the seasons if Earth's axis were not inclined relative to the plane of the ecliptic?
7. What would be the effect on the annual march of the seasons if the North Pole always leaned toward the Sun?
8. If Earth's axis were tilted only 20° from perpendicular, what would the latitudes of the Tropic of Cancer and Arctic Circle be?
9. Why are standard time zones 15° of longitude wide?
10. Most weather satellite images are "time-stamped" using UTC or "Zulu" time (UTC expressed using 24-hour, military time) instead of the local time of the region below. Why?

Exercises

1. Using formulas found in Appendix I (p. A-1), make the following conversions between the International System (S.I.) and English system of measurements:
   a. 21 centimeters ______ inches
   b. 130 kilometers ______ miles
   c. 18,000 feet ______ meters
   d. 7 quarts ______ liters
   e. 11 kilograms ______ pounds
   f. 20°C ______°F
2. Using a world map or globe, estimate the latitude and longitude of both Chicago, Illinois, and Shanghai, China. Be sure to specify whether these locations are north or south latitude, and east or west longitude.
3. The solar altitude (SA) can be calculated for any latitude on Earth for any day of the year, by using the formula $SA = 90° - AD$, where AD is the "arc distance" (the difference in latitude between the declination of the Sun and the latitude in question). Use Figure 1-25 to estimate the declination of the Sun at the following locations on the day given, and then calculate the solar altitude:
   a. Madrid, Spain (40° N), on November 1
   b. Nairobi, Kenya (1° S), on September 1
   c. Fairbanks, Alaska (65° N), on May 1
4. Using the map of North American time zones (Figure 1-28a) for reference, if it is 4:00 P.M. standard time on Thursday in Baltimore (39° N, 77° W), what are the day and time in Sacramento (39° N, 121° W)?
5. Using the map of world time zones (Figure 1-27) for reference, if it is 8:00 A.M. UTC, what is the standard time in Seattle (48° N, 122° W)?

Environmental **Analysis** THE JUNE AND DECEMBER SOLSTICES

Data
Soltices
https://media.pearsoncmg.com/ph/esm/esm_mcknight_physgeo_13/media/

The amount of solar energy received in a particular location during a day is mostly a consequence of the number of hours of Sun exposure and the intensity of radiation. In turn, that intensity is largely determined by the Sun angle—the height of the Sun above the horizon. (Cloud cover also plays a role.)

Activities

Refer to Table 1-3.

1. Should the equator or 30° N receive more solar energy on the June solstice? Why?
2. The North Pole has 24 hours of daylight on the June solstice, whereas the equator has only about 12. What helps explain why the North Pole is so much colder on that day?
3. In terms of Sun exposure, why is it "winter" at 30° S on the June solstice?

Go to suncalc.org and enter your location in the search box. Find the solar altitude (in degrees above the horizon) for your city on the June solstice and on the December solstice. Clicking on the date and time boxes allows you to change them.

4. What is the highest (noon) solar altitude in your city on each of those days?
5. What is the difference (in degrees) between the noon solar altitudes on those two days?

6. How does the difference in solar altitude relate to the difference in the Sun's declination (shown in Figure 1-25) on those days?
7. How much daylight does your city have on the June solstice and December solstice? What about for Anchorage, Alaska, and Key West, Florida? (If you live in either city, use Omaha, Nebraska, as your third city.)
8. Contrast and explain the differences in daylight on these two days in all three cities.

Clicking on the time box allows you to "Start animation . . . " to see the Sun's path across the sky from dawn to dusk.

9. Is the sun's noon location to the north of Key West on the solstices? What about for Panama Canal, Panama? Explain why the answers are different.

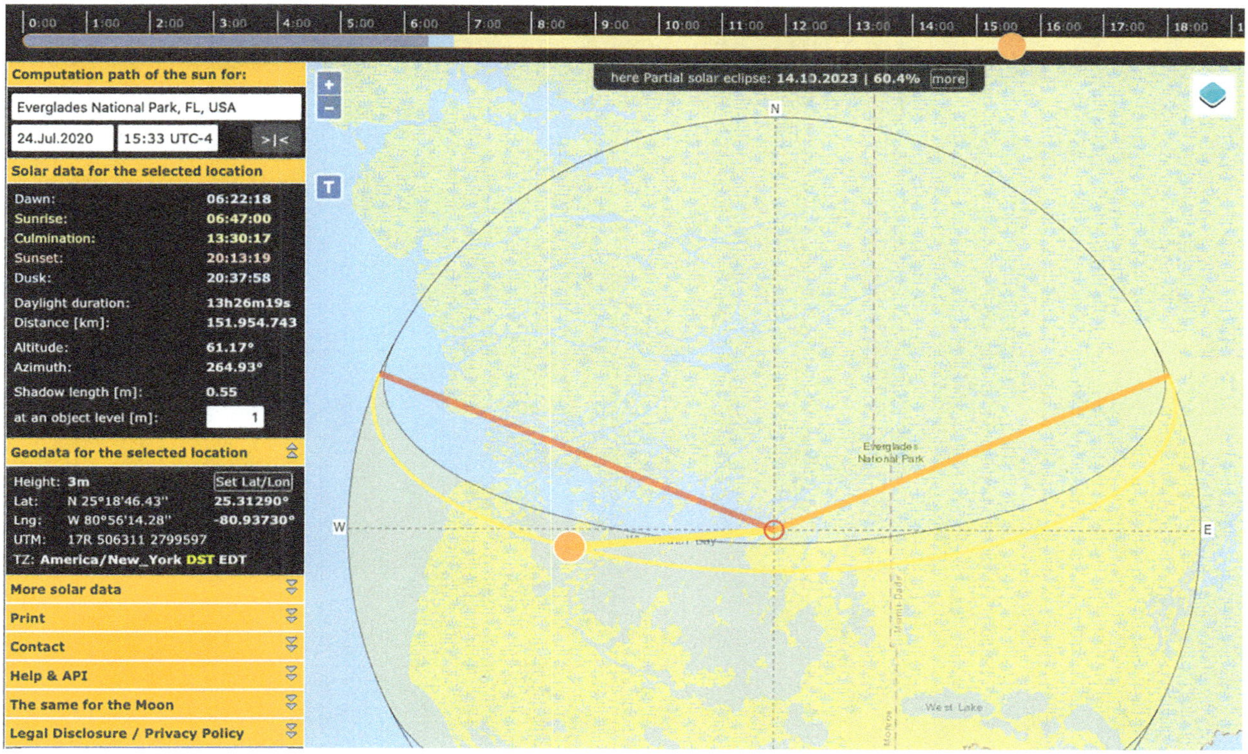

Seeing Geographically in Our National Parks

Look again at the photograph of Florida at the beginning of the chapter (p. 2). What examples of Earth's environmental spheres can you see? Using a globe or world map for reference, what are the approximate latitude and longitude of Lake Okeechobee, the dark patch in Florida at the center of the image? Florida is about 200 kilometers wide there—does it appear that the tops of the clouds are more than or less than 200 kilometers above the surface?

Mastering Geography®

Looking for additional review and test prep materials? Visit the Study Area in **Mastering Geography** to enhance your geographic literacy, spatial reasoning skills, and understanding of this chapter's content by accessing a variety of resources, including Map Master interactive maps, geoscience animations, *Mobile Field Trips*, videos, *Project Condor* Quadcopter videos, *In the News* RSS feeds, flashcards, web links, self-study quizzes, and an eText version of *McKnight's Physical Geography*.

Seeing Geographically in Our National Parks

This is a shaded relief map of the area of Grand Teton National Park in Wyoming. The Teton Range runs north to south down the left side of the map. Is the east side or west side of the Teton Range steeper? The map is based on precise elevation data processed to show topography. What kinds of features in the landscape are not shown clearly in this kind of map?

Grand Teton National Park, WY

Portraying Earth

2

HAVE YOU EVER WONDERED how the GPS receiver in your phone knows exactly where you are here in Grand Teton National Park? Or how it can tell you the fastest route through city traffic? Or how you can zoom in on a satellite image close enough to see your house? The technology, which many of us now take for granted, is rooted in the topic of this chapter: how we gather, convey, and analyze geographic information with maps, satellite images, and the Global Positioning System, and how we can make practical use of this information through geographic information systems.

The size and complexity of Earth's surface make it difficult to visualize and examine without special tools. The most basic tool for geographic studies is the map. The mapping of a geographic feature is often an essential first step toward understanding the spatial distribution and relationships of that feature. In this book you'll find hundreds of maps of various kinds, each included to further your understanding of some concept, fact, or relationship.

In this chapter we explain many of the tools of the modern geographer. We begin with the basic characteristics, attributes, and limitations of maps. We then describe the many ways that Earth's features can be portrayed and studied with photographs and remotely sensed imagery. We conclude with a discussion of how maps and computer databases have melded into powerful analytical systems that are now used for tasks ranging from scientific research to our personal shopping.

As you study this chapter, think about these **Key**Questions:

- **How is a map different from a globe?**
- **What is meant by the *scale* of a map?**
- **What are the differences between *equivalent* ("equal area") maps and *conformal* maps?**
- **Why are different map projections needed?**
- **How do *isolines* convey information on a map?**
- **How does a GPS unit know where we are?**
- **What is *remote sensing*?**
- **How does GIS help us analyze geographic data?**

Maps and Globes

A globe is a true representation of Earth (**Figure 2-1**). Not only does a globe convey the spherical shape of Earth, it can show, essentially without distortion, the spatial relationships of Earth's surface, maintaining correct size, shape, distance, and direction relationships of features around the planet.

The obvious limitation of a globe is that it cannot show much detail. In order for a globe to show as much detail as the map of Grand Teton National Park on the previous page, it would need to be about 45 meters (150 feet) in diameter! Because maps are much more portable and versatile than globes, literally billions of maps are in use all over the world, whereas globes are extremely limited in both number and variety.

Maps

In the simplest terms, a **map** is a flat representation of Earth, shown reduced in size with only selected features or data showing.

Maps portray distance, direction, size, and shape in their horizontal (that is, two-dimensional) spatial relationships. Most maps show other kinds of information as well. Most maps have a special purpose, and that purpose is usually to show the distribution of one or more phenomena. Such *thematic maps* may be designed to show street patterns, the distribution of Tasmanians, the ratio of sunshine to cloud, the biomass per cubic meter of rainforest, or any of an infinite number of other facts or combinations of facts—see the box *Featured National Park: Mount Rainier—Mapping the Many Faces of a Volcano.*

▲ **Figure 2-1** A model globe provides a splendid broad representation of Earth at a very small scale, but few details can be portrayed.

Because maps depict graphically "what is where" and because they are often helpful in providing clues as to "why" such a distribution occurs, maps are indispensable tools for geographers. Even so, it is important to realize that maps have limitations.

Map Distortions: Most of us understand that not every statement we may read in a book or on the Internet is necessarily correct (thus the somewhat cynical adage "Don't believe everything you read"). Yet we may uncritically accept all information portrayed on a map as being correct. However, no map can be perfectly accurate because it is impossible to portray the curved surface of Earth on a flat map without distortion. Imagine trying to flatten an orange peel. In order to do this, you must either stretch or tear the peel; effectively, the same thing must happen to Earth when we flatten its surface onto a map.

The extent to which the geometric impossibility of flattening a sphere without distortion becomes a problem on a map depends on two related variables: (1) how much of Earth is being shown on the map—these distortions are always significant on a world map, but they are less so on a map showing a very small region of Earth, and (2) the *scale* of the map—the topic to which we turn next.

LearningCheck 2-1 **Why can't a map represent Earth's surface as perfectly as a globe? (Answer on p. AK-1)**

Map Scale

Because a map is smaller than the portion of Earth's surface it represents, in order to understand the geographic relationships (distances or relative sizes, for example) depicted on that map, we must know how to use a **map scale**. The scale of a map describes the relationship between distance measured on the map and the actual distance it represents on Earth's surface. Knowing the scale of a map makes it possible to measure distance, determine area, and compare sizes.

Because Earth's surface is curved and a map's surface is flat, scale can never be perfectly correct over an entire map. In practice, if the map shows a small area, a single scale can be used across the whole map. However, if the map shows a large portion of Earth's surface (such as a world map), scale may vary significantly from one part of the map to the next. Such a map, for example, might need to list different scales for different latitudes.

Scale Types

Three ways of portraying map scale are widely used: the graphic scale, the fractional scale, and the verbal scale (**Figure 2-2**).

Graphic Map Scales: A **graphic map scale** uses a line marked off in distances to represent actual distance on Earth's surface. To use a graphic map scale, we measure the distance

Featured National Park

Mount Rainier

Mapping the Many Faces of a Volcano

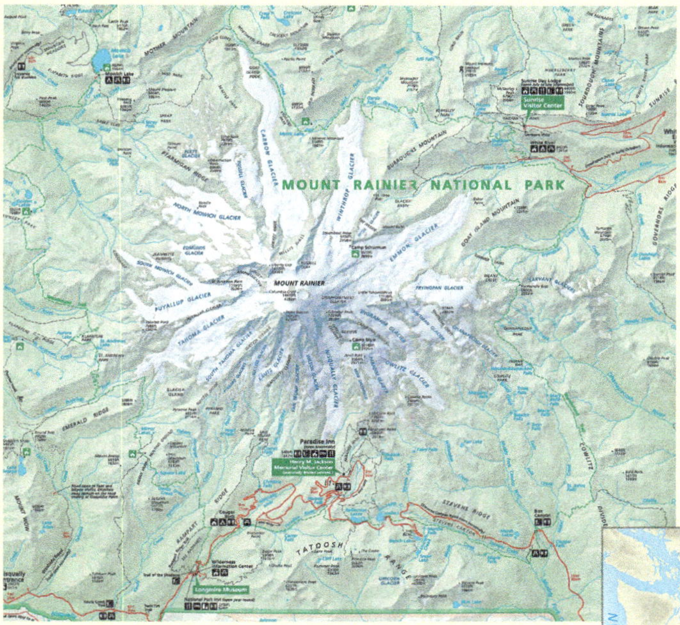

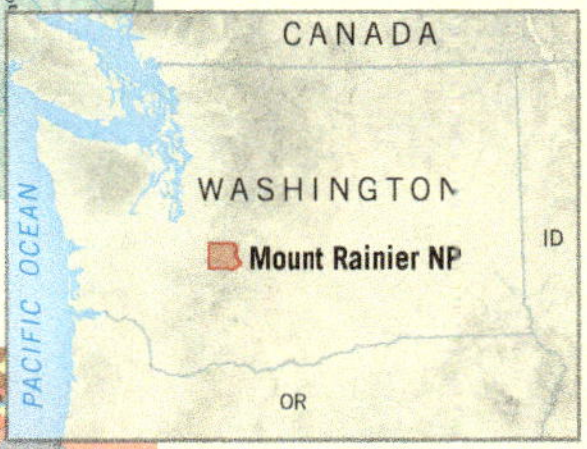

▲ **Figure 2-A** Park map.

Mount Rainier National Park in Washington is home to the highest active volcano in the Cascade Range. Standing 4392 meters (14,410 feet) above sea level, this peak towers over the surrounding landscape and includes the greatest number of active glaciers of any mountain in the conterminous United States.

Like many places geographers might study, Mount Rainier presents a challenge to map. Depending on which aspect of the environment we're interested in, we would need to use a different kind of map.

As a first-time visitor to the Park, we might want a general map showing roads and physical features (**Figure 2-A**). If we want to study the rocks, we'd use a geologic map (**Figure 2-B**). To gain insights into potential new eruptions, a map showing recent earthquakes would be useful (**Figure 2-C**). For studies of the mountain's glaciers, we could use a *Lidar* image (Lidar is described later in this chapter) showing topography and the extent of ice (**Figure 2-D**) along with a precipitation map to show us where on the mountain snowfall would be greatest (**Figure 2-E**).

Questions

1. How is the surface topography of the mountain shown in some of these maps?
2. Which of these maps might be useful if you were studying the distribution of vegetation on the mountain? Why?

▲ **Figure 2-B** Geologic map; white = ice; pink and red = recent lava flows; yellow = young stream deposits

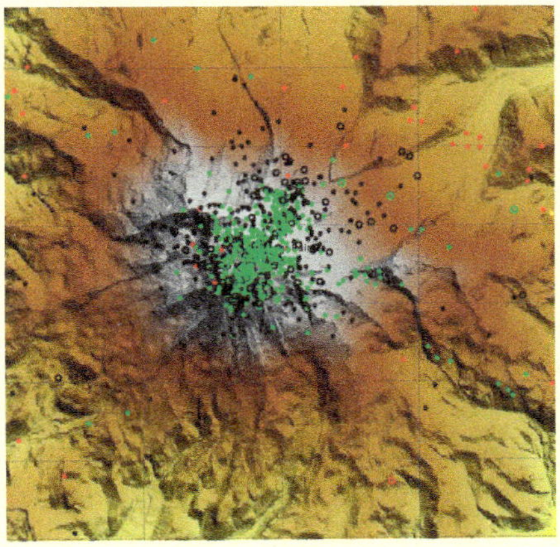

▲ **Figure 2-C** Earthquake map showing epicenters over a seven-year period.

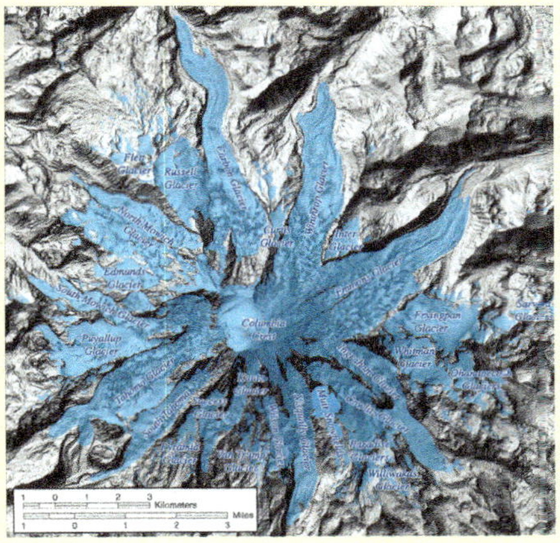

▲ **Figure 2-D** Lidar topography and glaciers; blue = current extent of glaciers.

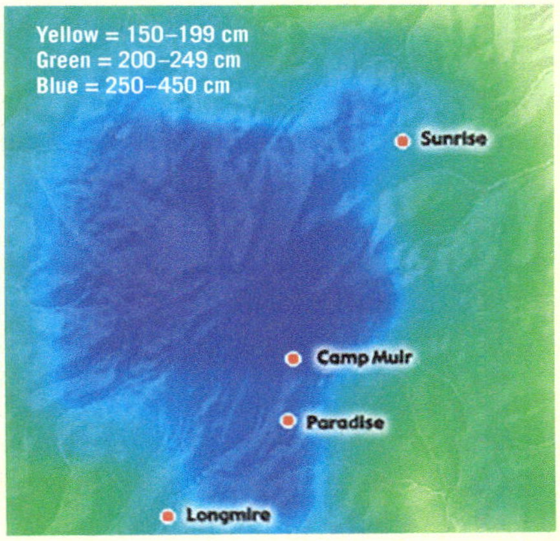

▲ **Figure 2-E** Average annual precipitation (dark blue areas get over 400 cm [155 in.] per year).

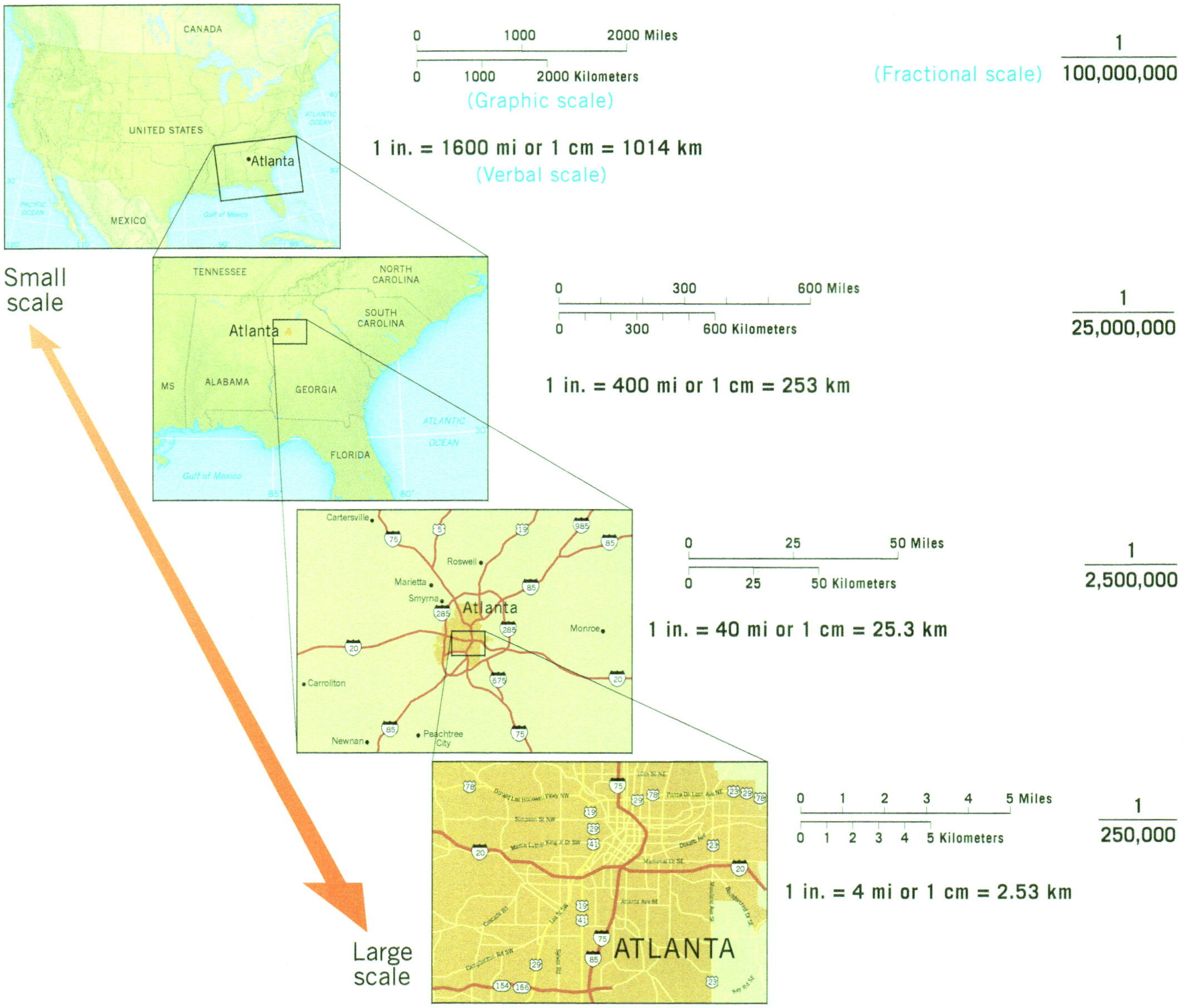

▲ **Figure 2-2** The three types of map scale, and comparisons of distance and area at various scales. A small-scale map portrays a large part of Earth's surface but depicts only the most important features, whereas a large-scale map shows only a small part of the surface but in considerably more detail.

between any two points on the map and then compare that measured distance with the graphic map scale, which gives you a direct reading of the actual distance. The advantage of a graphic scale is its simplicity; for example, you can quickly estimate travel distances on a road map with a graphic scale. Moreover, a graphic scale remains correct when the map is enlarged or reduced—such as you might do when looking at a map on your phone—because the length of the graphic scale line is enlarged or reduced along with the map.

Fractional Map Scales: A **fractional map scale** conveys the relationship between distance measured on a map and the actual distance that represents on Earth with a fraction or ratio called a *representative fraction*. For example, a common fractional scale uses the representative fraction 1/250,000 (often expressed as the ratio 1:250,000)—this means that *1 unit of measure on the map represents an actual distance of 250,000 units of measure on Earth*. The "units of measure" are the same on both sides of the fraction. So, 1 centimeter measured on the map represents an actual distance of 250,000 centimeters on Earth's surface, whereas 1 inch measured on the map represents a distance of 250,000 inches on Earth's surface. In the same way, on a map with a scale of 1/63,360, a measured distance of 1 inch represents an actual distance of 63,360 inches, and so forth.

Verbal Map Scales: A **verbal map scale** (or *word scale*) states in words the relationship between the distance on the map and the actual distance on Earth's surface, such as "one centimeter to ten kilometers" or "one inch equals five miles." A verbal scale is a mathematical manipulation of the fractional scale. For instance, there are 63,360 inches in 1 mile, so on a map with a fractional scale of 1:63,360 we can say "one inch represents one mile."

LearningCheck 2-2 On a map with a fractional scale of 1:10,000, one centimeter measured on the map represents what distance on Earth's surface?

Large-Scale and Small-Scale Maps

The adjectives *large* and *small* are comparative rather than absolute. In other words, scales are "large" or "small" only in comparison with other scales (see Figure 2-2). A **large-scale map** is one that has a relatively large representative fraction, which means that the denominator is small. Thus, 1/10,000 is a larger map scale than 1/1,000,000. Large-scale maps are used to show a small portion of Earth's surface in considerable detail. For example, if this page were covered with a map having a scale of 1:10,000, the map would be able to show just a small part of a single city, but that area would be depicted in great detail.

A **small-scale map** has a small representative fraction—that is, one with a large denominator, such as 1/10,000,000. Covered with a map of that scale, this page would be able to portray about one-third of the United States. Such small-scale maps are useful for showing geographical relationships over large areas but only in limited detail.

LearningCheck 2-3 Compare small-scale and large-scale maps.

Map Projections and Properties

Animation
Map Projections
https://media.pearsoncmg.com/ph/esm/esm_mcknight_physgeo_13/media/

The challenge to the *cartographer* (mapmaker) is to try to combine the geometric exactness of a globe with the convenience of a flat map. The fundamental problem is always the same: to transfer data from a spherical surface to a flat map with a minimum of distortion. This transfer is accomplished with a *map projection*.

Map Projections

A **map projection** is a system in which the spherical surface of Earth is transformed for display on a flat surface. Imagine a transparent globe on which are drawn meridians, parallels, and continental boundaries; also imagine a lightbulb in the center of this globe. A piece of paper, either held flat or rolled into a shape such as a cylinder or cone, is placed over the globe (Figure 2-3). When the bulb is lit, all the lines on the globe are projected outward onto the paper. These lines are then sketched on the paper. When the paper is laid out flat, a map projection has been produced. Few map projections have been made in this way by "optical" projection from a globe onto a piece of paper; instead, map projections are derived by mathematically transferring the features of a sphere onto a flat surface.

Because a flat surface cannot be closely fitted to a sphere without wrinkling or tearing, no matter how a map projection is made, data from a globe (parallels, meridians, continental boundaries, and so forth) cannot be transferred to a map without distortion of shape, relative area, distance, and/or direction. The cartographer can choose to control or reduce one or more of these distortions—although all distortions cannot be eliminated on a single map.

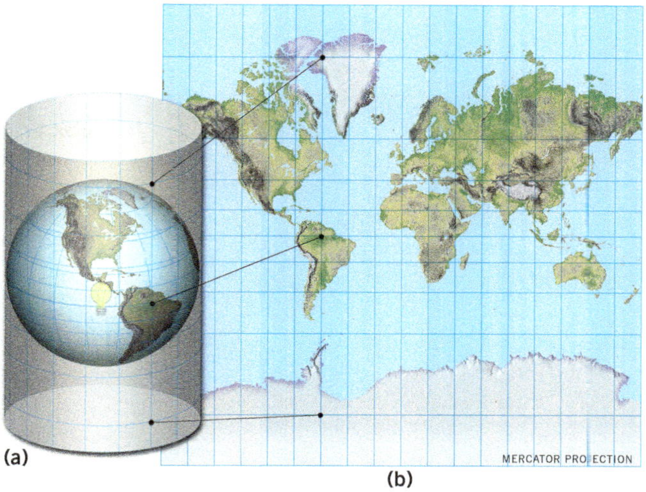

▲ Figure 2-3 The concept of map projection. (a) A cylinder is wrapped around a globe with a light at its center, and the features of the globe are projected onto the cylinder. (b) The resulting map is called a cylindrical projection.

LearningCheck 2-4 What is a map projection?

Map Properties

Cartographers often strive to maintain accuracy either of size or of shape—map properties known as *equivalence* and *conformality*, respectively (Figure 2-4).

Equivalence: In an **equivalent map projection** (also called an **equal area map projection**), the correct size ratio of area on the map to the corresponding actual area on Earth's surface is maintained over the entire map. For example, on an equivalent world map, if you were to place one coin on Brazil, one on Australia, one on Siberia, and one on South Africa, the area on Earth covered by each coin would be the same. Most of the world maps in this book are equivalent projections because they are so useful in showing distributions of the various geographic features we will study.

However, there are tradeoffs. Equivalence is difficult to achieve on small-scale maps because correct shapes must be sacrificed to maintain proper area relationships. Most equivalent world maps (which are necessarily small-scale maps) show distorted shapes of landmasses—especially in the high latitudes. For example, on equivalent maps the shapes of Greenland and Alaska are usually shown as more "squatty" than they actually are (see Figure 2-4a).

Conformality: In a **conformal map projection**, proper angular relationships are maintained across the entire map. Although it is impossible to depict true shapes for large areas such as a continent, in practice for small areas we can say that conformal maps show correct shapes. Conformal projections have meridians and parallels crossing each other at right angles, just as they do on a globe.

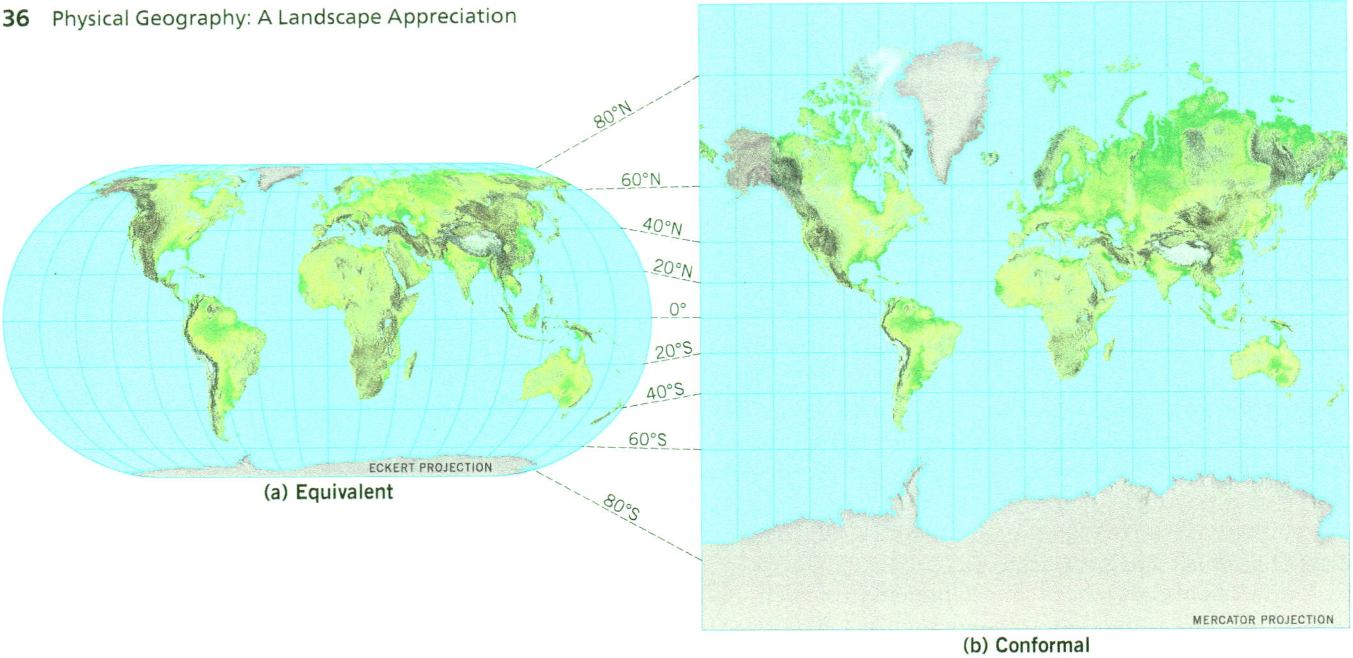

▲ **Figure 2-4** Equivalent and conformal maps. (a) An equivalent (equal area) projection (the Eckert) is accurate with regard to size, but shapes are badly distorted at high latitudes. (b) A conformal projection (the Mercator) depicts accurate shapes, but sizes are severely exaggerated at high latitudes. It is impossible to portray both correct size and correct shape on a world map. Compare the sizes and shapes of Antarctica, Alaska, and Greenland.

The main problem with conformal projections is that the size of an area must often be considerably distorted to depict the proper shape. Thus, the scale necessarily changes from one region to another. For example, a conformal map of the world normally greatly enlarges landmasses in the high latitudes (see Figure 2-4b).

Compromise Projections: Except for maps of very small areas (in other words, large-scale maps), where both properties can be closely approximated, equivalence and conformality cannot be maintained on the same projection. Thus the art of mapmaking, like politics, is often an art of compromise. The Robinson projection is one **compromise map projection**; it is neither equivalent nor conformal but instead balances reasonably accurate shapes with reasonably accurate areas (Figure 2-5). The Robinson projection is a popular choice as a general-purpose classroom map.

As a rule of thumb, some map projections are purely conformal, some are purely equivalent, none are both conformal and equivalent, and many are instead a compromise between the two.

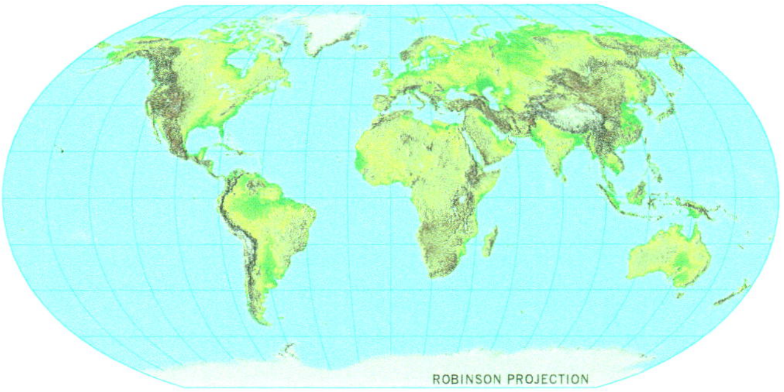

▲ **Figure 2-5** Many world maps are neither purely conformal nor purely equivalent but a compromise between the two. One of the most popular compromises is the Robinson projection.

LearningCheck 2-5 **What is the difference between an equivalent map and a conformal map?**

Families of Map Projections

Because distortion cannot be avoided completely, no map projection is ideal for all uses. Thus, hundreds of different map projections have been devised for one purpose or another. Most of them can be grouped into just a few families. Projections in the same family generally have similar properties and related distortion characteristics.

Cylindrical Projections

A **cylindrical projection** is obtained by, in effect, "wrapping" the globe with a cylinder of paper in such a way that the paper touches the globe only at the globe's equator (see Figure 2-3). We say that paper positioned this way is *tangent* to the globe at the equator, which is called the *circle of tangency* and becomes the *standard parallel* of the projection. (Some cylindrical projections choose a standard parallel other than the equator.) The curved parallels and meridians of the globe then form a perfectly rectangular grid on the map with meridians and parallels meeting at right angles. There is no size distortion at the circle of tangency, but size distortion increases progressively with increasing distance from this circle, a characteristic clearly exemplified by the *Mercator projection*.

Mercator—The Most Famous Projection: It is remarkable that the most famous of all map projections, the **Mercator projection**, which was devised in 1569 by Flemish geographer and cartographer Gerardus Mercator, is still commonly used today without significant modification (see Figure 2-4b).

The Mercator projection is a conformal map projection designed to facilitate oceanic navigation. The Mercator projection shows *loxodromes* as straight lines. A **loxodrome** (or *rhumb line*) is a line on the surface of a sphere that crosses all meridians at the same angle and represents a line of constant compass direction. A navigator first plots the shortest distance between origin and destination on a map projection in which great circles are shown as straight lines, such as a *gnomonic projection* (**Figure 2-6a**). (Great circle routes are discussed in Chapter 1.) That route is then transferred to a Mercator projection with straight-line loxodromes (**Figure 2-6b**). This procedure allows the navigator to take the shorter path of a great circle route by simply making periodic changes in the compass course of the airplane or ship (**Figure 2-6c**). Today, all such calculations are done by computer.

A Mercator map is relatively undistorted at the low latitudes. However, because the meridians do not converge at the poles but instead remain parallel to each other, size distortion increases rapidly in the mid- and high latitudes. Furthermore, to maintain conformality and the map's navigational virtues, Mercator compensated for the east–west stretching by spacing the parallels of latitude increasingly farther apart so that north–south stretching occurs at the same rate. This procedure allowed shapes to be approximated with reasonable accuracy—but at great expense to proper size relationships. Area is distorted by 4 times at the 60th parallel of latitude and by 36 times at the 80th parallel. If the North Pole could be shown on a Mercator projection, it would be a line as long as the equator rather than a single point!

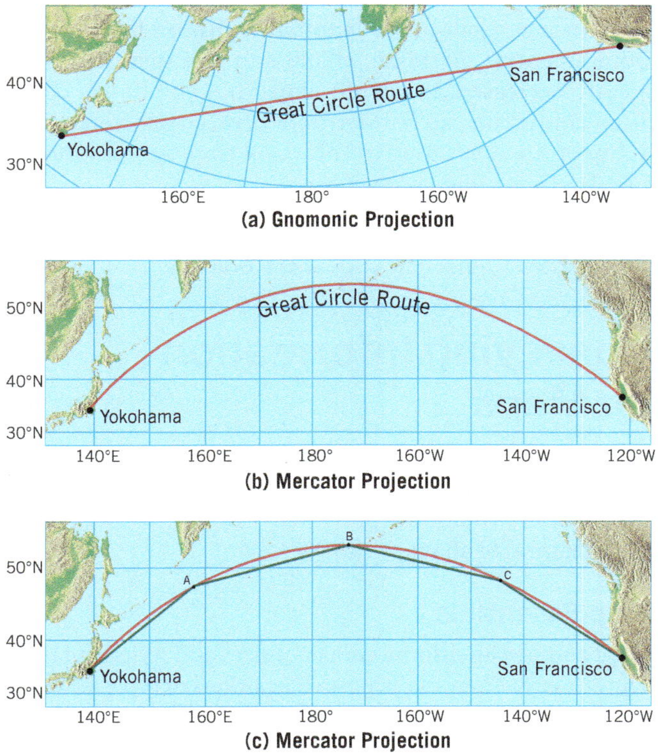

▲ **Figure 2-6** The prime virtue of the Mercator projection is its usefulness for straight-line navigation. (a) The shortest distance between two locations—here San Francisco and Yokohama—can be plotted on a *gnomonic* projection (on which great circles are shown as straight lines). (b) The great circle route can be transferred to a Mercator projection. (c) On the Mercator projection, straight-line loxodromes can then be substituted for the curved great circle. The loxodromes allow a navigator to maintain constant compass headings over small distances while still approximating the curve of the great circle.

The Mercator projection was a major leap forward in cartography when it was devised. Unfortunately, by the early twentieth century Mercator projections were widely used in American classrooms and atlases—several generations of American students passed through school with their principal view of the world provided by a Mercator map. This has created many misconceptions about the relative sizes of high-latitude landmasses: on a Mercator projection, the island of Greenland appears to be as large as or larger than Africa, Australia, and South America. In reality, Africa is 14 times larger than Greenland, South America is 9 times larger, and Australia is 3.5 times larger.

LearningCheck 2-6 Would a Mercator projection be a good choice for a map used to study the loss of forest cover around the world? Why or why not?

Planar Projections

A **planar projection** (also called a *plane, azimuthal,* or *zenithal projection*) is obtained by projecting the markings of a center-lit globe onto a flat piece of paper that is tangent to the globe at one point (**Figure 2-7**). That point is usually the North or South Pole, or some spot on the equator. There is no distortion immediately around the point of tangency, but distortion increases progressively away from this point.

Typically, planar projections show only one hemisphere. Some types can provide a perspective of Earth similar to the view we get when looking at a globe or that of an astronaut looking at Earth from space (called an *orthographic* projection). This half-view-only characteristic can be a drawback, just as it is with a globe, although planar projections can be useful for focusing attention on a specific region. They are commonly used for mapping the Arctic and Antarctic regions.

Conic Projections

A **conic projection** is obtained by projecting the markings of a center-lit globe onto a cone wrapped tangent to, or intersecting, a portion of the globe (**Figure 2-8**). Normally the apex of the cone is positioned above a pole, which means that the circle of tangency coincides with a parallel. Distortion is least near this standard parallel but increases progressively as distance from it increases. Consequently, conic projections are best suited for regions of east–west orientation in the midlatitudes, being particularly useful for maps of the United States, Europe, or China. It is impractical to use conic projections for more than one-fourth of Earth's

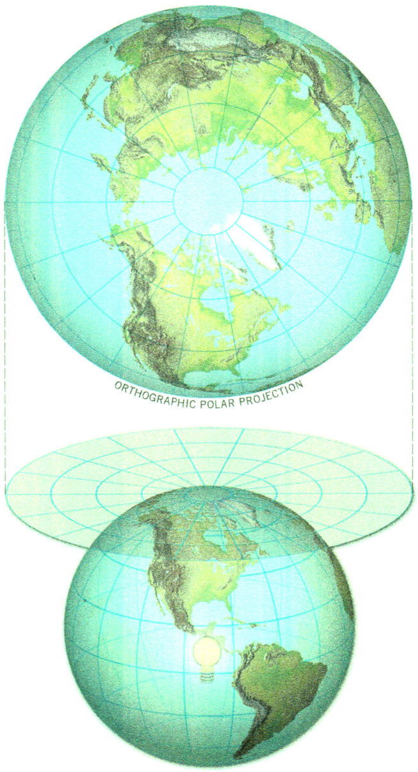

▲ Figure 2-7 The origin of a planar projection, as illustrated by a globe with a light in its center, projecting images onto an adjacent plane. This *orthographic* planar projection shows Earth as it would appear from space.

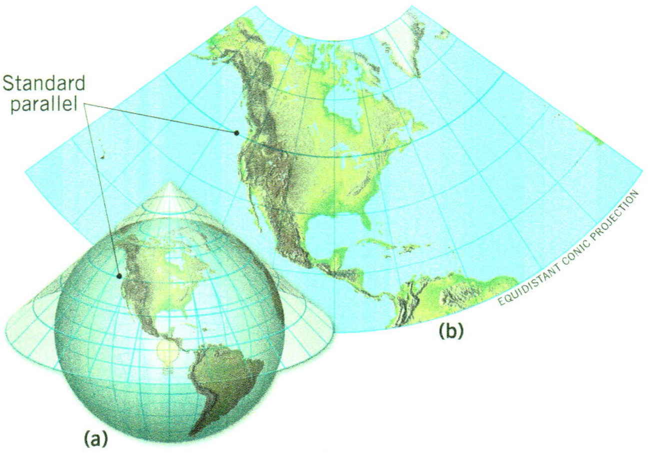

▲ Figure 2-8 (a) The origin of a conic projection, as illustrated by a globe with a light in its center, projecting images onto a cone. (b) The resulting map is a conic projection.

surface, but they are particularly well adapted for mapping relatively small areas, such as a state or county.

Pseudocylindrical Projections

A **pseudocylindrical projection** (also called an *elliptical* or *oval projection*) is roughly a football-shaped map. It is usually of the entire world (see the Eckert in Figure 2-4a and the Robinson in Figure 2-5), although sometimes only the central section of a pseudocylindrical projection is used for maps of lesser areas. A pseudocylindrical projection wraps around the equator like an ordinary cylindrical projection but further "curves in" toward the poles, effectively conveying some of the curvature of Earth.

In most pseudocylindrical projections, a central parallel (usually the equator) and a central meridian (often the prime meridian) cross at right angles in the middle of the map, which is a point of no distortion. Distortion in size and/or shape normally increases progressively as distance from this point increases in any direction. All of the parallels are drawn parallel to each other, whereas all meridians, except the central meridian, are shown as curved lines.

Interrupted Projections: One technique used with pseudocylindrical projections to minimize distortion of the continents is to "interrupt" oceanic regions. *Goode's interrupted homolosine equal-area projection* (Figure 2-9) is a popular example. Goode's projection is equivalent, and although it is impossible for this map to be conformal, the shapes of continental coastlines are very well maintained even at high latitudes.

When global distributions are mapped, the continents are often more important than the oceans, yet the oceans occupy most of the map space in a typical projection. A projection can be interrupted ("torn apart") in the Pacific, Atlantic, and Indian Oceans with central meridians that pass through each major landmass; with no land area far from a central meridian, shape and size distortions are greatly decreased. For world maps that emphasize ocean areas, continents can be interrupted instead of ocean basins. Many of the maps used in this book employ variations of Goode's interrupted projection.

LearningCheck 2-7 **What are the advantages of an "interrupted" projection, such as the Goode's?**

Conveying Information on Maps

Now that we have described the fundamentals of map scale, properties, and projections, let's think about some of the ways that information is presented on maps. We begin with basic features of all maps.

Map Essentials

Maps come in an infinite variety of sizes and styles and serve a limitless diversity of purposes. Regardless of type, however, every map should contain a few basic components to facilitate its use (Figure 2-10):

- **Title:** A brief summary of the map's content or purpose. It should identify the area covered and provide some indication of content.
- **Date:** The time span over which the information was collected, the date of publication, or both.
- **Legend:** An explanation of the symbols, colors, and shadings used to represent features or data.
- **Scale:** A graphic, fractional, or verbal scale.

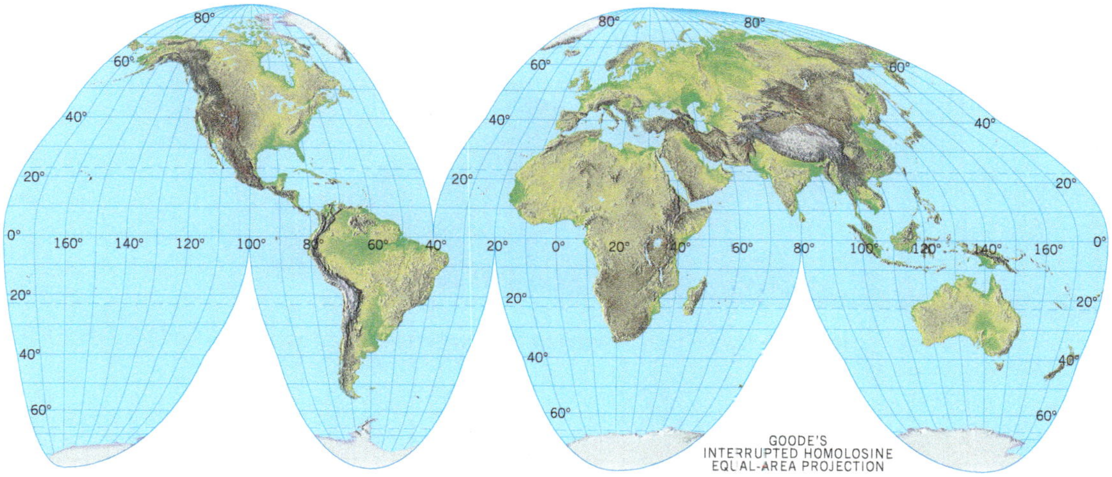

▲ **Figure 2-9** An interrupted projection of the world. With interruptions, the map portrays certain areas (usually continents) more accurately, at the expense of portions (usually oceans) that are not important to the map's theme. This map is a *Goode's interrupted homolosine equal-area projection*. We use a variation of that projection for many maps in this book.

- **Direction:** Typically shown by the grid of parallels and meridians or a straight arrow pointing toward the north geographic pole.
- **Location:** Most commonly, latitude and longitude. Other types of reference grids (such as on road maps) use a simple *x*- and *y*-coordinate grid for locating features. Some maps display more than one coordinate system.
- **Data Source:** The reference source that published the data.
- **Projection Type:** The name of the projection. It is particularly useful on small-scale maps to help the user assess inherent distortions.

Isolines

Maps can display data in a number of ways. One of the most widespread techniques for portraying the geographic distribution of some phenomenon is the *isoline* (from the Greek *isos*, meaning "equal"). An **isoline** is any line that joins points of equal value of something.

Some isolines represent tangible surfaces, such as **elevation contour lines**, which connect points of equal elevation on topographic maps (**Figure 2-11**). Most isolines, however, signify such intangible features as temperature or precipitation, and some express relative values such as ratios or proportions (**Figure 2-12**). More than 100 kinds of isolines

▶ **Figure 2-10** A typical thematic map containing all of the essentials.

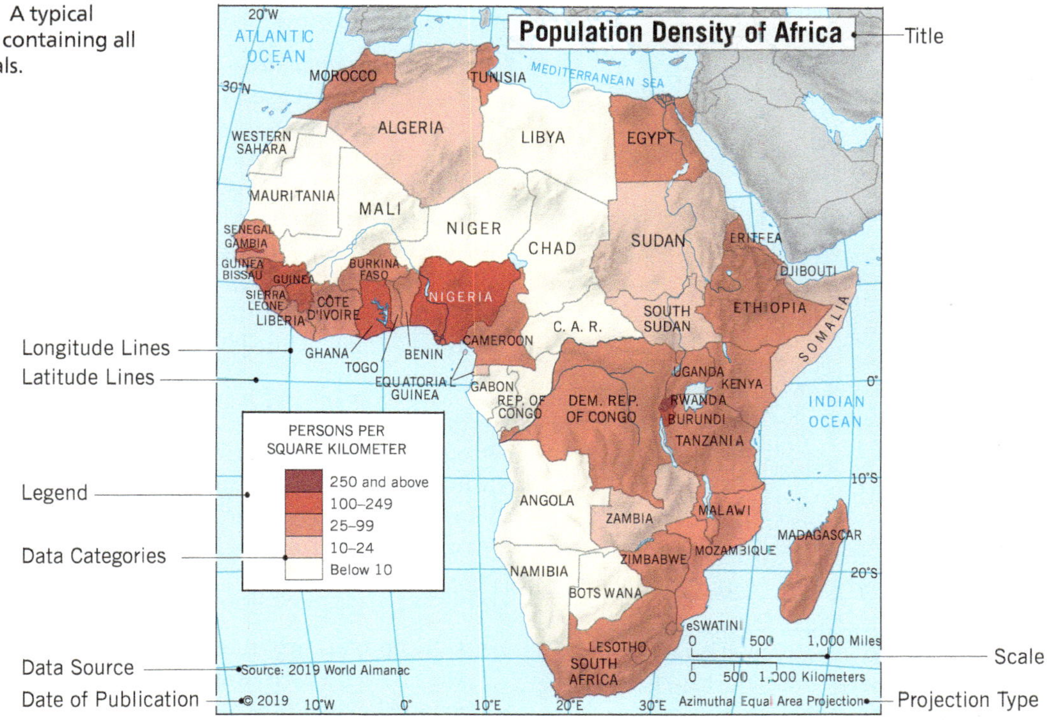

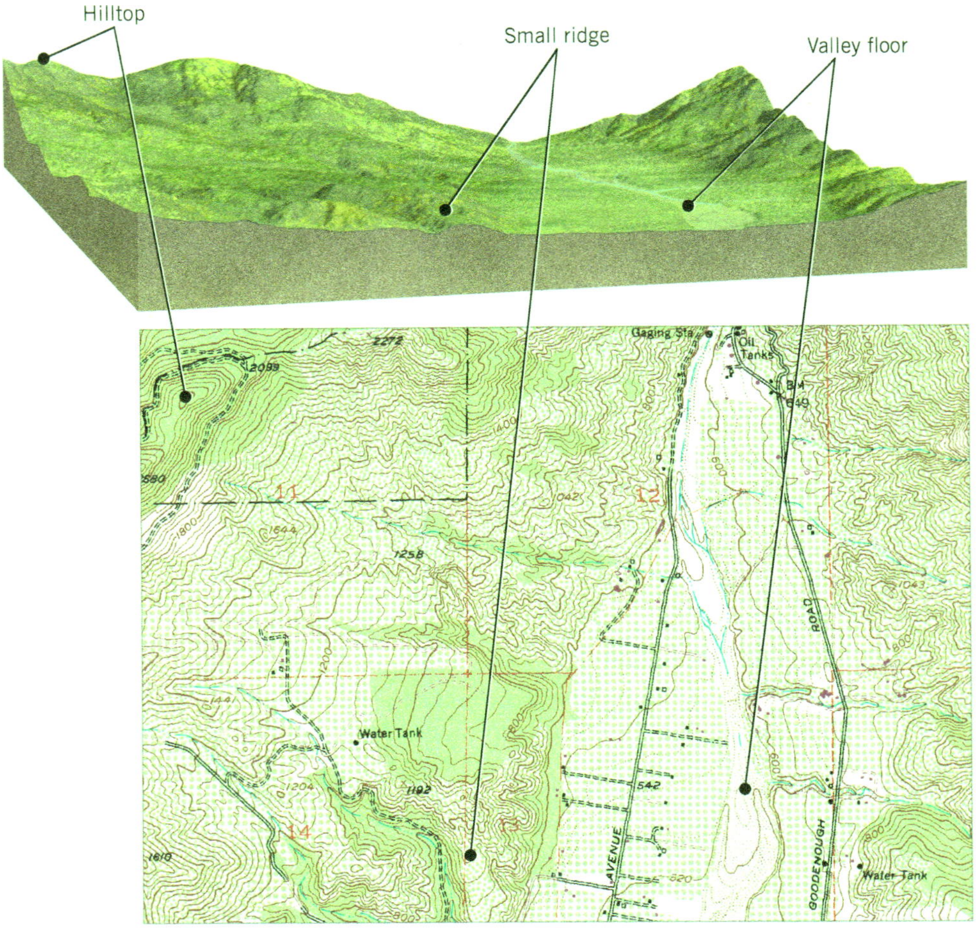

◀ **Figure 2-11** The use of elevation contour lines, shown with a matching landscape diagram and labeled features. This section of the Fillmore, California, quadrangle is part of a typical U.S. Geological Survey topographic map quadrangle. The original map scale was 1:24,000; the contour interval is 40 feet (12 meters).

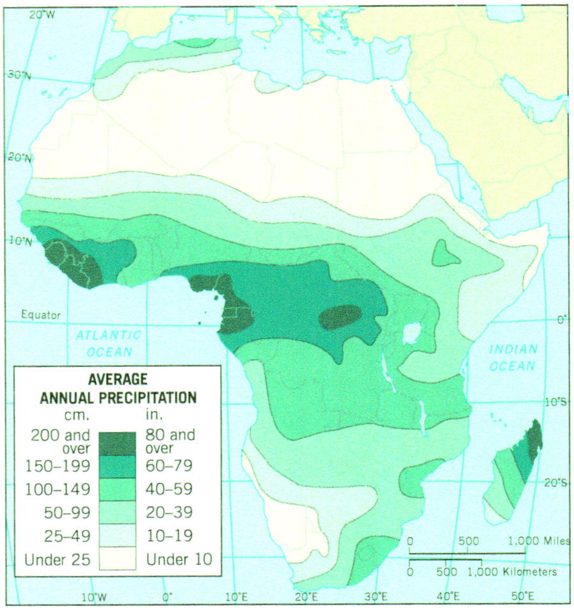

▲ **Figure 2-12** Isolines can be used to show the spatial variation of even intangible features, such as average annual precipitation of Africa. (On this map the areas between isolines have been shaded to clarify the pattern.)

have been identified by name, but only a few types are important in an introductory physical geography course:

- *Elevation contour line*—a line joining points of equal elevation (see Appendix II for a description of U.S. Geological Survey topographic maps)
- *Isotherm*—a line joining points of equal temperature
- *Isobar*—a line joining points of equal atmospheric pressure
- *Isohyet*—a line joining points of equal quantities of precipitation (*hyeto* is from the Greek, meaning "rain")
- *Isogonic line*—a line joining points of equal magnetic declination

Edmund Halley (1656–1742), an English astronomer and cartographer (for whom Halley's Comet is named), was not the first person to use isolines, but in 1701 he produced a map that was apparently the first published map to have isolines. This map showed isogonic lines in the Atlantic Ocean.

Characteristics of Isolines:

- Conceptually, isolines are always closed lines; that is, they have no ends. In practice, however, isolines often extend beyond the edge of a map, as in Figure 2-11.

- Because they represent gradations in a quantity, isolines cannot touch or cross one another, except under special circumstances (see Figure 2-12).
- The numerical difference between one isoline and the next is called the *interval*. Although intervals can vary according to the wishes of the mapmaker, it is normally useful to maintain a constant interval all over a given map.
- Isolines close together indicate a steep gradient (in other words, a rapid change); isolines far apart indicate a gentle gradient.

Drawing Isolines: When we draw an isoline on a map, we often must estimate values that are not available. **Figure 2-13** illustrates the basic steps in constructing an isoline map—in this case, an elevation contour line map. Each dot represents a data collection location (Figure 2-13a), and the number next to each dot is the elevation above sea level in meters.

We begin by drawing the 115-meter elevation contour: the 115-meter contour line passes between 114 and 116, and between 113 and 116 (Figure 2-13b). We repeat the process for other elevation contours (Figure 2-13c) and add shading to clarify the pattern (Figure 2-13d).

Isoline maps are now commonplace and are very useful to geographers even though an isoline is an artificial construct—that is, it does not occur in nature. For instance, an isoline map can reveal spatial relationships that might otherwise go undetected. Patterns that are too large, too abstract, or too detailed for ordinary comprehension are often significantly clarified by the use of isolines.

LearningCheck 2-8 **Define *isoline* and give one example of a kind of distribution pattern that can be mapped with isolines.**

Portraying the Three-Dimensional Landscape

Although many maps are simply flat representations of Earth, in physical geography the vertical aspect of the landscape is often an important component of study. In addition to raised-relief models of landforms, many other methods can be used to convey the three-dimensional aspect of the landscape on a two-dimensional map.

Elevation Contours: For many decades, topographic maps using elevation contour lines were a workhorse of landform study (see Figure 2-11). They remain so today, even as we transition from traditional paper maps to electronic maps such as those available from the U.S. Geological Survey (USGS) on its online *National Map* site (http://nationalmap.gov). Topographic maps and contour line rules are discussed in detail in Appendix II.

Digital Elevation Models: A remarkable recent advance in cartography has been the use of **digital elevation models (DEMs)** to convey topography. The starting point for creating a DEM image is a detailed database of precise elevations. For example, the USGS maintains such a database for the United States at several spatial resolutions. (*Spatial resolution* is the size of the smallest feature that can be identified.) One of the most commonly used is a 30-meter grid (meaning that elevation data are available at distance intervals of 30 meters, both north–south and east–west, across the entire country), but data have increasingly become available at a spatial resolution of 1 meter.

From digital elevation data, a computer generates a shaded-relief image of the landscape by portraying the landscape as if it were illuminated from the northwest by the Sun (**Figure 2-14**; also see the map at the start of this chapter). Although shaded relief maps were drafted by hand in the

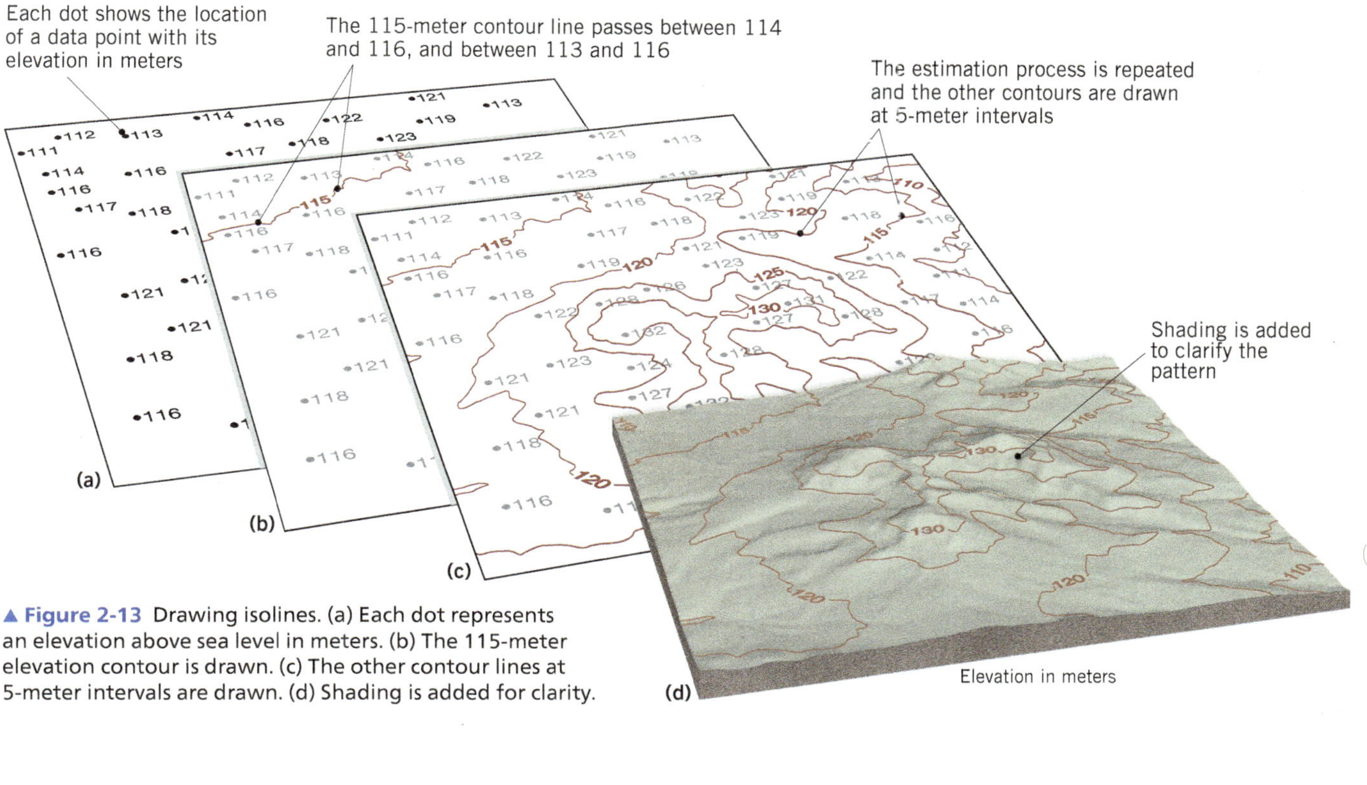

▲ **Figure 2-13** Drawing isolines. (a) Each dot represents an elevation above sea level in meters. (b) The 115-meter elevation contour is drawn. (c) The other contour lines at 5-meter intervals are drawn. (d) Shading is added for clarity.

▲ Figure 2-14 An oblique shaded-relief digital elevation model of post-1980 eruption Mount St. Helens in southwestern Washington.

past, one of the great virtues of a DEM is that the parameters of the image—such as its orientation, scale, and vertical exaggeration of the topography—can be readily manipulated. Furthermore, various kinds of information or images can be overlain on the topography to create maps that were once impossible to conceive (for example, see Figure 2-26).

LearningCheck 2-9 **How does a digital elevation model convey the topography of Earth's surface?**

GNSS—Global Navigation Satellite System

Since the 1970s, new technologies have transformed navigation and mapmaking. One such technology provides precise location data for points on or near Earth's surface. The **Global Navigation Satellite System (GNSS)** is such a system of satellite technologies. It includes the United States' **Global Positioning System (GPS)**, Russia's *Global Navigation Satellite System (GLONASS)*, Europe's *Galileo* GNSS, and China's *BeiDou* ("Compass") system.

GPS was developed in the 1970s and 1980s by the U.S. Department of Defense to aid in navigating aircraft, guiding missiles, and controlling ground troops. The first receivers were the size of a file cabinet, but continued technological improvement has reduced GPS receiver modules to the size of your fingertip. In fact, increasingly mobile devices such as tablets, cameras, and cell phones contain built-in GNSS receivers, revolutionizing both the way that data from field observations are gathered for use on maps and the way that map data are retrieved in the field.

The GPS system—formally called NAVSTAR GPS (Navigation Signal Timing and Ranging Global Positioning System)—is based on a constellation of at least 24 high-altitude satellites configured so that a minimum of four—preferably six—are in view from any position on Earth. (Currently, 30 satellites are active, with several older satellites still in orbit as backups.)

Each satellite continuously transmits both identification and positioning information that can be picked up by receivers on Earth (Figure 2-15). The distance between a given receiver and each of four or more satellites is calculated by comparing clocks stored in both the receiver and the satellite. Then the three-dimensional coordinates of the receiver's position are calculated through *trilateration*.

The more channels a GPS unit has (even inexpensive units now have 12), the more satellites that can be tracked and the better the accuracy. The simplest GPS units determine position to within about 5 meters (16 feet), and even mobile devices such as cell phones can attain an accuracy of better

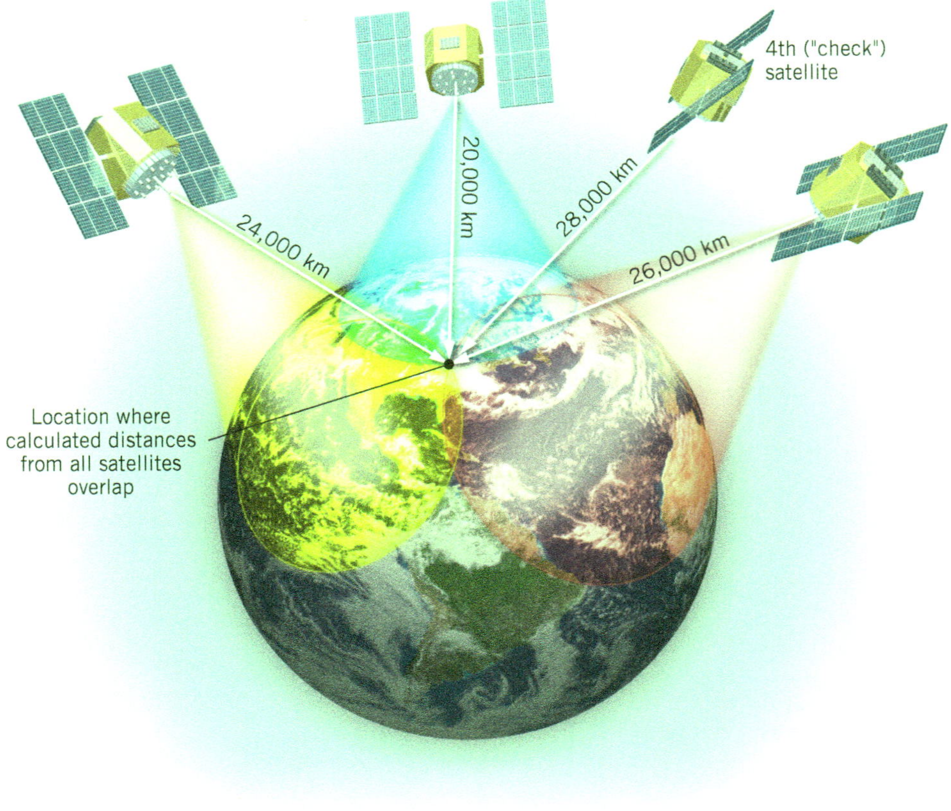

◄ Figure 2-15 Global Positioning System (GPS) satellites orbiting 20,200 kilometers (12,600 miles) above Earth broadcast timing signals that are picked up by a receiver on the surface. The time it takes the signal to travel from a satellite to the receiver is used to calculate a distance "circle" on the surface. Where the distance circles of at least three satellites overlap, the exact location of the receiver is determined through trilateration.

than 2 meters (6 feet) when satellite positioning is augmented by cell phone towers or another ground-based reference.

Wide Area Augmentation System (WAAS): The *Wide Area Augmentation System (WAAS)* was developed by the Federal Aviation Administration (FAA) to increase the accuracy of instrument-based flight approaches for airplanes. Several dozen ground-based reference stations across North America monitor GPS signals and then transmit correction messages to the satellites. With WAAS, GPS units should achieve a location accuracy of 7.6 meters (25 feet) 95 percent of the time; in practice, the actual location accuracy over most of the United States is typically better than 1.0 meters (3.2 feet) horizontally and 1.5 meters (4.9 feet) vertically. WAAS capability is built into nearly all new GPS receivers. Similar systems are being implemented in Asia (Japan's Multi-Functional Satellite Augmentation System) and Europe (the Euro Geostationary Navigation Overlay Service).

Continuously Operating GPS Reference Stations (CORS): The National Oceanic and Atmospheric Administration (NOAA) manages a system of permanent GPS receiving stations known as *Continuously Operating GPS Reference Stations (CORS)*. These highly accurate units can detect location differences of less than 1 centimeter of latitude, longitude, and elevation. Their uses include the long-term monitoring of slight changes in the ground surface caused by lithospheric plate movement or the bulging of magma below a volcano.

GPS Modernization Program: The United States has an ongoing modernization program for its GPS system. The upgrades include replacing older satellites with newer ones that provide greater accuracy, and a fourth civilian broadcast GPS signal (known as "L1C") that is compatible with the GNSS of other countries. The first new-generation GPS satellites (GPS III) were launched in 2018; all of the new satellites are scheduled to be in orbit by 2034.

GPS Applications: Since 1983, when access to GPS was made free to the public, growth has been astounding, and scientific and commercial applications now far outnumber military uses of the system. Practically everything that moves—airplane, truck, train, car, bus, ship, tablet, cell phone—can be equipped with a GPS receiver. When your cell phone location tracking is activated, it's a fairly simple matter for a commercial search database to direct you quickly to the closest coffee shop or to the nearest store with the item you want to buy.

GPS has been employed in earthquake forecasting, ocean floor mapping, volcano monitoring, and even tracking the COVID-19 pandemic. The Federal Emergency Management Agency (FEMA) uses GPS-enabled devices to collect data for damage assessment following such natural disasters as floods and hurricanes. Following Hurricane Sandy in October 2012, researchers using GPS data gathered from New York City taxis assessed the traffic patterns during this extreme event to give planners information that would better prepare the city for future disasters. What was born as a military system has become a scientific and economic resource.

It is important to remember that a number of factors can reduce the position accuracy of a GPS receiver: satellite signals can be blocked or reflected by buildings or topography, and radio interference or even solar storms can degrade the timing signal. Further, the mapping software in a GPS unit may be outdated or incomplete, occasionally leading unwary travelers toward roads that are poorly maintained or even nonexistent!

GPS Display of Latitude and Longitude: As we saw in Chapter 1, latitude and longitude traditionally are described in degrees, minutes, and seconds; for example, 40°45′10″N. Each 1″ (1 second) of latitude represents a distance of about 31 meters (101 feet). Even simple GPS units routinely achieve greater accuracy than that, so coordinates are frequently reported in decimal units, such as 40°45′10.4″N. (A difference of 0.1″ of latitude is about 3.1 meters [10 feet].)

Most GPS-enabled devices give users the option of displaying location coordinates without seconds by using "decimal minutes"—so 40°45′10.4″N would be 40°45.173′N. In some cases, even minutes are ignored, presenting latitude simply as 40.75288° N.

LearningCheck 2-10 How does GPS determine locations?

Remote Sensing

Throughout most of history, maps were the only tools available to depict anything more than a tiny portion of Earth's surface with any degree of accuracy. However, sophisticated technology developed in recent years permits precision recording instruments to operate from high-altitude vantage points, providing a remarkable new set of tools for the study of Earth. **Remote sensing** refers to any measurement or acquisition of information by a recording device that is not in physical contact with the object under study—in this case, Earth's surface.

Originally utilizing only airplanes, remote sensing was revolutionized by the use of satellites. Now more than 1200 Earth observation satellites, from dozens of countries, are perched high in the atmosphere. There they circle Earth either in a "low" orbit (an altitude of 20,000 kilometers [12,400 miles] or less) or in a lofty *geosynchronous orbit* (usually about 36,000 kilometers [22,400 miles] high) that allows a satellite to remain over the same spot on Earth at all times. These satellites gather data and produce images that provide communications, global positioning, weather data, and a variety of other information for a wide range of commercial and scientific applications—for example, see the box *Global Environmental Change: Growing a City in the Desert.*

Aerial Photographs

Video
Studying Fires Using Multiple Satellite Sensors
https://media.pearsoncmg.com/ph/esm/esm_mcknight_physgeo_13/media/

Until the 1960s, aerial photography was almost the only form of remote sensing. The earliest *aerial photographs* were taken from balloons in France in 1858. By World War II (1939–1945), black-and-white and color photographs taken from airplanes had become crucial to the military, and *photogrammetry*—the science of obtaining reliable measurements and mapping from aerial photographs—had developed. Although

Global Environmental **Change**

Growing a City in the Desert

• **Bradley Shellito,** Youngstown State University

As of 2020, the world population was an estimated 7.8 billion people, up from 5 billion in 1987. As a result of this tremendous growth, demand to provide housing, industries, and amenities for everyone across the planet is extensive—sometimes in the unlikeliest of environments. For example, in Saudi Arabia, crops are grown in the desert, while China's Pearl River Delta, which was mostly rural only 35 years ago, is now the world's largest urban area.

Viva Las Vegas: For several decades, Las Vegas, Nevada, has been one of the fastest growing cities in the United States. Its population grew from about 126,000 in 1970 to 258,000 in 1990; by 2019, the city's population was greater than 651,000. In addition, visitors in 2019 numbered more than 42 million, about double the number of visitors in 1990. That's a huge number of people living, working, and touring in a city built in the middle of a desert ecosystem. Las Vegas sits within a basin in the Mojave Desert, and sidewalks in new housing developments at the city's edges lead straight into the desert.

This level of urban development in a desert environment brings plenty of challenges and questions, particularly concerning water usage and sustainability. Water levels in a reservoir of the Colorado River at Lake Mead, the main source of water for the region, have been dropping. Water conservation efforts are now in place to aid sustainability, including returning indoor wastewater consumption to the lake, limits and prohibitions on the planting of turf grass, and watering restrictions in public places.

Geospatial technologies can be used to examine the "big picture" of the growth of Las Vegas and its environs and provide the monitoring needed to maintain sustainable growth measures. Satellite remote sensing allows us to view the expansion of urban development spatially, so we can see where the city is growing and at what rates. For example, the Landsat satellite archive, stretching back more than 40 years at intervals of every 16 days, allows us to keep a constant eye on urban growth within fragile ecosystems. Landsat imagery of Las Vegas in 1984 and 2011 (**Figure 2-F**) gives a dramatic look at the growth in urban developments (including houses, shops, utilities, and tourist locations) necessary to accommodate the growing population and visitors. With this monitoring, we can then use geographic information systems (GIS; discussed later in this chapter) to analyze different planning and water management strategies for the city.

(a) Las Vegas, 1984

(b) Las Vegas, 2011

▲ **Figure 2-F** Las Vegas in (a) 1984 and (b) 2011 as viewed by the Landsat 5 TM sensor.

Artificial Archipelago: Las Vegas is not the only site for which geospatial technologies can help in monitoring and planning. For example, the Palm Islands in the Persian Gulf, just off the coast of Dubai in the desert country of the United Arab Emirates, is an archipelago that was built for tourism. Building a series of islands out of sand and rocks to be used as resorts and hotels brings with it a series of environmental challenges, but the island's growth can be monitored through remote sensing technologies (**Figure 2-G**). Remote sensing is especially useful for monitoring environmental conditions such as water quality in the region as well as documenting urbanization and the environmental consequences of development. For example, see the growth of the Palm Islands in Dubai at https://earthobservatory.nasa.gov/world-of-change/Dubai, NASA's Earth Observatory. Similarly, the Timelapse app (http://world.time.com/timelapse/) allows you to view yearly Landsat imagery from 1984 until 2012 of locations around Earth (including Las Vegas and Dubai).

Looking Ahead: Earth's population is expected to climb to 9 billion people by 2040 and to continue to grow. This dramatic growth will carry with it a variety of environmental impacts all around the world. Remote sensing satellites will allow us to monitor these kinds of impacts and growth; GIS will let us analyze the patterns. By looking through the lens of geospatial technologies, we can understand global environmental impacts and plan for a sustainable future.

Questions

1. How can city officials use satellite imagery taken at regular intervals to make decisions for smart growth or sustainability strategies?
2. What other types of sustainability challenges do urban developments in desert and coastal ecosystems face? How can remote sensing be used to address them?

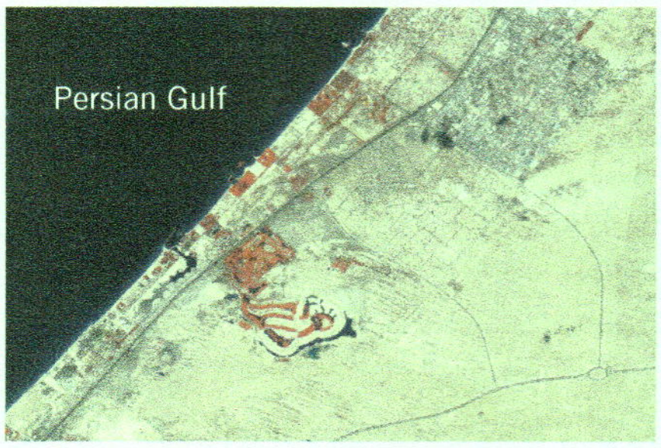

(a) Dubai, 2000

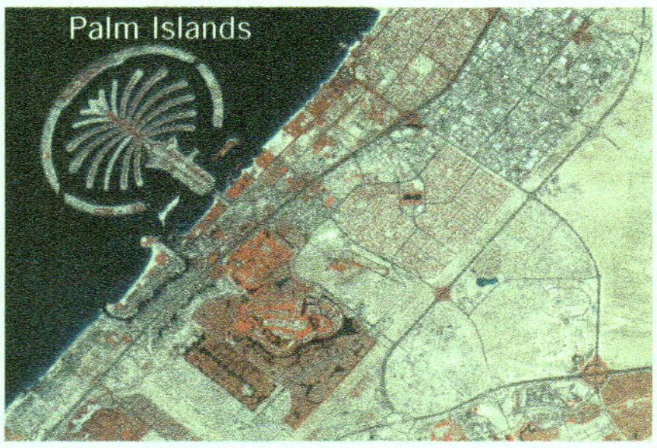

(b) Dubai, 2011

▲ **Figure 2-G** Satellite imagery of Dubai in (a) 2000 and (b) 2011. Here vegetation appears in red.

satellite imagery has taken over the role of aerial photography for some applications, aerial photographs available in digital form from agencies such as the U.S. Geological Survey remain an important source of large-scale geographic imagery.

Orthophoto Maps: An *orthophoto map* is a multicolored, low-distortion photographic map prepared from aerial photographs or digital images. Distortions caused by camera tilt or differences in terrain elevation are removed, giving orthophoto images the geometric characteristics of a map (**Figure 2-16**). An orthophoto can show the landscape in much greater detail than a conventional map can but, like a map, retains a common scale that allows precise measurement of distances. Orthophoto maps are particularly useful in flat-lying coastal areas because they can show subtle topographic detail in areas of very low relief.

▲ **Figure 2-16** Orthophoto map of Wilmington, North Carolina; original scale: 1:24,000.

Visible Light and Infrared Sensing

One of the most important advancements in remote sensing came when wavelengths of radiation other than visible light were first utilized. As we will see in Chapter 4, *electromagnetic radiation* includes a wide range of wavelengths of energy emitted by the Sun and other objects (**Figure 2-17**). The human eye (and traditional photographic film) is sensitive to only the narrow portion of the electromagnetic spectrum known as *visible light*—the colors seen in a rainbow. However, a wide range of other wavelengths of energy—such as *X-rays, ultraviolet radiation, infrared radiation*, and *radio waves*—are emitted, reflected, or absorbed by surfaces and can be detected by special films or instruments, yielding a wealth of information about the environment.

Color infrared (color IR) imagery uses electronic sensors or photographic film sensitive to radiation in the *near infrared* portion of the electromagnetic spectrum—wavelengths of radiation just longer than the human eye can see. With color IR imagery, sensitivity to visible blue light is replaced by sensitivity to near infrared wavelengths. The images produced in this way are "false-color"; living vegetation appears red instead of green. First widely used in World War II, color IR film was often called "camouflage-detection" film because it discriminates living vegetation from the withering vegetation used to hide objects during the war. Today, one of the major uses of color IR imagery remains the identification and evaluation of vegetation (**Figure 2-18**).

Thermal Infrared Sensing

None of the middle or far infrared part of the electromagnetic spectrum, called *thermal infrared* (thermal IR), can be sensed with conventional digital cameras or traditional photographic film. As a result, special supercooled scanners are needed. Thermal scanning measures the radiant temperature of objects and may be carried out day or night. The photograph-like images produced in this process are particularly useful for showing diurnal temperature differences between land and water, between bedrock and sediment for studying thermal water pollution, and for detecting forest fires.

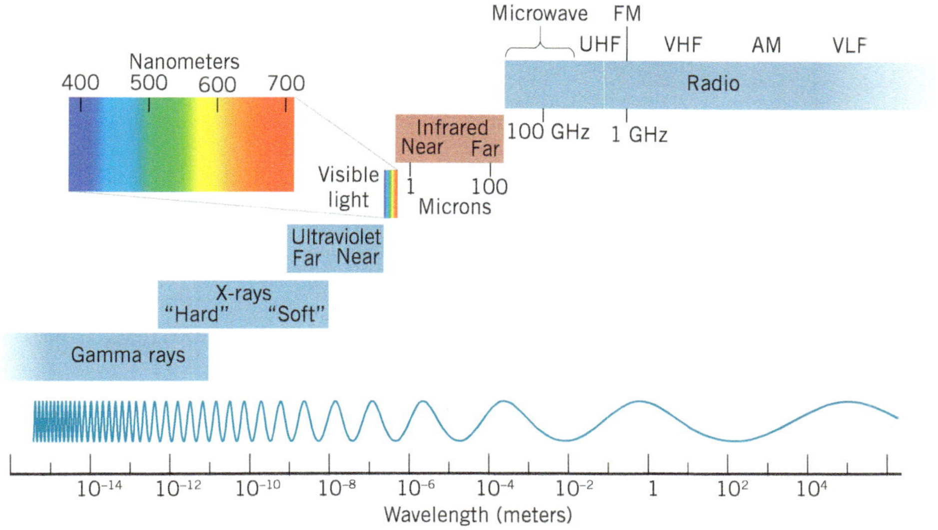

◄ **Figure 2-17** The electromagnetic spectrum. The human eye can sense radiation from the visible-light region only. Conventional photography also can use only a small portion of the total spectrum. Various specialized remote-sensing scanners are capable of "seeing" radiation from other parts of the spectrum.

▲ **Figure 2-18** Color infrared image from the Advanced Spaceborne Thermal Emission and Reflection Radiometer (ASTER) of the cities of Palm Springs, Cathedral City, and Palm Desert, California. In this false-color IR image, healthy vegetation is shown in red; bare ground, in gray-blue.

By far the greatest use of thermal IR scanning systems has been on meteorological satellites (for example, see "Focus: GOES Weather Satellites" in Chapter 6). Although the spatial resolution is not as high as for some other kinds of sensing systems, it is more than sufficient to make weather forecasting far more accurate and complete than ever before.

LearningCheck 2-11 **What are the differences between near infrared and thermal infrared images, and what kinds of features might be studied with each?**

Multispectral Remote Sensing

Today, most sophisticated remote sensing satellites are **multispectral**, or *multiband*. These instruments detect and record many *bands*, or regions, of the electromagnetic spectrum simultaneously. Thus, although traditional photographic film was sensitive to only a narrow band of visible light, a satellite equipped with a multiband instrument images the surface of Earth in several spectrum regions at once—visible light, near infrared, middle infrared, and thermal infrared—each useful for different applications.

A digital multispectral satellite image is conveyed through a matrix of numbers, with each number representing a single value for a specific *pixel* (picture element) and band. These data are stored in the satellite, eventually transmitted to a receiving station on Earth, numerically processed by a computer, and produced as a set of gray values and/or colors in the final image (**Figure 2-19**).

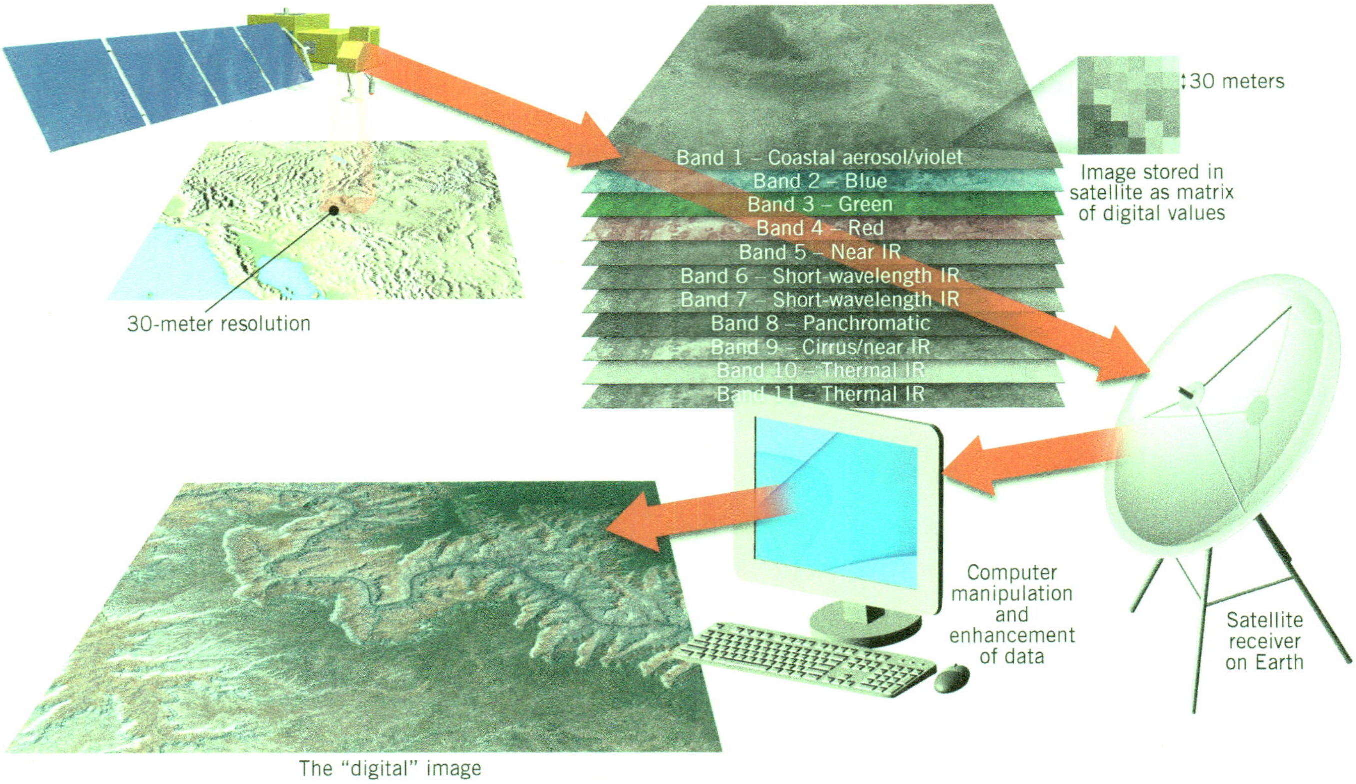

▲ **Figure 2-19** The sequence of events that takes place as multispectral satellite scans are converted to a digital image.

Landsat: The early NASA space missions (*Mercury*, *Gemini*, and *Apollo*) used multiband photography obtained through multicamera arrays. These imaging experiments were so successful that NASA then developed the *Earth Resources Technology Satellite (ERTS) series*, which was later renamed *Landsat*. The 1970s and 1980s saw the launch of five Landsat satellites carrying a variety of sensor systems.

Landsat 7, launched in 1999, carries an instrument array called the Enhanced Thematic Mapper Plus, which provides images in eight spectral bands with a resolution of 15 meters (49 feet) in the panchromatic band (sensitive to visible and near infrared wavelengths), 30 meters (98 feet) in the six narrow bands of visible and short infrared wavelengths, and 60 meters (197 feet) in thermal infrared. Although the satellite was originally designed for a life of less than 10 years, as of this writing Landsat 7 remains in active operation.

Landsat 8, launched in 2013, has 11 wavelength bands (Table 2-1) and includes improved imaging (Figure 2-20) and data collection with instruments such as the Operational Land Imager (OLI) and the Thermal Infrared Sensor (TIRS). A new Landsat 9 satellite is scheduled for launch in 2021.

▲ **Figure 2-20** Landsat 8 satellite image of Acadia National Park in Maine.

Earth Observing System Satellites: In 1999 NASA launched the first of its *Earth Observing System (EOS)* satellites known as *Terra*. The key instrument of these satellites is the *Moderate Resolution Imaging Spectroradiometer* (MODIS), which gathers data in 36 spectral bands and provides images covering the entire planet every one to two days (Figure 2-21). Other devices onboard Terra include the Clouds and the Earth's Radiant Energy System (CERES) instruments for monitoring the energy balance of Earth, and the Multiangle Image Spectroradiometer (MISR), capable of distinguishing various types of atmospheric particulates, land surfaces, and cloud forms. With special processing, three-dimensional models of image data are possible.

The more recently launched EOS satellite, *Aqua*, is designed to enhance our understanding of Earth's water cycle by monitoring water vapor, clouds, precipitation, glaciers, and soil wetness. In addition to instruments such as MODIS, Aqua includes the Atmospheric Infrared Sounder (AIRS), designed to permit very accurate temperature measurements throughout the atmosphere.

In 2011, NASA launched an Argentine-built satellite that included an instrument called *Aquarius* that enables scientists to monitor concentrations of dissolved salts near the surface of the ocean (see Figure 9-6). It is improving our understanding

TABLE 2-1 Bands of Landsat 8

Band Number	Bandwidth (micrometers)	Band Designation	Resolution (meters)	Typical Applications
1	0.43–0.45	Coastal aerosol	30	Coastal and aerosol studies
2	0.45–0.51	Blue	30	Bathymetric mapping; soil and vegetation analysis
3	0.53–0.59	Green	30	Vegetation analysis
4	0.64–0.67	Red	30	Vegetation slopes
5	0.85–0.88	Near IR	30	Biomass content and shorelines
6	1.57–1.65	Short-wavelength IR	30	Soil moisture; penetrate thin clouds
7	2.11–2.29	Short-wavelength IR	30	Soil and vegetation moisture content
8	0.50–0.68	Panchromatic	15	High-resolution images
9	1.36–1.38	Cirrus		Detection of cirrus clouds
10	10.60–11.19	TIRS 1	100	Thermal mapping; soil moisture
11	11.50–12.51	TIRS 2	100	Thermal mapping; soil moisture

◀ **Figure 2-21** Natural-color satellite image showing the Aleutian Islands extending to the southwest from the Alaska Peninsula. Katmai National Park is located at the northeastern end of the Alaska Peninsula. This image was taken with the MODIS (Moderate Resolution Imaging Spectroradiometer) instrument aboard NASA's Aqua satellite.

of the effects of long-term climate change and short-term phenomena such as *El Niño* (discussed in Chapter 5).

Many satellite images are now easily available for viewing and downloading via the Internet from NASA, NOAA, and the USGS. For example, you can visit the following sites: www.earthobservatory.nasa.gov, www.goes.noaa.gov, and http://eros.usgs.gov/imagegallery/.

Commercial High-Resolution Satellites: In addition to free or low-cost imagery that is available from government-operated satellites (such as the GOES satellites, Landsat, and the EOS satellites), a number of companies sell very high-resolution imagery (up to 50- to 60-centimeter [20- to 24-inch] resolution) for commercial applications, including *SPOT* (*Satellite Pour l'Observation de la Terre*), *GeoEye-1*, and *WorldView-3*. *SkySat* (originally called *Skybox*) developed small satellites that can be deployed at much lower cost than larger traditional satellites, promising economical high-quality images. The commercial market for remotely sensed images from such companies is growing rapidly.

LearningCheck 2-12 What is "multispectral" remote sensing?

Radar, Sonar, and Lidar: All the systems mentioned so far work by sensing the natural radiation emitted by or reflected from an object and are therefore characterized as *passive systems*. Another type of system, called an *active system*, has its own source of electromagnetic radiation. The most important active sensing system used in the Earth sciences is **radar**, the acronym for *radio detection and ranging*. Radar senses wavelengths longer than 1 millimeter, using the principle that the time it takes an emitted signal to reach a target and then return to the sender can be converted to distance information.

Radar is capable of operating by day or night, but it is unique in its ability to penetrate atmospheric moisture (Figure 2-22). Thus, some wet tropical areas that could never be sensed by other systems have now been imaged by radar. Radar imagery is particularly useful for terrain analysis in places of frequent cloud cover or thick vegetation and for meteorology—especially in the real-time study and mapping of rainfall and severe weather. (The use of specialized *Doppler radar* in meteorology is discussed in Chapter 7.)

Another active remote sensing system, **sonar** (*sound navigation and ranging*), permits underwater imaging so that scientists can determine the form of Earth's crust that is hidden by the world ocean.

Lidar (derived from the words *light* and *radar*) uses reflected laser light to measure distances. From ground-based or aerial platforms, lidar offers three-dimensional information about surface features with a precision previously unavailable through remote sensing (Figure 2-23).

Geographic Information Systems (GIS)

Cartography dates back to the days of the early Egyptians, but with the introduction of computers in the 1950s, incredible improvements in speed and image-handling ability have been possible. Of all the technological advances in cartography over the last few decades, however, one of the most revolutionary has been *geographic information systems*.

Geographic information systems—commonly called **GIS**—are computer systems designed to analyze and display spatial data. GIS software allows users to collect,

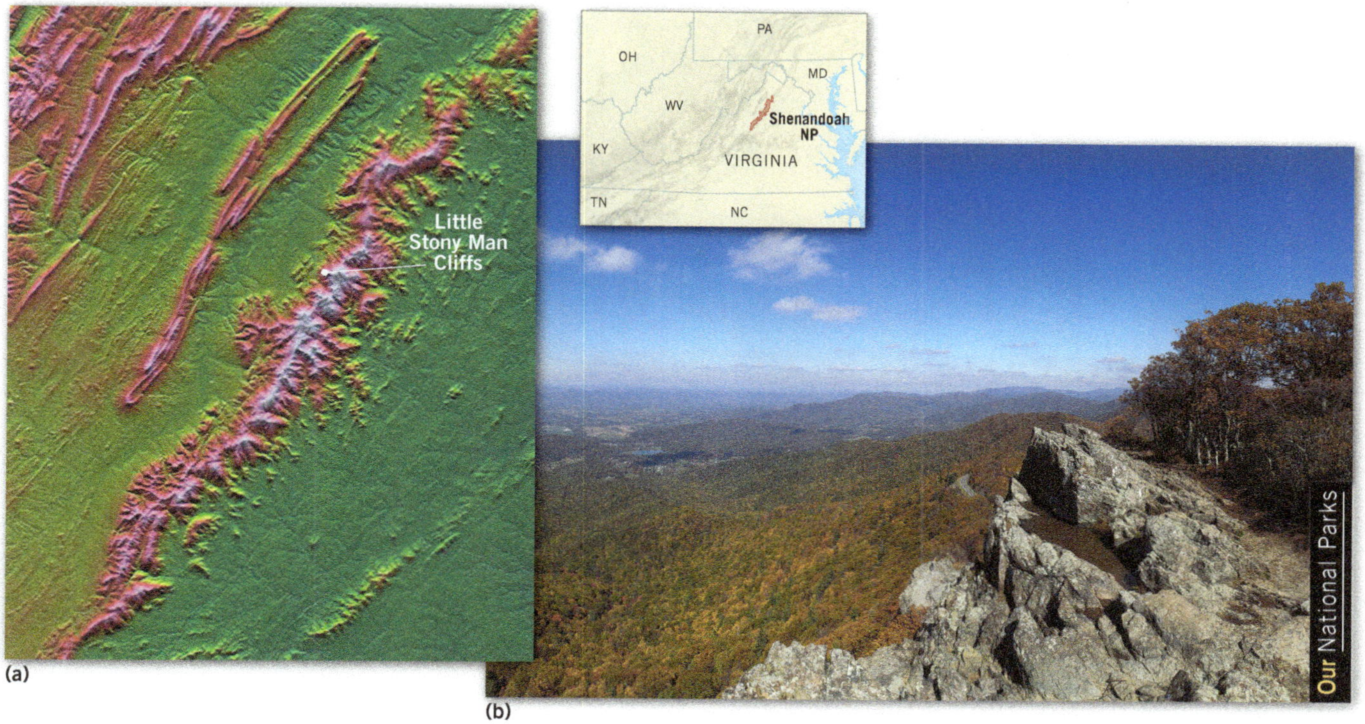

▲ **Figure 2-22** (a) Radar image showing the topography of Shenandoah National Park and the Blue Ridge Mountains in Virginia. The data were gathered from the Shuttle Radar Topography Mission aboard the Space Shuttle *Endeavour* in 2000. The data were processed with elevations represented by different colors, ranging from green for lowlands to white for high mountaintops. Shaded relief was added to highlight the topography. (b) View from Little Stony Man Cliffs.

▲ **Figure 2-23** (a) Lidar image of Great Sand Dunes National Park in Colorado. (b) View of sand dunes and Mount Herard.

store, retrieve, reorganize, analyze, and map geographic data from the real world (**Figure 2-24**).

Geographic information systems originally developed out of computer science, geography, and cartography, and they found their greatest early uses in surveying, photogrammetry, spatial statistics, and remote sensing. So common are they now in geographical analysis that GIS has become a science of spatial analysis by itself, known as *geographic information science* (*GISci*). The software has spun off a multibillion-dollar industry in spatial data and spatial information.

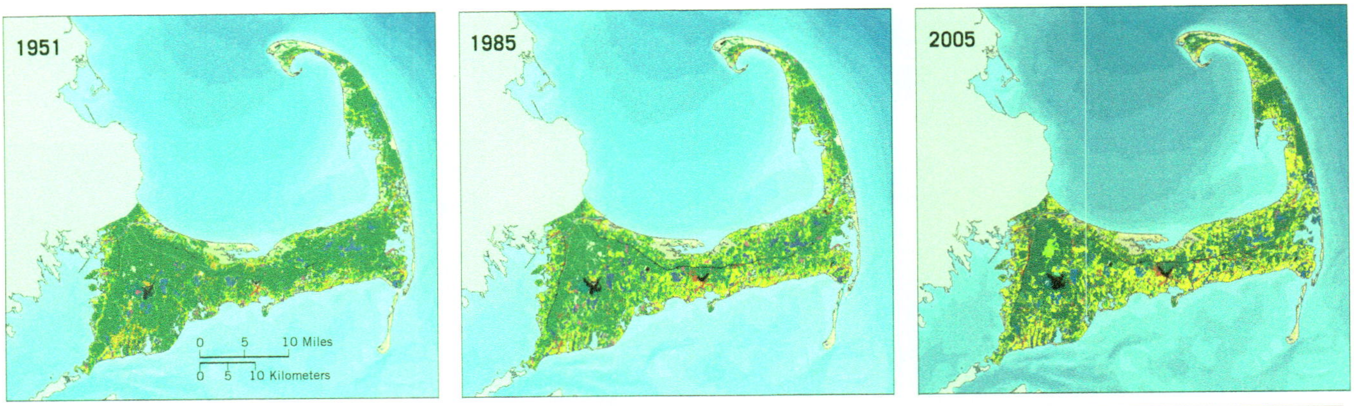

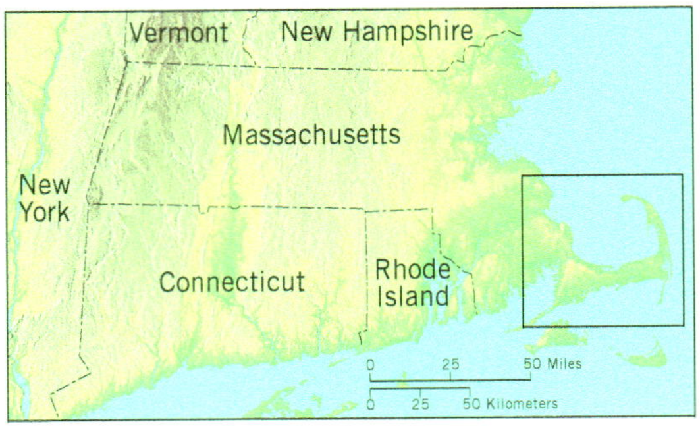

▲ **Figure 2-24** GIS was used to study land-use changes on Cape Cod from 1951 to 2005, showing the expansion of residential housing and the accompanying loss of forest. Dark green shows areas of forest, yellow shows areas of residential housing, and red shows areas of commercial and industrial development.

Geographic information systems are in effect libraries of information that use maps to organize, store, view, and analyze information in an intuitive, visual manner. Just as a computer database can be used to manipulate rows and columns of data, a GIS allows data management using the link between data and a map. The map and data are encoded, usually as numbers representing coordinates of locations at points on a grid covering the mapped area. The operator can then use the map to organize or search the data or use the data to search the map. An important attribute is the capability of GIS data from different maps and sources, such as field data, map data, and remotely sensed images, to be registered together within a common database, with a common map scale and map projection. In this way, one map layer, such as the locations of rivers, can then be cross-referenced to another, such as geology, soils, or slope.

Overlay Analysis

GIS is frequently used in *overlay analysis*, in which two or more layers of data are superimposed or integrated. GIS treats each spatially distributed variable as a particular layer in a sequence of overlays (Figure 2-25). Input layers combine such diverse elements as topography, vegetation, land use, land ownership, and land survey. Details of these various components are converted to digital data and are synthesized onto a reference map or data set. With GIS, particularly useful images for the study of physical geography can be developed when data or satellite images are overlain on topography generated with a digital elevation model, offering oblique views of the landscape that were previously impossible to obtain (Figure 2-26).

GIS in Decision-Making

Although once restricted to the domain of GIS specialists, in recent years GIS software has become increasingly user-friendly, and numerous open-source GIS platforms are now available. In addition, many GIS applications and data have migrated to the "Cloud," allowing the rapid sharing and mapping of real-time data. Live Web maps can be embedded in applications and accessed from mobile devices. Decisions can be made with tools never before available—everything from which route to take to avoid a traffic jam, to where to deploy firefighters on a blazing mountainside.

With GIS technology now available to the everyday decision-maker, it is rapidly becoming a necessary component of the policy-making structure of many organizations

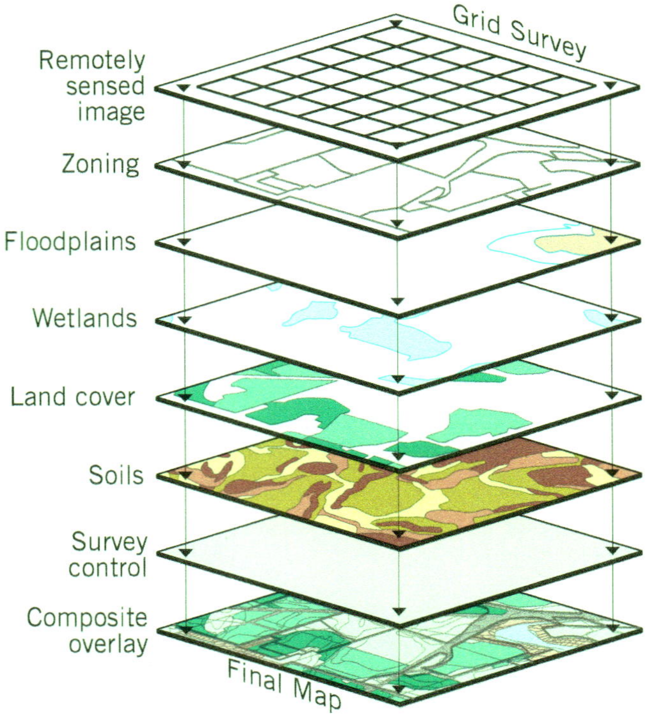

▲ **Figure 2-25** Much GIS work involves layers of spatial data superimposed on one another.

◄ **Figure 2-26** This oblique view of Katmai National Park in Alaska was created by configuring a natural color Landsat 8 image over a digital elevation model of the topography created from data gathered by NASA's Terra satellite.

(**Figure 2-27**). GIS can bring a new and more complete perspective to resource management, environmental monitoring, natural hazards assessment, and a host of other fields. The growth of GIS is so rapid today that few fields of academic study, sectors of the economy, and divisions of government do not use these powerful tools—for example, see the box *Focus: GIS for Geographic Decision Making.*

LearningCheck 2-13 **How is GIS different from GPS?**

Tools of the Geographer

The vast array of maps, remotely sensed imagery, satellite data, and GIS applications that are now available are making the tools of the geographer more widely used than ever before. The effective use of these tools, however, still entails thoughtful consideration. Although it is easy to download a satellite image or quickly print a map from the Internet, those images and maps may or may not be useful for geographical analysis—and might actually be deceptive—unless we choose an appropriate map projection, an appropriate scale, and an appropriate selection of data to depict.

Choosing Effective Maps and Imagery: Certain types of imagery are useful for particular purposes. For example, when we study major features of Earth's surface, high-altitude satellite imagery is especially valuable (**Figure 2-28**). For studying the hydrosphere, multiband satellite images

◄ **Figure 2-27** Scientists at the Indian National Centre for Ocean Information use GIS in their early warning system for tsunami and storm surges.

Focus

GIS for Geographic Decision-Making

• **Keith Clarke,** University of California–Santa Barbara

A coastal engineer wishes to know what will be the future impact of sea-level rise and increased storm surge on coastal wetlands. An emergency manager needs to know how best to evacuate residents during a hurricane. A city planner wants to know the future impact that land-use changes will have on the emission of carbon dioxide and other greenhouse gases. In each case, geographic information systems (GIS) can help bring together geographic facts that are relevant to decisions about natural and built environments. Such a system first assimilates and brings together spatial data—data that include geographic location or coordinates—from multiple sources, which may include government data clearinghouses, state agencies, and field measurements. It then overlays all of the data in a common reference frame of coordinates as layers (see Figure 2-25).

Bringing Data Together: The data can be retrieved for use in analysis and then displayed in a way that enables decision-makers to explore their options, usually by investigating future scenarios. For example, in some scenarios, the coastline of San Francisco Bay might face 1 meter (3 feet) of future sea-level rise (**Figure 2-H**). Using an online GIS at https://coast.noaa.gov/slr/, the coastal engineer can move the sea-level rise slider to see that almost all of the impacts on the Bay will occur in the low-lying fringes, now mostly salt evaporation basins. With GIS, the emergency manager can overlay the output from spatial analyses and models atop the geographic information to help make informed decisions about safe evacuation routes.

Modeling the Future for Better Decisions: GIS modeling is ideal for land-use studies. Compare an actual land-use map for Little Rock, Arkansas, as mapped by satellite in 1992 (**Figure 2-I-a**) with the land use as forecast by a land-use-change model for the year 2050 (**Figure 2-I-b**). This model is based on an Intergovernmental Panel on Climate Change (IPCC) scenario in which the world of the future has very rapid economic growth, global population peaks in mid-century and then declines, and we see rapid introduction of more efficient technologies, with a reduction in income disparity. (The findings of the IPCC are discussed in Chapter 8.) The U.S. Geological Survey used this model output to assess carbon and greenhouse gas (GHG) emissions under this scenario. The carbon and GHG impacts were then compared within the GIS among several IPCC scenarios as part of the National Assessment of Ecosystem Carbon and Greenhouse Gas Fluxes. Being able to see a variety of possible outcomes helps inform decision-makers and can lead to better policy.

Trends for GIS and Decision-Making: Today, decision-support tools are an integrated component of planning, business, industry, and research. They are used by governments at all levels, businesses, and planners to do everything from delivering packages by mail to planning for the next hurricane. Increasingly sophisticated decision-making tools are being integrated into GIS: mapping with multiple contributing factors, land suitability analysis, demographic and transportation planning tools, and many more. In addition, displays such as maps can be supplemented with information graphics, animations, and three-dimensional views and fed into broader decision-making tools such as optimization and scenario exploration software. You already employ many of these tools when you use your geographic position in searching the Internet.

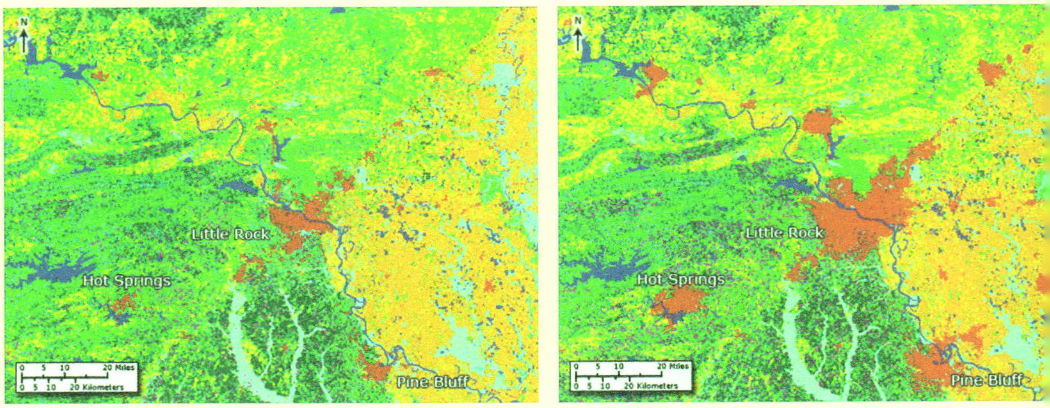

(a) 1992 (b) 2050

▲ **Figure 2-I** Land use in Little Rock, Arkansas, (a) in 1992; (b) as predicted for 2050 by GIS modeling. (Red = developed; orange = cropland; yellow = pasture; dark green = evergreen forest)

▲ **Figure 2-H** One meter of sea-level rise as shown by the NOAA Coastal exploration tool (coast.noaa.gov/slr).

Questions

1. What layers of imagery and data that you can see were brought together by the National Oceanic and Atmospheric Administration for the San Francisco Bay example?
2. The IPCC scenario described here is often called "business as usual." What will happen to land use in Little Rock if this scenario ensues by 2050?
3. How is GIS likely to change, in terms of decision support, in the future?

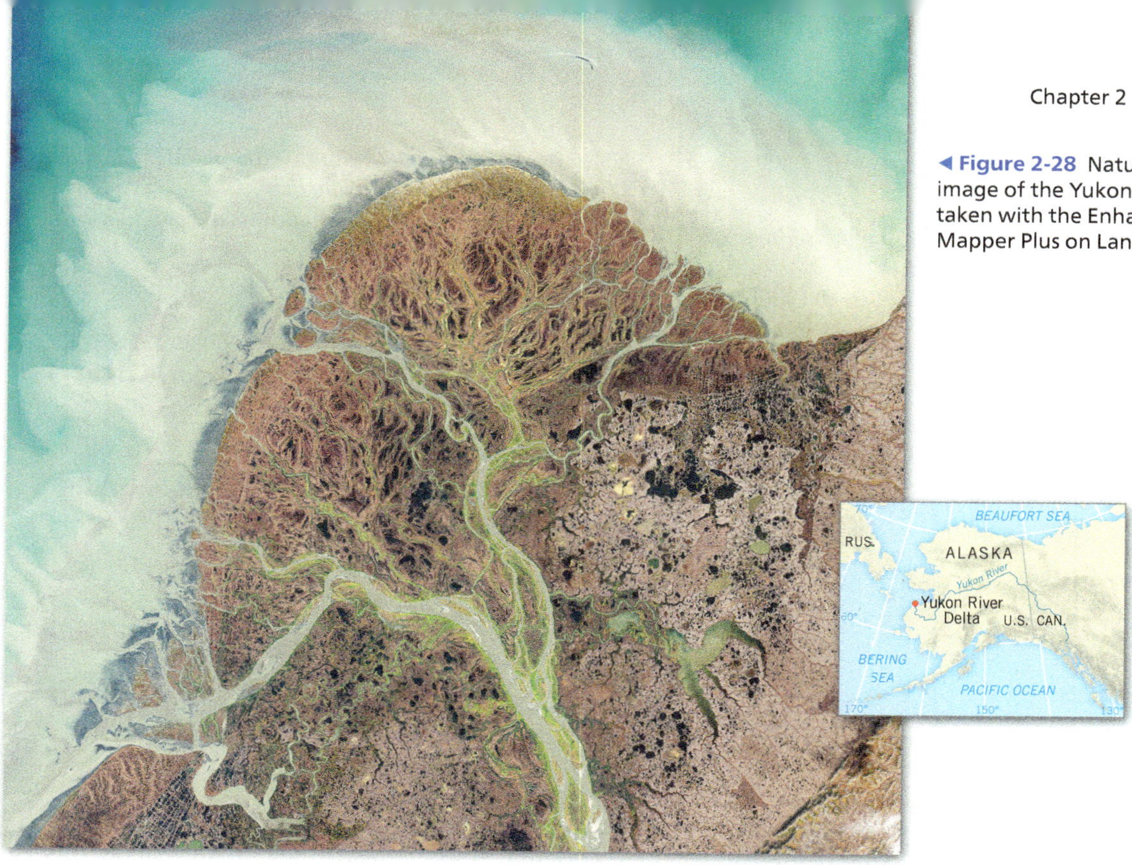

◀ **Figure 2-28** Natural color satellite image of the Yukon River Delta, Alaska, taken with the Enhanced Thematic Mapper Plus on Landsat 7.

of an entire hemisphere can tell us much about the water content in clouds, air masses, glaciers, and snowfields at a given time. Changes in vegetation patterns in the biosphere are often visible with time-series color infrared imagery. Cities or transportation networks may be best interpreted by using large-scale satellite imagery or aerial photographs. In all cases, GIS may be used to uncover or highlight geographic relationships that are not obvious when we employ any single source of data or kind of imagery.

In using these tools, the geographer should recognize that understanding does not come simply through the application of technology. Understanding comes from a carefully designed investigation, often using technology but frequently supported by such traditional sources of information as field study and observation.

After studying this chapter, you should be able to answer the following questions. Key terms from each text section are shown in **bold type.** Definitions for key terms are also found in the glossary at the back of the book.

Learning Review 2

Key Terms and Concepts

Maps and Globes (*p. 32*)

1. How is a **map** different from a globe?
2. Why is it impossible for a map of the world to portray Earth as accurately as can be done with a globe?

Map Scale (*p. 32*)

3. Describe and explain the concept of **map scale**.
4. Contrast **graphic map scales**, **fractional map scales**, and **verbal map scales**.
5. What is meant by a map scale with a representative fraction of 1/100,000 (also written 1:100,000)?
6. Explain the difference between **large-scale maps** and **small-scale maps**.

Map Projections and Properties (*p. 35*)

7. What is meant by a **map projection**?
8. Explain the differences between an **equivalent (equal area) map projection** and a **conformal map projection**.
9. Is it possible for a map to be both conformal and equivalent?
10. What is a **compromise map projection**?

Families of Map Projections (*p. 36*)

11. Briefly describe the four major families of map projections: **cylindrical projections**, **planar projections**, **conic projections**, and **pseudocylindrical projections**.
12. Why is a **Mercator projection** useful as a navigation map? Why is it not ideal for use as a general-purpose map?
13. What is a **loxodrome** (*rhumb line*)?

Conveying Information on Maps (*p. 38*)

14. Explain the concept of **isolines**.
15. What characteristics on maps are shown by *isotherms*, *isobars*, and **elevation contour lines**?
16. How does a **digital elevation model (DEM)** depict the landscape?

GNSS—Global Navigation Satellite System (*p. 42*)

17. Briefly explain how a **Global Navigation Satellite System (GNSS)**, such as the U.S. **Global Positioning System (GPS)**, works.

Remote Sensing (*p. 43*)

18. What is **remote sensing**?
19. Briefly define the following terms: *aerial photograph*, *photogrammetry*, *orthophoto map*.
20. What are some of the applications of *color infrared* imagery?
21. What are some of the applications of *thermal infrared* imagery?
22. Describe **multispectral** remote sensing.
23. Compare and contrast **radar**, **sonar**, and **lidar**.

Geographic Information Systems (GIS) (*p. 48*)

24. Distinguish between GPS and **GIS (geographic information systems)**.

Study Questions

1. Why are there so many types of map projections?
2. What kind of map projection would be best for studying changes in the amount of permafrost in the Arctic? Why? Consider both the general family of projection and its properties, such as equivalence and conformality.
3. Look at Figure 1-27, the world map of time zones shown in Chapter 1:
   a. Is this map an equivalent, conformal, or compromise projection? How can you tell?
   b. In which of the four families of map projections does it belong? How can you tell?
4. Isolines never start or stop on a map—every isoline must close on itself, either on or off the map. Why?
5. A GPS receiver in your car simply calculates your current latitude and longitude. How can it use this basic locational data to determine your *speed* and *direction* of travel?
6. Describe one kind of application in which radar imagery would be useful for geographical analysis. Explain the advantages of radar over other kinds of remote sensing in your example.

Exercises

1. On a map with a fractional scale of 1:24,000,
   a. One inch represents how many feet? _____
   b. One centimeter represents how many meters? _____
   c. If the map is 17 inches wide and 23 inches tall, how many square miles are shown on the map? _____
2. If we construct a globe at a scale of 1:1,000,000, what will be its diameter? (You may give your answer in either feet or meters.)
3. Convert the following latitude and longitude coordinates presented in decimal form (as might be shown on a GPS unit) to their conventional form of degrees/minutes/seconds:
   42.6700° N = _____°_____′_____″N
   105.2250° W = _____°_____′_____″W
4. Convert the following latitude and longitude coordinates from their conventional form of degrees/minutes/seconds to decimal form:
   22°20′15″N = _____° N
   137°30′45″E = _____° E
   22°20′15″N = 22°20._____′ N
   137°30′45″E = 137._____°E

Mastering Geography®

Looking for additional review and test prep materials? Visit the Study Area in **Mastering Geography** to enhance your geographic literacy, spatial reasoning skills, and understanding of this chapter's content by accessing a variety of resources, including Map Master interactive maps, geoscience animations, *Mobile Field Trips*, videos, *Project Condor* Quadcopter videos, *In the News* RSS feeds, flashcards, web links, self-study quizzes, and an eText version of *McKnight's Physical Geography*.

Environmental Analysis EXPLORING OUR WORLD WITH REMOTE SENSING

Data
Lake Erie
https://media.pearsoncmg.com/ph/esm/esm_mcknight_physgeo_13/media/

With remote sensing technologies, scientists can obtain and analyze large quantities of data without having to collect it by hand. How have land use and environmental conditions changed as a result of human activities? Remotely sensed data can help us answer that question.

Activities

Install and open Google Earth Pro™ (see https://www.google.com/earth/versions/). Use the Search function to go to Elwha River, Washington.

1. What is the site's elevation where the Elwha River drains from Lake Aldwell near the town of Elwha?
2. What is the condition of Lake Aldwell?

Use the historical imagery function (symbol that looks like a clock), and use the Polygon tool to make measurements.

3. Go to imagery for 2009. Lake Aldwell is located at approximately 48°04′39″ N and 123°34′25″ W. What is the site's elevation where the Elwha River drains from Lake Aldwell?
4. Estimate Lake Aldwell's surface area in square meters.
5. Reservoirs are never uniformly deep, but assume that the lake averaged 12 meters in depth. How many cubic meters of water did the lake hold in 2009?
6. Using Google Earth Pro, estimate when the Elwha Dam was removed. How did you determine this?

Return to the most recent imagery.

7. In river miles (the distance the water travels downstream), how far is the site of the old dam from the mouth of the Elwha River, where it empties into the Strait of Juan de Fuca? *(Use the Path tool.)*
8. Describe the mouth of the river. From where did the material there come?

Read the article of the "Restoration of the Elwha River" at https://earthobservatory.nasa.gov/images/78953/restoration-of-the-elwha-river.

9. Why was the Lake Aldwell dam removed?

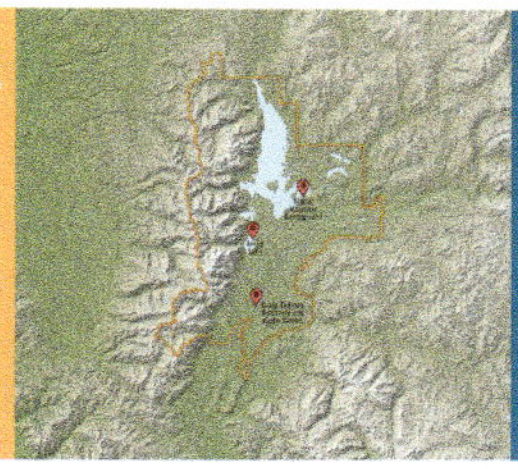

Seeing Geographically in Our National Parks

Look again at the image of Grand Teton National Park at the beginning of the chapter (p. 30). A shaded relief map uses shadows to create the illusion of three dimensions; from what direction is the Sun "shining" to create the shadows you see? What other types of map could we use to show the topography of this area?

Seeing Geographically in Our National Parks

The blue haze or "smoke" shown in this picture is what gives the Great Smoky Mountains their name. This haze is created by moisture and aerosols from the trees in this region. Humankind can modify the environment, but so can plants. How does this environment differ from where you live? What geographic feature do you think enhances the environment necessary to form haze? Do you think the season would influence haze formation?

Great Smoky Mountains National Park, NC, TN

3

Introduction to the Atmosphere

HAVE YOU EVER WONDERED why there are so many different environments on the surface of Earth? Or why the Great Smoky Mountains National Park has a warm, humid climate with a great diversity of plant and animal life, while the top of Denali in Denali National Park is bitter cold with relatively little plant or animal life? This is possible because of the way the atmosphere interacts with the solid earth and oceans. It is indeed our atmosphere that makes the great variety of life on Earth possible.

Our atmosphere supplies most of the oxygen that animals and plants must have to survive as well as the carbon dioxide that plants need. It helps maintain a water supply, which is essential to all living things. It insulates Earth's surface against temperature extremes and thus provides a livable environment over most of the planet. It also shields Earth from much of the Sun's ultraviolet radiation, which otherwise would be damaging to most life-forms.

The atmosphere is a complex and dynamic system. This chapter provides a foundation for understanding the atmosphere and its short-term patterns and processes, known as *weather*—as well as its long-term patterns, or *climate*. First, we describe the composition and structure of the atmosphere. Then we provide an overview of the basic elements or "ingredients" of weather and climate and the most important "controls" or influences of weather and climate. Finally, we discuss how human activity has altered the atmosphere.

As you study this chapter, think about these **Key**Questions:

- **What are the components of the atmosphere?**
- **What are the different layers of the atmosphere, and what are their properties?**
- **How is weather different from climate?**
- **How have humans changed the atmosphere?**

Size and Composition of the Atmosphere

Air—generally used as a synonym for atmosphere—is not a specific gas but rather a mixture of gases, mainly nitrogen and oxygen. It often contains small quantities of tiny solid and liquid particles held in suspension in the air as well as varying amounts of gaseous impurities.

Pure air is odorless, tasteless, and invisible. We can smell many impurities, however, and the air may even become visible if enough microscopic solid and liquid impurities coalesce (stick together) to form particles large enough to reflect or scatter sunlight. Clouds, by far the most conspicuous visible features of the atmosphere, represent the coalescing of water droplets or ice crystals around microscopic particles.

Size of Earth's Atmosphere

The atmosphere completely surrounds Earth and can be thought of as a vast ocean of air, with Earth at its bottom (Figure 3-1). It is held to Earth by gravity and therefore accompanies our planet in all its celestial motions. The attachment of Earth and atmosphere is a loose one, however. The atmosphere can move on its own much like the ocean, doing things that the solid Earth cannot do.

Density Decreases with Altitude: Although the atmosphere extends outward at least 10,000 kilometers (6000 miles), most of its mass is concentrated at very low altitudes. More than half of the mass of the atmosphere lies below the summit of North America's highest peak, Denali (formerly known as Mount McKinley) in Alaska, which reaches an elevation of 6.2 kilometers (3.8 miles). More than 98 percent of the atmosphere's mass lies within 26 kilometers (16 miles) of sea level (Figure 3-2). Therefore, relative to Earth's diameter of about 13,000 kilometers (8000 miles), the "ocean of air" we live in is very shallow.

In addition to reaching upward above Earth's surface, the atmosphere extends slightly downward. Because air expands to fill empty spaces, it penetrates into caves and crevices in rocks and soil. Moreover, it is dissolved in the waters of Earth and in the bloodstreams of organisms.

The atmosphere interacts with other components of Earth's environment, and it is instrumental in providing a hospitable setting for life. Whereas we often speak of human beings as creatures of Earth, it is perhaps more accurate to consider ourselves creatures of the atmosphere. As surely as a crab crawling on the sea bottom is a resident of the ocean, a person living at the bottom of the ocean of air is a resident of the atmosphere.

Learning Check 3-1 **What happens to the density of the atmosphere with increasing altitude? (Answer on p. AK-1)**

Development of Earth's Modern Atmosphere

The atmosphere today is very different from what it was during Earth's early history. Shortly after Earth formed about 4.6 billion years ago, the atmosphere probably consisted mostly

▼ Figure 3-1 Tall storm clouds are seen rising over northern Pakistan in this photograph taken from the International Space Station. Beyond the narrow blue band of the atmosphere is the blackness of outer space.

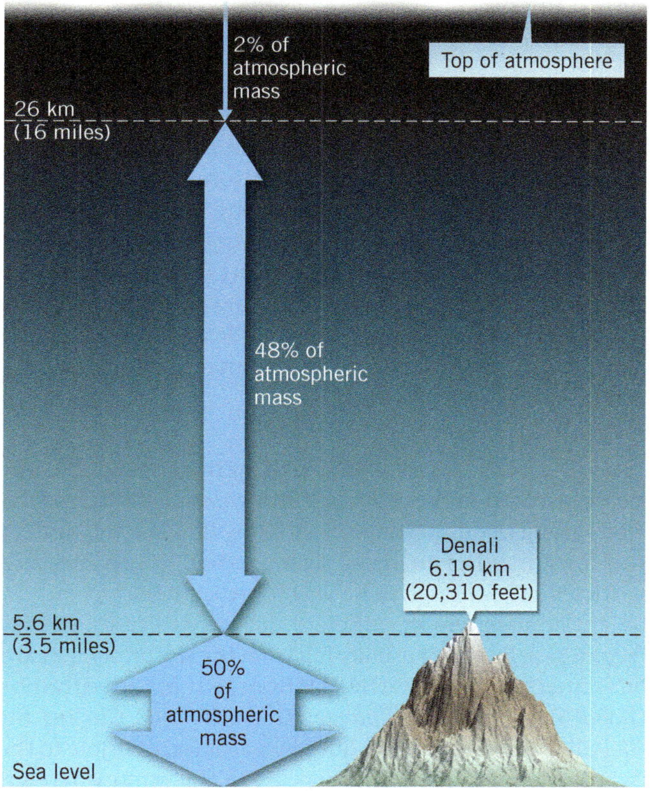

▲ **Figure 3-2** Most of the atmospheric mass is close to Earth's surface. More than half of the mass is below the highest point of Denali, North America's highest peak. About 98 percent of the atmosphere's mass lies below an altitude of 26 kilometers (16 miles, or 84,500 feet).

of light elements such as hydrogen and helium. By perhaps 4 billion years ago, those light gases were being lost, and outgassing from volcanic eruptions added large amounts of carbon dioxide and water vapor, along with small amounts of other gases, such as nitrogen. It is likely that arriving comets also contributed water to Earth's atmosphere. As ancient Earth cooled, most of the water vapor condensed out of the atmosphere, forming the world ocean.

By about 3.5 billion years ago, early forms of life—such as bacteria that could survive without oxygen—had begun to remove carbon dioxide and release oxygen into the atmosphere. Over time, as more complex forms of life evolved, algae and a variety of oceanic and terrestrial plants continued the transformation from a carbon dioxide-rich to an oxygen-rich atmosphere through *photosynthesis* (a process that is discussed in Chapter 10). Our modern atmosphere, therefore, was significantly influenced by life on Earth.

Composition of the Modern Atmosphere

The chemical composition of pure, dry air at lower altitudes (below about 80 kilometers or 50 miles) is simple and uniform, and the concentrations of the major components—the *permanent gases*—are now essentially unvarying. However, certain minor gases and nongaseous particles—the *variable gases* and *particulates*—vary significantly from place to place or from time to time, as does the amount of moisture in the air.

Permanent Gases

Nitrogen and Oxygen: The two most abundant gases in the atmosphere are nitrogen (N_2) and oxygen (O_2) (**Figure 3-3**). Nitrogen makes up more than 78 percent of the total and oxygen nearly 21 percent. Nitrogen is added to the air by the decay and burning of organic matter, volcanic eruptions, and the chemical breakdown of certain rocks. It is removed by certain biological processes and by lightning, which makes nitrogen easier to combine with other elements, thus enabling some of it to be washed out by rain or snow. Overall, the addition and removal of nitrogen gas are balanced, so the quantity present in the air remains constant over time. Oxygen is produced by vegetation and is removed by a variety of organic and inorganic processes; its total quantity is also relatively stable. The remaining 1 percent of the atmosphere's volume consists mostly of the inert gas argon. These three principal atmospheric gases—nitrogen, oxygen, and argon—have a minimal effect on weather and climate and therefore need no further consideration here.

Variable Gases

Several other gases occur in small but highly variable quantities in the atmosphere, but their influence on weather and climate is significant.

Water Vapor: Water in the form of a gas is known as water vapor. **Water vapor** (H_2O) is invisible—the visible forms of water in the atmosphere, such as clouds and precipitation, consist of water in its liquid or solid form (ice). Water vapor is most abundant in air overlying warm, moist surface areas such as tropical oceans, where water vapor may amount to as much as 4 percent of total volume. Over deserts and in polar regions, the amount of water vapor is but a tiny fraction of 1 percent. Thus, it is listed as a "variable gas" in Figure 3-3 because it varies by location.

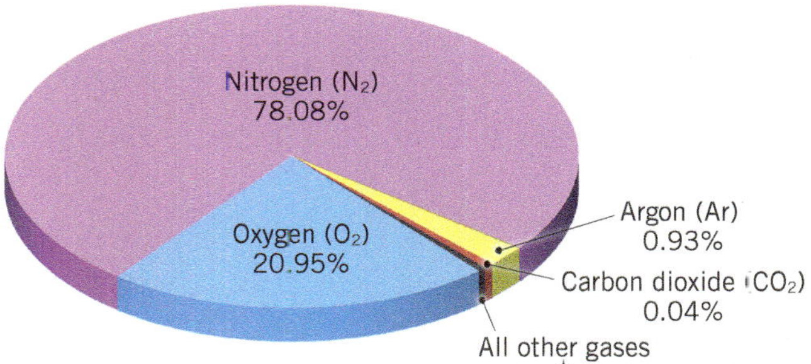

Permanent	Percent by volume	Concentration in parts per million
Neon (Ne)	0.00182	18.2
Helium (He)	0.00052	5.2
Krypton (Kr)	0.00011	1.1
Hydrogen (H_2)	0.00005	0.5
Variable		
Water vapor (H_2O)	0 to about 4	—
Methane (CH_4)	0.00018	1.8
Ozone (O_3)		<2

▲ **Figure 3-3** Percent by volume of the gases in the atmosphere. Nitrogen and oxygen are the dominant components. Although found in tiny amounts, some variable gases, such as carbon dioxide and water vapor, play important roles in atmospheric processes.

In the atmosphere as a whole, the total amount of water vapor remains nearly constant, though climate change may affect this. Water vapor has a significant effect on weather and climate: it is the source of all clouds and precipitation. It also absorbs certain wavelengths of radiation, so it plays an important role in regulating the temperature of the atmosphere. Water vapor also stores energy (called *latent heat*), so moving water vapor transports the energy needed for storms.

Carbon Dioxide: Another important atmospheric component is **carbon dioxide** (CO_2). Like water vapor, carbon dioxide has a significant influence on climate, primarily because it absorbs thermal infrared radiation and thereby helps warm the lower atmosphere. Carbon dioxide is distributed fairly uniformly in the lower layers of the atmosphere. Its concentration has been increasing steadily for the last century or so because of the increased burning of **fossil fuels**—naturally occurring fuels (such as coal, petroleum, and natural gas) that form over geologic time from organic materials. The proportion of carbon dioxide in the atmosphere has increased at a rate of more than 0.0002 percent (2 parts per million) per year and at present is more than 415 parts per million. Most atmospheric scientists conclude that the increased levels of (CO_2) are causing the lower atmosphere to warm enough to produce significant, although still somewhat unpredictable, global climatic changes. (The topic of *global warming* is presented in greater detail in Chapter 4.)

Ozone: Another minor but vital gas in the atmosphere is **ozone**, which is a molecule made up of three oxygen atoms (O_3) instead of the more common two oxygen atoms (O_2, or "diatomic" oxygen). Most of the ozone is concentrated in a region of the atmosphere called the ozone layer, which lies between 15 and 48 kilometers (9 and 30 miles) above Earth's surface. Ozone is an excellent absorber of ultraviolet solar radiation; it filters out enough of this radiation to protect life-forms from potentially deadly effects. (We discuss the recent thinning of the ozone layer later in this chapter.) A small amount of ozone is located near the surface and is considered a pollutant.

Other Variable Gases: Methane (CH_4), introduced into the atmosphere both naturally and through human activity, absorbs certain wavelengths of radiation and so plays a role in regulating the temperature of the atmosphere. Tiny amounts of other variable gases—carbon monoxide, sulfur dioxide, nitrogen oxides, and various hydrocarbons—are also increasingly being introduced into the atmosphere by emission from factories and automobiles. All of them can be hazardous to life, and many affect climate.

Particulates (Aerosols)

In addition to the gases that make up the atmosphere, the air holds both large and small nongaseous particles. The larger nongaseous particles are mainly liquid water and ice that form clouds, rain, snow, sleet, and hail. Sometimes dust particles large enough to be visible are kept aloft in the turbulent atmosphere in sufficient quantity to cloud the sky (Figure 3-4). Dust will usually fall from the sky pretty quickly, but sometimes it gets carried far from its source (see the box *Global Environmental Change: Aerosol Plumes Circling the Globe* later in the chapter). Smaller particles, invisible to the naked eye, may remain suspended in the atmosphere for months or even years.

The solid and liquid particles that are small enough to remain suspended in the atmosphere are collectively called **particulates** or **aerosols**. They have innumerable sources—some natural and some the result of human activities. Volcanic ash, windblown soil and pollen grains, meteor debris, smoke from wildfires, and salt spray from breaking waves are examples of particulates from natural sources. The Smoky Mountains pictured at the beginning of this chapter are well known for the "smoky" air caused by vapor released

◀ **Figure 3-4** Dust particles sometimes cloud the sky for a short time over a limited part of Earth's surface. On some occasions, as in this scene over the Kajaki Bazaar in Zeebruge, Afghanistan, the term *dust storm* is very appropriate, and the visual effect is imposing, if not menacing.

from vegetation. Particulates coming from human sources consist mostly of industrial and automotive emissions and smoke and soot from fires of human origin.

These tiny particles are most numerous near their places of origin—above cities, seacoasts, active volcanoes, and some desert regions. They may be carried great distances, however, both horizontally and vertically, by the restless atmosphere. They affect weather and climate in two major ways:

1. Many are *hygroscopic* (water absorbing), and water vapor condenses around such *condensation nuclei*. This accumulation of water molecules is a critical step in cloud formation, as we shall see in Chapter 6.
2. Some kinds of particulates absorb radiation, while others reflect it. Thus, the presence of particulates can influence the temperature of the atmosphere.

Learning Check 3-2 **What is the most abundant gas in the atmosphere? Does this gas play an important role in processes in the atmosphere? What role does ozone play in the atmosphere?**

Vertical Structure of the Atmosphere

The next five chapters of this book deal with atmospheric processes and their influence on climatic patterns. Our attention in these chapters is devoted primarily to the lower portion of the atmosphere, which is the zone in which most weather phenomena occur. Even though the upper layers of the atmosphere usually affect the environment of Earth's surface only minimally, it is still useful to have some understanding of the total atmosphere.

A given layer (altitude zone) of the atmosphere has different names, depending on the characteristic or feature under discussion. We begin with the layers of the atmosphere largely defined by temperature characteristics; later, we introduce other overlapping layers defined by different characteristics.

Thermal Layers

Most of us have had some personal experience with temperature differences associated with altitude. As we climb a mountain, for instance, we notice a decrease in temperature. Until about a century ago, it was generally assumed that temperature decreased with increasing altitude throughout the whole atmosphere, but now we know that is not the case. The vertical pattern of temperature is complex, consisting of a series of layers in which temperature alternately decreases and increases (Figure 3-5). From the surface of Earth upward, these thermal layers are called the *troposphere*, *stratosphere*, *mesosphere*, and *thermosphere*. Special terms are often used when we refer to the transition from one layer to another: the *tropopause* is the top of the troposphere; the *stratopause*, the top of the stratosphere; and the *mesopause*, the top of the mesosphere.

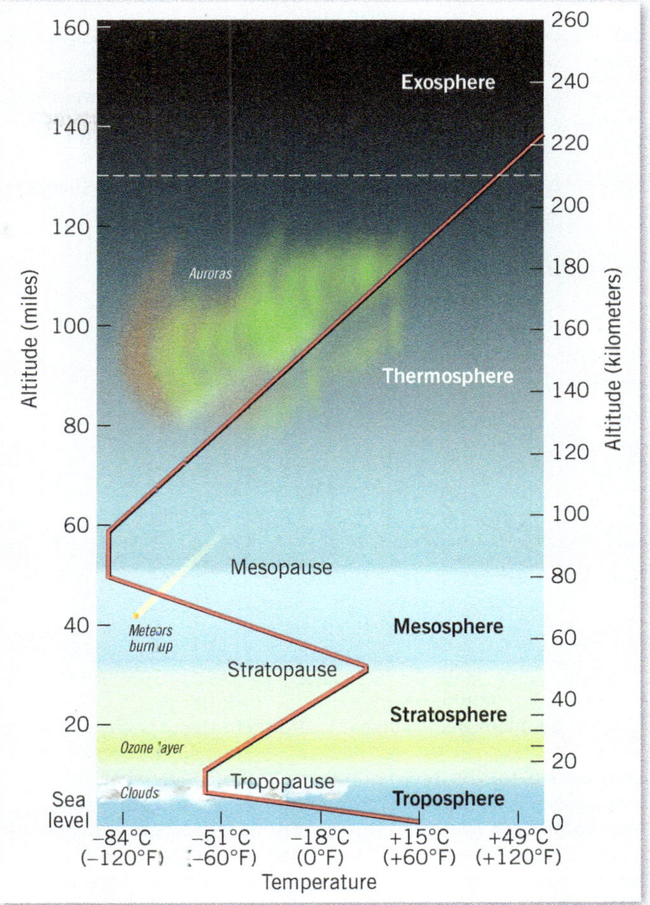

▲ **Figure 3-5** Average thermal structure of the atmosphere. Air temperature (see the red line) decreases with increasing altitude in the troposphere and mesosphere and increases with increasing altitude in the stratosphere and thermosphere. Nearly all weather processes occur within the troposphere.

Troposphere: The lowest layer of the atmosphere—the one in contact with Earth's surface—is known as the **troposphere**. The names *troposphere* and *tropopause* are derived from the Greek word *tropos* ("turn") and denote an overturning of the air in this zone. The depth of the troposphere varies in both time and place (Figure 3-6). It is deepest over warm tropical regions and shallowest over the cold poles, is deeper in summer than in winter, and varies with the passage of warm and cold air masses. On average, the top of the troposphere is about 18 kilometers (11 miles) above sea level at the equator and about 8 kilometers (5 miles) above sea level over the poles.

A number of characteristics of the troposphere are important to us:

- Within the troposphere, temperature usually decreases with increasing altitude (see Figure 3-5). Upward from sea level, where the average global temperature is about 15°C (59°F), temperature decreases steadily with increasing altitude, declining to an average temperature of about −57°C (−71°F) at the tropopause. The tropopause is the "cold spot" in the atmosphere until much higher altitudes. The source of warmth for the lower troposphere is the surface of Earth itself—solar energy warms the surface, and this energy is in turn transferred to the troposphere through a number of processes. (We discuss the warming of the troposphere in detail in Chapter 4.)

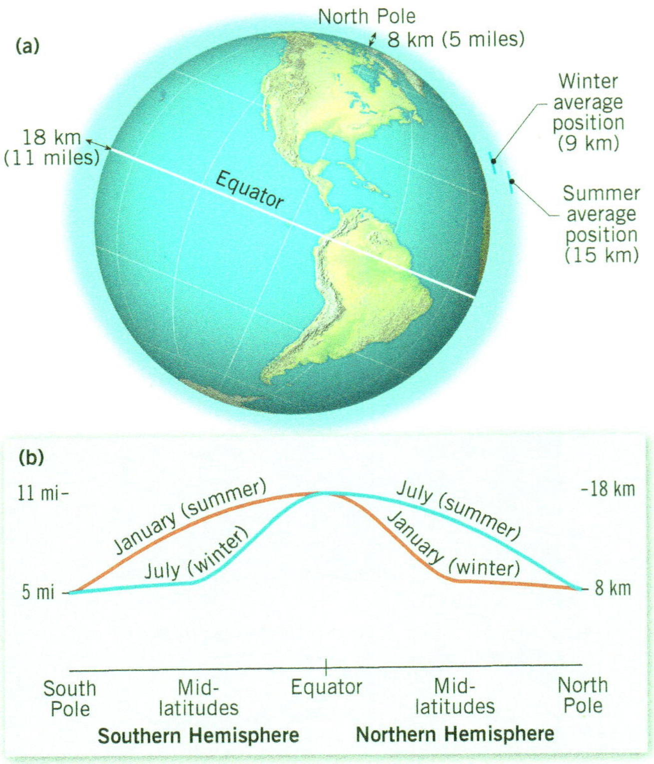

▲ **Figure 3-6** (a) The depth of the troposphere is variable. This thermal layer is deepest over the equator, where surface temperatures are warm and thermal mixing is greatest, and it is shallowest over the poles. It is deeper in summer than in winter. The thickness of the atmosphere is greatly exaggerated here. (b) The depth of the troposphere also changes with season. The depth is larger in the summer hemisphere than in the winter hemisphere.

- The tropopause represents the upper limit of surface-initiated turbulence in the atmosphere—such as the updrafts of air warmed by the surface and the tops of tall thunderstorm clouds.
- About 80 percent of the total mass of the atmosphere is within the troposphere.
- Nearly all water vapor and nearly all clouds are found within the troposphere. Above about 16 kilometers (10 miles), the temperature is so low that there is rarely enough moisture to provide even a wisp of a cloud. If you have ever flown, you may recall the remarkable sight of a cloudless sky overhead once the ascending plane breaks through the top of a solid cloud layer.

In our study of the atmosphere, we focus primarily on the troposphere. It is in this lowest layer of the atmosphere that nearly everything we call "weather" occurs. We do, however, need to discuss briefly the thermal layers above, especially the stratosphere, which overlies the troposphere.

Stratosphere: The names **stratosphere** and *stratopause* come from the Latin *stratum* ("a cover"), indicating a layered or stratified condition without vertical mixing. If we describe the air in the troposphere as being turbulent, we can describe the air in the stratosphere as being stagnant. As Figure 3-5 shows, temperature remains constant through the tropopause and for a short distance into the stratosphere. At an altitude of about 20 kilometers (12 miles), air temperature begins increasing with increasing altitude, reaching a maximum at 48 kilometers (30 miles) at the bottom of the mesosphere, where the temperature is about −2°C (28°F). The stratosphere extends from an altitude of about 18 kilometers (11 miles) above sea level to about 48 kilometers (30 miles).

The temperature increase in this part of the atmosphere is associated with the stratospheric *ozone layer*. Within the ozone layer, the gas ozone absorbs ultraviolet radiation from the Sun and thereby warms the atmosphere. (We discuss the ozone layer in detail later in this chapter.)

Upper Thermal Layers: Above the stratosphere, temperature decreases with increasing altitude through the *mesosphere* (from the Greek *meso*, for "middle"), beginning at 48 kilometers (30 miles) and extending to the top of this layer about 80 kilometers (50 miles) above sea level where temperature in the atmosphere reaches its minimum. Temperature decreases within the mesosphere simply because it lacks a source of warmth as is found in the lower troposphere or in the stratosphere.

Above the mesosphere is the *thermosphere* (from the Greek *therm*, meaning "heat"), which begins at an altitude of 80 kilometers (50 miles). Temperature increases until, at an altitude of 200 kilometers (125 miles), it is higher than the maximum temperature in the troposphere. The increase in temperature occurs because various atoms and molecules absorb ultraviolet energy from the Sun and are thus split and heated.

The thermosphere has no definite top. Instead, it merges gradually into interplanetary space, a region sometimes referred to as the *exosphere*. Traces of atmosphere extend for thousands of kilometers. Therefore, "top of the atmosphere" is a theoretical concept rather than a reality, with no true boundary between atmosphere and outer space.

Learning Check 3-3 **What generally happens to the temperature of the atmosphere from the surface of Earth to the tropopause? What happens to temperature above the tropopause in the stratosphere? Why do these changes in temperature occur?**

Although our attention in this book is directed almost entirely to the troposphere, from time to time we must consider variations in factors such as pressure and composition throughout the atmosphere.

Pressure

Atmospheric pressure is the force exerted per unit area by the molecules in the atmosphere. For simplicity's sake, we can think of atmospheric pressure as the "weight" of the overlying air. (In Chapter 5, we explore the concept of pressure in much greater detail.) The more air above an object, the greater the air pressure exerted on that object. Almost 50 percent of the atmosphere lies below the highest mountain peak in North America (see *Featured National Park: Denali*).

Featured National Park

Denali

Sitting on Top of the World

▲ **Figure 3-A** Hiking and camping are popular activities. This view from Wonder Lake Campground (near the end of the road) shows the subalpine woodland and tundra that surround Denali. Summers are short and cool; winters are long and extremely cold. Temperature decreases rapidly with height, so no vegetation grows at the top of Denali.

Denali National Park and Preserve is located north of Anchorage, Alaska. It covers 6 million acres and is larger than the states of New Jersey and Massachusetts combined. It is home to the tallest mountain in the United States. Only one road runs into the park, and most of the Park can only be accessed by riding tour bus or by hiking. See **Figures 3-A** to **3-D**.

Its pristine state makes it ideal for scientists and researchers to study the natural world. Scientists study geology, archeology, ecology, and climate change in the Park.

NPS Video
Science in the Park
https://media.pearsoncmg.com/ph/esm/esm_mcknight_physgeo_13/media/

▶ **Figure 3-B** Keeping the 600,000 Denali visitors company each year are a variety of animals suited to the rough terrain, including grizzly bears, wolves, moose, caribou, and Dall sheep. Dall sheep are a species of wild sheep that inhabit and are protected in the Denali National Park.

▼ **Figure 3-C** Denali (formerly known as Mount McKinley) is North America's highest mountain peak at 6.19 km (20,310 feet). It attracts climbers from around the world brave enough to test their skills. In addition to extreme weather, air pressure and density decrease with height. While respirators are not required for Denali, climbers must acclimate to the thinner air.

Climbers can expect to encounter 45 m/s (100 mph) winds, temperatures as low as −40°C (−40 °F), ice, heavy snow, and glaring sunlight.

▲ **Figure 3-D** Because weather conditions on Denali can change very rapidly, climbers must come prepared for extreme weather conditions. Altitude sickness (mountain sickness) occurs when the body hasn't adapted to lower pressure and lower oxygen levels. Symptoms include headache, confusion, shortness of breath even while resting, nausea, and insomnia. For mild cases, it may take a day or two to get used to the altitude. More severe cases can result in death.

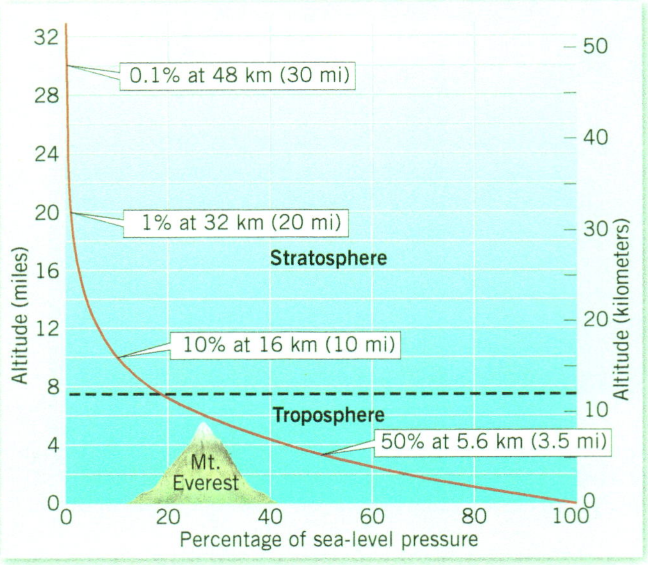

▲ **Figure 3-7** Air pressure decreases with increasing altitude, but not at a constant rate. Pressure decreases to 50 percent of sea level by an altitude of 5.6 kilometers (3.5 miles) and to just 1 percent of sea level at an altitude of 32 kilometers (20 miles).

Because air is highly compressible, the lower layers of the atmosphere are compressed by the air above, and this compression increases both the pressure exerted by the lower layers and the density (mass per unit volume) of these layers (see Figure 3-2).

Air pressure and density are normally highest at sea level and decrease rapidly with increasing altitude. The change of pressure with altitude is not constant, however. As a generalization, pressure decreases upward at a decreasing rate (**Figure 3-7**).

At an altitude of 5.6 kilometers (3.5 miles), atmospheric pressure has decreased to 50 percent of its sea-level value, and therefore its density is about half that of sea level. In other words, about half of all the gas molecules making up the atmosphere lie below 5.6 kilometers or 18,500 feet—this is why climbers on the highest mountains must carry oxygen tanks! About 90 percent of the gas molecules are in the first 16 kilometers (10 miles) above sea level (a typical altitude of the tropopause over the tropics). Pressure becomes so slight in the upper layers that, above about 80 kilometers (50 miles), there is not enough to register on an ordinary barometer, the instrument used to measure air pressure. Above this level, atmospheric molecules are so scarce that air pressure is less than that in the most perfect laboratory vacuum at sea level.

Composition

The principal gases of the atmosphere have a remarkably uniform vertical distribution throughout the lowest 80 kilometers (50 miles) or so of the atmosphere. This zone of homogenous composition is referred to as the *homosphere* (**Figure 3-8**). The sparser atmosphere above this zone does not display such uniformity. Rather, the gases are layered in accordance with their molecular or atomic weights—molecular nitrogen (N_2) below, with atomic oxygen (O), helium (He), and hydrogen (H) successively above. This higher zone is called the *heterosphere*.

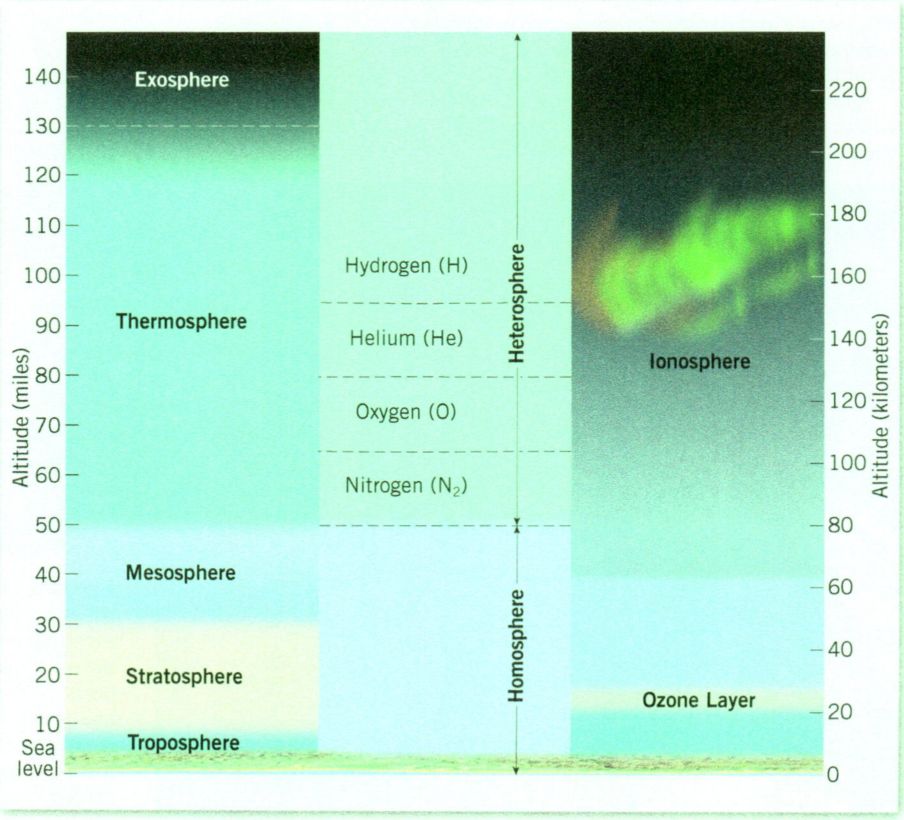

▶ **Figure 3-8** Relationships among the layers of the atmosphere. The homosphere is a zone of uniform vertical distribution of gases; in the heterosphere above, however, gases are distributed according to molecular or atomic weight—heavier below, lighter above. The ozone layer contains significant concentrations of ozone; the ionosphere is a deep layer of ions, which are electrically charged molecules and atoms.

▲ Figure 3-9 The Aurora Borealis (Northern Lights) in the ionosphere, as seen over Denali National Park.

Ozone Layer: The **ozone layer** lies between 15 and 48 kilometers (9 and 30 miles) up. It is sometimes referred to as the *stratospheric ozone layer* because it is centered in the lower stratosphere. Despite its name, the ozone layer is not composed primarily of ozone. It was so named because that is where the concentration of ozone relative to other gases is at its maximum. Even in the section of the ozone layer where the ozone attains its greatest concentration, at about 25 kilometers (15 miles) above sea level, this gas accounts for no more than about 15 parts per million of the atmosphere.

Ionosphere: The *ionosphere* is a deep layer of *ions* (electrically charged molecules or atoms) in the middle and upper mesosphere and the lower thermosphere, between about 60 and 400 kilometers (40 and 250 miles). The ionosphere is significant because it aids long-distance communication by reflecting radio waves back to Earth. It is also known for its auroral displays, such as the "Northern Lights" (Figure 3-9) that develop when charged atomic particles from the Sun are trapped by the magnetic field of Earth near the poles. In the ionosphere, these particles "excite" the nitrogen molecules and oxygen atoms, causing them to emit light, not unlike a neon lightbulb.

Learning Check 3-4 Describe the layers of the atmosphere in terms of composition.

Weather and Climate

Now that we have described the composition and structure of the atmosphere, we turn more specifically to the broad set of processes operating within this ocean of air.

Weather

Our atmosphere is energized by solar radiation, stimulated by earthly motions, and affected by contact with Earth's surface. The atmosphere reacts by producing an infinite variety of conditions and phenomena known collectively as *weather*; the study of weather is known as *meteorology*. The term **weather** refers to short-term atmospheric conditions that exist for a given time in a specific area. It is the combination of temperature, humidity, cloudiness, precipitation, pressure, winds, storms, and other atmospheric variables for a short period of time. Thus, we speak of the weather of the moment, the week, the season, or perhaps even of the year or the decade.

Climate

Weather is in an almost constant state of change, yet in the long view, we can generalize the variations into a composite pattern that is termed *climate*. **Climate** is the aggregate of day-to-day weather conditions over a long period of time. It encompasses not only the average characteristics but also the variations and extremes of weather. To describe the climate of an area requires weather information over an extended period, normally at least three decades.

Weather and climate, then, are related but not synonymous terms. The distinction between them is the difference between immediate specifics and long-term generalities. As a whimsical country philosopher once said, "Climate is what you expect; weather is what you get."

Weather and climate have direct and obvious influences on agriculture, transportation, and human life in general. Moreover, climate is a significant factor in the development of all major aspects of the physical landscape—soils, vegetation, animal life, hydrography (water bodies), and topography.

Our ultimate goal in studying the atmosphere is to understand the distribution and characteristics of climate around the world—the long-term patterns of the atmosphere. To achieve this understanding, however, we must spend the next four chapters gaining an appreciation for weather—the dynamics of the momentary state of the atmosphere.

Learning Check 3-5 What is the difference between weather and climate?

The Elements of Weather and Climate

The atmosphere is a complex medium, and its mechanisms and processes are sometimes very complicated. Its nature, however, is generally expressed in terms of only a few measurable variables.

These variables can be thought of as the **elements of weather and climate**. The most important are (1) temperature, (2) moisture content, (3) pressure, and (4) wind. These are the basic "ingredients" of weather and climate—the ones you hear about on weather reports. Measuring how they vary in time and space makes it possible to decipher at least partly the complexities of weather dynamics and climatic patterns.

The Controls of Weather and Climate

The nearly continuous variations in the elements of weather and climate are caused, or at least strongly

▲ Figure 3-10 Solar energy coming to Earth. As we will see in Chapter 4, the amount of solar energy received at Earth's surface varies by latitude.

influenced, by a collection of semipermanent attributes of our planet that are often referred to as the **controls of weather and climate**. The principal controls are briefly described in the paragraphs that follow but are explained in much more detail in later chapters. Although we discuss them individually here, we emphasize that there often is overlap and interaction among them, with widely varying effects.

Latitude: We noted in Chapter 1 that the continuously changing seasonal relationship between the Sun and Earth brings continuously changing amounts of sunlight, and therefore of radiant energy, to different parts of Earth's surface. Thus, the basic distribution of solar energy and the resulting warming/cooling over Earth is first and foremost a function of latitude (Figure 3-10). In terms of elements and controls, we say that the *control* latitude strongly influences the *element* temperature. Overall, latitude is the most fundamental control of climate.

Distribution of Land and Water: Probably the most fundamental distinction concerning the geography of climate is the distinction between continental climates and maritime (oceanic) climates. Oceans warm and cool more slowly and to a lesser degree than do landmasses. This means that maritime areas experience milder temperatures than continental areas in both summer and winter. For example, Seattle, Washington, and Fargo, North Dakota, are at approximately the same latitude (47°N), with Seattle on the western coast of the United States and Fargo deep in the interior (Figure 3-11). Seattle has an average January temperature of 6°C (42°F), whereas the January average in Fargo is −14°C (7°F). In the opposite season, Seattle has a July average temperature of 19°C (66°F), whereas in Fargo the July average is 22°C (71°F).

Oceans are also a much more abundant source of atmospheric moisture than land. Thus, maritime climates are normally more humid than continental climates. The uneven distribution of continents and oceans over the world, then, is a prominent control of the elements moisture content and temperature.

General Circulation of the Atmosphere: The atmosphere is in constant motion, with flows that range from temporary local breezes to vast regional wind regimes. At the planetary scale, a semipermanent pattern of major wind and pressure systems dominates the troposphere and greatly influences most elements of weather and climate.

As a simple example, in the tropics, most surface winds come from the east, whereas the middle latitudes are characterized by flows that are mostly from the west (Figure 3-12).

General Circulation of the Oceans: Somewhat analogous to atmospheric movements are the motions of the oceans (Figure 3-13). Like the atmosphere, the oceans have many minor motions, but they also have a broad general pattern of currents. These currents assist in energy transfer by moving warm water poleward and cool water equatorward. Although the influence of currents on climate is much less than that of atmospheric circulation, ocean currents are not inconsequential. For example, warm currents are found off the eastern coasts of continents, and cool currents occur off western coasts—a distinction that has a profound effect on coastal climates.

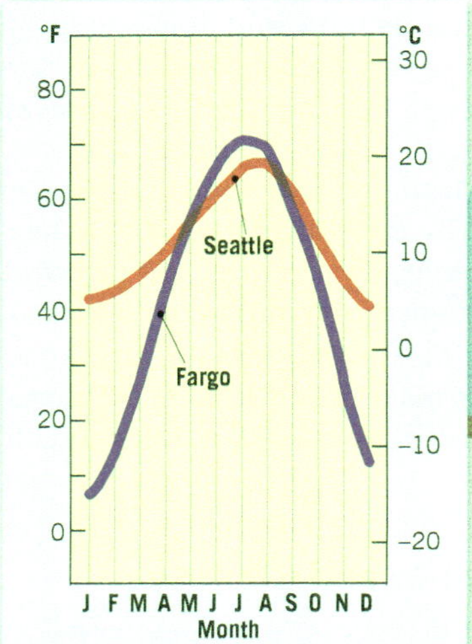

(a) Climograph: land versus water

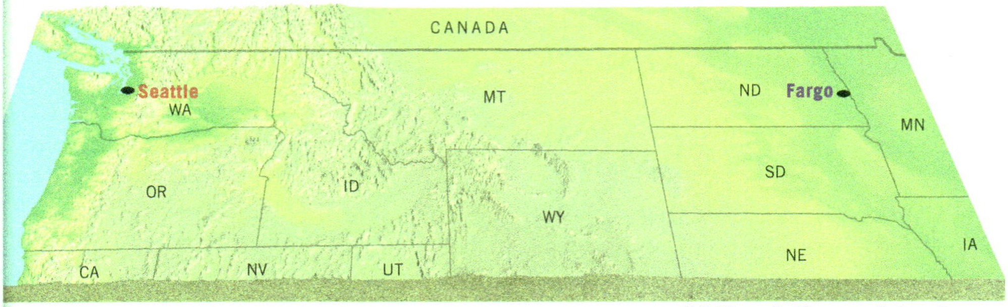

(b) Map of climograph locations

▲ Figure 3-11 Land–water contrasts. (a) A climograph showing the average monthly temperature in Fargo, North Dakota, and Seattle, Washington. (b) The inland city of Fargo experiences both hotter summers and colder winters than the coastal city of Seattle.

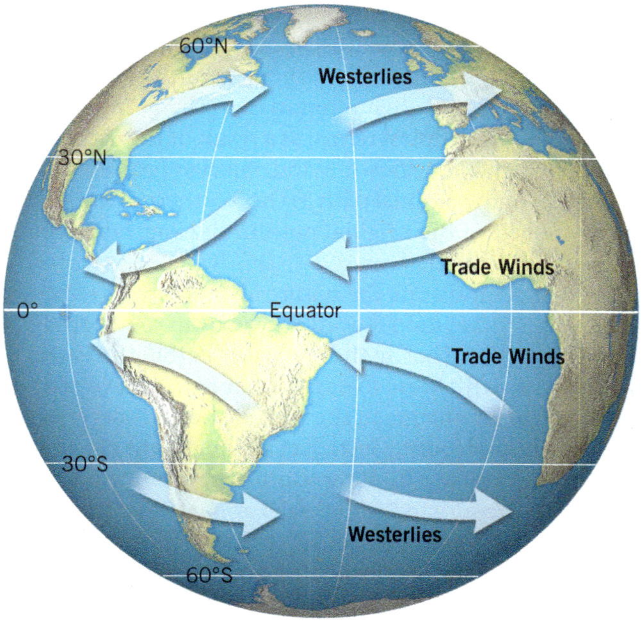

▲ **Figure 3-12** The general circulation of the atmosphere is an important climatic control. This highly simplified diagram shows that surface wind generally blows from the east in the tropics and from the west in the midlatitudes. The complete patterns of atmospheric wind and pressure are discussed in Chapter 5.

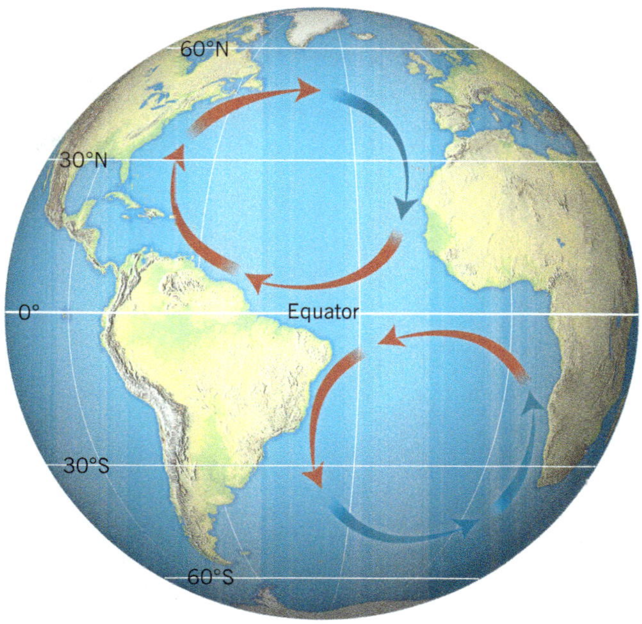

▲ **Figure 3-13** The general circulation of the oceans involves the movement of large amounts of warm water (red arrows) and cool water (blue arrows). These surface ocean currents have a significant climatic effect on neighboring landmasses.

Altitude: We have already noted that three of the four weather elements—temperature, pressure, and moisture content—generally decrease upward in the troposphere and are therefore under the influence of the control altitude. This simple relationship between the three elements and the control has significant ramifications for many climatic characteristics, particularly in mountainous regions (**Figure 3-14**).

▲ **Figure 3-14** Increasing altitude affects many components of the environment, as indicated by the variety of natural vegetation patterns on the slopes of Blanca Peak in south-central Colorado. The upper limit for trees (the treeline) is mainly due to low summer temperatures.

Topographic Barriers: Mountains and large hills sometimes have prominent effects on one or more elements of climate by diverting wind flow (**Figure 3-15**). The side

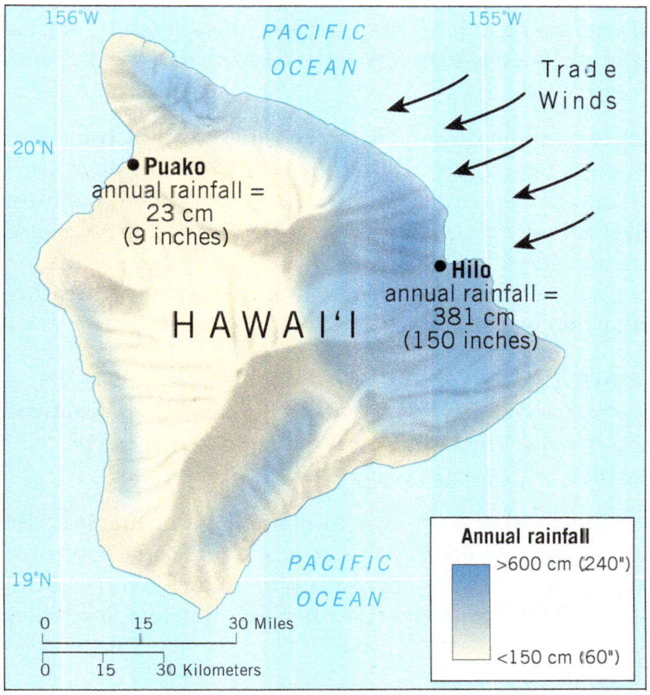

▲ **Figure 3-15** A topographic barrier as a control of climate. The difference in average annual rainfall in these two locations on the island of Hawai'i is caused by the mountain range separating them. Moisture-laden trade winds from the northeast drop their moisture when forced to rise by the eastern face of the mountains. The result is a very wet eastern side of the island and a very dry western side.

▲ **Figure 3-16** A prominent midlatitude cyclone storm system over the British Isles is counterpointed by localized thunderstorms over North Africa, Sicily, and Italy.

of a mountain range facing the wind (the *windward* side), for example, is likely to have a wetter climate than the sheltered (*leeward*) side.

Storms: Some storms have very widespread distribution, whereas others are localized (**Figure 3-16**). Although they often result from interactions among other climate controls, all storms create specialized weather circumstances and so are considered to be a control. Indeed, some storms are prominent and frequent enough to affect not only weather but climate as well.

Learning Check 3-6 **What is the relationship of the *controls* of weather and climate to the *elements* of weather and climate? What is the most fundamental control of weather and climate? Why?**

Before we discuss how humans have influenced the atmosphere, we need to introduce one additional control of weather and climate—or perhaps more correctly, a control of some of the controls of weather and climate: the rotation of Earth.

The Coriolis Effect

As a result of Earth's rotation, things moving over the surface of Earth appear to drift sideways. This deflection in the path of free-moving objects is known as the **Coriolis effect**, named in honor of Gaspard G. Coriolis (1792–1843), a French civil engineer and mathematician who quantitatively explained this phenomenon in the early 1800s.[1]

Stated in the simplest terms: as a result of the rotation of Earth, the path of any free-moving object appears to deflect to the right of the original path in the Northern Hemisphere and to the left of the original path in the Southern Hemisphere (**Figure 3-17a**).

We can demonstrate the nature of the Coriolis effect by imagining a rocket fired toward Boston from the North Pole (see Figure 3-17a). During the few minutes that the rocket is in the air, Earth's west-to-east rotation will have moved Boston a few kilometers to the east. If the Coriolis effect is not taken into account, the rocket would pass to the west of the city. To a person looking southward from the launch point, the uncorrected flight path appears to drift to the right. Although the Coriolis effect deflection is easier to envision when an object is moving north–south, the deflection occurs no matter which direction an object moves (**Figure 3-17b**). The Coriolis effect influences any freely moving object—a baseball, an automobile, even a person walking—but for these and other short-distance movements, the deflection is so minor as to be insignificant. Long-range movements, however, can be influenced considerably by the Coriolis effect.

Important Consequences of the Coriolis Effect: Although strictly speaking the Coriolis effect is an apparent deflection caused by Earth's rotation, the consequences of the Coriolis effect are very real. Ocean currents (Chapter 4) and wind belts (Chapter 5) deflect to the right in the Northern Hemisphere and to the left in the Southern Hemisphere (see Figures 3-12 and 3-13, for example). The circulation of wind within storms (Chapter 7) such as hurricanes is influenced by the Coriolis effect, and hurricanes never form on the equator because there is no Coriolis effect there. By examining the behavior of wind and currents, we can see the following aspects of the Coriolis effect:

1. Regardless of the initial direction of motion, any freely moving object appears to deflect to the right in the Northern Hemisphere and to the left in the Southern Hemisphere.
2. The apparent deflection is strongest at the poles and decreases progressively toward the equator, where the deflection is zero.
3. A fast-moving object is deflected more than a slower one—although the Coriolis effect influences direction of movement only; it does not change the speed of an object.

One phenomenon that the Coriolis effect does not influence is the circulation of water draining from a sink or bathtub. A folktale claims that Northern Hemisphere sinks drain

[1]The Coriolis effect is frequently referred to as the *Coriolis force*, especially when its consequences are calculated. In this book, we retain the more general term *Coriolis effect*.

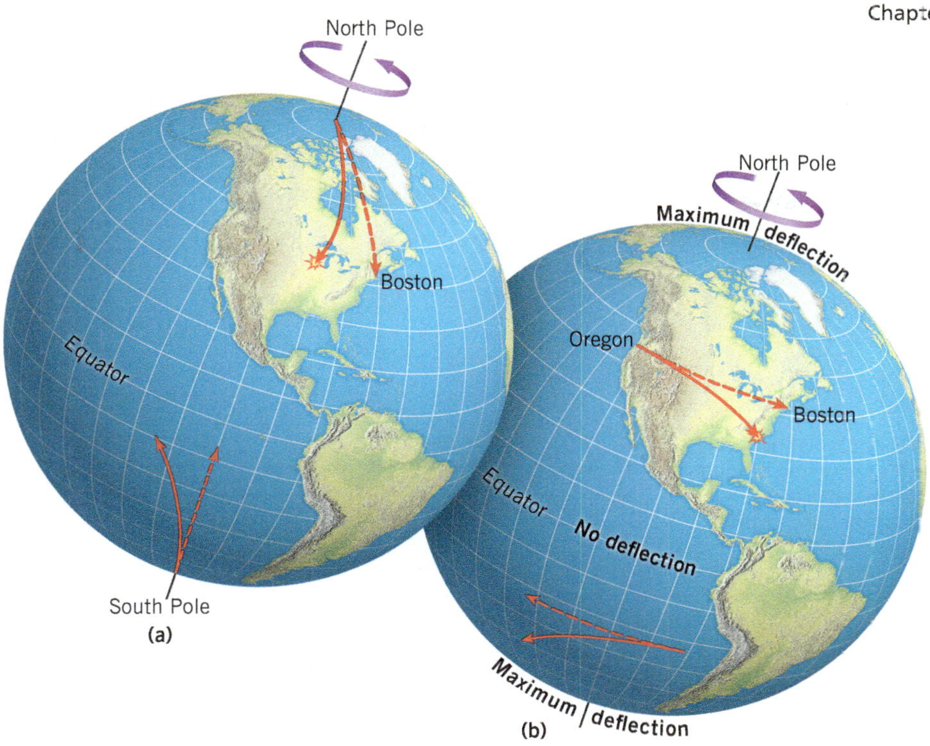

◀ **Figure 3-17** The deflection caused by the Coriolis effect is to the right in the Northern Hemisphere and to the left in the Southern Hemisphere. The dashed lines represent the planned route of a hypothetical rocket, and the solid lines represent what would be the actual movement. (a) A rocket launched from the North Pole toward Boston would land to the west of the target if the Coriolis effect is not considered when the flight path is computed. (b) A rocket fired toward Boston from a point at the same latitude in Oregon would also appear to curve to the right because of the Coriolis effect deflection.

Animation
Coriolis Effect
https://media.pearsoncmg.com/ph/esm/esm_mcknight_physgeo_13/media/

Video
Coriolis Effect Merry Go Round
https://media.pearsoncmg.com/ph/esm/esm_mcknight_physgeo_13/media/

clockwise and Southern Hemisphere sinks counterclockwise. The time involved is so short and the speed of the water so slow that the Coriolis effect cannot explain these movements—the characteristics of the plumbing system, the shape of the washbowl, and pure chance are more likely to determine the flow patterns.

Learning Check 3-7 **Describe the Coriolis effect and its consequences.**

Human-Caused Atmospheric Change

Over the past century, human activity has increasingly had unintended and uncontrolled effects on the atmosphere—effects seen around the globe. This human impact, in simplest terms, consists of the introduction of impurities into the atmosphere at a pace previously unknown—impurities capable of altering global climate and harming forms of life. The consequences of *anthropogenic* (human-produced) changes in the atmosphere, especially global climate change, have been a concern for atmospheric scientists for several decades. In recent years, however, those consequences have received international attention not only from the scientific community but from the general public as well.

In November 2018, the United States Global Change Research Program, a joint scientific effort involving more than a dozen federal agencies and the White House, released its fourth *National Climate Assessment*. The report offers a blunt conclusion regarding global climate change:

> "It is *extremely likely* that human influence has been the dominant cause of the observed warming since the mid-20th century. For the warming over the last century, there is no convincing alternative explanation supported by the extent of the observational evidence."

These conclusions echo the findings of other major research efforts, including those presented in the *Fifth Assessment Report* of the Intergovernmental Panel on Climate Change (discussed in later chapters).

In subsequent sections of the book, we highlight a number of aspects of human-caused environmental change and some of the steps that can be—and are being—taken to ameliorate them. The first major topic of global environmental change we spotlight is the depletion of the ozone layer.

Depletion of the Ozone Layer

Animation
Ozone Depletion
https://media.pearsoncmg.com/ph/esm/esm_mcknight_physgeo_13/media/

As we saw earlier, ozone is naturally produced in the stratosphere. It is an oxygen molecule consisting of three atoms of oxygen (O_3) rather than the more common two atoms (O_2). Ozone is created in the upper atmosphere by the action of ultraviolet solar radiation on *diatomic oxygen* (O_2) molecules.

Natural Formation of Ozone: Ultraviolet (UV) radiation from the Sun is divided into three bands (from least to most energetic wavelengths): *UV-A, UV-B,* and *UV-C*. (Radiation is discussed in more detail in Chapter 4.) In the stratosphere, under the influence of UV-C, O_2 molecules split into oxygen atoms; some of the free oxygen atoms combine with O_2 molecules and form O_3 (**Figure 3-18**). The natural breakdown of ozone in the stratosphere occurs when, under the influence of UV-B and UV-C, ozone breaks down into O_2 and a free

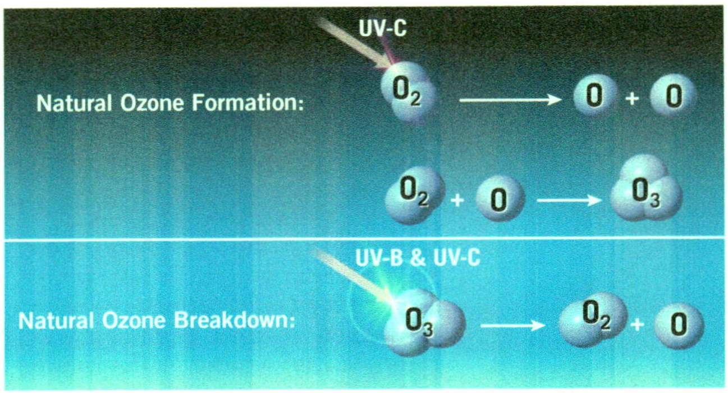

▲ **Figure 3-18** The natural formation and breakdown of ozone. Ultraviolet radiation splits oxygen molecules (O_2) into free oxygen atoms (O), some of which combine with other O_2 molecules, forming ozone (O_3). Also under the influence of UV light, ozone naturally breaks down into O_2 and a free oxygen atom.

oxygen atom. Through this ongoing natural process of ozone formation and breakdown, nearly all of UV-C and much of UV-B radiation is absorbed by the ozone layer. The absorption of UV radiation in this photochemical process also warms the stratosphere.

About 90 percent of all atmospheric ozone is found in the stratosphere, where it absorbs most of the potentially dangerous ultraviolet radiation from the Sun. Ultraviolet radiation can be biologically harmful in many ways. Prolonged exposure to UV radiation is linked to skin cancer; it is also linked to increased risk for cataracts; it can suppress the human immune system, diminish the yield of many crops, and disrupt the aquatic food chain by killing microorganisms such as phytoplankton on the ocean surface.

Ozone is also produced near Earth's surface in the troposphere through human activities. It forms one of the components of photochemical smog (discussed later in this chapter). However, it was a thinning of the stratospheric ozone layer first observed in the 1970s that triggered extensive research and monitoring.

Video
Ozone Hole
https://media.pearsoncmg.com/ph/esm/esm_mcknight_physgeo_13/media/

The "Hole" in the Ozone Layer: Although natural factors can alter the ozone layer, the consensus among atmospheric scientists today is that the dramatic thinning of the ozone layer observed since the 1970s is due primarily (if not entirely) to the release of human-produced chemicals. Pioneering research by atmospheric scientists Sherwood Rowland and Mario Molina in the 1970s showed that the most problematic of these chemicals are **chlorofluorocarbons (CFCs)**, but other ozone-depleting substances include halons (used in some kinds of fire extinguishers), methyl bromide (a pesticide), and nitrous oxide. (In 1995, Rowland, Molina, and fellow scientist Paul Crutzen received the Nobel Prize for Chemistry for their research on ozone depletion.)

CFCs are odorless, nonflammable, noncorrosive, and generally nonreactive. They were widely used in refrigeration and air-conditioning (the cooling liquid Freon™) is a CFC), in foam and plastic manufacturing, and in aerosol

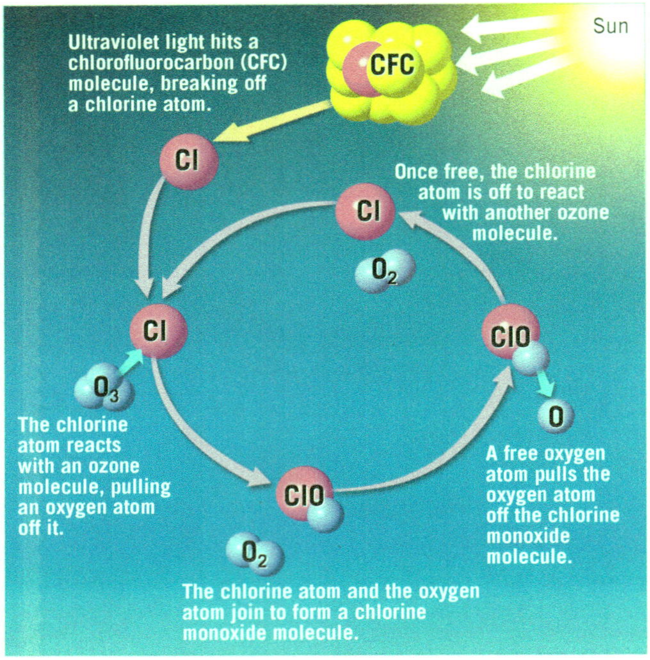

▲ **Figure 3-19** The destruction of ozone in the stratosphere by chlorine atoms derived from the breakdown of CFCs in the atmosphere. Chlorine atoms are unchanged by the reaction and can repeat the process. Thus, a single chlorine atom can destroy tens of thousands of ozone molecules.

sprays. Although extremely stable and inert in the lower atmosphere, CFCs are broken down by UV radiation once they reach the ozone layer. Under the influence of UV radiation, a chlorine atom is released from a CFC molecule (**Figure 3-19**). The chlorine atom then reacts with ozone, breaking it apart and forming one chlorine monoxide (ClO) molecule and one O_2 molecule. The ClO molecule can then react with a free atom of oxygen, forming an O_2 molecule while freeing the chlorine atom to react with another ozone molecule. As many as 100,000 ozone molecules can be destroyed for every chlorine atom released.

Not only was the ozone layer thinning, but in some places it had almost disappeared temporarily. Monitored today by the *Ozone Monitoring Instrument* onboard NASA's Aura satellite, the annual ebb and flow of the ozone layer has been continuously mapped since 1979, with a "hole" developing over Antarctica each year (**Figure 3-20**). By the late 1980s, ozone depletion was found over the Arctic as well. In March 2020 unusually quiet atmospheric motion and persistent cold temperatures created a hole over the Arctic that lasted for about six weeks.

Antarctic Polar Atmosphere: Why is ozone depletion more severe over the polar regions, particularly Antarctica? In part because of the extreme cooling of Antarctica during the winter, a whirling wind pattern known as the *polar vortex* develops, effectively isolating polar air from the atmosphere in lower latitudes. Within the stratosphere, ice crystals form thin *polar stratospheric clouds* (PSCs). The presence of these clouds can dramatically accelerate ozone destruction. The ice crystals in PSCs provide surfaces on which a number of

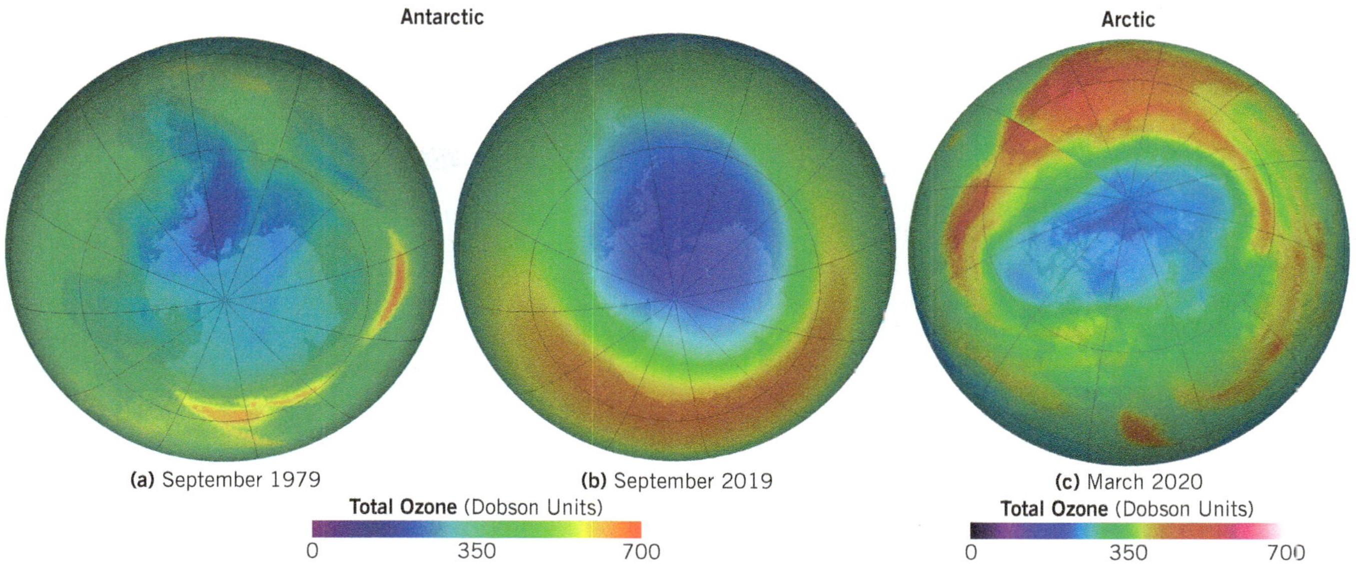

▲ **Figure 3-20** The ozone hole over the Antarctic in (a) 1979 and (b) 2019 and (c) the Arctic in 2020. The area over Antarctica, shown in dark blue and purple, has the lowest concentration of ozone. Ozone concentration is given in Dobson Units. (One Dobson Unit is the number of ozone molecules needed to create a 0.01-mm thick layer of pure ozone at sea level at 0°C.) As a result of international efforts to limit the production of ozone-depleting chemicals, after growing in size since the 1970s, the Antarctic ozone hole seems to be shrinking, as indicated by lighter blue colors and the absence of purple in 2019. An ozone hole over the Arctic is very unusual.

NASA video
Exploring Earth
https://media.pearsoncmg.com/ph/esm/esm_mcknight_physgeo_13/media/

reactions can take place, including the accumulation of chlorine-based molecules. With the return of sunlight in the polar spring (September in the Southern Hemisphere), UV radiation triggers the depletion of ozone. The thinning of ozone over the Arctic has been generally less severe than over the Antarctic. That is because over the North Pole, land masses usually disrupt the polar vortex.

The Montreal Protocol: The discovery of the effects of increasing levels of CFCs was sufficiently alarming that a number of countries, including the United States, banned their use in aerosol sprays in 1978. A major international treaty—the *Montreal Protocol on Substances That Deplete the Ozone Layer*—was negotiated in 1987 to set timetables for phasing out the production of the major ozone-depleting substances. One hundred ninety-six countries and the European Union have ratified the proposal. Following stipulations of the treaty and its more recent amendments, the industrialized countries of the world banned CFC production by 1996. Moreover, the protocol signatories pledged a fund of more than $2 billion through 2005 to help developing countries implement alternatives to CFCs.

As a result of the ban of CFCs, ozone depletion decreased 20 percent during the winter months between 2005 and 2016. Even with the Montreal Protocol fully implemented, the ozone layer will still take decades to recover because the reservoir of CFCs in the atmosphere may persist for 50 to 100 years—and because levels of nitrous oxide released by human activity remain high. The largest measurable Antarctic ozone hole was observed in 2006, and since then it appears that ozone is recovering. September 2019 had the smallest ozone hole since its discovery in the 1970s. However, some studies suggest that it may be 2050 before recovery is well under way.

Addressing the depletion of the ozone layer is considered by many scientists to be an example of an environmental success story: a human-produced problem was identified, and a global strategy was implemented to counteract it.

The UV Index: Stratospheric ozone depletion has been correlated with increased levels of ultraviolet radiation reaching ground level in Antarctica, Australia, mountainous regions of Europe, central Canada, and New Zealand. Ultraviolet radiation forecasts describe the expected level of UV radiation one day in advance. A *UV Index* (**Figure 3-21**), has been established to provide the public with information

UV Index Exposure Risk	Recommendations
0–2 Low	If you burn easily or are exposed to reflections off snow or water, wear sunglasses and apply broad spectrum SPF 30+ sunscreen.
3–5 Moderate	Limit midday Sun exposure. Wear protective clothing, hats, and sunglasses. Apply broad spectrum SPF 30+ sunscreen every 2 hours.
6–7 High	Stay in shade near midday. Use UV-blocking sunglasses and a wide-brimmed hat. Generously apply broad spectrum SPF 30+ suncreen every 2 hours, even on cloudy days.
8–10 Very High	Minimize Sun exposure between 10 A.M. and 4 P.M. Generously apply broad spectrum SPF 30+ sunscreen every 2 hours, even on cloudy days.
11+ Extreme	Avoid midday Sun exposure between 10 A.M. and 4 P.M. Generously apply broad spectrum SPF 30+ sunscreen every 2 hours, even on cloudy days, and after swimming.

Source: U.S. Environmental Protection Agency, SunWise Program.

▲ **Figure 3-21** UV Index scale.

about the intensity of UV radiation in an area in part because of the increased health risks posed by higher levels of UV radiation reaching the surface (Figure 3-22). They use a scale of 0 to 11+, with 1 representing low exposure risk and levels of 8 or greater representing high exposure risk. Each risk category of the UVI is accompanied by recommended precautions. The UV Index depends on cloud cover, altitude, and the amount of ozone in the air (Figure 3-23).

Unprotected exposure to ultraviolet (UV) radiation—from the Sun or tanning beds—is considered the most preventable risk factor for skin cancer. In the United States, skin cancer is the most commonly diagnosed form of cancer. Skin damage from exposure to UV radiation is cumulative. It is especially important that young children of all skin types use sunscreens of SPF (Sun Protection Factor) 30 or greater and take other measures, such as wearing hats and protective clothing. To find the UVI forecast for your location, go to www.epa.gov/sunwise/uvindex.html.

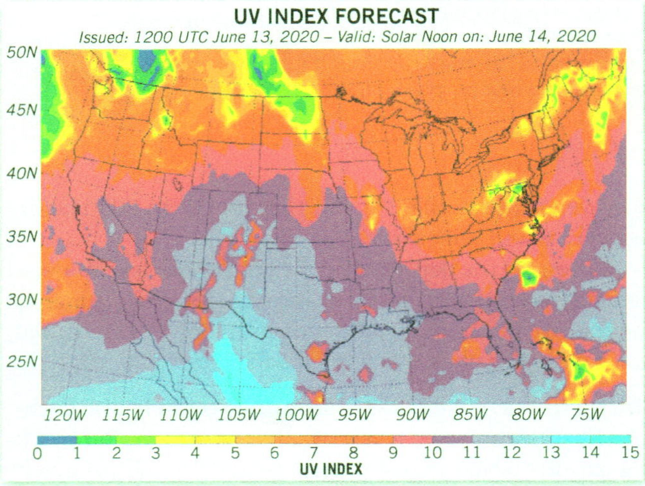

▲ Figure 3-22 UV Index forecast map.

▲ Figure 3-23 Visible satellite image showing the location of cloud cover over the United States at the time of the UVI forecast map in Figure 3-22.

Learning Check 3-8 What is the explanation for the thinning of the ozone layer that has been observed since the 1970s?

Air Pollution

In addition to diminishing stratospheric ozone, humans have altered the composition of the atmosphere in other ways. The atmosphere has never been without pollutants, but with the start of the Industrial Revolution in the 1700s, air pollution became widespread. By the twentieth century, it was recognized as a major problem.

By far the greatest difficulties are associated with cities, where people and activities are concentrated, especially from the pollutants produced by internal combustion engines and industry. The presence of pollutants in the air is most obviously manifested by reduced visibility due to fine particulate matter and photochemical smog. More critical, however, is the health hazard imposed by increasing concentrations of chemical impurities in the air.

Carbon Monoxide: The gas carbon monoxide (CO) is the most plentiful **primary pollutant**—a contaminant released directly into the air. It is formed by the incomplete combustion of carbon-based fuels, especially by motor vehicles. Because this gas is odorless and colorless, people exposed to carbon monoxide in confined spaces can be quickly overcome after CO enters the bloodstream and decreases the amount of oxygen available to their brain and other organs. There was a 59 percent decrease in CO between 2000 and 2018 because of legislation. Still, about two-thirds of the total weight of primary pollutants released by the United States each year is CO. More than half of the current CO comes from vehicles on the road.

Nitrogen Compounds: Nitric oxide (NO) is a gas that can form in water and soil as a natural by-product of biological processes; it generally breaks down quite quickly. However, NO may also form through combustion that takes place at high temperatures and pressures—such as in an automobile engine. Nitric oxide reacts in the atmosphere to form acid rain and nitrogen dioxide (NO_2), a corrosive gas that gives some polluted air its yellow or reddish-brown color. Although NO_2 itself tends to break down quickly, it reacts under the influence of sunlight and form several components of smog.

Sulfur Compounds: A large portion of the sulfur compounds found in the atmosphere are of natural origin, released from volcanoes or hydrothermal vents such as those in Yellowstone National Park. Hydrogen sulfide (H_2S) with its familiar "rotten egg" smell is one example. However, especially over the past century, human activity has increased the release of sulfur compounds into the atmosphere, primarily through the burning of fossil fuels such as coal and petroleum. Sulfur is a minor impurity in coal and oil; when those fuels are burned, sulfur compounds such as sulfur dioxide (SO_2) are released. Sulfur dioxide is a lung

Global Environmental Change

Aerosol Plumes Circling the Globe

Air pollution is not a local problem. Wind blowing across China, India, and Africa transports plumes of pollution and dust around the globe, affecting the environment thousands of kilometers from where they originated (**Figure 3-E**). Air samples show that on some days more than a quarter of the aerosols in California's air originated in Asia. Computer models estimate that these aerosol plumes circle Earth in only about 3 weeks.

Studying Plumes from Within: To study the impact of aerosol plumes, National Center for Atmospheric Research scientists use a modified Gulfstream V aircraft known as HIAPER (High-performance Instrumental Airborne Platform for Environmental Research). HIAPER is unique in that it can fly at very high altitudes (up to 15.5 kilometers [51,000 feet]) for very long distances (up to 11,000 kilometers [7000 miles]).

HIAPER collects data on the chemical composition and size distribution of aerosol particles within a plume. These data are then analyzed to determine the impact of plume aerosols on the local environment.

▲ **Figure 3-E** Satellite image showing aerosols blowing across the Sea of Japan (East Sea) into the Pacific Ocean. Pollution and dust originate in China, North Korea, South Korea, and Japan.

Impacts of Aerosol Plumes: Aerosols coming from Asia contain nutrients that replenish those lost due to rainfall runoff in the Amazon rainforest. We now know that these plumes also contain sulfate aerosols that are highly reflective to sunlight. These reflective aerosols likely mask some of the warming due to global climate change.

In addition, the plume aerosols enhance cloud formation. Cloud formation requires an abundant supply of *condensation nuclei*, and aerosols are effective condensation nuclei. The long-term impacts of aerosol plumes will likely include a modification of precipitation patterns and storm tracks. Evidence suggests that dust aerosols can diminish rain and snow events and may ultimately suppress precipitation altogether in some areas.

Questions

1. Why is HIAPER uniquely qualified to study dust plumes?
2. What are three known impacts of dust plumes?

---

irritant and corrosive, but it may react in the atmosphere and produce compounds such as sulfur trioxide (SO_3) or sulfuric acid (H_2SO_4)—a contributor to acid rain (discussed in Chapter 6). Legislation has led to a 79 percent decrease in the national average amount of SO_2, but it is still a problem for major cities.

Particulates: Particulates (or aerosols) are tiny solid particles or liquid droplets suspended in the air. Primary sources of particulates from human activities include smoke from combustion and dust emitted from industrial activities. The concentration of particulates may also increase through secondary processes, such as when small particles coalesce into larger particles or when liquid droplets develop around condensation nuclei. Many of the health hazards associated with particulates appear to be greatest when the particles are less than 2.5 micrometers (<2.5 millionths of a meter) in diameter, known as PM2.5. In 1997, the U.S. Environmental Protection Agency (EPA) revised its regulations to take into account the harmful effects of such fine particulates. However, regulating particulates is difficult because some travel great distances—see the box *Global Environmental Change: Aerosol Plumes Circling the Globe*.

Photochemical Smog: A number of gases react to ultraviolet radiation in strong sunlight, producing **secondary pollutants** (contaminants that form as a consequence of chemical reactions or other processes in the atmosphere). They make up what is known as **photochemical smog** (**Figure 3-24**). (The word *smog* was originally derived from a combination of the words *smoke* and *fog*; photochemical smog usually includes neither.) Nitrogen dioxide and hydrocarbons (also known as *volatile organic compounds*, or *VOC*)—both of which can result from the incomplete burning of fuels such as gasoline—are major contributors to photochemical smog. Nitrogen dioxide breaks down under UV radiation, forming nitric oxide. That may then react with VOC, forming peroxyacetyl nitrate (PAN), which has become a significant cause of crop and forest damage in some areas.

The breakdown of NO_2 into NO also frees an oxygen atom, which can react with O_2 molecules and form ozone—the main component of photochemical smog. Ozone has an acrid, biting odor that is distinctive to photochemical smog, causing damage to vegetation, corroding building materials (such as paint, rubber, and plastics), and damaging sensitive body tissues (eyes, lungs, and nose).

Atmospheric Conditions and Air Pollution: The state of the atmosphere is an important determinant of the level of air pollution, especially photochemical smog and particulates.

▲ **Figure 3-24** China established a four-color alert system in 2013—a red alert is the highest level of the warning system. On red alert days, heavy pollution industries and schools will close. Cars are ordered off the roads, and people are encouraged to stay home. Shown here is an orange-alert day in Harbin China with people traversing a bridge wearing masks to protect against the heavy smog.

If there is considerable air movement, pollutants can be quickly and widely dispersed. On the other hand, stagnant air allows for a rapid accumulation of pollutants. If the air is particularly stable, *temperature inversions* (cooler air below warmer) develop, functioning as "stability lids" that inhibit updrafts and general air movement. (Temperature inversions are discussed in Chapter 4.) Almost all the cities with persistent air pollution—such as Mexico City, Los Angeles, and Beijing—are characterized by a high frequency of temperature inversions. San Francisco had the worst air pollution in the world during parts of 2018 because of the California wildfires and temperature inversions.

Consequences of Anthropogenic Air Pollution: Carbon monoxide, sulfur dioxide, and particulates can contribute to cardiovascular disease, while prolonged exposure to some particulates may promote lung cancer. Nitrogen oxides and sulfur dioxide are the principal contributors to acid rain. Tropospheric ozone damages crops and trees; it is now the most widespread air pollutant. The U.S. EPA reports that perhaps one-fifth of all hospital cases involving respiratory illness in the summer are a consequence of exposure to ground-level ozone.

In the past few decades in the United States, there has been a prominent downward trend in the emission of nearly all pollutants except ozone and carbon dioxide, largely as a result of increasingly stringent emission standards imposed by the EPA starting in 1970. However, in many newly industrialized countries, pollutant loads are expanding, and an ongoing effort on the part of all countries will be required to curtail this global problem.

Learning Check 3-9 **How does photochemical smog form, and what are some of its effects on human health?**

Energy Production and the Environment

Although the release of many air pollutants has decreased in recent years, the release of the carbon dioxide through the burning of fossil fuels is still on the rise. The full significance of this increase is explored in subsequent chapters, but here we begin our look at a topic that underlies this and many of the environmental and economic challenges faced by the world today: our growing demand for energy to power our homes, automobiles, and industries.

In the chapters that follow, we explore a number of the methods used to generate power, from our long-standing use of fossil fuels such as coal and oil to renewable methods such as wind and tides, describing the technology as well as the virtues and drawbacks of each. There is no simple solution to our demand for energy. However, understanding how energy production fits in with physical geography and Earth's interconnected systems can help us assess the long-term consequences of our decisions.

We begin with the box *Energy for the 21st Century: Transitioning from Fossil Fuels*.

Energy for the 21st Century

Transitioning from Fossil Fuels

(a) India Gate December 6, 2019 (b) India Gate April 9, 2020

▲ **Figure 3-G** In late 2019, when the COVID-19 pandemic started to move across the world, many forms of fossil-fuel-powered transportation were greatly reduced in an attempt to reduce the virus's spread. In the months that followed, the world saw a dramatic decrease in air pollution concentrations worldwide, hinting at what a reduction in such fuel consumption might look like.

• **Michael E. Mann,** Penn State University

Fossil fuels (coal, oil, and natural gas) are the product of millions of years of accumulated energy from sunlight, absorbed in plant life and trapped as hydrocarbon matter beneath Earth's surface. Although fossil fuels have been the primary energy source powering human civilization since the dawn of the industrial revolution, a transition to newer, cleaner forms of energy is now underway.

Historical Significance of Fossil Fuels: Before the use of fossil fuels, people did most mechanical work by using their own muscle power and that of animals, both ultimately derived from the Sun's energy stored in plants through photosynthesis and plant-eating animals (discussed in Chapter 10). The shift to fossil fuels led to machinery, such as steam engines, that ran without the force of muscle power and eventually to electrical power generation and automobiles. This change allowed for dramatic gains in labor productivity and the growth of transportation networks. Moreover, the increasing reliance on fossil fuels freed up thousands of acres that had been used for energy, such as farmland used to grow feed for working animals and forest land that provided wood and charcoal for heating, cooking, and metal production.

The United States ranks second in energy consumption, with 81% of its energy from fossil fuels in 2019 (**Figure 3-F**). This is higher than the global average. Renewable energy continues to add more capacity with each passing year.

Consequences of Reliance on Fossil Fuels: Despite the historical benefits that fossil fuel energy has provided, it has come with considerable costs. First, the combustion of coal, oil, and gas produces enormous amounts of pollution. For example, sulfur dioxide (SO_2) from coal-fired power plants causes acid rain (discussed in Chapter 6). Passage of the Clean Air Acts in the 1970s, which required scrubbers in factory smoke stacks to remove SO_2 emissions, has largely alleviated that problem. However, a more fundamental environmental threat arises from the fact that all fossil fuel use emits carbon dioxide (CO_2), the primary human-produced greenhouse gas causing global climate change (**Figure 3-G**), discussed in Chapter 4. Moreover, the uneven distribution of fossil fuels creates geopolitical conflict over access to, and control over, energy resources. Although this problem is most visible in conflicts over oil (as in the Middle East), other areas of contention include natural gas pipelines in the Ukraine and North America and "fracking" (see Chapter 13) and mountaintop coal mining in the United States.

Alternative Energy: Support for switching to *alternative energy* has been growing among scientists, policymakers, and the public in recent decades. Most alternatives to fossil fuels generate electricity. Electricity is primarily generated by the combustion of coal or gas to create steam that turns a turbine. The transportation sector runs mostly on liquid fuels derived from crude oil. The only liquid fuel alternatives are biofuels (see Chapter 10), often derived from crops grown on farmland, thus competing with food production. The rapid increase in use of electric and plug-in hybrid vehicles, which can be powered off the electric grid, provides a means for reducing reliance on fossil fuels in the transportation sector.

Many alternative energy sources show promise, although each has some limitations. Nuclear power is the largest current alternative energy source, but it brings risks of calamitous accidents, such as the Fukushima disaster in Japan following the tsunami of 2011 (see Chapter 20) , and produces radioactive waste. Hydroelectric power uses dams and the power of falling water to generate electricity (see Chapter 16). Dams, however, degrade river ecosystems and often displace thousands of people. Some alternatives, such as geothermal (see Chapter 17) and tidal power (see Chapter 20) , may be limited to locations far from population centers.

Wind power recently has been the fastest growing alternative (see Chapter 5), but its capacity to generate electricity ultimately

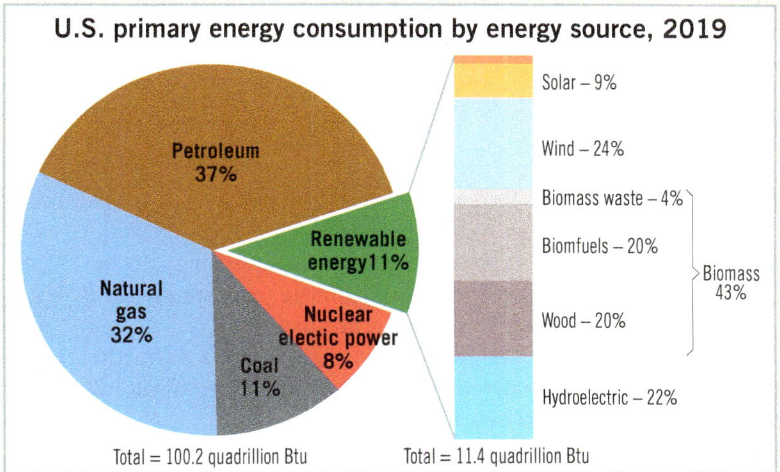

▲ **Figure 3-F** United States energy consumption by source in 2019.

continued on next page

depends on when the wind blows. Solar power harnesses direct sunlight to generate electricity through either photovoltaic cells or the boiling of water to create steam (see Chapter 4) . Like wind, solar is intermittent; it is not available when it is dark or cloudy. "Smart grids" that adaptively combine energy from a diverse mix of sources would alleviate these intermittency limitations.

The main barriers to an "energy transition" away from fossil fuels are political and economic (see Chapter 8). Historically, fossil fuels have been cheaper, making it difficult for alternative energy to compete when only short-term costs are considered. The current high profitability of the fossil fuel industry inhibits a shift toward initially less profitable alternative energy sources. Despite these obstacles, advances in technology are making renewable energy sources increasingly competitive with fossil fuels, and that trend should continue. More widespread support for pricing of carbon emissions to account for the damage done by the resulting changes in climate promises to accelerate this process.

Questions

1. An aphorism holds that "The Stone Age didn't end for want of stones." What relevance does this statement have for the transition to renewable energy?
2. How can putting a price on carbon emissions accelerate the transition away from fossil fuels to renewable energy sources?
3. Some advocates argue that cheap fossil fuels are the best way to assist economic development in poor nations. Why might this *not* be true?

3 Learning Review

After studying this chapter, you should be able to answer the following questions. Key terms from each text section are shown in **bold type.** Definitions for key terms are also found in the glossary at the back of the book.

Key Terms and Concepts

Size and Composition of the Atmosphere (*p. 58*)

1. What is meant by the terms *permanent gases* and *variable gases* in the atmosphere?
2. Describe the most important permanent gases of the atmosphere.
3. Briefly describe some of the roles that **water vapor**, **carbon dioxide**, **ozone**, and **particulates (aerosols)** play in atmospheric processes.
4. How has the burning of **fossil fuels** over the past 200 years changed the composition of the atmosphere?
5. Describe both the vertical distribution of water vapor in the atmosphere and its horizontal (geographic) distribution near Earth's surface.

Vertical Structure of the Atmosphere (*p. 61*)

6. Discuss the size and general temperature characteristics of the **troposphere** and **stratosphere**.
7. Describe how atmospheric pressure changes with increasing altitude.
8. What is the **ozone layer**, and where is it located?

Weather and Climate (*p. 65*)

9. What is the difference between **weather** and **climate**?
10. What are the four **elements of weather and climate**?
11. Briefly describe the seven dominant **controls of weather and climate**.
12. Describe the **Coriolis effect** and its cause.

Human-Caused Atmospheric Change (*p. 69*)

13. How is ozone formed, and why is it important in the atmosphere?
14. What is meant by the "hole" in the ozone layer, and what role have **chlorofluorocarbons (CFCs)** played in this?
15. Describe and contrast **primary pollutants** and **secondary pollutants** in the atmosphere.
16. Describe and explain the causes of **photochemical smog**.

Study Questions

1. What prevents the atmosphere from "escaping" into space?
2. Why is the question "How deep is the atmosphere?" difficult to answer?
3. In what ways was life on Earth responsible for the composition of our modern atmosphere?
4. Why are you likely to get out of breath more easily when hiking in the mountains than at sea level?
5. Why does the altitude of the tropopause vary from summer to winter and from the equator to the poles?
6. Why is it inaccurate to talk about a change in climate from last year to this year?
7. In our study of physical geography, why do we concentrate primarily on the troposphere rather than on other zones of the atmosphere?
8. Why does the Coriolis effect influence the direction of ocean currents but not the direction of water draining in a kitchen sink?
9. Why should humans be concerned about the depletion of the ozone layer?

Exercises

1. Use Figure 3-3: How much more nitrogen is in the atmosphere than oxygen?
2. Use Figure 3-3: How much more oxygen is in the atmosphere than carbon dioxide?
3. Use Figure 3-7: If you are on top of Pikes Peak in Colorado at an altitude of 4.3 kilometers (2.7 mi.; 14,110 ft.), what is the approximate percentage of surface atmospheric pressure?
4. Use Figure 3-7: If you were in an unpressurized balloon at an altitude of 10 kilometers (6.2 mi.; 33,000 ft.), what is the approximate percentage of surface atmospheric pressure?

Environmental **Analysis** CO_2 MONITORING IN YOUR NEIGHBORHOOD

Data
CO_2 Levels
https://media.pearsoncmg.com/ph/esm/esm_mcknight_physgeo_13/media/

With scientists now convinced that rising atmospheric CO_2 levels are largely responsible for the post-1900 increase in average global temperature, and with the potential for dire consequences, widespread monitoring of CO_2 concentrations is critical. The National Oceanic and Atmospheric Administration measures greenhouse gas concentrations in air samples taken continually by a network of more than 100 sites worldwide at various elevations.

Activities

Go to the Global Greenhouse Gas Reference Network at www.esrl.noaa.gov/gmd/ccgg/ggrn.php.

1. Describe the distribution of air sampling sites and the measurement methods used.

Click on "View map and listing of reference network sites." Using the map, select the site nearest to you that is not discontinued (yellow).

2. What is the site's elevation in meters above sea level (masl)?
3. What sampling methods are used at the site?

"Data visualization" allows you to plot the data from your site. Select "carbon cycle gases" and plot a time series. Using default settings, opt for a time series plot to graph values of the "carbon cycle gas" carbon dioxide (CO_2) for all available data (be sure to hit "Submit"). If no plot is generated, choose a different data type or measurement site.

4. What is the plotted time span?
5. What is the range of variation of CO_2 (shown by the circle symbols)?
6. Describe the data—how does CO_2 vary over time?
7. Is a seasonal cycle visible in the multiyear graph? Explain.

In a second browser window, select a second active measurement site in a distant location.

Repeat Activities 2 through 7, and describe any significant differences in the graphs you obtained for the two locations.

Seeing Geographically in Our National Parks

Look again at the Great Smoky Mountain picture at the beginning of the chapter (p. 56). Are the particulates (aerosols) that form the blue haze or "smoke" natural or human-made? The source of moisture needed for the "smoke" formation is not the ocean. Why not? What is the source of moisture?

Mastering Geography®

Looking for additional review and test prep materials? Visit the Study Area in **Mastering Geography** to enhance your geographic literacy, spatial reasoning skills, and understanding of this chapter's content by accessing a variety of resources, including Map Master interactive maps, geoscience animations, *Mobile Field Trips*, videos, *Project Condor* Quadcopter videos, *In the News* RSS feeds, flashcards, web links, self-study quizzes, and an eText version of *McKnight's Physical Geography*.

Seeing Geographically in Our National Parks

Kenai Fjords National Park, on the southern coast of Alaska, has nearly 40 glacier flows that melt into icy cold streams. Is the sun ever straight overhead here? What do you think the temperature was like when this picture was taken? What season is it? How do you know?

Kenai Fjords National Park, AK

Insolation and Temperature

4

HAVE YOU EVER WONDERED why the equator is warm and the poles are cold? Or why the glaciers in Kenai Fjords National Park don't completely melt in summer? The answers to both of these questions come with an understanding of the first element of weather and climate: temperature.

The warming of Earth's surface and atmosphere by solar energy is one of the most fundamental of all Earth systems. Many of the geographic patterns discussed in chapters that follow—including atmospheric pressure and wind, the distribution of precipitation, the characteristics of storms, the distribution of plants and animals, and the development of soils—are in part related to the arrival of solar energy, the warming and cooling that takes place on Earth's surface and in the atmosphere, and the transfer of this energy from one part of the planet to another.

In this chapter, we introduce some of the most important concepts in physical geography: how variations in temperature (by latitude) develop around the world; how land and water influence the atmosphere above them in different ways—and are influenced by the atmosphere in different ways; and how human activity can alter global climate.

As you study this chapter, think about these **Key**Questions:

- **How does energy from the Sun warm the atmosphere?**
- **Why are temperatures near the equator so different than at the poles?**
- **Why do continents experience greater temperature extremes than oceans?**
- **How is energy transferred around the world by the atmosphere and ocean currents?**
- **What factors explain global patterns of temperature?**
- **How has recent human activity changed global temperatures?**

The Impact of Temperature on the Landscape

Global temperature patterns leave a prominent mark on the landscapes of Earth. Many physical features of the landscape are affected by local temperature conditions. For example, temperature fluctuations are one cause of the breakdown of exposed bedrock and the rate of chemical processes associated with soil development. Furthermore, long-term temperature patterns influence the presence of many agents of erosion and deposition, such as streams and glaciers.

Animals and plants often evolve in response to hot or cold climates, so the types of fauna and flora in an area reflect the capability of the various species to withstand the long-term temperature conditions (**Figure 4-1**). Human activity may also be constrained by local temperature patterns, influencing everything from the type of buildings we construct to the type of economic activities that are possible.

Energy, Heat, and Temperature

In the most basic sense, the universe is made up of just two kinds of "things": *matter* and *energy*. The concept of matter is fairly easy to comprehend. Matter is the "stuff" of the universe: the solids, liquids, gases—and their atomic particles—from which all things are made. Matter has mass and volume; we can easily see and feel many kinds of matter. The concept of energy, on the other hand, may be more difficult to grasp.

Energy

Energy is what makes "stuff" move. For example, it takes energy to cause something to move faster, change direction, or break apart. The transfer of energy from one object to another or from one form to another causes changes to the condition of matter—whether that matter is a single molecule in the atmosphere or the entire volume of water in Earth's oceans.

There are many forms of energy: *kinetic energy*, *chemical energy*, *gravitational potential energy*, and *radiant energy*, among many others. Although energy can neither be created nor destroyed, it can—and frequently does—change from one form to another. In our discussion of the warming of the atmosphere, we look at just a few forms of energy and the transformations between those forms.

Work: Energy is commonly defined as the "ability to do work." *Work* refers to *force acting over distance*. When a force is involved in moving matter around, energy has been transformed from one form to another. The International System (S.I.) unit of energy is called the joule (J). How much energy does a joule represent?

- Lifting a 1-kilogram (2.2-pound) mass a distance of 1 meter (about 3 feet) increases the *gravitational potential energy* of that mass (the energy that object possesses because of its position in Earth's gravitational field) by about 10 joules.

(a) *Melanerpes uropygialis* (b) *Lynx canadensis*

▲ **Figure 4-1** Plants and animals are adapted to the climate of their habitat, a function of their location. (a)The gila woodpecker, found in the Saguaro National Park of Arizona, forages on taller desert vegetation, feeding on insects and cactus fruits. (b) The lynx, found in Alaska's Denali National Park, is well adapted to living in much colder temperatures, feeding on snowshoe hare and small rodents found in the Alaskan wilderness.

- The energy unit 1 *calorie* is the amount of energy needed to increase the temperature of 1 gram of water by 1°C (at 15°C). It takes 4.184 joules of energy to increase the temperature of 1 gram of water by 1°C. So, 1 calorie = about 4.184 joules, and 1 joule = about 0.239 calories.

Power: Another way to look at energy is to measure how much energy is *transferred per unit of time*. For example, later in this chapter we describe how much energy is delivered by the Sun to Earth. Energy per unit of time is defined as power. The S.I. unit of power is the watt (W): 1 watt is equal to 1 joule per second (and 1 W = 0.239 calories per second). The power consumption of lightbulbs and electrical appliances is commonly described in watts.

Internal Energy: To understand the warming of the atmosphere, or anything else, we first need to consider what is happening at the atomic or molecular level—a scale that we do not readily comprehend in everyday life. A great simplification can help us understand one of the relationships between energy and matter: all substances are composed of extraordinarily tiny, constantly "jiggling" *atoms*, commonly bonded together into combinations of atoms called *molecules*. The state of those substances—whether they are solid, liquid, or gas (or *plasma*, which is an ionized gas)—depends in part on how vigorously the molecules jiggle in place. Because of this constant movement, the molecules in all substances possess energy. This kind of *internal energy* is a form of **kinetic energy**—the energy of movement.

The average amount of kinetic energy possessed by the molecules in a substance is closely associated with a physical property that we *can* readily sense: how hot or cold something is. When a substance becomes warmer, it means that the average kinetic energy of the molecules in that substance has increased. In other words, energy has been added that has caused the molecules to jiggle back and forth more vigorously. Cooling results in a decrease in the average kinetic energy. This brings us to several important definitions.

Temperature and Heat

Temperature is a description of the average kinetic energy of the molecules in a substance (strictly speaking, the average *translational kinetic energy* of the molecules). The more vigorously the molecules jiggle—and therefore the greater the internal kinetic energy—the higher the temperature of a substance.

Heat refers to energy that transfers from one object or substance to another because of a *difference* in temperature. Sometimes the term *thermal energy* is used interchangeably with the term *heat*. A substance doesn't really "store" heat or "contain" heat—heat is simply the energy that is transferred from an object with a higher temperature to an object with a lower temperature, thereby increasing the internal energy of the cooler object and decreasing the internal energy of the warmer object.

In addition to heat transfer, the total internal energy of an object can be changed in other ways. For instance, work (the force applied to an object over some distance) can change an object's internal energy. When you strike a piece of metal with a hammer, you increase the internal energy of the molecules in the metal, so the metal is warmed. As with heat, an object doesn't "contain" work, but you can change its internal energy by doing work or having work done on it.

LearningCheck 4-1 **What is the difference between temperature and heat? (Answer on p. AK-1)**

Measuring Temperature

Thermometers are instruments used to measure temperature. The temperature of an object is usually relative to one of three common temperature scales (Figure 4-2). Each permits a precise measurement, but the existence of three scales creates an unfortunate degree of confusion.

Fahrenheit Scale (°F): The temperature scale that is most widely used in the United States—and announced in our public weather reports—is the Fahrenheit scale (named after Daniel Gabriel Fahrenheit, the eighteenth-century German physicist who devised it). On this scale, the sea-level freezing point of pure water is 32°, and its boiling point is 212°.

Celsius Scale (°C): In most other countries the Celsius scale (named for Anders Celsius, the eighteenth-century Swedish astronomer who devised it) is used either exclusively or predominantly. It is an accepted component of the International System of measurement (S.I.) because it is a decimal scale with 100 units (degrees) between the freezing and boiling points of water at sea level. The Celsius scale has long been used for scientific work but has not yet superseded the Fahrenheit scale in the United States.

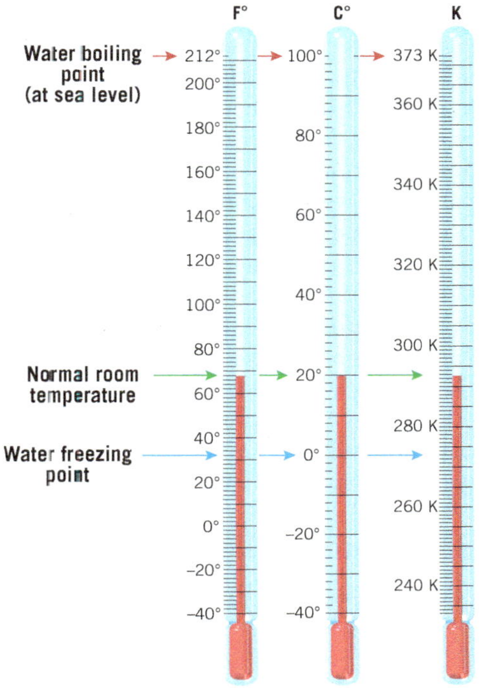

▲ Figure 4-2 The Fahrenheit, Celsius, and Kelvin temperature scales.

To convert from degrees Celsius to degrees Fahrenheit, we use the following formula:

$$\text{degrees Fahrenheit} = (\text{degrees Celsius} \times 1.8) + 32°$$

To convert from degrees Fahrenheit to degrees Celsius, we use the following formula:

$$\text{degrees Celsius} = (\text{degrees Fahrenheit} - 32°) \times 1.8$$

Kelvin Scale (K): For many scientific purposes, the Kelvin scale (named for nineteenth-century British physicist William Thomson, also known as Lord Kelvin) has long been used. It measures *absolute temperatures*, which means that the scale begins at *absolute zero* (0 K, or "zero kelvin"), the lowest possible temperature.[1] This scale maintains a 100-unit range between the boiling and freezing points of water. There are no negative values. Because this scale is not normally used by climatologists or meteorologists, we ignore it in this book except to compare it with the Fahrenheit and Celsius scales. On the Celsius scale, absolute zero is at about −273°, so the conversion is simple:

$$\text{degrees Celsius} = \text{kelvin} - 273$$
$$\text{kelvins} = \text{degrees Celsius} + 273$$

Solar Energy

The Sun is the only significant source of energy for Earth's atmosphere. Millions of other stars radiate energy, but they are too far away to affect Earth. Energy is also released from inside Earth, primarily from radioactive decay of elements such as uranium (^{238}U), thorium (^{232}Th), and potassium (^{40}K). This energy can be transferred through Earth and eventually released at the surface. For example, energy is released on the ocean floor through hydrothermal vents, although probably not enough to influence the atmosphere significantly. The Sun supplies essentially all of the energy that drives most atmospheric processes. Furthermore, we will see that it is the *unequal* warming of Earth by the Sun that ultimately puts the atmosphere in motion and is responsible for the most fundamental patterns of weather and climate.

The Sun is a prodigious generator of energy. In a single second, it produces more energy than the amount used by humankind since civilization began. The Sun functions as an enormous thermonuclear reactor, producing energy through *nuclear fusion*—under extremely high temperatures and pressures, nuclei of hydrogen fuse together, forming helium. This process utilizes only a very small portion of the Sun's mass but provides an immense and continuous flow of energy that is dispersed in all directions.

Electromagnetic Radiation

The Sun gives off energy in the form of **electromagnetic radiation**—sometimes referred to as **radiant energy**. (The Sun also gives off energy as streams of ionized particles called the *solar wind*, but we can ignore that kind of energy in our discussion here because its effect on weather is minimal.) We experience different kinds of electromagnetic radiation every day; visible light, microwaves, X-rays, and radio waves are all forms of electromagnetic radiation.

Electromagnetic radiation entails the flow of energy in the form of waves. These waves of energy move through space by way of rapidly oscillating electromagnetic fields. The electromagnetic fields oscillate at the same frequency as the vibrations of the electrical charges that form them. For example, it is the oscillation of electrons within an atom that can generate visible light. Electromagnetic radiation does not require a medium (the presence of matter) to pass through. The electromagnetic waves traverse the great voids of space in unchanging form. These waves travel outward from the Sun in straight lines at the speed of light—300,000 kilometers (186,000 miles) per second.

Earth intercepts only a tiny fraction of the Sun's total energy output. The waves travel through space without loss of energy, but their intensity (concentration) continuously diminishes because the same energy gets spread over a larger area as they get farther from the Sun (Figure 4-3). This is similar to the rubber of a balloon getting spread over a larger area as you blow the balloon up. The energy decreases according to the *inverse square law*: energy intensity decreases with the square of the distance from the source. For instance, twice the distance means only one-fourth ($1/2^2$) the intensity. As a result of this intensity decrease and the distance separating Earth from the Sun, less than one-two-billionth of total solar output reaches the outer limit of Earth's atmosphere, having traveled 150,000,000 kilometers (93,000,000 miles) in just over 8 minutes. Although it consists of only a minuscule portion of total solar output, in absolute terms the amount of solar energy Earth receives is enormous: the amount received in 1 minute is approximately equivalent to all the energy consumption by humans on Earth in a year.

Because Earth receives so much energy from the Sun, solar energy is increasingly viewed as an important source of renewable energy that can be used to generate electricity, as explained in the box *Energy for the 21st Century: Solar Power*.

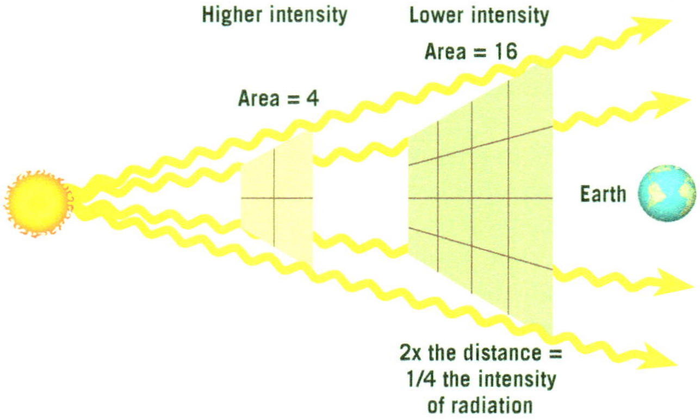

▲ Figure 4-3 The intensity of solar radiation decreases with distance. Following the inverse square law, at twice the distance solar energy is only one-fourth as intense.

[1]In classical physics, *absolute zero* is the temperature at which molecules have their *zero-point energy*—they have no kinetic energy to give up.

Energy for the 21st Century

Solar Power

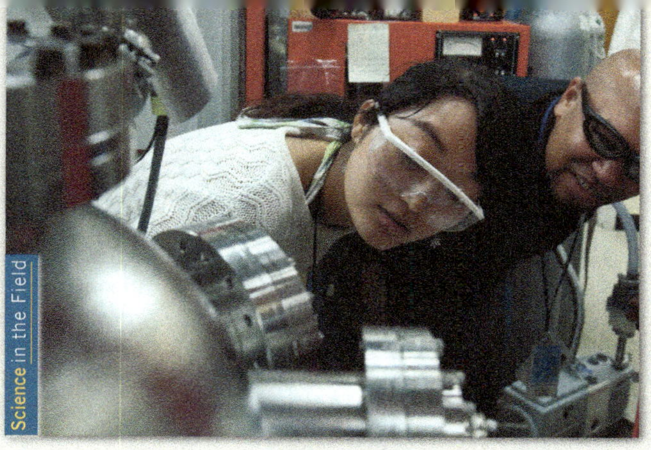

◀ **Figure 4-B** National Renewable Energy Laboratory scientists experimenting with ways to make solar cells more efficient.

• **Ryan Longman,** The University of Hawai'I at Mānoa

In one day, an average of 164 watts per square meter shines on Earth's surface, providing more than enough energy to meet the electrical generation needs of the entire planet. Yet humans make little use of this clean and relatively cheap form of energy. Although solar power provides only a fraction of a percent of global energy, the world's photovoltaic capacity has increased from 15 gigawatts (15 billion watts) in 2008 to approximately 633 gigawatts at the end of 2019.

How Photovoltaic Cells Work: *Photovoltaic (PV) cells* are devices that convert solar energy into electricity. They are constructed of inorganic materials such as silicon. However, silicon is a poor electrical conductor, so minute amounts of other elements are added to increase conductivity and to determine polarity. Alternate layers of positive and negative polarity are stacked and contacts are added, forming a "silicon sandwich" that comprises a complete electrical circuit. When photons (particles of light) strike the cell's surface, some electrons are displaced from the negative layer and are forced to flow into the positive layer (**Figure 4-A**). This electron flow creates an electrical current that can then be used as a power source. Although the electrical output of one PV cell is small, many cells are wired together to form a module, and many modules are connected to form an array.

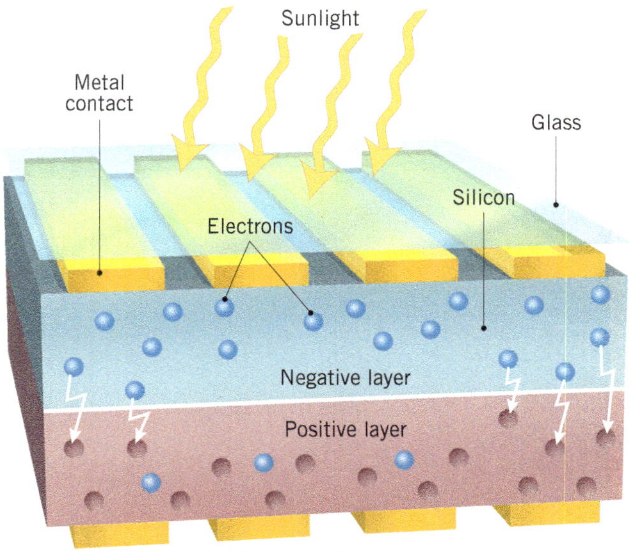

▲ **Figure 4-A** When sunlight strikes a photovoltaic cell, some electrons are displaced from the negative layer to the positive layer, generating an electrical current.

Advantages and Drawbacks of Solar Energy: Using photovoltaics for electrical generation offers several advantages. Compared with coal and nuclear power plants, a PV power plant has a low initial cost and requires little maintenance. No fuel is required! Perhaps one of the greatest advantages is the fact that solar power allows for decentralized electrical generation: a solar array does not have to be wired into a grid but instead can generate its own energy. Nevertheless, photovoltaic cells must first overcome several challenges, including the need for the improved efficiency of both PV cells and arrays.

The Geography of Solar Capacity: Solar capacity varies regionally, with cloud cover and latitude being the most significant limiting factors. Clouds, which typically have a high albedo (as high as 90 percent), reflect solar energy back into space, limiting the amount received by a solar array on the surface. Solar capacity also decreases as latitude increases because as distance from the equator increases, Earth's curved surface causes incoming solar radiation to be spread out over a larger area, thus diminishing beam intensity. For example, during the equinoxes, the equator receives a direct beam (that is, 90°), whereas at both 45° N and S latitudes, the Sun's rays strike at an angle of only 45°. This reduction in Sun angle translates to nearly a 71 percent reduction in energy on those days. Therefore, high-latitude regions are at a disadvantage in developing their solar potential.

Some countries that receive lower levels of solar radiation have increased their solar capacity by adopting policies that favor development of alternative energy. For example, Germany, which is located roughly at 50° N latitude, currently leads the world in PV solar capacity, with a 49-gigawatt peak—about 8 percent of Germany's total electricity. The United States added an additional 13.3 gigawatts in 2019, ranking fifth in the world, with a peak PV capacity of 77.7 gigawatts.

The Future of Solar Technology: Solar electricity generation is a low-carbon technology with the potential to grow to a very large scale. A massive expansion of global solar generating capacity is very likely an essential component of a workable strategy to mitigate the effects of climate change. In recent years, installed PV capacity has grown rapidly and great improvements have been made to technologies, price, and performance. For instance, concentrating solar power (CSP) plants use mirrors to concentrate the Sun's energy to drive traditional steam turbines or engines that create electricity. Other research (**Figure 4-B**) is underway to develop multispectral PV cells that can capture more of the electromagnetic spectrum—including infrared energy—and to incorporate organic elements that will integrate PV capability into thin, flexible, and even transparent materials. Such light-sensitive materials have the potential to allow everyday objects, such as paint, windows, curtains, and clothing, to generate electricity.

Questions

1. Is latitude the only factor that limits a country's ability to develop its PV capacity? Explain.
2. Why do you think PV capacity has increased so drastically in recent years, and how do you expect it to change over the next decade?

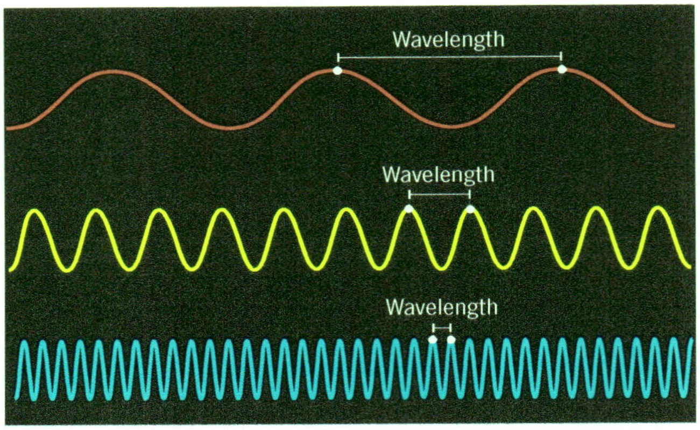

▲ **Figure 4-4** Electromagnetic waves can be of almost any length. The distance from one crest to the next is called the wavelength.

The Electromagnetic Spectrum: Electromagnetic radiation can be classified on the basis of *wavelength*—the distance between the crest of one wave and the crest of the next (Figure 4-4). Collectively, electromagnetic radiation of all wavelengths comprises what is called the **electromagnetic spectrum** (Figure 4-5). Electromagnetic radiation ranges from the exceedingly short wavelengths of gamma rays and X-rays (with some wavelengths less than one-billionth of a meter) to the exceedingly long wavelengths of television and radio waves (with some wavelengths measured in kilometers). For the physical geographer, however, only three areas or "bands" of the spectrum are important:

1. **Visible Light**. The human eye is sensitive to radiation wavelengths of only a fairly narrow band of the electromagnetic spectrum known as **visible light**—wavelengths between about 0.4 and 0.7 micrometers (μm; 1 micrometer = one-millionth of a meter). Visible light ranges from the shortest wavelength of radiation the human eye can sense—violet—through the progressively longer wavelengths of blue, green, yellow, orange, and finally red, the longest wavelength of radiation the human eye can see. This sequence of color is the same as what you see in a rainbow, from inner to outer (Figure 4-6). Although visible light makes up a narrow band of the electromagnetic spectrum, the peak intensity of electromagnetic radiation arriving from the Sun is in the visible portion of the spectrum, and approximately 47 percent of total solar energy arrives at Earth as visible light.
2. **Ultraviolet Radiation**. Wavelengths of radiation just shorter than the human eye can sense, from about 0.01 to 0.4 micrometers, make up the **ultraviolet (UV)** portion of the electromagnetic spectrum. The Sun is a prominent source of ultraviolet radiation (approximately 8 percent of the total solar energy). However, as we saw in Chapter 3, much of the UV radiation from the Sun is absorbed by the ozone layer. Thus, the shortest ultraviolet wavelengths do not reach Earth's surface, where they would cause considerable damage to most living organisms.
3. **Infrared Radiation**. Wavelengths of radiation just longer than the human eye can sense, between 0.7 and about 1000 micrometers (1 millimeter), make up the **infrared (IR)** portion of the electromagnetic spectrum. Infrared radiation ranges from the short or *near infrared* wavelengths emitted by the Sun to much longer wavelengths sometimes called **thermal infrared**. (Infrared "heat lamps" emit thermal infrared energy.) Approximately 45 percent of all solar energy comes to Earth as short infrared radiation, whereas radiation emitted by Earth is entirely thermal infrared.

Shortwave versus Longwave Radiation: Solar radiation is almost completely in the form of visible light, ultraviolet, and short infrared radiation, which as a group is referred to as **shortwave radiation** (Figure 4-7). Radiation emitted by Earth—or **terrestrial radiation**—is entirely in the thermal infrared portion of the spectrum and is referred to as **longwave radiation**. A wavelength of about 4 micrometers is considered the boundary on the spectrum separating longwave radiation from shortwave radiation. Thus, all terrestrial radiation is longwave radiation, whereas virtually all solar radiation is shortwave radiation. Shortwave

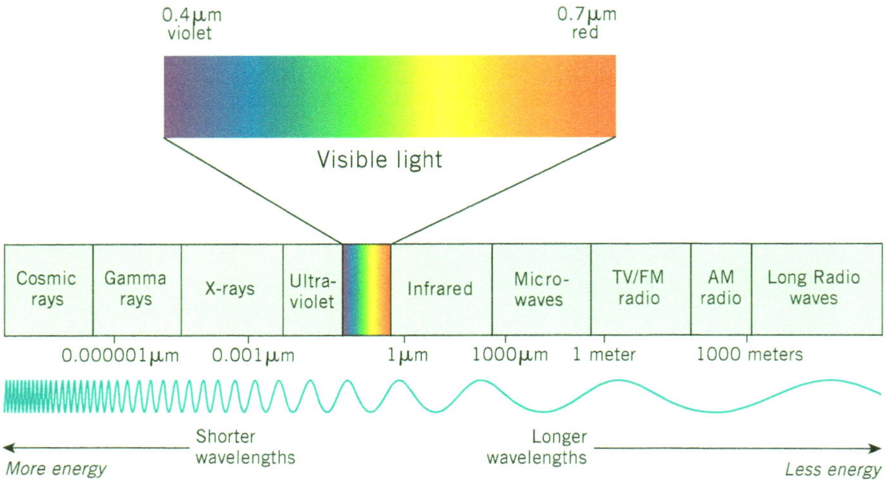

◀ **Figure 4-5** The electromagnetic spectrum.

▲ **Figure 4-6** The sequence of colors in a rainbow is determined by wavelength: from shortest to longest—violet, blue, green, yellow, orange, and finally red. A rainbow is seen in the sky opposite the Sun and forms when sunlight refracts (bends) as it enters a raindrop, reflects inside, and refracts again as it leaves the drop, separating the light into the colors of the spectrum.

radiation, such as UV, has more energy and does more damage than longwave (terrestrial) radiation. Sunscreen works by blocking the UV radiation that causes a sunburn.

LearningCheck 4-2 **Contrast shortwave radiation with longwave radiation.**

Insolation

The total **insolation** (*in*coming *sol*ar radi*ation*) received at the top of the atmosphere is believed to be constant when averaged over a year, although it may vary slightly over long periods of time with fluctuations in the Sun's temperature. This constant amount of incoming energy—referred to as the *solar constant*—is about 1370 watts per square meter (W/m^2). (Recall that 1 watt is equal to 1 joule per second.)

The entrance of insolation into the upper atmosphere is just the beginning of a complex series of events in the atmosphere and at Earth's surface. Some of the insolation is reflected off the atmosphere back out into space, where it is lost. The remaining insolation may pass through the atmosphere, where it can be transformed before or after reaching Earth's surface. We discuss this reception of solar energy—and the resulting energy cascade that ultimately warms Earth's surface and atmosphere—after we define a set of important concepts.

Basic Warming and Cooling Processes in the Atmosphere

Before looking at the events that occur after energy travels from the Sun to Earth, we need to examine the physical processes involved in the transfer of energy. Our goal is to provide practical explanations of the most important processes associated with the warming and cooling of the atmosphere. As such, in some cases we limit our discussion of a process to the aspects of direct importance to meteorology.

Radiation

Radiation—or **emission**—is the process by which electromagnetic energy is emitted from an object. So the term *radiation* refers to both the emission and the flow of electromagnetic energy. All objects emit electromagnetic energy, but in general, the hotter the object, the more intense its radiation. (Radiation *intensity* is commonly described in W/m^2—the amount of energy emitted or received in a given period of time in a given area.) Because the Sun is much

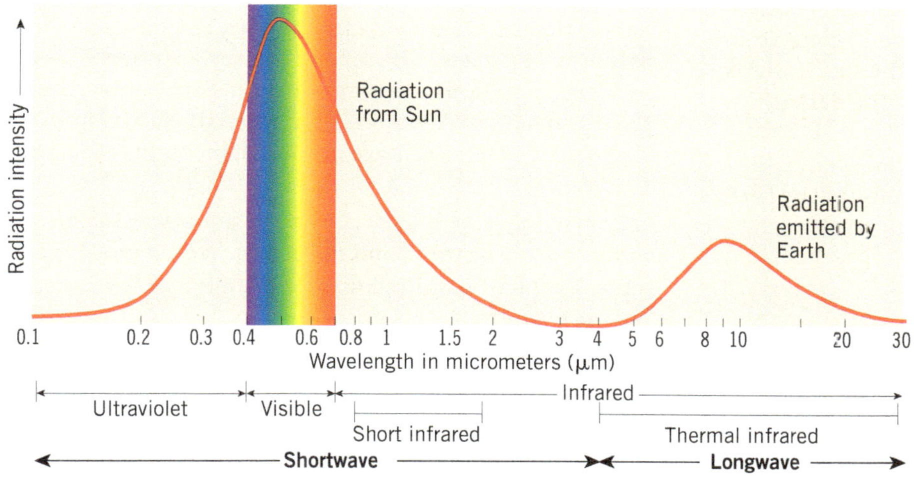

◀ **Figure 4-7** Solar versus terrestrial radiation. Shortwave radiation from the Sun consists of ultraviolet wavelengths (about 8 percent of total solar energy), visible light (about 47 percent of the total), and short infrared wavelengths (about 45 percent of the total). Longwave radiation emitted by Earth consists entirely of thermal infrared wavelengths. The wavelength scale (shown here in micrometers [μm]) is logarithmic—it compresses the scale dramatically as wavelengths increase to the right. If a linear wavelength scale had been used, the peak of radiation emitted by Earth would be 50 centimeters (20 inches) off the page!

hotter than Earth, it emits about two billion times more energy than Earth.

In addition, the hotter the object, the shorter the wavelengths of that radiation. Hot bodies radiate mostly short wavelengths of radiation, whereas cooler bodies radiate mostly long wavelengths. The Sun is the ultimate "hot" body of our solar system, so nearly all of its radiation is in the shortwave portion of the electromagnetic spectrum. Earth is cooler than the Sun, so Earth emits longer wavelengths of radiant energy (thermal infrared).

Temperature, however, is not the only control of radiation intensity. Objects at the same temperature may vary considerably in their radiating efficiency. A body that emits the maximum possible amount of radiation at all wavelengths is called a *blackbody radiator*. Both the Sun and Earth function very nearly as blackbodies—that is, as perfect radiators. They radiate with almost 100 percent efficiency for their respective temperatures. The atmosphere, on the other hand, is not as efficient a radiator as either the Sun or Earth's surface.

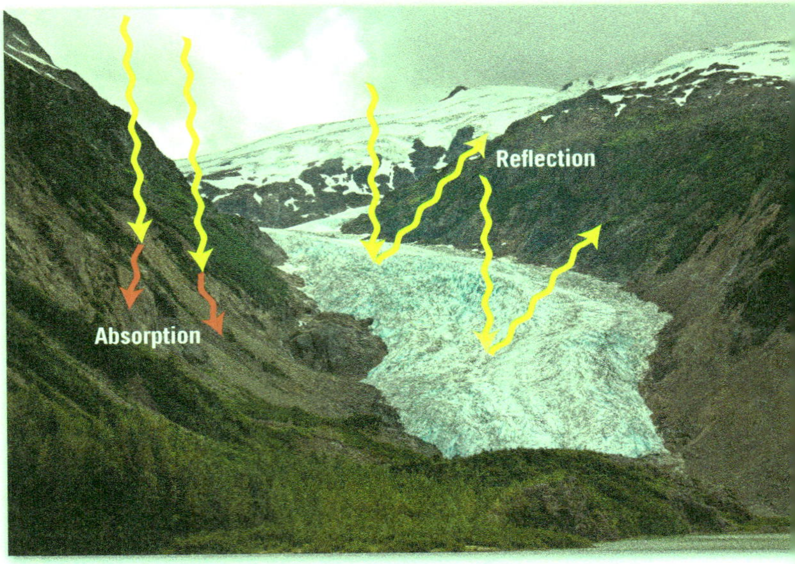

▲ Figure 4-8 Most solar radiation that reaches Earth's surface is either absorbed or reflected. The white snow of the Salmon Glacier reflects much of the solar energy, whereas the dark soil surrounding the glacier absorbs energy, thereby melting the glacier's edges. As the glacier melts, it exposes more soil, which melts more glacier. This is one consequence of climate change.

Absorption

Electromagnetic waves striking an object may be assimilated by that object; this process is called **absorption**. Different materials have different absorptive capabilities, with the variations depending in part on the wavelength of radiation involved. Although it is a great simplification, when electromagnetic waves strike a material, the electrons in atoms or even whole molecules in that material may be forced into vibration by the frequency of the incoming electromagnetic waves. (Since all wavelengths of electromagnetic radiation travel at the same speed [the speed of light], short wavelengths of radiation arrive with a higher frequency than do longer wavelengths.) The increased vibrations result in an increase in the internal energy of the absorbing material, and that leads to an increase in temperature—a typical response to the absorption of electromagnetic radiation. This also explains why UV radiation damages skin cells and can give you a sunburn.

In general, an object that is a good radiator is also a good absorber, and a poor radiator is a poor absorber. Mineral materials (rock, soil) are generally excellent absorbers; snow and ice are poor absorbers; water surfaces vary in their absorbing efficiency. One important distinction concerns color. Dark-colored surfaces are much more efficient absorbers of radiation in the visible portion of the spectrum than light-colored surfaces (as exemplified by the dark soil surrounding Salmon Glacier in Figure 4-8).

As we will see, both water vapor and carbon dioxide are efficient absorbers of longwave radiation emitted by Earth's surface, whereas nitrogen, the most abundant gas in the atmosphere, is not.

Reflection

If incoming solar radiation is not absorbed, it may be reflected. **Reflection** is the ability of an object to repel ("bounce back") electromagnetic waves that strike it. When insolation is reflected by the atmosphere or the surface of Earth, it is deflected back to space at the same angle and initial wavelength with which it arrived. A mirror, for example, is designed to be highly efficient in reflecting visible light—reflecting 90 percent or more of the incoming light.

In our context, reflection is the opposite of absorption. If the wave is reflected, it cannot be absorbed. Hence, an object that is a good absorber is a poor reflector, and vice versa (see Figure 4-8). A simple example of this principle is the existence of unmelted snow on a sunny day. Although the air temperature may be well above freezing, the snow does not melt rapidly because its white surface reflects rather than absorbs most of the solar energy that strikes it. Fresh snow is an excellent reflector, which makes you want to grab your sunglasses.

Albedo is the overall reflectivity of an object or surface, usually described as a percentage; the higher the albedo, the greater the amount of radiation reflected. Snow, for example, has a very high albedo (as much as 95 percent), whereas a dark surface, such as dense forest cover, can have an albedo as low as 14 percent. The texture of a surface affects albedo, as does the angle of incoming radiation. For example, overall the ocean has very low albedo, but when the sun is low in the sky, water can reflect a large portion of incoming radiation.

LearningCheck 4-3 **Compare the processes of radiation, absorption, and reflection. Would you expect a black asphalt parking lot to have a high or low albedo? Why?**

Scattering

Gas molecules and particulate matter in the air can deflect light waves and redirect them in a type of reflection known

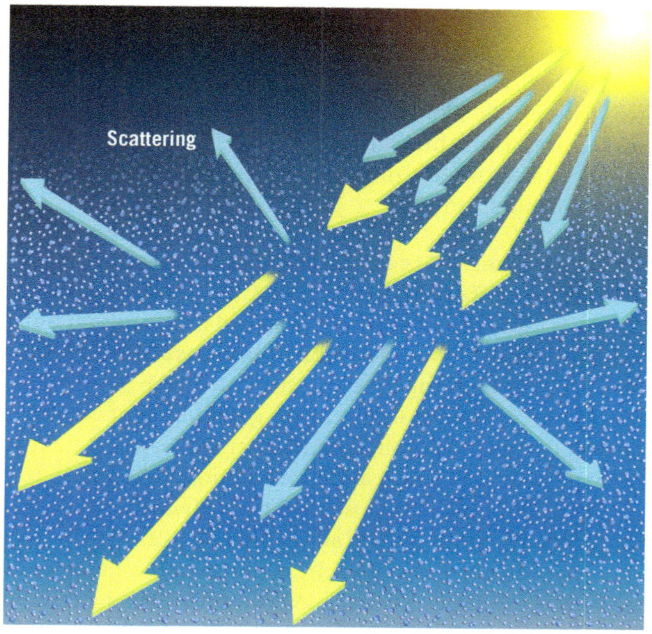

▲ **Figure 4-9** Gas molecules and impurities in the atmosphere scatter light waves. Some waves are scattered into space and therefore lost to Earth; others are scattered but continue through the atmosphere in altered directions. There is greater scattering of blue light than the longer wavelengths of visible light, resulting in a blue sky.

as **scattering** (**Figure 4-9**). This deflection involves a change in the direction of the light wave but no change in wavelength. Some of the waves are backscattered into space and thus are lost to Earth, but many of them continue through the atmosphere in altered but random directions, eventually striking the surface as *diffuse radiation*.

The amount of scattering that takes place depends on the wavelength of the light as well as on the size, shape, and composition of the molecule or particulate. In general, shorter wavelengths are more readily scattered than longer wavelengths by the gases in the atmosphere. One prominent kind of scattering, known as *Rayleigh scattering*, takes place when the shortest wavelengths of visible light—violet and blue—are scattered more easily in all directions by the gas molecules in the atmosphere than are the longer wavelengths of visible light—orange and red. This is why on a clear day the sky is blue. The sky is blue and not violet because of the greater prevalence of blue wavelengths in solar radiation and because our eyes are less sensitive to violet light. Were blue light not scattered by the atmosphere, the sky would appear black. When the Sun is low in the sky, the light has passed through so much atmosphere that nearly all of the blue wavelengths have been scattered away, leaving only the longest wavelengths of visible light, orange and red—the dominant colors of light we see at sunrise and sunset (**Figure 4-10**).

When the atmosphere contains great quantities of fairly large particles, such as suspended aerosols, all wavelengths of visible light are more equally scattered (a process known as *Mie scattering*). That makes the sky look gray rather than blue. It's also why the fog in the distance of the photo on the chapter opening looks gray.

In terms of atmospheric warming, one of the consequences of scattering—especially Rayleigh scattering—is that the intensity of solar radiation striking the surface of Earth is diminished, particularly when the Sun is low in the sky, because scattering redirects a portion of the insolation back out to space (**Figure 4-11**).

Transmission

If incoming solar radiation is neither absorbed nor reflected, it may pass through a surface or object. **Transmission** is the process whereby electromagnetic waves pass completely through a medium, as when light waves are transmitted through a pane of clear, colorless glass. Mediums vary considerably in their capacity to transmit electromagnetic waves.

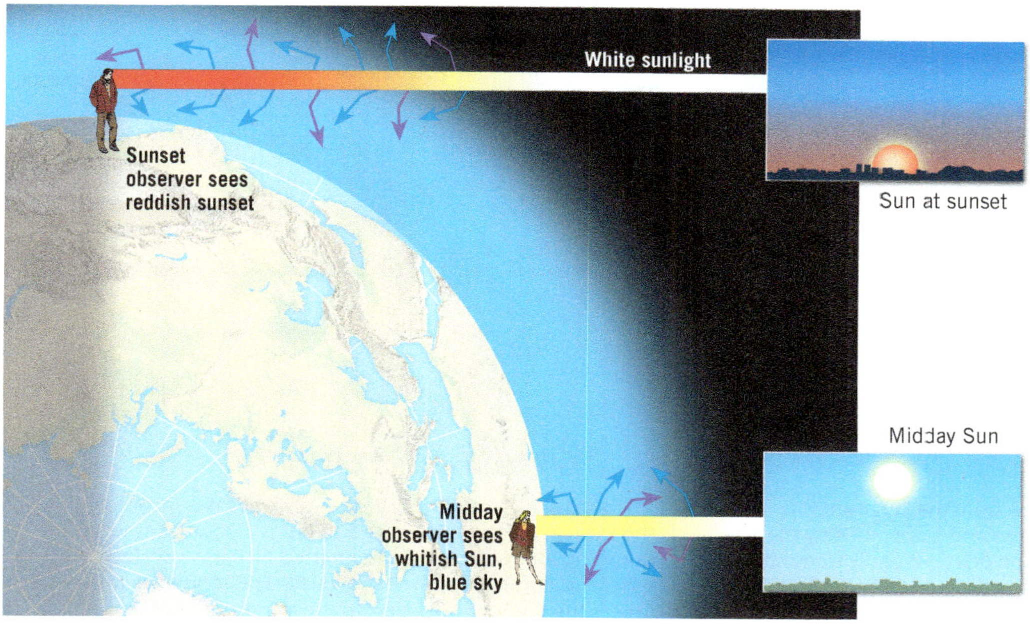

◄ **Figure 4-10** Short wavelengths of visible light (blue and violet) are scattered more easily than longer wavelengths. Consequently, when the Sun is overhead on a clear day, one can see blue sky in all directions because blue light was selectively scattered. At sunrise or sunset, however, the path that light must take through the atmosphere is much longer, with the result that most of the blue light is scattered out before the light waves reach Earth's surface. Thus, the Sun appears reddish.

▲ Figure 4-11 The predominant colors of a sunset are orange and red due to the scattering away of blue light, as in this scene of the Sun setting behind the Golden Gate Bridge in San Francisco.

Earth materials, for example, are typically opaque and therefore very poor transmitters of insolation; sunlight is absorbed at the surface of rock or soil and does not penetrate. Water, on the other hand, transmits sunlight well: even in very murky water, light penetrates some distance below the surface. In clear water, sunlight may illuminate to considerable depths.

The transmission ability of a medium generally depends on the wavelength of radiation. For example, glass has high transmission of shortwave radiation but not of longwave radiation. Temperature increases inside a closed automobile parked in the Sun in part because shortwave radiation transmits through the window glass and is absorbed by the upholstery, which gets warmer. The longwave radiation then emitted by the interior of the car does not readily transmit back through the glass, causing the inside of the car to warm up (Figure 4-12). This is commonly—but incorrectly—called the *greenhouse effect*.

As it turns out, actual greenhouses are not warmed simply by the greenhouse effect; the lack of mixing of warm inside air with cooler outside air (convection and advection) also plays a role. Nonetheless, the term *greenhouse effect* is used to describe the warming of the lower atmosphere caused by the differential transmission of shortwave and longwave radiation.

The Atmospheric Greenhouse Effect: The atmospheric **greenhouse effect** occurs because the atmosphere readily transmits incoming shortwave radiation from the Sun but a number of gases, known as **greenhouse gases**, do not easily transmit outgoing longwave terrestrial radiation (Figure 4-13) back to space. The most important greenhouse gas is water vapor, followed by carbon dioxide. Many other gases, such as methane, also play a role, as do some kinds of clouds.

In the simplest terms, incoming shortwave solar radiation transmits through the atmosphere to Earth's surface, where this energy is absorbed, increasing the temperature of the surface. However, the longwave radiation emitted by Earth's surface is inhibited from transmitting back through the atmosphere by the greenhouse gases. Much of this outgoing terrestrial radiation is absorbed by greenhouse gases and clouds. Some of the absorbed longwave radiation is then reradiated back toward the surface, warming the lower layers of the atmosphere.

The atmospheric greenhouse effect is one of the most important warming processes in the troposphere. The greenhouse effect keeps Earth's surface and lower troposphere much warmer than would be the case if there were no atmosphere. Without this greenhouse effect, the average temperature of Earth would be about −18°C (0°F) rather than the present average of 15°C (59°F).

Although the ongoing, natural greenhouse effect in the atmosphere makes life as we know it possible, over the past century or so a significant increase in greenhouse gas concentration—especially carbon dioxide—has been measured.

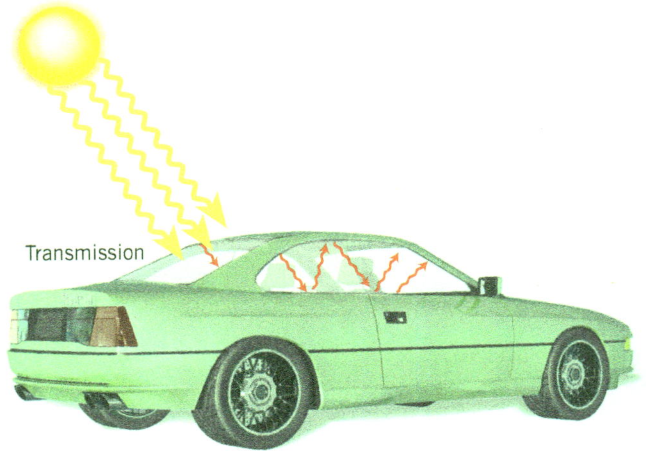

▲ Figure 4-12 Shortwave radiation from the Sun transmits through the glass of the car windows and is absorbed by the car interior. Because longwave radiation does not easily transmit back out through the glass and the warmed air inside doesn't mix with air outside the car, the temperature inside the car increases.

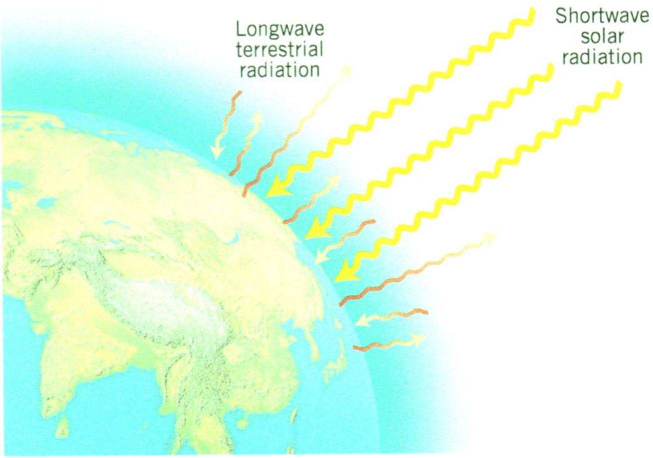

▲ Figure 4-13 The atmosphere easily transmits incoming shortwave radiation from the Sun but is a poor transmitter of outgoing longwave radiation emitted by Earth's surface. This differential transmission of shortwave and longwave radiation causes the greenhouse effect in the atmosphere.

This increase in atmospheric carbon dioxide is closely associated with human activity, especially the burning of fossil fuels such as petroleum and coal. (Carbon dioxide is one of the by-products of combustion.) The increase in greenhouse gas concentration has been accompanied by a slight yet significant increase in average global temperature. This important issue, commonly referred to as *global warming*, is addressed in more detail at the end of this chapter after we discuss atmospheric warming processes and patterns.

LearningCheck 4-4 **Explain the atmospheric greenhouse effect. What are the two most important greenhouse gases?**

Conduction

The transfer of heat from one molecule to another without changes in their relative positions is called **conduction**. This process enables energy to be transferred from one part of a stationary body to another or from one object to a second object when the two are in contact.

Conduction occurs through molecular collisions, as Figure 4-14 illustrates. A "cool" molecule becomes increasingly agitated as a "hotter" molecule collides with it, transferring some kinetic energy to it. In this manner, the energy is passed from one place to another. When two molecules of unequal temperatures are in contact, energy transfers from the hotter molecule to the cooler one until they attain the same temperature.

The ability of different substances to conduct heat is quite variable. For example, most metals are excellent conductors of heat. If you pour hot coffee into a metal cup, the warmth of the coffee is quickly conducted throughout the metal, which burns your lips. In contrast, hot coffee poured into a ceramic cup warms the cup very slowly because such earthy material is a poor conductor.

Earth's land surface warms rapidly during the day because it is a good absorber of incoming shortwave radiation, and some of that warmth is transferred away from the surface by conduction. Only a small portion is conducted deeper underground because Earth materials are not good conductors. Another portion of this absorbed energy is transferred to the lowest part of the atmosphere by conduction from the ground surface. Air, however, is a poor conductor, so only the air layer touching the ground (perhaps just the lowest few millimeters) is warmed very much; physical movement of the air is required to spread the warmth around. In contrast, when the ground surface is very cold, heat can transfer from the air to the ground through conduction, chilling the air above. Moist air is a slightly more efficient conductor than dry air. If you are outdoors on a winter day, you will stay warmer if there is less moisture in the air to conduct heat away from your body.

Convection

In the process of **convection**, energy is transferred from one point to another by the predominately vertical circulation of a fluid, such as air or water. Convection involves movement of the warmed molecules from one place to another. (In *convection*, molecules move from one place to another; in *conduction*, they vibrate back and forth and undergo molecular collisions.)

A convective pattern frequently develops in the atmosphere (Figure 4-15). For example, hot surface conditions may warm a parcel of air relative to the surrounding air, and the warm air parcel will rise. The warm air expands and moves upward in the direction of lower pressure. Cooler surrounding air then moves in from the sides, and air from above sinks down to replace it, thus establishing a convective circulation system, also called a **convection cell**. The prominent elements of the system are an updraft of warm air and a downdraft of air after it has cooled. Convection is common in each hemisphere during its summer and throughout the year in the tropics.

Advection

When the dominant direction of energy transfer in a moving fluid is horizontal (sideways), the term **advection** is applied. In the atmosphere, wind may transfer warm or cool air

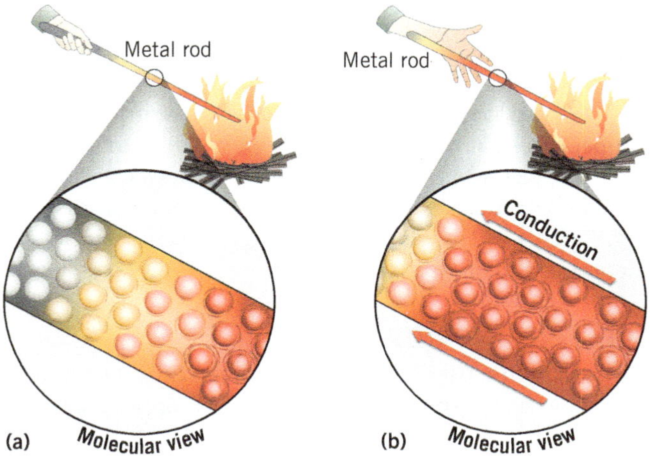

▲ Figure 4-14 Energy is conducted from one place to another by molecular agitation. (a) One end of the metal rod is in the flame and so becomes hot. (b) This energy then conducts the length of the rod and will burn the hand of the rod holder.

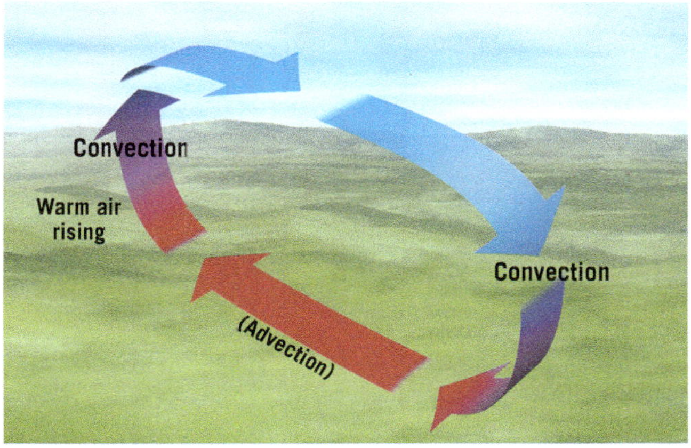

▲ Figure 4-15 Convection in the atmosphere. Air warmed by the surface rises and cooler air flows in to replace it, producing a convection cell. The horizontal movement of air in the convection cell is referred to as advection.

horizontally from one place to another through the process of advection. As we will learn in Chapter 5, some wind systems develop as part of large atmospheric convection cells; the horizontal component of air movement within such a convection cell is properly called advection (see Figure 4-15).

LearningCheck 4-5 Contrast the energy transfer processes of conduction, convection, and advection.

Adiabatic Cooling and Warming

Whenever air ascends or descends, its temperature changes. This invariable result of vertical air movement is due to the change in pressure. When air rises, it expands because there is less air above it, so less pressure is exerted on it (Figure 4-16). When air sinks, it is compressed because there is more air above it, so more pressure is exerted on it.

Expansion—Adiabatic Cooling: The expansion that occurs in rising air is a cooling process even though no energy is lost. As air rises and expands, the molecules spread through a greater volume of space. The "work" done by the molecules during expansion reduces their average kinetic energy, so the temperature decreases. This is called **adiabatic cooling**—cooling by expansion (*adiabatic* means without the gain or loss of energy). In the atmosphere, any time air rises, it cools adiabatically.

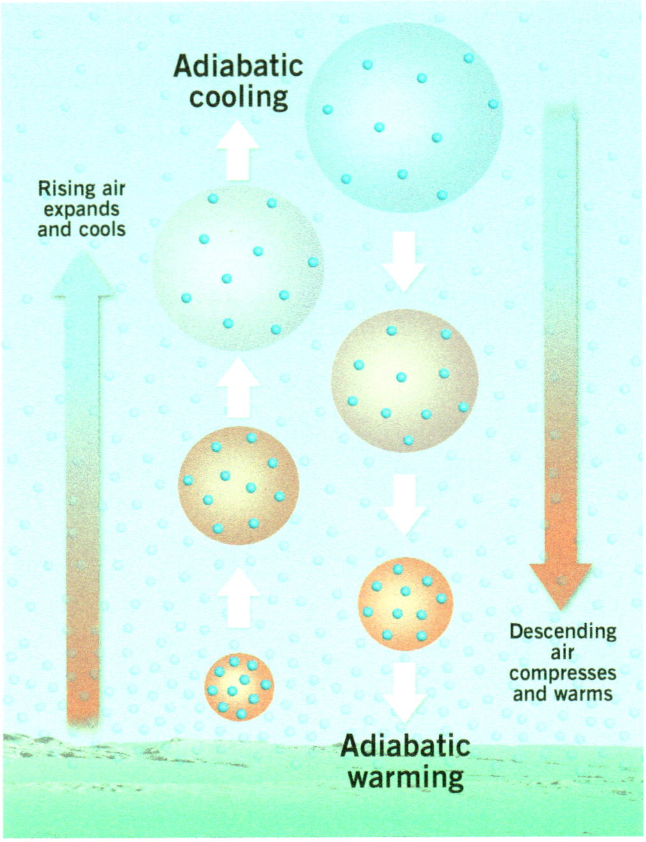

▲ Figure 4-16 Rising air (represented by an isolated "parcel" of air) expands and cools adiabatically, whereas descending air compresses and warms adiabatically. No energy is transferred in either process.

Compression—Adiabatic Warming: Conversely, when air descends, it becomes warmer. The descent causes compression as the air comes under increasing pressure. The work done on the molecules by compression increases their average kinetic energy, so the temperature increases even though no energy was added from external sources. This is called **adiabatic warming**—warming by compression. In the atmosphere, any time air descends, it warms adiabatically.

As we will see in Chapter 6, adiabatic cooling of rising air is one of the most important processes involved in cloud development and precipitation, whereas the adiabatic warming of descending air results in cloud dissipation.

LearningCheck 4-6 What happens to the temperature of air as it rises, and what happens to the temperature of air as it descends? Why does this occur?

Latent Heat

The physical state of water in the atmosphere frequently changes—ice changes to liquid water, liquid water changes to water vapor, and so forth. Any phase change involves an exchange of energy known as **latent heat** (*latent* is from the Latin, which means "lying hidden"). The two most common phase changes are **evaporation**, in which liquid water is converted to gaseous water vapor, and **condensation**, in which water vapor is converted to liquid water. During evaporation, latent heat is absorbed, or "stored," so evaporation cools the environment. On the other hand, during condensation, latent heat is released, so condensation warms the environment. We have all experienced latent heat. Think about the last time you were out by the pool and evaporation of water from your skin made you feel cold.

As we explore more fully in Chapters 6 and 7, a great deal of energy is transferred from one place to another in the atmosphere through the movement and phase changes of water vapor. Energy that is stored in one location through evaporation can be released in another location far away. We will also see that many storms, such as hurricanes, are fueled by the release of latent heat during condensation.

Earth's Solar Radiation Budget

We now turn to the specifics of atmospheric warming. What happens to solar radiation that enters Earth's atmosphere? How is it received and distributed? How does electromagnetic radiation warm the atmosphere? We begin by discussing Earth's solar radiation "budget"—the balance of incoming and outgoing radiation.

Long-Term Energy Balance

In the long run, there is a balance between the total amount of energy received by Earth and its atmosphere as insolation on one hand, and the total amount of energy returned to space on the other (Figure 4-17). (As we suggested earlier, humans are altering the energy balance of the atmosphere slightly through greenhouse gas emissions. For the purpose of understanding atmospheric warming processes, we will

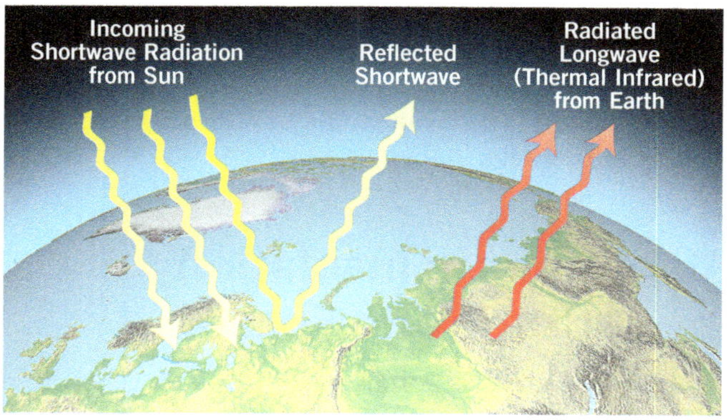

▲ **Figure 4-17** The big picture: Earth's energy budget in simplified form. Incoming shortwave radiation and outgoing longwave radiation are in a long-term balance.

ignore that for the moment.) Although there is an overall long-term balance between incoming and outgoing radiation, the details of the energy exchanges between Earth's surface and atmosphere are important for understanding basic weather processes.

Global Energy Budget

The annual balance between incoming and outgoing radiation is the *global energy budget*. We can illustrate it by using 100 "units" of energy to represent total insolation (100 percent of insolation) received at the outer edge of the atmosphere and tracing its dispersal (**Figure 4-18**). The values shown are approximate annual averages for the entire globe and do not apply to any specific location. Incoming energy can be broken down in to three parts: radiation that is reflected back to space, radiation that is absorbed by the atmosphere, and radiation that is absorbed by the ground.

Radiation Loss from Reflection: Most of the incoming solar radiation does not warm the atmosphere directly. About 31 units of total insolation are reflected (or scattered) back into space by the atmosphere and the surface. The *albedo* of Earth, therefore, is about 31 percent. Thus, nearly one-third of incoming solar radiation has no effect on atmospheric processes.

Atmospheric Absorption: Less than one-quarter of incoming solar radiation is absorbed directly by the

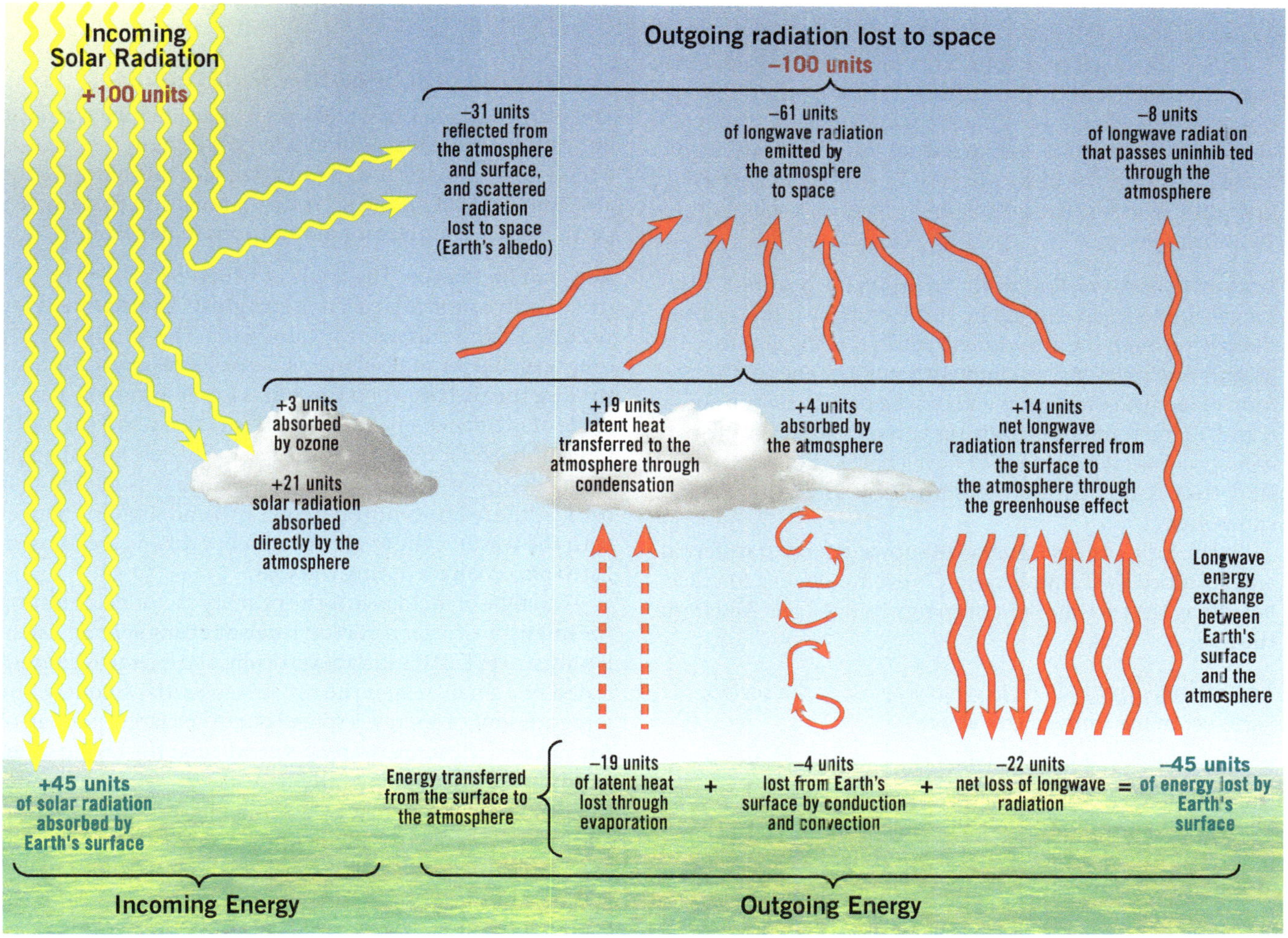

▲ **Figure 4-18** The generalized energy budget of Earth and its atmosphere, shown with 100 "units" of energy arriving from the Sun balanced by 100 units of energy eventually lost to space. About one-third of incoming shortwave radiation is reflected and scattered away, and nearly half of incoming radiation simply transmits through the atmosphere and is absorbed by the surface. The surface then warms the atmosphere above.

Animation
Atmospheric Energy Balance
https://media.pearsoncmg.com/ph/esm/esm_mcknight_physgeo_13/media/

atmosphere. About 3 units of radiation (in the ultraviolet portion of the spectrum) are absorbed by ozone and so warm the ozone layer. Another 21 units are absorbed by gases and clouds as incoming radiation passes through the rest of the atmosphere.

Surface Absorption: Nearly half of incoming radiation—about 45 units of energy—simply transmits through the atmosphere to Earth's surface, where it is absorbed, warming the surface. The warmed surface then in turn transfers energy to the atmosphere above in a number of ways.

Surface-to-Atmosphere Energy Transfer: Of the 45 units of incoming radiation that reach the ground, about 4 units of energy are conducted from Earth's surface back into the atmosphere, where it is dispersed by convection. Energy is also transferred from the surface to the atmosphere through the transport of latent heat in water vapor. About three-fourths of all sunshine falls on a water surface when it reaches Earth. Much of this energy is utilized in evaporating water from oceans, lakes, and other bodies of water—about 19 units of energy pass into the atmosphere as latent heat stored in water vapor and are eventually released through condensation.

Greenhouse gases absorb large amounts of longwave radiation emitted by the surface. In turn, they radiate much of this energy back to the surface where it may be absorbed—and then reemitted as longwave radiation again. Through the absorption of terrestrial radiation by greenhouse gases, the atmosphere receives a net gain of 14 units of energy.

Loss of Longwave Radiation to Space: A portion of the longwave radiation emitted by Earth's surface is transmitted directly through the atmosphere without being absorbed by greenhouse gases. Approximately 8 units of energy—in the form of longwave radiation with wavelengths between about 8 and 12 micrometers—transmit through what is called an *atmospheric window*, a range of wavelengths of infrared radiation that is not strongly absorbed by any atmospheric component.

In the long run, as Figure 4-18 shows, energy transferred and absorbed by the atmosphere is eventually lost to space, balancing the total amount of energy initially received from the Sun.

LearningCheck 4-7 **In what ways does the surface of Earth warm the troposphere above?**

Consequences of Indirect Warming of Atmosphere: For the most part, then, the atmosphere is warmed indirectly by the Sun: the Sun warms the surface, and the surface, in turn, warms the air above. This complicated sequence of atmospheric warming has many ramifications. Because the atmosphere is warmed mostly from below rather than from above, in the troposphere cold air overlies warm air. This "unstable" situation (explored further in Chapter 6) creates an environment of almost constant convective activity and vertical mixing. If the atmosphere were warmed directly by the Sun, resulting in warm air at the top of the atmosphere and cold air near Earth's surface, the situation would be stable, essentially without vertical air movements. The result would be a troposphere that is largely motionless.

Variations in Insolation by Latitude and Season

The energy budget we just discussed is broadly generalized. Many latitudinal and vertical imbalances are in this budget, and these are among the most fundamental causes of weather and climate variations.

In essence, we can trace a causal continuum wherein insolation absorption differences lead to temperature differences that lead to air-density differences that lead to pressure differences that lead to wind differences that often lead to moisture differences. It has already been noted that world weather and climate differences are fundamentally caused by the unequal warming of Earth and its atmosphere. This unequal warming is the result of latitudinal and seasonal variations in insolation.

Latitudinal and Seasonal Differences

There are only a few basic reasons for the unequal warming by latitude. These reasons include variations in the angle at which solar radiation strikes Earth, the influence of the atmosphere itself on the intensity of radiation that reaches Earth's surface, and seasonal variations in day length.

Angle of Incidence: The angle at which rays from the Sun strike Earth's surface is called the **angle of incidence**. A ray striking Earth's surface vertically, when the Sun is directly overhead, has an angle of incidence of 90° (Figure 4-19); a ray striking the surface when the Sun is lower in the sky has an angle of incidence smaller than 90°; and a ray striking Earth tangent to the surface (as at sunrise or sunset) has an angle of incidence of 0°. Because Earth's surface is curved and because the relationship between Earth and the Sun changes with the seasons, the angle of incidence for any location on Earth also changes during the year.

The angle of incidence is the primary factor determining the intensity of solar radiation received at any spot on Earth. If a ray strikes Earth's surface vertically, the energy is concentrated in a small area; if the ray strikes Earth obliquely, the same amount of energy is spread out over a larger portion of the surface. The more nearly perpendicular the ray (in other words, the closer angle of incidence is to 90°), the smaller the surface area warmed by a given amount of insolation and the more effective the heating. Averaged over the year as a whole, the insolation received by high-latitude regions is much less intense than that received by tropical areas (Figure 4-20).

Effect of the Atmosphere: Insolation does not travel through the atmosphere unaffected—it encounters various obstructions in the atmosphere. We have already noted that clouds, particulates, and gas molecules in the atmosphere

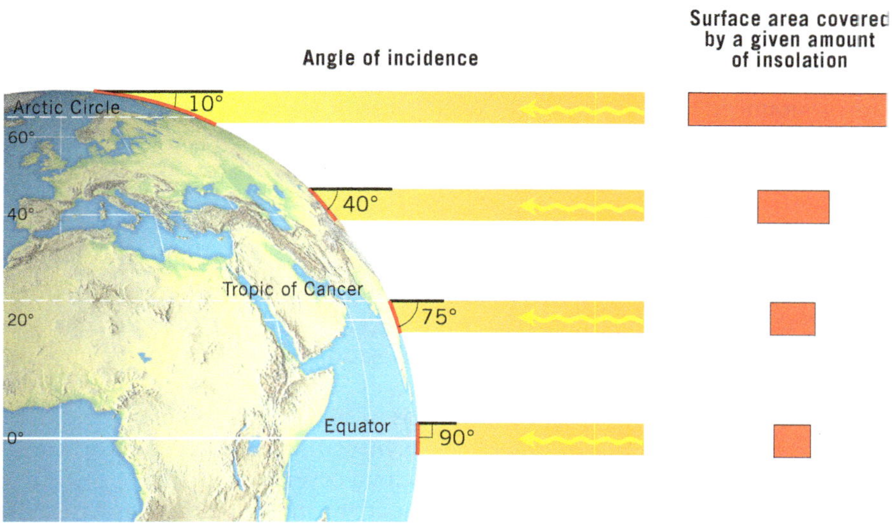

◀ **Figure 4-19** The angle at which solar rays hit Earth's surface varies with latitude. The higher the angle, the more concentrated the energy and therefore the more effective the warming. The day represented here is the March/September equinox.

may absorb, reflect, or scatter incoming solar radiation, reducing the intensity of this energy by the time it reaches Earth's surface. On average, sunlight received at Earth's surface is only about half as strong as it is at the top of Earth's atmosphere.

The attenuation (weakening) of radiation that passes through the atmosphere varies from time to time and from place to place depending on two factors: the amount of atmosphere through which the radiation passes and the transparency of the air. The distance a ray of sunlight travels through the atmosphere (commonly referred to as *path length*) is determined by the angle of incidence (Figure 4-21). A high-angle ray traverses a shorter course through the atmosphere than a low-angle one. A tangent ray (one with an incidence angle of 0°) must pass through nearly 20 times as much atmosphere as a vertical ray (one striking Earth at an angle of 90°).

The effect of atmospheric obstruction tends to reinforce the pattern of solar energy distribution at Earth's surface established by the angle of incidence. In high latitudes, the Sun has a lower angle of incidence and a greater path length through the atmosphere than in the tropics. Thus, there are smaller losses of energy in the tropical atmosphere than in the polar atmosphere. However, as we will see, this general pattern is complicated by patterns of cloud cover.

Day Length: The duration of sunlight is another important factor in explaining latitudinal differences in warming. Longer days allow more insolation to be received and thus more solar energy to be absorbed. In tropical regions, this factor is relatively unimportant because the number of

▼ **Figure 4-20** The Sun is low in the sky even during the summer in high latitudes. This time-lapse photograph shows the track of the "midnight Sun" on the day of the June solstice in Alaska.

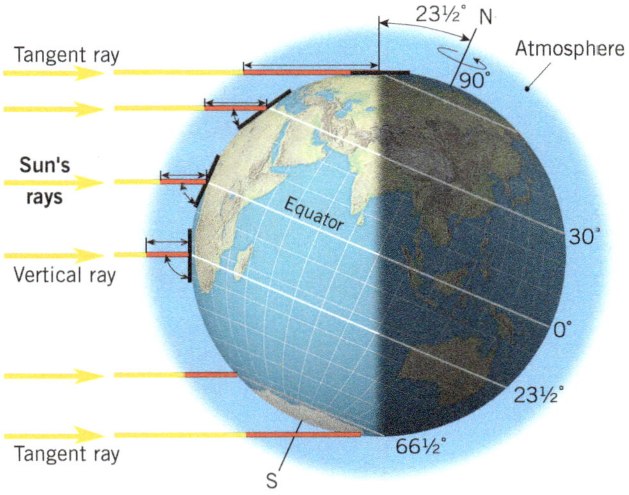

▲ **Figure 4-21** Atmospheric obstruction of sunlight. Low-angle rays (such as in the high latitudes) must pass through more atmosphere than high-angle rays do. The longer the path length (red line) through the atmosphere, the more the intensity of radiation is reduced through reflection, scattering, and absorption. The day represented here is the December solstice.

hours of daylight does not vary significantly from one month to another; at the equator, daylight and darkness are equal in length (12 hours each) every day of the year. In middle and high latitudes, however, there are pronounced seasonal variations in day length. The conspicuous buildup of warmth in summer in these regions is largely a consequence of the long hours of daylight, and the winter cold is a manifestation of limited insolation being received because of the short days.

LearningCheck 4-8 **Why does more solar energy reach the surface when the Sun is high in the sky than when it is low in the sky?**

Latitudinal Radiation Balance

As the vertical rays of the Sun shift northward and southward across the equator during the course of the year, the belt of maximum solar energy swings back and forth through the tropics. Thus, in the low latitudes, between about 38° N and 38° S, there is an energy surplus, with more incoming than outgoing radiation. In the latitudes north and south of these two parallels, there is an energy deficit, with more energy loss than gain. The surplus of energy in low latitudes is directly related to the consistently high angle of incidence, and the energy deficit in high latitudes is associated with low angles.

Figure 4-22 shows the distribution of average daily insolation at the surface around the world for December and June. The maps show the average daily insolation received in watts per square meter (W/m^2; 1 watt = 1 joule per second). The variations are largely latitudinal, as is to be expected. The major interruptions to the simple latitudinal pattern are due to the presence or absence of frequent cloud cover, from which insolation is reflected, diffused, and scattered. For example, notice that insolation in the United States in June is greatest in the Southwest, where clouds are consistently sparse. It is for this reason that Death Valley is one of the hottest places on Earth. Insolation is least in the Northwest and Northeast, where cloud cover is frequent.

Despite the variable pattern shown on the maps, there is a long-term balance between incoming and outgoing radiation for the Earth–atmosphere complex as a whole; in other words, the net radiation balance for Earth is zero. The mechanisms for exchanging energy between the surplus and deficit regions involve the general circulation patterns of the atmosphere and oceans, which are discussed later in the book.

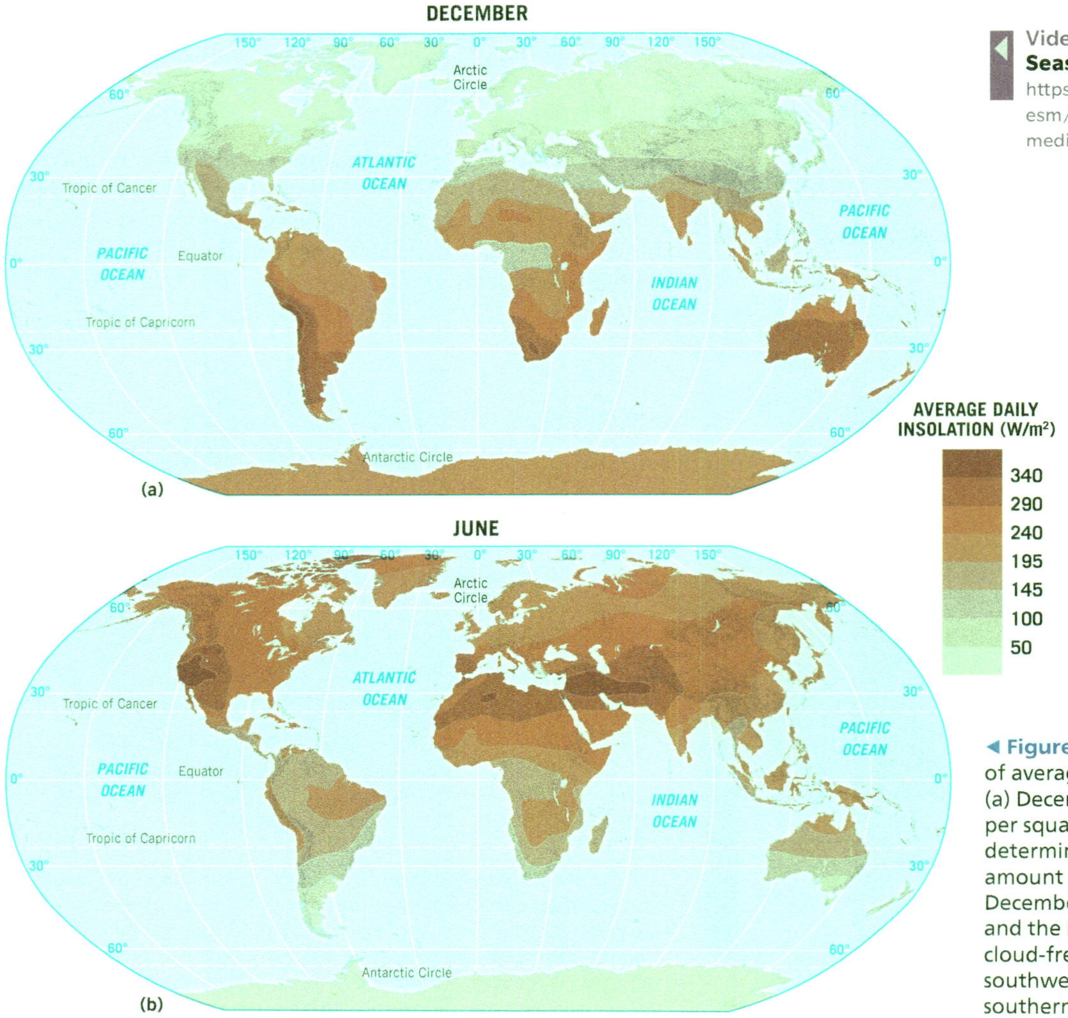

Video
Seasonal Radiation Patterns
https://media.pearsoncmg.com/ph/esm/esm_mcknight_physgeo_13/media/

◀ **Figure 4-22** World distribution of average daily insolation for (a) December and (b) June in watts per square meter. The pattern is determined mainly by latitude and amount of cloudiness. Note the low December levels in northern areas and the high levels in June in such cloud-free desert areas as the southwestern United States and southern Eurasia.

Land and Water Temperature Contrasts

As we've seen, the atmosphere is warmed mainly by energy reradiated and transferred from Earth's surface rather than by energy received directly from the Sun. So the warming of Earth's surface is a primary control of the warming of the air above it. To understand variations in air temperatures, it is useful to understand how different kinds of surfaces react to solar energy. There is considerable variation in the absorbing and reflecting capabilities of the almost limitless kinds of surfaces found on Earth—soil, water, grass, trees, cement, rooftops, and so forth. Their varying receptivity to insolation in turn causes differences in the temperature of the overlying air.

Although Earth has many kinds of surfaces, by far the most significant contrasts are those between land and water. As a generalization: land warms and cools faster and to a greater extent than water. You may have noticed this if you've been to the beach on a summer afternoon; the land will be much warmer than the water.

Warming of Land and Water

A land surface warms up more rapidly and reaches a higher temperature than a comparable water surface receiving the same amount of insolation. In essence, a thin layer of land is warmed to relatively high temperatures, whereas a thick layer of water is warmed more slowly to moderate temperatures. There are several significant reasons for this difference (Figure 4-23).

1. **Specific Heat**. Water has a higher *specific heat* than land. **Specific heat** (or **specific heat capacity**) is the amount of energy required to increase the temperature of 1 gram of a substance by 1°C. The specific heat of water is about five times as great as that of land, which means that water must absorb five times the amount of energy to show the same temperature increase as land.
2. **Transmission**. The Sun's rays penetrate water more deeply than they do land; that is, water is better than land at transmitting radiation. Thus, in water, solar energy is absorbed through a much greater volume of matter, spreading out the warming; on land, all of the warming is concentrated at the surface and maximum temperatures can be much higher.
3. **Mobility**. Water is highly mobile, so turbulent mixing and ocean currents disperse the energy broadly and deeply through convection—warm water mixes with cooler water, reducing the local temperature increase. Land is essentially immobile, so energy is dispersed by conduction only (and land is a relatively poor conductor of energy).
4. **Evaporative Cooling**. The unlimited availability of moisture on a water surface means that evaporation is much more prevalent over water than over land. The latent heat needed for this evaporation is drawn from the water and its immediate surroundings, causing a decrease in temperature. Thus, evaporative cooling (discussed more completely in Chapter 6) counteracts some of the warming of a water surface.

LearningCheck 4-9 How does the high specific heat of water influence how quickly water warms?

Cooling of Land and Water

Land cools more rapidly and to a lower temperature than a water surface for many of the same reasons it warms more rapidly. For example, during winter, the shallow, warmed layer of land radiates energy away quickly. Water loses its warmth more gradually because it has high specific heat and because the energy has been stored deeply and is brought only slowly to the surface for radiation. As the surface water cools, it sinks and is replaced by warmer water from below. The entire water body must be cooled before the surface temperatures decrease significantly.

Implications

The significance of these contrasts between the warming and cooling rates of land and water is that both the hottest and coldest areas of Earth are found in the interiors of

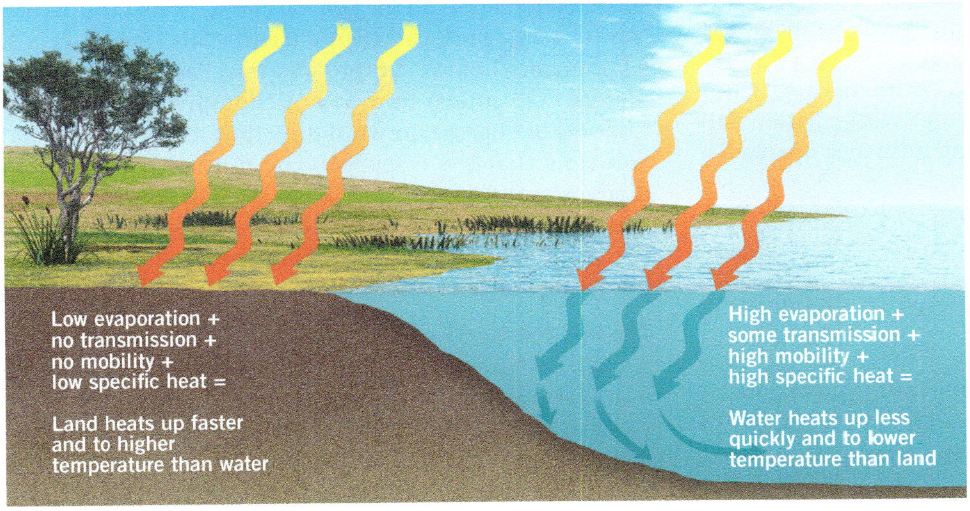

◀ **Figure 4-23** Some contrasting characteristics of the warming of land and water. In general, land warms faster and to a greater extent than water.

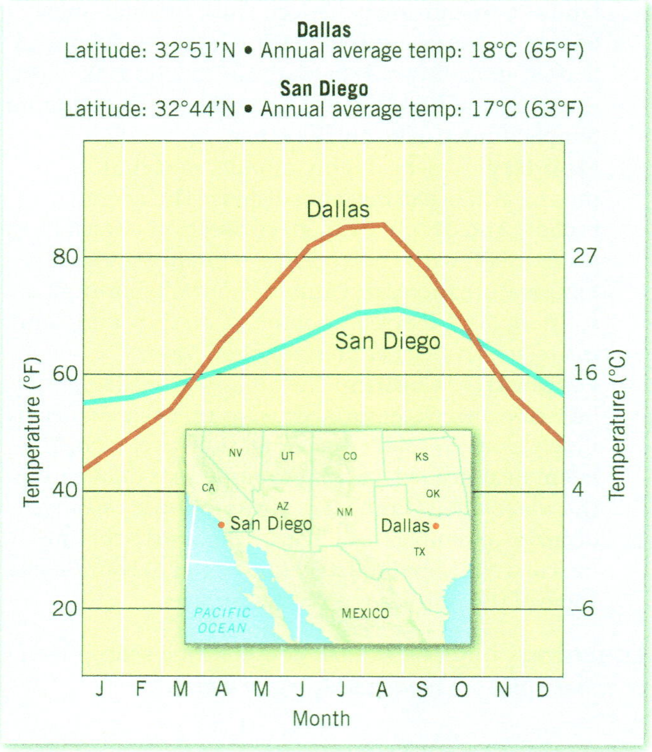

▲ **Figure 4-24** Annual temperature curves for San Diego, California, and Dallas, Texas. Although both cities have similar average annual temperatures, they have very different annual temperature regimes. In summer and winter, San Diego, situated on the coast, experiences milder temperatures than inland Dallas.

continents, distant from the influence of oceans. In the study of the atmosphere, probably no single geographic relationship is more important than the distinction between continental and maritime climates. A continental climate experiences greater seasonal extremes of temperature—hotter in summer, colder in winter—than a maritime climate.

These differences are shown in **Figure 4-24**, which portrays average monthly temperatures for San Diego and Dallas. These two cities are at approximately the same latitude and experience almost identical lengths of day and angles of incidence. Although their annual average temperatures are almost the same, the monthly averages vary significantly. Dallas, in the interior of the continent, experiences notably warmer summers and cooler winters than San Diego, which enjoys the moderating influence of an adjacent ocean.

The oceans, in a sense, act as great reservoirs of energy. In summer the oceans absorb solar energy and store it. In winter they release some of this energy and warm the air. Thus, they function as a sort of global thermostat, moderating temperature extremes.

LearningCheck 4-10 **Which midlatitude location will typically experience higher summer temperatures: a coastal area or a location deep in the interior of a continent? Why?**

Mechanisms of Global Energy Transfer

As we've seen, during the year more total solar energy arrives in the tropics than at the poles, and this is the basis for the broad difference in temperature observed between low latitudes and high latitudes. However, were it not for mechanisms of energy transfer, the tropics would be warmer than they actually are, and the poles would be colder.

Two important and related mechanisms of energy transfer—circulation patterns in the atmosphere and in the oceans—shift some of the warmth of the low latitudes toward the high latitudes. They thereby moderate both the warmth of low latitudes and the cold of the high latitudes. Both the atmosphere and oceans act as enormous thermal engines, with their latitudinal imbalance of energy driving the currents of air and water, which in turn transfer energy and somewhat modify the imbalance.

Atmospheric Circulation

Of the two mechanisms of global energy transfer, by far the more important is the general circulation of the atmosphere. Air moves in an almost infinite number of ways, but there is a broad planetary circulation pattern that serves as a general framework for moving warm air poleward and cool air equatorward. Some 75 to 80 percent of all horizontal energy transfer is accomplished by atmospheric circulation.

Our discussion of atmospheric circulation is withheld until Chapter 5, where we explain the fundamentals of atmospheric pressure and wind.

Oceanic Circulation

A close relationship exists between the general circulation patterns of the atmosphere and oceans. It is wind blowing over the surface of the water that is the principal force driving the major surface ocean currents. However, the influence works both ways: energy stored in the oceans has important effects on patterns of atmospheric circulation.

For purposes of understanding global energy transfer, we are concerned primarily with the broad-scale surface currents that make up the general circulation of the oceans (**Figure 4-25**). Surface ocean currents can flow as fast as 9 kilometers/hour (5.6 mph) and travel as far as 220 kilometers (135 miles) in a day. Although major currents respond to changes in wind direction, they are so broad and slow that the response time normally amounts to many months. In practice, ocean currents reflect average wind conditions over a period of several years. (We discuss the wind patterns responsible for ocean current movement in detail in Chapter 5.)

The Basic Pattern: All the oceans of the world are interconnected. Because of the location of landmasses and the pattern of atmospheric circulation, however, it is convenient to visualize five relatively separate ocean basins—North Pacific, South Pacific, North Atlantic, South Atlantic, and South Indian. Within each of these basins, there is a similar pattern

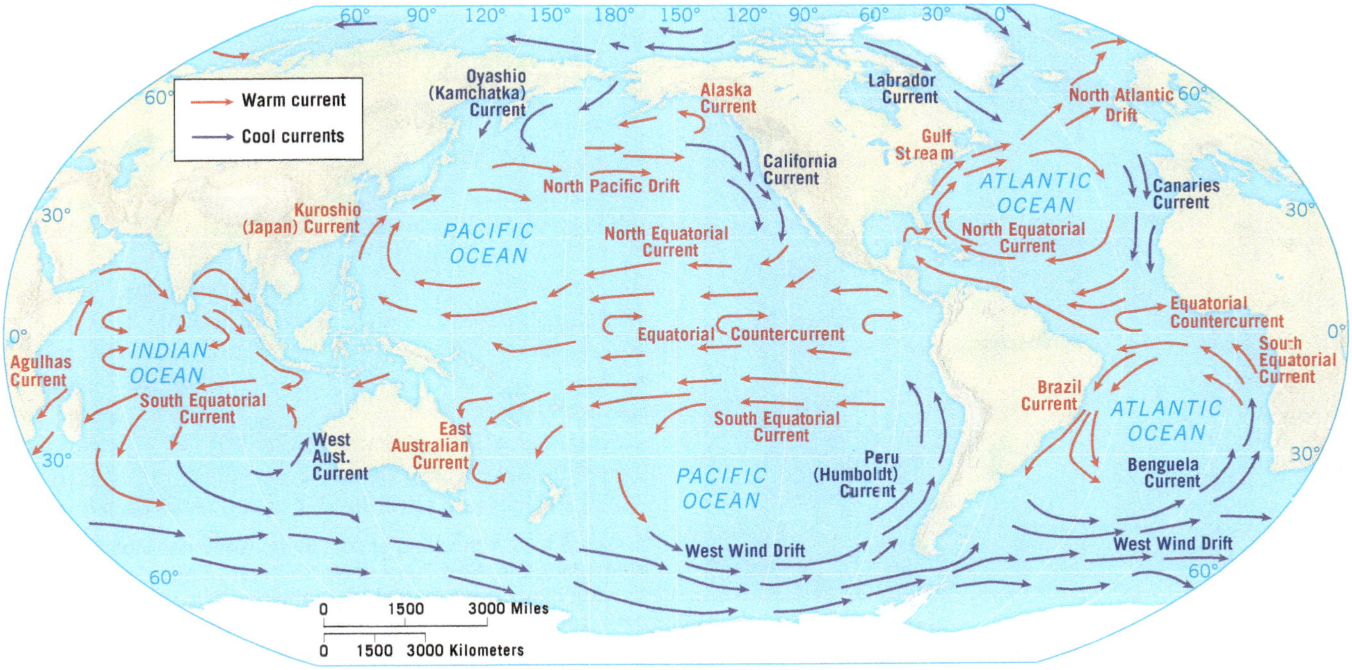

▲ **Figure 4-25** The major surface ocean currents. Warm currents are shown with red arrows and cool currents with blue arrows. Notice that in the midlatitudes, warm currents flow along the eastern coasts of continents, while cool currents flow along the western coasts.

Video
Ocean Circulation Patterns–Subtropical Gyres
https://media.pearsoncmg.com/ph/esm/esm_mcknight_physgeo_13/media/

of surface current flow based on a general similarity of prevailing wind patterns.

A single simple pattern of surface currents is characteristic of all the basins. It consists of a series of enormous elliptical loops elongated east–west and centered approximately at 30° of latitude (except in the Indian Ocean, where it is centered closer to the equator). These loops, called **subtropical gyres**, flow clockwise in the Northern Hemisphere and counterclockwise in the Southern Hemisphere (see Figure 4-25).

On the equatorward side of each subtropical gyre is an *Equatorial Current*, which moves steadily from east to west. The equatorial currents have an average position 5° to 10° north or south of the equator and, as we see in Chapter 5, are propelled by the dominant wind system of the tropics: the east-to-west blowing *trade winds*.

Near the western margin of each ocean basin, the general current curves poleward. As these currents approach the poleward margins of the ocean basins, they curve back to the east—here propelled by the west-to-east-blowing *westerly winds*. The currents curve back toward the equator as they reach the eastern edges of the basins, producing an incompletely closed loop in each basin.

The movement of these currents, although impelled by the wind, is influenced by the Coriolis effect (discussed in Chapter 3). A result of Earth's rotation, the Coriolis effect deflects the path of ocean currents to the right in the Northern Hemisphere and to the left in the Southern Hemisphere. A glance at the basic pattern shows that the current movement around the gyres responds precisely to the Coriolis effect.

Northern and Southern Variations: In the two Northern Hemisphere basins—North Pacific and North Atlantic—the bordering continents lie so close together at the northern basin margin that the bulk of the current flow is prevented from entering the Arctic Ocean. This effect is more pronounced in the Pacific than in the Atlantic. The North Pacific has very limited flow northward between Asia and North America, whereas in the North Atlantic a larger proportion of the flow escapes northward between Greenland and Europe.

In the Southern Hemisphere, the continents are far apart. Thus, the southern segments of the gyres in the South Pacific, South Atlantic, and South Indian Ocean basins are connected as one continuous flow in the uninterrupted belt of ocean that extends around the world in the vicinity of latitude 60° S. This circumpolar flow is called the *West Wind Drift*.

Current Temperatures: Of utmost importance to our understanding of global energy transfer are the temperatures of the various ocean currents. Each major current is classified as either *warm* or *cool* relative to the surrounding water at that latitude. Each "leg" of the subtropical gyres has its own general temperature characteristics:

- Low-latitude currents (Equatorial Currents) have relatively warm water.
- Poleward-moving currents on the western sides of ocean basins (off the east coast of continents) carry relatively warm water toward higher latitudes, such as the Gulf Stream off the east coast of North America (Figure 4-26).

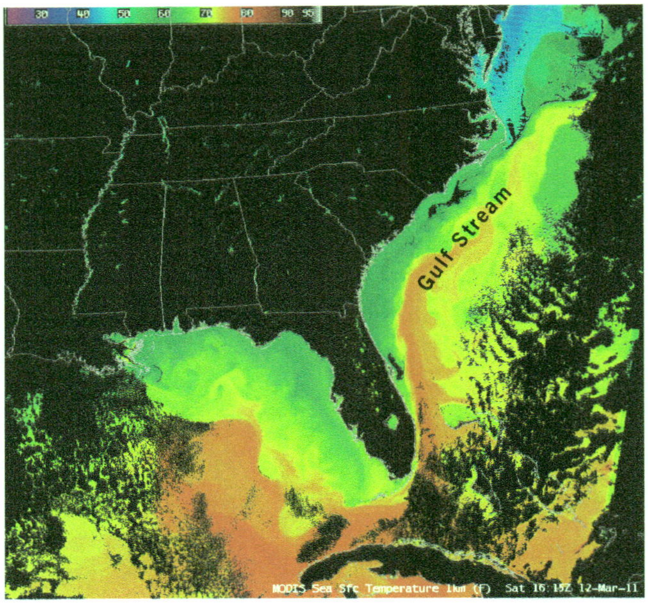

▲ **Figure 4-26** Multipass satellite image showing the Gulf Stream off the east coast of North America. Red represents relatively high water temperatures; blue and green represent relatively low water temperatures. (Vertical and horizontal bands are not real features but are caused by incomplete satellite data.)

Animation
Gulf Stream
https://media.pearsoncmg.com/ph/esm/esm_mcknight_physgeo_13/media/

- The high-latitude currents in the Northern Hemisphere gyres carry relatively warm water to the east, while the high-latitude currents in the Southern Hemisphere gyres (generally combined into the West Wind Drift) carry relatively cool water to the east.
- Equatorward-moving currents on the eastern sides of ocean basins (off the west coasts of continents) carry relatively cool water toward the equator.

In summary—from the perspective of the margins of the continents—we see a poleward flow of warm, tropical water along the east coasts of continents and an equatorward flow of cool, high-latitude water along the west coasts of continents.

LearningCheck 4-11 **What is the relative temperature of the ocean current flowing along the west coast of a midlatitude continent? Along the east coast?**

Western Intensification: The poleward warm currents off the east coast of continents tend to be narrower, deeper, and faster than the equatorward cool currents flowing off the west coast of continents. This phenomenon is called *western intensification* because it occurs on the western side of the subtropical gyres. (In other words, it occurs in the currents flowing poleward off the east coasts of continents in the midlatitudes.) This intensification of poleward warm currents arises for a number of reasons, including the Coriolis effect. Recall from Chapter 3 that the Coriolis effect is greater in higher latitudes, and so the eastward high-latitude current flow is deflected back toward the equator more strongly than the westward equatorial current flow is

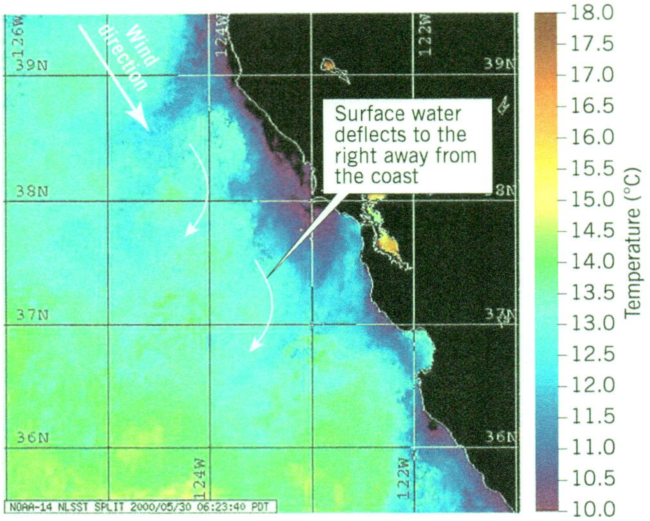

▲ **Figure 4-27** Upwelling of cold water (shown in violet and blue) along the West Coast of North America near San Francisco. As wind blows over the surface of the ocean, the surface water is deflected to the right by the Coriolis effect and veers away from the coast. That water is replaced by cold water from below.

deflected toward the poles. This means that cool water is slowly flowing back toward the equator across much of the eastward high-latitude currents, whereas the poleward warm currents are confined to a fairly narrow zone off the east coasts of continents.

Upwelling: Wherever an equatorward-flowing cool current pulls away from a subtropical western coast, a pronounced and persistent **upwelling** of cold water from below occurs. For example, if winds along the west coast of a Northern Hemisphere continent are blowing from the north or northwest, the Coriolis effect will deflect some of the surface water to the right, away from the coast. As surface water pulls away from the coast, it will be replaced with water from deeper below. Upwelling brings nutrient-rich water to the surface, making west-coast marine ecosystems highly productive. The upwelling also brings colder water to the surface, generally decreasing the surface temperature of the already cool currents off the western coast of continents (Figure 4-27). Upwelling is most striking off South America but is also notable off North America, northwestern Africa, and southwestern Africa. It is much less developed off the coast of western Australia.

Note that western intensification and upwelling tend to reinforce the water temperature contrasts off the eastern and western coasts of the continents in the midlatitudes.

Three final ocean current patterns are worth noting:

1. Notice in Figure 4-25 that the North and South Equatorial Currents are separated by an Equatorial Countercurrent—a west-to-east-moving warm current approximately along the equator in each ocean. The various equatorial currents feed the Equatorial Countercurrent near its western margin in each basin. Water from the Equatorial Countercurrent in turn drifts poleward to feed the equatorial current near the eastern end of its path.

2. The northwestern portions of Northern Hemisphere ocean basins receive an influx of cool water from the Arctic Ocean. For example, the Labrador Current is a prominent cool current flowing southward from the Arctic Ocean between Greenland and the Canadian coast (see Figure 4-25). A smaller flow of cold water issues from the Bering Sea southward along the coast of Siberia to Japan.
3. In addition to the surface ocean currents we have just described, there is a deep ocean circulation pattern—sometimes called the global conveyor-belt circulation—that influences global climate in subtle but important ways. The deep ocean conveyor-belt circulation is discussed in greater detail in Chapter 9.

Vertical Temperature Patterns

As we study the geography of weather and climate, most of our attention is directed to the horizontal dimension; in other words, we are concerned with the geographic distribution of atmospheric phenomena across the surface of Earth. However, to understand these distributional patterns fully, we must also pay attention to a number of important vertical patterns in the atmosphere. One such vertical pattern involves variations in temperature with increasing altitude in the troposphere.

Environmental Lapse Rate

Temperature change with increasing altitude within the troposphere is relatively predictable. As we learned in Chapter 3, throughout the troposphere, under typical conditions, a general decrease in temperature occurs with increasing altitude (Figure 4-28a). However, there are many exceptions. Indeed, the rate of vertical temperature decline can vary according to season, time of day, amount of cloud cover, and a host of other factors. In some cases, there is even an opposite trend, with the temperature increasing upward for a limited distance.

The observed trend of vertical temperature change in the atmosphere is called the **environmental lapse rate**. Determining the lapse rate of a "column" of air involves measuring air temperature at various altitudes in one location. Then a graph of temperature as a function of height is drawn to produce a temperature profile of that air column. When measuring such a temperature change, only the thermometer is moved; the air is at rest. (If the air is moving vertically, expansion or compression will cause an adiabatic temperature change. Such temperature changes are explored more fully in Chapter 6.)

Average Lapse Rate

Although the environmental lapse rate varies from place to place and from time to time, particularly in the lowest few hundred meters of the troposphere, the average rate of temperature change is about 6.5°C per 1000 meters (3.6°F per 1000 feet). This is called the **average lapse rate**, or *average vertical temperature gradient* within the troposphere. The average lapse rate tells us that if a thermometer measures the temperature 1000 meters above a previous measurement, the reading will be, on average, 6.5°C cooler. Conversely, if a second measurement is made 1000 meters lower than the first, the temperature will be about 6.5°C warmer.

Temperature Inversions

The most prominent exception to an average lapse-rate condition is a **temperature inversion**, a situation in which temperature in the troposphere *increases*, rather than decreases, with increasing altitude. Inversions are relatively common in the troposphere but are usually of brief duration and restricted depth. They can occur near Earth's surface, as in Figure 4-28b, or at higher levels, as in Figure 4-28c.

Inversions influence weather and climate. As we will see in Chapter 6, an inversion inhibits vertical air movements and greatly reduces the possibility of precipitation. Inversions also contribute significantly to increased air pollution because they create stagnant air conditions that greatly limit

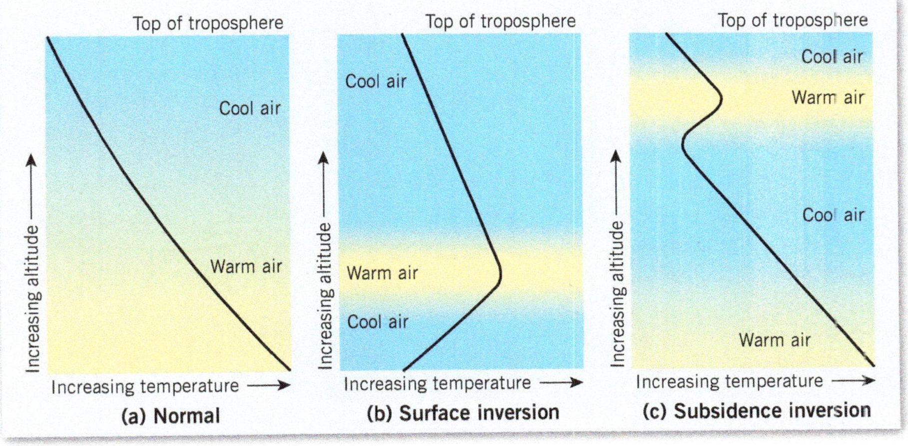

◀ **Figure 4-28** A comparison of normal and inverted lapse rates. (a) Tropospheric temperature normally decreases with increasing altitude. (b) In a surface inversion, temperature increases with increasing altitude from ground level to some distance above the ground. (c) In a subsidence inversion, temperature first decreases with increasing altitude as in a normal lapse rate, but then at some altitude well below the tropopause begins to increase with increasing altitude because sinking air warms.

▲ **Figure 4-29** Photochemical smog over the city of Santiago, Chile. The top of the temperature inversion layer trapping the smog is seen just below the peaks of the nearby mountains.

the natural upward dispersal of urban-industrial pollutants (**Figure 4-29**).

Surface Inversions: The most readily recognizable inversions are those found at ground level. These are often *radiation inversions*, which can develop on a long, cold winter night when a land surface rapidly emits longwave radiation into a clear, calm sky. The cold ground then cools the lowest few hundred meters of the air. Radiation inversions can occur on any night when the sky is clear and the wind is calm, but they are strongest in the winter because there is a short daylight period for solar warming and a long night for radiational cooling.

Advection inversions develop where wind brings cold air into an area. This condition is commonly produced by cool maritime air blowing into a coastal locale. Advection inversions are usually short-lived (typically overnight) and shallow. A similar kind of surface inversion develops when cold air slides down a slope into a valley, thereby displacing slightly warmer air, producing a *cold-air-drainage inversion*.

Subsidence Inversions: Upper-air *subsidence inversions* develop when air descends (or subsides) toward the surface, as is common in the subtropics, where high-pressure cells are found throughout the year. This descending air warms adiabatically, leaving a layer of warm air above cooler air closer to the surface.

LearningCheck 4-12 **What is a temperature inversion? Why is air pollution a bigger problem when there is a temperature inversion?**

Global Temperature Patterns

The goal of this chapter and the four succeeding ones is to examine the global pattern of climate. With the preceding pages as background, we turn our attention to the worldwide distribution of temperature—the first of the four elements of weather and climate.

Maps of global temperature patterns usually show seasonal extremes rather than annual averages. January and July are the months of lowest and highest temperatures for most places on Earth, and maps portraying the average temperatures of these two months provide a simple but meaningful expression of temperature conditions in winter and summer (**Figure 4-30** and **Figure 4-31**). Temperature

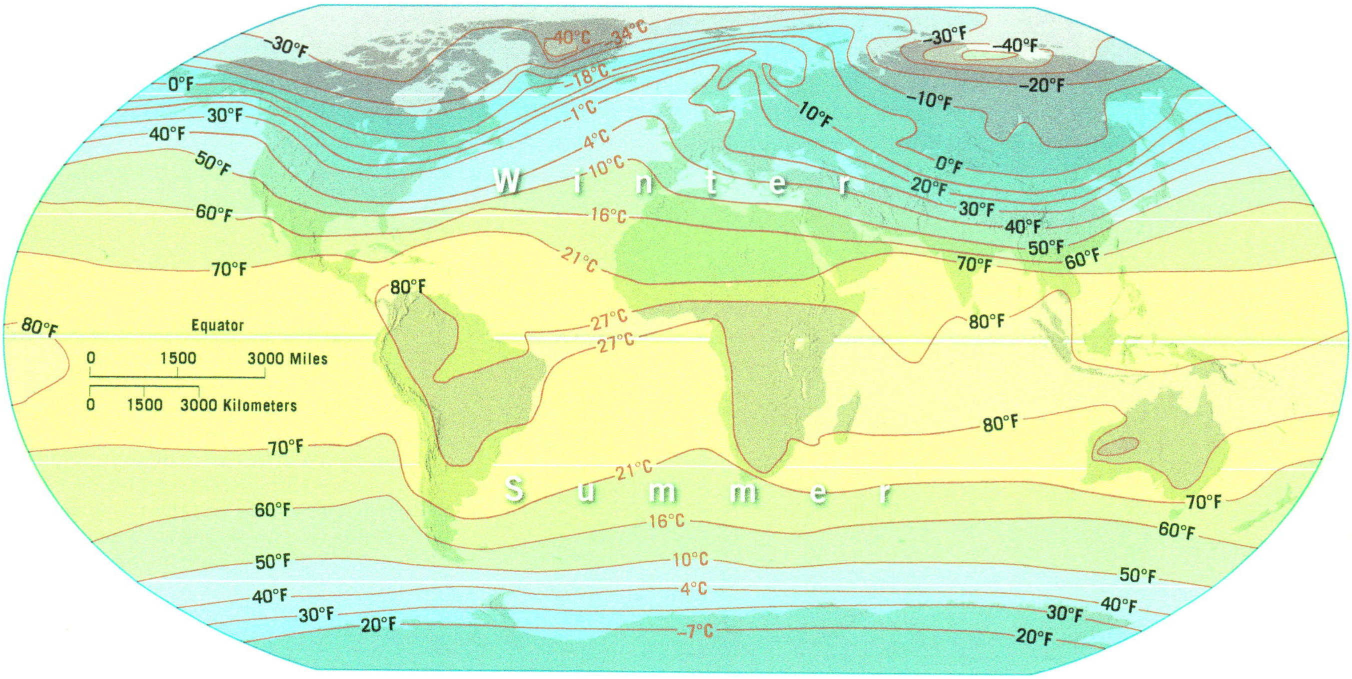

▲ **Figure 4-30** Average January sea-level temperatures.

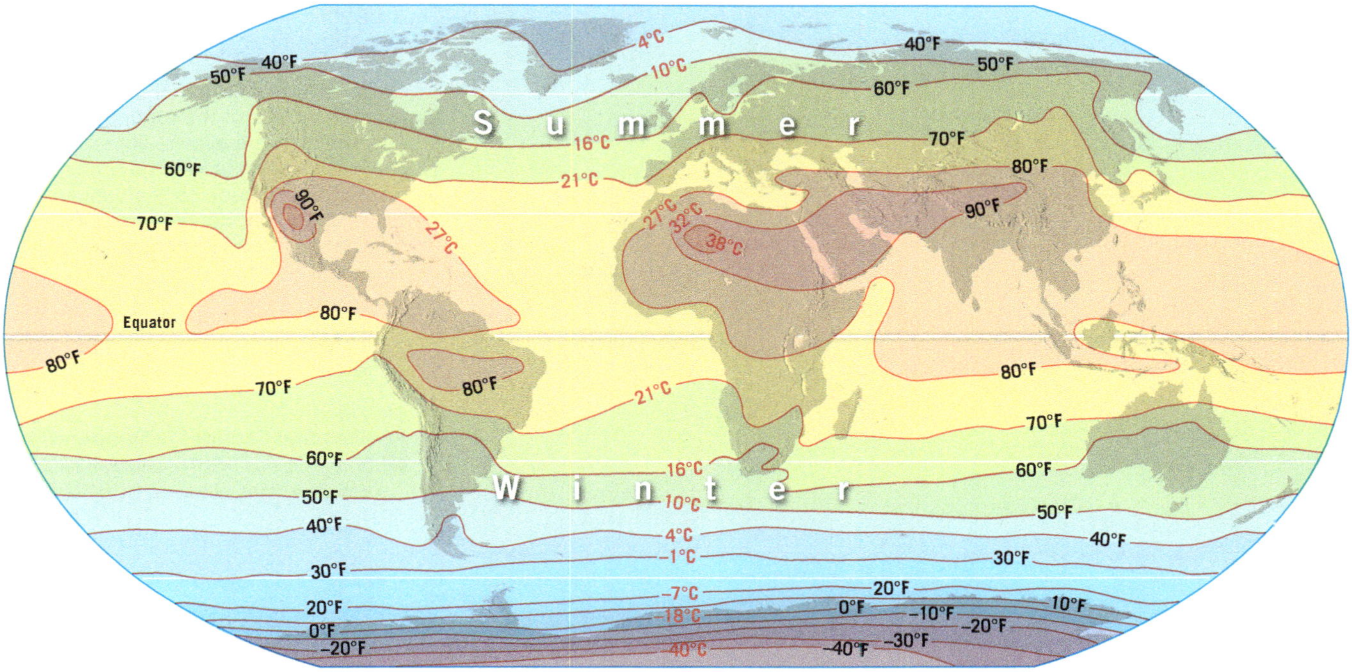

▲ **Figure 4-31** Average July sea-level temperatures.

distribution is shown by means of **isotherms**, lines joining points of equal temperature. Temperature maps are based on monthly averages, which are based on daily averages; the maps do not show the maximum daytime heating or the maximum nighttime cooling. Although the maps are on a very small scale, they permit a broad understanding of global temperature patterns.

Prominent Controls of Temperature

Patterns of temperature are controlled largely by four factors: latitude, altitude, land–water contrasts, and ocean currents.

Latitude: The most conspicuous feature of any world temperature map is the general east–west trend of the isotherms, roughly following the parallels of latitude. If Earth had a uniform surface without ocean currents and wind systems, the isotherms would probably coincide exactly with parallels, showing a progressive decrease of temperature poleward from the equator. However, temperature variations between land and water, and the circulation of ocean currents around the margins of ocean basins, complicate the actual temperature pattern significantly. Nonetheless, the fundamental cause of temperature variation around the world is insolation variation, which is governed primarily by latitude, so general global temperature patterns reflect this latitudinal control.

Altitude: Because temperature responds sharply to changes in altitude, it would be misleading to plot actual temperature values on global temperature maps such as Figures 4-30 and 4-31 because high-altitude stations would almost always be colder than low-altitude stations. Consequently, on these maps temperatures are adjusted to what they would be if the station were at sea level. This is done by using the average lapse rate, a method that eliminates the complication of terrain differences. Maps plotted in this way are useful in showing world patterns, but they are not satisfactory for indicating actual temperatures for locations that are not close to sea level.

Land–Water Contrasts: The differential warming and cooling of land and water are also reflected conspicuously on a temperature map. Summer temperatures are higher over the continents than over the oceans, as shown by the poleward curvature of the isotherms over continents in the respective hemispheres (July in the Northern Hemisphere, January in the Southern Hemisphere). Winter temperatures are lower over the continents than over the oceans; the isotherms bend equatorward over continents in this season (January in the Northern Hemisphere, July in the Southern Hemisphere). Thus, in both seasons, isotherms make greater north–south shifts over land than over water. Isotherms may migrate northward or southward more than 1550 kilometers (970 miles)—some 14° of latitude (Figure 4-32).

Ocean Currents: Some of the most obvious bends in the isotherms occur in near-coastal areas of the oceans, where prominent warm or cool currents reinforce the isothermal curves caused by land–water contrasts. Cool currents deflect isotherms equatorward, whereas warm currents deflect them poleward. Cool currents produce the greatest isothermal bends in the warm season: note the July conditions off the western coast of North America, or the January situation off the western coast of South America and the southwestern coast of Africa. Warm currents have their most prominent effects in the cool season: witness the isothermal pattern in the North Atlantic Ocean in January.

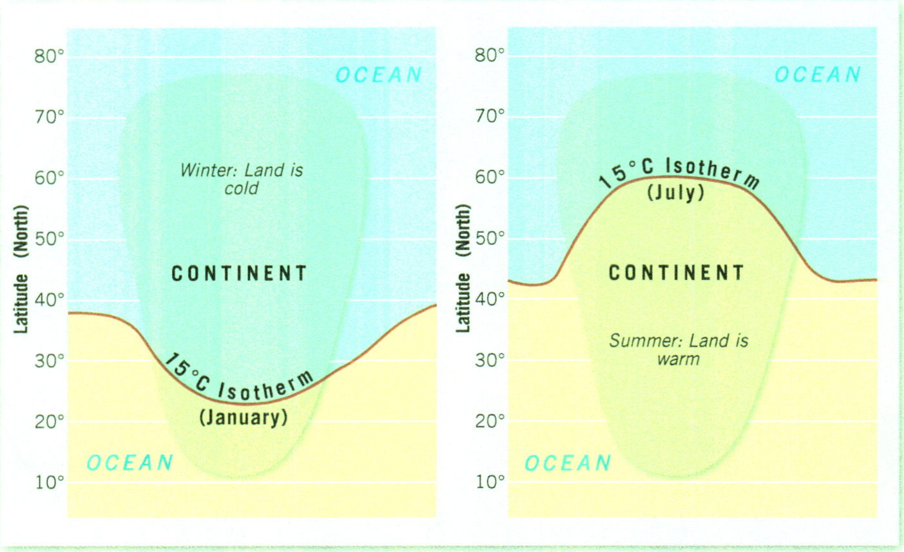

▲ **Figure 4-32** Idealized seasonal migration of the 15°C (59°F) isotherm over a hypothetical Northern Hemisphere continent. The latitudinal shift is greatest over the interior of the continent and least over the adjacent oceans. For example, over the western ocean, the isotherm moves from latitude 38°N in January to only 42°N in July, but over the continent the change is from 22°N in January all the way to 60°N in July.

Video
Seasonal Changes in Temperature
https://media.pearsoncmg.com/ph/esm/esm_mcknight_physgeo_13/media/

LearningCheck 4-13 **What generally happens to temperatures as you move from the equator toward the poles? What factors help explain this pattern?**

Seasonal Patterns

Apart from the general east–west trend of the isotherms, probably the most conspicuous feature of Figures 4-30 and 4-31 is the latitudinal shift of the isotherms from one map to the other. The isotherms follow the changing balance of insolation during the course of the year, moving poleward in summer and equatorward in winter. Note, for example, the 10°C (50°F) isotherm in southernmost South America: in January (midsummer), it is positioned at the southern tip of the continent, whereas in July (midwinter), it is shifted considerably to the north.

This isotherm shift is much more pronounced at high latitudes than at low ones and also much more pronounced over continents than over oceans. Thus, tropical areas, particularly tropical oceans, show relatively small displacement of isotherms from January to July.

Isotherms are also more tightly "packed" in winter. This close line spacing indicates that the temperature gradient (rate of temperature change with horizontal distance) is steeper in winter than in summer, which in turn reflects the greater contrast in radiation balance in winter. The temperature gradient is also steeper over continents than over oceans.

Coldest Winter Locations: The coldest places on Earth are over landmasses in the higher latitudes. During July, the polar region of Antarctica is the dominant area of coldness. In January, the coldest temperatures occur many hundreds of kilometers south of the North Pole, in subarctic portions of Siberia, Canada, and Greenland. Denali National Park, the featured National Park in Chapter 3, is the coldest place in the United States. The principle of greater cooling of land than water is clearly demonstrated.

Hottest Summer Locations: The highest temperatures also are found over the continents. The locations of the warmest areas in summer, however, are not equatorial. Rather, they are in subtropical latitudes, where descending air in high-pressure cells maintains clear skies most of the time, allowing for almost uninterrupted insolation. Frequent cloudiness prevents such a condition in the region around the equator. Thus, the highest July temperatures occur in northern Africa and in the southwestern portions of North America and Eurasia, whereas the principal areas of January warmth are in subtropical parts of Australia, southern Africa, and South America. Summer temperatures in these locations can become dangerously high.

Average annual temperatures are highest in equatorial regions, however, because these regions experience so little winter cooling. Subtropical locations cool substantially on winter nights, so their annual average temperatures are lower. The ice-covered portions of Earth—Antarctica and Greenland—remain quite cold throughout the year.

Annual Temperature Range

Another map useful in understanding the global pattern of air temperature is one that portrays the average annual range of temperatures (Figure 4-33). **Average annual temperature range** for a location is the difference between the average temperature of the warmest month and the average temperature of the coldest month—normally July and January. Enormous seasonal variations in temperature occur in the interiors of high-latitude continents, and continental areas in general experience much greater ranges than do equivalent oceanic latitudes. At the other extreme, the average temperature fluctuates only slightly from season to season in the tropics, particularly over tropical oceans.

Differences Between Hemispheres: The moderating influence of the oceans on annual temperature range is

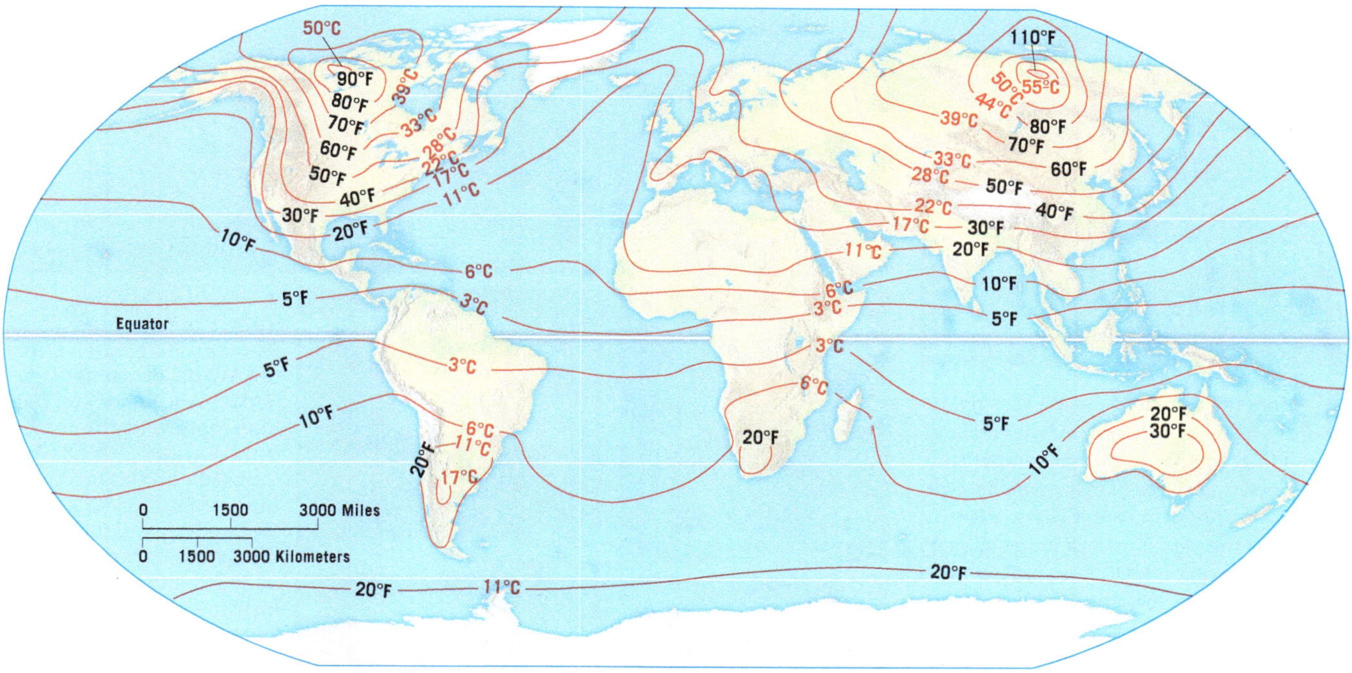

▲ **Figure 4-33** The world pattern of average annual temperature range. The largest ranges occur in the interior of high-latitude landmasses.

clearly seen in Figure 4-33. The Northern Hemisphere is often thought of as a "land hemisphere" because 39 percent of its area is land surface; the Southern Hemisphere is a "water hemisphere," with only 19 percent of its area as land. It is obvious that the land hemisphere has greater overall temperature variation during the year than the water hemisphere.

LearningCheck 4-14 **What kinds of locations around the world have a very small average annual temperature range, and what kinds of locations have a very large average annual temperature range?**

Measuring Global Temperatures

Until the late twentieth century, temperature was measured mostly by surface instruments. Thus, remote land locations as well as most of the ocean area of Earth had sparse coverage. In recent years, however, scientists have used satellites to gather surface temperature data around the globe, greatly enhancing the ability to study temperature patterns in the atmosphere and ocean. See the box *Focus: Measuring Earth's Surface Temperature by Satellite*.

Urban Heat Islands

Long-term temperature records indicate that urban areas tend to be warmer than rural areas. This often-observed increase in temperature associated with cities is known as the **urban heat island (UHI) effect**. The UHI effect is thought to result from reduced nighttime cooling because buildings inhibit the loss of longwave radiation to space and reduce the mixing of warmer surface air with cooler air above. Lower albedo (thus, greater absorption of insolation) in urban areas may also play a role.

The U.S. Environmental Protection Agency (EPA) estimates that temperatures within a city of one million people are typically 1–3°C (1.8–5.4°F) warmer than nearby rural areas, with nighttime temperature differences much greater than this. However, some recent research suggests that the UHI effect may be smaller than previously thought—perhaps as little as a 0.9°C (1.6°F) difference between cities and the surrounding rural areas. In addition to temperature, precipitation in urban areas (and downwind of urban areas) may be affected by the higher concentrations of aerosols released by industries or other human activity.

Climate Change and Global Warming

As we described earlier in this chapter, the "natural" greenhouse effect has been part of the basis of life on Earth since the early atmosphere formed. Without it, our planet would be a frozen mass, perhaps 33°C (59°F) colder than it is today. Over the past four decades, the scientific community has alerted the media and the general public to human-produced changes to the greenhouse effect. Data gathered from surface weather stations, ships, buoys, balloons, satellites, ice cores, and other paleoclimatological sources indicate that the climate of Earth is, on average, warming. This warming trend became known as **global warming**. Most

Focus

Measuring Earth's Surface Temperature by Satellite

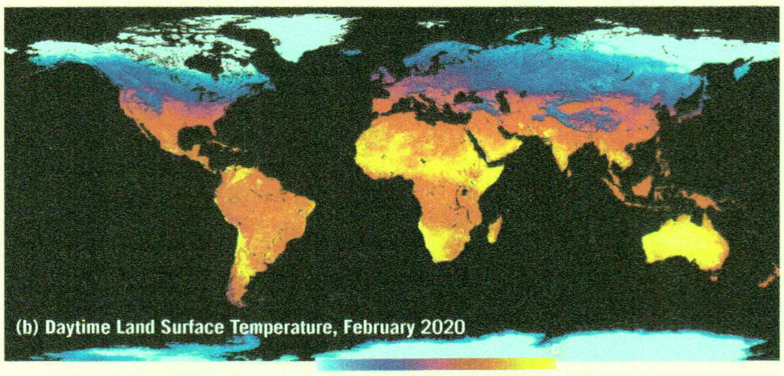

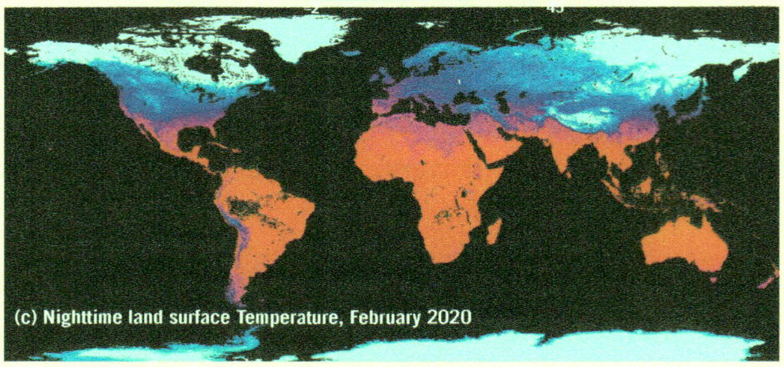

◄ **Figure 4-C** (a) Sea surface temperatures for February 2020 derived from the MODIS instruments on NASA's Aqua satellite. Dark blue represents the coldest surface water (−2°C); white (not seen on this image), the warmest (35°C). (b) Daytime land surface temperatures for February 2020 from the MODIS instruments on NASA's Terra satellite. Light blue represents the coldest land surface temperature (−25°C); bright yellow, the warmest (45°C). (c) Nighttime land surface temperature for February 2020 from Terra.

Before the advent of remote sensing, scientists relied on ships, buoys, and land-based instruments to gather temperature data. There were large gaps in coverage, especially over oceans. In recent years, scientists have used the Moderate Resolution Imaging Spectroradiometer (MODIS) on NASA's Aqua and Terra satellites to gather surface temperature data around the globe. Using computer algorithms to compensate for factors such as absorption and scattering in the atmosphere, scientists can estimate the "skin" temperature of ocean and land surfaces by measuring emitted thermal infrared radiation.

Sea Surface Temperature: Sea surface temperature (SST; Figure 4-C-a) influences not only the temperature of air masses that originate over ocean areas (thus the weather of continental areas over which these air masses may pass), but also the intensity of storms such as hurricanes (discussed in Chapter 7). Regular monitoring of SST also helps scientists anticipate the onset of El Niño events (discussed in Chapter 5) and provides important information about long-term changes to Earth's environment.

Land Temperature: Satellite-derived daytime and nighttime land temperatures (Figures 4-C-b and 4-C-c) measure the temperature of the surface itself, not the temperature of the air just above the surface. Notice the high daytime surface temperatures in the Northern Hemisphere subtropical and midlatitude deserts—a consequence of high Sun and sparse cloud cover. These same areas cool off significantly at night, as do high mountain areas, such as the Himalayas, where the thinner atmosphere allows rapid heat loss at night.

Questions

1. Are equatorial waters equally warm around the world? What might explain the pattern?
2. What area of South America cools off most dramatically from day to night? Why?
3. Compared with northern and southern Africa, equatorial Africa cools very little from day to night. Why?

climate scientists prefer the more general term *climate change* because it reflects the fact that not every place will warm and encompasses the many effects of warming, such as changes in precipitation patterns.

Temperature Change Over the Past Century

Since 1880, average global temperature has increased by about 0.99°C (1.78°F)—in the last quarter of the twentieth century alone temperatures increased 0.2–0.3°C (0.4°F; Figure 4-34). The temperature increase over the past 100 years is apparently greater than that of any other century in at least the past 1000 years. Overall, global temperatures are higher today than they have been in more than 1000 years, and the past two decades have been the hottest since widespread instrument readings began in 1880. One consequence of the warming trend can be seen in Isle Royale National Park, see *Featured National Park: Isle Royale*.

Because direct instrument measurements of Earth's temperature go back only a few centuries, past temperature patterns are calculated by using "proxy" measures—such as data deciphered from polar ice cores, oceanic sediments, or pollen analysis. (Proxy measures are discussed in Chapter 8.) Although there is a margin of error in these data, evidence clearly points to warming.

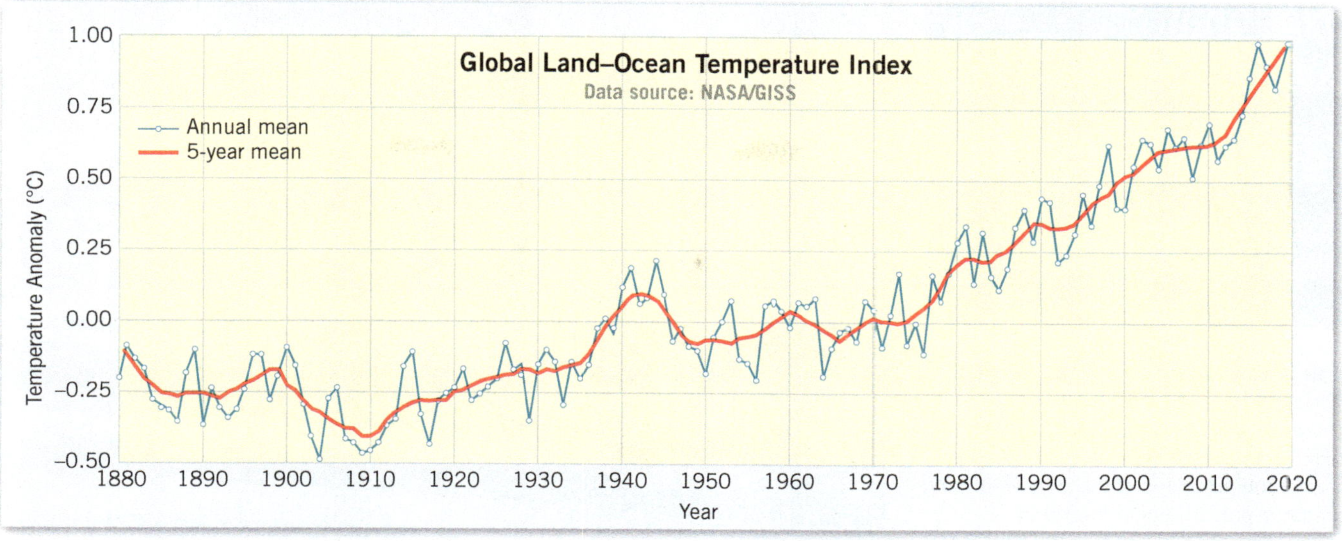

▲ **Figure 4-34** Global mean temperature over land and ocean, 1880–2019. The scale on the left shows the temperature difference relative to the 1951–1980 average. The dark red line shows the smoothed temperature trend.

Increasing Greenhouse Gas Concentrations

The cause of global climate change in the last 150 years is almost certainly *human-enhanced greenhouse effect*. Since the industrial era began, human activities have increased the concentrations of greenhouse gases, such as carbon dioxide, methane, tropospheric ozone, and chlorofluorocarbons in the atmosphere. As greenhouse gas concentrations in the atmosphere increase, more terrestrial radiation is retained in the lower atmosphere, thereby increasing global temperatures.

Carbon dioxide (CO_2) is responsible for about 65 percent of the human-enhanced greenhouse effect. CO_2 concentrations have been rising steadily since the Industrial Revolution began in the mid-1700s (**Figure 4-35**). Carbon dioxide is a principal by-product of combustion of anything containing carbon, such as coal and petroleum. Since 1750—when estimates show the concentration of CO_2 in the atmosphere was about 280 parts per million (ppm)—carbon dioxide levels in the atmosphere have increased by about 47 percent. The latest climatological data indicates that the current concentration of CO_2 in the atmosphere of over 415 ppm is greater than at any time in the past 800,000 years.

Many other greenhouse gases have been added to the atmosphere by human activity. Methane—produced by grazing livestock and rice paddies and as a by-product of the combustion of wood, natural gas, coal, and oil—has increased by 150 percent since 1750 and is about 25 times more potent as a greenhouse gas than CO_2 over a 100-year period. Nitrous oxide (N_2O), which comes from chemical fertilizers and automobile emissions, has increased by about 20 percent since 1750; adding one N_2O molecule to the atmosphere results in more greenhouse warming than 300 molecules of CO_2. Chlorofluorocarbons (CFCs) are synthetic chemicals that were widely used as refrigerants and as propellants in spray cans until the 1990s. (See Chapter 3 for a discussion of another consequence of CFCs and nitrous oxide in the atmosphere: ozone depletion.) Many of these gases, and others, are being released into the atmosphere at accelerating rates.

The increase in greenhouse gas concentrations, especially carbon dioxide, correlates closely with the observed increase in global temperature: as CO_2 has increased, so have average global temperatures.

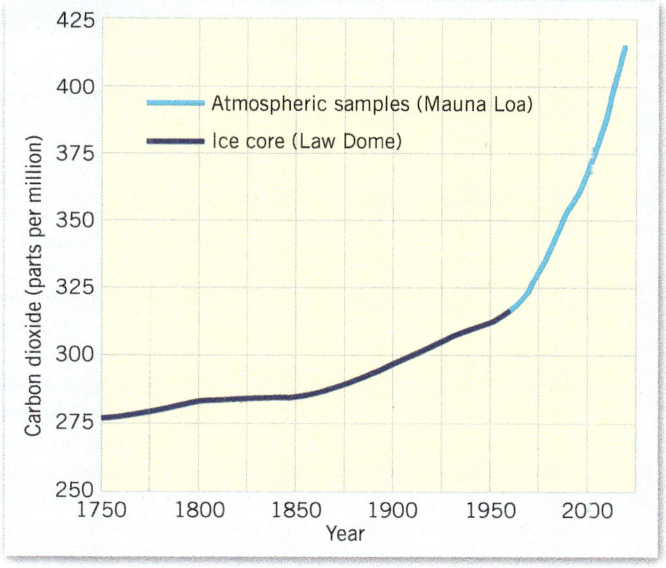

▲ **Figure 4-35** Change in atmospheric carbon dioxide concentration from 1750–2019. The dark blue line shows values derived from ice cores at Law Dome, Antarctica; the light blue line shows measured values from Mauna Loa, Hawai'i.

Animation
Global Warming
https://media.pearsoncmg.com/ph/esm/esm_mcknight_physgeo_13/media/

Featured National Park

Isle Royale, MI

Consequence of a Warmer Climate

Isle Royale National Park is made up of an island chain in northern Lake Superior. There are only two large terrestrial mammals that live there: wolves and moose. And they are the subject of one of the longest running predator-prey studies in the world.

The Wolves: Wolves living on the island use a 20-mile ice bridge to return to the mainland, a stretch of ice that also enables new wolves to find their way onto the island. Warmer temperatures over the past few decades have made ice bridges like the one seen in **Figure 4-D** rare, so much so that wolves can no longer travel back and forth to the island on a regular basis (**Figure 4-E**).

Wolves play a critical role in the ecosystem on Isle Royale by keeping the moose population in check. However, the wolf population on the island is dwindling. At the end of 2019, the population stood at 17, nine males and eight females. And those that remain are weakened by inbreeding. Wolf relocation efforts are currently underway to increase the number of wolves on the island (**Figure 4-F**), but it's a delicate process because wolves are very territorial and will attack one another.

The Moose: Over 2,000 moose live on Isle Royale (**Figure 4-G**), which has an area of only 553.5 km^2 (206 $miles^2$). With so few wolves to prey upon them, the population has grown unchecked, devastating the island's vegetation. This puts pressure on other populations of animals in the ecosystem that depend on that vegetation for food and shelter.

Having a large herd is not good for the health of the moose either. Overpopulation leads to competition for resources. The moose are smaller and live shorter lives than the moose that inhabited the island just 40 years ago. Wolf repopulation is one way to reestablish balance to the island's ecosystem. It takes only a few new wolves to reinvigorate the wolf population, so scientists are hopeful.

▶ **Figure 4-D** Rare ice bridge seen by satellite in January 2019.

▼ **Figure 4-E** Wolves crossing an ice bridge to get from the island to the mainland.

▶ **Figure 4-F** Wolf being released after relocation to Isle Royale.

▼ **Figure 4-G** A cow moose and her calf.

Questions

1. In what sense is this ecosystem "natural" give that researchers try to keep the herds healthy when they introduce new wolves?
2. Why does the moose population need to be kept in check?

NPS Video
Outside Science: Isle Royale
https://media.pearsoncmg.com/ph/esm/esm_mcknight_physgeo_13/media/

Intergovernmental Panel on Climate Change (IPCC)

It is well known that climate undergoes frequent natural fluctuations, warming or cooling regardless of human activities. A large body of evidence, however, indicates that *anthropogenic* (human-induced) factors are largely, if not wholly, responsible for the recent temperature increase. The Intergovernmental Panel on Climate Change (IPCC) is the most important international organization of atmospheric scientists and policy analysts assessing global climate change. (In 2007, in recognition of their many years of work on climate change, the IPCC was the co-recipient of the Nobel Peace Prize.)

The *Fifth Assessment Report* of the IPCC released in 2013 and 2014 concludes bluntly:

> Warming of the climate system is unequivocal, and since the 1950s, many of the observed changes are unprecedented over decades to millennia. The atmosphere and ocean have warmed, the amounts of snow and ice have diminished, and sea level has risen.

With regard to the causes of this climate change, the IPCC concluded:

> It is *extremely likely* [95% probability or higher] that more than half of the observed increase in global average surface temperature from 1951 to 2010 was caused by the anthropogenic increase in GHG [greenhouse gas] concentrations and other anthropogenic forcings together.

Because both the causes and implications of climate change are so complicated—due especially to the many feedback loops involved in climate systems—the preceding description of global warming serves only as our introduction to the topic. In subsequent chapters, after increasing our understanding of processes of weather and climate, we further explore both the natural and anthropogenic aspects of global environmental change, including global warming, in much greater detail.

LearningCheck 4-15 **How has human activity likely caused an increase in global temperature over the past century?**

After studying this chapter, you should be able to answer the following questions. Key terms from each text section are shown in **bold type.** Definitions for key terms are also found in the glossary at the back of the book.

Learning Review 4

Key Terms and Concepts

Energy, Heat, and Temperature (*p. 80*)

1. What is the difference between **heat** (thermal **energy**) and **temperature**?
2. What is the relationship between the internal **kinetic energy** of a substance and its temperature?

Solar Energy (*p. 82*)

3. Briefly describe the following bands of **electromagnetic radiation (radiant energy)**: **visible light**, **ultraviolet (UV)**, **infrared (IR)**, **thermal infrared**.
4. Describe and contrast the portions of the **electromagnetic spectrum** referred to as **shortwave radiation** and **longwave radiation (terrestrial radiation)**.
5. What is **insolation**?

Basic Warming and Cooling Processes in the Atmosphere (*p. 85*)

6. Describe and contrast the following processes associated with electromagnetic energy: **radiation (emission)**, **absorption**, **reflection**, and **transmission**.
7. What generally happens to the temperature of an object that absorbs electromagnetic radiation?
8. What is **albedo**?
9. How is **scattering** different from reflection?
10. Describe and explain the atmospheric **greenhouse effect**, noting the two most important natural **greenhouse gases**.
11. What is the difference between **conduction** and **convection**?
12. How and why does conduction influence the temperature of air above a warm surface? Above a cold surface?
13. Describe the pattern of air movement within a **convection cell**.
14. Briefly describe the process of **advection** in the atmosphere.
15. How does expansion lead to **adiabatic cooling** and compression to **adiabatic warming**?
16. What happens to the temperature of rising air? Of descending air? Why?
17. What is **latent heat**? What role does latent heat play in **condensation** and **evaporation**?

Earth's Solar Radiation Budget (*p. 90*)

18. Approximately what percentage of incoming solar radiation is reflected and scattered away from Earth?
19. Approximately what percentage of incoming solar radiation is transmitted through the atmosphere and absorbed by the surface?
20. Briefly describe how the troposphere is warmed by the Sun.

Variations in Insolation by Latitude and Season (*p. 92*)

21. What is meant by the **angle of incidence** of the Sun's rays?
22. Explain the reasons for the unequal warming (by latitude) of Earth by the Sun.

Land and Water Temperature Contrasts (*p. 95*)

23. How does the **specific heat** of a substance affect its rate of warming?
24. Explain why land warms faster and to a greater extent than water.
25. Explain why land cools faster and to a greater extent than water.

Mechanisms of Global Energy Transfer (*p. 96*)

26. What are the two dominant mechanisms of energy transfer around the world?
27. What is the relative temperature of the ocean current flowing along the west coast of a continent in the midlatitudes? Along the east coast?
28. Describe the basic pattern of ocean currents around the margins of a major ocean basin (the **subtropical gyres**). (Include the relative temperature—either "cool" or "warm"—of each current.) You should be able to sketch the direction of movement and note the relative temperature of major ocean currents on a blank map of an ocean basin.
29. Describe **upwelling** and its cause.

Vertical Temperature Patterns (*p. 99*)

30. What is meant by the **environmental lapse rate**?
31. What is the **average lapse rate** in the troposphere?
32. What is a **temperature inversion**?
33. What is the difference between a radiational inversion and an advectional inversion?

Global Temperature Patterns (*p. 100*)

34. What is an **isotherm**?
35. Where in the world do we find the greatest **average annual temperature ranges**, and where do we find the smallest average annual temperature ranges? Why?
36. What is the **urban heat island (UHI) effect**?

Climate Change and Global Warming (*p. 103*)

37. What is meant by the term **global warming**?
38. How might humans be enhancing the natural greenhouse effect?

Study Questions

1. Why is the sky blue? Why are sunsets orange and red?
2. Why do we say that **evaporation** is a cooling process and **condensation** is a warming process?
3. Why does temperature generally decrease with increasing altitude in the troposphere?
4. "The atmosphere is warmed mostly by Earth's surface rather than directly by the Sun." Comment on the validity of this statement.
5. If the concentration of greenhouse gases were to *decrease* significantly, what would likely happen to global temperatures? Why?
6. Why are seasonal temperature differences greater in the high latitudes than in the tropics?
7. Why are the hottest and coldest places on Earth over land, not over water?
8. How would global temperature patterns be different without the transfer of energy by the circulation of the atmosphere and ocean currents?
9. Using the isotherm maps of average January and July sea-level temperatures (Figures 4-30 and 4-31), explain the influence of latitude, season, land–water contrasts, and ocean currents on global temperature patterns. For example, explain why the following isotherms vary in latitude across the maps:
   - January −1°C (30°F) isotherm in the Northern Hemisphere
   - July 21°C (70°F) isotherm in the Northern Hemisphere

Exercises

1. Make the following conversions between the Celsius and Fahrenheit temperature scales:
   a. 15°C = _____°F
   b. −30°C = _____°F
   c. 102°F = _____°C
   d. 5°F = _____°C
2. If the temperature of the air is 25°C at sea level, use the average lapse rate to calculate the expected temperature at the top of a 5000-meter-high mountain: ______°C.
3. If the air temperature outside a jet airplane flying at 36,000 feet is −40°F, use the average lapse rate to estimate the temperature at sea level below: ______°F.
4. One gram of water and 1 gram of soil ("land") are at a temperature of 20°C (68°F). If 5 calories of heat are added to each, what will be the temperature of the water and the approximate temperature of the soil?

Environmental Analysis SURFACE TEMPERATURES SEEN FROM SPACE

Data Surface Temperatures
https://media.pearsoncmg.com/ph/esm/esm_mcknight_physgeo_13/media/

Polar-orbiting satellites used for climate research gather raw temperature data. With the NASA Worldview tool, for instance, we can use these data to examine mechanisms that control surface temperatures.

Activities

Go to https://worldview.earthdata.nasa.gov/. In the "Layers" tab turn on the "Place Labels" overlay, click "Add Layers" then add the "Latitude-Longitude Lines" overlay (located in "All"), and change the Timeline at the bottom of the map from "Days" to "Months". View several different months in various years.

1. The default display shows cloud cover and land color as if viewed from space. Notice that some regions of Earth are completely white. Where are these regions? Qualitatively, what are temperatures like there?
2. One region is always tan with almost no cloud cover. Where is this region, and what are temperatures like there?

Add the "Land Surface Temperature (Day) Terra/MODISs" overlay; hide the "Corrected Reflectance (True Color)" base layer. Temperature is given in kelvins (see Figure 4-2).

3. Zoom out to see the entire world, and alternate between the most recent February and August land-surface temperatures. Where is the difference between February and August the least? Why?
4. Where is the difference between February and August land-surface temperatures the greatest? Why?

Add the "Sea Surface Temperature (L3, Day, Monthly, Thermal, 4 km) Terra/MODIS" overlay; hide the "Land Surface Temperature (Day) Terra/Modis" overlay.

5. Alternate between the most recent February and August sea-surface temperatures. Are Northern Hemisphere oceans warmer in February or August? Why?
6. Are Southern Hemisphere oceans warmer in February or August? Why?

Add back the "Land Surface Temperature (Day) Terra/ Modis" overlay, and go to the most recent view on the Timeline. Add the "Land Surface Temperature (Night)" overlay. Zoom in to the West Coast of the United States.

7. Alternate between day and night land-surface temperatures. Is the ocean warmer or cooler than the land during the day? At night? Why?

Seeing Geographically in Our National Parks

Look again at the photograph of a meadow full of wildflowers in Kenai Fjords National Park at the beginning of the chapter (p. 78). How strong is solar heating at Kenai Fjords, compared to the Midwest or Florida? How long are the days when the picture was taken, more than 12 hours or less than 12 hours? How do you know?

Mastering Geography®

Looking for additional review and test prep materials? Visit the Study Area in **Mastering Geography** to enhance your geographic literacy, spatial reasoning skills, and understanding of this chapter's content by accessing a variety of resources, including Map Master interactive maps, geoscience animations, *Mobile Field Trips*, videos, *Project Condor* Quadcopter videos, *In the News* RSS feeds, flashcards, web links, self-study quizzes, and an eText version of *McKnight's Physical Geography*.

Seeing Geographically in Our National Parks

Strong winds send snow high into the air at Denali National Park, complicating these hikers' plans. What can you see that suggests that the wind here was exceptionally strong on this day? How is the mountain influencing the speed and direction of the wind?

National Park Name, State code

5

Atmospheric Pressure and Wind

HAVE YOU EVER WONDERED why the wind blows? Or how snow storms can rage with such ferocity, as seen at Denali National Park? In some ways, the answer is simple: wind is air moving from an area of high pressure toward an area of low pressure. However, as we see in this chapter, understanding why pressure differences develop and exactly how air moves between these different areas requires a lengthy explanation!

Atmospheric pressure is a difficult element of weather and climate for us to comprehend. We can more easily appreciate the other three elements—temperature, wind, and moisture—because we can feel warmth, air movement, and moisture. Our bodies, though, cannot sense the relatively small changes in atmospheric pressure that are responsible for significant changes in the atmosphere. We're usually aware of pressure changes only when we experience rapid vertical movement, as in an elevator or on an airplane, when the difference in pressure inside and outside our ears causes them to "pop."

Despite the difficulty of recognizing pressure changes, they are tied closely to the other weather elements, acting on them and responding to them. Because variations in pressure are responsible for wind, pressure and wind are often discussed together, as we do in this chapter. The movement of the atmosphere and the global circulation patterns of wind and pressure are components of several major Earth systems. Not only is the general circulation of the atmosphere a consequence of the receipt of solar energy, it is one of the key mechanisms of energy transfer itself. Furthermore, patterns of wind and pressure are key aspects of the hydrologic cycle—the systematic movement of water around the planet (discussed in Chapters 6 and 9).

As you study this chapter, think about these **Key**Questions:

- **How are areas of high and low atmospheric pressure produced?**
- **What determines the direction in which wind blows?**
- **What are the global patterns of wind and pressure, and how do they form?**
- **What are monsoons, and how do they form?**
- **What is El Niño?**

The Impact of Pressure and Wind on the Landscape

The influence of atmospheric pressure on the landscape is indirect but significant. Wind, which responds to pressure changes, has the energy to transport solid particles. Wind makes vegetation bend and shifts loose dust and sand from place to place. The results are usually temporary, but pressure and wind are major elements of weather and climate. We can't overestimate their interaction with other atmospheric components and processes.

The Nature of Atmospheric Pressure

Gas molecules, unlike those of a solid or liquid, are not strongly bound to one another. Instead, they are in continuous motion, colliding frequently with one another and with any surfaces to which they are exposed. Consider a container in which a gas is confined (**Figure 5-1**). The molecules of the gas zoom around inside the container and collide again and again with the walls. The pressure of the gas is the force the gas exerts on the container walls.

The atmosphere is made up of gases that have mass; the atmosphere has *weight* because this mass is pulled toward Earth by gravity. **Atmospheric pressure** is the force exerted by the weight of those gas molecules on a unit of area on Earth's surface or on any other body—including yours! At sea level, the pressure (the "weight") exerted on the ground by the atmosphere is about 14.7 pounds per square inch—in S.I. units, about 10 newtons (N) per square centimeter (1 newton is the force required to accelerate a 1-kilogram mass 1 meter per second per second).[1] This value decreases with increasing altitude because the farther away you get from Earth, the fewer gas molecules are present above a location.

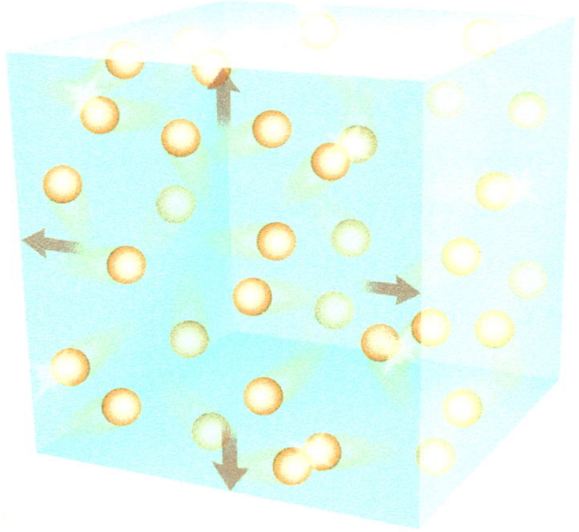

▲ **Figure 5-1** Gas molecules are always in motion. In this closed container, they bounce around, colliding with one another and with the container's walls. These collisions give rise to the pressure exerted by the gas.

▲ **Figure 5-2** The empty plastic bottle on the left was opened and then sealed tightly at an elevation of 3030 meters (9945 feet); when the bottle was brought down to sea level, the surrounding higher atmospheric pressure partially collapsed the bottle. The bottle on the right contains air at sea-level pressure.

The atmosphere exerts pressure on every surface it touches. Pressure is exerted equally in all directions—up, down, sideways, and diagonally. Every square centimeter of any exposed surface—animal, vegetable, or mineral—at sea level is subjected to atmospheric pressure (**Figure 5-2**). We are not sensitive to this ever-present burden of pressure because our bodies contain solids and liquids that are not significantly compressed and air spaces that are at the same pressure as the surrounding atmosphere. In other words, outward pressure and inward pressure within us balance exactly.

LearningCheck 5-1 **Explain why atmospheric pressure decreases with altitude. (Answer on p. AK-2)**

Factors Influencing Atmospheric Pressure

The pressure, temperature, and density of a gas are all related. If one of these variables changes, it can cause changes in the other two variables.

The Ideal Gas Law: The relationship among pressure, temperature, and density can be summarized by an equation called the *ideal gas law*:

$$P = \rho RT$$

where P is pressure, ρ ("rho") is density, R is a constant of proportionality, and T is temperature. In words, this equation says that pressure (P) increases if density remains constant but temperature (T) increases, and that pressure also increases if temperature remains constant but density (ρ) increases.

In a closed container (a sealed jar, for example), the relationship among pressure, temperature, and density is straightforward. However, the atmosphere is not a closed container; these variables cannot be held constant, so the

[1]The phrase "per second per second" may seem strange. However, a newton describes the force required to accelerate a mass: in other words, the force required to change its speed or direction.

cause-and-effect relationships are quite complex. It is still useful to look at how pressure varies with density, with temperature, and with the vertical movement of air.

Density and Pressure Relationships: Density is the mass of matter in a unit volume. For example, if you have a 10-kilogram cube of material with edges 1 meter long, the density of that material is 10 kilograms per cubic meter (10 kg/m^3). The density of solid material is the same on Earth or the Moon or in space. The density of liquids varies very slightly from one place to another, but that of gases varies greatly with location. Gas density changes easily because a gas is free to expand as far as the environmental pressure will allow.

For example, if 10 kilograms of gas in a container has a volume of 1 cubic meter, the gas density is 10 kg/m^3. If you transfer all the gas to a container with twice the volume (2 cubic meters), the gas expands to fill the larger volume. The same number of gas molecules are now spread through a volume twice as large, so the gas density is half what it was before, or 5 kg/m^3 (10 kg divided by 2 cubic meters).

The pressure exerted by a gas is proportional to its density, but in the atmosphere, we also have to consider temperature and air movement. The atmosphere is held to Earth by the force of gravity, which prevents the gas molecules from escaping into space. *Atmospheric pressure is simply the weight of the atmosphere above a location.* At lower altitudes, more molecules are above a location (Figure 5-3). Because more molecules are overhead, pressure is higher. At higher altitudes, fewer molecules are overhead, and pressure decreases correspondingly.

Temperature and Pressure Relationships: If air is warmed, the speed of molecules increases (see Chapter 4). This increase in speed produces a greater force to their collisions and results in higher pressure. Therefore, if other conditions remain the same (in particular, if volume is held constant), an increase in the temperature of a gas produces an increase in pressure, and a decrease in temperature produces a decrease in pressure.

You might conclude that the air pressure is high on warm days and low on cold days. Such is not usually the case, however; warm air can be associated with higher or lower atmospheric pressure. Although this seems contradictory, recall that we made the qualifying statement "if other conditions remain the same." Atmospheric pressure depends on more than just temperature and density.

Dynamic Influences on Air Pressure: Surface air pressure may also be influenced by "dynamic" factors. In other words, air pressure may be influenced by the vertical movement of the air, as well as convergence and divergence. As a generalization, descending air tends to be associated with relatively high pressure at the surface, whereas rising air tends to be associated with relatively low pressure at the surface.

In short, atmospheric pressure is affected by differences in air density, air temperature, and air movement.

▲ **Figure 5-3** In the upper atmosphere, gas molecules are far apart so density is lower, and fewer molecules are overhead so atmospheric pressure is lower. In the lower atmosphere, the molecules are closer together so density is higher, and more molecules exist above so pressure is higher.

It is important for us to be alert to these linkages, but it is often difficult to predict how a change in one variable will affect the others in a specific instance. Nevertheless, we can make some useful generalizations about the factors associated with areas of high pressure and low pressure near the surface:

- Strongly descending air is usually associated with high pressure at the surface—a **dynamic high**.
- Very cold surface conditions are often associated with high pressure at the surface—a **thermal high**.
- Strongly rising air is usually associated with low pressure at the surface—a **dynamic low**.
- Very warm surface conditions are often associated with relatively low pressure at the surface—a **thermal low**.

Surface pressure conditions usually can be traced to one of these factors being dominant.

LearningCheck 5-2 **Which is more likely to be associated with descending air: high or low atmospheric pressure at the surface? With rising air?**

Mapping Pressure with Isobars

Atmospheric pressure is measured with instruments called **barometers**. The first liquid-filled barometers date back to the 1600s, and measurement scales based on the height of a column of mercury are still in use (average sea-level pressure using a mercury barometer is 760 millimeters or 29.92 inches). For meteorologists in the United States, however, the most common unit of measure for atmospheric pressure is the **millibar**. The millibar (mb) is an expression of force per surface area. Average sea-level pressure is 1013.25 millibars. The S.I. unit used to describe pressure is the *pascal* (Pa; 1 Pa = 1 newton/m^2). One millibar equals 100 pascals or 1 *hectopascal* (hPa). In some countries the *kilopascal* is used (kPa; 1 kPa = 10 mb).

Highs and Lows: Pressure differences are shown on a weather map with isolines of equal pressure called **isobars** (Figure 5-4). The pattern of the isobars reveals the horizontal distribution of pressure in the region under consideration. Prominent on such maps are roughly circular or oval areas characterized as either "high pressure" or "low pressure." These **highs** and **lows** represent relative conditions: pressure that is higher or lower than that of the surrounding areas. In a similar way, a **ridge** is an elongated area of relatively high pressure, whereas a **trough** is an elongated area of relatively low pressure. It is important to keep in mind the relative nature of pressure centers. Notice in Figure 5-4, for example, that a pressure of 1008 millibars could be either "high" or "low," depending on the pressure of the surrounding areas. Notice also that pressure varies much less horizontally than it does vertically.

Surface Pressure Maps: On most maps showing surface air pressure, actual pressure readings are adjusted to represent pressures at a common elevation, usually sea level. This is done because, as we first saw in Chapter 3, pressure decreases rapidly with increasing altitude (Table 5-1). Significant variations in pressure readings are likely at different weather stations simply because of differences in elevation.

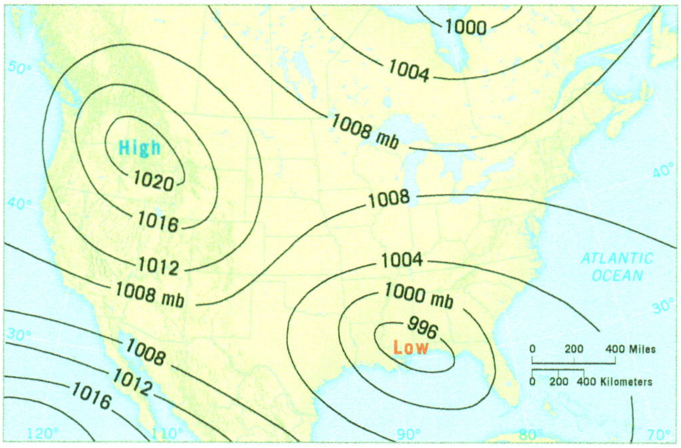

▲ **Figure 5-4** Isobars are lines connecting points of equal atmospheric pressure. When they have been drawn on a weather map, we can determine the location of high-pressure and low-pressure centers. This simplified weather map shows pressure in millibars.

TABLE 5-1 Atmospheric Pressure Variation with Altitude

	Altitude	
Pressure (millibars)	**Kilometers**	**Miles**
100	16	10
300	9.1	5.7
500	5.5	3.4
700	2.8	1.8
850	1.5	0.9
1013	0	0

This pressure change is most rapid at lower altitudes, and the rate of decrease diminishes significantly above about 3 kilometers (about 2 miles).

Pressure Gradients: As with other types of isolines, the relative closeness of isobars indicates the horizontal rate of pressure change, or **pressure gradient**. The pressure gradient can be thought of as representing the "steepness" of the pressure "slope" (or more correctly, the abruptness of the pressure change over a distance), a characteristic that has a direct influence on wind, which we discuss next.

The Nature of Wind

Animation
Development of Wind Patterns
https://media.pearsoncmg.com/ph/esm/esm_mcknight_physgeo_13/media/

The atmosphere is virtually always in motion. Air is free to move in any direction, its specific movements being shaped by a variety of factors. Some airflow is weak and brief; some is strong and persistent. Atmospheric motions often involve both horizontal and vertical movement.

Wind refers to horizontal air movement. Small-scale vertical motions are normally referred to as *updrafts* and *downdrafts*; large-scale vertical motions are *ascents* and *subsidences*. Although both vertical and horizontal motions are important in the atmosphere, much more air is involved in horizontal movements than in vertical ones. As we saw in Chapter 4, the horizontal movement of air in wind is a key process in the transfer of energy by means of *advection*.

Direction of Movement

Insolation is the ultimate cause of wind because all winds originate from the same basic sequence of events: unequal warming of Earth's surface produces temperature gradients that generate pressure gradients, and these pressure gradients set air into motion. Wind represents nature's attempt to even out the uneven distribution of air pressure across Earth's surface.

Air generally begins to flow from areas of higher pressure toward areas of lower pressure, but in practice wind rarely blows *directly* from high pressure to low pressure. If Earth did not rotate, that is precisely what would happen—a direct movement of air from a high-pressure region to a low-pressure region. However, the actual direction of wind is

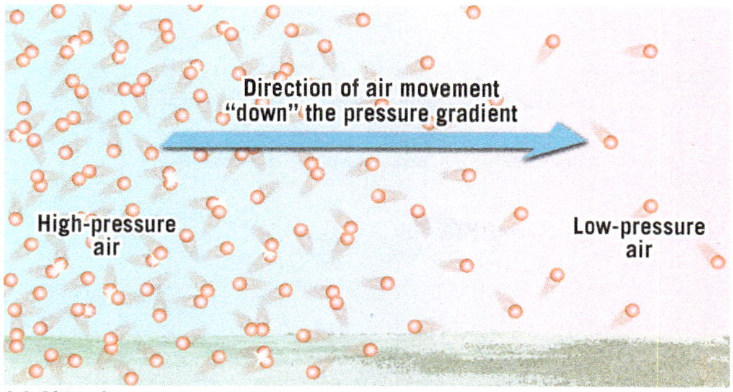

(a) Side view

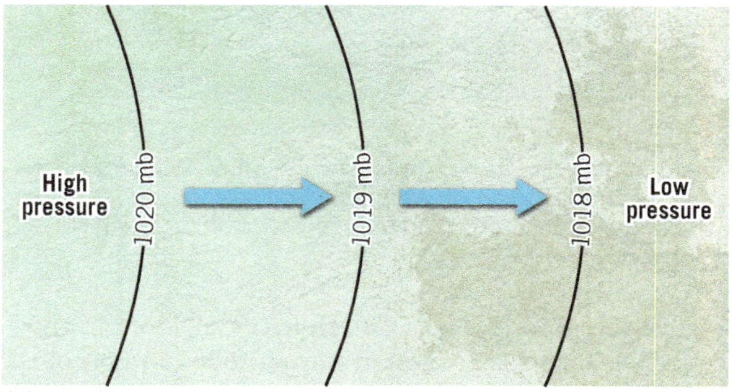

(b) Top view

▲ **Figure 5-5** (a) Air tends to move from areas of higher pressure toward areas of lower pressure. We say this movement is "down the pressure gradient." (b) If the pressure gradient force were the only force involved, air would flow perpendicular to the isobars (would cross the isobars at 90°).

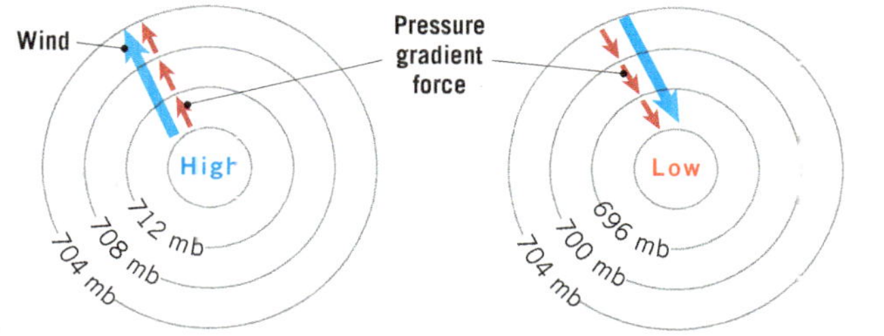

(a) If the pressure gradient force were the only factor, wind would blow "down" the pressure gradient away from higher pressure and toward lower pressure, crossing the isobars at an angle of 90°.

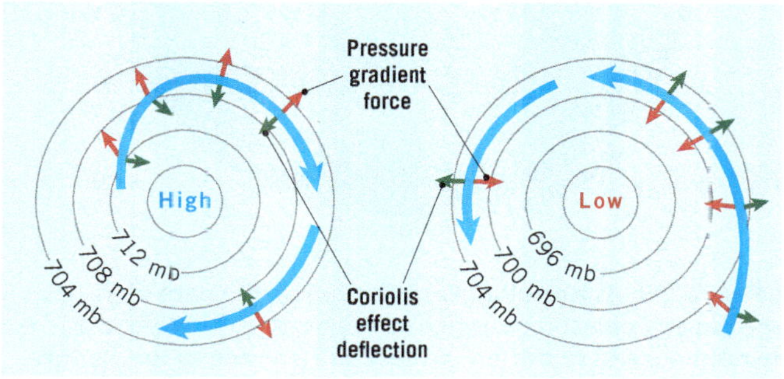

(b) In the upper atmosphere the balance between the pressure gradient force and the Coriolis effect results in geostrophic wind blowing parallel to the isobars.

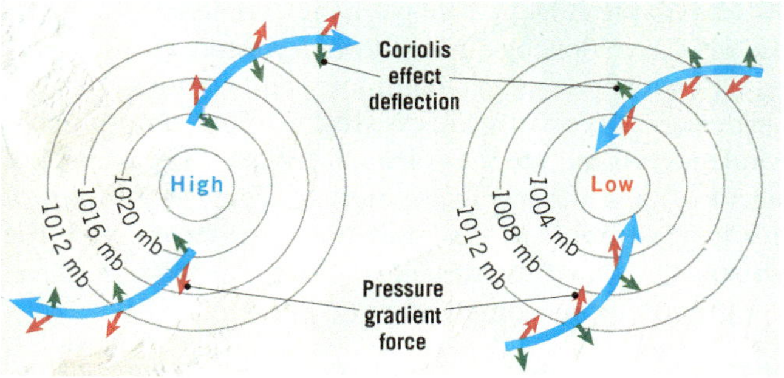

(c) In the lower atmosphere, friction slows the wind (which results in less Coriolis effect deflection) and so wind diverges clockwise out of a high and converges counterclockwise into a low in the Northern Hemisphere.

▲ **Figure 5-6** Wind direction is influenced by a combination of three factors: pressure gradient force, Coriolis effect, and friction. (a) Hypothetical pattern if pressure gradient force was the only factor. (b) Geostrophic winds in the upper atmosphere (above about 1000 m/3300 ft). (c) In the friction layer of the lower atmosphere, friction slows the wind (which results in less Coriolis effect deflection), so wind diverges out of a high and converges into a low.

determined by the interaction of three factors: the pressure gradient force, the Coriolis effect, and friction.

Pressure Gradient Force: If one area is at a higher pressure than another, air will begin to move from the higher pressure area toward the lower pressure area in response to the *pressure gradient force* (**Figure 5-5**). If you visualize a high-pressure area as a pressure "hill" and a low-pressure area as a pressure "valley," imagine air flowing "down" the pressure gradient in the same manner that water flows down a hill. (Keep in mind that these terms are metaphorical—air is not necessarily actually flowing downhill.)

The pressure gradient force acts at right angles to the isobars in the direction of the lower pressure. If no other factors needed to be considered, this is the way the air would move, crossing the isobars at 90° (**Figure 5-6a**). However, such a flow rarely occurs in the atmosphere.

The Coriolis Effect: Because Earth rotates, any object moving freely near Earth's surface is deflected by the *Coriolis effect* (**Figure 5-7**). Recall from Chapter 3 the significant aspects of the Coriolis effect:

Animation
Coriolis Effect
https://media.pearsoncmg.com/ph/esm/esm_mcknight_physgeo_13/media/

- Regardless of the initial direction of motion, any freely moving object appears to deflect to the right of its direction of motion in the Northern Hemisphere and to the left in the Southern Hemisphere.
- The apparent deflection is strongest at the poles and decreases progressively toward the equator, where the deflection is zero.
- A fast-moving object is deflected more than a slower one—although the Coriolis effect influences direction of movement only; it does not change an object's speed.

The Coriolis effect has an important influence on the direction of wind flow. The Coriolis effect deflection acts at 90° from the direction of movement—to the right in the Northern Hemisphere and to the left in the Southern Hemisphere. Once air begins to move, there is a "battle"

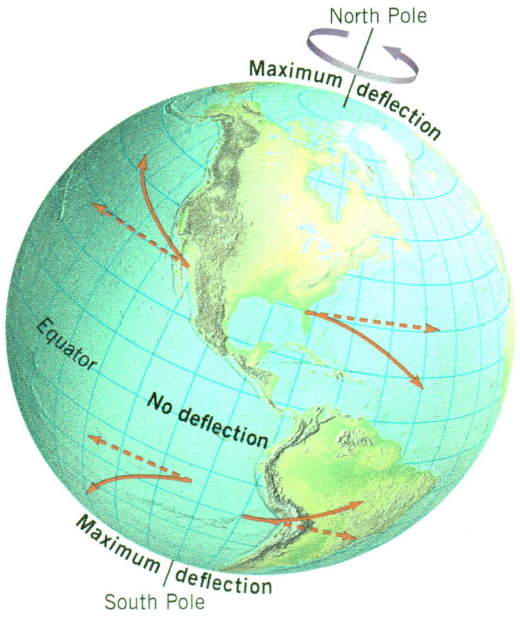

▲ Figure 5-7 As a result of Earth's rotation, the path of a free-moving object is deflected to the right in the Northern Hemisphere and to the left in the Southern Hemisphere by the Coriolis effect.

between the pressure gradient force moving air from higher toward lower pressure and the deflection of the Coriolis effect 90° from its pressure-gradient path: the Coriolis effect prevents the wind from flowing directly down a pressure gradient.

Where the pressure gradient force and Coriolis effect are balanced—as is usually the case in the upper atmosphere—wind moves parallel to the isobars and is called a **geostrophic wind**[2] (Figure 5-6b). Actually, most winds are geostrophic or nearly geostrophic in that they flow nearly parallel to the isobars. Only near the surface is another factor—friction—significant, further complicating the situation.

Friction: In the lowest portions of the troposphere, a third force influences wind direction: *friction*. The frictional drag of Earth's surface slows wind movement, so the influence of the Coriolis effect is reduced there. (Recall that rapidly moving objects are deflected more by the Coriolis effect than are slowly moving objects.) Instead of blowing perpendicular to the isobars (along the pressure gradient) or parallel to them (where the pressure gradient force and the Coriolis effect balance), the wind gradually crosses the isobars at angles between 0° and 90° (Figure 5-6c). In essence, friction reduces wind speed, which in turn reduces the Coriolis effect deflection. Although the Coriolis effect does introduce a deflection to the right (in the Northern Hemisphere), the pressure gradient "wins the battle," and air flows into an area of low pressure and away from an area of high pressure. As a general rule, the frictional influence is greatest near Earth's surface and diminishes progressively upward (Figure 5-8). Thus, the angle of wind flow across the isobars is greatest at low altitudes and becomes smaller at increasing elevations. The **friction layer** of the atmosphere extends to only about 1000 meters (approximately 3300 feet) above the surface. Higher than that, most winds follow a geostrophic or near-geostrophic course.

[2]Strictly speaking, geostrophic wind is found only in areas where the isobars are parallel and straight; gradient wind is a more general term used to describe wind flowing parallel to the isobars. In this book we use the term geostrophic to mean all wind blowing parallel to the isobars.

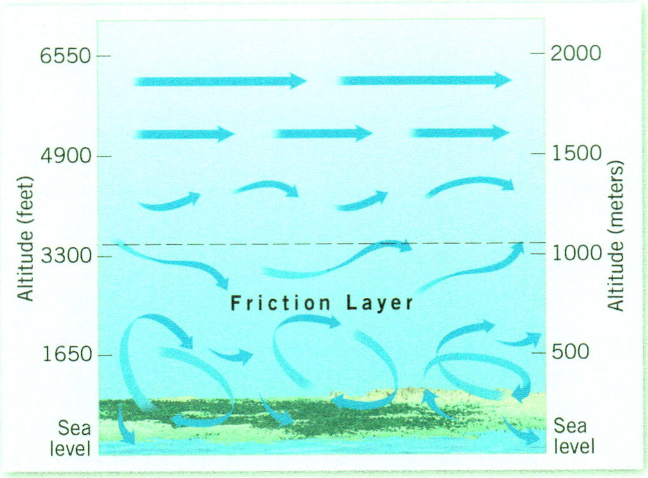

▲ Figure 5-8 Near Earth's surface friction causes wind flow to be turbulent and irregular. Above the friction layer (altitudes higher than about 1000 meters/3300 feet), the wind flow is generally smoother and faster.

LearningCheck 5-3 **Explain how friction influences the direction of wind flow.**

Wind Speed

Thus far, we have been considering the direction of wind movement and paying little attention to speed (other than to note that wind speed influences the amount of Coriolis effect deflection). Although some complications are introduced by factors such as *inertia* (the tendency of an object to resist changes in its motion), the speed of wind flow is determined primarily by the pressure gradient. If the gradient is steep, the air moves swiftly; if the gradient is gentle, the speed is slow. This relationship can be portrayed in a simple diagram (Figure 5-9). The closeness of the isobars indicates the steepness of the pressure gradient.

Describing Wind Speed: In meteorology, wind speed is frequently described in terms of knots (nautical miles per hour). Recall from Chapter 1 that a nautical mile is a bit longer than a "statute" mile; 1 knot is equivalent to 1.15 statute miles per hour, or 1.85 kilometers per hour. When

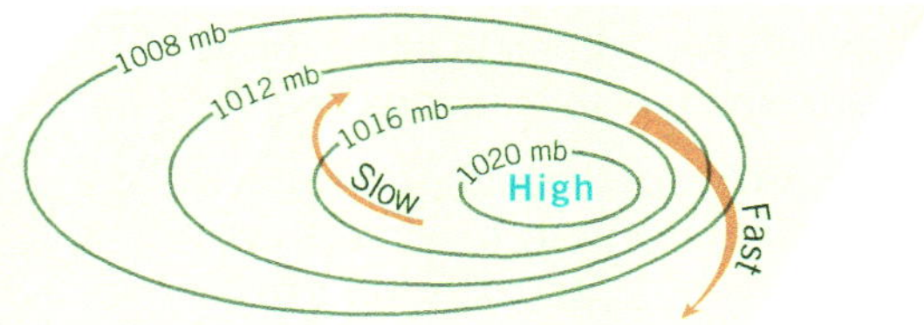

▲ Figure 5-9 Wind speed is determined by the pressure gradient, which is indicated by the spacing of isobars. Where isobars are close together, the pressure gradient is "steep," and wind speed is high; where isobars are far apart, the pressure gradient is "gentle," and wind speed is low.

speeds are given for ships and airplanes, the knot is the most common unit of measurement.

Global Variations in Wind Speed: Over most of the world most of the time, surface winds are relatively gentle. For example, annual average wind speeds across the United States are generally between 7 and 19 knots. Cape Denison in Antarctica holds the dubious distinction of being the windiest place on Earth, with an annual average wind speed of 38 knots. The highest surface wind speed ever recorded was on Barrow Island, Australia, during Typhoon Olivia in 1996: 220 knots (253 mph; 407 km/h).

The most persistent winds are typically in coastal areas or high mountains. In the United States, the relatively open terrain of the Great Plains results in high average wind speeds. Locations with persistent winds are often suitable for facilities that generate electricity by using wind power—see the box *Energy for the 21st Century: Wind Power*.

Wind speed is quite variable from time to time and from one altitude to another, usually increasing with height. Winds tend to move faster above the friction layer. As we shall see in subsequent sections, the strongest tropospheric winds are usually found in what are called *jet streams* at intermediate levels or in violent storms near Earth's surface.

Cyclones and Anticyclones

Distinct and predictable wind-flow patterns develop around all high-pressure and low-pressure centers—patterns determined by the pressure gradient, Coriolis effect, and friction. A total of eight circulation patterns are possible, four in the Northern Hemisphere and four in the Southern Hemisphere. Within each hemisphere, two patterns are associated with high-pressure centers and two with low-pressure centers (Figure 5-10).

Animation
Cyclones and Anticyclones
https://media.pearsoncmg.com/ph/esm/esm_mcknight_physgeo_13/media/

High-Pressure Wind Patterns: A high-pressure center is known as an **anticyclone**, and the flow of air associated with it is described as being *anticyclonic*. The four patterns of anticyclonic circulation are shown in Figure 5-10:

1. In the upper atmosphere of the Northern Hemisphere, the winds move clockwise in a geostrophic manner parallel to the isobars.
2. In the friction layer (lower altitudes) of the Northern Hemisphere, there is a divergent clockwise flow, with the air spiraling out away from the center of the anticyclone.
3. In the upper atmosphere of the Southern Hemisphere, there is a counterclockwise, geostrophic flow parallel to the isobars.
4. In the friction layer of the Southern Hemisphere, the pattern is a mirror image of the Northern Hemisphere, with air diverging in a counterclockwise pattern.

Low-Pressure Wind Patterns: Low-pressure centers are called **cyclones**, and the associated wind movement is said to be *cyclonic*. As with anticyclones, Northern Hemisphere cyclonic circulations are mirror images of their Southern Hemisphere counterparts:

5. In the upper atmosphere of the Northern Hemisphere, air moves counterclockwise in a geostrophic pattern parallel to the isobars.
6. In the friction layer of the Northern Hemisphere, a converging counterclockwise flow exists.
7. In the upper atmosphere of the Southern Hemisphere, a clockwise geostrophic flow occurs paralleling the isobars.
8. In the friction layer of the Southern Hemisphere, the winds converge in a clockwise spiral.

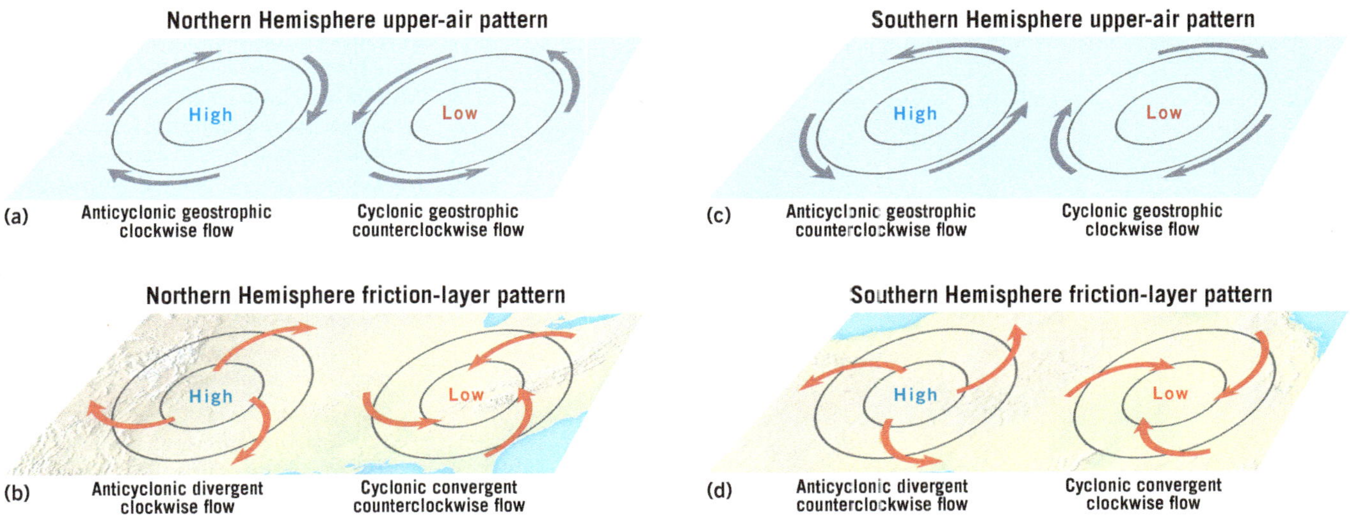

▲ **Figure 5-10** The eight basic patterns of air circulation around pressure cells. In the Northern Hemisphere (a) upper atmosphere (above the friction layer), the wind is geostrophic, blowing parallel to the isobars; (b) at the surface, wind diverges from a high and converges into a low. In the Southern Hemisphere (c) upper atmosphere and (d) surface, wind circulates in the opposite direction.

Energy for the 21st Century

Wind Power

• **Stephen Stadler,** Oklahoma State University

Electricity generated from wind power is here to stay. It has become a significant industry in a world looking toward the inevitable depletion of fossil fuel reserves and efforts to decrease carbon emissions. More than 80 countries produce wind power commercially. Worldwide, wind is among the fastest growing power source, increasing its capacity by double digits each year and approaching 651,000 megawatts (1 megawatt = 1 million watts). A 1.5-megawatt turbine running at capacity can power more than 750 average American homes. Wind power generates more than 5 percent of the world's electricity, and that amount is expected to increase significantly. For example, more than 47 percent of Denmark's electricity is wind generated.

Wind Geography: In wind power generation, it can be said that geography is everything. Physical geography is directly relevant because windier climates can help turbines generate at maximum capacity more of the time. Geographers have used GIS to pair landscapes (topography and vegetation) with spatial displays of the wind resource in order to pinpoint potential wind farm sites (**Figure 5-A**). The best land placements are on high ground away from trees and other obstructions to minimize friction with Earth's surface. Windy high mountain ranges are usually excluded because it is impractical to assemble components in these remote locations. Offshore locations almost always have better wind resources than nearby coastal areas, but the costs of development in a marine environment are greater.

Human geography plays a large role in the siting of turbines. Some windy places, such as lowly populated parts of the Great Plains of the United States and Canada, are lightly exploited because long-distance transmission lines are scarce.

Wind Turbines: Today's utility-scale turbines are massive machines costing millions of dollars per unit (**Figure 5-B**). Turbines are placed on tall towers because wind availability increases dramatically with elevation above the surface. A large turbine has three blades typically mounted 80 meters (260 feet) or more above the surface. The blades are more than 35 meters (115 feet) long, turn 12 to 15 revolutions per minute (rpm), and power a low-speed shaft attached to the hub. In a gearbox, low rpms are converted to high rpms on a shaft turning an electric generator. Typically, the electricity from large turbines goes directly to power transmission lines—battery storage is very expensive.

Recent Developments: Wind turbine technology is improving with worldwide contributions to engineering that decrease the cost of wind energy. Some new turbines have direct drives with no gearboxes to maintain. Blade materials keep getting lighter, so they can be longer and the towers higher, increasing efficiency. Improvements in resistance to the corrosive oceanic atmosphere have allowed reliable tapping of offshore winds. Size constraints of transportation and assembly have become greatly reduced, so turbines can be built larger.

▲ **Figure 5-B** A small portion of a wind farm in west central Oklahoma. Each of these turbines has a capacity of 1.5 million watts. The agricultural land use is relatively undisturbed.

Advantages of Wind Power: Electricity from wind power has notable advantages. "Fuel" is the wind, which is free. Wind power has no "carbon footprint" because no fuel is burned. Turbines require small amounts of land. Also, wind generation holds much promise for the developing world, where smaller turbines can serve rural populations not well connected to regional power transmission grids.

Downsides of Wind Power: The biggest drawback is that wind is not constant, even in the windiest areas. If an uninterrupted power supply is necessary, then wind generation must be mixed with other power sources. Nominally, the turning blades are a hazard to birds and bats, although the towering heights of modern turbines and the visibility of the large blades have minimized this problem. They are also cost-prohibitive in lower income areas and poorer countries. New advancements are helping, but all the cost is still up front. Finally, there is an aesthetic perspective: although the machines appear graceful to some, they are ugly to others.

Questions

1. The United States and Canada use only a small percentage of their huge wind resources. Explain why this is so.
2. Speculate on the relative environmental harm of wind turbines as compared with that of coal, natural gas, and nuclear sources of electrical generation.
3. Consider the rural landscape closest to your home. Explain why this area would or would not be suitable for commercial wind power development.

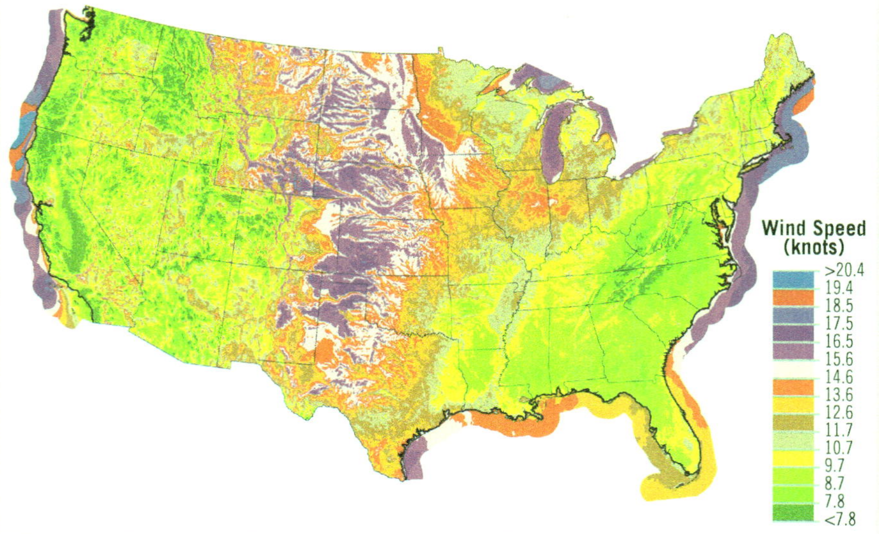

▲ **Figure 5-A** Average land and offshore wind speeds. The speeds are estimated for a height of 80 m (260 feet) above the surface. The Great Plains tend to have high-speed winds in all seasons; the strong coastal winds in California result from heating in the interior Central Valley.

Cyclonic patterns in the lower troposphere may be puzzling at first glance because the wind seems to defy the Coriolis effect: in the Northern Hemisphere, the wind seems to "bend" to the left instead of the right. Remember, however, that wind direction is the balance of several forces acting in different directions (see Figure 5-6c): as air begins to flow down the pressure gradient into a low, the Coriolis effect does indeed deflect the wind to the right; the combination of the pressure gradient "pulling" air into the low and the Coriolis effect nudging it to the right results in the counterclockwise flow into a Northern Hemisphere cyclone.

LearningCheck 5-4 **In the Northern Hemisphere, what is the pattern of wind circulation associated with a surface cyclone? With a surface anticyclone?**

Vertical Movement within Cyclones and Anticyclones: A prominent vertical component of air movement is also associated with cyclones and anticyclones. Air descends in anticyclones and rises in cyclones (Figure 5-11). Such motions are particularly notable in the lower troposphere. We can visualize the anticyclonic pattern as upper air sinking down into the center of the high and then diverging near the ground surface. Opposite conditions prevail in a low-pressure center, with the air converging horizontally into the cyclone and then rising. These patterns match our earlier generalizations about pressure: descending air is associated with high pressure at the surface, and rising air is associated with low pressure at the surface.

Notice in Figure 5-11 that cyclones and rising air are associated with clouds, whereas anticyclones and descending air are associated with clear conditions. We explain the reasons for this in Chapter 6.

Well-developed cyclones and anticyclones often "lean" with height—they are not absolutely vertical.

LearningCheck 5-5 **Describe the direction of vertical air movement within a cyclone and within an anticyclone.**

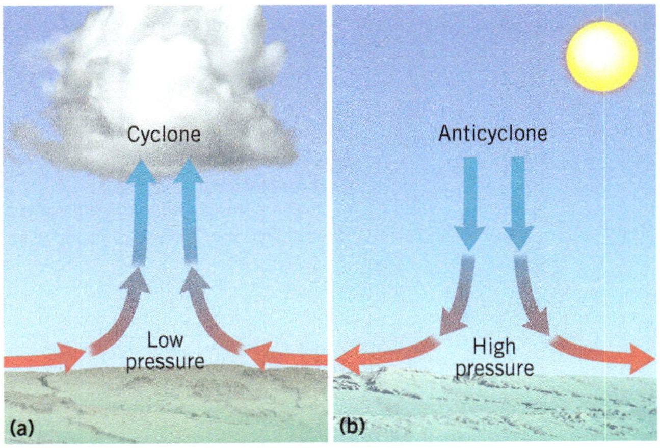

▲ Figure 5-11 (a) In a cyclone (low-pressure cell), air converges and rises. (b) In an anticyclone (high-pressure cell), air descends and diverges.

The General Circulation of the Atmosphere

Earth's atmosphere is an extraordinarily dynamic medium. It is constantly in motion, responding to the various forces described previously as well as to a variety of more localized conditions. Most important to an understanding of geography is the general pattern of circulation, which involves major semipermanent conditions of both wind and pressure. This circulation is the principal mechanism for both latitudinal and longitudinal energy transfer through *advection*. As a control of world climate patterns, the general circulation is exceeded by only the global pattern of insolation.

Idealized Circulation Patterns

Hypothetical Pattern of Nonrotating Earth: If Earth were a nonrotating sphere with a uniform surface, we could expect a very simple global atmospheric circulation pattern (Figure 5-12). The greater amount of solar warming in the equatorial region would produce a band of low pressure around the world, and radiational cooling at the poles would develop caps of high pressure in those areas. Surface winds in the Northern Hemisphere would flow directly "down the pressure gradient" from north to south, whereas those in the Southern Hemisphere would follow a similar gradient from south to north. Warm air would rise at the equator in a large convection cell and flow toward the poles (south to north in the Northern Hemisphere and north to south in the Southern Hemisphere), where cold air would subside into the polar highs.

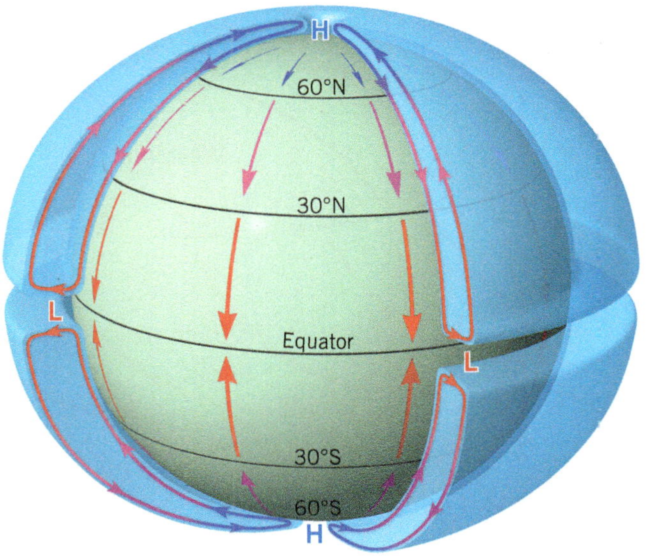

▲ Figure 5-12 Wind circulation patterns would be simple if Earth's surface were uniform (no distinction between continents and oceans) and if the planet did not rotate. High pressure at the poles and low pressure at the equator would produce northerly surface winds in the Northern Hemisphere and southerly surface winds in the Southern Hemisphere.

The Hadley Cells: Earth does rotate, however, and has an extremely varied surface as well. Consequently, the broad-scale circulation pattern of the atmosphere is much more complex than that shown in Figure 5-12. Only the tropical regions have a complete vertical convective circulation cell. Similar cells have been postulated for the middle and high latitudes, but observations indicate that the midlatitude and high-latitude cells either do not exist or are weakly and sporadically developed.

The low-latitude cells—one north and one south of the equator—are gigantic convection systems (Figure 5-13). These two prominent tropical convection cells are called Hadley cells, after George Hadley (1685–1768), an English meteorologist who, in 1735, first conceived the idea of enormous convective circulation cells.

Around the world in equatorial latitudes, warm air rises, producing a region of relative low pressure at the surface. This air ascends to great heights, mostly in thunderstorm updrafts. By the time this air reaches the upper troposphere at elevations of about 15 kilometers (50,000 feet), it has cooled. The air then spreads north and south, moving poleward, eventually descending at latitudes of about 30° N and 30° S, where it forms bands of high pressure at the surface (Figure 5-14). One portion of the air diverging from these surface high-pressure zones flows toward the poles, whereas another portion flows back toward the equator—where the Northern and Southern Hemisphere components converge and the warm air rises again.

LearningCheck 5-6 Describe and explain the pattern of air movement within the Hadley cells.

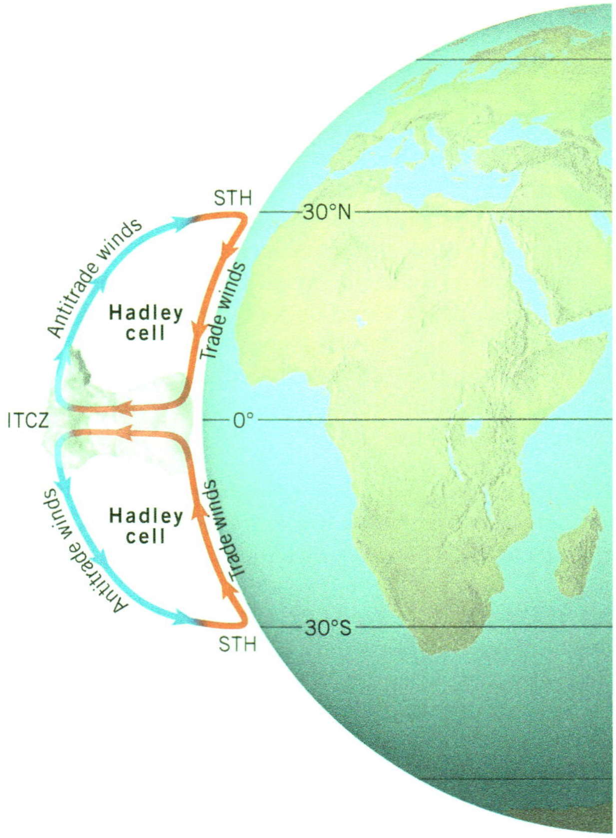

▲ **Figure 5-13** Hadley cells, distinct cells of vertical circulation, occur in tropical latitudes. The equatorial air rises to some 12 to 15 kilometers (40,000 to 50,000 feet) in the intertropical convergence zone (ITCZ) before spreading poleward. This air descends at about 30° N and 30° S into subtropical high-pressure (STH) cells. The vertical dimension of the Hadley cells is considerably exaggerated in this idealized diagram.

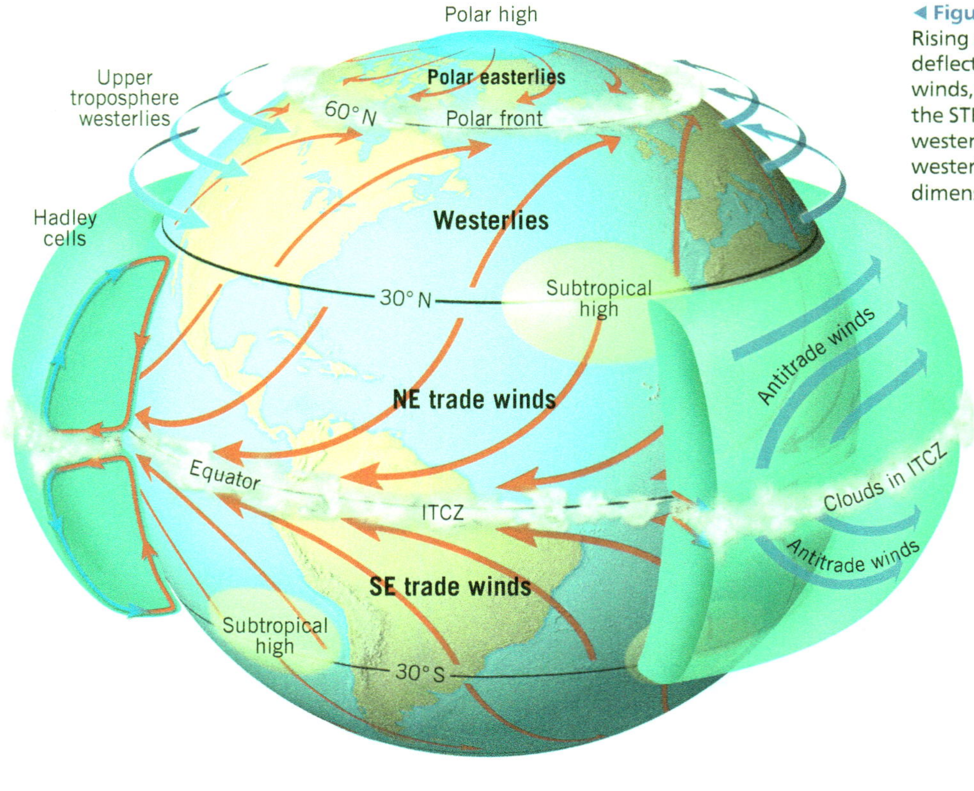

◀ **Figure 5-14** Idealized global circulation. Rising air of the ITCZ in the Hadley cell is deflected aloft and forms westerly antitrade winds, while surface winds diverging from the STH form easterly trade winds and the westerlies. The upper atmosphere flow of the westerlies is shown by blue arrows. (Vertical dimension exaggerated considerably.)

Animation
Global Atmospheric Circulation
https://media.pearsoncmg.com/ph/esm/esm_mcknight_physgeo_13/media/

Components of the General Circulation

Although the Hadley cell model is a simplification of reality, it is a useful starting point for understanding the main components of the general circulation of the atmosphere. The basic pattern has seven surface components of pressure and wind and one prominent upper troposphere wind component, intimately linked together. The Northern Hemisphere and Southern Hemisphere patterns are mirror images of each other. From the equator to the poles, the components are as follows:

1. Intertropical convergence zone (ITCZ)
2. Trade winds
3. Subtropical highs
4. Westerlies
5. Jet Streams (upper troposphere winds)
6. Polar front (subpolar lows)
7. Polar easterlies
8. Polar highs

The pattern of general circulation within the troposphere is essentially a closed system, with neither a beginning nor an end, so we could begin describing it almost anywhere. Rather than start at the equator or the poles, however, it is helpful to begin our discussion of the general circulation in the subtropical latitudes of the five major ocean basins—where the descending air of the Hadley cells becomes the "source" of two of the major surface wind systems: the *trade winds* and the *westerlies*. The third major surface wind system, the *polar easterlies*, develops out of the *polar highs*.

Subtropical Highs

Each ocean basin has a large semipermanent high-pressure cell centered at about 30° of latitude called a **subtropical high (STH)** (Figure 5-15). These gigantic anticyclones, with an average diameter of perhaps 3200 kilometers (2000 miles), develop from the descending air of the Hadley cells. The STHs represent intensified cells of high pressure in two general regions of high pressure, one in each hemisphere, that extend around the world at these latitudes. The high-pressure regions are significantly broken up over the continents, especially in summer, when inland temperatures produce thermal lows. The STHs normally persist over the ocean basins throughout the year because temperatures and pressures there remain nearly constant.

The STHs are usually elongated east–west and tend to be centered in the eastern portions of an ocean basin (in other words, just off the west coasts of continents). Their positions shift a few degrees poleward in summer and a few degrees equatorward in winter. The STHs are so persistent that some have been given proper names, such as the *Azores High* (also called the Bermuda High) in the North Atlantic and the *Hawaiian High* (also called the Pacific High) in the North Pacific (Figure 5-16).

Associated with these high-pressure cells is a general subsidence of air from higher altitudes in the form of a broad-scale, gentle downdraft. A permanent feature of the STHs is a *subsidence temperature inversion* that covers wide areas in the subtropics.

Weather of the Subtropical Highs: Within an STH, the weather is nearly always clear, warm, dry, and calm. We see in the next chapter that descending air does not form clouds and therefore cannot produce rain, so these areas are characterized by dry conditions and warm, subtropical sunshine. Thus, it comes as no surprise that the subtropical highs coincide with many of the world's major deserts.

Subtropical highs are also characterized by a general absence of wind: in the center of an STH, air is primarily subsiding; horizontal air movement and divergence begin toward the edges. These regions are sometimes called the **horse latitudes**, presumably because some sixteenth- and seventeenth-century sailing ships were becalmed there and their cargos of horses were thrown overboard to conserve drinking water.

The air circulation pattern around an STH is anticyclonic: diverging clockwise in the Northern Hemisphere and counterclockwise in the Southern Hemisphere. In essence, we can think of the STHs as gigantic "wind wheels" whirling in the lower troposphere, fed with air sinking from above (Figure 5-17). The winds are not dispersed uniformly around an STH, however, but are concentrated on the northern and southern sides.

Although the global flow of air is essentially a closed circulation from a viewpoint at Earth's surface, the STHs are the source of two of the world's three major surface wind systems: the *trade winds* and the *westerlies*.

LearningCheck 5-7 **Describe the general locations and the kind of weather associated with subtropical highs.**

Trade Winds

Diverging from the equatorward sides of the subtropical highs is the major wind system of the tropics—the **trade winds**. These winds cover most of Earth between about latitude 25° N and latitude 25° S (see Figure 5-17). They are particularly prominent over oceans but tend to be significantly interrupted and modified over landmasses. Because of the vastness of Earth in tropical latitudes and because most of this expanse is oceanic, the trade winds dominate more of the globe than do any other wind system. The trade winds are predominantly "easterly" winds—that is, they generally blow from east to west. In meteorology, winds are named for the direction *from which they blow*: an easterly wind blows from east to west, a westerly wind blows from the west, and so forth.

In the Northern Hemisphere, the trade winds usually blow from the northeast (and are sometimes called the *northeast trades*); south of the equator, they are from the southeast (the *southeast trades*). This general pattern has

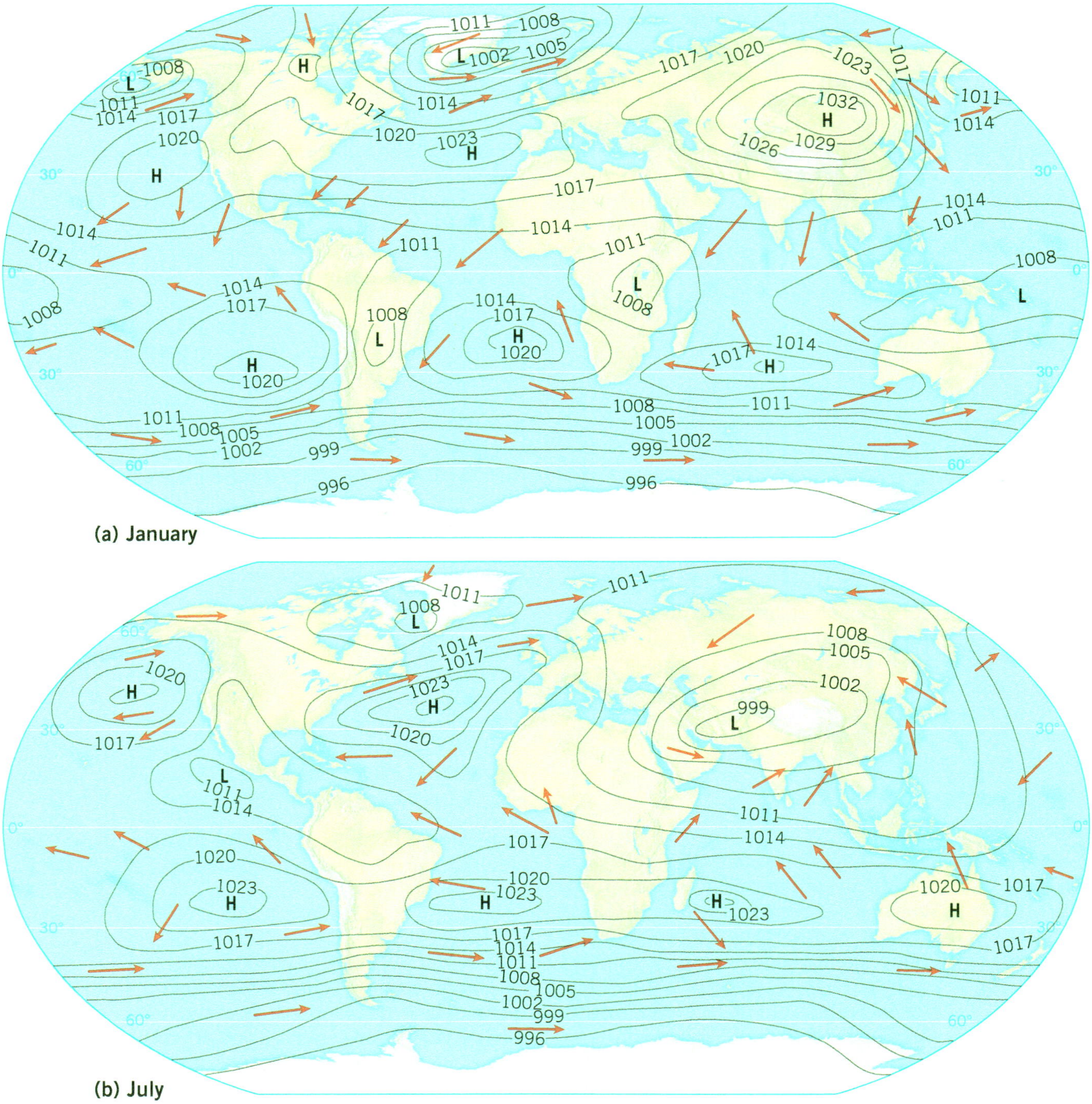

▲ **Figure 5-15** Average atmospheric pressure and wind direction in (a) January and (b) July. Pressure is reduced to sea-level values and shown in millibars. Arrows indicate generalized surface wind movements.

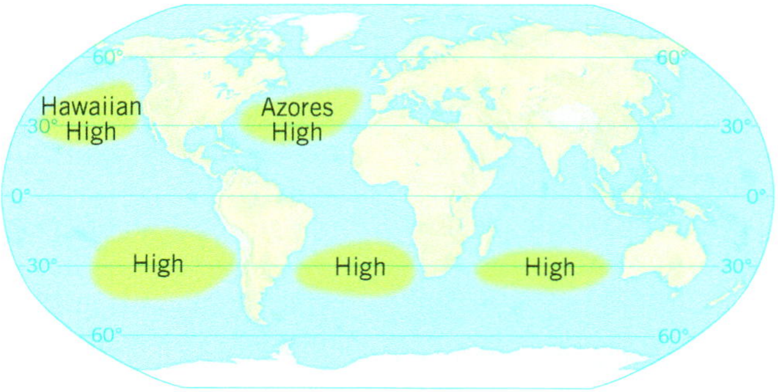

◀ **Figure 5-16** The subtropical highs are generally located over ocean basins at latitudes about 30° N and 30° S—called the horse latitudes.

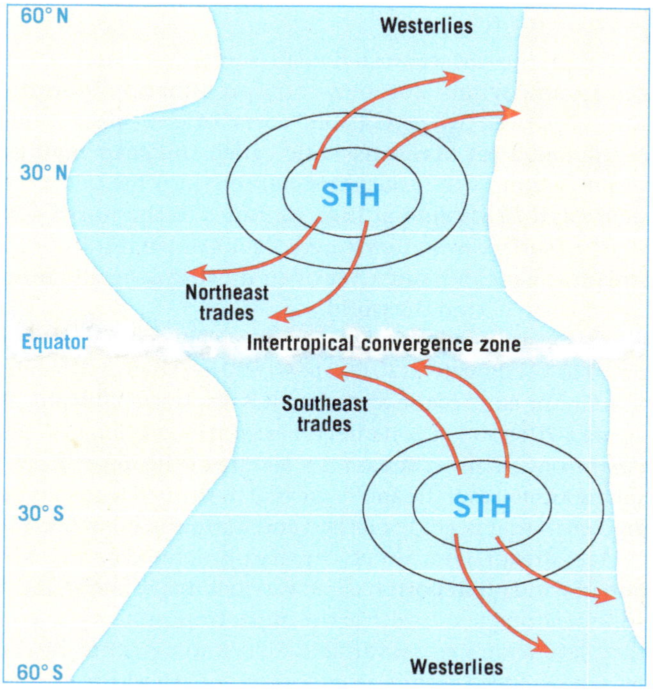

▲ **Figure 5-17** Air descending and diverging out of the subtropical highs (STH) is the source of the surface trade winds and westerlies. This map shows the generalized location of the intertropical convergence zone, trade winds, subtropical highs, and westerlies in hypothetical Northern and Southern Hemisphere ocean basins. Compare these wind patterns and the ocean current patterns of Figure 4-25.

(a) Trade winds uplifted on windward side of island

(b) Rainforest in Kīpahulu, Haleakalā National Park, Maui, Hawai'i

▲ **Figure 5-18** (a) Trade winds are usually heavily laden with moisture but rarely produce clouds and rain unless forced to rise. Thus, they may blow across a low-lying island with little or no visible effect. (b) An island of greater elevation, like Maui, Hawai'i, causes the air to rise up the side of the mountain. The result is heavy rain and a tropical wet climate in Haleakalā National Park.

exceptions, especially over the Indian Ocean, where westerly winds sometimes prevail, but for the most part, the flow is easterly over the tropical oceans.

Consistency of Trade Winds: The trade winds are by far the most "reliable" of all winds. They are extremely consistent in both direction and speed. They blow most of the time in the same direction at the same speed, day and night, summer and winter. This steadiness is reflected in their name: trade winds really means "winds of commerce." Mariners of the sixteenth century recognized early that the quickest and most reliable route for their sailing vessels from Europe to the Americas lay in the belt of northeasterly winds of the southern part of the North Atlantic Ocean. Similarly, the trade winds were used by Spanish galleons in the Pacific Ocean, and the name became applied generally to these tropical easterly winds.

The trades originate as warming, drying winds capable of holding an enormous amount of moisture. As the trades blow across the tropical oceans, they evaporate vast quantities of moisture and therefore have a tremendous potential for storminess and precipitation. They do not release the moisture, however, unless forced to do so when uplifted by a topographic barrier or some sort of pressure disturbance (the reasons for this are explained in Chapter 6). Low-lying islands in the trade-wind zone are often desert islands because the moisture-laden winds pass over them without dropping any rain. Even a slight topographic irregularity that forces air to rise may release abundant precipitation (**Figure 5-18**). Some of the wettest places in the world are windward slopes in the trade winds, such as in Hawai'i (see Figure 3-20).

Intertropical Convergence Zone (ITCZ)

The northeast and southeast trades come together in the general vicinity of the equator, although the latitudinal position shifts seasonally northward and southward following the Sun. This shift is greater over land than over sea because the land warms more. The zone where the air from the Northern Hemisphere and Southern Hemisphere meet is usually called the **intertropical convergence zone**, or simply the **ITCZ**, but it is also referred to as the **doldrums**. (This last name is attributed to the fact that sailing ships were often becalmed in these latitudes.)

Weather of the ITCZ: A zone of convergence and weak horizontal airflow, the ITCZ is characterized by feeble and erratic winds. It is a globe-girdling zone of warm surface conditions, low pressure associated with high rainfall, instability, and rising air in the Hadley cells (see Figure 5-14). It is not a region of continuously ascending air, however. Almost all the rising air of the tropics ascends in the updrafts that occur in thunderstorms in the ITCZ. These updrafts pump an enormous amount of sensible heat and latent heat of condensation into the upper troposphere, where much of it spreads poleward.

The ITCZ often appears as a well-defined, relatively narrow cloud band over the oceans near the equator (Figure 5-19). Over continents, however, it is likely to be more diffused and indistinct, although thunderstorm activity is common.

LearningCheck 5-8 Describe the general location and kind of weather associated with the ITCZ.

The Westerlies

The fourth component of the general atmospheric circulation is the great wind system of the midlatitudes, commonly called the **westerlies**. They are represented by the arrows that issue from the poleward sides of the STHs in Figure 5-17. These winds flow basically from west to east around the world in the latitudinal zones between about 30° N and 60° N and between about 30° S and 60° S. Because the circumference of Earth is smaller at these latitudes than in the tropics, the westerlies are less extensive than the trades; nevertheless, they cover much of Earth.

The surface westerlies are much less constant and persistent than the trades; in the midlatitudes, surface winds do not always flow from the west. The westerly flow gets interrupted and modified near the surface, which can be likened to eddies and countercurrents in a river. These interruptions are caused by surface friction, by topographic barriers, and especially by migratory pressure systems, which produce airflow that is not westerly.

LearningCheck 5-9 What is the relationship of the subtropical highs to the trade winds and the westerlies?

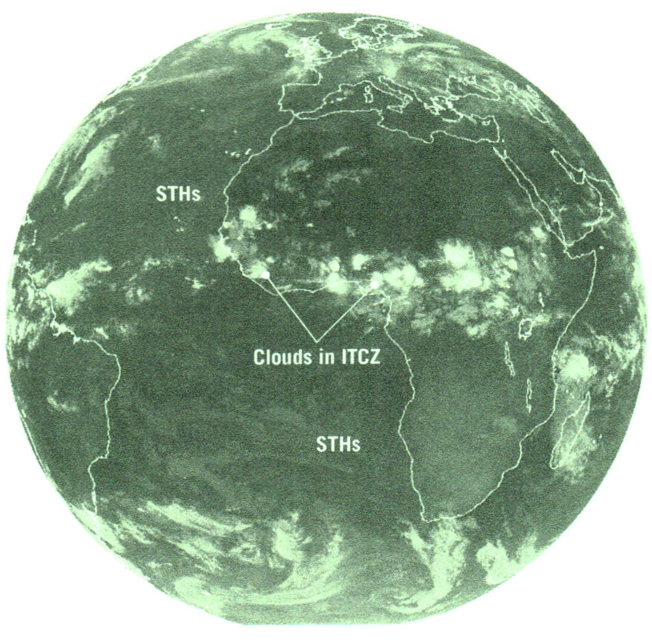

▲ Figure 5-19 A well-defined band of clouds marks the ITCZ over equatorial Africa in this infrared satellite image. (Darker grays indicate warmer surface temperatures.) This image was taken during the Northern Hemisphere summer, so the ITCZ has shifted slightly north of the equator. Note the generally clear skies off the west coasts of northern and southern Africa, corresponding to the areas of the subtropical highs, and the cloudiness and storms in the band of westerlies in the midlatitudes.

Jet Streams

Although the surface westerlies are somewhat variable, the geostrophic winds aloft blow very prominently from the west. Moreover, two remarkable "cores" of high-speed winds occur called **jet streams**—one called the *polar front jet stream* (or simply the *polar jet stream*) and the other called the *subtropical jet stream*—at high altitudes to the north and south of surface westerlies (Figure 5-20). We can think of a jet stream as a meandering river of air moving generally from west to east around the world.

The polar front jet stream, which usually occupies a position 9 to 12 kilometers (30,000 to 40,000 feet) high, is not centered in the band of the westerlies; it is displaced poleward, as Figure 5-20 shows. (The name comes from its location near the polar front.) This jet stream is a feature of the upper troposphere located over the area of greatest horizontal temperature gradient—that is, cold poleward and warm equatorward.

A jet stream is not always the sharply defined narrow ribbon of wind often portrayed on weather maps; rather, it is a zone of strong winds within the upper troposphere westerly flow. Jet stream speed is variable. Sixty knots is generally the minimum speed required for recognition as a jet stream, but wind speeds five times that fast have been recorded.

Commercial air travel can be significantly influenced by the high-speed flow of upper tropospheric winds. The cruising altitude of commercial jetliners is usually 9 to 12 kilometers (30,000 to 40,000 feet)—a typical elevation for the polar front jet stream. It generally takes longer to fly from east to west across North America than it does to fly from west to east. When you travel from the east, a "headwind" is likely to impede progress, whereas when you travel from the west, a "tailwind" may reduce travel time.

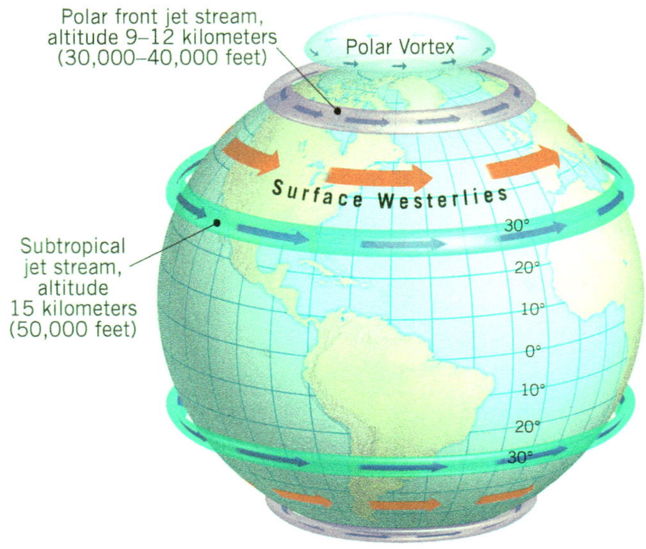

▲ Figure 5-20 Neither jet stream is centered in the band of the westerlies. The polar front jet stream is closer to the poleward boundary of this wind system, and the subtropical jet stream is closer to the equatorward boundary. The two jet streams are not at the same altitude—the subtropical jet stream is at a higher altitude than the polar front jet stream. The polar vortex is located in the stratosphere above the poles.

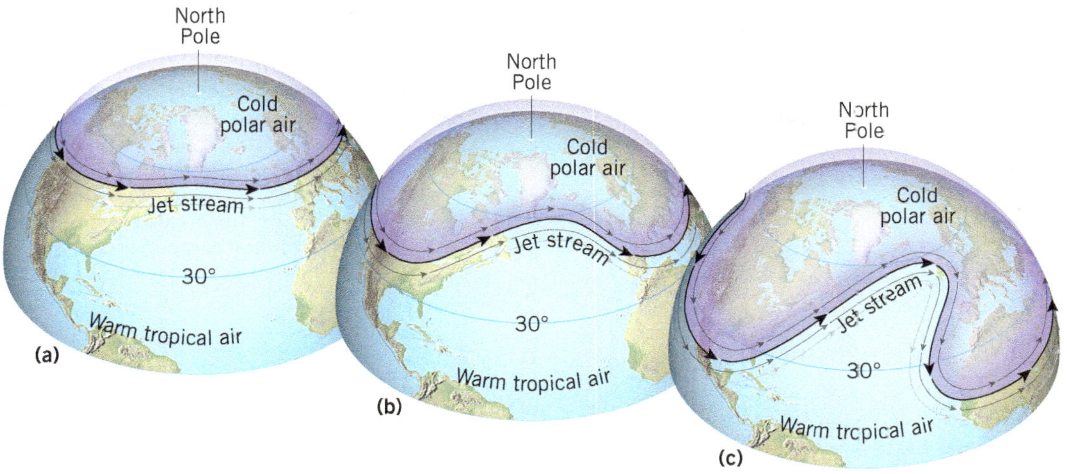

▲ **Figure 5-21** Rossby waves as part of the general flow (particularly in the upper air) of the westerlies. (a) When waves are few and their amplitude (north–south component of movement) is small, cold air usually remains poleward of warm air. (b) This distribution pattern begins to change as the Rossby waves grow. (c) When the waves have great amplitude, cold air pushes equatorward and warm air moves poleward.

Rossby Waves: The polar front jet stream often has a wavy path, and this change has considerable influence on the path of the wind. Although the basic direction of movement is west to east, frequently sweeping undulations develop in the westerlies and produce a meandering jet stream path that wanders widely north and south (Figure 5-21). These curves are very large and are generally referred to as **Rossby waves** (after Swedish-American meteorologist C. G. Rossby, who first explained their nature). They are also called "planetary waves."

At any given time, there are usually three to seven Rossby waves in the jet stream of each hemisphere. These waves in effect separate cold polar air from warmer tropical air. When the polar front jet stream path is more directly west-east, there is a *zonal flow* pattern in the weather, with cold air poleward of warm air. However, when the jet stream begins to oscillate and the Rossby waves develop significant amplitude (which means a prominent north–south component of movement), there is a *meridional flow*: cold air is brought equatorward, and warm air moves poleward, producing frequent and severe weather changes in the midlatitudes.

The subtropical jet stream is also located at high altitudes—just below the tropopause (Figure 5-22)—over the poleward margin of the subsiding air of the STH. It is weaker and has less influence on surface weather patterns because less temperature contrast is in the associated air streams. Sometimes, however, the polar front jet and the subtropical jet merge, producing a broad belt of high-speed winds in the upper troposphere—a condition that can intensify the weather conditions associated with either zonal or meridional flow of the Rossby waves.

Animation
The Jet Stream and Rossby Waves
https://media.pearsoncmg.com/ph/esm/esm_mcknight_physgeo_13/media/

All things considered, no other portion of Earth experiences such strong short-term variability of weather as the midlatitudes.

LearningCheck 5-10 **What are jet streams, and where are they usually found?**

Polar Highs

Situated over both polar regions are high-pressure cells called **polar highs** (see Figure 5-14). The Antarctic high, which forms over an extensive, high-elevation, very cold continent, is strong, persistent, and almost a permanent feature above the Antarctic continent. The Arctic high is much less pronounced and more transitory, particularly in winter.

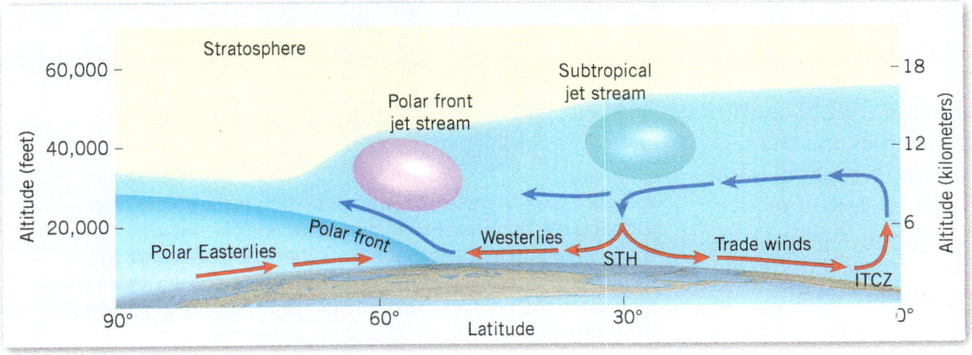

◀ **Figure 5-22** A vertical cross section of the atmosphere from the equator to the poles, showing the usual relative positions of the two jet streams. The jet streams sometimes merge, and the result is intensified weather conditions.

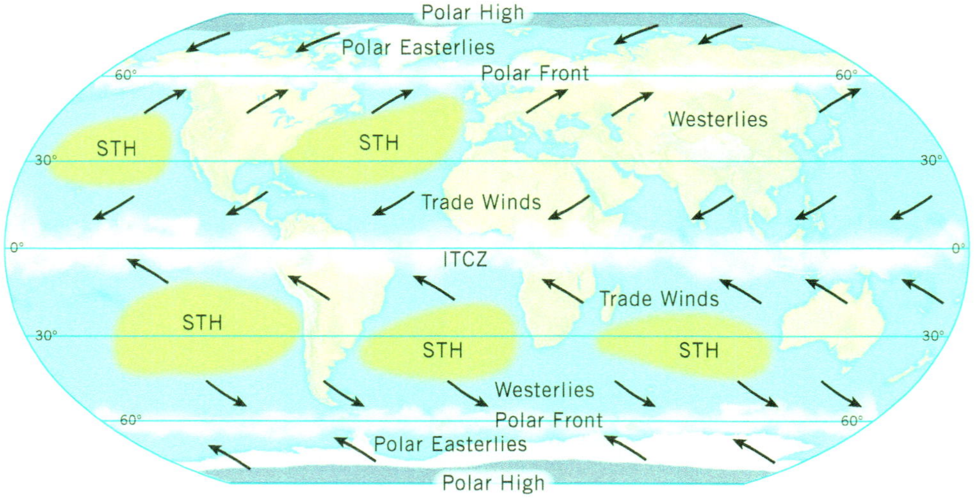

◀ **Figure 5-23** The generalized locations of the seven components of the general circulation patterns of the atmosphere: intertropical convergence zone (ITCZ), trade winds, subtropical highs (STH), westerlies, polar fronts, polar easterlies, and polar highs. (In this projection, the areal extent of the high-latitude components is considerably exaggerated.)

It tends to form over northern continental areas rather than over the Arctic Ocean. Air movement associated with these cells is typically anticyclonic. Air from above sinks down into the high and diverges horizontally near the surface (clockwise in the Northern Hemisphere and counterclockwise in the Southern Hemisphere), forming the third of the world's wind systems, the polar easterlies.

Polar Vortex: The **polar vortex** is a region of low pressure that exists in the winter stratosphere above the polar high (see Figure 5-20). This upper-level circulation is responsible for the record-breaking cold temperatures that grip the United States every few years. A stable polar vortex leads to a jet stream like the one shown in Figure 5-21(a) and keeps frigid air far north in the Arctic, while a weak polar vortex allows this incredibly cold air to move southward as shown in Figure 5-21(c).

Polar Easterlies

The third broadscale global wind system occupies most of the area between the polar highs and about 60° of latitude (**Figure 5-23**). The winds blow from the polar highs east to west and are called the **polar easterlies**. They are typically cold and dry but quite variable.

Polar Front

The final surface component of the general pattern of atmospheric circulation is a zone of low pressure at about 50° to 60° N and S. The zone is commonly called the **polar front**, although it is sometimes most clearly visible by the presence of semipermanent zones of low pressure called the **subpolar lows**.

The polar front is a meeting ground and zone of conflict between the cold winds of the polar easterlies and the relatively warmer westerlies (see Figure 5-23). The subpolar low of the Southern Hemisphere is nearly continuous over the uniform ocean surface of the cold seas surrounding Antarctica. In the Northern Hemisphere, however, the low-pressure zone is discontinuous, being interrupted by the continents. It is much more prominent in winter than in summer and is best developed over the northernmost reaches of the Pacific and Atlantic Oceans, forming the *Aleutian Low* and the *Icelandic Low*, respectively.

The polar front area is characterized by rising air, widespread cloudiness, precipitation, and generally unsettled or stormy weather conditions. Many of the migratory storms that travel with the westerlies have their origin in the conflict zone of the polar front.

Vertical Patterns of the General Circulation

As we have seen, over tropical regions, between the equator and 20° to 25° of latitude, surface winds generally blow from the east. In the midlatitudes, the surface winds are generally westerly, whereas in the highest latitudes, sur- face winds are again easterly. In the upper altitudes of the troposphere, however, the wind patterns are somewhat different (**Figure 5-24**).

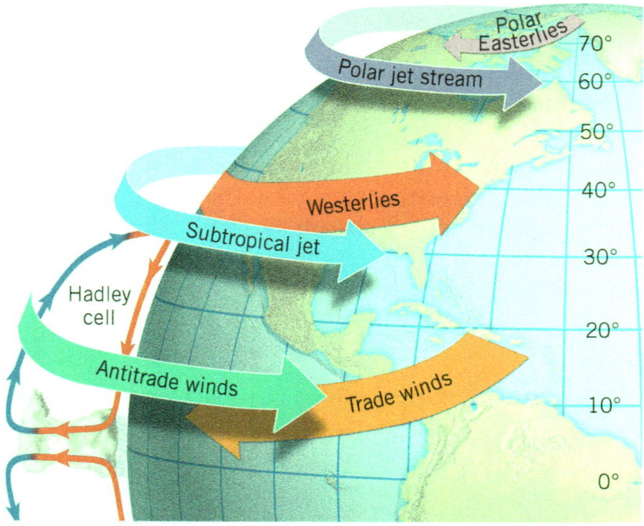

▲ **Figure 5-24** A generalized cross section through the troposphere, showing the dominant wind directions at different latitudes near the surface and in the upper troposphere. Surface winds in the tropics are generally easterly, but high above, the antitrade winds blow from the west, as are the upper troposphere jet streams of the westerlies (vertical dimension is exaggerated).

Antitrade Winds: The most dramatic difference is seen over the tropics. After equatorial air has risen in the ITCZ, the high-elevation poleward flow of air in the Hadley cell is deflected by the Coriolis effect (see Figure 5-14). As a result, upper-elevation winds called the **antitrade winds** blow from the southwest in the Northern Hemisphere and from the northwest in the Southern Hemisphere. This flow eventually becomes more westerly and encompasses the subtropical jet stream. Thus, at the surface within the tropics, winds are generally from the east, whereas high above, the antitrade winds blow from the west.

Modifications of the General Circulation

Many variations to the pattern are discussed on the preceding pages, and all features of the general circulation may appear in altered form, much different from the idealized description. Indeed, components sometimes disappear from sizable parts of the atmosphere where they are expected to exist. Even the tropopause sometimes "disappears"; for example, during a high-latitude winter with very cold surface temperatures, the atmospheric temperature may steadily increase with height into the stratosphere. In such cases the tropopause cannot be identified using temperature.

Nevertheless, the generalized pattern of global wind and pressure systems comprises the seven components described previously. To understand how real-world weather and climate differ from this general picture, we must discuss two important modifications of the generalized scheme.

Seasonal Variations in Location

The seven surface components of the general circulation shift north or south with the changing seasons. When sunlight, and therefore surface warming, is concentrated in the Northern Hemisphere (Northern Hemisphere summer), all components are displaced northward; during the opposite season (Southern Hemisphere summer), everything is shifted southward. The displacement is greatest in the low latitudes and least in the polar regions. The ITCZ, for example, can be found as far as 25° N in July and 20° S in January (Figure 5-25), while the polar highs experience little or no latitudinal displacement from season to season.

Weather is affected by shifts in the general circulation components only minimally in polar regions, but the effects can be quite significant in the tropics and midlatitudes. For example, as we see in Chapter 8, regions of *mediterranean climate* found along the west coasts of continents at about 35° N and 35° S have warm, rainless summers while under the influence of the STH; in winter, however, the belt of the westerlies shifts equatorward, bringing changeable and frequently stormy weather to these regions (see Figure 5-15). Also, as we will see, the shift of the ITCZ is closely tied to seasonal rainfall patterns in large areas of the tropics.

LearningCheck 5-11 **Why does the location of the ITCZ shift north during the Northern Hemisphere summer?**

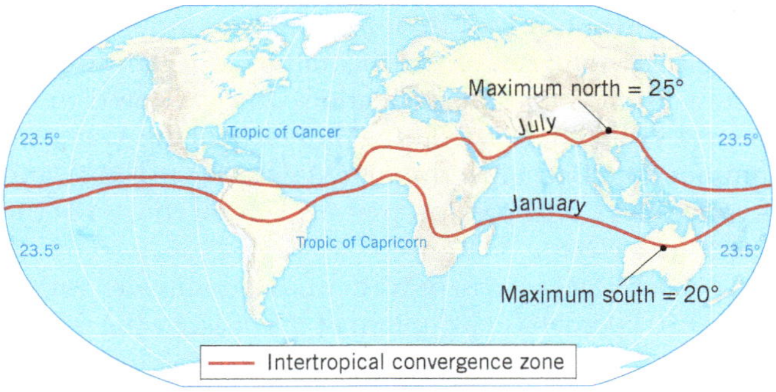

▲ **Figure 5-25** Typical maximum poleward positions of the intertropical convergence zone at its seasonal extremes. The greatest variation in location is associated with monsoon activity in Eurasia and Australia.

Monsoons

By far the most significant deviation from the general circulation pattern is the development of **monsoons** in certain parts of the world, particularly southern and eastern Eurasia (Figure 5-26). The word *monsoon* is derived from the Arabic *mawsim* (meaning "season"). It has come to mean a seasonal reversal of winds—a general sea-to-land movement, called *onshore flow*, in summer and a general land-to-sea movement, called *offshore flow*, in winter. Associated with the monsoon wind pattern is a distinctive seasonal precipitation regime: heavy summer rains derived from the moist maritime air of the onshore flow and a pronounced winter dry season when continental air moving seaward dominates the circulation.

Causes of Monsoons: It is easiest to explain monsoon circulation simply on the basis of the unequal warming of continents and oceans. A strong thermal (warm surface) low-pressure cell generated over a continent in summer pulls oceanic air onshore; similarly, a prominent thermal anticyclone over a continent in winter produces an offshore circulation. These thermally induced pressure differences contribute to monsoon development (see Figure 5-15) but are not the whole story.

Monsoon winds essentially represent unusually large latitudinal migrations of the trade winds associated with the large seasonal shifts of the ITCZ over southeastern Eurasia. The Himalayas evidently also play a role. This significant

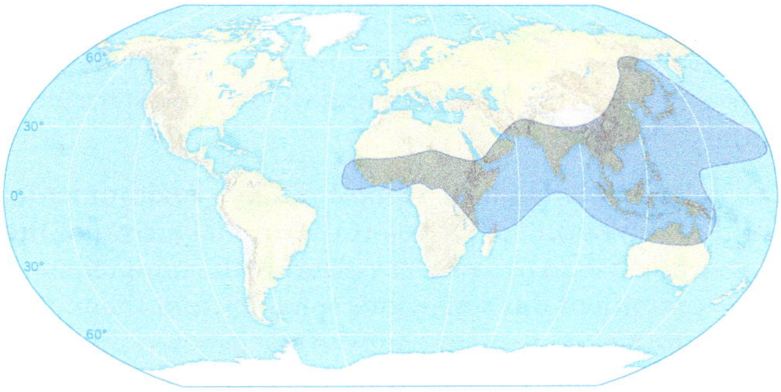

▲ **Figure 5-26** The principal monsoon areas of the world.

(a) Jharkhand, India, September 2016

(b) North Queensland, Australia, February 2019

▲ Figure 5-27 (a) The monsoon rain is necessary for the rice crops throughout much of India, but (b) too much rain, like we see in North Queensland, Australia, can be destructive.

topographic barrier allows greater winter temperature contrasts between South Asia and the interior of the continent to the north, which in turn may influence the location and persistence of the subtropical jet stream in this region.

Significance of Monsoons: We cannot overestimate the importance of monsoon circulation. More than half of the world's population inhabits regions with monsoon-controlled climates. Generally, the majority of the populace in these regions depends on agriculture for its livelihood (Figure 5-27a). Their lives are intricately bound up with the reality of monsoon rains, which are essential for food production and cash crops. The failure or even late arrival of monsoon moisture causes widespread hunger and economic disaster. Monsoons also can be destructive to built environments, similar in effect to large-scale flooding (Figure 5-27b).

Although the causes are complex, we can describe the monsoon patterns with some precision. There are two major monsoon systems (in South Asia and East Asia), two minor monsoon systems (in Australia and West Africa), and several other regions where monsoon patterns develop (especially in Central America and the southwestern areas of the United States).

South Asian Monsoon: The most notable environmental event each year in South Asia is the annual arrival of the summer monsoon (Figure 5-28a). During the summer monsoon, prominent onshore winds spiral in from the Indian Ocean, bringing life-giving rains to the parched subcontinent. The late arrival of the 2015 monsoon caused an extended heat wave and resulted in 2300 deaths in India alone. In winter, South Asia is dominated by outblowing dry air diverging from the northeast (Figure 5-28b). Recent research suggests that monsoon patterns may be changing—see the box *Global Environmental Change: Changes in the South Asian Monsoon*.

East Asian Monsoon: Winter is the more prominent season in the East Asian monsoon system, which primarily affects China, Korea, and Japan. A strong outflow of dry continental air, largely from the northwest, is associated with the Siberian High, anticyclonic circulation around the massive thermal high-pressure cell over eastern Eurasia (Figure 5-28d). The onshore flow of maritime air in summer is not as notable as that in South Asia, but it does bring southerly and southeasterly winds, as well as considerable moisture, to the region (Figure 5-28c).

Other Monsoon Areas: In one of the two minor systems, the northern quarter of the Australian continent experiences a distinct monsoon circulation. Onshore flow of moist air arrives from the north during the height of the Australian summer from December through March (Figure 5-29a), and dry, southerly, offshore flow prevails most of the rest of the year (Figure 5-29b).

The south-facing coast of West Africa is dominated within about 650 kilometers (400 miles) of the coast by the second minor monsoonal circulation (Figure 5-29c). Moist oceanic air flows onshore from the south and southwest during summer, and dry, northerly, continental flow prevails in the opposite season. Central Africa is the lightning capital of the world in part because of the monsoon circulation (Figure 5-29d).

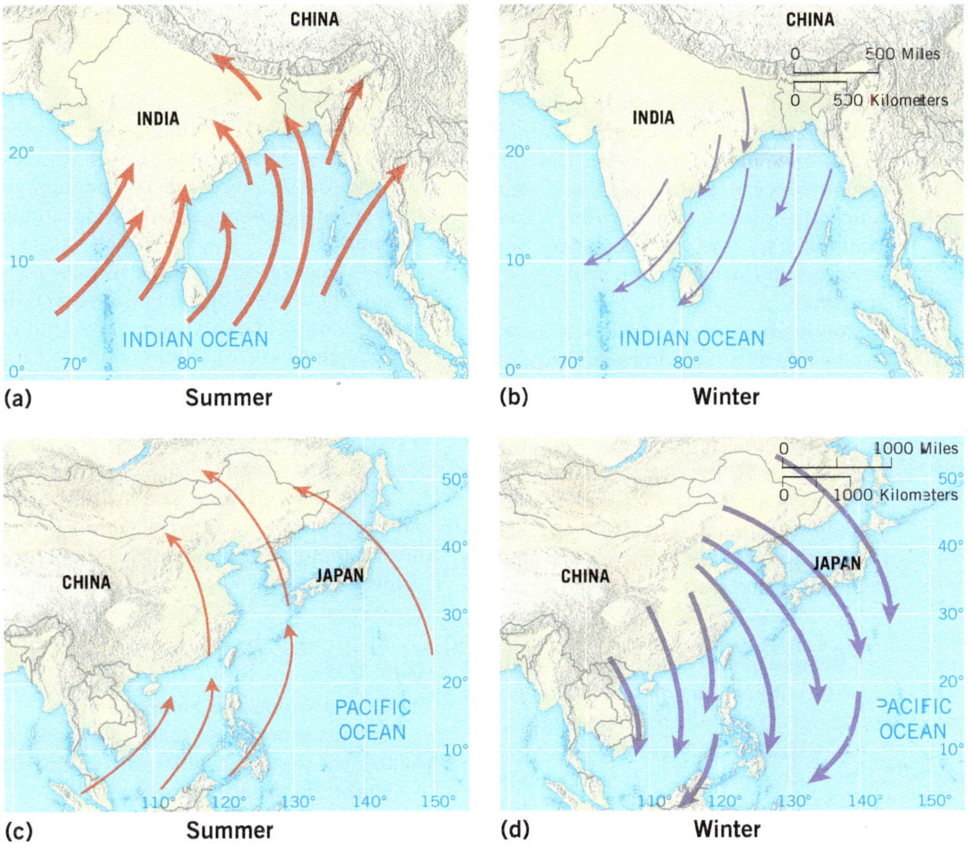

◀ **Figure 5-28** The two major monsoon systems. The South Asian monsoon is characterized by (a) a strong onshore flow in summer (rainy season) and (b) a somewhat less pronounced offshore flow in winter (dry season). In East Asia, (c) the inblowing summer monsoon is weaker than (d) the outblowing winter monsoon.

Animation
Seasonal Pressure and Precipitation Patterns
https://media.pearsoncmg.com/ph/esm/esm_mcknight_physgeo_13/media/

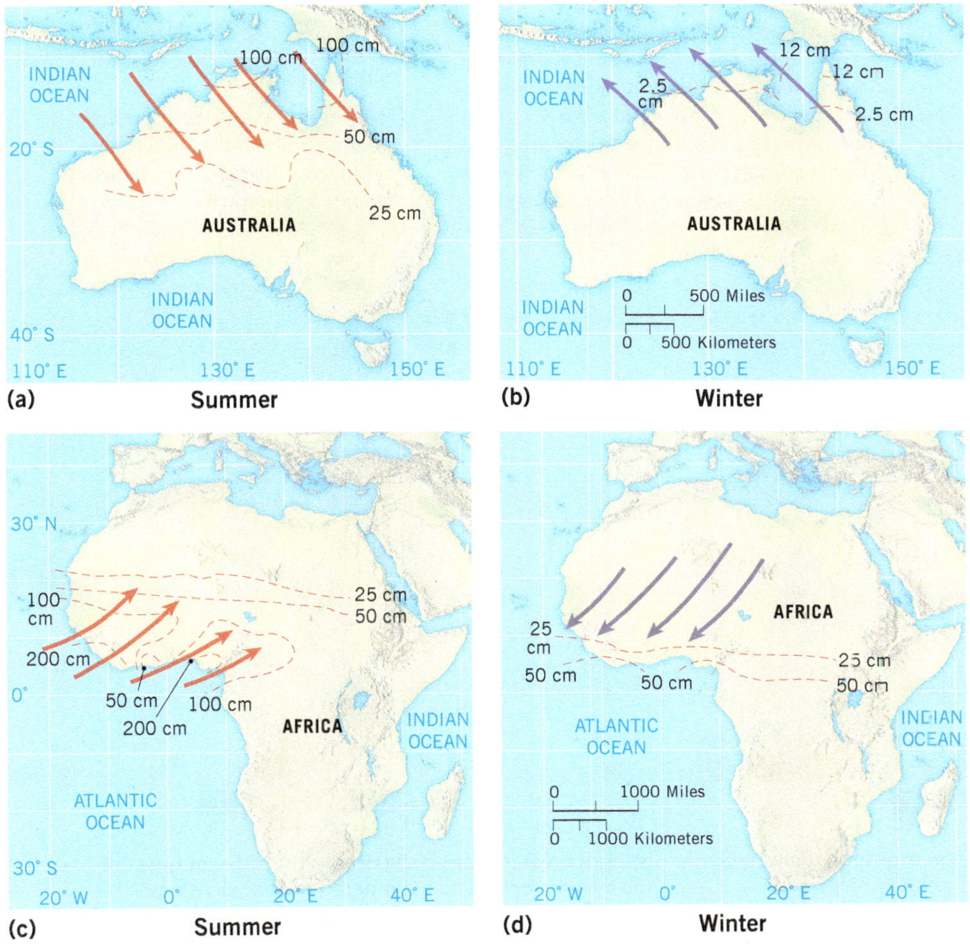

◀ **Figure 5-29** The two minor monsoon systems, showing three-month seasonal lines of equal rainfall. In Australia, (a) northwesterly summer winds bring the wet season to northern Australia; (b) dry southeasterly flow dominates in winter. In West Africa, (c) summer winds are from the southwest, and (d) winter winds are from the northeast.

Global Environmental Change

Changes in the South Asian Monsoon

• **Stephen Stadler,** Oklahoma State University

In India, the summer monsoon is critical to agriculture. Normally, this South Asian monsoon first bursts onto the west coast of India in June, moves eastward and northward across the subcontinent, and is over by the beginning of October. When the rains are late, crops can fail because seeds don't sprout. If rain is superabundant, crops can drown. Nowhere else on the planet do so many people depend on the nature of the rainy season.

In their *Fifth Assessment Report*, the Intergovernmental Panel on Climate Change concludes that the risks and impacts of climate change on human populations will be unevenly distributed around the world. Accumulating evidence strongly suggests that global changes will substantially alter the South Asian summer monsoon. Although the South Asian monsoon is notoriously fickle from year to year, precipitation is usually within 10 percent of the historic averages for India as a whole. Regionally, however, within a single monsoon season, it is common for parts of India to be too wet or too dry.

South Asian Monsoon Research: Scientists, including geographers and climatologists, have researched monsoon variability for more than a century. They have discovered teleconnections with the strength and variability of El Niño–Southern Oscillation. (This coupled pattern of oceans and atmosphere is discussed later in the chapter.) This discovery has led to contemporary research forecasting changes in the South Asian monsoon many years into the future.

Key to a better forecast has been increased availability of surface and satellite (particularly oceanic) weather observations. These observations are incorporated into supercomputer models that predict changes decades and even centuries in advance. We have confidence that the computer models are correct in their principal findings because we are able to use historic data and the models have correctly predicted the monsoons of today.

A Glimpse of the Future: India is a large country with an area of 3.29 million square kilometers (1.27 million square miles). Given what we know about regionalities of climate change, changes to the summer monsoon are likely to vary in magnitude across various parts of the country. Models show that over most of India the monsoon rains will decrease (**Figure 5-C**) and the monsoon's onset will be delayed as compared with the present onset dates (**Figure 5-D**). Breaks in precipitation will become more common. Disturbances will move more slowly, ironically increasing the potential for flooding while seasonal precipitation totals decrease. Alarmingly, in a changed climate scenario, the summer rains are projected to fail (defined as 40–70 percent of today's normal) every five years.

The Human Factor: Of India's population of over 1.35 billion people, two-thirds work the land. The security of India's food supply is threatened by the projected changes in the monsoon.

The government of India is quite concerned about the monsoon's future, and its Meteorological Department closely monitors the monsoon. Mitigation policies have been proposed, but it is difficult to plan far ahead when *this year's* food supply is always a top priority of the government. In 2015, the Indian Ministry of Earth Sciences and the U.S. National Oceanic and Atmospheric Administration agreed to collaborate in predicting features of the monsoon, hopefully leading to the increased understanding of its changes.

Questions

1. Given that much of the population of India is projected to be vulnerable to climate change, what sorts of things might the Indian government do to mitigate adverse effects of the monsoon changes that are predicted to occur?
2. Explain how increased rainfall from the summer monsoon might be harmful to Indian agriculture.
3. Why are rural Indians thought to be more vulnerable than their urban counterparts to monsoon climate change?

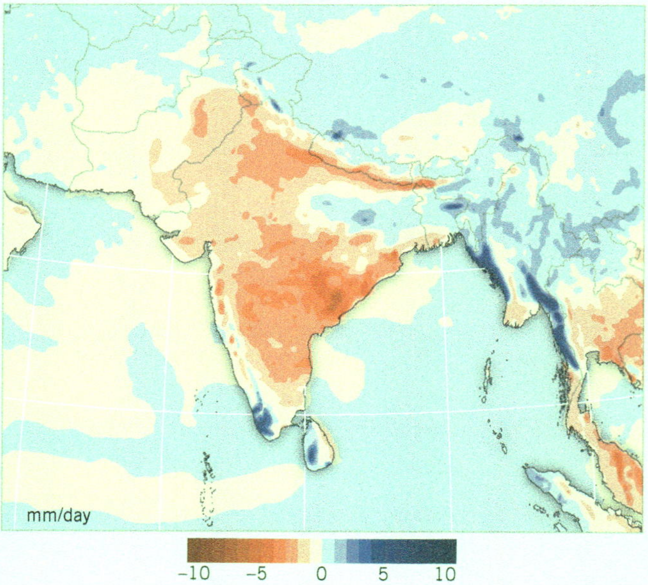

▲ **Figure 5-C** Modeled changes in summer monsoon daily precipitation compared with present amounts.

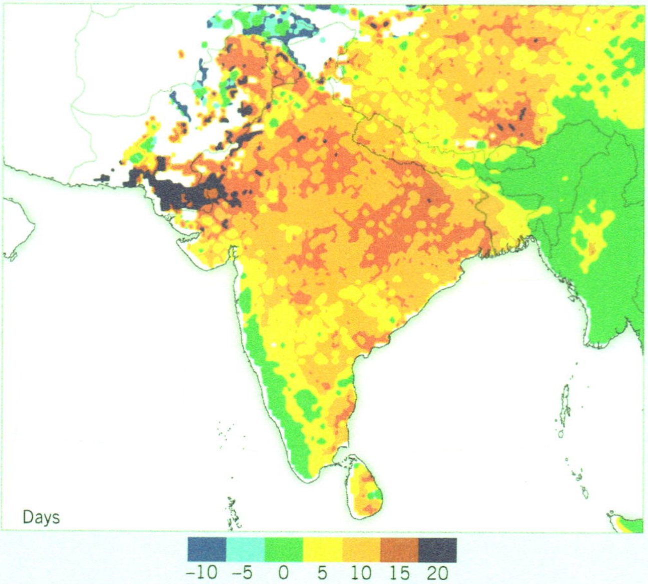

▲ **Figure 5-D** Modeled changes in the dates of onset of monsoon precipitation compared with present dates.

The "Arizona Monsoon" of the southwestern United States is part of a broader minor monsoon pattern called the North American Monsoon. These onshore winds in summer carry moisture from the Gulf of California and the Gulf of Mexico into New Mexico, Arizona, and northwestern Mexico, bringing bursts of thunderstorm activity.

LearningCheck 5-12 **How does wind direction in South Asia differ from summer to winter?**

Localized Wind Systems

The preceding sections have dealt with only the broadscale wind systems that make up the global circulation and influence the climatic pattern of the world. Many kinds of lesser winds, however, significantly affect weather and climate at a localized scale. Such winds are the result of local pressure gradients that develop in response to topographic configurations or contrasting thermal conditions in the immediate area.

Sea and Land Breezes

A common local wind system along tropical coastlines, and to a lesser extent during the summer in midlatitude coastal areas, is the cycle of **sea breezes** during the day and **land breezes** at night (Figure 5-30). (As is usual with winds, the name tells the direction from which the wind comes: a sea breeze blows from sea to land, and a land breeze blows from land to sea.) This is essentially a convectional circulation caused by the differential warming of land and water surfaces. The land warms up rapidly during the day, causing the air to expand and rise. That creates low pressure, attracting surface breezes from over the adjacent water body. Because the onshore flow is relatively cool and moist, it holds down daytime temperatures in the coastal zone and provides moisture for afternoon showers. Sea breezes are sometimes strong, but they rarely are influential for more than 15 to 30 kilometers (10 to 20 miles) inland.

The reverse flow at night is normally considerably weaker than the daytime wind. The land and the air above it cool more quickly than the adjacent water body, producing relatively higher pressure over land. Thus, air flows offshore in a land breeze.

Valley and Mountain Breezes

Another notable daily cycle of airflow is characteristic of many hill and mountain areas. During the day, conduction and reradiation from the land surface cause air near the mountain slopes to warm more than air over the valley floor (Figure 5-31). The warmed air rises, creating a low-pressure area, and then air from the valley floor flows upslope from the high-pressure area to the low-pressure area. This upslope flow is called a **valley breeze**. The rising air often causes clouds to form around the peaks, so afternoon showers are common in the high country. After dark, the pattern is reversed. The mountain slopes lose warmth rapidly through radiation, which chills the adjacent air, causing it to slip downslope as a **mountain breeze**.

Valley breezes are particularly prominent in summer, when solar warming is most intense. Mountain breezes are often weakly developed in summer and likely to be more prominent in winter. Indeed, a frequent winter phenomenon in areas of even gentle slope is *cold air drainage*—the nighttime sliding of cold air downslope, collecting in the lowest spots. This is a modified form of mountain breeze.

LearningCheck 5-13 **How do sea breezes form?**

Katabatic Winds

Related to simple air drainage is the more general and powerful spilling of air downslope in the form of **katabatic winds**

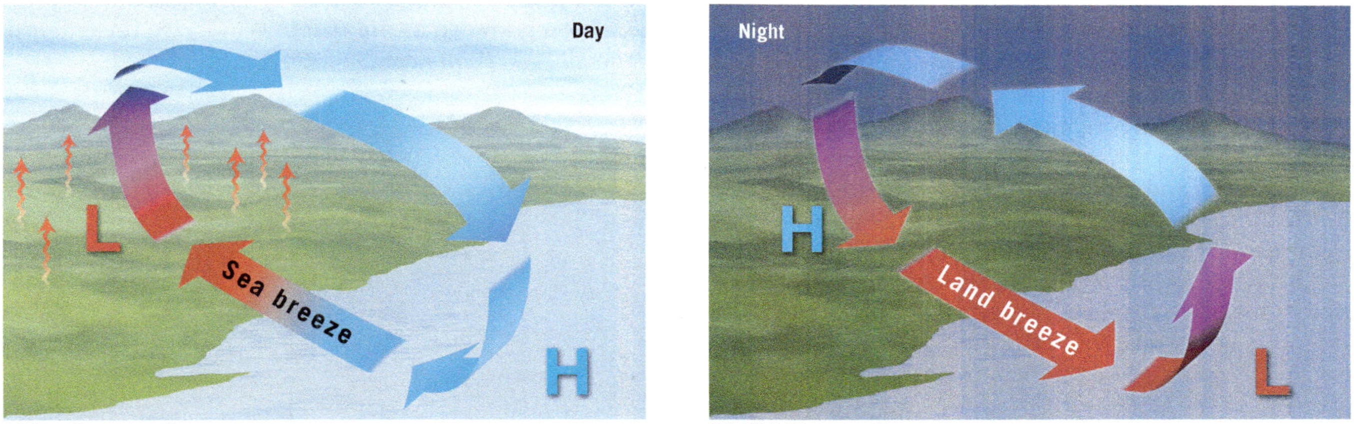

(a) Daytime warming (b) Nighttime cooling

▲ Figure 5-30 In a typical sea–land breeze cycle, (a) daytime warming over the land produces relatively low pressure there; this low-pressure center attracts an onshore flow of air from the sea. (b) Nighttime cooling over the land causes high pressure there, creating an offshore flow of air.

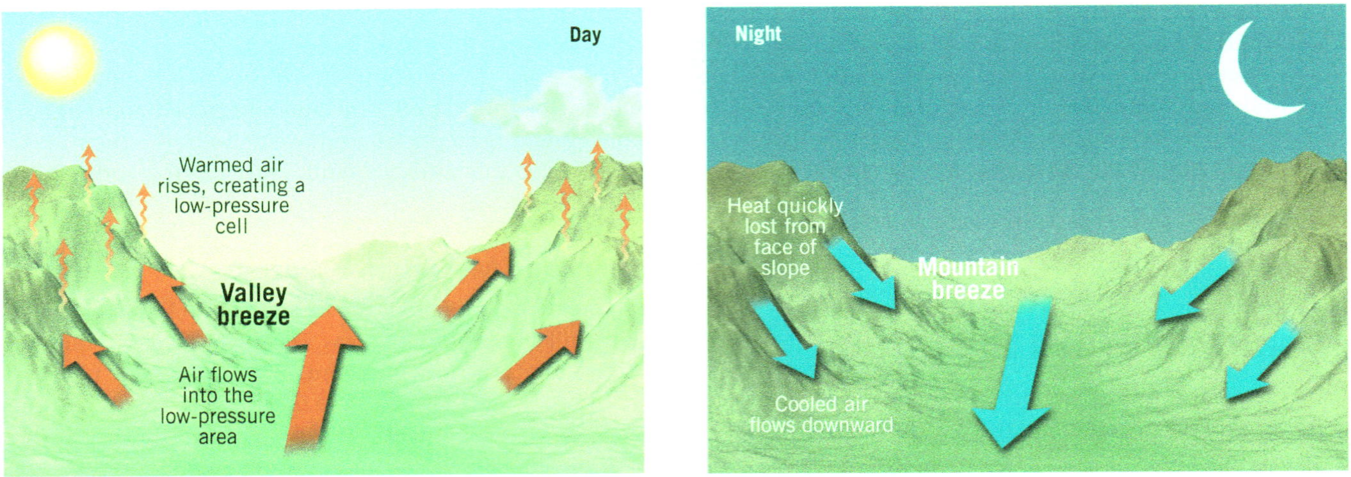

(a) Daytime warming

(b) Nighttime cooling

▲ Figure 5-31 (a) Daytime warming of the mountain slopes causes the air above the slopes to warm and rise, creating lower air pressure than over the valley. Cooler valley air then flows along the pressure gradient (up the mountain slope) as a valley breeze. (b) At night, the slopes radiate their warmth away; as a result, the air just above them cools. This cooler, denser air flows down into the valley as a mountain breeze.

(from the Greek *katabatos*, which means "descending"). These winds originate in cold upland areas and cascade toward lower elevations under the influence of gravity; they are sometimes referred to as *gravityflow* winds. The air in them is dense and cold. Although warmed adiabatically as it descends, it is usually colder than the air it displaces in its downslope flow.

Katabatic winds are particularly common in Greenland and Antarctica, especially where they whip off the edge of the high, cold ice sheets. Sometimes a katabatic wind will become channeled through a narrow valley, where it may develop high speed and considerable destructive power. An infamous example is the *mistral*, which sometimes surges down France's Rhône Valley from the Alps to the Mediterranean Sea. Similar winds are called *bora* in the Adriatic region and *taku* in southeastern Alaska.

Foehn and Chinook Winds

Another downslope wind is called a **foehn** (pronounced as in "fern" with a silent "r"), also called a **chinook** in the Rocky Mountains. It originates only when a steep pressure gradient develops with high pressure on the windward side of a mountain and a lower pressure on the leeward side. Air moves down the pressure gradient, from the windward side to the leeward side (Figure 5-32).

The downflowing air on the leeward side is dry: it has lost its moisture through precipitation on the windward side. It is warm relative to the air on the windward side because it contains all the latent heat of condensation given up by the condensing of snow or rain that fell at the peak. As the wind blows down the leeward slope, it is further warmed adiabatically, so it arrives at the base of the range as a warming, drying wind. It can produce a remarkable rise of temperature leeward of the mountains in just a few minutes. It is known along the Rocky Mountains front as a "snow-eater" because it melts the snow rapidly and quickly dries the resulting mud.

Santa Ana Winds

Similar drying winds in California, known as **Santa Ana winds**, develop when a cell of high pressure persists over the interior of the western United States for several days. The wind diverges clockwise out of the high, bringing dry, warm northerly or easterly winds to the coast (instead of the more typical cool, moist air off the ocean from the westerlies). The Santa Anas are noted for high speed, high temperature, and extreme dryness. Their presence provides ideal conditions for wildfires. Virtually every year they fan large brush fires that destroy dozens of homes in late summer, fall, and occasionally spring. A similar phenomenon is known as the Diablo Winds; see the *Featured National Park: Point Reyes.*

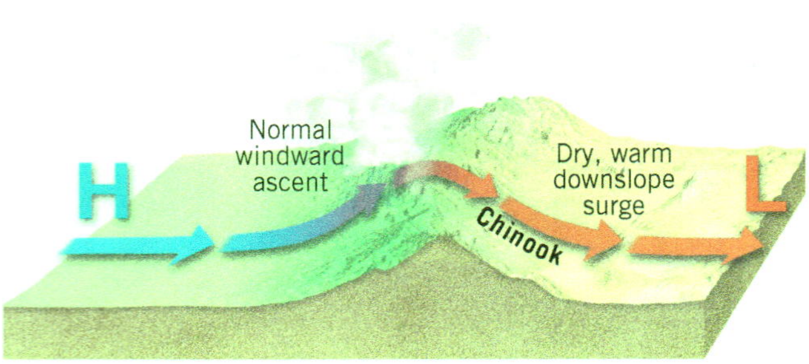

▲ Figure 5-32 A chinook is a rapid downslope movement of relatively warm air. It is caused by a pressure gradient on a mountain's two faces.

Featured National Park

Point Reyes

Fire and Wind

▲ **Figure 5-H** The Woodward Fire at Point Reyes, seen here from from Bolinas Ridge on August 18, 2020. This was one of 25 major wildfires burning in California at the time.

Point Reyes National Seashore is located on the west coast of California, just north of San Francisco. The "point" of Point Reyes is covered by scrubby brush and Douglas firs. For much of the year, strong winds come off the ocean with frequent fog in summer, but sometimes during the fall and winter, the wind pattern changes for a few days at a time (**Figure 5-E**).

The Diablo Wind: Similar to Santa Ana winds, Diablo winds blow from the east, carrying hot air that dries vegetation. These winds make their way to the coast and contribute to an ecosystem that is naturally characterized by regular periods of wildfire.

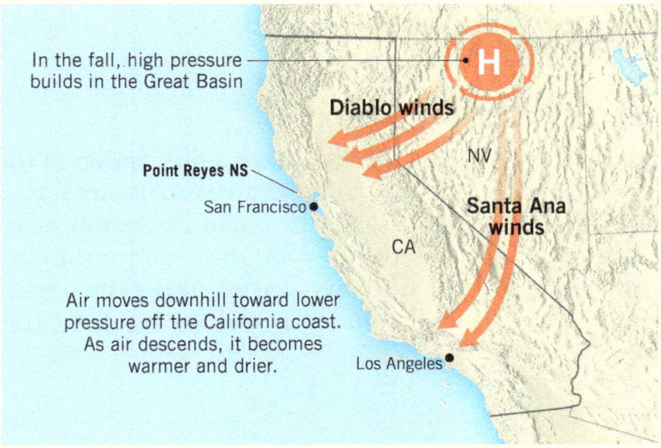

▲ **Figure 5-E** Relationship between the Point Reyes National Seashore and Diablo/Santa Ana winds.

▲ **Figure 5-F** Satellite image of smoke coming from a Diablo fire—the Kincade Fire, October 2019.

"Fighting Fires": The Coast Miwok, an indigenous people who have lived along this coast for thousands of years, used to burn brush yearly to manage vegetation and increase food availability. This action helped to maintain a healthy landscape and control the spread of wildfire. In the early days of the National Park Service, the conventional wisdom was exactly opposite to that of the Coast Miwok. We needed to "fight fires" to preserve our natural landscape and the U.S. Forest Service decided to adopt a policy of putting out all fires on state and federal lands.

Over time this practice lead to a build-up of forest litter and made people feel safe building in wild landscapes. The result was a series of devastating wildfires, starting in the1980s (**Figure 5-F**). We now recognize that fires are an important part of the Point Reyes ecosystem (**Figure 5-G**). Fire suppression is predominately done to protect neighboring communities.

Taken to Extremes: Climate change has made fires more common and more destructive. In 2020, record-breaking heat contributed to fires along the West Coast. Hotter temperatures mean vegetation is drier and burns more readily. The Woodward Fire, caused by lightning, closed Point Reyes temporarily (**Figure 5-H**).

Questions

1. Based on what you know about the Santa Ana winds, explain how the Diablo wind may contribute to wildfires.
2. Why would the Diablo wind that blows in during the fall be more dangerous than that that comes in the spring?

▲ **Figure 5-G** Controlled burning at Point Reyes National Seashore. This practice removes fuel to prevent larger fires. Controlled burning is directed at the base of the trees and brush to clear out grass and forest litter, without damaging the older growth.

NPS Video
Point Reyes Wilderness
https://media.pearsoncmg.com/ph/esm/esm_mcknight_physgeo_13/media/

El Niño–Southern Oscillation

A basic tenet of physical geography is the interrelatedness of the various elements of the environment. So far in our look at weather and climate, we have introduced patterns of temperature, ocean circulation, wind, and pressure without fully exploring important feedback mechanisms that are a part of such interactions. Nor have we explored the complexity introduced by cyclical variations in oceanic and atmospheric patterns, occurring over periods of years or even decades.

A prime example of this interrelatedness and complexity is **El Niño**, an episodic atmospheric and oceanic phenomenon of the equatorial Pacific Ocean, particularly prominent along the west coast of South America.

Effects of El Niño

During an El Niño event, abnormally warm water appears at the surface of the ocean off the west coast of equatorial South America, replacing the cold, nutrient-rich water that usually prevails. We now know that El Niño, once thought to be a local phenomenon, is associated with changes in pressure, wind, precipitation, and ocean conditions over large regions of Earth. During a strong El Niño event, productive Pacific fisheries off South America are disrupted, and heavy rains pound some parts of the world while drought comes to others.

A slight periodic warming of Pacific coastal waters has been noticed by South American fishermen for many generations. It occurs every two to seven years and typically begins around Christmastime—hence the name El Niño (Spanish for "the boy" in reference to the Christ child). Less frequently, however, the warming of the ocean is much greater, and fishing is likely to be much poorer. Historical records have documented the effects of El Niño for several hundred years, with archeological and paleoclimatological evidence pushing the record back several thousand years.

Not until the major 1982–1983 event, however, did El Niño receive worldwide attention. Over a period of several months, droughts crippled Australia, India, Indonesia, the Philippines, Mexico, Central America, and southern Africa; floods devastated the western and southeastern United States, Cuba, and northwestern South America; tropical cyclones destroyed parts of the Pacific (such as in Tahiti and Hawai'i) where they are normally rare; and a vast sweep of ocean water as much as 8°C (14°F) warmer than normal stretched over 13,000 kilometers (8000 miles) of the equatorial Pacific, causing massive die-offs of fish, seabirds, and coral. Directly attributable to these events were more than 1300 human deaths, damage estimated at nearly $9 billion, and vast ecological changes. An example of the temperature and precipitation changes associated with the stronger 1997–1998 El Nino are shown in Figure 5-33.

In 2014–2016, the strongest El Niño cycle on record took place. This time El Niño affected 60 million people worldwide, caused a record-breaking year for tropical cyclones in the central Pacific, and led to the most severe coral bleaching event for the Great Barrier Reef. El Niño caused the biggest one-year jump in carbon dioxide levels because of widespread drought (fewer plants to remove CO_2) and fires (more sources of CO_2.)

What happens during El Niño events that correlates with such widespread changes in the weather?

Normal Pattern

To understand El Niño, we begin with a description of the normal conditions in the Pacific Ocean basin (Figure 5-34a). As we saw in Chapter 4, usually the waters off the west coast of equatorial South America are cool. The wind and pressure patterns in this region are dominated by the persistent trade winds associated with convergence along the ITCZ (see Figure 5-14).

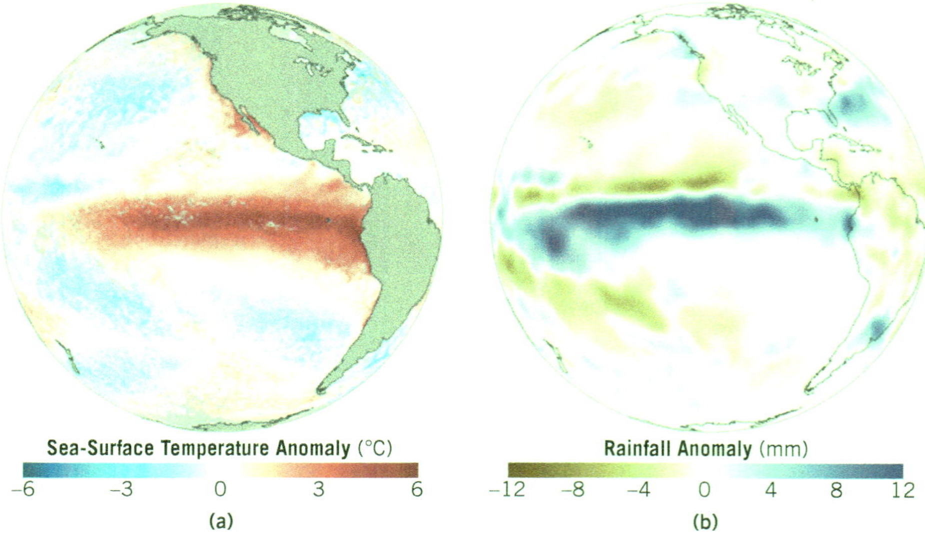

Video
El Niño
https://media.pearsoncmg.com/ph/esm/esm_mcknight_physgeo_13/media/

◄ **Figure 5-33** The 1997–1998 El Niño event. (a) Sea-surface temperature anomalies and (b) precipitation anomalies for December 1997. Ocean water along the west coast of South and North America was abnormally warm. Higher than average precipitation fell in parts of the eastern Pacific (and in the southwestern United States in the spring of 1998); drought was experienced in parts of the tropical western Pacific.

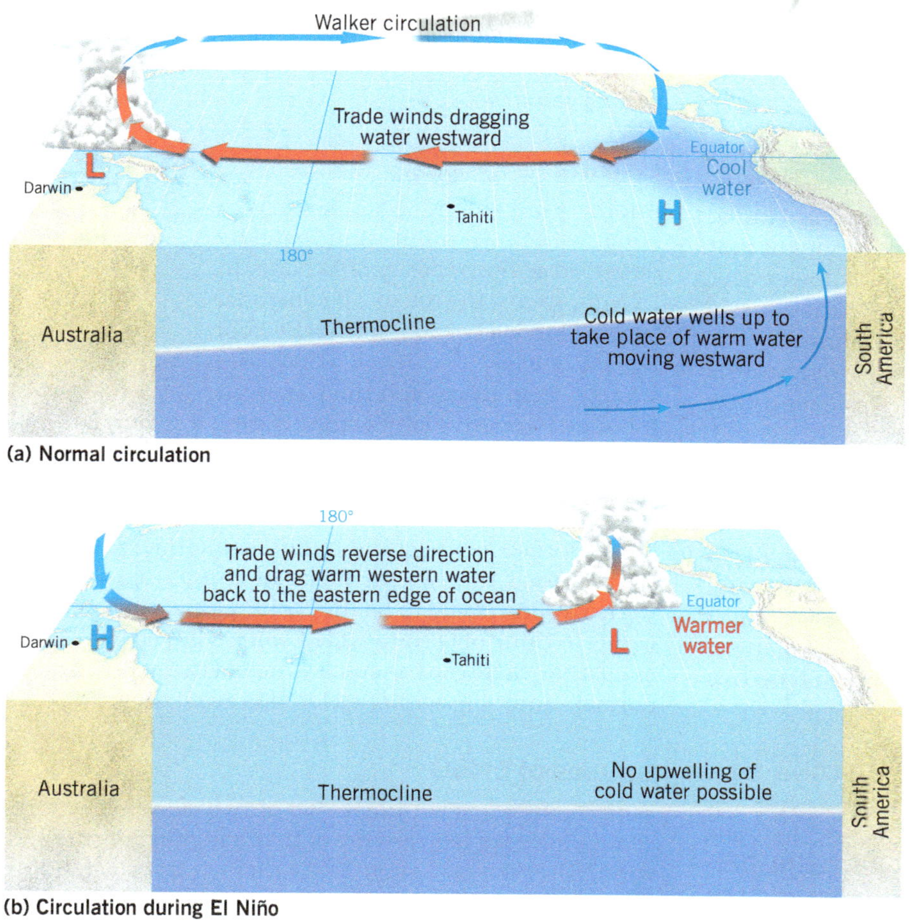

(a) Normal circulation

(b) Circulation during El Niño

◀ **Figure 5-34** (a) Normal conditions in the South Pacific. The trade winds carry warm equatorial water across the Pacific from east to west. (b) These conditions either weaken or reverse during an El Niño event. The upwelling of cold water off the coast of South America diminishes, the thermocline—the thermal boundary between near-surface and cold deep water—lowers, and much warmer water than usual is present at the surface.

Animation
El Niño
https://media.pearsoncmg.com/ph/esm/esm_mcknight_physgeo_13/media/

Video
El Niño Triggers Disease
https://media.pearsoncmg.com/ph/esm/esm_mcknight_physgeo_13/media/

As the trade winds blow across South America, they are slowed by friction, but the wind speeds up again when it reaches the Pacific Ocean. This causes a net surface divergence off the coast of equatorial South America during normal conditions because the air is moving away from the coast faster than the air is arriving from the land. This divergence causes sinking air, which creates a slightly higher pressure and is often depicted with an H on a map. It's not a high pressure cell, in the sense of an STH, but it is a *higher* pressure than over Indonesia where the opposite happens as the fast-moving trade winds encounter land again. As the winds slow down, air is forced to rise. This is often depicted with an L on a map because it creates a slightly lower pressure.

Cool Water off West Coast: As the trade winds diverge from the STH, they flow from east to west across the Pacific—dragging surface ocean water westward across the Pacific basin in the warm Equatorial Currents (introduced in Chapter 4; see Figure 4-25). As surface water pulls away from the coast of South America, an upwelling of cold, nutrient-rich ocean water rises into the already cool Peru Current. This combination of cool water and higher pressure (sinking air) results in relatively dry conditions along much of the west coast of South America.

In contrast to the cold water and sinking air near South America, in a normal year on the other side of the Pacific Ocean, things are quite different. The trade winds and Equatorial Currents pile up warm water, raising the sea level near Indonesia as much as 60 centimeters (about 2 feet) higher than near South America, turning the tropical western Pacific into an immense storehouse of energy and moisture.

The Walker Circulation: Warm water and persistent lower pressure (rising air) prevail around northern Australia and Indonesia; local convective thunderstorms develop in the ITCZ, producing high annual rainfall in this region. After this air rises in the ITCZ, it begins to flow back to the east to form the upper-atmosphere westerly antitrade winds. Some of this airflow aloft eventually subsides on the other side of the Pacific (see Figure 5-34a). This general circuit of airflow is called the Walker Circulation, after British meteorologist Gilbert Walker (1868–1958) who first described it. Although Figure 5-34a shows the Walker Circulation as a closed cell, recent studies suggest that although the upper atmosphere is generally flowing from west to east, a closed "loop" of airflow probably does not exist.

El Niño Pattern

Every few years, the normal-pressure patterns in the Pacific change (**Figure 5-34b**). Trade winds weaken or reverse. Higher pressure develops over northern Australia, and lower pressure develops to the east near Tahiti. This "seesaw" of pressure is known as the **Southern Oscillation**. It was first recognized by Gilbert Walker in the 1920s. Years before, when Walker was director of the Meteorological Service in colonial India, he searched for a method to predict the monsoon. When the life-giving South Asian monsoon failed to develop, drought and famine ravaged India. In the global meteorological records, Walker thought that he saw a pattern: in most years, pressure is lower over northern Australia (specifically, Darwin, Australia) and higher over Tahiti, and in these years, the monsoon usually comes as expected. However, in some years pressure is higher in Darwin and lower in Tahiti, and the monsoon would often fail.

As it turned out, Walker's observed correlation between the monsoons in India and the Southern Oscillation of pressure was not reliable enough to predict the monsoons. However, by the 1960s meteorologists recognized a connection between Walker's Southern Oscillation and the occurrence of strong El Niño warming near South America. This overall coupled ocean–atmosphere pattern is now known as the **El Niño–Southern Oscillation (ENSO)**.

Onset of El Niño Event: Although no two ENSO events are exactly alike, we can describe a typical El Niño cycle. For many months before the onset of an El Niño, the trade winds pile up warm water in the western Pacific near Indonesia. A bulge of warm equatorial water perhaps 25 centimeters (10 inches) high, known as a Kelvin wave, then begins to move slowly to the east across the Pacific. A Kelvin wave might take two or three months to arrive off the coast of South America. The bulge of warm water in a Kelvin wave is deflected little as it moves across the ocean because the Coriolis effect is zero on the equator. Rising air and thunderstorms form above the leading edge of the warm oceanic water.

Arrival of El Niño Conditions: When a Kelvin wave reaches South America, sea level rises as the warm water pools. The usual higher pressure in the subtropics has weakened. Upwelling no longer brings cold water to the surface, so ocean temperature increases still further—an El Niño is underway. Thunderstorms and flooding occur in the normally dry equatorial South America.

The *thermocline*, the thermal boundary between near-surface and cold deep ocean waters, lowers. Pressure increases over Indonesia, and the most active portion of the ITCZ in the Pacific shifts toward the now-warmer central and eastern Pacific basin. Drought strikes northern Australia and Indonesia; the South Asian monsoon may fail or develop weakly. The path of the subtropical jet stream over the eastern Pacific shifts, guiding winter storms into the southwestern United States. California and Arizona experience more powerful winter storms than usual, resulting in high precipitation and flooding. The upper Midwest, on the other hand, enjoys milder winters.

LearningCheck 5-14 **How do the trade winds differ during an El Niño event compared with the normal pattern?**

La Niña

Video
La Niña
https://media.pearsoncmg.com/ph/esm/esm_mcknight_physgeo_13/media/

Another aspect of the ENSO pattern is La Niña, which is often simplistically described as the opposite of El Niño. In La Niña events, the waters off equatorial South America become unusually cool (Figure 5-35a), the trade winds are stronger than usual, and the waters off Indonesia are unusually warm. During a strong La Niña event, the southwestern United States is drier than usual, whereas Southeast Asia and northern Australia are wetter.

El Niño and La Niña conditions are generally identified by sea-surface temperature trends; sometimes El Niño is referred to as the "warm" phase of ENSO (Figure 5-35b), while La Niña is said to be the "cold" phase. However, it is wrong to think of El Niño and La Niña as linked in a simple oscillation: there can be several swings between El Niño and normal conditions without a La Niña event occurring.

Causes of ENSO

Which comes first with the onset of an El Niño event—the change in ocean temperature or the change in pressure and wind? The "trigger" of an ENSO event is not clear, but it

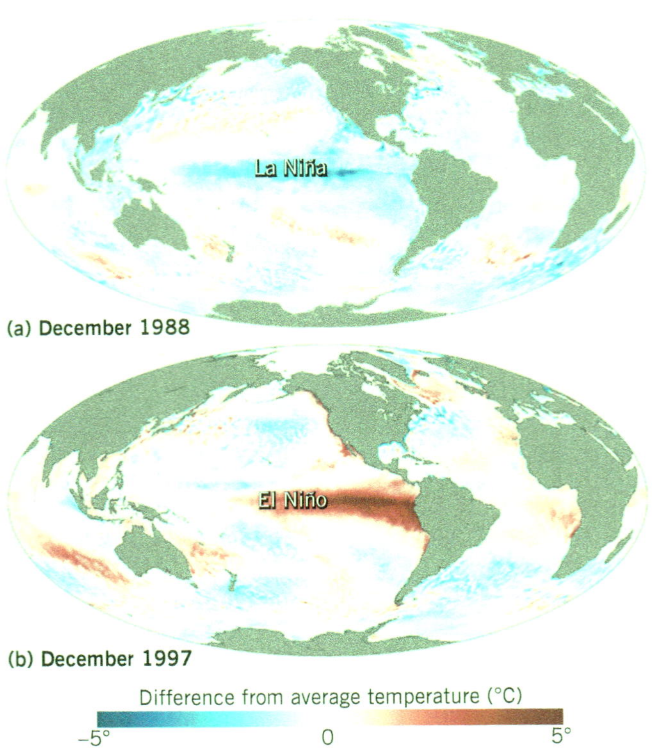

▲ **Figure 5-35** In contrast to the warm ocean conditions in the eastern equatorial Pacific during an El Niño event, ocean conditions are much cooler during a La Niña. These images show the temperature anomalies for (a) the 1988 La Niña and (b) the 1997 El Niño.

appears to have something to do with the location and strength of the Pacific STH regions. Atmospheric pressure and wind patterns are tied together with the ocean in a complex feedback loop with no clear starting point: if atmospheric pressure changes, wind changes; when wind changes, ocean currents and ocean temperature may change; when ocean temperature changes, atmospheric pressure changes—and that in turn may change wind patterns still more. In short, we don't fully understand the causes of ENSO.

Not only are the causes of ENSO elusive, but even the effects are not completely predictable. While a strong El Niño generally brings high precipitation to the southwestern United States (1982–1983, for example), the strong El Niño that occurred in 2015 did not bring high precipitation. El Niño did not even bring enough precipitation for California to recover from its drought. Furthermore, although we can generalize that during a strong El Niño year high rainfall is more likely in California than in La Niña years, very wet winters can occur in any year no matter what ENSO is doing. In other words, El Niño might "open the storm door" every few years, but there is no guarantee that the storms will come.

Teleconnections

As we learn more about ENSO, its connections with oceanic and atmospheric conditions inside and outside the Pacific basin are increasingly being recognized (**Figure 5-36**). Drought in Brazil, cold winters in the southeastern United States, high temperatures in the Sahel, a weak monsoon in India, tornadoes in Florida, fewer hurricanes in the North Atlantic—all seem to correlate quite well with a strong El Niño event. Such couplings of weather and oceanic events in one part of the world with those in another are termed **teleconnections**. Adding to the complexity of these teleconnections is growing evidence that long-term ENSO patterns may be influenced by other ocean–atmosphere

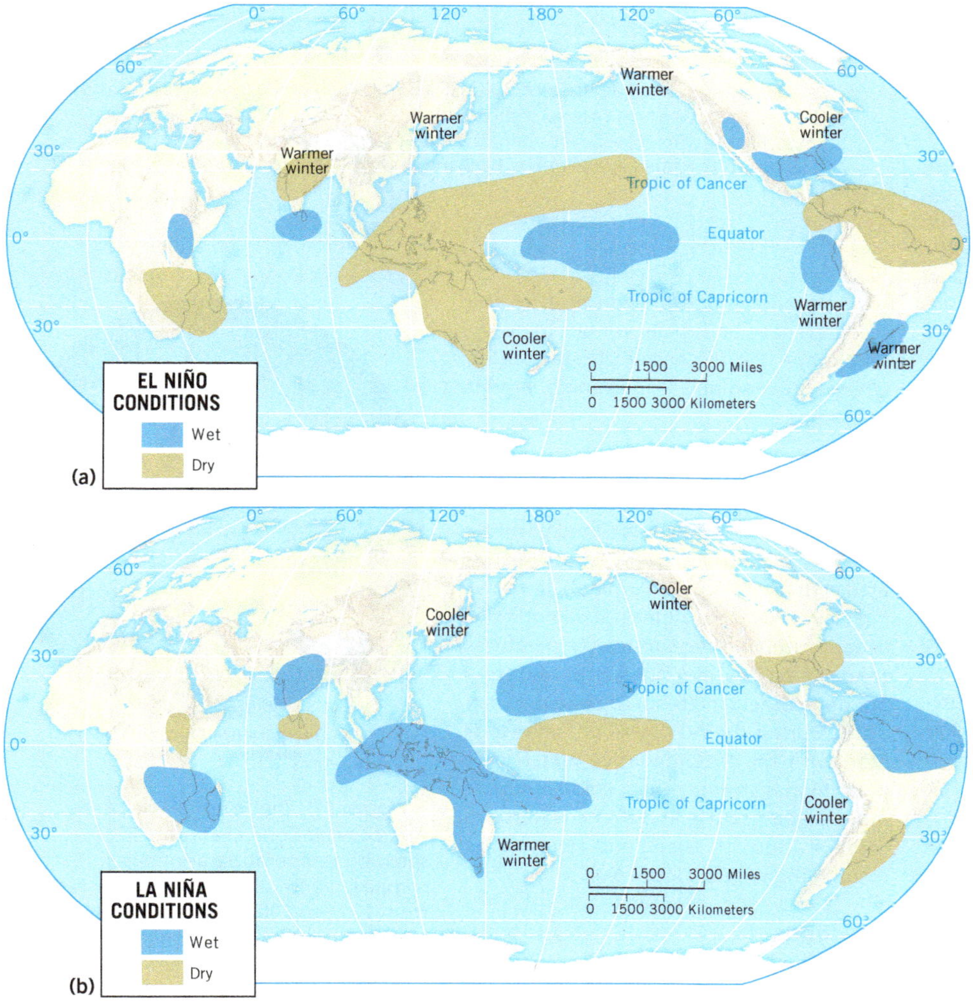

▲ **Figure 5-36** Generalized precipitation and temperature patterns during (a) El Niño conditions and (b) La Niña conditions.

Mobile Field Trip
El Niño
https://media.pearsoncmg.com/ph/esm/esm_mcknight_physgeo_13/media/

Focus

Multiyear Atmospheric and Oceanic Cycles

In addition to El Niño, several other long-term atmospheric and oceanic cycles have been identified. Their origins and the full extent of their teleconnections are not completely understood.

Pacific Decadal Oscillation: The *Pacific Decadal Oscillation* (*PDO*) is a long-term cycle of sea-surface temperature change in the Pacific Ocean. PDO cycles last up to 30 years but can shift abruptly. During the PDO "positive" or "warm" phase, the east tropical Pacific is relatively warm and the northern/west tropical Pacific is relatively cool. With the onset of the PDO "negative" or "cool" phase, the pattern abruptly switches, leaving cooler sea-surface temperatures in the east tropical Pacific and warmer conditions in the northern/west tropical Pacific (**Figure 5-I**). Although the causes of the PDO are not well understood, the change in ocean water temperature seems to influence the location of the jet stream and in turn affects storm tracks across North America. During a warm phase, temperatures in Alaska are often warmer than average, whereas during a cool phase, they are cooler. The PDO can also influence the intensity of El Niño events: when the PDO is in a positive phase with warmer water in the eastern tropical Pacific, the influence of El Niño on regional weather patterns appears to be more significant.

North Atlantic Oscillation: The *North Atlantic Oscillation* (*NAO*) is an irregular "seesaw" of pressure differences between the Icelandic Low and the Azores High (the STH in the North Atlantic; see Figures 5-16 and 5-17). In the "positive" phase of the NAO, the Icelandic Low exhibits lower-than-usual pressure and the Azores High exhibits higher-than-usual pressure; winter storms tend to take a more northerly track across the Atlantic, bringing mild, wet winters to Europe and the eastern United States but colder, drier conditions to Greenland.

In the "negative" phase of the NAO, both the Icelandic Low and the Azores High are weaker. Winter storms tend to bring higher-than-average precipitation to the Mediterranean and colder winters to northern Europe and the eastern United States, whereas Greenland experiences milder conditions.

Arctic Oscillation: The *Arctic Oscillation* (*AO*) alternates between warm and cold phases that are closely associated with the NAO. During the AO "warm" phase (associated with the NAO positive phase), the polar high is weaker. Cold air masses do not move as far south, and sea-surface temperatures tend to be warmer in Arctic waters—although Greenland tends to be colder than usual. During the AO "cold" phase (associated with the NAO negative phase), the polar high is strengthened, cold air masses move farther south, and Arctic waters are colder—although Greenland tends to be warmer than usual (**Figure 5-J**).

Questions

1. Why can the "positive" phase of the PDO intensify an El Niño event?
2. How does the NAO influence the Icelandic Low and the Azores High?

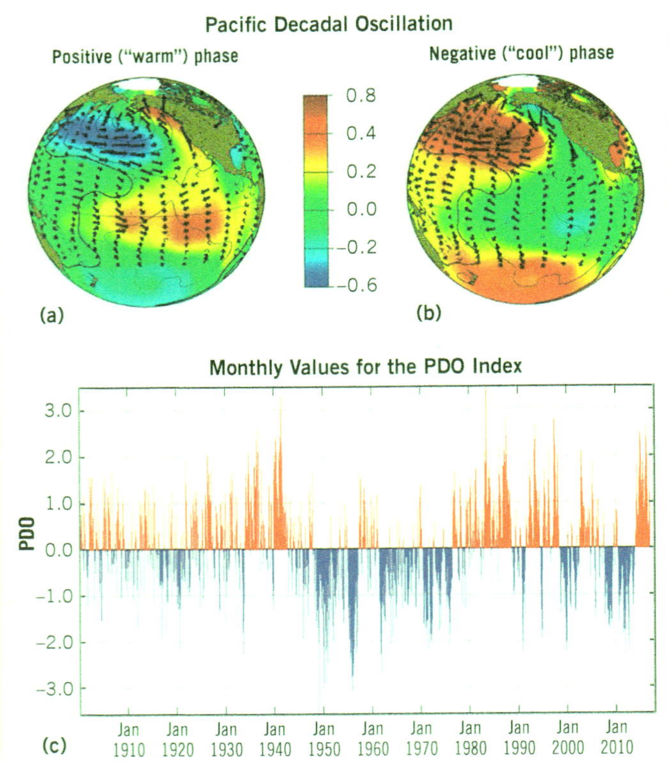

▲ **Figure 5-I** The Pacific Decadal Oscillation (PDO). (a) Positive values (in units of degrees Celsius) indicate warm water in the eastern tropical Pacific, (b) whereas negative values indicate relatively cool water in the same area. (c) The PDO index from 1900 to 2015.

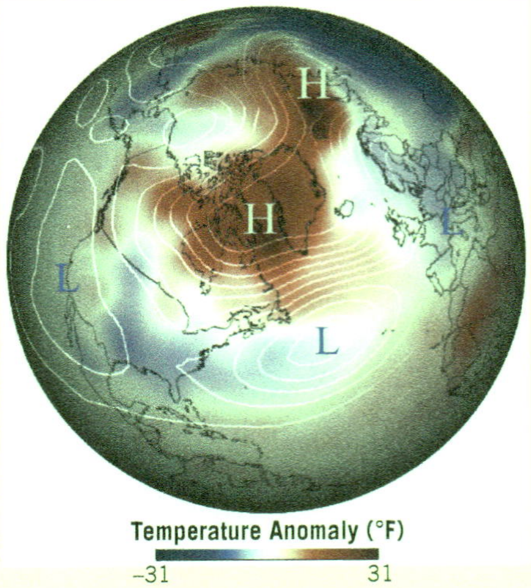

▲ **Figure 5-J** The "cold" phase of the Arctic Oscillation in February 2010, showing temperature anomalies compared with the average temperatures from 1971 to 2000. The polar high was stronger than usual, and the midlatitudes were generally colder; Greenland was warmer.

cycles, such as the *Pacific Decadal Oscillation* (see the box *Focus: Multiyear Atmospheric and Oceanic Cycles*).

Over the last century, El Niño events have occurred on average once every two to seven years. It appears that El Niños have been becoming more frequent and progressively warmer in recent decades. The 2014–2016 event was probably the strongest El Niño of the last 200 years, and the second-strongest event, 1997–1998, developed more rapidly than any in the last 50 years—although we don't yet know why. Some scientists speculate that global warming may be influencing the intensity of the El Niño cycle, but a clear connection has not yet been found.

In the last 30 years, great strides have been made toward forecasting El Niño events months in advance—largely because of better satellite monitoring of ocean conditions and the establishment of TAO/TRITON, an array of oceanic buoys in the tropical Pacific. Although much remains to be learned, a clearer understanding of cause and effect, as well as countercause and countereffect of ENSO, is gradually emerging.

LearningCheck 5-15 **What is a teleconnection? Describe one example of a teleconnection.**

After studying this chapter, you should be able to answer the following questions. Key terms from each text section are shown in **bold type.** Definitions for key terms are also found in the glossary at the back of the book.

Learning Review 5

Key Terms and Concepts

The Nature of Atmospheric Pressure (*p. 112*)

1. What generally happens to **atmospheric pressure** with increasing altitude?
2. Explain how atmospheric pressure is related to air density and air temperature.
3. What causes a **thermal high** near the surface? A **thermal low**?
4. What causes a **dynamic high** near the surface? A **dynamic low**?
5. Define the following terms: **barometer**, **millibar**, **isobar**.
6. In reference to air pressure, what is a **high**, a **low**, a **ridge**, and a **trough**?
7. What is a **pressure gradient**?

The Nature of Wind (*p. 114*)

8. What three factors influence the direction of wind flow?
9. How and why are **friction layer** (surface) winds different from upper-atmosphere **geostrophic winds**?
10. Describe the relationship between the "steepness" of a pressure gradient and the speed of the wind along that gradient. Describe the general wind speed associated with a gentle (gradual) pressure gradient and a steep (abrupt) pressure gradient.

Cyclones and Anticyclones (*p. 117*)

11. Describe and explain the pattern of wind flow in the Northern Hemisphere around the following:
    - a surface high
    - a surface low
    - an upper atmosphere high
    - an upper atmosphere low

    (You should be able to sketch the wind direction on isobar maps of highs and lows near the surface and in the upper atmosphere for both the Northern and Southern Hemispheres.)
12. What explains the difference in wind flow patterns in the Northern Hemisphere and the Southern Hemisphere?
13. What is a **cyclone**? An **anticyclone**?
14. Describe the pattern of vertical air movement within a cyclone and within an anticyclone.

The General Circulation of the Atmosphere (*p. 119*)

15. What are the **Hadley cells**, and what generally causes them?
16. Describe the general location and characteristics of the following atmospheric circulation components:
    - **intertropical convergence zone (ITCZ)**
    - **trade winds**
    - **subtropical highs (STHs)**
    - **westerlies**

    (You should be able to sketch the location of these four components on a blank map of an ocean basin.)
17. Discuss the characteristic weather associated with the ITCZ and the characteristic weather associated with subtropical highs.
18. What are the **horse latitudes** and the **doldrums**?
19. Describe the general location and characteristics of the **jet streams** of the westerlies.
20. What are **Rossby waves**?

21. Briefly describe the location and general characteristics of the high-latitude components of the general circulation patterns of the atmosphere:
    - **polar front (subpolar lows)**
    - **polar easterlies**
    - **polar highs**
22. Differentiate between trade winds and **antitrade winds**.

Modifications of the General Circulation (*p. 127*)

23. Describe and explain the seasonal shifts of the general circulation patterns; note the significance of the seasonal shifts of the ITCZ and the subtropical highs.
24. Describe and explain the South Asian **monsoon**.

Localized Wind Systems (*p. 131*)

25. Explain the origin of **sea breezes** and **land breezes**.
26. In what ways are sea breezes and land breezes similar to **valley breezes** and **mountain breezes**?
27. What is a **katabatic wind**, and where are such winds commonly found?
28. In what ways are **Santa Ana winds** similar to **foehn** and **chinook** winds?

El Niño–Southern Oscillation (*p. 134*)

29. What is the **Walker Circulation**?
30. Why is **El Niño** commonly referred to as **El Niño–Southern Oscillation (ENSO)**?
31. Contrast the oceanic and atmospheric conditions in the tropical Pacific Ocean basin during an El Niño event with those of the normal pattern.
32. Contrast the oceanic and atmospheric conditions during an El Niño event with those of a **La Niña** event.
33. What are **teleconnections**?

Study Questions

1. Why does atmospheric pressure decrease with height?
2. Why is it misleading to describe atmospheric pressure as simply the weight of only the air pressing down on a surface?
3. When wind moves over any significant distance, why doesn't it flow straight "down" a pressure gradient?
4. Why are upper-atmosphere winds usually faster than surface winds?
5. Why do the trade winds cover so much of the globe?
6. Why are the subtropical highs and the ITCZ characterized by little wind?
7. Explain why the trade winds and the antitrade winds blow in opposite directions.
8. Why does the ITCZ generally shift north of the equator in the Northern Hemisphere summer and south of the equator in winter?

Exercises

1. Using Table 5.1, estimate the altitude where atmospheric pressure has decreased to about half of its surface value. ______ kilometers
2. Using Table 5.1, estimate the atmospheric pressure at the cruising altitude of a jet airliner (about 34,000 feet). ______ mb
3. When the winds in a hurricane are blowing at 150 knots, what is the wind speed in
   a. Miles per hour: _____ b. Kilometers per hour: _____
4. If the atmospheric pressure is 1010 millibars, what is the equivalent pressure in hectopascals? ______ hPa
5. If the atmospheric pressure is 99,000 pascals, what is the equivalent pressure in millibars? ______ mb

Mastering Geography®

Looking for additional review and test prep materials? Visit the Study Area in **Mastering Geography** to enhance your geographic literacy, spatial reasoning skills, and understanding of this chapter's content by accessing a variety of resources, including Map Master interactive maps, geoscience animations, *Mobile Field Trips*, videos, *Project Condor* Quadcopter videos, *In the News* RSS feeds, flashcards, web links, self-study quizzes, and an eText version of *McKnight's Physical Geography*.

Environmental Analysis MEAN SEA-LEVEL PRESSURE

Data Sea-Level Pressure https://media.pearsoncmg.com/ph/esm/esm_mcknight_physgeo_13/media/

Measurements of atmospheric pressure on Earth's surface are converted to mean sea-level pressure (MSLP) values, which can then be used to track weather systems. Low MSLP is generally associated with cloudiness and precipitation; high MSLP with fair weather.

Activities

Go to www.nnvl.noaa.gov/view, the NOAA View Data Exploration Tool. (You can watch the Video Tour.) Click on "Add Data"; choose "Weather Models" and then "Mean Sea Level Pressure." Check the "Data Values" box.

1. Pressure varies greatly in the vertical direction but not horizontally. What is the lowest MSLP in North America? The highest?
2. What is the lowest MSLP globally? The highest?
3. What is the MSLP at your location?
4. If MSLP varies greatly over a short distance, then a region will experience strong winds. Where are the winds strongest in North America?
5. If MSLP varies minimally over a great distance, then a region will experience weak or calm winds. Where are the winds weakest in North America?

The initial map shows today's MSLP. To view MSLP predictions for the next nine days, drag the slider bar to the right.

6. Based on the variation of MSLP, how is weather expected to change at your location over the next nine days? Indicate any changes in wind speed, cloud cover, and precipitation. How did you determine these changes?

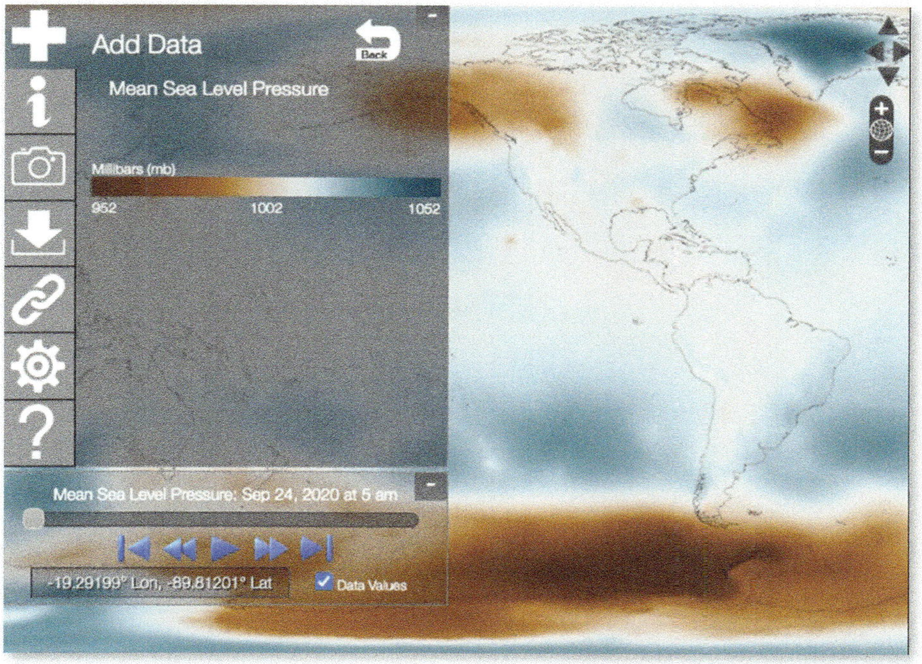

Seeing Geographically in Our National Parks

Look again at the photograph of the distant snow storm shown in Denali National Park at the beginning of the chapter (p. 110). Why is one side of the mountain in whiteout conditions, but the remainder of the landscape is mostly clear? This national park is at latitude 63° N; which global wind system most frequently blows over Denali National Park?

Seeing Geographically in Our National Parks

A sunset in Everglades National Park, Florida. What does the landscape suggest about the general climate of this region? What do the clouds look like?

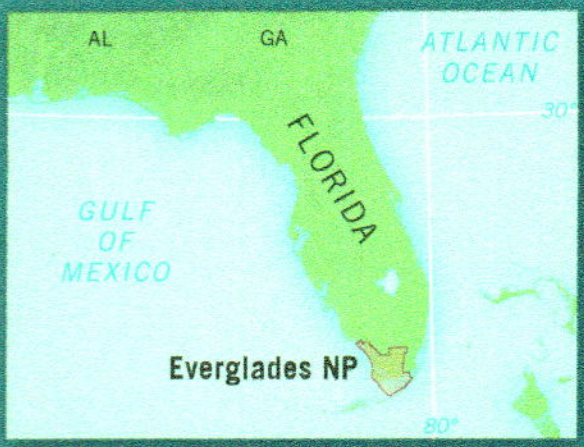

Everglades National Park, FL

Atmospheric Moisture

6

HAVE YOU EVER WONDERED why some places on Earth are deserts while other places are soaked by rain almost every day? Or why Death Valley National Park gets less than 3 inches of rain per year whereas the Everglades National Park gets 60 inches? Such vast differences in rainfall around the world are the consequences of the variations in temperature, pressure, and wind that we learned about in previous chapters, as well as those of the fourth element of weather and climate: *moisture*.

Although we're familiar with water, it is one of the most difficult components of the atmosphere to understand fully. This complexity arises partly because water occurs in the atmosphere in three physical states: as a *solid* (such as snow, hail, and sleet), as a *liquid* (such as rain and water droplets in clouds), and as a *gas* in the form of water vapor. Of these three states, water as a gas is the most important in terms of the dynamics of the atmosphere. Before we can interpret global precipitation patterns, we must understand the behavior of water vapor.

Water vapor in the atmosphere is a component of several key Earth systems. Most obviously, the movement of water vapor is part of the *hydrologic cycle*—a key to understanding weather and climate, the biosphere, and many processes shaping Earth's surface. Also, the evaporation of water, the transport of water vapor from one location to the next, and the eventual condensation of water vapor to form clouds and precipitation is a key component of Earth's *energy cycle*.

In this chapter, we limit our focus to the role of water in weather processes in Earth's atmosphere. In the chapters that follow, we broaden our discussion of water to include the dynamics of the oceans, lakes, streams, and underground water.

As you study this chapter, think about these **Key**Questions:

- **What properties of water influence its effects in the atmosphere?**
- **What is latent heat?**
- **What causes relative humidity to change?**
- **Why is rising air so important to cloud formation and precipitation?**
- **What factors are generally responsible for the wet regions of Earth, and what factors are responsible for dry regions?**
- **What are the causes and effects of acid rain?**

The Impact of Atmospheric Moisture on the Landscape

When the atmosphere contains enough moisture, water vapor condenses into haze, fog, clouds, rain, sleet, hail, or snow, producing a visible, tangible skyscape. Through precipitation, water can produce dramatic short-term changes in the landscape, such as when puddles form, streams and rivers flood, or snow and ice blanket the ground. The long-term effects of atmospheric moisture are even more fundamental. Water vapor stores energy that can incite the atmosphere into action. Rainfall and snowmelt on soil and rock are integral to weathering and erosion. In addition, precipitation is critical to the survival of most plants.

The Nature of Water: Commonplace But Unique

Water is the most widespread substance on Earth's surface, occupying more than 70 percent of the surface area of the planet. Water is also perhaps the most distinctive substance found on Earth. Water set the stage for the evolution of life and is an essential ingredient of all life today. In order to understand the significance of the properties of water, we begin with an introduction to water at two different scales: the *hydrologic cycle* and the *water molecule*.

The Hydrologic Cycle

The widespread distribution of water vapor in the atmosphere reflects the ease with which moisture can change from one state to another at the pressures and temperatures found in the lower troposphere.

Moisture can leave Earth's surface as a gas and return as a liquid or solid. Indeed, a continuous interchange of moisture occurs between Earth and the atmosphere (Figure 6-1). This unending circulation of our planet's water supply is referred to as the **hydrologic cycle**. Its essential feature is that liquid water (primarily from the oceans) evaporates into the air, condenses to the liquid (or solid) state, and returns to Earth as some form of precipitation. Water is predominately stored in water bodies (oceans, rivers and lakes), but water also exists in the land (soil moisture), the atmosphere (clouds and humidity), the biosphere (plants and animals), and polar ice.

The movement of moisture through the cycle is intricately related to many atmospheric processes and is an important determinant of climate. We discuss the complete hydrologic cycle in greater detail in Chapter 9, but here we begin with the basic dynamics of water transfer to and from the atmosphere.

Before we look specifically at the role of water in weather and climate, we must look at the water molecule itself. The characteristics of the water molecule help explain many of water's unique properties. What we learn about the nature of water over the next few pages will help us understand many processes discussed in the remaining chapters.

The Water Molecule

As we discussed in Chapter 4, the fundamental building blocks of matter are *atoms*. Atoms are almost incomprehensibly small—a thimble filled with 1 gram of water holds about 100,000,000,000,000,000,000,000 atoms! Atoms are composed of still smaller *subatomic particles*: positively charged *protons* and uncharged *neutrons* comprise the *nucleus* of the atom, surrounded by negatively charged *electrons*.[1]

[1]Although it was once thought that the smallest particles of matter were electrons, protons, and neutrons, we know now that some subatomic particles are made of still smaller particles such as *quarks*.

▼ Figure 6-1 The hydrologic cycle is a continuous interchange of moisture between the atmosphere and Earth.

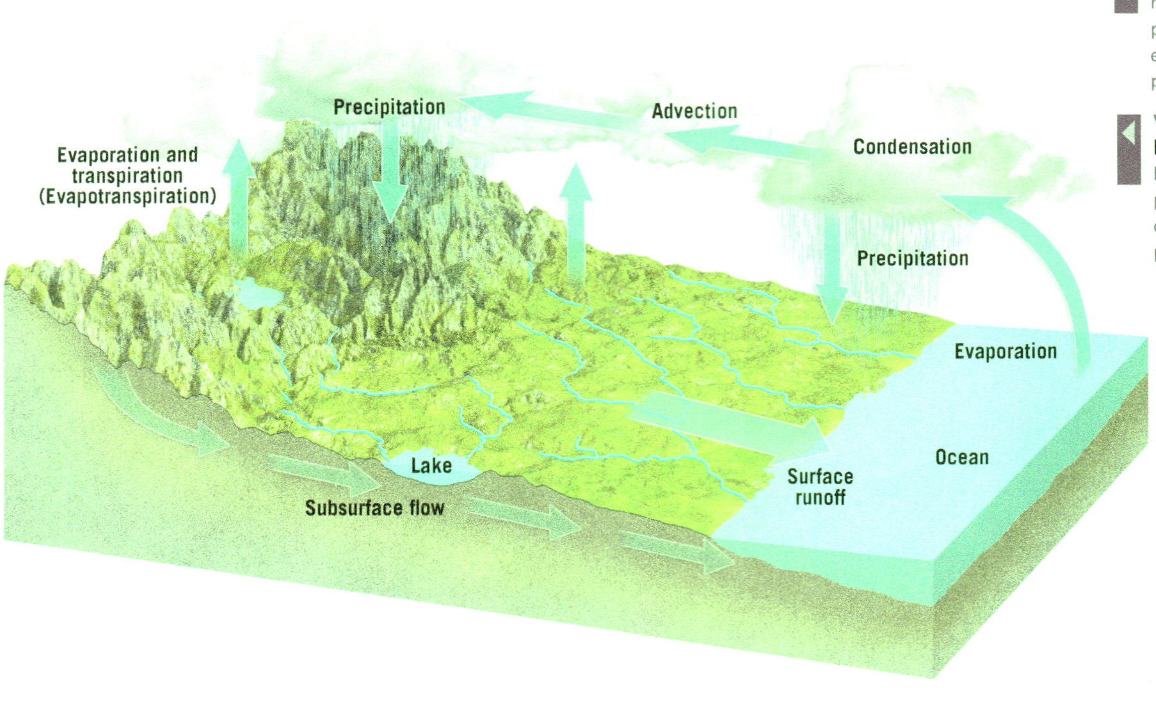

Animation
Hydrologic Cycle
https://media.pearsoncmg.com/ph/esm/esm_mcknight_physgeo_13/media/

Video
Hydrological Cycle
https://media.pearsoncmg.com/ph/esm/esm_mcknight_physgeo_13/media/

Electrons are held in atoms by electrical attraction to the protons and are found in "shells" surrounding the nucleus. In many atoms, the numbers of electrons, protons, and neutrons are normally equal—for example, an atom of the element oxygen typically has eight protons and eight electrons, along with eight electrically neutral neutrons. The number of neutrons in an element's individual atoms can vary without affecting the chemical behavior of the atom. Atoms with different numbers of neutrons are called *isotopes*. If the number of electrons in an atom is different from the number of protons, the neutral atom becomes an electrically charged *ion*.

Hydrogen Bonds: Two or more atoms form a *molecule* when they are held together in a specific geometric arrangement. In the water molecule, two atoms of hydrogen and one atom of oxygen (H_2O) are held together by *covalent bonds*, in which the atoms share electrons (some of the electrons are found between energy shells of two atoms; **Figure 6-2a**). The hydrogen atoms are separated by an angle of 105°. As a consequence of this geometry, a water molecule has a weak electrical *polarity*: the oxygen side of the molecule has a slight negative charge, whereas the hydrogen side has a slight positive charge (**Figure 6-2b**).

The weak electrical polarity of the molecules gives water many of its interesting properties. For example, water molecules tend to orient themselves toward each other so that the negatively charged oxygen side of one water molecule is next to the positively charged hydrogen side of another. This attraction forms a **hydrogen bond** between adjacent water molecules (see Figure 6-2b). Although hydrogen bonds are relatively weak (compared with the covalent bonds that hold together the atoms within a water molecule), water molecules tend to "stick" to each other.

LearningCheck 6-1 **What is a hydrogen bond?** **(Answer on p. AK-2)**

Important Properties of Water

Water has a number of properties that are important in our study of physical geography.

Liquidity: One of the most striking properties of water is that it exists as a liquid at the temperatures found at most places on Earth's surface. The liquidity of water greatly enhances its versatility as an active agent in the atmosphere, lithosphere, and biosphere.

Ice Expansion: Most substances contract as they get colder, no matter what the change in temperature. When freshwater becomes colder, however, it contracts only until it reaches 4°C (39°F) and then expands (as much as 9 percent) as it cools from 4°C to its freezing point of 0°C (32°F). As water is cooled from 4°C to its freezing point, water molecules begin to form hexagonal structures, held together by hydrogen bonds. When frozen, water is made entirely of these structures—hexagonal snowflakes reflect the internal structure of these ice crystals. As we see in Chapter 15, this expansion of ice can break apart rocks; it is an important process of *weathering* (the disintegration of rock exposed to the atmosphere).

Because water expands as it approaches freezing, ice is less dense than liquid water. The density of liquid water at 4°C is 1 gram per cubic centimeter (1 g/cm^3), but the density of ice is only 0.92 g/cm^3. As a result, ice floats on and near the surface of water. If it were denser than water, ice would sink to the bottom of lakes and oceans, where melting would be virtually impossible, and eventually many water bodies would become ice-choked. In fact, because freshwater becomes less dense as it approaches its freezing point, water that is ready to freeze rises to the tops of lakes. Hence all lakes freeze from the top down. (*Note*: In ocean water the high salinity prevents hexagonal structures from forming until the water is completely frozen.)

Surface Tension: Because of its electrical polarity, liquid water molecules tend to stick together (*cohesion*), giving water extremely high **surface tension**. A thin "skin" of molecules forms on the surface of liquid water, causing it to "bead." Some insects use the stickiness of water to stride atop the surface of a body of water—the insect's weight, spread over the surface area of the water, is less than the force of the hydrogen bonds holding the water molecules together

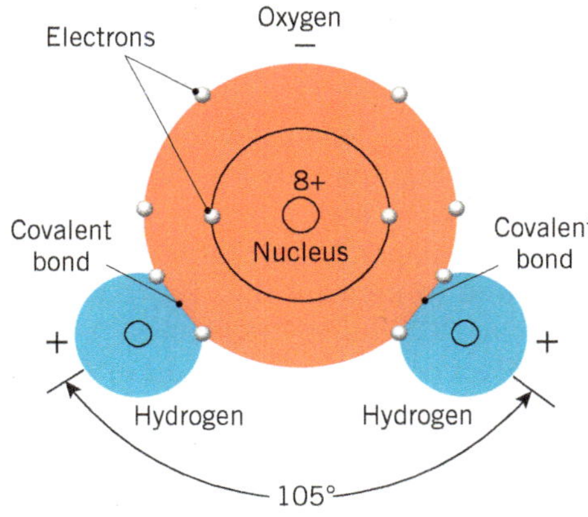

(a) The oxygen side of the water molecule has a slight negative charge, and the hydrogen side has a slight positive charge.

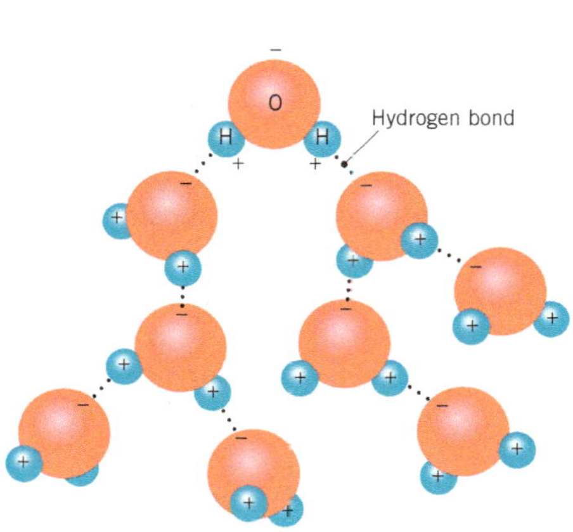

(b) Hydrogen bonds form between water molecules because the negatively charged oxygen side of one molecule is attracted to the positively charged hydrogen side of another molecule.

▲ **Figure 6-2** (a) A water molecule consists of two hydrogen atoms and one oxygen atom. (b) Hydrogen bonds between water molecules.

▲ Figure 6-3 The weight of this raft spider (*Dolomedes fimbriatus*) is supported by the surface tension of the water surface.

(Figure 6-3). Later in this chapter we see that surface tension influences the way that water droplets grow within a cloud.

LearningCheck 6-2 **What happens to the volume of water when it freezes?**

Capillarity: Water molecules also "stick" easily to many other substances—a characteristic known as *adhesion*. Surface tension combined with adhesion allows water to climb upward in narrow openings for many centimeters or even meters. This upward action under confinement is called **capillarity**. Capillarity enables water to circulate upward through rock cracks, soil, and the roots and stems of plants.

Solvent Ability: Water can dissolve almost any substance; it is sometimes referred to as the "universal *solvent*." Because of their polarity, water molecules are attracted not only to each other but also to other "polar" chemical compounds. Water molecules attach themselves quickly to the ions that constitute the outer layers of solid polar materials and in some cases can overcome the strength of their bonds, tearing the ions out of the solid and eventually dissolving the material. As a result, water in nature is nearly always impure—it contains various other chemicals in addition to its hydrogen and oxygen atoms. As water moves through the atmosphere, on the surface of Earth, and in soil, rocks, plants, and animals, it carries a remarkable diversity of dissolved minerals and nutrients as well as tiny solid particles in suspension.

Specific Heat Capacity: Another environmentally important characteristic of water is its large heat capacity. As we saw in Chapter 4, *specific heat capacity* (also called *specific heat* or *heat capacity*) is defined as the amount of energy required to raise the temperature of 1 gram of a substance by 1°C. When water is warmed, it can absorb an enormous amount of energy with only a small increase in temperature. Water's specific heat capacity (1 calorie/gram or about 4184 joules/kg) is exceeded by no other common substance except ammonia and is about five times greater than that of materials such as soil and rock. In other words, it takes five times as much energy to increase the temperature of water by 1°C as it does to increase the temperature of soil or rock by the same amount.

The high specific heat capacity of water is a consequence of the relatively large amount of kinetic energy required to overcome the hydrogen bonds between water molecules. The practical result, as we saw in Chapter 4, is that bodies of water are very slow to warm up during the day or in summer, and very slow to cool off during the night or in winter. Thus, water bodies have a moderating effect on the temperature of the overlying atmosphere by serving as reservoirs of warmth during winter and having a cooling influence in summer.

LearningCheck 6-3 **How does the specific heat capacity of water influence how rapidly the water warms in summer?**

Phase Changes of Water

Matter is found in three states or phases—gas, liquid, and solid—which can be converted from one to the other. These conversions, or *phases changes*, are as follows:

- *Sublimation* from solid to gas; *deposition* from gas to solid
- *Melting* from solid to liquid; *freezing* from liquid to a solid
- *Condensation* from gas to liquid; *vaporization* from liquid to gas

Earth's water is found naturally in all three states: as a liquid, as a solid, and as a gas. The great majority of the world's moisture is in the form of liquid water, which is converted to water vapor by **evaporation** or to ice by **freezing**. Water vapor can be converted to liquid water by **condensation** or directly to ice by **deposition**. Ice can be converted to liquid water by **melting** or to water vapor by **sublimation**.

In each of these phase changes, an exchange of *latent heat* takes place (Figure 6-4), a concept we introduced in Chapter 4. Latent means hidden, so this is heat that is hidden in phase changes. We must understand phase changes and latent heat to comprehend several atmospheric processes we describe later in this chapter.

Latent Heat

If you insert a thermometer into a pan of cool water and then warm the water on a stove, you will notice an interesting pattern of temperature change. At first, the temperature of the water increases, as you might expect. However, once the water begins to boil, the temperature of the water does not increase above 100°C (212°F; at sea level)—even if you turn up the flame! This result is just one of several important observations we can make about water as it changes states.

Figure 6-5 is a temperature graph showing the amount of energy (in calories) required to melt a 1-gram block of ice and then convert it to water vapor. Recall that 1 calorie is the amount of heat required to increase the temperature of 1 gram of liquid water by 1°C—about 4.184 joules (J).

We begin with our block of ice at a temperature of −40°C. As we add energy, the temperature of the ice increases quickly; only 20 calories (84 J) of energy are needed to increase the temperature of the ice to its melting point of 0°C. Once the ice begins to melt, it absorbs 80 calories (335 J)

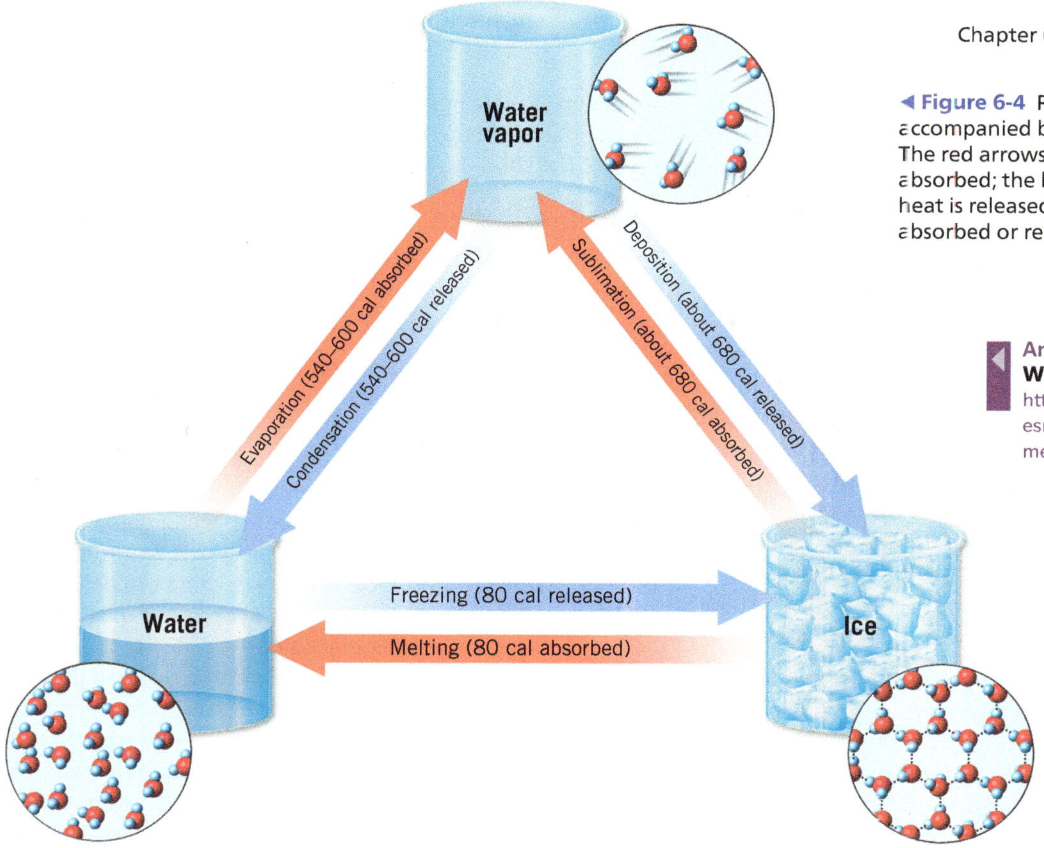

◄ **Figure 6-4** Phase changes of water are accompanied by the exchange of latent heat. The red arrows indicate that latent heat is absorbed; the blue arrows indicate that latent heat is released. The values given for heat absorbed or released are for 1 gram of water.

Animation
Water Phase Changes
https://media.pearsoncmg.com/ph/esm/esm_mcknight_physgeo_13/media/

of energy per gram, but the *temperature of the ice water does not increase* above 0°C until all of the ice has melted.

We then add another 100 calories (418 J) of energy to increase the temperature of the liquid water from 0°C to 100°C, its boiling point at sea level. The boiling water absorbs 540 calories (2260 J) per gram, but once again, the *temperature does not increase* above 100°C until all of the liquid water is converted to water vapor.

The key point is that even as energy was added, the temperature of the water did not increase while it was undergoing a *phase change*. Here's why: in order for ice to melt, energy must be added to "agitate" water molecules enough to break some of the hydrogen bonds that hold these molecules together as ice crystals. The energy added does not increase the temperature of the ice, but it does increase the *internal structural energy* of the water molecules. They can break free and become liquid.[2]

As we saw in Chapter 4, this energy exchanged during a phase change is called **latent heat**. *Latent heat* is different from *specific heat capacity*. It may be helpful to think of *latent heat* as the energy used to break or form the bonds (in rearranging molecular structure between two phases), whereas *specific heat capacity* is the energy used to raise the temperature (the "speed" of the molecules in one phase).

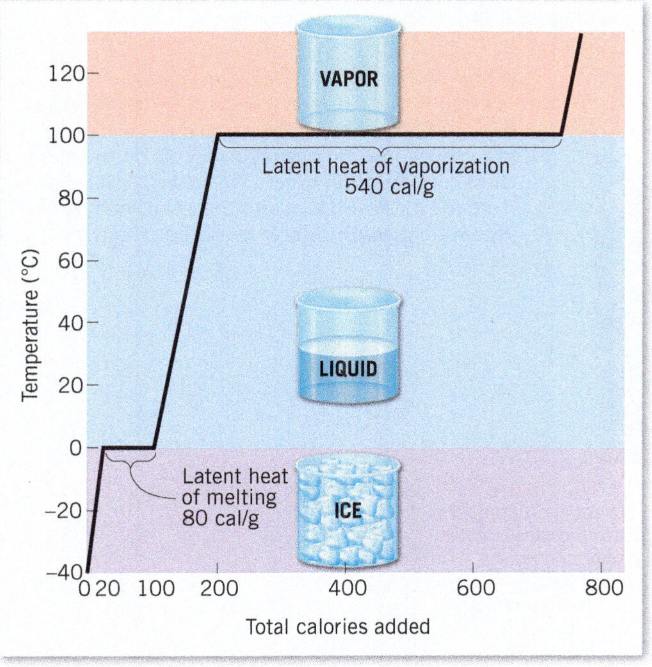

▲ **Figure 6-5** The energy input (in calories) and the associated temperature changes as 1 gram of ice starting at a temperature of −40°C is melted and then converted to water vapor. The latent heat of vaporization is much greater than the latent heat of melting.

Melting and Freezing: Energy is required to melt ice; that energy is called the *latent heat of melting*. The opposite is also true: when water freezes, the liquid water molecules must give up some of their internal structural energy in order to revert to a less agitated condition in which ice can form. The energy released as water freezes is called the *latent heat of fusion*. For each gram of ice, 80 calories (335 J) of energy are absorbed when ice melts, and 80 calories of energy are released when water freezes.

[2]The *internal structural energy* of water molecules includes the *translational kinetic energy* associated with temperature (the "jiggling" of molecules) as well as energy associated with the rotation and vibration of the molecules and the *potential energy* associated with forces between molecules, such as bonds.

Evaporation and Condensation: In a similar way, energy must be added to agitate liquid water molecules enough for them to escape into the surrounding air as water vapor. The energy added does not increase the temperature of the liquid water but instead increases the energy of water molecules; they can then break free and become vapor. The energy required to convert liquid water to water vapor is called the *latent heat of vaporization*.

Once again, the opposite is also true: when water vapor condenses back to liquid water, the highly agitated water vapor molecules must give up some of their internal structural energy in order to revert to a less agitated liquid state. The energy released during condensation is called the **latent heat of condensation**. For each gram of liquid water at a temperature of 100°C, 540 calories (2260 J) of energy are absorbed as water vaporizes, and 540 calories of energy are released when water vapor condenses.

Evaporation versus Vaporization: The value given above for the latent heat of vaporization refers to circumstances in which water is boiling. *Boiling* occurs when vaporization takes place beneath the liquid surface—not just at the surface. In nature, however, most water vapor is added to the atmosphere through simple *evaporation* off the surface of water bodies at temperatures below 100°C. In this case, the energy required for vaporization is greater than it is for boiling water. The **latent heat of evaporation** ranges from 540 calories to about 600 calories, depending on the temperature of the water. (It is approximately 585 calories [2450 J] when the liquid water is at a temperature of 20°C [68°F].)

Notice in Figures 6-4 and 6-5 that about seven times more energy is needed to evaporate 1 gram of liquid water than is needed to melt 1 gram of ice. Notice also that when sublimation takes place, the latent heat exchange is simply the total of the solid–liquid and liquid–gas exchanges.

LearningCheck 6-4 **Why doesn't the temperature of a block of ice increase while the ice melts?**

Importance of Latent Heat in the Atmosphere

The significance of the exchange of latent heat during phase changes—especially between liquid water and water vapor—is straightforward. Whenever evaporation takes place, energy is removed from the liquid to vaporize some of the water, so the temperature of the remaining liquid is reduced. Because latent heat is energy "stored" in water vapor during evaporation, *evaporation is a cooling process.* A swimmer who leaves a swimming pool on a dry, warm day experiences such evaporative cooling: as water evaporates off the wet skin, the skin is chilled (Figure 6-6a).

Conversely, because latent heat must be released during condensation, *condensation is a warming process.* Water vapor represents a "reservoir" of energy that can be transferred from one location to another through the movement of air masses by wind. Whenever and wherever that water vapor condenses, this energy is added back to the atmosphere. When the wind moves water vapor, the wind is moving the energy necessary for storms. As we begin to see later in this chapter, the release of latent heat during condensation plays important roles in the stability of the atmosphere and in the power of many storms. On a smaller scale, your glass of iced lemonade cools the air near the glass so that condensation occurs. Condensation of moisture on the outside of your glass releases latent heat and causes the ice cubes to melt *faster* (Figure 6-6b).

(a) For the skin, evaporation is a cooling process.

(b) For the liquid in the glass, condensation is a warming process.

◀ **Figure 6-6** (a) Swimmers feel cold when they get out of the water on a warm, dry day because evaporation cools. (b) The air next to an ice-cold glass cools until condensation occurs on the glass. Once condensation occurs, the water molecules from the air slow down and the excess energy (latent heat) warms a cold drink, helping the ice melt.

Water Vapor and Evaporation

Although the vast majority of water on Earth is in liquid form, for the moment we direct our attention to water vapor, the source of moisture for clouds and precipitation.

Water vapor is a colorless, odorless, tasteless, invisible gas that mixes freely with the other gases of the atmosphere. We are only likely to become aware of water vapor (or lack thereof) when humidity is very high or very low. As we saw in Chapter 3, the amount of water vapor in the atmosphere varies by place and time. It is virtually absent in some places but constitutes as much as 4 percent of the total atmospheric volume in others. Essentially, water vapor is restricted to the lower troposphere: more than half of all water vapor is found within 1.5 kilometers (about 1 mile) of Earth's surface, and only a tiny fraction exists above 6 kilometers (about 4 miles).

Evaporation and Rates of Evaporation

The process that puts water vapor into the atmosphere is evaporation. In describing the phase changes of water, we simplified things a bit. As it turns out, evaporation and condensation may take place at the same time. Strictly speaking, water vapor is added to the air when the *rate* of evaporation exceeds the *rate* of condensation; in other words, when there is *net evaporation* (Figure 6-7). The rate of evaporation from a water surface, and therefore net evaporation, depends on several factors: the temperature (of air and water), the amount of water vapor already in the air, and air movement.

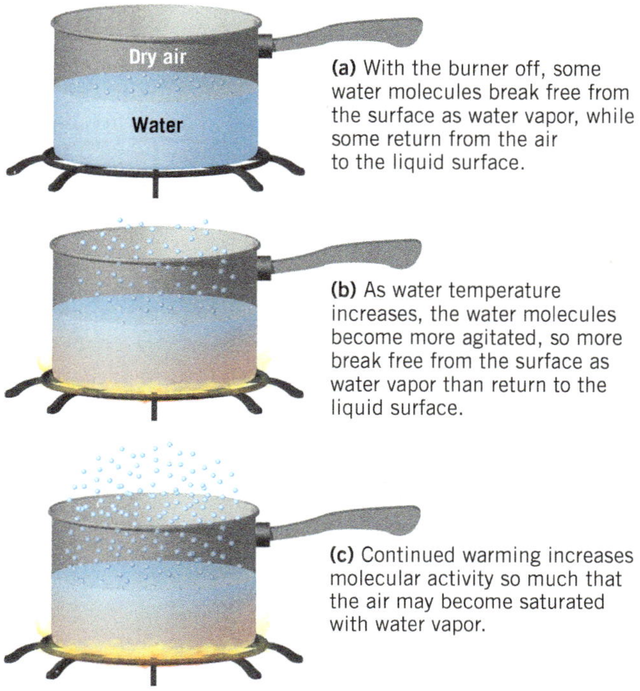

▲ Figure 6-7 Evaporation involves the escape of water molecules from a liquid surface into the air as water vapor. It can take place at any temperature, but high temperatures increase the energy of the molecules in the liquid and therefore increase the rate of evaporation.

Temperature: The water molecules in warm water are more "agitated" than those in cool water, so there tends to be more evaporation from warm water than from cold. Warm air also promotes evaporation: the more "energetic" gas molecules in warm air may collide with the liquid water surface and impart enough kinetic energy for some of the liquid water molecules to break their hydrogen bonds and enter the air above as vapor.

Water Vapor Content of Air: Water molecules cannot enter the air indefinitely, however. As we learned in Chapter 5, each gas in the atmosphere exerts pressure. Total atmospheric pressure is the sum of the pressures exerted by all of the individual gases in the atmosphere. The pressure exerted by water vapor is called the **vapor pressure**.

At any temperature, there is a maximum vapor pressure: the higher the temperature, the higher the maximum vapor pressure. In other words, warm air can hold more water vapor than in cold air can.

When water molecules in the air exert the maximum possible vapor pressure at a given temperature, the air is said to reach **saturation**, and a balance exists between the rates of evaporation and condensation. If this maximum vapor pressure is exceeded, more water vapor molecules will leave the air through condensation than are added through evaporation. *Net condensation* will occur until the rates of evaporation and condensation are again matched and the air is again saturated.

In practice, this means that evaporation tends to take place more rapidly when relatively little water vapor is in the air and that the rate of evaporation drops off as the air gets closer to saturation. On a hot, humid day you may have noticed that the air feels "heavy." Humid air is actually lighter than dry air, but it feels "heavy" because the sweat on your skin is evaporating very slowly (see footnote 6 on page 155.)

Wind: If the air overlying a water surface is almost saturated with water vapor, the rate of evaporation is about the same as the rate of condensation, and very little further evaporation can take place. If the air is in motion (as wind or turbulence), however, the water vapor molecules in it are dispersed more widely and the rate of evaporation increases.

LearningCheck 6-5 What happens to the rate of evaporation as air approaches saturation?

Evapotranspiration

Although most of the water that evaporates into the air comes from bodies of water, a relatively small amount comes from the land. Evaporation from land has two sources: (1) soil and other inanimate surfaces and (2) plants. The amount of moisture that evaporates from soil is minor compared with the land-derived moisture that comes from plants. Plants give up moisture through their leaves by a process called *transpiration*; the combined process of water vapor entering the air from land sources is called **evapotranspiration** (see Figure 6-1). Thus, water vapor in the atmosphere is added through evaporation from bodies of water and through evapotranspiration from sources on land.

Potential Evapotranspiration: Whether a given land location is wet or dry depends on the rates of evapotranspiration and precipitation. To analyze these rates, we use the concept of *potential evapotranspiration*: the amount of evapotranspiration that *could* occur if the ground in that location were always sopping wet. Potential evapotranspiration is estimated with a formula that takes into account a location's temperature, vegetation, soil, and actual evapotranspiration characteristics.

Wherever the annual precipitation is greater than the potential evapotranspiration, a water surplus accumulates in the ground. Conversely, in places where the annual potential evapotranspiration is greater than the actual precipitation, no water is available for storage in soil and in plants; dry soil and sparse vegetation are the result.

Measures of Humidity

The amount of water vapor in air is referred to as the *humidity*. Measures of humidity can express either the actual amount of water vapor in the air or the relative amount.

Actual Water Vapor Content

The actual amount of water vapor in the air can be described in several ways.

Absolute Humidity: One direct measure of the water vapor content of air is **absolute humidity**—the mass of water vapor in a given *volume* of air. Absolute humidity is usually expressed in grams of water vapor per cubic meter of air (g/m^3; 1 gram is approximately 0.035 ounces and 1 cubic meter is about 35 cubic feet). For example, if a cubic meter of air contains 12 grams of water vapor, its absolute humidity would be 12 g/m^3.

If the volume of air changes (as happens when air expands or compresses as it moves vertically), the value of the absolute humidity also changes, even though the total amount of water vapor remains unchanged. For this reason, absolute humidity is generally not used to describe moisture in air that is rising or descending.

Specific Humidity: The mass of water vapor in a given *mass* of air is called the **specific humidity** and is usually expressed in grams of water vapor per kilogram of air (g/kg). For example, if 1 kilogram of air contains 15 grams of water vapor, the specific humidity is 15 g/kg.

Specific humidity changes only as the quantity of water vapor varies; it is not affected by variations in air volume, as is absolute humidity.[3] Specific humidity is particularly useful in studying the characteristics and movements of air masses (discussed in Chapter 7).

The maximum specific humidity of air is determined by the temperature: cold air has a small maximum specific humidity, and warm air has a large maximum specific humidity. The maximum specific humidity at a given temperature

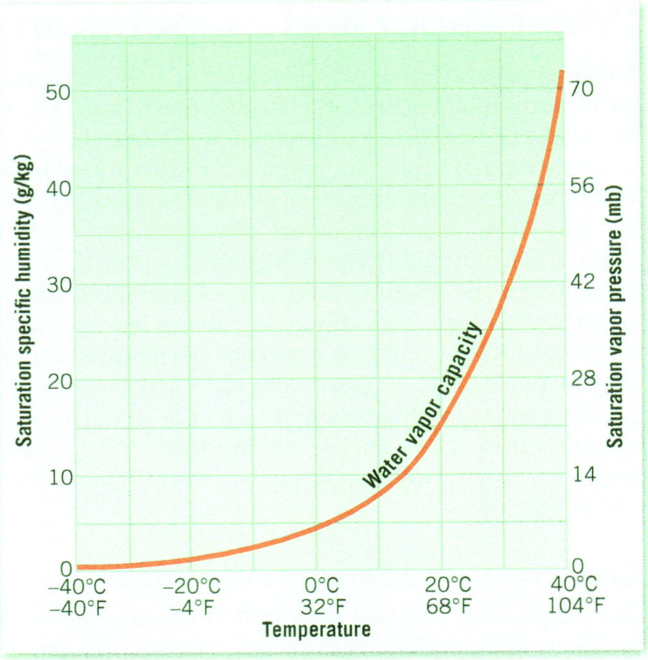

▲ Figure 6-8 Water vapor capacity (the maximum amount of water vapor that can be in the air) increases as temperature increases. This graph shows *saturation specific humidity* (in g/kg) and *saturation vapor pressure* (in millibars).

is called the *saturation specific humidity*. At a temperature of 20°C (68°F), the air can contain about 15g/kg of water vapor (its *water vapor capacity*), whereas at a temperature of 10°C (50°F), it can contain only about 8 g/kg (Figure 6-8).

Vapor Pressure: As we saw earlier, the contribution of water vapor to the total pressure of the atmosphere is called the *vapor pressure*. We can express vapor pressure in the same way as total atmospheric pressure, in millibars (mb). The maximum vapor pressure at a given temperature is called the *saturation vapor pressure*. Notice in Figure 6-8 that at a temperature of 10°C the saturation vapor pressure is a little more than 10 mb, whereas at 30°C it is about 40 mb.

Absolute humidity, specific humidity, and vapor pressure are all ways of expressing the actual amount of water vapor in the air—and as such are indications of the quantity of water that could be extracted by condensation and precipitation. However, before we discuss condensation and precipitation, we need to introduce the important concept of *relative humidity*.

LearningCheck 6-6 **What does it mean when the specific humidity of the air is 10 g/kg?**

Relative Humidity

The most familiar of humidity measures is **relative humidity**. Unlike absolute humidity, specific humidity, and vapor pressure, however, relative humidity does *not* describe the actual water vapor content of the air. Rather, relative humidity describes how close the air is to saturation with water vapor. Relative humidity is a ratio (expressed as a percentage) that

[3]For comparison with absolute humidity, 1 cubic meter of air at sea level has a mass of about 1.4 kilograms at room temperature.

compares the actual amount of water vapor in the air to the *water vapor capacity* of the air.

Water Vapor Capacity: **Water vapor capacity** is the maximum amount of water vapor that air can contain at a given temperature. Capacity is conceptually equivalent to *saturation absolute humidity*, *saturation specific humidity*, and *saturation vapor pressure*. The proper term depends on which measure of actual water vapor content is used.

As Figure 6-8 shows, cold air has a low water vapor capacity, and warm air has a high water vapor capacity. It is sometimes said that warm air can "hold" more water vapor than cold air, but this is somewhat misleading. The air does not actually hold water vapor as if it were in a sponge. Water vapor is simply one component of the atmosphere. The water vapor capacity of the air is determined by the temperature, which determines the rate of vaporization of water.

Relative humidity is calculated with a simple formula:

$$\text{Relative humidity} = \frac{\text{Actual water vapor in air}}{\text{Capacity}} \times 100$$

For example, suppose that 1 kilogram of air contains 10 grams of water vapor (the specific humidity is 10 g/kg). If the temperature is 24°C (75.2°F), the capacity of the air is about 20 g/kg (see Figure 6-8), so the relative humidity is

$$\frac{10\text{ g}}{20\text{ g}} \times 100 = 50\%$$

A relative humidity of 50 percent means that the air contains half of the maximum water vapor at that temperature. In other words, the air is 50 percent of the way to saturation.

Learning Check 6-7 **If the water vapor content of air is 5 g/kg and the capacity is 20 g/kg, what is the relative humidity?**

Factors Changing Relative Humidity: Relative humidity changes when the water vapor content of the air changes or the water vapor capacity of the air changes. If, in our previous example, 5 grams of water vapor are added to the air through evaporation while the temperature remains constant (so the capacity doesn't change), relative humidity increases:

$$\frac{15\text{ g}}{20\text{ g}} \times 100 = 75\%$$

Conversely, if water vapor is removed from the air by condensation or dispersal, the relative humidity decreases.

Relative humidity will also change when the temperature changes—*even if the actual amount of water vapor in the air remains the same*. If the temperature increases from our initial 24°C to 32°C (89.6°F), the water vapor capacity increases from 20 grams to 30 grams, so relative humidity decreases:

$$\text{at } 24°\text{C } (75.2°\text{F}): \frac{10\text{ g}}{20\text{ g}} \times 100 = 50\%$$

$$\text{at } 32°\text{C } (89.6°\text{F}): \frac{10\text{ g}}{30\text{ g}} \times 100 = 33\%$$

If instead the temperature *decreases* from 24°C to 15°C (59.0°F), the capacity decreases from 20 to 10 grams (see Figure 6-8), so relative humidity increases:

$$\text{at } 24°\text{C } (75.2°\text{F}): \frac{10\text{ g}}{20\text{ g}} \times 100 = 50\%$$

$$\text{at } 20°\text{C } (68.0°\text{F}): \frac{10\text{ g}}{15\text{ g}} \times 100 = 67\%$$

$$\text{at } 15°\text{C } (59.0°\text{F}): \frac{10\text{ g}}{10\text{ g}} \times 100 = 100\%$$

Notice that air can be brought to saturation (100 percent relative humidity) through a decrease in temperature alone—with no water vapor added. As we will see, the most common way that air is brought to the point of saturation and condensation is through cooling. It is unusual for air to reach saturation through an increase in water vapor content via evaporation; remember, when air already contains a lot of water vapor, the rate of evaporation is very low.

Temperature–Relative Humidity Relationship: The relationship between temperature and relative humidity is one of the most important in all of meteorology: as temperature increases, relative humidity decreases; as temperature decreases, relative humidity increases (until condensation begins).

This inverse relationship is portrayed in Figure 6-9, which demonstrates the fluctuation in temperature and relative humidity during a typical day (assuming no variation in the amount of water vapor in the air). In the early morning, the temperature is low, and the relative humidity is high because the air's water vapor capacity is low. As the air warms up during the day, the relative humidity decreases because warm air has a higher water vapor capacity than cool air. As evening approaches, air temperature decreases, the air's water vapor capacity diminishes, and relative humidity increases.

See Appendix III for a description of how relative humidity can be determined using a simple instrument known as a *psychrometer*.

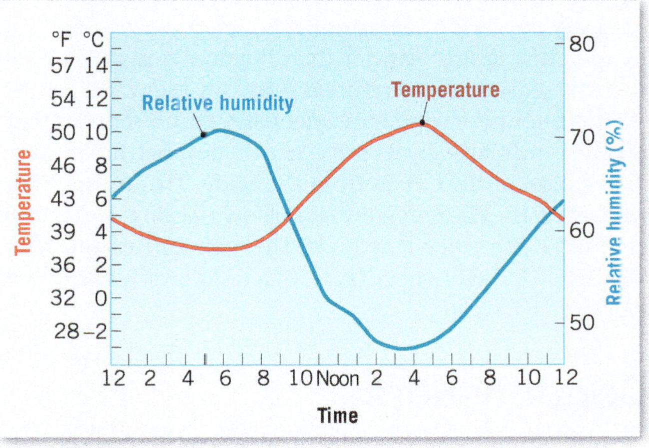

▲ **Figure 6-9** Typically there is an inverse relationship between temperature and relative humidity on any given day. As the temperature increases, the relative humidity decreases. Thus, relative humidity tends to be lowest in midafternoon and highest just before dawn.

LearningCheck 6-8 What happens to the relative humidity of unsaturated air when the temperature decreases? Why?

Related Humidity Concepts

Two other concepts related to relative humidity are useful in a study of physical geography: *dew point temperature* and *sensible temperature*.

Dew Point Temperature: As we have seen, when air cools, the water vapor capacity decreases and relative humidity increases. Cooling can bring unsaturated air to the saturation point. The temperature to which air must cool in order to saturate is called the **dew point temperature**, or **dew point**. The dew point temperature varies with the moisture content of air. In our previous example, air containing 10 grams of water vapor per kilogram of air reaches its dew point when chilled to about 15°C (59.0°F); air containing 20 grams of water vapor per kilogram of air reaches its dew point at about 24°C (75.2°F).

Although dew point is a temperature, in practice it can also describe the actual water vapor content of the air. Using our example, if the water vapor content of the air is 10 g/kg, its dew point temperature will be 15°C. Therefore, if the air reaches its dew point at 15°C, its water vapor content must be 10 g/kg.

Sensible Temperature: The term **sensible temperature** refers to the temperature that a person's body feels. In addition to the actual air temperature, relative humidity, dew point, and wind influence our perception of warmth and cold.

On a warm, humid day, air seems hotter than the thermometer indicates, and the sensible temperature is said to be high. This is because the air is near saturation. Perspiration on our skin does not evaporate readily—thus, there is little evaporative cooling, and the air seems warmer than it actually is. On a warm, dry day, evaporative cooling is effective, so the air seems cooler than it actually is. In this case, we say that the sensible temperature is low.

On a cold, humid day, the coldness seems more piercing because body heat is conducted away more rapidly in damp air; the sensible temperature is again described as low. On a cold, dry day, body heat is not conducted away as fast. The temperature seems warmer than it actually is, and we say that the sensible temperature is relatively high.

The amount of wind movement also affects sensible temperature, primarily by its influence on evaporation and the convection of heat away from the body. This is especially true when the air temperature is below freezing—on windy cold days, the wind may lower the apparent temperature significantly. See Appendix III for a description of the *heat index* and *wind chill*.

Condensation

Condensation is the opposite of evaporation. It is the process whereby water vapor is converted to liquid water. In other words, it is a change in state from gas to liquid. In order for condensation to take place, the air must be saturated. In theory, this saturated state can come about through the addition of water vapor to the air, but in practice it is usually the result of the air being cooled to a temperature below the dew point.

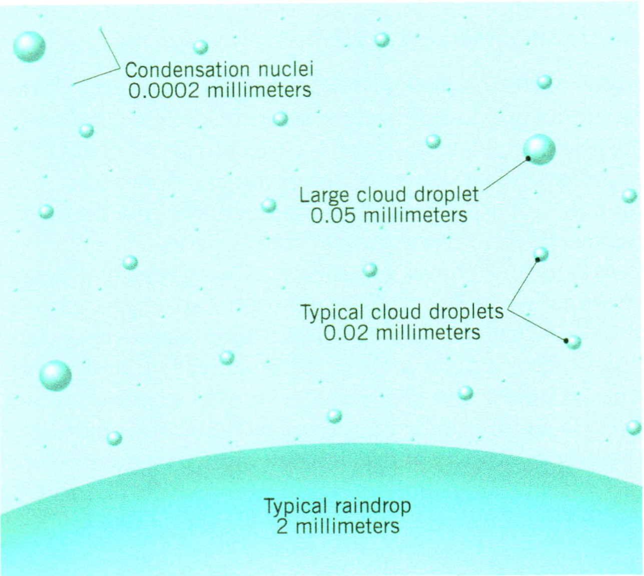

▲ **Figure 6-10** Comparative sizes of condensation nuclei and droplets. Condensation nuclei can be particles of dust, smoke, salt, pollen, bacteria, or any other microscopic matter found in the air.

The Condensation Process

Saturation alone is not enough to cause condensation. Surface tension makes it virtually impossible to grow liquid droplets of pure water from water vapor. Surface tension inhibits an increase in surface area, so it is very difficult for additional water molecules to enter or form a droplet. (Conversely, molecules can easily leave a small droplet by evaporation, thereby decreasing its area.) Thus, it is necessary to have a surface on which condensation can take place. If no such surface is available, condensation can only occur under extreme conditions. If cooling continues in such a situation, the air becomes **supersaturated** (has a relative humidity greater than 100 percent[4]).

Condensation Nuclei: Normally, plenty of surfaces are available for condensation. At ground level, availability of a surface is obviously no problem. The air above the ground usually has an abundance of "surfaces" also, as represented by tiny particles of dust, smoke, salt, pollen, bacteria, and other compounds. Most of these particles are microscopic—invisible to the naked eye (Figure 6-10). They are most concentrated over cities, seacoasts, and volcanoes, which are the source of much particulate matter, but they are present in lesser amounts throughout the troposphere. Referred to as

[4]This is often the case in the upper troposphere on a cloud-free day. If an airplane flies through and leaves a condensation trail that lasts for a long time, then the air was supersaturated.

hygroscopic particles or **condensation nuclei**, they serve as collection points for water molecules during condensation.

As soon as the air temperature cools to the dew point, water vapor molecules begin to condense around condensation nuclei. The droplets grow rapidly as more and more water vapor molecules stick to them. As they become larger, they bump into one another and coalesce as the colliding droplets stick together. Continued growth can make them large enough to be visible, forming haze or cloud particles. A single raindrop may contain a million or more condensation nuclei plus all of their associated moisture.

LearningCheck 6-9 Name the two conditions necessary for condensation to occur.

Supercooled Water: Clouds may be composed of liquid water droplets even when their temperature is below freezing. Although water in large quantity freezes at 0°C (32°F), if it is dispersed as fine droplets, it can remain in liquid form down to −40°C (−40°F). Water that persists in liquid form at temperatures below freezing is said to be "supercooled." **Supercooled water** droplets are important to condensation because they promote the growth of ice particles in cold clouds by freezing around the particles or by evaporating into vapor from which water molecules are readily added to the ice crystals.

Adiabatic Processes

One of the most significant facts in physical geography is that the only way in which large masses of air above the surface can be cooled to the dew point temperature is by expansion as the air masses rise. Thus, the only prominent mechanism for the development of clouds and the production of rain is *adiabatic cooling*. As we noted in Chapter 4, when air rises, its pressure decreases, so it expands and cools adiabatically.

Dry and Saturated Adiabatic Rates

Dry Adiabatic Rate: As a parcel of unsaturated air rises, it cools at the relatively steady rate of 10°C per 1000 meters (5.5°F per 1000 feet). This is known as the **dry adiabatic rate** (or the *dry adiabatic lapse rate*). This term is a misnomer: the air is not necessarily "dry"; it is simply unsaturated (its relative humidity is less than 100 percent).

If the air mass rises high enough, it cools to the dew point temperature, the air saturates, condensation begins, and clouds form. The altitude at which this occurs is known as the **lifting condensation level (LCL)**. Under many circumstances, the LCL is clearly visible as the often flat base of the clouds that form (Figure 6-11).

▲ **Figure 6-11** The flat bottom of this cumulus cloud represents the lifting condensation level.

Saturated Adiabatic Rate: As soon as condensation begins, latent heat is released (this energy was absorbed originally as the latent heat of evaporation). If the air continues to rise, cooling due to expansion continues, but release of the latent heat during condensation counteracts some of the adiabatic cooling and lessens the rate of cooling. This diminished rate of cooling is called the **saturated adiabatic rate** (also called the *wet* or *saturated adiabatic lapse rate*, or the *moist adiabatic rate*) (Figure 6-12). It depends on temperature, moisture, and pressure but averages about 6°C per 1000 meters (3.3°F per 1000 feet).

Adiabatic Warming of Descending Air: Adiabatic warming occurs when air descends. Typically, descending air will warm at the dry adiabatic rate of 10°C/1000 meters (5.5°F/1000 feet).[5] The increasing temperature of descending air increases the water vapor capacity of the air and thus

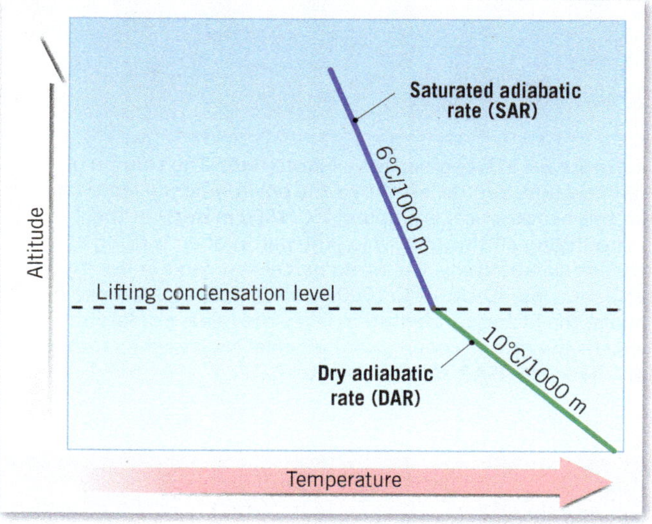

▲ **Figure 6-12** Unsaturated rising air cools at the *dry adiabatic rate* (DAR). Above the lifting condensation level, rising saturated air cools at the *saturated adiabatic rate* (SAR).

[5]Although descending air typically warms at the dry adiabatic rate, this is not always the case. When air descends through a cloud, some water droplets may evaporate, and the evaporative cooling will counteract some of the adiabatic warming. As a result, such descending air can warm at a rate close to the saturated adiabatic rate. As soon as evaporation of water droplets ceases, this descending air will warm at the dry adiabatic rate.

causes saturated air to become unsaturated. In short, this is why descending air cannot make clouds—and why the subtropical highs we discussed in Chapter 5 are characterized by dry weather.

Adiabatic Rates versus Environmental Lapse Rate: In any consideration of adiabatic temperature changes, remember that we are dealing with a parcel of air that is rising or descending. Do not confuse the adiabatic rates with the *environmental lapse rate* (or its average value of 6.5°C/1000 m), discussed in Chapter 4; the environmental lapse rate describes the temperature of *still* air at different altitudes in the atmosphere (Figure 6-13).

LearningCheck 6-10 **Why does air rising above the lifting condensation level cool at a lesser rate than air rising below that level?**

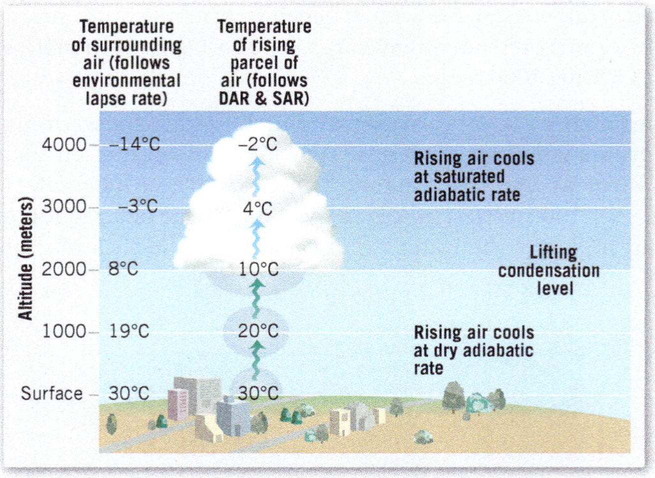

▲ **Figure 6-13** A comparison of lapse rates. The column of temperatures on the left shows the *environmental lapse rate* (in this hypothetical example, 11°C/1000 meters) of the surrounding air through which the parcel of air is rising and cooling adiabatically. The rising parcel first cools at the *dry adiabatic rate* (DAR; 10°C/1000 meters). As the parcel of air rises above the lifting condensation level, the release of latent heat lessens the rate of cooling, so the parcel cools at the *saturated adiabatic rate* (SAR; 6°C/1000 meters).

Significance of Adiabatic Temperature Changes

Some of the implications of adiabatic temperature changes can be seen in Figure 6-14, showing the temperature changes associated with a hypothetical parcel of air moving up and over a mountain range.

The parcel is unsaturated when it begins to rise over the mountain, so it cools at the dry adiabatic rate. Once the lifting condensation level is reached, the air continues to rise and cool at the saturated adiabatic rate as condensation forms a cloud. Once the air reaches the summit, it begins to descend down the leeward side of the mountain. The descending air warms at the dry adiabatic rate, so by the time the air has reached sea level again, it is significantly warmer and significantly drier—in both relative and absolute terms—than when it started. In this hypothetical example, we assume that moisture condensed out of the rising air is left as precipitation or clouds on the windward side of the mountain and that no evaporation takes place as the air descends. In most cases, essentially the same thing happens in the real world. This circumstance is one way in which deserts form.

Atmospheric Stability

Because most condensation and precipitation result from the cooling of rising air, conditions that promote or hinder the upward movement of air are of great importance to weather and climate. Depending on how "buoyant" it is, air rises more freely under some circumstances than under others.

Buoyancy

The tendency of any object to rise or sink in a fluid under the influence of gravity is called the *buoyancy* of that object. In general:

- If an object is less dense than the surrounding fluid, the object will float or rise.
- If it is denser than the surrounding fluid, it will sink.
- If it is the same density as the fluid, it will neither rise nor sink.

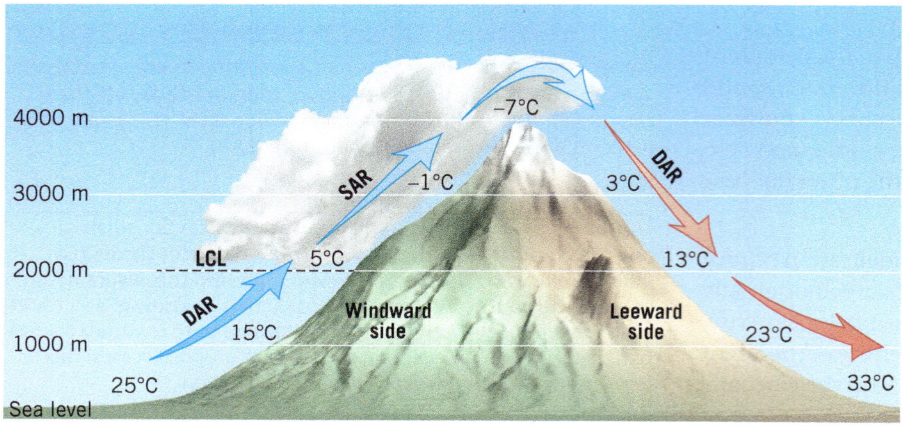

◄ **Figure 6-14** Temperature changes in a hypothetical parcel of air passing over a mountain 4000 meters (13,100 feet) high. The lifting condensation level (LCL) of the parcel is 2000 meters, the dry adiabatic rate (DAR) is 10°C/1000 meters, and the saturated adiabatic rate (SAR) is 6°C/1000 meters. Because latent heat is released during condensation on the windward side of the mountain, by the time the air has descended to sea level on the leeward side, it is warmer and drier than before it started up the windward side. (This example assumes that no evaporation takes place as the air descends.)

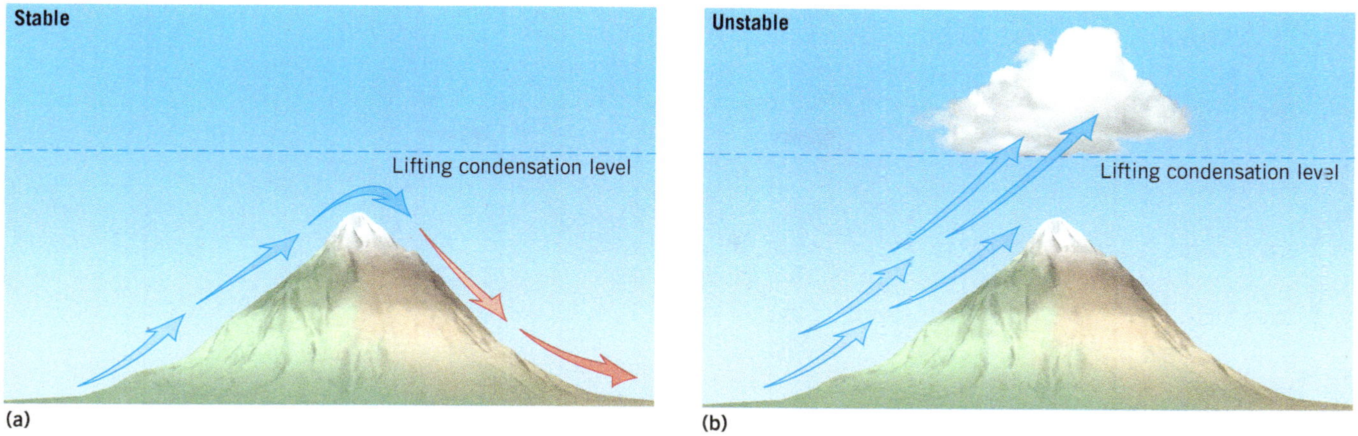

▲ **Figure 6-15** (a) As stable air blows over a mountain, the air rises only as long as it is forced to do so. On the leeward side, it moves downslope. (b) When forced up a mountain slope, unstable air is likely to continue rising until it reaches surrounding air of similar temperature and density. If it rises to the lifting condensation level, clouds form.

The Stability of Air

In the case of air, imagine a "parcel"—a volume of air with boundaries, such as a balloon—as being an "object" and the surrounding air as being the fluid. As with other gases (and liquids, too), an air parcel tends to seek its *equilibrium level*. This means that a parcel of air moves up or down until it reaches an altitude at which the surrounding air is of equal density. In practice, if a parcel of air is warmer, and thus less dense, than the surrounding air, the parcel tends to rise. If a parcel is cooler, and therefore denser, than the surrounding air, it tends to sink or at least resist uplift. Thus, we say that warm air is more buoyant than cool air.[6]

Stable Air: If a parcel of air resists uplift, it is said to be **stable** (**Figure 6-15**). If stable air is forced to rise, perhaps by wind forcing it up a mountain slope, it does so only as long as the force is applied. In other words, stable air is nonbuoyant. However, when unstable air comes up against the same mountain, it continues to rise once it has passed the peak.

[6]Water vapor content may also slightly influence air buoyancy. Because water vapor molecules have a lower molecular weight than nitrogen (N_2) or oxygen (O_2) molecules, when air has a high water vapor content, heavier N_2 and O_2 molecules have been displaced by lighter H_2O molecules. Thus, such moist air is just slightly "lighter" than dry air. Overall, however, air temperature is by far the most important determinant of air buoyancy.

In the atmosphere, high stability is promoted when cold air is beneath warm air, a condition most frequently observed during a *temperature inversion* (discussed in Chapter 4). With colder, denser air below warmer, lighter air, upward movement is unlikely. A cold winter night is typically a highly stable situation, although high stability can also occur in the daytime. Because it does not rise, highly stable air provides little opportunity for adiabatic cooling unless there is some sort of forced uplift. Highly stable air is not associated with cloud formation and precipitation unless it is forced to rise over a front or up a mountain.

Unstable Air: Air is said to be **unstable** if it rises without any external force other than the buoyant force or if it continues to rise after such an external force has ended. In other words, unstable air is buoyant. When a mass of air is warmer than the surrounding air, it becomes unstable. This is a typical condition on a warm summer afternoon (**Figure 6-16**). The unstable air rises until it reaches its equilibrium level, the altitude where the surrounding air has similar temperature and density. While ascending, it will be cooled adiabatically. In this situation, clouds are likely to form if sufficient moisture is in the air.

Latent Heat and Instability: Between stability and instability is an intermediate condition, sometimes called *conditional instability*. Near the surface, such a parcel of air is cooler than the surrounding air, so it is stable. If it is forced

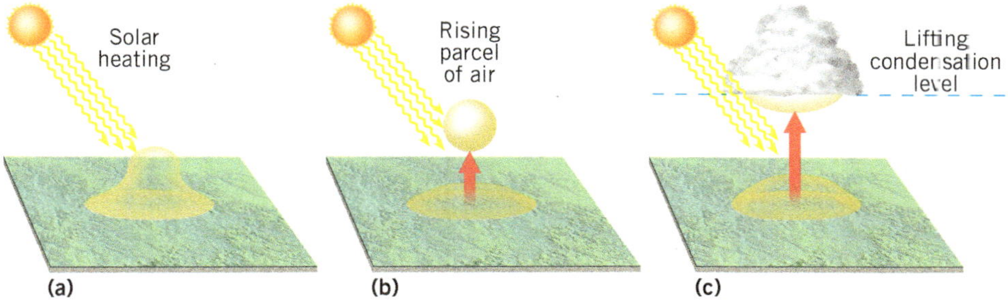

◀ **Figure 6-16** (a) Localized heating of Earth's surface warms a parcel of air. (b) Because this parcel of air is warmer than the surrounding air, it is unstable and begins to rise. (c) If this air rises high enough, it may reach the lifting condensation level, forming a cloud.

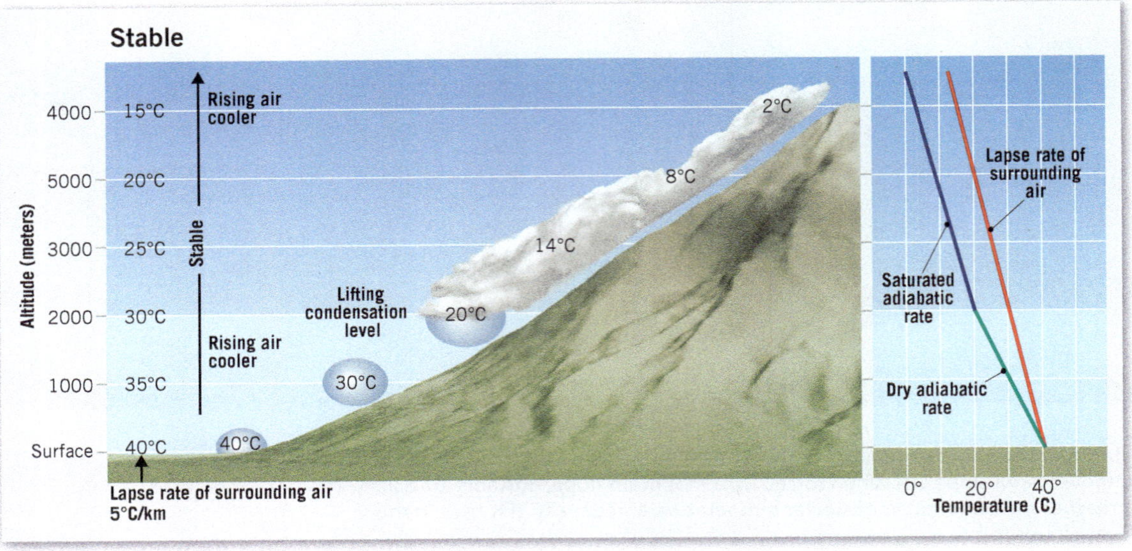

▲ **Figure 6-17** Rising stable air, shown with a diagram and a graph. At all altitudes, the rising parcel of air is cooler than the surrounding air, so the parcel is stable and will rise only if forced. (Dry adiabatic rate = 10°C/1000 meters; saturated adiabatic rate = 6°C/1000 meters; lifting condensation level = 2000 meters; lapse rate of surrounding air [environmental lapse rate] = 5°C/1000 meters.)

Animation
Adiabatic Processes and Atmospheric Stability
https://media.pearsoncmg.com/ph/esm/esm_mcknight_physgeo_13/media/

to rise above the lifting condensation level, the release of latent heat during condensation may warm the air enough to make the parcel unstable. It will then rise until it reaches an altitude where the surrounding air has density and temperature similar to its own.

LearningCheck 6-12 What makes a parcel of air unstable? Stable?

Determining Atmospheric Stability

An accurate determination of the stability of any mass of air generally depends on temperature measurements.

Temperature, Lapse Rate, and Stability: The temperature of a parcel of rising air can be compared with the temperature of surrounding (nonrising) air by a series of thermometer readings at different elevations. The rising air cools (at least initially) at the dry adiabatic rate of 10°C per 1000 meters (5.5°F per 1000 feet). The *environmental lapse rate* of the surrounding (nonrising) air depends on many things and is likely different from the dry adiabatic rate.

For example, if the environmental lapse rate of the surrounding air is less than the dry adiabatic rate of the rising air (Figure 6-17), at every elevation the rising air is *cooler* than the surrounding air and therefore stable. Under such conditions, the air rises only when forced to do so.

If instead the environmental lapse rate of the surrounding air is greater than the dry adiabatic rate of the rising air, at every elevation the rising air is *warmer* than the surrounding air, so it is unstable. The unstable air rises until it reaches an elevation where the surrounding air is of similar temperature and density (Figure 6-18).

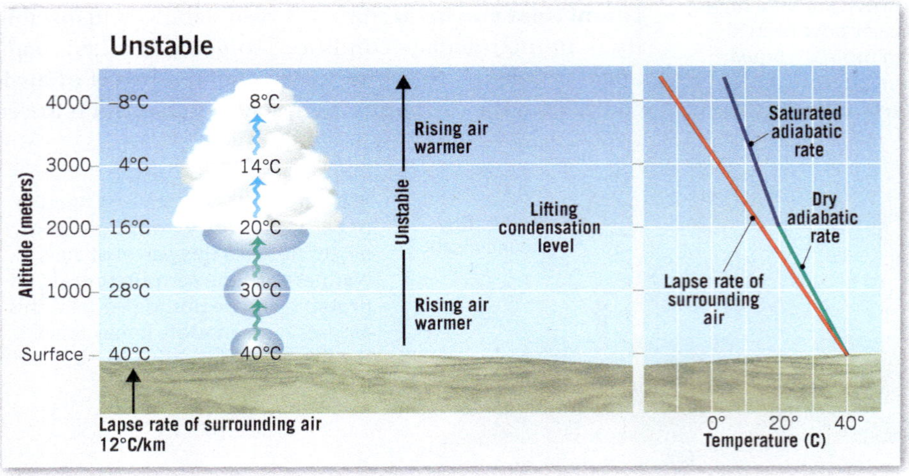

◀ **Figure 6-18** Rising unstable air. At all altitudes, the rising parcel of air is warmer than the surrounding air, so the parcel is unstable and will rise because of its buoyancy. (Dry adiabatic rate = 10°C/1000 meters; saturated adiabatic rate = 6°C/1000 meters; lifting condensation level = 2000 meters; lapse rate of surrounding air = 12°/1000 meters.)

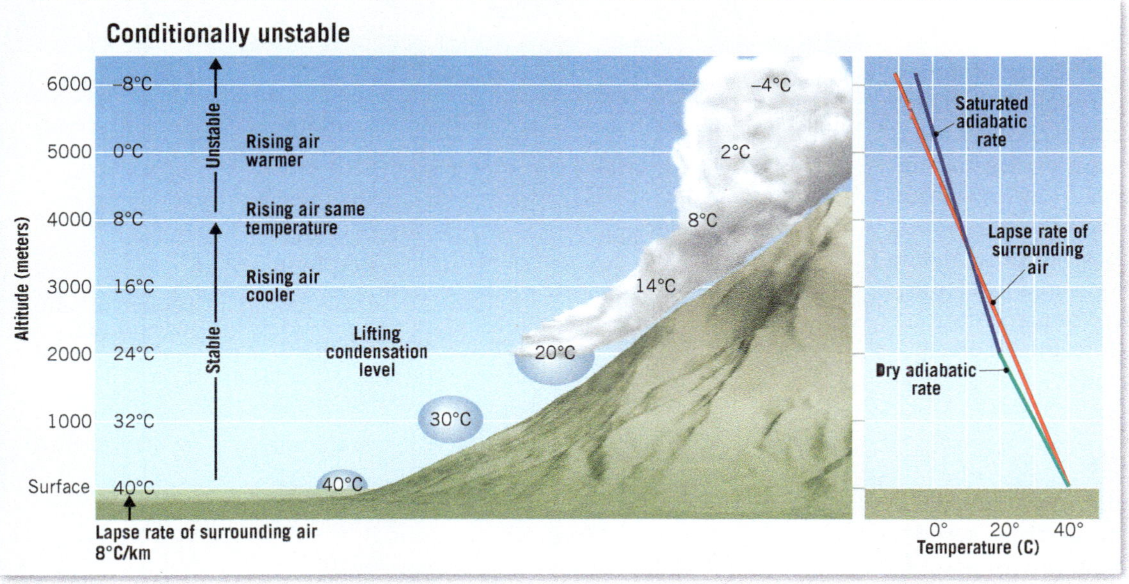

▲ **Figure 6-19** Conditionally unstable air. In this case, the rising parcel of air is cooler than the surrounding air and is stable up to an altitude of 4000 meters. However, above that altitude, the release of latent heat during condensation warms the rising air enough to make it unstable—thus, the rising air is conditionally unstable. (Dry adiabatic rate = 10°C/1000 meters; saturated adiabatic rate = 6°C/1000 meters; lifting condensation level = 2000 meters; lapse rate of surrounding air = 8°C/1000 meters.)

In a third situation—conditional instability—rising stable air is cooled to its dew point temperature at the lifting condensation level. The release of latent heat during condensation can then cause this rising air to become unstable (Figure 6-19).

Visual Determination of Stability: The cloud pattern in the sky is often indicative of air stability. Unstable air is associated with distinct updrafts, vertical air currents that are likely to produce vertical clouds (Figure 6-20). Thus, the presence of cumulus (puffy) clouds suggests instability, and a towering cumulonimbus (thunderstorm) cloud is an indicator of pronounced instability. Horizontally developed clouds, most notably stratiform (flat), are characteristic of stable air that has been forced to rise, and a cloudless sky may indicate stable air that is immobile.

Regardless of stability conditions, no clouds form unless the air is cooled to the dew point temperature, so the mere absence of clouds is not certain evidence of stability; it is only an indication. The general features of stable and unstable air are summarized in Table 6-1.

▲ **Figure 6-20** A large cumulonimbus cloud as seen from an airplane. Such clouds develop in association with highly unstable air. The anvil-shaped top is characteristic of cumulonimbus clouds; it is formed of ice crystals carried by high winds near the tropopause.

TABLE 6-1 Characteristics of Stable and Unstable Air

Stable Air	Unstable Air
Nonbuoyant; remains immobile unless forced to rise	Buoyant; rises without outside force
If clouds develop, tend to be flat (stratiform) but can be thin or thick	If clouds develop, tend to be puffy and have vertical development (cumuliform)
If precipitation occurs, tends to be drizzly or light steady rain	If precipitation occurs, tends to be showery or a downpour

Clouds

Clouds are collections of minute droplets of liquid water or tiny crystals of ice. They are the visible expression of condensation and other processes occurring in the atmosphere. At a glance, they reflect the present weather and can be harbingers of things to come. At any given time, about two-thirds of Earth is covered by clouds, and they are the source of precipitation. Not all clouds precipitate, but all precipitation comes from clouds.

TABLE 6-2 The International Classification Scheme for Clouds

Family	Type	Symbol	Form	Characteristics
High	Cirrus	Ci	Cirriform	Thin, white, icy
	Cirrocumulus	Cc	Cirri and cumuliform	
	Cirrostratus	Cs	Cirri and stratiform	
Middle	Altocumulus	Ac	Cumuliform	Layered or puffy; mixture of liquid and ice particles
	Altostratus	As	Stratiform	
Low	Stratus	St	Stratiform	General overcast
	Stratocumulus	Sc	Strati and cumuliform	
	Nimbostratus	Ns	Stratiform	
Vertical	Cumulus	Cu	Cumuliform	Tall, narrow, puffy
	Cumulonimbus	Cb	Cumuliform	

Classifying Clouds

Although clouds occur in an almost infinite variety of shapes and sizes, certain general forms recur commonly. Moreover, the various cloud forms are normally found only at certain generalized altitudes. It is on the basis of these two factors—form and altitude—that clouds are classified (Table 6-2).

Cloud Form: The international classification scheme for clouds recognizes three forms:

1. *Cirriform* clouds (Latin *cirrus*, meaning "a lock of hair") are thin and high and composed of ice crystals rather than water droplets.
2. *Stratiform* clouds (Latin *stratus*, meaning "spread out") appear as grayish sheets that cover most or all of the sky, rarely broken up into individual cloud units.
3. *Cumuliform* clouds (Latin *cumulus*, meaning "heap" or "pile") are massive and rounded, usually with a flat base and limited horizontal extent but often billowing upward to great heights.

These three cloud forms are subclassified into 10 types based on shape and altitude (Figure 6-21). The types overlap, and cloud development frequently is in a state of change, so one type may evolve into another. Three of the 10 types are purely of one form, and these are called **cirrus clouds** (cirriform), **stratus clouds** (stratiform), and **cumulus clouds** (cumuliform). The other types may be combinations of these three (Figure 6-22). Cirrocumulus clouds, for example, have the thin, iciness of cirrus clouds and the puffiness of cumulus clouds. Cirrostratus clouds often look like clear skies, but they can cause optical effects such as sun pillars (see Figure 6-22d).

Clouds that produce precipitation have "nimb" in their name—specifically, *nimbostratus* or *cumulonimbus*. These clouds produce precipitation that reaches the ground. Normally these types develop from other types; that is, cumulonimbus clouds develop from cumulus clouds, and nimbostratus clouds develop from stratus clouds.

Cloud Families: The 10 cloud types are divided into four families on the basis of altitude.

1. *High clouds* are generally found above 6 kilometers (20,000 feet). Because of the small amount of water vapor and low temperature at such altitudes, these clouds are thin, white, and composed of ice crystals. Included in this family are *cirrus*, *cirrocumulus*, and *cirrostratus*. High clouds often are harbingers of an approaching weather system or storm.
2. *Middle clouds* normally occur between about 2 and 6 kilometers (6500 and 20,000 feet). They may be either stratiform or cumuliform and are composed of liquid water. Included types are puffy *altocumulus* clouds, which usually indicate settled weather conditions, and lengthy *altostratus* clouds, which are often associated with changing weather.
3. *Low clouds* usually are below 2 kilometers (6500 feet). They sometimes occur as individual clouds but more often appear as a general overcast. Low cloud types include *stratus*, *stratocumulus*, and *nimbostratus*. Low clouds often are widespread and are associated with somber skies and drizzly rain.
4. *Clouds of vertical development* grow upward from low bases to heights of as much as 18 kilometers (60,000 feet). Their horizontal spread is usually very restricted. They indicate very active vertical movements in the air. The relevant types are *cumulus* clouds, which usually indicate fair weather, and **cumulonimbus clouds**, which are thunderstorm clouds.

LearningCheck 6-11 **Name the three main forms by which clouds are classified.**

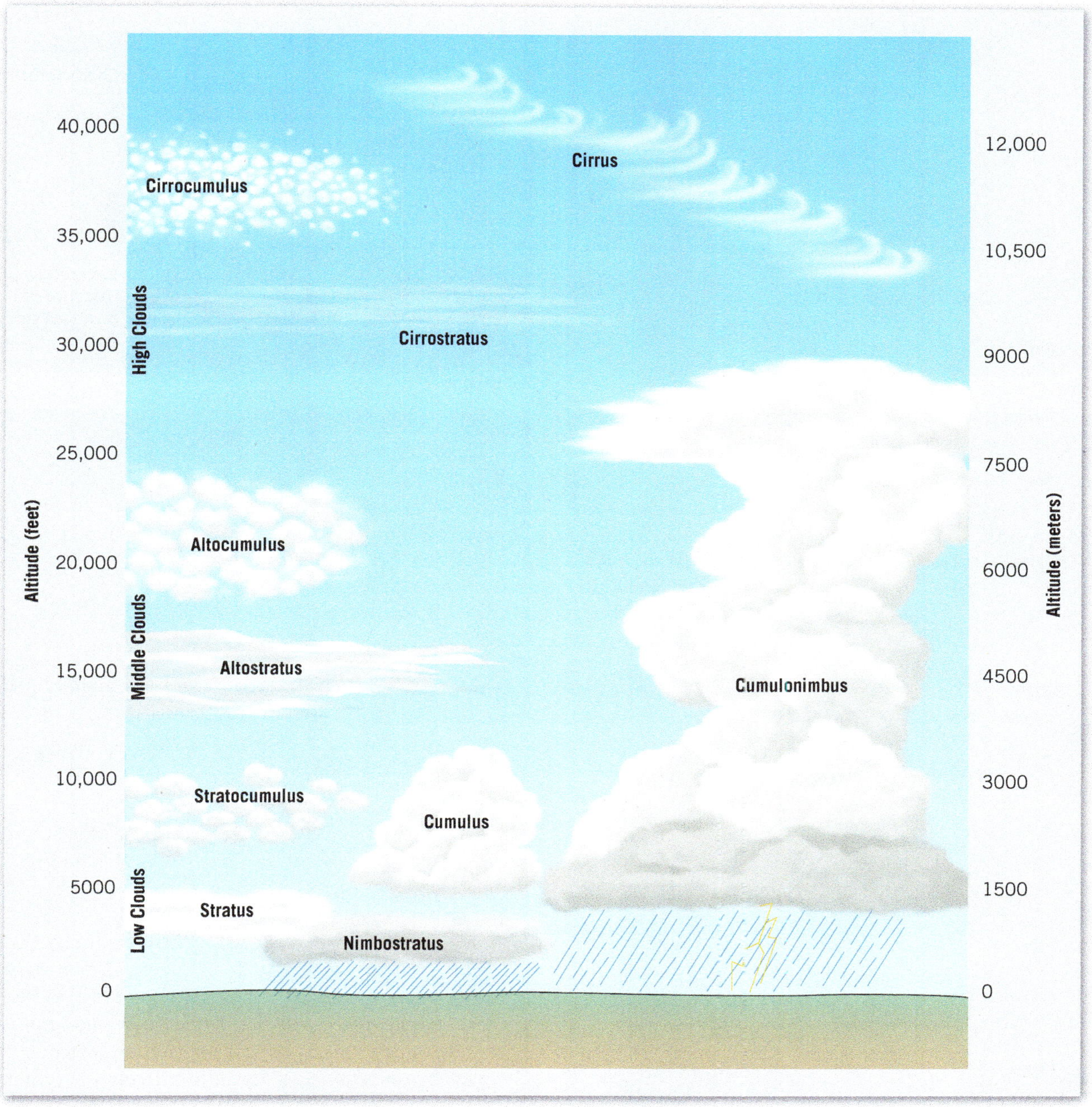

▲ **Figure 6-21** Typical shapes and altitudes of the 10 principal cloud types.

Fog

From a global standpoint, fog represents a minor form of condensation. Its importance to humans is disproportionately high, however, because it can hinder visibility enough to make surface transportation hazardous or even impossible (**Figure 6-23**). **Fog** is simply a cloud on the ground. There is no physical difference between a cloud and fog, but there are important differences in how each forms. Most clouds develop as a result of adiabatic cooling in rising air, but only rarely is uplift involved in fog formation. Instead, most fogs form when air at Earth's surface cools to below its dew point temperature or when enough water vapor is added to the air to saturate it.

Four types of fog are generally recognized (**Figure 6-24**):

1. A *radiation fog* results when the ground radiates away heat, usually at night. The air closest to the ground cools as heat flows conductively from it to the relatively cool ground, and fog condenses in the cooled air at the dew point, often collecting in low areas.
2. An *advection fog* develops when warm, moist air moves horizontally over a cold surface, such as snow-covered ground or a cold ocean current. Air moving from sea to land is the most common source of advection fogs.

(a) Cirrus
(b) Stratus
(c) Cumulus
(d) Cirrostratus
(e) Altocumulus
(f) Nimbostratus

▲ **Figure 6-22** Common types of clouds.

3. An *upslope fog*, or *orographic fog* (from the Greek *oro*, meaning "mountain"), is created by adiabatic cooling when humid air climbs a topographic slope.
4. An *evaporation fog* or *steam fog* results when water vapor is added to cold air that is already near saturation. This usually occurs when cold air flows over a warm body of water.

Areas of heavy fog in North America are mostly coastal (Figure 6-25). The western-mountain and Appalachian fogs are mostly radiation fogs. Areas of minimal fog are in the Southwest, Mexico, and the Great Plains in both the United States and Canada, where available atmospheric moisture is limited and winds are strong.

Because of the great variety of local environments, almost any kind of cloud might appear somewhere within the National Park System—see *Snapshots from Our National Parks: From Mount Rainier to the Everglades: A Diversity of Clouds*.

Dew

Dew usually originates from terrestrial radiation. Nighttime radiation cools objects (grass, pavement, automobiles, or whatever) at Earth's surface, and the adjacent air is in turn

▲ **Figure 6-23** Advection fog enveloping the harbor at Ketchikan, Alaska.

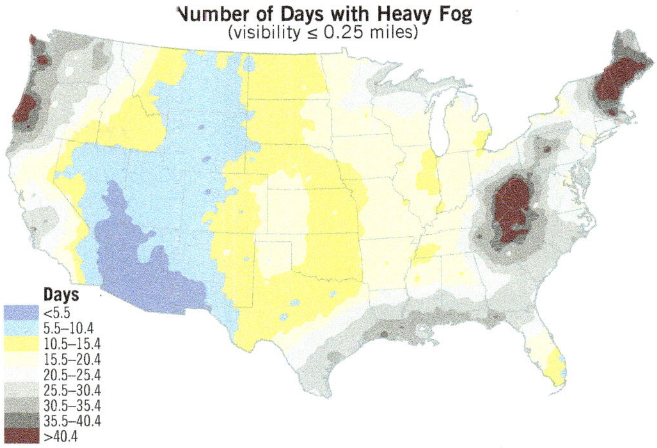

▲ **Figure 6-25** The annual number of heavy fog days in the United States (visibility less than or equal to 0.25 mile).

cooled by conduction. If the air is cooled enough to reach saturation, tiny beads of water collect on the cold surface of the object (**Figure 6-26**). If the temperature is below freezing, ice crystals (*frost*) rather than water droplets form. The dew point temperature is often a good indicator of the night time low temperature because condensational heating prevents the temperature from dropping further.

Clouds and Climate Change

In addition to their obvious role in precipitation, clouds are important because of their influence on radiant energy. They receive both insolation from above and terrestrial radiation from below, and then they may absorb, reflect, scatter, or reradiate this energy. Thus understanding the function of clouds in the global energy budget is important, and we must take clouds into account when we try to anticipate the causes or consequences of climate change.

Precipitation

All **precipitation**—droplets of liquid or solid—originates in clouds, but most clouds do not yield precipitation. Exhaustive experiments have demonstrated that condensation alone is not enough to produce raindrops. The tiny water

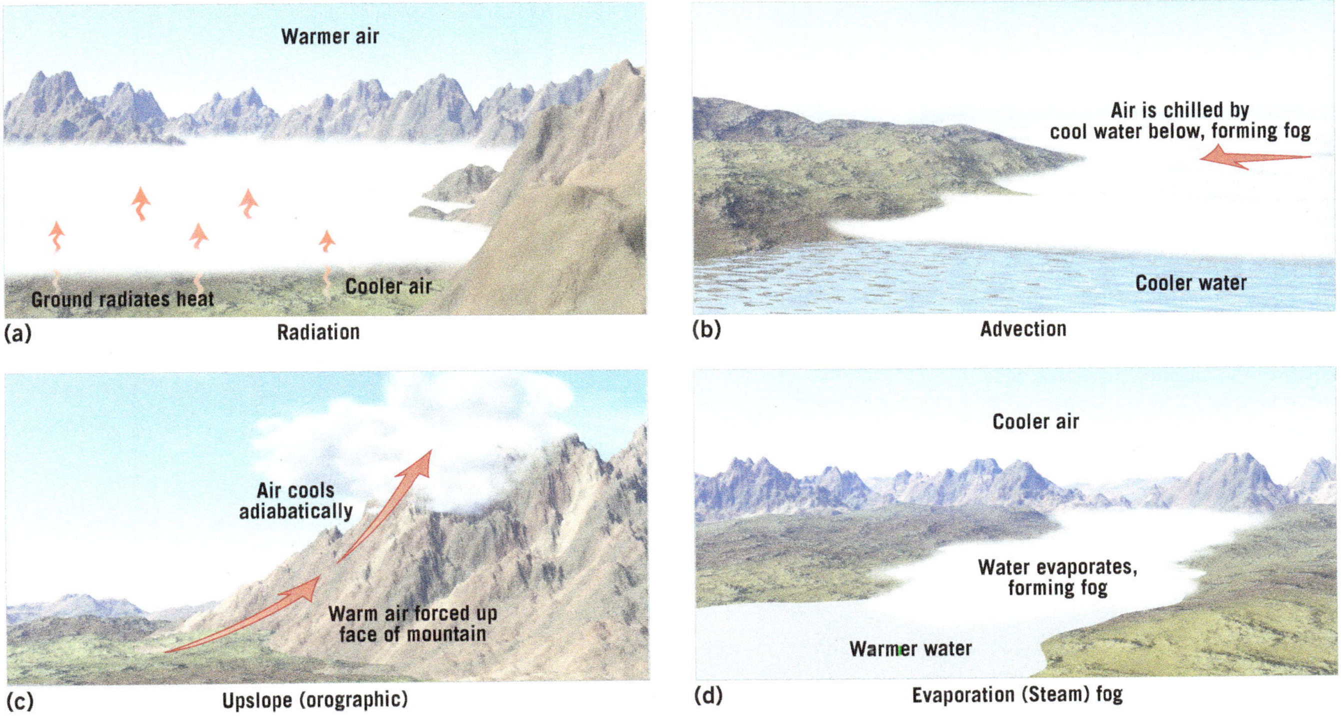

▲ **Figure 6-24** The four principal types of fog: (a) radiation, (b) advection, (c) upslope (orographic), (d) evaporation (steam).

Snapshots From Our National Parks

From Mount Rainier to the Everglades

A Diversity of Clouds

Clouds form when air rises and cools to the dew point temperature. The geography and weather patterns in our National Parks provide a variety of mechanisms for different types of clouds to form, including some rather unique ones (**Figures 6-A** to **6-E**). The parks also provide the perfect platform for viewing these atmospheric displays.

▲ **Figure 6-A** Bison stand in the fog at Yellowstone National Park (Idaho, Montana, and Wyoming). This evaporation (or steam) fog forms when cold air draining into the valley flows over a warmer body of water. Moisture gets added to the air and condenses because the air is cool.

▶ **Figure 6-B** A lenticular clouds forms over Mount Rainier in Mount Rainier National Park (Washington). Lenticular clouds form when stable, moist air flows over a mountain, forming a lens-shaped cloud. As the air rises, moisture condenses at the top of the wave. Multiple lenticular clouds may appear on the downwind side of mountains. Sometimes they are mistaken for UFOs.

▲ **Figure 6-C** Mammatus clouds form over Lake Superior near Isle Royale National Park, Michigan. In the Midwest, mammatus clouds form on the bottom side of very strong thunderstorms. Mammatus clouds indicate that the atmosphere is very unstable, so if you see them, you should seek shelter.

▶ **Figure 6-D** Satellite image of Hurricane Irma as it passes over the Florida Keys. Hurricanes are very organized groups of thunderstorms with a huge expanse of cloud. Hurricane Irma did an enormous amount of damage to Everglades National Park on September 10, 2017. An estimated 60 percent of the mangroves were heavily damaged by the hurricane.

◀ **Figure 6-E** A rainbow forms over the alpine tundra of Denali National Park and Preserve in Alaska. You can see a rainbow when the sun is behind you and you are facing the rain. A tundra is a large treeless region where the subsoil is permanently frozen.

▲ **Figure 6-26** Dewdrops on a wild daisy.

droplets that make up clouds cannot fall as rain because their size makes them very buoyant and the typical turbulence of the atmosphere keeps them aloft. Even in still air, their descent would be so slow that it would take many days for them to reach the ground from even a low cloud. Besides that, most droplets would evaporate in the drier air below the cloud before they got far.

Despite these difficulties, rain and other forms of precipitation are commonplace in the troposphere. What, then, produces precipitation in its various forms?

The Processes

An average-sized raindrop contains several million times as much water as the average-sized water droplet found in any cloud. Consequently, great multitudes of droplets must join together and form a drop large enough to overcome both turbulence and evaporation. Then the drop can fall to Earth under the influence of gravity.

Two mechanisms are believed to be principally responsible for producing precipitation particles: (1) collision and coalescence of water droplets and (2) ice-crystal formation.

Collision/Coalescence: In many cases, particularly in the tropics, cloud temperatures are greater than 0°C (32°F); these are known as *warm clouds*. In such clouds, rain is produced by the collision and coalescence (merging) of water droplets. Condensation alone cannot yield rain because condensation produces a lot of small droplets but no large drops. Thus, tiny condensation droplets must coalesce into drops large enough to fall as precipitation. Larger water droplets fall faster, overtaking and often coalescing with smaller ones, which are swept along in the descent (Figure 6-27). This sequence of events favors the continued growth of the larger drops.

Not all collisions result in coalescence, however. The air displaced around a falling large drop can push very tiny droplets out of its path. In addition, evidence suggests that differences in the electrical charges of droplets may influence coalescence as well.

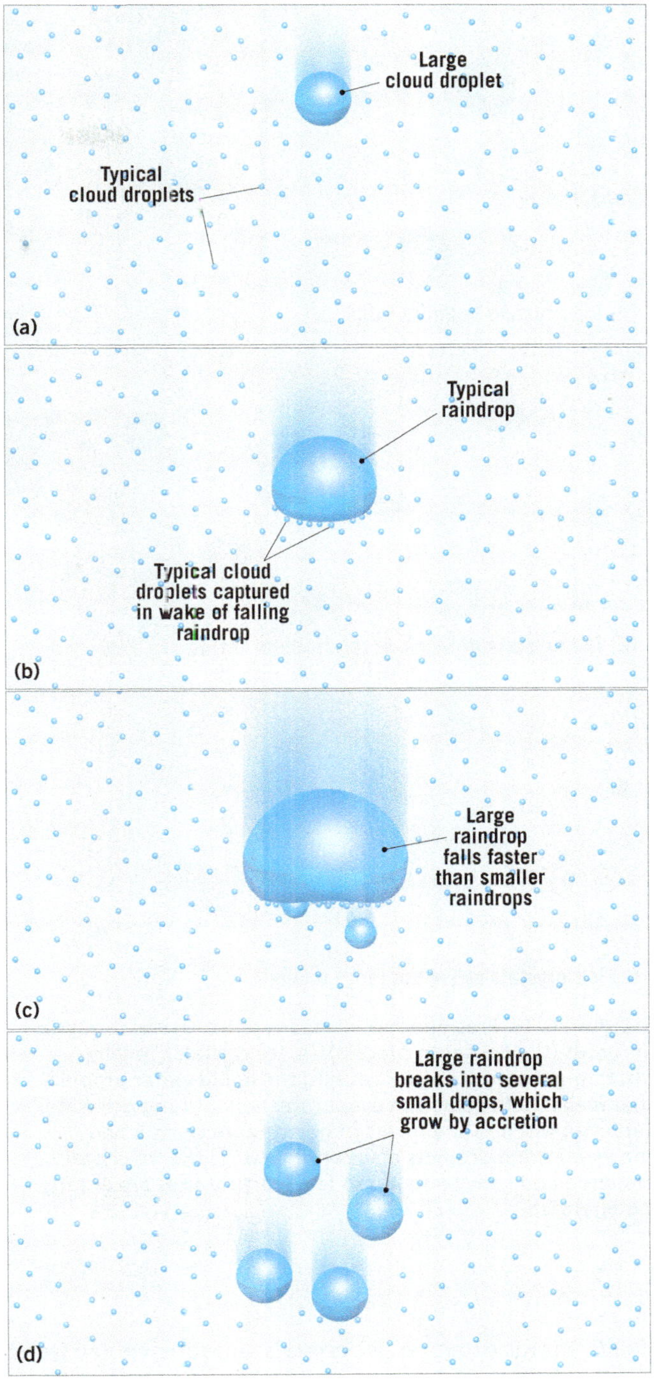

▲ **Figure 6-27** Raindrops forming by collision and coalescence. (a) Large droplets fall more rapidly than small ones, (b) coalescing with some and sweeping others along in their descending path. (c) As droplets become larger during descent, (d) they sometimes break apart. (Cloud droplets and raindrops are not drawn at the same scale.)

Collision/coalescence is the process most responsible for precipitation in the tropics, and it also produces some of the precipitation in the middle latitudes.

Ice-Crystal Formation: Clouds or portions of clouds that extend high enough to have temperatures well below the freezing point of liquid water are known as *cold* or *cool*

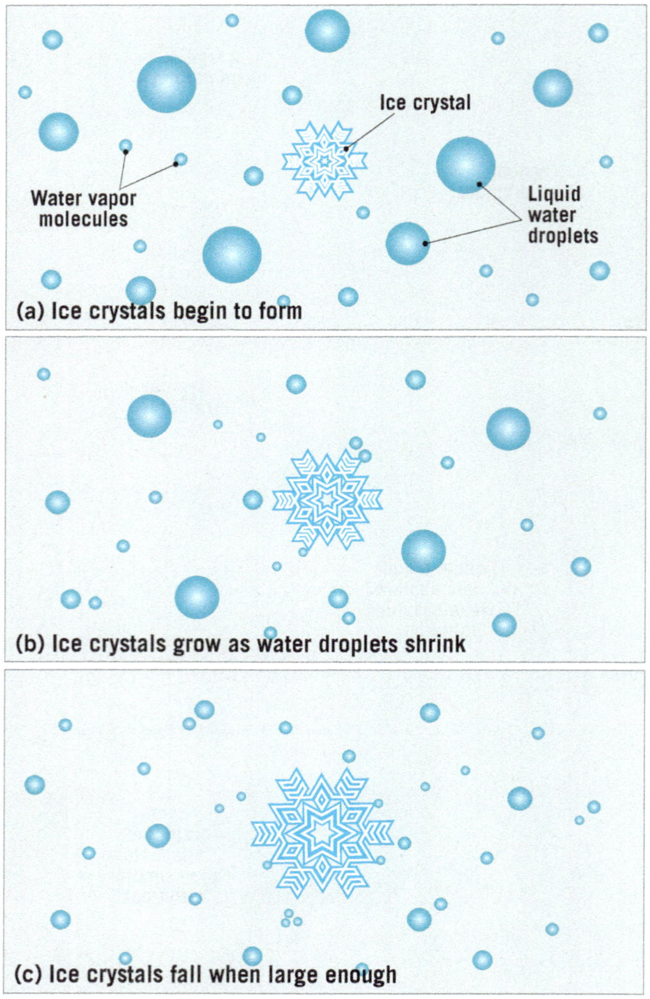

▲ **Figure 6-28** Precipitation by means of ice-crystal formation in clouds (the Bergeron process). (a) Ice crystals grow by attracting water vapor, (b) causing the liquid water droplets that make up the cloud to evaporate, replenishing the water vapor supply. (c) The process of growing ice crystals and shrinking cloud droplets may continue until the ice crystals are large and heavy enough to fall. (Particle sizes are greatly exaggerated.)

clouds. In this situation, ice crystals and supercooled water droplets often coexist in the cloud. These two types of particles are in direct "competition" for the water vapor that has not yet condensed.

Saturation vapor pressure is lower around the ice crystals than around liquid water droplets. If the air around a liquid water droplet is saturated (100 percent relative humidity), that same air is *supersaturated* around an ice crystal. Thus, the ice crystals attract most of the water vapor, and the liquid water droplets, in turn, evaporate, replenishing the diminishing supply of vapor (**Figure 6-28**). Therefore, the ice crystals grow at the expense of the water droplets until the crystals are large enough to fall. As they fall through the lower, warmer portions of the cloud, they pick up more moisture and grow still larger. They may then

▲ **Figure 6-29** A towering cumulonimbus cloud produces a small but intense thunderstorm over Gateway Arch National Park, Missouri.

either precipitate from the cloud as snowflakes or melt and precipitate as raindrops.

Precipitation by ice-crystal formation was first proposed by Swedish meteorologist Tor Bergeron more than half a century ago. It is now known as the *Bergeron process* and is believed to account for the majority of precipitation outside of tropical regions.

Forms of Precipitation

Several forms of precipitation can result from the processes just described, depending on air temperature and turbulence.

Rain: By far the most common and widespread form of precipitation is **rain**, which consists of drops of liquid water (**Figure 6-29**). Most rain is the result of condensation and precipitation in rising air that has a temperature above freezing, but some results from the melting of ice crystals as they descend through warmer air.

Meteorologists often make a distinction among *rain*, which goes on for a relatively long time; *showers*, which are relatively brief and involve large drops; and *drizzle*, which consists of very small drops and usually lasts for some time.

Snow: The general name given to solid precipitation in the form of ice crystals, small pellets, or flakes is **snow**. It forms when water vapor is converted directly to ice through deposition without an intermediate liquid stage. (The water vapor may have evaporated from supercooled liquid cloud droplets inside cold clouds.)

Sleet: In the United States, *sleet* refers to small raindrops that freeze during descent and reach the ground as small pellets of ice. In other countries, the term is often applied to a mixture of rain and snow.

▲ Figure 6-30 The largest documented hailstone fell in Vivian, South Dakota, on July 23, 2010. It weighed 879 g (1.94 lb.).

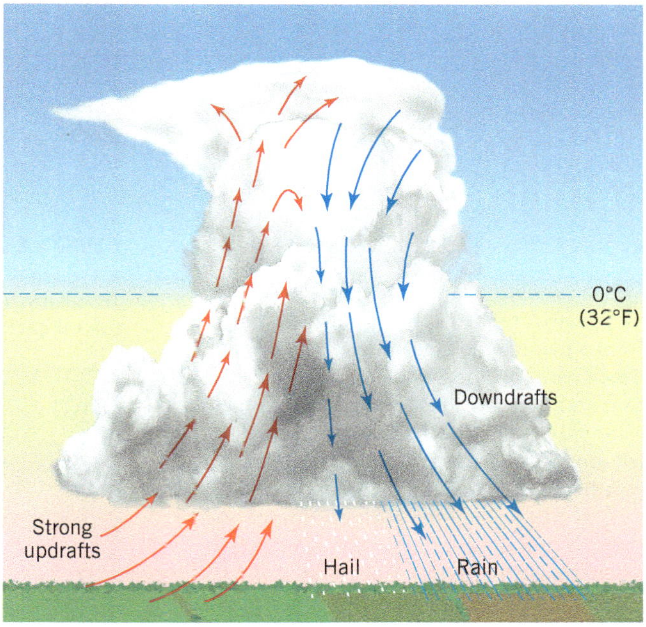

▲ Figure 6-31 Hail is produced in cumulonimbus clouds with strong updrafts that are partly at a temperature above the freezing point of water and partly at a temperature below the freezing point. Hail grows as it passes through supercooled droplets.

Glaze: *Glaze* (or *freezing rain*) is rain that turns to ice the instant it collides with a solid object. Raindrops fall through a shallow layer of subfreezing air near the ground. Although the drops do not freeze in the air (in other words, they do not turn to sleet), they become supercooled while in this cold layer and instantly become icy when they land. The result can be a thick coating of ice that makes travel hazardous and breaks tree limbs and transmission lines.

Hail: The precipitation form with the most complex origin is **hail**, which consists of either small pellets or larger lumps of ice (Figure 6-30). Hailstones are usually composed of roughly concentric layers of clear and cloudy ice. The cloudy portions contain numerous tiny air bubbles among small crystals of ice, whereas the clear parts are made up of large ice crystals.

Hail is produced in cumulonimbus clouds as a result of great instability and strong updrafts and downdrafts (Figure 6-31). Because highly unstable air is needed for hail to form, it tends to be more common in spring and summer than in winter. In order for hail to form, the lower part of a cloud must be warmer than the freezing point (0°C; 32°F) and the upper part must be colder. Updrafts carry water droplets from the warm layer of the cloud (or small ice pellets from the lowest part of the below-freezing layer) upward; ice then grows by collecting moisture from the surrounding supercooled cloud droplets. When the ice particles become too large to be supported in the air, they fall, gathering more moisture on the way down (see Figure 6-31).

A hailstone normally continues to grow whether it rises or falls in the below-freezing layer, providing it passes through portions of the cloud that contain supercooled droplets. If considerable supercooled moisture is available, the hailstone becomes surrounded by a wet layer that freezes slowly, producing large ice crystals and forcing air out of the water. The result is clear-ice rings. If the supply of supercooled droplets is more limited, the water may freeze almost instantly around the hailstone. This fast freezing produces small crystals with tiny air bubbles trapped among them, forming opaque rings of ice.

The eventual size of a hailstone depends on the amount of supercooled water in the cloud, the strength of the updrafts, and the total length (up, down, and sideways) of the path it takes through the cloud. The largest documented hailstone fell in South Dakota on July 23, 2010 (see Figure 6-30). It was 20 centimeters (8 inches) in diameter and weighed 879 grams (1.9375 pounds).

LearningCheck 6-13 **Why is hail associated with highly unstable air?**

Virga: If the relative humidity of the air below a precipitating cloud is quite low, falling precipitation may evaporate before reaching the surface. These streaks of rain that disappear before hitting the ground are called *virga*.

Among the most important tools for meteorologists studying precipitation are weather satellites (Figure 6-32). By providing such information as the moisture content of air masses, temperature profiles of the atmosphere, and the location and characteristics of clouds, weather satellites help forecasters predict rain, cloud cover, and other weather events. For example, see the box *Focus: GOES Weather Satellites*.

Atmospheric Lifting and Precipitation

The role of rising air and adiabatic cooling has been stressed in this chapter. Only through these events can any significant amount of precipitation originate. What causes air to

Focus

GOES Weather Satellites

▲ **Figure 6-F** Visible light satellite image taken by NOAA's GOES-East satellite. Tropical Storm Cristobal is in the Gulf of Mexiso. Notice that it is still night in the far east.

▲ **Figure 6-G** Infrared satellite image taken by GOES-East.

▲ **Figure 6-H** Water vapor satellite image taken at the same time as Figures 6-F and 6-G. Note the relatively dry area (shown in yellow and orange) across Mexico and Brazil and the moist areas (green and white) of thunderstorms and Tropical Storm Cristobal.

The satellite images commonly seen on weather reports in North America come from a pair of satellites known as GOES, or *Geostationary Operational Environmental Satellites*. The satellites are operated by the National Oceanic and Atmospheric Administration (NOAA) and are essential tools for weather forecasting.

The GOES satellites are geostationary, orbiting at a distance of 35,800 kilometers (22,300 miles) in fixed locations above Earth's surface. GOES-East (GOES-16) orbits above the equator in South America (75° W), where it can see the conterminous United States as well as much of the North and South Atlantic Ocean. GOES-West (GOES-17) orbits above the equator in the Pacific (137° W), where it can see most of the Pacific Ocean from Alaska to New Zealand. Similar weather satellites are maintained by other countries, providing complete global coverage.

The GOES satellites have instruments to measure vertical temperature and moisture variations, ozone distribution, and both reflected and radiated electromagnetic energy in several wavelength bands. These satellites send back several images each hour, from which time-lapse satellite "movie" loops are produced.

Viewing GOES Images: GOES satellite images are readily available through many Internet sites, such as www.weather.gov/satellite.

Visible Light Images: Visible light ("VIS") satellite images show sunlight that has been reflected off the surface of Earth or by clouds in the atmosphere (**Figure 6-F**). Brightness depends on the *albedo* (reflectance) of a surface, as well as the *thickness* of a cloud layer and the angle of the light striking it. The brightest surfaces in visible light are typically thick clouds and snow- or ice-covered surfaces. The darkest surfaces are typically land areas (especially unvegetated land surfaces) and the oceans—which are usually the darkest surfaces seen on visible light satellite images.

Infrared Images: The infrared ("IR") images of Earth and its atmosphere are produced day and night and are among the most widely used in meteorology. Infrared images show the longwave ("thermal infrared") radiation emitted by Earth's surface or by clouds in the atmosphere (**Figure 6-G**). Warm objects emit more longwave radiation than cold objects, so infrared images show us, in effect, differences in temperature. On IR images, cooler surfaces are shown in color, whereas warmer surfaces are shown in gray. The tops of high clouds, such as massive cumulonimbus clouds, are much colder than low clouds and fog, so high clouds appear yellow or orange on infrared images. Low clouds are warmer, so they appear nearly the same shade of gray as the surface below.

Water Vapor Images: By measuring the wavelengths of longwave infrared radiation that is strongly absorbed and reemitted by water vapor (at wavelengths of 6.7 μm and 7.3 μm), meteorologists can estimate the quantity of water vapor in the atmosphere. Satellite water vapor ("WV") images show regions of dry air (with low emission at 6.7 μm and 7.3 μm) and moist air (with high emission at 6.7 μm and 7.3 μm)—even if these areas are cloud free and thus will not show clearly on VIS or IR weather satellite images (**Figure 6-H**). Areas with high moisture content appear as white or cool colors, while dry regions appear red and yellow.

Questions

1. Why are both visible light and infrared satellite images useful to meteorologists?
2. In Figure 6-G, how can you tell whether the clouds in Tropical Storm Cristobal in the Gulf of Mexico are high clouds or low clouds?

◄ Figure 6-32 A National Weather Service meteorologist in Hastings, Nebraska, analyzes satellite images and weather maps to prepare a forecast.

rise? There are four principal types of atmospheric lifting. One type is spontaneous, and the other three require the presence of an external force. Frequently, however, the various types operate together.

Convective Lifting

Because of unequal heating of different surface areas, a parcel of air close to the ground may be warmed by conduction. The density of the warmed air is reduced as the air expands, so the parcel rises by **convective lifting** (Figure 6-33a). As the warm air rises, it expands and cools adiabatically to the dew point temperature. Condensation begins and a cumulus cloud forms. With the proper humidity, temperature, and stability conditions, the cloud may grow into a towering cumulonimbus thunderhead, with a downpour of showery raindrops and/or hailstones accompanied by lightning and thunder (see Figure 6-29).

An individual convective cell is likely to cover only a small area, although multiple cells may form very close to

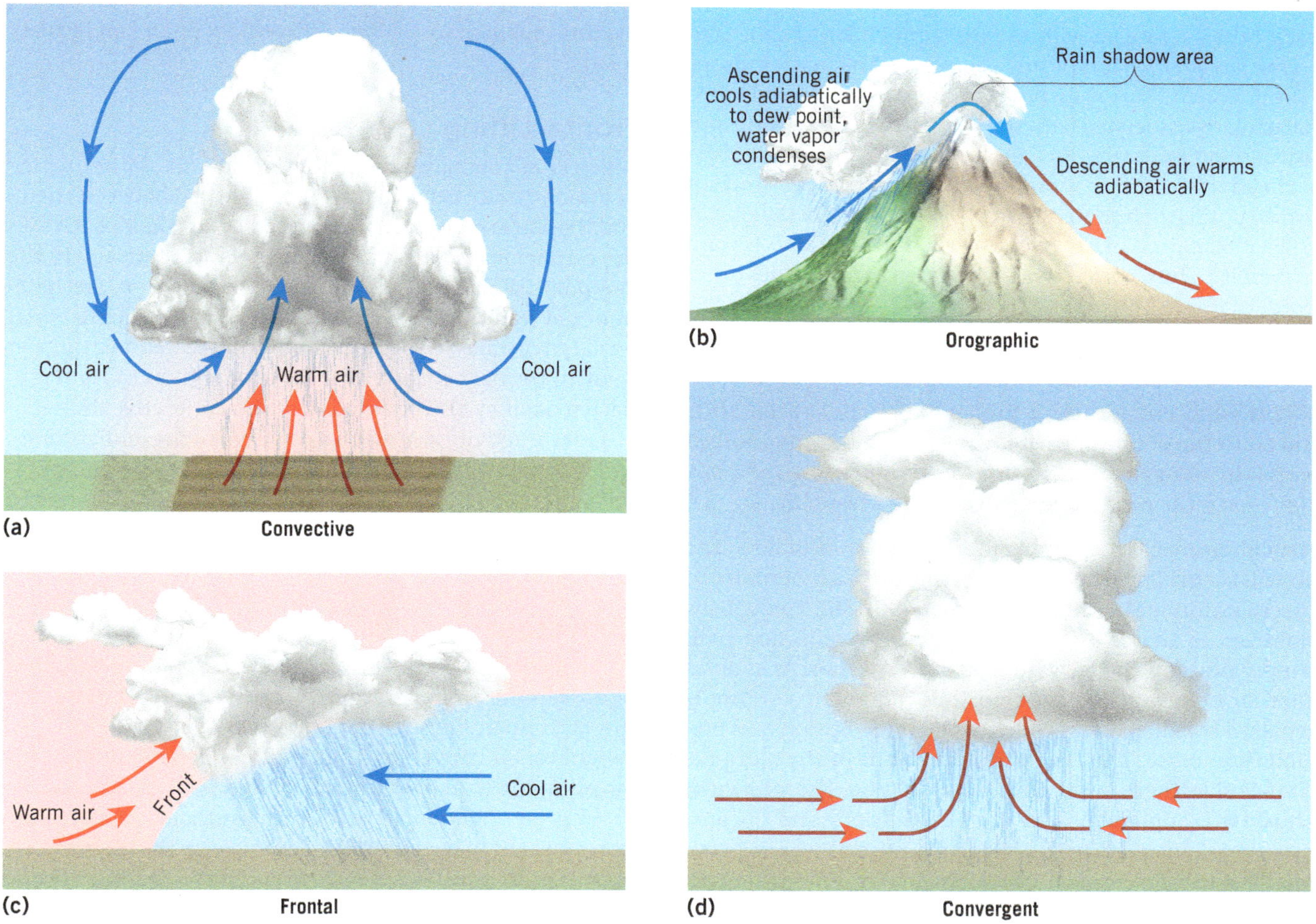

▲ Figure 6-33 The four basic types of atmospheric lifting and precipitation: (a) convective, (b) orographic, (c) frontal, and (d) convergent.

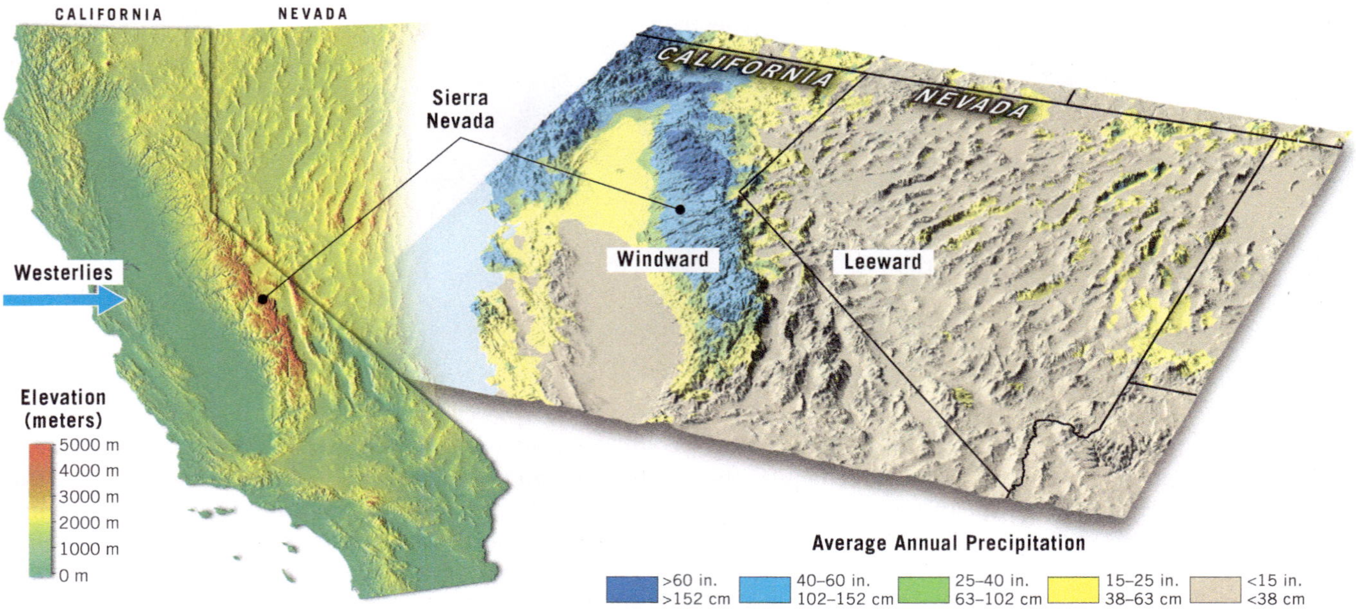

▲ **Figure 6-34** Nevada is in the rain shadow of the Sierra Nevada Mountains of California. The west slope of the Sierra has high rainfall because of the orographic lifting of the moist westerly winds coming from the Pacific Ocean. East of the Sierra, the climate is arid.

each other—close enough to form a much larger cell. *Convective precipitation* is typically showery, with large raindrops falling fast and furiously but for a short duration only. It is particularly associated with the warm parts of the world and warm seasons.

In addition to this spontaneous lifting, various kinds of forced uplift, such as air moving over a mountain range, can trigger formation of a convective cell if the air tends toward instability. Thus, convective lifting often accompanies other kinds of uplift.

Orographic Lifting

When wind encounters a topographic barrier, the air is forced to travel upslope (**Figure 6-33b**). This kind of forced ascent by **orographic lifting** can produce *orographic precipitation* if the rising air is cooled to the dew point.

Rain Shadows: If orographic lifting triggers instability, the air will keep rising when it reaches the top of a mountain, so precipitation may continue (see Figure 6-15b). Frequently, however, air that is forced to rise up a mountain slope will simply descend the leeward side of the barrier. As soon as that air begins to move downslope, adiabatic cooling is replaced by adiabatic warming, and condensation and precipitation cease. Thus the windward slope of the barrier is the wet side; the leeward slope is the dry side. The term **rain shadow** is applied to both the leeward slope and the area beyond as far as the drying influence extends (**Figure 6-34**). Many of the desert areas of the world develop in the rain shadows of mountain ranges (**Figure 6-35**).

Orographic lifting can occur at any latitude, any season, any time of day. Orographic precipitation is likely to be prolonged because the upslope flow of air is relatively steady.

Frontal Lifting

When unlike air masses meet, they do not mix. Rather, a zone of discontinuity called a *front* is established between them, and the warmer air rises over the cooler air (**Figure 6-33c**). As the warmer air is forced to rise, it cools adiabatically to the dew point, with resulting clouds and precipitation. Precipitation that results from such **frontal lifting** is referred to as *frontal precipitation*. We discuss frontal precipitation in greater detail in the next chapter. It tends to be widespread and protracted but may also be associated with convective showers.

Frontal activity is most characteristic of the midlatitudes, which are meeting grounds of cold polar air and warm tropical air. It is less significant in the high latitudes and rare in the tropics; those regions contain air masses that tend to be like one another.

Convergent Lifting

Less common than the other three types, but nevertheless significant in some situations, is **convergent lifting** and the accompanying *convergent precipitation* (**Figure 6-33d**). Whenever air converges, the result is a general uplift. This forced uplift enhances instability and is likely to produce showery precipitation. It is frequently associated with cyclonic storm systems. It is common, for example, in the intertropical convergence zone (ITCZ; discussed in Chapter 5) and is notable in such tropical disturbances as hurricanes and easterly waves.

LearningCheck 6-14 What causes a rain shadow?

▲ **Figure 6-35** Clouds over the Amargosa Range in Death Valley, California. Air that is forced to rise over a mountain may produce clouds and precipitation, leaving the leeward side in a dry rain shadow.

Mobile Field Trip
Clouds: Earth's Dynamic Atmosphere
https://media.pearsoncmg.com/ph/esm/esm_mcknight_physgeo_13/media/

Global Distribution of Precipitation

The most important geographic aspect of atmospheric moisture is the spatial distribution of precipitation. The broadest pattern is based on latitude, but many other factors are involved and the overall pattern is complex. We focus on a series of maps that illustrate worldwide and U.S. precipitation distribution. To show variations in precipitation, these maps use **isohyets**, lines joining points of equal quantities of precipitation.

The amount of precipitation on any part of Earth's surface is determined by the nature of the air mass involved and the degree to which that air is uplifted. The moisture content, temperature, and stability of the air mass depend mostly on where the air originated (over land or water; in high or low latitudes) and on the path it has followed. The amount of uplift of that air mass is determined largely by global pressure patterns, topographic barriers, storms, and other atmospheric disturbances. The combination of these factors produces the annual distribution of precipitation (Figure 6-36).

Regions of High Annual Precipitation

High annual precipitation is generally found in three types of locations.

Region of the ITCZ and Trade-Wind Uplift: The worldwide annual precipitation pattern clearly shows that the tropics contain most of the wettest areas. The warm easterly trade winds are capable of carrying enormous amounts of moisture, and where they are forced to rise by topographic obstacles, very heavy rainfall is usually produced. Equatorial regions particularly reflect these conditions where warm ocean water easily vaporizes and warm, moist, unstable air is uplifted in the ITCZ. Because the trades are easterly winds, it is the eastern coasts of tropical landmasses—for example, the east coast of Central America, northeastern South America, and Madagascar—where the orographic effect is most pronounced.

Tropical Monsoon Regions: Where the normal trade wind pattern is modified by monsoons, the onshore trade-wind flow may occur on the western coasts of tropical landmasses. Thus, the wet areas on the western coast of southeastern Asia, India, and what is called the Guinea Coast of West Africa are caused by the onshore flow of southwesterly winds that are nothing more than trade winds diverted from a "normal" pattern by the South Asian and West African monsoons.

Coastal Areas in Westerlies: The only other regions of high annual precipitation shown in Figure 6-36 are narrow zones along the western coasts of North and South America between 40° and 60° of latitude. These areas reflect a combination of frequent onshore westerly airflow, considerable storminess, and mountain barriers running perpendicular to the direction of the prevailing westerly winds. The presence of these north–south mountain ranges near the coast restricts the precipitation to a relatively small area and creates a pronounced rain shadow effect to the east of the ranges.

LearningCheck 6-15 **Why do equatorial regions generally have high annual precipitation?**

Animation
Seasonal Pressure and Precipitation Patterns
https://media.pearsoncmg.com/ph/esm/esm_mcknight_physgeo_13/media/

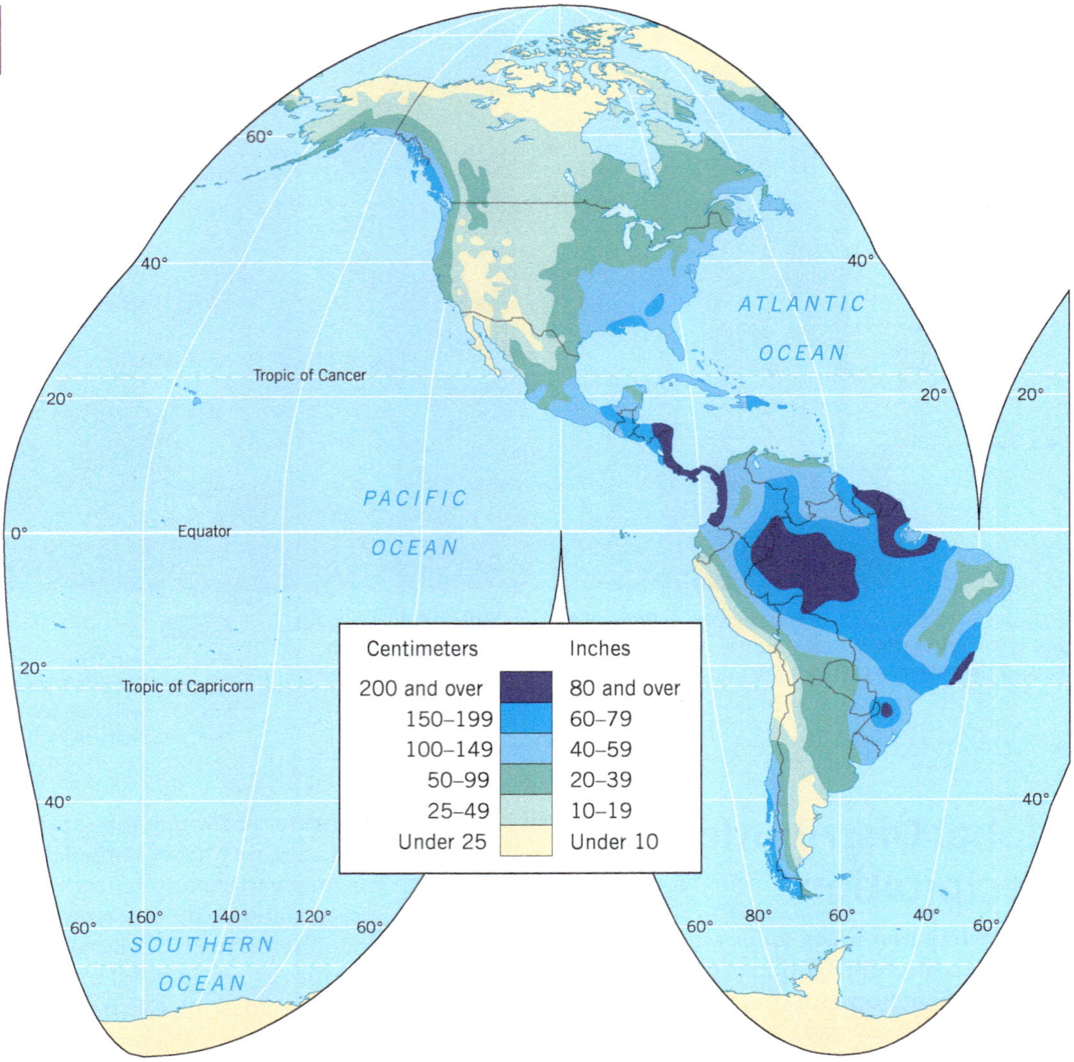

▲ **Figure 6-36** Average annual precipitation over the land areas of the world.

Regions of Low Annual Precipitation

The principal regions of sparse annual precipitation are found in three types of locations (see Figure 6-36).

Areas of Subtropical Highs: Dry lands are most prominent on the western sides of continents in the subtropics (centered at 25° to 30°). The subtropical highs dominate at these latitudes, particularly on the western sides of continents. High pressure is associated with descending air, which is not conducive to condensation and precipitation.

The presence of cool ocean currents also contributes to the atmospheric stability and dryness of these regions. These dry zones are most extensive in North Africa and Australia primarily because of the blocking effect of landmasses or highlands to the east. (The presence of such landmasses prevents moisture from coming in from the east.)

Interiors of Continents: Dry regions in the midlatitudes are most extensive in central Eurasia, but they also occur in western North America and southeastern South America. In each case, the dryness is due to lack of access to moist air masses. In Eurasia, much of the land surface is far from any ocean. In North and South America, rain shadow situations occur in regions of predominantly westerly airflow.

High-Latitude Regions: In the very high latitudes, not much precipitation falls anywhere. Water surfaces are scarce and cold, so little opportunity exists for moisture to evaporate into the air. As a result, polar air masses have low water vapor content and precipitation is slight. These regions are referred to accurately as cold deserts.

LearningCheck 6-16 **Why do west coast locations at about 25° to 30° N and S typically have low annual precipitation?**

Seasonal Precipitation Patterns

A geographic understanding of climate requires knowledge of seasonal as well as annual precipitation patterns. Over most of the globe, the amount of precipitation received in

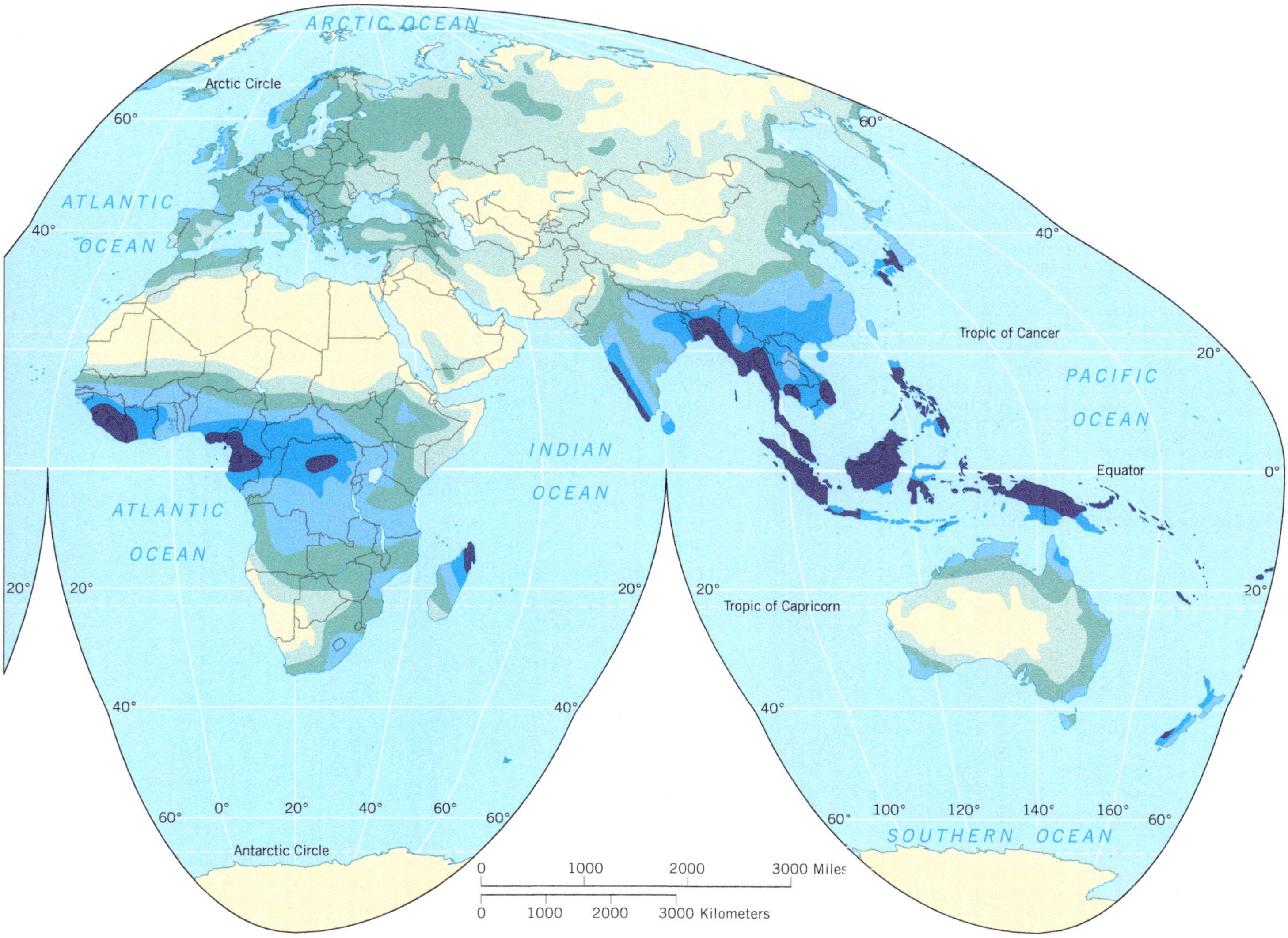

▲ **Figure 6-36** *(Continued)*

summer is considerably different from the amount received in winter. This variation is most pronounced over continental interiors, where strong summer warming at the surface causes greater instability and the potential for greater convective activity. Thus, in interior areas, most of the year's precipitation occurs during summer, and winter is generally a time of anticyclonic conditions with diverging airflow. Coastal areas often have a more balanced seasonal precipitation regime, which is again a reflection of their nearness to moisture sources.

Maps of average January and July precipitation show the contrasts between winter and summer precipitation around the world (Figure 6-37). Several seasonal patterns are important to note.

ITCZ Shifts: The seasonal shifts of major pressure and wind systems—shifts that follow the Sun (north by July and south by January)—are mirrored in the seasonal shifts of wet and dry zones. This is seen most clearly in tropical regions, where the heavy rainfall belt of the ITCZ migrates north and south in different seasons.

Subtropical High Shifts: Summer is the time of maximum precipitation over most of the world. The Northern Hemisphere experiences the heaviest rainfall in July, and the Southern Hemisphere receives the most precipitation in January. The only important exceptions occur in relatively narrow zones along western coasts between about 30° and 45° of latitude, including the United States (Figure 6-38). The same trend occurs in South America, New Zealand, and southernmost Australia. These regions experience the summer dryness associated with the seasonal shift of the subtropical highs.

Monsoon Regions: The most conspicuous seasonal variation in precipitation is found in the regions of tropical monsoons (principally southern and eastern Eurasia, northern Australia, and West Africa). Summer in these regions tends to be very wet, and winter is generally dry.

Precipitation Variability

All of the maps considered thus far portray long-term average conditions. The data on which they are based were gathered over decades; thus the maps represent abstraction

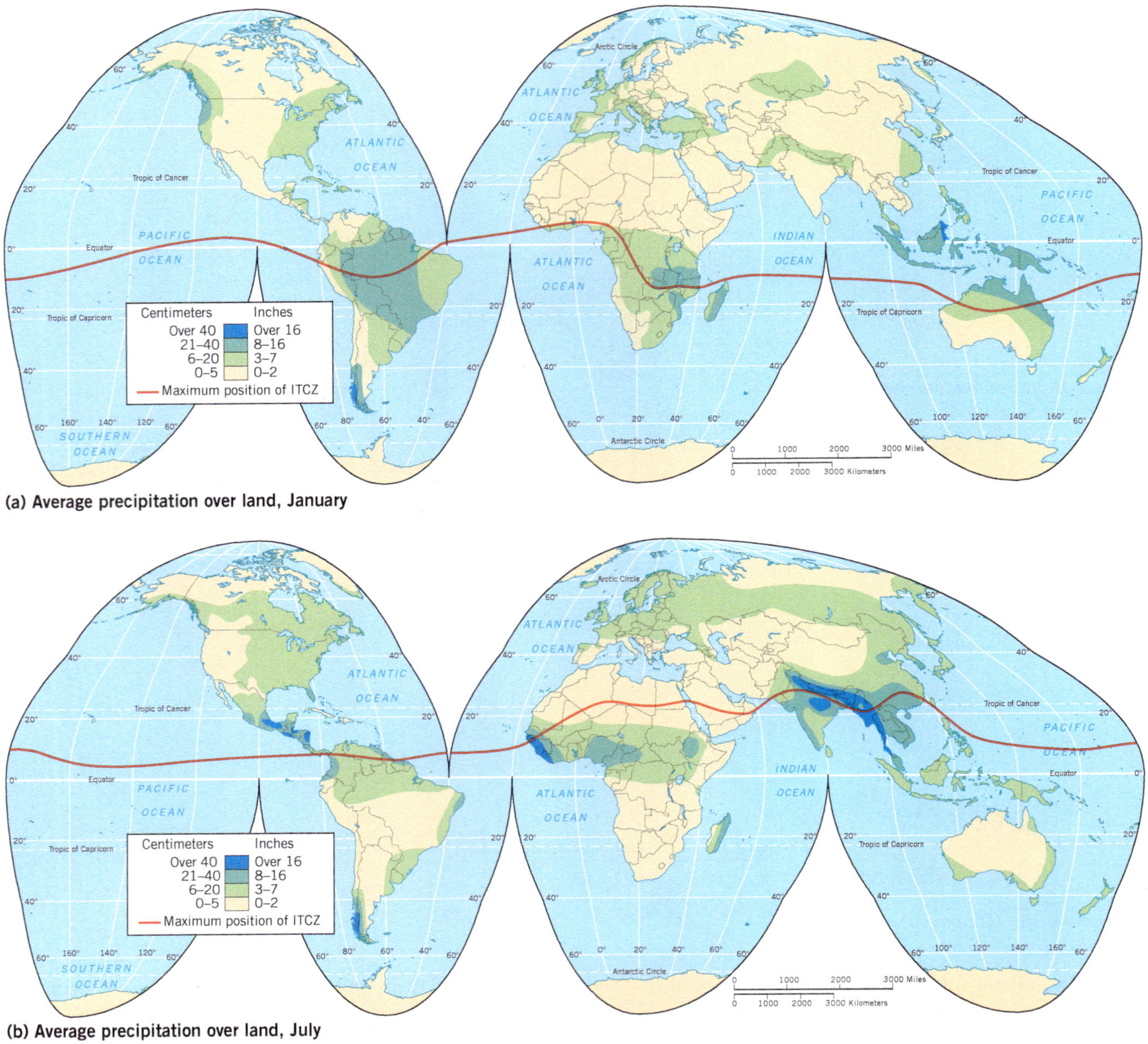

(a) Average precipitation over land, January

(b) Average precipitation over land, July

▲ **Figure 6-37** (a) Average January and (b) average July precipitation over the land areas of the world. The red line marks the typical maximum poleward position of the intertropical convergence zone (ITCZ).

rather than reality. In any year or season, the amount of precipitation may be quite different from the long-term average.

Precipitation variability is the expected departure from average precipitation in any given year, expressed as a percentage above or below average. For example, a precipitation variability of 20 percent means that a location expects to receive either 20 percent more or 20 percent less precipitation than average in any given year. If a location has a long-term average annual precipitation of 50 cm (20 in.) and a precipitation variability of 20 percent, the "normal" rainfall for a year would be either 40 cm (16 in.) *or* 60 cm (24 in.).

Regions of normally heavy precipitation experience the least variability; normally dry regions experience the most (Figure 6-39). Said another way, dry regions experience great fluctuations in precipitation from one year to the next.

LearningCheck 6-17 Is a desert likely to have high precipitation variability or low precipitation variability?

Acid Rain

One of the most troublesome environmental problems of the last half-century is **acid rain**—more generally called *acid precipitation* or *acid deposition*. This term refers to the deposition of either wet or dry acidic materials from the atmosphere on Earth's surface. Although most conspicuously associated with rainfall, the pollutants may fall to Earth with snow, sleet, hail, or fog or in the dry form of gases or particulate matter.

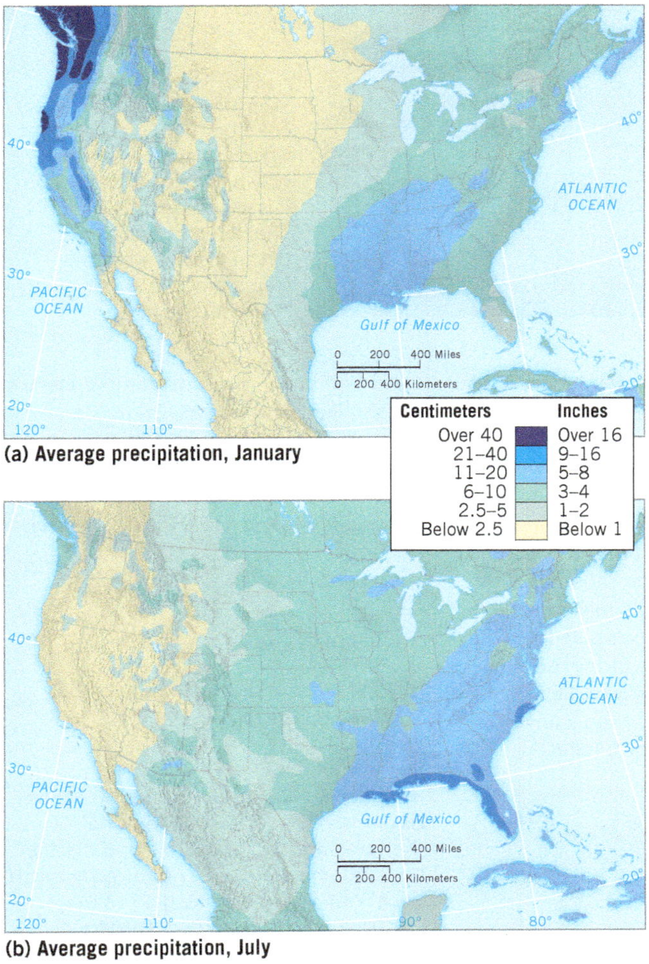

▲ **Figure 6-38** (a) Average January and (b) average July precipitation in southern Canada, the conterminous United States, and northern Mexico. Winter precipitation is heaviest in the Pacific Northwest; summer rainfall is greatest in the Southeast.

Sources of Acid Precipitation

Sulfuric and nitric acids are the principal culprits that have been recognized thus far. Evidence indicates that the principal human-induced sources are sulfur dioxide (SO_2) emissions from smokestacks (such as electric utility companies in the United States and the smelting of metal ores in Canada) and nitrogen oxides (NO_x) from motor vehicle exhaust. When emissions of sulfur and nitrogen compounds are expelled into the air, they may drift hundreds or even thousands of kilometers by winds. While adrift, they may mix with atmospheric moisture, forming the sulfuric and nitric acids that are later precipitated.

Measuring Acidity: Acidity is measured on a *pH scale*, based on the relative concentration of hydrogen ions (H^+) (**Figure 6-40**). The scale ranges from 0 to 14. The lower end represents extreme *acidity* (battery acid has a pH of 1), and the upper end extreme *alkalinity* (lye has a pH of 13). Alkalinity is the opposite of acidity. The pH scale is a logarithmic scale, which means that a difference of one whole number on the scale reflects a 10-fold change in absolute values.

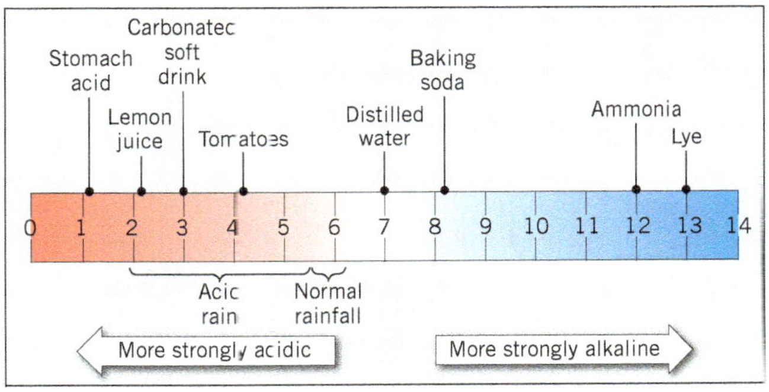

▲ **Figure 6-40** The pH scale. Rainfall in clean air has a pH of about 5.6 (slightly acidic). Acid rain can have a pH as low as 2.0.

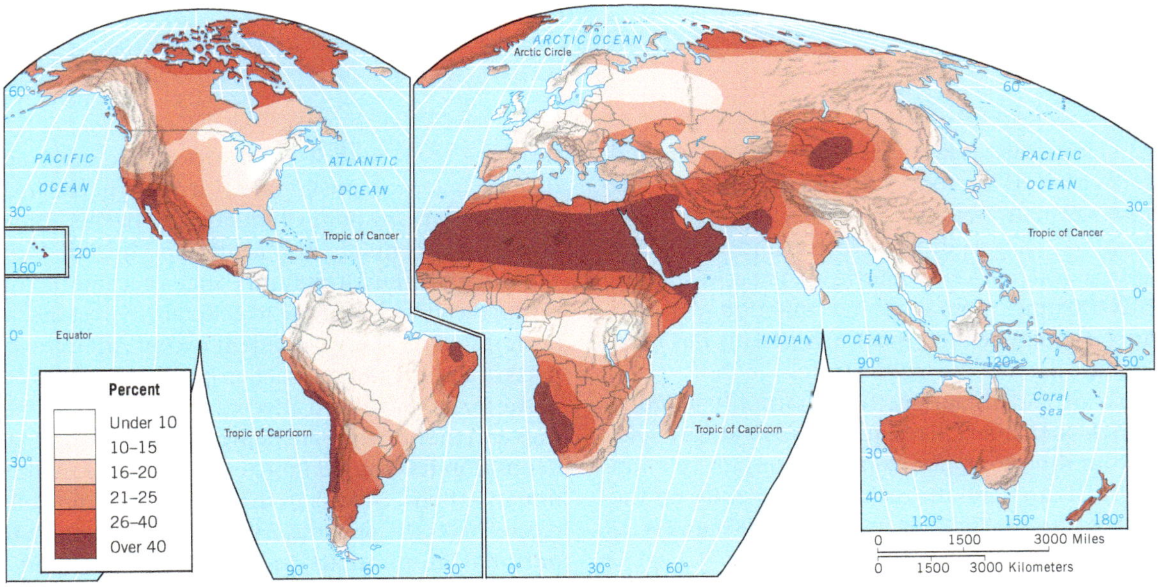

▲ **Figure 6-39** Precipitation variability is the expected departure from average precipitation in a given year, expressed as a percentage above or below average. Dry regions (such as northern Africa, the Arabian peninsula, southwestern Africa, central Asia, and much of Australia) experience greater variability than humid areas (such as the eastern United States, northern South America, central Africa, and western Europe).

Featured National Park

The National Mall

Acid Rain's "Slow Dissolve"

◀ **Figure 6-J** The light-green color on the Ulysses S. Grant Memorial is typical of the way bronze weathers when it is exposed to acidic rain. The rain dissolves some of the metal and causes staining of the white marble bases.

The National Mall and Memorial Parks, located in Washington, DC, receive approximately 24 million visitors each year and is a symbol of our nation's history. It has eight memorials and numerous statues, many made of limestone and marble (**Figure 6-I**). Both of these building materials are susceptible to acid rain and what the Park Service described as the "slow dissolve" in one of its publications.

In 1997, the average pH of the rain in Washington, DC, was about 4.2, which is almost as acidic as a carbonated soda. This level of acidity has weakened structures and pitted the surfaces of most of the limestone and marble structures on the National Mall. In addition to the acid rain damage, a biofilm has taken up residence in the cracks and crevices, giving the monuments a dark, dirty look. Repairs are currently underway to fix any damage and remove the biofilm.

Acid rain also causes damage to bronze statues, like the Ulysses S. Grant Memorial seen in **Figure 6-J**, causing it to have green streaks. To protect from further damage, the National Park Service washes the bronze statues periodically and applies a wax coating.

Questions

1. Why is it significant that buildings were built using limestone and marble?
2. What kinds of damage has acid rain caused on the National Mall?

◀ **Figure 6-I** Cleaning the volutes of the columns in the Jefferson Memorial. Even sheltered areas of limestone and marble buildings that rain does not directly touch are at risk. Sulfur dioxide gas in the air reacts with calcite grains in the stone causing the edges to dissolve. You can see from the detail of this column capital volute how the surface has a rough "sugary" texture.

Rainfall in clean, dust-free air has a pH of about 5.6, and precipitation with a pH less than 5.6 is considered to be acid precipitation. Normally, rain is just slightly acidic (because slight amounts of carbon dioxide dissolve in raindrops, forming *carbonic acid* [H_2CO_3], a mild acid), but acid rain can be as much as 100 times more acidic. Precipitation with a pH of less than 4.5 (below which most fish perish) has been recorded in some parts of the United States.

Many parts of Earth's surface have naturally alkaline soil or bedrock that neutralizes acid precipitation. Soils developed from limestone, for example, contain calcium carbonate, which can neutralize acid. Granitic soils, on the other hand, have no neutralizing component (**Figure 6-41**).

LearningCheck 6-18 What are the main sources of acid rain?

Damage from Acid Precipitation: The most noticeable damage from acid precipitation, a major hazard to the environment, is being done to aquatic ecosystems. Thousands of lakes and streams are now acidic, and hundreds of lakes in the eastern United States and Canada became biological deserts in recent decades due to acid rain. Forest diebacks have been noted on every continent except Antarctica; in some parts of eastern and central Europe, 30 percent to 50 percent of the forests have been affected or killed by acid rain. Buildings and monuments are being damaged or destroyed; acid deposition has caused more erosion on the marble Parthenon in Athens, Greece, in the last 24 years than all forms of erosion over the previous 24 years. See how acid rain affects national monuments in Washington, DC, in the box *Featured National Park: The National Mall: Acid Rain's "Slow Dissolve."*

▲ **Figure 6-41** Areas in the United States and Canada that are particularly sensitive to acid rain because of a scarcity of natural buffers.

A great complexity of this situation is that much of the acid precipitation is deposited great distances from its source. Downwind locations receive unwanted acid deposition from upwind origins. Thus Scandinavians and Germans complain about British pollution; Canadians blame U.S. sources; New Englanders accuse the Midwest.

One of the thorniest issues in North American international relations during the 1980s was Canadian dissatisfaction with U.S government efforts to mitigate acid rain. Canadians viewed acid rain as a grave environmental concern, and at the time perhaps half the acid rain falling on Canada came from U.S. sources, particularly older coal-burning power plants in the Ohio and Tennessee River valleys.

Action to Reduce Acid Precipitation: In the United States, significant progress toward reducing emissions that cause acid rain has been made since Title IV of the 1990 Clean Air Act Amendments was implemented, creating the Acid Rain Program monitored by the Environmental Protection Agency (EPA). The program requires major reductions in SO_2 and NO_x, especially from coal-burning electricity-generating plants that annually produced about 70 percent of SO_2 emissions and 20 percent of NO_x emissions in the United States.

In 1991, Canada and the United States signed the bilateral Air Quality Agreement, which addressed the issue of acid rain and transboundary air pollution. In 2000, the United States and Canada signed Annex 3 to the Agreement, with the goal of reducing emissions of NO_x and volatile organic compounds (VOC). In the years that followed, initiatives also addressed reducing particulate matter.

These programs have been quite successful: the EPA reports that the national average of SO_2 saw a 88 percent decrease between 1980 and 2017. NO_x emissions had been reduced to 2 percent of 1990 levels. Such reductions in emissions are central to the sustained effort needed to reduce the effects of acid rain.

After studying this chapter, you should be able to answer the following questions. Key terms from each text section are shown in **bold type.** Definitions for key terms are also found in the glossary at the back of the book.

Learning Review 6

Key Terms and Concepts

The Nature of Water: Commonplace but Unique (*p. 144*)

1. Briefly describe how water moves through the **hydrologic cycle**.
2. What is a **hydrogen bond** between water molecules?
3. Describe what happens to the density of water as it freezes.
4. What is meant by **surface tension** of water?
5. What is **capillarity**?

Phase Changes of Water (*p. 146*)

6. Briefly define the following terms: **evaporation**, **condensation**, **sublimation**.
7. How is energy exchanged during phase changes of water? (In other words, explain **latent heat**.)
8. What is meant by the **latent heat of condensation**? **Latent heat of evaporation**?

Water Vapor and Evaporation (*p. 149*)

9. Describe the conditions associated with relatively high rates of evaporation and the conditions associated with relatively low rates of evaporation.
10. What is meant by the **vapor pressure** of water in the atmosphere?
11. What is meant by **saturation** with water vapor?
12. What is **evapotranspiration**?

Measures of Humidity (*p. 150*)

13. What is **absolute humidity**? **Specific humidity**?
14. What is meant by saturation vapor pressure?
15. What determines the **water vapor capacity** of air?
16. Describe **relative humidity**, and explain what is meant by a relative humidity of 50 percent.
17. What is the **dew point**? The **dew point temperature**?

18. Explain **sensible temperature**.

Condensation (p. 152)

19. Under what circumstances can air become **supersaturated**?
20. Explain the role of **condensation nuclei** in the condensation process.
21. What are **supercooled water** droplets?

Adiabatic Processes (p. 153)

22. Which cooling process in the atmosphere is responsible for the formation of most clouds (and nearly all clouds that produce precipitation)?
23. What happens to the relative humidity of an unsaturated parcel of air as it rises? Why?
24. How is the dew point temperature of a parcel of air related to its **lifting condensation level**?
25. Contrast the **dry adiabatic rate** and a **saturated adiabatic rate**.

Atmospheric Stability (p. 154)

26. What is the difference between **stable** air and **unstable** air?
27. What conditions make a parcel of air unstable?

Clouds (p. 158)

28. Are stratus clouds associated with stable or unstable air? Are cumulus clouds associated with stable or unstable air?
29. Briefly describe the three main forms of **clouds**: **cirrus clouds**, **stratus clouds**, and **cumulus clouds**; describe **cumulonimbus clouds**.
30. Identify the four families of clouds.
31. Describe the four principal types of **fog**.
32. How and where does **dew** form?

Precipitation (p. 161)

33. Briefly describe the following kinds of **precipitation**: **rain**, **snow**, **hail**.
34. How is hail related to atmospheric instability?

Atmospheric Lifting and Precipitation (p. 165)

35. Describe the four main lifting mechanisms of air: **convective**, **orographic**, **frontal**, and **convergent lifting**.
36. What is a **rain shadow**?

Global Distribution of Precipitation (p. 169)

37. What is an **isohyet**?
38. What is **precipitation variability**?
39. How does precipitation variability relate to average annual precipitation?

Acid Rain (p. 172)

40. What are some of the circumstances that cause **acid rain**?

Study Questions

1. Why does ice float on liquid water?
2. Why is evaporation a "cooling" process and condensation a "warming" process?
3. What happens to the relative humidity of an unsaturated parcel of air when the temperature decreases? Why?
4. What happens to the relative humidity of an unsaturated parcel of air when the temperature increases? Why?
5. Why does a rising parcel of unsaturated air cool at a greater rate than a rising parcel of saturated air (in which condensation is taking place)?
6. Why can't descending air form clouds?
7. Why does the dew point temperature of an air parcel indicate its actual water vapor content?
8. How can rising stable air become unstable above the lifting condensation level?
9. Explain the role of adiabatic temperature changes, as well as changes in both the relative humidity and the actual water vapor content of the air, in the formation of rain shadows.
10. Using the global map of average annual precipitation (Figure 6-36), explain the causes of
    a. wet regions within the tropics.
    b. wet regions along the west coasts of continents in the midlatitudes (between about 40 and 60° N and S).
    c. dry regions along the west coasts of continents in the subtropics (at about 20 to 30° N and S).
    d. dry areas within the midlatitudes.
11. Using the maps of average January and July precipitation (Figure 6-37), contrast and explain the seasonal rainfall patterns in central Africa.

Exercises

1. Calculate the relative humidity of a parcel of air for which the specific humidity is 5 g/kg and the water vapor capacity is 25 g/kg: _____%
2. Calculate the relative humidity of a parcel of air for which the specific humidity is 30 g/kg and the water vapor capacity is 40 g/kg: _____%
3. Use Figure 6-8 to estimate the water vapor capacity (the saturation specific humidity in g/kg) of air at the following temperatures:
    a. 0°C (32°F): _____ g/kg
    b. 30°C (86°F): _____ g/kg

4. Using your answers for Exercise 3, calculate the relative humidity of the following parcels of air at the temperature given:
   a. If the specific humidity is 4 g/kg at a temperature of 0°C: _____%
   b. If the specific humidity is 4 g/kg at a temperature of 30°C: _____%
5. Assume that a parcel of unsaturated air is at a temperature of 24°C at sea level before it rises up a mountain slope, and that the lifting condensation level of this parcel is 3000 meters:
   a. What is the temperature of this parcel after it has risen to 2000 meters? _____ °C
   b. To 5000 meters? _____ °C
6. Assume that 1 gram of liquid water is at a temperature of 20°C. How many calories (or joules) of energy must be added for the water to warm to 40°C? _____ calories or _____ joules

Environmental **Analysis** CLOUD CLIMATOLOGY

Data
Cloud Climatology
https://media.pearsoncmg.com/ph/esm/esm_mcknight_physgeo_13/media/

The International Satellite Cloud Climatology Project (ISCCP) collects cloud data from weather satellites of several nations to help us understand the role of clouds in climate.

Activities

Go to http://isccp.giss.nasa.gov/products/browsed2.html, the ISCCP website. Retain the variable "Total Cloud Amount (%)" and time period "Mean Annual"; then click "View."

1. The map indicates the average annual percentage of cloud-covered sky. What is the range of cloud cover amounts in the far north? In the far south?
2. In general, is cloud cover higher over oceans or land? What would cause this?

Go back and select the variable "Mean Precipitable Water for 1000–680 mb"; then click "View."

3. The map indicates how much moisture is available for precipitation in the lower half of the troposphere. How much precipitable water is available in the far north?
4. Notice that precipitable water amounts are high in equatorial regions and decrease poleward, as do temperatures. Why are precipitable water and temperature related in this way?

Go back and select the variable "Mean Precipitable Water for 680–310 mb"; then click "View."

5. The map indicates how much moisture is available for precipitation in the upper half of the troposphere. Again, precipitable water is most abundant in equatorial regions. What type of cloud is likely to form there?
6. Recall the patterns of cloud cover (Activity 1) and precipitable water (Activity 3) in the far north. What types of clouds are most likely to form in the far north? It may help to refer to Figure 6-21.

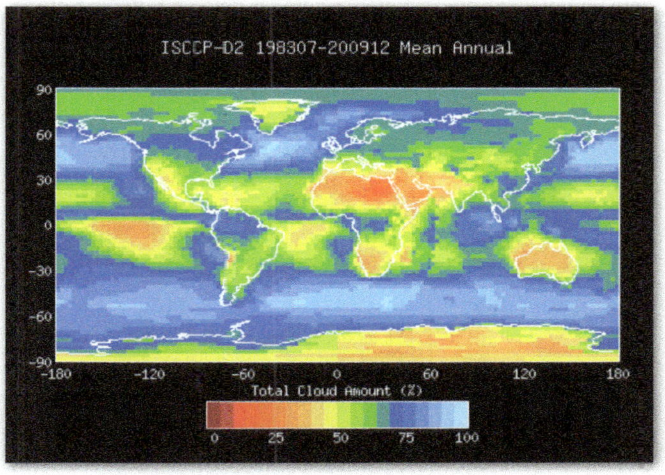

Seeing Geographically in Our National Parks

Look again at the photograph of the sunset in Everglades National Park at the beginning of the chapter (p. 142). What type of clouds fill the sky? What does the cloud type suggest about the stability of the air?

Mastering Geography®

Looking for additional review and test prep materials? Visit the Study Area in **Mastering Geography** to enhance your geographic literacy, spatial reasoning skills, and understanding of this chapter's content by accessing a variety of resources, including Map Master interactive maps, geoscience animations, *Mobile Field Trips*, videos, *Project Condor* Quadcopter videos, *In the News* RSS feeds, flashcards, web links, self-study quizzes, and an eText version of *McKnight's Physical Geography*.

Seeing Geographically in Our National Parks

This springtime thunderstorm formed over the Badlands National Park, South Dakota. Describe the general topography of this region. Do you think the Badlands gets an abundance of rainfall from thunderstorms? How do you determine this?

Badlands National Park, SD

Atmospheric Disturbances

7

HAVE YOU EVER WONDERED why springtime tornadoes are a worry in Oklahoma City but not in Seattle? Or why a summer thunderstorm is more likely at the Badlands National Park than in San Francisco? Or why hurricanes regularly strike the East Coast of the United States but not the West Coast? Regional differences in storm activity are directly tied to the global patterns of temperature, pressure, wind, and precipitation that we explored in the previous three chapters. As we'll see, the Great Plains of Oklahoma are the ideal breeding ground for tornadoes: relatively flat terrain with strong springtime air-mass contrasts and instability. The atmosphere around Seattle is stabilized by the cool water surrounding it, making tornadoes rare. Hurricanes, as we'll also see, can develop and survive only in areas of warm ocean water; cool water off the West Coast would quickly weaken a hurricane.

In this chapter, we turn from the broad patterns and conditions of weather that we discussed earlier to events in the atmosphere that are smaller in scale and shorter in duration. These more limited events include the passing of air masses and fronts, which affect large portions of the country. They also include a variety of atmospheric disturbances usually referred to as *storms*, which may cause more damage, but affect smaller regions. Particularly in the midlatitudes, such disturbances move with the general circulation of the atmosphere, persisting from several days (hurricanes, air masses, and fronts) to only a few seconds or minutes (lightning and tornadoes), before dissipating. Although air masses, fronts, and storms are migratory and temporary, in some parts of the world they are so frequent and dominating that their interactions are major determinants of weather and even climate.

As you study this chapter, think about these **Key**Questions:

- **How does an air mass form?**
- **What is a front?**
- **How do midlatitude cyclones influence weather?**
- **How do hurricanes form, move, and cause damage?**
- **How do thunderstorms and tornadoes form?**

The Impact of Storms on the Landscape

Storms can be very dramatic events: expansive clouds, swirling winds, heavy rain, and perhaps thunder and lightning. The landscape may be quickly and significantly transformed by a storm—uprooted trees, accelerated erosion, flooded valleys, damaged buildings, and ruined crops can result. Most storms also have a positive long-term effect on the landscape, however. They promote diversity in the vegetative cover, increase the size of lakes and ponds, and stimulate plant growth through the moisture they add to the ground.

Air Masses

The troposphere is by no means a uniform blanket of air. Instead, it is composed of many large distinct parcels of air referred to as **air masses**.

Characteristics

To be recognized as an air mass, a parcel of air:

1. *Must be large.* A typical air mass is more than 1600 kilometers (1000 miles) across and several kilometers deep (from its top to Earth's surface).
2. *Must be uniform in the horizontal dimension.* At any given altitude in the air mass, its physical characteristics—primarily temperature, humidity, and stability—are relatively homogeneous.
3. *Must travel as a unit.* It must be distinct from the surrounding air; when it moves it must retain its original characteristics and not be torn apart by differences in airflow.

Origin and Source Regions

An air mass develops its characteristics when it remains over a uniform land or sea surface long enough to acquire the temperature, humidity, and stability characteristics of the surface below. Stable air is more likely than unstable air to remain stagnant for days, so regions with anticyclonic (high pressure) conditions commonly form air masses.

The formation of air masses is usually associated with what are called *source regions*: regions of Earth's surface that are particularly well suited to generate air masses. Such regions are large, physically uniform, and associated with stationary or anticyclonic air. Ideal source regions are ocean surfaces and extensive flat land areas that have a uniform covering of snow, forest, or desert. Air masses rarely form over the irregular terrain of mountain ranges. Figure 7-1 portrays the principal source regions for air masses that affect North America.

It may well be that the concept of source regions is of more theoretical value than actual value, however. A broader view, to which many atmospheric scientists subscribe, holds that air masses can originate almost anywhere in the low or high latitudes but rarely in the midlatitudes, where the persistent prevailing westerlies would prevent air-mass formation.

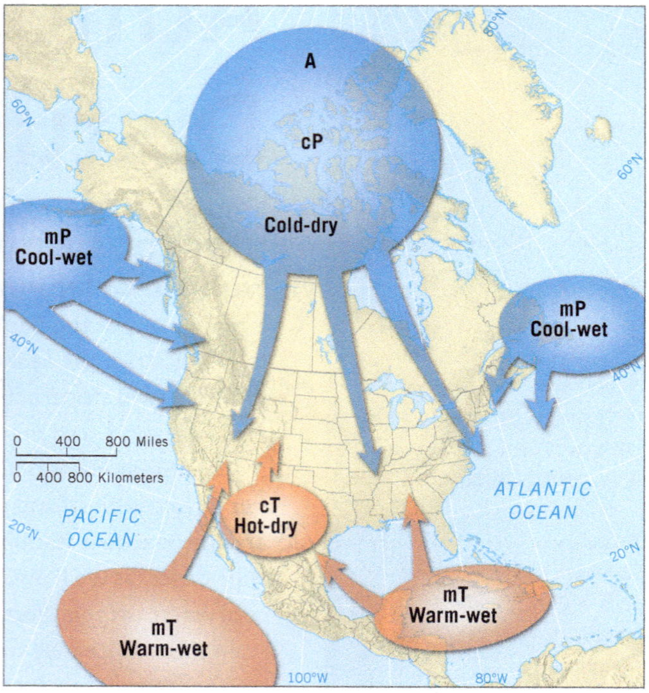

▲ Figure 7-1 Major air masses that affect North America and their generalized paths. The tropics and subtropics are important source regions as are the high latitudes. Air masses do not originate in the middle latitudes except under unusual circumstances. (For an explanation of the air-mass codes A, cP, mP, cT, and mT, see Table 7-1.)

Classification

Air masses are classified on the basis of source region. The latitude of the source region correlates directly with the temperature of the air mass, and the nature of the surface strongly influences the humidity within the air mass. Thus, a low-latitude air mass is warm or hot and therefore more unstable; a high-latitude one is cool or cold and therefore more stable. If the air mass develops over a continental surface, it is likely to be dry; if it originates over an ocean, it is usually moist.

A one- or two-letter code is generally used to identify air masses. They are classified into six basic types (Table 7-1) based on temperature and humidity.

LearningCheck 7-1 **How do air masses form? Why do air masses rarely originate in the midlatitudes? (Answer on p. AK-2)**

Movement and Modification

Some air masses remain in their source regions for long periods. Our interest, however, is in masses that leave their source region and move into other areas, particularly into the midlatitudes. As an air mass moves, temperature, moisture, and stability change.

TABLE 7-1 Simplified Classification of Air Masses

Type	Code	Source Regions	Source Region Properties
Arctic/Antarctic	A	Antarctica, Arctic Ocean and fringes, and Greenland	Very cold, very dry, very stable
Continental polar	cP	High-latitude plains of Eurasia and North America	Cold, dry, very stable
Maritime polar	mP	Oceans in vicinity of 50°–60° N and S latitude	Cold, moist, relatively unstable
Continental tropical	cT	Low-latitude deserts	Hot, very dry, unstable
Maritime tropical	mT	Tropical and subtropical oceans	Warm, moist, often unstable

The air mass itself is modified, and it modifies the weather of the regions into which it moves. In a classic example of this modification (Figure 7-2), a midwinter cold air outbreak of continental polar (cP) air from northern Canada sweeps down across the central part of North America. Starting with a source-region temperature of −46°C (−50°F) around Great Slave Lake, the air mass warms to −34°C (−30°F) by the time it reaches Winnipeg, Manitoba, and continues to warm as it moves southward. In its southward course, the air mass also brings some of the coldest weather that each of these places will receive all winter. Thus, the air mass is modified, but it also modifies the weather in regions through which it passes.

North American Air Masses

The North American continent is a prominent area of air-mass interaction. The lack of mountains trending east to west permits polar air to sweep southward and tropical air to flow northward unhindered by terrain, particularly over the eastern two-thirds of the continent (see Figure 7-1). In the western part of the continent, though, air masses moving off the Pacific are impeded by the prominent north–south trending mountain ranges.

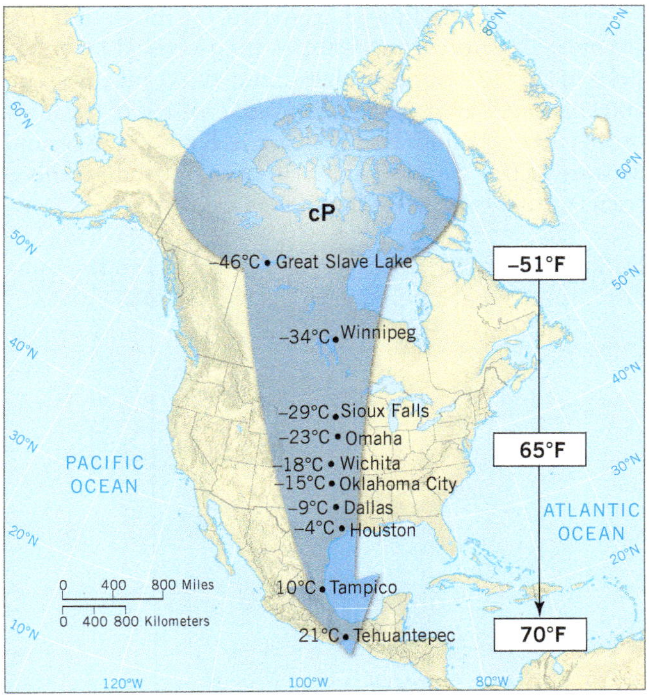

▲ Figure 7-2 An example of temperatures resulting from a strong midwinter outburst of cP air from Canada. The cP air mass starts with a temperature of −46°C (−51°F) in Great Slave Lake and warms to 21°C (70°F) as it moves south to Tehuantepec.

Continental polar (cP) air masses develop in central and northern Canada, whereas *Arctic* (A) air masses originate farther north and are colder and drier than cP air masses. Both are dominant features in winter with their cold, dry, stable nature.

Maritime polar (mP) air from the Pacific in winter can bring cloudiness and heavy precipitation to the mountainous west coastal regions. In summer, cool Pacific mP air produces fog and low stratus clouds along the coast. North Atlantic mP air masses are also cool, moist, and unstable. These air masses sometimes affect the New England area but are usually kept out to sea by the prevailing westerlies.

Maritime tropical (mT) air from the Atlantic/Caribbean/Gulf of Mexico is warm, moist, and unstable. It strongly influences weather and climate east of the Rockies in the United States, southern Canada, and much of Mexico and is the principal precipitation source in this broad region. It is more prevalent in summer than in winter, bringing periods of uncomfortable humid heat.

Pacific mT air originates over water in areas of anticyclonic subsidence, so it tends to be cooler, drier, and more stable than Atlantic mT air. It is a source of summer rains in the southwestern interior and may produce coastal fog.

Continental tropical (cT) air originates in the desert southwest region of North America. In summer, hot, very dry, unstable cT air surges into the southern Great Plains area on occasion, bringing heat waves and dry conditions.

LearningCheck 7-2 What are the temperature and moisture characteristics of a maritime polar (mP) air mass? A continental tropical (cT) air mass?

Fronts

When unlike air masses meet, they do not mix readily; instead, a boundary zone called a **front** develops between them. A front is not a very sharp boundary. A typical front is a narrow transition zone several kilometers or even tens of kilometers wide. Within this zone, the properties of the air change rapidly from one air mass to the other.

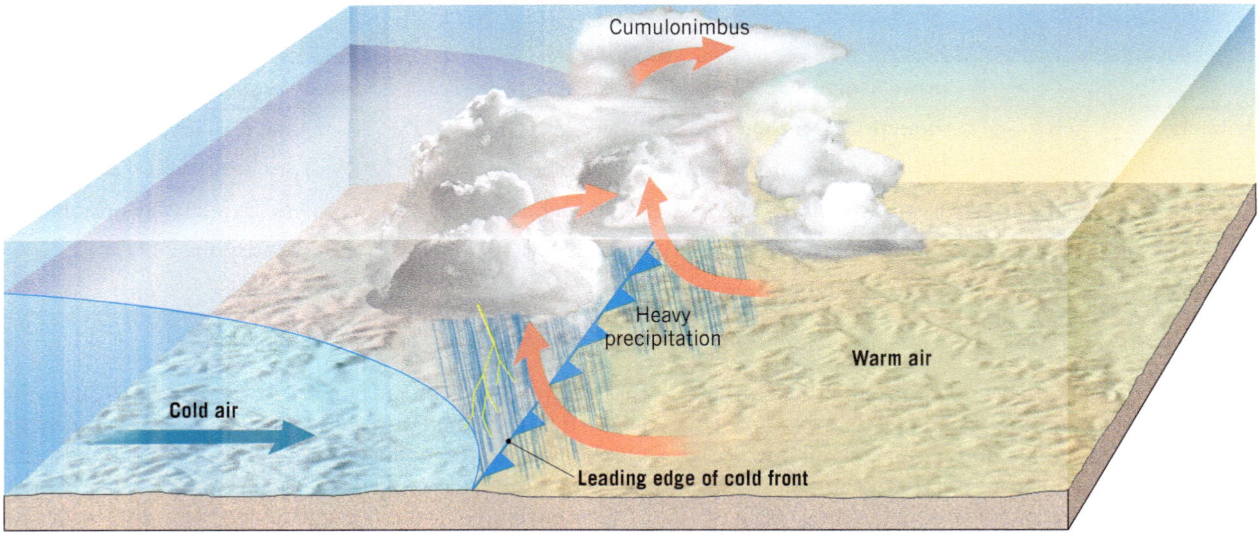

▲ **Figure 7-3** A cold front forms when a cold air mass actively underrides a warm air mass. As a cold front advances, the warm air ahead of it is forced to rise. This uplift often creates cloudiness and relatively heavy precipitation along and immediately behind the ground-level position of the front. (In this diagram, the vertical scale has been exaggerated.)

Animation
Cold Fronts
https://media.pearsoncmg.com/ph/esm/esm_mcknight_physgeo_13/media/

The front concept was developed during World War I by Norwegian meteorologists, who coined the term *front* because they considered the clash between unlike air masses to be analogous to a confrontation between opposing armies along a battlefront. As the more "aggressive" air mass advances at the expense of the other mass, a small amount of mixing of the two occurs within the frontal zone, but for the most part the air masses retain their separate identities as one displaces the other.

The most conspicuous difference between air masses is usually temperature. A **cold front** forms where an advancing cold air mass meets and displaces warmer air (**Figure 7-3**), whereas a **warm front** forms where an advancing warm air mass meets colder air (**Figure 7-4**). In both cases, warm air is on one side of the front and cool air on the other, with a fairly abrupt temperature gradient between. Air masses may also have different densities, humidity levels, wind patterns, and stabilities, and these factors can have a steep gradient through the front as well.

Regardless of which air mass is advancing, the cooler, denser air functions as a wedge that forces the warmer, lighter air aloft. As Figures 7-3 and 7-4 show, fronts "lean" or slope upward from the surface. Along this slope, the warmer air rises and cools adiabatically, forming clouds and often precipitation. Indeed, fronts lean so much that they are more like horizontal features than vertical ones. The slope of a typical front averages about 1:200, meaning that 200 kilometers away from the surface position of the front, the height of the front is only 1 kilometer above the ground. Because of this very low angle of slope, most diagrams of fronts greatly exaggerate their steepness.

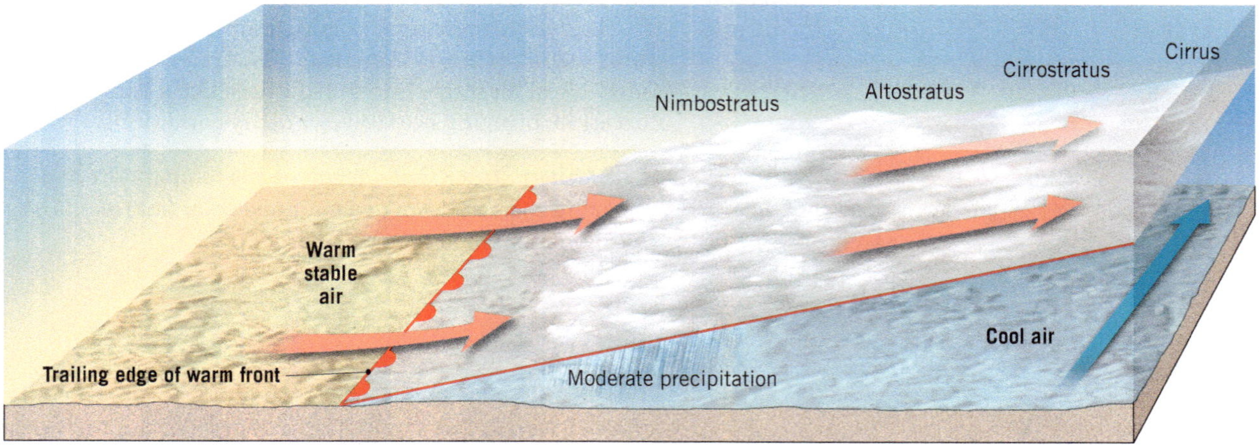

▲ **Figure 7-4** A warm front forms when a warm air mass actively overrides a cooler air mass. As warm air rises above cooler air, widespread cloudiness and precipitation develop along and in advance of the ground-level position of the warm front. Higher and less dense clouds are often dozens or hundreds of kilometers ahead of the ground-level position of the front. (In this diagram, the vertical scale has been exaggerated.)

Animation
Warm Fronts
https://media.pearsoncmg.com/ph/esm/esm_mcknight_physgeo_13/media/

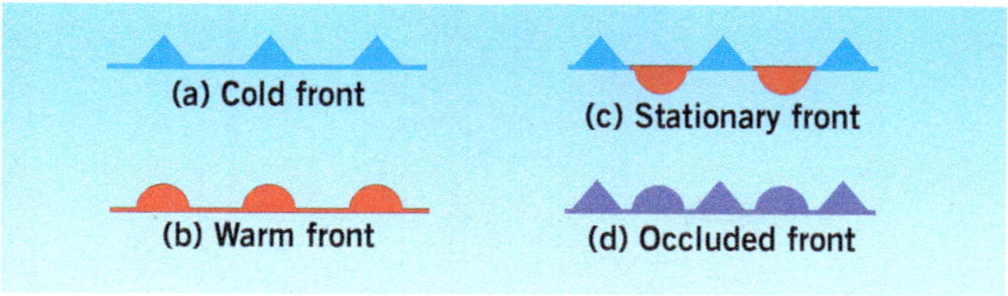

▲ **Figure 7-5** Weather map symbols for fronts. (a) Cold front: The solid triangles extend in the direction toward which the front moves. (b) Warm front: The solid semicircles extend in the direction toward which the front moves. (c) Stationary front: Cold air is opposite the triangles, and warm air opposite the half circles. (d) Occluded front: Warm and cold front symbols alternate on the same side of the line.

Notice that the "leading edge" of a cold front precedes its higher-altitude "trailing edge," whereas a warm front leans "forward" so the higher-altitude part of the front is ahead of its lower-altitude trailing edge. The symbols that are used on a weather map indicate the ground-level position of a front (Figure 7-5).

Cold Fronts

Because of friction with the ground, the advance of the lower portion of a cold air mass is slowed relative to the upper portion. As a result, a cold front tends to become steeper as it moves forward and usually develops a protruding "nose" a few hundred meters above the ground (see Figure 7-3). The average cold front is twice as steep as the average warm front. Moreover, cold fronts normally move faster than warm fronts because the dense, cold air mass easily displaces the lighter, warm air.

This combination of steeper slope and faster advance leads to rapid lifting and adiabatic cooling of the warm air ahead of the cold front. The rapid lifting often makes the warm air very unstable, and the result is blustery, violent weather along the cold front. Vertically developed clouds, such as cumulonimbus clouds, are common, along with turbulence and showery precipitation. Both clouds and precipitation tend to be concentrated along and immediately behind the ground-level position of the front. Precipitation is usually more intense but shorter-lived than that associated with a warm front.

Warm Fronts

The slope of a typical warm front is gentler than that of a cold front, averaging about 1:200 (see Figure 7-4). As the warm air pushes against and rises over the retreating cold air, it cools adiabatically, usually resulting in clouds and precipitation. Because the frontal uplift is very gradual, clouds form slowly, and turbulence is limited. High-flying cirrus clouds may signal the approaching front many hours before it arrives. As the front comes closer, the clouds become lower, thicker, and more extensive, typically developing into altocumulus or altostratus clouds. Precipitation frequently occurs broadly; if the rising air is inherently unstable, however, precipitation can be showery or even violent. Most precipitation falls ahead of the ground-level position of the moving front.

LearningCheck 7-3 **Compare the characteristics of cold fronts and warm fronts.**

Stationary Fronts

When neither air mass displaces the other—or when a cold front or warm front "stalls"—the boundary between air masses is called a **stationary front**. It is difficult to generalize about the weather along such a front, but often gently rising warm air produces limited precipitation similar to that along a warm front. Long-lasting stationary fronts can lead to flooding.

Occluded Fronts

A fourth type of front, called an *occluded front*, forms when a cold front overtakes a warm front. In winter, occluded fronts are sometimes associated with blizzards. The development of occluded fronts is discussed later in this chapter.

Air Masses, Fronts, and Major Atmospheric Disturbances

We now turn our attention to the major kinds of atmospheric disturbances that are discussed in this chapter. Most of these disturbances involve unsettled and sometimes violent atmospheric conditions and are referred to as *storms*. Some, however, produce calm, clear, quiet weather that is the opposite of stormy. Some of these disturbances involve air-mass contrasts or fronts, and many are associated with migrating pressure cells.

In general, atmospheric disturbances

- are smaller than the components of the general circulation, although they are extremely variable in size;
- are migratory;
- are brief in duration, persisting for only a few days, hours, or minutes; and
- produce relatively predictable weather conditions.

Midlatitude Disturbances: The midlatitudes are the principal "battleground" of tropospheric phenomena: where polar and tropical air masses meet, where most fronts occur, and where weather is most dynamic and changeable from season to season and from day to day. Many kinds of atmospheric disturbances are associated with the midlatitudes, but *midlatitude cyclones* and *midlatitude anticyclones* are much bigger and more prevalent.

Tropical Disturbances: The low latitudes are characterized by monotony—the same weather day after day, week after week, month after month. Almost the only breaks are provided by transient atmospheric disturbances, of which by far the most significant are *tropical cyclones* (locally known as *hurricanes* or *typhoons* when they intensify).

Localized Severe Weather: Other localized atmospheric disturbances occur in many parts of the world. Short-lived but sometimes severe atmospheric disturbances such as *thunderstorms* and *tornadoes* often develop in conjunction with other kinds of storms. (See the box *Snapshots from our National Parks* on *Weather Observations* for some types of localized disturbances that the National Park Service copes with.)

Snapshots From Our National Parks

Grand Teton, Rocky Mountains, and Death Valley

Weather Observations

▲ **Figure 7-A** This Grand Teton National Park weather observation station was installed in the Surprise Lake/Disappointment Peak area. It is predominately used to forecast avalanche conditions. This is a great help to those who want to do backcountry skiing, snowshoeing, and hiking in the winter.

For many national parks, monitoring and predicting weather conditions is a matter of visitor safety and not just gathering scientific data. In fact, the Grand Canyon National Park website provides a separate landing page for "Weather Dangers," which include lightning, rock falls, and flash floods. Other parks must anticipate avalanches (**Figure 7-A**), heavy snow fall (**Figure 7-B**), and temperatures that can kill (**Figure 7-C**).

NPS Video
Lightening Safety at Grand Canyon
https://media.pearsoncmg.com/ph/esm/esm_mcknight_physgeo_13/media/

▶ **Figure 7-B** The Rocky Mountains National Park weather stations are used to monitor changing weather conditions during both summer and winter. Winter conditions change rapidly, which could lead to sudden heavy snowfall. Summer conditions often lead to thunderstorms near the top of the mountain, and the heavy rains that result can travel down through valleys and be very dangerous.

▲ **Figure 7-C** The Death Valley weather stations point to a landscape that is very hot and very dry most of the time. These hot, dry conditions are dangerous and not for the faint of heart. The weather stations can also pick up rare rainfall, which can lead to dangerous flash flooding.

Midlatitude Cyclones

Probably most significant of all atmospheric disturbances are **midlatitude cyclones**. Throughout the midlatitudes, cyclones dominate weather maps, are responsible for most day-to-day weather changes, and bring precipitation to much of the populated portions of the planet. These large, migratory low-pressure cells are usually called *depressions* in Europe and *lows* or *low-pressure systems*, *extratropical cyclones*, or simply (although not very precisely) *storms* in the United States.

Midlatitude cyclones are associated with air-mass movement in regions between about 30° and 70° of latitude. Thus, they are found almost entirely within the band of westerly winds. Their general path of movement is toward the east, which explains why weather forecasting in the midlatitudes is essentially a west-facing vocation.

Each midlatitude cyclone is different. The discussion that follows describes the "typical" or idealized conditions of these storms. Moreover, these conditions are presented as Northern Hemisphere phenomena. For the Southern Hemisphere, the patterns of isobars, fronts, and wind flow are mirror images of the Northern Hemisphere patterns.

Characteristics

A typical mature midlatitude cyclone has a diameter of 2000 kilometers (about 1200 miles) or so. It is essentially a vast cell of low-pressure air, with ground-level pressure in the center typically between 990 and 1000 millibars. The system (shown by closed isobars on a weather map, as in **Figure 7-6a**) usually tends toward an oval shape, with the long axis trending northeast–southwest. Usually a clear-cut trough of low pressure extends along the cold front from the center to the southwest.

Formation of Fronts: Midlatitude cyclones are low pressure cells, so they have a converging counterclockwise circulation pattern in the Northern Hemisphere. This wind flow pattern brings cool air down from the north and warm air up from the south (see Figure 7-6a). The convergence of these unlike air masses characteristically creates two fronts: (1) a cold front that extends to the southwest from the center of the cyclone and runs along the pressure trough extending from the center of the storm and (2) a warm front extending eastward from the center and running along another, usually weaker, pressure trough.

Sectors: The two fronts divide the cyclone into a *cool sector* north and west of the center where the cold air mass is in contact with the ground, and a *warm sector* to the south and east where the warm air mass is in contact with the ground. At the surface, the cool sector is the larger of the two, but aloft the warm sector is more extensive. This size relationship exists because both fronts "lean" over the cool air. Thus, the cold front slopes upward toward the northwest, and the warm front slopes upward toward the northeast (**Figure 7-6b**).

LearningCheck 7-4 **What causes fronts to develop within a midlatitude cyclone?**

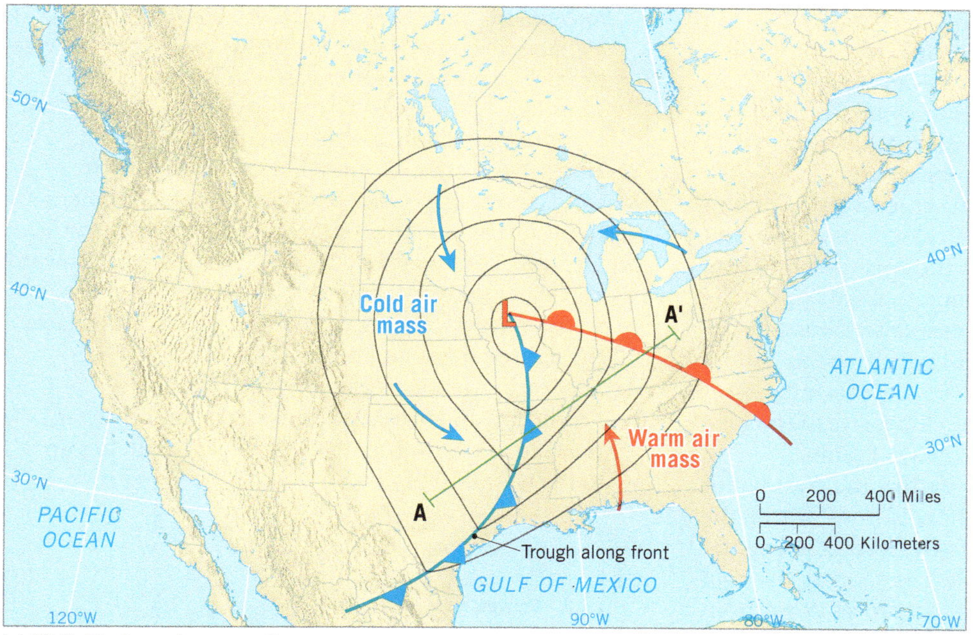

(a) Midlatitude cyclone: weather map

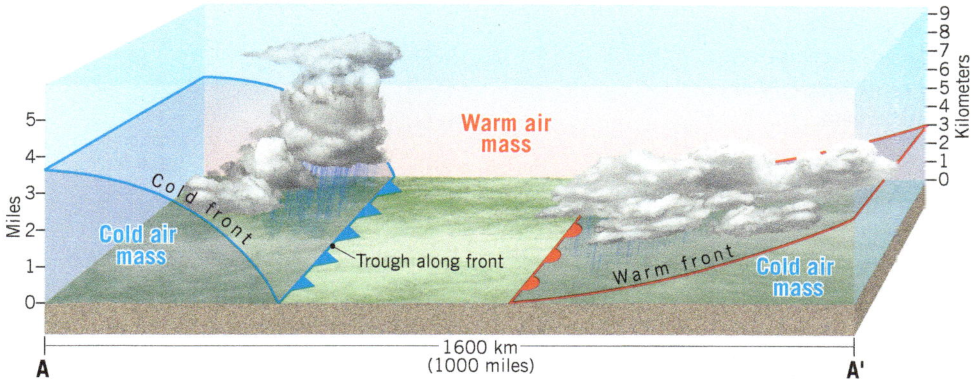

(b) Midlatitude cyclone cross section

Animation
Midlatitude Cyclones
https://media.pearsoncmg.com/ph/esm/esm_mcknight_physgeo_13/media/

◀ **Figure 7-6** (a) A map and (b) a cross section of a typical mature midlatitude cyclone. In the Northern Hemisphere, a cold front usually extends to the southwest, and a warm front extends toward the east. A well-developed trough of low pressure usually accompanies the surface position of the cold front.

◀ **Figure 7-7** A large midlatitude cyclone centered near Lake Michigan. A cold front extends across the southern states to the southwest.

Clouds and Precipitation: Clouds and precipitation develop in the zones within a midlatitude cyclone where air is rising and cooling adiabatically. Because warm air rises along both fronts, the typical result is two zones of cloudiness and precipitation that overlap around the center of the storm (where air is rising in the center of the low-pressure cell).

Along and immediately behind the ground-level position of the cold front (the steeper of the two fronts), a band of cumuliform clouds usually yields showery precipitation (**Figure 7-7**). The air rising along the more gradual slope of the warm front produces a more extensive expanse of horizontally developed clouds and typically lower-intensity precipitation. In both cases, most of the moisture for the precipitation is derived from the warm air mass. These clouds give the cyclone the characteristic "comma" shape shown in Figure 7-7.

Most frontal precipitation falls within the cool sector north of the warm front and along the cold front, within the low pressure cell. The general area to the west of the center of the cyclone is frequently cloudless as soon as the cold front has passed because cold, stable air moves down from the north. The clouds to the north and west of the low clear shortly after the center of the low has passed. In contrast, the air of the warm sector is often moist and tending toward instability, so thermal convection may produce sporadic thunderstorms. Also, sometimes one or more *squall lines* of intense thunderstorms develop in the warm sector in advance of the cold front.

Movements

Midlatitude cyclones are transient features, on the move throughout their existence. Four kinds of movement are involved (**Figure 7-8**):

1. The whole storm moves with the westerlies, traversing the central United States and Canada. The rate of movement averages 30 to 45 kilometers (about 20 to 30 miles) per hour, so the storm can cross North America in three to four days (often faster in winter than in summer). The route of a cyclone is generally from west to east, often following the path of the jet stream.
2. The system has a cyclonic wind circulation, with wind converging counterclockwise (in the Northern Hemisphere) into the center of the storm at the surface.
3. The cold front is created by cold air moving southward and usually advances faster than the center of the storm. (The advancing dense, cold air easily displaces the lighter, warm air ahead of the front.)
4. The warm front is created by warm air moving northward and usually advances more slowly than the cold front. Eventually the cold front catches up to the warm front.

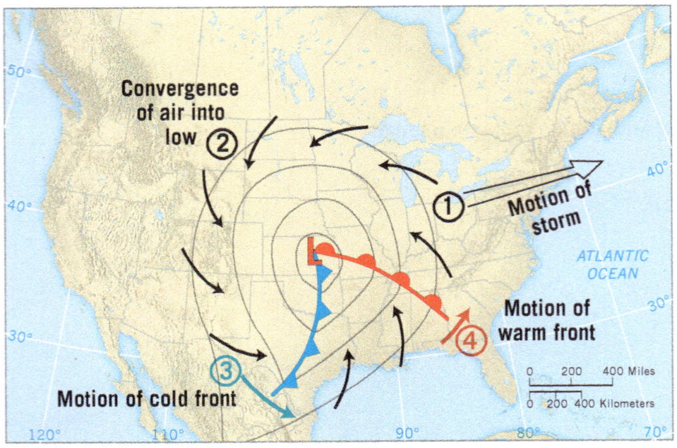

▲ **Figure 7-8** Four components of movement occur in a typical midlatitude cyclone: (1) The entire storm moves west to east in the general flow of the westerlies; (2) airflow is cyclonic, converging counterclockwise (in the Northern Hemisphere); (3) the cold front advances; (4) the warm front advances.

LearningCheck 7-5 **Explain where and why precipitation develops within a midlatitude cyclone.**

Life Cycle

Cyclogenesis: A typical midlatitude cyclone progresses from origin to maturity, and then to dissipation, in about three to ten days. The most common cause of **cyclogenesis**, or "birth" of cyclones, is upper troposphere conditions in the vicinity of the polar front jet stream. Most midlatitude cyclones begin as "waves" along the polar front. Recall from Chapter 5 that these waves are undulations or curves that develop in the path taken by upper-level winds such as a jet stream, and that the polar front is the contact zone between the relatively cold polar easterlies and the relatively warm westerlies. The opposing airflows normally have a relatively smooth linear motion on either side of the polar front (Figure 7-9a). On occasion, however, the smooth frontal surface may be distorted into a wave shape (Figure 7-9b).

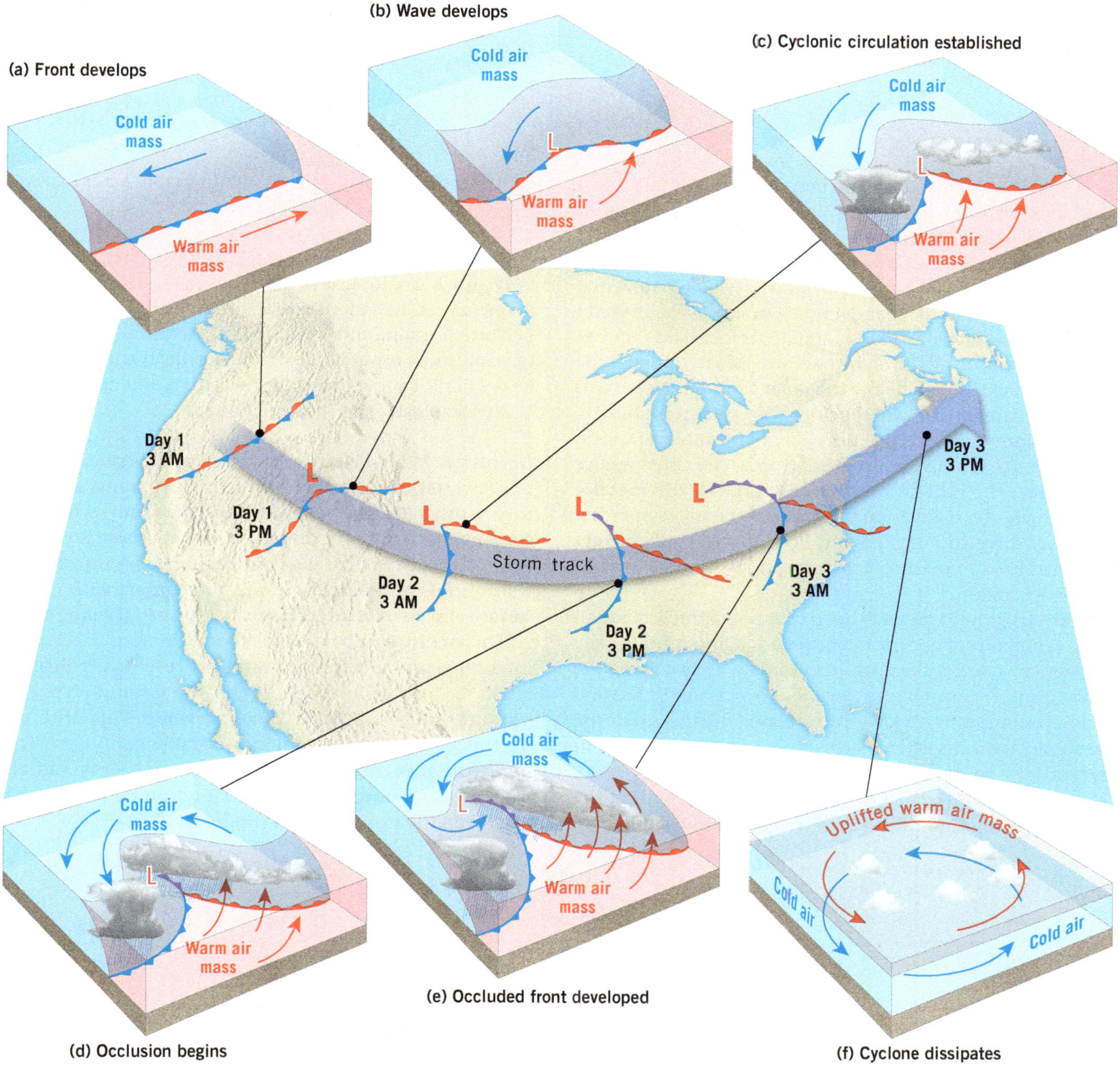

▲ **Figure 7-9** Schematic representation of the life cycle of an idealized midlatitude cyclone passing over North America during a three-day period. (a) Front develops between unlike air masses. (b) Wave appears along front. (c) Cyclonic circulation is well developed around a low. (d) Occlusion begins. (e) Occluded front is fully developed. (f) Cyclone dissipates after all warm surface air has been lifted and cooled.

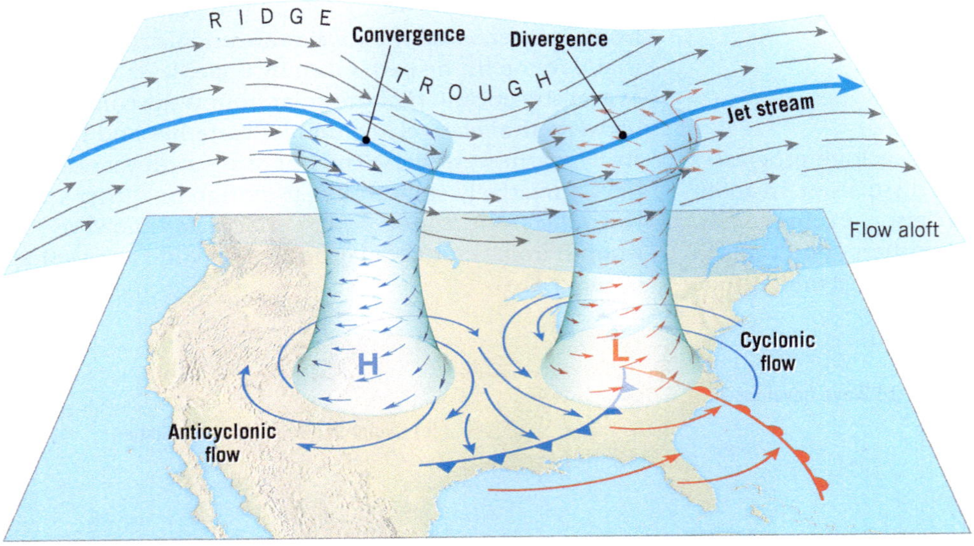

◄ **Figure 7-10** A typical weather situation in which the upper-level airflow, such as the path of the jet stream, is meridional (meandering north and south). Convergence and divergence aloft support anticyclonic and cyclonic circulation at ground level.

There is a close relationship between upper-level airflow and ground-level cyclones. When the upper airflow is *zonal*—by which we mean the wind blows relatively straight from west to east—ground-level cyclonic activity is unlikely. When the winds aloft become *meridional* (Figure 7-10)—that is, the wind begins to meander north to south—large waves of alternating pressure troughs and ridges form and cyclonic activity at ground level intensifies. Most midlatitude cyclones are centered below the polar front jet stream axis and downstream from an upper-level pressure trough.

A cyclone is unlikely to develop at ground level unless a divergence is above it. In other words, the convergence of air near the ground must be supported by divergence aloft. Such divergence can be related to changes in either speed or direction of the wind flow, but it nearly always involves broad north-to-south meanders (Rossby waves) in the jet stream.

Various ground factors—such as topographic irregularities, temperature contrasts between sea and land, or the influence of ocean currents—can initiate a wave along the polar front. For example, cyclogenesis occurs on the leeward side of mountains. A low-pressure area drifting with the westerlies weakens when it crosses a mountain range. As air ascends the range, the air column compresses and spreads, slowing its counterclockwise spin. When descending the leeward side, the air column stretches vertically and contracts horizontally. This change in shape causes it to spin faster and initiates cyclonic development.

This chain of events happens with some frequency in winter on the eastern flanks of the Rocky Mountains, particularly in Colorado and Alberta, Canada, and with lesser frequency on the eastern side of the Appalachian Mountains, in North Carolina and Virginia. Cyclones formed in this way typically move toward the east and northeast and often bring heavy rain or snowstorms to the central and northeastern United States and southeastern Canada.

Occlusion: Ultimately, the storm dissipates because the cold front overtakes the warm front. As the two fronts get closer to each other (Figures 7-9c, 7-9d, 7-9e), the warm sector at the ground is increasingly displaced, forcing more and more warm air aloft. Where the cold front catches up with the warm front, warm air is no longer in contact with Earth's surface and an **occluded front** has formed (Figure 7-11).

This **occlusion** process may result in a short period of intensified precipitation and wind until eventually all of the warm air mass is forced aloft and the ground-level low-pressure center is "filled in" by convergence. The entire area on the ground is now cool air—a stable condition. This sequence of events shuts off the storm's energy and air-lifting mechanism, and thus its cloud-producing mechanism, and the storm dies out (Figure 7-9f).

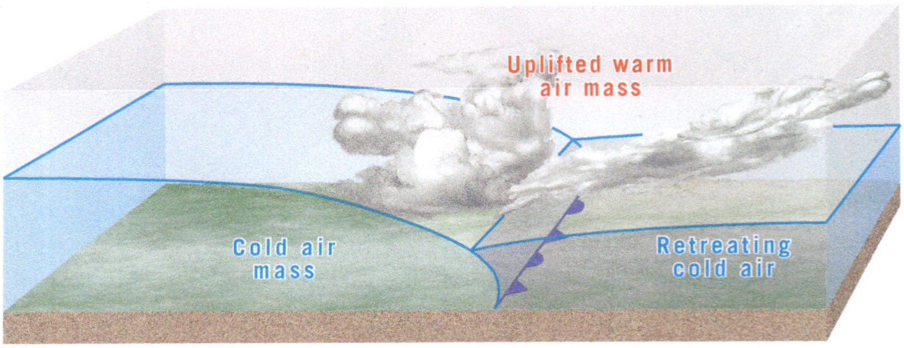

◄ **Figure 7-11** An occluded front develops when the leading edge of a cold front catches up with the trailing edge of a warm front, lifting all of the warm air off the ground. Once lifted, moisture condenses into thick clouds.

Conveyor Belt Model of Midlatitude Cyclones: The description of midlatitude cyclones we've just provided is sometimes called the "Norwegian" model because it was first presented by meteorologists in Norway in the 1920s. Although this explanation of midlatitude cyclones remains useful today, new data have provided a more complete explanation of these storms, especially airflow in the upper troposphere, with moisture moving in *atmospheric rivers*. The *conveyor belt model* now offers a better explanation of the three-dimensional aspects of these storms—see the box *Focus: Conveyor Belt Model of Midlatitude Cyclones*.

LearningCheck 7-6 **Describe the process that forms an occluded front.**

Weather Changes with the Passing of a Midlatitude Cyclone

Although the exact details vary from storm to storm, the basic structure and movements of a midlatitude cyclone that we just described can help us understand the often abrupt weather changes we experience on the ground when such a storm occurs. This is especially true when the cold front of a midlatitude cyclone passes through in winter.

For example, imagine we're in the warm sector of a midlatitude cyclone—the situation just before the cold front moves through (see Figure 7-6). Remember, the whole storm moves from west to east, so the cold front is moving closer to us hour by hour. When the cold front passes, all four elements of weather will likely change:

Temperature: As the cold front passes, temperature drops abruptly because the cold front is the boundary between the cold and warm air masses of the storm.

Pressure: The trough of low pressure associated with the cold front extends south from the heart of the storm. As the trough along the front approaches, pressure decreases, reaching its lowest point at the front. Then, as the cold front passes and the trough travels with it, pressure begins to increase steadily.

Wind: Due to the overall converging counterclockwise wind pattern (in the Northern Hemisphere), winds in the warm sector come from the south (ahead of the cold front). Once the front passes, winds shift and come from the west or northwest.

Clouds and Precipitation: The generally clear skies ahead of the cold front are replaced by cloudiness and precipitation at the front. They are generated by the adiabatic cooling of the warm air as it is lifted along the front—and are in turn replaced hours later by clear skies in the cold air mass behind the cold front.

Similar changes, although of lesser magnitude, occur with the passage of a warm front.

LearningCheck 7-7 **Why does pressure fall as a cold front approaches and rise as a cold front moves away?**

Occurrence and Distribution

At any given time, five to ten midlatitude cyclones exist in the Northern Hemisphere midlatitudes and a similar number in the Southern Hemisphere. They occur at scattered but irregular intervals throughout the zone of the westerlies.

In part because temperature contrasts are greater during the winter, these migratory disturbances are more numerous, better developed, and faster moving in winter than in summer. They also follow much more equatorward tracks in winter. In the Southern Hemisphere, Antarctica provides a prominent year-round source of cold air, so vigorous cyclones are almost as numerous in summer as in winter. The summer storms are farther poleward than their winter cousins, however, and are mostly over the Southern Ocean. Thus, they have little effect on land areas.

Midlatitude Anticyclones

Another major disturbance in the general flow of the westerlies is the **midlatitude anticyclone**, frequently referred to simply as a "high" (H). This is an extensive, migratory high-pressure cell of the midlatitudes (Figure 7-12). Typically, a high-pressure cell is larger than a midlatitude cyclone and generally moves west to east with the westerlies (distinguishing them from the semipermanent subtropical highs).

Characteristics

As with any other high-pressure center, a midlatitude anticyclone has air converging into it from above, subsiding, and diverging at the surface—clockwise in the Northern Hemisphere and counterclockwise in the Southern Hemisphere. Because unlike air masses are not being brought together, anticyclones contain no fronts. (The fronts shown in Figure 7-12 are outside the high-pressure system.) The weather is typically clear and dry with little or no opportunity for cloud formation. Wind movement is limited near

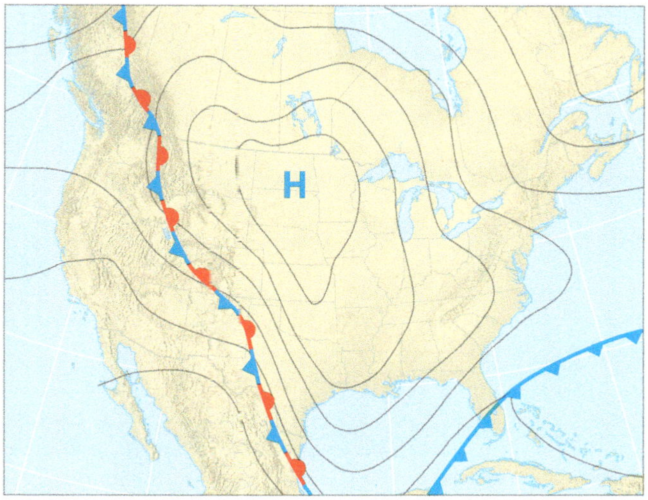

▲ **Figure 7-12** A typical well-developed midlatitude anticyclone centered over the Dakotas. Both fronts shown here are outside the high-pressure system

Focus

Conveyor Belt Model of Midlatitude Cyclones

• **Ted Eckmann,** University of Portland

Satellite and weather balloon measurements have revealed that midlatitude cyclones involve more than just surface fronts and a low-pressure center; they also tend to include several well-defined channels of air called *conveyor belts* (**Figure 7-D**). Our discussion uses examples from the Northern Hemisphere, but the conveyor belt model fits many Southern Hemisphere midlatitude cyclones as well. The conveyor belt model has improved our understanding of midlatitude cyclones and forecasting of their effects by explaining interactions between surface and upper-level winds.

The Warm Conveyor Belt: A midlatitude cyclone's surface low draws air northward from the southeastern portion of the cyclone. The air to the southeast tends to be warm and moist because those are characteristics of the air mass where it originates. A *warm conveyor belt* (also called an atmospheric river) develops from this air. It starts at the surface but eventually rises up and over the cooler air to the north of the warm front because air in the warm conveyor belt is less dense. As the warm conveyor belt reaches higher altitudes, it can turn eastward by joining the prevailing westerly winds of the upper troposphere (see Figure 7-D).

North of the warm front, the warm conveyor belt produces mostly stratus-type clouds because the air is rising too slowly to form clouds with more vertical development. These clouds tend to produce light but steady precipitation over a large area. However, the warm conveyor belt also delivers plenty of moisture just ahead of the cold front. The cold front then lifts this air rapidly, forming convective clouds such as cumulonimbus. These clouds produce showers that are more intense, but more sporadic, than those from the stratus-type clouds north of the warm front.

The Cold Conveyor Belt: Just north of the warm front, cooler, drier surface air moves westward toward the cyclone's central low, forming the *cold conveyor belt*. Like the warm conveyor belt, some of this cold air can rise and merge with the general westerly flow at upper levels (see Figure 7-D). However, the cold conveyor belt can also split, with the rest of the air turning cyclonically and rising as it moves toward the low-pressure center. While the cold conveyor belt starts out relatively dry, it can gain moisture as precipitation falls into it from the warm conveyor belt above, and thus be moist enough to support significant snow in winter by the time it reaches the area northwest of the low.

The Dry Conveyor Belt: On the western side of a typical midlatitude cyclone, convergence in the upper troposphere produces descending air, some of which swirls counterclockwise into the cyclone's low and forms the *dry conveyor belt*. This air from the upper troposphere is much drier than air in the other conveyor belts because it is farther from the surface and thus farther from sources of moisture. Few clouds, if any, can form in this dry air, which often results in a "dry slot" of air, with limited cloud cover, behind the cold front (**Figure 7-E**).

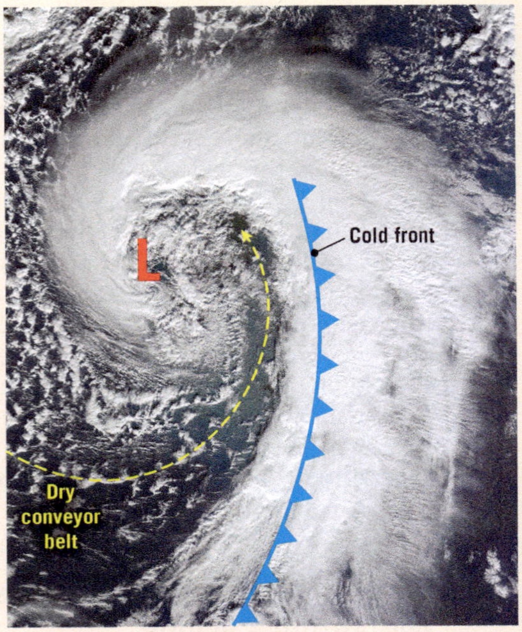

▲ **Figure 7-E** The dry conveyor belt spirals toward the low-pressure center, producing a dry slot behind the cold front that has significantly less cloud cover than adjacent areas (where the warm and cold conveyor belts produce thick clouds and precipitation).

Questions

1. The key ingredients for intense snow are cold air at the surface and rising, moist air. Use the conveyor belt model to explain why the heaviest snowfall in winter occurs just northwest of the low of a midlatitude cyclone.
2. How would the conveyor belt model differ for a midlatitude cyclone in the Southern Hemisphere? Why?

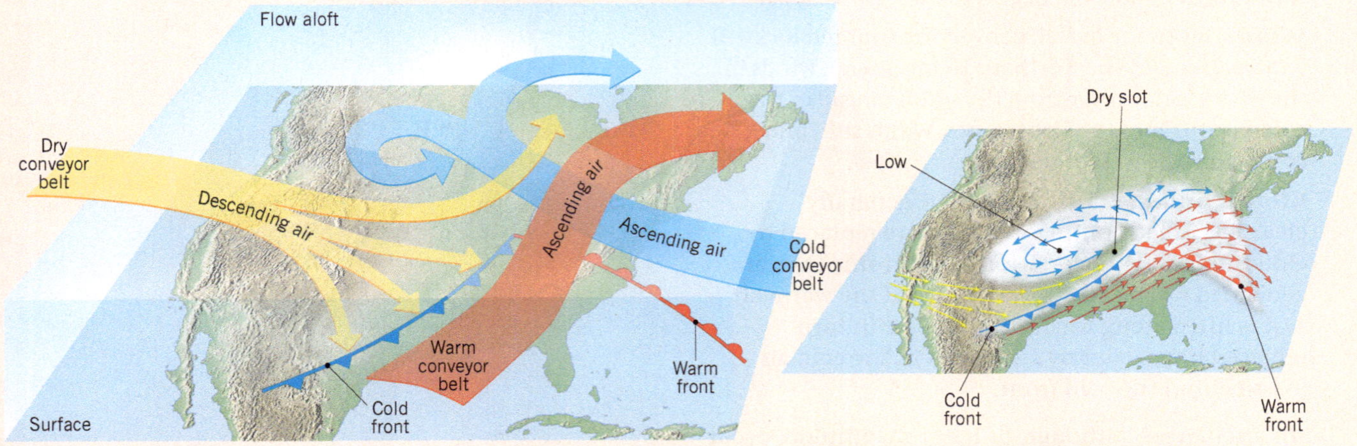

(a) Midlatitude cyclone **(b) Cloud cover of midlatitude cyclone**

▲ **Figure 7-D** (a) A typical midlatitude cyclone in the Northern Hemisphere with the warm conveyor belt, the cold conveyor belt, and the dry conveyor belt. (b) The typical cloud coverage produced by a midlatitude cyclone such as this.

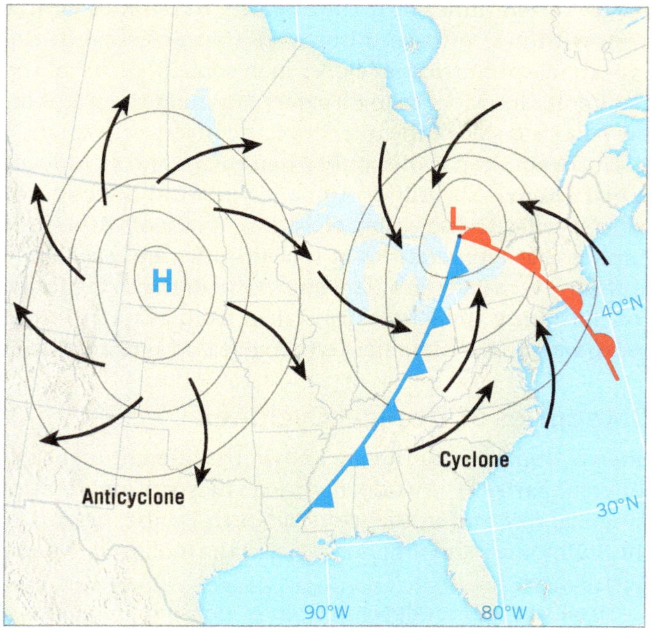

▲ **Figure 7-13** Midlatitude cyclones and anticyclones often are linked in the midlatitudes, with the anticyclone pumping cold air into the cyclone.

the center of an anticyclone but increases progressively outward. Particularly along the eastern margin (the leading edge) of the system, winds may be strong. In winter, anticyclones are characterized by very low temperatures. (Recall that high-pressure cells may be associated with cold surface conditions.) In summer, anticyclones may be characterized by heat waves.

Anticyclones move toward the east either at the same rate as or a little slower than midlatitude cyclones. Unlike cyclones, however, anticyclones occasionally stagnate and remain over the same region for several days. This stalling brings clear, stable, dry weather to the affected region, which enhances the likelihood that air pollutants will become concentrated under a subsidence temperature inversion. Such stagnation may block the eastward movement of cyclonic storms, causing precipitation in another region while the anticyclonic region remains dry.

Relationships of Cyclones and Anticyclones

Midlatitude cyclones and anticyclones often alternate with one another in (irregular) sequence around the midlatitudes globally. Each can occur independently of the other, but there is often a functional relationship between them. This relationship can be seen when an anticyclone closely follows a cyclone (Figure 7-13). The winds diverging from the eastern margin of the high fit into the flow of air converging into the western side of the low. It is easy to visualize this anticyclone as a polar air mass that has the cold front of the cyclone as its leading edge.

LearningCheck 7-8 **Why are midlatitude anticyclones associated with dry weather?**

Tropical Cyclones: Hurricanes

Tropical cyclones are intense, low-pressure disturbances that develop in the tropics and move poleward, occasionally as far as the midlatitudes. Tropical cyclones are considerably smaller than midlatitude cyclones, typically between 160 and 1000 kilometers (100 and 600 miles) in diameter (Figure 7-14).

Origin and Environment

Many tropical cyclones originate in a tropical disturbance called an easterly wave. An **easterly wave** is a long but weak migratory, low-pressure system that can occur almost anywhere between 5° and 30° of latitude (Figure 7-15).

Easterly waves drift slowly westward in the flow of the trade winds. Ahead of the wave is fair weather with divergent airflow at the surface. Behind the wave, convergent conditions prevail, and moist air is uplifted, yielding widespread cloudiness and sometimes convective thunderstorms. Little or no temperature change occurs with the passage of easterly waves. These waves are usually several hundred kilometers long and nearly always oriented north–south.

▲ **Figure 7-14** Even strong hurricanes come in a wide range of sizes. (a) Super Typhoon Tip measured 2221 kilometers (1380 miles) across and set a record for the largest storm on Earth shortly before it made landfall in Japan on October 19, 1979. (b) Hurricane Michael measured just 563 kilometers (350 miles) across and caused devastation along the Florida panhandle on October 10, 2018.

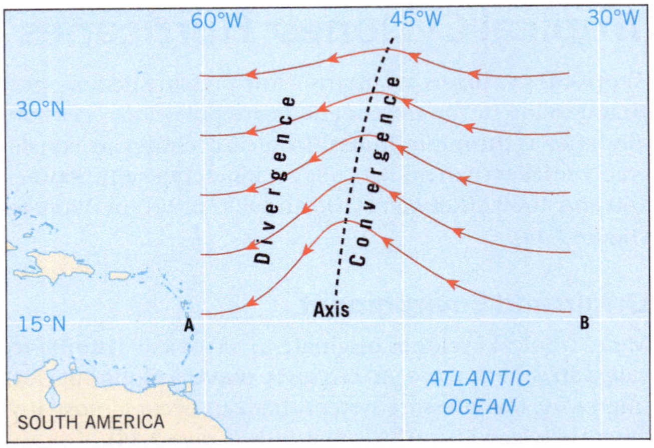

(a) Easterly wave, map

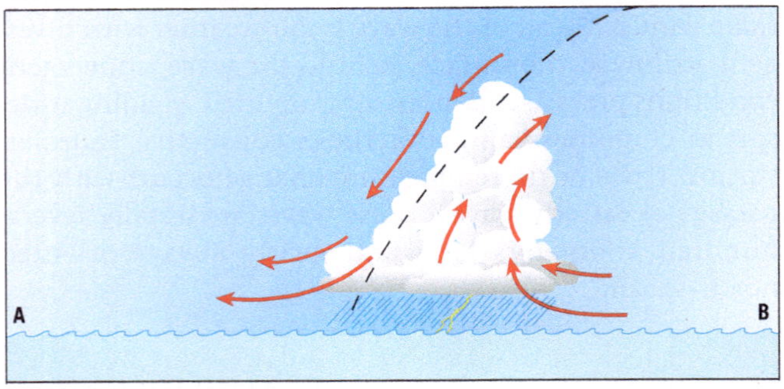

(b) Easterly wave, cross section

▲ **Figure 7-15** (a) Map view and (b) cross section of an easterly wave. The arrows indicate the general direction of airflow.

Most of the easterly waves that move across the North Atlantic originate over North Africa. The vast majority of easterly waves weaken and die out over the ocean, but a small percentage intensify into more powerful tropical cyclones.

Tropical cyclones that may become hurricanes form only over warm oceans in the tropics and at least a few degrees north or south of the equator (Figure 7-16). The ocean water temperature generally needs to be at least 26.5°C (80°F) to a depth of 50 meters (164 feet) or more. Because the Coriolis effect is so minimal near the equator, hurricanes do not form within 5° of it, and hurricanes do not cross it; the appearance of hurricanes closer than some 8° or 10° of the equator is rare. More than 80 percent originate in or just on the poleward side of the intertropical convergence zone.

Hurricanes can evolve only when there is no significant **wind shear**—a significant change in wind direction or wind speed with increasing elevation. A lack of wind shear implies that temperatures at low altitudes are reasonably uniform over a wide area. This allows the air to rise vertically, concentrating the latent heat taken from evaporation of ocean water. This latent heat is then released as clouds form.

Categories of Tropical Cyclones

Intense tropical cyclones are known by different names in different parts of the world: *hurricanes* in North and Central America, *typhoons* in the western North Pacific, *baguios* in the Philippines, and simply *cyclones* in the Indian Ocean and Australia.

Tropical cyclones develop from minor low-pressure disturbances in the trade winds (such as easterly waves). Generally called *tropical disturbances* by the U.S. National Weather Service, about 100 of these disturbances are identified each year over the tropical North Atlantic. Only a few strengthen into hurricanes. Three categories of tropical disturbances are recognized on the basis of wind speed:

1. A **tropical depression** has wind speeds up to 62 kilometers per hour (33 knots; 38 mph) but has developed a closed wind circulation pattern.
2. A **tropical storm** has winds between 63 and 118 kilometers per hour (34-63 knots; 39–73 mph).
3. A **hurricane** has winds that reach or exceed 119 kilometers per hour (64 knots; 74 mph).

Named Storms: When a tropical disturbance intensifies into a tropical storm, it is assigned a name from an alphabetical list that the World Meteorological Organization (WMO) has prepared. The storm keeps the same name if it becomes a hurricane. The WMO network of Regional Specialized Meteorological Centers and Tropical Cyclone

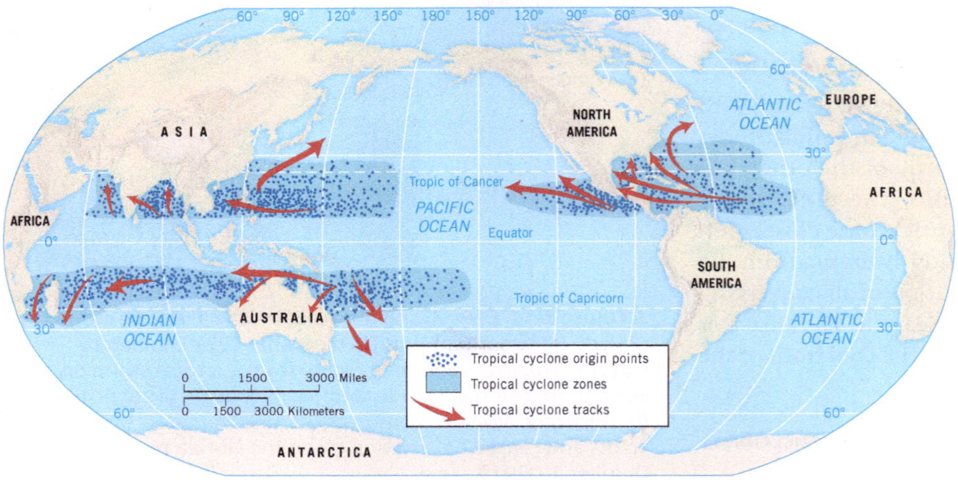

◀ **Figure 7-16** Generalized tracks of tropical cyclones (hurricanes and typhoons). Dots within the major tropical cyclone zones show origin points of tropical cyclones over a 19-year period.

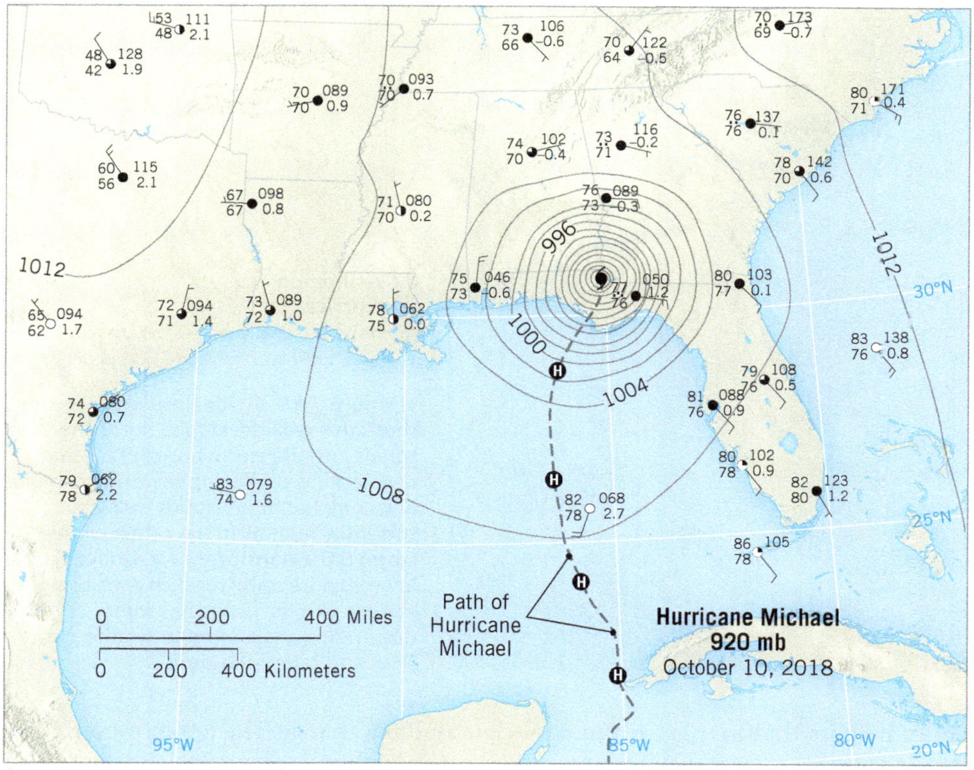

◀ **Figure 7-17** Weather map of Hurricane Michael on October 10, 2019. The isobars show a very steep gradient around the low-pressure center. The dashed line shows the position of Michael beginning on October 7.

Warning Centers monitors tropical storms around the world. For example, the National Hurricane Center in Miami is responsible for North Atlantic and northeastern Pacific storms, the Central Pacific Hurricane Center in Hawai'i is responsible for north-central Pacific storms, and the Japan Meteorological Agency in Tokyo is responsible for northwestern Pacific storms.

Each region uses a different name list. After a few years, hurricane and typhoon names are used again unless the storm was especially notable—such as Florence and Michael—in which case the name is retired from the list.

Characteristics

Hurricanes (as we generally refer to tropical cyclones in this chapter) consist of prominent low-pressure centers that are essentially circular, with a steep pressure gradient outward from the center (**Figure 7-17**). As a result, strong winds spiral inward. Winds must reach a speed of 119 kilometers per hour (64 knots; 74 mph) for the storm to be officially classified as a hurricane, although winds in a well-developed hurricane more than double that speed.

The converging cyclonic wind pattern of a hurricane pulls in warm, moist air—the "fuel" that powers the storm. As warm, water vapor–laden air spirals into the storm, it rises in intense updrafts within towering cumulonimbus clouds. As the air rises, it cools adiabatically, bringing the air to saturation. Condensation then releases vast amounts of liquid water that builds up the huge clouds and feeds the heavy rain. Condensation also releases latent heat, powering and strengthening a storm by increasing the instability of the air. In a short period of time and in a relatively small area, a hurricane releases an enormous amount of energy into the atmosphere. An average mature hurricane in one day releases approximately as much energy as is generated by all electric utility plants in the United States in one year.

Hurricanes are not characterized by fronts. All of the air within a hurricane is warm and moist, so unlike air masses do not converge as in a midlatitude cyclone.

LearningCheck 7-9 **Why is warm, moist air the "fuel" for a hurricane?**

Eye of a Hurricane: A remarkable feature of a well-developed hurricane is the nonstormy **eye** in the center of the storm (**Figure 7-18**). The winds do not converge to a central point but rather reach their highest speed at the *eye wall*, which is the edge of the eye. The eye has a diameter of from 30 to 65 kilometers (20 to 40 miles) and is a singular area of calmness in the maelstrom that whirls around it (see Figure 7-14).

The weather pattern within a hurricane is relatively symmetrical around the eye. Bands of dense cumulus and cumulonimbus clouds (called *spiral rain bands*) curve in from the edge of the storm to the eye wall, producing heavy rain in the band that generally increases in intensity inward. Updrafts are common throughout the hurricane, becoming most prominent around the eye wall. Near the top of the storm, most of this air diverges out clockwise into the upper troposphere. However, a small portion of the air descends back down into the eye as a gentle downdraft.

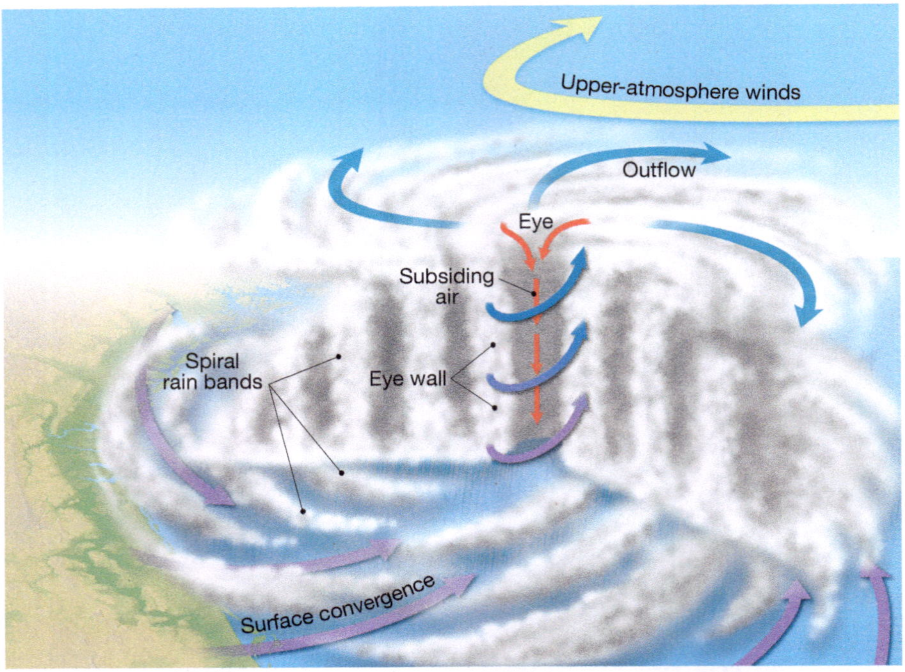

Animation
Hurricanes
https://media.pearsoncmg.com/ph/esm/esm_mcknight_physgeo_13/media/

◀ **Figure 7-18** An idealized cross section through a well-developed hurricane. Air spirals into the storm horizontally and rises rapidly, producing towering cumulus and cumulonimbus clouds that yield torrential rainfall. In the center of the storm is the nonstormy eye. Unlike typical low-pressure cells, in which air is rising, the eye of a hurricane has some descending air, leading to relative clear skies and calm conditions.

The clouds of the eye wall tower to heights that may exceed 16 kilometers (10 miles). Within the eye, there is no rain and almost no low clouds; adiabatic warming of descending air inhibits cloud formation there. In the eye, scattered high clouds may part, letting in intermittent sunlight. The wall of thunderstorms circling the eye is sometimes surrounded by a new wall of thunderstorms. The inner wall disintegrates and is replaced by the outer wall. The process, called *eye-wall replacement*, usually lasts less than 24 hours and tends to weaken the storm.

Movement

Hurricanes occur in a half-dozen low-latitude regions (see Figure 7-16). They are most common in the North Pacific basin, originating largely in two areas: east of the Philippines and west of southern Mexico and Central America. The third most notable region of hurricane development is in the west-central portion of the North Atlantic basin, extending into the Caribbean Sea and Gulf of Mexico.

These ferocious storms are also found in the western portion of the South Pacific and all across the South Indian Ocean, as well as in the North Indian Ocean both east and west of the Indian peninsula. They are very rare in the South Atlantic and in the southeastern part of the Pacific, apparently because the water is too cold and there is too much wind shear. The strongest and largest hurricanes are typically those of the western Pacific Ocean. Only a few storms in other parts of the world have attained the size and intensity of the large East Asian "super" typhoons (see Figure 7-14).

Hurricane Tracks: Once formed, hurricanes follow irregular tracks within the general flow of the trade winds. A specific path is very difficult to predict many days in advance, but the general pattern of movement is highly predictable. Roughly one-third of all hurricanes travel east to west without much latitudinal change. The rest, however, begin on an east–west path and then curve prominently poleward, where they either dissipate over the adjacent continent or become enmeshed in the general flow of the midlatitude westerlies (**Figure 7-19**).

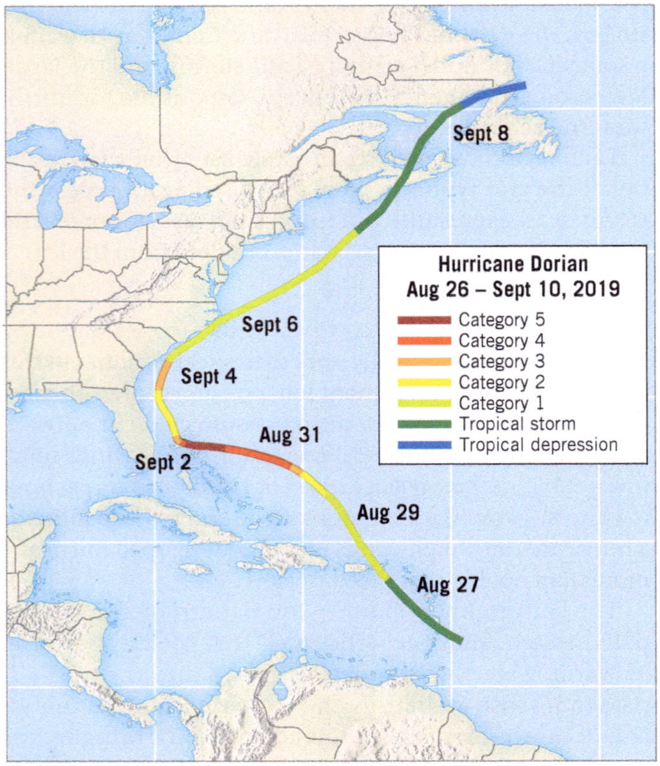

▲ **Figure 7-19** Path of Hurricane Dorian from August 26 to September 10, 2019. Its strength increased as the storm passed over the warm water in the Gulf Stream but diminished after moving north.

Hurricanes sometimes survive (with diminished intensity) off the east coasts of continents in the midlatitudes because of the warm ocean currents there. Hurricanes do not survive in the midlatitudes off the west coasts of continents because cool ocean currents are present.

The southwestern Pacific Ocean north and northeast of New Zealand has a marked variation from this general flow pattern. Hurricanes in this part of the Pacific usually move erratically from northwest to southeast. Therefore, when they strike an island such as Fiji or Tonga, they approach from the west, a situation not replicated anywhere else in the world at such a low latitude. These seemingly aberrant tracks result from the fact that the tropospheric westerlies extend quite far equatorward in the Southwest Pacific; hurricanes there are basically steered by the general circulation pattern.

LearningCheck 7-10 **Why can hurricanes move up into the midlatitudes along the East Coast of North America but not along the West Coast?**

Life Span: Whatever their trajectory, hurricanes do not last long. The average hurricane exists for only about a week, with four weeks as the maximum duration. As soon as a hurricane leaves the ocean and moves over land, it begins to dissipate because its energy source (warm, moist air) is cut off. If it stays over the ocean but moves into the midlatitudes, it dissipates as it penetrates the cooler environment. It is not unusual for a tropical hurricane that moves into the midlatitudes to diminish in intensity but grow in area until it develops into a midlatitude cyclone that travels with the westerlies.

In most regions, hurricanes have a marked seasonality. They are largely restricted to late summer and fall, peaking in early September in the Northern Hemisphere, because this is the time that ocean temperatures are highest and the intertropical convergence zone is shifted farthest poleward.

Damage and Destruction

Hurricanes are known for their destructive capabilities. Some of the destruction comes from high winds and torrential rain, and hurricanes may spawn tornadoes. However, the overwhelming cause of damage and loss of life is the flooding brought by high seas.

Hurricane Strength: Although the amount of damage caused by a hurricane depends in part on the physical configuration of the landscape, speed of the storm, and the population size and density of the affected area, storm strength is the most important factor. In the United States, the **Saffir-Simpson Hurricane Scale** is used to rank the relative intensity of hurricanes (based on wind speed), ranging from 1 to 5, with 5 being the most severe (Table 7-2). (See the box *Featured National Park: Everglades: Hurricane Impacts* for how a Category 4 storm affected this national park.)

Storm Surge: Wind-driven waves increase water height, producing a **storm surge** of water as much as 7.5 meters (25 feet) above normal tide level when the hurricane pounds into a shoreline (Figure 7-20). The low pressure in the center of the storm allows the ocean surface near the eye to bulge up as much as an additional 1 meter (3 feet). Thus, a low-lying coastal area can be severely inundated (Figure 7-21), and 90 percent of hurricane-related deaths are caused by water.

The greatest hurricane disaster in U.S. history occurred in 1900 when Galveston Island, Texas, was overwhelmed by a 6-meter (20-foot) storm surge that killed as many as 8000 people, nearly one-sixth of Galveston's population.

TABLE 7-2 Saffir-Simpson Hurricane Scale

	Wind Speed			
Category	km/hr	mph	knots	Damage
1	119–153	74–95	64–82	Dangerous winds will produce some damage.
2	154–177	96–110	83–95	Very dangerous winds will cause extensive damage.
3	178–208	111–129	96–112	Devastating damage will occur.
4	209–251	130–156	113–136	Catastrophic damage will occur.
5	>252	>157	>137	Catastrophic damage will occur.

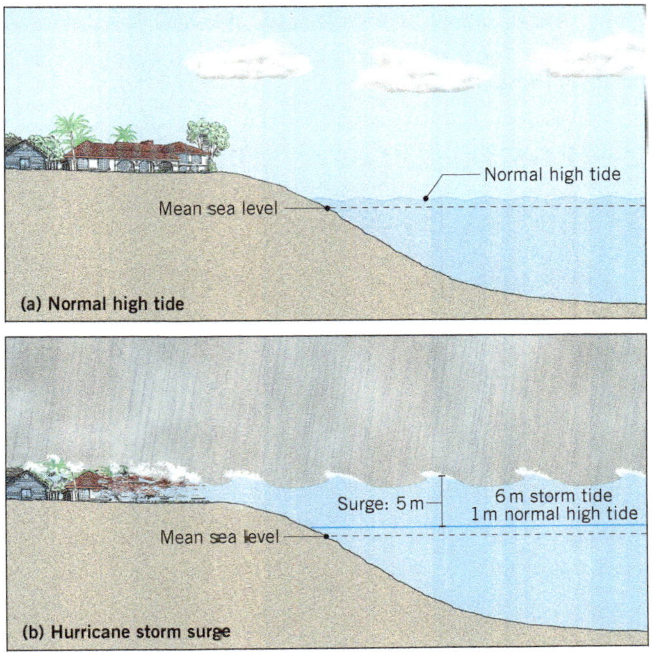

▲ **Figure 7-20** When (a) a normal high tide is accompanied by (b) a hurricane storm surge, flooding can overwhelm a coastal area.

Featured National Park

Everglades

Hurricane Impacts

(a) March 28, 2017

(b) December 1, 2017

▲ **Figure 7-G** Images taken from an airplane on (a) March 28, 2017, and (b) December 1, 2017, show what a small section of the Everglades looked like before Hurricane Irma, then after.

The Everglades National Park encompasses 20,200 square kilometers (7800 square miles) of mangroves, pine forest, and marshland, which line the southern edge of Florida. This picturesque swampland is home to a wide variety of birds, reptiles, fish, and other animals. The mangrove forests are a necessary component of the south Florida wetlands. They stabilize the soil and reduce erosion as well as provide a home for wildlife (**Figure 7-F**).

Hurricane Irma made landfall on September 10, 2017, as a Category 4 storm. Irma is the fifth costliest hurricane to hit the mainland. The damage caused in Everglades National Park was barely counted in the monetary total, but it was devastating nonetheless. The high wind and 5-foot storm surge stripped the leaves off of the mangroves, snapped branches, and tore the plants from the ground. More than 60 percent of the mangrove forest was severely damaged (**Figure 7-G**).

By the end of 2018, there were early signs of recovery. Seedlings were starting to sprout and live plants put out leaves, but it will likely take 10–15 years for the mangrove forest to fully recover.

▲ **Figure 7-F** The mangroves support a wide variety of wildlife in the Everglades, including these white ibis.

Question

1. Roughly 7 million people live along the coastal areas buffered by the Everglades. What benefits do you think they derive from these swamplands?

NPS Video
Outside Science: Keeping an Eye on the Everglades
https://media.pearsoncmg.com/ph/esm/esm_mcknight_physgeo_13/media/

NPS Video
Climate Change: Everglades
https://media.pearsoncmg.com/ph/esm/esm_mcknight_physgeo_13/media/

▲ **Figure 7-21** A neighborhood in Houston, Texas, after the landfall of Hurricane Harvey on August 25, 2017.

In other regions of the world, hurricane devastation has been much greater. The flat deltas of the Ganges and Brahmaputra rivers in Bangladesh have been subjected to enormous losses of human life from Indian Ocean cyclones, killing 500,000 in 1970 and 175,000 in 1991. Cyclone Nargis struck Myanmar (Burma) in 2008, killing at least 130,000. In 2013, Super Typhoon Haiyan swept over the Philippines, killing more than 7000.

Heavy Rain and Flooding: Strong hurricanes can inflict heavy damage from flooding even after they weaken and move inland. For example, on August 25, 2017, Hurricane Harvey brought extensive flooding to southeastern Texas. Harvey made landfall as a Category 4 hurricane near Houston, Texas, then weakened but remained over the area for 5 days. Harvey dumped a record-breaking 1538 mm (60.58 inches) of rain in Nederland, Texas, which is 15 inches more than the average annual rainfall for the Texas coast. More than 17,000 people had to be rescued

Video
2005 Hurricane Season
https://media.pearsoncmg.com/ph/esm/esm_mcknight_physgeo_13/media/

◀ **Figure 7-22** Satellite image of Hurricane Katrina as it made landfall on August 29, 2005.

from their homes or flooded highways. Flood waters covered southeast Texas over an area the size of the state of New Jersey (see Figure 7-21). Hurricane Harvey is the second costliest hurricane on record for the United States, following Hurricane Katrina.

Hurricane Katrina: In August 2005, Hurricane Katrina created one of the greatest natural disasters in U.S. history. After developing in the southeastern Bahamas, Katrina moved over south Florida, weakened slightly, then intensified when it moved over the warm waters of the Gulf of Mexico. On August 28, it became a Category 5 storm with winds of more than 275 kilometers per hour (170 mph; 150 knots) and a central pressure of 902 mb. Katrina made its second landfall southeast of New Orleans on August 29 as a strong Category 3 storm (**Figure 7-22**).

Although the hurricane spawned 62 tornadoes, the greatest devastation came from Katrina's storm surge. New Orleans first flooded at about 7:00 A.M. when a 3.0- to 6.0-meter (10- to 20-foot) storm surge rolled up the Intercoastal Waterway from the Gulf into the city's Industrial Canal, overtopping and scouring levees and flooding the eastern parts of the city. Three hours later, a second flood hit the city when the London Avenue and 17th Street canals failed, sending water pouring into the central parts of the city from Lake Pontchartrain. By August 31, 80 percent of New Orleans was underwater—in some places 6 meters (20 feet) deep (**Figure 7-23**). The loss of life from Hurricane Katrina may never be known with certainty, but it is likely to have been more than 1500.

Many factors contributed to the Katrina disaster. New Orleans sits in a shallow "bowl" alongside the Mississippi River. Draining this once-swampy area over the last few hundred years has led to soil compaction and subsidence that left parts of the city below sea level, protected only by a series of levees and pumps. The wetlands of the Mississippi River delta once slowed and absorbed the punch from hurricane storm surges, but flood control efforts upstream have changed the sediment load of the river. Over the last few decades these marshlands on the lower delta have been sinking and eroding; they now offer little protection from storms. Finally, the levees themselves were inadequate.

Levees, floodwalls, and pumps have since been repaired and improved, but some scientists and local officials still wonder what will happen when the next great storm arrives.

▶ **Figure 7-23** New Orleans on August 30, 2005, after Hurricane Katrina.

Hurricane Michael: Hurricane Michael brought catastrophic destruction to the Florida Gulf Coast (Figure 7-24) on October 10, 2018. The maximum wind speed measured was 234 kilometers per hour (139 mph), before the sensor broke. Based on damage, winds were estimated at 257 kilometers per hour (160 mph), and the storm was updated to a Category 5 rating. Michael was the first Category 5 hurricane to make U.S. landfall since Andrew in 1992 and the first ever to strike the Florida panhandle. Wind and storm surge made Michael the most intense (based on level of destruction) hurricane to strike the United States since 1969, with more than $25 billion in damages. Part of the reason the damage total wasn't much higher is that the Florida panhandle coast does not have any big cities.

▲ **Figure 7-24** Destruction in Mexico Beach, Florida, (a) before Hurricane Michael and (b) after Michael's wind and storm surge destroyed houses and relocated boats. Arrows mark the same location in both photographs.

NASA Video
Hurricane Michael as Seen from Space
https://media.pearsoncmg.com/ph/esm/esm_mcknight_physgeo_13/media/

Hurricane Patricia: On October 23, 2015, after intensifying rapidly from the previous day, Hurricane Patricia made landfall in southwestern Mexico as a Category 5 storm. The strongest documented hurricane in the Western Hemisphere, it had a central pressure of 872 mb and winds reaching 345 kilometers per hour (215 mph). It weakened rapidly after landfall but brought flooding to central Mexico.

Destruction and tragedy are not the only legacies of hurricanes, however. Such regions as northwestern Mexico, northern Australia, and southeastern Asia rely on tropical storms for much of their water supply. Hurricane-induced rainfall is often a critical source of moisture for agriculture.

LearningCheck 7-12 **What causes a hurricane's storm surge?**

Hurricanes and Climate Change

The 2005 hurricane season in the North Atlantic was the most active on record, with 28 named storms. Three of the most powerful hurricanes ever measured in terms of minimum atmospheric pressure in the eye occurred that year: Katrina, Rita, and Wilma (Table 7-3). The 2012 season was

TABLE 7-3 Most Intense Hurricanes in the North Atlantic Basin[1] from 1851 to 2019 (based on lowest central pressure)

	Hurricane	Year	Minimum Pressure
1	Hurricane Wilma	2005	882 mb
2	Hurricane Gilbert	1988	888 mb
3	1935 Labor Day Hurricane	1935	892 mb
4	Hurricane Rita	2005	895 mb
5	Hurricane Allen	1980	899 mb
6	Hurricane Katrina	2005	902 mb
7	Hurricane Camille	1969	905 mb
	Hurricane Mitch	1998	905 mb
	Hurricane Dean	2007	905 mb
10	Hurricane Maria	2017	908 mb

[1]Typhoon Tip (1979; Western North Pacific) was the world's most intense documented storm, with a pressure of 872 mb. Hurricane Patricia (2015; Eastern Pacific) was the Western Hemisphere's most intense documented storm, with a pressure of 872 mb.

(Data source: U.S. Department of Commerce/NOAA.)

another very active year, tied with 1887, 1995, 2010, and 2011 for third overall, with 19 named storms. (The annual average is 11.)

The 2013–2014 *Fifth Assessment Report* (AR5) of the IPCC concluded that "[o]cean warming dominates the increase in energy stored in the climate system . . . " and that it is "virtually certain" (99–100% probability) that the upper 700 meters (2200 feet) of the ocean warmed between 1971 and 2010. Given the connection between high ocean temperature and hurricane formation, we could ask whether the recent increase in hurricane activity (the number of storms or their intensity) is tied to global climate change.

Number of Hurricanes: Over the last 25 years or so, the annual number of hurricanes in the North Atlantic has generally increased. However, many meteorologists think that this increase in frequency is part of the *Atlantic Multi-Decadal Signal*, a cycle of hurricane activity that has been well documented since the early 1900s. This pattern includes such factors as higher sea-surface temperatures (SST), lower vertical wind shear, and an expanded upper-level westward flow of the atmosphere off North Africa. Although the underlying causes of all components of the Atlantic Multi-Decadal Signal are not completely understood, the recent upswing in hurricane frequency can likely be explained without tying it to global warming. Meteorologists at NOAA's Climate Prediction Center have generally been forecasting a higher-than-average number of North Atlantic tropical storms since the early 2000s.

Intensity of Hurricanes: The IPCC's AR5 concluded that it is "virtually certain" (99 percent to 100 percent probability) that the activity of intense North Atlantic hurricanes has increased since 1970 and that further changes in the Western North Pacific and North Atlantic Ocean basins by the end of this century are "more likely than not" (greater than 50 percent to 100 percent probability). That said, the AR5 reports "low confidence" that there is a "human contribution" to these changes. In other words, hurricane activity and intensity has been increasing, but no definitive evidence ties them to anthropogenic climate change.

Further research may clarify the relationship between climate change and hurricane activity. However, given the clear trend toward warmer oceans and IPCC projections that the rate of sea-level rise this century will very likely exceed that of 1971–2010, the possibility of increased hurricane damage from storm surges in the future is a real worry.

Thunderstorms and Tornadoes

Several kinds of smaller atmospheric disturbances, such as *thunderstorms* and *tornadoes*, are common in some parts of the world. Although these events generally affect a more restricted area than tropical cyclones or midlatitude cyclones, such localized severe weather is frequently associated with both kinds of larger storms.

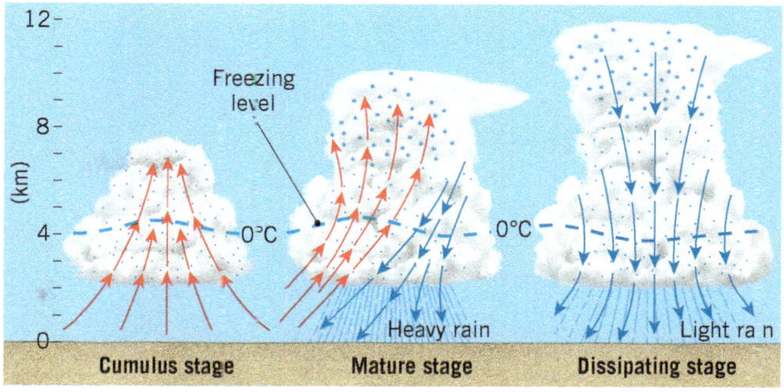

▲ **Figure 7-25** Sequential development of a thunderstorm cell. Red arrows show updrafts and blue arrows show downdrafts.

Thunderstorms

A **thunderstorm** is a localized, short-lived, and sometimes violent convective storm accompanied by thunder and lightning. It is always associated with vertical air motion, considerable humidity, and instability—a combination that produces a towering cumulonimbus cloud and nearly always intense precipitation.

Thunderstorms can occur as individual clouds, produced by nothing more than thermal convection—such developments are commonplace in the tropics and during summer in much of the midlatitudes. Thunderstorms also occur with larger kinds of storms or as a result of instability triggered by uplift. Thus, they often accompany hurricanes, fronts (especially cold fronts) in midlatitude cyclones, and orographic lifting that may produce instability.

Development: The uplift of warm, moist air releases latent heat of condensation that helps sustain rising air. In the early stage of a thunderstorm (**Figure 7-25**), called the *cumulus stage*, updrafts prevail and the cloud grows. As the rising air reaches the tropopause it spreads out to form an "anvil" top (see Figure 6-20). Above the freezing level, supercooled water droplets and ice crystals coalesce; when they become too large to be supported by the updrafts, they fall and melt. These falling particles drag air with them, initiating a downdraft. When the downdraft with its precipitation leaves the bottom of the cloud, the thunderstorm enters the *mature stage*, in which updrafts and downdrafts coexist as the cloud continues to enlarge. This stage is the most active, with heavy rain often accompanied by hail, blustery winds, lightning, and thunder. Eventually downdrafts dominate and the *dissipating stage* is reached, with light rain ending and turbulence ceasing. Ordinary (nonsevere) thunderstorms only last about an hour. Severe thunderstorms can last several hours.

Thunderstorms are most common where there are high temperatures, high humidity, and high instability—a combination typical of the intertropical convergence zone. Thunderstorm frequency generally decreases away from the equator and is virtually zero poleward of 60° of latitude (**Figure 7-26**). Thunderstorm frequency is much greater over land than water because temperatures are higher over land and most thunderstorms occur when it's warm. In the

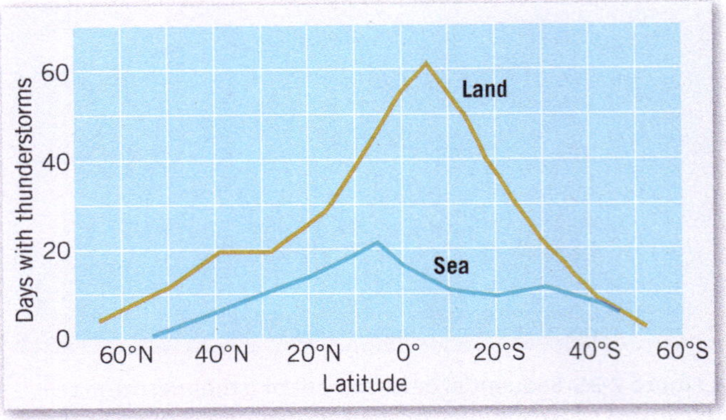

▲ Figure 7-26 Average number of days per year with thunderstorms, by latitude. Most thunderstorms are in the tropics. Land areas experience many more thunderstorms than ocean areas because land warms up much more in summer than water does.

▲ Figure 7-28 Lightning over Tucson, Arizona.

United States, the greatest annual thunderstorm activity occurs in Florida and the Gulf Coast, where moist, unstable air often prevails in spring and summer (Figure 7-27). The fewest thunderstorms occur along the Pacific coast, where cool water and subsidence from the subtropical high lead to stable conditions.

Downbursts: If the air beneath a thunderstorm is relatively dry when precipitation starts, then a *downburst* may form. Precipitation in the falling air evaporates and cools the air, which then falls very rapidly. When a downburst hits the ground, the wind spreads out, flattening everything in its path if the wind is strong enough. Downbursts include localized *microbursts* that are small in area but especially dangerous to aircraft during takeoff and landing. Much larger and long-lived wind storms known as *derechos* may also develop from the downburst of a rapidly moving thunderstorm. Derechos-producing storms can travel across several states, with winds in excess of 92 kilometers per hour (57 mph). Straight-line winds created by downbursts are not the same as the disorganized winds of a tornado, but they can be as dangerous.

LearningCheck 7-13 **Describe the sequence of development and dissipation of a typical thunderstorm.**

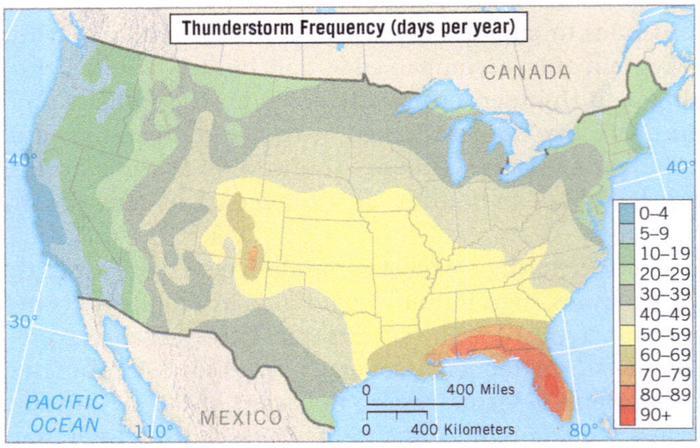

▲ Figure 7-27 Average number of days per year with thunderstorms in the conterminous United States.

Lightning: At any moment some 2000 thunderstorms exist over Earth. These storms produce about 6000 flashes of **lightning** every minute, or more than 8.5 million lightning bolts daily. A lightning flash heats the air along its path to as much as 28,000°C (50,000°F) and can develop 100,000 times the amperage used in household electricity (Figure 7-28). With such frequency and power, lightning clearly poses a significant potential danger for humanity. In the United States, on average about 27 deaths are blamed on lightning annually.

The first step in the sequence of events that leads to lightning discharge is the development of a large cumulonimbus cloud. Within the cloud, a separation of electrical charges occurs. Updrafts carry small positively charged water droplets or crystals upward while ice pellets and large drops fall, gathering negative charges and transporting them downward (Figure 7-29). The growing negative charge in the lower part of the cloud attracts a growing positive charge on

▲ Figure 7-29 Typical arrangement of electrical charges in a thunderstorm cloud. Positively charged particles are mostly high in the cloud, whereas negatively charged particles tend to be concentrated near the base.

◀ **Figure 7-30** This EF-2 tornado near McCook, Nebraska, on May 17, 2019, travelled through farmland for 6 miles.

Earth's surface immediately below. The contrast (the *electric potential*) between the cloud base and ground surface builds to tens of millions of volts before the insulating barrier of air that separates the charges is overcome.

Finally, a finger of negative current flicks down from the cloud and meets a positive charge darting upward from the ground. This makes an electrical connection of ionized air from cloud to ground, and the first lightning flash surges downward. Other flashes may follow until all or most of the negative charges have been drained out of the cloud base.

In addition to such ground-to-cloud discharges, less spectacular but more frequent lightning is exchanged between adjacent clouds or between the upper and lower portions of the same cloud.

Thunder: A lightning bolt abruptly heats the air, which expands fast enough to break the sound barrier and produce **thunder**. The lightning and thunder occur simultaneously, but we perceive them at different times. Lightning is seen at essentially the instant it occurs because its image travels at the speed of light. Thunder, however, travels at the much slower speed of sound. Thus, you can estimate the distance of a lightning bolt by timing the interval between sight and sound: a 3-second delay means the lightning strike was about 1 kilometer away; a 5-second interval indicates that the lightning flash was about 1 mile away. Rumbling thunder is indicative of a long lightning trace some distance away, with one portion being nearer to you than another. If you hear no thunder, the lightning is far away—probably more than 20 kilometers (a dozen miles).

Tornadoes

Although very small and localized, the **tornado** is one of the most destructive of all atmospheric disturbances. It is the most intense vortex in nature: a deep low-pressure cell surrounded by a violently whirling cylinder of wind (**Figure 7-30**). Primarily due to better forecasting and emergency broadcasts, the average annual death toll from violent tornadoes in the United States has declined over the last 50 years. The 551 deaths in 2011 were a tragic exception (**Figure 7-31**)—see the box *Global Environmental Change: Are Tornado Patterns Changing?*

Tornadoes are tiny, concentrated rotations caused by severe thunderstorms. Generally less than 400 meters (a quarter of a mile) in diameter, they have the most extreme pressure gradients known—as much as a 100-millibar difference from the center of the tornado to the air immediately outside the funnel. This extreme pressure difference produces winds of extraordinary speed, up to 480 kilometers (300 miles) per hour. Air sucked into the vortex also rises at an inordinately fast rate. These storms are of such incredible power, it is no wonder that one was invoked by author Frank Baum, and later by Hollywood, to transport a little girl and her dog to the Land of Oz.

▲ **Figure 7-31** Damage in Joplin, Missouri, two days after an EF-5 tornado devastated the city in May 2011.

Global Environmental Change

Are Tornado Patterns Changing?

• **Ted Eckmann**, University of Portland

Tornado patterns may be changing. Tornado numbers have recently shattered long-standing records. However, some years have lots of tornadoes, and others have very few (**Figure 7-H**). The deadliest tornado outbreak since modern recordkeeping began occurred in 2011, with 343 confirmed tornadoes in just a few days (**Figure 7-I**), more than the total number sometimes reported for an entire year! But 2012 was one of the weakest tornado seasons in history.

How We Study Tornado Trends: Post-storm damage surveys provide the main evidence of tornadoes and their strengths, although reports from trained storm spotters and the public provide important information as well. Doppler weather radar alone does not confirm tornadoes, but installation of the U.S. Doppler network in the 1990s, plus advances since that deployment, dramatically improved tornado warnings and detection. Prior to these radar advances, some small tornadoes in unpopulated areas likely went unnoticed. However, the radar network now covers most of the United States, and tornado recording has likely improved. Increases in population and in the numbers of storm spotters have also made tornado detection more likely, so it's not surprising that the number of *recorded* tornadoes has increased dramatically over the last few decades. To see whether the *actual* number of tornadoes has increased, researchers tend to analyze trends in only larger tornadoes, which probably would have been noticed and recorded even before the Doppler network was installed.

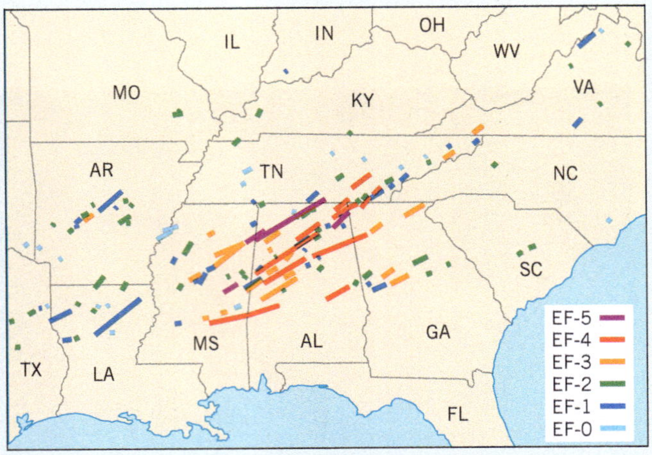

▲ **Figure 7-I** Approximate paths of tornadoes from April 25–28, 2011, color-coded by maximum intensity reached along the path.

More Outbreaks: The overall number of large tornadoes per decade has not changed significantly, but outbreaks, such as the one in 2011, are becoming more common. These gigantic events overwhelm the resources of emergency responders. Outbreaks don't happen every year, but when they do, they're deadly: by the end of 2011, tornadoes had killed 551 people in the United States—the largest annual total in 66 years of modern records. More than half of those deaths occurred during the April outbreak. Tornado patterns are thus changing in ways that make outbreaks increasingly common and increasingly deadly.

What Is Causing the Changes? Thunderstorms tend to be more frequent and more powerful when it's hot and humid at the surface. Due to global climate changes, warming surface temperatures are increasing evaporation rates, so the air is becoming hotter and more humid in many tornado-producing regions. Global climate changes also appear to be decreasing the number of days with adequate wind shear for tornadic formation. (Thunderstorms need wind shear to produce tornadoes.) This fits observed changes in tornado patterns: fewer days with tornadoes, but on days when there is enough wind shear, increased tendency for outbreaks.

Although around three-quarters of all tornadoes occur in the United States, tornadic activity is also changing elsewhere, such as in Australia, Europe, and Japan. However, tornado records from other regions are not as thorough as in the United States. Researchers are still working to understand the changes in tornado patterns, along with the causes of these changes.

Questions

1. Use Figure 7-H to explain why stronger tornadoes were more likely to be observed than the weakest ones even before Doppler weather radar existed.
2. How does the change in tornado patterns explain the recent rise in the death rate from tornadoes in the United States?

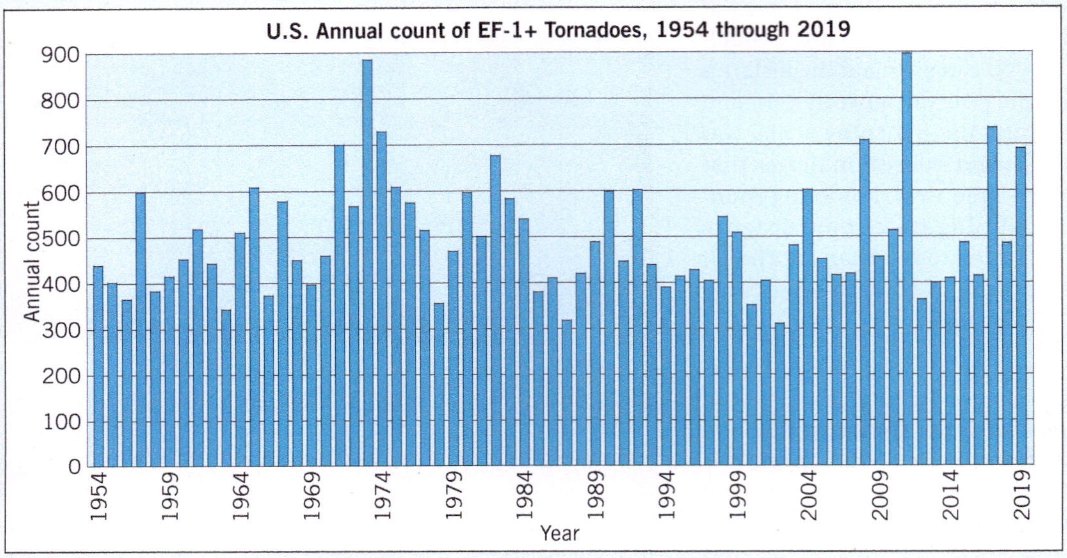

◀ **Figure 7-H** Annual number of moderate or stronger (EF-1+) tornadoes in the United States, 1954–2014. (See Table 7-4 for a description of the EF Scale for tornadoes.)

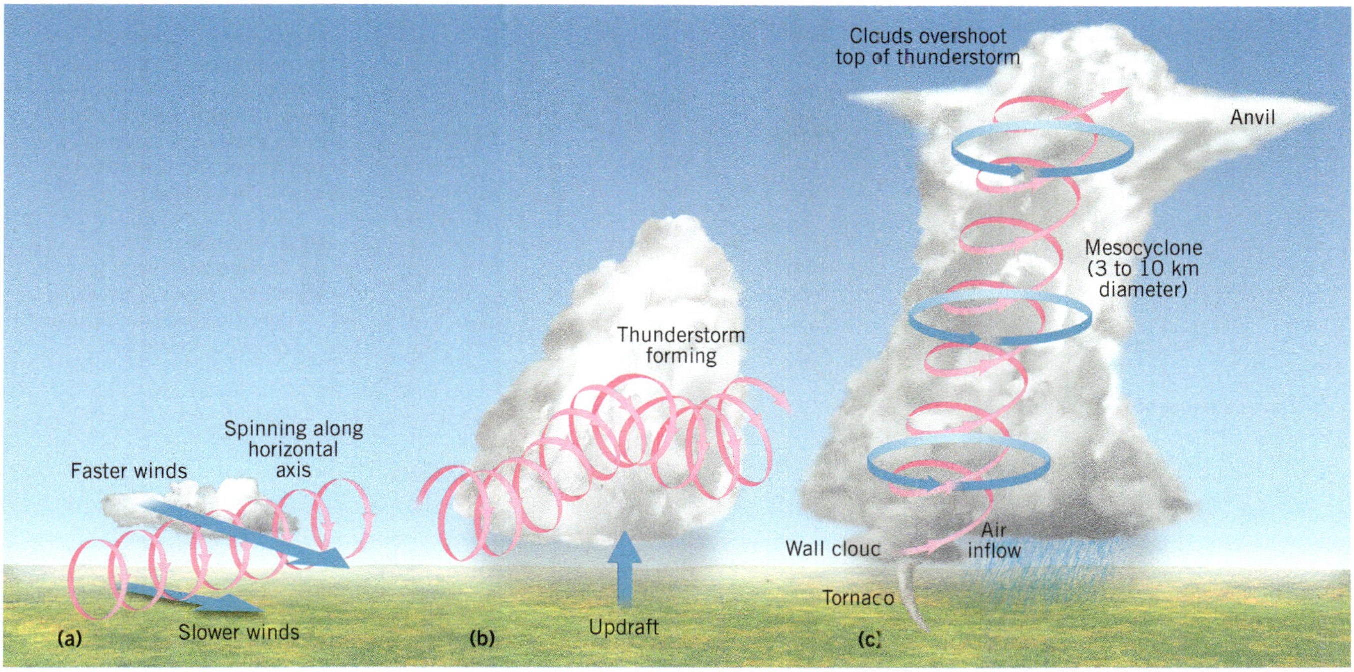

▲ **Figure 7-32** The formation of a mesocyclone often precedes tornado formation. (a) Winds are stronger aloft than at the surface; the wind shear produces a rolling motion around a horizontal axis. (b) Strong thunderstorm updrafts tilt the horizontally rotating air to a nearly vertical alignment. (c) The mesocyclone, a vertical cylinder of rotating air, is established. If a tornado develops it will descend from a slowly rotating wall cloud in the lower portion of the mesocyclone.

Funnel Clouds: Tornadoes usually originate a few hundred meters above the ground, the rotating vortex becoming visible when water vapor condenses into a rotating **funnel cloud**. Once rotation reaches the ground, it is a tornado. The tornado advances along an irregular track that generally extends from southwest to northeast in the United States. Sometimes the tornado sweeps along the ground, devastating everything in its path, but its trajectory is usually twisting and dodging; at frequent intervals the tornado may lift off the ground and then touch down again nearby.

Typically, tornadoes have damage paths 50 meters (about 150 feet) wide, move at about 56 kilometers (35 miles) per hour, and last for only a few minutes. Extremely destructive ones may be more than 1.5 kilometers (1 mile) wide, travel at 95 kilometers (60 miles) per hour, and remain on the ground for more than an hour, with maximum longevity recorded at 3.5 hours.

The dark, twisting funnel of a tornado contains not only cloud but also sucked-in dust and debris. Damage is caused largely by the strong winds, flying debris, and swirling updraft. The old advice of opening a window when faced with an approaching tornado (supposedly to reduce an abrupt drop in pressure when the center of the storm passes over a closed building) is no longer given—and in fact opening a window may *increase* the chance of injury from flying debris. Likewise, the notion to seek shelter under an overpass is wrong. The overpass actually increases the wind speed, making it more dangerous for shelter-seekers.

Tornado Formation: As with many other kinds of storms, the exact mechanism of tornado formation is not well understood. Tornadic thunderstorms can develop in the warm, moist, unstable air associated with a midlatitude cyclone, such as precedes a rapidly advancing cold front or along the cold front itself. They can also develop in association with tropical cyclones. Virtually all tornadoes are generated by severe thunderstorms. The basic requirement is vertical wind shear—a significant change in wind speed or direction from the bottom to the top of the storm.

On a tornado day, low-level winds are southerly, and jet stream winds are southwesterly. This difference in direction (directional shear) causes the thunderstorm updraft to rotate. The conventional updrafts that become thunderstorms reach several kilometers up into the atmosphere, and an increase in wind speed with height can cause air to roll along a horizontal axis (**Figure 7-32**). Strong updrafts in such a rapidly maturing *supercell* thunderstorm may then tilt this rotating air vertically, developing into a **mesocyclone** and *wall cloud*, with a diameter of 3 to 10 kilometers (2 to 6 miles). Less than one quarter of all mesocyclones result in a tornado.

Spring and early summer are favorable for tornado development because of the sharp air-mass contrasts present in the midlatitudes at that time; a tornado can form in any month, however. Most occur in midafternoon, at the time of maximum heating.

Tornadoes do occur in the midlatitudes and subtropics, but more than 75 percent are reported in the United States

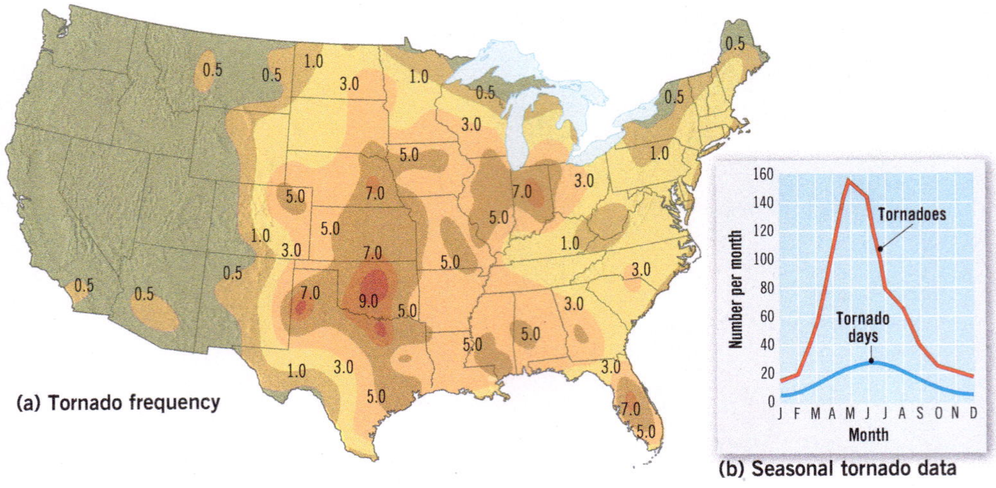

◀ **Figure 7-33** (a) Map showing the average number of tornadoes per 26,000 square kilometers (10,000 square miles) over a 27-year period. (b) The chart shows the seasonality of tornadoes and tornado days in the United States.

Animation
Tornadoes
https://media.pearsoncmg.com/ph/esm/esm_mcknight_physgeo_13/media/

(Figure 7-33). Such concentration reflects optimum environmental conditions, with the relatively flat terrain of the central and southeastern United States providing an unhindered zone of interaction between prolific source regions for Canadian cP and Gulf mT air masses. Although between 800 and 1200 tornadoes are recorded annually in the United States, the actual total may be considerably higher because many small tornadoes that occur briefly in uninhabited areas are not reported.

LearningCheck 7-14 **Why is wind shear required for tornado formation?**

Strength: The strength of a tornado is commonly described using the **Enhanced Fujita Scale** (**EF Scale**; Table 7-4), named after the late University of Chicago meteorologist Theodore Fujita. The EF Scale is based on estimates of 3-second gust wind speeds as determined by observed damage after a tornado. About 69 percent of all tornadoes in the United States are classified as *light* or *moderate* (EF-0 or EF-1), about 29 percent are classified as *strong* or *severe* (EF-2 or EF-3), and 2 percent are classified as *devastating* or *incredible* (EF-4 or EF-5).

Waterspouts: Similar-appearing funnels over the ocean, called **waterspouts**, have a lesser pressure gradient, gentler winds, and reduced destructive capability. Waterspouts can form on a clear, sunny day.

Severe Storm Watches and Warnings

Forecasting severe thunderstorms and tornadoes has improved remarkably over the last few decades, largely due to a better understanding of these events, better computer simulations, and more sophisticated observing technology—see the box *Focus: Weather Radar.*

TABLE 7-4 Enhanced Fujita Scale (EF Scale) for Tornadoes

	Enhanced F Scale		
EF Scale	**3-second gust km/hr**	**3-second gust mph**	**Expected Level of Damage**
0	105–137	65–85	Light: Broken tree branches; uprooted small trees; billboards and chimneys damaged.
1	138–177	86–110	Moderate: Roof surfaces peeled off; mobile homes overturned or pushed off their foundations.
2	178–217	111–135	Strong: Mobile homes destroyed; roofs blown off wooden-frame houses; uprooted large trees; lightweight objects become "missiles."
3	218–266	136–165	Severe: Trains derailed or overturned; walls and roofs torn from well-constructed houses; heavy cars thrown off ground.
4	267–322	166–200	Devastating: Structures with weak foundations blown for some distance; well-built houses destroyed; large objects become missiles.
5	Over 322	Over 200	Incredible: Well-built houses lifted off foundations and carried considerable distance before complete destruction; tree bark removed; automobile-sized missiles carried more than 100 meters.

Focus

Weather Radar

• **Steve Stadler,** Oklahoma State University

Radar was developed shortly before World War II to detect ships and aircraft. Radar operators noticed that liquid water and ice reflect and refract these microwave transmissions. The strength of these returns and changes over short times provide a remotely sensed look at important weather characteristics, such as type and rate of precipitation and motions in weather disturbances.

The use and understanding of radar have enabled timely warnings of weather events. The United States is overlain by the radar beams from more than 150 governmental radars known as NEXRAD (NEXt-generation RADar).

NEXRAD: NEXRAD transmits microwave pulses through the atmosphere from inside a geodesic dome mounted on a tower (**Figure 7-J**). This network uses the *Doppler effect* to sense motion toward and away from the radar site and thereby determine a storm's internal structure. Returns from parts of the storm moving toward the radar have shorter wavelengths than the transmitted wavelengths; returns from parts moving away from the radar have longer wavelengths. You have experienced the audio version of the Doppler effect if you have ever stopped at a railroad track as a speeding train passes. As the train approaches, the pitch of its whistle seems to get higher. After the train passes, the whistle's pitch seems to get lower. In reality, the pitch does not change.

▲ **Figure 7-J** NEXRAD tower and radar dome.

NEXRAD has undergone improvements since its deployment in the 1990s. For instance, the radar now has *dual polarization*, transmitting simultaneous horizontal and vertical pulses. This allows us to distinguish among rain, sleet, hail, and ice crystals for better short-term forecasting.

Tracking Severe Storms—Thunderstorms and Tornadoes: NEXRAD can slice through storms at various angles and allows a storm to be considered in three dimensions. A crucial use of NEXRAD is tornado detection. In 2013 a thunderstorm containing an EF-5 tornado passed over Moore, Oklahoma. Twenty-four people were killed, and many dozens injured. A reflectivity scan from the NEXRAD radar site shows how severe this tornado was (**Figure 7-K**). More intense colors represent higher reflectivity from very large raindrops and hail.

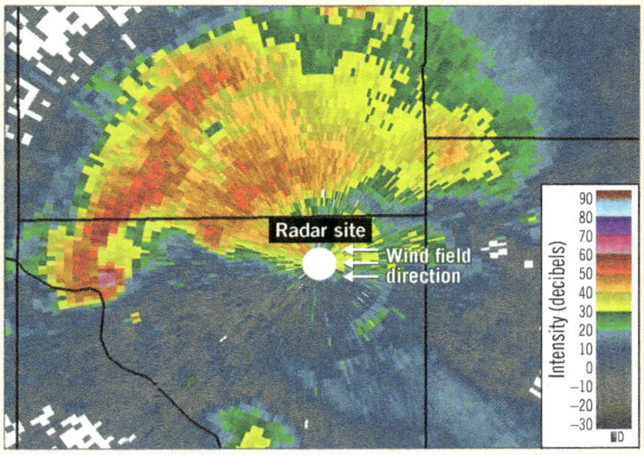

▲ **Figure 7-K** NEXRAD reflectivity display as a tornado nears Moore, Oklahoma, 25 kilometers (15 miles) west of the radar tower. The reflectivity intensities correspond to sizes of water particles and, in this case, debris. The tornado itself is not visible. Data courtesy of NWS; displayed on Oklahoma Mesonet software.

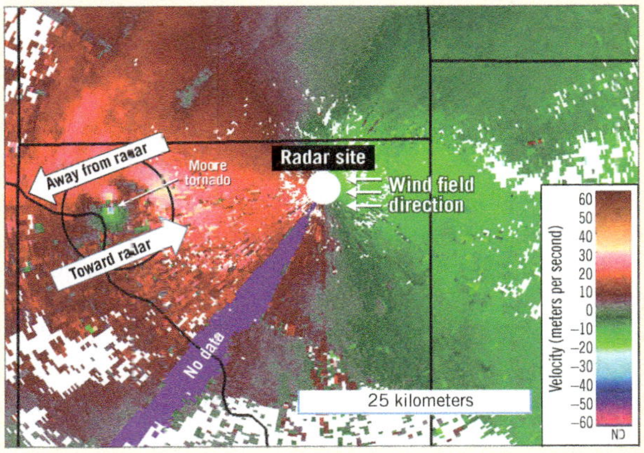

▲ **Figure 7-L** Image taken at the same time as Figure 7-K. Color indicates the velocity of motion (1 meter per second = 2.2 miles per hour). The strongest velocities are within the circle and about to rip into Moore. The closeness of reds and greens implies a strong counterclockwise mesocyclone, parent to the tornado below. Data courtesy of NWS; displayed on Oklahoma Mesonet software.

Tornado funnels are usually only a few hundred meters across. They cannot be detected far from the tower because radar beams are transmitted in straight paths and Earth's surface is curved; thus the radar overshoots the funnel itself. However, NEXRAD can peer into tall thunderstorms to detect horizontal motion (**Figure 7-L**). Like Figure 7-K, Figure 7-L shows the Moore storm—but also uses NEXRAD's motion-detection capability. Greens indicate motion toward the radar; reds, motion away from it. The proximity of intense reds and greens implies strong counterclockwise rotation. This large area of rotation is a *mesocyclone*, the most frequent immediate parent to a tornado. Mesocyclones form before tornadoes can touch down. We cannot forecast a tornado's exact path, but based on the Figure 7-K display, the National Weather Service issued a tornado warning 16 minutes before this tornado touched down, doubtless saving many lives.

Questions

1. Why can't radar predict tornadoes a day in advance?
2. How does Figure 7-L better capture a tornado's probable position more precisely than Figure 7-K?
3. In Oklahoma, trained storm spotters are deployed on days with a chance of tornadoes. Why are human eyes still necessary even though the state is well covered by radar?

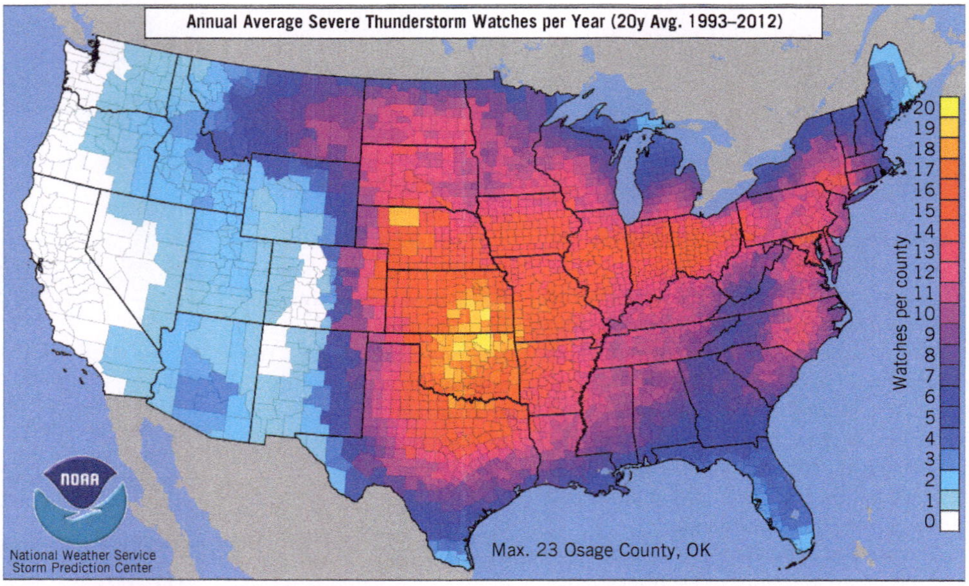

▶ **Figure 7-34** Map showing the average number of severe thunderstorm watches issued from 1993 to 2012.

The U.S. National Weather Service provides two types of advisories when severe weather, such as a strong thunderstorm or tornado, is likely: *watches* and *warnings*.

A severe **storm watch** is an advisory issued for a region (perhaps the area of one or more counties) where over the next four to six hours the conditions are favorable for the development of severe weather (Figure 7-34). A watch means only that the conditions are right for severe weather, not that such weather has developed or is impending.

In contrast, a severe **storm warning** is issued by a local weather forecasting office when a severe thunderstorm or tornado has actually been observed. Such warnings are often accompanied by civil defense sirens and emergency broadcasts on radio, cell phones, and television. When a warning is given, the public is directed to take immediate action, such as seeking safety in a storm shelter.

Weather forecasting: Data come from weather satellites (see *Focus: GOES Weather Satellites*, page 166), Doppler radar (see *Focus: Weather Radar*), surface observation stations, and weather balloons. More than 900 Automated Surface Observing Systems (ASOS) collect surface weather data across the United States. Radiosondes (weather balloons) are launched twice a day from almost 900 sites worldwide.

The data values, along with the previous forecast, are then entered into a computer simulation that is comprised of a set of complicated equations to predict the future state of the atmosphere. Forecasts are usually pretty good out to a couple days. As we look further into the future, the chaotic nature of the atmosphere, the presence of atmospheric processes we do not understand, and the presence of small variations that are not captured in the data make forecasts less reliable.

7 Learning Review

After studying this chapter, you should be able to answer the following questions. Key terms from each text section are shown in **bold type.** Definitions for key terms are also found in the glossary at the back of the book.

Key Terms and Concepts

Air Masses (*p. 180*)

1. What is an **air mass**, and what conditions are necessary for one to form?
2. What regions of Earth are least likely to produce air masses? Why?
3. Contrast and explain the moisture and temperature characteristics of a mT (maritime tropical) air mass with that of a cP (continental polar) air mass.

Fronts (*p. 181*)

4. What is the relationship of air masses to a **front**?
5. What is a **cold front**? What is a **warm front**?
6. What is a **stationary front**?

Midlatitude Cyclones (*p. 184*)

7. Describe the pressure and wind patterns of a **midlatitude cyclone**.

8. Describe the locations of fronts and the surface "sectors" of a mature midlatitude cyclone.
9. Describe and explain the regions of cloud development and precipitation within a midlatitude cyclone.
10. Discuss the four components of movement of a midlatitude cyclone.
11. Explain the process of **occlusion**.
12. Why does an **occluded front** usually indicate the "death" of a midlatitude cyclone?
13. Discuss the **cyclogenesis** of midlatitude cyclones. What is the relationship between upper-level airflow and the formation of surface disturbances in the midlatitudes?
14. Describe and explain the changes in wind direction, atmospheric pressure, sky conditions (such as clouds and precipitation), and temperature with the passing of a cold front of a midlatitude cyclone.

Midlatitude Anticyclones (*p. 189*)

15. Describe the pressure pattern, wind direction, and general weather associated with a **midlatitude anticyclone**.
16. How are midlatitude anticyclones often associated with midlatitude cyclones?

Tropical Cyclones: Hurricanes (*p. 191*)

17. What is an **easterly wave**?
18. Distinguish among a **tropical depression**, a **tropical storm**, and a **hurricane**.
19. Describe and explain the pressure and wind patterns of a **tropical cyclone** (hurricane).
20. Discuss the characteristics of the **eye** of a hurricane.
21. What is **wind shear**?
22. Discuss the conditions necessary for a hurricane to form.
23. Describe and explain the typical paths taken by hurricanes in the North Atlantic Ocean basin.
24. Why do hurricanes weaken when they move over land?
25. Briefly explain the **Saffir-Simpson Hurricane Scale**.
26. What is a hurricane **storm surge**, and what causes one?
27. Do hurricanes have any beneficial effects? Explain.

Thunderstorms and Tornadoes (*p. 199*)

28. Discuss the general sequence of **thunderstorm** development and dissipation.
29. What is the relationship of **thunder** to **lightning**?
30. Describe the wind and pressure characteristics of a **tornado**.
31. What is a **funnel cloud**?
32. Discuss the general formation of a tornado from a supercell thunderstorm and **mesocyclone**.
33. Briefly explain the **Enhanced Fujita Scale (EF Scale)** for tornadoes.
34. What is the difference between a tornado and a **waterspout**?
35. What is the difference between a **storm watch** and a **storm warning**?

Study Questions

1. Why is an air mass unlikely to form over the Rocky Mountains of North America?
2. Why are maritime polar (mP) air masses from the Atlantic Ocean less important to the United States than mP air masses from the Pacific Ocean?
3. Explain why clouds develop along cold fronts and warm fronts.
4. Why do midlatitude cyclones develop in the midlatitudes but not in the tropics?
5. Why are there no fronts in a midlatitude anticyclone?
6. Why are there no fronts in a hurricane?
7. Why are tropical cyclones common along the east coasts of continents in the midlatitudes but not along the west coasts?
8. Why is it very unlikely that a hurricane could cross the equator?
9. Why are thunderstorms more common over land than over water?
10. Why are thunderstorms sometimes called "convective storms"?

Exercises

1. If you see a flash of lightning and you hear thunder 20 seconds later, how far away are you from the lightning in the thunderstorm? ______ miles
2. You see a flash of lightning, and you hear thunder 20 seconds later. Four minutes after the first flash of lightning, you see another flash from the same storm, but the thunder arrives in only 15 seconds. How fast is the thunderstorm moving toward you? ______ miles per hour
3. Look at the map of thunderstorm activity in the United States (Figure 7-27). Explain why the west coast of California has so little thunderstorm activity, while Florida has so much.

Environmental Analysis TORNADO CLIMATOLOGY

Data
Tornado Climatology
https://media.pearsoncmg.com/ph/esm/esm_mcknight_physgeo_13/media/

The National Oceanic and Atmospheric Administration's National Centers for Environment Information collect tornado data to create a climatology of tornadoes. This climatology includes information about location and strength as well as information about the deadliest tornadoes.

Activities

Go to www.windows2universe.org, the Windows to the Universe site. Type "world map of tornadoes" into the "Search" field, then select "World Map of Tornado Occurrence and Agriculture Production."

1. Tornadoes tend to form in agricultural regions. What conditions make agricultural regions favorable for tornadoes?

Go to the NOAA site at www.ncdc.noaa.gov. Select "Climate Information"; click on "Extreme Events" and then "Tornado Climatology."

2. Globally, which latitudes are most likely to have tornadoes? Why do tornadoes occur at these latitudes?
3. Which parts of the world are most prone to tornadoes?
4. Which country has the largest number of tornadoes per year?
5. Which U.S. state has the largest number of tornadoes per year? Which state has the fewest?

Click on "Recent Tornado Reports and Information."

6. How many tornadoes occurred in the United States during the most recent month? How does this number compare with the average number of tornadoes for that month?

Go back to NOAA's U.S. Tornado Climatology website and click on "Historical Records and Trends."

7. In your region, at what time of day do most tornadoes occur?

Go back to NOAA's U.S. Tornado Climatology website and click on "Deadliest Tornadoes."

8. With the exception of 2011, none of the deadliest tornadoes occurred in the past 50 years. Why are more recent tornadoes less deadly than those of the past?

Go to https://www.weather.gov/media/publications/assessments/Joplin_tornado.pdf, the NWS summary of the May 22, 2011, Joplin tornado. Read the Executive Summary section of this report.

9. How did most residents of Joplin, Missouri, respond to the tornado warning? How should they have responded?
10. What suggestions did the report make that could help reduce fatalities during strong and violent tornadoes in the future?

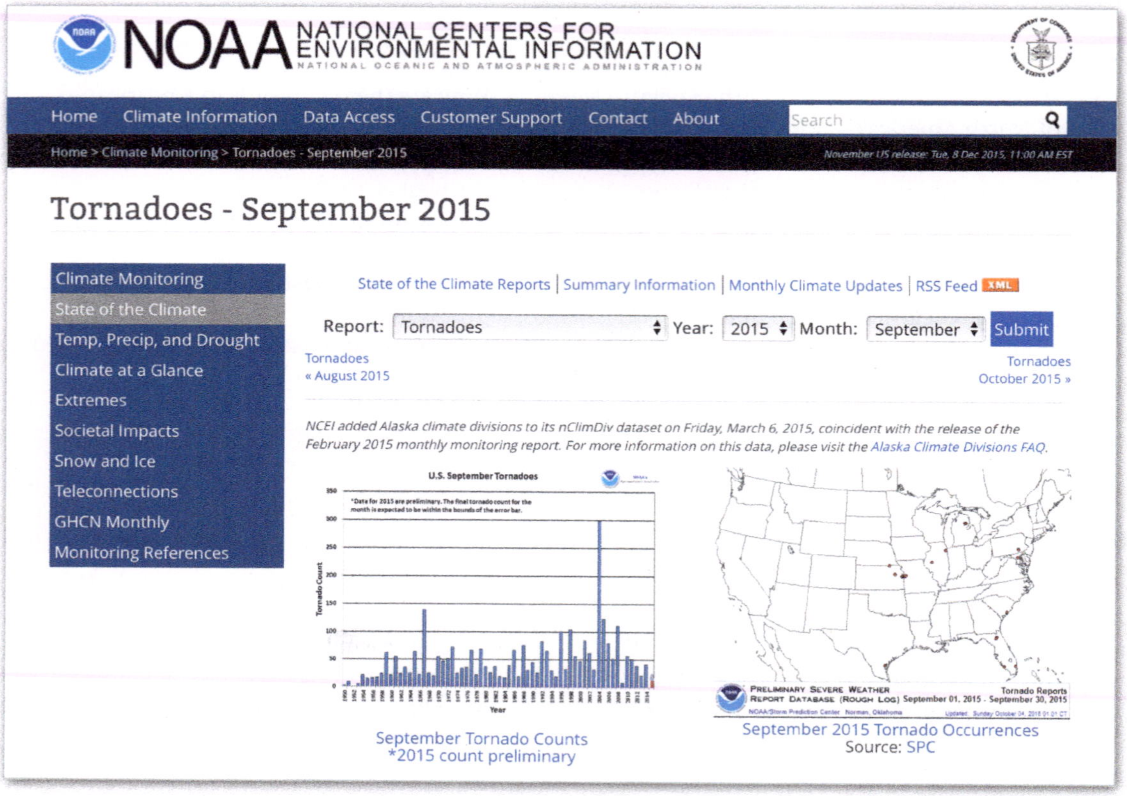

Seeing Geographically in Our National Parks

Look again at the photograph of the thunderstorm at the beginning of the chapter (p. 178). How might the topography of this region influence the likelihood of thunderstorms? Why are spring and early summer the most common times for thunderstorms? What additional ingredient would need to be added to increase the likelihood of thunderstorms in the Badlands?

Mastering Geography®

Looking for additional review and test prep materials? Visit the Study Area in **Mastering Geography** to enhance your geographic literacy, spatial reasoning skills, and understanding of this chapter's content by accessing a variety of resources, including Map Master interactive maps, geoscience animations, *Mobile Field Trips*, videos, *Project Condor* Quadcopter videos, *In the News* RSS feeds, flashcards, web links, self-study quizzes, and an eText version of *McKnight's Physical Geography*.

Seeing Geographically in Our National Parks

Joshua trees have roots that can extend up to 11 meters (36 feet). How might these unusual trees, native to Joshua Tree National Park in California, provide information about the region's climate? Does this area have abundant rainfall? What do you see that supports your answer? What environmental factors might account for this?

Joshua Tree National Park, CA

Climate and Climate Change

8

HAVE YOU EVER WONDERED why the landscape and vegetation in Joshua Tree National Park are so different from other locations? Or whether we will always have these beautiful places to enjoy or will climate change take them away? The answers to these questions are at the heart of one of the most important environmental issues of our day and the topic of this culminating chapter in our study of the atmosphere: understanding climate and climate change.

We now turn from our focus of the last few chapters on *weather*, which consists of the short-term patterns of the atmosphere, to *climate*, which describes the long-term patterns of weather. Recall that the climate of a location doesn't refer to only the "average" weather conditions over several decades, but also takes into account weather extremes that we expect to happen from time to time. This expected variation in weather from year to year complicates our tasks of both defining the climate of a region and recognizing when that climate is changing.

In this chapter we begin by describing the classification and distribution of global climate types—discussing climate largely as it has been over the last few thousands of years. We conclude the chapter with a discussion of climate change: recognizing, explaining, and anticipating the consequences of a climate system in a state of flux.

As you study this chapter, think about these **Key**Questions:

- **What information do you need to classify the climate of a location?**
- **Why are the tropics so warm and rainy?**
- **Why do deserts form where they do?**
- **Why do some midlatitude locations have mild winters, whereas others have severe, cold winters?**
- **Why are polar regions so cold and dry?**
- **How do we know what climate was like in the past?**
- **What can cause climate change?**
- **What are some of the possible consequences of present-day climate change?**

Climate Classification

Describing the geographic distribution of many things—population density, voting patterns, or a host of other tangible things or relationships—is relatively easy. Climate, however, is the product of several elements that are, for the most part, continuously and independently variable. This makes the task of describing climate distribution quite complex.

Temperature, for example, is one of the simplest climatic elements to describe, yet in this book we use more than 40 maps and diagrams to convey various temperature patterns around the world: daily, monthly, and annual averages, as well as temperature ranges and extremes. Climate involves almost continuous variation from place to place, not only of temperature but also many other factors.

Adding to the complexity of understanding climate distribution is the fact that climate changes over time. The geologic record reveals times in both the distant and recent past when Earth's climate was very different from today—such as the great ice ages of the Pleistocene epoch, which ended about 10,000 years ago. (The Pleistocene is discussed in Chapter 19.) More important for us, however, are indications that climate change is under way today as a result of human activity.

Our first task is to explore a system of climate classification that helps us simplify, organize, and generalize a vast array of data. To illustrate how a classification system can help us understand global climate patterns, suppose that a geography student in Atlanta, Georgia, is asked to describe the climate of southeastern China. A world map showing the distribution of climatic regions (such as the map in Figure 8-3) reveals that southeastern China has the same climate as the southeastern United States. So, the student's familiarity with the climate in Georgia provides a basis for understanding the general characteristics of the climate of southeastern China.

The Köppen Climate Classification System

The **Köppen climate classification system** is by far the most widely used modern system. Wladimir Köppen (1846–1940; pronounced "kur-pin" with a silent r) was a Russian-born German climatologist who was also an amateur botanist. The first version of his climate classification scheme appeared in 1884. He continued to modify and refine it for the rest of his life, the last version being published in 1936.

The system is based on only the average annual and average monthly values of temperature and precipitation, combined and compared in a variety of ways. Consequently, the necessary statistics are easily acquired. Data for any location (called a *station*) on Earth can be used to determine the precise classification of that place, and the geographical extent of any recognized climatic type can be determined and mapped. This means that the classification system is functional at both the local and the global scale.

Köppen defined four of his five major climatic groups primarily by temperature characteristics, and the fifth (the B group) on the basis of moisture. He then subdivided each group into climate types according to various temperature and precipitation relationships.

Köppen was unsatisfied with his last version and did not consider it a finished product. Many geographers and climatologists have used the Köppen system as a springboard to devise systems of their own or to modify Köppen's classifications. The system of climate classification used in this book is properly called the *modified Köppen system*. It encompasses the basic design of the Köppen system but with a variety of minor modifications. Many such changes follow the lead of the late University of Wisconsin geographer Glen Trewartha.

The modified Köppen system describes five major climate groups (groups A, B, C, D, and E), which are subdivided into a total of 14 individual climate types, along with the special category of highland (H) climate (Figure 8-1). It is important

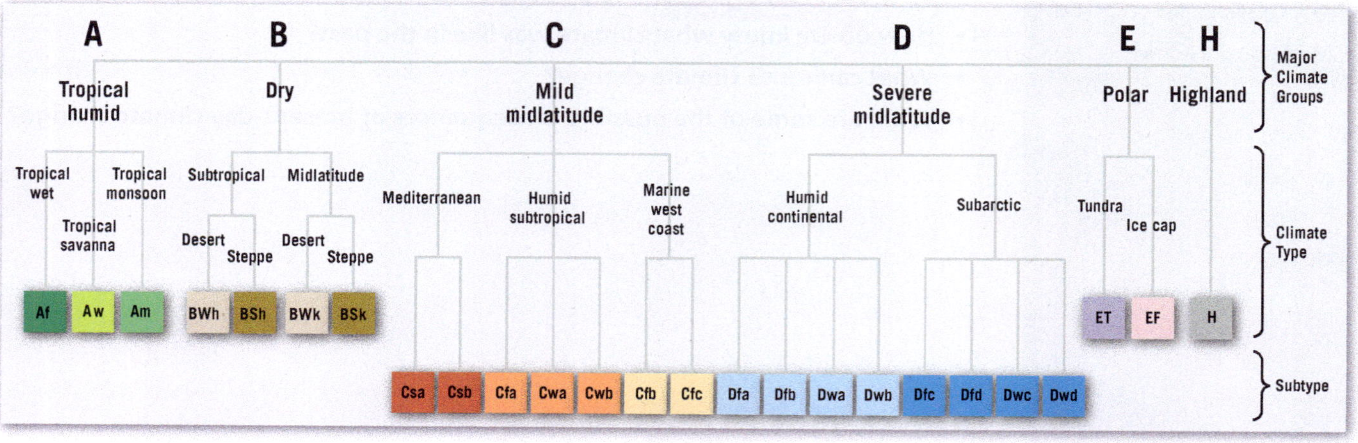

▲ Figure 8-1 The modified Köppen climatic classification. There are five major climate groups (and the special category of Highland) and 15 individual climate types.

to keep in mind that there are rarely sharp boundaries between climate types; instead, one type transitions into another. We use the Köppen system here simply as a tool to help us comprehend the general pattern of world climates.

LearningCheck 8-1 **What kinds of data are used to classify climates in the Köppen system? (Answer on p. AK-3)**

Köppen Letter Code System: In the modified Köppen system, each climate type is designated by a descriptive name and by a series of letters defined by specific temperature and/or precipitation values. The first letter designates the major-climate group, the second letter usually describes precipitation patterns, and the third letter (if any) describes temperature patterns (Table 8-1). (See Appendix V for the exact definitions used for classification.)

Although the letter code system may seem complicated, it provides a shorthand method for summarizing key characteristics of each climate. For example, if we look for the definitions of the letters in *Csa*, one of the letter code combinations for a *mediterranean climate*, we see that:

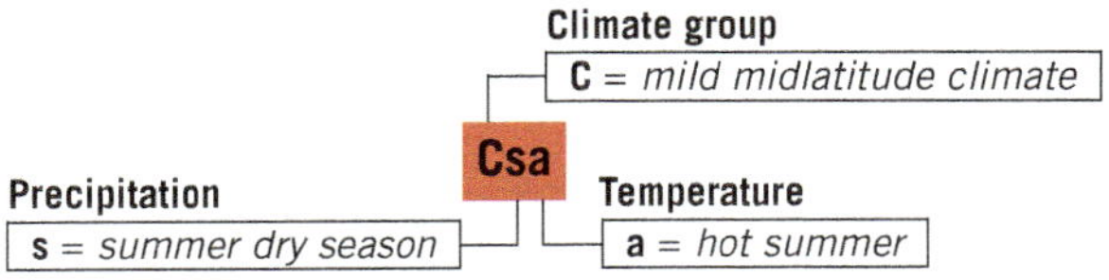

As we see later in this chapter, this is a very good synopsis of the distinctive characteristics of this climate type.

Climographs

Probably the most useful tool in a general study of world climatic classification is a simple graphical representation of monthly temperature and precipitation for a specific weather station. Such a graph is called a **climograph** or *climatic diagram* (Figure 8-2). A climograph has 12 columns, one for each month, with a temperature scale on the left side and a precipitation scale on the right side. Average monthly temperatures are connected by a curved line in the upper portion of the diagram, and average monthly precipitation is represented by bars extending upward from the bottom.

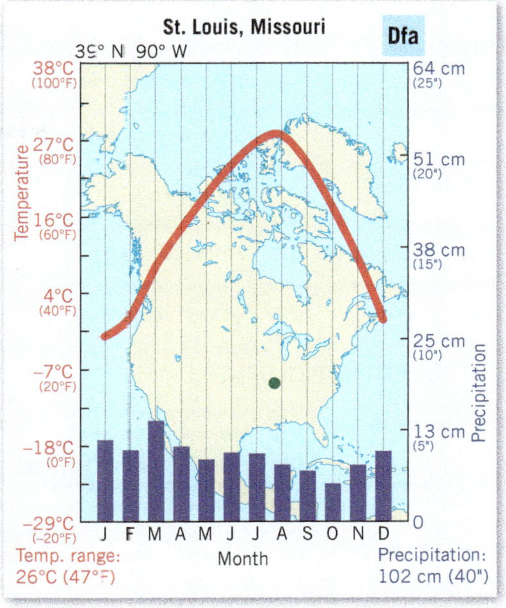

▲ **Figure 8-2** A typical climograph. Average monthly temperature is shown with a solid red line (scale on the left side), and average monthly precipitation is shown with dark blue bars extending from the bottom (scale on the right side).

The value of a climograph is twofold: (1) it displays precise details of important aspects of the climate of a specific place, and (2) it can be used to recognize and classify the climate of that place.

LearningCheck 8-2 **What information is conveyed on a climograph?**

TABLE 8-1 Generalized Meaning of Köppen Letter Codes

First Letter: Major Climate Group			Second Letter		Third Letter	
A	Tropical Humid	*low latitude; warm and wet*	A, C, & D Climate Precipitation		C & D Climate Temperature	
B	Dry	*deserts and steppes*	f	*wet all year*	a	*hot summers*
C	Mild Midlatitude	*mild winters*	m	*monsoonal pattern*	b	*warm summers*
D	Severe Midlatitude	*severe, cold winters*	w	*winter dry season*	c	*cool summers*
E	Polar	*high-latitude cold climates*	s	*summer dry season*	d	*very cold winters*
H	Highland	*altitude is dominant control*	B Climate Precipitation		B Climate Temperature	
			W	*desert*	h	*hot desert or steppe*
			S	*steppe*	k	*cold desert or steppe*
			E Climate Temperature			
			T	*tundra*		
			F	*ice cap*		

A TROPICAL HUMID CLIMATES

Tropical wet	Af	(Wet all year)
Tropical savanna	Aw	(Dry winter; wet summer)
Tropical monsoon	Am	(Dry winter; very wet summer)

B DRY CLIMATES

Subtropical desert	BWh	("Hot" desert)
Subtropical steppe	BSh	("Hot" semiarid)
Midlatitude desert	BWk	("Cold" desert)
Midlatitude steppe	BSk	("Cold" semiarid)

C MILD MIDLATITUDE CLIMATES

Mediterranean	Csa	(Hot, dry summer)
	Csb	(Warm, dry summer)
Humid subtropical	Cfa	(Wet all year; hot summer)
	Cwa	(Dry winter; hot summer)
	Cwb	(Dry winter; warm summer)
Marine west coast	Cfb	(Wet all year; warm summer)
	Cfc	(Wet all year; cool summer)

D SEVERE MIDLATITUDE CLIMATES

Humid continental	Dfa	(Cold winter; wet all year; hot summer)
	Dfb	(Cold winter; wet all year; warm summer)
	Dwa	(Cold, dry winter; hot summer)
	Dwb	(Cold, dry winter; warm summer)
Subarctic	Dfc	(Cold winter; no dry season; cool summer)
	Dfd	(Very cold winter; no dry season)
	Dwc	(Cold, dry winter; cool summer)
	Dwd	(Very cold, dry winter)

E POLAR CLIMATES

Tundra	ET	(Polar tundra; no true summer)
Ice cap	EF	(Polar ice cap)

H HIGHLAND CLIMATES

	H	(High elevation climates)

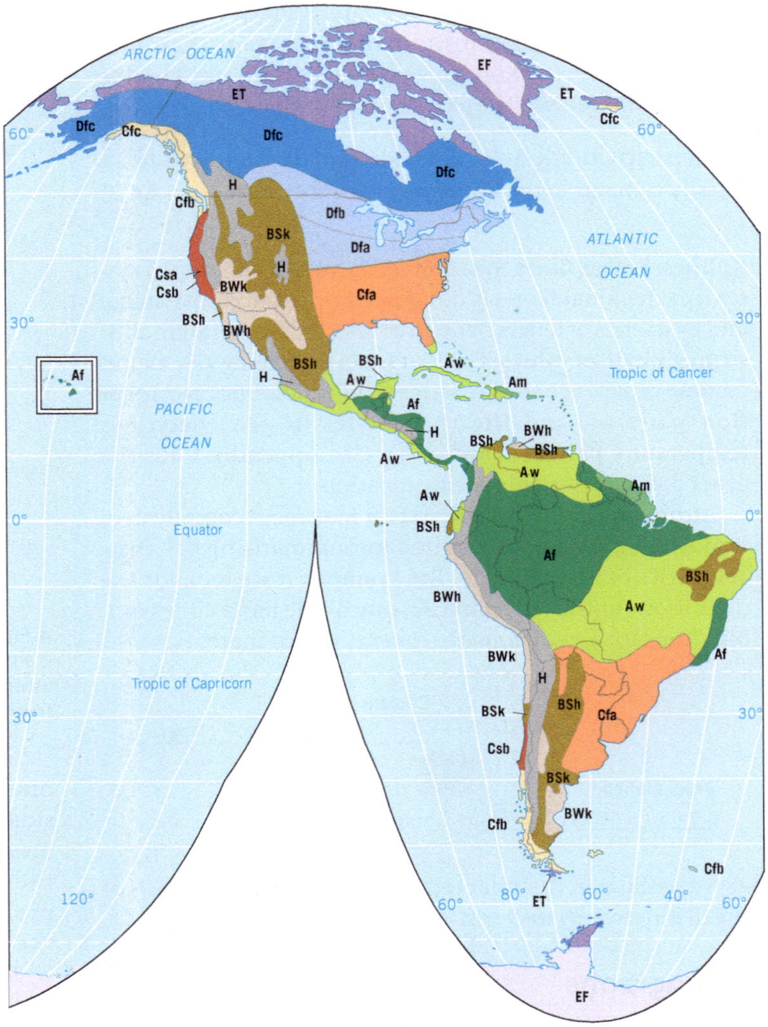

▲ **Figure 8-3** Climatic regions over land areas (modified Köppen system).

Animation
Seasonal Pressure and Precipitation Patterns
https://media.pearsoncmg.com/ph/esm/esm_mcknight_physgeo_13/media/

World Distribution of Major Climate Types

Figure 8-3 is a map showing the global distribution of climate. It also provides a general description of each climate type.

In the discussion of climatic types that follows, our attention is focused primarily on three questions:

1. *Where are the various climate types located?* We pay special attention to the typical latitude and position on a continent—such as east coast, west coast, or interior—of each climate.
2. *What are the characteristics of each climate?* We are especially interested in the annual and seasonal temperature and precipitation patterns of each climate.
3. *What are the main controls of these climates?* Recall from Chapter 3 that the major controls of climate include latitude, altitude, distribution of land and water, the general circulation patterns of the atmosphere, ocean currents, topography, and storms. These factors interact, producing the characteristic patterns of temperature, pressure, wind, and moisture of the different climates.

As we begin our look at climates of the world—a seemingly daunting task at first glance—keep in mind that very few new concepts are introduced in these next sections. Rather, this discussion of climate distribution lets us systematically organize what we have *already* learned about weather and climate in previous chapters.

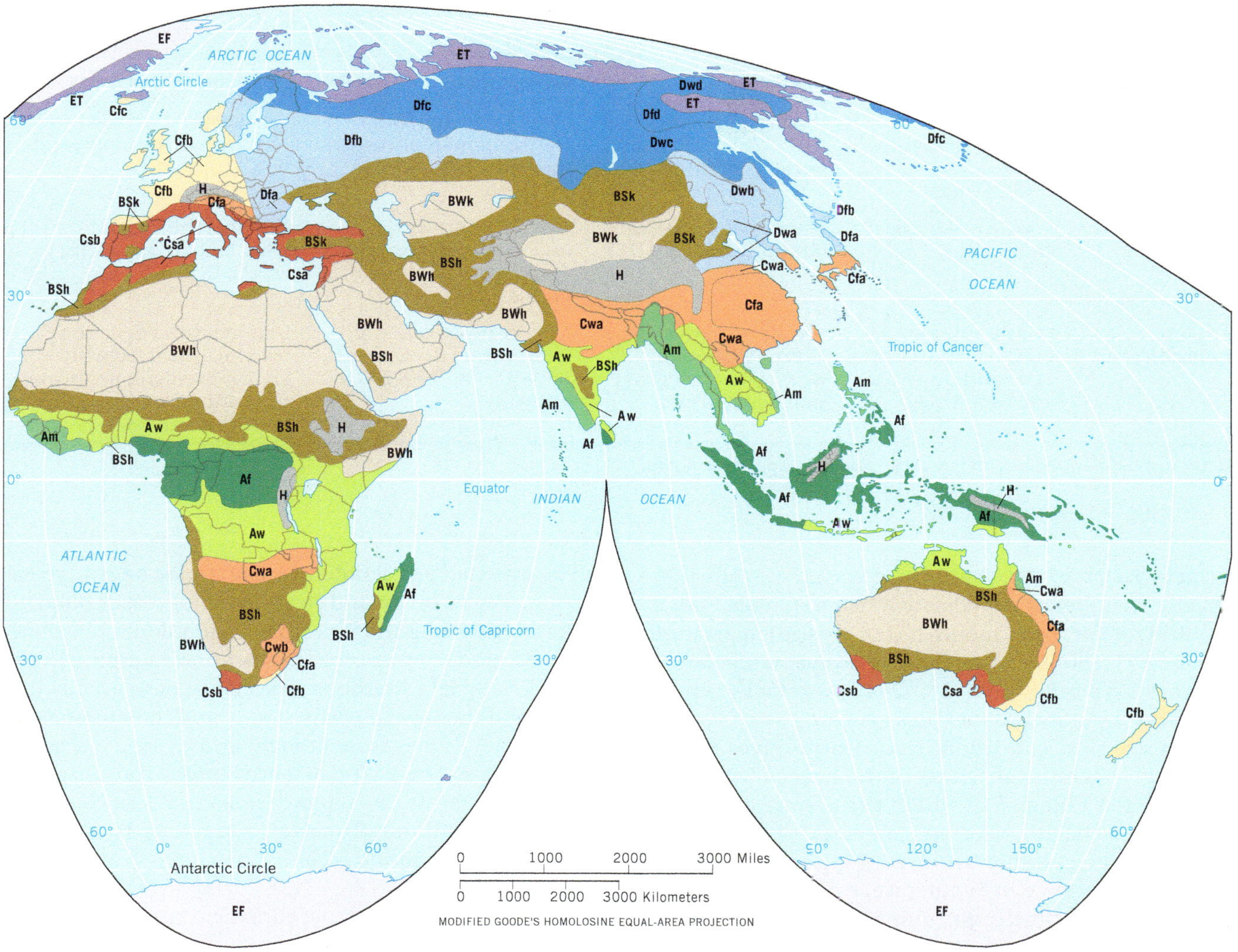

▲ **Figure 8-3** *(Continued)*

Take advantage of the climate "summary" table for each major climate group (such as **Table 8-2**). This series of tables concisely describes the general location, temperature and precipitation characteristics, and dominant controls for each climate in the Köppen system. When a climate type includes several letter code variations (such as Cfb and Cfc for *marine west coast* climate), we usually emphasize the most widespread of these variations (Cfb, in this case).

Tropical Humid Climates (Group A)

The *tropical humid climates* (Group A) occupy almost all of the land area within the tropics (**Figure 8-4**), interrupted only here and there by mountains or small arid regions. The A climates, like Hawai'i, are noted not so much for warmth as for lack of coldness. Interestingly though, they do not experience the world's highest temperatures.

These are winterless climates with an abundance of moisture. However, we sometimes use the terms *winter* to refer to the time of year when the Sun is lowest in the sky (the "low-Sun season") and *summer* to refer to the time of year when the Sun is highest (the "high-Sun season")—such as in the "winter monsoon" or "summer monsoon."

The tropical humid climates are classified into three types on the basis of annual rainfall (Table 8-2). The *tropical wet* has abundant rainfall every month of the year. The *tropical savanna* is characterized by a low-Sun dry season and a moderately rainy high-Sun wet season. The *tropical monsoon* has a dry season and a distinct very rainy high-Sun wet season with some of the highest rainfall totals in the world.

Tropical Wet Climate (Af)

The **tropical wet climate** is predominantly equatorial, found in an east–west sprawl astride the equator. The Af climate typically extends to about 10° N and S (see Figure 8-4), but in a few east-coast locations is as much as 25° poleward of the equator.

TABLE 8-2 Summary of Group A Climates: Tropical Humid

Type	Location	Temperature	Precipitation	Dominant Controls of Climate
Tropical wet (Af)	Within 5°–10° of equator; farther poleward on eastern coasts	Warm all year; very small ATR; small DTR; high sensible temperature	No dry season; 150 cm (60 in.) or more annually; many thunderstorms	Latitude; ITCZ; trade wind convergence; onshore wind flow
Tropical savanna (Aw)	Fringing Af between 25° N and S	Warm to hot all year; moderate ATR and DTR	Distinct summer wet and winter dry seasons; 90–180 cm (35–70 in.) annually	Seasonal shifting of tropical wind and pressure belts, especially ITCZ
Tropical monsoon (Am)	Windward tropical coasts of Asia, Central and South America, and west Africa	Similar to Af with slightly larger ATR; hottest weather just before summer monsoon	Very heavy in summer; short winter dry season; 250–500 cm (100–200 in.) annually	Seasonal wind direction reversal associated with ITCZ movement; jet stream fluctuation; continental pressure changes

ATR = annual temperature range; DTR = daily temperature range.

Characteristics of Af: The most descriptive word that applies to the tropical wet climate is "monotonous." It is a seasonless climate. Every month has an average temperature of about 27°C (80°F) and an average annual temperature range that is typically only 1–2°C (2–4°F)—by far the smallest range of any climatic type (Figure 8-5).

This seasonless condition in Af regions gives rise to the saying, "Night is the winter of the tropics." The Af is one of the few climates in which the *average daily temperature range* (the difference in temperature between day and night) is greater than the *average annual temperature range* (the difference in temperature between summer and winter). Daytime highs climb into the low 30s°C (high 80s°F) but fall only into the low 20s°C (low 70s°F) just before dawn.

Because both absolute and relative humidities are high, the air feels warm regardless of the thermometer reading. Short-lived but heavy rain from convective thunderstorms can be expected just about every day (Figure 8-6). A typical morning dawns bright and clear. Cumulus clouds build up in the late morning and develop into cumulonimbus thunderheads, producing a furious convectional rainstorm in the afternoon. By late afternoon, the clouds have usually dispersed, and the sky is a partly cloudy and the sunset glorious. The clouds may recur at night and create a nocturnal thunderstorm. The next day dawns bright and clear, and this pattern repeats. Each month receives at least moderate precipitation, with annual totals of 150 centimeters (60 inches) or more (Figure 8-5b and 8-5c).

Controls of Af: The principal climatic control is relatively straightforward: latitude. The high Sun throughout the year and the little or no variation in day length lead to uniform insolation, so seasonal temperature variation is low. This extensive warming produces considerable thermal convection, which accounts for a portion of the raininess. More

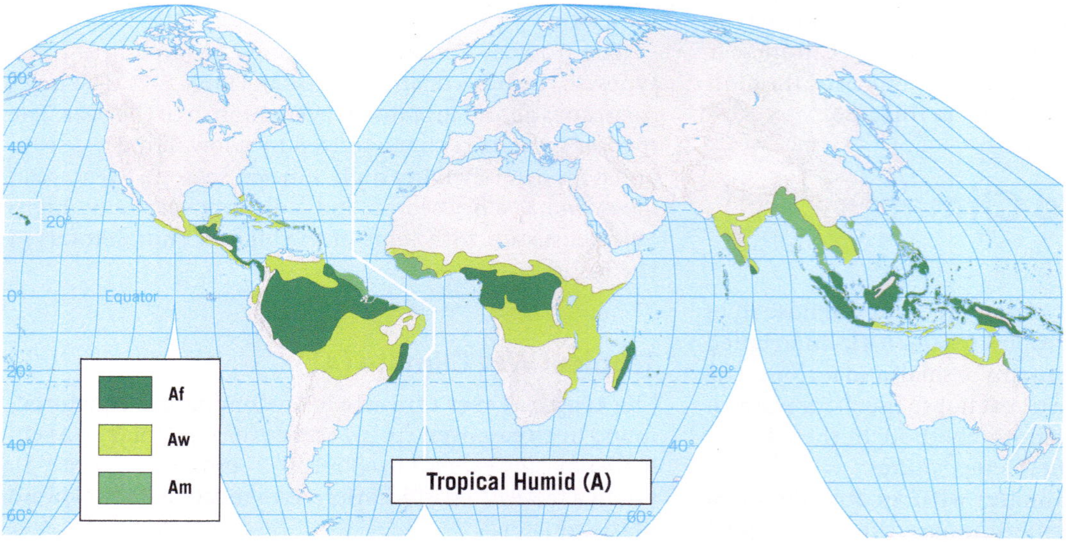

◀ **Figure 8-4** The global distribution of tropical humid (A) climates.

(a) Tropical wet climate

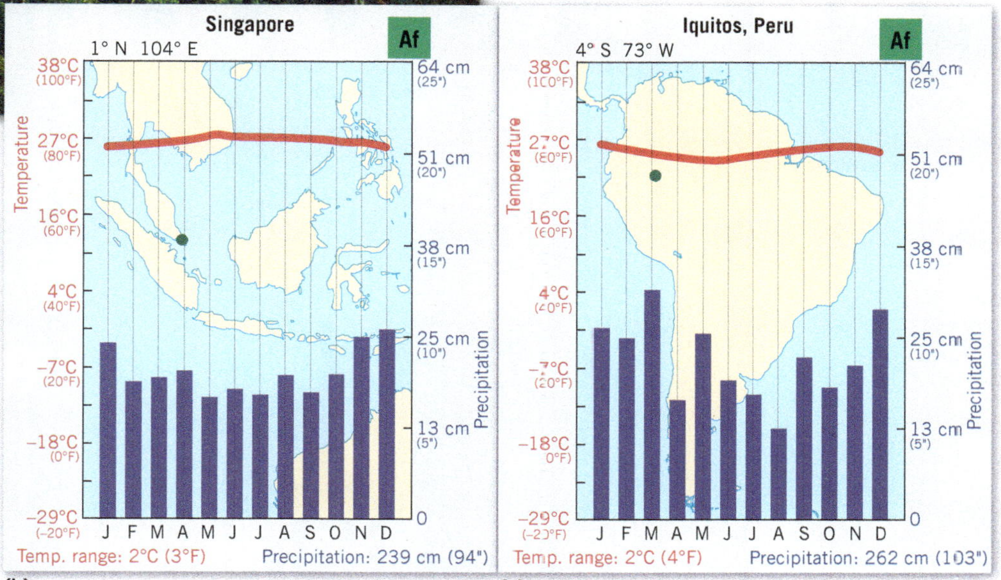

(b) (c)

◀ **Figure 8-5** Tropical wet climate. (a) The Amazon rainforest in Peru's Manu National Park. Climographs for representative tropical wet stations in (b) Singapore and (c) Iquitos, Peru.

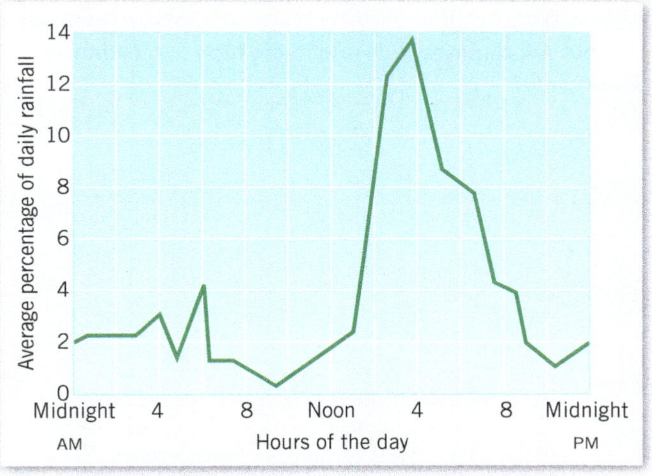

▲ **Figure 8-6** A typical daily pattern of rainfall in an Af climate. This example, from Malaysia, shows a heavy concentration in midafternoon with a small secondary peak at or shortly before dawn.

important, the influence of the intertropical convergence zone (ITCZ) for most or all of the year leads to widespread uplift of warm, humid, unstable air. In addition, onshore winds along trade-wind (east-facing) coasts provide a consistent source of moisture and add another mechanism for precipitation: orographic uplift. Interior areas, however, typically lack a great deal of wind—a consequence of the persistent influence of the ITCZ and its ongoing convective updrafts.

Tropical Savanna Climate (Aw)

The most extensive of the A climates, the **tropical savanna climate**, generally lies to the north and south of Af climates (see Figure 8-4). In a few places it extends to 25° N and S.

Characteristics of Aw: The Aw climate is distinguished by its clear-cut seasonal summer-wet and winter-dry periods (**Figure 8-7**). Typical Aw annual rainfall averages are 90–180 centimeters (35–70 inches), almost all coming during the high-Sun season.

The average annual temperature range in Aw regions is typically 3–8°C (5–15°F), slightly greater than that of

(a) Tropical savanna climate

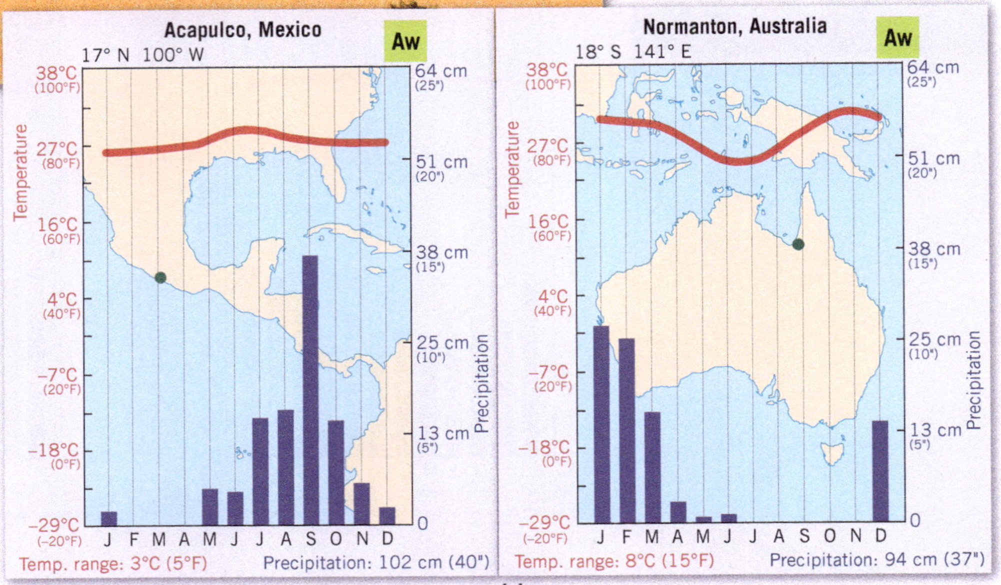

(b) (c)

◀ **Figure 8-7** Tropical savanna climate. (a) Giraffe walking across the savanna grassland during the dry season in Masai Mara National Reserve in Kenya. Climographs for representative tropical savanna stations in (b) Acapulco, Mexico, and (c) Normanton, Australia.

Af regions. The higher annual variations occur in locations farther from the equator. The hottest time of the year is likely to be late spring, just before the onset of the summer rains.

Controls of Aw: The rainfall characteristic of the Aw climates is explained by their location between the unstable, converging air of the ITCZ (which dominates the Af climates all year) on their equatorial side and stable, subsiding air of the subtropical high on their poleward side.

During the low-Sun season (winter), when pressure belts shift equatorward, savanna regions are dominated by the dry conditions and clear skies associated with the subtropical highs. In summer, the pressure systems "follow the Sun," shifting poleward and bringing the thunderstorms and convective rain of the ITCZ into the Aw region. (These seasonal precipitation patterns are clearly seen in Figure 6-37.) The poleward limits of Aw climate are approximately equivalent to the poleward maximum migration of the ITCZ (**Figure 8-8**).

LearningCheck 8-3 Why do Af climates have rain all year but Aw climates have rain in the high-Sun (summer) season only?

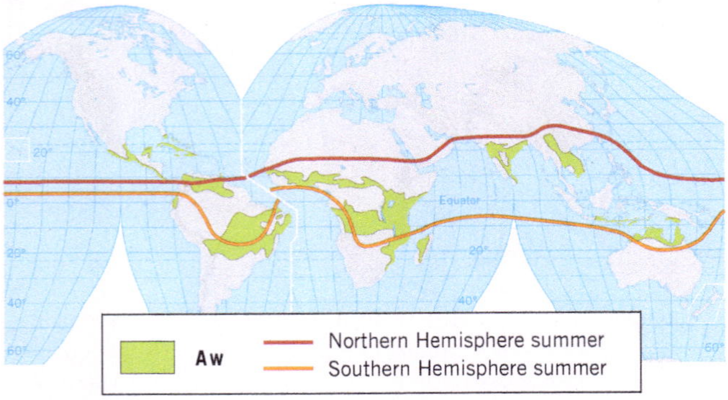

▲ **Figure 8-8** The intertropical convergence zone (ITCZ) migrates widely during the year. The red and orange lines show its typical northern and southern boundaries, respectively. These lines coincide approximately with the poleward limits of Aw climate.

Tropical Monsoon Climate (Am)

The **tropical monsoon climate** is found in tropical regions with a prominent monsoon wind pattern. It is most extensive on the windward (west-facing) coasts of southeastern Eurasia (primarily India, Bangladesh, Myanmar [formerly Burma], and Thailand), but also occurs in more restricted coastal regions of western Africa, northeastern South America, the Philippines, and northeastern Australia (see Figure 8-4).

Characteristics of Am: The Am climate is distinctive primarily in its rainfall pattern (**Figure 8-9**). During the high-Sun season, an enormous amount of rain falls in association with the "summer" monsoon. It is not unusual to have more than 75 centimeters (30 inches) of rain in each of two or three months. The annual total for a typical Am station is 250–500 centimeters (100–200 inches). An extreme example is Cherrapunji (in the Khasi hills of Assam in eastern India), with an annual average of 1143 centimeters (450 inches). Cherrapunji has been inundated with 210 centimeters (84 inches) in three days, with 930 centimeters (366 inches) in one month, and with a memorable 2647 centimeters (1042 inches) in its record year (**Figure 8-10**).

The highest Am temperatures normally occur in late spring just before the onset of the summer monsoon. The heavy cloud cover of the wet monsoon period shields out some insolation, resulting in slightly lower temperatures in summer than in spring.

Controls of Am: The extremely wet high-Sun season of the Am climate is explained by the onset of the summer monsoon that brings moist onshore winds and thunderstorms. During the low-Sun season, Am climates are dominated by offshore winds. The "winter" monsoon during this season produces little precipitation, and one or two months may be rainless.

(a) Tropical monsoon climate

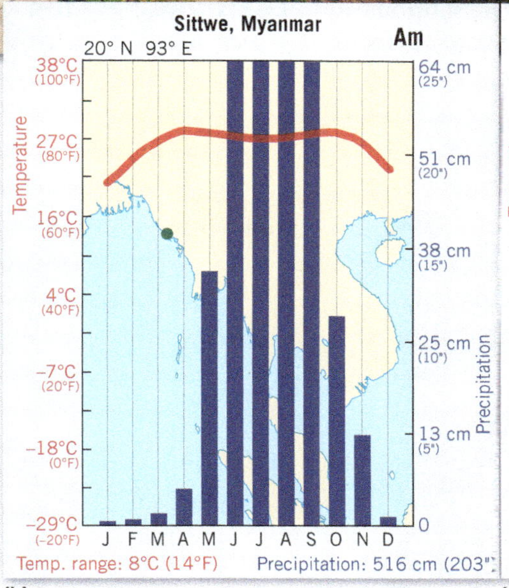

(b)

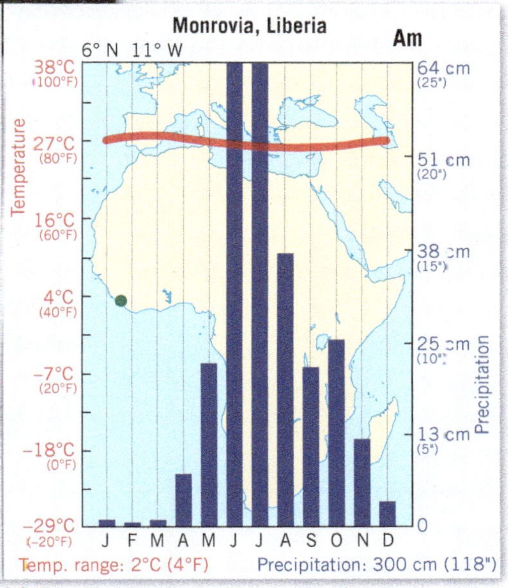

(c)

◀ **Figure 8-9** Tropical monsoon climate. (a) Flooding from monsoon rains in the city of Prayagraj (formerly Allahabad), India, in 2016. Climographs for representative tropical monsoon stations in (b) Sittwe, Myanmar, and (c) Monrovia, Liberia.

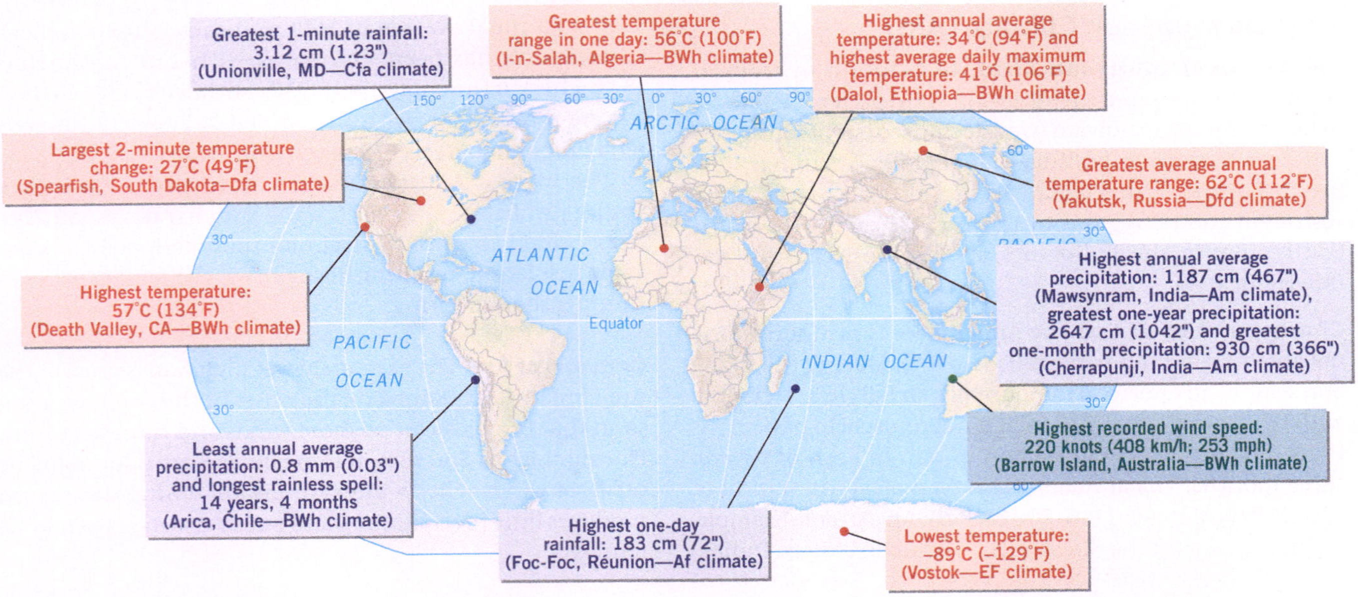

▲ **Figure 8-10** Locations of some extreme weather records.

Dry Climates (Group B)

The *dry climates* (Group B) cover about 30 percent of the world's land area (**Figure 8-11**)—more than any other climatic group. Although at first glance their distribution pattern appears erratic and complex, it actually has a considerable degree of predictability: arid regions of the world develop as a result of (1) the lack of air uplift necessary for cloud formation or (2) the lack of moisture in the air—in some cases, both.

The concept of a dry climate is complex because it involves the balance between precipitation and evapotranspiration, so it depends on temperature as well as rainfall. Generally, higher temperature produces greater potential evapotranspiration, so hot regions can receive more precipitation than cool regions and yet be classified as dry.

The largest expanses of dry areas are in subtropical latitudes, especially in the western and central portions of continents, where subsidence associated with the subtropical highs (STH) and cool ocean currents serve to increase atmospheric stability. Desert conditions also occur over extensive ocean areas; it is quite reasonable to refer to oceanic deserts.

In the midlatitudes, particularly in central Eurasia, arid climates are found in areas that are far from sources of moisture or are located in the rain shadow of mountain ranges.

The two main categories of B climates are *desert* and *steppe* (**Table 8-3**). Deserts are extremely arid, whereas steppes are semiarid. Most deserts are large core areas of aridity surrounded by a transitional fringe of steppe that is slightly less dry. The two B climates are further classified based on temperature into "hot" *subtropical desert* and *subtropical steppe*, and "cold" *midlatitude desert* and *midlatitude steppe*. Here we focus on deserts because they represent the epitome of dry conditions—the arid extreme. Most of what is stated about deserts applies to steppes but in modified intensity.

LearningCheck 8-4 Explain the two main causes of dry climates.

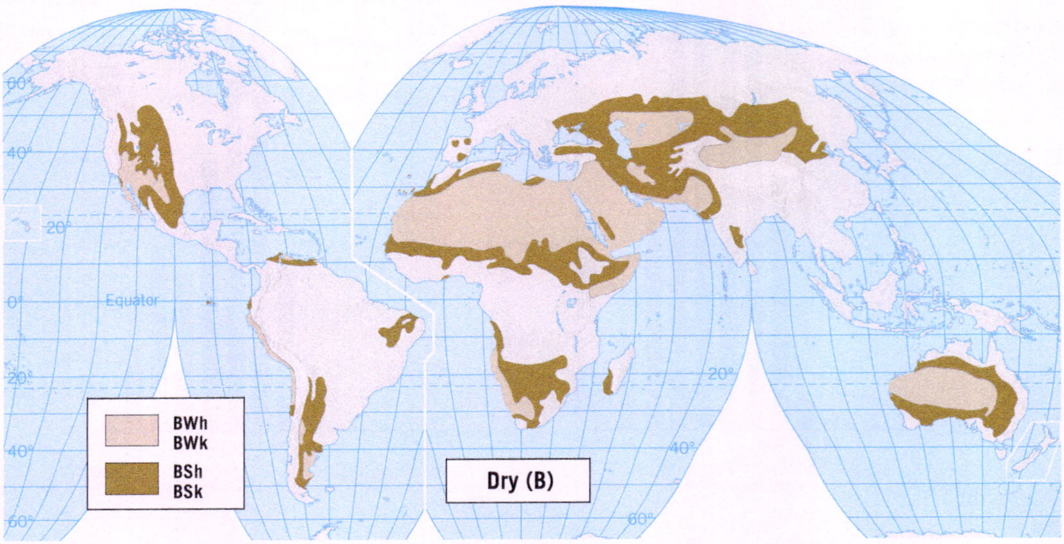

◀ **Figure 8-11** The global distribution of dry (B) climates.

TABLE 8-3 Summary of Group B Climates: Dry

Type	Location	Temperature	Precipitation	Dominant Controls of Climate
Subtropical desert (BWh)	Centered at latitudes 25°–30° on western sides of continents, extending into interiors; most extensive in northern Africa and southwestern Eurasia	Very hot summers, relatively mild winters; enormous DTR, moderate ATR	Rainfall scarce, typically less than 30 cm (12 in.); unreliable; intense; little cloudiness	Subsidence from STH; cool ocean currents; may be extended by rain shadow of mountains
Subtropical steppe (BSh)	Fringing BWh except on west	Similar to BWh but more moderate	Semiarid	Similar to BWh
Midlatitude desert (BWk)	Central Asia; western interior of United States; Patagonia	Hot summers, cold winters; very large ATR, large DTR	Meager: typically, less than 25 cm (10 in.); erratic, mostly showery; some winter snow	Distant from sources of moisture; rain shadow of mountains
Midlatitude steppe (BSk)	Peripheral to BWk; transitional to more humid climates	Similar to BWk but slightly more moderate	Semiarid; some winter snow	Similar to BWk

ATR = annual temperature range; DTR = daily temperature range.

Subtropical Desert Climate (BWh)

In both the Northern and Southern Hemispheres, **subtropical desert climates** lie in or very near the band of the STH (Figure 8-12). They are centered between 25° and 30° N and S latitude, especially along the west coasts of continents.

The enormous expanse of BWh climate in northern Africa (the Sahara) and southwestern Eurasia (the Arabian Desert) represents more desert area than is found in the rest of the world combined. Here, desert extends across the continent to the east coast, primarily because the massive Eurasian continent eliminates maritime moisture sources from the northeast. Subtropical desert climate is also expansive in Australia (nearly 20 percent of the land area) because the mountains that parallel the east coast are just high enough to prevent Pacific winds from penetrating.

Characteristics of BWh: Three adjectives describe the precipitation in subtropical deserts (Figure 8-13): scarce, unreliable, and intense.

1. *Scarce*—Subtropical deserts are among the most nearly rainless regions on Earth. Although some experience several consecutive years without a drop of rain, most average 5–20 centimeters (2–8 inches), and some receive as much as 38 centimeters (15 inches).
2. *Unreliable*—Recall from Chapter 6 that the lower the average annual precipitation, the greater its year-to-year variability, so the concept of an "average" annual rainfall in a BWh location is misleading. Yuma, Arizona, for example, has a long-term average rainfall of 8.4 centimeters (3.3 inches), but over the last two decades, it has received as little as 0.5 centimeters (0.19 inches) and as much as 18.4 centimeters (7.3 inches) in a given year.
3. *Intense*—Most precipitation in these regions falls in vigorous convective showers that are localized and brief. Rare rains may bring short-lived floods to regions that have been deprived of surface moisture for months.

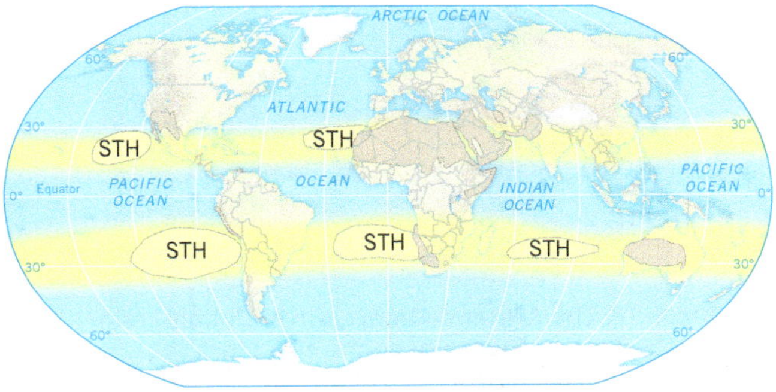

▲ Figure 8-12 The coincidence of the subtropical high-pressure zones (STH) and BWh climates is striking.

Temperatures in BWh regions are also distinctive. The combination of low-latitude location (where the vertical or near-vertical rays of the Sun strike in summer) and lack of cloudiness permits a great deal of insolation to reach the surface. Summers are long and blisteringly hot, with monthly averages in the middle to high 30s°C (high 90s°F)—significantly hotter than most equatorial regions. Midwinter months have average temperatures in the high teens °C (60s°F), which gives moderate annual temperature ranges.

Daily temperature ranges, however, are sometimes astounding. Summer days are so hot that the nights do not have time to cool off significantly, but during the transition seasons of spring and fall, a 28°C (50°F) fluctuation between the heat of the afternoon and the cool of the following dawn is not unusual. Generally clear skies and low water vapor content permit rapid nighttime cooling because longwave radiation transmission is unimpeded through the atmosphere (less local "greenhouse effect").

Controls of BWh: Subtropical deserts extend inland from the west coasts of continents in the subtropics, where

(a) Subtropical desert climate

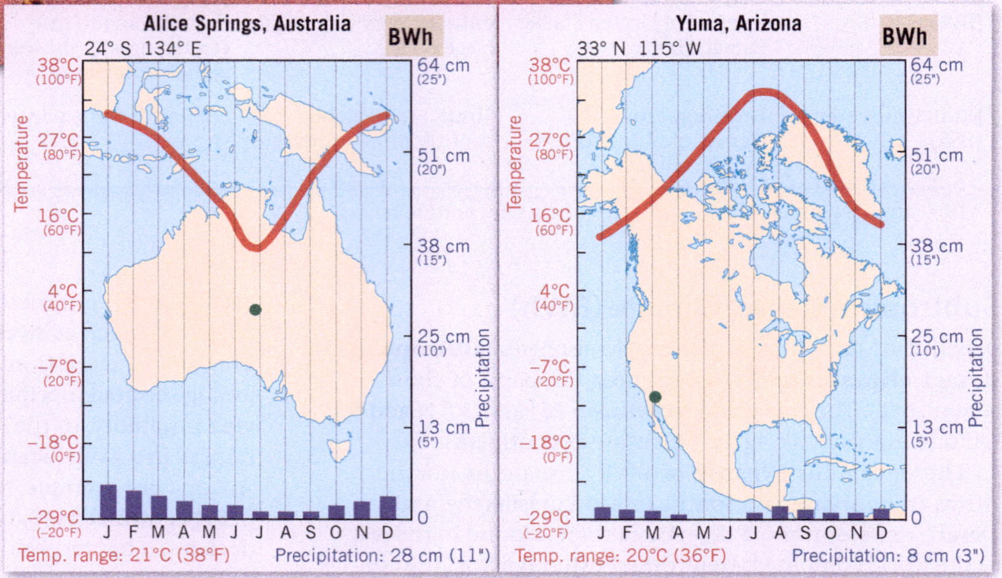

▲ **Figure 8-13** Subtropical desert climate. (a) A subtropical desert landscape (here, in Atacama Desert, Chile) sometimes contains an abundance of sand. Climographs for representative subtropical desert stations in (b) Alice Springs, Australia, and (c) Yuma, Arizona.

subsidence from the STH is stronger and where cool ocean currents help stabilize the air.

Subtropical deserts are restricted to coastal regions in southwest Africa, South America, and North America. The greatest north–south elongation occurs along the western side of South America, where the Atacama Desert is also the driest of the dry lands. The Atacama is sandwiched in a "double" rain shadow position (**Figure 8-14**): moist winds from the east are kept out of this region by the Andes Mountains, and Pacific air is thoroughly chilled and stabilized as it passes over the world's most prominent cool ocean current (the Peru Current, also known as the "Humboldt" Current).

▲ **Figure 8-14** The Atacama Desert is in a "double" rain shadow: the Andes Mountains to the east block the movement of moist air from the Atlantic, and the cold Peru (Humboldt) Current to the west stabilizes the moist Pacific air, inhibiting uplift.

West-Coast "Foggy" Deserts: Special conditions prevail along western coasts in subtropical deserts (**Figure 8-15**). The cold waters offshore (the result of cool currents and upwelling of cold deeper ocean water) chill any air that moves across them. This cooling produces frequent fog and low stratus clouds. Precipitation almost never results from this advective cooling, however, and the influence normally extends only a few kilometers inland. The immediate coastal region is characterized by such abnormal desert conditions as relatively low summer temperatures (typical summer averages in the low 20s°C or low 70s°F), continuously high

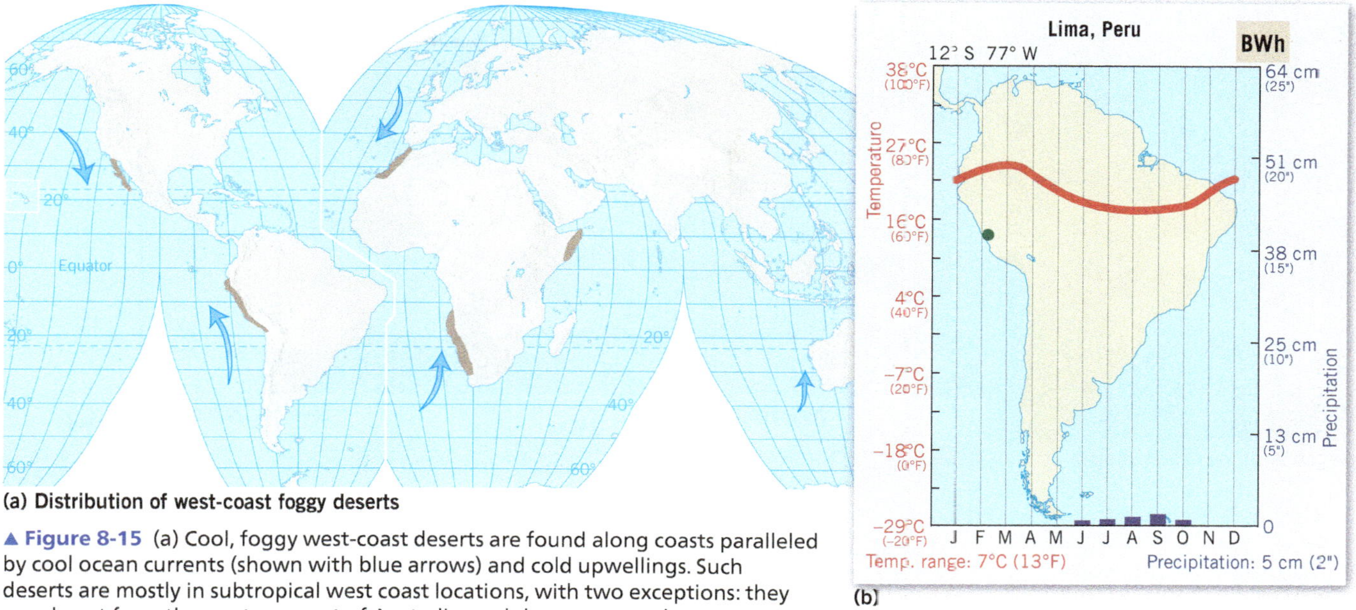

(a) Distribution of west-coast foggy deserts

(b)

▲ **Figure 8-15** (a) Cool, foggy west-coast deserts are found along coasts paralleled by cool ocean currents (shown with blue arrows) and cold upwellings. Such deserts are mostly in subtropical west coast locations, with two exceptions: they are absent from the western coast of Australia, and they occur on the eastern coast of the "horn" of Africa (Somalia). (b) Climograph for a cool west-coast desert station in Lima, Peru.

relative humidity, and greatly reduced annual and daily temperature ranges. (Compare Figures 8-13b and 8-15b.)

Subtropical Steppe Climate (BSh): The *subtropical steppe climates* characteristically surround the BWh climates (except on the western side where desert extends to the ocean), separating the deserts from the more humid climates. Temperature and precipitation conditions are similar to those for BWh regions, except that the extremes are more muted in the steppes (**Figure 8-16**).

LearningCheck 8-5 **Why are subtropical desert (BWh) climates located along the west coast of continents?**

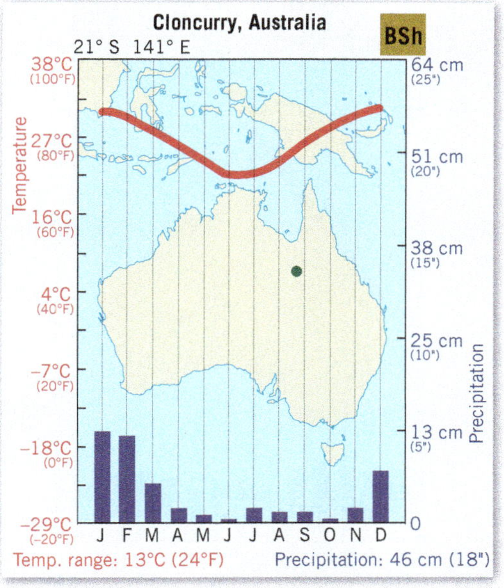

▲ **Figure 8-16** Climograph for representative subtropical steppe station in Cloncurry, Australia.

Midlatitude Desert Climate (BWk)

The **midlatitude desert climates** occur primarily in the deep interiors of continents (see Figure 8-11). The largest expanse of midlatitude dry climates, in central Eurasia, is both distant from any ocean and protected by massive mountains on the south (especially the Himalayas) from any contact with the South Asian summer monsoon. In North America, high mountains closely parallel the western coast; as a result, the dry climates begin just east of these mountains. The only other significant BWk region is in southern South America, where the desert reaches the eastern coast of Patagonia (southern Argentina).

Characteristics of BWk: Precipitation in midlatitude deserts is much like that of subtropical deserts—meager and erratic. Differences lie in two aspects: seasonality and intensity. Most BWk regions receive the bulk of their precipitation in summer (**Figure 8-17**), when continental warming and instability are common.

Unlike subtropical deserts, BWk regions have severely cold winters and when precipitation falls, it may come as snow. The average cold-month temperature is normally below freezing (some BWk stations have 6 months with below-freezing averages), producing much greater average annual temperature ranges than BWh regions. Some BWk locations have a 30°C (54°F) difference between winter and summer months.

Controls of BWk: Midlatitude deserts develop in the band of the westerlies in locations either far removed geographically from the ocean or in the rain shadows of mountain ranges. The low winter temperature and large annual temperature range reflect these midlatitude, continental locations.

LearningCheck 8-6 **Explain the differences in temperature patterns of the BWh and BWk climates.**

(a) Midlatitude desert climate

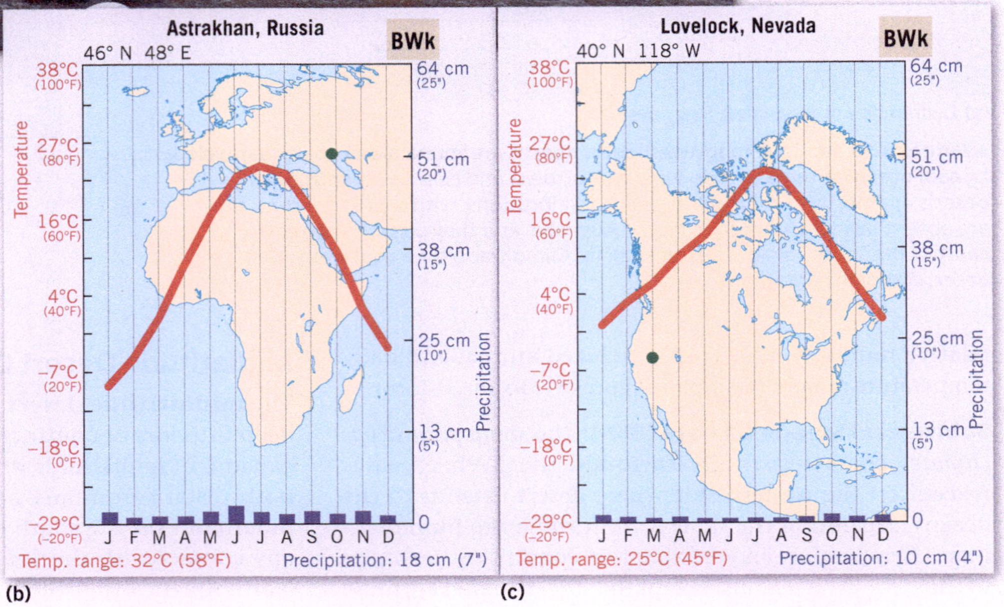

▲ **Figure 8-17** Midlatitude desert climate. (a) Midlatitude deserts are characterized by cold winters. Here a fresh snowfall decorates the desert landscape at Capital Reef National Park, Utah. Climographs for representative midlatitude desert stations in (b) Astrakhan, Russia, and (c) Lovelock, Nevada.

Midlatitude Steppe Climate (BSk): As in the subtropics, *midlatitude steppe climates* generally occupy transitional positions between deserts and humid climates. Typically, midlatitude steppes have more precipitation and lesser temperature extremes than midlatitude deserts (Figure 8-18). In western North America, the steppe climate is extensive. Only in the interior southwest of the United States is the climate dry enough to be a desert.

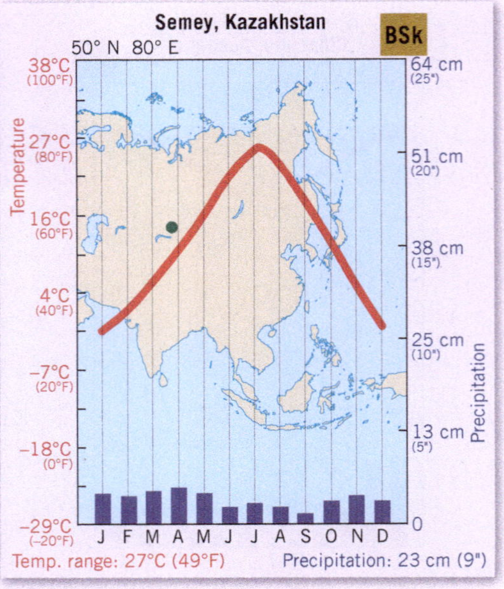

▲ **Figure 8-18** Climograph for representative midlatitude steppe station in Semey, Kazakhstan.

Mild Midlatitude Climates (Group C)

The *mild midlatitude climates* (Group C) occupy the equatorward margin of the midlatitudes, extending farther poleward along the west coasts of continents than along the east (Figure 8-19). They constitute a transition between warmer tropical climates and colder severe midlatitude climates.

The midlatitudes are a region of air mass contrast, creating a kaleidoscope of atmospheric disturbances and weather variability. The seasonal rhythm of temperature is usually more prominent than that of precipitation. Whereas in the tropics the seasons are more likely to be characterized as

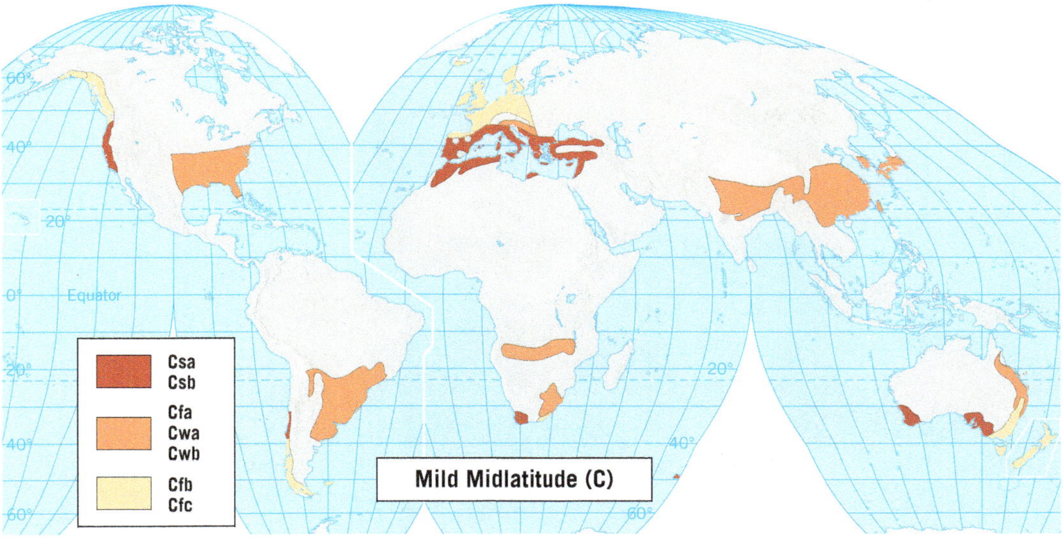

◄ Figure 8-19 The global distribution of mild midlatitude (C) climates.

"wet" and "dry," in the midlatitudes they are clearly "summer" and "winter."

Summers in the C climates are long and sometimes hot; winters are short and relatively mild. Most C zones experience winter frosts and therefore do not have a year-round growing season. Precipitation is highly variable with regard to both total amount and seasonal distribution.

The C climates are subdivided into three types primarily on the basis of precipitation seasonality and secondarily on the basis of summer temperatures: *mediterranean*, *humid subtropical*, and *marine west coast* (Table 8-4).

Mediterranean Climate (Csa, Csb)

The two Cs climates are sometimes referred to as *dry summer subtropical* or, more commonly, **mediterranean climate**.[1] Cs climates are found in five parts of the world on the western side of continents, centered at about 35° N and 35° S. The Cs regions are restricted to the coasts in California, central Chile, the southern tip of Africa, and the two southwestern "corners" of Australia; the only extensive area of mediterranean climate is around the borderlands of the Mediterranean Sea.

Characteristics of Cs: Cs climates (Figure 8-20) have three distinctive characteristics:

1. The modest annual precipitation falls in winter; summers are virtually rainless.
2. Winter temperatures are unusually mild for the midlatitudes, and summers vary from warm to hot.
3. Clear skies and abundant sunshine are typical, especially in summer.

Average annual precipitation is modest—about 25–40 centimeters (10–15 inches)—with the two or three midsummer months totally dry. Only one other climatic type, *marine west coast*, has such a concentration of precipitation in winter.

Most mediterranean climate is classified as "hot summer" Csa, with midsummer monthly averages of 24–29°C (75–85°F) and frequent high temperatures above 38°C (100°F). Coastal "warm summer" Csb areas have much

[1]The proper terminology for a type of climate is lowercase; Mediterranean, with a capital M, refers to a specific region around the Mediterranean Sea.

TABLE 8-4 Summary of Group C Climates: Mild Midlatitude

Type	Location	Temperature	Precipitation	Dominant Controls of Climate
Mediterranean (Csa, Csb)	Centered at 35° latitude on western sides of continents; limited east–west extent except in Mediterranean Sea area	Warm/hot summers; mild winters; year-round mildness in coastal areas	Summer dry season; moderate rain: 25–40 cm (10–15 in.) annually, nearly all in winter; much sunshine, some coastal fog	Subtropical high subsidence and stability in summer; westerly winds and cyclonic storms in winter
Humid subtropical (Cfa, Cwa, Cwb)	Centered at 30° latitude on eastern sides of continents; considerable east–west extent	Summers warm/hot, sultry; winters mild to cold	Abundant: 100–165 cm (40–65 in.) annually, mostly rain; summer maxima but no true dry season	Westerly winds and storms in winter; moist onshore flow in summer; monsoons in Asia
Marine west coast (Cfb, Cfc)	Latitudes 40°–60° on western sides of continents; limited inland extent except in Europe	Very mild winters for the latitude; generally mild summers; moderate ATR	No dry season; moderate to abundant: 75–125 cm (30–50 in.) mostly in winter; many days with rain; much cloudiness	Westerly flow and oceanic influence year-round

ATR = annual temperature range.

(a) Mediterranean climate

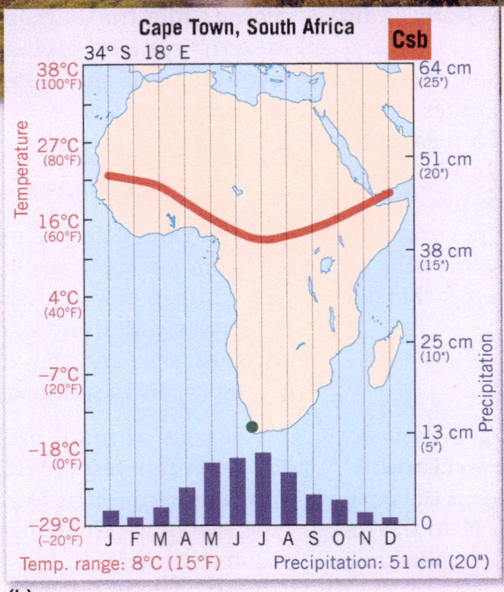

(b)

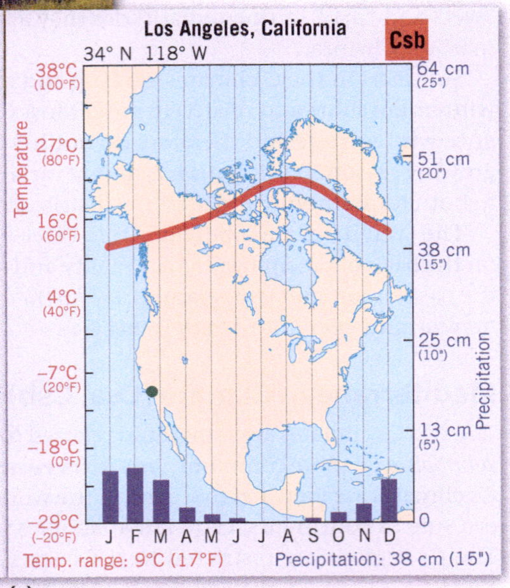

(c)

▲ **Figure 8-20** Mediterranean climate. (a) Sage scrub and chaparral are a typical mediterranean landscape at Santa Monica Mountains National Recreation Area. Climographs for representative mediterranean stations in (b) Cape Town, South Africa, and (c) Los Angeles, California.

milder summers and slightly milder winters than inland mediterranean areas (Figure 8-21). Only rarely do temperatures fall below freezing.

Controls of Cs: The origin of mediterranean climates is clear-cut: in summer, these regions are dominated by dry, stable, subsiding air from the eastern portions of STHs; in winter, the wind and pressure belts shift equatorward, and mediterranean regions come under the influence of the westerlies with their migratory midlatitude cyclones and associated fronts. Almost all precipitation comes from these cyclonic storms.

Humid Subtropical Climate (Cfa, Cwa, Cwb)

Whereas mediterranean climates are found on the western side of continents, **humid subtropical climates** are found on the eastern side, centered at about the same latitude. However, humid subtropical climates extend farther inland and over a greater range of latitude than do mediterranean climates—especially in North America, South America, and Eurasia. Cfa is the most common variation.

Characteristics of Cfa: The humid subtropical climates differ from mediterranean climates in several important respects (Figure 8-22):

- Summer temperatures in humid subtropical regions are generally warm to hot, but with higher humidity than Cs areas. Cfa days tend to be hot and sultry, and often night brings little relief.
- Precipitation tends to have a summer maximum; winter is a time of diminished precipitation, but it is not really a dry season (except in the Cwa areas of China, where dry winter monsoon conditions dominate). In the North American and Eurasian coastal areas, a late summer–autumn bulge in the precipitation curve is due to rainfall from tropical cyclones. Annual precipitation is generally abundant, averaging 100–165 centimeters (40–65 inches), with a general decrease inland.
- Winter temperatures in Cfa regions are mild but a little cooler than in mediterranean regions. Winter is punctuated by cold waves that can bring severe weather and killing frosts for a few days at a time.

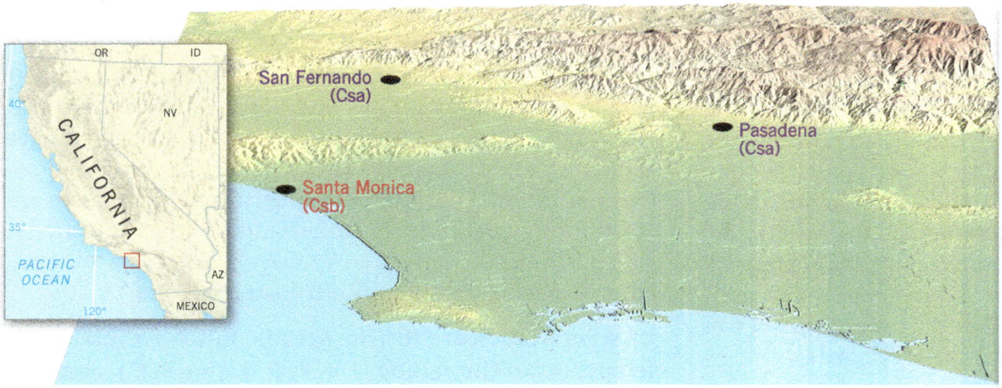

(a) Mediterranean areas: coastal versus inland

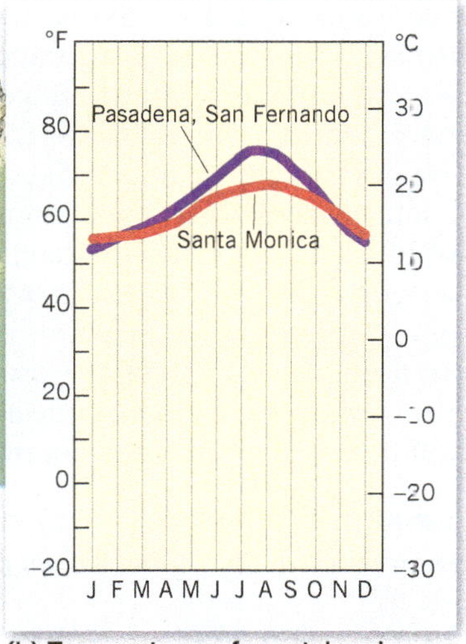

(b) Temperatures of coastal and inland stations

▲ **Figure 8-21** There are often significant temperature differences between coastal and inland mediterranean areas. (a) The physiographic map shows the topographic and coastal relationships of these stations. (b) This climograph shows the annual temperature curves for three Southern California stations: Santa Monica (a coastal station; Csb climate), and Pasadena and San Fernando (inland stations having exactly the same average temperature each month; Csa climate).

(a) Humid subtropical climate

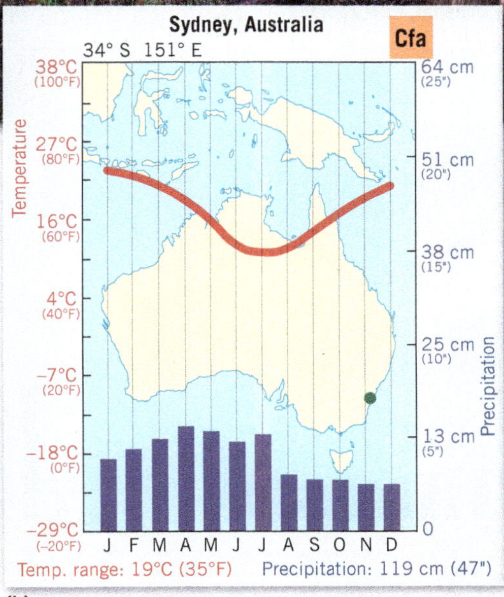

(b)

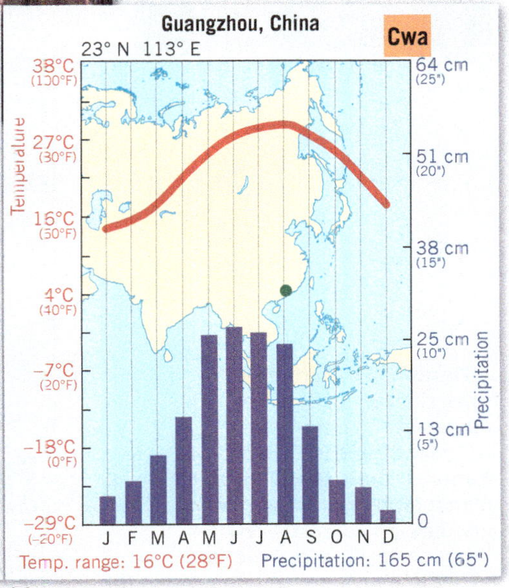

(c)

▲ **Figure 8-22** Humid subtropical climate. (a) Cypress trees and cypress knees, Congaree National Park, South Carolina. Climographs for representative humid subtropical stations in (b) Sydney, Australia, and (c) Guangzhou, China.

Controls of Cfa: To understand the controls of humid subtropical climates, again a comparison with mediterranean regions is useful because both generally lie in the same latitude. During winter, westerly winds bring midlatitude cyclones and precipitation to both regions. In summer, the relatively cool ocean water along west coasts and the poleward migration of the STH stabilize the atmosphere in mediterranean regions, bringing dry weather; on the eastern side of continents, however, where humid subtropical climates are found, no such stability exists. Summer rain comes from the onshore flow of maritime air and frequent convective uplift, as well as the influence of tropical cyclones.

LearningCheck 8-7 **Why do Cs climates have dry summers, but Cfa climates have precipitation all year?**

Marine West Coast Climate (Cfb, Cfc)

Marine west coast climates are situated on the western side of continents between about 40° and 60° N and S; this is a windward location in the band of the westerlies. Cfb is the most common variation.

The most extensive area of marine west coast climate is in western and central Europe. The North American region is much more restricted by the presence of mountain ranges that run perpendicular to the direction of onshore flow. Only in the Southern Hemisphere, where landmasses are small in these latitudes (New Zealand, southeast Australia, and southernmost South America), does this oceanic climate extend across to eastern coasts (see Figure 8-19).

Characteristics of Cfb: The oceanic influence moderates temperatures throughout the year (**Figure 8-23**). Summer months average 16–21°C (60–70°F), and winter months average 2–7°C (35–45°F). Prolonged summer heat waves are rare, and winter frosts are relatively infrequent. The growing season is abnormally long for the latitude. For instance, the growing season in Vancouver, British Columbia, is the same length as that in Birmingham, Alabama, which is 16° of latitude closer to the equator.

(a) Marine west coast climate

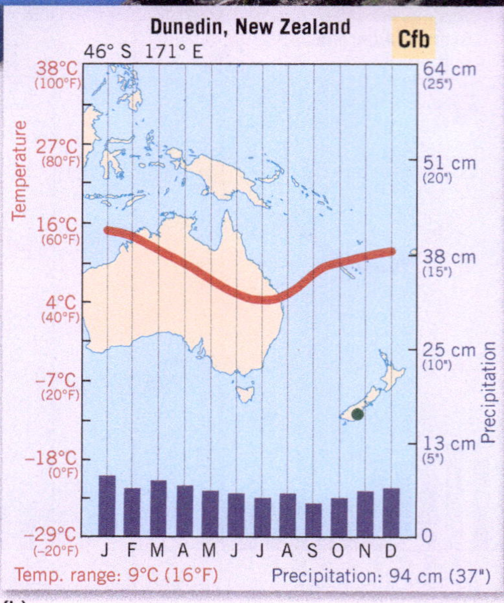

(b)

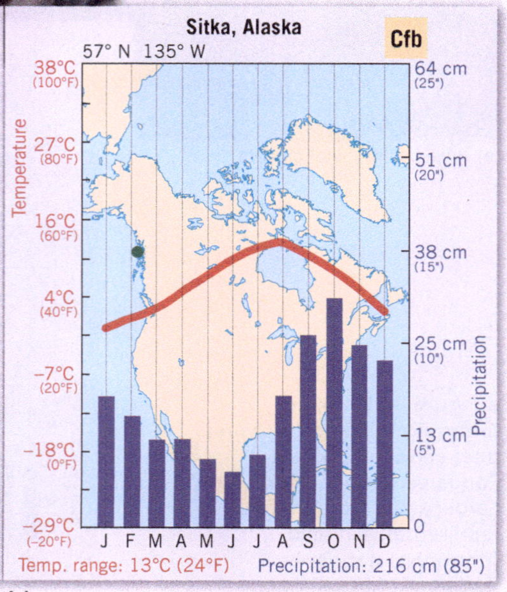

(c)

▲ **Figure 8-23** Marine west coast climate. (a) Big trees dominate this marine west coast landscape in Olympic National Forest, Washington. Climographs for representative marine west coast stations in (b) Dunedin, New Zealand, and (c) Sitka, Alaska.

Marine west coast climates are among the wettest of the midlatitudes. Annual totals of 75–125 centimeters (30–50 inches) are typical, with much higher totals recorded on exposed mountain slopes. Snow is uncommon in the lowlands, but higher, west-facing slopes receive some of the heaviest snowfalls in the world.

Perhaps a more important characteristic of the marine west coast climate than total precipitation is precipitation frequency. Vancouver, British Columbia, for example, receives only 41 percent of the total possible sunshine each year, in contrast to 73 percent in Los Angeles. London has experienced as many as 89 consecutive days with rain; is it any wonder that the umbrella is that city's civic symbol? Some places on the western coast of New Zealand's South Island have recorded 325 rainy days in a single year!

Controls of Cfb: The dominant climate control for these regions is the year-round cool maritime influence brought by the onshore flow of the westerlies. That influence leads to frequent cloudiness and high proportions of days with some precipitation, as well as an extraordinarily temperate climate considering the latitude.

Severe Midlatitude Climates (Group D)

The *severe midlatitude climates* (Group D) occur in the Northern Hemisphere only (Figure 8-24) because the Southern Hemisphere has limited landmasses at the appropriate latitudes—between 40° and 70°. This climatic group extends broadly across North America and Eurasia.

Continentality, or remoteness from oceans, is a keynote of D climates. Landmasses are broader at these latitudes than anywhere else. Even though these climates extend to the eastern coasts of the two continents, they experience little maritime influence because the general flow of the westerlies brings air from the interior of the continents to the east coasts.

These climates have four clearly recognizable seasons: a long, cold winter; a relatively short summer that varies from warm to hot; and transition periods in spring and fall. Annual temperature ranges are very large, particularly at more northerly locations, where winters are most severe. Summer is the time of precipitation maximum, but winter is by no means completely dry, and snow cover lasts for many weeks or months.

The severe midlatitude climates are subdivided into two types on the basis of temperature. The *humid continental* type has long, warm summers, whereas the *subarctic* type is characterized by short summers and very cold winters (Table 8-5).

Humid Continental Climate (Dfa, Dfb, Dwa, Dwb)

The **humid continental climate** (Figure 8-25) is found over a large area of east-central North America and northern and northeastern Eurasia, between 35° and 55° N. Dfa is the most common variation.

Characteristics of Dfa: Day-to-day variability and dramatic changes are prominent features of the weather pattern of humid continental climates. These are regions of cold waves, heat waves, blizzards, thunderstorms, tornadoes, and other dynamic atmospheric phenomena.

Summer temperatures are warm, generally averaging in the mid-20s°C (mid-70s°F)—comparable to, although shorter than, those of the humid subtropical climate to the south. The average winter-month temperature is usually between −12°C and −4°C (10–25°F), however, with one to five months averaging below freezing. Winter temperatures decrease rapidly northward in the Dfa climates (compare Figures 4-30 and 4-31), and the growing season diminishes from about 200 days on the southern margin to about 100 days on the northern edge.

Despite their name, humid continental climates do not experience high precipitation. Annual totals average 50–125 centimeters (20–50 inches), with the highest values

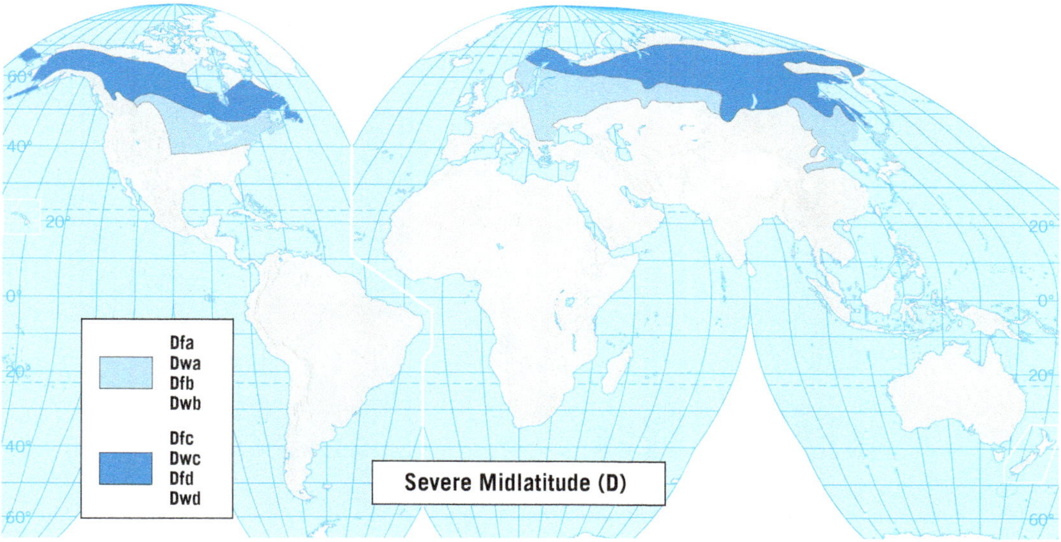

◀ Figure 8-24 The global distribution of severe midlatitude (D) climates.

TABLE 8-5 Summary of Group D Climates: Severe Midlatitude

Type	Location	Temperature	Precipitation	Dominant Controls of Climate
Humid continental (Dfa, Dfb, Dwa, Dwb)	Northern Hemisphere only; latitudes 35°–55° on eastern sides of continents	Warm/hot summers; cold winters; much day-to-day variation; large ATR	Moderate to abundant: 50–125 cm (20–50 in.) annually with summer maxima; diminishes interiorward and poleward	Westerly winds, storms, and continental air masses, especially in winter; monsoons in Asia
Subarctic (Dfc, Dfd, Dwc, Dwd)	Northern Hemisphere only, latitudes 50°–70° across North America and Eurasia	Long, dark, very cold winters; brief, mild summers; enormous ATR	Meager: 13–50 cm (5–20 in.) annually with summer maxima; light snow in winter but little melting	Pronounced continentality; westerlies and cyclonic storms alternating with prominent anticyclonic conditions

ATR = annual temperature range.

(a) Humid continental climate

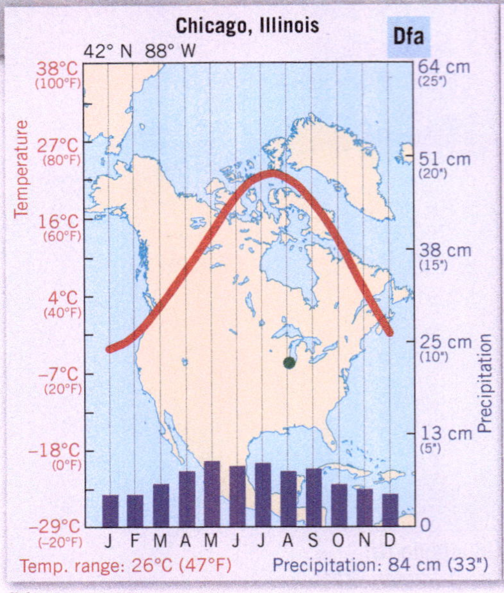

(b)

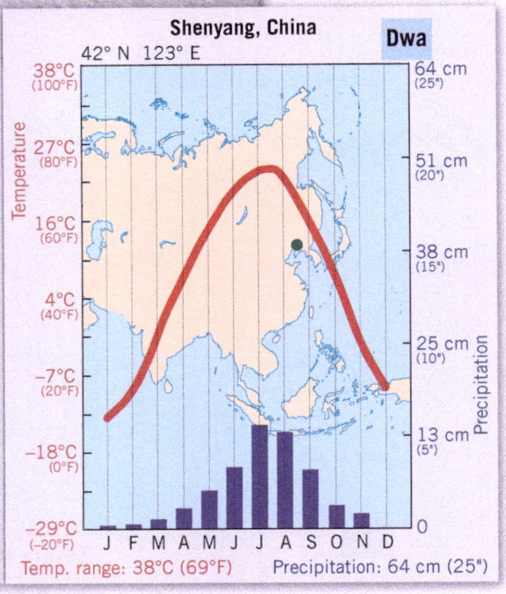

(c)

▲ **Figure 8-25** Humid continental climate. (a) The snow-covered landscape of winter in a humid continental climate in Voyageurs National Park, Minnesota. Climographs for representative humid continental stations in (b) Chicago, Illinois, and (c) Shenyang, China.

on the coast and a general decrease inland and from south to north (see Figure 6-36). Both trends reflect increasing distance from warm, moist air masses.

Controls of Dfa: This climate is dominated by the westerlies throughout the year, resulting in frequent weather changes associated with the passage of midlatitude cyclones and anticyclones, especially in winter. Although large areas of humid continental climate are located along the coastlines of eastern North America and Eurasia—at much the same latitude as marine west coast climates—the temperature pattern is continental. Westerlies bring maritime influence to marine west coast locations throughout the year, but westerly winds bring continental air to east coast locations in Dfa climates, especially cold continental air masses in winter.

Summer rain is mostly convective or monsoonal and/or frontal in origin. Winter precipitation is associated with midlatitude cyclones, and much of it falls as snow. During a typical winter, snow covers the ground for only two or three weeks in southern parts of Dfa regions but for as long as eight months in northern portions.

LearningCheck 8-8 Why do coastal Dfa climates—such as in New York City and Boston—have such cold winters?

Subarctic Climate (Dfc, Dfd, Dwc, Dwd)

The **subarctic climate** occupies the higher midlatitudes generally between 50° and 70° N. This climate occurs as two vast, uninterrupted expanses across the broad northern landmasses: from western Alaska across Canada, and across Eurasia from Scandinavia to easternmost Siberia (see Figure 8-24). The name *boreal* (which means "northern" and comes from *Boreas*, the mythological Greek god of the north wind) is sometimes applied to this climatic type in the Yukon-Charley Rivers National Preserve, Alaska (Figure 8-26). In Eurasia it is often called *taiga*, after the Russian name for the forest in the region where this climate occurs. Dfc is the most common variation.

(a) Subarctic Climate

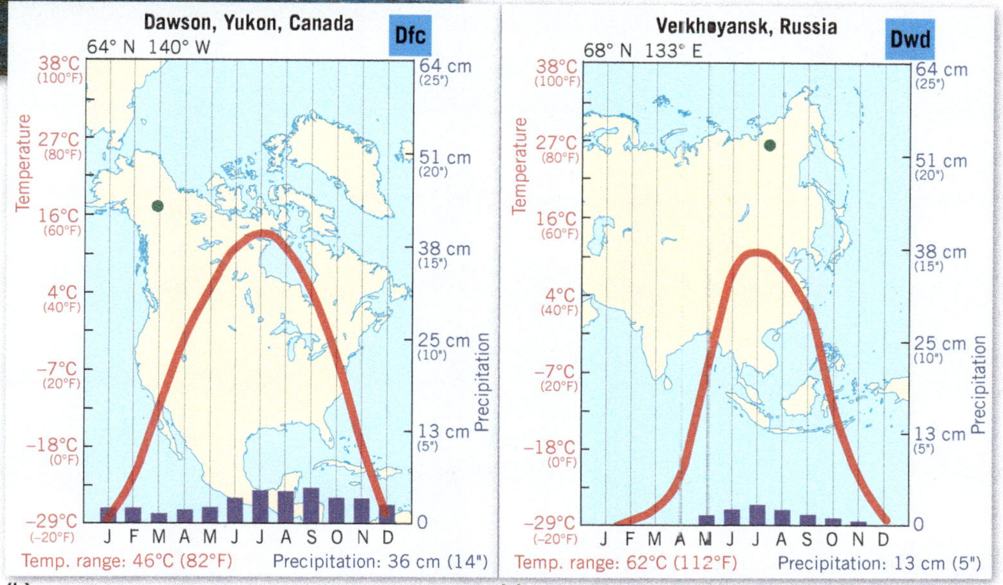

(b)

(c)

◀ **Figure 8-26** Subarctic climate. (a) Coniferous forest in Yukon-Charley Rivers National Preserve, Alaska. Climographs for representative subarctic stations in (b) Dawson, Yukon, Canada, and (c) Verkhoyansk, Russia. Note: The January temperature for Verkhoyansk is off the scale of the climograph at –47°C (–53°F).

Characteristics of Dfc: The keyword of the subarctic climate is "winter," which is long, dark, and bitterly cold. Summers are short, and fall and spring slip by rapidly. In most places, ice begins to form on the lakes in September or October and doesn't thaw until May or later. For six or seven months, the average temperature is below freezing, and the coldest months have averages below −38°C (−36°F). The world's coldest temperatures, apart from the Antarctic and Greenland ice caps, are found in the subarctic climate. (The records are −62°C [−80°F] in Alaska and −68°C [−90°F] in Siberia.)

Summer warms up remarkably despite its short duration. Although the intensity of sunlight is low (because the angle of incidence is small), summer days are very long and nights are too short to permit much radiational cooling. Average summer temperatures are typically in the midteens °C (high 50s°F), but frosts may occur in any month.

Winter-to-summer annual temperature ranges in Dfc climates are the largest in the world, frequently exceeding 45°C (80°F). The *absolute annual temperature variation* (the difference in temperature from the coldest to the hottest ever recorded) can be unbelievable, especially deep inland. The world record is −68°C to +37°C (−90°F to +98°F), a range of 105°C (188°F) in Verkhoyansk, Russia!

Average annual precipitation is usually low, perhaps 13–50 centimeters (5–20 inches), with the higher values occurring in coastal areas. Summer is the wet season, and most precipitation comes from scattered convective showers. Winter experiences only modest snowfalls (except near the coasts), perhaps 60–90 centimeters (2–3 feet). A continuous thin snow cover remains into spring despite the sparseness of snowfall.

Controls of Dfc: Subarctic climates are largely the consequence of their high-midlatitude continental location, which removes them from both the moisture and temperature-moderating effects of the ocean. The westerlies bring in cyclonic storms that provide their meager precipitation. The low winter temperatures allow for little moisture in the air, and anticyclonic conditions predominate. Despite these sparse totals, the evaporation rate is low, and the soil is frozen for much of the year, so enough moisture is present to support a forest.

Polar and Highland Climates (Groups E and H)

Being farthest from the equator, the *polar climates* (Group E) receive too little insolation for significant warming (Figure 8-27). By definition, in a polar climate no month has an average temperature above 10°C (50°F). If the wet tropics represent conditions of monotonous warmth, the polar climates are known for their enduring cold. They have the coldest summers and the lowest annual and absolute temperatures. They are also extraordinarily dry, but evaporation is so minuscule that Group E is classified as nonarid in the Köppen system.

Indications are that high-latitude climates are changing more rapidly than midlatitude and tropical areas. As the Arctic warms, both natural and human-built environments are changing. Over the last few decades, average temperatures in the Arctic have been increasing at about twice the rate observed in lower latitudes (Figure 8-28). Climate change will be more fully discussed in the second half of this chapter.

The two types of polar climates are distinguished by summer temperature. The *tundra* climate has at least one month

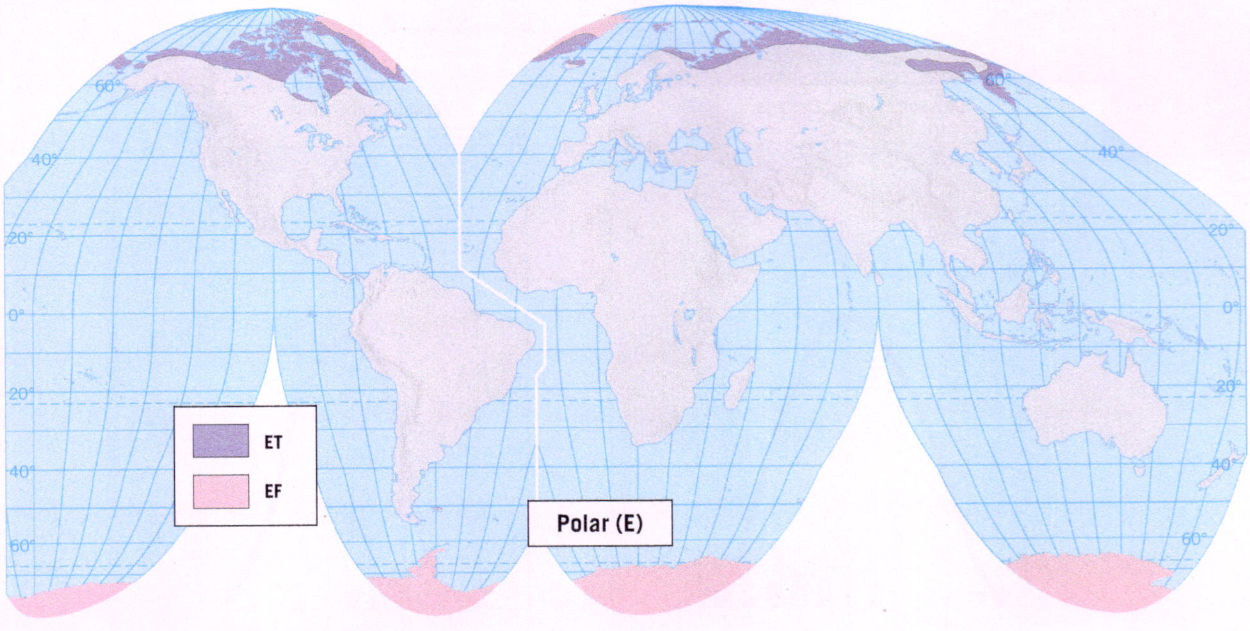

▲ Figure 8-27 The global distribution of polar (E) climates.

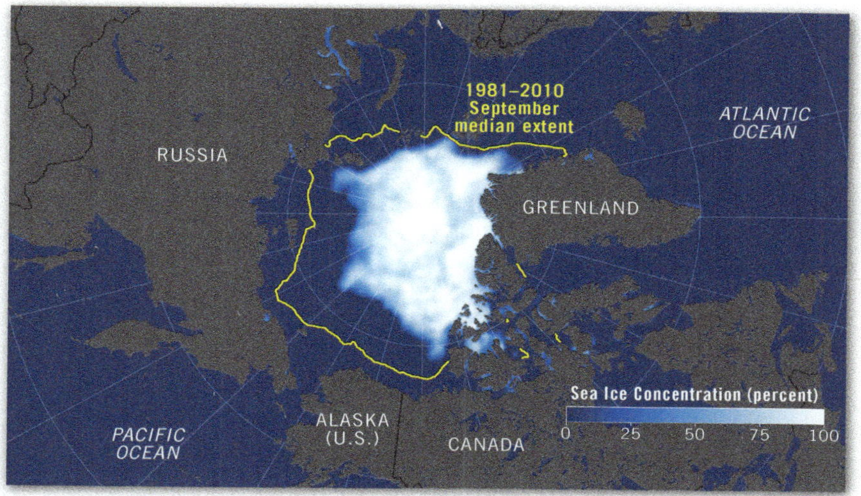

(a) The sea ice cap shrank to 4.15 million square kilometers (1.60 million square miles) in 2019.

(b) Polar bears, such as these on the Chuckchi Sea, are losing their habitat to climate change.

▲ **Figure 8-28** (a) As the extent of the summer sea ice decreases it changes the living conditions of Arctic communities and (b) puts habitats at risk, including the habitat of these Alaskan polar bears.

Mobile Field Trip
Changing Arctic
https://media.pearsoncmg.com/ph/esm/esm_mcknight_physgeo_13/media/

with an average temperature exceeding the freezing point; the *ice cap* climate does not (Table 8-6).

Tundra Climate (ET)

The name *tundra* originally referred to the low, ground-hugging vegetation of high-latitude and high-altitude regions, but the term has been adopted to refer to the climate of the high-latitude regions as well. The generally accepted equatorward edge of the **tundra climate** is the 10°C (50°F) isotherm for the average temperature of the warmest month. This isotherm corresponds approximately with the poleward limit of trees, so the boundary between D and E climates is the *tree line*. Much of the area of ET climate is underlain by *permafrost*, permanently frozen ground (discussed in Chapter 9).

At the poleward margin, the ET climate is bounded by the isotherm of 0°C (32°F) for the warmest month, which coincides approximately with the extreme limit for growth of *any* plant cover. More than for any other climatic type, the delimitation of the tundra climate demonstrates Köppen's contention that climate is best delimited in terms of plant communities (Figure 8-29).

Characteristics of ET: Long, cold, dark winters and brief, cool summers characterize the tundra. Only one to four months experience average temperatures above freezing, and the average summer temperature is 4–10°C (40–50°F). Freezing temperatures can occur any time of year, and frosts are likely every night except in midsummer. Although an ET winter is bitterly cold—winter months might average −18°C (0°F) in coastal areas and −35°C (−30°F) in interior regions—it may not be as severe as in the more-continental subarctic climate just equatorward. Annual temperature ranges are fairly large, commonly more than 30°C (54°F).

Annual total precipitation is typically less than 35 centimeters (14 inches). Generally, more precipitation falls in the warm season than in winter, although the total amount

TABLE 8-6 Summary of Group E Climates: Polar

Type	Location	Temperature	Precipitation	Dominant Controls of Climate
Tundra (ET)	Fringes of Arctic Ocean; small coastal areas in Antarctica	Long, cold, dark winters; brief, cool summers; large ATR, small DTR	Sparse: less than 35 cm (14 in.) annually, mostly snow	Latitude; distance from sources of warmth and moisture; extreme seasonal contrasts in sunlight and darkness
Ice cap (EF)	Antarctica and Greenland	Long, dark, windy, bitterly cold winters; cold, windy summers; no monthly average above 0°C (32°F); large ATR, small DTR	Very sparse: less than 13 cm (5 in.) annually, all snow	Latitude; distance from sources of heat and moisture; extreme seasonal contrasts in sunlight and darkness; polar anticyclones

ATR = annual temperature range; DTR = daily temperature range.

(a) Tundra climate

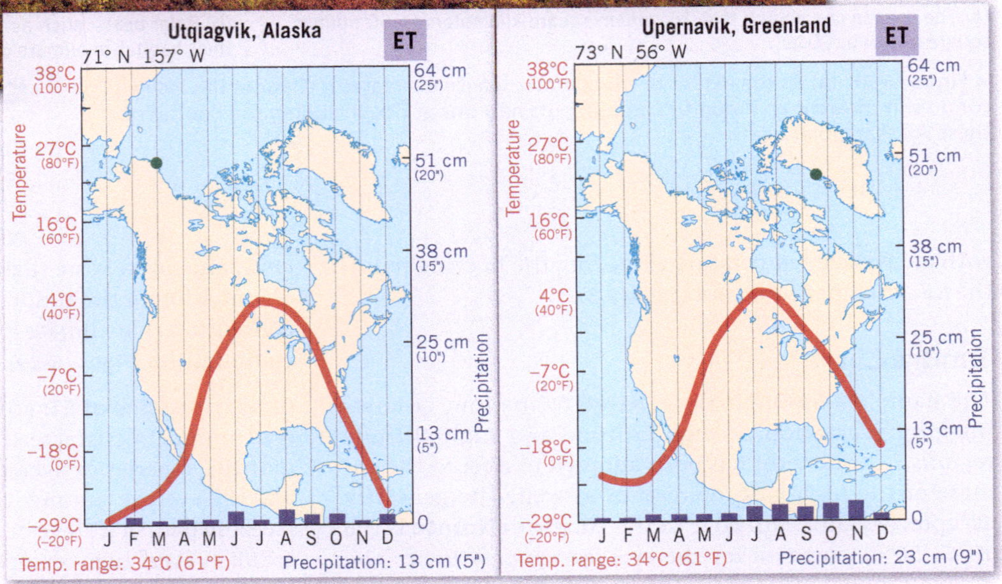

(b)

(c)

▲ **Figure 8-29** Tundra climate. (a) Autumn colors of the tundra in western Denali National Park, Alaska. Climographs for representative tundra stations in (b) Utqiagvik (formerly Barrow), Alaska, and (c) Upernavik, Greenland.

in any month is small and the month-to-month variation is minor. Winter snow appears to be more extensive than it actually is because no melting occurs and because winds swirl it horizontally even when no snow is falling. Radiation fogs are fairly common throughout ET regions, and sea fogs are sometimes prevalent for days along the coast.

Controls of ET: Daily temperature ranges are small because the Sun is above the horizon for most of the time in summer and below the horizon for most of the time in winter. Thus, nocturnal cooling is limited in summer, and daytime warming is almost nonexistent in winter. Moisture availability is very restricted in ET regions despite the proximity of an ocean—the air is too cold to contain much water vapor, so the absolute humidity is almost always very low. Moreover, anticyclonic conditions are common, with little uplift to encourage condensation.

Ice Cap Climate (EF)

The most severe of climates, **ice cap climate**, is restricted to Greenland (all but the coastal fringe) and most of Antarctica. The combined extent of these two regions is more than 9 percent of all land area (see Figure 8-27).

Characteristics of EF: The EF climate is one of perpetual frost where vegetation cannot grow, and the landscape consists of a permanent cover of ice and snow (**Figure 8-30**). The extraordinary severity of EF temperatures is increased because both Antarctica and Greenland are ice plateaus, so relatively high altitude adds to high latitude as a thermal factor. All months have average temperatures below freezing, and in the most extreme locations the average temperature of even the warmest month is below −18°C (0°F). Winter temperatures average between −34 and −51°C (−30°F and −60°F), and extremes well below −73°C (−100°F) have been recorded at interior Antarctic weather stations.

The air is chilled so intensely from the underlying ice that strong surface temperature inversions prevail most of the time. Heavy, cold air often flows downslope as a vigorous *katabatic wind* (discussed in Chapter 5). Strong winds and blowing snow characterize the ice cap climate, particularly in Antarctica.

(a) Ice cap climate

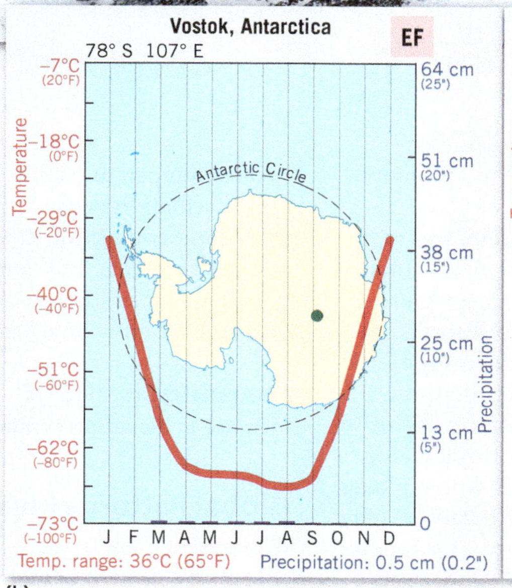

(b)

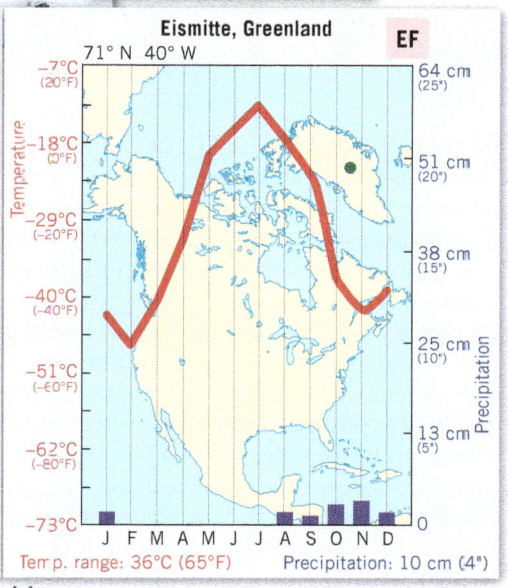

(c)

▲ **Figure 8-30** Ice cap climate. (a) Long-tailed Gentoo penguins congregate on Cuverville Island, Antarctica. Climographs for representative ice cap stations in (b) Vostok, Antarctica, and (c) Eismitte, Greenland. Notice that Vostok receives almost no measurable precipitation in a typical year. (The temperature scales are different from those of the other climographs in this chapter. Also, the climograph for Elsmitte is based on less than 30 years of data.)

Precipitation is very limited, but the snow that falls doesn't readily melt. These regions are essentially polar deserts; most places receive less than 13 centimeters (5 inches) of moisture annually.

Controls of EF: The EF climates are brutally cold as a result of their high latitude—even when the Sun is above the horizon, it is so low in the sky that little warming is possible. The extreme dryness of these areas is largely due to the low temperatures: the air is so cold that it has almost no water vapor capacity. The cold conditions lead to stability with too little likelihood of uplift to permit much precipitation.

LearningCheck 8-9 Why are polar climates so dry?

Highland Climate (Group H)

Highland climate (Group H) is not defined in the same sense as all the other groups. Climatic conditions in mountainous areas have almost infinite variations from place to place, and many of the differences extend over very limited horizontal distances. Köppen did not recognize highland climate as a separate group, but most researchers who have modified his system have added such a category. Highland climates are found in relatively high uplands (mountains and plateaus) having complex local climate variation in small areas (Figure 8-31). The climate of any highland location is usually closely related to that of the adjacent lowland, particularly with regard to seasonality of precipitation.

Characteristics and Controls of H Climates: Altitude variations influence all four elements of weather and climate. As we learned in Chapters 4, 5, and 6, with increasing altitude, temperature and pressure generally decrease; wind is less predictable but tends to be brisk and abrupt with many local wind systems. As a result of orographic lifting, precipitation is characteristically heavier in highlands than it is in the surrounding lowlands. Therefore, mountains usually stand out as moist islands on a rainfall map (Figure 8-32).

Altitude is more significant than latitude in determining climate in highland areas. Steep vertical gradients of climatic change are expressed as horizontal bands along the slopes. An increase of a few hundred meters in elevation may be equivalent to a journey of several hundred kilometers poleward in terms of temperature and related environmental characteristics. This phenomenon, called *vertical zonation*, is particularly prominent in tropical highlands

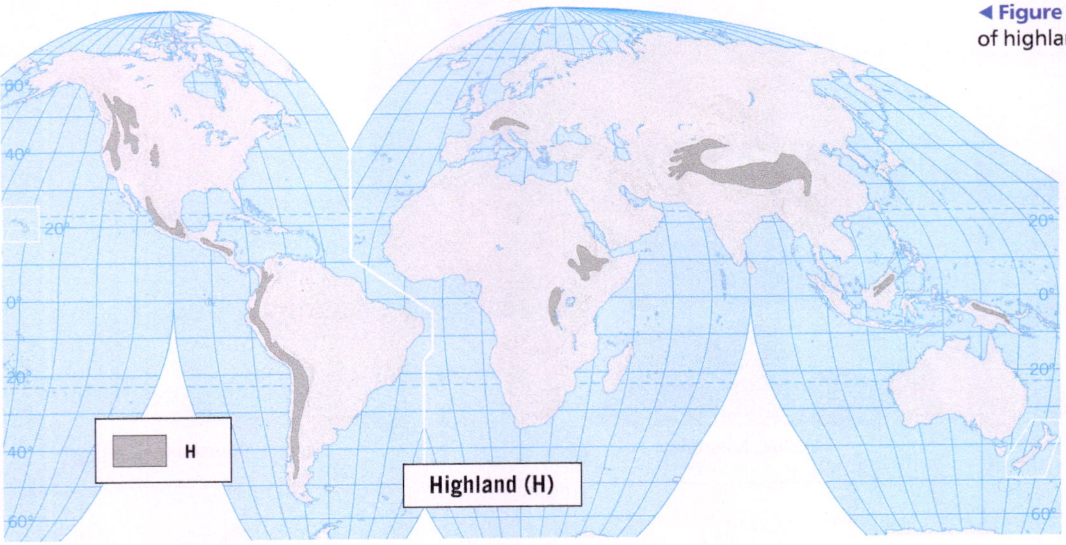

◀ **Figure 8-31** The global distribution of highland (H) climate.

(**Figure 8-33**). (We explore this topic again in Chapter 11 when we discuss patterns of *biogeography*.)

Exposure—whether a slope, peak, or valley faces windward or leeward—has a profound influence on climate. Ascending air on a windward face brings a strong likelihood of heavy precipitation, whereas a leeward location is sheltered from moisture or has predominantly downslope wind movement with limited opportunity for precipitation. The angle of exposure to sunlight is also a significant factor in determining climate, especially outside the tropics. Slopes that face equatorward receive direct sunlight, which makes them warm and dry; adjacent slopes facing poleward may be much cooler and moister simply because of a smaller angle of solar incidence and more shading.

Changeability is perhaps the most conspicuous characteristic of highland climate. The thin, dry air permits rapid influx of insolation by day and rapid loss of radiant energy at night, so daily temperature ranges are very large, with frequent and rapid oscillation between freeze and thaw. Daytime upslope winds and convection cause rapid cloud development and abrupt storminess. Travelers in highland areas are well advised to prepare for sudden changes from hot to cold, from wet to dry, from clear to cloudy, from quiet to windy, and vice versa.

LearningCheck 8-10 **In terms of climate, how is an increase in elevation similar to an increase in latitude?**

Global Patterns Idealized

It should be clear to you by now that there is a fairly predictable global pattern of climatic types and that this pattern is based primarily on latitude, position on a continent, and the general circulation of the atmosphere and oceans. **Figure 8-34** summarizes the idealized distribution of the mild

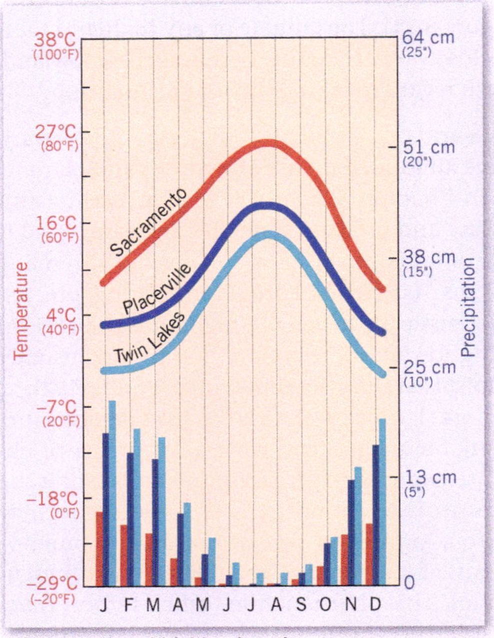

(a) Lowland and highland stations

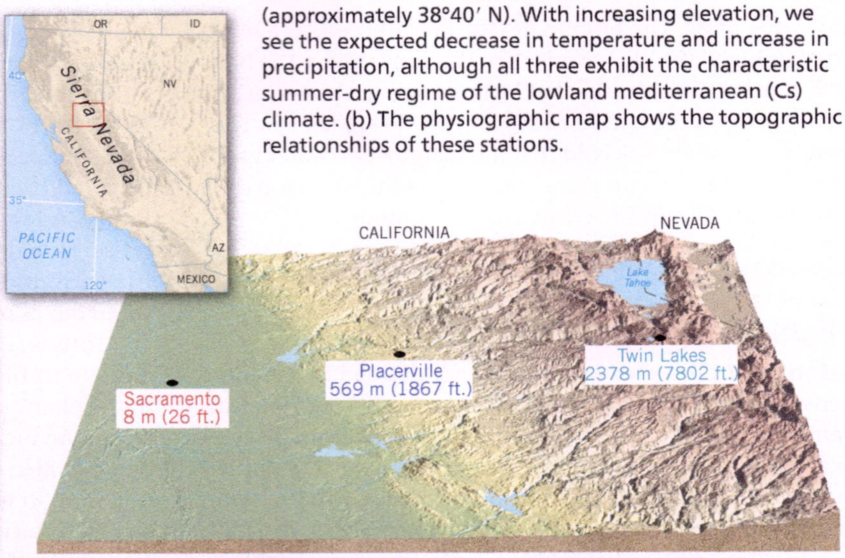

(b) Elevation differences

◀ **Figure 8-32** (a) Climate of a lowland station (Sacramento) and two nearby highland localities (Placerville and Twin Lakes) at about the same latitude in California (approximately 38°40′ N). With increasing elevation, we see the expected decrease in temperature and increase in precipitation, although all three exhibit the characteristic summer-dry regime of the lowland mediterranean (Cs) climate. (b) The physiographic map shows the topographic relationships of these stations.

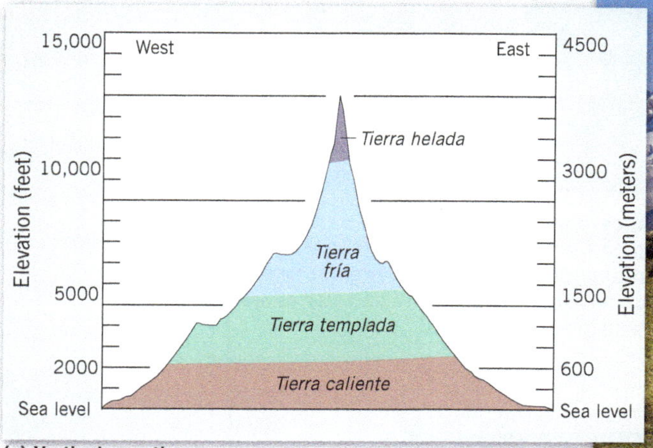

(a) Vertical zonation

(b) Highland climate

▲ **Figure 8-33** Vertical climate zonation in the tropics. (a) Idealized diagram. *Tierra caliente* ("hot land") is a zone of high temperatures, dense vegetation, and tropical agriculture. *Tierra templada* ("temperate land") is an intermediate zone of slopes and plateaus and moderate temperatures. *Tierra fria* ("cold land") is characterized by warm days and cold nights; its agriculture is limited to hardy crops. *Tierra helada* ("frozen land") has cold weather throughout the year. (b) The tierra fria zone in Ecuador. The Chimborazo Volcano rises into the tierra helada zone.

(A, B, and C) climates along the west coasts of continents, where the distribution pattern is slightly more regular than along the east coasts. Notice especially the relationship of climate types to the seasonal shifts of the intertropical convergence zone and the STHs.

Global Climate Change

So far in this chapter we have discussed global climate patterns as we observe them in the present day. We now explore how these patterns can change over time.

For the remainder of this chapter, we look specifically at global climate change. We address questions first raised in Chapter 4: Is human activity causing global climate change, and if so, what are the likely long-term consequences? In providing a thoughtful answer to these questions, we look at how past climates are identified, how climate change is recognized, what kinds of factors may lead to climate change, and what evidence indicates that climate is changing today.

As we first saw in Chapter 3, within any climate system, fluctuations in weather occur from year to year. We can think of this random year-to-year variation in weather as the background "noise" in the long-term climate record. Within the climate record, however, other kinds of variation may be present: episodic events and natural cycles with periods ranging from a few years to many decades (examples include El Niño and the Pacific Decadal Oscillation, discussed in Chapter 5), as well as long-term global climate change itself. Long-term climate change influences plant and animal biomes. See the box *Snapshots from Our National Parks: Impacts of Climate Change* for more on how climate change affects the plants and animals living in our national parks.

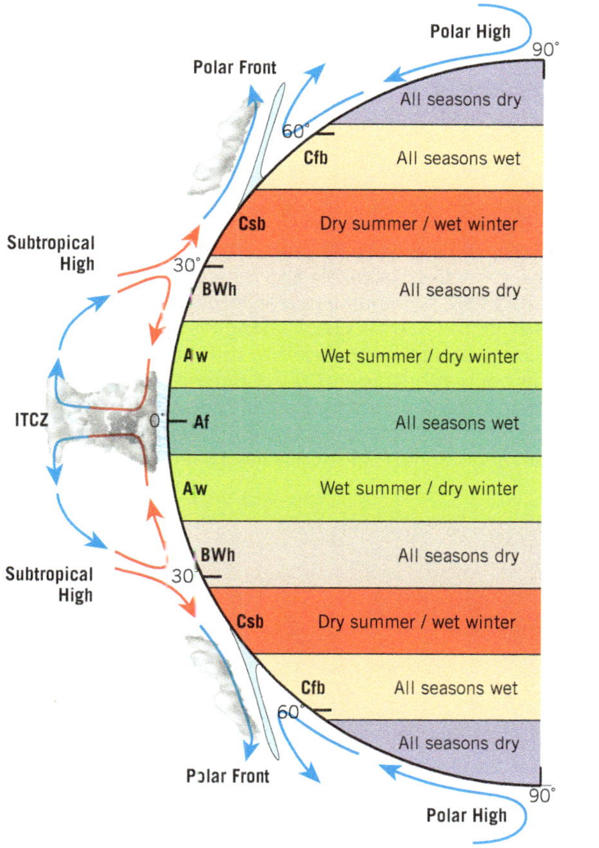

▲ **Figure 8-34** Idealized seasonal precipitation patterns and climates along the west coast of continents. Notice that the progressions north and south of the equator are mirror images. Much of this pattern is due to the seasonal shifts of the ITCZ and the subtropical highs

Snapshots From Our National Parks

Impact of Climate Change

Ecosystems under Stress

▲ **Figure 8-B** Wild Assateague ponies roam Assateague Island National Seashore. Assateague Island is a 37-mile-long barrier island off the coasts of Maryland and Virginia and includes the Chincoteague National Wildlife Refuge in Virginia. Sea-level rise alone is likely to put much of the island underwater. Sea-level rise and storms are moving the island westward toward the mainland in a process called "island rollover."

Climate change causes a shift in temperature and precipitation properties for a location. These properties influence the plants and animals that are capable of living in that place. As climates change, ecosystems (communities of living organisms) migrate, but often not at the same rate. Take for example is the symbiotic relationship that exists between the yucca moth and Joshua trees (shown in the chapter opener image). Yucca moths pollinate Joshua trees, and Joshua tree flowers provide a home for yucca moth larvae. As climate changes, the overlap between the habitats of Joshua trees and yucca moths is getting smaller. By 2100, scientists fear that there won't be any more Joshua trees in Joshua Tree National Park. For other examples of park ecosystems under stress, see **Figures 8-A** to **8-D**.

▶ **Figure 8-C** Corals are slow-growing invertebrates that attach to hard surfaces on the sea floor. Colonies of coral grow into reefs that support diverse food chains and shelter many marine animals. Coral bleaching, like this seen at the National Marine Sanctuary of American Samoa, occurs when the algae that give coral their color die as ocean temperatures rise. The algae normally provide nourishment and a protective coating for the delicate corals.

▼ **Figure 8-A** The Haleakalā silversword of Haleakalā National Park can live up to 90 years. It blooms once in its lifetime, then dies. Haleakalā and Hawai'i Volcanoes National Parks are both home to a wide variety of rare plants with limited habitat space. Changing trade wind patterns are causing these habitats to shift faster than their plant populations. Scientists use climate change predictions to develop propagation and relocation strategies to ensure these species can survive in a wider ecological range. So far 135 species have been "resettled."

▲ **Figure 8-D** Yosemite, one of this country's most iconic national parks, is experiencing a series of climate change impacts. Rising temperatures result in smaller snowpack, shrinking glaciers, and dry waterfalls. Persistent drought has made trees weaker and susceptible to invasive insects that can kill the trees. All these changes affect the Park's habitats and the diverse communities of plants and animals they support. Affected too are the natural cycles of wildfire that kept the Park's forests healthy. The length of the fire season is increasing and large-scale fires have become more common. Here, visitors look out on Yosemite Valley through the haze of smoke caused by the 2018 Ferguson fire that swept through three national forests and burned more than 95,000 acres.

Time Scales of Climate Change

The time scale of observations determines the kinds of patterns that stand out in the climate record. For example, in the case of global temperature:

- Over the last 70 million years, a clear global cooling trend is visible (**Figure 8-35a**).
- Over the last 150,000 years, temperature fluctuated significantly until about 10,000 years ago, when it increased sharply at the end of the last glacial advance of the Pleistocene epoch; temperature has remained warm and fairly steady since (**Figure 8-35b**).
- Over the last 150 years, a warming trend appears underway—a clear departure from at least the previous 1000 years (**Figure 8-35c**).

The long-term temperature trends are even more complicated than might first appear: these temperature changes weren't uniform around the world.

Determining Climates of the Past

Before we can assess the possibility of climate change today, we must have some understanding of past climates and the causes of climate change, a field of study known as **paleoclimatology**.

Detailed instrument records of weather go back only a few hundred years, so it is necessary to use *proxy* ("substitute") measures of climate to reconstruct conditions in the past. Information about past climates comes from many sources: tree rings, ice cores, oceanic sediments, coral reefs, relic soils, and pollen, among others. No one method for determining past climate conditions is ideal—each has strengths and weaknesses. However, by correlating the results obtained from one method with those from another, paleoclimatologists can construct a detailed history of Earth's climate.

Dendrochronology

Most trees growing in temperate areas increase trunk diameter by adding one concentric *tree ring* per year of growth.

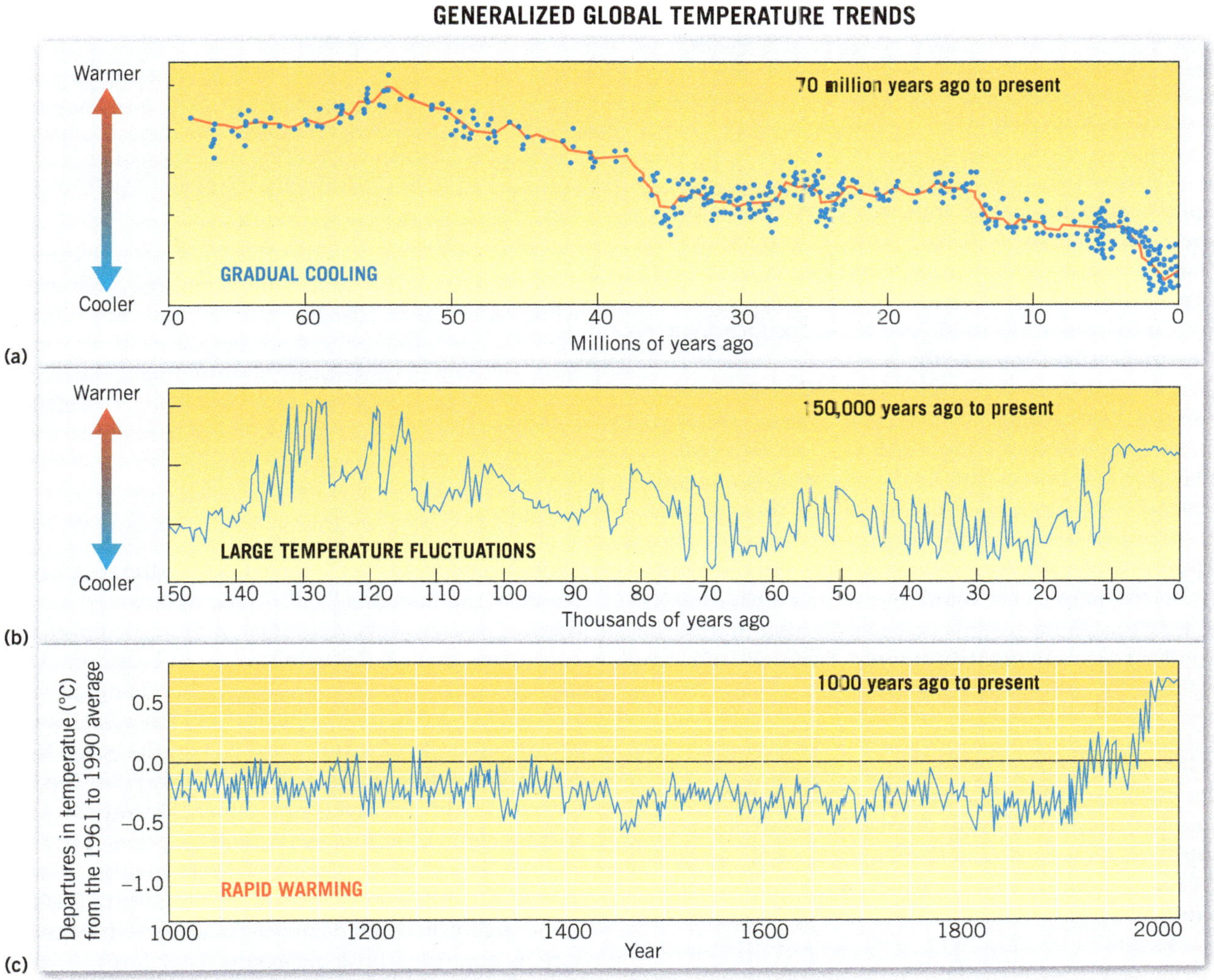

▲ **Figure 8-35** Generalized global temperature trends at three time scales. (a) 70 million years ago to present. (Curve was generalized from ice volume and ocean temperature proxy measures.) (b) 150,000 years ago to present. (Curve was derived from North Atlantic Ocean proxy measures.) (c) 1000 years ago to present. (Curve was derived from proxy measures and the historical record after 1900.)

Animation
End of the Last Ice Age
https://media.pearsoncmg.com/ph/esm/esm_mcknight_physgeo_13/media/

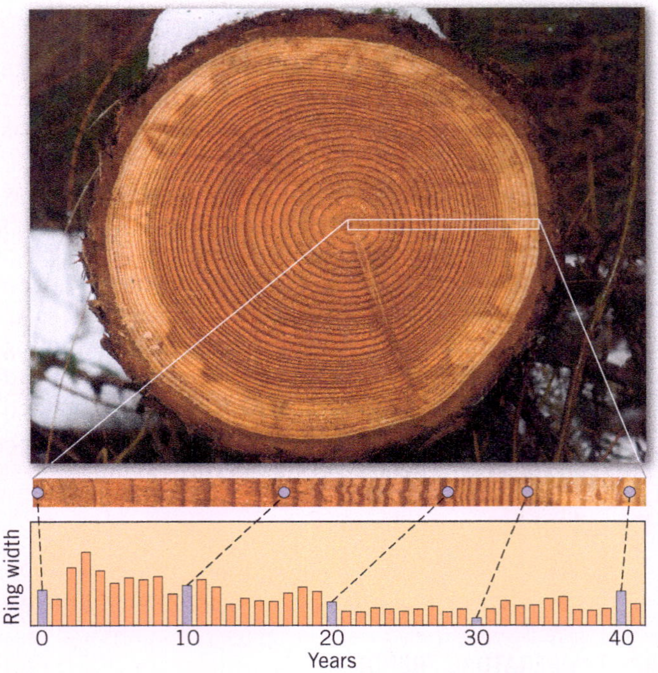

▲ **Figure 8-36** Dendrochronology uses tree rings to identify past climate. Each ring represents one growing season. Wider rings often indicate favorable growing conditions, whereas narrow rings indicate harsher conditions.

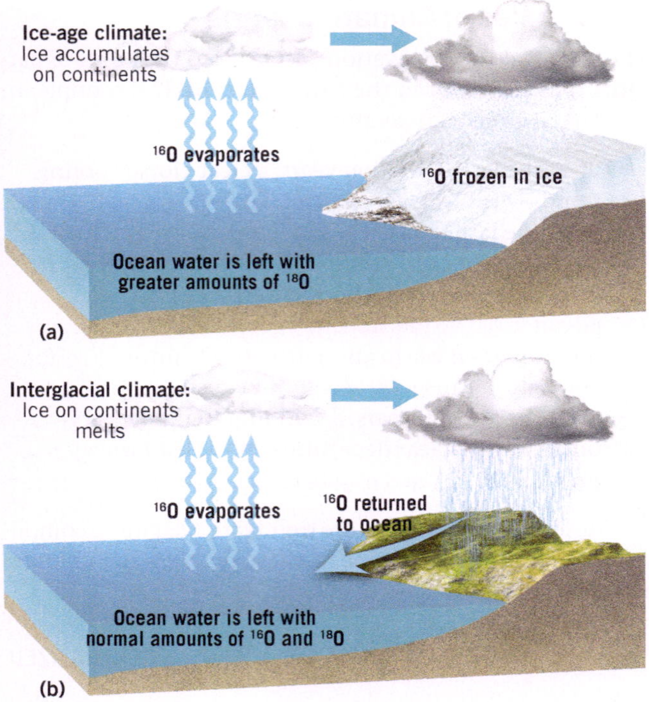

▲ **Figure 8-37** The concentration of oxygen isotopes in ocean water varies with climate. (a) During an ice age, water with ^{16}O evaporates from the ocean and is locked-up in continental ice, so the relative concentration of ocean water with ^{18}O increases. (b) When glaciers melt, ^{16}O is returned to the ocean.

By counting the number of rings, we can determine the age of a tree. Tree rings also provide information about climate: during years when the growing conditions are favorable (such as mild temperatures and/or ample precipitation), tree rings tend to be wider than in years when the growing conditions are harsher (Figure 8-36).

Living or dead trees can be used in **dendrochronology** (the study of dating past events by analyzing tree rings). Small cores are taken so that living trees are not harmed, and fallen trees, buried trunks, and timber from archeological sites all may be used to extend tree-ring chronologies. In addition to determining the age of trees, dendrochronologists can, by comparing and correlating the tree-ring patterns of many trees in an area, establish dates for catastrophic events (such as floods and fires that have killed trees) and identify periods of drought and lower or higher temperatures.

Oxygen Isotope Analysis

In the nucleus of many atoms, the number of protons and neutrons is the same. For example, most oxygen atoms have 8 protons and 8 neutrons, giving them an atomic weight of 16 (^{16}O or "oxygen 16"). However, a small percentage of oxygen atoms have two extra neutrons, giving these atoms an atomic weight of 18 (^{18}O). ^{16}O and ^{18}O are known as *isotopes* of oxygen. Both isotopes of oxygen are found in common molecules such as water (H_2O) and calcium carbonate ($CaCO_3$).

Through **oxygen isotope analysis**, the ratio of $^{18}O/^{16}O$ in the molecules of substances such as water and calcium carbonate can tell us something about the environment in which those molecules formed. Because they contain the lighter oxygen isotope, water molecules with ^{16}O evaporate more easily than those with ^{18}O. Thus, precipitation such as rain and snow tends to be relatively rich in ^{16}O. During an ice age, great quantities of ^{16}O are locked up in glacial ice on the continents, leaving a greater concentration of ^{18}O in the oceans (Figure 8-37a). During a warmer, interglacial period, the glacial ice melts, returning ^{16}O to the oceans (Figure 8-37b).

Oceanic Sediments: Ocean floor sediments provide a record of the changing $^{18}O/^{16}O$ ratio in seawater over periods of thousands of years. A number of common marine microorganisms, such as *foraminifera*, secrete external shells of calcium carbonate that retain the $^{18}O/^{16}O$ ratio of the water in which these microorganisms live. The remains of these organisms build up layer upon layer on the ocean floor (see Figure 9-8). By comparing the $^{18}O/^{16}O$ ratio of the calcium carbonate in these layers of sediment, scientists can determine when periods of glaciation took place. If the $^{18}O/^{16}O$ ratio in the calcium carbonate is high (in other words, relatively high amounts of ^{18}O are present in the shells), then those organisms lived during a glacial period; if the $^{18}O/^{16}O$ ratio is low, then those organisms lived during an interglacial period.

Coral Reefs: Coral reefs are built up in tropical ocean areas by massive colonies of tiny coral polyps that excrete calcium carbonate exoskeletons. (Coral is discussed in more detail in

Chapter 20.) Because this $CaCO_3$ was extracted from the seawater, the $^{18}O/^{16}O$ ratio of the coral provides information about the climate at the time the reef was forming. Furthermore, because coral reefs develop in shallow water, the relative height of old reefs may help us determine past fluctuations in sea level.

LearningCheck 8-11 **How can oxygen isotope analysis tell us about temperature in the past?**

Ice Cores

The analysis of ice cores can also give us information about past climates. By drilling down into glaciers, we can obtain a record of snowfall going back hundreds of thousands of years in some locations (Figure 8-38). Such ice cores offer several kinds of information about past conditions. A greater number of water molecules with ^{18}O evaporate from the oceans when temperatures are high than when temperatures are low, thus the $^{18}O/^{16}O$ ratio found in a layer of ice serves as a "thermometer" for the climate at the time that snow fell. Oxygen isotope analysis of ice cores has become one of the most important ways scientists have been able to construct a record of past global temperature.

Ice cores also provide direct information about the composition of the atmosphere in the past. As snow accumulates and is compacted into glacial ice, air bubbles from the atmosphere become frozen into the ice. These tiny air bubbles deep in glaciers are preserved samples of the ancient atmosphere, allowing direct measurements of the concentration of CO_2 and other gases. In addition, particulates in the ice can offer information about past cataclysmic events such as major volcanic eruptions.

Dome C in Antarctica: Of all the locations where ice cores have been extracted from glaciers, the most significant is probably Dome C in Antarctica. Dome C is located about 1670 kilometers (1040 miles) from the South Pole at a latitude of 75° S and a longitude 123° E—where the Antarctic ice cap is thickest (Figure 8-39a). The coring was undertaken by the European Project for Ice Coring in Antarctica (EPICA). A multinational team of scientists extracted a continuous 10-centimeter (4-inch) diameter ice core record to a depth of more than 3 kilometers (2 miles). The analysis has provided climate and atmospheric composition data back 800,000 years—far longer than any other ice core—providing climate information on eight complete glacial/interglacial cycles (Figure 8-39b).

The findings at Dome C closely match the proxy climate records derived from both the nearby Vostok, Antarctica, ice core and the oxygen isotope analysis of the calcium carbonate in the shells of foraminifera found in oceanic sediments, adding to the scientists' confidence in the soundness of the Dome C data. Scientists are currently searching for possible drilling sites that can extend the ice core record back even further.

The Dome C climate record shows that the present concentration of CO_2 in the atmosphere is greater than at any time in the last 800,000 years and that increases and decreases in global temperature closely correlate with changes in greenhouse gas concentration (especially carbon dioxide and methane). Lower concentrations of greenhouse gases are present during glacial episodes, whereas higher concentrations of greenhouses gases are present during interglacial (warm) periods. However, research shows that atmospheric CO_2 concentrations usually increased several hundred years after an interglacial temperature increase began. This suggests that the increase in greenhouse gas concentration was not the trigger for the interglacial temperature increase; increasing CO_2 levels may have simply amplified the warming climate. Today, however, an increase in temperature closely follows an increase in greenhouse gases, indicating that the temperature increase has been caused by the increase in greenhouse gases.

LearningCheck 8-12 **How can ice cores give us information about past concentrations of greenhouse gases in the atmosphere?**

◄ **Figure 8-38** Scientists in the Arctic Circle gather ice cores for an ongoing ice characterization project. The cores will be tested for salinity and ice density.

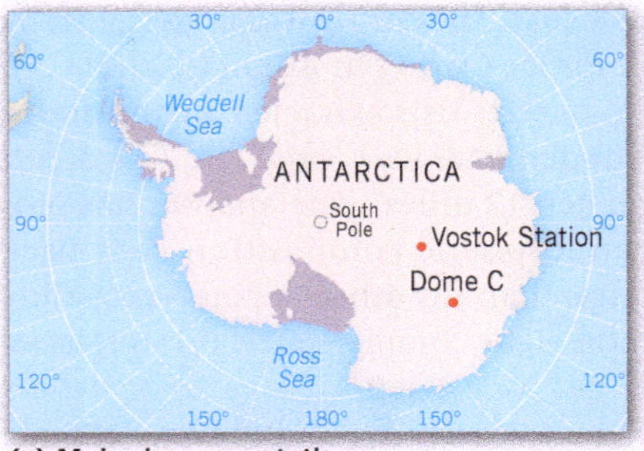

(a) Major ice core stations

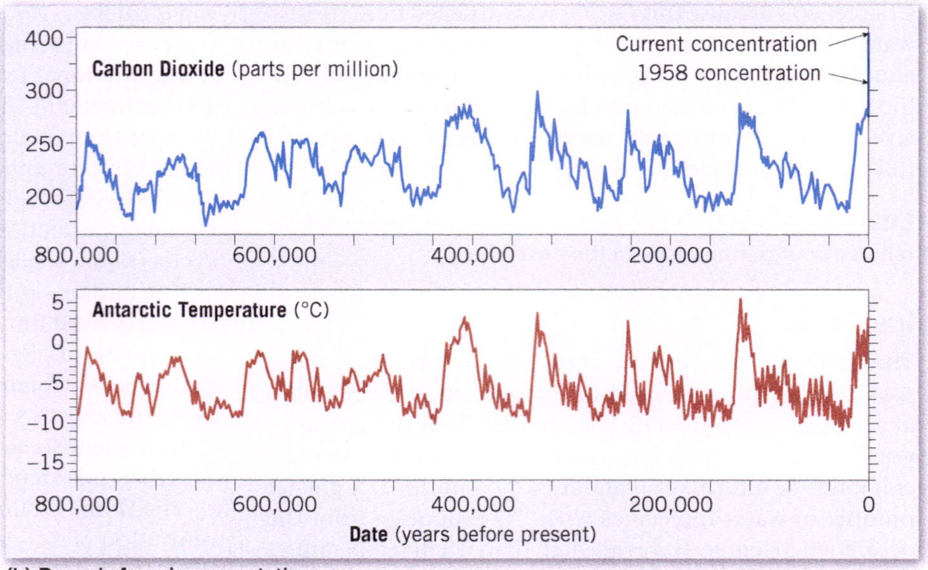

(b) Records from ice core stations

▲ **Figure 8-39** (a) Map of Antarctica showing the location of EPICA Dome C and Vostok Station. (b) CO_2 record over the last 800,000 years from EPICA Dome C ice core and other locations, and the corresponding temperature record derived from Dome C. Temperature scale is relative to average temperature over the last 1000 years.

Pollen Analysis

Video
18,000 Years of Pine Pollen
https://media.pearsoncmg.com/ph/esm/esm_mcknight_physgeo_13/media/

Another important method used for determining past climates comes from the field of *palynology*, or *pollen analysis*. Airborne pollen from trees and other plants can be preserved in sediment layers on lake bottoms and in bogs. Cores of these sediment layers are taken and analyzed.

Radiocarbon dating of material in each layer provides an estimate of the age of organic material younger than about 50,000 years. In radiocarbon dating, the relative amounts of two isotopes of carbon found in organic material are compared. Radioactive ^{14}C decays over time at a known rate into ^{14}N, a stable isotope of nitrogen; the more common stable isotope of carbon, ^{12}C, remains constant no matter how old the sample is. So the $^{14}C/^{12}C$ ratio serves as an indication of how long ago the plant or animal material was alive. Once the age of a sediment layer is known, palynologists look at the types of pollen present to determine which plants and trees were living at the time that layer was forming. Since plant species change as the environment changes—some species are better adapted for colder conditions, for example—the prevalence of one plant community over another acts as a proxy for the climate of the time.

Remnant Glacial Landforms

As we see in Chapter 19, glaciers leave distinctive marks on the landscape. As glaciers move slowly, they abrade the adjacent terrain. Striated (scratched) or polished bedrock on valley walls can show the height of past glaciers. Mounds of unsorted glacial rock deposits called *moraines* accumulate at the edges of glaciers and are used to determine the original extent of the ice and the amount of glacier retreat over the last century.

Speleothems

Deposits of calcium carbonate in limestone caverns provide further clues to past climates. *Speleothems* are deposits that form slowly as mineral-rich water drips into cave openings. The most familiar types are *stalactites*, which grow downward, and *stalagmites*, which grow upward (see Figure 17-4). As calcium carbonate precipitates out of solution on the speleothem, it leaves a growth ring, somewhat like a tree ring. The size of the ring, plus the isotopes of carbon and oxygen present, provide information about the precipitation and temperature conditions of the environment when the speleothem formed.

Causes of Climate Change

A quick glance back to Chapter 4 and the diagram showing Earth's solar radiation budget (see Figure 4-18) will remind you of the many variables that influence patterns of temperature: the quantity of incoming solar radiation, the albedo of the atmosphere and surface, the concentration of greenhouse gases, the transfer of energy through oceanic and atmospheric circulations, and many others. A change in any one of those variables has the potential to change temperature, wind, pressure, and moisture patterns.

Not surprisingly, then, a number of mechanisms are likely to have had a role in past climate change and are operating in the present day as well. In the following sections we describe several kinds of climate-change mechanisms.

Atmospheric Aerosols

Large quantities of particulates ejected into the atmosphere by volcanic eruptions can alter global temperatures (**Figure 8-40**). Fine volcanic ash and other aerosols can reach the stratosphere, where they may circle the globe for years, blocking incoming solar radiation and lowering

▲ **Figure 8-40** Eruption of the Calbuco volcano, south of Santiago, Chile, in April 2015. The plume of ash reached an altitude of more than 15 kilometers (9.3 miles).

temperatures. For example, the 1991 eruption of Mount Pinatubo in the Philippines increased aerosol concentrations (especially sulfur dioxide) in the atmosphere enough to lower global temperatures by about 0.5°C (0.9°F) for the next year.

Large asteroid impacts also can inject enough dust into the atmosphere to alter global climate significantly. It is widely accepted by scientists that the impact of a 10-kilometer-wide asteroid 65 million years ago contributed to the extinction of the dinosaurs. (Environmental change associated with massive *flood basalt* eruptions occurring in India at the same time may have been a more significant factor; see Chapter 14.)

Anthropogenic aerosols, especially sulfates and black carbon, may influence climate. Sulfates released by the burning of fossil fuels tend to scatter incoming solar radiation, so they have a net cooling effect. Conversely, black carbon emissions from the burning of diesel and biofuels tend to absorb solar radiation, leading to warming.

Over the last 40 years, stricter environmental regulations in industrialized countries have cut sulfate emissions in half while emissions of black carbon from industrializing countries in the Northern Hemisphere, especially in Asia, have risen. A 2009 NASA study suggested that much of the warming in the Arctic since 1976 may be linked to changes in aerosols—less cooling from sulfates combined with more warming from black carbon.

Fluctuations in Solar Output

A link between past climates and variations in the energy output of the Sun has been pursued for many decades. For example, researchers have looked for a connection between climate and sunspot activity. *Sunspots* appear as dark spots on the surface of the Sun. They are slightly cooler regions of the Sun's *photosphere* (outer layer) and are associated with intense magnetic storms on the surface of the Sun, as well as the ejection of charged particles. (Some of the charged particles reach Earth as part of *solar wind* and produce auroral displays in the upper atmosphere; see Figure 3-9.)

Some scientists have noted that several past climate events, such as the *Little Ice Age*—a period of unusually cold weather from about 1400 to 1850—roughly corresponded with the *Maunder Minimum*, a period of greatly reduced sunspot activity between 1645 and 1715. Recent satellite measurements show that the energy striking Earth's upper atmosphere, or *total solar irradiance* (TSI), varies by about 0.1 percent during the well-documented 11-year cycle of sunspot activity—with less energy received during periods of low sunspot activity. Ultraviolet radiation seems to vary much more than this, so fluctuations in TSI may affect the ozone layer and perhaps other aspects of energy flow through the atmosphere.

Although some of the relationships between sunspot cycles and climate appear at first to be strong statistically, many scientists remain unconvinced—in part because not all episodes of unusual sunspot activity correlate with changes in climate. Furthermore, although variation in TSI may well play a role in climate change, the measured fluctuations do not seem to be nearly enough to account for the global warming trend observed over the last century.

LearningCheck 8-13 **How can differences in atmospheric aerosols influence climate?**

Variations in Earth–Sun Relations

As we saw in Chapter 1, the change of seasons is largely a consequence of the changing orientation of Earth's rotational axis to the Sun during the year (see Figure 1-24). Recall that Earth's axis maintains an inclination of 23.5° at all times, that Earth's axis always points toward the North Star, Polaris, and that Earth is closest to the Sun on January 3. As it turns out, all of these aspects of Earth–Sun relations change over periods of many thousands of years in a series of predictable, well-documented cycles (**Figure 8-41**).

Eccentricity of Orbit: The "shape" of Earth's elliptical orbit, or *eccentricity*, varies in a series of cycles lasting about 100,000 years. Sometimes Earth's orbit is nearly circular, whereas at other times the orbit is more elliptical, thus altering the distance between the Sun and Earth. The present difference in distance between the Sun and Earth at aphelion and perihelion is only about 3 percent (about a 5,000,000-kilometer difference), but over the last 600,000 years or so, the difference has varied from 1 percent to 11 percent. The greater the difference in distance, the greater the difference in insolation at various points in Earth's orbit.

Obliquity of Rotational Axis: The inclination of Earth's axis, also called *obliquity* or *tilt*, varies from 22.1° to 24.5° in a cycle lasting about 41,000 years. When inclination is greater, seasonal variation between the low latitudes and high latitudes tends to be greater; when inclination is lesser, seasonal contrasts are less significant.

Precession: Earth's axis "wobbles" like a spinning top, so over time it points in different directions relative to the stars in a 25,800-year cycle called *precession*. Precession alters the timing of the seasons relative to Earth's position in its orbit around the Sun.

Milankovitch Cycles: These long-term cycles "overlap"; as a result, there are intervals when significantly less radiation reaches Earth's surface (especially in the high latitudes) and intervals of greater or lesser seasonal contrasts. For example, at times when there is less "seasonality"—in other words, when there are smaller contrasts between winter and summer—more snow can accumulate in high latitudes (due to greater snowfall from milder winters) and less melting will take place (due to lower summer temperatures).

These long-term cycles are known as **Milankovitch cycles**, after Milutin Milankovitch, an early twentieth-century Yugoslavian astronomer. Milankovitch looked at the combined effects of these cycles and concluded that they were responsible for an approximately 100,000-year cycle of insolation variation that has taken Earth into and out of periods of glaciation over the last few millions of years. Although his ideas were mostly rejected by other scientists during his lifetime, recent paleoclimatological work has shown that cyclical variation in Earth–Sun relations is one key to the timing of glacial and interglacial events.

However, these astronomical cycles present some problems as well—for example, the effects of precession that produce cooler conditions in one hemisphere should produce warmer conditions in the other, but this does not appear to be the case. Other factors certainly play a part in the onset and end of glacial periods.

LearningCheck 8-14 What are Milankovitch cycles?

Greenhouse Gas Concentrations

As we first saw in Chapter 4, *greenhouse gases*, such as water vapor, carbon dioxide (CO_2), and methane, play an important role in regulating the temperature of the troposphere. It is also well known—most notably through ice-core analysis—that past fluctuations in global temperature have been accompanied by fluctuations in greenhouse gas concentrations. Warmer periods in the past are generally associated

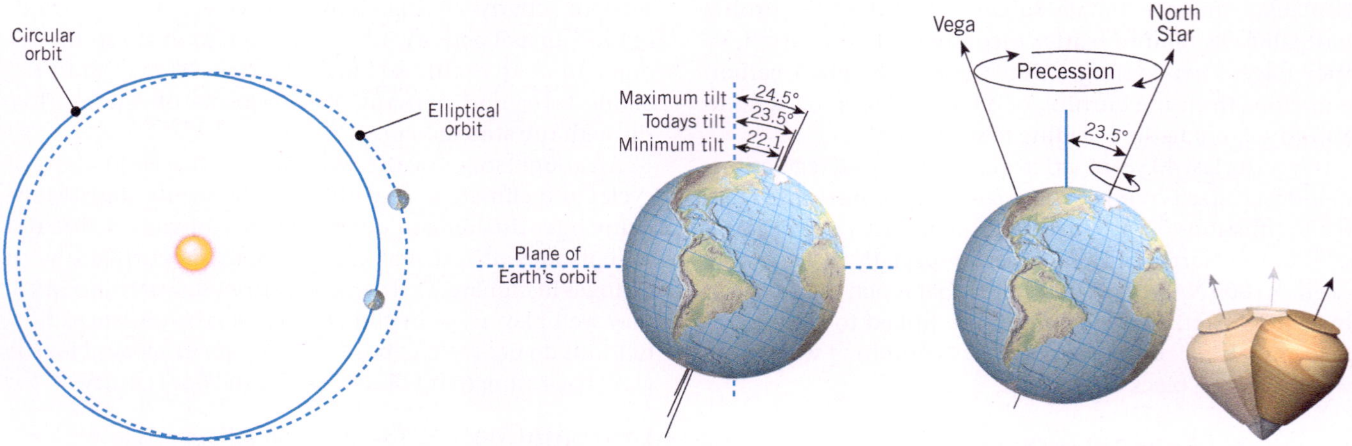

(a) The shape of Earth's elliptical orbit, or *eccentricity*, varies in a cycle lasting about 100,000 years.

(b) The inclination of Earth's axis, or *obliquity*, varies in a cycle lasting 41,000 years.

(c) The orientation of Earth's axis varies in the 25,800-year cycle of *precession*.

▲ **Figure 8-41** Orbital variation cycles in Earth–Sun relations. (a) The shape of Earth's elliptical orbit, or eccentricity. (b) The inclination of Earth's axis, or obliquity. (c) The orientation of Earth's axis, or precession. Currently, Earth's axis points toward the North Star, Polaris; in the future, it will point toward the star Vega.

Animation
Orbital Variations and Climate Change
https://media.pearsoncmg.com/ph/esm/esm_mcknight_physgeo_13/media/

with higher concentrations of gases such as CO_2, whereas cooler periods are associated with lower concentrations of CO_2 (see Figure 8-39b).

How do we know if the sharp increase in atmospheric CO_2 over the last century is the result of human activity or the result of natural processes? Several lines of evidence support the conclusion that human activity is responsible. For example:

- The increase in atmospheric CO_2 correlates well with fossil fuel use since the industrial revolution began in the mid-1700s—although only about 40 percent of the human-released CO_2 has remained in the atmosphere. The balance was either absorbed by the oceans or fixed by plants through photosynthesis.
- The radioactive isotope ^{14}C forms naturally in the upper atmosphere and exists along with the stable ^{12}C isotope. Measurements of the $^{14}C/^{12}C$ ratio derived from tree cores show that the relative level of radioactive ^{14}C in the atmosphere has been decreasing since the industrial revolution. This trend indicates that the additional CO_2 in the atmosphere during that time frame was released from old "dead" sources of carbon, such as coal and petroleum. These sources formed so long ago that most of the radioactive ^{14}C has decayed, leaving relatively more of the stable ^{12}C isotope.

Greenhouse Gas Potency and Residence Time: A complication in trying to understand the role of greenhouse gases in climate change is that the relative potency and the residence time of these gases vary greatly by the gas. For example, methane has much shorter residence time in the atmosphere than does CO_2—after about 12 years in the atmosphere, methane oxidizes to CO_2. However, methane is about 72 times more "efficient" as a greenhouse gas than CO_2 over the first 20 years, and about 25 times more efficient over a 100-year time period. In other words, the effects of methane in the short term may not be the same as in the long term.

Feedback Mechanisms

Adding to the complexity of the relationship between greenhouse gas concentrations and global temperatures are a number of important *feedback mechanisms*. (We first described feedback mechanisms in Chapter 1 where we introduced Earth systems.)

Sea ice and continental ice sheets have high *albedos*—in other words, these surfaces reflect much of the shortwave radiation that strikes them. If global climate were to cool, ice would expand and in turn reflect more incoming radiation, which would lead to more cooling—which in turn would lead to greater expansion of ice. This is a *positive feedback mechanism*. If climate were to warm instead, the ice cover would diminish, which would reduce the albedo and therefore increase absorption by the surface, which would lead to higher temperatures and therefore still less ice cover (see Figure 4.8). This too is a positive feedback mechanism.

One of the most complicated series of feedback mechanisms associated with climate change has to do with water vapor. As climate warms, evaporation increases—thereby increasing the concentration of water vapor in the atmosphere. As we have learned, water vapor is a key greenhouse gas. As warming increases evaporation, the greater concentration of water vapor in the atmosphere leads to still more warming—another positive feedback mechanism. However, here is where things become complicated.

An increase in water vapor would also likely lead to increased cloud cover. Clouds can act as efficient reflectors of solar radiation (leading to cooling), as well as efficient absorbers of terrestrial radiation (contributing to warming). In other words, increased cloud cover can lead to both cooling and warming!

The net effect—cooling or warming—depends on the type of clouds that form. For example, a deep layer of cumulus clouds does not increase albedo as much as it increases absorption of longwave radiation from the surface, thus leading to warming. On the other hand, the expansion of stratus cloud cover increases albedo more than it increases longwave absorption, thus leading to cooling. The effects of clouds remain one of the great complications of modeling future climate change.

LearningCheck 8-15 **How can rising temperature lead to a feedback mechanism that further increases global temperature?**

The Roles of the Ocean

Many important variables in climate change involve the oceans. They absorb a great deal of CO_2 from the atmosphere—by some estimates, the oceans have absorbed about 30 percent of the carbon that human activities have released into the atmosphere since the industrial revolution. Marine plants and animals then remove carbon from ocean water through photosynthesis and the extraction of calcium carbonate to form shells and exoskeletons. Much of this carbon is eventually deposited in sediment layers on the ocean floor.

In addition, some oceanic sediments contain vast quantities of carbon in the form of *methane hydrates* (a kind of ice made from the greenhouse gas methane and water). Should these methane hydrates be destabilized by higher temperatures and the trapped methane be released into the atmosphere, a sudden increase in atmospheric temperature could result. (Some evidence suggests that this happened about 55 million years ago, resulting in a prominent warming period.)

Finally, the oceans obviously play an important role in transferring energy from low latitudes to high latitudes. Should patterns of ocean circulation change, climate could be significantly altered. (We consider this possibility in Chapter 9.)

Anthropogenic Climate Change

The paleoclimatological record provides the backdrop with which we can compare both the rate and the magnitude of climate change over the last century. We're especially interested in knowing whether human activity is responsible for climate change—what is sometimes called *anthropogenic forcing*.

Observed Current Climate Change

As we saw in Chapter 4, the Intergovernmental Panel on Climate Change (IPCC) is the most authoritative international body providing information about climate change to global leaders. In evaluating evidence of climate change over the last century, the *Climate Change 2014: Synthesis Report* of the IPCC's *Fifth Assessment Report* (AR5), released in 2014, concluded that the warming of global climate is "unequivocal."

Some of the specific findings presented in the AR5[2] and studies since include the following:

Changes in Air Temperature

- Based on the longest data set available, the total increase in global temperature between the average of the period 1850–1900 and the average of the period 2003–2012 was 0.78°C (1.4°F). **Figure 8-42** shows the global *temperature anomalies*, or departures from average, from 1850 to 2019.
- Calculated as a linear trend, the averaged land and sea surface temperature between 1880 and 2012 increased by a little more than 1°C (2°F). (Notice that estimates of temperature changes over the last 150 years vary slightly depending on how the calculation is made.)
- Based on instrument data of temperature since 1880, 19 of the 20 warmest years on record occurred since 2001 (1998 is the only year in the twentieth century that is in the "top 20"). The warmest year on record is now 2016.
- The period between 1983 and 2012 was *likely* warmer than any 30-year period in the last 1400 years in the Northern Hemisphere.

Changes in the Ocean

- It is *virtually certain* that the upper 700 meters (2300 feet) of the ocean warmed between 1971 and 2010.
- There is *high confidence* that more than 90 percent of the energy accumulated in the climate system between 1971 and 2010 was stored in the ocean. Only about 1 percent was stored in the atmosphere.
- Because of absorption of CO_2, the ocean has become 28 percent more acidic (an increase in pH of 0.1) since the beginning of the industrial era, according to recent studies.
- In part because of thermal expansion of seawater, between 1901 and 2010, global sea level rose by about 0.19 meters (0.6 feet).
- It is *very likely* that between 1901 and 2010, sea level rise averaged 1.7 millimeters (0.07 inches) per year, but according to recent studies, between 2006 and 2015 the rate had increased to 3.6 millimeters (0.14 inches) per year.

Changes in Polar Regions

- In the Arctic, there is *high confidence* that surface temperature warming exceeded that of the global average.

[2] The IPCC uses the following terms to express the level of consensus, confidence, or probability of their estimates and projections: *likely* (probability of 66 percent or greater), *very likely* (probability of 90 percent or greater), and *virtually certain* (probability of 99–100 percent); also, *medium*, *high*, and *very high confidence* (various degrees of certainty, based on the quality of the evidence and the level of agreement among researchers).

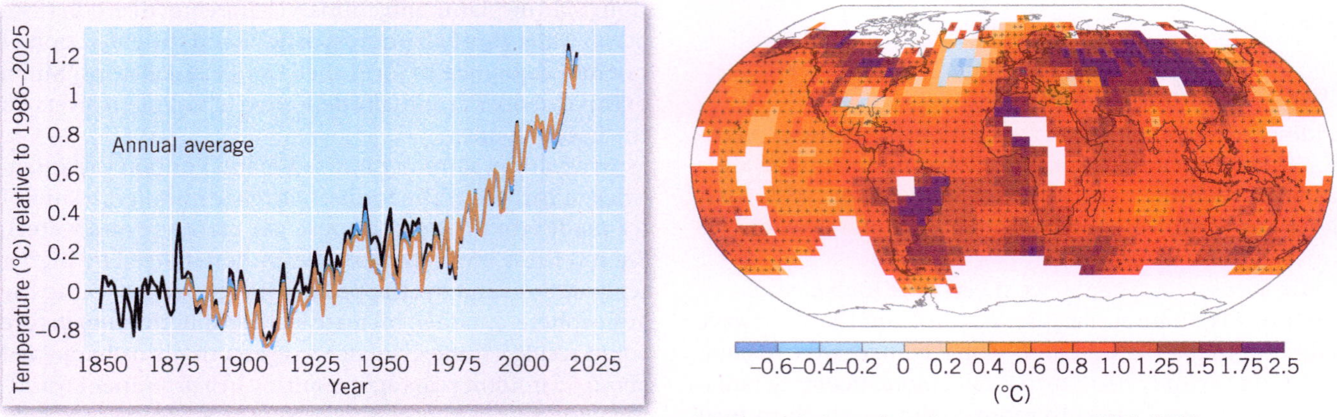

(a) Observed globally averaged combined land and ocean surface temperature anomaly, 1850–2019

(b) Observed change in surface temperature, 1901–2012

▲ **Figure 8-42** (a) Temperature anomalies of global air temperature from 1850 to 2019 derived from multiple data sets. (b) Observed change in temperature from 1901 to 2012; notice that the greatest increase was generally in the high latitudes of the Northern Hemisphere. (Blank areas are regions with insufficient data.)

- There is *high confidence* that between 1992 and 2011 the ice sheets on Antarctica and Greenland lost mass and that glaciers around the world continued to shrink. (Recent data from NASA's Grace satellites show that since 2002 the ice sheet of Antarctica has lost 147 billion metric tons per year and the ice sheet of Greenland has lost 280 billion metric tons per year.)
- Between 1979 and 2012, it is *very likely* that the annual average Arctic sea ice extent decreased by 3.5–4.1 percent per decade. Over that same period, it is *very likely* that average Antarctic sea ice extent increased by 1.2–1.8 percent per decade; however, there is *high confidence* that strong regional differences exist, with some regions around Antarctica increasing and others decreasing.
- It is *very likely* that over the period 1979–2020 the summer sea ice minimum in the Arctic decreased at a rate of 13.1 percent per decade. By the end of the summer in 2012, the extent of Arctic sea ice was the smallest measured since regular satellite monitoring of the ice pack began in 1979. (Data through 2019 indicate that the Arctic summer minimum sea ice extent has decreased 35 percent since 1981. See Figure 8-28a.)
- There is *high confidence* that since 1980 temperatures at the top of the *permafrost* layer have increased in most areas. Warming by as much as 3°C (5.4°F) was observed in parts of northern Alaska.
- There is *very high confidence* that since the mid-twentieth century, the snow cover in the Northern Hemisphere has decreased.

Changes in Weather Patterns

- Extreme weather events are becoming more common. It is *very likely* that the number of both warm days and nights and cold days and nights has increased globally.
- It is *likely* that more land areas have experienced increased numbers of heavy precipitation events than have experienced decreased numbers.

Changes in the Biosphere

- There is *high confidence* that shifts in the geographical ranges, migration patterns, and abundances have occurred in many terrestrial, marine, and freshwater species.
- There is *high confidence* that in many areas crop yields have shown more negative impacts from climate change than positive ones.

LearningCheck 8-16 **What are temperature anomalies? Describe the trend in land–ocean temperature anomalies over the last century.**

Natural or Anthropogenic Climate Change?

So, is it possible that the changes in climate observed over the last century have natural causes and are not the result of anthropogenic increases in greenhouses gases? Although we must always leave open that possibility, the preponderance of evidence leads atmospheric scientists to conclude that the observed changes in global climate over the last century cannot be explained through natural causes alone.

The observed increases in global temperature and the secondary effects of this warming correlate very closely with the increase in greenhouse gas concentrations tied to human activity. In AR5 the IPCC concludes:

> Historical emissions have driven atmospheric concentrations of carbon dioxide, methane and nitrous oxide to levels that are unprecedented in at least the last 800,000 years, leading to an uptake in energy by the climate system.

Carbon dioxide in the atmosphere, the most important anthropogenic greenhouse gas, had increased in concentration from a preindustrial level of about 280 parts per million (ppm) to over 417 ppm by May 2020. Methane, another key anthropogenic greenhouse gas, increased from a preindustrial level of about 715 parts per billion (ppb) to 1872 ppb in 2020. It is likely that this increase in greenhouse gases would have caused even more warming than was observed if that effect was not offset by slight cooling from anthropogenic and volcanic aerosols. The IPCC reports *high confidence* that cooling from aerosols counteracted a "substantial portion" of the warming from greenhouse gases.

A broad consensus exists within the scientific community about global climate change. A few examples of that consensus follow.

In 2019 the American Geophysical Union (AGU), an international organization of research scientists, concluded:

> Human activities are changing Earth's climate, causing increasingly disruptive societal and ecological impacts. Such impacts are creating hardships and suffering now, and they will continue to do so into the future—in ways expected as well as potentially unforeseen. To limit these impacts, the world's nations have agreed to hold the increase in global average temperature to well below 2°C (3.6°F) above pre-industrial levels. To achieve this goal, global society must promptly reduce its greenhouse gas emissions. . . . Society must also prepare to cope with and adapt to the adverse impacts of climate change. Done strategically, efficiently, and equitably, the needed transformations provide a pathway toward greater prosperity and well-being, while inaction will prove very costly for humans and other life on the planet.

Source: *American Geophysical Union*, AGU Position Statement: Human-Induced Climate Change Requires Urgent Action, *adopted by AGU December 2003, revised and reaffirmed August 2013 and November 2019.*

The report *Global Climate Change Impacts in the United States*, released by U.S. Global Change Research Program (USGCRP) in 2014, states succinctly:

> Climate change is happening now. The United States and the world are warming, global sea level is rising, and some types of extreme weather events are becoming more frequent and severe. These changes have already resulted in a wide range of impacts across every region of the country and many sectors of the economy.
>
> **Source:** *www.globalchange.gov/climate-change*

Finally, the *Climate Change 2014: Synthesis Report* of AR5 picks up where the IPCC's earlier reports left off. It summarizes:

> The evidence for human influence on the climate system has grown since AR4 [IPCC's *Fourth Assessment Report* of 2007]. Human influence has been detected in warming of the atmosphere and the ocean, in changes in the global water cycle, in reductions in snow and ice, and in global mean sea level rise; and it is *extremely likely* to have been the dominant cause of the observed warming since the mid-20th century.

In assessing the credibility of these statements and the quality of the science behind them, keep in mind that IPCC reports are the result of the work of hundreds of scientists. The conclusions of these reports are refereed by a separate group of reviewers, and the final edited work is approved by delegates from more than 100 countries. Because IPCC's reports are consensus views, the agreed-upon conclusions often involve compromise on the part of scientists. Climate change researcher and biologist Tim Flannery has described these conclusions as "lowest-common-denominator science" and states that they "carry great weight . . . precisely because they represent a consensus view."[3]

Future Climate Change

Even though the conclusion of atmospheric scientists is that human activity is indeed altering global climate, predicting the extent of future climate change and its consequences are less certain. How do atmospheric scientists predict what climate will be like in the decades ahead?

Using Models to Predict Future Climate

Projections of future temperature and precipitation patterns are mostly the product of sophisticated computer simulations known as *general circulation models (GCMs)*. In short, a GCM is a mathematical model of Earth's climate system. In such a computer-simulated world, scientists make assumptions about the amount of radiation striking the surface, the extent of cloud cover, variations in ocean temperature, changes in wind and pressure patterns, changes in greenhouse gas concentration, variations in surface albedo, and many other parameters.

More recently, *Earth system models (ESMs)* have been used to study climate change. These models, in effect, couple a GCM with an ocean circulation model. They also take into account land and sea ice and the effect of the biosphere on the carbon cycle.

Both GCMs and ESMs create a three-dimensional grid of "boxes"—dividing the atmosphere, land surface, and water surfaces into blocks. Within each box, variables can be changed and the results observed (**Figure 8-43**). Each box, whether it consists of air, land, or water, behaves according to known physical laws. As conditions such as the amount of CO_2 in an atmospheric box change, equations adjust the temperature based on, say, the amount of longwave radiation from the surface absorbed by the CO_2 in that box. The change in temperature can in turn alter the amount of water vapor in this atmospheric box, which can vary the amount of cloud cover, which will affect the albedo, which can change the energy balance still more, and so on. Once the computer simulation starts running, changes in one box can modify the conditions in adjacent boxes, and all of these changes are tracked over time.

By adjusting the initial conditions—such as the concentration of greenhouse gases or the amount of solar radiation—scientists can see how the atmosphere, land, and oceans respond to different "scenarios."

Testing Climate Models with Past Climates: A GCM or ESM must pass several "tests" before it can predict future climate. First, the model must be able to simulate present atmospheric processes and climate—in other words, it must reliably replicate processes and patterns in today's atmosphere. Next, the GCM or ESM must be able to simulate past climate change, so data on past climate conditions are input into the program, and the simulation is run "forward" to the

▲ **Figure 8-43** An Earth system model divides the atmosphere, land surface, and ocean into a grid of "boxes." The model then calculates changes in conditions, such as temperature and moisture, within each box as the computer simulation progresses through time.

[3] Tim Flannery, *The Weather Makers: How Man Is Changing the Climate and What It Means for Life on Earth* (New York: Atlantic Monthly Press, 2005), 246.

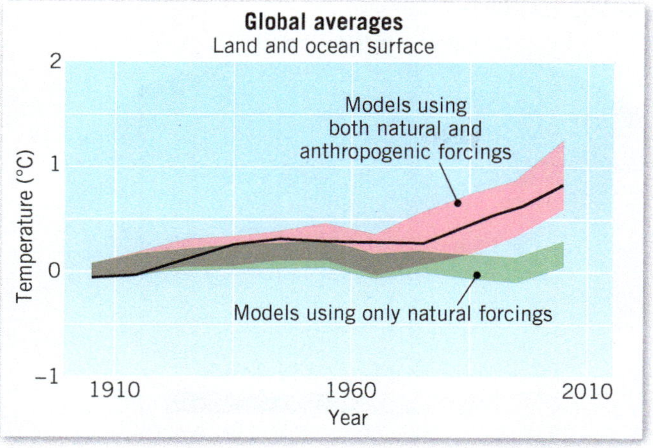

▲ **Figure 8-44** A comparison of computer models and actual temperature change. The black line shows observed temperature anomalies over the twentieth century. The green shaded area shows the range of model predictions of temperature change assuming only natural climate forcings (such as volcanic eruptions and variations in solar activity). The pink shaded area shows the range of model predictions of temperature change assuming anthropogenic plus natural forcings. Natural factors alone do not explain the temperature change over the last century.

present day. If the model cannot "predict" what has actually happened to climate, adjustments must be made. Only then can the model be run forward to make projections about future climate.

How well do GCMs and ESMs "predict" the observed climate changes over the last century? Quite well (**Figure 8-44**).

Because of the enormous complexities in the atmosphere, different GCMs and ESMs use different assumptions about how the atmosphere works. Hence, each model comes up with slightly different projections about climate in the future. Among the most important uses of GCMs and ESMs today is to anticipate how climate will change given various scenarios of greenhouse gas increase in the atmosphere.

Projections of Future Climate

Video
Temperature and Agriculture
https://media.pearsoncmg.com/ph/esm/esm_mcknight_physgeo_13/media/

Projections of the temperature increase and other changes in climate expected by the middle or end of this century are difficult to calculate. In addition to the great complexity of the global climate system itself, how levels of greenhouse gases will change in coming decades is uncertain.

The IPCC's AR5 defines four *Representative Concentration Pathways (RCPs)*—each describing a different global greenhouse gas (GHG) emission scenario through the end of this century, including a "best-case" (lowest emission) scenario and a "worst-case" (highest emission) scenario. Which scenario does the IPCC think is most likely? They don't try to answer that question. It is impossible to know how government policies, population growth, per capita economic growth, and technology will change in the decades to come. Instead, these scenarios give the climate models a range of possibilities to test.

The IPCC summarizes their projections of future climate changes, risks, and impacts presented in the *Fifth Assessment Report* as follows:

> Continued emission of greenhouse gases will cause further warming and long-lasting changes in all components of the climate system, increasing the likelihood of severe, pervasive and irreversible impacts for people and ecosystems.

Specific projections presented in the IPCC's AR5 include the following:

Temperature Change Projections

- Over the next two decades, the projected temperature increase is similar in all RCP scenarios. There is *medium confidence* that average global surface temperature increase by 2035 will *likely* be in the range of 0.3–0.7°C (0.5–1.3°F).
- Relative to the period 1850–1900, there is *high confidence* that the projected global surface temperature change by the interval 2081–2100 is *likely* to be greater than 1.5°C (2.7°F) in the intermediate and highest RCP emission scenarios. There is *medium confidence* that it is *unlikely* to exceed 2.0°C (3.6°F) in the lowest emission scenario (**Figure 8-45**).
- It is *virtually certain* that hot temperature extremes will become more frequent and that cold temperature extremes will become less frequent.

Ocean Change Projections

- Throughout the current century, the ocean will continue to warm.
- Due to thermal expansion and increased rates of ice flow from Antarctica and Greenland, sea level will continue to rise during this century. It is *very likely* that the rate of sea level rise will exceed the observed 1971–2010 rate of about 2.0 millimeters (0.08 inches) per year. It is *virtually certain* that global sea level will continue to rise for many centuries beyond 2100.

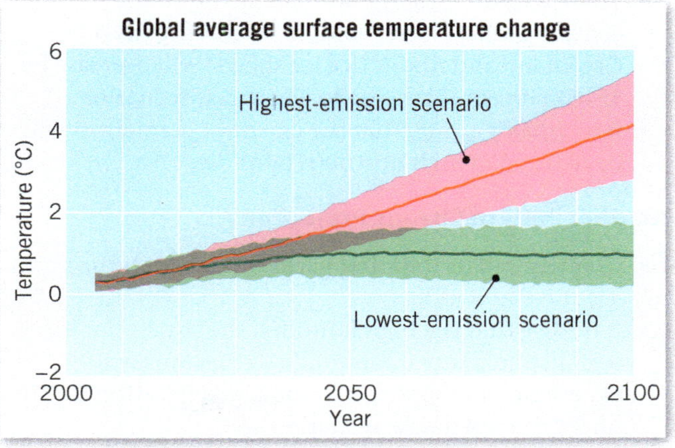

▲ **Figure 8-45** Projected temperature changes by 2100 (relative to 1986–2005 average) under two IPCC AR5 emission scenarios. The shaded areas show the range of projected changes.

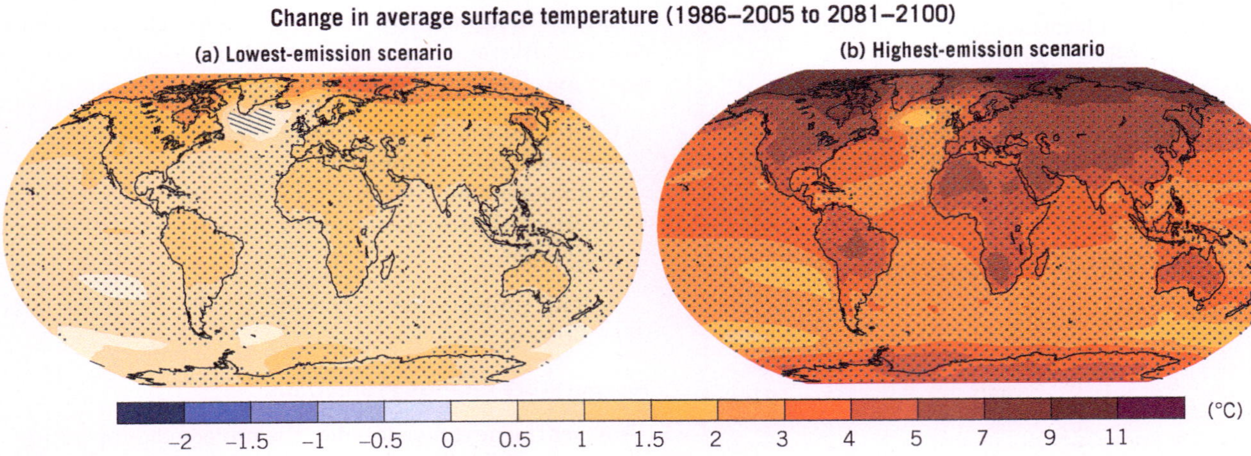

▲ **Figure 8-46** Maps showing projected change in surface temperatures by 2100 under the IPCC AR5's (a) best-case scenario and (b) worst-case scenario. In dotted areas, the projected change is large compared with the region's natural variability; in hatched areas, the projected change is close to natural variability.

- The rise in sea level will not be uniform around the world, but it is *very likely* to rise in more than 95 percent of the ocean area. Even a modest increase in sea level will cause widespread retreat of shorelines, inundating many heavily populated coastal areas and increasing vulnerability to storms such as hurricanes.
- It is *likely* that by midcentury, the total average sea-level rise will be 0.17–0.32 meters (0.6–1.0 feet) under the RCP best-case scenario and 0.22–0.38 meters (0.7–1.2 feet) under the worst-case scenario. These projections assume that the marine-based sectors of the Antarctic ice sheet *do not* collapse.

Polar Region Change Projections

- There is *very high confidence* that the Arctic region will continue warming more rapidly than the world as a whole (**Figure 8-46**).
- In all RCP scenarios, year-round Arctic sea ice is projected to be reduced by 2100. By the year 2100, it is *likely* that Northern Hemisphere spring snow cover will be reduced by 7–25 percent.
- There is *medium confidence* that by the end of the century global glacier volume (excluding the Greenland and Antarctica ice sheets) will decrease by 15–85 percent (best-case to worst-case scenarios).
- It is *virtually certain* that the extent of permafrost will be reduced in high northern latitudes.

Weather Pattern Change Projections

- Changes in precipitation patterns will not be uniform around the world.
- It is *likely* that the high latitudes, the equatorial Pacific region, and many midlatitude wet regions will experience an increase in annual precipitation by 2100 under the worst-case scenario.
- It is *likely* that many midlatitude and subtropical dry regions will experience a decrease in precipitation under the worst-case scenario.
- It is *very likely* that extreme precipitation events will become more intense and frequent over most midlatitude land areas and in wet tropical regions.

Biosphere Change Projections

- There is *high confidence* that a large fraction of terrestrial, marine, and freshwater species face increased risk of extinction during and beyond the current century due to climate change.
- There is *medium confidence* that many animal and plant species will be unable to adjust their ranges fast enough to adapt to climate changes during this century under the moderate and highest emission scenarios.
- There is *high confidence* that throughout the current century ill health among human populations will increase, especially in low-income countries.
- There is *high confidence* that human food security will be affected this century.

According to the IPCC, global CO_2 emissions must reach net-zero by around 2070 to have a good chance at limiting the warming to 2°C. As climate changes, entirely new climate types may emerge and others may disappear—see the box *Global Environmental Change: Disappearing and Novel Climates*.

LearningCheck 8-17 **Describe and explain at least one possible change in global climate projected by the IPCC to occur during this century.**

Addressing Climate Change

As we've just seen, the evidence that global climate change is happening now is strong, and the projections of the changes to come are stark but clear. It is primarily in the arenas of politics, economics, and public policy that disagreement about global climate change remains. This disagreement centers especially on what should be done to address global warming—see the box *Energy for the 21st Century: Strategies for Reducing Greenhouse Gas Emissions*.

Global Environmental Change

Disappearing and Novel Climates

• **Michael E. Mann,** Penn State University

We have seen in this chapter that the diverse climates found around the world can be classified through schemes such as the Köppen classification system, based on seasonal patterns of rainfall and temperature. We have also learned how human activity, in particular the burning of fossil fuels, is leading to climate change. Can we bring these concepts together and investigate how climate change is changing the distribution of climate zones themselves?

As human-caused climate change alters seasonal patterns of temperature and precipitation around the world, *novel climates*—new climates currently not found in the Köppen scheme—are predicted to emerge. Meanwhile, some that exist today may disappear. Novel climates are particularly likely in lower latitude regions where temperatures are likely to exceed the range currently found on Earth (**Figure 8-E-a**). Disappearing climates are most likely in alpine or boreal environments where the very cold temperatures that prevail today will no longer be encountered on any of Earth's continents (**Figure 8-E-b**).

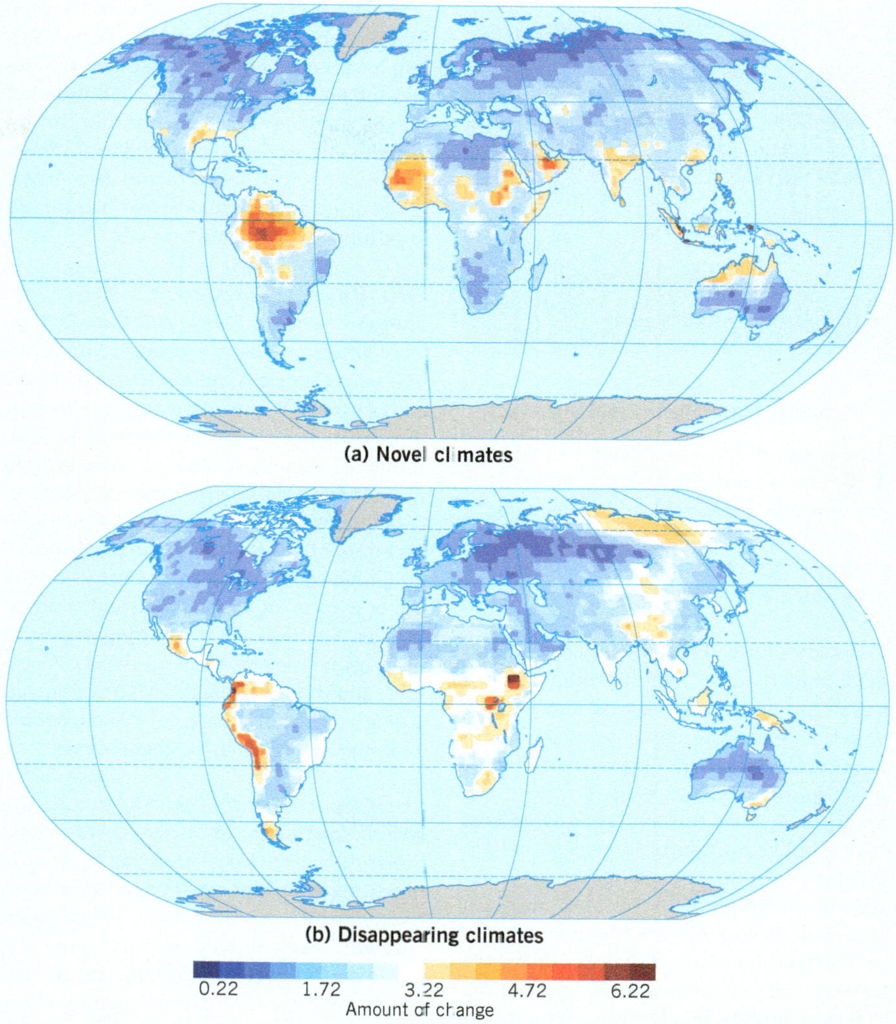

▲ **Figure 8-E** Maps showing areas where (a) novel climates (which fit no Köppen category) are expected to appear and (b) present climates are expected to disappear by the year 2100 if carbon emissions are not reduced. The "Amount of change" scale for each map is relative; change is likely for values above 3.22.

Threats to Biodiversity: The loss of climate zones poses a threat to biological diversity. The loss of the Arctic sea ice environment, for example, places animals such as the polar bear and walrus at risk. The ongoing retreat of the alpine tundra environment in North America threatens the American pika, a rabbit relative familiar to those who have hiked in the Rocky Mountains.
Tropical species are especially threatened because temperatures vary over a smaller range, and the natural ranges of animal species tend to be smaller nearer the equator. Even modest shifts in temperature may thus pose a greater challenge to adaptation of those species. The Amazon basin is especially at risk due to the added threats of increased wildfires and forest loss due to the predicted drying in that region.

Of significant concern is the fact that many of the areas where existing climates are predicted to disappear (the Andes, Mexico and Central America, southern and eastern Africa, the Himalayas, and the Philippines) also happen to be regions of great biological diversity. The potential for species loss is thus particularly great in these regions.

Questions

1. *Cloud forests* are found in the tropics at high elevations, where the atmosphere is cool enough for condensation of moisture. How might global warming threaten species that are unique to that habitat?
2. Where in the United States do we see a particularly high likelihood of novel climates appearing? Can you speculate as to why?

Energy for the 21st Century

Strategies for Reducing Greenhouse Gas Emissions

• **Michael E. Mann,** Penn State University

New and different strategies will be required to address the energy needs of the world in a way that addresses human-caused climate change. Just a few of the possibilities for reducing greenhouse gas emissions are described here.

Conservation: Perhaps the simplest and most immediate strategy to reduce greenhouse gas emissions is to use less energy. Together, electrical power generation and transportation account for almost 40 percent of all greenhouse gas emissions. In those areas, small steps can add up: driving less, taking public transportation, using energy-efficient lighting, improving home insulation, and turning off lights and low-drain electronic devices when not in use.

Being aware of our own resource use can be a helpful first step toward conservation. You can calculate your "carbon footprint" at several websites, such as http://www3.epa.gov/carbon-footprint-calculator/.

Putting a Price on Carbon: Reluctance on the part of the United States and some other industrialized countries to cut back immediately and significantly on greenhouse gas emissions largely centers on short-term costs. Many economists argue that any pragmatic approach to reducing carbon emissions must involve market incentives. There are two main approaches to this. One approach, a *cap-and-trade* system, places declining limits (a "cap") on industrial carbon emissions but allows industries that emit less than their carbon allotment to sell or trade their remainder to industries that emit too much. This system provides financial incentives to innovate and conserve, while offering more time for industries making a slower transition to cleaner energy technology.

The second approach, a *carbon tax*, penalizes industries that emit more than an established emissions cap. To a certain extent, a carbon tax acts as a financial "stick" to reduce emissions, while a cap-and-trade system offers somewhat of a "carrot." In both cases, the key to greenhouse gas reductions is setting meaningful emissions caps and timetables and ensuring a price signal in the market for the damages done by emitting carbon into the atmosphere. Some politicians favor "revenue-neutral" forms of the carbon tax, which offset any new taxation with decreases in existing (e.g., income) taxes.

"No Regrets" Opportunities: The Intergovernmental Panel on Climate Change highlights a number of "no regrets" opportunities—greenhouse gas emissions practices with costs that are offset by the savings through direct or indirect benefits to individuals or society. A few examples include the adoption of energy-efficient lighting, building insulation, and heating, which lowers energy bills; the reduction of air pollution, which reduces health care costs; and the use of carpooling, public transportation, or bicycles, which lessens traffic congestion and saves commuters money (Figure 8-F). Voluntarily adopting such practices is often advantageous at an individual level, making people healthier as well as saving them money.

The Clean Energy Transition: As the world continues to move toward less dependence on fossil fuels for its energy needs, an important question remains unanswered: which companies and which countries will lead the way by developing the cost-effective technologies and infrastructure to deliver so-called green energy? Many economists argue that those who have the foresight to invest in clean energy technology today may reap enormous financial benefits in the future. Developing nations such as China are leading the way through massive investment in renewable (e.g., solar) energy technologies and by introducing a price on carbon at the national level. Countries such as Germany that have introduced market incentives for renewable energy now provide nearly a third of their power production from renewables. In the United States, states accounting for nearly 30 percent of the nation's population now belong to regional climate pricing consortiums; executive actions have been implemented requiring increased automotive fuel efficiency standards and requiring states to reduce carbon emissions from power generation. These efforts are already making a difference at the global level. In 2014, for the first time in recent history, the global economy increased without an increase in global carbon emissions, and renewable energy outpaced fossil fuels in added global power generation capacity. An international agreement to reduce carbon emissions would accelerate the transition that is already underway.

Questions

1. What could you do today to help lower global carbon emissions? Think about both individual actions and your ability to influence policy decisions.
2. Consider two possible futures, 30 years from now: in one, carbon emissions have been dramatically lowered; in another, society has continued a trajectory of ever-increasing carbon emissions. Describe how those futures might look environmentally, economically, and societally.

▼ **Figure 8-F** Bicycling commuters in Portland, Oregon.

International Climate Change Agreements

One of the first international responses to the threat of global climate change took place in 1992 at the *Rio Earth Summit*. The adoption of the "Rio Convention" established the *United Nations Framework Convention on Climate Change* (UNFCC) and produced a framework that is working toward stabilizing greenhouse gas concentrations.

The Kyoto Protocol: An important global-scale step to cut carbon emissions was the Kyoto Protocol. The participants in the *Kyoto Protocol* to the UNFCC in 1997 agreed that all emitters of carbon dioxide (the top six are China, the United States, Russia, Japan, Germany, and India) would cut greenhouse gas emissions by an average of 5 percent below 1990 levels by the year 2012. In 2012, the *Doha Amendment to the Kyoto Protocol* was adopted with the goal of reducing greenhouse gas emissions 18 percent below 1990 levels over the commitment period of 2013 to 2020.

Although more than 190 countries (as well as the European Union as a regional economic organization) ratified the original agreement, the United States did not. Fewer countries agreed to the cutbacks called for in the second commitment period. So far, the United States and China have been unwilling to sign a treaty that would legally bind them to reduce greenhouse gas emissions.

By even the most optimistic accounts, the modest greenhouse gas reductions of the Kyoto Protocol would at best curtail global warming slightly. Nonetheless, the agreement was a historic first step.

U.N. Conference of Parties: The *United Nations Climate Action Summit* held in 2019 in Madrid, Spain, was the 25th annual "Conference of Parties" (COP25). (The 1997 Kyoto Protocol was adopted at COP3.) The goal of the conference was to establish a legally binding agreement to address greenhouse gas emissions and climate change, with the goal of keeping global warming at less than 2.0°C (3.6°F) over this century and carbon neutral by 2050. Seventy-three countries and 1214 groups (regions, cities, and businesses) committed to net-zero emissions by 2050.

Newly Industrialized Countries

As they expand their industrial infrastructures, many developing countries increase greenhouse gas emissions. Requiring all countries to meet the same emission reductions seems unfair to leaders in some newly industrializing countries. These countries argue that the rich industrialized nations created the problem of global warming through their own unfettered use of fossil fuel and "dirty" technologies. Should developing countries be asked to cut back emissions as much as rich countries?

The United States makes up about 4 percent of the world's population but emits about 15 percent of the world's total carbon dioxide. (CO_2 emissions started to stabilize over the last few years, however, primarily because of an increasing shift from coal to natural gas to generate electricity—burning of natural gas releases less CO_2 than burning of other fossil fuels does.) China recently surpassed the United States as the world leader in CO_2 emissions (about 26 percent of the world total), but on a per capita (per person) basis, the United States still far exceeds China. Clearly, all countries of the world need to participate in greenhouse gas reductions, but just how the economic needs of all countries can be balanced is not yet clear.

Mitigating and Adapting

Because some greenhouse gases have long residence times in the atmosphere, even if the goals of CO_2 reduction specified under the Kyoto Protocol—or still more ambitious goals proposed for new negotiations—are achieved, we can expect global temperatures to continue rising for some time. The IPCC's *Fifth Assessment Report* concludes that temperature increases, sea-level increases, and rainfall changes are "effectively irreversible" for perhaps several centuries after CO_2 emissions are stabilized.

The long-lasting effects of climate change raise several immediate questions. In addition to efforts to prevent greater global warming by emissions reductions, what should we do to adapt to the consequences of climate change? For example, should we raise coastal dikes or seawalls to compensate for higher sea levels and storm surge potential? Should we expand drinking-water storage reservoirs to compensate for reductions in winter snow pack? The costs and environmental consequences of these and many other strategies need to be thoughtfully weighed.

Some questions are even more difficult to answer. Should we attempt "geoengineering"—large-scale engineering projects to change the energy balance of the atmosphere, perhaps by introducing aerosols to block incoming radiation or by trying to remove CO_2 from the atmosphere? We must consider the costs, the practicality—and perhaps most of all, the likelihood of unintended consequences—of such actions.

Poorest, Most Vulnerable Countries: The findings of the IPCC point to a related problem: the consequences of global climate change will not be shared equally around the world. Many of the populations most vulnerable to sea-level rise, changes in precipitation for agriculture and drinking water, and the expanding ranges of insect pests and disease are found in the poorest parts of the world. They are the populations with the least financial means to adapt. What can and should the richest countries of the world do to assist the poorest to adapt to climate change?

It remains to be seen how the people of the world—as individuals and as countries—will choose to address global climate change, a problem without quick or easy solutions.

8 Learning Review

After studying this chapter, you should be able to answer the following questions. Key terms from each text section are shown in **bold type.** Definitions for key terms are also found in the glossary at the back of the book.

Key Terms and Concepts

Climate Classification (*p. 212*)

1. Explain the basic concept of the **Köppen climate classification system.**
2. In the Köppen climate classification letter code system, what information is given by the first letter, the second letter, and the third letter?
3. Briefly describe the major climate groups of the modified Köppen climate classification system: A, B, C, D, E, and H.
4. What information is conveyed in a **climograph**?

Tropical Humid Climates (Group A) (*p. 215*)

5. Describe the general location, temperature characteristics, precipitation characteristics, and main controls of the following climates. You should be able to recognize these climates from a climograph:
   - Af **Tropical wet**
   - Aw **Tropical savanna**
   - Am **Tropical monsoon**
6. Why do Af (tropical wet) climates receive rain all year, whereas Aw (tropical savanna) climates receive rain only in the summer (the high-Sun season)?

Dry Climates (Group B) (*p. 220*)

7. Describe the general location, temperature characteristics, precipitation characteristics, and main controls of the following climates. You should be able to recognize these climates from a climograph:
   - BWh **Subtropical desert**
   - BWk **Midlatitude desert**
8. What is the general difference between a desert climate and a steppe climate?
9. What are the main differences in controls of BWh (subtropical desert) and BWk (midlatitude desert) climates?

Mild Midlatitude Climates (Group C) (*p. 224*)

10. Describe the general location, temperature characteristics, precipitation characteristics, and main controls of the following climates. You should be able to recognize these climates from a climograph:
    - Cs **Mediterranean**
    - Cfa **Humid subtropical**
    - Cfb **Marine west coast**
11. Why do Cs (mediterranean) climates have dry summers and wet winters?
12. What causes the relatively mild temperatures of marine west coast climates?

Severe Midlatitude Climates (Group D) (*p. 229*)

13. Describe the general location, temperature characteristics, precipitation characteristics, and main controls of the following climates. You should be able to recognize these climates from a climograph:
    - Dfa **Humid continental**
    - Dfc **Subarctic**
14. What is meant by the phrase "continentality is a keynote of D climates"?
15. Why do subarctic (Dfc) climates have such a wide annual temperature range?

Polar and Highland Climates (Groups E and H) (*p. 232*)

16. What general temperature characteristic distinguishes **tundra climate** from **ice cap climate**?
17. Why are polar climates so dry?
18. In determining a **highland climate**, in what ways is altitude more important than latitude?

Global Climate Change (*p. 237*)

19. What is meant by **paleoclimatology**?
20. How is **dendrochronology** used in studies of past climate?
21. How does the **oxygen isotope analysis** of ocean floor sediments and glacial ice tell us about past temperatures?
22. How can ice-core analysis provide information about the gas composition of the atmosphere in the past?

Causes of Climate Change (*p. 242*)

23. What are **Milankovitch cycles**, and in what ways might they help explain past climate change?
24. Describe and explain at least one feedback mechanism that would further increase global temperatures once a warming trend has started.
25. What kinds of clouds tend to cool the surface of Earth, and what kinds of clouds tend to warm it?
26. What roles do the oceans play in influencing the carbon-dioxide concentrations in the atmosphere?

Anthropogenic Climate Change (*p. 246*)

27. What is the overall trend in global temperature since 1880?
28. What is the cause of the observed changes in climate over the last century?

Future Climate Change (*p. 248*)

29. What is a climate model?
30. How do climate models project how climate will change in the future?

Addressing Climate Change (*p. 250*)

31. What is the Kyoto Protocol?

Study Questions

1. Why are subtropical desert (BWh) climates generally hotter in summer than tropical humid (A) climates?
2. Why are subtropical desert (BWh) climates usually displaced toward the western sides of continents?
3. Why are dry climates much more extensive in North Africa than in any other subtropical location?
4. Although both cities are coastal, New York City has a continental climate whereas Seattle, Washington, has a maritime climate. Why?
5. Why don't severe midlatitude (D) climates exist in the Southern Hemisphere?
6. What is the annual temperature pattern likely to be at high elevation on the equator?
7. What explains the alternating bands of wet and dry seasons in Figure 8-34?
8. How does the loss of sea ice in the Arctic contribute to a positive feedback mechanism and rising temperatures?

Exercises

1. Estimate the average monthly precipitation in Iquitos, Peru, by using the climograph in Figure 8-5c: _____ centimeters (inches)
2. Estimate the average monthly precipitation in Yuma, Arizona, by using the climograph in Figure 8-13c: _____ centimeters (inches)
3. Carbon-14 (^{14}C) is a radioactive isotope of carbon with a half-life of 5730 years—meaning that half of the ^{14}C in a sample decays into ^{14}N during that time. Suppose a scientist analyzes a sample of ancient plant material from a bog and finds that it contains one-eighth as much ^{14}C as when it formed. Roughly how old is the sample? _____ years

Environmental Analysis SEA ICE TRENDS

Data
Sea Ice Trends
https://media.pearsoncmg.com/ph/esm/esm_mcknight_physgeo_13/media/

The National Snow and Ice Data Center monitors sea ice coverage and concentration. The Sea Ice Index is calculated from this coverage to assess the change in sea ice compared with the 30-year average.

Activities

Go to http://nsidc.org/data/seaice_index. Use the default map parameters of "Monthly" and "Extent" for the Arctic.

1. When did Sea Ice Index data collection start?
2. The "median ice edge" line on the map represents the 30-year median ice extent. Is the current ice cover extent more than or less than the 30-year median?

View the graph entitled "Northern Hemisphere Extent Anomalies" for the most recent month, and click on the "i" (Information icon) to get more information about the graph.

3. How is the monthly ice extent anomaly calculated?
4. Given the 1981–2010 mean sea ice extent labeled near the bottom of the graph, calculate the current sea ice extent.
5. How does the extent anomaly graph compare with the temperature anomaly graph of Figure 4-34?

Click on "Animate Monthly Images".

6. Which month has the lowest sea ice extent? Which month has the greatest sea ice extent?
7. According to the map, how does the current month's sea ice extent compare with the 1981–2010 average?

Seeing Geographically in Our National Parks

Look again at the photograph of a Joshua tree at the beginning of the chapter (p. 210). How would plants that adapt to extremely harsh environments help scientists study the impact of climate change? As climate changes in the Mojave Desert, plant and animal species must shift to higher elevations for adequate rainfall. How might this influence the Joshua tree, which can live for more than 150 years?

Mastering Geography®

Looking for additional review and test prep materials? Visit the Study Area in **Mastering Geography** to enhance your geographic literacy, spatial reasoning skills, and understanding of this chapter's content by accessing a variety of resources, including Map Master interactive maps, geoscience animations, *Mobile Field Trips*, videos, *Project Condor* Quadcopter videos, *In the News* RSS feeds, flashcards, web links, self-study quizzes, and an eText version of *McKnight's Physical Geography*.

Seeing Geographically in Our National Parks

Sandstone Falls, in New River Gorge National River, West Virginia. Does the local environment appear to be arid or moist? How can you tell? Which season of the year does this appear to be? What do you observe that makes you say this? How many different forms of water are shown in this photograph?

New River Gorge National River, WV

The Hydrosphere

9

HAVE YOU EVER WONDERED what happens to rainwater once it reaches the ground? The answer might seem obvious: it ends up in a stream such as New River, here in New River Gorge National River, West Virginia, and eventually flows into a lake or the ocean. However, the complete answer is more complicated. Some of the water may indeed flow in streams, but some of it may evaporate back into the atmosphere as water vapor, and some may soak into the surface and become *groundwater*—or even become frozen underground as *permafrost*.

The *hydrosphere* is the most pervasive, but in some ways the least well defined, of the four "spheres" of Earth's physical environment that we introduced in Chapter 1. It includes the surface water in oceans, lakes, rivers, and swamps; all underground water; frozen water in the form of ice, snow, and high-cloud crystals; water vapor in the atmosphere; and moisture temporarily stored in plants and animals.

The hydrosphere overlaps significantly with the other three spheres. Liquid water, ice, and even water vapor occur in the soil and rocks of the *lithosphere*. Water vapor and cloud droplets of liquid water and ice are important constituents of the *atmosphere*, and water is a critical component of every living organism of the *biosphere*. Life as we know it is impossible without water. Water solutions in organisms dissolve or disperse nutrients for nourishment. Most waste products are carried away in solutions. Indeed, the total mass of every living thing is more than half water—from about 60 percent for some animals to more than 95 percent for some plants.

Through water, then, the interrelationships of the four spheres are most conspicuous. In Chapter 6 we introduced many of the physical properties of water and the roles of water in weather and climate. In this chapter, we examine the geography of water more broadly and explore one of the most important of all Earth systems.

As you study this chapter, think about these **Key**Questions:

- **How does water move through the *hydrologic cycle*?**
- **How do the characteristics of the ocean differ around the world?**
- **What causes *tides*?**
- **What is *permafrost*?**
- **How do lakes and reservoirs change over time?**
- **What factors influence the availability of *groundwater*?**

The Hydrologic Cycle

Animation
Hydrologic Cycle
https://media.pearsoncmg.com/ph/esm/esm_mcknight_physgeo_13/media/

We begin this chapter with a more detailed look at the movement of water—in all its forms—through the hydrosphere. Water is distributed very unevenly on, in, and above Earth. More than 99 percent of Earth's water is in "storage"—in oceans, lakes, and streams, locked up as glacial ice, or held in rocks below Earth's surface (Figure 9-1). Water frozen as ice in glaciers and continental ice sheets represents about three-fourths of all of the freshwater on the planet.

The proportional amount of moisture in these various storage "reservoirs" is relatively constant over thousands of years. These components only change notably during an ice age. As we see in Chapter 19, during periods of glaciation, the volume of the oceans decreases as the ice sheets grow and the amount of atmospheric water vapor diminishes; then during deglaciation, the ice melts, the volume of the oceans increases as the meltwater flows into them, and atmospheric water vapor increases.

The remaining small fraction of Earth's water that is not in storage—less than 1 percent of the total—is involved in an almost continuous sequence of movement and change. This tiny portion of Earth's water supply moves from one storage area to another—from ocean to air, from air to ground, and so on—in the **hydrologic cycle**. We first introduced the hydrologic cycle in Chapter 6. In this chapter, we examine the complexities of this cycle in detail.

We can view the hydrologic cycle as a series of storage areas interconnected by various transfer processes, in which there is a ceaseless interchange of moisture in terms of both its geographic location and its physical state (Figure 9-2). Liquid water on Earth's surface evaporates, becoming water vapor in the atmosphere. That vapor then condenses and precipitates, either as liquid water or as ice, back onto the surface. This precipitated water then runs off into storage areas and later evaporates into the atmosphere again. Because this is effectively a closed, circular system, we can begin the discussion at any point. It is perhaps clearest to start with the movement of moisture from Earth's surface into the atmosphere.

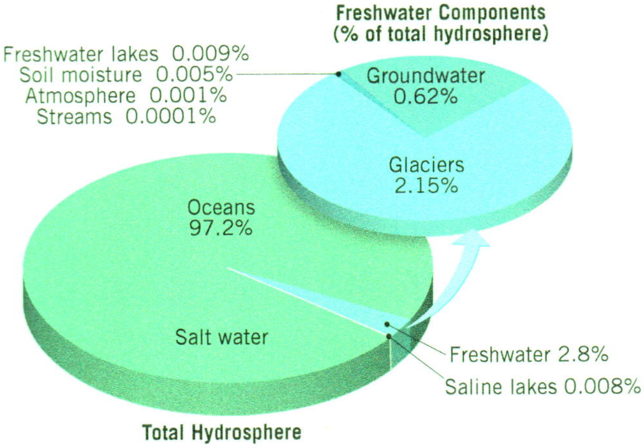

▲ Figure 9-1 Moisture inventory of Earth. More than 97 percent of all water is contained in the oceans. Less than 3 percent of all water is freshwater.

Surface-to-Air Water Movement

Most of the moisture that enters the atmosphere from Earth's surface does so through *evaporation*. (Transpiration from plants is the source of the rest.) The oceans, of course, are the principal source of water for evaporation. They occupy 71 percent of Earth's surface and are extensive in low latitudes, where high temperatures and wind facilitate evaporation. As a result, an estimated 86 percent of all

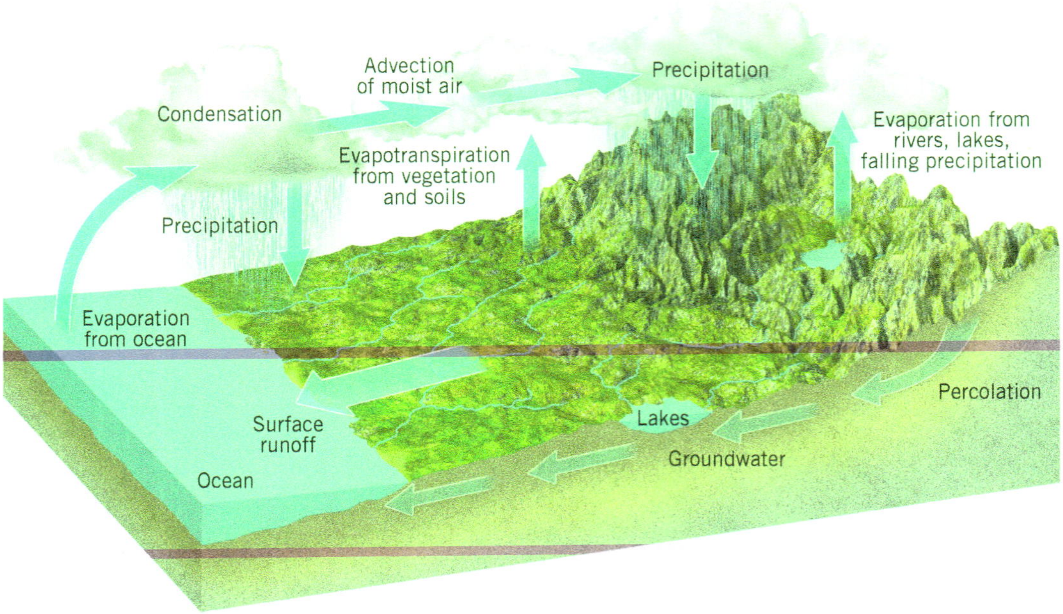

◄ Figure 9-2 The hydrologic cycle. The two major components are evaporation from surface to air and precipitation from air to surface. Other important elements include transpiration of moisture from vegetation to atmosphere; surface runoff and subsurface flow of groundwater from land to sea; condensation of water vapor, forming clouds from which precipitation may fall; and advection of moisture from one place to another.

Video
Hydrological Cycle
https://media.pearsoncmg.com/ph/esm/esm_mcknight_physgeo_13/media/

TABLE 9-1 Moisture Balance of Continents and Oceans			
	Total of World Surface Area	Total of World Precipitation Received	Total of World Water Vapor from Surface
Oceans	71%	78%	86%
Continents	29%	22%	14%

evaporated moisture is derived from ocean surfaces (Table 9-1). The 14 percent that comes from land surfaces includes the twin processes of evaporation and transpiration, referred to as *evapotranspiration*.

Water vapor from evaporation remains in the atmosphere a relatively short time—sometimes just a few hours, but typically about 10 days. During that interval, however, it may move a considerable distance, either vertically through convection or horizontally through advection driven by wind.

Air-to-Surface Water Movement

Sooner or later, water vapor in the atmosphere condenses to liquid water or deposits directly as ice and forms clouds. As we saw in Chapter 6, under the proper circumstances clouds may drop precipitation in the form of rain, snow, sleet, or hail. As Table 9-1 shows, 78 percent of this precipitation falls into the oceans and 22 percent falls onto land.

Over the span of several years, the total worldwide precipitation is approximately equal to the total worldwide evaporation/transpiration. Although the amounts of precipitation and evaporation/transpiration balance in time, they do not balance in place: evaporation exceeds precipitation over the oceans, whereas the opposite is true over the continents. This imbalance is explained by the advection of moist maritime air onto land areas. Except for coastal spray and storm waves, the only route by which moisture moves from sea to land is via the atmosphere.

Movement on and beneath Earth's Surface

The "surplus" precipitation over the continents is effectively returned to the ocean through surface **runoff**, water draining off the land and back into the sea in streams.

The 78 percent of total global precipitation that falls on the ocean is incorporated immediately into the water already there; the 22 percent that falls on land goes through a more complicated series of events. Rain falling on a land surface collects in lakes, runs off slopes, or infiltrates the ground. Any water that pools on the surface eventually either evaporates or sinks into the ground; runoff water eventually ends up in the ocean; and infiltrated water is either stored temporarily as soil moisture or percolates farther down, becoming part of the underground water supply.

Much of the soil moisture eventually evaporates or transpires back into the atmosphere, and much of the underground water eventually reappears at the surface in *springs*

NPS Video
Hidden Waters
https://media.pearsoncmg.com/ph/esm/esm_mcknight_physgeo_13/media/

▲ **Figure 9-3** Roaring Springs in Grand Canyon National Park, Arizona.

(Figure 9-3). Most of the water that reaches the surface evaporates again, and the rest is incorporated into streams and rivers and becomes runoff flowing into the oceans. This runoff water from continents to oceans amounts to 8 percent of all moisture circulating in the global hydrologic cycle. Runoff balances the excess of precipitation over evaporation taking place on the continents and keeps the oceans from drying up and the land from flooding.

LearningCheck 9-1 Is the amount of precipitation and evapotranspiration over the continents the same? Explain. (Answer on p. AK-3)

▲ **Figure 9-4** Aurora Lake in Mount Rainier National Park, Washington

Residence Times

Although the hydrologic cycle is a closed system, *residence times* vary enormously for individual molecules of water as they move through different parts of the cycle. For example, a molecule of water may be stored in an ocean or deep lake, or stashed away as glacial ice, for thousands of years without moving through the cycle. It can even be trapped in rocks buried deep beneath Earth's surface and excluded from the cycle for thousands or even millions of years.

However, water that *is* moving through the cycle is in almost continuous motion (**Figure 9-4**). Runoff water can travel hundreds of kilometers to the sea in a few days, and moisture evaporated into the atmosphere may remain there for only a few minutes or hours before it is precipitated back to Earth. At any given moment, the atmosphere contains only a few days' potential precipitation.

Energy Transfer in the Hydrologic Cycle

As we've seen, the hydrologic cycle is powered by the Sun. Recall from our discussion of Earth's energy cycle in Chapters 4 and 6 that water vapor represents not only a reservoir of moisture for precipitation but also a reservoir of energy. The latent heat "stored" in water vapor is released during condensation—acting as the fuel for storms such as hurricanes—and serving as one means of energy transfer from the tropics toward the poles.

The Oceans

Although most of Earth's surface is oceanic and the vast majority of all water is in the oceans (see Figure 9-1), our knowledge of the seas was fairly limited until recently. Only within the last six decades or so has sophisticated equipment been available to catalog and measure details of the maritime environment.

How Many Oceans?

From the broadest viewpoint, there is only one interconnected ocean. This "world ocean" has a surface area of 362 million square kilometers (139 million square miles) and contains 1.34 billion cubic kilometers (321 million cubic miles) of saltwater. It spreads over almost three-fourths of Earth's surface, interrupted here and there by continents and islands. Although tens of thousands of small pieces of land rise above its waters, the world ocean is so vast that half a dozen continent-sized portions of it are totally devoid of islands. We are usually referring to one or more of these large expanses of water when we use the term *ocean*.

The world ocean is divided into five principal parts (**Figure 9-5**):

1. The *Pacific Ocean* (**Figure 9-6a**) occupies about one-third of the total area of Earth, more than all the world's land surfaces combined. It contains the greatest average depth of any ocean as well as the deepest known oceanic trenches. Although the Pacific extends almost to the Arctic and Antarctic Circles, it is largely a tropical ocean.
2. The *Atlantic Ocean* is slightly less than half the size of the Pacific (**Figure 9-6b**). Its average depth is a little less than that of the Pacific.
3. The *Indian Ocean* (**Figure 9-6c**) is a little smaller and on average slightly less deep than the Atlantic. Nine-tenths of the Indian Ocean's area is south of the equator.
4. The *Arctic Ocean* (**Figure 9-6d**) is much smaller and shallower than the Pacific, Atlantic, or Indian and is mostly covered with sea ice.
5. The more-recently designated *Southern Ocean* (Figures 9-6a–c) surrounds Antarctica, extending to a latitude of 60° S.

As we saw in Chapter 4, when describing ocean current patterns we further subdivide some of these oceans into major basins: the North and South Pacific, the North and South Atlantic, and the South Indian.

Around the margins of the ocean are many partly landlocked smaller bodies of water called *seas*, *gulfs*, and *bays*. This nomenclature is clouded by the term *sea*, which is often used synonymously with *ocean*, sometimes to denote a specific smaller body of water around the edge of an ocean, and occasionally to denote an inland body of water.

Characteristics of Ocean Waters

Wherever they are found, the waters of the world ocean have many similar characteristics, but they also show significant differences from place to place. The differences are particularly notable in the surface layers, down to a depth of about 200 meters (about 660 feet).

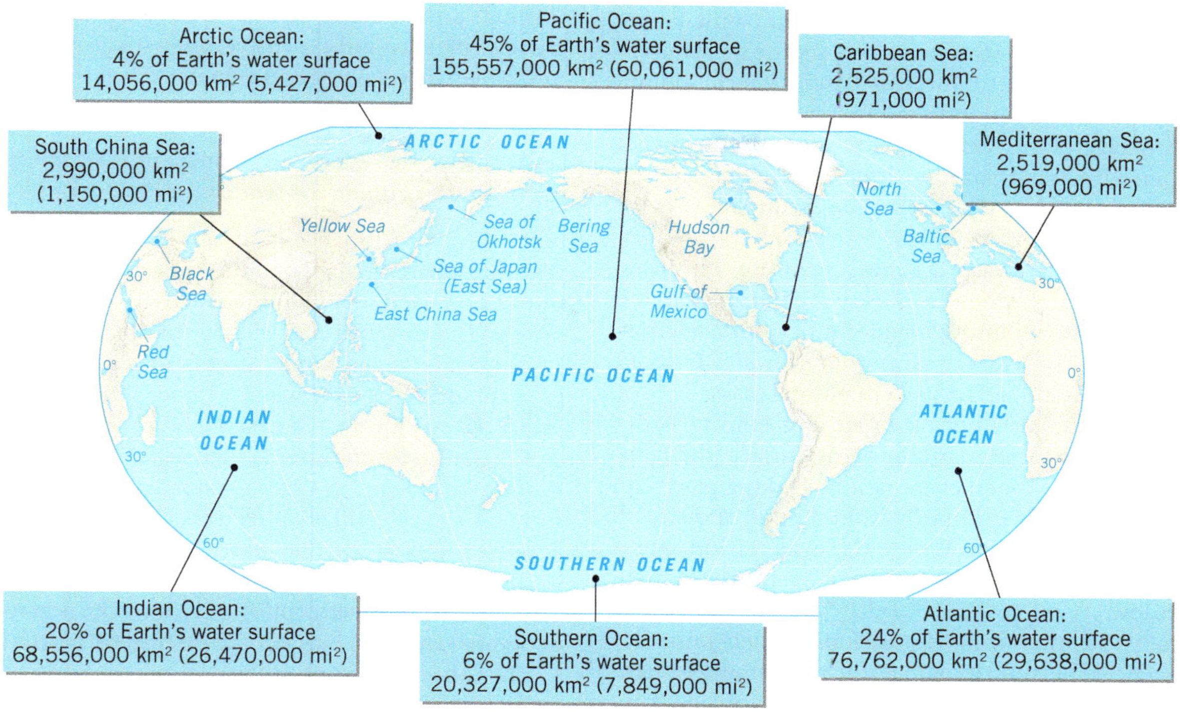

▲ **Figure 9-5** The five principal oceans and the three major seas of the world.

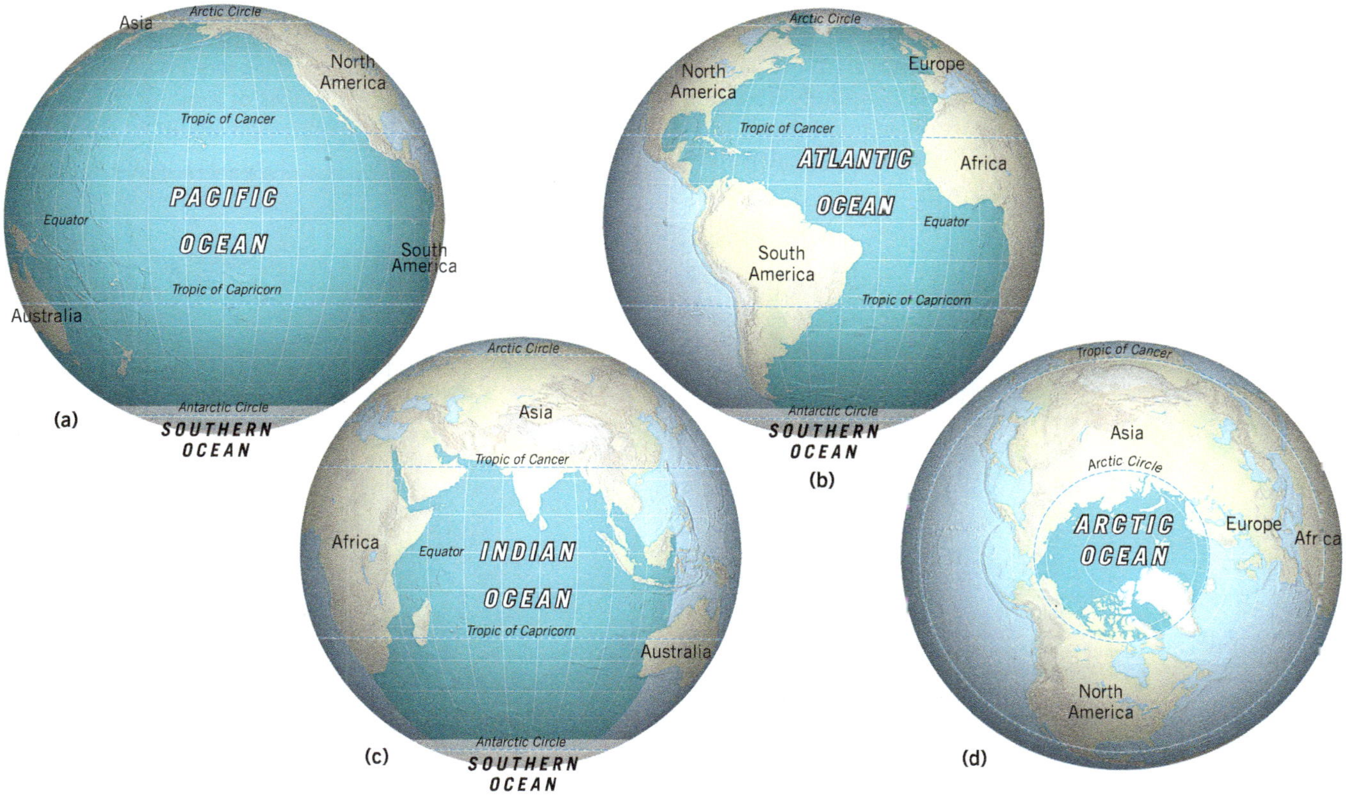

▲ **Figure 9-6** The five major parts of the world ocean: (a) the Pacific Ocean, (b) the Atlantic Ocean, (c) the Indian Ocean, (d) the Arctic Ocean—and (a)–(c) the Southern Ocean, which is composed of the southerly portions of the South Atlantic, South Pacific, and South Indian Ocean basins to a latitude of 60° S.

Chemical Composition: Almost all known elements are found to some extent in seawater, but by far the most important are sodium (Na) and chlorine (Cl), which form sodium chloride (NaCl)—the common salt we know as "table salt." In the language of chemistry, "salts" are substances that result when a *base* neutralizes an *acid*. For instance, sodium chloride is formed when the base sodium hydroxide (NaOH) neutralizes hydrochloric acid (HCl).

The **salinity** of seawater is a measure of the concentration of dissolved salts, which are mostly sodium chloride but also include salts containing magnesium, sulfur, calcium, and potassium. The average salinity of seawater is about 35 parts per thousand, or 3.5 percent of total mass.

The geographic distribution of surface salinity varies (Figure 9-7). At any location on the ocean surface, the salinity depends on how much evaporation is taking place and how much freshwater (primarily from rainfall and stream discharge) is being added. Where the evaporation rate is high, so is salinity; where the inflow of freshwater is high, salinity is low.

Typically the lowest salinities are found where rainfall is heavy and near the mouths of major rivers. Salinity is highest in partly landlocked seas in dry, hot regions because there the evaporation rate is high and stream discharge is minimal. Salinity is lowest in equatorial regions because of heavy rainfall, cloudiness, and humidity, all of which inhibit evaporation, and because of considerable river discharge. Salinity is at a maximum in the subtropics, where precipitation is low and evaporation extensive, and at a minimum in the polar regions, where evaporation is lowest and inflow of freshwater from rivers and ice caps is considerable.

It is possible to convert seawater into drinking water through *desalination*. Common methods include distillation (heating, then condensing water vapor evaporated from saltwater) and reverse osmosis (forcing saltwater through a membrane to filter out salts). Desalination is energy intensive and expensive, but with growing needs for freshwater, especially in arid regions such as the Middle East, its use is projected to grow rapidly over the next few years.

LearningCheck 9-2 **How and why does the salinity of seawater vary around the world?**

Increasing Acidity: The oceans absorb carbon dioxide from the atmosphere. Perhaps one-third of the excess CO_2 released into the air each year by human activity is absorbed by the oceans. The absorbed CO_2 then forms *carbonic acid* (H_2CO_3), a weak acid. Research now suggests that as a result of the great quantities of CO_2 absorbed since the beginning of the industrial revolution, the ocean is becoming more acidic.

Currently, ocean water is slightly alkaline, with a pH of 8.1 (see "Acid Rain" in Chapter 6 for a description of the pH scale). Although still alkaline, this value is estimated to be about 0.1 lower—in other words, more acidic—than it was in the preindustrial era. Given the current rate of fossil fuel use and continued absorption of CO_2 by the oceans, the pH of ocean water could drop to 7.7 by the end of this century.

The consequences of a slightly more acidic ocean are not completely known, but it will likely affect the growth of organisms such as coral polyps and microscopic creatures such as *foraminifera* (Figure 9-8) that build their shells or exoskeletons from calcium carbonate ($CaCO_3$) extracted from seawater. As the oceans become more acidic, fewer calcium ions are left in seawater, so the growth of calcium carbonate shells is inhibited. It is not clear whether these creatures will be able to adapt to the changing chemistry of the ocean.

Foraminifera are at the bottom of the oceanic food web. Among the potentially important consequences of a decline in their numbers would be the loss of food for fish such as mackerel and salmon. If the increased acidity of the oceans

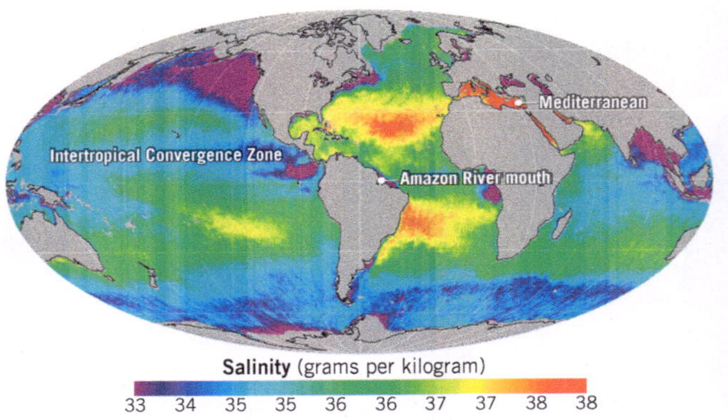

▲ Figure 9-7 Average salinity of oceans in March 2020. Lowest salinity (violet and blue areas) is found in areas of freshwater runoff, such as the mouths of rivers, and where rainfall is high (as in the intertropical convergence zone, the ITCZ); highest salinity (red and yellow) is found where evaporation rates are highest. The data were gathered by NASA's *Soil Moisture Active Passive* (SMAP) satellite mission.

▲ Figure 9-8 Scanning electron micrograph (magnified approximately 100 times) of *Elphidium crispum*, one of many kinds of single-celled foraminifera with exoskeletons made of calcium carbonate.

Animation
The Carbonate Buffering System
https://media.pearsoncmg.com/ph/esm/esm_mcknight_physgeo_13/media/

reduces the growth of coral polyps, coral reefs—already under stress worldwide from higher temperatures—might degrade even further. (A more complete discussion of coral reefs is in Chapter 20.)

Temperature: As expected, surface seawater temperatures generally decrease with increasing latitude. The temperature often exceeds 26°C (80°F) in equatorial locations and decreases to −2°C (28°F), the average freezing point for seawater, in Arctic and Antarctic seas. (Dissolved salts lower the freezing point of water below the 0°C (32°F) of pure water.) The western sides of oceans are nearly always warmer than the eastern margins because of the movement of major ocean currents (see Figure 4-25 in Chapter 4). This pattern of warmer western parts is due to the contrasting effects of poleward warm currents on the west side of ocean basins and equatorward cool currents on the east side.

Density: Seawater density varies with temperature, salinity, and depth. High temperature produces low density, and high salinity produces high density. Deep water has high density because its temperature is low and the pressure of the overlying water is high.

Surface layers of seawater tend to contract and sink in cold regions, whereas in warmer areas deeper waters tend to rise to the surface. Surface currents also affect this situation, particularly by producing an upwelling of colder, denser water in some localities. As we see later in this chapter, differences in density are partially responsible for a vast, slow circulation of deep ocean water.

Movement of Ocean Waters

The waters of the ocean are in near-continuous motion. We group this motion under three headings: tides, currents, and waves. The movement of almost anything over the surface—the wind, a boat, a swimmer—can set the water surface into motion, so the ocean surface is almost always ridged with swells and waves. Disturbances of the ocean floor also trigger significant movements in the water (see Chapter 20 for a discussion of *tsunami*) . The gravitational attraction of the Moon and Sun causes the greatest movements of all: the tides.

Tides

Animation
Tides
https://media.pearsoncmg.com/ph/esm/esm_mcknight_physgeo_13/media/

On the shores of the world ocean, almost everywhere, sea level fluctuates regularly. For about six hours each day, the water rises; then for about six hours it falls. **Tides** are essentially "bulges" in the sea surface in some places that are compensated by lower areas or "sinks" in the surface elsewhere. Thus, tides are primarily vertical motions. In shallow-water areas around ocean margins, however, the vertical oscillations of the tides produce significant horizontal water movements as well, when tides cause ocean water to advance and retreat along gently sloping coastal plains.

Causes of Tides: Every object exerts an attractive gravitational force on every other object. Thus, Earth exerts an attractive force on the Moon, and the Moon exerts an attractive force on Earth. The same is true of Earth and the Sun. It is the gravitational attraction between the Moon and Earth, and between the Sun and Earth, that causes tides.

The strength of the force of gravity between two bodies is inversely proportional to the square of the distance between them. Because Earth is much farther from the Sun than from the Moon—150,000,000 kilometers (93,000,000 miles) versus 385,000 kilometers (239,000 miles)—*lunar tides* are about twice as strong as *solar tides*. To keep things simple, let us ignore solar tides for the moment.

Gravitational attraction pulls ocean water toward the Moon. The side of Earth facing the Moon (closest to the Moon) experiences more gravitational attraction than on the opposite side of Earth. The difference in force slightly elongates the shape of the global ocean, so two bulges of ocean water develop—one on the side of Earth facing the Moon, the other on the opposite side. As Earth rotates, coastlines move into and out of these bulges, producing simultaneous high tides on the opposite sides of Earth and low tides halfway between. As Earth rotates to the east, these tidal bulges progress westward around the world.

Because the Moon revolves around Earth in the same direction that Earth is rotating, a lunar "day" lasts about 50 minutes longer than a 24-hour "solar" day. This means that Earth rotates through the pair of tidal bulges once every 24 hours and 50 minutes. As a result, nearly all oceanic coastlines experience two high tides and two low tides about every 25 hours.

The magnitude of tidal fluctuation is quite variable in time and place, but the sequence of the cycle is generally similar everywhere. From its lowest point, the water rises gradually for about 6 hours and 13 minutes. It moves up along the coast in what is called a **flood tide**, reaching the maximum water level, or *high tide*. Soon the water level begins to drop, and for the next 6 hours and 13 minutes water gradually moves away from the coast, a movement called an **ebb tide**. When the minimum water level (*low tide*) is reached, the cycle begins again.

Monthly Tidal Cycle: The difference in elevation between high and low tide is called the **tidal range**. Changes in the relative positions of Earth, Moon, and Sun produce periodic variations in tidal ranges. Normally, there is no special alignment of these bodies (Figure 9-9a). The greatest range (that is, the highest tide) occurs when the three bodies are positioned in a straight line, which usually occurs twice a month near the times of the full and new Moons. Then the joint gravitational pull of the Sun and Moon is along the same line, so the combined pull is at a maximum. This is true both when the Moon is between Earth and the Sun and when Earth is between the Moon and Sun. In either case, this is a time of higher than usual tides, called **spring tides** (Figure 9-9b). (The name has nothing to do with the season; think of water "springing" up to a very high level.)

When the Sun and Moon are located at right angles to one another with respect to Earth, their individual gravitational pulls are diminished, resulting in a lower than

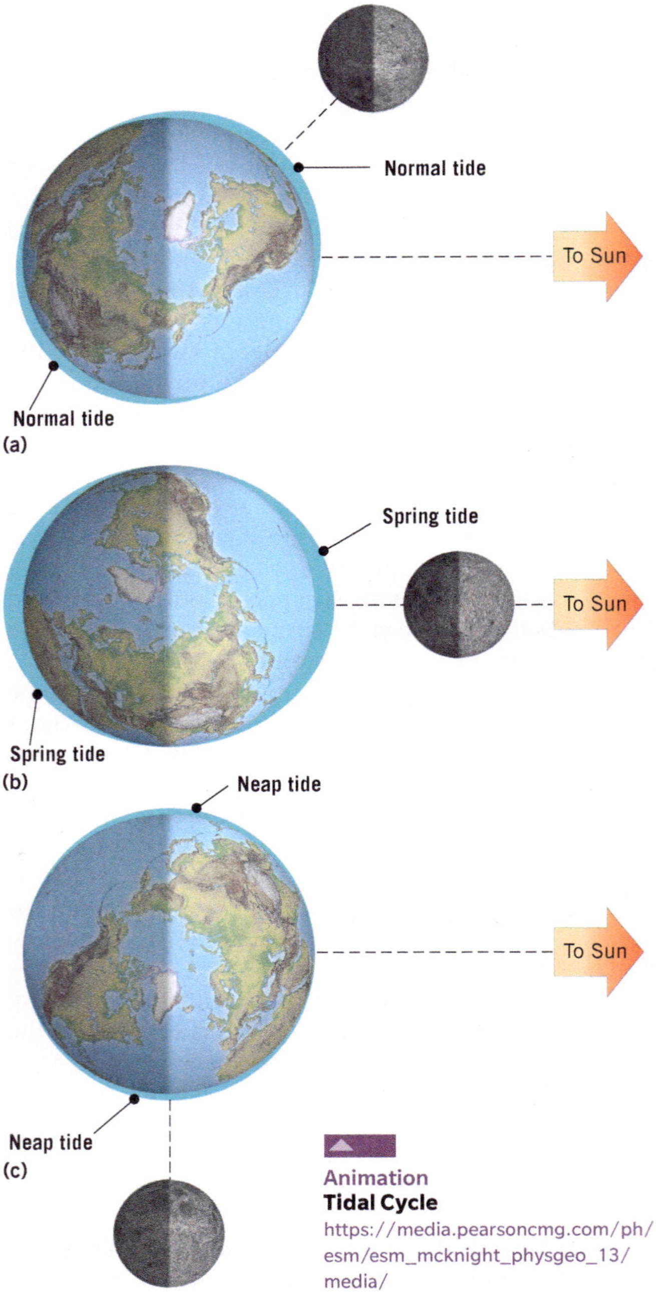

Animation
Tidal Cycle
https://media.pearsoncmg.com/ph/esm/esm_mcknight_physgeo_13/media/

▲ **Figure 9-9** The monthly tidal cycle. The juxtaposition of the Sun, Moon, and Earth accounts for variations in Earth's tidal range. (a) When the Moon and Sun are neither aligned nor at right angles to each other, levels of high tides on both sides of Earth are normal. (b) When the Sun, Earth, and Moon are positioned along the same line, *spring tides* (the highest high tides) are produced. (c) When the line joining Earth and the Moon forms a right angle with the line joining Earth and the Sun, *neap tides* (the lowest high tides) result.

normal tidal range called a **neap tide** (Figure 9-9c). The Sun–Moon alignment that causes neap tides generally takes place twice a month at about the time of first-quarter and third-quarter moons.

Tidal range is also affected by the Moon's nearness to Earth. The Moon follows an elliptical orbit in its revolution around Earth; the nearest point (called *perigee*) is about 45,000 kilometers (27,900 miles), or 11 percent, closer than the farthest point (*apogee*). Three or four times a year, when the Moon's perigee coincides with a spring-tide alignment of the Moon and Sun, a higher than usual tidal range occurs. Such a *perigean spring tide* (also called a "king tide") is even greater when this alignment occurs at the time of Earth's early-January perigee in its orbit around the Sun.

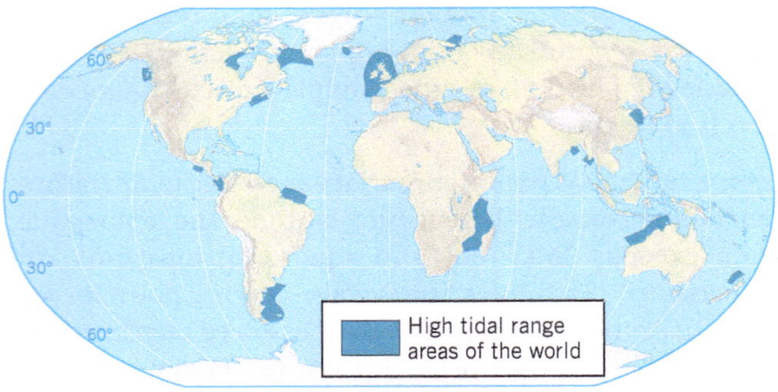

▲ **Figure 9-10** Areas with tidal ranges exceeding 4 meters (13 feet). The pattern is not predictable because it depends on a variety of unrelated factors, particularly shoreline and sea bottom configuration.

LearningCheck 9-3 **Describe the positions of the Sun, Moon, and Earth that produce the highest high tides and the lowest low tides.**

Global Variations in Tidal Range: Tidal range fluctuates all over the world at the same times of the month. However, enormous variations in range occur along different coastlines (Figure 9-10). Midocean islands may experience tides of only 1 meter (3 feet) or less, whereas continental seacoasts have greater tidal ranges because the amplitude is greatly influenced by the shape of the coastline and the configuration of the sea bottom beneath coastal waters. Along most coasts, the tidal range is 1.5–3 meters (5–10 feet). Some partly landlocked seas, such as the Mediterranean, have almost negligible tides. Other places, such as the northwestern coast of Australia, experience enormous tides of 10 meters (35 feet) or so.

The greatest tidal range is found at the upper end of the Bay of Fundy in eastern Canada (Figure 9-11). A 15-meter (50-foot) water-level fluctuation twice a day is not uncommon there, and a wall of seawater—called a **tidal bore**—several centimeters to more than a meter in height rushes up the Petitcodiac River in New Brunswick for many kilometers.

Tidal variation is exceedingly small in inland bodies of water. Even the largest lakes usually experience a tidal rise and fall of no more than 5 centimeters (2 inches). Effectively, then, tides are important only in the world ocean, and we normally notice them only around its shorelines.

Ocean Currents

As we learned in Chapter 4, the world ocean contains a variety of currents that shift vast quantities of water both horizontally and vertically. Surface currents are caused primarily

▼ **Figure 9-11** (a) The world's maximum tides are in the Bay of Fundy, where ocean water moves long distances up many of the coastal rivers twice a day. (b) A tidal bore arriving at Truro, Nova Scotia, Canada.

(a) Bay of Fundy

(b) Tidal bore

by winds, whereas other currents are set in motion by contrasts in temperature and salinity. All currents may be influenced by the size and shape of the particular ocean, the configuration and depth of the sea bottom, and the Coriolis effect. Some currents involve subsidence of surface waters downward; other vertical flows bring an upwelling of deeper water to the surface.

Animation
Ocean Circulation Patterns—Subtropical Gyres
https://media.pearsoncmg.com/ph/esm/esm_mcknight_physgeo_13/media/

Geographically speaking, the most prominent currents are the major horizontal flows that make up the general circulation of the various oceans. The dominant surface currents introduced in Chapter 4 are generally referred to as *subtropical gyres* (see Figure 4-25). They are set up by the action of the dominant surface wind systems in the tropics and midlatitudes: the trade winds and the westerlies.

Ocean currents can transport floating debris, both natural and human-produced, across large expanses of ocean. They also can concentrate floating trash in several areas of the Pacific Ocean—see the box *People & the Environment: The Great Pacific Garbage Patch.*

Deep Ocean Circulation: In addition to the major surface ocean currents, an important system of deep ocean circulation exists. This deep-water circulation occurs because of differences in water density that arise from differences in salinity and temperature. Thus, this water movement is referred to as **thermohaline circulation**. Ocean water becomes denser and sinks if its salinity increases or its temperature decreases. This happens predominantly in high-latitude ocean areas where the water is cold and salinity increases when sea ice develops. (When water freezes, it expels dissolved salts, so the salinity of the remaining water increases.)

The combination of deep ocean water movement through thermohaline circulation and influences from surface ocean currents establishes an overall **global conveyer-belt circulation** pattern (**Figure 9-12**). Cold, dense water in the North Atlantic sinks and slowly flows south deep below the surface; it eventually joins the eastward deep, cold, high-salinity water circling Antarctica. Some of this deep water flows north into the Indian and Pacific Oceans, where it rises and forms a shallow, warm current that flows back to the North Atlantic. There it sinks and begins its long journey once again. These deep ocean currents travel so slowly they may require many centuries to complete a single circuit.

Animation
Ocean Circulation Patterns—Global Conveyor-Belt Circulation
https://media.pearsoncmg.com/ph/esm/esm_mcknight_physgeo_13/media/

The global conveyor-belt circulation does not have as immediate an effect on weather and climate as the subtropical gyres and other surface currents discussed in Chapter 4 do. Nonetheless, this circulation plays a role in energy transfer around the world—and therefore long-term climate patterns. Research suggests that a connection exists between the global conveyer-belt circulation and global climate. For example, if global climate becomes warmer, freshwater runoff from the melting of Greenland's glaciers could form a

People
& the
Environment

The Great Pacific Garbage Patch

• **Jennifer Rahn,** Samford University

Characteristics that make plastics useful—their light weight, strength, and durability—make them a problem in marine environments. Each year more than 270 billion kilograms (300 million tons) of plastic is produced worldwide, and about 10 percent of it ends up in the ocean.

The ocean carries many kinds of plastic. Certain plastics, including microbeads in some face scrubs and toothpastes, are manufactured to be very small. Also, some tiny plastics are shards of larger items. Instead of biodegrading, most plastics very slowly break up by *photodegrading* from sunlight and other environmental factors, such as waves and big storms, into ever smaller pieces (microplastics). Plastics never really go away.

Location: Trash tends to accumulate where winds and currents are weak in the center of *subtropical gyres*—especially in the area of the subtropical high (STH). These patches of garbage and flotsam in the Pacific (and all other ocean basins) are not new; they have always been there. What is new is that now most of the junk is nonbiodegradable plastic; formerly it consisted of materials such as vegetation, wood, glass bottles, and glass fishing floats. The best-studied patch is the *Great Pacific Garbage Patch* between Hawai'i and California (**Figure 9-A**).

Some of the trash sinks, but most of it floats in the upper 10 meters (33 feet) of the ocean (**Figure 9-B**). Estimates of the size of the trash patch now in the Pacific Ocean vary from the size of Texas to 20 times that. Estimates of the total mass of the trash also vary greatly, from about 80 million kilograms (88 thousand tons) to 3.2 billion kilograms (3.5 million tons) of trash, and 80 percent of it is plastic from East Asia and the United States. A 2012 study by the Scripps Institution of Oceanography estimates that the trash accumulating in the patch increased by 100 times in the last four decades.

Hazards to Marine Life: Marine animals often mistake the floating plastic for food. To sea turtles or birds, a plastic bag floating in the water looks like a jellyfish—a favorite food—but they cannot digest plastic. If animals eat too much plastic, it fills their stomachs, and they starve to death.

Marine animals can become entangled in plastic bags or rings that hold drink cans; unable to move freely, many of the animals die. More than 600 marine species have been harmed by the debris; about 100,000 whales, seals, turtles, birds, dolphins, and other marine animals are killed each year by plastic-induced asphyxiation, strangulation, contamination, or entanglement.

▲ **Figure 9-B** Floating debris in the Great Pacific Garbage Patch.

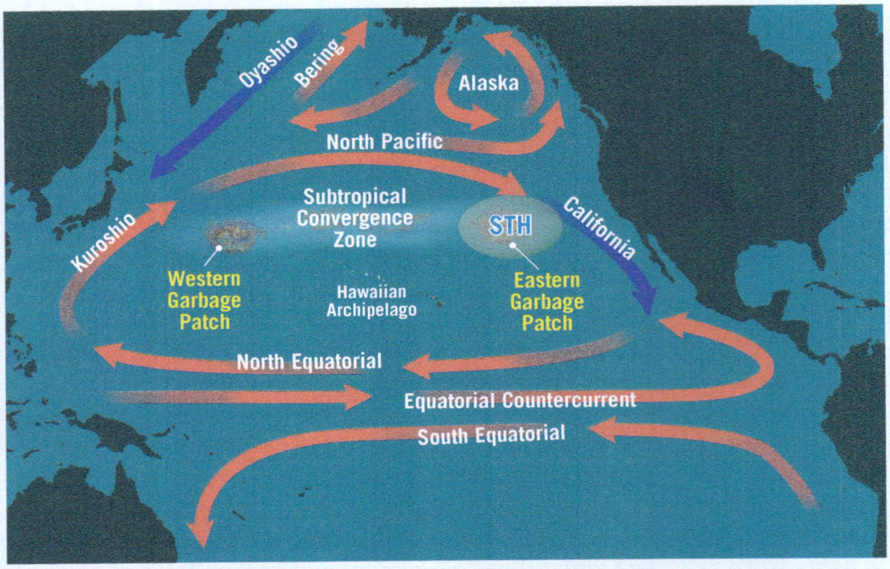

▲ **Figure 9-A** The location of the Great Pacific Garbage Patch, in the area of the subtropical high (STH). Garbage also accumulates in the western portion of the North Pacific Subtropical Convergence Zone.

Fish consume small plastic particles, thinking they are plankton, their main food source. The Great Pacific Garbage Patch now contains six times more plastic than plankton. Toxin levels in fish there have reached a million times the amount in surrounding water. The fish may then be consumed by whales, dolphins, or humans—so the toxins are concentrated at the top of the food chain. As we eat large fish, we receive concentrated doses because the effects increase up the food chain.

In 2017, the United States banned microplastics in cosmetics. The European Union is moving to ban the use of microplastics in most products in 2020.

Questions

1. Describe how trash from the West Coast of the United States ends up in the middle of the Pacific Ocean.
2. How could plastics be harmful to you and to marine life?

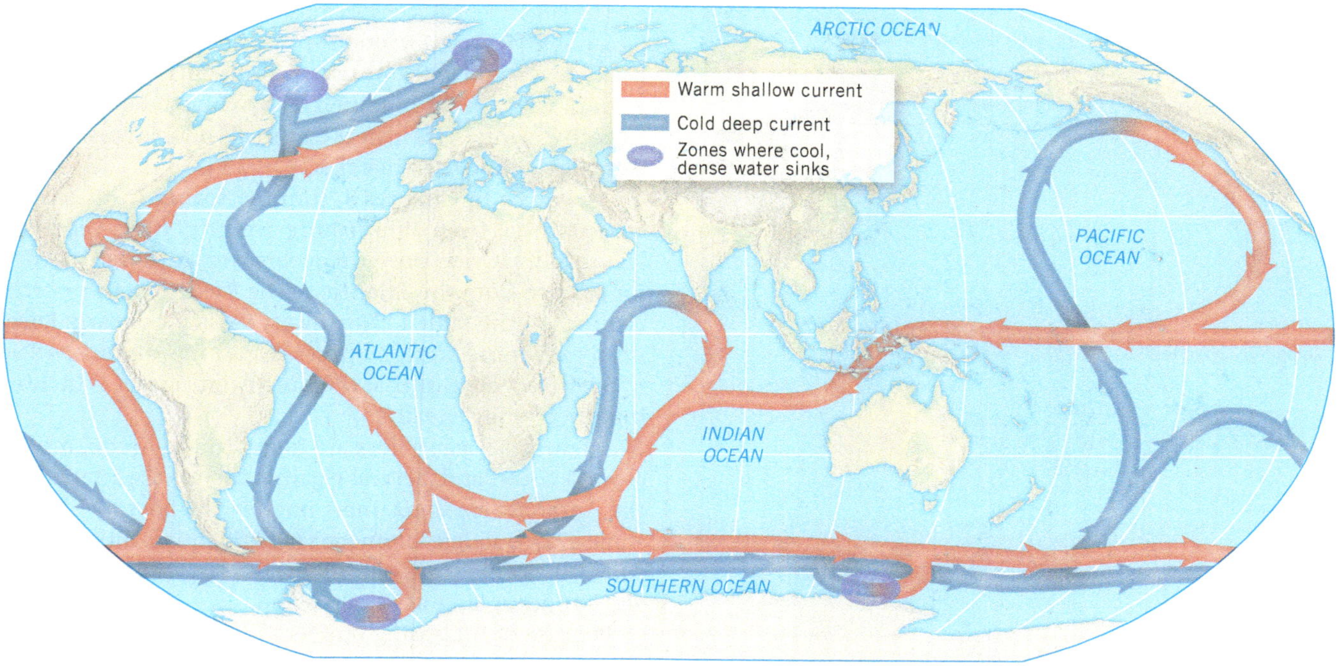

▲ **Figure 9-12** Idealized global conveyor-belt circulation. In the North Atlantic, cool, dense water sinks and moves south as a deep subsurface flow. It joins cold, deep water near Antarctica, eventually moving into the Indian and North Pacific oceans, where the water rises slowly, eventually flowing back to the North Atlantic and again sinks. The violet ovals show major locations where cool, dense surface water feeds the flow of deep water. One circuit may take many hundreds of years to complete.

Animation
North Atlantic Deep Water Circulation
https://media.pearsoncmg.com/ph/esm/esm_mcknight_physgeo_13/media/

pool of lower-density water in the North Atlantic that could disrupt the downwelling of water in that region. Such disruption would alter the redistribution of heat—and thereby climate—around the world.

LearningCheck 9-4 **Explain what causes the movement of water in the global conveyor-belt circulation.**

Waves To the casual observer, the most conspicuous motion of the ocean is provided by waves. Most of the sea surface is in a state of constant agitation, with wave crests and troughs bobbing up and down. Moreover, around the margins of the ocean, waves break on the shore in endless procession.

From the water's point of view, most of this movement is like "running in place" with little forward progress. The movement of a wave across the surface of the open ocean is a movement of form rather than substance—a transfer of energy rather than a transfer of matter. Individual water particles make only small oscillating movements as the form of the wave passes. Only when a wave "breaks" does any significant shifting of water take place. Waves are discussed in detail in Chapter 20.

Permanent Ice—The Cryosphere

Second only to the world ocean as a storage reservoir for water is the solid portion of the hydrosphere—the ice of the world, or *cryosphere*. Although tiny in comparison with the amount of water in the oceans, the moisture content of ice at any given time is more than twice as large as the combined total of all other types of storage (groundwater, surface waters, soil moisture, atmospheric moisture, and biological water).

The cryosphere is divided between ice on land and ice floating in the ocean, with the land portion being the larger. Ice on land is found as mountain glaciers, ice sheets, and ice caps, all of which are studied in Chapter 19. Approximately 10 percent of the land surface of Earth is covered by ice (**Figures 9-13** and **9-14**). It is estimated that enough water is locked up in this ice to feed all the rivers of the world at their present rate of flow for nearly 900 years.

Oceanic ice has various names, depending on size:

- **Ice pack:** An extensive and cohesive mass of floating ice.
- **Ice shelf:** A massive portion of a continental ice sheet that projects out over the sea.
- **Ice floe:** A large, flattish mass of ice that breaks off from larger ice bodies and floats independently.
- **Iceberg:** A chunk of floating ice that breaks off from an ice shelf or glacier.

Because ice is less dense than liquid water, only about 14 percent of the mass of an iceberg is exposed above the water, with about 86 percent below (**Figure 9-15**).

Some oceanic ice freezes directly from seawater. All forms of oceanic ice are composed almost entirely of freshwater, however, because the salts present in seawater are not incorporated into ice crystals when that water freezes.

▲ **Figure 9-13** Winter sea ice covers the entire Arctic Ocean. Although it has diminished over the last few decades, summer sea ice still covers much of the Arctic Ocean.

Animation
Arctic Sea Ice Minimum
https://media.pearsoncmg.com/ph/esm/esm_mcknight_physgeo_13/media/

The largest ice pack covers most of the surface of the Arctic Ocean (see Figure 9-13); on the other side of the globe, an ice pack fringes most of the Antarctic continent (see Figure 9-14). Both of these packs become greatly enlarged during their respective winters, their areas essentially doubling as freezing increases around their margins.

As we saw in Chapter 8, summer sea ice in the Arctic especially has been diminishing over the last 45 years (see Figure 8-28). Once, powerful icebreaker ships were needed for passage from the Atlantic to the Pacific via this northern route—the fabled "Northwest Passage" of early European explorers. Arctic summer sea ice has diminished so much in recent years that the Northwest Passage is increasingly becoming a reality.

A few small ice shelves are found in the Arctic, mostly around Greenland. Several gigantic shelves are attached to the Antarctic ice sheet, most notably the Ross Ice Shelf of some 472,00 square kilometers (182,00 square miles). Some Antarctic ice floes are enormous; the largest ever observed was 4 times as large as Rhode Island.

Because of increasing temperatures, formerly stable ice shelves in Antarctica have broken apart. Since the early 1990s, as much as 8000 square kilometers (more than 3000 square miles) of Antarctic ice shelves have disintegrated. In 2002, the Larsen-B Ice Shelf on the Antarctic Peninsula disintegrated in less than a month, and the much larger Larsen-C shelf just to the south is showing signs that its mass is being reduced because of increasing water temperatures below it. In 2008, the Wilkins Ice Shelf in Antarctica also began to disintegrate. (See Chapter 19 for further discussion of changes in glaciers and ice sheets around the world.)

LearningCheck 9-5 **Where is most of the ice in the cryosphere found?**

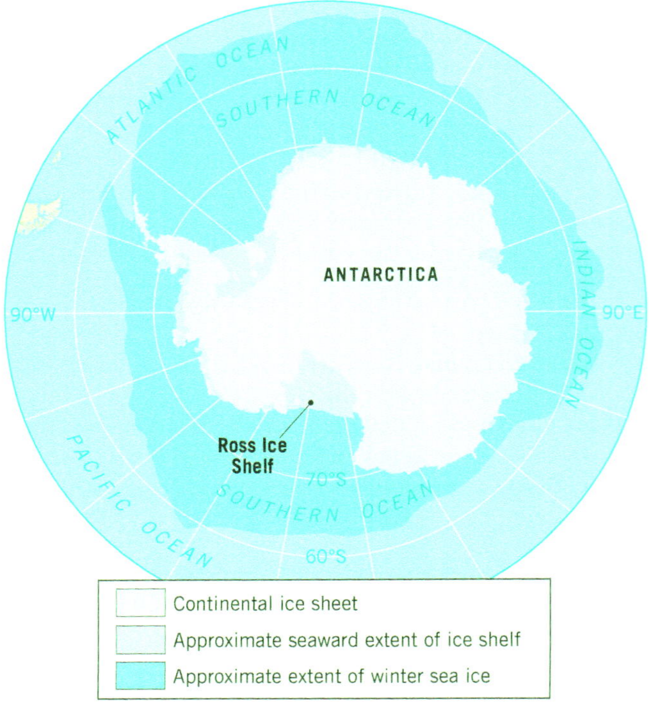

▲ **Figure 9-14** Maximum extent of sea ice in Antarctica. The ice sheet covers land, and the ice shelf and ice pack are oceanic ice.

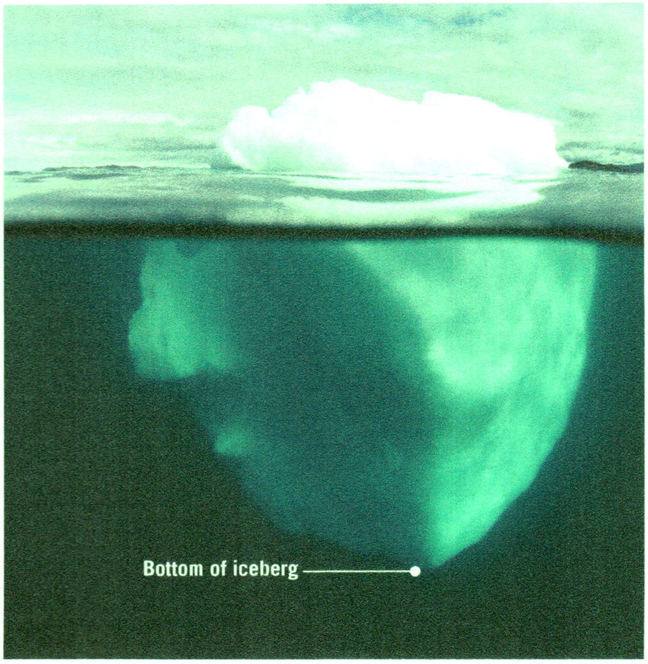

▲ **Figure 9-15** Most of an iceberg is underwater, as shown here in the waters of Disko Bay, Greenland.

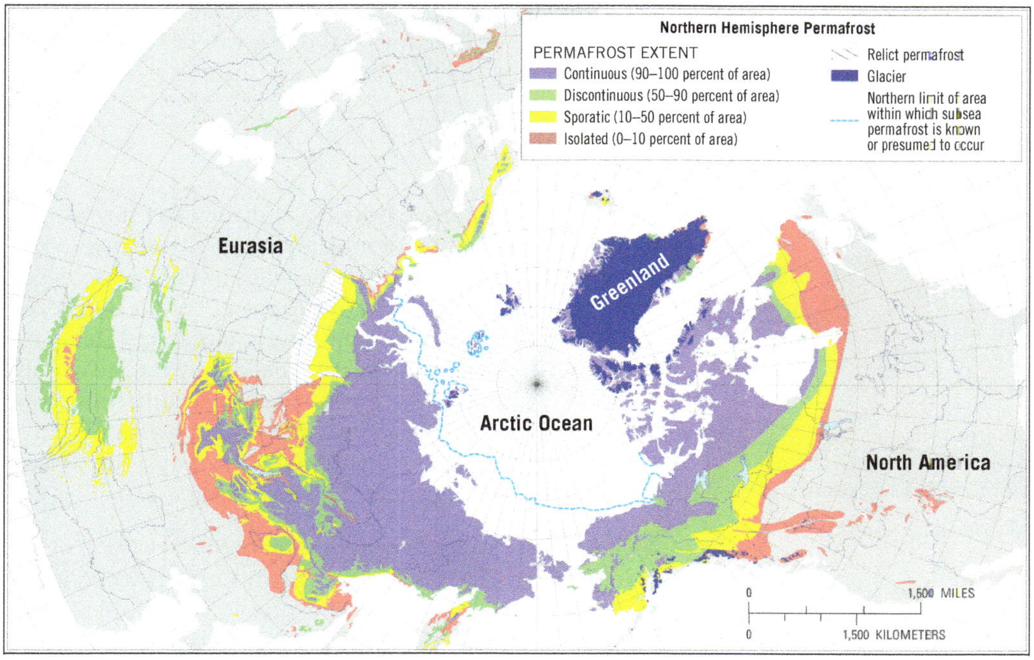

◀ **Figure 9-16** Extent of permafrost in the Northern Hemisphere. All of the high-latitude land areas and some of the adjacent midlatitude land areas are underlain by permafrost. Climate change is slowly reducing the extent of permafrost. (Source: USGS.)

Permafrost

A relatively small proportion of the world's ice occurs beneath the land surface as ground ice. This type of ice occurs only in areas where the temperature is continuously below the freezing point, so it is restricted to high-latitude and high-elevation regions (Figure 9-16). Most permanent ground ice is **permafrost**, which is permanently frozen subsoil. It is widespread in northern Canada, Alaska, and Siberia and found in small patches in many high mountain areas. Some ground ice is aggregated as veins of frozen water, but most of it develops as ice crystals in the spaces between soil particles.

Thawing of Permafrost: In locations such as the region around Fairbanks, Alaska, permafrost is widespread just below the surface. During the summer, only the upper 30 to 100 centimeters (12 to 40 inches) of soil thaws in what is called the *active layer*; below that is a layer of permanently frozen ground perhaps 50 meters (165 feet) thick. Much of the permafrost in the high-latitude areas of the world has been frozen for at least the last few thousand years, but in response to higher average temperatures, it is beginning to thaw. In the last 45 years, a warming trend has been observed (Figure 9-17), bringing the ground temperature in some areas above the melting point of the permafrost. Deep in the

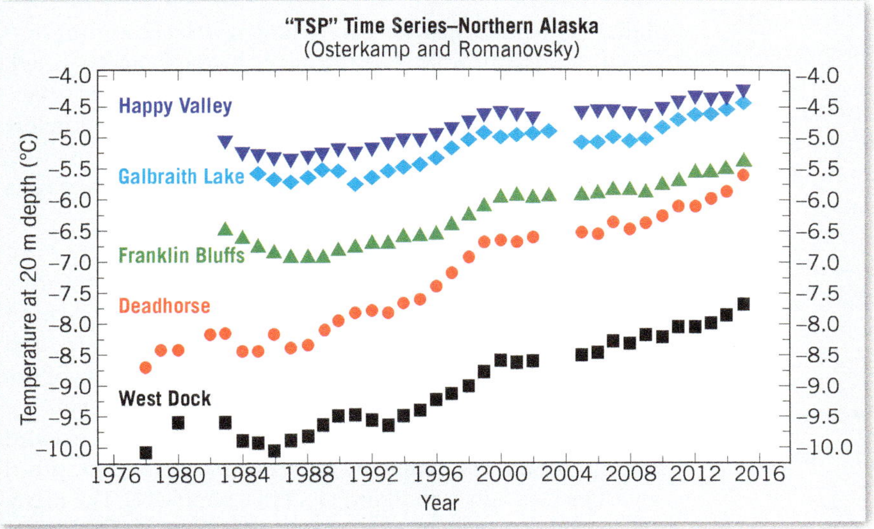

(a) Permafrost temperatures

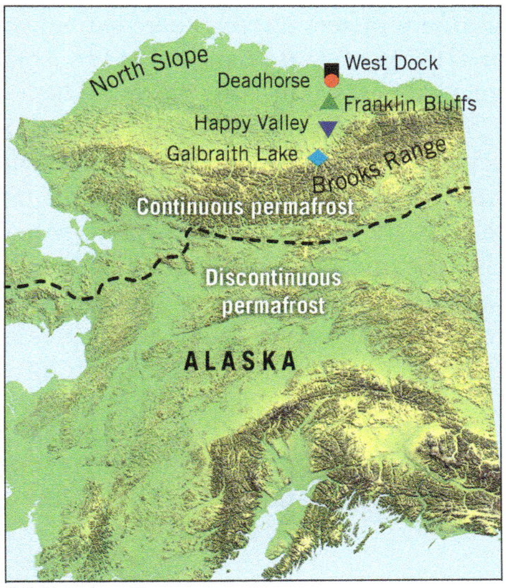

(b) Data locations

▲ **Figure 9-17** Thermal State of Permafrost (TPS) Time Series. (a) Changes in permafrost temperature at a depth of 20 meters (65 feet) from 1976 to 2015 in Alaska. (b) Sites ranged from the Brooks Range to the North Slope.

▲ Figure 9-18 House in Alaska collapsed as a result of melting permafrost.

permafrost layer where ground still remains frozen, temperatures are rising also.

For people accustomed to temperate environments, it might seem that having the ground thaw would not be a problem, but such is not the case. As the ground thaws, buildings, roads, pipelines, and airport runways are increasingly destabilized, and transportation and business are likely to be disrupted (Figure 9-18). In areas with poor surface drainage, the degradation of permafrost can lead to what is called *wet thermokarst* conditions, in which the surface subsides and the ground becomes oversaturated with water. In some cases, unpaved roads become impassible. In the last four decades, the number of days that the Alaska Department of Natural Resources permits oil exploration activity in areas of tundra has been cut in half due to the increasingly soft ground.

Along the Beaufort Sea, rising temperatures are thawing permafrost in the coastal bluffs and contributing to more rapid erosion of the coastline. From an average rate of erosion of 6 meters (20 feet) per year between the mid-1950s and 1970s, the rate had jumped to as much as 20 meters (65 feet) per year in some places by 2010.

The thawing of frozen soils leads to an increase in the activity of microorganisms in the soil. This in turn increases the rate of decomposition of organic matter long sequestered in the frozen ground. As microorganisms decompose this organic matter, carbon dioxide or methane can be released, contributing to increasing greenhouse gas concentrations in the atmosphere. In 2019, researchers estimated that 300 to 600 million metric tons of carbon are released into the atmosphere each year from thawing permafrost.

LearningCheck 9-6 **What are some of the consequences of thawing permafrost around the Arctic?**

Surface Waters

Surface waters represent only about 0.02 percent of the world's total moisture supply (see Figure 9-1), but to humans they are of incalculable value. Lakes, wetlands, swamps, and marshes abound in many parts of the world, and all but the driest parts of the continents are seamed by rivers and streams.

Lakes

In the simplest terms, a **lake** is a body of water surrounded by land. No minimum or maximum size is attached to this definition, although the word *pond* is often used to designate a very small lake. Well over 90 percent of the nonfrozen surface water of the continents is contained in lakes. Lake Baykal (often spelled Baikal) in Siberia is by far the world's largest freshwater lake in terms of volume, containing considerably more water than the combined contents of all five Great Lakes in central North America. It is also the world's deepest lake—1642 meters (5387 feet) deep.

Saline Lakes: More than 40 percent of the lake water of the planet is salty. The lake that we call the Caspian Sea contains more than three-quarters of the total volume of the world's nonoceanic saline water. (Utah's famous Great Salt Lake contains less than 1/2500 the volume of the Caspian.) Any lake that has no natural drainage outlet, either as a surface stream or as a sustained subsurface flow, will become saline: virtually all freshwater contains tiny quantities of salts and minerals dissolved from rocks on land. This water flows into a closed basin and evaporates, leaving these soluble minerals to accumulate over time.

Most small salt lakes and some large ones are *ephemeral*, which means that they contain water only sporadically and are dry much of the time. They are in dry regions with insufficient inflow to maintain them on a permanent basis. We discuss ephemeral lakes in desert regions in greater detail in Chapter 18.

Formation of Lakes: Most lakes are fed and drained by streams, but how do lakes originate? Two conditions are necessary for the formation and continued existence of a lake: (1) a natural basin having a restricted outlet and (2) enough inflow of water to keep the basin partly filled. The water balance of most lakes is maintained by surface inflow, sometimes combined with springs and seeps below the lake surface. A few lakes are fed entirely by springs. Most freshwater lakes have only one stream that serves as a drainage outlet.

Lakes are distributed very unevenly over the land (Figure 9-19). They are very common in regions that were glaciated in the recent geologic past because glacial erosion and deposition disturbed the normal drainage patterns and created innumerable basins (Figure 9-20). Some parts of the world notable for lakes were not glaciated, however. For example, a remarkable series of large lakes in eastern and central Africa was created by faulting as Earth's crust spread apart tectonically (see Figure 14-14 in Chapter 14). The many thousands of small lakes in Florida were formed by sinkhole collapse when rainwater dissolved calcium carbonate from the limestone bedrock (for example, see Figure 17-10 in Chapter 17).

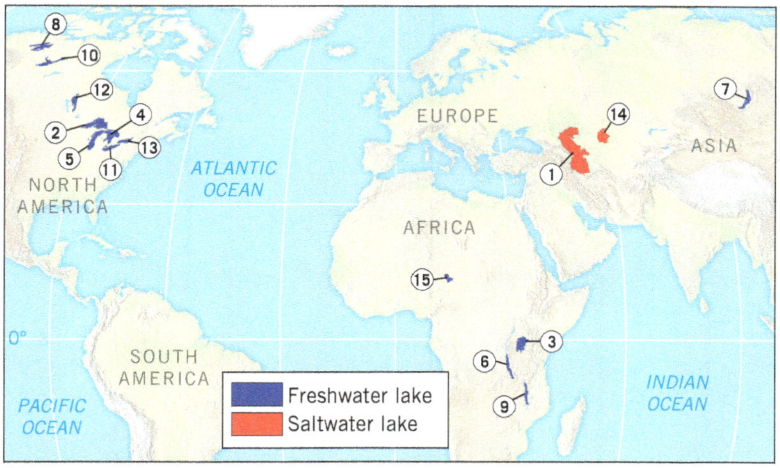

(a) Largest lakes: locations

	World's Largest Lakes by Surface Area	Square km	Square mi
(1)	Caspian Sea	372,450	143,250
(2)	Lake Superior	82,420	31,700
(3)	Lake Victoria	69,400	26,700
(4)	Lake Huron	59,800	23,000
(5)	Lake Michigan	58,000	22,300
(6)	Lake Tanganyika	33,000	12,650
(7)	Lake Baikal	31,700	12,200
(8)	Great Bear Lake	31,500	12,100
(9)	Lake Nyasa (L. Malawi)	30,000	11,550
(10)	Great Slave Lake	29,400	11,300
(11)	Lake Erie	25,700	9900
(12)	Lake Winnipeg	23,500	9100
(13)	Lake Ontario	19,500	7500
(14)	Aral Sea	*	*
(15)	Lake Chad	*	*

* Greatly reduced in size from that shown here

(b) Surface areas

▲ **Figure 9-19** (a) The largest lakes of the world, (b) ranked by surface area.

Natural Disappearance of Lakes: On the grand scale of geologic time, most lakes are relatively temporary features of the landscape. Few have been in existence for more than a few thousand years. Inflowing streams bring sediment that fills lakes up; outflowing streams cut channels progressively deeper and drains them. As lakes becomes shallower, a continuous increase in plant growth accelerates the infilling (a process discussed in Chapter 10). Thus, the destiny of most lakes is to disappear naturally.

Human Alteration of Natural Lakes: Human activity also plays a part in the disappearance of lakes. For example, the diversion of streams flowing into California's Mono Lake (east of Yosemite National Park) has reduced its volume by one-half since the 1940s. However, in 1994 environmental activists won a court case mandating that diversions from the lake be reduced, and lake level has stabilized; it should rise slightly in years to come.

▲ **Figure 9-20** Glaciation is responsible for the formation of Convict Lake in California. The lake formed behind a natural dam of rocks deposited by glaciers.

(a) 1973 (b) 2018

(c) Aral "Sea" floor

▲ **Figure 9-21** In 1960, the Aral Sea was the world's fourth-largest lake. As irrigation needs forced farmers to drain water from the rivers feeding the Aral, its size has decreased by 90 percent in the last 60 years. (a) The Aral Sea in 1973 and (b) 2018. (c) Where fish once swam, boats are left stranded on dry land. If present trends continue, the Aral may cease to exist.

More dramatically, the Aral Sea was once the world's fourth-largest lake in terms of surface area, but beginning in the 1960s, irrigation projects designed to boost agricultural production in Soviet Central Asia cut off much of the water flowing into the lake. This sea is now less than 10 percent of its original size (**Figure 9-21**). The once-viable commercial fishing industry is gone, and winds now carry away a cloud of choking clay and salt dust lifted from the exposed lake bottom. The sea has split into several pieces. A new dam on the Syr Darya River is allowing the northern remnant of the Aral Sea to recover slightly, but the southern remnants are likely to stay dry.

In some cases, both human and natural changes are responsible for the loss of a lake. Fifty years ago, Lake Chad was one of the largest lakes in Africa, but ongoing drought has reduced it to about 10 percent of its original size (**Figure 9-22**). Nearly all of the lake's water comes from the Chari River flowing in from the south. The lake is surrounded by an extensive wetlands area—once the second largest in Africa. The lake is also shallow, so it responds quite quickly to changes in inflow. Although water diversion projects along the Chari River have contributed to the reduction of Lake Chad, climate change in the region is likely responsible for much of its ongoing decline.

LearningCheck 9-7 **How and why has the Aral Sea changed over the last few decades?**

Reservoirs: One notable thing we have done to alter the natural landscape is to produce artificial lakes, or *reservoirs*. They have been created largely by the construction of dams, ranging from small earth mounds across a gully to immense concrete structures blocking major rivers. Some reservoirs are as large as medium-sized natural lakes. See the box *Snapshots from Our National Parks: Glen Canyon and Lake Mead.*

Mobile Field Trip
Moving Water Across California
https://media.pearsoncmg.com/ph/esm/esm_mcknight_physgeo_13/media/

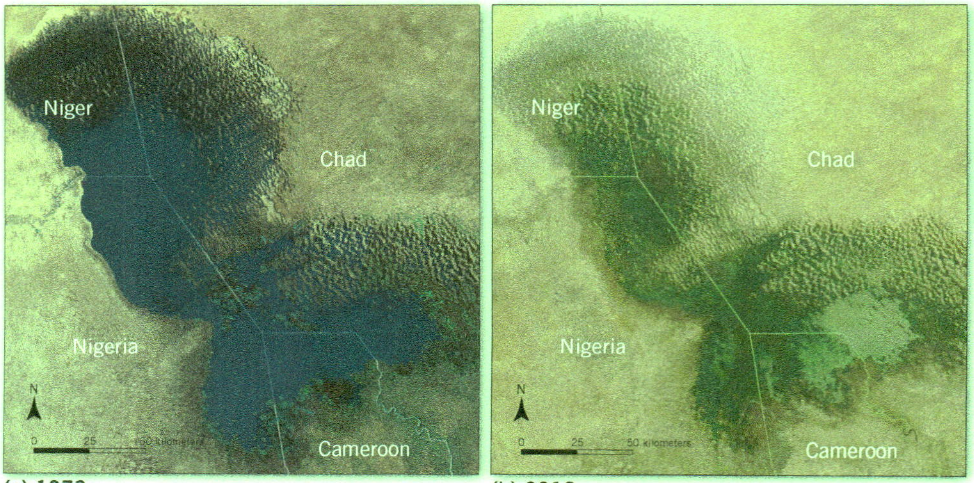

(a) 1973 (b) 2018

◀ **Figure 9-22** Lake Chad in (a) 1973 and (b) 2018.

Snapshots From Our National Parks

Glen Canyon and Lake Mead

Bookends of the Grand Canyon

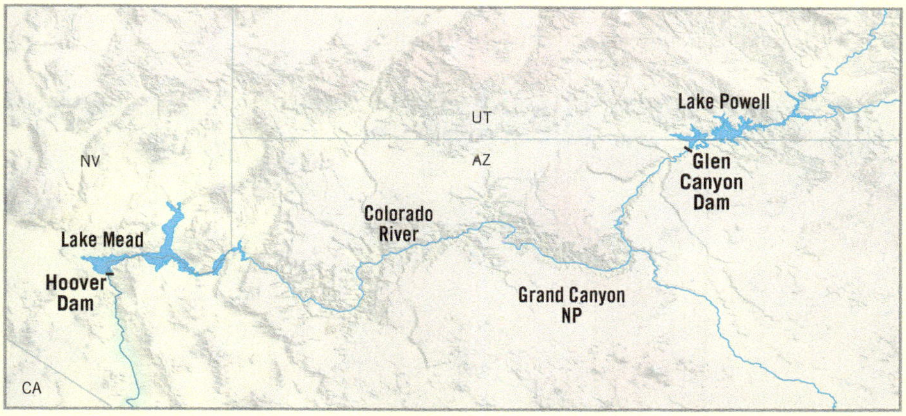

▲ **Figure 9-C** Map of the Colorado River connecting the Hoover and Glen Canyon dams.

Some of the most heated environmental battles of the twentieth century centered on the construction of dams in the western United States. Dams were built by different local, state, and federal agencies, but the U.S. Bureau of Reclamation was responsible for many of the largest projects. These dams were constructed for many reasons: flood control, agricultural and municipal water supply, and hydroelectric power generation.

Today, reservoirs of two of the most famous dams in the West—Hoover Dam and Glen Canyon Dam on the Colorado River—are overseen by the National Park Service as National Recreation Areas and act as "bookends" to Grand Canyon National Park in Arizona (**Figure 9-C**).

Hoover Dam: Hoover Dam, downstream of the Grand Canyon, was built at the height of the Great Depression (**Figure 9-D**). When it was completed in 1936, it was the world's tallest dam. Behind the dam, Lake Mead became the largest reservoir in the United States.

Most of the water flowing into Lake Mead from the Colorado River originally falls as snow in Colorado, Wyoming, Utah, and New Mexico. Each year, a minimum amount of Colorado River water is allotted to Lake Mead, and a specified amount of water is released for use by California, Arizona, Nevada, and Mexico. In a typical year, more water is released than flows into the reservoir.

◄ **Figure 9-D** Hoover Dam and Lake Mead on the Colorado River. The "bathtub ring" around the margin of the reservoir marks the water level when Lake Mead is at full capacity.

Glen Canyon Dam: Glen Canyon Dam, upstream of the Grand Canyon, was quite controversial. Although first considered as a dam site in the 1920s, the Glen Canyon project moved forward in the late 1950s. The growing population in the desert southwest United States convinced the Bureau of Reclamation that new reservoirs were needed along the Colorado River. Several new dams were proposed, including two within Grand Canyon National Park and another at Echo Park within Dinosaur National Monument.

The Sierra Club, led by David Brower, was strongly opposed to the new dams. After a long fight, plans for the Echo Park Dam were scrapped by the Bureau of Reclamation, but by 1956 a dam was being constructed across Glen Canyon. When completed in 1966, Lake Powell (named in honor of John Wesley Powell who led the first U.S. Geological Survey expedition through the Grand Canyon in 1869) began to fill the spectacular narrow sandstone canyons that many considered to be among the most beautiful in the West (**Figure 9-E**). The famous Sierra Club book of Eliot Porter, *The Place No One Knew*, contains photographs taken before the canyon was flooded and mourns the loss of this landscape. With the flooding of Glen Canyon, Brower and the Sierra Club mounted another campaign and ultimately stopped the dam projects within Grand Canyon National Park.

Consequences of the Dams: Hoover Dam and Glen Canyon Dam, as well as many others, had immense ecological and economic consequences, some unforeseen at the time of construction. In addition to the loss of tribal and scenic land that has been inundated by reservoir waters, downstream ecosystems are altered by restricted stream flows; in some locations, rapid sedimentation may limit the useful life of a reservoir. (In Chapter 16, we consider the implications of flood control through the use of dams and river levees.)

▲ **Figure 9-E** Lake Powell in Glen Canyon National Recreation Area.

Especially here in the arid southwest, precipitation from year to year is quite variable (see Figure 6-38), and so annual surface runoff to refill these reservoirs can differ significantly. One visible consequence of this varying runoff is the "bathtub rings" seen around many reservoirs when a few years of lower-than-average rainfall causes a large drawdown in water level (see Figure 9-D).

In 1983, the opposite problem occurred. Unusually heavy upstream precipitation filled Lake Powell to its brim. Only emergency releases of water through its spillways prevented what might have been a catastrophic dam failure.

Questions

1. Why do the water levels of Lake Mead and Lake Powell vary from year to year?
2. What are some of the negative consequences of dam construction?

▲ Figure 9-23 Marshes are particularly numerous along the poorly drained South Atlantic and Gulf of Mexico coasts of the United States. This marshland is part of Everglades National Park in southern Florida.

Wetlands

Closely related to lakes but less numerous and containing much less water are **wetlands**—broadly defined as land areas where saturation with water is the overriding factor that influences soil development and plant and animal communities. As we see in Chapter 11, not only do wetland areas play important roles in local ecosystems as sources of food and nutrients, but also they act as "filters" for surface runoff and thus control the water quality of many lakes, streams, and coastal waters. Furthermore, as we saw in Chapter 7, coastal saltwater wetlands can act as buffers, reducing the immediate impact of hurricane storm surges.

Swamps and Marshes: Swamps and marshes are flattish places that are submerged in water at least part of the time but are shallow enough to permit the growth of water-tolerant plants (Figure 9-23). A **swamp** has a plant growth that is dominantly trees, whereas a **marsh** is vegetated primarily with grasses and rushes. Both are usually associated with coastal plains, broad river valleys, or recently glaciated areas. Sometimes they represent an intermediate stage in the infilling of a lake.

Rivers and Streams

Although containing only a small proportion of the world's water at any given time, rivers and streams are an extremely dynamic component of the hydrologic cycle. (In common usage, a *stream* is smaller than a *river*; geographers, however, call any flowing water a "stream," no matter what its size.) Streams provide the means by which the land surface drains and by which water, sediment, and dissolved chemicals are moved ever seaward. The occurrence of rivers and streams is closely, but not absolutely, related to precipitation patterns. Humid lands have many rivers and streams, most of which flow year-round. Dry lands have fewer, almost all of which are ephemeral (they are dry most of the time).

Table 9-2 lists the world's largest rivers by discharge volume and length, while Figure 9-24 shows the world's largest river drainage basins (a *drainage basin* is all the land area drained by a river and its tributaries). A mere two dozen great rivers produce half of the world's total stream discharge. The mighty Amazon yields nearly 20 percent of the total, more than five times the discharge of the second-ranking river, the Congo River. Indeed, the discharge of the Amazon is three times as great as the total combined discharge of all rivers in the United States. The Mississippi is North America's largest river by far, with a drainage basin that encompasses about 40 percent of the total area of the 48 conterminous states and a flow that amounts to about one-third of the total discharge from all other U.S. rivers.

We explore the ways in which streams shape the landscape of the continents in Chapter 16.

Groundwater

Beneath the land surface is another important component of the hydrosphere: underground water. As Figure 9-1 shows, the total amount of underground water is many times that contained in lakes and streams. Moreover, underground water is much more widely distributed than surface water. Whereas lakes and rivers are found only in restricted locations, underground water is almost ubiquitous, occurring beneath the land surface throughout the world. Its quantity is sometimes limited, its quality is sometimes poor, and it may exist at great depth, but almost anywhere on Earth, if we dig deep enough, we can find water. Strictly speaking, **groundwater** is underground water in the subsurface zone where the pore spaces are completely filled with water (the *zone of saturation*), but the term *groundwater* often refers to all underground water, as we use it in this chapter.

More than half of the world's groundwater is found within 800 meters (about half a mile) of the surface. Below

TABLE 9-2 The World's Largest Rivers by Discharge and Length

Name	Rank by Discharge Volume	Rank by Length	Approximate Discharge (cubic meters per second)	Approximate Length (kilometers)	Approximate Drainage Area (square kilometers)	Continent
Amazon	1	2	210,000	6400	5,800,000	South America
Congo	2	9	40,000	4700	4,000,000	Africa
Ganges-Brahmaputra	3	23	39,000	2900	1,730,000	Eurasia
Yangtze	4	3	21,000	6300	1,900,000	Eurasia
Paranà-La Plata	5	8	19,000	4900	2,200,000	South America
Yenisey	6	5	17,000	5550	2,600,000	Eurasia
Mississippi-Missouri	7	4	17,000	6000	3,200,000	North America
Orinoco	8	27	17,000	2700	880,000	South America
Nile	25	1	5000	6650	2,870,000	Africa

Note: Estimates of river lengths and discharges are approximate; the relative rankings of some rivers vary from one data source to another.

▲ **Figure 9-24** The world's largest drainage basins are scattered over the four largest continents in all latitudes. (Dark green lines mark divides between adjacent drainage basins.)

that depth, the amount of water generally decreases gradually and erratically. Although water has been found at depths below 10 kilometers (6 miles), it is almost immobilized because the pressure exerted by overlying rocks is so great and openings are so few and small.

Movement and Storage of Groundwater

Virtually all groundwater comes originally from above. Its source is precipitation that either percolates directly into the soil or eventually seeps downward from lakes and streams.

Porosity: Once water gets underground, what happens next depends largely on the nature of the soil and rocks it infiltrates. The quantity of water that can be held in subsurface material (rock or soil) depends on the **porosity** of the material, which is the percentage of the total volume of the material that consists of voids (pore spaces or cracks) that can fill with water. The more porous a material is, the more open space it contains, and the more water it can hold.

Permeability: Porosity is not the only factor affecting groundwater flow. If water is to move through rock or soil, the pores must be connected to one another and be large enough for the water to move through them. The ability to transport groundwater (as opposed to just hold it) is termed **permeability**, and this property of subsurface matter is determined by the size of pores and by their degree of interconnectedness. The water twists and turns through these small, interconnected openings. The smaller and less connected the pore spaces, the less permeable the material and the slower the water moves.

The rate at which water moves through rock depends on both porosity and permeability. For example, clay has a great many *interstices* (openings) among the minute flakes that make up the sediment, but the interstices are so tiny that molecular attraction binds the water to the clay flakes and holds it in place. Thus, clay is typically very porous but relatively impermeable. Consequently it can trap large amounts of water and prevent drainage.

Aquifers: Underground water is stored in, and moves slowly through, moderately to highly permeable rocks called **aquifers** (from the Latin, *aqua*, meaning "water," and *ferre*, meaning "to bear"). The rate of movement of the water varies with the situation. In some aquifers, the flow rate is only a few centimeters a day; in others, it may be several hundred meters per day. A "rapid" rate of flow would be 12–15 meters (40–50 feet) per day.

Impermeable materials composed of components such as clay or very dense unfractured rock, which hinder or prevent water movement, are called **aquicludes** (Figure 9-25).

We can understand the general distribution of groundwater by visualizing a vertical subsurface cross section containing three main hydrologic zones. From top to bottom, these layers are called the *zone of aeration*, the *zone of saturation*, and the *waterless zone*.

Zone of Aeration

The topmost band, the **zone of aeration**, is a mixture of solids, water, and air. Its depth can vary from a few centimeters to hundreds of meters. The interstices in this zone are filled partly with water and partly with air. The amount of water fluctuates considerably with time. After a rain, the pore spaces may be saturated with water, but the water may drain away rapidly. Some of the water evaporates, but much is absorbed by plants, which later return it to the atmosphere by transpiration. Water that molecular attraction cannot hold seeps downward into the next zone.

LearningCheck 9-8 What is an aquifer?

Zone of Saturation

Immediately below the zone of aeration is the **zone of saturation**, in which all pore spaces in the soil and cracks in the rocks are fully saturated with water. Groundwater seeps slowly through the ground following the pull of gravity and guided by rock structure.

The top of the saturated zone is referred to as the **water table**. The orientation and slope of the water table usually conform roughly to the slope of the land surface above—generally closer to the surface in valley bottoms and farther from it beneath a ridge or hill. Where the water table intersects Earth's surface, water flows out, forming a spring. A lake, swamp, marsh, or permanent stream is almost always an indication that the water table reaches the surface there. In humid regions, the water table is higher (that is, the zone of saturation is nearer the surface) than in arid regions. Some desert areas have no saturated zone.

A localized zone of saturation may develop above an aquiclude. That configuration forms a *perched water table*.

Cones of Depression: A well dug into the zone of saturation fills with water up to the level of the water table. When water is taken from the well faster than it can flow in from the saturated rock, the water table drops in the immediate vicinity of the well in the approximate shape of an inverted cone. This striking feature is called a **cone of depression**

Animation
Groundwater Cone of Depression
https://media.pearsoncmg.com/ph/esm/esm_mcknight_physgeo_13/media/

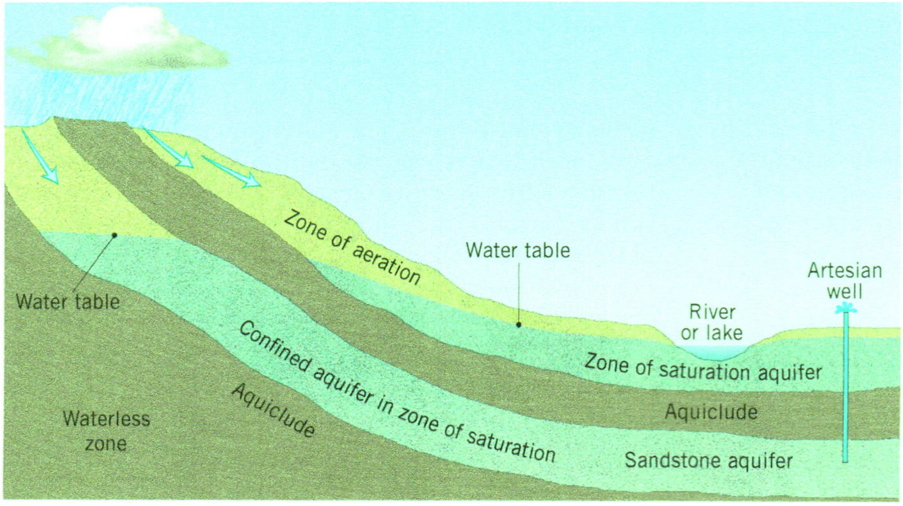

Mobile Field Trip
Mammoth Cave
https://media.pearsoncmg.com/ph/esm/esm_mcknight_physgeo_13/media/

Animation
The Water Table
https://media.pearsoncmg.com/ph/esm/esm_mcknight_physgeo_13/media/

◀ **Figure 9-25** An aquifer is a rock structure that is permeable or porous enough to hold water, whereas the structure of an aquiclude is too dense to allow water to penetrate.

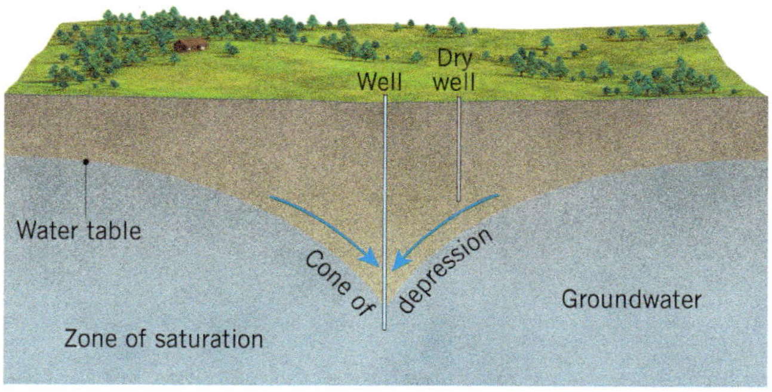

▲ **Figure 9-26** If water is withdrawn from a well faster than it can be replenished, a cone of depression will develop, effectively lowering the water table over a large area. Nearby shallow wells may run dry because they lie above the lowered water table.

(Figure 9-26). If many wells withdraw water faster than it is being replenished naturally, the water table may be significantly depressed over a large area, causing shallower wells to go dry.

Water percolates slowly through the saturated zone. Gravity supplies much of the energy for groundwater percolation, leading it from areas where the water table is high toward areas where it is lower—that is, toward surface streams or lakes. Percolation flow is not always downward, however. Often the flow follows a curving path and then turns upward (against the force of gravity), entering the stream or lake from below. This trajectory is possible because saturated-zone water at any given height is under greater pressure beneath a hill than beneath a stream valley. The water moves toward points where the pressure is least.

LearningCheck 9-9 How does a cone of depression form, and how does it affect an aquifer?

Confined Aquifers: In many parts of the world, some aquifers in the zone of saturation are confined by aquicludes above and below and so are known as **confined aquifers** (see Figure 9-25). Sometimes such aquifers alternate with aquicludes. Water cannot penetrate a confined aquifer directly by infiltration from above because of the impermeable barrier, so any water the zone contains must have percolated along the aquifer from a distant area where no aquiclude interfered. Characteristically, then, a confined aquifer is a sloping or dipping layer that reaches to, or almost to, the surface at some location, where it can absorb infiltrating water. The water works its way down the sloping aquifer from the catchment area, building up considerable pressure in its confined situation.

Artesian Wells: If a well is drilled from the surface down into the confined aquifer, the pressure forces water to rise in the well. The elevation to which the water rises is known as the **piezometric surface**. In some cases, the pressure is enough to allow the water to rise above the ground. This free flow of water is called an **artesian well** (Figure 9-27). If the confining pressure is sufficient to push the water only partway to the surface and it must be pumped the rest of the way, the well is *subartesian*.

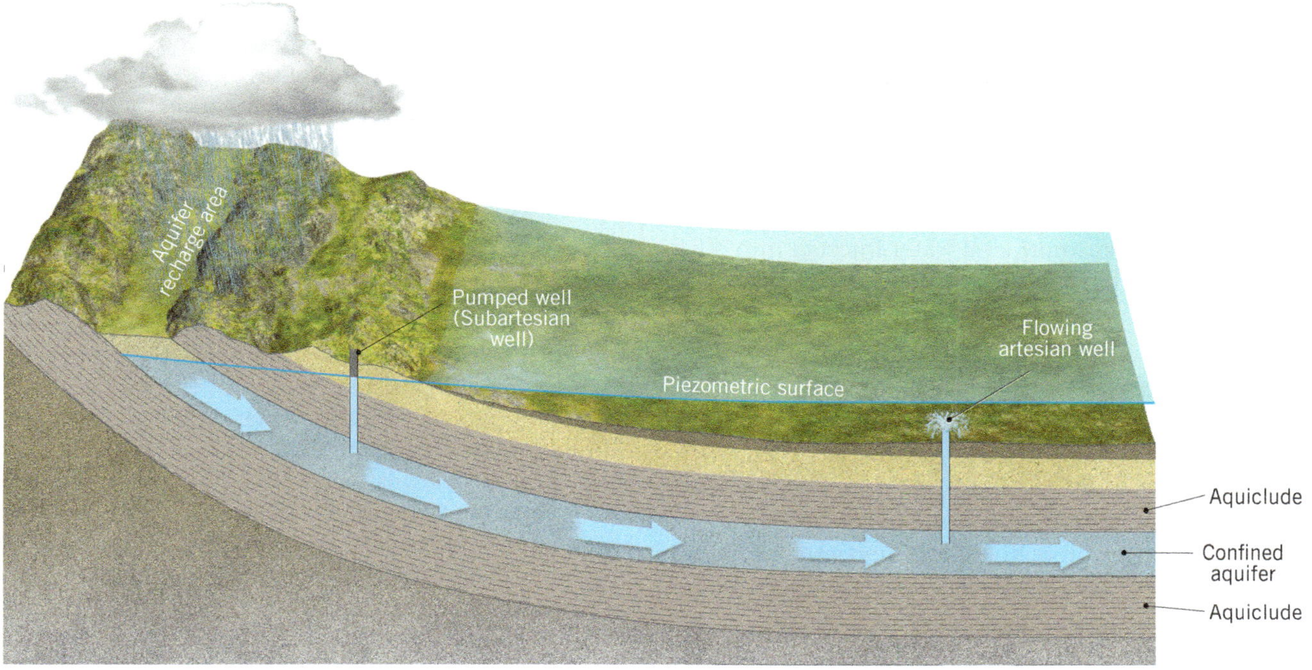

▲ **Figure 9-27** An artesian system. Surface water penetrates the aquifer in the recharge (replenishment) area and infiltrates downward. It is confined to the aquifer by (impermeable) aquicludes above and below. If a well is dug through the upper aquiclude into the aquifer, the confining pressure forces the water to rise in the well to the level of the piezometric surface. In an artesian well, the pressure forces the water to the surface; in a subartesian well, the water is forced only partway to the surface and must be pumped the rest of the way.

▲ **Figure 9-28** An artesian well in Australia's Great Artesian Basin, New South Wales. It is the largest and most productive source of confined water in the world.

Unlike the distribution of groundwater, which is closely related to precipitation, the global distribution of confined aquifers is quite erratic. Confined water underlies many arid or semiarid regions that are poor in surface water or groundwater, thus providing a critical resource for these dry lands (**Figure 9-28**).

Groundwater Pollution: When polluted surface water seeps into an aquifer, that contamination may remain in the groundwater indefinitely. Sources of such contamination vary widely: leaking industrial waste storage sites, seepage from underground service station gasoline tanks, agricultural pesticide runoff, and water injected into oil and gas wells during hydraulic fracturing, or "fracking" (see Chapter 13).

Waterless Zone

The lower limit of the zone of saturation lacks pore spaces and therefore lacks water. This boundary may be a single layer of impermeable rock, or the overlying pressure on the rock may be so great that there are effectively no pore spaces—the rock here cannot hold or transmit groundwater. This *waterless zone* generally begins several kilometers beneath the land surface.

Groundwater Mining

In most parts of the world where groundwater occurs, it has been accumulating for a long time. Rainfall and snowmelt seep and percolate downward into aquifers, where the water may be stored for decades or centuries or millennia. Only recently have most of these aquifers been discovered and tapped. They represent valuable sources of water that can supplement surface water resources. Farmers in particular use groundwater to irrigate in areas that contain insufficient surface water.

The accumulation of groundwater is tediously slow. Its use by humans, however, can be distressingly rapid. In many parts of the U.S. Southwest, for example, the recharge (replenishment) rate averages only 0.5 centimeter (0.2 inch) per year, but it is not uncommon for a well to pump 75 centimeters (30 inches) per year. Thus, yearly pumpage is equivalent to 150 years' recharge. We can liken this rate of groundwater use to mining because a finite resource is being removed with no hope of replenishment. For this reason, the water in some aquifers is referred to as *fossil water*. Almost everywhere in the world that underground water is being utilized on a large scale, the water table is dropping steadily and often precipitously.

In addition to depleting the water supply, groundwater mining causes a variety of problems. In some places, the compaction of sediments that takes place when groundwater is extracted faster than it is recharged leads to subsidence of the surface. For example, in the southern part of California's Central Valley, groundwater pumping resulted in 8.5 meters (about 29 feet) of subsidence in the mid-twentieth century. During the 1990s, groundwater pumping caused the Las Vegas Valley of Nevada to sink as much as 20 centimeters (8 inches). Because many parts of the world are becoming more dependent on groundwater, monitoring the state of this vital resource is becoming ever more critical—see the box *Global Environmental Change: Monitoring Groundwater Resources from Space.*

Global Environmental Change

Monitoring Groundwater Resources from Space

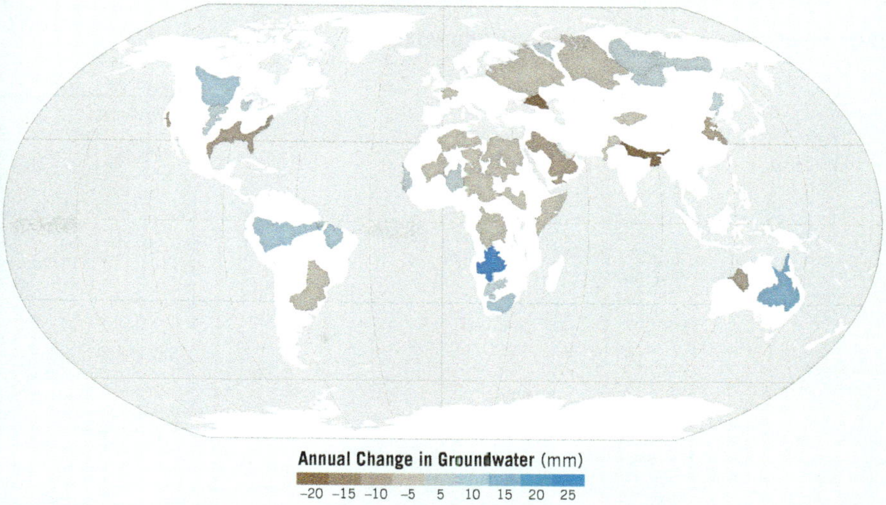

▲ **Figure 9-G** Annual change in water table in the 37 largest aquifers, January 2003–December 2013. Many of the most overstressed aquifers (shown in brown) are found in the driest parts of the world.

The United Nations Educational, Scientific and Cultural Organization (UNESCO) estimates that about 2.5 billion people around the world obtain all of their drinking water from groundwater, including large populations in relatively poor parts of the world. UNESCO also estimates that about 40 percent of all irrigation water comes from groundwater. The demand for groundwater is so high in many regions that extraction rates far exceed the natural recharge rates, leading to depletion of this critical resource. Monitoring changes in groundwater is difficult in many parts of the world. Innovative technology, however, is allowing scientists to monitor changes in the status of groundwater from space.

GRACE Satellites: In 2002, the German Aerospace Center in partnership with NASA launched a pair of polar orbiting satellites that circle the planet about 220 kilometers (137 miles) apart. Called the *Gravity Recovery and Climate Experiment* (GRACE), from 2002 to 2017 the satellites measured tiny differences in the distance between them that are caused by slight variations in Earth's gravity field. These local differences in gravity are caused by differences in the mass of Earth below. In 2018, the *GRACE Follow On* (GRACE-FO) mission continued this work.

The GRACE system is so sensitive that it can detect differences in the mass of water and ice on and within Earth, allowing scientists to track changes in the exchange of water between ice sheets and the ocean, and between groundwater and the surface.

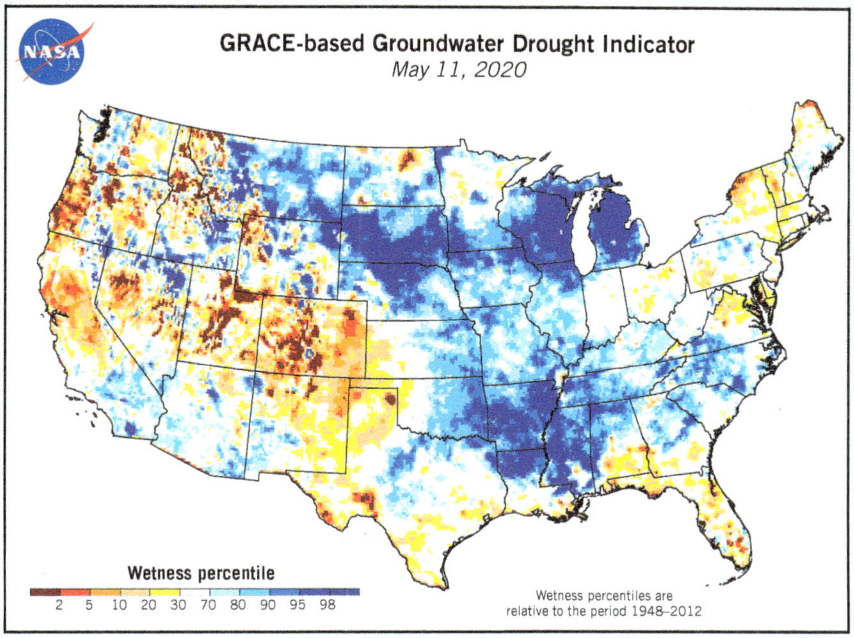

▲ **Figure 9-F** Estimated wetness percentiles of groundwater storage on May 11, 2020, compared with the average moisture content of the ground between 1948 and 2012. The dark maroon areas show dryness in the northwest and wetness in the northern Great Plains.

Groundwater depletion and the lowering of the water table of aquifers can result from overpumping as well as from drought.

GRACE data are used to measure short-term differences in soil moisture and groundwater levels caused by weather variations such as droughts. For example, **Figure 9-F** shows the wetness percentile of groundwater storage in May 2020 relative to the average of wetness from 1948 to 2012. The decline in groundwater associated with dry conditions in the northwestern United States is clearly shown.

Global Decline in Groundwater Storage: In 2015, a multidisciplinary research project using data from GRACE, as well as from ground-based observations and models, found that 13 of the 37 largest aquifers on Earth are being depleted, receiving little or no natural recharge (**Figure 9-G**). The most overstressed aquifer is reported to be the Arabian Aquifer System, followed by the Indus Basin Aquifer of northwestern India and Pakistan.

Although the problem of groundwater depletion is now well documented, the actual amount of water stored in many of the world's aquifers is only roughly known to researchers, making "time to depletion" estimates uncertain. Climate change will likely make such estimates even more difficult.

Questions

1. How does the GRACE satellite system measure variations in groundwater quantities over time?
2. Which parts of the world face the greatest challenges from groundwater decline? Why?

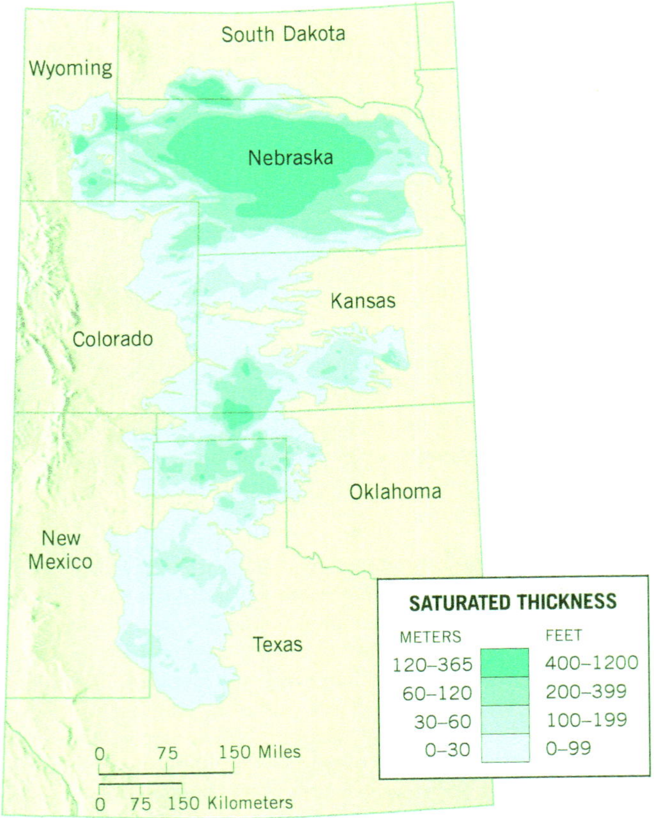

▲ Figure 9-29 The Ogallala or High Plains Aquifer. Darker areas indicate greater thickness of the water-bearing strata.

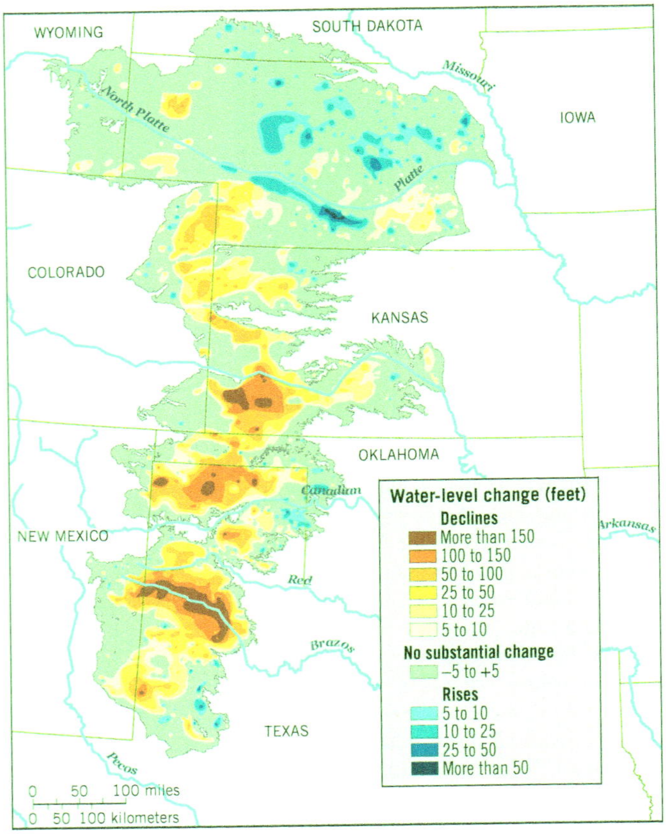

▲ Figure 9-30 Water table change in the Ogallala Aquifer from the early twentieth century through 2018. Increases in the water table depth occurred in places where surface water irrigation has been extensive.

The Ogallala Aquifer: A classic example of groundwater mining is seen in the southern and central parts of the Great Plains, where the largest U.S. aquifer, the Ogallala or High Plains Aquifer, underlies 450,000 square kilometers (174,000 square miles) of eight states. The Ogallala formation consists of a series of limey and sandy layers that function as a gigantic underground reservoir ranging in thickness from a few centimeters in parts of Texas to more than 300 meters (1000 feet) under the Nebraska Sandhills (Figure 9-29). Water has been accumulating in this aquifer for some 30,000 years. At the midpoint of the twentieth century, it was estimated to contain 1.4 billion acre-feet (1.7 quadrillion liters or 456 trillion gallons) of water, an amount roughly equivalent to the volume of one of the larger Great Lakes.

Farmers began to tap the Ogallala in the early 1930s. Before the end of that decade, the water table was already dropping. After World War II, the development of high-capacity pumps, sophisticated sprinklers, and other technological innovations encouraged the rapid expansion of crop irrigation based on Ogallala water. Water use in the region has almost quintupled since 1950. The results of this accelerated usage have been spectacular. Above ground, high-yield farming has spread rapidly into areas never before cultivated (especially in Nebraska), and irrigation has soared in all eight Ogallala states. Beneath the surface, however, the water table is sinking ever deeper—dropping more than 45 meters (150 feet) over large areas (Figure 9-30). Farmers who once obtained water from 15-meter (50-foot) wells now must bore to 45 or 75 meters (150 or 250 feet). Some 170,000 wells tapped the Ogallala in the 1970s, but thousands of those have now been abandoned as a result of the high cost of pumping.

Some farmers have shifted to crops that require less water. Others are adopting water- and energy-conserving measures that range from the decision to irrigate less frequently to the installation of sophisticated machinery that uses water in the most efficient fashion (Figure 9-31). Many farmers have faced or will soon face the prospect of abandoning irrigation. During the next four decades, it is estimated that 2 million hectares (5 million acres) now irrigated will revert to dry-land production. Other farmers concentrate on high-value crops before it is too late, hoping to make a large profit and then get out of farming.

Water conservation is further complicated by the fact that groundwater is no respecter of property boundaries. A farmer who is very conservative in his or her water use must face the reality that careless neighbors who are pumping

▲ **Figure 9-31** Fields using center pivot irrigation in northern Texas, near Dimitt.

from the same aquifer may seriously diminish the water available to everyone.

The situation varies from place to place. The Nebraska Sandhills have the most favorable conditions. The aquifer is deepest there, previous water use was minimal, and the recharge rate is relatively rapid. Indeed, in parts of Nebraska, the water table has been *rising* in places where natural recharge is high and withdrawal rates have changed. In contrast, the 13 counties of southwestern Kansas have a withdrawal rate more than 20 times the recharge rate—a clearly unsustainable situation.

LearningCheck 9-10 **How has groundwater mining affected the Ogallala Aquifer since the 1930s?**

Learning Review 9

After studying this chapter, you should be able to answer the following questions. Key terms from each text section are shown in **bold type.** Definitions for key terms are also found in the glossary at the back of the book.

Key Terms and Concepts

The Hydrologic Cycle (*p. 258*)

1. Where is most of the world's freshwater located?
2. Explain the role of evaporation in the **hydrologic cycle**.
3. What is the relationship between transpiration and evaporation?
4. Describe the roles of advection and **runoff** in the hydrologic cycle.

The Oceans (*p. 260*)

5. Is the Pacific Ocean significantly different from other oceans? Explain.
6. Why does **salinity** vary in different parts of the world ocean?
7. Why are the oceans becoming slightly more acidic?

Movement of Ocean Waters (*p. 263*)

8. Why do most oceanic areas experience two high **tides** and two low tides each day?
9. What is meant by **tidal range**?
10. Distinguish between **flood tide** and **ebb tide**.
11. Describe and explain **spring tides** and **neap tides**.
12. What is a **tidal bore**?
13. What is **thermohaline circulation**?
14. Explain the **global conveyor-belt circulation**.

Permanent Ice—The Cryosphere (*p. 267*)

15. Where is most of the ice in the cryosphere located?
16. Distinguish among an **ice pack**, **ice shelf**, **ice floe**, and **iceberg**.
17. Why does all sea ice consist of freshwater?
18. Describe the characteristics and global distribution of **permafrost**.

Surface Waters (*p. 270*)

19. Distinguish among a **lake**, **wetlands**, a **swamp**, and a **marsh**.

Groundwater (*p. 274*)

20. What is the difference between **porosity** and **permeability**?
21. Contrast an **aquifer** with an **aquiclude**.
22. Briefly define the following terms: **groundwater**, **zone of aeration**, **zone of saturation**.
23. Explain the concept of a **water table**.
24. Describe and explain the cause of a **cone of depression**.
25. Under what circumstances can a **confined aquifer** develop?
26. What is the **piezometric surface**?
27. Distinguish between an **artesian well** and a *subartesian well*.

Study Questions

1. In what part of the hydrologic cycle is water most likely to stay for a very short time? A very long time? Why?
2. "How many oceans are there?" Why is this a difficult question to answer?
3. How can an increase in the acidity of ocean water affect corals and other ocean creatures with calcite skeletons?
4. What are some of the consequences of melting permafrost?
5. Explain why and how the Aral Sea has changed in recent decades.
6. Why are most lakes considered to be "temporary" features of the landscape?
7. Why is the water from some aquifers referred to as fossil water?
8. How has groundwater mining affected the Ogallala Aquifer since the 1930s?

Exercises

1. If a floating iceberg has a surface area of 150 square meters and is 10 meters high, what is the total volume of ice, including both the exposed portion above water and the portion below the water? _____ cubic meters
2. If the natural recharge rate of an aquifer is 1 centimeter per year but the rate of groundwater pumping is 10 centimeters per year, how far will the water table drop in 10 years? _____ centimeters
3. Assuming a natural recharge rate of 0.5 centimeters per year, if groundwater pumping lowers the water table by 40 centimeters, how many years of fossil water have been extracted? _____ years

Mastering Geography®

Looking for additional review and test prep materials? Visit the Study Area in **Mastering Geography** to enhance your geographic literacy, spatial reasoning skills, and understanding of this chapter's content by accessing a variety of resources, including MapMaster interactive maps, geoscience animations, *Mobile Field Trips*, videos, *Project Condor* Quadcopter videos, *In the News* RSS feeds, flashcards, web links, self-study quizzes, and an eText version of *McKnight's Physical Geography*.

Environmental Analysis OCEAN SALINITY

Data
Ocean Salinity
https://media.pearsoncmg.com/ph/esm/esm_mcknight_physgeo_13/media/

Salinity varies both spatially across the ocean and with depth in the ocean. Many factors determine salinity in a particular location, including mixing with freshwater, ocean currents, and temperature.

Activities

Go to www.nnvl.noaa.gov/view, the NOAA Data Exploration Tool. (You can watch the Video Tour.) Click on "Add Data"; choose "Ocean" followed by "Chemistry" and then "Salinity." Choose "0 Meters" and "Monthly" data.

1. In which general latitudinal regions are salinity values high? Low?
2. If you were to graph average salinity against latitude from the South Pole to the North Pole, what letter of the alphabet would the graph resemble?
3. Where do you find the absolute highest salinity values?
4. Consult the map inside the front cover or Figure 9-26. Is salinity higher or lower where rivers empty into the ocean? The Mississippi River and Amazon River are good examples.

"Back" up to examine the salinity at various depths. Look at "Yearly" data.

5. How does salinity vary with depth?
6. Why are there so little data at 5000 meters?

Go back to monthly salinity at 0 meters. In another window, choose "Ocean" followed by "Temperature" and then "At the Surface." Choose "Monthly" data.

7. Is the current that travels up the east coast of the United States a warm or cold current?
8. Is the current that travels down the west coast of the United States a warm or cold current?
9. Compare the temperature and salinity maps. How are temperature and salinity related away from the equator?

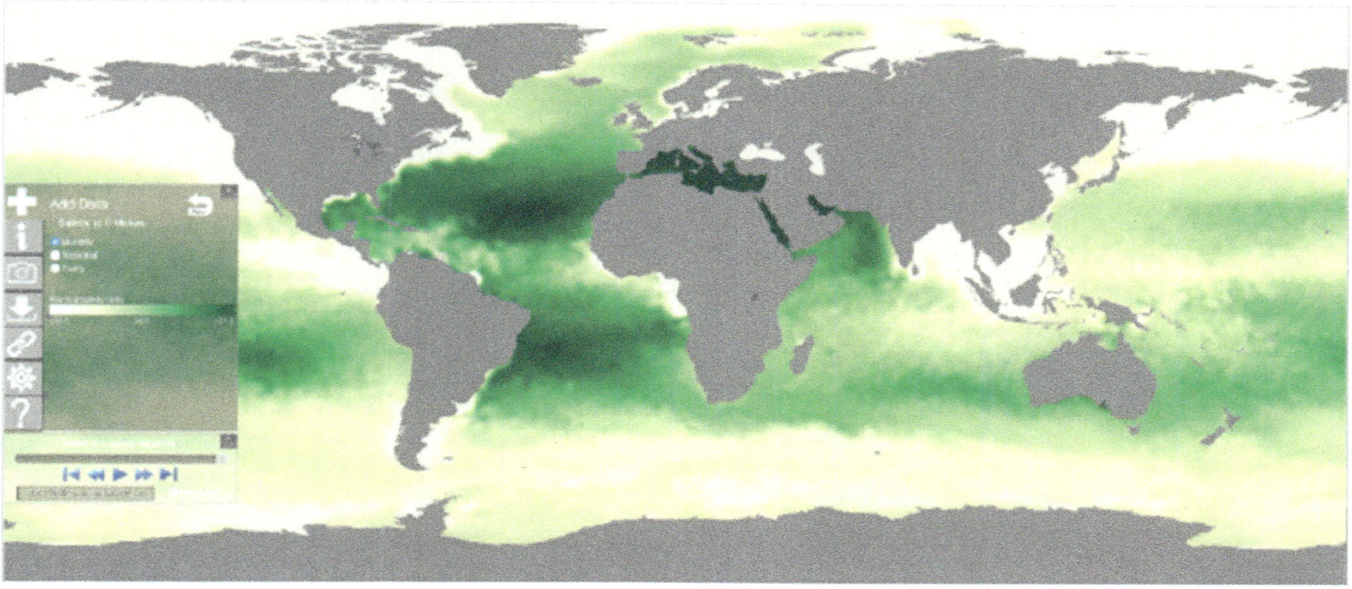

Seeing Geographically in Our National Parks

Look again at the photograph of Sandstone Falls along New River, West Virginia, at the beginning of the chapter (p. 256). What aspects of the hydrologic cycle can you see in this photograph? In what ways is water in this area being transferred back to the atmosphere as water vapor? Based on the climate of this location (38° N, 81° W), is there an annual dry season to greatly reduce seasonal runoff into this river?

Seeing Geographically in Our National Parks

Temperate forest in Olympic National Park, Washington. Describe the kinds of plants you see in this photograph and their overall abundance. What do you see that suggests the relative amount of sunlight and moisture that are available on this forest floor?

Olympic National Park, WA

Cycles and Patterns in the Biosphere

10

HAVE YOU EVER WONDERED why plants and animals that live here in Olympic National Park are so different from those in Grand Canyon National Park? The answer seems simple: each species is adapted to a different environment. Plant and animal species typically have characteristics that limit them to habitats with a specific range of temperature, moisture, topographic, seasonal, and other factors. But it's actually not quite that simple. These relationships work both ways: the physical environment of an area influences what lives there, but the plants and animals in that area in turn influence the physical environment itself. It is to these interrelationships of the biosphere that we turn in this chapter.

Of the four environmental spheres, the biosphere's boundaries are hardest to pin down. The atmosphere consists of the envelope of air that surrounds the planet, the lithosphere is the solid portion, and the hydrosphere encompasses the various forms of water. These three spheres are conceptually distinct from one another and easy to visualize. The biosphere, however, consists of the numerous and diverse array of organisms that populate our planet—most obviously plants and animals, but also less obvious types of organisms such as bacteria and fungi.[1] Most of these organisms exist at the interface between atmosphere and lithosphere, but some live largely or entirely within the hydrosphere or the lithosphere, and others move relatively freely from one sphere to another.

As we begin our study of the biosphere, look for connections to Earth systems and processes that we discussed in earlier chapters. For example, consider the close relationships of Earth's solar radiation budget (Chapter 4) and hydrologic cycle (Chapters 6 and 9) to the *biogeochemical cycles* that we present in this chapter. Note especially the relationship between the distribution patterns of organisms and the distribution of climate that we presented in Chapter 8.

As you study this chapter, think about these **Key**Questions:

- **What is biogeography?**
- **How do water, carbon, oxygen, and nitrogen move through the biosphere and other environmental spheres?**
- **What is a food chain or a food pyramid?**
- **What factors influence the distribution patterns of plants and animals?**

[1]Over the years, biologists have employed different systems for classifying the diverse forms of life on Earth. For more on the subject of biological classification, see Appendix VI.

The Impact of Plants and Animals on the Landscape

Thousands of years ago, when Earth was only thinly populated by humans, vegetation grew in abundance wherever the land surface was not too dry or too cold. Today, native plants are still widespread in the sparsely populated parts of the world (**Figure 10-1a**). However, much of the natural vegetation in populated areas has been removed or modified by the introduction of crops, weeds, and ornamental plants (**Figure 10-1b**). In a similar way, the distribution of animals has been modified by human activity. The significance of plants and animals extends well beyond their presence in the landscape: they can be important influences in the development and evolution of other landscape components, such as soil, landforms, and water.

The Geographic Approach to the Study of Organisms

Even the simplest living organism is an extraordinarily complex entity. An organism differs in many ways from other aspects of the environment, but most obviously in that it is alive and its survival depends on an intricate set of life processes. As beneficial as a complete understanding of the world's organisms might be for us, as with every other feature of the world, the geographer focuses on certain aspects of the landscape only.

Biogeography

As we saw in earlier chapters, the geographic viewpoint is one of broad understanding. This does not mean that when we study the biosphere we ignore individual organisms; rather, it means that we seek generalizations and patterns. Here, as elsewhere, the geographer is interested in distributions and relationships. The subfield of geography that looks at the biosphere is known as **biogeography**, the study of the distribution patterns of living organisms and how these patterns change over time. Although biogeography focuses on the *biotic* (living) systems of Earth, it also entails an understanding of the ways that *abiotic* (nonliving) systems fit in with the biosphere.

Biodiversity: As geographers look for and explain patterns in the biosphere, *biodiversity* is one of the most significant components of these patterns. **Biodiversity** refers to the number of different kinds of organisms present in a location. An area of *high biodiversity* has many types of organisms, whereas an area of *low biodiversity* has just a few kinds. Declining biodiversity in a location is often an indication that the overall health of the natural environment is deteriorating.

(a) Intact native vegetation

(b) Displacement by crops

▲ **Figure 10-1** (a) Colorful autumn leaves fill this forest scene in Shenandoah National Park, Virginia. Many parts of Earth are still covered with native vegetation like as this, with little or no human impact in evidence. (b) A small natural depression of wetland remains in this field near Tosterup, Sweden, but most of the natural vegetation has been displaced by crops.

Flora and Fauna: Approximately 1.5 million known species of plants and animals live on Earth, and most scientists think that the actual number is considerably higher than this. The term **biota** refers to this total complex of plant and animal life. The basic subdivision of biota separates **flora**, or plants, from **fauna**, or animals. In this book, we recognize a further fundamental distinction—between *oceanic biota* and *terrestrial biota* (living on land; *terrenus* is Latin for "earth").

The inhabitants of the oceans are generally divided into three groups: *plankton* (floating plants and animals), *nekton* (animals such as fish and marine mammals that swim freely), and *benthos* (animals and plants that live on or in the ocean bottom). Although marine life-forms are fascinating and 70 percent of Earth's surface is oceanic, in this book we give limited attention to oceanic biota primarily because of constraints of time and space. The terrestrial biota is our primary focus for most of our study of the biosphere.

LearningCheck 10-1 What is biodiversity? (Answer on p. AK-3)

With such an overwhelming diversity of organisms, how can we study their distributions and relationships in any meaningful manner?

The Search for a Meaningful Classification Scheme

When a geographer studies any phenomena, she or he attempts to classify specific observations in such a way that broad patterns may be recognized. In some cases, the geographer borrows classification schemes from other disciplines, but sometimes these schemes are not ideally suited for geographic studies.

Classifying Organisms: A widely used classification scheme, adopted from the discipline of biology, is the *binomial* (two name or "scientific-name") *system* of nomenclature originally developed by Swedish botanist and zoologist Carolus Linnaeus in the eighteenth century (see Appendix VI). The first term in a scientific name identifies the genus, or group of closely related organisms. The second term identifies the species, or group of organisms that is genetically distinct from other groups—even from closely related ones. As an example: *Ursus americanus* is the scientific name for the American black bear, whereas *Ursus arctos* is a related but different species, the grizzly bear (or brown bear). Biologists historically have also grouped organisms into a hierarchy of increasingly inclusive categories that include genera and species (again, see Appendix VI).

In the past, biologists classified organisms largely by morphology (structure and form) and grouped them on the basis of structural similarity. However, today biologists are more concerned with evolutionary relationships among organisms. Biological classification schemes, including the Linnaean binomial system, are widely used by geographers, but because we are also concerned with distribution patterns and environmental preferences, we also use other ways of classifying organisms.

Habitats and Niches: Biogeographers are especially interested in the relationships among the different organisms in a location and in the relationships between those organisms and the surrounding environment (**Figure 10-2**). Each species has a characteristic *habitat*—the kind of environment in which it is adapted to live. (For example, most woodpeckers live in forests or woodland habitats.) Further, each species can be thought of as having a *niche*—a specific role or function in that habitat. (For example, a woodpecker eats wood-boring insects—thereby helping control insect populations.) Any given habitat is likely to contain many species, each with its own niche. If two species have exactly the same niche in a habitat—for example, two kinds of animals that eat the same food or try to live in the same space—over time one of those species will outcompete the other.

Ecosystems and Biomes: The term *ecosystem* is used to describe all of the organisms in an area and their interactions with the immediate environment. At the global scale, biogeographers recognize broader groupings of plants and

◀ **Figure 10-2** Researcher studying the effects of drought on a Giant Sequoia tree in Sequoia National Park, California.

animals known as *biomes*—large, recognizable assemblages of plants and animals living in a functional relationship with the environment.

Ecosystems and biomes are explored in much greater detail in Chapter 11. Here, we set the stage for that discussion by describing fundamental cycles and relationships in the biosphere. We begin with a look at the broadest patterns and cycles in the biosphere—*biogeochemical cycles*.

Biogeochemical Cycles

The web of life comprises a great variety of organisms coexisting in a diversity of ecosystems. Organisms are sustained by flows of energy, water, and nutrients. These flows are different in different parts of the world, in different seasons of the year, and under various local circumstances.

It is generally believed that, for the last billion years or so, Earth's atmosphere and hydrosphere have been composed of approximately the same balance of chemical components we live with today. This constancy is the result of a planet-wide condition in which the various chemical elements have been maintained by cyclic passage through the tissues of plants and animals—first absorbed by an organism and then returned to the air, water, or soil through decomposition. These grand cycles are collectively called **biogeochemical cycles**.

If the biosphere is to function properly, its chemical substances must be recycled continually through these biogeochemical cycles. In other words, after one organism uses a substance, that substance must be converted, with the expenditure of some energy, to a reusable form. For some components, this conversion can be accomplished in less than a decade; for others, it may require hundreds of millions of years. In recent years, however, human activity and the accelerating rate at which we consume Earth's resources have disrupted or harmed many of these cycles.

Although we describe biogeochemical cycles separately in the following subsections, you will see that many of these cycles are closely interrelated. It is appropriate, then, for us to begin with the most fundamental of these cycles: the flow of energy through the biosphere.

The Flow of Energy

The Sun is the basic energy source on which nearly all life ultimately depends. (Forms of life utilizing geochemical energy from hydrothermal vents on the ocean floor are a well-known exception.) Solar energy drives life processes in the biosphere through *photosynthesis*, the production of organic matter by chlorophyll-containing plants and bacteria.

Only about 1 percent of the solar energy that reaches Earth is captured for use in photosynthesis. More than half of that total is used immediately in the plant's own *respiration*, and the remainder is temporarily stored. Eventually this remainder enters a *food chain*.

Photosynthesis and Respiration: The biosphere is a temporary recipient of a small fraction of the solar energy that reaches Earth, which is "fixed" (made stable and usable) by green plants through the process of **photosynthesis** (Figure 10-3). The key to photosynthesis is a light-sensitive pigment known as *chlorophyll*, which is found within organelles called *chloroplasts* in leaf cells. Chlorophyll absorbs certain wavelengths of visible light, while prominently reflecting green light. (That is why leaves look green.)

In the presence of sunlight and chlorophyll, a photochemical reaction takes carbon dioxide (CO_2) from the air, combines it with water (H_2O), and forms the energy-rich *carbohydrate* compounds we know as sugars while releasing molecular oxygen. In this process, the energy from sunlight is stored as chemical energy in the sugars. In simplified form, the chemical equation for photosynthesis is:

Video **Global Carbon Uptake by Plants** https://media.pearsoncmg.com/ph/esm/esm_mcknight_physgeo_13/media/

$$CO_2 + H_2O \xrightarrow{\text{light}} \text{Carbohydrates} + O_2$$

In turn, plants can use simple sugars to build more complex carbohydrates, such as starches. The stored chemical energy in the form of carbohydrates is utilized in the biosphere primarily in two ways. First, some of the stored energy cycles through the biosphere when animals eat either the photosynthesizing plants or other animals that have eaten plants. Second, the other portion of the stored energy in carbohydrates is consumed directly by the plant itself in a process known as **plant respiration**. In the process of respiration, the stored energy in carbohydrates is *oxidized*, releasing water, carbon dioxide, and heat. The simplified chemical equation for plant respiration is:

$$\text{Carbohydrates} + O_2 \rightarrow CO_2 + H_2O + \text{Energy (heat)}$$

LearningCheck 10-2 Why is photosynthesis so critical to the existence of the biosphere?

Net Primary Productivity: Plant growth depends on a surplus of carbohydrate production. The *net photosynthesis* of a plant is the difference between the amount of carbohydrate produced in photosynthesis and that lost in plant respiration. When we discuss an ecosystem as a whole, the term **net primary productivity** describes the total amount of chemical energy stored (or carbon "fixed" through photosynthesis) in a plant community. Net primary productivity is reflected in the dry weight of organic material, or **biomass**, of that community.

Annual net primary productivity is the net primary productivity of a plant community over a period of one year, usually measured in the amount of fixed carbon per unit area (grams of carbon per square meter per year). Monthly or seasonal variations in productivity can also be determined by measuring net photosynthesis.

Net primary productivity varies widely around the world. It tends to be highest on land within the tropics, where both high precipitation and high insolation are available for plant growth, but generally diminishes poleward, especially in extremely arid and extremely cold environments (Figure 10-4). In the oceans, productivity is

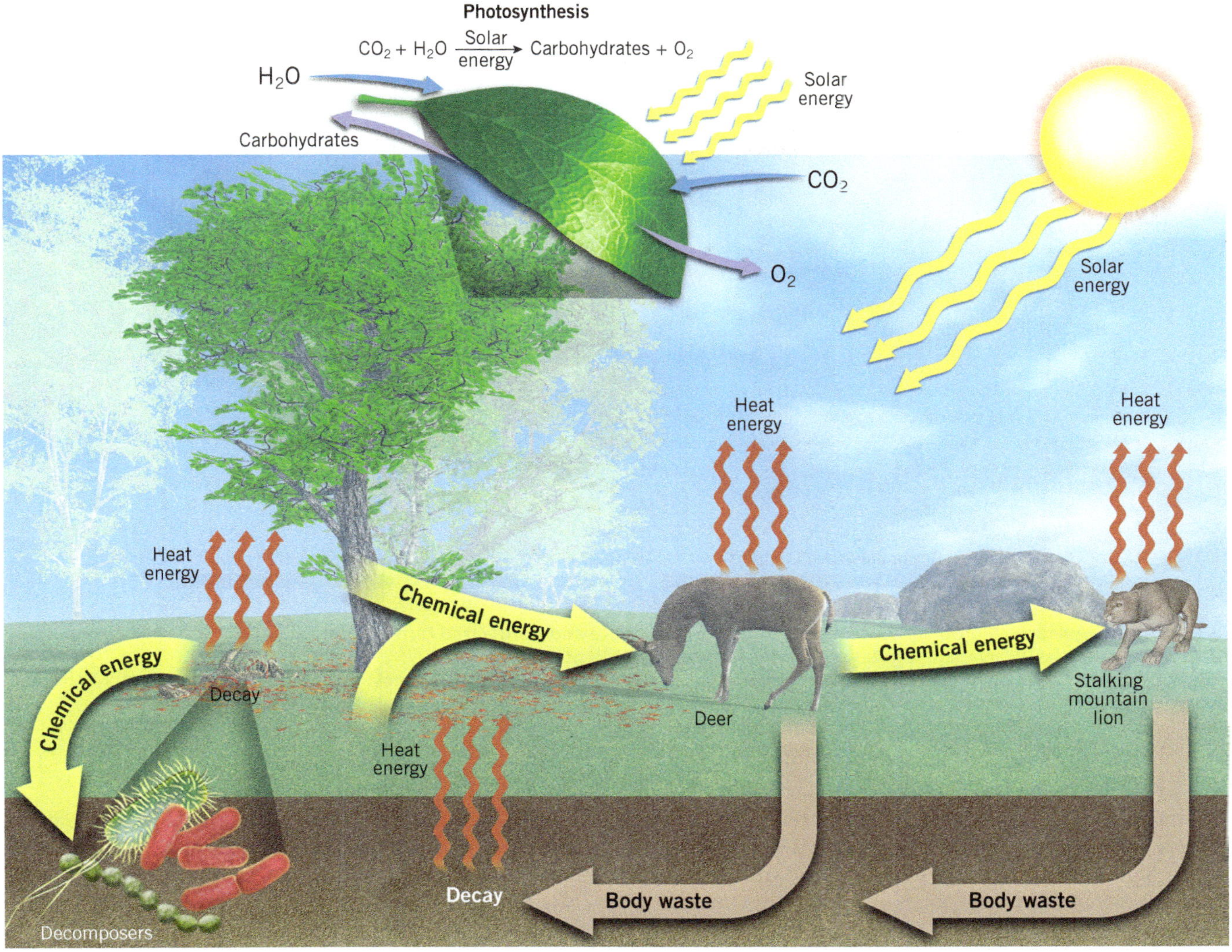

▲ **Figure 10-3** Energy flow in the terrestrial biosphere. Plants use solar energy in photosynthesis, storing that energy in the sugar molecules they manufacture. Grazing and browsing animals acquire that energy when they eat the plants. Other animals eat the grazers/browsers and thereby acquire some of the energy originally in the plants. Body wastes from the animals, the bodies of dead animals, and dead plant matter return energy to the soil. As the waste and dead matter decay, they give off energy in the form of heat. The arrows represent energy pathways but do not reflect the relative amounts of energy moving along any one pathway.

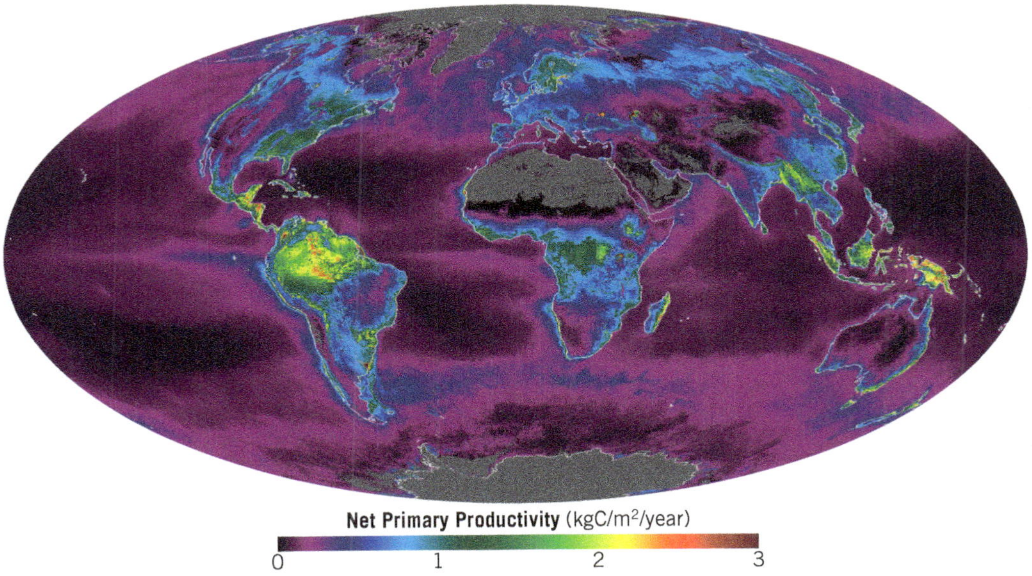

◀ **Figure 10-4** Annual net primary productivity. This composite satellite image shows net primary productivity, based on the rate at which plants absorb carbon dioxide from the air (the mass of carbon absorbed per square meter per year). The areas of highest annual average productivity (yellow and red) are most notably in the tropical rainforests of the world.

Animation
Net Primary Productivity
https://media.pearsoncmg.com/ph/esm/esm_mcknight_physgeo_13/media/

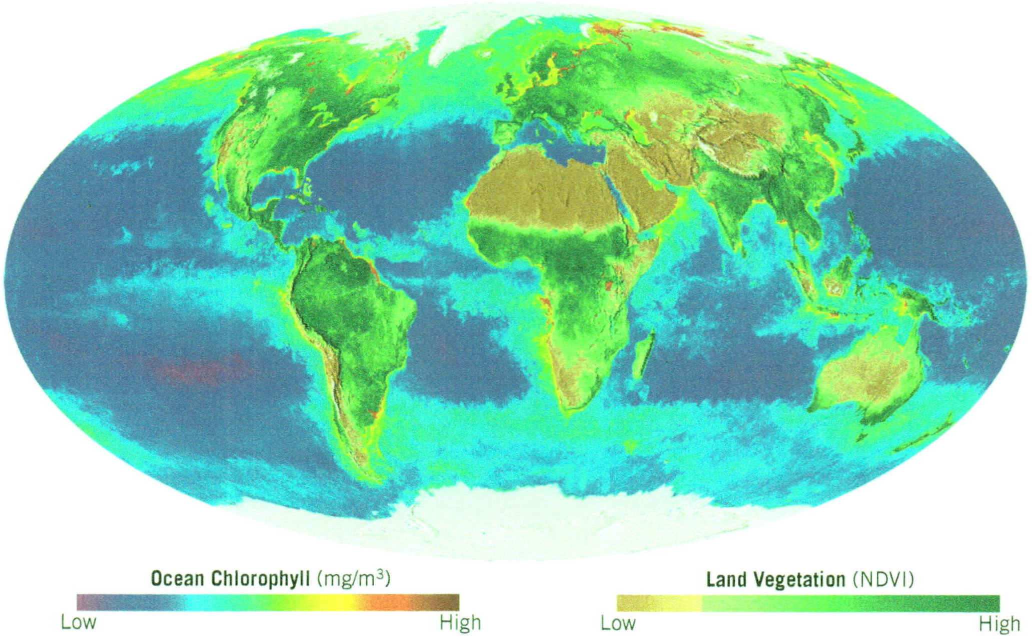

▲ **Figure 10-5** Average chlorophyll concentration at the ocean surface and Normalized Difference Vegetation Index (NDVI; average vegetation density on land) during September of 2017. Chlorophyll concentration in the ocean reflects net primary productivity. Chlorophyll is greatest where phytoplankton is most concentrated—in the cold, productive polar waters and along coastlines where upwelling brings nutrients to the surface. In the center of ocean basins, productivity is generally low. Data were collected by NOAA satellites and the SeaWiFS satellite.

Animation
Biological Productivity in Midlatitude Oceans
https://media.pearsoncmg.com/ph/esm/esm_mcknight_physgeo_13/media/

strongly influenced by the nutrient content of the water (**Figure 10-5**). For example, off the west coasts of continents in the midlatitudes, the upwelling of cold, nutrient-rich water results in high net primary productivity. (Upwelling is discussed in Chapter 4.)

The carbohydrates incorporated into plant tissue are in turn consumed by animals or decomposed by microorganisms. Plant-eating animals convert some of the consumed carbohydrates back to CO_2 and exhale it into the air (*animal respiration*); the remainder is decomposed by microorganisms after the animal dies. The carbohydrates acted upon by microorganisms are ultimately oxidized into CO_2 and returned to the atmosphere (*soil respiration*).

Because photosynthesis captures and stores solar energy, we can utilize this energy reservoir in a number of ways. Burning wood for cooking or heating, for example, takes advantage of this stored solar energy, as does our use of other kinds of *biofuels*—see the box *Energy for the 21st Century: Biofuels*.

The Hydrologic Cycle

The most abundant single substance in the biosphere, by far, is water. As we learned in Chapter 9, the movement of water from one sphere to another is called the *hydrologic cycle* (see Figure 9-2). Water occurs in the biosphere in two ways as part of the hydrologic cycle: (1) *in residence*, with its hydrogen chemically bound into plant and animal tissues, and (2) *in transit*, as part of the transpiration–respiration pathway in which water moves back and forth between the environment and living organisms.

All living things require water to carry out their life processes. Through chemical reactions that take place in a water solution, an organism converts nutrients to energy or to the materials it needs to grow and repair itself. In addition, the organism needs water to carry away waste products.

Most organisms contain considerably more water in their mass than anything else (**Table 10-1**). Every living thing depends on keeping its water supply within a narrow range. For example, humans can survive without food for two months or more, but they can live without water for only about a week.

The Carbon Cycle

Carbon is one of the basic elements of life and a part of all living things. The biosphere contains a complex mixture of carbon compounds, more than half a million in total. These compounds are in continuous states of creation, transformation, and decomposition.

TABLE 10-1 Water Content in Some Plants and Animals

Organism	Percentage Water in Body Mass
Human	65
Elephant	70
Earthworm	80
Ear of corn	70
Tomato	95

Energy for the 21st Century

Biofuels

• **Rob Bailis,** Stockholm Environment Institute

Heavy reliance on fossil fuels causes air pollution, contributes to climate change, and reduces energy security, motivating policies that support alternative energy sources. For transportation, such policies encourage the development of liquid *biofuels*—fuels made of or from biological materials, which are thought to reduce pollution and increase energy security. Between 2000 and 2018, global biofuel production grew 10-fold, driven by fuel-blending mandates in the United States, Brazil, Europe, and a score of other countries. Mandates also support domestic agriculture, which reinforces political support for these policies.

However, as production increased, negative impacts became apparent. For example, biofuels are not completely clean; they can be polluting. Increased demand for crops such as corn and soybeans can affect international markets, raise prices, and contribute to food insecurity. In addition, farmers may expand crop production into previously uncultivated regions, causing undesirable land-use change, such as deforestation. These and other impacts have raised serious concerns. Many countries designed policies to address these issues, but there is disagreement over how successful they've been.

Types of Biofuels: Ethanol and biodiesel are the two main categories of biofuels. Ethanol is distilled from starch- or sugar-based crops, such as corn or sugarcane. Biodiesel is a common diesel substitute made from plant oils, waste oil, or animal fat. Through a different process, the same raw materials are used to make "renewable gas" (RG) or "renewable diesel" (RD).

Greenhouse Gas Emissions and Other Pollution: Fossil fuels emit carbon dioxide (CO_2) that has been buried for millions of years. In contrast, biofuels emit CO_2 taken out of the atmosphere during recent plant growth, which is fixed by future growth. If the biofuel crops are harvested sustainably, there are no net CO_2 emissions. However, although the cycle of harvesting, combustion, and regrowth emits no net CO_2, other stages of biofuel production require fertilizers and other inputs that emit greenhouse gases (GHGs) and other pollution. Most biofuels emit fewer GHGs than the fossil fuels they replace, but they may result in higher emissions of other pollutants. For example, fertilizers applied to crops release nitrogen, which can damage lakes and rivers.

If biofuel production leads to land-use change, then biofuels may emit more GHGs than the fossil fuels they replace. Some policies are designed to minimize the risk of land-use change by discouraging the use of crops such as palm oil from Indonesia or soybeans from Brazil (Figure 10-A). However, global commodity markets are interconnected, so even biofuel crops that are not *directly* associated with land-use change may influence land use *indirectly*.

Biofuels and Food Security: A major concern is food security, which the United Nations defines as regular access to sufficient food that is safe and nutritious. In the mid-2000s, when biofuel consumption began to increase, the prices of many staple foods also rose. Researchers think biofuel policies contributed to this, which led many to question the use of crops for fuel. In response, some countries forbade the use of food crops for biofuels and incentivized the development of alternatives. However, the vast majority of biofuels are still derived from corn, sugarcane, soybeans, canola, or palm oil.

Nevertheless, the tradeoff between food and fuel is not simple. Many biofuel crops have multiple uses. In the United States, for instance, when corn is refined into ethanol, or soy oil into biodiesel, the refineries also make co-products that are used to feed livestock (Figure 10-B). In Brazil, while refineries process sugarcane into both sugar and ethanol, they simultaneously burn sugarcane waste to produce electricity, which they sell to local utilities.

▲ **Figure 10-B** An ethanol refinery in Wisconsin.

Future Directions: So far, biofuels have failed to deliver the full range of benefits that many people hoped for. However, alternative pathways are under development. For example, ethanol can be made from cellulose-based materials such as grasses, crop waste, or wood—requiring less energy, emitting fewer pollutants, and delivering higher yields than corn-based ethanol. (At present, however, cellulosic ethanol is more expensive.) Biodiesel derived from algae would be more efficient than soy, canola, or palm oil, with lower risk of food/fuel conflict and land-use change. Renewable gasoline and diesel are other pathways receiving attention. Many options exist. Today's most popular pathways carry benefits and risks. Alternatives may be less risky but require more research and development.

Questions

1. What are the main reasons that fuel-blending policies have been introduced in many nations?
2. What are some of the negative impacts that have been associated with increased biofuel production?

▼ **Figure 10-A** A soybean plantation in cleared tropical rainforest, Rondônia State, western Brazil.

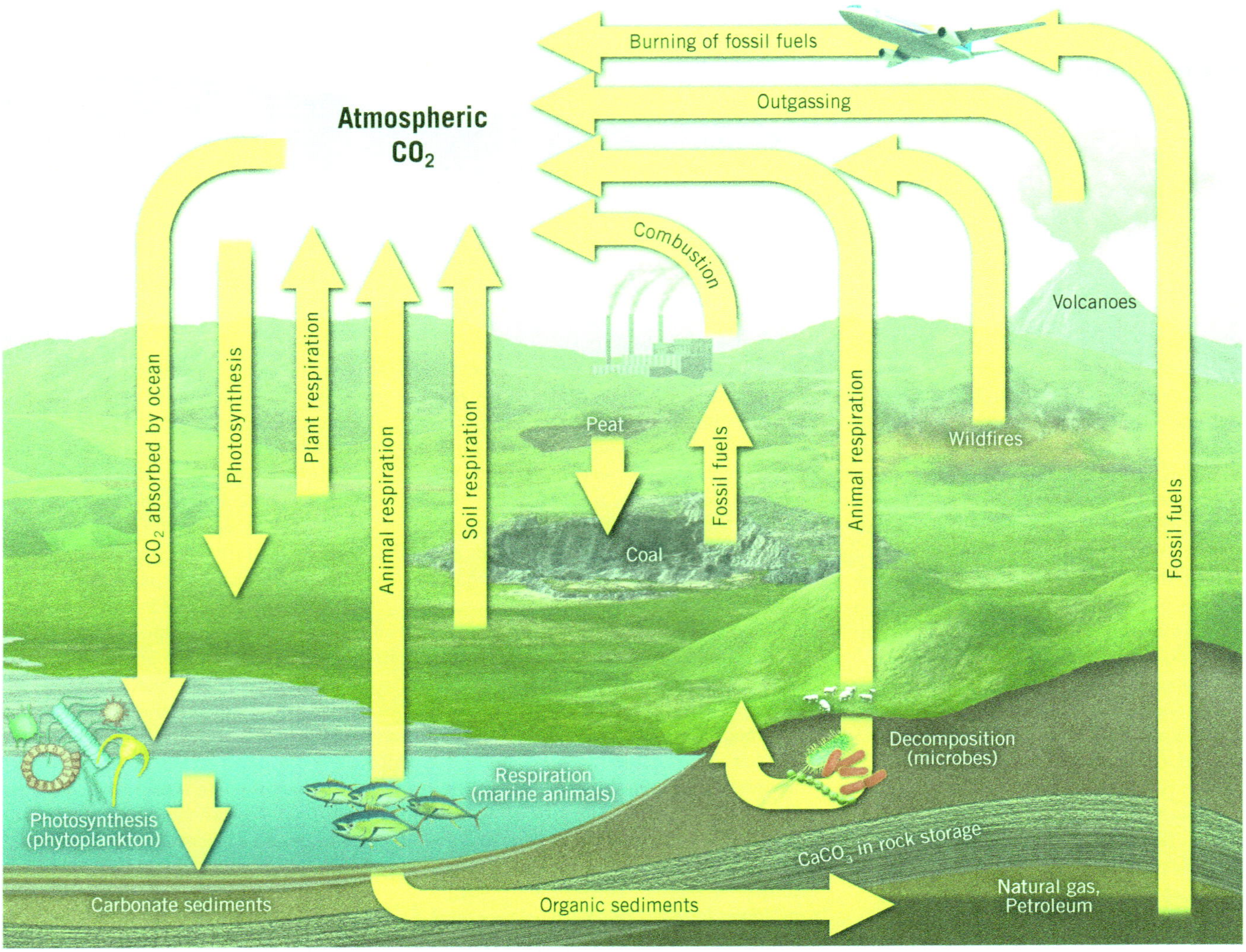

▲ **Figure 10-6** The carbon cycle. Carbon from the CO_2 in the atmosphere is used by plants to make carbon-containing sugars during photosynthesis. Some organic compounds are stored in rocks or petroleum. Over time, these compounds are eventually converted back to CO_2 and returned to the atmosphere. The arrows represent pathways but do not reflect the relative amounts of carbon moving along any one pathway.

In the **carbon cycle**, carbon is transferred from CO_2 to living matter and then back to CO_2 (**Figure 10-6**). This transfer is initiated when CO_2 from the atmosphere is converted into carbohydrate compounds during photosynthesis. In turn, plant respiration and soil respiration return carbon to the atmosphere in the form of CO_2. Think of the carbon cycle as a complex of interlocking cycles in which carbon moves constantly from the inorganic reservoir to the living system and back again. A similar cycle takes place in the ocean.

Different components of the carbon cycle operate at different rates. For example, some carbon cycles through quickly via photosynthesis, whereas other carbon cycles through much more slowly by way of geologic processes that can take thousands of years. The overwhelming bulk of near-surface carbon has been concentrated over millions of years in geologic deposits—such as coal, petroleum, and carbonate rocks, including limestone. These deposits are composed of dead organic matter that accumulated mostly on sea bottoms and was later buried. Carbon from this reservoir is normally incorporated into the cycle very gradually, mostly by rock "weathering" (discussed in Chapter 13).

In the last century and a half, humans have added large amounts of CO_2 to the atmosphere by extracting and burning fossil fuels (especially coal, oil, and natural gas) that contain carbon fixed by photosynthesis many millions of years ago. As we discussed in Chapter 8, this rapid acceleration of the rate at which carbon is freed and converted to CO_2 is having far-reaching effects on both the atmosphere and the biosphere. For example, increasing concentrations of atmospheric CO_2 has led to changes in the natural distribution patterns of both plants and animals.

LearningCheck 10-3 Over long periods of time, how can a molecule of carbon dioxide move from the atmosphere to the biosphere to the lithosphere and back to the atmosphere?

The Nitrogen Cycle

Although nitrogen gas (N_2) is an apparently inexhaustible component of the atmosphere (air is about 78 percent N_2), only a limited number of organisms can use this essential nutrient in its gaseous form. The movement of nitrogen in and out of the biosphere—from forms of nitrogen that can be used by organisms and back to forms that cannot—is called the **nitrogen cycle**.

For the vast majority of living organisms, atmospheric nitrogen is usable only after it has been converted to nitrogen compounds (*nitrates*) that can be used by plants (Figure 10-7). This conversion process is called **nitrogen fixation**. Some nitrogen is fixed in the atmosphere by lightning and cosmic radiation, and some is fixed in the ocean by marine organisms, but the amount involved in these processes is modest. Nitrogen-fixing bacteria living in the soil and associated plant root nodules provide most of the usable nitrogen to Earth's biosphere.

Once atmospheric nitrogen has been fixed into an available form (nitrates), it is absorbed by green plants, some of which are eaten by animals. The animals then excrete nitrogenous wastes in their urine. These wastes, as well as the dead animal and plant material, are attacked by bacteria, and *nitrite* compounds are released as a further waste

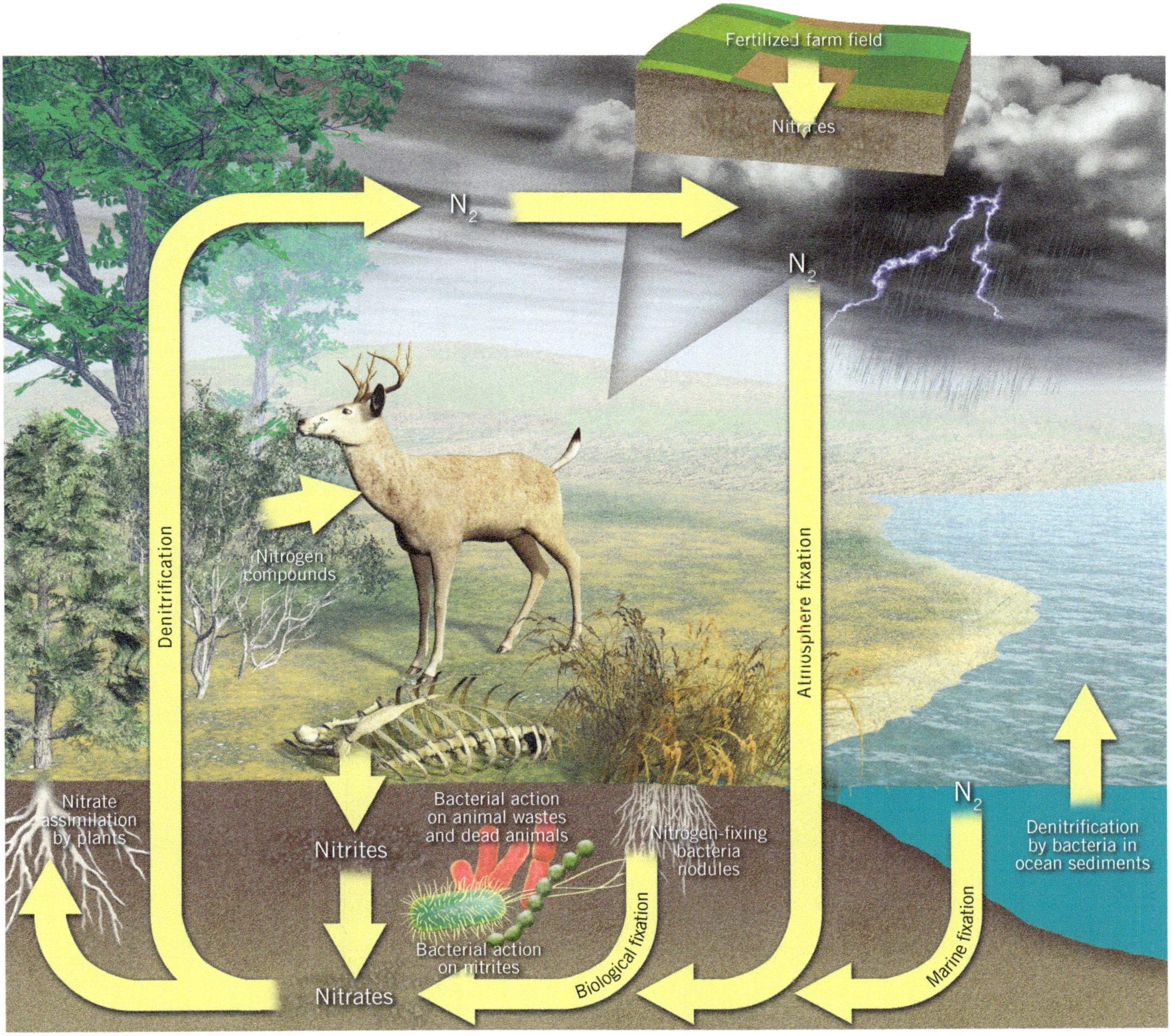

▲ Figure 10-7 The nitrogen cycle. Atmospheric nitrogen is fixed into nitrates, which are then taken up by green plants, some of which are eaten by animals. Dead plant and animal materials, as well as animal wastes, contain various nitrogen compounds. Bacteria modify these compounds, producing nitrites. The nitrites are then converted by different bacteria to nitrates; thus the cycle continues. Still other bacteria denitrify some of the nitrates, releasing free nitrogen into the air. The arrows represent pathways but do not reflect the relative amounts of nitrogen moving along any one pathway.

product. Different bacteria convert the nitrites to nitrates, making them available again to green plants. Still other bacteria convert some of the nitrates to nitrogen gas in a process called **denitrification**, and the gas becomes part of the atmosphere. Rain carries this atmospheric nitrogen back to Earth, where it re-enters the soil–plant portion of the cycle.

Eutrophication: Human activities have produced a major modification in the natural nitrogen cycle. The synthetic manufacture of nitrogenous fertilizers and widespread introduction of nitrogen-fixing crops (such as alfalfa, clover, and soybeans) have significantly changed the balance between fixation and denitrification. One short-term result has been an excessive accumulation of nitrogen compounds in many lakes and streams. Such an oversupply of nutrients is called **eutrophication** and can lead to a "bloom" of plants or algae in lakes or estuaries. Such a bloom can deplete the water of oxygen, leading to the death of fish and other aquatic organisms (discussed further in the next section). Once the plants and algae die, their decomposition releases CO_2, which makes the water more acidic and so slows fish and shellfish growth.

LearningCheck 10-4 **How is nitrogen made available to living organisms through the nitrogen cycle?**

The Oxygen Cycle

Oxygen is a building block in most organic molecules and consequently makes up a significant proportion of the atoms in living matter. As part of the **oxygen cycle**, oxygen released into the atmosphere through photosynthesis may be taken up by respiring organisms or may react chemically with rocks. The oxygen cycle (Figure 10-8) is extremely complicated and is summarized only briefly here.

Oxygen occurs in many chemical forms and is released into the atmosphere in a variety of ways. Most of the oxygen in the atmosphere is molecular oxygen (O_2) produced when plants decompose water molecules in photosynthesis. Some atmospheric oxygen is bound up in water molecules that came from evaporation or plant transpiration, and some is bound up in the CO_2 released during animal respiration. Photosynthesis eventually recycles much of this CO_2 and water through the biosphere.

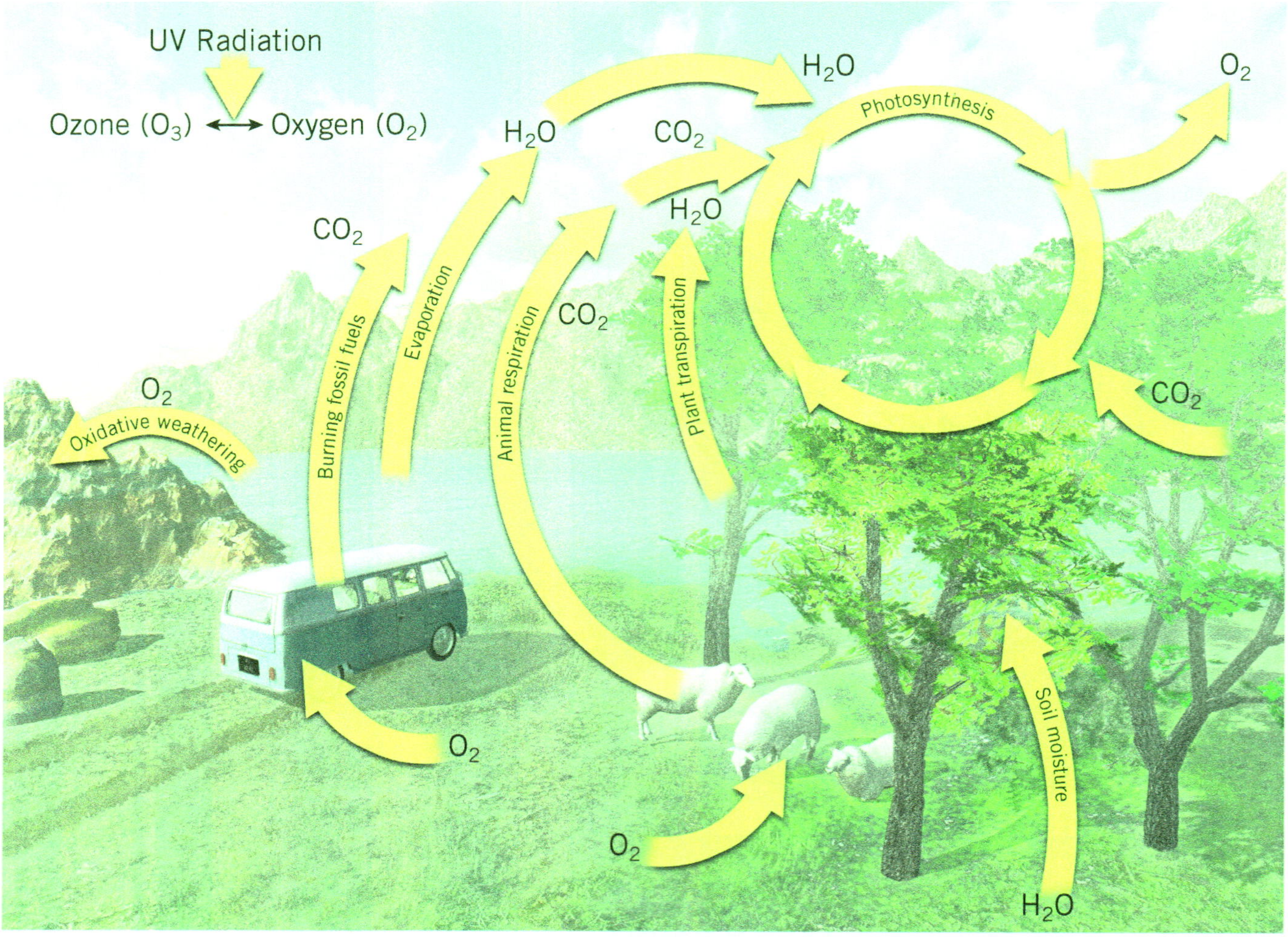

▲ Figure 10-8 The oxygen cycle. Molecular oxygen is essential for almost all forms of life. It is made available to the air through a variety of processes such as photosynthesis and is recycled in a variety of ways. The arrows represent pathways but do not reflect the relative amounts of oxygen moving along any one pathway.

Other sources of oxygen for the oxygen cycle include atmospheric ozone (O_3), oxygen involved in the oxidative weathering of rocks, oxygen stored in and sometimes released from carbonate rocks, and some processes (such as the burning of fossil fuels, in which oxygen is combined with carbon in CO_2) that are human induced.

Although our atmosphere is now oxygen rich, it was not always so. Earth's earliest atmosphere was oxygen-poor; indeed, in the early days of life on this planet, about 3.4 billion years ago, oxygen was poisonous to living cells. Evolving life had to develop mechanisms to neutralize or, better still, exploit its presence. This exploitation was so successful that most life now cannot function without oxygen. The oxygen now in the atmosphere is largely a by-product of plants. Once life could sustain itself in the presence of high amounts of oxygen, primitive plants made possible the evolution of more advanced plants and animals by providing molecular oxygen for their metabolism.

Aquatic Dead Zones: Aquatic ecosystems depend on oxygen just as much as terrestrial ones do. Fish, for example, extract dissolved oxygen by forcing water over the capillary-dense tissues in their gills. If the amount of dissolved oxygen in the water becomes too low, fish cannot survive. As we saw, this happens in lakes and marine coastal areas where nitrogen-rich runoff containing agricultural fertilizers or sewage leads to eutrophication and a surge in algae growth. Once the algae die, bacteria decompose them and deplete the water of oxygen. In recent decades, large *dead zones*—tracts of ocean and lakes with nearly no aquatic life—have developed in oxygen-starved water.

Mineral Cycles

Although carbon, nitrogen, and oxygen—along with hydrogen—are the principal chemical components of the biosphere, many minerals are critical nutrients for plants and animals. Most notable among these trace minerals are phosphorus, sulfur, and calcium, but more than a dozen others are occasionally significant.

Some nutrients are cycled in part along gaseous pathways, as we saw in the carbon, oxygen, and nitrogen cycles. Other nutrients—such as calcium, phosphorus, sulfur, copper, and zinc—follow primarily sedimentary pathways. For example, an element is weathered from bedrock into the soil. Some of it is then washed downslope with surface runoff or percolated into the groundwater supply. Much of it reaches the ocean, where it may be deposited in a sedimentary rock or is ingested by aquatic organisms. Later it is released into the cycle again through waste products and dead organisms.

Food Chains

The unending flows of energy, water, and nutrients through the biosphere are channeled in significant part by direct passage from one organism to another in pathways referred to as *food chains*. A **food chain** is a simple concept (Figure 10-9): Organism A is eaten by Organism B, which thereby absorbs A's energy and nutrients; Organism B is eaten by Organism C, with similar results; Organism C is eaten by D; and so on.

In nature, however, the matter of who eats whom may be extraordinarily complex, with a bewildering number of

▲ Figure 10-9 A simple food chain showing a producer (green plant), a primary consumer (caterpillar), and secondary, tertiary, and quaternary consumers (frog, snake, and eagle).

interlaced strands. Therefore, *chain* is a misleading word in this context because it implies an orderly linkage of equivalent units. It is more accurate to think of this energy transfer process as a "web" with interconnected parts or links. Each link acts as an energy transformer that ingests some of the energy of the preceding link, uses some of that energy for its own sustenance, and then passes some of the balance on to the next link.

The fundamental units in any food chain are the **producers**: the *autotrophs*, or "self-feeders"—in other words, plants. Plants fix carbon and effectively store solar energy through photosynthesis. Plants may then be eaten by **consumers**, or *heterotrophs*. Plant-eating animals are called *herbivores* (*herba* is Latin for "plant"; *vorare*, "to devour") and are referred to as **primary consumers**. Herbivores become food for other animals. Meat-eating animals are called *carnivores* (*carne* is Latin for "meat") and are referred to as **secondary consumers** or *predators*. A food chain may have many levels: secondary, tertiary, quaternary, and so on (see Figure 10-9). *Omnivores*, such as humans, are animals that eat both plants and other animals and thus may have several roles within a food chain.

LearningCheck 10-5 **Describe producers, primary consumers, and secondary consumers.**

Food Pyramids

A food chain can also be conceptualized as a **food pyramid**. The number of energy-storing organisms is much, much larger than the number of primary consumers; the number of primary consumers is larger than the number of secondary consumers; and so on up the pyramid (**Figure 10-10**). There are usually several levels of carnivorous consumers, each succeeding level consisting of fewer and usually larger animals. The final consumers at the top of the pyramid are usually the largest and most powerful *apex predators* in the area (**Figure 10-11**). We say that organisms share the same **trophic level** when they occupy the same general position in a food chain and thus consume the same general types of food in a food pyramid.

The consumers at the apex of the pyramid do not constitute the final link in the food chain, however. When they die, they are fed on by scavenging animals and by tiny (mostly microscopic) organisms that function as **decomposers**, returning the nutrients to the soil to be recycled into yet another food pyramid.

Why are so many producers needed to support so few apex predators in a food pyramid? The answer has to do with the relative inefficiency with which stored energy is transferred

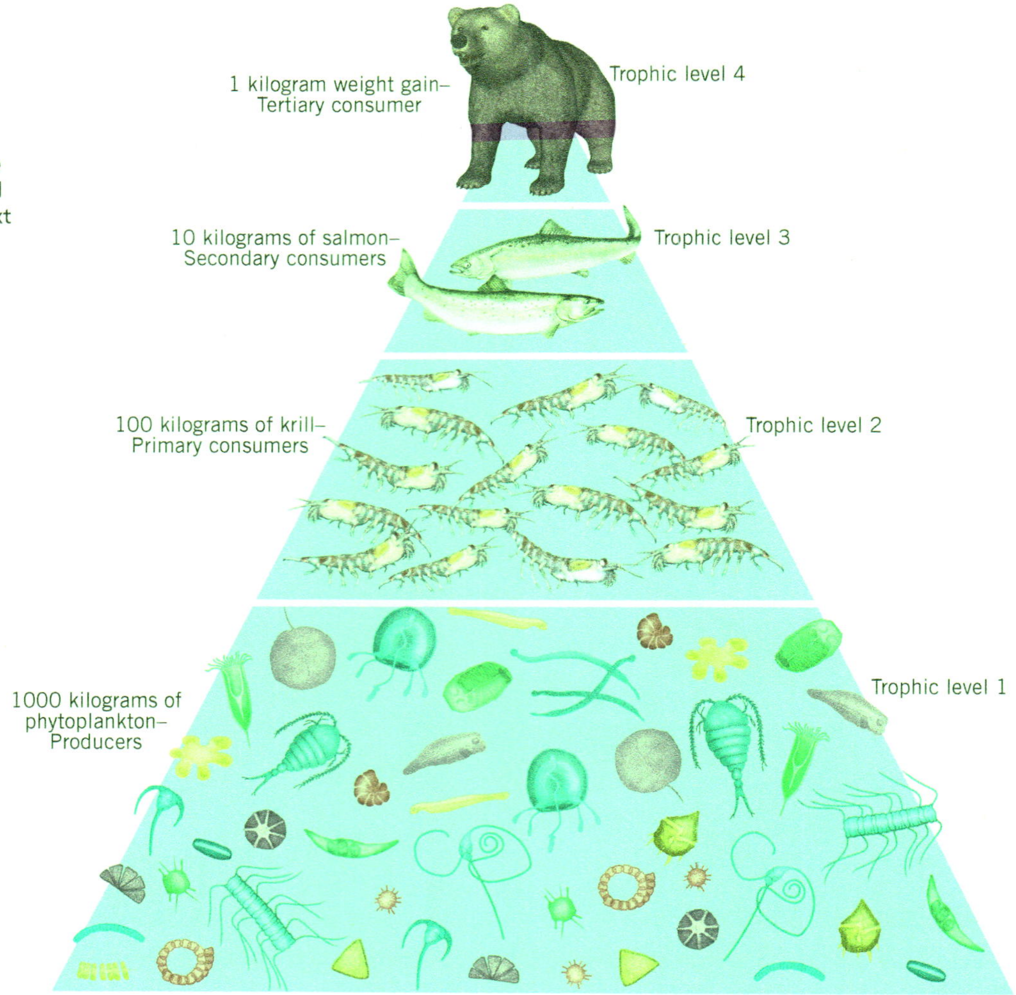

▶ **Figure 10-10** A food pyramid. It takes 1000 kilograms (about 1 ton) of phytoplankton (microscopic marine plants) to provide a 1-kilogram (2-pound) weight gain for a bear. Notice that only about 10 percent of the stored energy in one trophic level is eventually passed on to the next trophic level.

▲ **Figure 10-11** A grizzly bear fishing for salmon along the Brooks River in Katmai National Park, Alaska. The bear is an apex predator.

from one trophic level to the next. When a primary consumer eats a plant, perhaps only 10 percent of the total energy stored in the plant can be effectively stored by that primary consumer. In turn, when a secondary consumer eats the primary consumer, there is further energy transfer inefficiency. In short, only a portion of the stored energy is transferred from the organisms of one trophic level to those in the next trophic level. (In terms of human diet and food supply, this is why it is much more energy efficient to feed a population directly with grains than with meat from grain-fed animals.)

LearningCheck 10-6 **Why does the amount of biomass change at each trophic level of a food pyramid?**

Pollutants in the Food Chain

Although stored energy is not efficiently passed on from one organism to the next in a food chain, some chemical pollutants can be. An increasing concern is evidence that some chemical pollutants can become concentrated within a food chain—a process referred to as *biomagnification*. For example, although some chemical pesticides released into the air, water, or soil degrade relatively quickly into harmless substances, others, such as DDT (dichlorodiphenyl-trichloroethane), are quite stable and may become concentrated in the fatty tissues of organisms at higher levels of a food chain. The concentration of pesticides, as well as the concentration of some heavy metals such as mercury and lead, has resulted in harmful effects and even death in the animal (and human) consumers at the top of the food chain (Figure 10-12).

A somewhat similar problem was discovered in 1982 at the Kesterson National Wildlife Refuge amid the farmland of California's southern Central Valley. This artificial wetlands area was created by using runoff from local agricultural irrigation. Minute quantities of natural selenium dissolved in the irrigation water were concentrated through evaporation in the water entering the Wildlife Refuge, resulting in deformities and high mortality for waterfowl using the wetlands. This discovery prompted the U.S. Department of the Interior to create the National Irrigation Water Quality Program to reduce the likelihood of similar problems elsewhere.

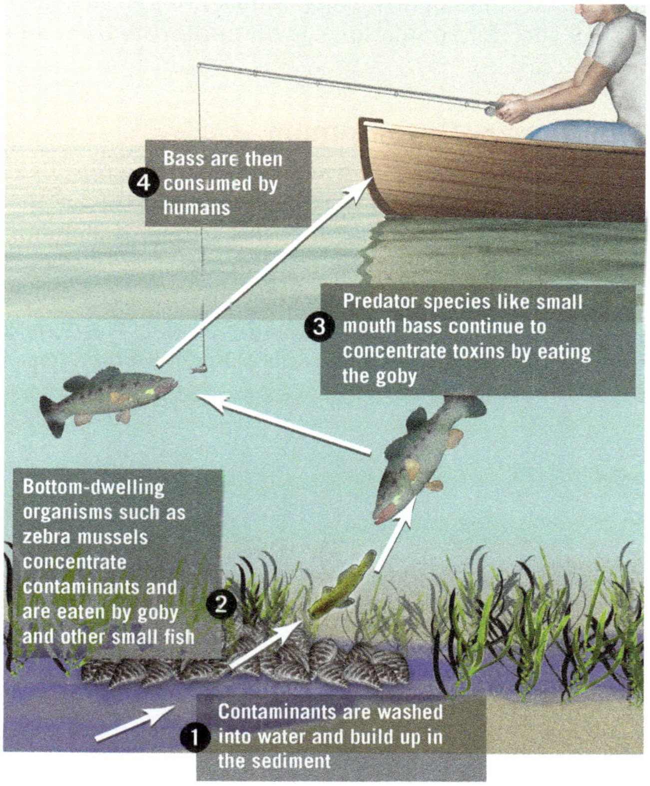

▲ **Figure 10-12** Biomagnification. Toxins such as mercury or some pesticides are washed into a body of water and build up in the sediments. Bottom-dwelling organisms, which are primary consumers, concentrate the toxins in their bodies and are eaten by secondary consumers, such as small fish, that further concentrate the toxins. Small fish are in turn eaten by larger fish and eventually by humans.

The 2010 explosion on the *Deepwater Horizon* drilling platform killed 11 oil workers and released about 4.9 million barrels (775 million liters; 205 million gallons) of petroleum into the Gulf of Mexico. The spilled oil floating on the water surface—and the large plume of oil that remained out of sight below the surface—had immediate and dramatic impacts on Gulf Coast animal and marine life. Most visible were oil-soaked birds and beached dolphins, but the most extensive and long-lasting effects of the spill will likely be the circulation of oil and chemical pollution through the marine food chain.

Biological Factors and Natural Distributions

The study of biogeochemical cycles and food chains shows how the biosphere is part of the flow of materials through Earth systems. But the most basic studies of organisms by geographers are usually concerned with distribution. Biogeography deals with such questions as, "What is the range of a certain species or group of plants or animals?" "What are the reasons behind this distribution pattern?" and "What is the significance of the distribution?"

The natural distribution of any species or group of organisms is determined by several primarily biological factors: evolutionary development, migration and dispersal, reproductive success, population die-off and extinction, and plant succession.

Evolutionary Development

The Darwinian theory of *natural selection* explains the origin of a species as the outcome of an ongoing process of descent, with modification, from parent forms. Our understanding of how this happens has come from increased knowledge of genes. A *gene* is a segment of a long strand of DNA (*deoxyribonucleic acid*). Called a *chromosome*, this strand occurs in pairs and is found in the nucleus of a cell. Genes contain the genetic information of traits that is replicated and passed on from organisms to their offspring.

Over time, the genes of an individual organism acquire slight mutations (errors) when DNA strands are not replicated (copied) perfectly during cell division. Such mutations are random events, but other types of mutations are caused by external factors that affect a cell. Because of mutations, new versions of a gene (called *alleles*) develop.

Many genetic mutations have no effect—positive or negative—on an organism; they are simply passed on from one generation to the next. Other mutations are harmful to an organism and might lead to death or otherwise prevent that individual from producing offspring. Still other mutations in the genes are beneficial, at least in some circumstances, and give an individual an advantage in survival or reproduction. These "advantageous" mutations would be passed on to offspring. If mutations accumulate in a population over time, the individuals in that population become genetically distinct from other members of their species, and a new species may form. It is this long, slow, and essentially endless process that accounts for the development of all species.

Acacias and Eucalyptus: Because the local environment influences which traits of an individual are advantageous to survival, to understand the distribution of any species or genus, we must consider where the species evolved. For example, consider the contrast in apparent origin of two important groups of plants: *acacias* and *eucalyptus*. Acacias are an extensive genus of shrubs and low-growing trees, represented by numerous species found in low-latitude portions of every continent that extends into the tropics or subtropics (Figure 10-13). Eucalyptus, on the other hand, is a genus of trees native to Australia and a few adjacent islands

▼ **Figure 10-13** Hundreds of species of acacias grow in semiarid and subhumid portions of the tropics. This scene from southern Kenya shows an acacia in tree form, although lower shrub forms are more common.

◀ **Figure 10-14** The original forests of Australia were composed almost entirely of species of eucalyptus. This scene is near Kalgoorlie, Western Australia.

only (**Figure 10-14**). Acacias apparently evolved prior to the separation of the continents, whereas the genus Eucalyptus evolved after the Australian continent was isolated. (We discuss the changing locations of the continents over geologic time when we introduce the theory of plate tectonics in Chapter 14.)

Endemic Species: When a variety of plant or animal is found in one location or region but nowhere else, we describe that species as **endemic**. Endemic species are especially common in relatively isolated environments, such as on islands or remote mountain areas, where evolution proceeds without an influx of new genetic material.

Migration and Dispersal

Throughout Earth's history, organisms have moved from one place to another. Animals possess active mechanisms for locomotion—legs, wings, fins, and so on—and their possibilities for migration are obvious. Plants are also mobile, however. Although most individual plants become rooted and therefore fixed in location for most of their lives, many opportunities for passive migration exist, particularly during the seed stage. Wind, water, and animals are the principal natural mechanisms of seed dispersal, although human activity can influence dispersal patterns—see the box *Global Environmental Change: Honey Bees at Risk.*

The contemporary distribution pattern of many organisms is often the result of natural migration or dispersal from an original center of development. Next, we use just two of thousands of examples to illustrate this process.

Coconut Palms: The coconut palm (*Cocos nucifera*) is believed to have originated in southeastern Asia and adjacent Melanesian islands. It is now extraordinarily widespread along the coasts of tropical continents and islands all over the world. Most of this dispersal apparently has come about because coconuts, the large hard-shelled seeds of the plant, can float in the ocean for years without losing their fertility. Thus, they wash up on beaches throughout the world and colonize successfully if environmental conditions are right (**Figure 10-15**). Humans significantly augmented

▲ **Figure 10-15** Coconuts have a worldwide distribution in tropical coastal areas, in part because the nut can float long distances and then take root if it finds a favorable environment. This sprouting coconut is from the island of Vanua Levu in Fiji.

Global Environmental Change

Honey Bees at Risk

- **Sandra L. Arlinghaus,** University of Michigan; **Diana Sammataro,** U.S. Department of Agriculture (retired); **Diana Brand,** Honey Bee Research Services

Honey bees (*Apis mellifera*) are important pollinators of about one-third of the world's crops, including numerous fruits and vegetables. Beeswax, produced from the honeycomb, is even more valuable; it is used in food, cosmetics, and pharmaceuticals. Any pest that threatens the honey bee is a threat to our global economy.

Beekeepers deal with a variety of pathogens that threaten the honey bee population; these may contribute, particularly in combination, to *colony collapse disorder*, or a dead colony. Varroa mites (*Varroa destructor*), parasites of honey bee colonies, have become a major, common threat to colonies since the 1980s. Generally, these tick-like mites directly reduce the bee's lifespan by feeding on its circulatory system fluid. They reduce it indirectly by serving as "vectors" that can spread up to 20 viruses. Mature female Varroa mites, about the size of the head of a pin, live on immature and adult honey bees. The smaller tan male mites do not feed directly on bees. The mite is much smaller than the bee (Figure 10-C). Small pests are difficult to detect; contemporary scientists are working assertively to manage the problem.

Many of the viruses transmitted by these mites are deadly. Controlling Varroa populations in a hive often controls the associated viruses. Viruses are an emerging frontier in the science of understanding honey bee diseases. For many reasons, these mites are now one of the primary causes of honey bee decline. Left untreated, an infested colony will likely die in one to three years.

Dispersal of the Varroa Mite:

The global dispersal of the Varroa mite is similar to other dispersal patterns, such as the spread of an invasive species of plant along the U.S. Interstate Highway System or the spread of contagious diseases via the proliferation of commercial air travel. All dispersals can gravely impact the delicate environmental fabric; a tiny mite is a good model to study!

In the early twentieth century in Asia, the Varroa mite was found only on the island of Java. By the middle of the twentieth century, the mite had reached Sumatra and the USSR (Figure 10-D).

By the second half of the twentieth century, the mite was found worldwide. Although improved observation techniques and added interest from trained observers might account for some of the change in distribution pattern, a significant increase has been documented. One theory relates the mite's spread to improvements in transportation after World War II, associated with both natural and commercial shipment of honey bees.

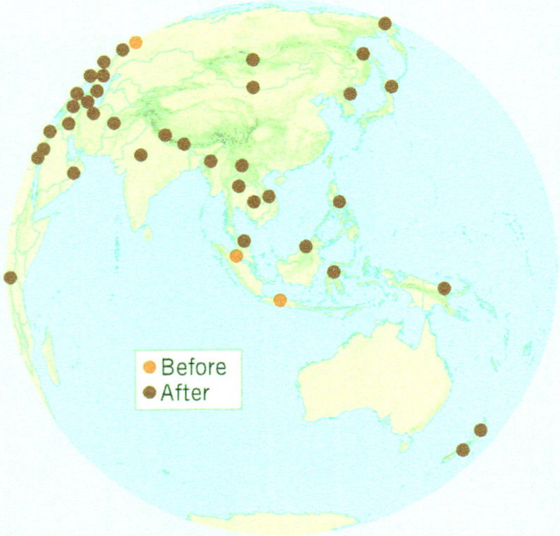

▲ Figure 10-D The distribution of Varroa mite observations in the first half of the twentieth century and in the entire twentieth century. Notice the increase in observations over time.

Mite observation has continued into the twenty-first century. Mites have been found in St. Kitts and Nevis, Panama, New Zealand, and Kenya. Varroa mites arrived in Australia in 2018, but so far, their damage to commercial beehives has been contained.

Future Impacts on Bees?

The 2009 discovery of mites in the Eastern Rift Valley in Kenya, home to a diverse and unusual population of wild animals, is alarming in yet another way. Bees and honey are an integral part of subsistence-level farming there, where honey is an important source of income. Beyond important impacts on the global economy, the removal of an entire species could have long-range, and perhaps unforeseen, consequences at all levels—from the local to the global.

Scientists are seeking creative solutions to the decline of honey bee populations. Some researchers, for instance, are studying certain honey bees that defend themselves by performing grooming behaviors that brush off the parasitic mites or by biting the mites' legs. If we can isolate the genes responsible for these defensive actions, we can selectively breed honey bees that display those behaviors and give the bees a fighting chance.

▲ Figure 10-C Honey bee with a Varroa mite.

Questions

1. Explain the statement, "Any pest that threatens the honey bee is a threat to our global economy."
2. Based on the Varroa mite's pattern of dispersal through the twentieth century and into the twenty-first century, where might the mite take hold next? Why?

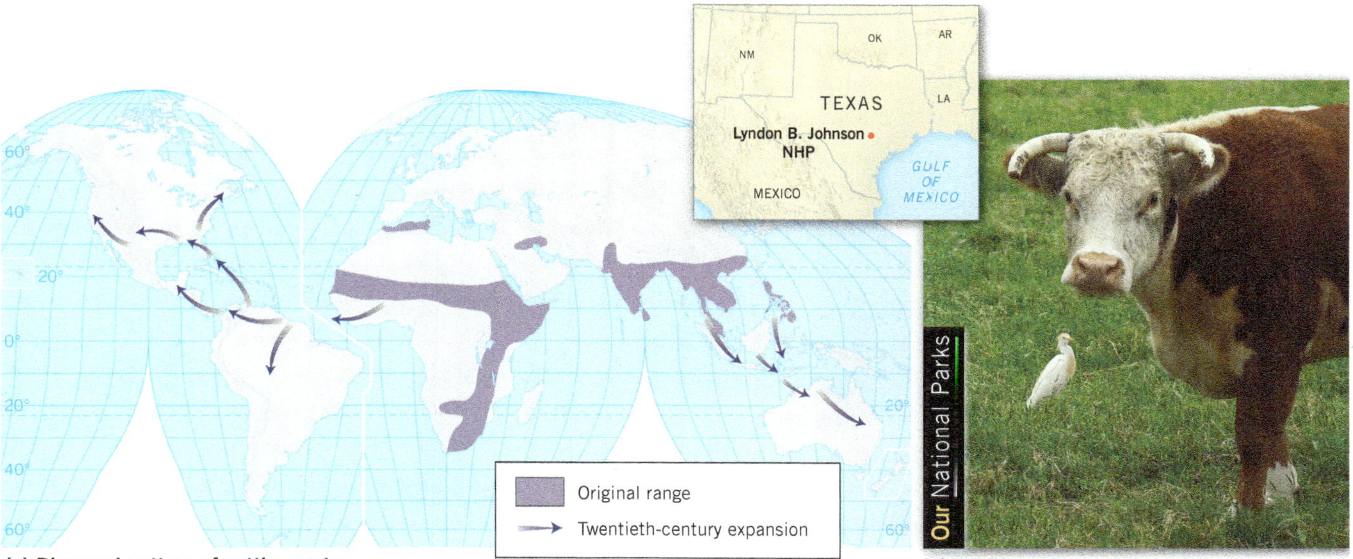

(a) Dispersal pattern of cattle egrets

(b) Habitat

▲ **Figure 10-16** Natural dispersal of an organism. (a) During the twentieth century, cattle egrets expanded from their Asian–African range into new habitats in the Americas and Australia. (b) Cattle egrets have found a happy home among the cattle herds of Texas, here at the Lyndon B. Johnson National Historical Park.

this natural dispersion in the Atlantic region, particularly by the deliberate transport of coconuts from the Indian and Pacific Ocean areas to the West Indies.

Cattle Egrets: The cattle egret (*Bubulcus ibis*) apparently originated in southern Asia but during the last few centuries has spread to other warm areas, particularly Africa (**Figure 10-16**). In recent decades, a change in land use in South America has caused a dramatic expansion of the cattle egret's range. At least as early as the nineteenth century, some cattle egrets crossed the Atlantic from West Africa to Brazil, but they were unable to find suitable ecological conditions and thus did not become established. The twentieth-century introduction of extensive cattle ranching in tropical South America provided the missing ingredient, and egrets quickly adapted to the newly suitable habitat. Their descendants spread northward throughout the subtropics and now breed in the Gulf coastal plain of the southeastern United States. Today they are at least seasonal residents in nearly all of the conterminous 48 states and Hawai'i. Also within the twentieth century, cattle egrets dispersed at the other end of their "normal" range to enter northern Australia and spread across that continent.

LearningCheck 10-7 **What factors account for the distribution of acacias and cattle egrets?**

Reproductive Success

A key factor in the continued survival of any biotic population is reproductive success. Poor reproductive success can come about for a number of reasons: heavy predation (a fox eats quail eggs from a nest); climatic change (heavy-furred animals perishing in a climate that was once cold but has warmed up); failure of food supply (a string of unusually cold winters keeps plants from setting seed); and so on.

Reproductive success is usually the limiting factor that allows one competing population to flourish while another languishes. For example, when waters on the fringes of the Arctic Ocean warmed in recent decades, cod expanded their range at the expense of several other types of fish. The American bison was nearly exterminated by overhunting in the late 1800s, but when allowed to live in a suitable prairie habitat, bison exhibit high reproductive success and can be reestablished as a primary grazing species (**Figure 10-17**).

Population Die-off and Extinction

The range of a species can diminish when some or all of a population dies out. The history of the biosphere has many such examples, varying from minor adjustments in a small area to extinction over the entire planet. Evolution is a continuing process. No species is likely to be a permanent inhabitant of Earth, and during the period of its ascendancy there is apt to be a great deal of distributional variation within a species, part of which is caused by local die-offs.

Extinction: An extinct species has been eliminated forever from the landscape of Earth. Extinction has taken place many times in Earth's history. Indeed, it is estimated that half a billion species have become extinct during the several-billion-year life of our planet. Probably the most dramatic example is the disappearance of the dinosaurs about 65 million years ago. For many millions of years, those gigantic reptiles were the dominant terrestrial life-forms, yet all were wiped out in a relatively short period of geologic time. The fact remains that the history of the world has recorded innumerable such natural extinctions of entire species.

As we discuss in Chapter 11, human activities are currently leading to the extinctions of plant and animal species at an alarming rate, primarily through the destruction of natural habitat.

◀ **Figure 10-17** Once nearly exterminated, American bison (*Bison bison*) now occur in large numbers in areas of suitable habitat, as here in Yellowstone National Park. Under natural conditions, they have a high level of reproductive success.

Plant Succession

One of the simplest and most localized examples of species change over time is **plant succession**, in which one type of vegetation is replaced naturally by another. Plant succession is a normal occurrence in a host of situations; a very common one involves the infilling of a pond (Figure 10-18). As the pond gradually fills with sediments and organic debris, the aquatic plants at the bottom are slowly choked out. The sedges, reeds, and mosses of shallow waters along the edges become more numerous and extensive, forming a marsh. Continued infilling further diminishes the aquatic habitat and allows for the increasing encroachment of low-growing land plants such as grasses and shrubs, yielding a meadow. As the process continues, trees move in and colonize the site, replacing the grasses and shrubs and completing the transition from pond to marsh to meadow to forest.

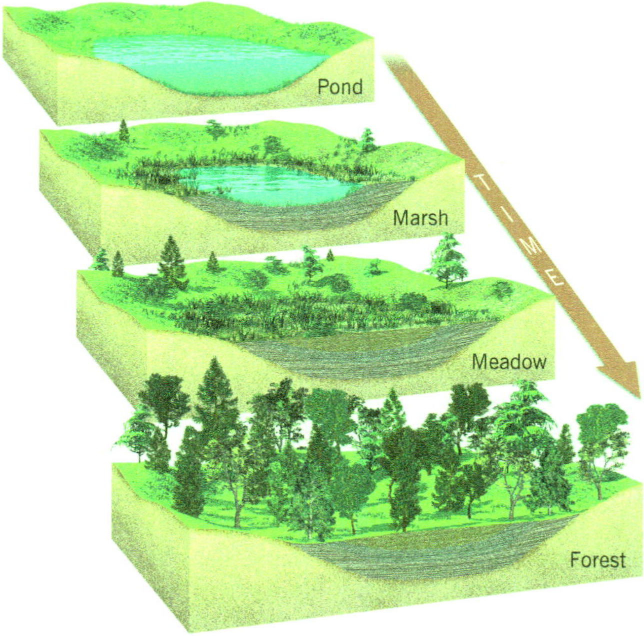

▲ **Figure 10-18** A simple example of plant succession: infilling of a small pond with sediment and organic material. Over time, successional colonization by different kinds of plants changes the area from pond to marsh to meadow to forest.

A similar series of local animal replacements would accompany the plant succession because of the significant habitat changes. Pond animals would be replaced by marsh animals, which in turn would be replaced by meadow and then forest animals.

Plant and animal successions occur after catastrophic natural events as well as gradual ones. For example, as an immediate consequence of the major 1980 eruption of Mount St. Helens, thousands of hectares of forest were covered by volcanic mudflows and pyroclastic flows or were blown down by the large lateral blast (see Chapter 14). In areas where all remnants of the original ecosystem were effectively covered and the soil obliterated, the first species of plants and animals to return comprised what is known as a *pioneer community* in a process called *primary succession*. Over time, one plant community gives way to another, and then another. Each succeeding association alters the local environment, making possible the establishment of the next association. The general sequential trend is toward taller plants and greater stability in species composition. The longer plant succession continues, the more slowly change takes place because more advanced associations usually contain species that live a relatively long time.

In other areas—and more common in general—*secondary succession* is taking place, in which remnants of the original plant communities and soil become the starting point for a succession sequence (Figure 10-19).

LearningCheck 10-8 **Explain the long-term process of plant succession in an infilling lake.**

Climax Vegetation: Eventually a plant community may stabilize, and each succeeding generation of the community is much like its predecessor. This stable association is generally referred to as **climax vegetation**, and the

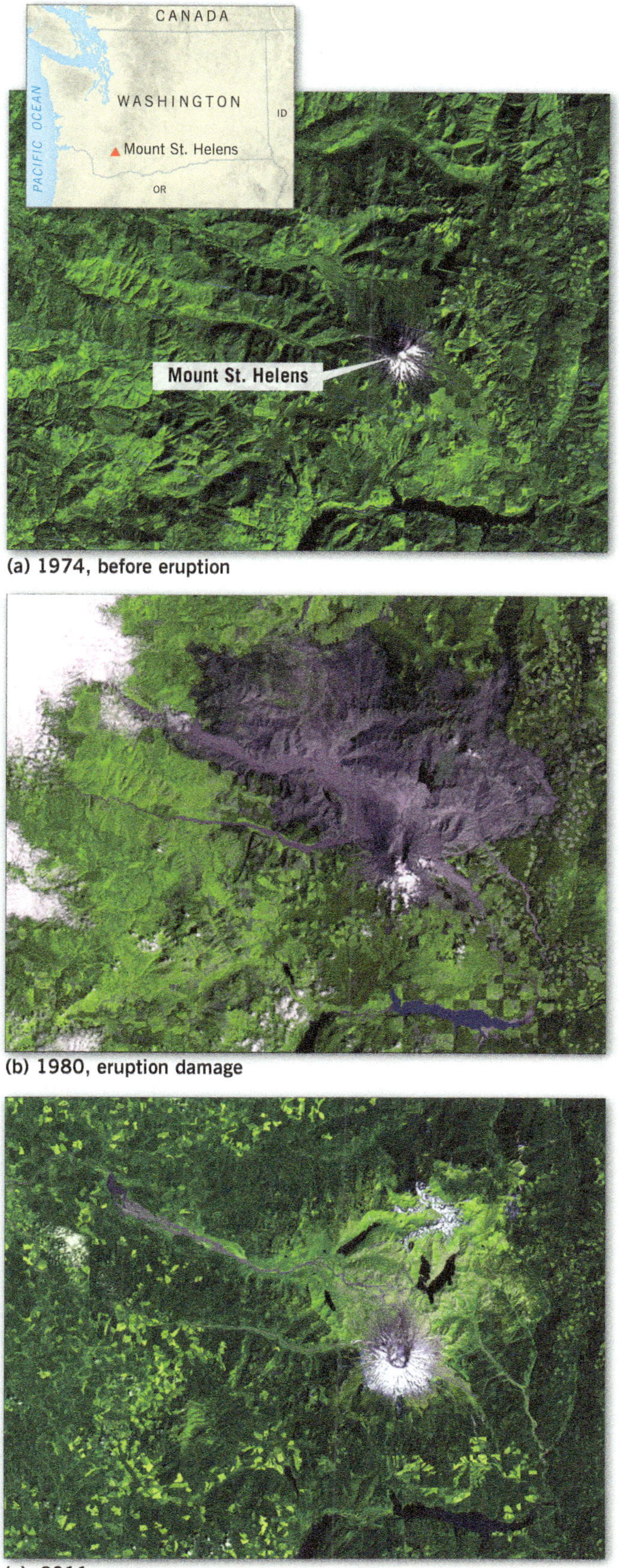

(a) 1974, before eruption

(b) 1980, eruption damage

(c) 2011, recovery

▲ **Figure 10-19** Landsat images showing ecosystem recovery after the Mount St. Helens eruption. (a) October 1974—before the eruption; (b) August 1980—devastation north of the volcano caused by the May 1980 eruption; (c) July 2011—vegetation had regrown in the area razed by the eruption.

various associations leading up to it are called *seral stages*. The implication of the term *climax vegetation* is that the dominant plants of a climax vegetation have demonstrated that, of all possibilities for that particular situation, they can compete the most successfully. Thus, they represent the "optimal" floristic cover for that environmental context.

However, it is important to emphasize that the idealized climax vegetation for a location is rarely realized for long in many environments. Nearly all ecosystems are continuously changing—even if subtly—in response to patterns of disturbance and environmental change. When these conditions change, the climax stage is disturbed, and another succession sequence is initiated.

Don't confuse plant succession with true extinction. Extinction is permanent, but species succession is not. Although a particular plant species may not grow at a given time in a given location, it may reappear quickly if environmental conditions change and if a seed source is available.

LearningCheck 10-9 **Explain the concept of climax vegetation.**

Environmental Factors

The survival of plants and animals depends on a set of environmental factors, including light, moisture, temperature, wind, soil, topography, and wildfire. The specific influence of these factors, of course, varies by species.

We can discuss environmental factors at various scales. For example, if we consider global or continental patterns of biotic distribution, we are concerned primarily with gross generalizations that deal with average conditions, seasonal characteristics, latitudinal extent, zonal winds, and other broad-scale factors. If our interest is instead a small area, such as an individual valley or a single hillside, we are more concerned with such localized environmental factors as degree of slope, direction of exposure, and permeability of topsoil.

Whatever the scale, the generalizations nearly always have exceptions, and the larger the area, the more numerous the exceptions. Thus, a region that is generally humid probably includes many localized sites that are dry, such as cliffs or sand dunes. Even a very dry desert is likely to have several places that are always damp, such as an oasis or a spring.

Limiting Factors: Throughout the following discussion of environmental factors, keep in mind that both *intraspecific competition* (among members of the same species) and *interspecific competition* (among members of different species) are at work. Both plants and animals compete with one another as they seek light, water, nutrients, and shelter in a dynamic environment. We use the term **limiting factor** to describe the variable that is most important in determining an organism's survival.

Human activities can influence the environmental factors affecting the success of some species—see the box *Featured National Park: Rocky Mountain—What's Killing the Park's Forests?*

Featured National Park

Rocky Mountain

What's Killing the Park's Forest?

▲ **Figure 10-E** Pine forest in Rocky Mountain National Park, Colorado, dying from the infestation of the mountain pine beetle.

Across Rocky Mountain National Park, thousands of hectares of pine forest are dying. Year by year, green pine forest is replaced by expanses of brown dead and dying trees caused by an infestation of a bark beetle (**Figure 10-E**). The mountain pine beetle (*Dendroctonus ponderosae*) has affected more than 18 million hectares (44 million acres) of forest in the western United States, and about the same area in Canada—and still growing (**Figure 10-F**). It may be the largest insect infestation in North America during historic times.

The mountain pine beetle is about the size of a grain of rice. It drills into wood to lay its eggs and then injects a fungus that stops the tree from excreting sap, which would kill the beetle larvae. The fungus causes a stain that turns the wood blue. The tree responds by emitting a waxy resin that can plug the holes and kill the beetles, but usually a tree is ultimately overwhelmed as more beetles arrive and drill still more holes.

Causes of Infestation: This extensive beetle infestation in Rocky Mountain National Park has several causes. First, fire suppression over the last century prevented large sections of the forests—especially the expanses of lodgepole pines (*Pinus contorta*)—from burning regularly. Thus, a large proportion of the trees are about the same age—and old enough to be susceptible to the bark beetle. (Younger trees are generally not targeted.) Second, extended drought that began in the late 1990s weakened many of the trees, making them more susceptible. Finally, climate change in western North America is bringing less frequent severely cold winters, allowing the beetle to extend its range. Today, the absolute winter low temperatures are about 3.5–5.5°C (6–10°F) higher than in the mid-twentieth century. With fewer extremely cold days in winter, more beetle larvae are surviving.

Consequences of Infestation: The consequences of the infestation are troubling. Outside the Park, resort towns in the Rocky Mountains worry about the loss of tourism in areas where forests are dying. The threat of rapidly moving "crown fires" that can jump from tree to tree has increased. (The likelihood of crown fires is reduced, however, once a dead tree's needles drop to the ground.) With many dead (or burned) trees, watersheds are more susceptible to flash floods and debris flows. Finally, falling dead trees are becoming a problem along highways, so large strips have been cleared to reduce road hazards.

Future Prospects: The mountain pine beetle is a native insect that has been around for thousands of years, playing its role in western pine forest ecosystems—as is a native spruce beetle also spreading in these same forests. Lodgepole pines are well adapted to major "stand-replacement events" such as this during which large areas of old forest are killed. The forest then regenerates as nutrients are cycled through the soil; grasses and other low vegetation return first, to be replaced in time by a new forest—a process already taking place in parts of the Park. The wild card in this infestation is climate change: milder winters mean fewer checks on the beetles, so they are spreading almost as a destructive introduced "invasive" species would.

Nothing suggests that this infestation will end any time soon. There is little Rocky Mountain National Park can—or should—do to address the spread of the mountain pine beetle: susceptible stands of old trees can be thinned; the bark of trees in heavily visited scenic tourist spots can be treated to kill the larvae; dead trees in high traffic areas can be removed to prevent hazards from falling branches. In the end, however, the infestation will need to run its course.

Questions

1. How does the mountain pine beetle weaken a tree?
2. How is climate change contributing to the beetle infestation?

▼ **Figure 10-F** Landsat satellite images of the area near Grand Lake, Colorado, just east of Rocky Mountain National Park, showing the expansion of the mountain pine beetle infestation (brown areas) from (a) September 2005 to (b) September 2011.

(a) Before beetle infestation

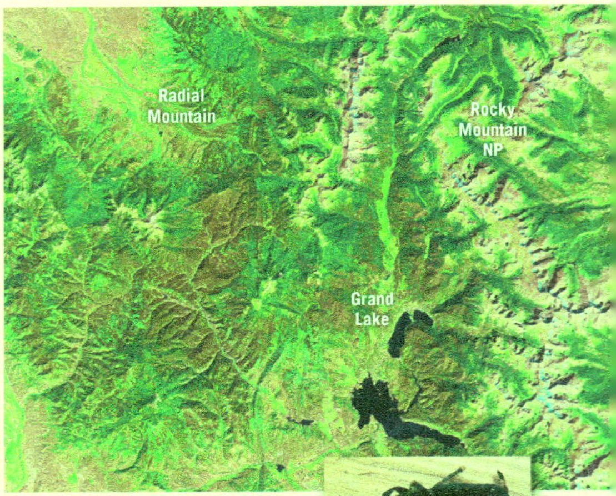

(b) Aftermath of beetle infestation

Mountain pine beetle

The Influence of Climate

At almost any scale, the most prominent environmental constraints on biota are exerted by various climatic factors.

Light: No green plant can survive without light. We have already discussed the basic process—photosynthesis—whereby plants produce stored chemical energy; this process is activated by light. Essentially this is why photosynthetic vegetation is absent from deeper ocean areas, where light does not penetrate.

Light can have a significant effect on plant shape. Where the amount of light is restricted, such as in a dense forest, trees are likely to be very tall but have limited lateral growth (Figure 10-20). In areas that have less dense vegetation, more light is available; as a result, trees are likely to be expansive in lateral spread but truncated vertically.

Another important relationship involves how much light an organism receives during any 24-hour period. This relationship is called **photoperiodism**. Except around the equator, the seasonal variation in the photoperiod becomes greater with increasing latitude. Fluctuation in the photoperiod stimulates seasonal behavior—such as flowering, leaf fall, mating, and migration—in plants and animals.

▲ Figure 10-20 In an open stand (left), light is abundant all around a tree, which responds by broad lateral growth rather than vertical growth. Under crowded conditions (right), less light reaches a tree, which elongates upward rather than spreading laterally.

Moisture: The broad distribution patterns of biota are governed more significantly by the availability of moisture than by any other environmental factor besides light. A prominent trend throughout biotic evolution has been the adaptation of plants and animals to either excesses or deficiencies in moisture availability (Figure 10-21).

Temperature: The temperature of the air and the soil is also important to biotic distributions. Fewer species of plants and animals can survive in cold regions than in areas of more moderate temperatures. Plants, in particular, have a limited tolerance for low temperatures because they are continuously exposed to the weather, and they experience tissue damage and other physical disruption when their cellular water freezes. Animals in some instances are able to avoid the bitterest cold by moving to seek shelter. Even so, the cold-weather areas of high latitudes and high elevations have a limited variety of animals as well as plants.

Wind: The influence of wind on biotic distributions is more limited than that of the other climatic factors. Where winds are persistent, however, they often serve as a constraint. The principal negative effect of wind is that it causes excessive

◄ Figure 10-21 Even the most stressful environments often contain distinctive and conspicuous plants, as in this scene from Saguaro National Park, Arizona, in the Sonoran Desert.

▲ **Figure 10-22** Some plants are remarkably adaptable to environmental stress. In this timberline scene from north-central Colorado, persistent wind from the right has so desiccated this subalpine fir that only branches growing toward the left survive. This preferred growth direction places the trunk of the tree between those branches and the wind.

▲ **Figure 10-23** Wildfires are commonplace in many parts of the world. The Forks Complex Fire burned this forest southeast of Hayfork, California, in August 2015.

Mobile Field Trip
Forest Fires
https://media.pearsoncmg.com/ph/esm/esm_mcknight_physgeo_13/media/

drying by increasing evaporation from exposed surfaces, thus causing a moisture deficiency. In cold regions, wind escalates the rate at which animals lose body warmth.

The sheer physical force of wind can also be influential: a strong wind can uproot trees, increase the heat intensity of wildfires, and modify plant forms (Figure 10-22). On the positive side, wind sometimes aids in dispersal by carrying pollen, seeds, lightweight organisms, and flying creatures.

LearningCheck 10-10 **Explain how the climate-related limiting factors of light and wind can influence plant distributions.**

Edaphic Influences

Soil characteristics, known as **edaphic factors** (from *edaphos*, Greek for "ground" or "soil"), also influence biotic distributions. These factors are direct and immediate in their effect on flora but usually indirect in their effect on fauna. Because soil is a major component of the habitat of any vegetation, its characteristics significantly affect rooting capabilities and nutrient supply. Especially significant are soil texture, soil structure, humus content, chemical composition, and the relative abundance of soil organisms. We discuss soil in much greater detail in Chapter 12.

Topographic Influences

In global distribution patterns of plants and animals, general topographic characteristics are the most important factor affecting distribution. For example, the assemblage of plants and animals in a plains region is very different from that in a mountainous region. At a more localized scale, the factors of slope and drainage are likely to be significant—primarily the steepness of the slope, its orientation with regard to sunlight, and the porosity of the soil on the slope.

Wildfire

Most environmental factors that affect the distribution of plants and animals are passive, and their influences are slow and gradual. Occasionally, however, abrupt and catastrophic events—such as floods, earthquakes, volcanic eruptions, landslides, insect infestations, and droughts—also play a significant role. By far the most important of these is wildfire (Figure 10-23). Except for always-wet regions where fire cannot start and always-dry regions where there is little combustible vegetation, uncontrolled natural fires occur with surprising frequency. Fires generally result in complete or partial devastation of the plant life and the killing or driving-away of all or most of the

▲ **Figure 10-24** Regrowth after the Yellowstone wildfires. (a) Firefighter mopping up some of the last remnants of fires in 1988. (b) The same location five years later, showing the regrowth of vegetation on the forest floor. (c) Regrowth on the Blacktail Deer Plateau 23 years after the fires.

animals. These results are temporary; sooner or later, vegetation sprouts, and animals return. At least in the short run, however, the composition of the biota is changed, and if fires occur with sufficient frequency, the change may be more than temporary.

Wildfire can be very helpful to the seeding or sprouting of certain plants and the maintenance of certain types of forest. For example, the forest fires that burned nearly half of Yellowstone National Park in 1988 triggered the extensive regrowth of understory plants and young trees (**Figure 10-24**). The lodgepole pine (*Pinus contorta*) is especially well adapted to rapid regeneration after "stand-replacement" events such as fire. In some cases, grasslands are sustained by relatively frequent natural fires, which inhibit the encroachment of tree seedlings. Moreover, many plant species, particularly trees such as the giant sequoia (*Sequoiadendron giganteum*), scatter their seeds only after the heat of a fire has caused the cones or other types of seedpods to open.

Wildland-Urban Interface: As urban populations encroach ever further into forest and woodland areas around the world, the problem of wildfires is increasing in the *wildland-urban interface*. Not only is a larger population directly exposed to the natural cycle of wildfire, but human activities are frequent triggers of these fires. With climate change leading to higher temperatures and decades of fire suppression having left a large fuel supply in many woodland areas, the conditions have been ripe in recent years for catastrophic conflagrations (**Figure 10-25**).

LearningCheck 10-11 **How can a wildfire be beneficial to a forest?**

Environmental Correlations

One of the most important themes in physical geography is the intertwining relationships of the various components of the environment. Time after time, we note situations in which one aspect of the environment affects another—sometimes conspicuously, sometimes subtly. In terms of broad distribution patterns, climate, vegetation, and soil have a particularly close correlation. Before we discuss some of the details of biogeographical patterns in the following chapter, it may be helpful to consider an example of some of these correlations by examining the distribution of tropical rainforest.

The Example of Tropical Rainforest

On any map showing the world distribution of major *plant associations* (a grouping of plant species typically found

◀ **Figure 10-25** In November 2018, at least 85 people were killed when the Camp Fire swept through the rural California town of Paradise.

together in a particular environment), one of the conspicuous units is the *tropical rainforest*, or *selva* (the Portuguese-Spanish word for "forest"). A tropical rainforest's plant association consists of many species of trees and other plants, such as vines and orchids that grow on the trees.

A vast extent of selva exists in northern South America (primarily within the Amazon River watershed) and central Africa (mostly within the Congo River watershed). More limited patches are found in Central America, Colombia, West Africa, Madagascar, Southeast Asia, and northeastern Australia (**Figure 10-26**).

Climate: With very limited exceptions, the tropical rainforest occurs wherever precipitation is relatively abundant and temperatures are uniformly warm throughout the year, especially in areas of tropical wet (Af) climate (see Chapter 8). It is tempting to state that the tropical wet climate "creates" the conditions necessary for tropical

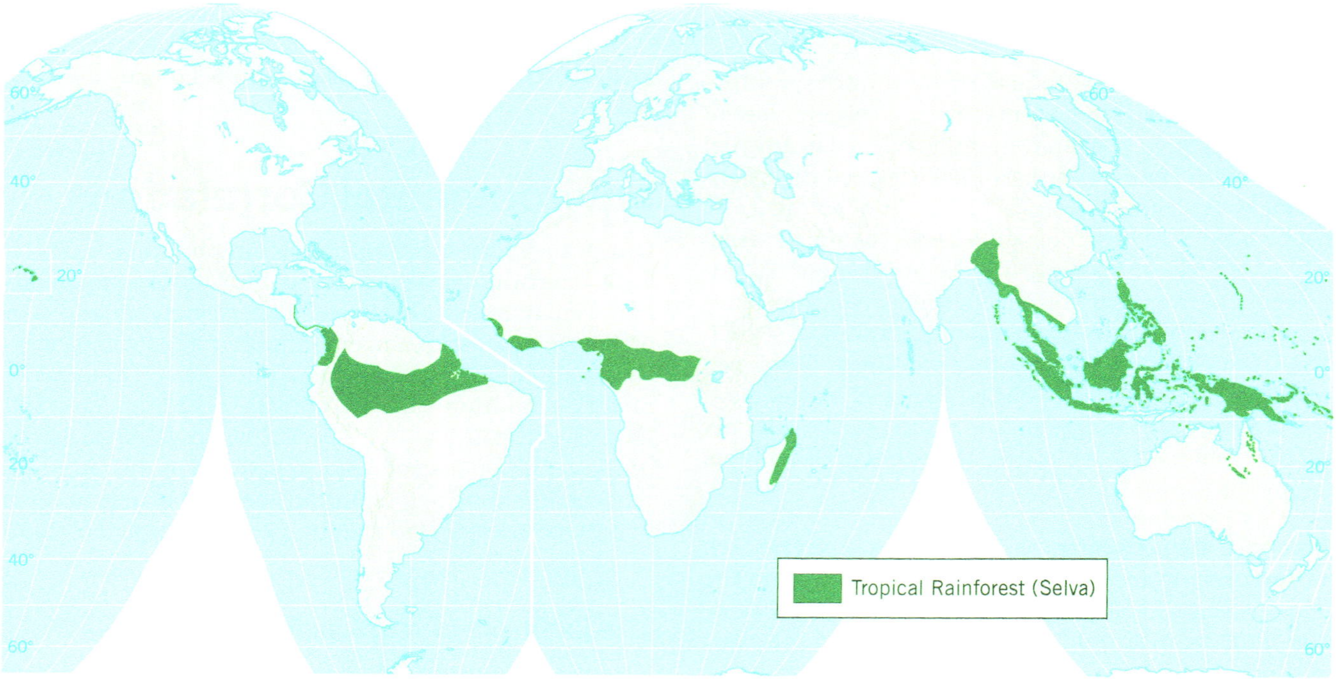

▲ **Figure 10-26** Generalized distribution of tropical rainforest, or *selva*.

rainforest, but the cause-and-effect relationship is not that simple. For example, transpiration from the local vegetation is very much a part of the hydrologic cycle in tropical rainforest regions.

Flora: Because of the high temperatures and high humidity, regions with a tropical wet climate are normally covered with natural vegetation that is unexcelled in luxuriance and variety. Tropical rainforest is a broadleaf evergreen forest with numerous tree species. Many of the trees are very tall, and their intertwining tops form an essentially continuous canopy that prohibits direct sunlight from shining on the forest floor. Often shorter trees form a second and even a third partial canopy at lower elevations.

Most of the trees have smooth bark and have no low limbs, although a profusion of vines and hanging plants entangle the trunks and dangle from higher limbs (Figure 10-27). The dimly lit forest floor is relatively clear of growth because lack of sunlight inhibits survival of bushes and shrubs. Where much sunlight reaches the ground, as along the edge of a clearing or banks of a stream, a maze of undergrowth can prosper. This sometimes impenetrable tangle of bushes, shrubs, vines, and small trees is called a *jungle*.

Fauna: Regions of tropical rainforest are the realm of flyers, crawlers, creepers, and climbers. Larger species, particularly hoofed animals, are rare. Birds and monkeys inhabit the forest canopy, often in great quantity and diversity. Snakes and lizards are common on the forest floor and in the trees. Rodents are sometimes numerous at ground level, but the sparser population of larger mammals is typically secretive and nocturnal. Aquatic life, particularly fish and amphibians, is usually abundant. Invertebrates, especially insects and arthropods, are superabundant.

Soil: The copious, warm, year-round rains provide an almost continuous infiltration of water downward, so soils are usually deep but highly leached and infertile. Leaves, twigs, flowers, and branches frequently fall from the trees to the ground, where they are rapidly decomposed by the abundant earthworms, ants, bacteria, and soil microfauna. The accumulated litter is continuously incorporated into the soil, where some of the nutrients are taken up by plants and the remainder is carried away by the infiltrating water.

Laterization (rapid weathering of mineral matter and speedy decomposition of organic matter) is the principal soil-forming process. It produces a thin layer of fertile topsoil that is rapidly used by plants and a deep subsoil that is largely an infertile mixture of such insoluble constituents as iron, aluminum, and magnesium compounds. These minerals typically impart a reddish color to the soil. River floodplains tend to develop soils of higher fertility because of the deposition of silt from a flood.

▲ Figure 10-27 A tropical rainforest scene in Ecuador.

LearningCheck 10-12 **How is it possible for relatively infertile tropical soils to support the huge biomass of a tropical rainforest?**

Hydrography: The abundance of runoff water on the surface feeds well-established drainage systems. There is usually a dense network of streams, most of which carry a great deal of water and a heavy load of sediment. Perhaps surprisingly, lakes are not common. Over time, lake basins tend to drain by stream erosion or fill with sediment, although if the land is very flat, swamps may be present due to slow drainage and the rapid growth of vegetation.

In the next chapter, we explore many other correlations among the patterns in the biosphere, lithosphere, atmosphere, and hydrosphere.

10 Learning Review

After studying this chapter, you should be able to answer the following questions. Key terms from each text section are shown in **bold type.** Definitions for key terms are also found in the glossary at the back of the book.

Key Terms and Concepts

The Geographic Approach to the Study of Organisms (p. 286)

1. Briefly define the following terms: **biogeography**, **biota**, **flora**, **fauna**.
2. What is meant by **biodiversity**?

Biogeochemical Cycles (p. 288)

3. What is a **biogeochemical cycle**?
4. What is the primary source of energy for the biosphere?
5. Describe and explain the process of **photosynthesis**.
6. Describe and explain the process of **plant respiration**.
7. What is meant by **net primary productivity**?
8. What is the relationship of **biomass** to net primary productivity?
9. Describe the basic steps in the **carbon cycle**.
10. Why is it difficult to integrate nitrogen gas from the atmosphere into the **nitrogen cycle** of the biosphere?
11. Explain the differences between **nitrogen fixation** and **denitrification**.
12. What causes **eutrophication**?
13. Briefly describe some of the components of the **oxygen cycle**.

Food Chains (p. 295)

14. What is the relationship of a **food chain** to a **food pyramid**?
15. Explain the roles of **producers**, **consumers**, and **decomposers** in the food chain.
16. What is the difference between **primary consumers** and **secondary consumers**?
17. Explain the concept of **trophic levels** in a food pyramid.

Biological Factors and Natural Distributions (p. 298)

18. Describe one mechanism through which plant seeds can be dispersed over great distances.
19. What is an **endemic** species?
20. Explain the concept of **plant succession**.
21. Explain the concept of **climax vegetation**.

Environmental Factors (p. 303)

22. What is meant by the term **limiting factor**?
23. Explain how both photosynthesis and **photoperiodism** depend on sunlight.
24. What is meant by **edaphic factors**?
25. What are the beneficial effects of wildfire?

Environmental Correlations (p. 307)

26. Describe the general environmental conditions associated with tropical rainforest locations.

Study Questions

1. How is solar energy “stored” in the biosphere?
2. Most of the carbon at or near Earth’s surface is not involved in any short-term cycling. Explain.
3. What is the primary source of carbon that humans have added to the atmosphere?
4. What is the importance of photosynthesis to the flow of energy, water, oxygen, and carbon through the biosphere?
5. See the food pyramid in Figure 10-10. Why does it take 1000 kilograms of phytoplankton to produce only 10 kilograms of fish?
6. Explain *biomagnification*.
7. Why are trees in dense forests likely to be tall with narrow tops?

Exercises

1. Using Figure 10-10, estimate how many kilograms of phytoplankton it takes to produce 25 kilograms of invertebrates. _____ kilograms of phytoplankton
2. Using Figure 10-10, estimate how many kilograms of invertebrates it takes to produce 10 kilograms of fish. _____ kilograms of invertebrates
3. Using the map of tropical rainforest distribution (Figure 10-26), as well as climate maps in Chapter 8, explain which natural factors likely inhibit the presence of tropical rainforest in the following areas:
   a. Equatorial East Africa
   b. Along the west coast of equatorial South America

Environmental Analysis NET PRIMARY PRODUCTIVITY

Data
Net Primary Productivity
https://media.pearsoncmg.com/ph/esm/esm_mcknight_physgeo_13/media/

Plants play an important role in Earth's carbon dioxide budget. Plants take in carbon dioxide via photosynthesis during the day and give off carbon dioxide via respiration at night. The difference between carbon uptake and release by plants is the net primary production. The Moderate Resolution Imaging Spectroradiometer (MODIS) instrument on NASA's Terra satellite monitors this carbon dioxide use.

Activities

Go to https://earthobservatory.nasa.gov, then to "Global Maps" (via the upper navigation menu). Select the "Net Primary Productivity" map.

1. What do negative net primary productivity values mean?
2. The map is missing data in northern Africa. Would you expect the net primary productivity to be high or low there? Why?

Use the arrow below the map to play the animation.

3. The northern half of South America has high positive net primary productivity year round. Why?
4. In the United States, the net primary productivity is lower in the Southwest than in the Northeast year round. Why?

Look for "Select and Compare" and click on "Show All Maps" below. Choose "Vegetation" and play the animation.

5. What is displayed on the vegetation map (lower map)? What factors control this?
6. In which months are vegetation values high but net primary productivity (upper map) is near or below zero in the United States? What season(s) is this in the Northern Hemisphere?
7. What does this indicate about the relationship between net primary productivity and day length? Net primary productivity and temperature?
8. Look at the winter months in the Southern Hemisphere. Day length is comparable to those in the Northern Hemisphere winter. What factor helps moderate temperatures in the Southern Hemisphere?

Go to www.esrl.noaa.gov/gmd/ccgg/trends to see Trends in Atmospheric Carbon Dioxide from NOAA's Earth System Research Laboratories.

9. How do monthly mean CO_2 concentrations change over the course of one year at Mauna Loa Observatory in Hawai'i? Explain this pattern.

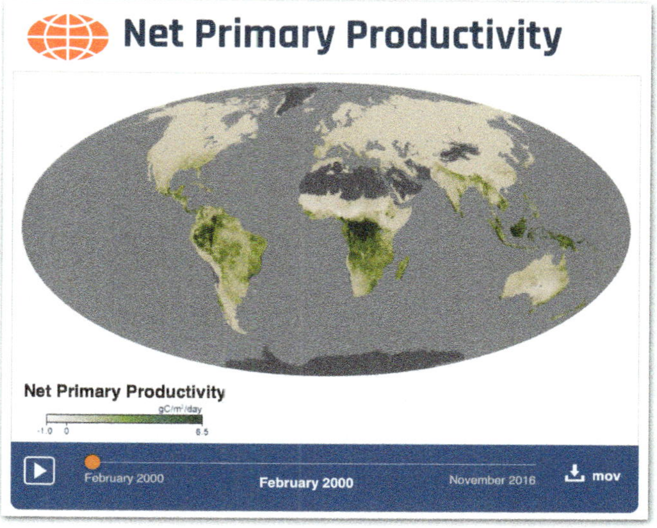

Seeing Geographically in Our National Parks

Look again at the photograph of the forest in Olympic National Park at the beginning of the chapter (p. 284). Does this forest appear to have relatively high or low species diversity? Why do you say so? Which environmental factors appear most significant in supporting this ecosystem? Do those factors correlate with the climate type you expect here (48° N, 124° W)?

Mastering Geography®

Looking for additional review and test prep materials? Visit the Study Area in **Mastering Geography** to enhance your geographic literacy, spatial reasoning skills, and understanding of this chapter's content by accessing a variety of resources, including Map Master interactive maps, geoscience animations, *Mobile Field Trips*, videos, *Project Condor* Quadcopter videos, *In the News* RSS feeds, flashcards, web links, self-study quizzes, and an eText version of *McKnight's Physical Geography*.

Seeing Geographically in Our National Parks

American Bison (*Bison bison*) at Wind Cave National Park, South Dakota. Describe the general topography and vegetation you see here. Does this appear to be a fairly humid or dry environment? What makes you say this?

Wind Cave National Park, South Dakota

Terrestrial Flora and Fauna

11

HAVE YOU EVER WONDERED why the wildlife in North America—such as here in Wind Cave National Park in South Dakota—is so different from that in Australia? Having looked at the fundamental patterns and processes in the biosphere in the previous chapter, we now turn more specifically to questions such as this. As we explore the geographical distribution of plants and animals, the relationships among the biosphere, atmosphere, hydrosphere, and lithosphere will become more conspicuous.

For example, we see that the relationship of North American and Australian animals and plants—as well as countless other biogeographical patterns around the world—isn't always a simple one. These particular patterns involve the movement of continents, evolution on a long-isolated landmass, similarities (and differences) in climate, and the deliberate efforts on the part of humans to move organisms from one part of the world to another.

We begin with the basics of plants and animals, as well as concepts that help us study groups of organisms. We then look at global biogeographical patterns.

As you study this chapter, think about these **Key**Questions:

- **How is a biome different from an ecosystem?**
- **How can plants adapt to very dry and very wet environments?**
- **How do animals adapt to very cold and very hot environments?**
- **What are the characteristics and distribution of Earth's major biomes?**
- **How have humans modified the natural distribution patterns of plants and animals?**

Ecosystems and Biomes

As we saw in Chapter 10, *biogeography* is the study of the distribution patterns of living organisms and how these patterns change over time. We saw also that in our search for organizing principles to help us comprehend the biosphere, the concepts of *ecosystem* and *biome* are of particular value. We now explore these concepts in greater detail.

Ecosystem: A Concept for All Scales

The term **ecosystem** is a contraction of the phrase *ecological system*. An ecosystem includes all organisms in a given area, but it is more than just a community of plants and animals existing together. An ecosystem encompasses the totality of interactions among all of the organisms, as well as interactions between organisms and the abiotic (nonliving) portion of the environment—the rocks, soil, water, sunlight, and atmosphere. The concept is built around the flow of energy and nutrients among the various components of the ecosystem, which are the essential determinants of how a biological community functions (Figure 11-1).

This functional ecosystem concept is very attractive as an organizing principle for the geographic study of the biosphere. We must approach it with caution, however, because of the various scales at which it can be applied. There is an almost infinite variety in the geographic size of ecosystems we might study. At one extreme of the scale might be a global ecosystem that encompasses the entire biosphere; at the other end of the scale might be the ecosystem of a fallen log, the underside of a rock, or even a drop of water.

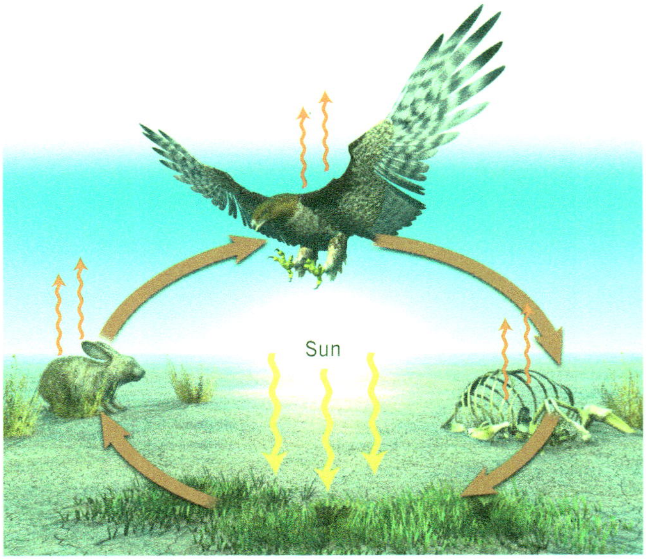

▲ Figure 11-1 The flow of energy in a simple ecosystem. Energy from the Sun is fixed by the grass during photosynthesis. The grass is then eaten by a rabbit, which is eaten by a hawk, which eventually dies. The energy originally contained in the photosynthetic products made by the grass passes through stages during which it is bound up in the molecules of the rabbit's and hawk's bodies. Ultimately that energy becomes heat lost from the live animals and from the decaying dead matter.

If we are going to identify and understand broad global patterns in the biosphere, we must focus on only ecosystems that can be recognized at a useful scale.

Biome: A Scale for All Biogeographers

Among terrestrial ecosystems, the type that provides the most appropriate scale for understanding world distribution patterns is the *biome*. A **biome** is defined as any large, recognizable assemblage of plants and animals in functional interaction with its environment—in other words, a collection of plants and animals over a large area that have broadly similar adaptations and relationships with the environment and climate. A biome is usually identified and named on the basis of its dominant vegetation. That vegetation normally constitutes the bulk of the *biomass* (the total weight of all organisms—plant and animal) in the biome, as well as being the most obvious and conspicuous visible component of the landscape.

There is no universally recognized classification system of the world's terrestrial biomes (we don't consider aquatic biomes in this book), but scholars commonly accept 10 major types:

1. Tropical rainforest
2. Tropical deciduous forest
3. Tropical scrub
4. Tropical savanna
5. Desert
6. Mediterranean woodland and shrub
7. Midlatitude grassland
8. Midlatitude deciduous forest
9. Boreal forest
10. Tundra

A biome comprises much more than merely the *plant association*—the grouping of plant species typically found together in a particular environment—for which it is named. Other kinds of vegetation usually grow among, under, and occasionally over the dominant plants. Diverse animal species also occupy the area. Often, as we saw in Chapter 10, significant and even predictable relationships exist between the biota (particularly the flora) of a biome and the associated climate and soil types.

Ecotones: On any map showing the major biome types of the world (such as Figure 11-22 later in this chapter), the regional boundaries are highly generalized. Biomes rarely occupy sharply defined areas in nature, no matter how sharp the demarcations may appear on a map. Normally the communities merge more or less imperceptibly with one another through **ecotones**: transition zones of competition in which the typical species of one biome intermingle with those of another (Figure 11-2).

LearningCheck 11-1 Explain the concept of a biome. (Answer on p. AK-3)

Before we describe the flora and fauna of the world's terrestrial biomes, we need to say something about plants and animals themselves.

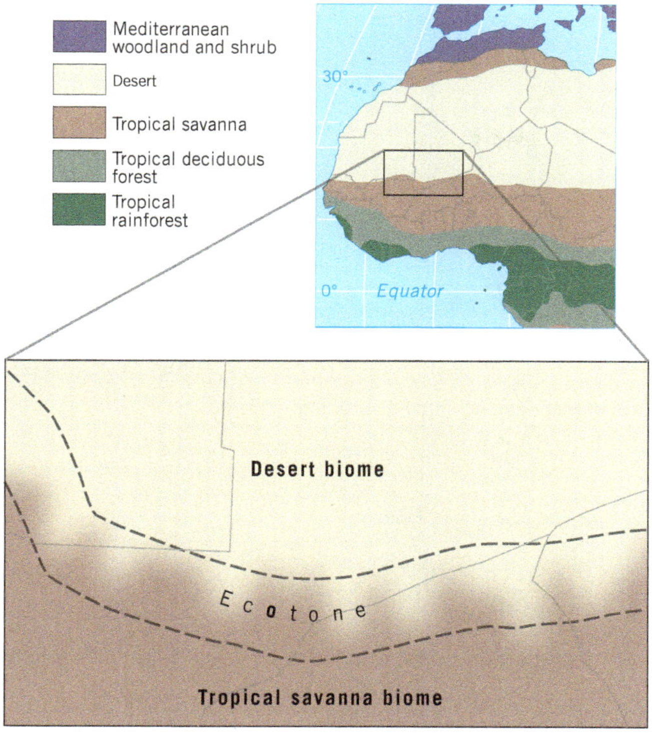

▲ **Figure 11-2** A hypothetical boundary between two biomes. The irregular boundary between the two shows much "interfingering." This transition zone is called an ecotone.

Terrestrial Flora

The natural vegetation of the land surfaces of Earth is important to geographers for three reasons. First, over much of the planet, terrestrial flora is the most significant visual component of the landscape. Second, vegetation is a sensitive indicator of other environmental attributes, reflecting subtle variations in sunlight, temperature, precipitation, evaporation, drainage, slope, soil conditions, and other natural factors. Third, vegetation often has a prominent and tangible influence on human settlement and activities.

Characteristics of Plants

Plant life varies remarkably in form, from microscopic algae to gigantic trees. Most plants, however, have common characteristics: roots to gather nutrients and moisture and anchor the plant; stems, branches, or both for support and for nutrient transportation from roots to leaves; leaves to absorb and convert solar energy for sustenance and to exchange gases and transpire water; and reproductive organs for regeneration. Despite the fragile appearance of many plants, most varieties are remarkably hardy. They survive, and often flourish, in the wettest, driest, hottest, coldest, and windiest places on Earth.

Reproductive Mechanisms: Plants can be broadly classified according to their reproductive strategies. Plants that survive seasonal fluctuations from year to year are called **perennials**, whereas those that perish during times of climatic stress (such as winter) but leave behind a reservoir of seeds to germinate during the next favorable period are called **annuals**.

Plants are further divided into two major categories based on reproductive mechanism: plants that reproduce through *spores* and those that reproduce through *seeds*. Spore producers include the *bryophytes* (such as true mosses, peat mosses, and liverworts) and the *pteridophytes* (such as ferns, horsetails, and club mosses). During much of geologic history, great forests of tree ferns, giant horsetails, and tall club mosses dominated continental vegetation, but they are less dominant today.

Seed producers also include two broad groups: *gymnosperms* and *angiosperms*. The more primitive are the **gymnosperms** ("naked seeds"), which carry their seeds in cones. When the cones open, the seeds fall out. For this reason, gymnosperms are sometimes called **conifers** (Figure 11-3). Gymnosperms were largely dominant in the geologic past; today, the only large surviving gymnosperms are cone-bearing trees such as pines and redwoods.

Much more common today are the **angiosperms** ("vessel seeds"), or *flowering plants*. Their seeds are encased in a protective body, such as a fruit, nut, or pod. The vast majority of modern trees, shrubs, grasses, crops, weeds, and garden flowers are angiosperms. Along with a few conifers angiosperms have dominated the vegetation of the planet for the last 50 or 60 million years.

LearningCheck 11-2 **Explain the difference between an annual plant and a perennial plant.**

▲ **Figure 11-3** Terminology used in describing plants. There are conspicuous differences between hardwood (angiosperm) and softwood (gymnosperm) trees. The most obvious difference is in general appearance.

Structural Features: Plants are also classified according to their stems and leaf patterns. *Woody plants*, such as trees and shrubs, have stems composed of hard fibrous material, whereas *herbaceous plants*, such as grasses, forbs (broadleaf herbaceous plants), and lichens, have soft stems.

An **evergreen tree** sheds its leaves on a sporadic or successive basis but always appears to be fully leaved. A **deciduous tree** experiences an annual period in which all leaves die and usually fall from the tree, due to either a cold season or a dry season.

Broadleaf trees have leaves that are flat and expansive, whereas **needleleaf trees** are adorned with thin slivers of tough, leathery, waxy needles rather than typical leaves. Almost all needleleaf trees are evergreen. The great majority of broadleaf trees are deciduous, except in the rainy tropics, where everything is evergreen.

Finally, the term *hardwood* is frequently used to refer to broadleaved and deciduous angiosperm trees; their wood has a relatively complex structure but is not always hard. *Softwoods* are gymnosperms. Nearly all are needleleaf and evergreen; their wood has a simple cellular structure but is not always soft.

LearningCheck 11-3 Contrast evergreen trees and deciduous trees.

Environmental Adaptations

Despite the hardiness of most plants, definite tolerance limits govern their survival, distribution, and dispersal. During hundreds of millions of years of development, plants have evolved a variety of protective mechanisms to shield against harsh environmental conditions and to expand their tolerance limits. Two prominent adaptations to environmental stress are associated with low water availability and high water availability.

Xerophytic Adaptations: Plants that are structurally adapted to withstand protracted dry conditions are called *xerophytic* (*xero* is Greek for "dry"; *phyt-* comes from *phuto-*, Greek for "plant"). **Xerophytic adaptations** include the following:

- Roots are modified to enable plants to seek widely for moisture. *Taproots* can extend to extraordinary depths to reach subterranean moisture, and numerous thin, hairlike rootlets penetrate tiny pore spaces in soil (Figure 11-4).
- Stems are modified into fleshy, spongy structures that can store moisture. Plants such as cacti with such fleshy stems are called *succulents*.
- Leaves are modified to decrease transpiration. A leaf surface may be hard and waxy to inhibit water loss or white and shiny to reflect insolation and thus reduce evaporation. Still more effective are tiny leaves or no leaves at all; in many dry-land shrubs, leaves have been replaced by spines, from which there is virtually no transpiration (Figure 11-5).
- Changes occurred in reproductive cycle. Many xerophytic plants lie dormant for years without perishing. When rain eventually arrives, these plants promptly initiate and pass through an entire annual cycle of germination, flowering, fruiting, and seed dispersal in only a few days, then lapse into dormancy again if the drought resumes.

▲ Figure 11-4 Desert plants have evolved various mechanisms for survival in an arid climate. (a) Some plants produce long taproots that penetrate deeply in search of the water table. (b) More common are plants that have no deep roots but rather have myriad small roots and rootlets that seek any near-surface moisture over a broad area.

Hygrophytic Adaptations: Plants with **hygrophytic adaptations** are suited to a wet terrestrial environment. A distinction is sometimes made between *hydrophytes*

▲ Figure 11-5 Cacti mostly or completely lack leaves but instead have spines from which there is no water loss through transpiration. This saguaro cactus is in Saguaro National Park, Arizona.

(a) Hydrophytes: lily pads

(b) Hygrophytes: bald cypress trees

▲ **Figure 11-6** Some types of plants flourish in a totally aqueous environment. (a) These lily pads virtually cover the lake surface in Pictured Rocks National Lakeshore, Michigan. (b) Many hygrophytic trees, such as these cypress in Big Thicket National Preserve, Texas, have wide, flaring trunks near the ground that provide firm footing in the wet environment.

(species living more or less permanently immersed in water, such as the water lilies in **Figure 11-6a**) and *hygrophytes* (moisture-loving plants that generally require frequent soakings with water, as do many ferns, mosses, and rushes). Both groups, however, are often called hygrophytes.

Hygrophytes are likely to have extensive root systems to anchor them in the soft ground, and hygrophytic trees often develop a widened, flaring trunk near the ground to provide better support (**Figure 11-6b**). Many hygrophytic plants that grow in standing or moving water have weak, pliable stems that can withstand the ebb and flow of currents rather than standing erect against them. The buoyancy of the water, rather than the stem, provides support for the plant.

LearningCheck 11-4 **How are xerophytic adaptations different from hygrophytic adaptations?**

Global Distribution of Plant Associations

A geographer attempting to recognize spatial groupings of plants faces some significant difficulties. Plant associations that are similar in appearance and in environmental relationships can occur in widely separated localities and are likely to contain different species. At the other extreme, very different plant associations are often detected within a very small area. When associations are portrayed on maps, therefore, their boundaries represent approximations in nearly all cases.

Another problem is that in many areas of the world, the natural vegetation has been completely removed or replaced through human interference. Forests have been cut, crops planted, pastures seeded, and urban areas paved. Over extensive areas of Earth's surface, therefore, *climax vegetation* (discussed in Chapter 10) is the exception rather than the rule. Most world maps that show natural vegetation ignore human interference and are actually maps of theoretical natural vegetation, in which the mapmaker makes assumptions about what the natural vegetation would be if it had not been modified by human activity.

The Major Plant Associations: For broad geographical purposes, we classify major plant associations largely based on the structure and appearance of the dominant plants (**Figure 11-7**):

- **Forests** consist of trees growing so close together that their individual leaf canopies generally overlap. Because trees must transport nutrients a relatively great distance from their roots to their leaves, forests typically require considerable precipitation throughout the growing season. Other plant associations can flourish where precipitation is relatively high, but they rarely become dominant because they are shaded out by trees. Thus, forests are likely to become the climax vegetation association in any area.
- **Woodlands** are tree-dominated plant associations in which the trees are spaced more widely apart than in forests and do not have interlacing canopies. Ground cover may be either dense or sparse, but it is not inhibited by lack of sunlight. Woodland environments are generally drier than forest environments.
- **Shrublands** are dominated by relatively short woody plants generally called *shrubs* or *bushes*. Shrubs take a variety of forms, but most have several stems branching near the ground and leafy foliage that begins very close to ground level. Trees and grasses may be interspersed with the shrubs but are less prominent in the landscape. Shrublands range widely but are generally restricted to semiarid or arid locales.
- **Grasslands** may contain scattered trees and shrubs, but the landscape is dominated by grasses and forbs. Prominent types of grassland include *savanna*, low-latitude grassland characterized by tall grasses; *prairie*, midlatitude grassland characterized by tall grasses; and *steppe*, midlatitude grassland characterized by short grasses and bunchgrasses. Grasslands are associated with semiarid and subhumid climates.

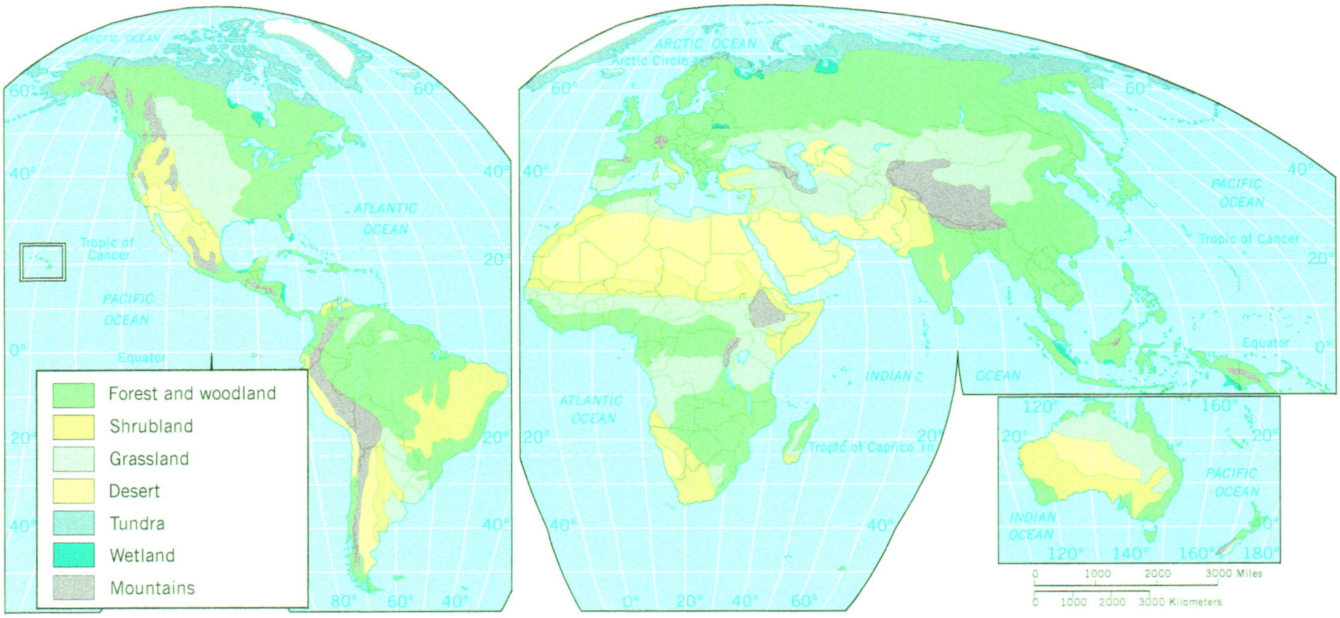

▲ **Figure 11-7** The major natural vegetation associations.

- **Deserts** are typified by widely scattered plants with much bare ground interspersed. As we saw in Chapter 8, *desert* is actually a climatic term, and desert areas may have a great variety of vegetation, including grasses, succulent herbs, shrubs, and straggly trees (**Figure 11-8**). Some extensive desert areas comprise loose sand, bare rock, or extensive gravel, with virtually no plant growth.
- **Tundra**, as we noted in Chapter 8, consists of a complex mix of very low plants, including grasses, forbs, dwarf shrubs, mosses, and lichens, but no trees. Tundra occurs in only the perennially cold climates of high latitudes or high altitudes.
- **Wetlands** have a much more limited geographic extent than other plant associations. They are characterized by shallow standing water all or most of the year, with vegetation rising above the water level. The most widely distributed wetlands are *swamps* (with trees as the dominant plant forms; see Figure 11-6b) and *marshes* (with grasses and other herbaceous plants dominant).

▶ **Figure 11-8** Some deserts contain an abundance of plants, but they are usually spindly and always xerophytic. This scene is from the Sonoran Desert near Phoenix, Arizona.

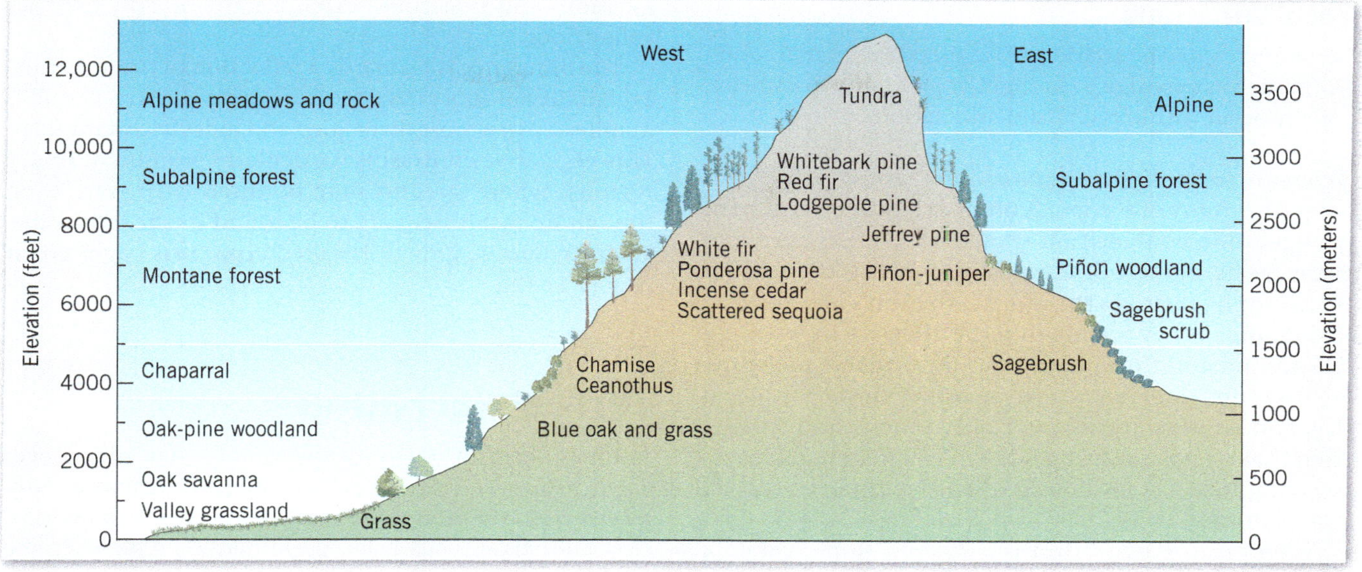

▲ **Figure 11-9** A west–east profile of California's Sierra Nevada, indicating the principal vegetation at different elevations on the western (wet) and eastern (dry) sides of the range. Characteristic plants within each vegetation zone are labeled along the mountain slope.

Vertical Zonation

As we first saw in Chapter 8, mountainous areas often have a distinct pattern of **vertical zonation** of climatic conditions and vegetation patterns (**Figure 11-9**). Significant changes in elevation over a short horizontal distance cause different plant associations to exist in relatively narrow zones on mountain slopes. This zonation is largely due to the effects of elevation on temperature and precipitation (**Figure 11-10**).

Elevation changes are the counterpart of latitude changes. In other words, to travel from sea level to the top of a tall tropical peak is roughly equivalent environmentally to a journey from the equator to the Arctic. This elevation-latitude relationship is shown most clearly by variations in the **tree line**—the elevation above which trees cannot survive, especially due to low summer temperatures and moisture availability. Near the equator, the tree line can be more than 4000 meters (13,000 feet), but at a latitude of 70° N, the tree line has dropped to sea level.

Interestingly, the latitude–tree line elevation relationship is different in the Northern and Southern Hemispheres. For reasons that are not completely understood, the tree line in the midlatitudes of the Southern Hemisphere tends to be much lower than at the same latitude in the Northern Hemisphere.

LearningCheck 11-5 **What is meant by vertical zonation? What determines the tree line for a location?**

◄ **Figure 11-10** Vertical zonation of vegetation patterns on the east slope of the Sierra Nevada in California. Sagebrush in the lowest elevations (about 1525 m; 5000 ft.) gives way to piñon woodland and pockets of subalpine forest in higher elevations; above the treeline, alpine tundra is found. The high peak in the center is Mount Whitney (4421 m; 14,505 ft.).

Local Variations

Variations caused by a variety of local environmental conditions, such as exposure to sunlight or proximity to a stream, usually occur within each plant association.

Exposure to Sunlight: The direction that a sloped surface faces (the *aspect* of a slope) is often a critical determinant of vegetation composition (Figure 11-11). Exposure has many facets, but one of the most pervasive is the angle at which sunlight strikes the slope. If the Sun's rays arrive at a high angle, they are much more effective in warming the ground and thus in evaporating available moisture. Such a sunny slope, also called an **adret slope**, is hot and dry, and its vegetation is not only sparser and smaller than that on adjacent slopes having a different exposure to sunlight, but is also likely to have a different species composition.

The opposite condition is a shaded slope, called a **ubac slope**. Sunlight strikes it at a low angle and is thus much less effective in heating and evaporating. This cooler condition produces more luxuriant, more varied vegetation.

Valley-Bottom Location: In mountainous areas where a river runs through a valley, the vegetation associations growing on either side of the river are significantly different from the association found upslope. This floral gradient is sometimes restricted to immediate streamside locations and sometimes extends more broadly over the valley floor.

The difference is primarily a reflection of the perennial availability of subsurface moisture near the stream and is manifested in a more diversified and more luxuriant flora. This vegetative contrast is particularly prominent in dry regions, where streams may be lined with trees even though no other trees are to be found in the landscape. Such streamside growth is called **riparian vegetation** (Figure 11-12).

Terrestrial Fauna

Animals occur in much greater variety than plants over Earth. However, in geographical studies animals usually receive less attention than plants for at least two reasons. First, animals are usually less prominent than plants in the landscape. Second, because animals are mobile, they can adjust to environmental variability more easily than plants, so environmental interrelationships are often less clearly demonstrated by animals than by plants.

Nonetheless, wildlife can be a prominent element of physical geography. In some places it is an important resource for humans, and in others a significant hindrance to human activity. Moreover, it is increasingly clear that animals are sometimes more sensitive indicators than plants of the health of a particular ecosystem.

(a) Adret and ubac slopes

(b) Vegetation differences on adret and ubac slopes

▲ Figure 11-11 A typical adret–ubac situation. (a) The noon Sun's rays strike the adret (sunny) slope at approximately a 90° angle, resulting in maximum heating. When the same rays strike the ubac (shaded) slope at approximately a 40° angle, the heating is spread over a large area and is therefore less intense. (b) Vegetation differences on south-facing slopes (left sides of ridges) and the cooler north-facing slopes (right sides of ridges) near the central California town of Parkfield.

◀ **Figure 11-12** Riparian vegetation is particularly prominent in dry lands. This stream in northern Queensland, Australia, would be inconspicuous if it were not for the trees growing along it.

Characteristics of Animals

Most people do not realize how diverse animal life-forms are. We commonly think of animals as being relatively large and conspicuous creatures. Actually, the term *animal* encompasses not only the larger, more complex forms, but also hundreds of thousands of species of smaller organisms that may be inconspicuous or even invisible to us.

Although the vast majority of animals are tiny, seemingly inconsequential animals sometimes play exaggerated roles in the biosphere—as carriers of disease or as hosts of parasites, sources of infection, or providers of scarce nutrients. For example, no geographic assessment of Africa can ignore the presence and distribution of the small tsetse fly (genus *Glossina*), which is responsible for the transmission of trypanosomiasis (sleeping sickness), a widespread and deadly disease for humans and livestock over much of the continent.

The variety of animal life is so great—such as the contrast between an enormous elephant and a tiny insect—that it is difficult to find many unifying characteristics. Animals have only two truly universal characteristics: animals are capable of self-generated movement, and they must eat plants or other animals for sustenance.

Kinds of Animals

An understanding of the principal kinds of animals is useful in the general study of physical geography.

Invertebrates: Animals without backbones are called **invertebrates**. More than 90 percent of all animal species are encompassed within this broad grouping, which includes worms, sponges, mollusks, various marine animals, and microscopic or near-microscopic creatures. Very prominent among invertebrates are the *arthropods*, a group that includes insects (Figure 11-13), spiders, centipedes, millipedes, and crustaceans (shellfish). With some 350,000 recognized species, beetles are the most numerous of animals, comprising about 40 percent of all insect species and more than one-fourth of all known animals.

Vertebrates: Vertebrates are animals that have a backbone (spine), which protects the main nerve (or spinal cord).

▲ **Figure 11-13** Insects, such as this silver spotted skipper moth in Kennesaw Mountain National Battlefield Park, Georgia, are arthropods, the largest group of invertebrate animals.

Vertebrates are divided into five principal groups:

1. *Fishes* are the only vertebrates that can breathe under water. (A few fish species are also capable of breathing in air.) Most fishes inhabit either freshwater or saltwater only, but some species (such as the bull shark, *Carcharhinus leucas*) can live in both environments. Several species, most notably salmon, spawn in freshwater streams but spend most of their lives in the ocean.
2. *Amphibians*, such as frogs and salamanders, are semiaquatic animals. When born they are fully aquatic and breathe through gills; as adults, they are air-breathers by means of lungs and through their glandular skin.
3. Most *reptiles* are totally land based. Ninety-five percent of all reptile species are either snakes or lizards. Most of the other species are turtles or crocodilians. Reptiles are cold-blooded.
4. *Birds* are believed to have evolved from reptiles; indeed, they have so many reptilian characteristics that they have been called "feathered reptiles." The more than 9000 species of birds all reproduce by means of eggs. Birds are so adaptable that some species can live almost anywhere on Earth's surface. However, they are warm-blooded.
5. *Mammals* are distinguished from all other animals in that only mammals produce milk with which they feed their young, and only mammals possess true hair. The great majority of all mammals are *placentals*, which means that their young grow and develop in the mother's body, nourished by the *placenta*, an organ that forms a link with the mother's bloodstream (**Figure 11-14**). A small group of mammals (about 334 species) are *marsupials*, whose females have pouches in which the very underdeveloped young live for several weeks or months after birth (**Figure 11-15a**). The most primitive of all mammals are the egg-laying *monotremes*, of which only two families and five species exist (**Figures 11-15b** and **11-15c**).

Environmental Adaptations

Evolution through eons of time has made it possible for animals to diverge remarkably in adjusting to different environments. Just about every existing environmental extreme has been met by evolutionary adaptations that make it feasible for some animal species to survive and even flourish. Recent climate changes, however, challenge the adaptability of many species, as you can see in the box *Snapshots from Our National Parks: Shenandoah, Badlands, and Denali—Changing Climate Affects Bird Populations*.

Physiological Adaptations: The majority of animal adaptations to environmental conditions have been physiological—that is, they are anatomical or metabolic changes. Among the most important adaptations are those associated with temperature.

Both mammals and birds are **endotherms** ("warm blooded"): their body temperature stays about the same under any climatic conditions. This adaptation enables species of mammals and birds to live in almost all parts of the world. Animals that cannot regulate their temperature internally, such as reptiles and amphibians, are *ectotherms* ("cold blooded"). This is why snakes and lizards tend to be sluggish in the morning, until they can warm up in the Sun. Some fish species are endotherms, but most are ectotherms.

Several hundred species of mammals have skin completely covered with a fine, soft, thick, hairy coat referred to

◀ **Figure 11-14** Most large land animals are placental mammals, such as this moose and her young calves in Denali National Park in Alaska.

(a) Marsupials: red kangaroos

(b) Monotreme: short-beaked echidna

(c) Monotreme: platypus

▲ **Figure 11-15** Marsupials and monotremes. (a) A red kangaroo joey (baby) peers out from its mother's pouch. As with other marsupials, baby kangaroos live for a time in their mother's pouch after birth. (b) Only two kinds of monotremes, or egg-laying mammals, exist: the echidna, which is found in Australia and New Guinea only, and (c) the duckbill platypus, which is primarily aquatic.

as *fur*. These mammals range in size from tiny mice and moles to the largest bears. Many fur-bearing species live in high-latitude or high-elevation locations, where winters are long and cold. A number of fur-bearing species live in aquatic environments; the remainder are widely scattered over the continents, occupying a considerable diversity of habitats. Many fur-bearing species, including those with the heaviest, thickest fur, live in regions where cold temperatures are common, but the climatic correlation in this case is partial because many other factors are involved.

In furred animals, ears are prime conduits for body heat, as they provide a relatively bare surface for its loss. Arctic foxes (Figure 11-16a) have unusually small ears, which minimizes heat loss; desert foxes (Figure 11-16b) possess remarkably large ears, which are a great advantage for radiating heat during the blistering desert summers.

The catalog of similar adaptations is almost endless: webbing between toes to make swimming easier; broad feet that will not sink in soft snow; increase in size and number of sweat glands to aid in evaporative cooling; and a host of others.

LearningCheck 11-6 Which major groups of animals are endothermic, and what advantages do endotherms have over ectotherms?

Behavioral Adaptations: An important advantage that animals have over plants, in terms of adjustment to environmental stress, is that animals can seek shelter from heat, cold, flood, or fire; travel far in search of relief from drought or famine; and shift from daytime (diurnal) to nighttime (nocturnal) activities to minimize water loss during hot seasons. Such techniques as *migration* (periodic movement from one region to another), *hibernation* (spending winter in a dormant condition), and *estivation* (spending a dry-hot period in an inactive state) are behavioral adaptations employed regularly by many species of animals.

Reproductive Adaptations: Harsh environmental conditions are particularly dangerous to newborns, so many species have evolved specialized reproductive cycles or have developed modified techniques of baby care. During lengthy periods of bad weather, for example, some species delay

Snapshots From Our National Parks

Shenandoah, Badlands, and Denali

Changing Climate Affects Bird Populations

▲ **Figure 11-A** Black-capped Chickadee (*Poecile atricapillus*).

Bird populations can be sensitive indicators of climate change. Many species have specific environmental requirements for feeding, nesting, and migration. Because birds are generally quite mobile, a population may respond quickly to local environmental changes. As climate changes, the birds found in our National Parks will likely also change.

Bird Research in National Parks: A 2018 study led by the National Park Service and the National Audubon Society concluded that one-quarter of the bird species seen in any given National Park could be different by the year 2050 if greenhouse gas emissions and climate change continue at their present rates. In some cases, the climate conditions for a given bird species may worsen; in other cases, the conditions may improve. For example, with warmer summers, some birds will need to follow the cooler temperatures into higher elevation areas or to higher latitudes.

By looking at the range of climatic and seasonal conditions that birds need for survival, the study mapped the changing suitability of National Parks for 500 different species of birds. One of the study's conclusions: with ongoing climate change and an increasingly urbanized American landscape, our National Parks will play an ever more important role as habitat for birds seeking new places to live, breed, or feed.

Many examples are cited in the study.

Shenandoah National Park In the forests of Shenandoah National Park, Virginia, warming summers may reduce the number of some birds such as the Black-capped Chickadee (**Figure 11-A**) and Cedar Waxwing. However, the warmer winters may allow new species to thrive here—birds such as the Orange-crowned Warbler and Common Yellowthroat may decide to winter in the Park.

Badlands National Park Badlands National Park in South Dakota has long been a desirable environment for many species of grassland birds that breed during the summer months. Higher temperatures by 2050 may lead to reductions in these birds, such as the Burrowing Owl (**Figure 11-B**) and the Mountain Bluebird. However, the changing climate may make the Park suitable for birds adapted to the arid conditions of the Great Plains to the south. If so, the Park may then see the arrival of birds like the Northern Bobwhite and the Mississippi Kite.

Denali National Park Like far northern areas in general, over the last half century Alaska's Denali National Park has warmed twice as fast as the rest of the country. As this warming continues, the Park may see the arrival of birds looking for cooler forests than they will find in their present range to the south. As many as 40 new bird species could find Denali National Park a suitable summer environment, including the Rose-breasted Grosbeak and Western Tanager (**Figure 11-C**).

Questions

1. Why are bird populations sensitive indicators of climate change?
2. Why are some bird population ranges shifting poleward in North America?

◄ **Figure 11-B** Burrowing Owl (*Athene cunicularia*).

▲ **Figure 11-C** Western Tanager (*Piranga ludoviciana*).

(a) Arctic fox

(b) Desert fox

▲ **Figure 11-16** (a) An Arctic fox near Cape Churchill in the Canadian province of Manitoba. Its tiny ears are an adaptation to conserve body heat in a cold environment. (b) Desert foxes have large ears, the better to radiate away body heat. This is a bat-eared fox in northern Kenya.

mating or postpone nest building. If fertilization has already taken place, some animal reproductive cycles are capable of almost indefinite delay, resulting in a protracted egg or larval stage or even total suspension of embryo development until the weather improves.

If the young have already been born, they sometimes remain longer than usual in the nest, den, or pouch, and the adults may feed them for a longer time. When good weather finally returns, some species are capable of hastened *estrus* (the period of heightened sexual receptivity by the female), nest building, den preparation, and so on, and more progeny than normal may be produced.

LearningCheck 11-7 **What advantage is there for an animal to hibernate during the winter?**

Examples of Animal Adaptations to Desert Life: The desert provides many examples of animal adaptations.

Often animals follow the rains in nomadic fashion. Many species demonstrate remarkable instincts for knowing where precipitation has occurred and will travel tens or even hundreds of kilometers to take advantage of locally improved conditions. Some African antelopes and Australian kangaroos have been observed to double in numbers almost overnight after a good rain, a feat impossible except by rapid, large-scale movement from one area to another.

To avoid excessive heat, most desert rodents and reptiles live in underground burrows, and ants and termites live in underground nests. Desert frogs, crayfish, and freshwater crabs often survive long dry spells by burying themselves. Most desert animals except birds, *ungulates* (hooved mammals), and many flying insects are almost completely nocturnal. A few species of rodents, most notably kangaroo rats (Figure 11-17a), can exist from birth to death without ever taking a drink, surviving exclusively on moisture ingested with their food.

Of all large animals, perhaps the *dromedary*, or one-humped camel (*Camelus dromedarius*), developed the most remarkable series of adjustments to the desert environment (Figure 11-17b). The upper lip of the dromedary is deeply

(a) Desert dweller: Ord kangaroo rat

(b) Desert dwellers: camels and handlers

▲ **Figure 11-17** Desert animals exhibit a number of remarkable adaptations. (a) The Ord kangaroo rat in Arizona can live its entire life without taking a drink, receiving all necessary moisture from its food. (b) Camels are ideal pack animals in arid environments, as here in Sudan.

cleft, with a groove that extends to each nostril, so any moisture expelled from the nostrils can be caught in the mouth rather than being wasted. The nostrils consist of horizontal slits that can be closed tightly to keep out blowing dust and sand. The eyes are set beneath shaggy brows, which help shield them from the Sun's glare; the eyes are further protected from blowing sand by a complex double set of eyelids. Dromedaries can tolerate extreme dehydration without their body temperatures rising to a fatal level; whereas humans and most other large animals have body temperatures that fluctuate by only 1°–2°C (2°–4°F) on hot days, the range for dromedaries is 7°C (12°F). The dromedary sweats little except during the very hottest hours, thereby conserving body water. When drinking water is available, they are capable of completely rehydrating quickly.

LearningCheck 11-8 **Describe and explain one kind of adaptation of desert animals.**

Competition among Animals

Competition among animals is even more intense than that among plants. Animals with similar dietary habits compete for food and occasionally for territory. Animals in the same area also sometimes compete for water. And animals of the same species often compete for territory and for mates. In addition to these ecological rivalries, animals are also involved in predator–prey relationships.

Competition among animals is a major part of the general struggle for existence that characterizes natural relationships in the biosphere. Individual animals are concerned with their own survival—and sometimes with that of their mates or their own young—in response to instincts. Maternal (and, much more rarely, paternal) instinct is by no means universal among animals, however. For the most part, animal survival is a matter of every creature for itself.

Cooperation among Animals

Despite the intensity of competition among animals, many kinds of animals also exhibit cooperative behaviors. Many animals live together in social groups of varying sizes, generally referred to as herds, flocks, or colonies (Figure 11-18). Fewer species show individual concern for their group, as represented in colonies of ants or prides of lions. Sometimes a social group encompasses several species in a communal relationship, such as zebras, wildebeest, and impalas living together on an East African savanna. Within such groupings, there may be a certain amount of cooperation among unrelated animals, but competition for both space and resources is likely to be prominent as well.

▲ **Figure 11-18** Seals and sea lions are among the most gregarious of mammals. This "hauling-out" beach at Año Nuevo State Park on the central coast of California is crowded with elephant seals.

Symbiosis: Symbiosis is the arrangement in which two dissimilar organisms live together. There are three principal forms:

1. *Mutualism* involves a mutually beneficial relationship between the two organisms, as exemplified by the tickbirds that are constant companions of many African ungulates. The birds remove ticks and other insects that infest the mammals' skin, and the birds have a readily available supply of food (Figure 11-19).
2. *Commensalism* involves two dissimilar organisms living together with one species benefiting without harming or helping the other, as represented by burrowing owls sharing the underground home of prairie dogs.
3. *Parasitism* involves one organism living on or in another, obtaining nourishment from the host, which is usually weakened and sometimes killed by the actions of the parasite. (Mistletoe is an example of a plant parasite of forest trees in North America and Europe.)

LearningCheck 11-9 **What is the difference between mutualism and parasitism?**

▲ **Figure 11-19** Hoofed animals and insectivorous birds often have a symbiotic relationship. In this scene from South Africa's Kruger National Park, two red-billed oxpeckers are riding on a zebra, whose back can provide tasty morsels.

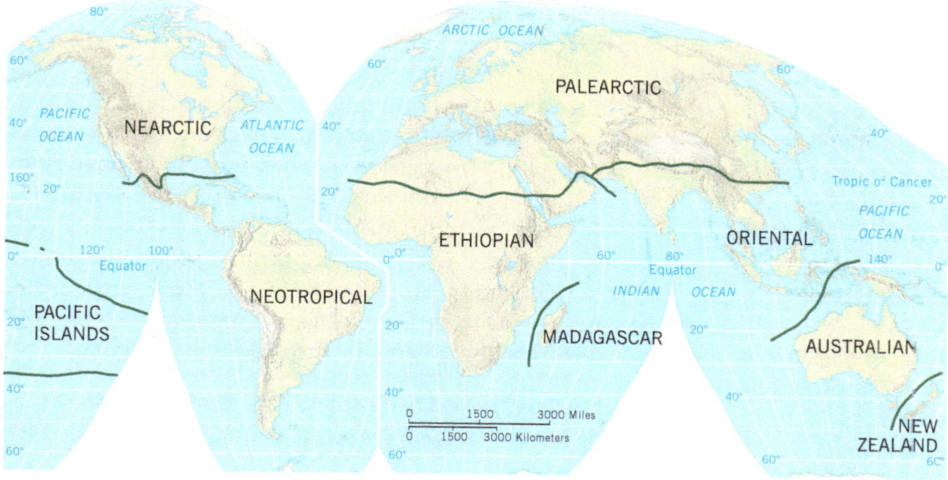

◄ **Figure 11-20** Zoogeographic regions of the world.

Zoogeographic Regions

As a result of animal mobility, the distribution of animals over the world is much more complex and irregular than that of plants. The richest faunal assemblages are found in the diverse, energy-rich environment of the humid tropics, and the dry lands and cold lands have the sparsest representations of both species and individuals.

The first significant attempt to define geographical regions based on the distribution of animals is credited to nineteenth-century British naturalist Alfred R. Wallace, whose scheme is based on the work of P. L. Sclater. Wallace originally recognized six **zoogeographic regions**, although today most biogeographers describe additional *biogeographical realms* or *ecozones* (**Figure 11-20**). Within each region we find a broadly similar collection of animals, thought to have evolved at least partially isolated from other regions by oceans, mountain ranges, or climate barriers.

- **Ethiopian Region:** The *Ethiopian Region* is primarily tropical or subtropical, separated from other regions by an oceanic barrier on three sides and a broad desert on the fourth. Its vertebrate fauna is the most diverse of all the zoogeographic regions and includes the greatest number of mammalian families.
- **Oriental Region:** The *Oriental Region* is similar to but less diverse than the Ethiopian Region, although it is rich in brilliantly colored birds and reptiles (particularly venomous snakes). Separated from the rest of Eurasia by mountains, it has some endemic groups but also shares some groups with the Palearctic and Australian regions.
- **Palearctic Region:** The *Palearctic Region* includes the rest of Eurasia and most of North Africa. Its fauna as a whole is much poorer than that of the Ethiopian and Oriental Regions. It contains only two minor families of endemic mammal (both rodents), and almost all of its birds belong to families that have a very wide distribution.
- **Nearctic Region:** The *Nearctic Region* consists of the nontropical portions of North America. Its faunal assemblage is relatively poor (other than reptiles and freshwater fishes) and is largely a transitional mixture of Palearctic and Neotropical groups. The similarities between Palearctic and Nearctic fauna have persuaded some zoologists to group them into a single superregion, the *Holarctic*, which had a land connection in the recent geologic past when glaciation lowered global sea level—the Bering land bridge across which considerable faunal dispersal took place (**Figure 11-21**).
- **Neotropical Region:** The *Neotropical Region* encompasses all of South America and the tropical portion of North America. Its fauna is rich and distinctive, reflecting both a variety of habitats and a considerable degree of isolation from other regions. Neotropical faunal evolution often followed a path different from that in other regions. It contains a larger number of endemic mammal families than any other region, and its bird fauna is exceedingly diverse.

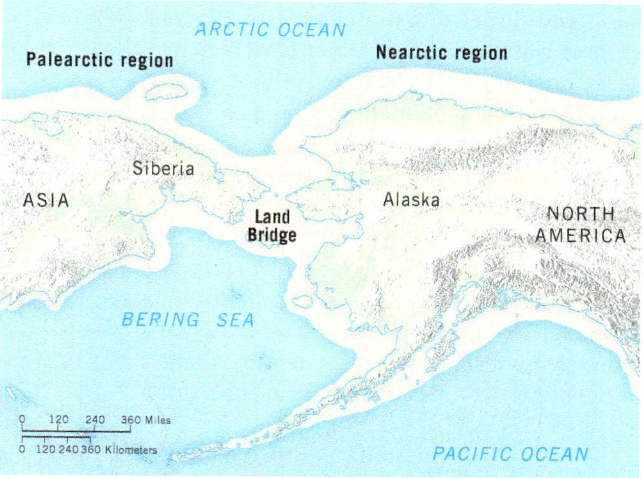

▲ **Figure 11-21** The Bering land bridge facilitated the interchange of animals between the Palearctic and Nearctic regions when sea level was lower during the Pleistocene glaciations. Most of the dispersal was from Asia (Palearctic) to North America (Nearctic).

- **Madagascar Region:** The *Madagascar Region*, restricted to the island of that name, has a very different fauna from that of nearby Africa. The Madagascan fauna is dominated by a relic assemblage of unusual forms in which primitive primates (lemurs) are notable.
- **New Zealand Region:** The *New Zealand Region* has a unique fauna dominated by birds, with a remarkable proportion of flightless types. It has almost no terrestrial vertebrates (no mammals; only a few reptiles and amphibians).
- **Pacific Islands Region:** The *Pacific Islands Region* includes a great many far-flung islands, mostly quite small. Its faunal assemblage is very limited.
- **Australian Region:** The *Australian Region* is restricted to the continent of Australia and some adjacent islands, particularly New Guinea. Its fauna is by far the most distinctive of any major region, due primarily to its lengthy isolation from other principal landmasses. The region has only nine families of terrestrial mammals, but eight of them are unique to it. The bird fauna is varied, and both pigeons and parrots reach their greatest diversity here. Freshwater fishes and amphibians are scarce. The Australian Region has a moderate number of reptiles, mostly snakes and lizards, including the largest lizard of all: the Komodo dragon. Within the region, there are many significant differences between the fauna of Australia and that of New Guinea.

Why is the Australian Region so different from the other zoogeographic regions? The answer becomes clearer when we discuss plate tectonics in Chapter 14, but for now, a quick explanation.

The Unique Biota of Australia: The unusual fauna and flora of Australia are largely the result of isolation. All continents were connected until about two hundred million years ago. When they separated, Australia became isolated from other continents, and evolution continued there with little genetic influence from the outside world.

More than 90 percent of all native trees in Australia are members of a single genus, *Eucalyptus*. Moreover, the eucalyptus (see Figure 10-14), of which there are more than 400 species, are native to no other continent. The shrubs and bushes of Australia are also dominated by a single genus, *Acacia*.

If the flora of Australia is unusual, the fauna is absolutely unique. This, too, is primarily the result of isolation. Since the time the continents separated, Australia has functioned for millions of years as a faunal "museum" where rare and vulnerable species were able to flourish in relative isolation from the competitive and predatory pressures that influenced animal evolution in other parts of the world.

Unlike all other continents, the Australian fauna is dominated by a single primitive mammalian group, the marsupials (see Figure 11-15a)—a group that has long disappeared from most other parts of the world (for example, North America has only the opossum). Australia also provides the only continental home for an even more primitive group, the monotremes (see Figures 11-15b and 11-15c). More remarkable, perhaps, is what is lacking: placental mammals, so notable elsewhere, are limited and inconspicuous in Australia.

LearningCheck 11-10 **Why does the Australian zoogeographic region have so many unique animal species?**

The Major Biomes

We now turn to a description of the major biomes of the world (Figure 11-22). Most biomes are named for their dominant vegetation association, but the biome concept also encompasses fauna as well as interrelationships with soil, climate, and topography.

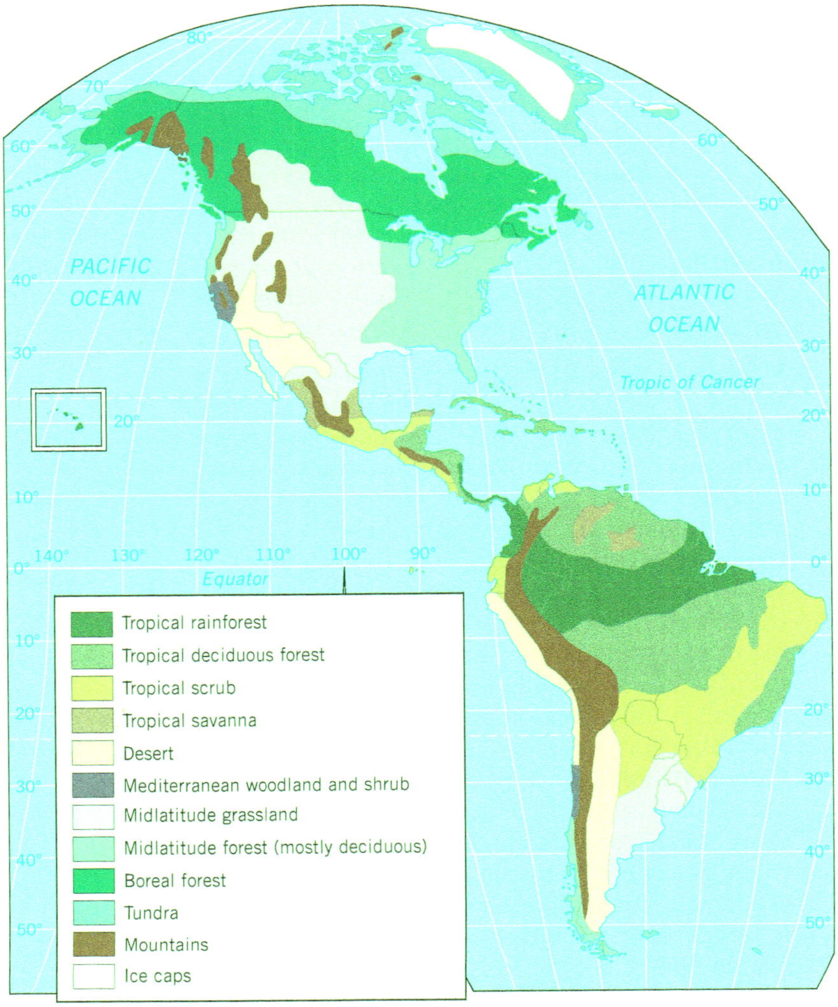

▲ Figure 11-22 Major biomes of the world.

Tropical Rainforest

As we saw in Chapter 10, the distribution of the **tropical rainforest** (or *selva*) biome is closely related to climate—high rainfall and relatively high temperatures. Thus, there is an obvious correlation with the location of tropical wet (Af) and some tropical monsoon (Am) climatic regions (Figure 11-23).

The rainforest is probably the most complex of all terrestrial ecosystems and the one with the greatest *biodiversity*. (Many species are found in an area, but often relatively few individuals of each species are present.) Most of the great variety of trees here are tall, high-crowned, broadleaf evergreen species (Figure 11-23a). The tropical rainforest has a layered structure; the second layer down from the top usually forms a complete canopy of interlaced branches that provides continuous shade to the forest floor. Bursting through the canopy to form the top layer are the forest giants—tall trees called *emergents* that often grow to great heights above the canopy level. Beneath the canopy is an erratic third understory layer of lower trees, palms and tree ferns, able to survive in the shade.

Undergrowth is relatively sparse in the tropical rainforest because sunlight is lacking. Only where there are gaps in the canopy, as alongside a river, does light reach the ground, resulting in the dense undergrowth associated with a jungle. *Epiphytes*, such as orchids and bromeliads, which get most of their moisture directly from the air, hang from or perch on tree trunks and branches. Vines often dangle from the arching limbs or wind their way up the trunks of trees.

The interior of the rainforest is a region of heavy shade, high humidity, windless air, continuous warmth, and an aroma of mold and decomposition. As plant litter accumulates on the forest floor, plant and animal decomposers act on it very rapidly. The upper layers of the forest are areas of high productivity, and nutrients are much more concentrated in the vegetation than in the soil. Indeed, most tropical soil is surprisingly infertile.

Rainforest fauna is largely *arboreal* (tree dwelling) because the principal food sources are in the canopy rather than on the ground (Figure 11-24). Large animals are generally scarce on the forest floor. The animal life of this biome is characterized by monkeys, arboreal rodents, birds, tree snakes and lizards, and multitudes of invertebrates.

Tropical Deciduous Forest

The distribution of the **tropical deciduous forest** biome with specific climatic types is irregular and fragmented

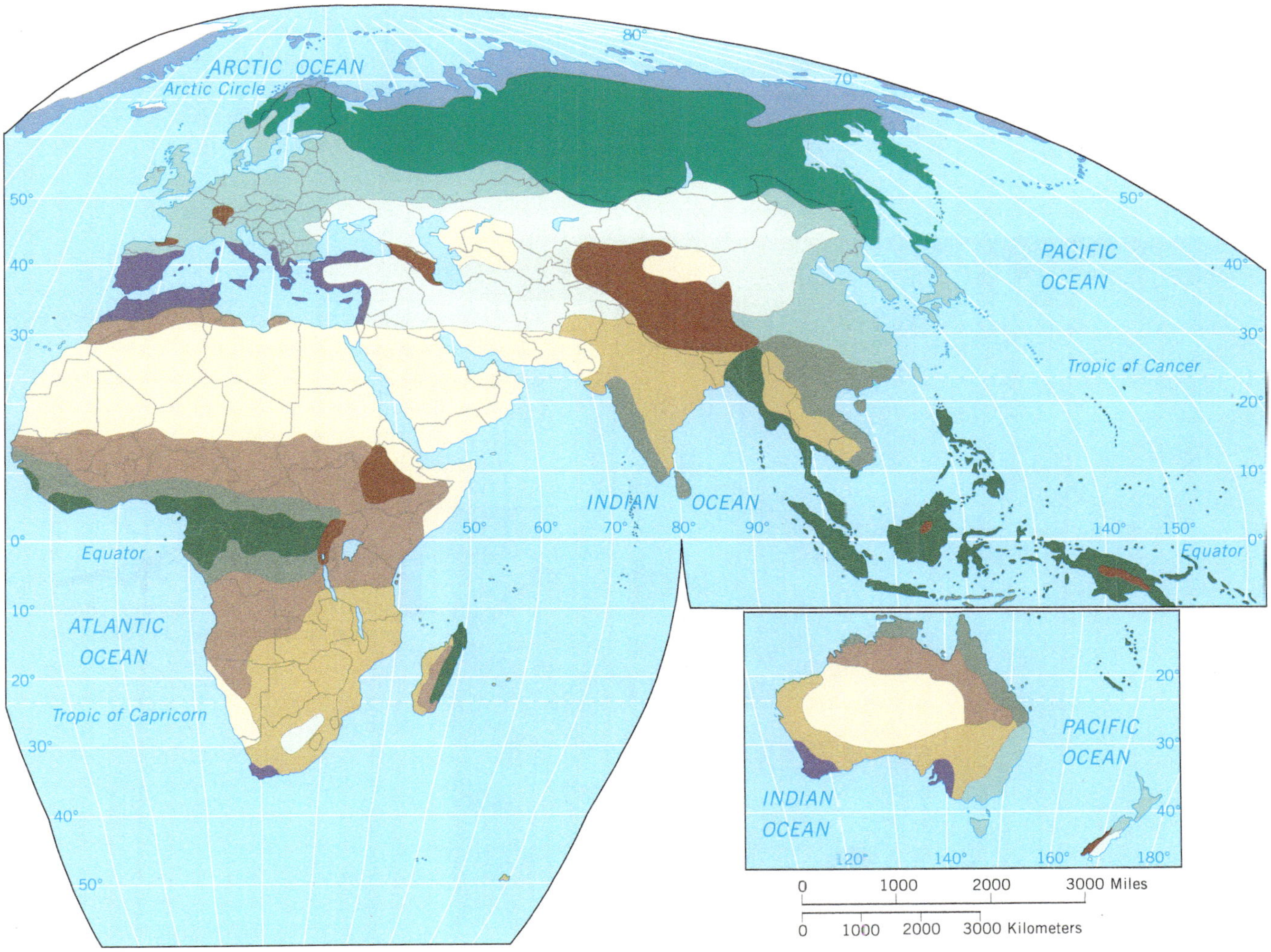

▲ Figure 11-22 (*Continued*)

(a) Tropical rainforest biome

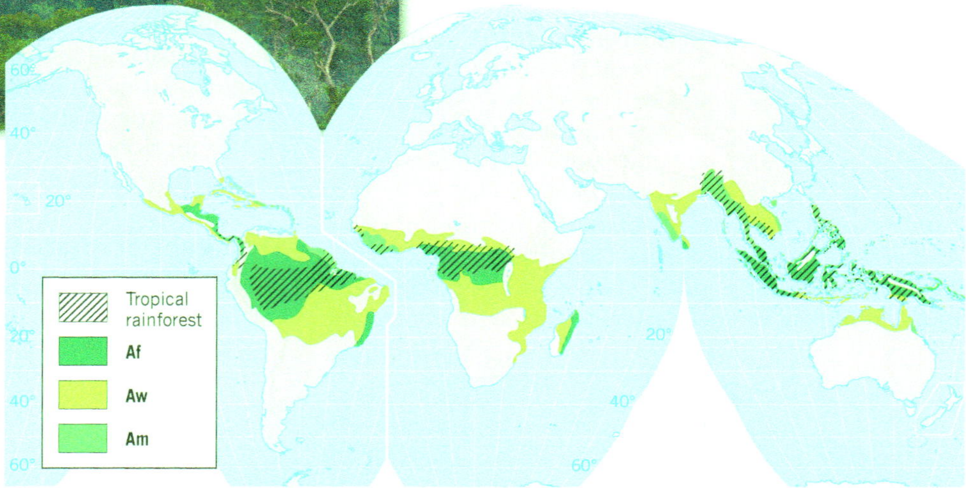

(b) Distribution of tropical rainforest

▲ **Figure 11-23** Tropical rainforest biome. (a) Mist rises from the forest canopy after heavy rains in the Amazon rainforest of Madidi National Park in Bolivia. (b) World distribution of tropical rainforest. The colors show the three A climates from Figure 8-3; the diagonal lines indicate the rainforest, which occurs in tropical wet (Af) and tropical monsoon (Am) climates.

▲ **Figure 11-24** A violet sabrewing hummingbird (*Campylopterus hemileucurus*) in the Monteverde cloud forest of Costa Rica.

Mobile Field Trip
Cloud Forest
https://media.pearsoncmg.com/ph/esm/esm_mcknight_physgeo_13/media/

(a) Tropical deciduous forest biome

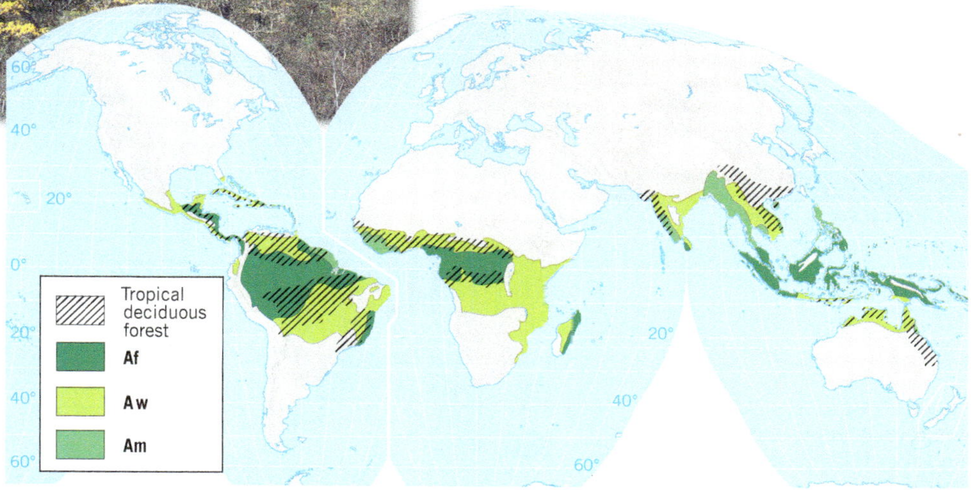

(b) Distribution of tropical deciduous forest

▲ **Figure 11-25** Tropical deciduous forest biome. (a) A tropical deciduous forest in Santa Rosa National Park, Costa Rica. (b) World distribution of tropical deciduous forest (diagonal lines) compared with distribution of A climates.

(**Figure 11-25**), although many tropical deciduous forests are found in the transition zone between tropical wet (Af) and tropical savanna (Aw) climates. Such regions have high temperatures all year but generally less—or more seasonal—rainfall patterns than in the tropical rainforest biome.

Tropical rainforest and tropical deciduous forest have structural similarity, but as a response to less precipitation, the deciduous forest canopy is less dense, the trees are somewhat shorter, and layers are fewer (Figure 11-25a). As a result of a pronounced dry period that lasts for several weeks or months, many of the trees shed their leaves at the same time, allowing light to reach the forest floor. This light produces an understory of lesser plants that often grow in such density as to produce classic jungle conditions. The diversity of tree species is not as great in this biome as in tropical rainforest, but there is a greater variety of shrubs and other smaller plants.

The faunal assemblage of the tropical deciduous forest is similar to that of the rainforest. Although there are more ground-level vertebrates than in the rainforest, arboreal species such as monkeys, birds, bats, and lizards are particularly conspicuous in both biomes.

LearningCheck 11-11 **Explain what it means to say that the tropical rainforest biome typically has high biodiversity.**

Tropical Scrub

The **tropical scrub** biome is widespread in drier portions of tropical savanna (Aw) climate and in some areas of subtropical steppe (BSh) climate (**Figure 11-26**). It is dominated by low-growing, scraggly trees and tall bushes, usually with an extensive understory of grasses (Figure 11-26a). The trees are 3–9 meters (10–30 feet) in height in variable density. Biodiversity is much less than in the tropical rainforest and tropical deciduous forest biomes; frequently just a few species comprise the bulk of the taller growth over vast areas. In the more tropical and wetter portions of the tropical scrub biome, most of the trees and shrubs are evergreen. Elsewhere, most species are deciduous. In some areas, a high proportion of the shrubs are thorny.

The fauna of tropical scrub regions is notably different from that of the tropical rainforest and tropical deciduous forest. There is a moderately rich assemblage of ground-dwelling mammals and reptiles and of birds and insects.

Tropical Savanna

There is an incomplete correlation between the distribution of the **tropical savanna** biome (or *tropical grassland* biome) and that of the tropical savanna (Aw) climate (**Figure 11-27**). The correlation tends to be most noticeable where seasonal rainfall contrasts are greatest, a condition particularly associated with the broad-scale annual shifting of the intertropical convergence zone (ITCZ).

Savanna lands are dominated by tall grasses (Figure 11-27a). Sometimes the grasses form a complete ground cover, but occasionally tufts of grass are dispersed over bare ground in a bunchgrass pattern. The name "savanna" usually refers to areas that are virtually without shrubs or trees, but that type of savanna is not the most common. In most cases, a wide scattering of both types dots the grass-covered terrain, and this mixture of plant forms is often referred to as *parkland* or *park savanna*.

(a) Tropical scrub biome

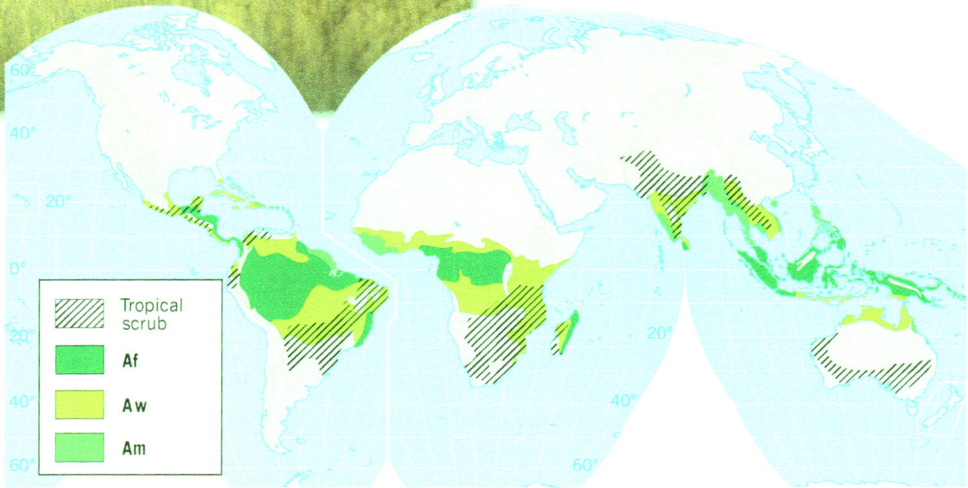

(b) Distribution of tropical scrub

▲ **Figure 11-26** Tropical scrub biome. (a) A tropical scrub scene in northern Namibia. (b) World distribution of tropical scrub (diagonal lines) compared with distribution of A climates. Much tropical scrub is found in Aw climates.

(a) Tropical savanna biome

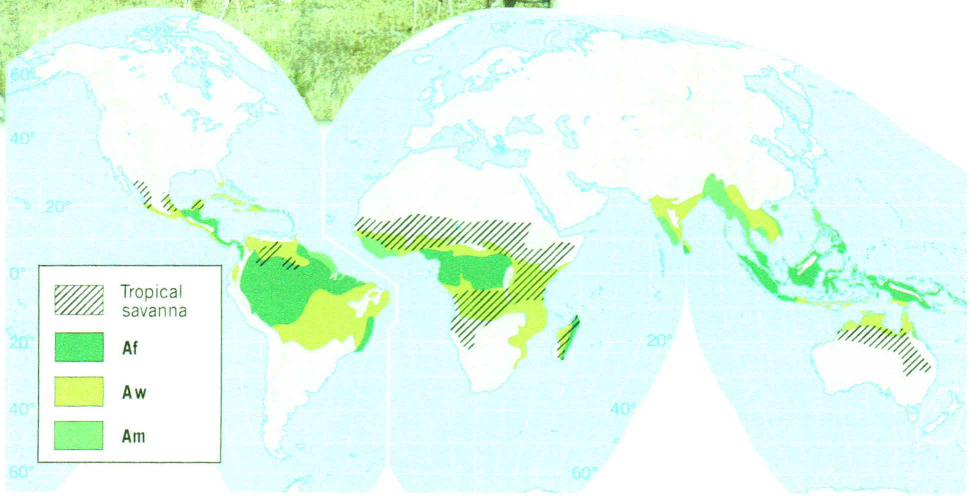

(b) Distribution of tropical savanna

▲ **Figure 11-27** Tropical savanna biome. (a) A mixed array of ungulates—zebra, kudu, and springbok—surrounds a waterhole in a typical savanna landscape in Namibia's Etosha National Park. (b) World distribution of tropical savanna (diagonal lines) compared with distribution of A climates.

In much of the savanna, the vegetation has been altered by humans. A considerable area of tropical deciduous forest and tropical scrub, and perhaps some tropical rainforest, has been converted to savanna over thousands of years through fires set by humans and through the grazing and browsing of domestic animals.

The savanna biome has a very pronounced seasonal rhythm. During the wet season, the grass grows tall, green, and luxuriant. During the dry season, the grass withers; before long, the above-ground portion is dead and brown. At this time, many of the trees and shrubs shed their leaves. The third "season" is the time of wildfires. The accumulation of dry grass provides abundant fuel, and most parts of the savannas experience natural burning every year or so. The recurrent fires burn away the unpalatable portion of the grass without causing significant damage to shrubs and trees. When the rains of the next wet season arrive, the grasses spring into growth with renewed vigor.

Savanna fauna varies from continent to continent. The African savannas are the premier "big game" lands of the world, with an unmatched richness of large animals, particularly ungulates and carnivores, but also including a remarkable diversity of other fauna. The Latin American savannas, conversely, have only a sparse population of large wildlife, and Eurasian and Australian areas are intermediate between these two extremes.

LearningCheck 11-12 **Describe the typical climate in regions of the tropical savanna biome.**

Desert

In previous chapters, we noted a general decrease in precipitation as distance from the equator in the low latitudes increases. This progression is matched by a gradation from the tropical rainforest biome of the equator to the **desert** biome of the subtropics. The desert biome also occurs extensively in midlatitude locations in Eurasia, North America, and South America with a fairly close correlation to subtropical desert (BWh) and midlatitude desert (BWk) climates (**Figure 11-28**).

Desert vegetation is surprisingly variable (Figure 11-28a). It consists largely of xerophytic plants, such as succulents and other drought-resisting plants with structural modifications that allow them to conserve moisture, and drought-evading plants capable of hasty reproduction during brief rainy times. The plant cover is usually sparse. Typically the plants are shrubs, which occur in considerable variety, each with its own mechanism to combat the stress of limited moisture. Succulents are common in the drier parts of most

(a) Desert biome

(b) Distribution of desert

▲ **Figure 11-28** Desert biome. (a) Natural vegetation typically is spindly and thorny in deserts, as in this scene from Joshua Tree National Park in California. (b) World distribution of desert (diagonal lines) compared with distribution of desert climates.

▲ **Figure 11-29** This desert collared lizard is sunning itself in Grand Canyon National Park.

desert areas, and many desert plants have tiny leaves or no leaves at all as a moisture-conserving strategy. Grasses and other herbaceous plants are widespread but sparse. Despite the dryness, trees can be found sporadically in the desert, especially in Australia.

Animal life is inconspicuous in most desert areas, yet most deserts have a moderately diverse faunal assemblage but with a limited variety of large mammals. A large proportion of desert animals avoid the principal periods of heat (daylight in general and the hot season in particular) by resting in burrows or crevices during the day and prowling at night (Figure 11-29).

Generally speaking, life in the desert biome is characterized by an appearance of stillness. In favorable times and in favored places (around water holes and oases), however, biotic activities abound, and sometimes the total biomass is of remarkable proportions. Favorable times are at night, and particularly after rains. For example, a heavy rain might trigger the germination of wildflower seeds that had remained dormant for decades.

Mediterranean Woodland and Shrub

The **mediterranean woodland and shrub** biome is found in six widely scattered and relatively small areas of the midlatitudes (Figure 11-30), all with the pronounced dry summer–wet winter precipitation typical of mediterranean (Cs) climates. In this biome, the dominant vegetation associations are physically very similar but taxonomically quite varied. The biome is dominated mostly by a dense growth of woody shrubs known in North America as a *chaparral* (Figure 11-30a); it has other names elsewhere. Chaparral includes many species of *sclerophyllous* plants, adapted to the prominent summer dry season by the presence of small, hard leaves that inhibit moisture loss. A second significant plant association of mediterranean regions is an open grassy woodland, in which the ground is almost completely grass covered but has a considerable scattering of trees (Figures 11-30b and 11-30c).

The plant species vary by region. Oaks of various kinds are by far the most significant genus in the Northern Hemisphere mediterranean lands, sometimes occurring as prominent medium-sized trees but also appearing as a more stunted, shrubby growth. In all areas, the trees and shrubs are primarily broadleaf evergreens. Their leaves are mostly small and have a leathery texture or waxy coating, which inhibits water loss during the long dry season. Moreover, most plants have deep roots. Summer is virtually rainless in mediterranean climates, so summer fires are relatively common. Many of the plants are adapted to rapid recovery after wildfires. Indeed, as noted in Chapter 10, some species have seeds that are released for germination only after the heat of a fire has caused their seedpods to open. Winter floods sometimes follow summer fires, as slopes left unprotected by the burning-away of grass and lower shrubs are susceptible to abrupt erosive runoff if the winter rains arrive before the vegetation can resprout.

The fauna of this biome is not particularly distinctive. Seed-eating, burrowing rodents are common, as are some bird and reptile groups. Animals in this biome and adjacent ones generally overlap.

LearningCheck 11-13 Which climate characteristic is most closely associated with the mediterranean woodland and shrub biome?

Midlatitude Grassland

Vast **midlatitude grasslands** occur widely in North America and Eurasia (Figure 11-31). In the Northern Hemisphere, this biome and the steppe (BSh and BSk) climatic type are well coordinated spatially. The smaller Southern Hemisphere areas (mostly the *pampa* of Argentina and the *veldt* of South Africa) have less distinct climatic correlations.

In the wetter areas of a grassland biome, the grasses grow tall, and the term *prairie* is often applied in North America (Figure 11-31a). In drier regions, the grasses are shorter; such growth is often referred to as *steppe*. Sometimes a continuous ground cover is missing, and the grasses grow in discrete tufts as bunchgrass or tussock grass.

Most of the grass species are perennials, lying dormant during the winter and sprouting anew the following summer. Trees are mostly restricted to riparian locations, whereas shrubs and bushes occur sporadically on rocky sites. Grass fires are fairly common in summer, which helps explain the relative scarcity of shrubs. The woody plants cannot tolerate fires and can generally survive only on dry slopes where there is little grass cover to fuel a fire.

Grasslands provide extensive pastures for grazing animals. Before encroachment by humans drastically changed population sizes, the grassland fauna included large numbers of relatively few species, such as the American bison in North America (see the photo at the beginning of this chapter). The larger herbivores were often migratory prior to human settlement. Many of the smaller animals spend time underground, where they find some protection from heat, cold, and fire.

(a) Mediterranean woodland and shrub biome: chaparral

(b) Mediterranean woodland and shrub biome, winter

(c) Mediterranean woodland and shrub biome, summer

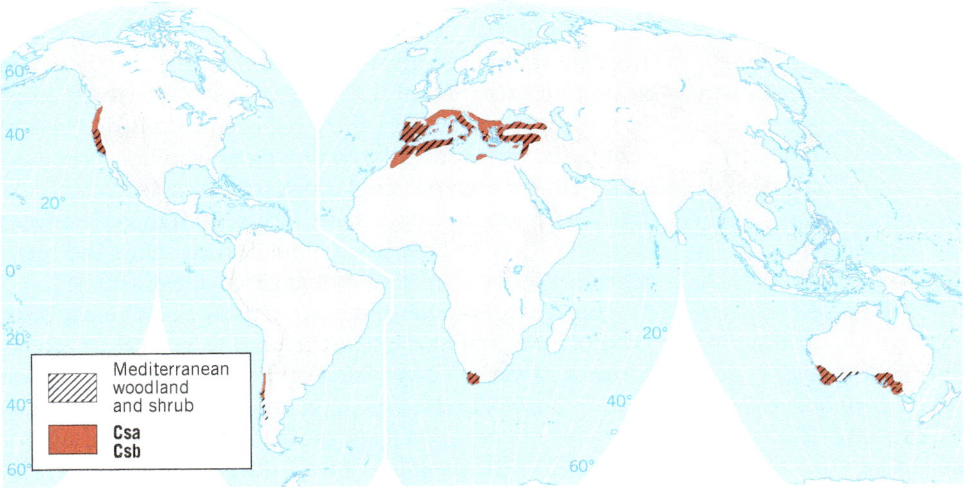

(d) Distribution of mediterranean woodland and shrub

▲ **Figure 11-30** Mediterranean woodland and shrub biome. (a) Dense mediterranean shrub biome chaparral vegetation in the San Gabriel Mountains of California. (b) Open mediterranean grassy woodland vegetation in the Mount Lofty Ranges of South Australia. Winter is moist, mild, and green, while (c) early summer is hot, dry, and brown; this is followed by the fire season. (d) World distribution of mediterranean woodland and shrub (diagonal lines) compared with distribution of Cs climates.

Midlatitude Deciduous Forest

Extensive areas on all Northern Hemisphere continents, as well as more limited tracts in the Southern Hemisphere, were originally covered with a forest of largely broadleaf deciduous trees (**Figure 11-32**). Except in hilly country, a large proportion of this **midlatitude deciduous forest** has been cleared for agriculture and other types of human use, so very little of the original natural vegetation remains.

The forest is characterized by a fairly dense growth of tall broadleaf trees with interwoven branches that provide a complete canopy in summer. Smaller trees and shrubs exist at lower levels, but for the most part, the forest floor is relatively barren of undergrowth. In winter, the appearance of the forest changes dramatically, owing to the seasonal fall of leaves (Figure 11-32a).

Tree species vary considerably from region to region, although most are broadleaf and deciduous. The principal exception is in eastern Australia, where the forest is composed almost entirely of varieties of eucalyptus, which are broadleaf evergreens. Northern Hemisphere regions have a

(a) Midlattitude grassland biome

(b) Distribution of midlatitude grassland

▲ **Figure 11-31** Midlatitude grassland biome. (a) The open grassland of Tallgrass Prairie National Preserve, Kansas. (b) World distribution of midlatitude grassland (diagonal lines) compared with distribution of BS climates.

northward gradational mixture with needleleaf evergreen species. In an unusual situation in the southeastern United States, extensive stands of pines (needleleaf evergreens) rather than deciduous species occupy most of the well-drained sites above the valley bottoms. In the Pacific Northwest of the United States, the forest association is primarily evergreen coniferous rather than broadleaf deciduous.

This biome generally has the richest assemblage of fauna in the midlatitudes. It has (or had) a considerable variety of birds and mammals, and in some areas reptiles and amphibians are well represented. Summer brings a diverse and active population of insects and other arthropods. All animal life is less numerous (partly due to migrations and hibernation) and less conspicuous in winter.

Boreal Forest

One of the most extensive biomes is the **boreal forest**, sometimes called *taiga* after the Russian word for the northern fringe of the boreal forest climate in that country. The boreal forest occupies a vast expanse of northern North America and Eurasia (**Figure 11-33**). The correlation between the location of the boreal forest biome and the subarctic (Dfc) climatic type is very close, with a similar correlation between the locations of the tundra climate and the tundra biome.

This great northern forest contains perhaps the simplest assemblage of plants of any biome (Figure 11-33a). Most of the trees are conifers, nearly all needleleaf evergreens, with the important exception of the tamarack, or larch, which drops its needles in winter. The variety of species is limited to mostly pines, firs, and spruces extending broadly in homogeneous stands. In some places, the coniferous cover is interrupted by areas of deciduous trees. These deciduous stands are also of limited variety (mostly birch, poplar, and aspen) and often represent a seral (intermediate) stage following a forest fire.

The trees grow taller and more densely near the southern margins of this biome, where the summer growing season is longer and warmer. Near the northern margins, the trees are spindly, short, and more openly spaced. Undergrowth is normally not dense beneath the forest canopy, but a layer of deciduous shrubs sometimes grows in profusion. The ground is usually covered with a complete growth of mosses and lichens, with some grasses in the south and a considerable accumulation of decaying needles overall.

Poor drainage is typical in summer, due partially to permanently frozen subsoil (*permafrost*), which prevents downward percolation of water, and partially to the disrupted drainage left by glaciers during the Pleistocene ice age. (Permafrost was discussed in Chapter 9.) Thus, bogs and swamps are numerous, and the ground is generally spongy in summer. During the long winters, all is frozen.

Harsh climate, floral homogeneity, and slow plant growth produce only a limited food supply for animals. Faunal species diversity is limited, although the number of individuals of

(a) Midlatitude forest biome

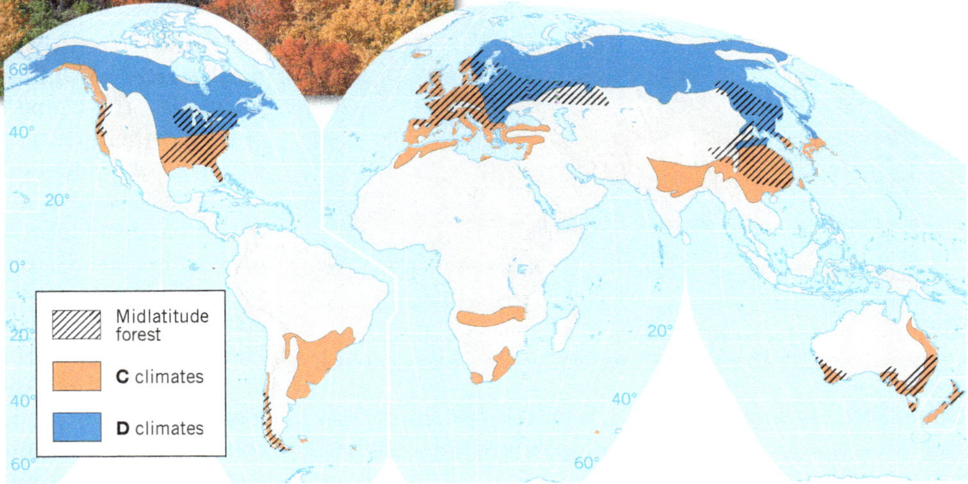

(b) Distribution of midlatitude forest

▲ **Figure 11-32** Midlatitude deciduous forest biome. (a) A midlatitude forest scene in the mountains in Hokkaido, Japan. As is common in some midlatitude locations, this forest consists of both deciduous and evergreen trees. (b) World distribution of midlatitude forest (diagonal lines) before human activities made significant changes in this biome. The colors indicate distribution of C and D climates.

(a) Boreal forest biome

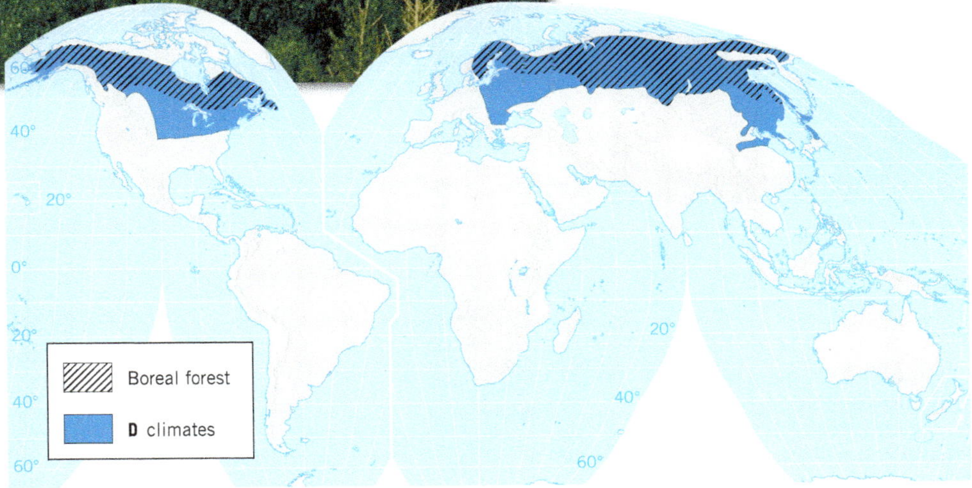

(b) Distribution of boreal forest

▲ **Figure 11-33** Boreal forest biome. (a) The boreal forest contains trees that generally are short, close growing, and of uniform species composition. This spruce forest surrounds Shady Lake near Prince Albert in Saskatchewan, Canada. (b) World distribution of boreal forest (diagonal lines) compared with distribution of D climates.

▲ **Figure 11-34** Bears live in diverse habitats, but mostly in cold climates. This grizzly bear inhabits Katmai National Park in Alaska.

some species is astounding. With relatively few animal species in such a vast biome, populations sometimes fluctuate enormously in the span of only a year or so. Mammals are represented prominently by species that have been traditionally hunted for their fur (**Figure 11-34**) and by a few species of ungulates. Birds are numerous and fairly diverse in summer, but nearly all migrate to milder latitudes in winter. Insects are absent in winter but superabundant during the brief summer.

LearningCheck 11-14 How does the biodiversity of the boreal forest biome differ from that of the tropical rainforest biome?

Tundra

The **tundra** is essentially a cold, arid grassland in which moisture is scarce and summers are so short and cool that trees cannot survive. This biome is distributed along the northern edge of the Northern Hemisphere continents, correlating closely to the distribution of tundra (ET) climate (**Figure 11-35**). The plant cover consists of a considerable mixture of species, many of them in dwarf forms (Figure 11-35a). Included are grasses, mosses, lichens, flowering herbs, and a scattering of low shrubs. These plants often occur in a dense, ground-hugging arrangement, although some places have more sporadic cover. The plants complete their annual cycles hastily during the brief summer, when the ground is often moist and waterlogged because of inadequate surface drainage—often as a consequence of permafrost just below the surface.

Animal life is dominated by birds and insects during the summer. Extraordinary numbers of birds flock to the tundra for summer nesting, migrating southward as winter

(a) Tundra biome

(b) Distribution of tundra

▲ **Figure 11-35** Tundra biome. (a) Tundra vegetation, colored in the brilliant reds of autumn, in Denali National Park, Alaska. (b) World distribution of tundra (diagonal lines) compared with distribution of E climates.

◀ **Figure 11-36** Alpine tundra vegetation high in the Rocky Mountains in Rocky Mountain National Park, Colorado.

approaches. Mosquitoes, flies, and other insects proliferate astoundingly during the short warm season, laying eggs that can survive the bitter winter. Other forms of animal life are scarcer—a few species of mammals and freshwater fishes but almost no reptiles or amphibians.

Alpine Tundra: An alpine version of tundra is found in many high-elevation areas. Many mountain areas above the timberline exhibit areas with a sparse cover of vegetation consisting of herbaceous plants, grasses, and low shrubs (**Figure 11-36**).

Human Modification of the Biosphere

Thus far in our discussion of the biosphere, most of our attention has been focused on "natural" conditions—that is, events and processes that take place in nature without the aid or interference of human activities. Such natural processes have gone on for millennia, and their effects on floral and faunal distribution patterns have normally been very slow and gradual. The pristine environment, uninfluenced by humankind, experiences its share of abrupt and dramatic events, such as fires and floods, but environmental changes generally proceed at a gradual pace. When humans appear, however, the tempo can change dramatically.

People are capable of exerting extraordinary influences on the distribution of plants and animals. Not only is the magnitude of the changes likely to be great, but also the speed with which they occur is sometimes exceedingly rapid. In broadest perspective, humankind exerts three types of direct influences on biotic distributions: physical removal of organisms, habitat modification, and the introduction of exotic species—all of which can lead to a loss of biodiversity.

Physical Removal of Organisms

As human population increases and spreads over the globe, there is often a wholesale removal of native plants and animals to make way for the severely modified landscape that is thought necessary for civilization. The natural plant and animal inhabitants are cut down, plowed up, paved over, burned, poisoned, shot, trapped, or otherwise eradicated in actions that have far-reaching effects on overall distribution patterns.

Habitat Modification

Habitat modification is another activity in which humankind excels. Farming changes the soil and vegetative environment; the construction of roads and buildings removes native vegetation, disrupts wildlife migration routes, and alters patterns of water runoff and infiltration; the grazing of livestock can change both the quantity and composition of vegetation (**Figure 11-37**); the

Video
Climate, Crops, and Bees
https://media.pearsoncmg.com/ph/esm/esm_mcknight_physgeo_13/media/

▼ **Figure 11-37** An overgrazed range in the Colorado high country illustrates human-induced habitat modification caused by overstocking of livestock in the area to the right of the fence.

introduction of impurities of various kinds degrades the atmosphere; the waters of the planet are impounded, diverted, and polluted. All such deeds influence the native plants and animals in the affected areas.

Tropical Rainforest Removal: Among the most dramatic recent human-initiated changes to global habitat is the removal of vast areas of tropical rainforest. Throughout much of history, most rainforests of the world were only modestly populated and as a consequence were affected by human activities in limited ways. By the twentieth century, however, rainforests were being exploited and devastated at an accelerating pace, and over the last 40 years, tropical deforestation became one of Earth's most serious environmental problems.

The total extent of forest lost and the exact rates of deforestation around the world—in both the tropics and temperate regions—are not precisely known. The cloud cover often found over the tropics complicates measurements by satellites, and data released by countries can vary in reliability.

Recent estimates suggest that about 10 percent of the world's tropical forest cover has been lost since the year 2000. Researchers at the University of Maryland estimated that 11.9 million hectares of tropical forest, an area about the size of Switzerland, were lost in 2019 alone, and of that total, 3.8 million hectares were primary tropical rainforest (Figure 11-38). See the box *Global Environmental Change: Rainforest Loss in the Amazon.*

In terms of the total percentage of rainforest cleared in each country, the picture is somewhat different. The United Nations Food and Agriculture Organization (FAO) reported in 2019 that in the tiny African island country of Comoros more than half of the land has been cleared and is now degraded. In 2017, the FAO reported that Burundi in central equatorial Africa had lost nearly all of its natural forest cover.

In South and Southeast Asia, where commercial exploration, especially for teak and mahogany, is important, about 45 percent of the original forest no longer exists. Approximately 40 percent of Latin America's rainforest has been cleared. Much of the very rapid deforestation in Central America has been due to expanded cattle ranching. Deforestation of the Amazon region as a percentage of the total area of rainforest has been moderate (perhaps 20 percent of the total has been cleared).

Rainforest Species Loss: As the forest goes, so goes its habitability for both indigenous peoples and native animal life. In the mid-1980s, it was calculated that tropical deforestation was responsible for the extermination of one species per day; by the mid-1990s, it was estimated that the rate was two species per hour. Moreover, loss of the forests contributes to accelerated soil erosion, drought, flooding, water quality degradation, declining agricultural productivity, and greater poverty for rural inhabitants. In addition, atmospheric carbon dioxide continues to be increased because burning trees as a way of clearing forest releases carbon to the air.

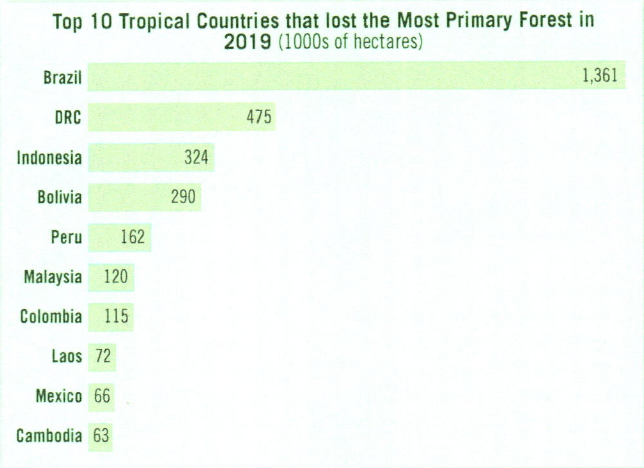

▲ Figure 11-38 Estimated loss of primary tropical rainforest (in hectares) in 2019.

The irony of tropical deforestation is that the anticipated economic benefits are usually temporary. Much of the forest clearing is in response to the social pressure of overcrowding and poverty in societies where many people are landless. The governments open "new lands" for settlement in the rainforest. The settlers clear the land for crop or livestock. The result is almost always an initial nutrient pulse of high soil productivity, followed in only two or three years by a pronounced fertility decline as the nutrients are quickly leached and cropped out of the soil, weed species rapidly invade, and erosion becomes rampant (Figure 11-39). Ongoing commercial agriculture can generally be expected only

▼ Figure 11-39 When rainforests are cleared for agriculture, the results are often small yields and accelerated soil erosion. This scene is from central Thailand.

Global Environmental Change

Rainforest Loss in the Amazon

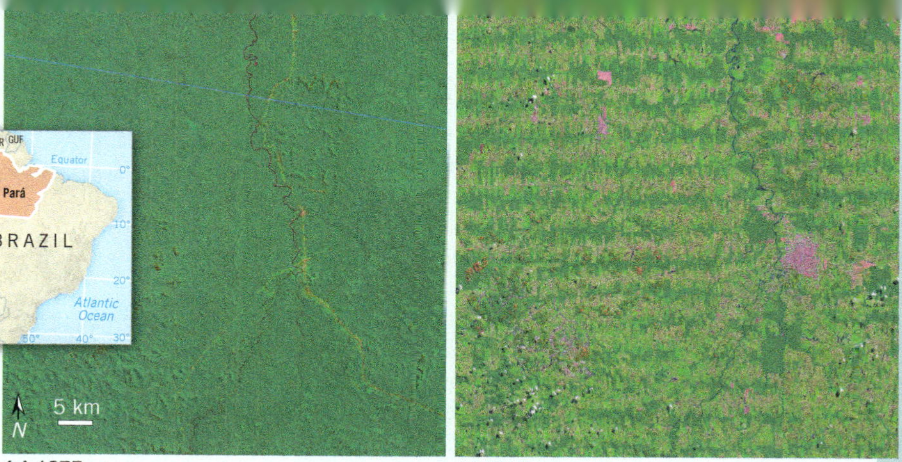

(a) 1975 (b) 2012

▲ **Figure 11-D** Deforestation in Rondônia, Brazil, (a) in 1975 and (b) in 2012. Images taken by Landsat satellites jointly managed by NASA and the U.S. Geological Survey.

Widespread clearing of the Amazon rainforest first became an environmental concern back in the 1980s. Today, about 20 percent of the total original rainforest has been lost. The peak rate of rainforest loss in the Amazon was reached in 2004, after which the annual rate of deforestation declined—until 2012. Since then, the rate of forest loss has increased again, most alarmingly in 2019, when Brazil's National Institute for Space Research showed that annual rate of deforestation in the Brazilian Amazon increased at least 30 percent over the previous year.

Rondônia: One of the earliest regions of concern in the Amazon was the southwestern state of Rondônia in Brazil. Its rainforest was still intact 15 years after completion of a highway across the province in 1960 (**Figure 11-D-a**), but by the 1990s, a "fishbone" pattern had emerged as forest was cleared for settlements, agriculture, ranching, and logging. By 2012, the extent of deforestation was extraordinary (**Figure 11-D-b**). In Rondônia, more than 90 percent of agricultural land is used for cattle grazing or annual crops—which rapidly deplete soil nutrients—rather than for sustainable perennial crops, such as cacao or coffee.

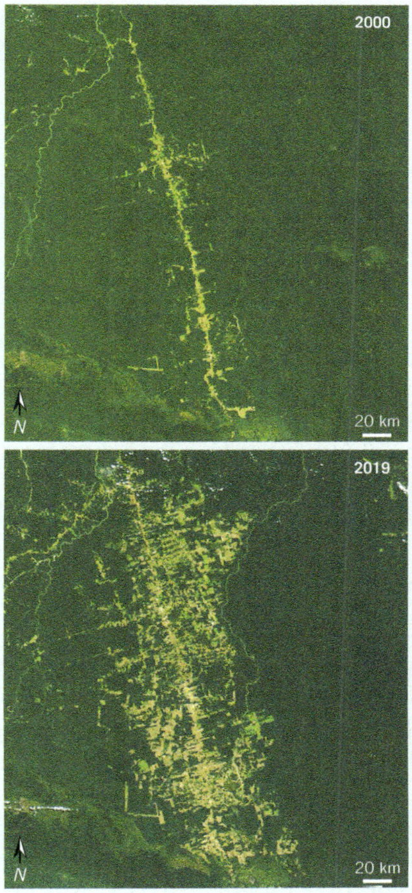

▲ **Figure 11-E** Landsat images of tropical rainforest loss in Pará, Brazil, (a) in 2000 and (b) in 2019.

Pará: In recent years, the area of greatest deforestation in Brazil has shifted east and north to Mato Grosso and Pará. In Pará, a gold rush in the 1980s fueled the local economy. Since then, the paving of a highway through the region linking the soy-growing areas in the south to an ocean port in the north has accelerated the economic growth and deforestation here (**Figure 11-E**).

Other Countries: Over the last 15 years, Amazon deforestation in Brazil's neighboring countries has also left a troubling mark By 2008, rainforest loss was accelerating in Peru and Columbia. In addition to cattle ranching, the expansion of palm oil agriculture in Peru contributed to an increase in forest clearing (**Figure 11-F**). After the end of the civil war in Columbia in 2015, renewed economic growth led to rainforest loss to cocoa and timber production.

Do Local Actions Cause Global Change? The effects of tropical deforestation are global in scale. With the burning of forest, Brazil is a significant emitter of greenhouse gases and a source of widespread air pollution. Forests help stabilize global climate, protect against soil erosion, and provide freshwater. Landsat imagery has tied the loss of tropical forest to changes in mid- and high-latitude rainfall patterns as sensible heat and latent heat are redistributed. In addition, the effects of biodiversity loss could be monumental; the potential value of now-threatened species as sources of medicines or foods is inestimable.

Questions

1. Briefly describe the causes and effects of rainforest loss.
2. Because deforestation causes wide-ranging devastation, why has it not been stopped in areas such as Brazil? Use examples from this box.

▲ **Figure 11-F** Rainforest in Peru cleared for cattle ranching.

with continuous heavy fertilization, a costly procedure, while sustainable land-extensive forms of traditional agriculture are slowly disappearing.

If left alone, forests can regenerate, providing seed trees are in the vicinity and the soil has not been stripped of all its nutrients—but the increasing fragmentation of tropical rainforest in some regions means that the original species composition may not return. This loss of biodiversity from tropical rainforests is an increasing concern. Valuable potential resources—pharmaceutical products, new food crops, natural insecticides, industrial materials—may disappear before they are even discovered. Wild plants and animals that could be bred with domesticated cousins to impart resistance to disease, insects, parasites, and other environmental stresses may also be lost.

LearningCheck 11-15 **Why is tropical rainforest loss a major environmental concern today?**

Addressing Forest Loss: Some steps have been taken to address tropical deforestation. The development of *agroforestry* (planting crops with trees, rather than cutting down the trees and replacing them with crops) is being fostered in many areas.

The United Nations Educational, Scientific, and Cultural Organization (UNESCO) administers the *Man and the Biosphere Programme*, which coordinates an international effort to establish and protect biosphere reserves. The goal of this project is to set aside tracts of largely pristine land—including regions of tropical rainforest—to preserve biodiversity before it is lost to development. At present, 701 biosphere reserves have been established in 124 countries.

Although deforestation remains high in many locations, the *rate* of forest loss has slowed in some parts of the world. In addition, global efforts to replant cleared forest with new trees are expanding. The FAO estimates that the *net forest loss* (the difference between forest loss and forest expansion through replanting) since 1990 has been 178 million hectares (of that, 81 million hectares was primary forest consisting of native tree species). The annual *rate* of forest loss has decreased from 7.8 million hectares per year between 1990 and 2000 to 4.7 million hectares per year between 2010 and 2020—and some regions, such as in parts of Asia, are actually experiencing a net gain in forest cover (Figure 11-40). The highest rates of net forest loss remain in Africa (3.9 million ha per year), followed by South America (2.6 million ha per year).

It is important to note, however, that both planted forests (which account for about 7 percent of the world forest cover today) and naturally regenerated forests (which account for about 57 percent of the world total) frequently do not contain the same biodiversity as the original forest—and in many parts of the world, replanted forests consist of introduced tree species. Thus, even in areas where forest cover is expanding, the original biodiversity has been lost. Overall, only about one-third of the world's forest cover is *primary forest*—forest with its natural species composition and little signs of human activity.

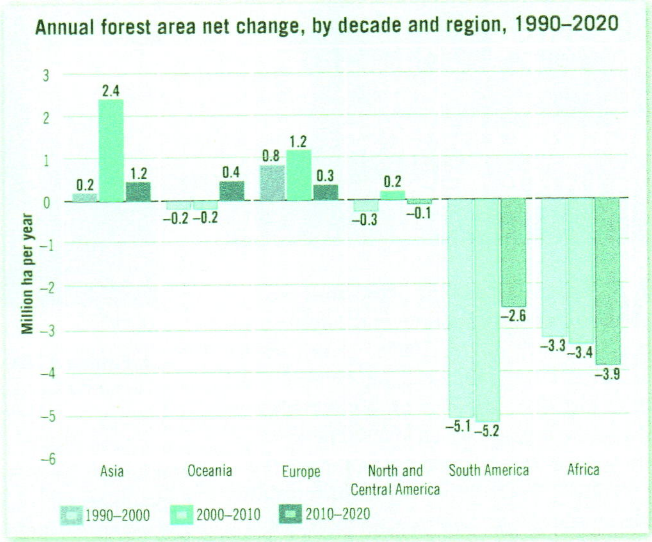

▲ Figure 11-40 Annual net forest area change by region, 1990 to 2020. Because of replanting, some countries experienced a net gain in forest cover.

Introduction of Exotic Species

People are capable of elaborate rearrangement of the natural complement of plants and animals in almost every part of the world. This is shown most clearly with domesticated species—crops, livestock, pets. There is now, for example, more corn than native grasses in Iowa, more cattle than native gazelles in Sudan, and more canaries than native thrushes in Detroit. More importantly for our discussion here, humans have also accounted for many introductions of wild plants and animals into "new" habitats; such organisms are called **exotic species** in their new homelands.

In some cases, the introduction of exotic species was deliberate. A few examples among a great many include taking prickly-pear cactus from Arizona to Australia; eucalyptus trees from Australia to California; crested wheat grass from Russia to Kansas; European boar from Germany to Tennessee; pronghorn antelope from Oregon to Hawai'i; and red fox from England to New Zealand. Frequently, however, the introduction of an exotic species was an accidental result of human carelessness (Figure 11-41). The European flea, for example, is one of the most widespread creatures on Earth because it has been an unseen accompaniment to human migrations all over the world. Similarly, the English sparrow and European brown rat have inadvertently been introduced to all inhabited continents by traveling as stowaways on ships.

One other type of human-induced introduction of animals involves the deliberate release or accidental escape of livestock to become established as a "wild"—properly termed *feral*—population. This has happened in

▲ **Figure 11-41** A ranger who is a member of the Invasive Plant Management Team sprays to prevent the spread of invasive fountain grass (*Pennisetum setaceum*) in Joshua Tree National Park.

NPS Video
Managing Invasive Species
https://media.pearsoncmg.com/ph/esm/esm_mcknight_physgeo_13/media/

▼ **Figure 11-42** Burros are prominent feral animals in much of the southwestern desert area of the United States. This group is in the Panamint Range of California.

many parts of the world, such as in North America with feral donkeys and horses and in Australia with feral pigs (**Figure 11-42**).

Invasive Species: When an exotic species is released in a new area, the results are often one of two extremes. Either it dies out in a short time because of environmental hazards or competitive-predatory pressures, or the introduced species finds both a benign climate and an unfilled ecologic *niche* and is able to flourish extraordinarily. When an exotic plant or animal thrives by rapidly filling such a local niche or by outcompeting native species, it is termed an **invasive species**.

The results of invasive species introduction can be disastrous. The world abounds with examples: sparrows and starlings, rabbits and pigs, mongooses and mynas, zebra mussels and Asian carp, lantana and prickly pear, mesquite and Scotch broom, kudzu and purple star thistle, Australia and New Zealand, Hawai'i and Mauritius. The list of species and places is virtually endless.

An argument could be made, however, that the recent and contemporary biotic history of Florida represents one of the worst examples, made all the more frightening because the cumulative impact will almost surely be much worse in the near future than it is already—see the box *Featured National Park: Everglades—Invaders of the Everglades*.

LearningCheck 11-16 What is an exotic species? An invasive species?

Loss of Biodiversity

One of the most troubling consequences of human modification of ecosystems is an ongoing loss of species and therefore a loss of biodiversity around the world.

In its history, Earth has undergone several *mass extinctions*—abrupt episodes when a large portion of life-forms die off. At the end of the Permian Period (252 million years ago), 95 percent of marine life and 70 percent of terrestrial species died off. At the end of the Cretaceous Period (66 million years ago), the dinosaurs, along with about 75 percent of all other species, became extinct. Many scientists think that Earth is currently undergoing another mass extinction event.

Earth's past mass extinctions were caused by natural events, such as major asteroid impacts and periods of major volcanic activity. The current extinctions have a human cause: habitat loss (such as clearing of tropical rainforests); the introduction of invasive species; unregulated hunting and fishing; and pollution of water, air, and soil. The *Fifth Assessment Report* of the Intergovernmental Panel on Climate Change reports *high confidence* that a few recent species extinctions are attributed to anthropogenic climate change and that during and beyond this century "... a large fraction of terrestrial, freshwater and marine species faces increased extinction risk due to climate change . . . ." Research published by the National Academy of Sciences in 2020 shows that the rate of extinction of terrestrial vertebrates is much higher than previously estimated—in *each year* of the last century, Earth lost as many species as would have been typically lost naturally over a period of 100 years.

As we've seen, change in natural systems is inevitable. However, the current loss of biodiversity at the hands of humans will likely have negative consequences we can't anticipate.

Featured National Park

Everglades

Invaders of the Everglades

Everglades National Park is a vast subtropical wetlands area that includes such diverse ecosystems as stands of tall cypress trees rising above slowly flowing water, freshwater marl prairie bordering deep marshy sloughs, and mangrove forest lining the sinuous coastal channels along the southern end of the state.

The Park is home to many rare and endangered animals, including the American Crocodile, the Florida Panther, and the Manatee. Unfortunately, the Park has also become the home of a large number of invasive species that have proven extremely difficult to eradicate.

How did they get here? Many factors help explain the proliferation of invasive species in the Everglades. Florida's mild subtropical climate is suitable for many plants and animals. During the last 50 years, the state has experienced a massive in-migration of people and extensive changes in land use—especially the modification of drainage systems. As a result, natural ecosystems were destabilized, making them susceptible to invasive species. A steady supply of exotic plants and animals arrived over the last few decades. Some are escaped pets, some species were deliberately turned loose for sport, and others have been brought in accidently in the holds of ships or in the baggage of travelers.

Invasive Plants: Dozens of species of exotic plants are now widespread in the Everglades. Prominent among them is the Melaleuca tree (*Melaleuca quinquenervia*) from Australia (**Figure 11-G**). Seeds from these "paperbark" trees were dispersed by airplane in the 1930s in the hope of developing a timber industry in the swamplands of southwestern Florida. The spread of Melaleuca changed swamp to forest, radically altering the entire regional ecosystem, but the lumber potential turned out to be negligible.

▲ **Figure 11-G** "Paperbark" Melaleuca trees growing in the Florida Everglades.

The Brazilian Pepper (*Schinus terebinthifolius*) is a widespread invasive evergreen bush that grows into a tangled mass with multiple stems. It was originally brought to Florida as an ornamental plant in the mid-1800s, but since then has taken over large areas, ranging from farmland to mangrove forests. It is both salt tolerant and fire resistant, and its leaves can irritate the skin. Currently, there are no good controls on its spread.

Invasive Animals: Exotic animals are sometimes less obvious in their occurrence than plants. Many species, however, are already well established and spreading rapidly throughout Florida, including the Mexican Nine-banded Armadillo, the Southeast Asian "Walking" Catfish, and the South American Cane Toad. Inside the Park itself, exotic snakes are perhaps the biggest problem.

The most widespread and troublesome snake invader in the Everglades is the Burmese Python (*Python molurus bivittatus*). These pythons are believed to be responsible for the observed decrease in mammals throughout the Everglades (**Figure 11-H**). After extensive study and effort, the Park Service has yet to find a way to mitigate the python problem.

A more recent invader to the Everglades water ecosystem is the Red Lionfish (*Pterois volitans*). This venomous predator fish, introduced from the tropical Pacific Ocean, first appeared in Florida waters in the 1980s and is now becoming increasingly common in Florida Bay in the southern end of the Park and also in nearby Biscayne National Park. Lionfish have already decimated native fish populations in some parts of the Caribbean, and Park personnel are trying find a mitigation solution before it becomes more problematic here.

In Everglades National Park—as is true in all of our National Parks—it is much better to prevent an invasive species problem than to solve it. As the Park Service reminds all visitors, "Don't let it loose!"

Questions

1. Why is Florida's environment so susceptible to invasive species?
2. Explain one example of an invasive species in Everglades National Park.

▲ **Figure 11-H** Burmese Python in Florida.

NPS Video
Outside Science: Hunting Lionfish in Biscayne National Park
https://media.pearsoncmg.com/ph/esm/esm_mcknight_physgeo_13/media/

After studying this chapter, you should be able to answer the following questions. Key terms from each text section are shown in **bold type.** Definitions for key terms are also found in the glossary at the back of the book.

Learning Review 11

Key Terms and Concepts

Ecosystems and Biomes (p. 314)

1. Contrast and explain the concepts of **ecosystem** and **biome**.
2. What is an **ecotone**?

Terrestrial Flora (p. 315)

3. What is the difference between a **perennial** and an **annual plant**?
4. Explain the difference between a **gymnosperm (conifer)** and an **angiosperm**. Name examples of each.
5. Explain the difference between a **deciduous tree** and an **evergreen tree**. Name examples of each.
6. Explain the difference between a **broadleaf tree** and a **needleleaf tree**. Name examples of each.
7. Describe some typical **xerophytic adaptations** of plants.
8. Describe some typical **hygrophytic adaptations** of plants.
9. What are the differences among **forests**, **woodlands**, and **shrublands**?
10. What are the similarities and differences among the **grasslands**: savanna, prairie, and steppe?
11. Briefly describe the **desert**, **tundra**, and **wetland** plant associations.
12. Explain what is meant by **vertical zonation** of vegetation patterns.
13. Define and explain what causes the **tree line**.
14. What is the difference between an **adret slope** and a **ubac slope**? How and why is vegetation likely to be different on an adret slope and a ubac slope?
15. What is **riparian vegetation**?

Terrestrial Fauna (p. 320)

16. What are some basic characteristics that distinguish plants from animals?
17. Contrast **invertebrates** with **vertebrates**. Provide one example of each.
18. Describe the distinguishing characteristics of mammals.
19. What are the advantages of animals that are **endotherms**?
20. Distinguish among the three ways (physiological, behavioral, reproductive) animals adapt to the environment.
21. What is meant by **symbiosis**?
22. Distinguish between *mutualism* and *parasitism*.

Zoogeographic Regions (p. 327)

23. Explain the concept of **zoogeographic regions**.

The Major Biomes (p. 328)

24. What climate characteristics are most closely associated with the tropical rainforest biome?
25. Contrast the general characteristics of the **tropical rainforest**, **tropical deciduous forest**, and **tropical scrub** biomes.
26. Describe the seasonal patterns of the **tropical savanna** biome.
27. Describe and explain the global distribution of the **desert** biome.
28. Discuss the general climate characteristics and types of vegetation associated with the **mediterranean woodland and shrub** biome.
29. What are the general differences in climate associated with the tropical savanna and **midlatitude grassland** biomes?
30. What are the general differences in climate associated with the **midlatitude deciduous forest** and **boreal forest** biomes?
31. Contrast the general species diversity in the tropical rainforest and boreal forest biomes.
32. Describe the general vegetation cover found in the **tundra** biome.

Human Modification of the Biosphere (p. 339)

33. What is an **exotic species**? An **invasive species**?
34. What is a *feral* animal population? Provide an example.

Study Questions

1. Why do trees require so much more moisture to survive than grass?
2. In what ways is an increase in altitude similar to an increase in latitude?
3. What generally happens to the elevation of the treeline going from the equator toward the poles? Why?
4. Describe and explain at least one animal adaptation to desert life.
5. Why are the flora and fauna of the Australian Region so distinctive?
6. Why is it usually difficult to maintain productive agriculture year after year in a cleared area of tropical rainforest?
7. Describe and explain one example of an exotic species that has disrupted a region's natural ecosystem.

Exercises

1. Using Figure 11-9 and the average lapse rate (discussed in Chapter 4) within the troposphere, estimate the typical temperature difference between the near-sea-level valley grassland ecosystem on the west side of the Sierra Nevada and the alpine tundra ecosystem near the summit of the range at an elevation of about 3500 meters (11,500 feet): _____°C (°F).
2. Refer to the climate map (Figure 8-3) and the biomes map in this chapter (Figure 11-22). The boreal forest biome generally corresponds to the distribution of which climate type?
3. Given a rate of world forest loss of 13 million hectares (32 million acres) per year, how much forest cover was lost around the world between 2010 and 2020? ______ million hectares

Environmental **Analysis** CHANGES IN GLOBAL FOREST COVER

Data
Forest Cover
https://media.pearsoncmg.com/ph/esm/esm_mcknight_physgeo_13/media/

Satellite imagery allows us to monitor forests without having to visit them. These images can detect forest extent. By monitoring the extent over time, we can determine forest loss or gain.

Activities

Go to http://earthenginepartners.appspot.com to see the Global Forest Change map.

1. How are trees defined there?
2. What is meant by "Forest Cover Loss"? By "Forest Cover Gain"?

In the "Data Products" drop-down menu, choose "Loss/Extent/Gain (Red/Green/Blue)" from its menu.

3. In what ranges of latitudes are most forests found? Why? (Think about the general circulation of air on Earth. Refer to Figures 5-22 and 5-23.)
4. Which areas show the greatest amount of forest loss? What major biomes are located in those areas?

In the "Example Locations" menu, choose "Forest Fires in Yakutsk" and click on "Zoom to area." Use any of these "Data Products" to complete the Activities that follow: "Loss/Extent/Gain (Red/Green/Blue)"; "Forest Cover Loss 2000–2019"; "Forest Cover Gain 2000–2012"; "2000 Percent Tree Cover."

5. How does forest loss compare with forest gain here? Estimate a percentage of forest loss relative to forest gain.
6. How does forest loss compare with forest extent? Estimate a percentage of forest loss relative to forest extent.

Go to? http://earthobservatory.nasa.gov and type "86414" into the "Search" field. Select "Fires in Siberia" and read the article.

7. Burning cropland is a common practice in Siberia, but the fires often get out of control. How many square kilometers had burned as of August 12, 2015?
8. Why did the smoke move toward North America?
9. What is a pyrocumulonimbus cloud?
10. Are more or fewer fires expected to occur in the Yakutsk area in the future? Why?
11. How might these fires impact climate change?

Seeing Geographically in Our National Parks

Look again at the photograph of the bison in Wind Cave National Park at the beginning of the chapter (p. 312). What kind of climate and biome(s) are likely found in this location (43° N, 104° W)? What adaptations might make bison suitable for this environment? In which zoogeographic region are these bison found?

Mastering Geography®

Looking for additional review and test prep materials? Visit the Study Area in **Mastering Geography** to enhance your geographic literacy, spatial reasoning skills, and understanding of this chapter's content by accessing a variety of resources, including Map Master interactive maps, geoscience animations, *Mobile Field Trips*, videos, *Project Condor* Quadcopter videos, *In the News* RSS feeds, flashcards, web links, self-study quizzes, and an eText version of *McKnight's Physical Geography*.

Seeing Geographically in Our National Parks

The Little Missouri River in Theodore Roosevelt National Park, North Dakota. Describe the general topography of this area. Is the river here fairly deep or quite shallow? How do you know? What helps explain why the steep slopes have little plant cover whereas the area along the river has many trees?

Theodore Roosevelt National Park, ND

12

Soils

HAVE YOU EVER WONDERED why some soils are better for agriculture than other soils? Or why the land alongside a river has such fertile soil—like along the Little Missouri River here in Theodore Roosevelt National Park in North Dakota? The qualities of a soil, such as its fertility, are the outcome of an intricate set of processes that can produce an almost infinite variety of characteristics. Soil may be the most underappreciated of all natural resources, and unfortunately, it's a resource that we have badly degraded or simply wasted in many places.

With this chapter and our look at soils, we begin our study of Earth's fourth environmental sphere, the *lithosphere*—the solid part of Earth. The lithosphere contrasts with the atmosphere, hydrosphere, and biosphere in its enormity and particularly in its seeming stability. The dynamics of the lithosphere, with a few spectacular exceptions, such as earthquakes and volcanic eruptions, operate with incredible slowness. Most people consider the phrase "the everlasting hills" to be a literal expression that describes the permanence of Earth's topography. In reality, the phrase is hyperbole that fails to recognize the remarkable changes that take place over time, largely unrecognizable to the casual observer.

Our goal in the remaining chapters of this book is to understand the contemporary character of Earth's surface and to explain the processes that are at work shaping it. In this chapter, we begin with the aspect of Earth that perhaps most dramatically links the lithosphere, the atmosphere, the hydrosphere, and the biosphere: the thin veneer of soil.

As you study this chapter, think about these **Key**Questions:

- **How is soil different from bedrock?**
- **What factors influence soil formation?**
- **What are the components of soil?**
- **What influences soil fertility?**
- **What is a soil profile?**
- **How does climate influence soil development?**
- **How are soils classified and mapped?**

Soil and Regolith

Despite the implication of the well-known phrase "as common as dirt," soil is remarkably diverse. **Soil** is a relatively thin surface layer made up of a mixture of weathered mineral particles, decaying organic matter, living organisms, gases, and liquid solutions. Because the bulk of most soil is inorganic, soil is classified as part of the lithosphere,[1] but it acts as a fundamental interface where atmosphere, hydrosphere, biosphere, and lithosphere meet.

Soil as a Component of the Landscape

The surface of the lithosphere is almost always covered by soil. Although extremely pervasive, soil is often an inconspicuous component of the landscape because it is normally masked by vegetation or roads, parking lots, buildings, and so on.

We recognize soil in the landscape mostly by its color. The depth of the soil layer becomes obvious only where it is exposed by gully erosion or a road cut.

Almost all land plants sprout from this precious medium, spread so thinly across the continental surfaces that fertile topsoil may have an average worldwide depth of only about 15 centimeters (6 inches). It occupies that part of the outer "skin" of Earth that generally extends from the surface down to the maximum depth to which plant roots can penetrate. Soil is characterized by its ability to produce and store plant nutrients, an ability made possible by the interactions of such diverse factors as water, air, sunlight, rocks, plants, and animals.

From Regolith to Soil

Soil development begins with the physical and chemical disintegration of rock exposed to the atmosphere and to the action of water percolating down from the surface. This disintegration is called *weathering*. As we shall learn in Chapter 15, the basic result of weathering is the weakening and breakdown of solid rock and the fragmentation of rock masses.

The principal product of weathering is a layer of loose inorganic material called **regolith** ("blanket rock"), which lies like a blanket over the solid, unfragmented *bedrock* below (Figure 12-1). Typically, regolith consists of material that has weathered from the underlying rock, with the largest and least fragmented pieces at the bottom, immediately adjacent to the unweathered, intact bedrock. Sometimes, however, regolith includes material that was transported from elsewhere, so it may vary significantly from the bedrock immediately below.

The uppermost layer of regolith—the soil—typically differs from the material below in several ways. The soil layer is composed mostly of finely fragmented mineral particles and is the ultimate product of weathering. It normally also contains an abundance of living plant roots, dead and rotting plant parts, microscopic plants and animals both living and dead, and a variable amount of air and water. Soil, however, is not the end product of a process but rather a stage in a never-ending continuum of physical, chemical, and biological processes (Figure 12-2).

[1]In this chapter we use the word *lithosphere* as a general term for the solid part of Earth. In Chapter 13 we see that the term has a more limited definition in the context of plate tectonics.

▲ Figure 12-1 Vertical cross section from surface to bedrock, showing the relationship between soil and regolith.

LearningCheck 12-1 **Describe the first step in the formation of soil. (Answer on p. AK-4)**

Soil-Forming Factors

Soil is an ever-evolving material. Metaphorically, soil acts like a sponge—taking in inputs and being acted on by the local environment, so it changes over time and when the inputs or local environment change. Five principal soil-forming factors are responsible for soil development: geology, climate, topography, biology, and time.

The Geologic Factor

The source of the rock fragments that make up soil is **parent material**, which may be either bedrock or loose sediments transported from elsewhere by water, wind, or ice. The nature of the parent material often influences the characteristics of the soil that develop from it; this factor sometimes dominates all others, particularly in the early stages of soil formation. Both the chemical composition and physical characteristics of the parent material may be influential in soil development, particularly in terms of texture and structure. Bedrock that weathers into large particles (as does sandstone, for example) normally produces a coarse-textured soil, one easily penetrated by air and water to some depth. Bedrock that weathers into minute particles (shale, for example) yields fine-textured soils with many small pores, inhibiting air and water from easily penetrating the surface.

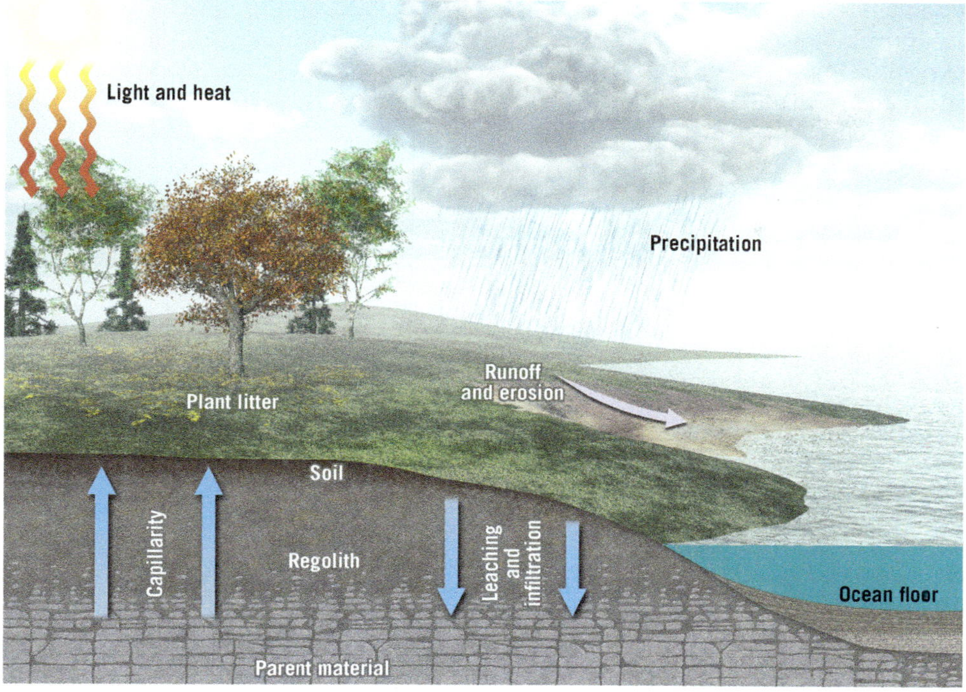

◀ **Figure 12-2** Soil develops through a complex interaction of physical, chemical, and biological processes. Parent-material bedrock weathers to regolith; then plant litter combines with the regolith, forming soil. Some of that soil washes to the ocean floor, where, over geologic time, it is transformed to sedimentary rock. Someday that ocean floor may be uplifted above sea level and the exposed sedimentary rock will again be weathered into soil.

Mobile Field Trip
The Critical Zone
https://media.pearsoncmg.com/ph/esm/esm_mcknight_physgeo_13/media/

Young soils are likely to be very reflective of the rocks or sediments from which they were derived. With the passage of time, however, other soil-forming factors become increasingly important, and the significance of the parent material diminishes. Eventually the influence of the parent material may be obliterated, and it is sometimes impossible to ascertain the nature of the rock from which the soil was derived.

The Climatic Factor

Temperature and moisture are the climatic variables of greatest significance to soil formation. As a basic generalization, both the chemical and biological processes in soil are usually accelerated by high temperatures and abundant moisture and are slowed by low temperatures and lack of moisture, so soils tend to be deepest in warm, humid regions and shallowest in cold, dry regions.

It is difficult to overemphasize the role of moisture moving through the soil. The flow is mostly downward because of the pull of gravity, but it is sometimes sideways (in response to drainage opportunities) or, in special circumstances, even upward. Water carries dissolved chemicals in solution and usually also carries tiny particles of matter in suspension. Thus, moving water is ever engaged in rearranging the chemical and physical components of the soil, as well as contributing to the variety and availability of plant nutrients.

In terms of general soil characteristics, climate is likely to be the most influential factor in the long run.

The Topographic Factor

Slope and drainage are the two main features of topography that influence soil characteristics. Wherever soil develops, its vertical extent undergoes continuous, usually very slow, change through a lowering of both the bottom and top of the soil layer. The bottom slowly gets deeper as weathering penetrates into the regolith and parent material and as plant roots extend to greater depths. At the same time, the soil surface is being lowered by sporadic removal of its uppermost layer through erosion—the removal of individual soil particles by running water, wind, and gravity.

Where the land is flat, soil tends to develop at the bottom more rapidly than it is eroded away at the top; surface erosion is extraordinarily slow. Thus, the deepest soils are usually on flat land. Where slopes are relatively steep, surface erosion is more rapid than soil deepening, with the result that such soils are nearly always thin and immaturely developed (**Figure 12-3**).

If soils are well drained, moisture relationships may be relatively unimportant factors in soil development. If soils have poor natural drainage (such as where water is standing on the surface), however, significantly different characteristics may

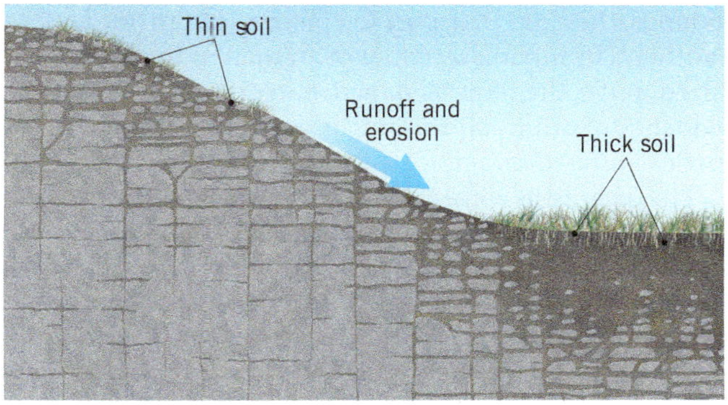

▲ **Figure 12-3** Slope is a critical determinant of soil depth. On flat land, soil normally develops more deeply with the passage of time because very little erosion is washing away the topmost soil. On a slope, the rate of erosion is equal to or greater than the rate at which soil forms at the bottom of the soil layer, so the soil remains shallow there.

develop. For example, a waterlogged soil tends to contain a high proportion of organic matter, and the biological and chemical processes that require free oxygen are impeded. (Air is the source of the needed oxygen, and a waterlogged soil contains essentially no air.) Most poorly drained soils are in valley bottoms or in some other flat locale because soil drainage is usually related to slope.

In some cases, such subsurface factors as permeability and the presence or absence of impermeable layers are more influential than slope.

LearningCheck 12-2 **What factors are of greatest importance in soil formation?**

The Biological Factor

By volume, soil is about half mineral matter and about half air and water, with only a small fraction of organic matter. However, this organic fraction, consisting of both living and dead plants and animals, is of utmost importance. The biological factor gives life to the soil and makes it more than just "dirt." Every soil contains (sometimes an enormous quantity of) living organisms, and every soil incorporates (sometimes a vast amount of) dead and decaying organic matter.

Vegetation of various kinds growing in soil performs certain vital functions. Plant roots, for instance, work their way down and around, providing passageways for drainage and aeration, as well as being the vital link between soil nutrients and the growing plants.

Many kinds of animals contribute to soil development as well. Even such large surface-dwelling creatures as elephants and bison affect soil formation by compacting soil with their hooves, rolling in the dirt, grazing on the vegetation, and dropping excreta. Ants, worms, and all other land animals fertilize the soil with their waste products and contribute their carcasses for eventual decomposition and incorporation into the soil.

Many small animals spend most or all of their lives in the soil layer, tunneling here and there, moving soil particles upward and downward, and providing passageways for water and air (Figure 12-4). Mixing and plowing by soil fauna can be remarkably extensive. As ants and termites build mounds, they also transport soil materials from one layer to another. The mixing activities of animals in the soil, generalized under the term *bioturbation*, tend to counteract the tendency of other soil-forming processes to accentuate the vertical differences among soil layers.

The organisms within a soil range from microscopic protozoans to larger animals that may accidentally alter certain soil characteristics. Of all creatures, however, the earthworm is probably the most important to soil formation and development.

Earthworms: The cultivating and mixing activities of earthworms are of great value in improving the structure, increasing the fertility, lessening the danger of accelerated erosion, and deepening the profile of the soil. The mere presence of earthworms, however, does not guarantee that a soil will be highly productive, as other kinds of inhibiting factors may

▲ **Figure 12-4** Like other burrowing animals, these prairie dogs at Badlands National Park in South Dakota contribute to soil development by bringing subsoil to the surface and providing passageways for air and moisture to get underground.

be present, such as a high water table. Nevertheless, an earthworm-rich soil has a higher potential productivity than similar soils lacking earthworms. In various controlled experiments, the addition of earthworms to wormless soil enhanced plant productivity by several hundred percent.

At least seven beneficial functions have been attributed to earthworms:

1. Their innumerable tunnels facilitate drainage and aeration and deepen the soil profile.
2. Their continual movement beneath the surface tends to produce a crumbly structure, which is generally favorable for plant growth.
3. The soil is further mixed as material is carried and washed downward into their holes from the surface. This is notably in the form of leaf litter dragged downward by the worms, which fertilizes the subsoil.
4. Their digestive actions and tunneling form aggregate soil particles that increase porosity and resist the impact of raindrops, helping deter erosion.
5. They release nutrients into the soil by the excretion of *casts* (expelled mineral material bound together with decomposed organic material). Earthworm casts are 5 times richer in available nitrogen, 7 times richer in available phosphates, and 11 times richer in available potash than the surrounding soil.
6. They rearrange material in the soil, especially by bringing deeper matter to the surface, where it can be weathered more rapidly. Where earthworms are numerous, they may deposit as much as 9000 kg/hectare (25 tons/acre) of casts on the surface in a year.
7. Their presence promotes nitrification, due to increased aeration, alkaline fluids in their digestive tracts, and the decomposition of earthworm carcasses.

Global Environmental Change

Invasive Earthworms Change Soils as We Know Them!

• **Randall Schaetzl,** Michigan State University; **Kyungsoo Yoo,** University of Minnesota

Invasive species are everywhere. They remind us how fragile our ecosystems are. Even the earthworm, an invader in many areas, is causing considerable changes under our feet.

For example, after glaciers left the upper Midwest 20,000–12,000 years ago, the debris left behind was almost devoid of macroscopic life-forms. No beetles, no badgers—and no earthworms. Invasive earthworms arrived when European settlers brought them to North America on ships and were further dispersed by loggers, farmers, and people using worms as fish bait. Earthworms migrate slowly, less than 10 meters (33 feet) per year. Evidence of these invaders is most dramatic at an "invasion front." They occur behind the front but not ahead of it. Their migration across the northern Great Lakes region, New England, northern Europe, and much of Canada has taken millennia and is still ongoing.

How Earthworms Change Soils: Invading earthworms can change soils dramatically and quickly. As they live in (and consume) soil, they also mix it. Such burrowing is called *bioturbation*. As earthworms bioturbate the soil, they consume soil microbes, mixing their own wastes into the soil as *fras*, or earthworm casts (**Figure 12-A**). This material is so rich in organic matter and nutrients that it is sold as a soil amendment.

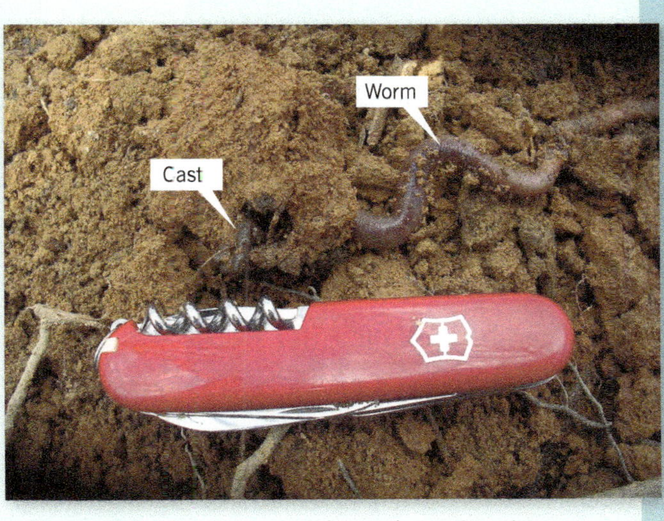

▲ **Figure 12-A** The common nightcrawler, an invasive earthworm. Notice the dark color of the excreted cast, rich in organic matter.

Studies of soils at invasion fronts have shown that bioturbation by introduced worms mainly affects the topsoil, where they primarily live and burrow. Forest leaves and litter are quickly consumed. In Minnesota, the layer of leaves at the soil surface is thinning or even disappearing due to invasive earthworms. This brings on a cascade of effects. For example, salamanders and ground-nesting songbirds are negatively affected by the loss of forest litter, whereas deer thrive on seedlings that used to be protected by the litter. The topsoil just below thickens and gains organic matter from the surface litter; this transference of carbon is of much interest to scientists studying the global carbon cycle. Do invasive earthworms deplete soils of carbon or enrich them? The jury is still out.

Questions

1. In what ways, other than those mentioned, could earthworms be transported to new ecosystems?
2. Some earthworms burrow vertically to great depths; others burrow mainly horizontally within the surface soil layer. How differently might these two burrowing behaviors affect soils?

In many parts of the world, earthworms are lacking. They are, for example, almost absent from arid and semiarid regions. In these dry lands, some of the earthworm's soil-enhancing functions are carried out by ants and earth-dwelling termites, but much less effectively. The migration of earthworms into new areas is occurring in many places around the world, and their arrival leads to significant changes—see the box *Global Environmental Change: Invasive Earthworms Change Soils as We Know Them!*

Microorganisms in the Soil: Another important aspect of the biological factor is microorganisms, both plant and animal, that occur in uncountable billions. An estimated three-quarters of a soil's metabolic activity is generated by microorganisms. These microbes help release nutrients from dead organisms by decomposing organic matter and by converting nutrients to forms that are usable by plants. Algae, fungi, protozoans, actinomycetes, and other minuscule organisms all play a role in soil development. Bacteria probably make the greatest contribution overall, although through the decomposition and decay of dead plant and animal material and the consequent release of nutrients into the soil. In some environments, cyanobacteria play an important role in the development of the soil surface—see the box *Featured National Park: Canyonlands—The Desert Soil Is Alive.*

LearningCheck 12-3 What are several ways that earthworms tend to increase the productivity of soil?

The Time Factor

For soil to develop on a newly exposed land surface requires time. How much time? That depends on the nature of the exposed parent material and the characteristics of the environment. Soil-forming processes are generally very slow, and many centuries may be required for a thin layer of soil to form on a newly exposed surface. A warm, moist environment is conducive to soil development. Normally of much greater importance, however, are the attributes of the parent material. For example, soil develops from sediments relatively quickly and from bedrock relatively slowly.

Soil Erosion: Most soil develops with geologic slowness—so slowly that changes are almost imperceptible within a human life span. It is possible, however, for a soil to be degraded, either

Featured National Park

Canyonlands

The Desert Soil Is Alive

In Canyonlands National Park, as in many other locations in the high desert of the Colorado Plateau, we find *biological soil crusts* (**Figure 12-B**). This knobby black surface is indeed alive and may account for as much as three-quarters of the ground cover biomass in some areas.

Cyanobacteria: The biological soil crusts in Canyonlands National Park are made up of bacteria, mosses, fungi, and lichen, but the most important component is *cyanobacteria* (once called blue-green algae). Cyanobacteria were among the earliest forms of life on Earth and one of the first organisms to live on land.

When the biological soil crust is dry, the cyanobacteria are dormant. However, when moistened, the cyanobacteria grow and move down through the soil, leaving sticky fibers that hold particles together in a living mat. With time, moss and lichen begin to grow in the mat along with the cyanobacteria. Such a cover is quite resistant to erosion by water and wind.

Biological soil crusts have a number of important functions. When wet, the fibers of the cyanobacteria swell and can hold moisture that can be used by plants. In addition, the cyanobacteria also fix nitrogen and so provide a vital nutrient to plants.

A Fragile Cover: Although a biological soil crust armors the desert surface in some ways, it is a relatively fragile armor. The crust is easily broken by vehicle tires and even by footprints. Once the crust is disrupted, the desert surface below is much more susceptible to erosion, and it may take decades—in some cases, centuries—for the crust to reform.

Where biological soil crusts are present, visitors should be sure to stay on established roadways and walking paths. As the Park Service reminds us, "Don't bust the crust!"

▲ **Figure 12-B** Dark, knobby biological soil crust in Canyonlands National Park.

Questions

1. How do biological soil crusts develop?
2. How can biological soil crusts be damaged by people?

NPS Video
Inside Canyonlands—Biological Soil Crust
https://media.pearsoncmg.com/ph/esm/esm_mcknight_physgeo_13/media/

through the physical removal associated with accelerated erosion or through depletion of nutrients, in only a few years.

Fine-textured soils—especially those with low rates of rainwater infiltration—tend to be soils that are most easily eroded by rainwater runoff and wind; steep slopes and lack of a vegetation cover also increase the likelihood of erosion. In regions where single-crop agriculture ("monoculture") is practiced, fields are often left bare and unplanted for several months each year, increasing the likelihood of erosion.

In the United States, some researchers estimate that nearly 40 percent of the productive soil in the wheat-growing Palouse region of Washington and Idaho and as much as 50 percent of the topsoil of Iowa have been lost to erosion over the last 150 years. Globally, perhaps 7 million hectares of cropland are lost each year to soil erosion—10 to 40 times faster than the rate at which productive soil can develop. It is important to realize that in the grand scale of geologic time, soil can be formed and reformed, but in the dimension of human time, it is a mostly nonrenewable resource.

Soil Components

The natural components of soil are classified into just a few main groups: inorganics, organics, air, and water.

Inorganic Materials

The bulk of most soils is mineral matter, mostly in the form of small but macroscopic particles. Inorganic material also occurs as microscopic clay particles and as dissolved minerals in solution.

About half the volume of an average soil is small, granular mineral matter called *sand* and *silt*. These particles may consist of a great variety of minerals, depending on the nature of the parent material from which they were derived, and are simply fragments of the wasting rock. Most common are bits of quartz, which are composed of silica (SiO_2) and appear in soil as very resistant grains of sand. Other prominent minerals making up sand and silt are some feldspars and micas.

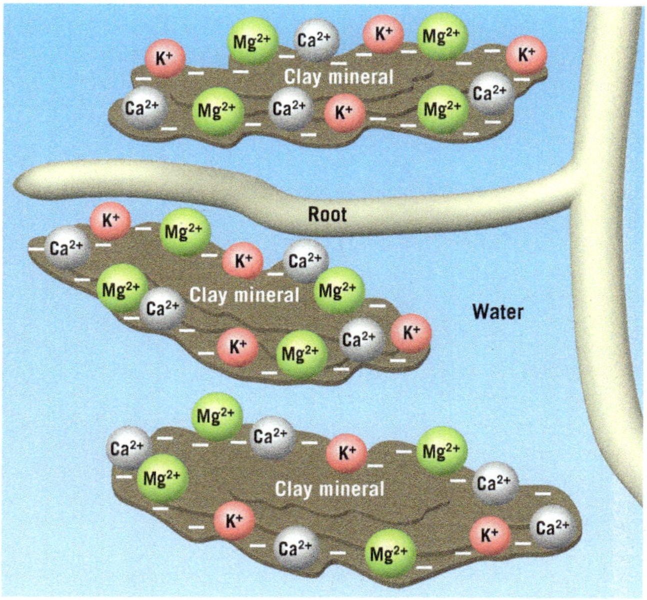

▲ **Figure 12-5** Clay particles offer a large surface area on which substances dissolved in soil water can cling. The particles are negatively charged and therefore attract cations (positively charged ions) from the water. These cations held by the clay are then absorbed by plant roots and become nutrients for the plant.

The smallest particles in soil are **clay**—usually a combination of silica and oxides of aluminum and iron found only in the soil and not in the parent material. Clay has properties significantly different from those of larger (sand or silt) fragments. Most clay particles are *colloidal* in size, which means they are larger than molecules but too small to be seen with the naked eye. They are usually flat platelets (Figure 12-5) and therefore relatively large in surface area.

Many chemical reactions occur at the surfaces of soil particles. The platelets group together in loose, sheet-like assemblages through which water moves easily. Substances dissolved in the water are attracted to and held by the sheets. Since the sheets are negatively charged, they attract positively charged *ions* called **cations**.[2] Many essential plant nutrients occur in soil solutions as cations. As a result, clay is an important reservoir for plant nutrients, just as it is for soil water. We discuss the role of cations in soils in greater detail later in this chapter.

Organic Matter

Although organic matter generally constitutes less than 5 percent of total soil volume, it has an enormous influence on soil characteristics and plays a fundamental role in the biochemical processes that make soil an effective medium of plant growth. Some of the organic matter is living organisms, some is dead but undecomposed plant parts and animal carcasses, some has decomposed to *humus*, and some is in an intermediate stage of decomposition.

Apart from plant roots, evidence of the variety and bounty of organisms living in soil may not be obvious, but most soils are seething with life. A half hectare (about 1 acre) may contain a million earthworms, and the total number of organisms in 30 grams (about 1 ounce) of soil likely exceeds 100 trillion. Microorganisms far exceed larger life-forms, both in total numbers and in cumulative mass. They are active in rearranging and aerating the soil and in yielding waste products that are links in the chain of nutrient cycling. Some make major contributions to the decay and decomposition of dead organic matter, and others make nitrogen available for plant use.

Litter: Leaves, twigs, stalks, and other dead plant parts accumulate at the soil surface, where they are referred to collectively as **litter**. The eventual fate of most litter is decomposition, in which the solid parts are broken down into chemical components, which are then absorbed into the soil or washed away. In cold, dry areas, litter may remain undecomposed for a long time; where the climate is warm and moist, however, decomposition may take place almost as rapidly as litter accumulates.

Humus: After most of the residues have been decomposed, a brown or black, gelatinous, chemically stable organic matter known as **humus** remains. This "black gold" is of utmost importance to agriculture because it loosens the structure and lessens the density of the soil, thereby facilitating root development. Moreover, humus, like clay, is a catalyst for chemical reactions and a reservoir for plant nutrients and soil water.

Soil Air

Nearly half the volume of an average soil is made up of pore spaces. These spaces provide a labyrinth of interconnecting passageways, called *interstices*, for air and water among the soil particles. On average, the pore spaces are about half filled with air and half with water, but at any given time and place, the amounts of air and water are quite variable (Figure 12-6).

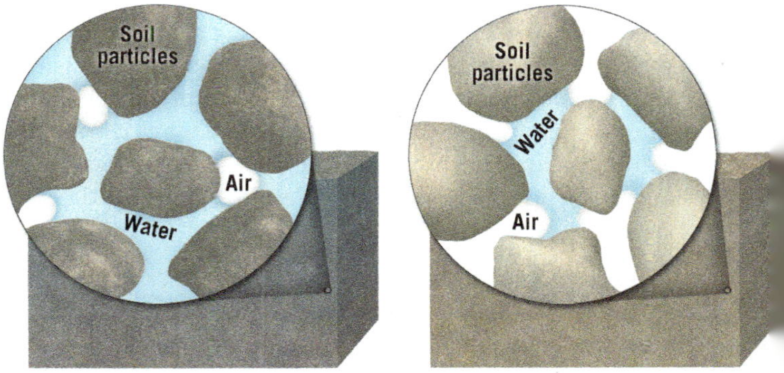

▲ **Figure 12-6** The relative amounts of water and air in soil pores vary from place to place and from time to time. (a) The interstices of wet soil contain much water and little air. (b) Dry soil has much air and little water.

[2]An *ion* is an electrically charged atom or molecule. An atom or molecule that has fewer electrons than protons has a positive charge and is called a *cation*; if it has more electrons than protons, it has a negative charge and is called an *anion*.

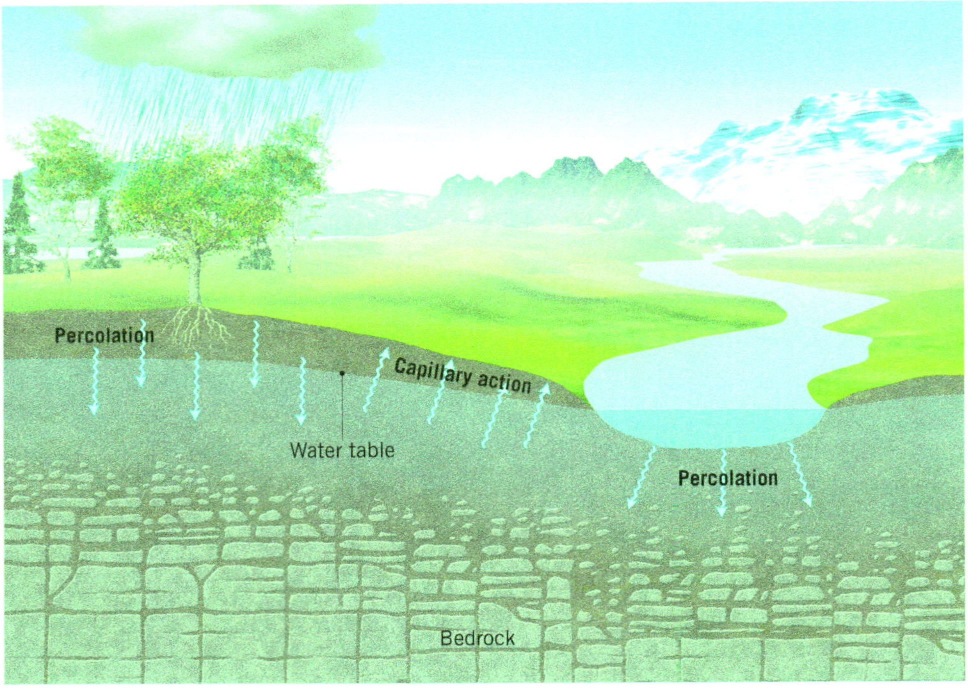

▶ Figure 12-7 Water is added to the soil layer by the percolation of rainwater and snowmelt from above. Additional moisture enters the soil from below as capillary action pulls groundwater upward above the water table.

Video
Maps of Soil Moisture
https://media.pearsoncmg.com/ph/esm/esm_mcknight_physgeo_13/media/

Video
California Drought
https://media.pearsoncmg.com/ph/esm/esm_mcknight_physgeo_13/media/

The characteristics of air in the soil are significantly different from those of atmospheric air. Soil air is found in openings generally lined with a film of water, and it is not exposed to moving air currents. Thus, it is saturated with water vapor. Soil air is also very rich in carbon dioxide and poor in oxygen because plant roots and soil organisms remove oxygen and respire carbon dioxide into the pore spaces. The carbon dioxide then slowly escapes into the atmosphere.

LearningCheck 12-4 Why are clay particles and humus important ingredients of soil?

Soil Water

Water comes into the soil largely by percolation of rainfall and snowmelt, but some is also added from below when groundwater is pulled up above the water table by *capillary action* (Figure 12-7; see Chapter 6 for a discussion of capillarity). Once it has penetrated the soil, water envelops each solid particle in a film and wholly or partially fills the pore spaces. Water can be lost from the soil by percolation down into the groundwater, by upward capillary movement to the surface followed by evaporation, or by plant use (transpiration).

Four forms of soil moisture play roles in soil processes (Figure 12-8):

1. **Gravitational Water:** *Gravitational water* results from prolonged infiltration from above and sinks through the interstices toward the groundwater zone. This water stays in the soil for a short time only and is not very effective in supplying plants because it drains away rapidly. However, gravitational water is the principal agent of *eluviation* and *illuviation* (discussed below) and thus makes the topsoil coarser and more open-textured and the subsoil denser and more compact.
2. **Capillary Water (Water of Cohesion):** *Capillary water*, which remains after gravitational water has drained away, consists of moisture held at the surface of soil particles by *surface tension* (the force exerted along the surface, caused by *cohesion*, or the attraction of water molecules to each other; see Chapter 6). Capillary water is by far the principal source of moisture for plants. This water is free to move equally in all directions in response to capillary action. It tends to move from wetter areas toward drier ones, which accounts for the upward movement of capillary water when no gravitational water is percolating downward.
3. **Hygroscopic Water (Water of Adhesion):** *Hygroscopic water* consists of a microscopically thin film of moisture bound rigidly to all soil particles by *adhesion* (the attraction of water molecules to solid surfaces). Hygroscopic water adheres so tightly to the particles that it is normally unavailable to plants.
4. **Combined Water:** *Combined water* is the least available of all. It is held in chemical combination with various soil minerals and is freed only if the compound is altered.

After gravitational water has drained away, the volume of water that remains, filling the pore spaces, is the soil's **field capacity**. If drought conditions prevail and the capillary water is used up by plants or evaporated, the plants are no longer able to extract moisture from the soil; then, the **wilting point** is reached.

Leaching: Water performs a number of important functions in the soil. It is an effective solvent, dissolving essential soil nutrients and making them available to plant roots. These dissolved nutrients are carried downward in solution and partly redeposited at lower levels. This process, called **leaching**, tends to deplete the topsoil of soluble nutrients.

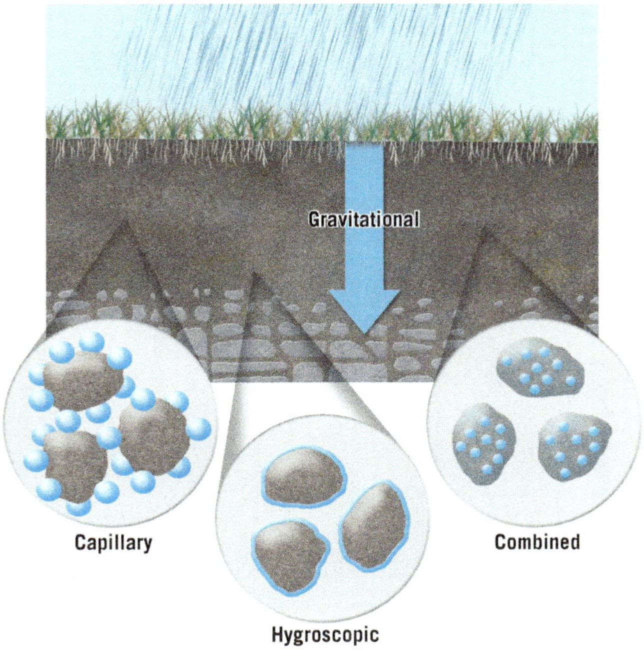

▲ Figure 12-8 The four forms of soil water.

Water is also required for many of the chemical reactions of clay and by the microorganisms that produce humus.

Eluviation and Illuviation: Water can have considerable influence on the physical characteristics of soil by moving particles around and depositing them elsewhere in the soil. For example, as water percolates into the soil, it picks up fine particles of mineral matter from the upper layers and carries them downward in a process called **eluviation** (Figure 12-9).

▲ Figure 12-9 In eluviation, fine particles in upper soil layers are picked up by percolating water and carried deeper into the soil. In illuviation, these particles are deposited in a lower soil layer.

These particles are eventually deposited at a lower level; this deposition process is known as **illuviation**.

LearningCheck 12-5 Which form of soil moisture is most important for plants? Why?

Soil-Water Budget: The moisture added to soil by percolation of rainfall or snowmelt is diminished largely through evapotranspiration. The dynamic relationship between these two processes is referred to as the **soil-water balance**. It is influenced by a variety of factors, including soil and vegetation characteristics, but it is primarily determined by temperature and humidity.

The amount of water available to plants is much more important to an ecosystem than is the amount of precipitation. Much of the water from rainfall or snowmelt becomes unavailable to plants because of evaporation, runoff, or deep infiltration. At any given time and place, the soil is likely to have either a surplus or a deficit of water. Generally speaking, warm weather causes increased evapotranspiration, which diminishes the soil-water supply, and cool weather slows evapotranspiration, allowing more moisture to be retained in the soil.

In a hypothetical Northern Hemisphere midlatitude location, January is a time of surplus water in the soil because low temperatures inhibit evaporation and there is little or no transpiration from plants. The soil is likely to be at or near field capacity. With the arrival of spring, temperatures rise, and plant growth accelerates, so both evaporation and transpiration increase. The soil-water balance tips from a water surplus to a water deficit. This deficit peaks in middle or late summer, as temperatures reach their greatest heights and plants need maximum water. Heavy use and diminished precipitation may deplete all of the moisture available to plants, which reach the wilting point. Thus, the amount of soil moisture available for plant use is essentially the difference between field capacity and wilting point.

In late summer and fall, as air temperature decreases and plant growth slackens, evapotranspiration diminishes rapidly. The soil-water balance shifts once again to a water surplus, which continues through the winter. Then the cycle begins again. Such variation in the soil-water balance through time is called a **soil-water budget** (Figure 12-10).

Soil Properties

We use the physical and chemical characteristics of color, texture, and structure to describe and classify soils (Figure 12-11).

Color

The most conspicuous property of a soil is usually its color, but color is by no means the most definitive property. Soil color can provide clues about a soil's nature and capabilities, but the clues can be misleading. Soil scientists recognize 175 gradations of color. The standard colors are generally shades

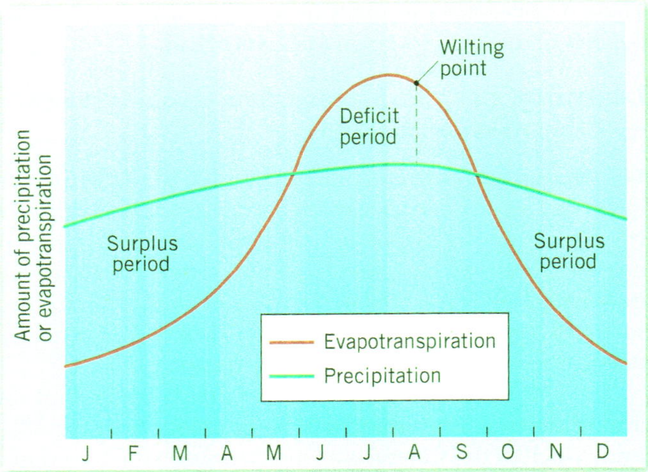

▲ **Figure 12-10** Hypothetical soil-water budget for a Northern Hemisphere midlatitude location. From January through early May, precipitation exceeds evapotranspiration, and the soil contains more water than any plant needs. From mid-May to mid-September, the evapotranspiration curve rises above the precipitation curve, indicating that more water leaves the soil than enters it. About August 1, so much water has been removed from the soil that plants begin to wilt. After mid-September, the evapotranspiration curve again dips below the precipitation curve and soil again has a surplus of water.

▲ **Figure 12-11** Park researcher sampling soil at Golden Gate National Recreation Area, California.

of black, brown, red, yellow, gray, or white. In most cases, color is the result of stains, imparted by metallic oxides or organic matter, on the surface of the soil particles.

Black or dark brown usually indicates a high humus content; the blacker the soil, the more humus it contains. Color gives a strong hint about fertility because humus is a key catalyst in releasing nutrients to plants. Dark color is not definitive, however, because it may be due to other factors, such as poor drainage or high carbonate content.

Reddish and yellowish colors generally indicate iron oxide stains on soil particles. These colors are most common in tropical and subtropical regions, where many minerals are leached away, leaving behind insoluble iron compounds. In such situations, a red color bespeaks good drainage, and a yellowish hue suggests imperfect drainage. Red soils are also common in desert environments, where the color is carried over intact from reddish parent materials rather than representing a surface stain.

Gray and bluish colors typically indicate poor drainage, whereas mottling indicates saturated conditions for part of the year. In humid areas, a light color implies so much leaching that even the iron has been removed, but in dry climates, it indicates an accumulation of salts. It may also indicate simply a lack of organic matter.

Texture

All soils are composed of myriad particles of various sizes (Figure 12-12), although smaller particles usually predominate. Rolling a sample of soil between your fingers can give you a feel for the principal particle sizes. Table 12-1 shows the standard classification scheme for particle sizes; in this scheme, the size groups are called **separates**. The gravel,

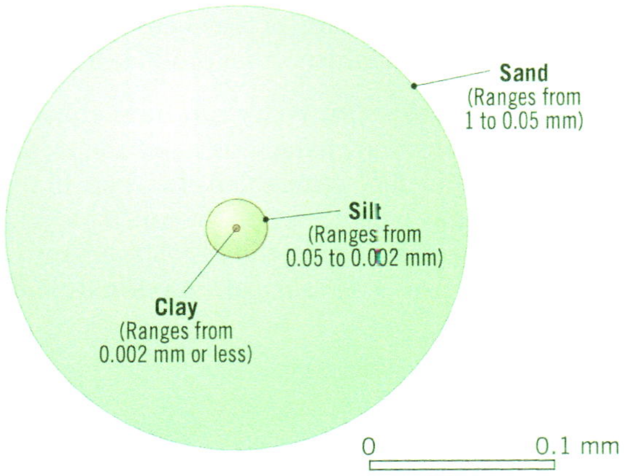

▲ **Figure 12-12** The relative sizes of sand, silt, and clay particles.

TABLE 12-1 Standard U.S. Classification of Soil Particle Size

Separate	Diameter
Gravel	Greater than 2 mm (0.08 in.)
Very coarse sand	1–2 mm (0.04–0.08 in.)
Coarse sand	0.5–1 mm (0.02–0.04 in.)
Medium sand	0.25–0.5 mm (0.01–0.02 in.)
Fine sand	0.1–0.25 mm (0.004–0.01 in.)
Very fine sand	0.05–0.1 mm (0.002–0.004 in.)
Coarse silt	0.02–0.05 mm (0.0008–0.002 in.)
Medium silt	0.006–0.02 mm (0.00024–0.0008 in.)
Fine silt	0.002–0.006 mm (0.00008–0.00024 in.)
Clay	0.002 mm (less than 0.00008 in.)

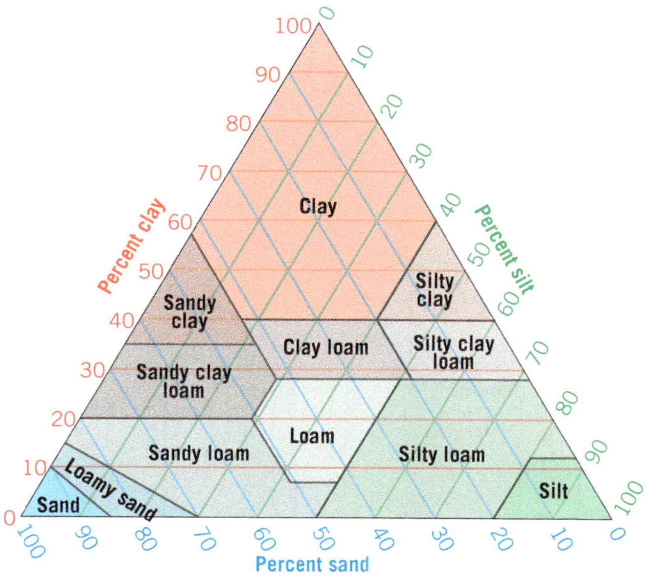

▲ **Figure 12-13** The standard soil-texture triangle. The texture of a soil is determined by the relative proportions of sand, silt, and clay particles. To identify the proportion for a given soil texture, follow each of the three colored lines to its scale on the side of the triangle.

sand, and silt separates are fragments of the weathered parent material and are mostly the grains of common minerals found in rocks, especially quartz, feldspars, and micas. These coarser particles are the inert materials of the soil mass—only the clay particles take part in the intricate chemical activities that occur in the soil.

Because no soil is made up of particles of uniform size, the texture of any soil is determined by the relative amounts of the various separates present. The *texture triangle* (**Figure 12-13**) shows the standard classification scheme for soil texture, based on the percentage of each separate by weight. Below the center of the triangle is **loam**, the name given to a texture in which none of the three principal separates dominates the other two. This fairly even-textured mix is generally the most productive for plants.

Structure

The individual particles of most soils tend to aggregate into clumps called **peds**, and these clumps determine soil structure. The size, shape, and stability of peds have a marked influence on how easily water, air, and organisms (including plant roots) move through the soil, and consequently on soil fertility. Peds are classified on the basis of shape as spheroidal, platy, blocky, or prismatic. These four shapes give rise to seven generally recognized soil structure types (**Figure 12-14**). Aeration and drainage are usually facilitated by peds of intermediate size; both massive and fine structures tend to inhibit these processes.

Some soils, particularly those composed largely of sand, do not develop a true structure—the individual grains do not aggregate into peds. Silt and clay particles readily aggregate in most instances. Other things being equal, aggregation is greatest in moist soils and least in dry ones.

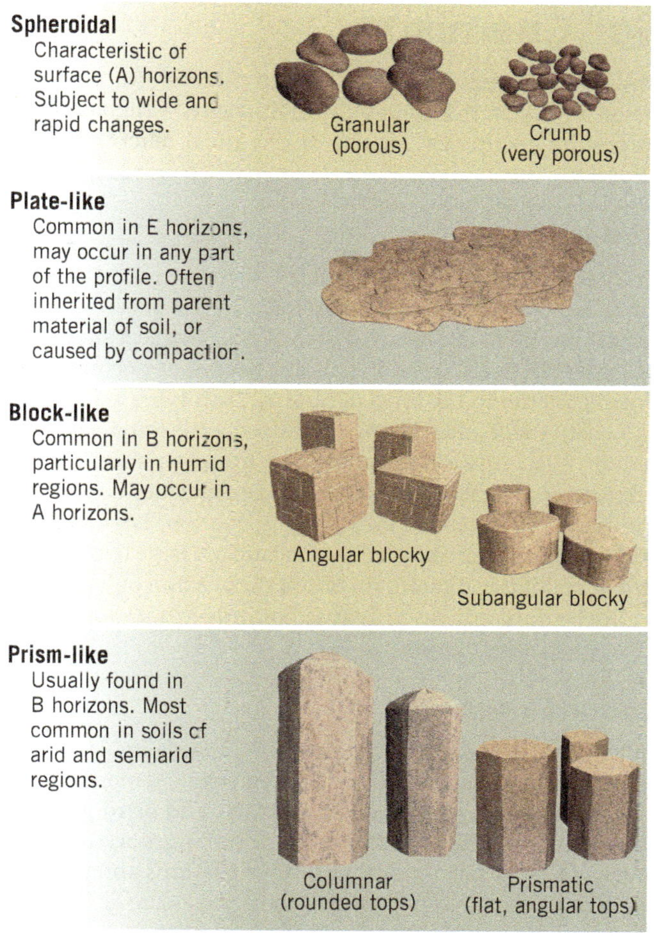

▲ **Figure 12-14** Various structure types of soil particle clumps, or peds, found in mineral soils. (Soil *horizons* are layers with distinct characteristics.)

Structure is an important determinant of a soil's *porosity* and *permeability*. As we learned in Chapter 9, porosity is the amount of pore space between soil particles or between peds, defined as

$$\text{Porosity} = \frac{\text{Volume of pore spaces}}{\text{Total volume}}$$

Porosity is usually expressed as a percentage or a decimal fraction. It is a measure of a soil's capacity to hold water and air.

Permeability is the ability of a soil to transmit water through it. This quantity is determined by the degree of interconnectedness of the pore spaces.

The relationship between porosity and permeability is not simple; the most porous materials are not necessarily the most permeable. Clay, for example, is the most porous separate, but it is the least permeable because the pores are too small for water to pass through easily.

LearningCheck 12-6 **Distinguish between soil porosity and permeability.**

Soil Chemistry

The effectiveness of soil as a growth medium for plants is based largely on the availability of nutrients, which is determined by an intricate series of chemical reactions. Soil chemistry involves microscopic soil particles and ions.

Colloids

Soil particles smaller than about 0.1 micrometer in diameter are called **colloids**. Inorganic colloids consist of clay in thin, crystalline, plate-like forms created by the chemical alteration of larger particles; organic colloids represent decomposed organic matter in the form of humus. Both types are chemically active soil particles. When mixed with water, colloids remain suspended indefinitely as a homogeneous, murky mixture. Some have remarkable storage capacities, so colloids influence the water-holding capacity of a soil. They soak up water, whereas the soil particles that are too large to be classified as colloids can maintain only a surface film of water.

Both inorganic and organic colloids attract and hold great quantities of ions.

Cation Exchange

As we saw earlier, cations are positively charged ions. Elements that form them include calcium, potassium, and magnesium, all of which are essential for soil fertility and plant growth. Colloids carry mostly negative electrical charges on their surfaces, which attract swarms of nutrient cations and prevent them from being leached from the soil (see Figure 12-5).

The combination of a colloid and attached cations is called a *colloidal complex*, and it maintains a delicate balance. If it holds the nutrients strongly, they will not be leached away, yet if the bonds that form the complex are too strong, the plants cannot absorb the nutrients.

Some types of cations are bound more tightly than others. Cations that tend to bond strongly in the colloidal complex may replace those that bond less strongly. For example, cations such as calcium (Ca^{2+}) and magnesium (Mg^{2+}) are fairly easily replaced by metal ions or hydrogen ions in a process called *cation exchange*. The capability of a soil to attract and exchange cations is known as its **cation exchange capacity (CEC)**. In general, the higher the CEC, the more fertile the soil. Soils with a high clay content have a higher CEC than more coarsely grained soils because the former have more colloids. Humus is a particularly rich source of high-CEC activity because humus colloids have a much higher CEC than inorganic clay minerals. The most fertile soils, then, tend to be those that are rich in clay and humus.

LearningCheck 12-7 **What is cation exchange capacity, and how does it relate to soil fertility?**

Acidity/Alkalinity

We can characterize chemical solutions, including those in soil, on the basis of acidity or alkalinity. An *acid* is a chemical compound that when dissolved in water produces hydrogen ions (H^+) or hydronium ions (H_3O^+), whereas a *base* is a chemical compound that when dissolved in water produces hydroxide ions (OH^-). When an acid reacts with a base, a *salt* forms. Solutions that contain dissolved acids are described as *acidic*. Those that contain dissolved bases are called either *basic* or *alkaline* solutions. Any chemical solution can be characterized by its acidity or alkalinity.

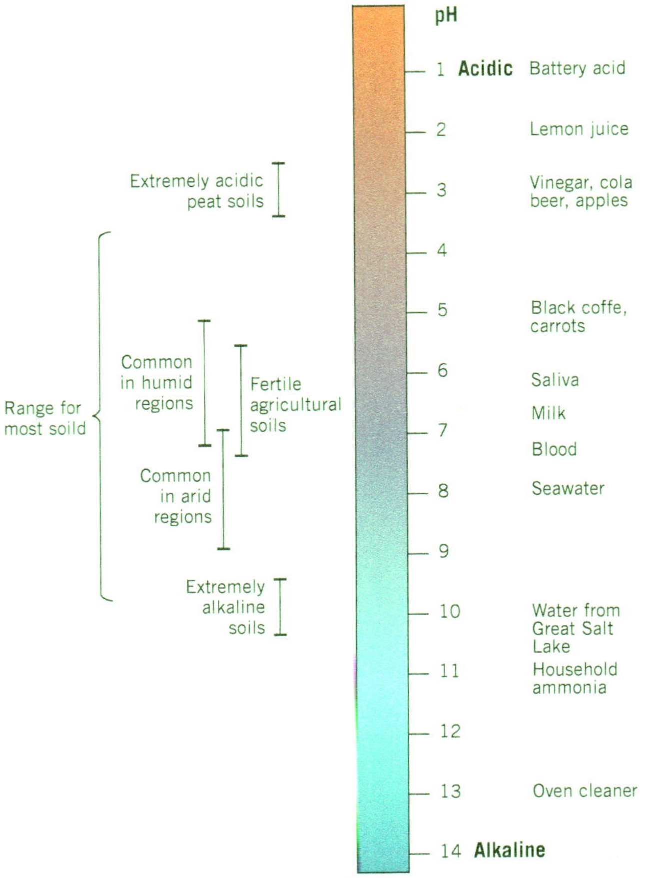

▲ **Figure 12-15** The standard pH scale. The most fertile soils tend to be neither too acidic nor too alkaline.

Nearly all nutrients are provided to plants in solution. An overly alkaline soil solution is inefficient in dissolving minerals and releasing their nutrients. However, if the solution is highly acidic, the nutrients are likely to be dissolved and leached away too rapidly for plant roots to absorb them. The optimum situation, then, is for the soil solution to be neutral: neither too alkaline nor too acidic. The acidity/alkalinity of a soil is determined to a large extent by its CEC.

As we first saw in Chapter 6, chemists describe the acidity/alkalinity of a solution by its pH, which is determined by the relative concentration of hydrogen ions (H^+) in the solution. The scale ranges from 0 to 14. The lower end represents acidic conditions; higher numbers indicate alkaline conditions (**Figure 12-15**). Neutral conditions are represented by a value of 7, and soil having a pH of about 7 is most suitable for the great majority of plants and microorganisms.

Soil Profiles

The development of soil is expressed in two dimensions: depth and time. Depth and age do not have a straight-line relationship; some soils deepen and develop much more rapidly than others.

Four processes deepen and age soils (**Figure 12-16**): *addition* (ingredients added to the soil), *loss* (ingredients lost

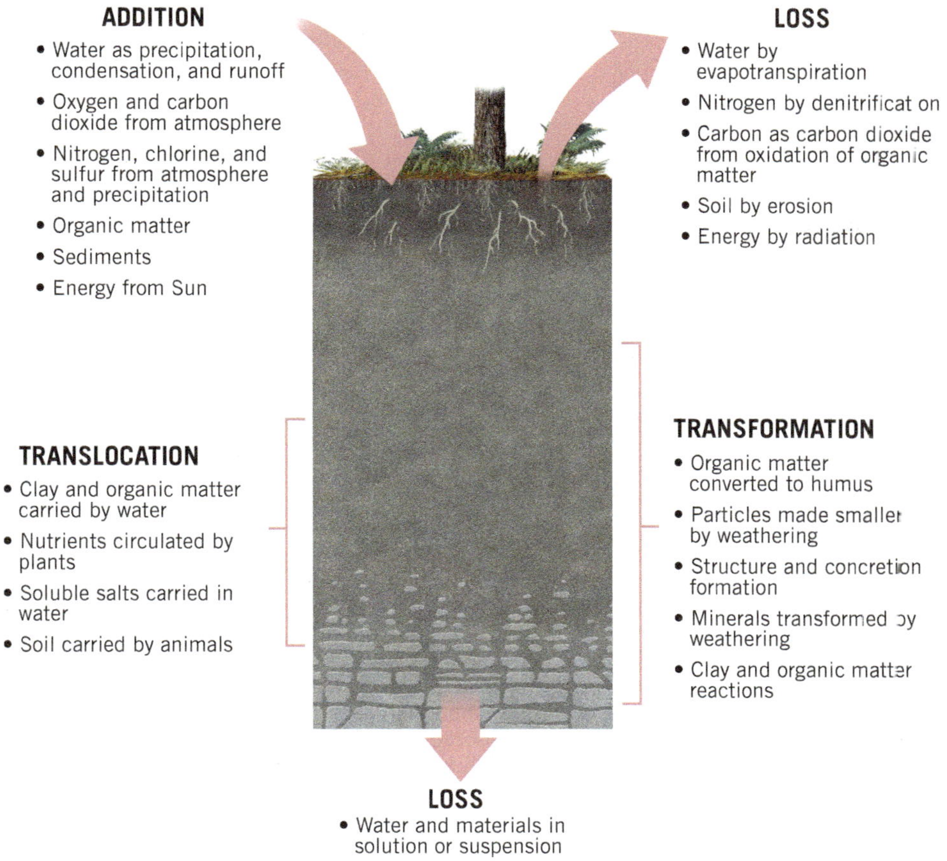

◀ **Figure 12-16** The four soil-forming processes: addition, loss, translocation, and transformation. Geologic, climatic, topographic, biological, and chronological (time) soil-forming factors influence the rate at which these four processes occur and therefore the rate at which soil forms.

from the soil), *translocation* (ingredients moved within the soil), and *transformation* (ingredients altered within the soil). The five soil-forming factors discussed earlier—geologic, climatic, topographic, biological, and time—influence the rate of these four processes, the result being the development of various soil *horizons* and the *soil profile*.

Soil Horizons

Soil tends to have more or less distinctly recognizable layers, called **horizons**, each with different characteristics. The horizons are positioned approximately parallel with the land surface, one above the other, normally separated by a transition zone rather than a sharp line. A vertical cross section (as might be seen in a road cut or the side of a trench) from Earth's surface down through the soil layers and into the parent material is referred to as a **soil profile**. The different varieties of soils in the world are usually grouped and classified on the basis of differences exhibited in their profiles.

In a well-developed soil profile, six horizons are distinguished (Figure 12-17):

- The **O horizon** is a surface layer of organic matter, both fresh and decaying. This horizon results essentially from litter derived from dead plants and animals. It is common in forests and generally absent in grasslands. Soils typically do not have an O horizon; the surface horizon of most soils is the A horizon.
- The **A horizon**, colloquially referred to as *topsoil*, is a mineral horizon that also contains considerable organic matter. It is formed either at the surface or immediately below an O horizon. A horizons generally contain enough organic matter to give the soil a darker color than underlying horizons. A horizons are also normally coarser in texture, having lost some of the finer materials by erosion and eluviation. Seeds germinate mostly in this horizon.
- The **E horizon** is normally lighter in color than either the overlying A or the underlying B horizon. It is essentially an eluvial layer from which clay, iron, and

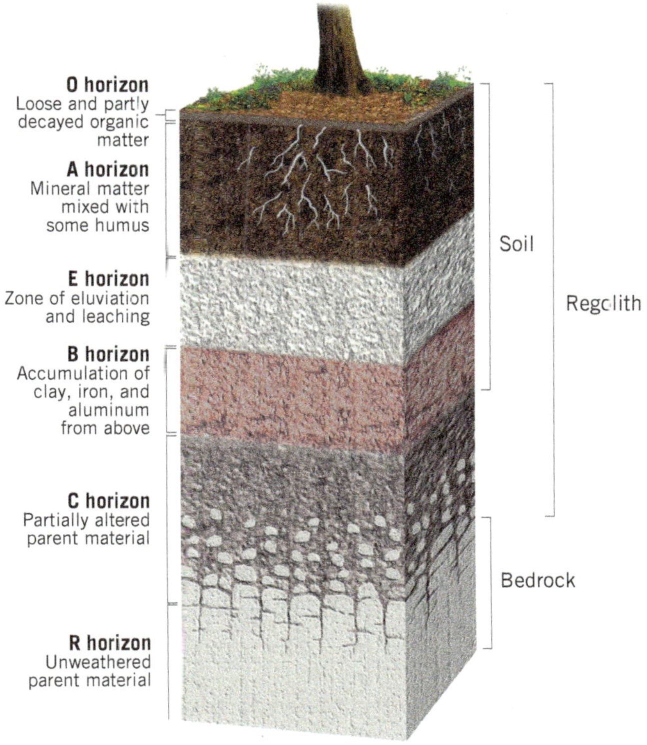

▲ **Figure 12-17** Idealized soil profile. The true soil, or solum, consists of the O, A, E, and B horizons.

aluminum have been removed, leaving a concentration of abrasion-resistant sand or silt particles.

- The **B horizon**, usually called *subsoil*, is a mineral horizon of illuviation where most of the materials removed from above have been deposited. A collecting zone for clay, iron, and aluminum, this horizon is usually of heavier texture, greater density, and relatively greater clay content than the A horizon.
- The **C horizon** is unconsolidated parent material (regolith) beyond the reach of plant roots and soil-forming processes except weathering. It lacks organic matter.
- The **R horizon** is bedrock with little evidence of weathering.

Strictly speaking, a true soil, which is called **solum**, extends down through only the B horizon.

Time is the critical passive factor in profile development, but surface water is the vital active factor. If there is no surface water, from rainfall or snowmelt or some other source, to infiltrate the soil, there can be no profile development. Descending water carries material from the surface downward, from topsoil into subsoil, by eluviation and leaching. Most of this transported material is deposited a few tens of centimeters (a few feet) below the surface. In the usual pattern, topsoil (A) becomes a somewhat depleted horizon through eluviation and leaching, and subsoil (B) develops as a layer of accumulation due to illuviation.

A profile that contains all horizons is typical of a humid area on well-drained but gentle slopes in an environment that has been undisturbed for a long time. In many parts of the world, however, such idealized conditions do not exist. A typical soil profile may have one horizon particularly well developed, one missing altogether, a *fossil horizon* formed under a different past climate, an accumulation of a hardpan (a very dense and impermeable layer), surface layers removed through accelerated erosion, or some other variation.

A soil containing only an A horizon atop partially altered parent material (C horizon) is said to be *immature*. The formation of an illuvial B horizon is normally an indication of a *mature* soil.

LearningCheck 12-8 Describe the soil horizons found in solum.

Pedogenic Regimes

Soil-forming factors and processes interact in almost limitless variations to produce soils of all descriptions. However, we can identify five major **pedogenic regimes** ("soil-forming" regimes): *laterization, podzolization, gleization, calcification*, and *salinization*. These regimes are environmental settings in which distinctive physical—chemical—biological processes prevail.

Laterization

Laterization is named for the brick-red color of the soil it produces (*later:* Latin, "brick"). The processes associated with this regime are typical of the warm, moist regions of

▲ **Figure 12-18** In humid tropical environments, laterization is the dominant soil-forming regime, and most soils are reddish as a result of the prominence of iron and aluminum compounds. The reddish layers exposed here in the Painted Hills in John Day Fossil Beds National Monument, Oregon, are bands of ancient lateritic soil that formed about 35 million years ago when the climate was tropical.

the world. The soil formed by laterization is most prominent, then, in the tropics and subtropics, in regions dominated by forest, shrub, or savanna vegetation.

Laterization is characterized by rapid weathering of parent material, dissolution of nearly all minerals, and speedy decomposition of organic matter. Probably the most distinctive feature of laterization is the leaching-away of silica, the most common constituent of most rock and soil and a constituent that is usually highly resistant to being dissolved. Most other minerals are also leached out rapidly, leaving behind primarily iron and aluminum oxides and barren grains of quartz sand. This residue normally imparts to the resulting soil the reddish color that gives this regime its name (Figure 12-18). The A horizon is highly eluviated and leached, whereas the B horizon has a considerable concentration of illuviated materials.

Because plant litter is rapidly decomposed wherever laterization is the predominant regime, little humus is incorporated into the soil. Even so, plant nutrients are not totally removed by leaching because the natural vegetation, particularly in a forest, quickly absorbs many of the nutrients in solution. If the vegetation is relatively undisturbed by human activities, this regime has the most rapid of nutrient cycles, and the soil is not totally impoverished by the speed of mineral decomposition and leaching. Where the forest is cleared for agriculture or some other purpose, however, most base nutrients are likely to be lost from the cycle because the tree roots that would bring them up are gone. The soil then rapidly becomes impoverished, and hard crusts of iron and aluminum compounds are likely to form.

Latosols: Soils produced by laterization are called *latosols*. They sometimes develop to depths of several meters because of the strong weathering activities and the fact that laterization continues year-round in these climates. Most latosols have little to offer as agricultural soils, but laterization often produces concentrations of iron and aluminum oxides that can be mined profitably.

Podzolization

Podzolization is another regime named after the color of the soil it produces; in this case, gray (*podzol:* Russian for "under ashes"). It also occurs in regions having a positive moisture balance and involves considerable leaching. However, podzolization occurs primarily in areas where the vegetation has limited nutrient requirements and where the plant litter is acidic, such as in mid- and high-latitude locales with a coniferous forest cover. The typical location for podzolization is under a boreal forest in the subarctic climates found only in the Northern Hemisphere.

In these cool regions, chemical weathering is slow, but the abundance of acids, plus adequate precipitation, makes leaching very effective. Mechanical weathering from frost action is relatively rapid during the unfrozen part of the year. Bedrock here consists mostly of ancient crystalline rocks (some transported by Pleistocene glaciers) that are rich in quartz and aluminum silicates and poor in the alkaline mineral cations important in plant nutrition. The boreal forests, dominated by conifers, require little in the way of soil nutrients, and their litter returns few nutrient minerals when it decays. The litter is largely needles and twigs, which accumulate on the soil surface and decompose slowly. Microorganisms do not thrive in this environment, so humus production is retarded. Moisture is relatively abundant in summer, so leaching of whatever nutrient cations are present in the topsoil—along with iron oxides, aluminum oxides, and colloidal clays—is fairly complete.

Podzols: Soils produced by podzolization are referred to as *podzols*. They are shallow and acidic and have a fairly distinctive profile, usually including an O horizon. The upper part of the A horizon is eluviated to a silty or sandy texture and is so leached as to appear bleached. It is usually an ashy, light gray color, imparted by its high silica content. The illuviated B horizon is a receptacle for the iron-aluminum oxides and clay minerals leached from above and has a sharply contrasting darker color (sometimes with an orange or yellow tinge). Soil fertility is generally low, and a crumbly structure makes the soil very susceptible to accelerated erosion if the vegetation cover is disturbed.

LearningCheck 12-9 **Contrast laterization and podzolization. What are the characteristics of soils produced by each process?**

Gleization

Gleization is a regime restricted to waterlogged areas, normally in a cool climate. (The name comes from the Slavic word *glej*, for "muddy ground.") The poor drainage that produces a waterlogged environment can be associated with flat land, a topographic depression, a high water table, or various other conditions. In North America, gleization is prominent around the Great Lakes, where recent glacial deposition disrupted previously established drainage patterns.

Gley Soils: Soils produced by gleization are known as *gley soils*. They characteristically have a highly organic A horizon, where decomposition proceeds slowly because bacteria are inhibited by the lack of oxygen in waterlogged sites. In such *anaerobic* conditions, chemical reduction takes place: notably, ferric iron (Fe^{3+}) compounds are reduced to ferrous iron (Fe^{2+}) compounds. (*Reduction* occurs when a substance gains an electron—the opposite of *oxidation*.) Reduced iron is more easily carried away than oxidized iron, so over time the soil tends to become iron poor and turn gray.

Gley soils are usually too acidic and oxygen poor to be productive for anything but water-tolerant vegetation. If drained artificially, however, and fertilized with lime to counteract the acid, their fertility can be greatly enhanced.

Calcification

In semiarid and arid climates, where precipitation is less than potential evapotranspiration, leaching is short-lived or absent. Natural vegetation consists of grasses or shrubs. **Calcification** (so called because calcium salts are produced in this regime) is the dominant pedogenic process in these regions, such as the drier prairies of North America, steppes of Eurasia, and savannas and steppes of the subtropics.

Calcic Hardpans: Eluviation and leaching are restricted by the absence of percolating water, so materials that would be carried downward in other regimes become concentrated in the soil where calcification occurs. Moreover, there is considerable upward movement of water by capillary action in dry periods. Calcium carbonate ($CaCO_3$), the most important chemical compound active in this regime, is carried downward by limited leaching after a rain. $CaCO_3$ is often concentrated in the B horizon, forming a dense layer of *hardpan*, then brought upward by capillary water and grass roots, and finally returned to the soil when the grass dies. Little clay is formed because chemical weathering is limited. Organic colloidal material, however, is often abundant.

Where calcification takes place under undisturbed grassland, the resulting soils are likely to be remarkably productive. Humus from decaying grass yields abundant organic colloidal material, giving the soil a dark color and contributing to a structure that can retain nutrients and soil moisture. Grass roots tend to bring calcium up from the B horizon sufficiently to inhibit or delay the formation of calcic hardpans. Where shrubs are the dominant vegetation, roots are fewer but deeper, so nutrients are brought up from deeper layers. Little material accumulates at the surface, with less humus incorporated into the soil

Where calcification is operative in true deserts, the soils tend to be shallower and sandier, calcic hardpans may form near the surface, little organic matter accumulates on or in the soil, and the soils resemble the parent material.

Salinization

In arid and semiarid regions, it is fairly common to find areas with poor drainage, particularly in enclosed valleys and basins. Intense evaporation draws moisture upward and into the atmosphere. The evaporating water leaves behind various salts in or on the soil surface, sometimes enough to make the surface a brilliant white, and the pedogenic regime is called **salinization** (*sal:* Latin for "salt"). These salts, which are mostly chlorides and sulfates of calcium and sodium, are toxic to most plants and soil organisms. The resulting soil can support very little life other than a few salt-tolerant grasses and shrubs.

Soils developed in this regime can sometimes be made productive through careful water management, but artificial drainage is equally necessary or else salt accumulation will intensify. Indeed, human-induced salinization has ruined good agricultural land in various parts of the world many times (Figure 12-19).

LearningCheck 12-10 **Under what conditions can salinization lead to soils that are unsuitable for agriculture?**

Climate and Pedogenic Regimes

The pedogenic regimes are distinguished primarily on the basis of climate, as reflected in temperature and moisture availability (Figure 12-20), and secondarily on the basis of vegetation cover. In regions where there is normally a surplus of moisture—that is, annual precipitation exceeds annual evapotranspiration—water movement in the soil is predominantly downward, and leaching is prominent. In such areas where temperatures are relatively high throughout the year, laterization is the dominant regime; where winters are long and cold, podzolization predominates; and where the soil is saturated most of the time due to poor drainage, gleization is notable.

A broadly valid generalization is that gleization can occur in many environments, as it is more dependent on local topography than on macroclimate. In regions with a moisture deficit, the principal soil moisture movement is upward (through capillarity), and leaching is limited. Calcification and salinization are the principal pedogenic regimes under these conditions.

Although climate is the primary factor influencing the pedogenic regimes, local variations in soil and soil-forming processes are frequently found. For example, see the box *Focus: Soil Differences—They're All about Scale.*

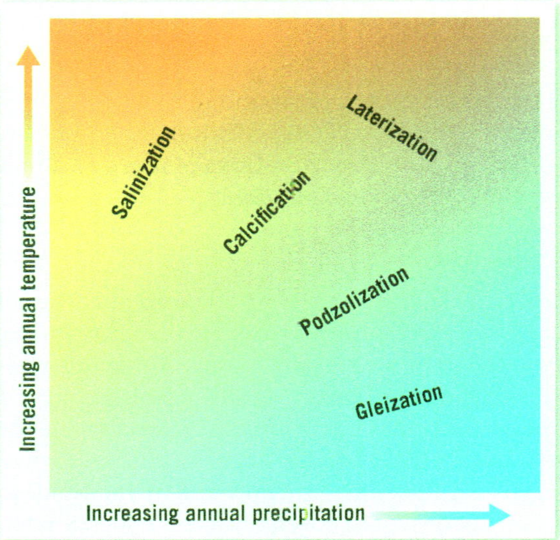

▲ **Figure 12-20** Relative temperature–moisture relationships for the five principal pedogenic regimes. Where temperature is high but precipitation low, salinization is the main method of soil formation. Where temperature is low but precipitation high, gleization predominates. Laterization occurs where both temperature and precipitation are high. Calcification and podzolization take place in less extreme environments.

▼ **Figure 12-19** Soil salinization in a vineyard near San Juan, Argentina.

Soil Classification

In no other subdiscipline of physical geography is the matter of classification more complicated than with soil science.

The Soil Taxonomy

Over the past century, scientists have devised various soil classifications. As the knowledge of soil characteristics and processes has become greater, so have the efforts at soil classification become more sophisticated. Several systems have been developed, particularly in Canada, the United Kingdom, Russia, France, and Australia. Moreover, United Nations agencies have their own classification schemes. The system that is presently in use in the United States was developed by the Natural Resources Conservation Service (NRCS) and is called the **Soil Taxonomy**.

The basic characteristic that sets Soil Taxonomy apart from previous systems is that it is *generic*, which means it is organized on the basis of observable soil characteristics. The focus is on the existing properties of a soil rather than on the environment, processes of development, or properties it would possess under virgin conditions. The logic of such a generic system is theoretically impeccable: soil has certain properties that can be observed, measured, and at least partly quantified.

Like other logical generic systems, the Soil Taxonomy is hierarchical, which means that it has several levels of

Focus

Soil Differences—They're All about Scale

• **Randall Schaetzl,** Michigan State University

Why do soils vary so much from place to place? The answer often depends on the *scale* of observation. Across large expanses, soils vary mainly because of changes in climate and vegetation. Early scientists knew that as we travel from the forests of the eastern United States to the grasslands of the central states, the soils also changed. This observation led many early soil scientists to identify major soil "zones" across the world, which not coincidentally correlate well to climate and vegetation belts. This early work remains valid today, although geographers have become increasingly focused on soil variation across smaller tracts of land. Knowing how soils vary across a field or subdivision, for example, helps us determine the most appropriate land use for that tract.

At local scales, most soil variation is a function of topography. Lowland areas are wet, and uplands have drier soils. For most agricultural and construction applications, drier sites—upland soils—are better. Dryness and aeration are manifested in soils by red or brown colors. Wet and waterlogged soils are typically darker and usually gray.

▼ **Figure 12-C** The root plate of a recently uprooted tree, and the pit remaining where the roots formerly grew.

▲ **Figure 12-D** After complete decay of the tree and its roots, a pit-and-mound pair, like this one, commonly forms.

◄ **Figure 12-E** After centuries of uprooting disturbance, a pock-marked landscape (now cleared of trees) is evident.

Across very short distances, soils can also vary tremendously. Here, the factors that impact the soils and cause major changes are often related to *disturbance*: an anthill, a gully, a toppled tree.

Tree Uprooting—A Major Disturbance Factor in Forests: Tree uprooting is particularly important in forested landscapes (**Figure 12-C**). As trees uproot, they tear up a considerable amount of soil. This soil eventually slumps off the roots, forming a conspicuous mound. A pit is left behind where the tree once stood. Thus, pits and mounds usually come in pairs (**Figure 12-D**). Over time, many of the soils across a forested landscape may become disturbed in this manner (**Figure 12-E**). Some researchers have estimated that every soil in a forest can be "plowed' in this way every few thousand years.

Pit-and-mound topography dramatically affects the soils below, leading to complex spatial patterns and tremendous spatial variability over short distances. Forest litter collects in the pits, and excess water funnels through them, leaving them enriched in organic matter, more leached, and generally better developed. Conversely, water and litter are shed off mounds, and soils there are weakly developed. In cold climates, snow accumulates in pits; mounds are blown free of snow, leaving them to freeze. In the end, the pit-and-mound microtopography created by uprooting leads to short-distance variation in soils that can be quite impressive. This complex spatial variability is a boon for the ecosystem, as it provides habitats for a variety of organisms. If they don't like the soil where they are, the soil just next door may be different—even better!

Questions

1. What other types of small-scale disturbances might create short-distance spatial variation in forest soils?
2. In a grassland ecosystem, what are the main types of disturbances that affect soils and create spatial variability?

TABLE 12-2 Name Derivations of Soil Orders

Order	Derivation
Alfisols	"Al" for aluminum, "f" for iron (chemical symbol Fe); soils rich in those elements
Andisols	Andesite, rock formed from type of magma in Andes Mountains volcanoes; soils high in volcanic ash
Aridisols	Latin *aridus*, "dry"; dry soils
Entisols	Last three letters in "recent"; recently formed soils
Gelisols	Latin *gelatio*, "freezing"; soils in areas of permafrost
Histosols	Greek *histos*, "living tissue"; soils containing mostly organic matter
Inceptisols	Latin *inceptum*, "beginning"; young soils at the beginning of their "life"
Mollisols	Latin *mollis*, "soft"; soft soils
Oxisols	Oxygen; soils rich in oxygen-containing compounds
Spodosols	Greek *spodos*, "wood ash"; ashy soils
Ultisols	Latin *ultimus*, "last"; soils that are leached of all nutrient bases
Vertisols	Latin *verto*, "turn"; soils in which the usual horizon order is inverted

generalization, with each higher level encompassing several members of the level immediately below it. The number of similarities among the members of a category increases with each step downward in the hierarchy; in the lowest-level category, all members have mostly the same properties. (See Appendix VII for details.)

Soil Order: At the highest level of the Soil Taxonomy is **soil order**. Only 12 orders are recognized worldwide (Table 12-2). The soil orders, and many of the lower-level categories as well, are distinguished from one another largely on the basis of certain diagnostic properties, which are often expressed in combination to form *diagnostic horizons*. The two basic types of diagnostic horizon are the *epipedon* (from the Greek word *epi*, meaning "over" or "upon"), which is essentially the A horizon (or the combined O/A horizon), and the *subsurface horizon*, which is roughly equivalent to the B horizon. (Note that all A and B horizons are not necessarily diagnostic, so the terms and concepts are not synonymous.)

Soil orders are subdivided into *suborders*, of which about 50 are recognized in the United States. The third level consists of *great groups*, which number about 250 in the United States. Successively lower levels in the classification are subgroups, families, and series. About 20,000 soil series have been identified in the United States to date, and the list will undoubtedly be expanded in the future. To comprehend general world distribution patterns, however, we need to concern ourselves with orders only (Figure 12-21).

LearningCheck 12-11 What does it mean to say that the Soil Taxonomy is a generic system of classification?

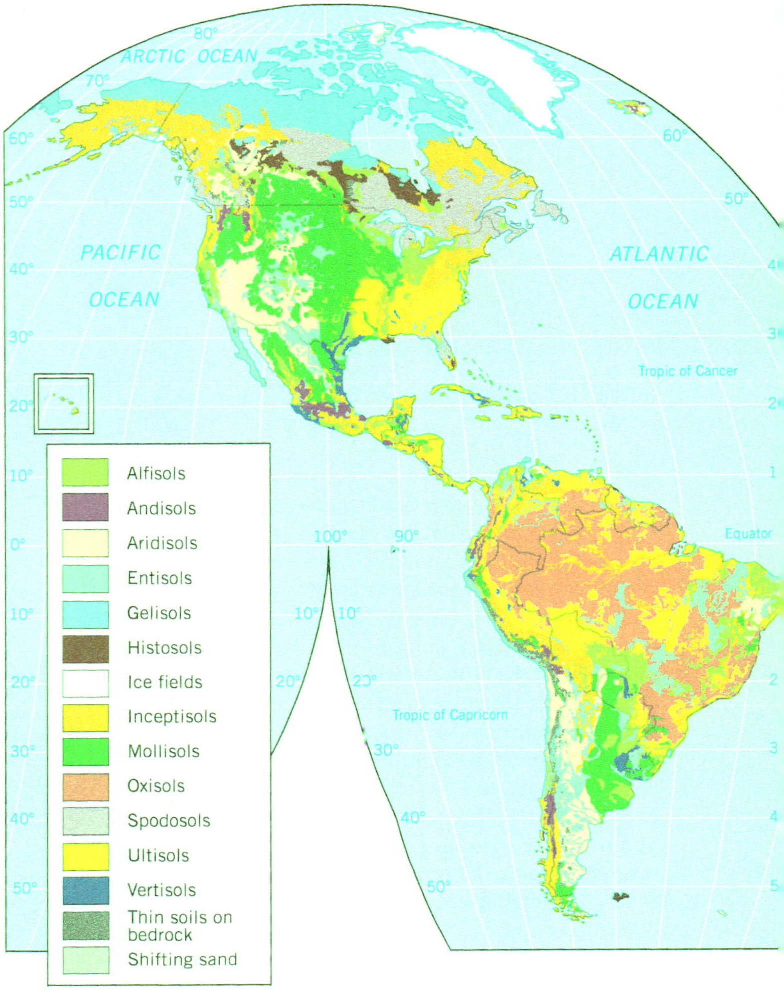

▲ **Figure 12-21** Soils of the world.

The Mapping Question

The higher levels of the Soil Taxonomy are generalized abstractions of average or typical conditions over broad areas, so the selection of an appropriate mapping technique can significantly influence our understanding of soil distribution. In most soil maps, if more than one soil type is present in a given area, that area is classified by the prevailing soil type. Such a map is useful in portraying the general distribution of the major soil types—the smaller the scale, the greater the generalization needed. The many variations and exceptions found at the local level can be shown on more detailed large-scale soil maps only.

Global Distribution of Major Soils

We recognize 12 orders of soils. They are distinguished largely on the basis of properties that reflect a major course of development, with considerable emphasis on the presence or absence of diagnostic horizons. We consider each of the 12 orders, beginning with those characterized by little profile development and progressing to those with the most highly weathered profiles, as generally represented from top to bottom in Figure 12-22.

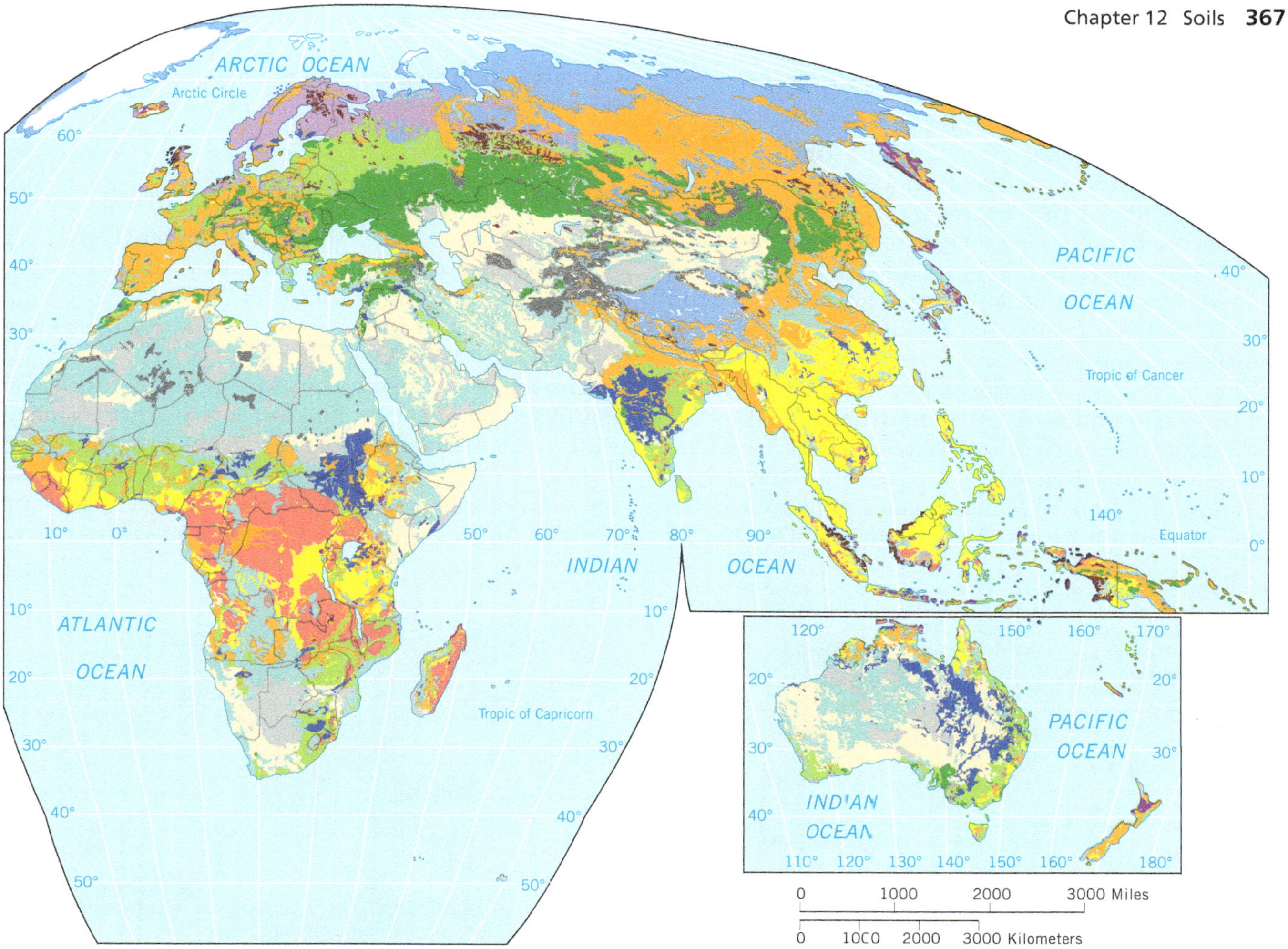

▲ **Figure 12-21** (*Continued*)

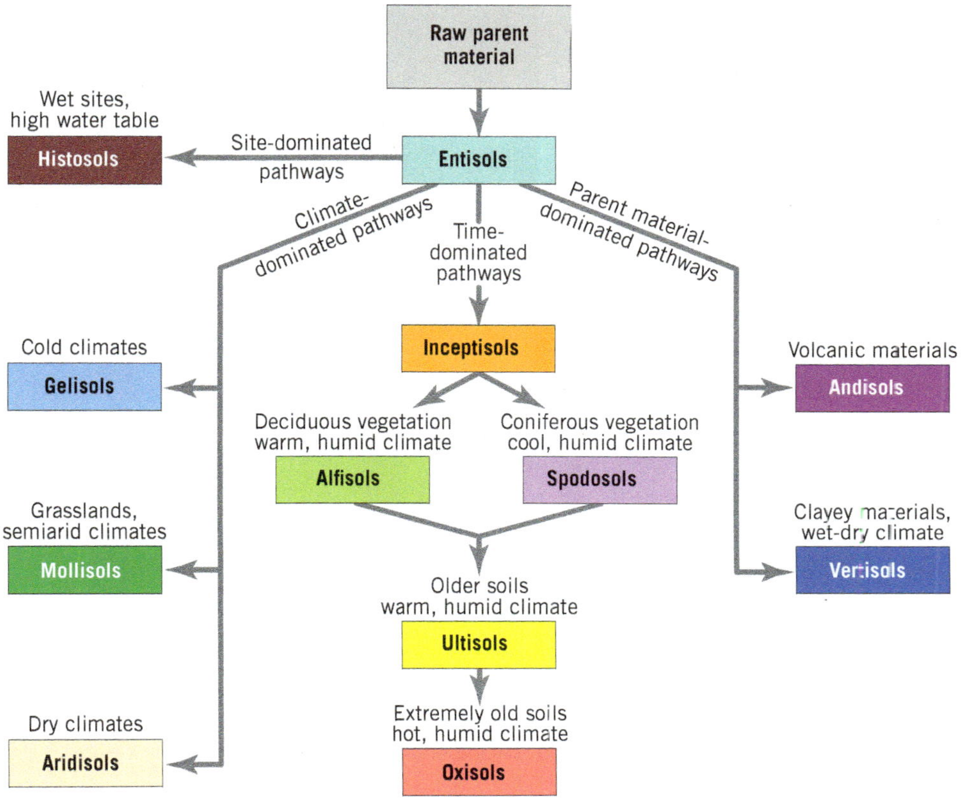

◀ **Figure 12-22** Theoretical soil order development pathways. Soils develop along "pathways" in which different factors—such as parent material, climate, local site conditions, or length of time—may dominate. (The colors for each soil order match those on the soil distribution maps in this chapter.)

Entisols (Very Little Profile Development)

The least well developed of all soils, **Entisols** have experienced little mineral alteration and are virtually without pedogenic horizons. Their undeveloped state is usually a function of time: most Entisols are surface deposits that have not been in place long enough for pedogenic processes to have had much effect. Some, however, are very old, and in these soils the lack of horizon development is due to a mineral content that does not alter readily, to a very cold climate, or to some other factor.

The distribution of Entisols is very widespread and cannot be specifically correlated with particular moisture or temperature conditions or with certain types of vegetation or parent materials (Figure 12-23). In the United States, Entisols are most prominent in the dry lands of the West but are found in most other parts of the country as well. They are commonly thin and/or sandy and have limited productivity, although those developed on recent alluvial deposits tend to be quite fertile.

Inceptisols (Few Diagnostic Features)

Another immature order of soils is the **Inceptisols**. They have not developed sufficiently to produce diagnostic horizons (Figure 12-24). If the Entisols can be called "youthful," the Inceptisols are "adolescent." They are primarily eluvial soils and lack illuvial layers.

Like Entisols, Inceptisols are widespread over the world in various environments. Also like Entisols, they include a variety of fairly dissimilar soils that share a lack of maturity. They are most common in tundra and mountain areas but are also notable in older valley floodplains. Their world distribution pattern is very irregular. This is also true in

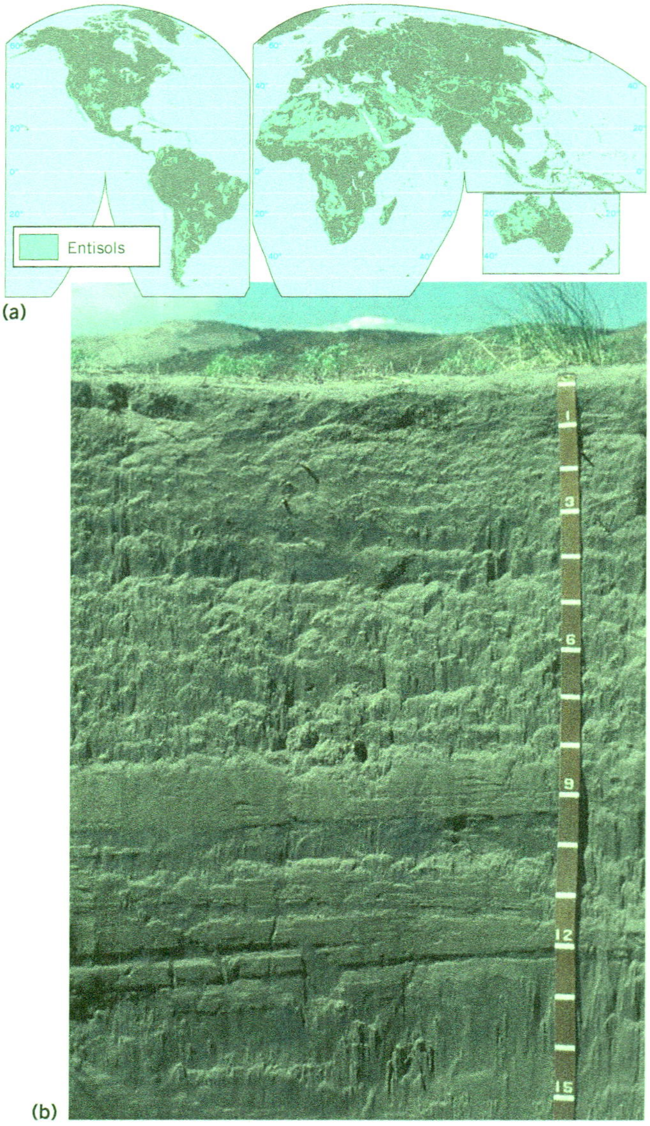

▲ Figure 12-23 Entisols. (a) World distribution of Entisols. (b) Profile of an Entisol in Idaho. This weakly developed soil formed from wind-blown sand deposits.

▲ Figure 12-24 Inceptisols. (a) World distribution of Inceptisols. (b) Profile of a New Zealand Inceptisol with a distinctive B horizon of white pebbly material. The scale is in meters.

the United States, where they are most typical of the Appalachian Mountains, the Pacific Northwest, and the lower Mississippi Valley.

LearningCheck 12-12 **Describe and explain the soil horizon development of Entisols.**

Andisols (Volcanic Ash Soils)

Having developed from volcanic ash, **Andisols** have been deposited in relatively recent geological time. They are not highly weathered, therefore, and there has been little downward translocation of their colloids. Profile development is minimal, and the upper layers are dark (Figure 12-25). Their inherent fertility is relatively high.

Andisols are found primarily in volcanic regions of Japan, Indonesia, and South America, as well as in the very productive wheat lands of Washington, Oregon, and Idaho.

Gelisols (Cold Soils with Permafrost)

Gelisols are young soils with minimal profile development (Figure 12-26). They develop only slowly because of cold temperatures and frozen conditions. These soils typically have a permafrost layer. Also commonly found in Gelisols is *cryoturbation*, or frost churning, which is the physical disruption and displacement of soil material by freeze–thaw action. Most of the soil-forming processes in Gelisols take place above the permafrost in the active layer that thaws every year or so.

▲ Figure 12-25 Andisols. (a) World distribution of Andisols. (b) Profile of an Andisol in Washington state. The various horizons are very clearly delineated.

▲ Figure 12-26 Gelisols. (a) World distribution of Gelisols. (b) Profile of a Gelisol in Alaska. The scale is in centimeters.

Gelisols are the dominant soils of Arctic and subarctic regions. They occur in association with boreal forest and tundra vegetation; thus, they are primarily found in Russia, Canada, and Alaska and are prominent in the Himalaya Mountain country of central Asia. Nearly 9 percent of Earth's land area has a Gelisol soil cover.

LearningCheck 12-13 **In which kind of climate are Gelisols typically found?**

Histosols (Organic Soils on Very Wet Sites)

The **Histosols** occupy only a small fraction of Earth's land surface, a much smaller area than any other order. These are organic rather than mineral soils, and they are saturated with water all or most of the time. They may occur in any waterlogged environment but are most characteristic in mid- and high-latitude regions that experienced Pleistocene glaciation. In the United States, they are most common around the Great Lakes, but they also occur in southern Florida and Louisiana. Nowhere, however, is their occurrence extensive (Figure 12-27).

Some Histosols are composed largely of undecayed or only partly decayed plant material, whereas others consist of a thoroughly decomposed mass of muck. The lack of oxygen in the waterlogged soil slows down the rate of bacterial action. The soil becomes deeper mostly by growing upward—that is, by more organic material being added from above.

Histosols are usually black, acidic, and fertile for water-tolerant plants only. If drained, they can be very productive agriculturally for a short while. Before long, however, they are likely to dry out, shrink, and oxidize. This series of steps leads to compaction, susceptibility to wind erosion, and danger of fire.

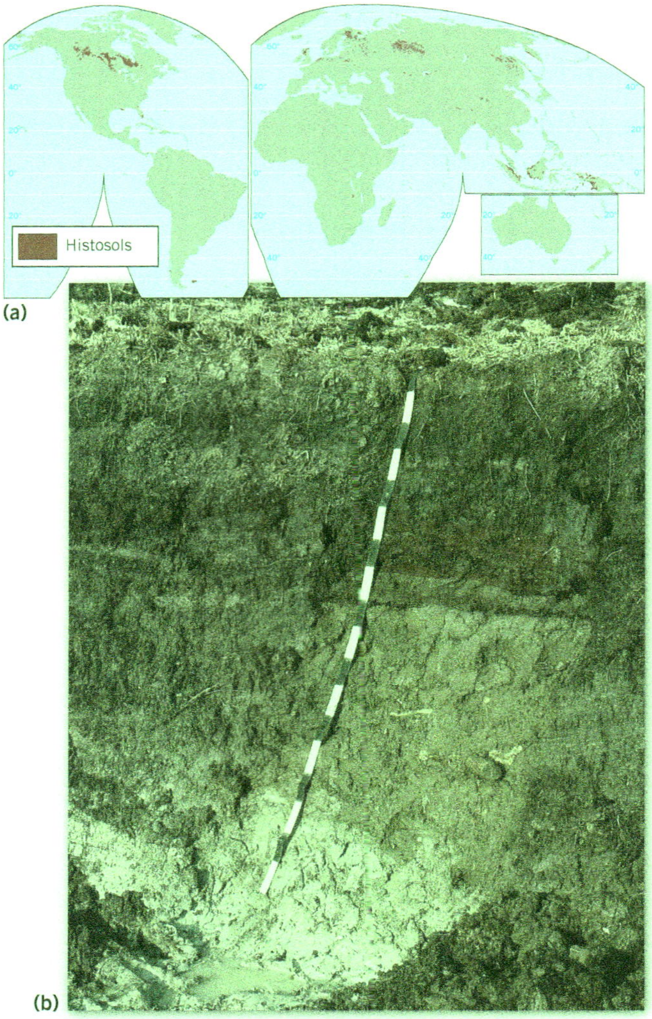

▲ Figure 12-27 Histosols. (a) World distribution of Histosols. (b) Histosols are characteristically dark and composed principally of organic matter.

Aridisols (Soils of Dry Climates)

Nearly one-eighth of Earth's land surface is covered with **Aridisols**, one of the most extensive soil orders (Figure 12-28). They are preeminently soils of the dry lands, occupying environments that do not have enough water to remove soluble minerals from the soil. Thus, their distribution pattern is largely correlated with that of desert and semidesert climates.

Aridisols are typified by a thin profile that is sandy and lacking in organic matter, characteristics clearly associated with a dry climate and a scarcity of penetrating moisture. The epipedon is almost invariably light in color. There are various kinds of diagnostic subsurface horizons, nearly all distinctly alkaline. Most Aridisols are unproductive, particularly because of lack of moisture; if irrigated, however, some display remarkable fertility. The threat of salt accumulation is nonetheless ever present.

LearningCheck 12-14 **In what kind of environment are Aridisols typically found?**

Vertisols (Swelling and Cracking Clays)

Vertisols contain a large quantity of clay that becomes a dominant factor in the soil's development. The clay of Vertisols is described as "swelling" or "cracking" clay. This clay-type soil has an exceptional capacity for absorbing water: when moistened, it swells and expands. As it dries, deep, wide cracks form, sometimes 2.5 centimeters (1 inch) wide and as much as 1 meter (3 feet) deep. Some surface material falls into the cracks, and more is washed in when it rains. When the soil is wetted again, more swelling takes place and the cracks close. This alternation of wetting and drying with expansion and contraction produces a churning effect that mixes the soil constituents ("Vertisol" connotes an inverted condition), inhibits the development of horizons, and may cause minor irregularities in the land surface (Figure 12-29).

The sequence of swelling and contraction needed for Vertisol formation requires an alternating wet and dry

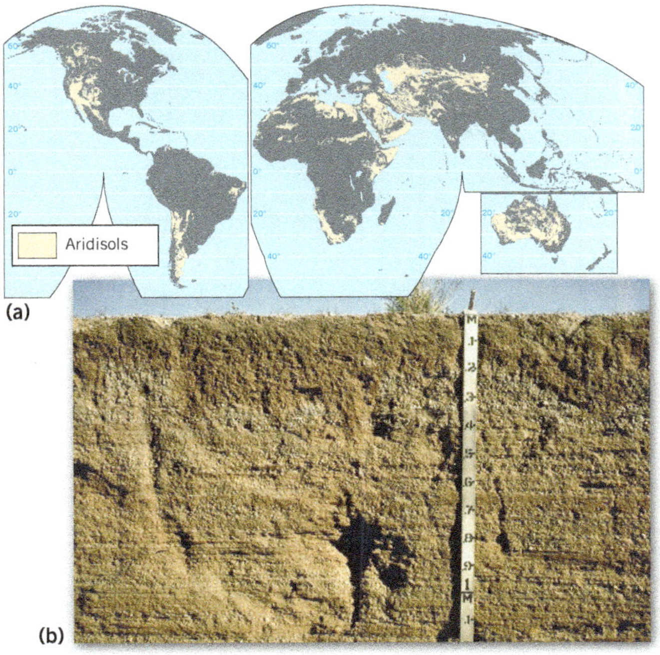

▲ **Figure 12-28** Aridisols. (a) World distribution of Aridisols. (b) The typical sandy profile of an Aridisol, in this case from New Mexico.

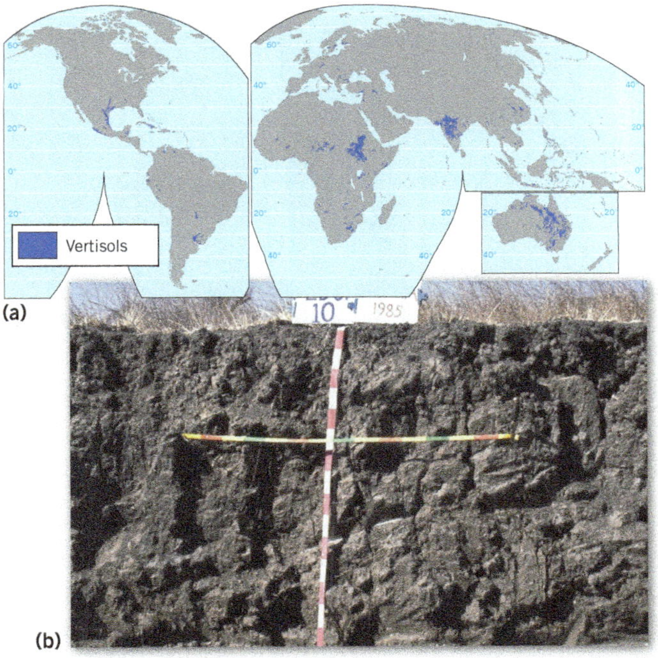

▲ **Figure 12-29** Vertisols. (a) World distribution of Vertisols. (b) A dark Vertisol profile from Zambia. Many cracks are typically found in Vertisols.

climate. Thus, the wet–dry climate of tropical and subtropical savannas is ideal where the proper parent material to yield the clay minerals exists. Consequently, Vertisols are widespread in distribution but very limited in extent. The principal occurrences are in eastern Australia, India, and a small part of East Africa. They are uncommon in the

▲ **Figure 12-30** Mollisols. (a) World distribution of Mollisols. (b) A Mollisol profile from Nebraska. It has a typical mollic epipedon, a surface horizon that is dark and replete with humus.

United States, although prominent in some parts of Texas and California.

The fertility of Vertisols is relatively high, as they tend to be rich in nutrient bases. They are difficult to till, however, because of their sticky plasticity, so they are often uncultivated.

Mollisols (Dark, Soft Soils of Grasslands)

Mollisols are characterized by the presence of a *mollic* epipedon, which is a mineral surface horizon that is dark and thick, contains abundant humus and basic cations, and remains soft (rather than becoming hard and crusty) when it dries out (Figure 12-30). Mollisols can be thought of as transition soils that develop in regions not dominated by either humid or arid conditions. They are typical of midlatitude grasslands and are thus most common in central Eurasia, the North American Great Plains, and the pampas of Argentina.

The grassland environment generally maintains a rich clay–humus content in a Mollisol soil. The dense, fibrous mass of grass roots permeates uniformly through the epipedon and to a lesser extent into the subsurface layers. Decay of plant parts is almost continuous, producing nutrient-rich humus for the living grass.

Mollisols are probably the most productive soil order. They are generally derived from loose parent material rather than from bedrock and tend to have favorable structure and texture for cultivation. Because they are not overly leached, nutrients are generally retained within reach of plant roots. Moreover, Mollisols provide a favored habitat for earthworms, which contribute to softening and mixing of the soil.

Alfisols (Clay-Rich B Horizons, High Base Status)

The most wide-ranging of the mature soils, **Alfisols** occur extensively in low and middle latitudes (Figure 12-31). They are found in a variety of temperature and moisture conditions and under diverse vegetation associations. They tend to be associated with transitional environments and are less characteristic of regions that are particularly hot, cold, wet, or dry. Their global distribution is extremely varied. They are widespread in the United States, with particular concentrations in the Midwest.

Alfisols are distinguished by a subsurface clay horizon and a medium to generous supply of basic cations, plant nutrients, and water. The epipedon is ochric (light-colored), but it has no other particularly diagnostic characteristics and can be considered an ordinary eluviated horizon. The moderate conditions under which Alfisols develop tend to produce balanced soils that are reasonably fertile. Alfisols rank second only to Mollisols in agricultural productivity.

▲ Figure 12-31 Alfisols. (a) World distribution of Alfisols. (b) Profile of an Alfisol in east-central Illinois. The soil has a reddened, clay-rich B horizon. It formed under forest vegetation and is heavily cropped with corn and soybeans.

LearningCheck 12-15 **Compare the relative suitability for agriculture of Vertisols, Mollisols, and Alfisols.**

Ultisols (Clay-Rich B Horizons, Low Base Status)

Ultisols are roughly similar to Alfisols but are more thoroughly weathered and more completely leached of nutrient bases. Ultisols have experienced greater mineral alteration than any other soil in the midlatitudes, although they also occur in the low latitudes. Many pedologists believe that the ultimate fate of Alfisols is to degenerate into Ultisols.

Typically, the A horizon of Ultisols is reddish as a result of a significant proportion of iron; aluminum is abundant as well. These soils usually have a fairly distinct layer of subsurface clay accumulation. The principal properties of Ultisols have been imparted by a great deal of weathering and leaching (Figure 12-32). Indeed, the name (derived from the Latin *ultimos*) suggests that these soils represent the ultimate stage of weathering. The result is a fairly deep soil that is acidic, lacks humus, and has a relatively low fertility due to the lack of bases.

Ultisols have a fairly simple world distribution pattern. They are mostly confined to humid subtropical climates and to some relatively youthful tropical land surfaces. In the United States, they are restricted largely to the Southeast and to a narrow strip along the northern Pacific Coast.

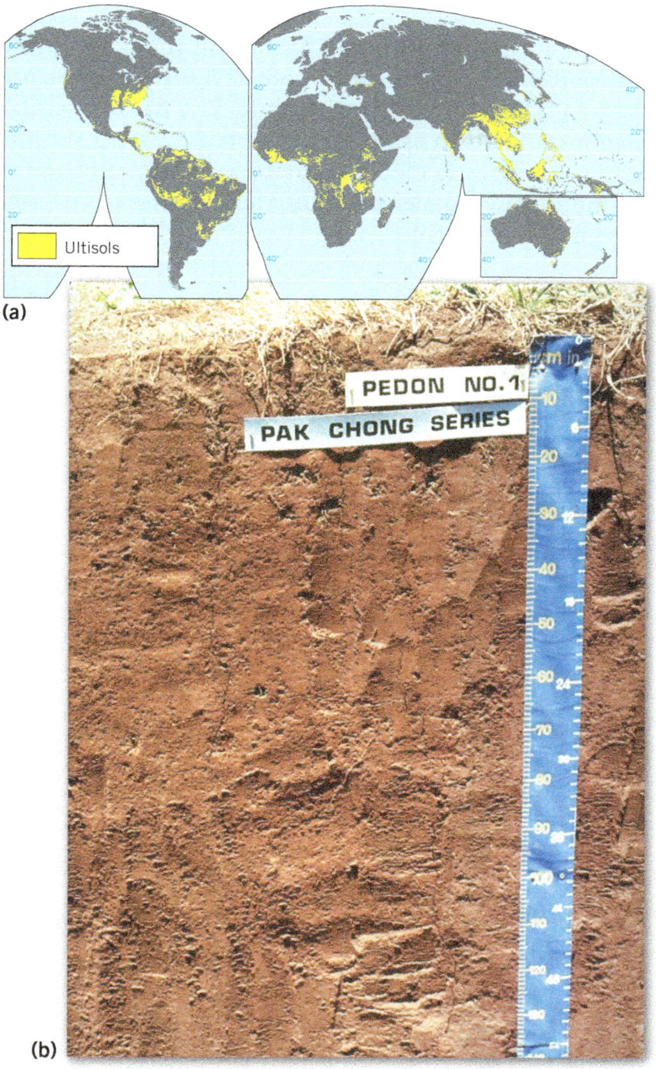

▲ **Figure 12-32** Ultisols. (a) World distribution of Ultisols. (b) A tropical Ultisol from Thailand. It is reddish throughout its profile, indicative of much leaching and weathering.

▲ **Figure 12-33** Spodosols. (a) World distribution of Spodosols. (b) Profile of a Spodosol in northern Michigan. This weakly developed soil was formed in sandy material under coniferous forest vegetation and contains few nutrients. The scale is in centimeters.

Spodosols (Soils of Cool, Forested Zones)

The key diagnostic feature of **Spodosols** is a *spodic* subsurface horizon, an illuvial dark or reddish layer where organic matter, iron, and aluminum accumulate. The upper layers are light-colored and heavily leached (Figure 12-33). At the top of the profile is usually an O horizon of organic litter. Such a soil is a typical result of podzolization.

Spodosols are notoriously infertile. They have been leached of useful nutrients and are acidic throughout. They do not retain moisture well and lack humus and, typically, clay. Spodosols are most widespread in areas of coniferous forest where there is a subarctic climate. Alfisols, Histosols, and Inceptisols also occupy these regions, however, and Spodosols are sometimes found in other environments, such as poorly drained portions of Florida.

Oxisols (Highly Weathered and Leached)

The most thoroughly weathered and leached of all soils are the **Oxisols**, which invariably display a high degree of mineral alteration and profile development. They occur mostly on ancient landscapes in the humid tropics, particularly in Brazil and equatorial Africa, and to a lesser extent in Southeast Asia (Figure 12-34). The distribution pattern is often spotty, with Oxisols mixed with less developed Entisols, Vertisols, and Ultisols. Oxisols are absent from the United States except for Hawai'i, where they are common.

Oxisols are essentially the products of laterization (and in fact were called Latosols in older classification systems). They form in warm, moist climates, although some are now found in drier regions, an indication of climatic change since the soils developed. The diagnostic

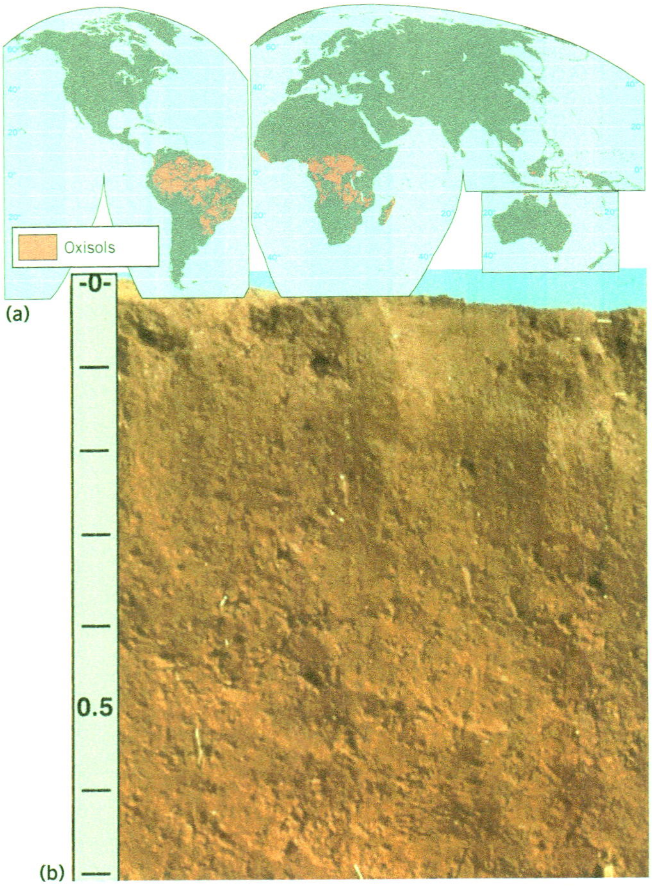

▲ **Figure 12-34** Oxisols. (a) World distribution of Oxisols. (b) Oxisols are impoverished tropical soils that usually are heavily leached and have indistinct horizons. This sample is from Brazil. The scale is in meters.

horizon for Oxisols is a subsurface dominated by oxides of iron and aluminum and with a minimal supply of nutrient bases (this is called an *oxic horizon*). These are deep soils but not inherently fertile. The natural vegetation is efficient in cycling the limited nutrient supply, but if the flora is cleared (to attempt agriculture, for example), the nutrients rapidly leach out and the soil becomes impoverished.

LearningCheck 12-16 **What characteristic do Spodosols and Oxisols have in common, and how does this affect their fertility?**

Distribution of Soils in the United States

The distribution of soil orders in the United States is quite different from that of the world as a whole (**Figure 12-35**). This difference is due to many factors, the most important being that the United States is essentially a midlatitude country that lacks significant expanses of area in the low and high latitudes.

Mollisols are much more common in the United States than in the world as a whole; they are the most prevalent soil order throughout the Great Plains, the prairies of the Midwest, and much of the West. Also significantly more abundant in the United States than elsewhere are Inceptisols and Ultisols. Aridisols and Entisols are proportionally less extensive. Almost totally lacking are Oxisols.

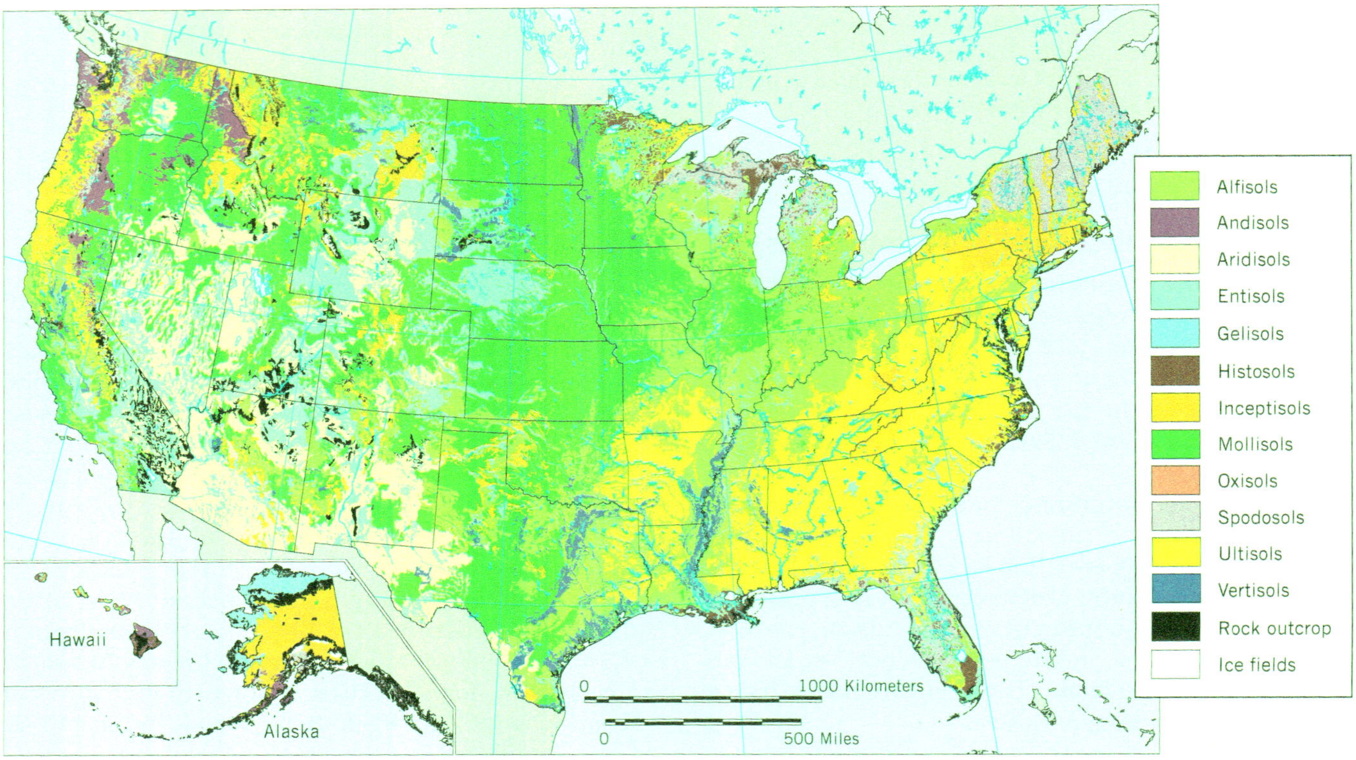

▲ **Figure 12-35** Distribution of predominant soil orders in the United States.

After studying this chapter, you should be able to answer the following questions. Key terms from each text section are shown in **bold type**. Definitions for key terms are also found in the glossary at the back of the book.

Learning Review 12

Key Terms and Concepts

Soil and Regolith (p. 350)

1. What is the relationship between weathering and regolith?
2. What is the difference between **soil** and **regolith**?

Soil-Forming Factors (p. 350)

3. Briefly describe the five principal soil-forming factors.
4. Explain the importance of **parent material** to the nature of the overlying soil.
5. What are some of the roles of animals in soil formation?
6. What roles do microorganisms play in the soil?

Soil Components (p. 354)

7. What are the importance of **clay** and **cations** to plant nutrients in the soil?
8. What is the difference between **litter** and **humus**?
9. Describe and explain the four forms of soil moisture.
10. Distinguish between **field capacity** and **wilting point**.
11. What is **leaching**?
12. Explain the processes of **eluviation** and **illuviation**.
13. What is the **soil-water balance**?
14. What is the role of temperature in the **soil-water budget**?

Soil Properties (p. 357)

15. Distinguish between soil texture and soil structure.
16. What are soil **separates**?
17. What is **loam**?
18. Explain the difference between porosity and permeability.
19. How do **peds** influence soil porosity?

Soil Chemistry (p. 360)

20. What is a **colloid**?
21. Explain what is meant by the **cation exchange capacity (CEC)** of a soil.
22. Which soil is more likely to be fertile: acidic soil or neutral soil? Why?

Soil Profiles (p. 360)

23. What is a soil **horizon**?
24. What is a **soil profile**?
25. Briefly describe the six possible soil horizons:
    - **O horizon**
    - **A horizon**
    - **E horizon**
    - **B horizon**
    - **C horizon**
    - **R horizon**
26. What is **solum**?

Pedogenic Regimes (p. 362)

27. Briefly describe and explain the five major **pedogenic regimes** and the soils they produce:
    - **Laterization**
    - **Podzolization**
    - **Gleization**
    - **Calcification**
    - **Salinization**

Soil Classification (p. 364)

28. How does the **Soil Taxonomy** differ from previous soil classification schemes?
29. What is a **soil order**?

Global Distribution of Major Soils (p. 366)

30. Briefly describe the most distinguishing characteristics of each of the 12 soil orders:
    - **Entisols**
    - **Inceptisols**
    - **Andisols**
    - **Gelisols**
    - **Histosols**
    - **Aridisols**
    - **Vertisols**
    - **Mollisols**
    - **Alfisols**
    - **Ultisols**
    - **Spodosols**
    - **Oxisols**

Study Questions

1. Why does soil tend to be deepest on flat land?
2. Why are earthworms generally considered beneficial to humans?
3. Explain the importance of clay as a constituent of soil.
4. What is a colloidal complex, and how does it relate to soil fertility?
5. What can you learn about a soil from its color?
6. Why do tropical rainforests usually have poor soils?
7. Why is it so difficult to portray soil distribution with reasonable accuracy on a small-scale map?
8. For each of the following locations, indicate which soil order is mostly likely to be found there and why: (a) regions of permafrost; (b) areas with coniferous forest; (c) deserts.
9. Assuming a warm climate with moderate rainfall, how would a soil developing on a sloping, sandstone surface compare with a soil developing on a flat, shale surface?

Exercises

1. Using the map of soil distribution (Figure 12-21 or 12-35), choose one of the soil orders that is found nearby your home and describe that order's general distribution and characteristics.
2. Use Figure 12-13 to determine the soil texture for a combination of 70 percent clay, 10 percent silt, 20 percent sand. ______
3. Use Figure 12-13 to determine the soil texture for a combination of 20 percent clay, 40 percent silt, 40 percent sand. ______
4. Use Figure 12-13 to determine the soil texture for a combination of 50 percent clay, 50 percent silt, 0 percent sand. ______

Environmental **Analysis** SOIL TYPES WHERE YOU LIVE

Data
Soil Types
https://media.pearsoncmg.com/ph/esm/esm_mcknight_physgeo_13/media/

Soil surveys have been completed for 95 percent of the counties in the United States. Farmers and both urban and rural planners depend on soil information to determine how the land can be used.

Activities

Go outside and collect a sample of about 1/4 cup of soil from your location.

1. What color is the soil?
2. Is the soil texture coarse, like gravel; medium, like sand; or fine, like silt or clay? Or is there a combination of textures? (Refer to Table 12-1 for common particle sizes.)
3. Does your soil form peds (clumps)? If so, what is the most common shape for the peds in your sample? (Refer to Figure 12-14 for common shapes.)

Go to http://websoilsurvey.sc.egov.usda.gov to see the USDA's Web Soil Survey. Click on the "Start WSS" button. Under "Quick Navigation," select "State and County"; enter your state and county, then click "View." Click on the data point nearest your location to navigate there. Click on the "i" tool and then your location to display location information.

4. What are the latitude and longitude of your location?
5. When was this aerial photograph taken?

In the Web Soil Survey, create an Area of Interest (AOI) around your location: click the AOI rectangle tool, then drag the mouse to create a space around your location that encompasses at least a few streets. When a striped area appears, click on the "Soil Map" tab to see the soil types in your area. Click on the Map Unit Legend to see soil types and on the soil type name under "Map Unit Name" for details.

6. What type(s) of soil is(are) present at your location?
7. What is the percentage range in your soil for sand? Silt? Clay? (Refer to Figure 12-13.)
8. Do these percentages agree with what you discovered in Activity 4? Why or why not? What might be an explanation if the percentages don't agree with what you observed?

Seeing Geographically in Our National Parks

Look again at the photograph of the Little Missouri River in Theodore Roosevelt National Park, North Dakota, at the beginning of the chapter (p. 348). Using Figure 12-35 for reference, notice that most of this part of North Dakota (47° N, 103° W) is dominated by Mollisols. What helps explain why this soil order is prevalent in this region? Notice also that a narrow band of Entisols is present in the area shown in the photograph. What might explain the existence of Entisols here?

Mastering Geography®

Looking for additional review and test prep materials? Visit the Study Area in **Mastering Geography** to enhance your geographic literacy, spatial reasoning skills, and understanding of this chapter's content by accessing a variety of resources, including Map Master interactive maps, geoscience animations, *Mobile Field Trips*, videos, *Project Condor* Quadcopter videos, *In the News* RSS feeds, flashcards, web links, self-study quizzes, and an eText version of *McKnight's Physical Geography*.

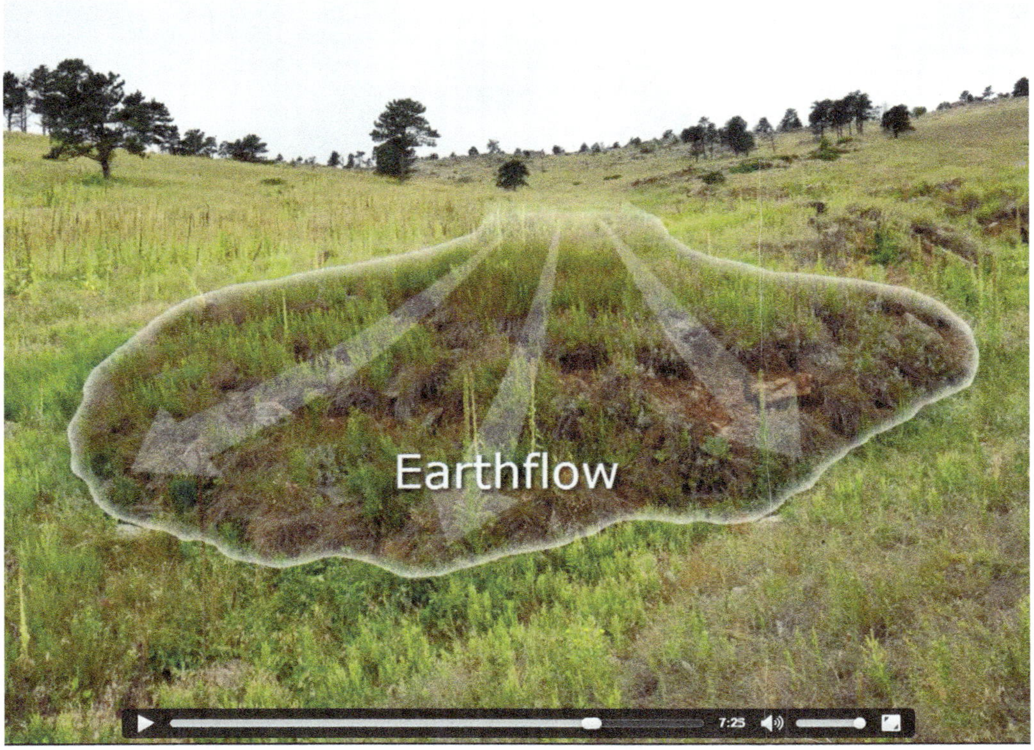

Seeing Geographically in Our National Parks

Mummy Cliff in Capitol Reef National Park, Utah. Describe the general topography of this landscape. How many different layers of rock can you see? What helps you recognize these different layers? What type of climate appears to exist here? What suggests this? Is an extensive layer of soil present? What might explain this?

Capitol Reef National Park, UT

Introduction to Landform Study

13

HAVE YOU EVER WONDERED how we know what Earth was like in the past? Or how we know that some of the rocks in Capitol Reef National Park formed 200 million years ago? Or how we know that the continents were in different positions millions of years ago and that mountain ranges rose, only to be worn down over time? Earth scientists are like detectives. Whether they are studying past climates or deciphering the history of Earth's surface, they look for clues, follow leads, and come to conclusions based on the evidence.

With this chapter, as we turn our attention directly to the solid part of Earth, we face an enormous object of study that is well beyond our usual scale of thinking. As geographers, we largely focus on the surface. However, in order to understand some of the processes that shape Earth's surface, we must also understand what happens in Earth's interior, and that is where we begin.

Our knowledge of the interior of Earth is somewhat limited and based largely on indirect evidence. The deepest existing mine shaft extends down only 3.9 kilometers (2.4 miles); the deepest drill holes from which sample cores have been brought up have penetrated only 12.3 kilometers (7.6 miles) into Earth. Earth scientists, in the colorful imagery of writer John McPhee, "are like dermatologists: they study, for the most part, the outermost two per cent of the earth. They crawl around like fleas on the world's tough hide, exploring every wrinkle and crease, and try to figure out what makes the animal move."[1]

After we describe the structure of Earth as a whole, we turn to a discussion of rocks, the solid material from which the planet is made. We conclude with fundamental concepts that set the stage for our study of the basic Earth systems operating within and on the surface of the lithosphere—especially the flow of energy and rock material within Earth and the interaction of the atmosphere and hydrosphere with the lithosphere—that together shape Earth's surface topography.

As you study this chapter, think about these **Key**Questions:

- **What is the internal structure of Earth?**
- **What are the three classes of rocks, and how do they form?**
- **What is *geomorphology*?**
- **What is *plate tectonics*?**
- **How do the concepts of *uniformitarianism* and *geologic time* help us understand the processes that shape Earth's surface?**
- **Why does the scale of study influence how we study geomorphology?**

[1]John McPhee, *Assembling California* (New York: Farrar, Straus and Giroux, 1993), p. 36.

The Structure of Earth

Most of what we know about Earth's interior has been collected by geophysical means, especially from the behavior of earthquake waves or vibrations from human-made explosions. The speed and direction of seismic waves change whenever these waves cross a boundary from one type of material to another. Analysis of such changes, along with related data on Earth's magnetism and gravitational attraction, has enabled Earth scientists to develop a model of Earth's internal structure.

Earth's Hot Interior

In general, temperature and pressure increase with depth inside Earth, with the highest temperatures and pressures at the center. The source of the warmth is largely from the release of energy from the decay of radioactive elements (in much the same way that the decay of radioactive material supplies the warmth to run a nuclear power plant). The transfer of heat from Earth's interior drives many Earth processes, such as plate tectonics and volcanism (discussed in Chapter 14).

Figure 13-1 shows Earth's internal structure. Earth consists of three basic layers, defined by differences in composition: the *crust, mantle,* and *core*. In addition, there are a number of functional layers, defined by differences in physical properties, that subdivide or overlap the compositional layers.

The Crust

The **crust**, Earth's outermost shell, consists of a broad mixture of rock types. Beneath the oceans, the crust has an average thickness of only about 7 kilometers (4 miles), whereas beneath the continents the thickness averages more than five times as much and in places exceeds 70 kilometers (40 miles). Oceanic crust is thinner but is composed of denser ("heavier") rocks than continental crust. In general, within the crust density gradually increases with depth. Altogether, the crust makes up less than 1 percent of Earth's volume and about 0.4 percent of Earth's mass.

At the base of the crust, a significant change in mineral composition occurs. This relatively narrow zone of change is called the **Mohorovičić discontinuity**, or simply the **Moho**, named for Croatian seismologist Andrija Mohorovičić (1857–1936), who discovered it.

The Mantle

Beneath the Moho is the **mantle**, composed of high-density rock that extends down to a depth of approximately

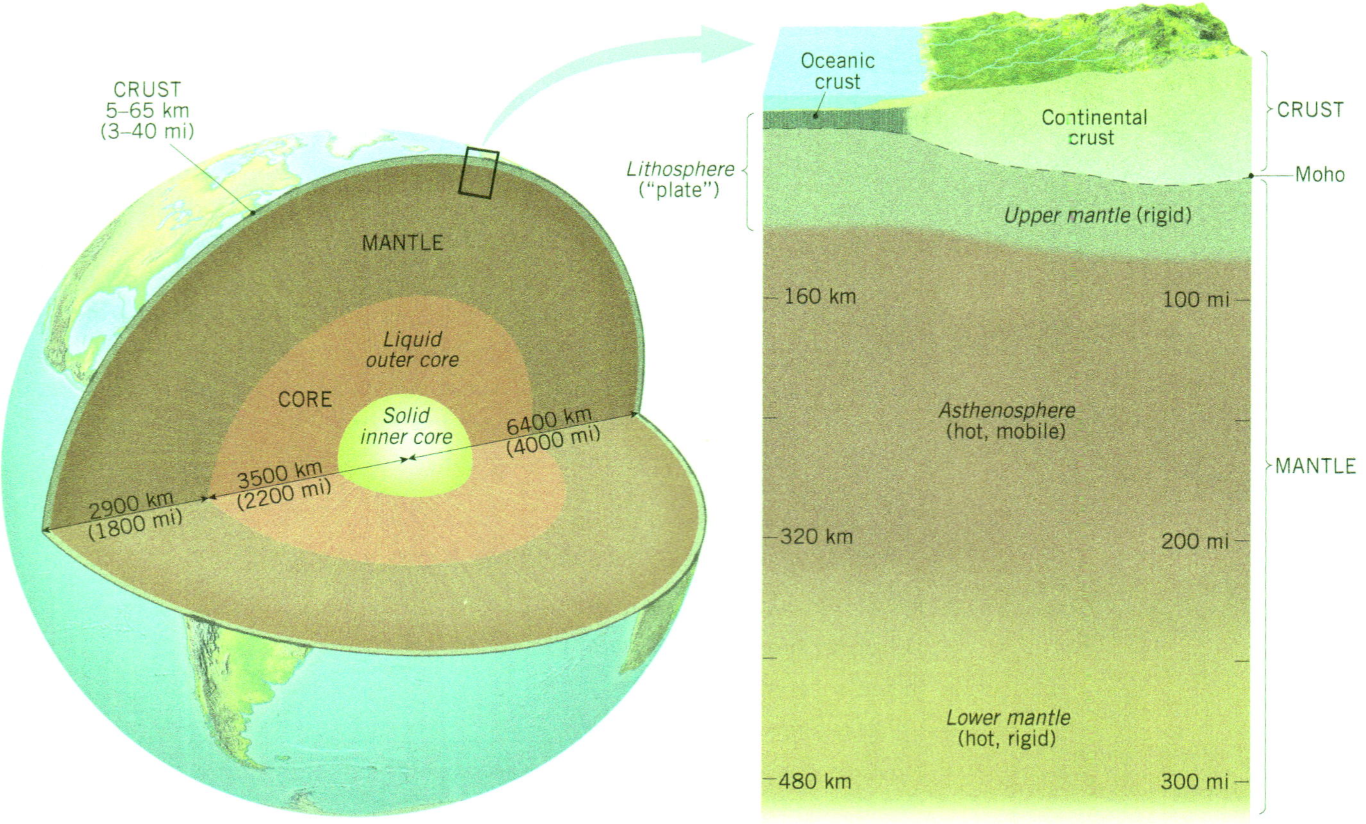

(a) Cross section of Earth's interior

(b) Cross section from crust into mantle

▲ Figure 13-1 The vertical structure of Earth's interior. (a) Below the thin outer crust of Earth is the broad zone of the mantle; below the mantle are the liquid outer core and solid inner core. (b) Idealized cross section through Earth's crust and part of the mantle. Together, the crust and uppermost mantle, both rigid zones, are called the lithosphere—the "plates" of plate tectonics. In the asthenosphere, the mantle is hot and therefore weak and easily deformed. In the lower mantle, the rock is generally rigid.

2900 kilometers (1800 miles). In terms of volume, the mantle is by far the largest of the four layers. It makes up 84 percent of the total volume of Earth and about two-thirds of Earth's total mass.

Upper Mantle: The uppermost mantle is relatively thin and rigid, extending to a depth of 65–100 kilometers (40–60 miles)—somewhat deeper under the continents than under the ocean floors. This uppermost mantle, together with the overlying oceanic or continental crust, is called the **lithosphere** (Figure 13-1b). For the remainder of this book as we study landforms, the term *lithosphere* has a more restricted meaning than when we introduced Earth's environmental "spheres" in Chapter 1. For our purposes now, *lithosphere* refers specifically to the combination of the crust and upper rigid mantle. As we see shortly, large pieces of the lithosphere are the "plates" of plate tectonics.

Asthenosphere: Beneath the rigid layer of the lithosphere, and extending to a depth of as much as 350 kilometers (200 miles), is a mantle zone in which the rocks are hot enough that they lose much of their strength and become "plastic"—they are easily deformed (somewhat like tar). This region is called the **asthenosphere** ("weak sphere").

Lower Mantle: Below the asthenosphere is the *lower mantle*, where the rocks are very hot but largely rigid because of higher pressures.

The Inner and Outer Cores

Beneath the mantle is the molten (liquid) **outer core** (Figure 13-1a), extending to a depth of about 5000 kilometers (3100 miles).

The innermost portion of Earth is the **inner core**, a solid (because of extremely high pressure) and very dense mass having a radius of about 1450 kilometers (900 miles). The inner and outer cores are thought to be made of iron/nickel or iron/silicate. Together, both cores make up about 15 percent of Earth's volume and 32 percent of its mass.

A common misconception is that the liquid outer core of Earth is the source of molten rock ("magma" and "lava") that is ejected by volcanoes, but this is not the case: the near-surface mantle is the source of magma. Earth's cores are instead the source of energy that drives the slow movement of hot rock through the mantle toward the surface (through the process of *convection*). This rising hot rock in the mantle is under so much pressure that it remains essentially solid. Only when this rising mantle material is very close to the surface is pressure low enough for it to melt.

Earth's Magnetic Field: Earth's *magnetic field* is generated in the outer core. Convective circulation within the conductive liquid iron and nickel outer core, spiraling in line with Earth's rotational axis, induces the magnetic field of our planet through what is sometimes called a *geodynamo*.[2]

[2]Most geophysicists think that for such a geodynamo to work, the presence of an initial weak magnetic field was needed, most likely coming from the Sun.

Interestingly, the strength and orientation of the magnetic field change over time, and the location of the north magnetic pole does not exactly match the true geographic North Pole. The position of the north magnetic pole slowly but continuously drifts several tens of kilometers each year—although in recent decades, it's been shifting faster. It is currently located at about 86° N, 170° E—but the position of the north magnetic pole can change significantly during even a single day! In addition, for reasons that are not completely understood, at irregular intervals of thousands to millions of years, the polarity of Earth's magnetic field reverses, with the north magnetic pole becoming the south. As we discuss in Chapter 14, a record of these magnetic polarity reversals has been recorded in the iron-rich rocks of the ocean floor.

LearningCheck 13-1 **Compare the characteristics of Earth's crust, mantle, inner core, and outer core.** **(Answer on p. AK-4)**

Plate Tectonics and the Structure of Earth

Although the generalized model of Earth's interior shown in Figure 13-1 is a useful starting point for understanding physical geography, we still have much to learn about the processes taking place within Earth's interior. Geophysicists continue to improve our model of the interior of Earth by using computer simulations, high-pressure lab experiments, analysis of earthquakes, variations in gravity, heat-flow maps, and other physical data.

Despite the difficulties of studying the interior of our planet, our understanding of how Earth works dramatically changed in the 1960s. The notion of "continental drift," first proposed in the early 1900s but rejected by most scientists for the next half-century, was then revived and expanded into the theory of *plate tectonics*—a theory that is now accepted by all Earth scientists. Geological, paleontological, seismic, and magnetic evidence makes it clear that the lithosphere of Earth is broken into large, sometimes continent-sized slabs, commonly called "plates," that float on and slowly move over the hot, soft asthenosphere below. These enormous plates pull apart, collide, and slide past each other—driven by the convective heat flow within Earth. Many internal processes, such as faulting, folding, and volcanic activity, are directly linked to the interactions taking place along the boundaries of these plates.

We explore the dynamics of plate tectonics in greater detail in Chapter 14. The remainder of this chapter sets the stage for that discussion by introducing some key concepts in the study of Earth's surface features, such as rocks, geologic time, and the doctrine of uniformitarianism.

LearningCheck 13-2 **Describe the characteristics of Earth's lithosphere and asthenosphere.**

The Composition of Earth

About 100 natural chemical elements are found in Earth's crust, mantle, and core, occasionally as discrete elements but usually bonded with one or more other elements to form compounds. Some of these compounds and elements form *minerals*—the building blocks of rocks, which are in turn the building blocks of the landscape itself.

Minerals

For a substance to be considered a **mineral**, it must:

- Be solid
- Be naturally found in nature
- Be inorganic (nonliving)
- Have a specific chemical composition that varies only within certain limits
- Contain atoms arranged in a regular pattern, forming solid crystals

Only about one-fourth of all elements are involved in the formation of minerals to any significant extent. Just eight elements account for more than 98 percent of the mass of Earth's crust—oxygen and silicon alone make up more than three-quarters of the mass of the crust. Approximately 4400 minerals have been identified, with new types identified almost every year. The majority of known minerals are found in the crust; a more limited number are found within the mantle or have been identified in extraterrestrial rocks such as meteorites or rocks brought back from the Moon. Of the roughly 4400 recognized minerals, only a few dozen are important constituents of crustal rocks. Some of the most common rock-forming minerals are listed in Table 13-1. Mineral nomenclature is very unsystematic. Some names reflect the mineral's chemical composition or a physical property, some names are based on a person or a place, and some appear to have been chosen at random!

The rock-forming minerals are grouped into seven principal "families"—categories based on their chemical properties and internal crystal structure.

TABLE 13-1 Some Common Rock-Forming Minerals

Group	Mineral	Chemical Composition	Common Characteristics
Ferromagnesian ("dark") Silicate	Olivine	$(Mg, Fe)_2 SiO_4$	Green to brown; glassy luster; gem peridot
	Pyroxene group (Augite)	$(Mg, Fe) SiO_3$	Green to black; commonly has two cleavage planes at 90°
	Amphibole group (Hornblende)	$Ca_2(Fe, Mg)_5Si_8O_{22}(OH)_2$	Black or dark green; often appear as elongated rod-shaped crystals
	Biotite mica	$K(Mg, Fe)_3AlSi_3O_{10}(OH)_2$	Black or brown; often appear as hexagonal crystals in thin sheets
Nonferromagnesian ("light") Silicate	Muscovite mica	$KAl_2(AlSi_3O_{10})(OH)_2$	Colorless or brown; splits into thin, translucent sheets
	Potassium feldspar (Orthoclase)	$KAlSi_3O_8$	White to gray or pink; good cleavage in two planes at 90°
	Plagioclase feldspar	$(Ca, Na)AlSi_3O_8$	White to gray; striations often appear along cleavage planes
	Quartz	SiO_2	Commonly colorless; forms six-sided elongated crystals
Oxide	Hematite	Fe_2O_3	Red to silver-gray; type of iron ore
	Magnetite	Fe_3O_4	Black with metallic luster; magnetic
	Corundum	Al_2O_3	Brown or blue; gems sapphires and rubies
Sulfide	Galena	PbS	Black to silver metallic luster; lead ore
	Pyrite	FeS_2	Brassy or golden yellow; often seen in well-formed cubes; sometimes called fool's gold
	Chalcopyrite	$CuFeS_2$	Brassy or yellow metallic luster; copper ore
Sulfate	Gypsum	$CaSO_4 \cdot 2H_2O$	White or transparent; used in plaster and wallboard
Carbonate	Calcite	$CaCO_3$	White or colorless; cleavage in three planes forms rhombohedra; fizzes in dilute acid
	Dolomite	$CaMg (CO_3)_2$	White or transparent; cleaves into rhombohedra; powdered form fizzes in dilute acid
Halide	Halite	$NaCl$	Transparent or white; table salt
	Fluorite	CaF_2	Transparent purple, green, or yellow; fluorine ore
Native Elements	Gold	Au	Bright yellow; highly malleable
	Silver	Ag	Brilliant white metallic luster when polished

(a) Quartz: Silicate family (SiO_2) (b) Pyrite: Sulfide family (FeS_2) (c) Fluorite: Halide family (CaF_2)

▲ Figure 13-2 (a) Quartz, a common silicate mineral. (b) Pyrite, a common sulfide mineral. (c) Fluorite, a common halide mineral.

Silicates: The largest and most important mineral family consists of the **silicates**, which combine the two most abundant chemical elements in the lithosphere: oxygen (O) and silicon (Si). The bulk of crustal rocks are composed of silicate minerals. Most silicates are hard and durable. The major subcategories of this group are *ferromagnesian silicates* (called the "dark" silicates) and *nonferromagnesian silicates* (called the "light" silicates), which are distinguished from one another by the presence or absence of iron and magnesium. Feldspars and quartz (Figure 13-2a) are the most abundant silicate minerals. Quartz is composed of pure *silica* (SiO_2).

Oxides: An *oxide* is an element combined with oxygen. The most widespread oxides are those that combine with iron—particularly hematite, magnetite, and limonite, all three of which are major sources of iron ore and are common rock-forming minerals. (Although quartz has the chemical composition of an oxide, it is classified as a silicate because of its internal structure.)

Sulfides: *Sulfides* are composed of reduced sulfur in some combination with one or more other elements. Pyrite (Figure 13-2b), for example, is a combination of iron and sulfur (FeS_2). Many of the most important ore minerals—such as galena (lead), sphalerite (zinc), and chalcopyrite (copper)—are sulfides. This group is common in many types of rock and may be massive or abundant in veins.

Sulfates: The *sulfate* group includes minerals such as gypsum that contain sulfur and oxygen in combination with some other element. Calcium is the principal combining element. The sulfate minerals are usually light-colored, and most are found in sedimentary rocks.

Carbonates: Carbonates are light-colored (or colorless) minerals that are common constituents of sedimentary rocks such as limestone (largely made of calcium carbonate [$CaCO_3$] in its mineral form, calcite). Carbonates are composed of one or more elements in combination with carbon and oxygen.

Halides: The *halide* group is the least widespread. The name is derived from a word meaning "salt." Halide minerals include halite, or common table salt (NaCl), and fluorite (CaF_2) (Figure 13-2c).

Native Elements: A few minerals may occur in nature as discrete elements (not combined chemically with another element). These are referred to as native elements. Included are such precious metals as gold, silver, and platinum.

LearningCheck 13-3 **Which family of minerals is most abundant in Earth's crust, and what most-abundant elements do these minerals contain?**

Rocks

Rocks are consolidated combinations of mineral material—sometimes just one kind of mineral but usually several different minerals. Rocks occur in great variety and complexity in the lithosphere, although fewer than 20 minerals account for more than 95 percent of the composition of all continental and oceanic crustal rocks.

Solid rock that is found right at the surface is called an **outcrop** (Figure 13-3). Over most of Earth's land area, though, solid rock exists as a buried layer of *bedrock* and is

▲ Figure 13-3 Bedrock is usually buried under a layer of regolith (weathered rock and soil) but occasionally appears as an outcrop.

covered by a layer of broken rock called *regolith*. Soil, when present, comprises the upper portion of the regolith (see Chapter 12).

The enormous variety of rocks can be classified systematically. A detailed knowledge of *petrology* (the study of the origin and characteristics of rocks) is unnecessary for our purposes, however, so we restrict our coverage to a survey of the three major rock groups or classes—*igneous*, *sedimentary*, and *metamorphic*—and their basic attributes (Table 13-2 and Figure 13-4).

Igneous Rocks

The word *igneous* is derived from the Latin *igneus* ("fire"). **Igneous rocks** are formed by the cooling and solidification of molten rock (Figure 13-4a). **Magma** is a general term for

TABLE 13-2 Rock Classification

Class	Subclass	Rock	General Characteristics
Igneous	Plutonic (Intrusive)	Granite	Coarse-grained; "salt and pepper" appearance; from high-silica felsic magma; plutonic equivalent of rhyolite
		Diorite	Coarse-grained; from intermediate silica magma; plutonic equivalent of andesite
		Gabbro	Coarse-grained; black or dark gray; from low-silica mafic magma; plutonic equivalent of basalt
		Peridotite	Common mantle rock consisting primarily of olivine and/or pyroxene
	Volcanic (Extrusive)	Rhyolite	Light color; from high-silica felsic magma; volcanic equivalent of granite
		Andesite	Typically gray; from intermediate silica magma; volcanic equivalent of diorite
		Basalt	Usually black; from low-silica mafic magma; volcanic equivalent of gabbro
		Obsidian	Volcanic glass; typically black and rhyolitic in composition
		Pumice	Volcanic glass with frothy texture; often rhyolitic in composition
		Tuff	Rock made from volcanic ash or pyroclastic flow deposits
Sedimentary	Detrital (Clastic)	Shale	Composed of very fine-grained sediments; typically thinly bedded
		Sandstone	Composed of sand-sized sediments
		Conglomerate	Composed of rounded, pebble-sized sediments in a fine-textured matrix
		Breccia	Composed of coarse-grained, angular sediments; typically thinly bedded, poorly sorted fragments
	Chemical and Organic	Limestone	Composed of calcite; may contain shells or shell fragments
		Travertine	Limestone deposited in caves or around hot springs; often deposited in banded layers
		Chert	Common chemical rock composed of microcrystalline quartz
		Bituminous coal	Composed of compacted plant material
Metamorphic	Foliated	Slate	Fine-grained; smooth surfaces; typically forms from the low-grade metamorphism of shale
		Phyllite	Fine-grained; glossy, wavy surfaces
		Schist	Thin, flaky layers of platy minerals
		Gneiss	Coarse-grained; granular texture; distinct mineral layers; high-grade metamorphism
	Nonfoliated	Quartzite	Composed primarily of quartz; often derived from sandstone
		Marble	Composed principally of calcite; typically derived from limestone
		Serpentinite	Greenish-black; slippery feel; from hydrothermal alteration of peridotite
		Anthracite	High-grade coal often derived from bituminous coal

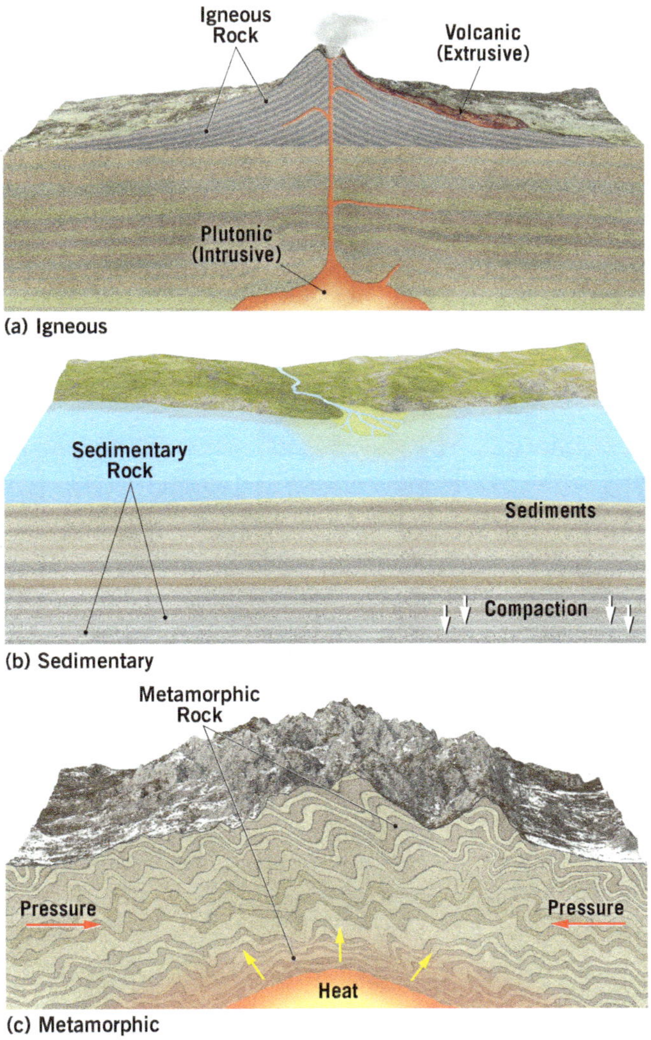

▲ **Figure 13-4** The three rock classes. (a) Igneous rocks form when magma or lava cools. (b) Sedimentary rocks result from consolidation of deposited particles. (c) Metamorphic rocks are produced when heat and/or pressure act on existing igneous or sedimentary rocks.

molten rock beneath the surface of Earth, whereas **lava** is molten rock that flows out on or is squeezed up onto the surface. Most igneous rocks form directly from the cooling of magma or lava; however, some igneous rocks develop from the "welding" of tiny pieces of solid volcanic rock, called **pyroclastics**, that have been explosively ejected onto the surface by a volcanic eruption.

The classification of igneous rocks is based largely on mineral composition and texture. The mineral composition of an igneous rock is determined by the "chemistry" of the magma—in other words, by the particular combination of molten mineral material in the magma. As we discuss in Chapter 14, one of the most important variables of magma composition is the relative amount of *silica* (SiO_2) present. Magmas with relatively large amounts of silica generally cool into *felsic* igneous rocks, which contain large portions of light-colored silicate minerals such as quartz and feldspar. The word *felsic* comes from *fel*dspar and *si*lica (quartz). Felsic minerals tend to have lower densities and lower melting temperatures than mafic minerals. Magmas with relatively low amounts of silica generally cool into *mafic* igneous rocks, which contain large portions of dark-colored, magnesium- and iron-rich silicate minerals such as olivine and pyroxene. The word *mafic* comes from *ma*gnesium and *f*errum (Latin for "iron").

The texture of an igneous rock is determined primarily by where and how the molten material cools. Magma beneath the surface cools slowly, leading to a coarse-grained texture, whereas lava on the surface cools more rapidly, leading to a fine-grained texture. Thus, one magma can produce a number of quite different igneous rocks, depending on whether the material cools below the surface, on the surface as a lava flow, or on the surface as an accumulation of pyroclastics.

Igneous rocks are divided into two main categories, based on where the rocks form. **Plutonic rocks**, or **intrusive igneous rocks**, form from the cooling of magma below the surface. **Volcanic rocks**, or **extrusive igneous rocks**, form from the cooling of lava or the bonding of pyroclastic materials on the surface (Figure 13-5).

LearningCheck 13-4 What is the difference between magma and lava? How are igneous rocks classified according to their composition and texture?

Plutonic (Intrusive) Rocks: Plutonic rocks cool and solidify beneath Earth's surface where surrounding rocks serve as insulation around the intrusion of magma, greatly slowing the rate of cooling. Because the magma may require many thousands of years for complete cooling, the individual mineral crystals in a plutonic rock can grow fairly large—large enough to see with the naked eye—giving the rock a very coarse-grained texture. Although originally buried, plutonic rocks may become important in surface topography by being pushed upward or by being exposed by erosion.

The most common and well-known plutonic rock is **granite**, a generally light-colored, coarse-grained igneous rock. Granite is made of a combination of light- and dark-colored minerals such as quartz, plagioclase feldspar, potassium feldspar, hornblende, and biotite. (Granite is a felsic igneous rock; see Figure 13-5.[3]) Granite and similar plutonic rocks, such as *granodiorite*, make up the ancient interior of many continents and the core of many mountain ranges, including the Front Range in Colorado and the Sierra Nevada in California—see the *Featured National Park* box on *The Granites of Yosemite*.

[3]Sometimes the term *granitic rock* is used broadly to refer to granite as well as closely related plutonic rocks, such as granodiorite and tonalite, which have varying amounts of quartz and slightly different combinations of potassium feldspar and plagioclase feldspar from true granite.

▲ **Figure 13-5** Common igneous rocks. Magma chemistry and the cooling rate help determine the final igneous rock formed. Felsic magmas contain relatively high amounts of silica and cool to form either a plutonic rock such as granite or the volcanic rock rhyolite. Mafic magmas contain relatively low amounts of silica and cool to form either a plutonic rock such as gabbro or the volcanic rock basalt. Magmas of intermediate composition cool to form either a plutonic rock such as diorite or the volcanic rock andesite.

Volcanic (Extrusive) Rocks: Volcanic rocks form on the surface of Earth, from either the cooling of lava or the accumulation of pyroclastic material, such as volcanic ash and cinders. When lava cools rapidly on Earth's surface, the lava may solidify within hours, so the mineral crystals in many volcanic rocks are so small as to be invisible without a microscope. On the other hand, volcanic rocks that form by the accumulation of pyroclastics may clearly show tiny fragments of shattered rock that were explosively ejected from a volcano.

Of the many kinds of volcanic rocks, by far the most common is the black or dark gray, fine-grained rock called **basalt** (Figure 13-6). Basalt forms from the cooling of lava and is composed of only dark-colored minerals such as plagioclase feldspar, pyroxene, and olivine. (Basalt is a mafic igneous rock; see Figure 13-5.) Basalt is the most common volcanic rock in Hawai'i (see the *Featured National Park* box on Hawaiian volcanoes in Chapter 14). Basalt is also widespread in parts of some continents, such as in the Columbia Plateau of the northwestern United States. Basalt also makes up the bulk of the ocean floor crust.

There are many other well-known volcanic rocks. *Obsidian*—typically black—is a type of volcanic "glass" (meaning it has no crystal structure) that forms from extremely rapid cooling of lava. *Pumice* forms from the rapid cooling of frothy, gas-rich molten material (and is sometimes light enough to float on water!). *Tuff* is a volcanic rock consisting of welded pyroclastic fragments (see Figure 13-12b).

In Chapter 14 we discuss volcanism and some of the topographic features associated with plutonic and volcanic activities.

LearningCheck 13-5 In what ways are granite and basalt different?

Featured National Park

Yosemite

The Granites of Yosemite

Yosemite National Park in California's Sierra Nevada Mountains is one of the best places in the world to see exposures of granite. Many "pulses" of magma intruded into the ancestral Sierra between about 105 million and 85 million years ago. These individual *plutons* cooled slowly deep below the surface, forming the 650-kilometer (400-mile) long Sierra Nevada *batholith* that was later uplifted and exposed by erosion (**Figure 13-A**).

▲ **Figure 13-A** Massive outcrops of granitic rock in California's Yosemite National Park.

Granitic Rocks: Although the plutonic rock in Yosemite is often simply called "granite," several kinds of granite, or more properly *granitic rock*, are exposed in the Park. The differences among them lie in their exact mineral composition.

Minerals in Granitic Rocks: Five common minerals are found in the granitic rocks of Yosemite. They are often easy to identify with a small magnifying glass (**Figure 13-B**):

- ***Quartz***: Clear or light gray with irregular or curved glassy surfaces.
- ***Plagioclase feldspar***: Chalky white or gray with flat "cleavage" surfaces that reflect sunlight, often with distinctive long striations.
- ***Potassium feldspar***: Chalky white or pink, also with flat cleavage surfaces that reflect sunlight but without striations.
- ***Hornblende***: Black or dark green long, rectangular crystals.
- ***Biotite***: Black hexagonal crystals that form shiny, thin flakes.

Classification of Granitic Rocks: Granitic rocks are classified according to the relative abundance of quartz, plagioclase, and potassium feldspar. True *granite* contains about equal amounts of quartz, plagioclase, and potassium feldspar (or more potassium feldspar than plagioclase). The most common granitic rock in Yosemite is *granodiorite*, which contains the same amount of quartz as granite but has a greater proportion of plagioclase than potassium feldspar. In the field, it can be very difficult to tell the difference.

▲ **Figure 13-B** Close-up of granite showing its common minerals.

The granitic rocks in Yosemite vary from pluton to pluton. Some, such as the Half Dome Granodiorite, contain large, obvious hornblende crystals—as well as crystals of the distinctive honey-colored mineral *sphene*. The El Capitan Granite appears nearly white and contains very little hornblende. Perhaps most distinctive of all is the Cathedral Peak Granodiorite exposed in the Yosemite high country along the Tioga Road through Tuolumne Meadows. It has large *phenocrysts* of potassium feldspar—some more than 5 centimeters (2 inches) long (**Figure 13-C**)

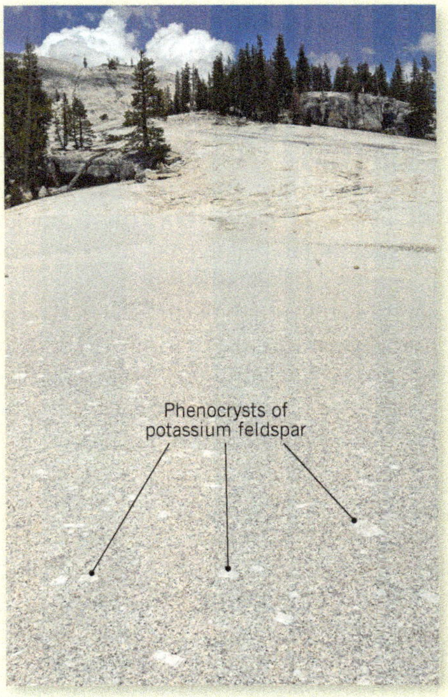

▲ **Figure 13-C** Large phenocrysts of potassium feldspar in the Cathedral Peak Granodiorite exposed near Tuolumne Meadows in Yosemite National Park.

Inclusions: In some places, darker inclusions appear in the granitic rock. These inclusions formed either from blobs of darker magma mixed in with the granitic magma or from pieces of older, solid rock that broke off and dropped down into the still-molten magma.

Transportation by Glaciers: Because many of the granitic rocks in Yosemite have a distinctive appearance or mineral composition, scientists can trace back the original source of boulders transported many tens of kilometers by glaciers during the Pleistocene ice ages. Such out-of-place boulders, called *erratics*, are discussed in Chapter 19.

Mobile Field Trip
Yosemite
https://media.pearsoncmg.com/ph/esm/esm_mcknight_physgeo_13/media/

Questions

1. In what ways do the granitic rocks in Yosemite National Park vary?
2. How can looking for different mineral compositions in granitic rock be helpful when studying the glacial history of Yosemite?

▲ **Figure 13-6** (a) Solidified lava flows of basalt create an other-worldly landscape in Craters of the Moon National Monument in south central Idaho. The smooth lava surface in the foreground is called pāhoehoe, whereas the rougher, blocky lava in the background is called 'a'ā. (b) Closeup of black basalt.

Sedimentary Rocks

External processes, mechanical and chemical, operating on rocks cause them to disintegrate. (Such rock "weathering" is discussed in Chapter 15.) This disintegration produces fragmented mineral material—called **sediment**—some of which is removed by water, wind, ice, gravity, or a combination of these agents. Much of this material, along with ions of compounds in solution, is transported by water moving in rivers or streams and is eventually deposited somewhere in a quiet body of water, particularly on an ocean floor (see Figure 13-4b).

Over long periods of time, sedimentary deposits can build to a remarkable thickness—many thousands of meters. The sheer weight of this massive overburden exerts enormous pressure, which causes individual particles of sediment to adhere to each other and to interlock through compaction. In addition, chemical cementation normally takes place. Various cementing agents—especially silica (SiO_2), calcium carbonate ($CaCO_3$), and iron oxide (Fe_2O_3)—precipitate from the water into the pore spaces between sediment grains (Figure 13-7). This combination of compaction and cementation consolidates and transforms the sediments to **sedimentary rock**.

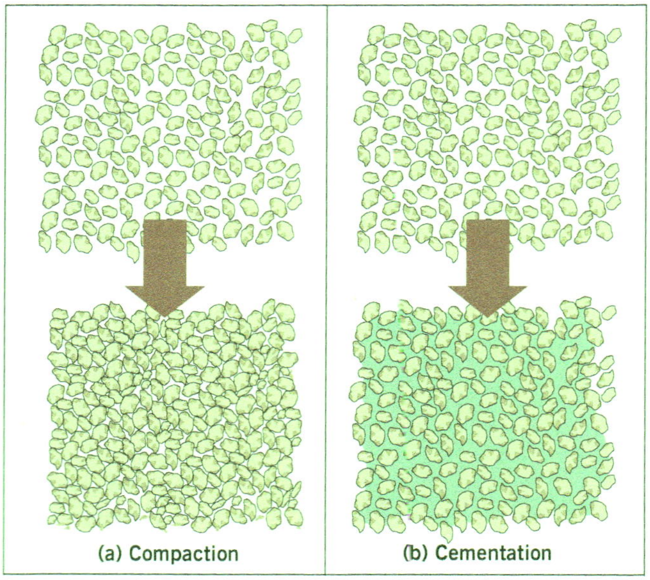

▲ **Figure 13-7** Sedimentary rocks are typically composed of small particles of sediment deposited by water or wind in layers and then consolidated by compaction and/or cementation. (a) Compaction consists of the packing of the particles as a result of the weight of overlying material. (b) Cementation involves infilling pore spaces among the particles by a cementing agent, such as silica, calcium carbonate, or iron oxide.

In some cases, buried organic material—such as the remains of plants or algae—accumulate along with inorganic sediments. This organic material may in time also become sedimentary rock.

As we discuss in Chapter 16, during transportation and deposition, sediments may be sorted roughly by size. Thus, many sedimentary rocks are composed of particles of fairly uniform size. Other variations in the composition of the sediments may be due to factors such as different rates of deposition, changes in climatic conditions, patterns of sediment movement in the oceans, or the composition of the original source material. Most sedimentary deposits are built up in more or less distinct horizontal layers called **strata** (Figure 13-8), which vary in thickness and composition. (Wind-deposited sediments are a notable exception to the horizontal layers of most sedimentary rocks.) The resulting parallel structure, or *stratification*, is a characteristic feature of most sedimentary rocks. Some sedimentary rocks exhibit distinct **bedding planes**—flat surfaces separating one layer from the next. Although originally deposited and formed in horizontal orientation, the strata may later be uplifted, tilted, and deformed by pressures from within Earth (Figure 13-9).

▶ **Figure 13-8** Gently tilted strata of limestone and shale along the cliffs of Kilve Beach, in Somerset, England.

(b) Close-up of limestone

(a) Sedimentary layers

▲ **Figure 13-9** (a) Sedimentary strata (mostly limestone and shale) that have been folded and tilted into an almost vertical orientation on Mount Angeles in the Olympic Mountains of Washington. (b) A sample of limestone containing small fossil mollusks.

(b) Close-up of sandstone

(a) Sandstone cliffs

▲ **Figure 13-10** (a) Nearly flat lying layers of Estrada Sandstone in Cathedral Valley, Capitol Reef National Park, Utah. The structure of the rock in these abrupt cliffs is easy to see. (b) Closeup of sandstone.

Sedimentary rocks are generally classified into three categories based on how they formed: *detrital, chemical,* or *organic.*

Detrital (Clastic) Sedimentary Rocks: Rocks composed of fragments of preexisting rocks in the form of cobbles, gravel, sand, silt, or clay are known as *detrital* or *clastic* sedimentary rocks. By far the most common are *shale* (or *mudstone*), which is composed of very fine silt and clay particles, and *sandstone,* which is made up of compacted, sand-size grains (**Figure 13-10**). When the rock consists of rounded, pebble-size fragments, it is called a *conglomerate.*

Chemical Sedimentary Rocks: Chemically accumulated sedimentary rocks usually form by the precipitation of solids from ions in solution but sometimes by more complicated chemical reactions. Calcium carbonate ($CaCO_3$) is a common component of such rocks, and *limestone* is the most widespread result (see Figure 13-9). Limestone can also form from the accumulated skeletal remains of coral or other lime-secreting sea animals. *Chert* forms in a similar way to limestone but is composed of silica (SiO_2) instead of calcium carbonate. Rock salt is an example of an *evaporite,* rock that forms as water evaporates and leaves behind dissolved minerals such as common table salt (NaCl; see Figure 18-32).

Organic Sedimentary Rocks: Organically accumulated sedimentary rocks, including *lignite* (soft brown coal) and *bituminous coal* (soft black coal), form from the compacted remains of dead plant material.

There is much overlap among these formation methods. As a result, in addition to there being many kinds of sedimentary rocks, there are many gradations among them. The majority of sedimentary rocks are shale (45 percent of total), sandstone (32 percent), or limestone (22 percent).

LearningCheck 13-6 **Describe and explain the formation of detrital (clastic), chemical, and organic sedimentary rocks. Give one example of each.**

Petroleum and Natural Gas: Although petroleum and natural gas are not rocks, they form by the accumulation of organic material in sedimentary deposits. The complex formation processes involve the accrual of hydrocarbons derived from marine organisms, often in association with heat (**Figure 13-11**). See the box *Energy for the 21st Century: Unconventional Hydrocarbons and the Fracking Revolution.*

Metamorphic Rocks

Any preexisting rock can become a **metamorphic rock** when it is physically or chemically altered (or both) by heat, pressure, and/or chemically active fluids. Metamorphic rocks are associated with conditions in the lithosphere, where the pressures and temperatures are greater than those that form sedimentary rocks but less than those that melt rocks into magma.

The effects of heat and pressure on rocks are complex, being strongly influenced by such things as the amount and composition of fluids in the rocks, as well as the length of time the rocks are heated and/or subjected to high pressure. Metamorphism can be an almost dry "baking" process that heats the rock, causing its mineral components to be recrystallized

Mobile Field Trip
Oil Sands
https://media.pearsoncmg.com/ph/esm/esm_mcknight_physgeo_13/media/

▶**Figure 13-11** Oil sands are a type of unconventional hydrocarbon, shown here being processed after extraction from the Murray oil sands in Alberta, Canada.

and rearranged (**Figure 13-12**), or it can be a more active, wet process, such as the hydrothermal metamorphism happening at "black smokers" on the ocean floor. In any case, the metamorphic result is often quite different from the original rock because metamorphism can alter the original rock's composition, texture, and structure.

Video
Black Smokers
https://media.pearsoncmg.com/ph/esm/esm_mcknight_physgeo_13/media/

Causes of Metamorphism: Metamorphism can take place in a number of environments. For example, **contact metamorphism** occurs beneath Earth's surface where the rocks in contact with magma are altered by heat and pressure. As a result, it is common to find exposed metamorphic rocks adjacent to plutonic rocks such as granite. **Regional metamorphism** takes place where large volumes of rock deep within the crust are subjected to heat and/or pressure over long periods of time, such as happens in areas of mountain building (see Figure 13-4c) or in *subduction zones,* where the edge of one lithospheric plate slides below another plate. **Hydrothermal metamorphism** occurs where hot, mineral-rich fluids circulate through cracks in rocks. As we discuss in Chapter 14, these and other kinds of metamorphism are common along many of the boundaries between lithospheric plates.

(a) Outcrop of basalt overlying tuff

(b) Close-up of tuff

▲ **Figure 13-12** (a) This bedrock exposure in northeastern California, near Alturas, shows a light brown basalt overlying a colorful layer of tuff (a volcanic rock formed by consolidation of volcanic ash). The basalt was extruded onto the tuff in molten form, and its great heat "baked," or metamorphosed, the upper portion of the tuff. The color difference between the metamorphosed and unmetamorphosed part of the tuff stratum is visual evidence of the metamorphism. (b) A representative sample of tuff.

Energy for the 21st Century

Unconventional Hydrocarbons and the Fracking Revolution

• **Matthew Fry,** University of North Texas

In 1982, an energy revolution was set in motion when commercial-scale shale gas production via *hydraulic fracturing*, or *fracking*, began on the Barnett Shale in Texas. After 2002, horizontal drilling, record-high natural gas prices, and relaxed environmental regulations combined to create a boom in "unconventional" hydrocarbon production. The expansion of unconventional production since then, in turn, has sparked controversy over fracking's costs and benefits.

Unconventional Hydrocarbons: Hydrocarbons are carbon- and hydrogen-based organic compounds, such as natural gas, petroleum, and bitumen, that form from decomposed organic matter in sedimentary deposits subjected to heat and pressure. Conventional methods to extract hydrocarbons involve drilling a well into a reservoir of crude oil and/or natural gas. Unconventional technology is employed to free trapped hydrocarbons from dense formations that have poor permeability (Figure 13-D).

Fracturing Process: To reach the hydrocarbons trapped in shale (referred to as shale gas and tight oil), first a vertical wellbore is drilled to the formation. The wellbore is then turned horizontally at depth, enabling operators to access greater area within the shale layer. Next, a mixture of water, chemicals, and "proppants" (often, sand) is injected at high pressure. The pressure breaks apart the shale, the chemicals free up the hydrocarbons, and the sand props open the fractures. The fractures serve as pathways for the movement of shale gas and tight oil to the surface, where gas, oil, and flowback (including the fracking fluid) are collected, separated, and transported.

Shale Production: In 2015, only the United States, Canada, China, and Argentina produced commercial volumes of shale gas or tight oil. In the United States, although the Barnett deposit has produced the most shale gas to date, the Marcellus deposit has the largest production potential (Figure 13-E). Tight oil production mostly comes from the Eagle Ford and Bakken Shales.

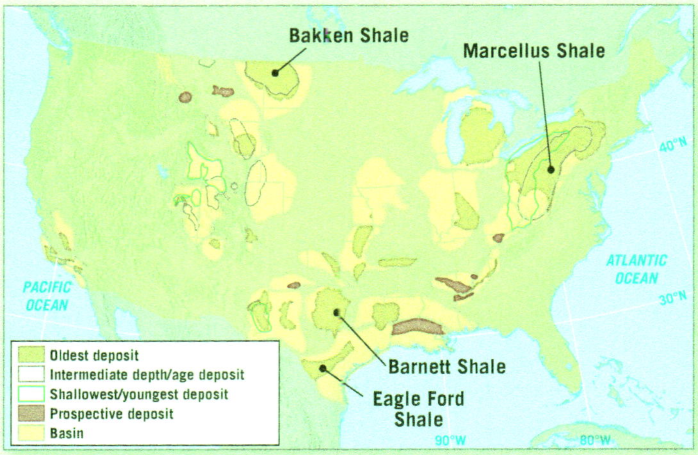

▲ Figure 13-E Shale gas deposits in the United States.

According to the U.S. Energy Information Administration, shale gas increased from 8 percent of U.S. natural gas production in 2007 to 75 percent in 2019. Production of tight oils also increased since 2008.

Controversy: Despite (and due to) increased shale gas and tight oil production, researchers have documented negative impacts of fracking. The use of large volumes of freshwater in fracking fluids can severely reduce local water supplies, especially in arid or semiarid regions, and local and regional air quality can be degraded from emissions associated with drilling activities. Chemicals used in fracking fluids also pose risks. Well-water studies in areas of the Barnett and Marcellus Shales have detected elevated levels of dissolved methane, heavy metals, and volatile organic carbon compounds. Safe disposal of the hazardous and often radioactive flowback fluids is an issue. The high-pressure injection of flowback fluids into deep porous deposits is linked to induced seismic activity (earthquakes) in Oklahoma, Texas, and Colorado. With more than 200,000 new U.S. wells since 2000, the fracking revolution also has caused widespread land-use change, displacement of native vegetation and habitat, and land-surface scarring in drilling areas.

Questions

1. Why are more risks and impacts associated with unconventional drilling than with conventional drilling?
2. How can policymakers balance the costs of hydraulic fracturing with its economic benefits?

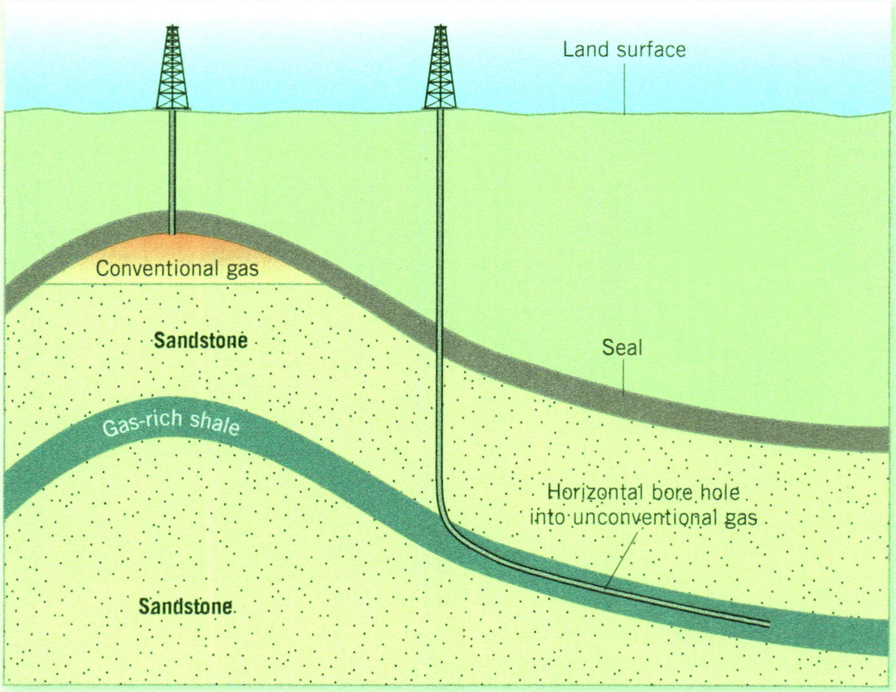

▲ Figure 13-D Conventional extraction targets trapped hydrocarbons that migrated from a source deposit. Unconventional drilling goes right to the source deposit.

(a) Mt. Rushmore: outcrop of granite overlying schist

Animation
Metamorphic Rock Formation
https://media.pearsoncmg.com/ph/esm/esm_mcknight_physgeo_13/media/

(b) Close-up of mica schist

▲ **Figure 13-13** (a) The presidents' heads at Mount Rushmore, South Dakota, are carved out of granite. Just below the granite is a much older metamorphic rock, a mica schist, into which the granite-forming magma intruded. (b) A sample of mica schist.

For some metamorphic rocks, it is possible to know what the original, unaltered rock was. In other cases, the metamorphism is so great that it is difficult to know with certainty the nature of the original rock.

Foliated Metamorphic Rocks: If the minerals in a metamorphic rock show a prominent alignment or orientation, we say that the rock is *foliated*. Dynamic geologic environments, such as the margins of colliding plates, create the directed stresses necessary to produce foliated metamorphic rocks. Such rocks may have a platy, wavy, or banded texture. Ranging from the least ("low-grade") metamorphism to the most ("high-grade") metamorphism, some common foliated metamorphic rocks are fine-grained *slate* (shale that has undergone low-grade metamorphism), narrowly foliated/medium-grained *schist* (**Figure 13-13**), and broadly banded *gneiss* (**Figure 13-14**).

(b) Close-up of gneiss

(a) Banded gneiss outcrop

▲ **Figure 13-14** (a) An outcrop of banded gneiss in western Greenland. This formation, of age 3.8 billion years, is one of the oldest on Earth. (b) A typical piece of gneiss, showing foliation.

Nonfoliated Metamorphic Rocks: If the original rock was dominated by a single mineral (as in some sandstones or limestones), foliation is less common. When limestone undergoes metamorphism it usually becomes *marble*, whereas sandstone is frequently metamorphosed into *quartzite*. When the organic sedimentary rock bituminous coal undergoes metamorphism, it becomes high-grade *anthracite* coal.

LearningCheck 13-7 Define *foliation*, and explain under what conditions metamorphic rocks are likely to be foliated.

The Rock Cycle

As the preceding description suggests, over long periods of time, the minerals in one rock might well end up as part of a different rock: igneous rocks can be broken down into sediments, which might then form sedimentary rocks, which in turn might undergo metamorphism, only to be eroded back into sediments. This ongoing "recycling" of lithospheric material is referred to as the **rock cycle** (Figure 13-15). For example, as we just saw, the quartz in a piece of granite might over time be broken down into sediments; grains of quartz sand from the granite might then be compacted and cemented into sandstone; sandstone might then be metamorphosed into quartzite.

Continental and Ocean Floor Rocks

The distribution of the three principal rock classes on Earth's surface is uneven and may at first glance appear chaotic. However, this pattern is not random. The distribution of rocks in Earth's crust is a consequence of processes operating on the surface and in the planet's interior.

On the continents, sedimentary rocks compose the most commonly exposed bedrock—perhaps as much as 75 percent. The sedimentary cover is not thick, however, averaging less than 2.5 kilometers (1.5 miles), and sedimentary rocks accordingly constitute only a very small proportion (perhaps 4 percent) of the total volume of the crust. The bulk of the continents consist of granite, along with an unknown proportion of metamorphic rocks such as gneiss and schist. The ocean floor crust, however, is composed almost entirely of basalt and *gabbro* (the plutonic equivalent of basalt), covered by a relatively thin veneer of oceanic sediments.

The distinction between the dominant rocks of the ocean floors and continents is important. Basalt is denser than granite (about 2.9 g/cm^3 versus about 2.7 g/cm^3, respectively). For this reason, oceanic lithosphere is more dense than continental lithosphere. Continental lithosphere "floats" quite easily on the denser asthenosphere below, whereas oceanic lithosphere is dense enough that it can be pushed down, or

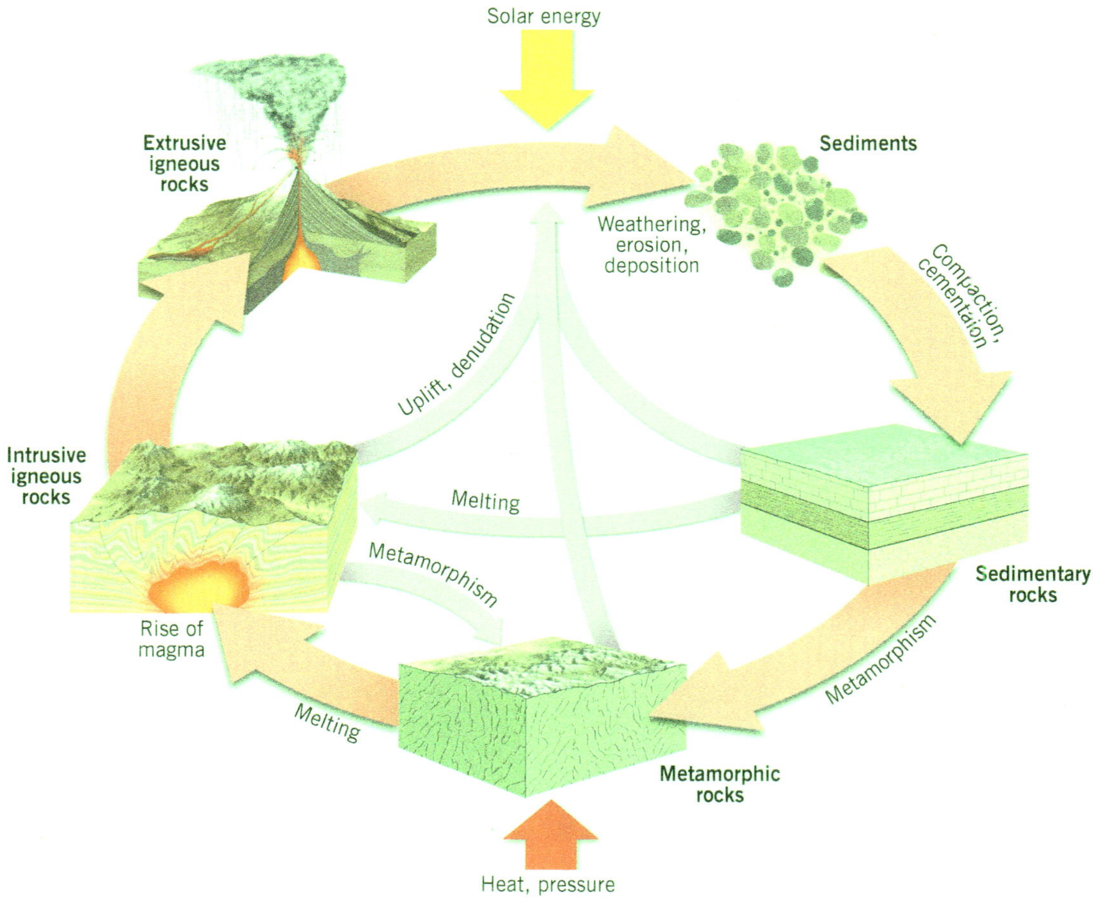

▲ Figure 13-15 The rock cycle, showing some of the relationships among the three classes of rocks.

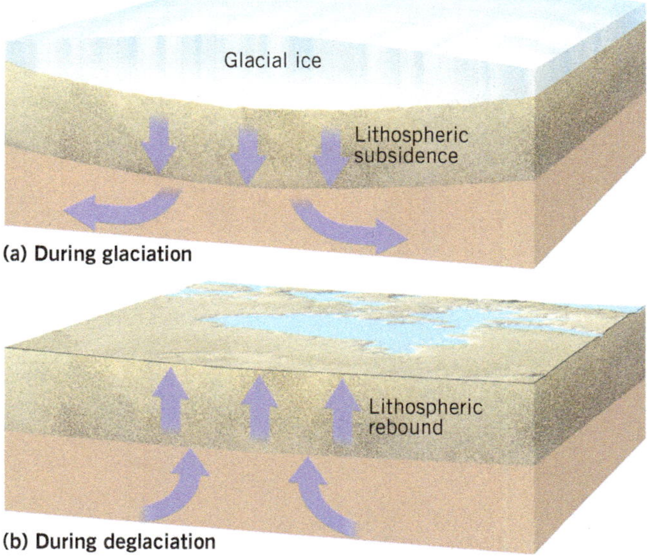

▲ **Figure 13-16** Isostatic adjustment. (a) In a glacial period, the heavy weight of accumulated ice depresses the lithosphere. (b) Deglaciation removes the weighty overburden, and as the ice melts, the lithosphere rises or "rebounds."

subducted, into the asthenosphere. The consequences of this are discussed at length in the following chapter.

Isostasy

Animation
Isostasy
https://media.pearsoncmg.com/ph/esm/esm_mcknight_physgeo_13/media/

Related to differences among oceanic crust, continental crust, and the mantle is the principle of *isostasy*. In simplest terms, the lithosphere "floats" on the denser, deformable asthenosphere below. Added weight on a portion of the lithosphere causes it to sink, whereas removal of a large weight allows the lithosphere to rise. This adjustment of Earth's crust due to the amount of load is **isostasy**.

Isostatic adjustment can have a variety of causes. The surface may be depressed, for example, by deposition of a large amount of sediment on a continental shelf, by the accumulation of a great mass of glacial ice on a landmass (**Figure 13-16**), or even by the weight of water trapped behind a large dam.

Depressed crust may rebound to a higher elevation as material on top erodes, as an ice sheet melts, or as a large body of water drains. Florida, for example, has experienced recent isostatic uplift because of the mass removed as groundwater dissolved the extensive limestone bedrock underlying the state. In central Canada around Hudson Bay, the region has uplifted more than 300 meters (almost 1000 feet) since the last of the Pleistocene ice sheets melted 8000 years ago. More recently, parts of southern Alaska have been uplifting at the astonishing rate of as much as 25 mm (1.0 in.) a year—most likely at least partly in response to the retreat of glaciers over the past century or so.

LearningCheck 13-8 **Is oceanic lithosphere or continental lithosphere denser? Why?**

The Study of Landforms

Our attention for the remainder of this book is directed primarily to **topography**: the surface configuration of Earth. A **landform** is an individual topographic feature. Thus, the term could refer to something as minor as a cliff or sand dune as well as to something as major as a peninsula or mountain range. The plural—*landforms*—is less restrictive and is generally considered synonymous with *topography*. Our focus on surface features is a field of study known as **geomorphology**, the study of the characteristics and development of landforms.

To organize our thinking, it may be helpful to take an analytic approach by isolating certain basic elements of the landscape: structure, process, slope, and drainage.

Structure: *Structure* refers to the nature, arrangement, and orientation of the materials making up a landform—the geologic underpinning of the landform. Is it composed of bedrock or not? If so, what kind of bedrock and in what configuration? If not made of bedrock, what are the nature and orientation of the sediments or other deposits? With a structure as clearly visible as that shown in Figure 13-9 or 13-10, these questions are easily answered, but this is not always the case.

Process: *Process* considers the actions that have combined to produce the landform. A variety of processes—geologic, hydrologic, atmospheric, and biotic—may be at work shaping the features of the lithospheric surface, and their interaction is critical to the formation of the features. In some cases, a landscape results primarily from one process, such as *glaciation* (**Figure 13-17**).

Slope: *Slope* is the fundamental aspect of shape for any landform. The angular relationship between a surface and the surrounding landscape is essentially a reflection of the contemporary balance among the various components of structure and process (see Figure 13-17).

Drainage: *Drainage* refers to the movement of water (from rainfall and snowmelt), either over Earth's surface or down into the soil and bedrock. Although moving water is also a "process" in landform development, the ramifications of streamflow, stream patterns, and other aspects of drainage are so significant that the general topic of drainage is considered a basic element in landform analysis (a topic discussed in detail in Chapter 16).

Once these basic elements have been recognized and identified, the geographer is prepared to analyze the topography by answering the fundamental questions at the heart of any geographic inquiry (**Figure 13-18**):

- What? The form of the feature or features
- Where? The distribution and pattern of the landform assemblage
- Why? An explanation of origin and development
- So what? The significance of the topography in relationship to other elements of the environment and to human life and activities

▶ **Figure 13-17** The spectacular terrain of Montana's Glacier National Park was produced by a variety of interacting processes, but most recently glaciation dominated. This view is across St. Mary Lake. Slope is a conspicuous visual element of some landscapes. Here the steep cliff on the left is Citadel Mountain.

LearningCheck 13-9 Explain what is meant by the "structure" of a landform.

Some Critical Concepts

One term used frequently in the following pages is **relief**, which is the difference in elevation between the highest and lowest points in an area. The term can be used at any scale. Thus, as we saw in Figure 1-7, the maximum world relief is approximately 20 kilometers (12 miles), which is the difference in elevation between the top of Mount Everest and the bottom of the Mariana Trench. At the other extreme, the local relief of the topography in places such as Florida can be a matter of merely a few meters.

Internal and External Geomorphic Processes

The relief we see in a landscape is temporary. It represents the momentary balance of two largely opposing sets of processes that shape and reshape the surface of Earth: *internal processes* and *external processes* (**Figure 13-19**).

◀ **Figure 13-18** Geologist from the Hawaiian Volcano Observatory sampling molten lava flow during the recent eruption of Kīlauea Volcano, Hawai'i Volcanoes National Park, Hawai'i.

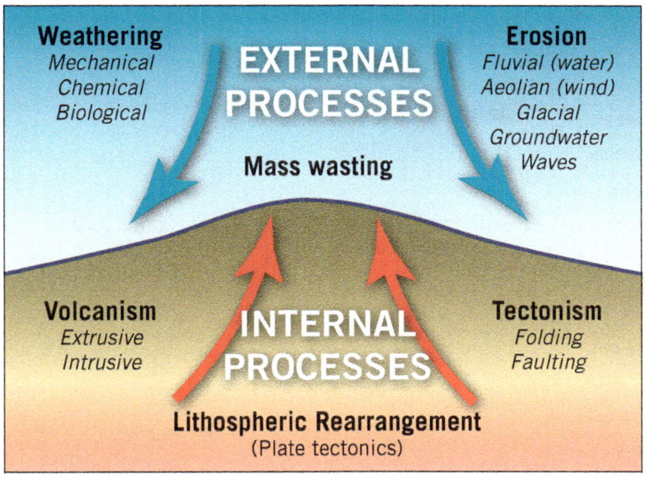

▲ **Figure 13-19** Schematic relationship between external and internal geomorphic processes. Earth's surface is uplifted by internal processes and worn away by external processes.

Internal Processes: The **internal processes** originate from within Earth. These processes result in crustal movements through folding, faulting, and volcanic activity of various kinds. In general, they are "constructive": uplifting, building processes that tend to increase the relief of the land surface. These processes are considered in detail in the next chapter.

External Processes: In contrast, the **external processes** are largely "subaerial" (they operate at the base of the atmosphere) and draw their energy mostly from sources above the lithosphere, either in the atmosphere or in the oceans. The external processes of *weathering, mass wasting,* and *erosion* may be thought of generally as "wearing-down" or "destructive" processes: eventually diminishing topographic irregularities and decreasing surface relief—broadly called *denudation.*

Thus, internal and external processes work in more or less direct opposition to one another. In some landscapes, both operate at the same time; we see the momentary balance between them. Their battleground is Earth's surface, the interface among all of the Earth-system spheres—the lithosphere, hydrosphere, atmosphere, and biosphere.

In succeeding chapters, we consider these various processes—their nature, dynamics, and effects—in detail, but it is useful to summarize them so that we can view them in totality before we treat them one at a time (see Figure 13-19). Note, however, that our classification scheme is a great simplification; whereas some items are clearly separate and discrete, others overlap with each other.

LearningCheck 13-10 Contrast the internal and external processes that shape Earth's surface.

Uniformitarianism

Fundamental to an understanding of internal and external processes and topographic development is familiarity with the doctrine of **uniformitarianism**, which holds that "the present is the key to the past." This concept, first put forth by geologist James Hutton in 1795, means that the processes that shape the landscape today are the same processes that formed the topography of the past—and are the same processes that will shape the topography of the future.

In the centuries before Hutton, most scientists accepted the doctrine of *catastrophism*—the idea that Earth's major features, such as large canyons and mountain ranges, were produced by sudden upheavals and catastrophic events in the past. Uniformitarianism, instead, says that by understanding the geomorphic processes we see at work today—and by taking into account long periods of time—we can study a landscape and begin to comprehend the history of its development. The development of landforms is an ongoing process, with the topography at any given time representing one moment in a continuum of change.

It is important to understand that uniformitarianism doesn't say that all geomorphic processes have operated at the same rate or to the same extent throughout Earth's history. For example, at some periods of time in the past, glaciation was more important than it is today. Furthermore, it doesn't mean that landscape change always takes place through slow, grain-by-grain wearing away of the rock. Change often comes rather episodically and abruptly through "local catastrophes"—such as an earthquake, a volcanic eruption, a flood, or a landslide.

Overall, however, because many of the internal and external processes operate very slowly by human standards, they may be difficult for us to grasp. For example, enormous lithospheric plates (which we discuss with plate tectonics in Chapter 14) move a few centimeters a year, on average—about as fast as your fingernails grow. To understand how sometimes imperceptibly slow processes alter the landscape, we must stretch our concept of time.

Geologic Time

Probably the most mind-boggling concept in physical geography is the length of geologic time (**Figure 13-20**). In our daily lives, we deal with such brief intervals of time as hours, months, years, and sometimes centuries, which does not prepare us for the scale of Earth's history. The sweep of geologic time encompasses epochs of millions or hundreds of millions of years.

The concept of *geologic time* refers to the vast periods of time over which geologic processes operate. For uniformitarianism to work as a basic premise for interpreting Earth history, that history must have occurred over a span of time long enough to allow considerable change to be accomplished by the internal and external processes operating at infinitesimally slow rates when measured on a human time scale.

Age of Earth: Scientists estimate that Earth is nearly 4.6 billion years old—a length of time that is almost unfathomable. It is difficult to grasp such an enormous period of time in which the Age of Dinosaurs lasted about 180 million years, or one in which the Rocky Mountains were initially uplifted approximately 65 million years ago.

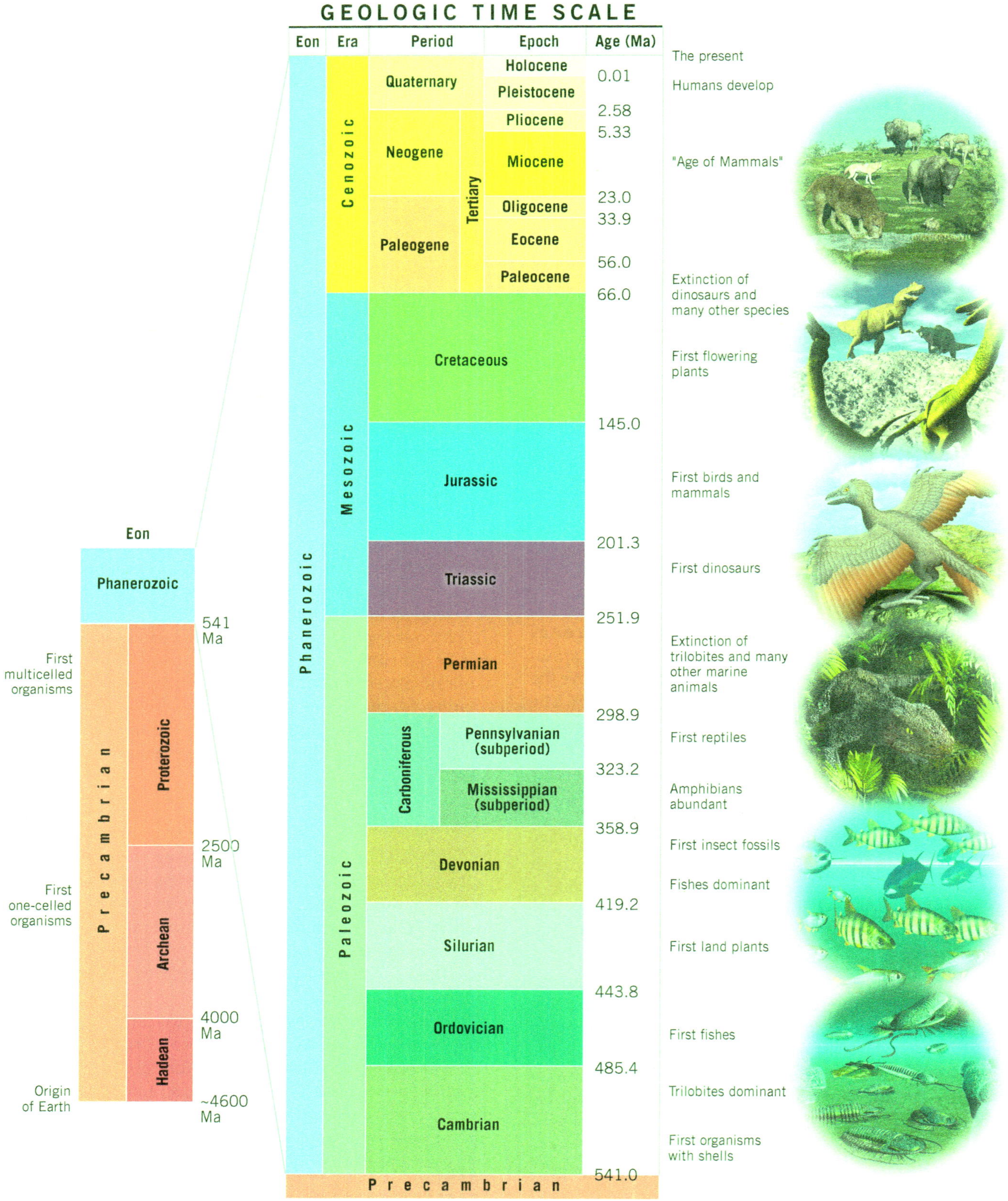

▲ **Figure 13-20** The geologic time scale. The divisions of Earth's history were originally based on relative dating—the sequence of major changes and extinctions seen in the fossil record. The absolute dates, shown as millions of years ago (Ma), were established more recently but are subject to revision. This current time scale is based on the International Commission on Stratigraphy's 2020 International Chronostratigraphic Chart v2020/01.

An analogy may help us begin to comprehend the almost incomprehensible length of geologic time: Envision the entire 4.6-billion-year history of Earth compressed into a single calendar year. Each day would be equivalent to 12.6 million years, each hour to 525,000 years, each minute to 8750 years, and each second to 146 years. On such a time scale, the planet was lifeless for more than two months, with primal forms of one-celled life not appearing until early March. These primitive algae and bacteria had the world to themselves until June, when the first multi-celled organisms began to evolve. The first vertebrate animals, fishes, appeared about November 21; before the end of the month, amphibians began establishing themselves as the first terrestrial vertebrates. Vascular plants, mostly tree ferns, club mosses, and horsetails, appeared about November 27, and reptiles began their era of dominance about December 7. Mammals arrived about December 14, with birds on the following day. Flowering plants would first bloom about December 21, and on December 24 both the first grasses and the first primates appeared. The first hominids walked upright in midafternoon on New Year's Eve, and *Homo sapiens* came on the scene about an hour before midnight. The age of written history encompassed the last minute of the year!

This extraordinary time scale gives credibility to the doctrine of uniformitarianism, allowing us to see the Grand Canyon as a "youthful" feature carved by that relatively small river seen deep in its inner gorge and to envision how Africa and South America were once joined but have drifted 3200 kilometers (2000 miles) apart. The geomorphic processes generally operate with slowness, but the vastness of geologic history provides a suitable time frame for their accomplishments.

Geologic Time Scale: The geologic time scale shown in Figure 13-20 is divided into units of time that reflect major events in Earth's history, especially changes in life on Earth. The largest units are called *eons*: the *Phanerozoic* (the time of "visible life"), which began about 541 million years ago, and the *Precambrian*, which includes about 88 percent of Earth's history. The Phanerozoic Eon is divided into three major *eras*: the *Cenozoic* (Age of Recent Life), *Mesozoic* (Age of Middle Life), and *Paleozoic* (Age of Ancient Life). Each era is divided into *periods*, which are in turn subdivided into epochs (shown here for the Cenozoic Era only).

Geologists in the nineteenth century identified the original divisions of the geologic time scale by using relative dating methods—largely based on the observed sequence of major transitions (especially mass extinctions) in the fossil record. The absolute dates in the scale were established beginning in the twentieth century through *radiometric dating*—measuring the decay of radioactive elements found in some rocks (see the *Featured National Park* box on the Grand Canyon: *Traveling through Geologic Time*).

The dates in the geologic time scale are subject to periodic revision as new data become available. For example, in 2009 the *International Commission on Stratigraphy* recommended that the boundary between the Pleistocene and Pliocene Epochs be changed from 1.81 million years ago to 2.58 million years ago.

Because of the enormous impact humans have had on the planet in recent centuries, some scholars propose that we call the current epoch of geologic time the *Anthropocene*, although this designation has yet to be universally accepted.

LearningCheck 13-11 **Explain the importance of geologic time to the doctrine of uniformitarianism.**

Scale and Pattern

Before we proceed with our systematic study of geomorphic processes, keep in mind two final concepts: scale and pattern.

An Example of Scale

The question of scale is fundamental in geography. Regardless of the subject of geographic inquiry, recognizable features and associations are likely to vary depending on the scale of observation. The aspects of the landscape observed in a close-up view are different from those observed at a more distant view.

As an example of the complexity and significance of scale, let us view a particular place on Earth's surface from different perspectives. The location is north-central Colorado within the boundaries of Rocky Mountain National Park, some 13 kilometers (8 miles) due west of the town of Estes Park. It includes a small valley called Horseshoe Park, with the steep slopes of Bighorn Mountain to the north and Mt. Chapin to the west; through it flows a clear mountain stream named Fall River (**Figure 13-21**).

- Figure 13-21a: To illustrate the largest or closest scale of ordinary human experience, we hike northward from the center of Horseshoe Park up the side of Bighorn Mountain. At this scale, we can observe such effects as streamflow and local erosion. The first topographic feature of note is a smooth stretch of Fall River that we must cross. We walk over a small sandbar at the south edge of the river, wade for a few steps, and tread up a low bank onto the mountainside, noting the dry bed of a small intermittent pond on our left. After 20 minutes of steep uphill scrambling, we reach a rugged granite outcrop called Hazels Cone, which presents us with an almost vertical cliff face to climb.
- Figure 13-21b: For a significantly different scale of observation, we travel by car through Horseshoe Park along U.S. Highway 34. After 20 minutes, we reach a magnificent viewpoint high on the mountain to the southwest of Horseshoe Park. From this vantage point, our view of the country through which we hiked is significantly widened; here we observe the relationships among rock type, vegetation, and slope. We can no longer recognize the sandbar, the bank, or the dry pond, and even the rugged cliff of Hazel Cone appears as little more than a pimple on the vast slope of Bighorn Mountain. Instead, we see that Fall River is a broadly meandering stream in a flat valley and that Bighorn Mountain is an impressive peak rising high above the valley.

Featured National Park

Grand Canyon

Traveling through Geologic Time

In carving the Grand Canyon, the Colorado River cut down through layers of rock that span nearly 2 billion years of Earth's history (**Figure 13-F**). As you travel down into the canyon, you are, in a sense, going back in time, making the Grand Canyon an ideal place to envision the great expanse of geologic time.

Earth scientists use both *absolute dating* and *relative dating* to study the geologic history of a landscape such as the Grand Canyon.

▲ **Figure 13-F** The Grand Canyon exposes rock layers dating back 1.7 billion years.

Absolute Age: *Absolute age* describes the actual number of years since a rock layer formed or a geologic event took place. These ages, such as those shown on the geologic time scale (see Figure 13-20), have been determined primarily though *radiometric dating* based on the steady, predictable decay of radioactive elements found in some rocks.

Relative Age: Relative dating is based on several principles that allow the sequence—but not necessarily the absolute age—of geologic events to be determined.

Principle of Original Horizontality: Because most sediments are transported by streams and deposited in basins or bodies of water, they accumulate in flat, horizontal layers. If those layers have been tilted or folded, we know that this deformation took place *after* deposition. Notice in **Figure 13-G** that most of the upper rock layers in the Grand Canyon retain their original horizontal orientation.

Principle of Superposition: If a sequence of sedimentary layers has not been disturbed, we know that the top layer is younger than the layer below. For example, the Kaibab Limestone at the rim of the Grand Canyon is younger than the limestone of the Toroweap Formation just below—which in turn is younger than the Coconino Sandstone below it.

Principle of Cross-Cutting Relationships: If a feature such as a fault or igneous intrusion cuts through a rock layer, we know that this cross cutting occurred *after* that rock layer formed.

Unconformities: Over time, layers of rock can be exposed to both tectonic and erosional processes. So, rather than finding uninterrupted sequences of rocks dating from the present day back to the earliest formation of Earth, we find gaps in the geologic sequence. For example, erosion may remove a sedimentary layer, exposing older rocks below—only to have those older rocks covered again with newer sedimentary layers. Such interruptions in the rock record are called *unconformities*.

The Grand Canyon has several major unconformities—hundreds of millions of years of the rock record are missing. Most dramatically, just above the 1.7 billion-year-old Vishnu Schist exposed in the inner gorge of the Grand Canyon is the 525 million-year-old Tapeats Sandstone (see Figure 13-G). This Great Unconformity represents a gap of more than 1 billion years in the geologic record!

▲ **Figure 13-G** Major rock layers in the Grand Canyon. Gaps in the rock record are known as *unconformities*. (Ma = rock age in millions of years.)

Questions

1. How might the presence of a fault help determine the relative ages of sedimentary rock layers?
2. How do unconformities in the rock record come about?

- Figure 13-21c: Our third observation of this area takes place from an airplane flying over this region at 12 kilometers (39,000 feet). From this elevation, Fall River is nearly invisible, and only careful observation reveals Horseshoe Park. Bighorn Mountain is now merged indistinguishably as part of the Mummy Range, which is seen as a minor offshoot of a much larger and more impressive mountain system called the Front Range. Here we can discern broader stream patterns and begin to see the connection of Bighorn Mountain to the rest of the local mountain range.
- Figure 13-21d: A fourth level of observation is available when we hitch a ride on the International Space Station, orbiting 390 kilometers (240 miles) above Earth. Our brief glimpse of northern Colorado is probably inadequate to distinguish the Mummy Range, and even the 400-kilometer (250-mile) long Front Range appears as only a component of the mighty Rocky Mountain cordillera, which extends from New Mexico to northern Canada. From this vantage point, we begin to recognize the relationship of the Front Range to the regional tectonic context of the Rocky Mountains as a whole.
- Figure 13-21e: At the smallest scale, the final viewpoint possible to humans is from a spacecraft rocketing away from Earth toward the Moon. Looking back in the direction of Horseshoe Park, we might recognize the Rocky Mountains, but the only conspicuous feature in this small-scale view is the North American continent. From here, we can place the present-day Rocky Mountains in the context of the sequence of events that "assembled" North America over billions of years.

For more on Rocky Mountain National Park see the feature in Chapter 10.

LearningCheck 13-12 **Which level of scale in Figure 13-21 is most suitable for studying an individual landform? The structure of a continent?**

Pattern and Process in Geomorphology

The prime goals of any geographic study are to recognize the distribution pattern of some phenomenon and to understand the processes that produced that pattern. In previous portions of this book, we saw that there is broad geographic predictability to many patterns of weather, climate, ecosystems, biomes, and soils based on latitude or location on a continent. We now enter into a part of physical geography in which such orderly patterns of distribution are more difficult to discern.

There are a few aspects of predictability; for example, we can anticipate that in desert areas, certain geomorphic processes are more conspicuous than others and certain landform features are likely to be found. Overall, however, the global distribution of topography appears irregular. Largely for this reason, the geomorphology portion of this book concentrates less on distribution and more on process. Comprehending the processes associated with topographic development is more important to an understanding of systematic physical geography than any amount of detailed study of landform distribution. We begin this study of landform-shaping processes in the chapter that follows.

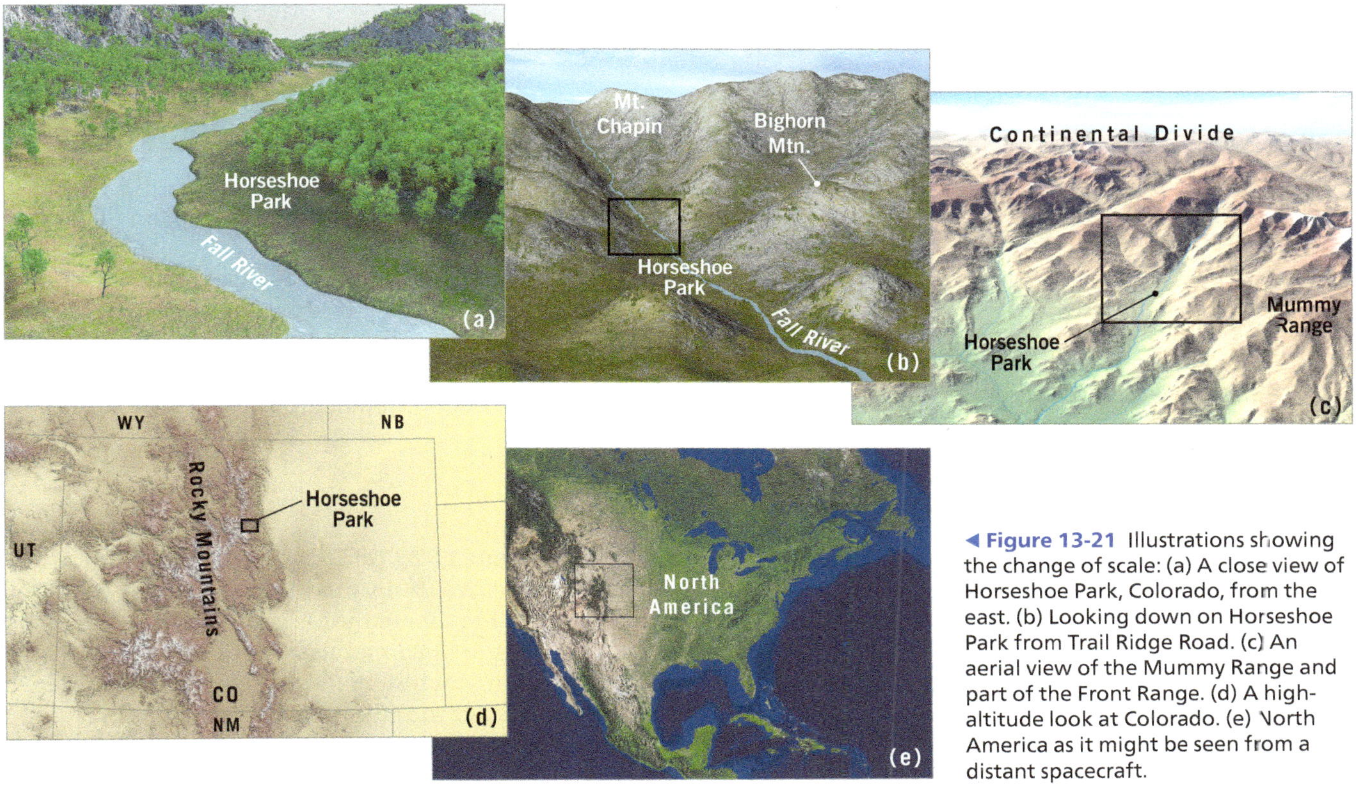

◀ **Figure 13-21** Illustrations showing the change of scale: (a) A close view of Horseshoe Park, Colorado, from the east. (b) Looking down on Horseshoe Park from Trail Ridge Road. (c) An aerial view of the Mummy Range and part of the Front Range. (d) A high-altitude look at Colorado. (e) North America as it might be seen from a distant spacecraft.

13 Learning Review

After studying this chapter, you should be able to answer the following questions. Key terms from each text section are shown in **bold type.** Definitions for key terms are also found in the glossary at the back of the book.

Key Terms and Concepts

The Structure of Earth (*p. 380*)

1. Briefly describe the overall structure of Earth, noting the four main layers: **crust**, **mantle**, **outer core**, and **inner core**.
2. What are the differences between the **lithosphere** ("plates") and the **asthenosphere**?
3. What is the **Moho (Mohorovičić discontinuity)**?

The Composition of Earth (*p. 381*)

4. How is a **mineral** different from a **rock**?
5. Provide a general description of the **silicate** mineral family. Name at least one common silicate mineral.
6. What is an **outcrop**?
7. Describe the general differences among **igneous**, **sedimentary**, and **metamorphic rocks**.
8. Briefly define the following terms: **magma**, **lava**, **pyroclastics**.
9. What is the difference between a **plutonic (intrusive igneous) rock** and a **volcanic (extrusive igneous) rock**?
10. What are the main differences between **granite** and **basalt**?
11. How is **sediment** formed?
12. Why do most sedimentary rocks form in flat, horizontal layers called **strata**?
13. What is a **bedding plane**?
14. Briefly contrast **contact metamorphism**, **regional metamorphism**, and **hydrothermal metamorphism**.
15. Over long periods of time, how can the minerals in one rock end up as part of a different rock (or a different kind of rock)? In other words, explain the **rock cycle**.
16. Explain the concept of **isostasy**.

The Study of Landforms (*p. 395*)

17. Briefly define the terms **topography**, **landform**, and **geomorphology**.
18. What is meant by the *structure* of a landform?

Some Critical Concepts (*p. 396*)

19. In the context of geomorphology, what is meant by the term **relief**?
20. Contrast the concepts of **internal processes** and **external processes** in geomorphology.
21. How does the doctrine of **uniformitarianism** help us understand the history of Earth?
22. What is generally meant by the term *geologic time*?

Scale and Pattern (*p. 399*)

23. In the study of geomorphology, why do we primarily concentrate on processes rather than on distribution patterns by latitude?

Study Questions

1. How can one kind of magma (one kind of magma chemistry) produce two or more different kinds of igneous rock?
2. Why are metamorphic rocks often found in contact with plutonic rocks such as granite?
3. Contrast the composition and characteristics of oceanic lithosphere with those of continental lithosphere.
4. Why are sedimentary rocks so common on the surface of the continents?
5. Describe the general structure, slope, drainage, and processes involved in the development of the landscape in Figure 13-17.
6. What is the importance of geologic time to the doctrine of uniformitarianism?

Exercises

1. If sediments are deposited in a body of water at a rate of 2 centimeters every 1000 years, how long does it take to accumulate a layer of sediment 8 meters thick? ______ years
2. Using Figure 13-20, calculate the percentage of Earth's 4.6-billion-year history in which fishes have been present (since the beginning of the Ordovician Period). ______%
3. Using Figure 13-20, calculate the percentage of Earth's 4.6-billion-year history in which dinosaurs were present (throughout the Mesozoic Era). ______%
4. Using Figure 13-20, calculate the percentage of Earth's 4.6-billion-year history in which humans have been present (since the beginning of the Pleistocene Epoch). ______%

Environmental **Analysis** ROCK AND MINERAL RESOURCES IN YOUR AREA

Data
Mineral Resources
https://media.pearsoncmg.com/ph/esm/esm_mcknight_physgeo_13/media/

Minerals are the building blocks of the rocks that make up the landscape. Minerals are mined throughout the world, including the United States. The evolution of the landscape determines which minerals are close enough to the surface for us to access.

Activities

Go to Mineral Resources Online Spatial Data at http://mrdata.usgs.gov, click on "Mineral Resources Data System (MRDS)", and choose the "by record quality" map. In the menu, check A through D under "Mineral Resources."

1. What information is displayed on this map?
2. Which areas of the conterminous United States have the largest concentration of mines?
3. Which states have the smallest concentration of mines?

Zoom in to your county (or a nearby county if yours has no mineral resource sites) and click on a single mineral resource site. Click on the first "Record ID" to open a new page with site information.

4. Which commodities are available from this site?
5. Click the link under "Nearby scientific data" to open a new page with geologic information. During what geologic age did this region form? Estimate the age in thousands or millions of years.
6. What major changes or extinctions occurred during this period based on Figure 13-20?
7. What are the major and minor rock types in this region? Are these rocks igneous, sedimentary, or metamorphic?

Go to the USGS State Minerals Statistics and Information map: http://minerals.usgs.gov/minerals/pubs/state/. Click on your state, then click on the most recent year's Minerals Yearbook Volume II in PDF format. Scroll down to the state map.

8. What is one mineral resource in your county (or a nearby county if yours does not list a resource)?
9. Do any of the surrounding counties have the same resource? Which resources are found in surrounding counties but not in your chosen county?

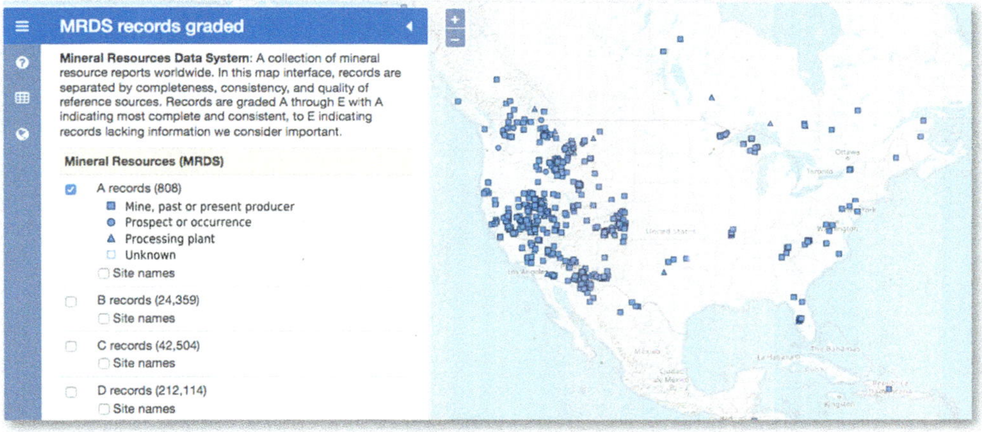

Seeing Geographically in Our National Parks

Look again at the photograph of Capitol Reef at the beginning of the chapter (p. 378). Which of the three classes of rock likely makes up most of what you see? Why do you say this? What is the structure of these exposed rocks? Describe a general sequence of internal and external processes that could have led to the present landscape.

Mastering Geography®

Looking for additional review and test prep materials? Visit the Study Area in **Mastering Geography** to enhance your geographic literacy, spatial reasoning skills, and understanding of this chapter's content by accessing a variety of resources, including Map Master interactive maps, geoscience animations, *Mobile Field Trips*, videos, *Project Condor* Quadcopter videos, *In the News* RSS feeds, flashcards, web links, self-study quizzes, and an eText version of *McKnight's Physical Geography*.

Seeing Geographically in Our National Parks

Eruption of Kīlauea Volcano on the island of Hawai'i, Hawai'i Volcanoes National Park. In addition to the molten lava flow, what other kind of material is being vented from the volcano? Does it appear that the lava is flowing down a steep slope or gentle slope? What does that tell you about how easily the lava is moving?

Hawai'i Volcanoes National Park, HI

The Internal Processes

14

HAVE YOU EVER WONDERED why volcanoes are common in Hawai'i? Or why they're not found in Connecticut? Interestingly, Connecticut has lots of volcanic rocks—just no active volcanoes. The explanation for these and many other intriguing patterns on Earth's surface comes with an understanding of the internal processes.

In the previous chapter, we learned about the interior of Earth and how different kinds of rocks form. We also gained an appreciation for changes in landscapes over the immense span of geologic time. In this chapter, we begin to explain how much of that change can take place through the internal processes. It is the slow movement of hot—but largely solid—rock through the mantle that drives plate tectonics and is responsible for nearly all volcanism, folding, and faulting—the processes that create mountain ranges and increase the relief of Earth's surface. We leave it to the chapters that follow to explain how this relief can be worn down.

As you study this chapter, think about these **Key**Questions:

- **How does the theory of plate tectonics explain the changing positions of continents and ocean basins?**
- **What evidence helps verify plate tectonics?**
- **What happens at divergent, convergent, and transform plate boundaries?**
- **How does the Hawaiian hot spot fit in with plate tectonics?**
- **What are the different kinds of volcanoes?**
- **What are synclines and anticlines, and how do they form?**
- **What landforms develop from normal faults and strike-slip faults?**
- **How does the nature of the bedrock influence the severity of ground shaking during an earthquake?**

The Impact of Internal Processes on the Landscape

Internal processes actively shape and reshape Earth's surface. The crust is buckled and bent, land is raised and lowered, rocks are fractured and folded, solid material is melted, and molten material is solidified. These actions are fundamentally responsible for the overall shape of the lithospheric landscape at any given time. The internal processes do not always act independently of each other, but in this chapter we isolate them to simplify our analysis.

From Rigid Earth to Plate Tectonics

The shapes and positions of the continents may seem fixed at the time scale of human experience, but at the geologic time scale, measured in millions or tens of millions of years, continents are quite mobile. Continents have moved, collided and merged, and then been torn apart again; ocean basins have formed, widened, and eventually closed off. These changes on the surface of Earth continue today, so the contemporary configuration of the ocean basins and continents is by no means final. Only in the last half-century, however, have Earth scientists come to understand how all of this could happen.

Until the mid-twentieth century, most Earth scientists assumed that the planet's crust was static, with continents and ocean basins fixed in position and significantly modified by changes in sea level and periods of mountain building only. The uneven shapes and irregular distribution of the continents were puzzling, but it was generally accepted that the present arrangement was emplaced in the distant past when Earth's crust cooled from its original molten state.

Although not widely accepted, the idea that the continents have changed position over time, or that a single "supercontinent" once existed before separating into large fragments, has been around for a long time. Various naturalists, physicists, astronomers, geologists, botanists, and geographers have been putting forth this idea since the days of geographer Abraham Ortelius in the 1590s and philosopher Francis Bacon in 1620. Until fairly recently, however, the idea was generally unacceptable to the scientific community at large.

Wegener's Continental Drift

Early in the twentieth century, the notion of **continental drift** was revived, most notably by German meteorologist and geophysicist Alfred Wegener. Wegener put together the first comprehensive theory to describe and partially explain the phenomenon, publishing his landmark book *Die Entstehung der Kontinente und Ozeane* (*The Origin of Continents and Oceans*) in 1915. Wegener postulated the existence of a massive supercontinent, which he called **Pangaea** (Greek for "whole land"), that began to break apart long ago into the present-day continents—and that have continued to move away from one another to this day (Figure 14-1).

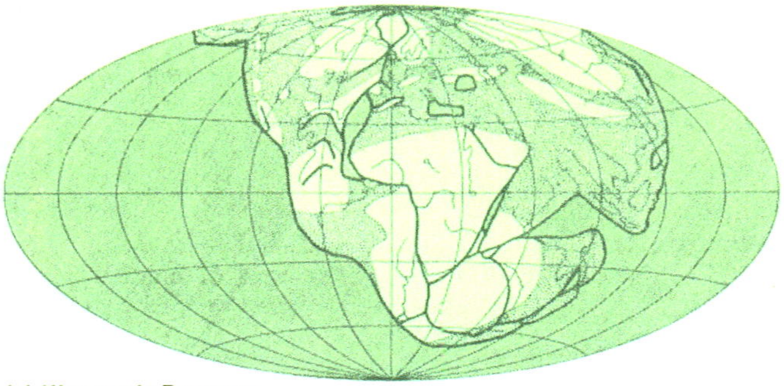

(a) Wegener's Pangaea

(b) Modern reconstruction of Pangaea

▲ **Figure 14-1** The supercontinent Pangaea. (a) Wegener's reconstruction from 1925. (b) Contemporary reconstruction of Pangaea as it probably appeared about 200 million years ago.

Wegener's Evidence for Continental Drift: Wegener accumulated a great deal of evidence to support his hypothesis, most notably the remarkable number of close affinities of geologic features on both sides of the Atlantic Ocean. He found that the continental margins of Africa and South America fit together with jigsaw-puzzle-like precision (Figure 14-2). He also determined that the

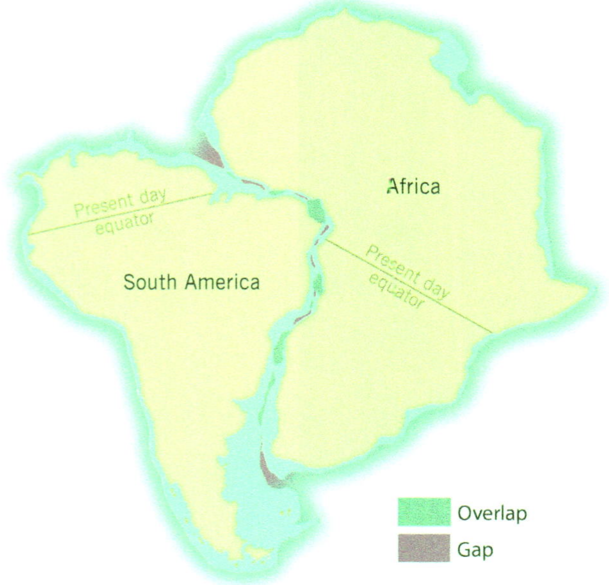

▲ **Figure 14-2** When the shapes of Africa and South America are compared at a depth of about 900 meters (500 fathoms) along their continental slopes, the match is convincing—there are few large gaps or overlaps.

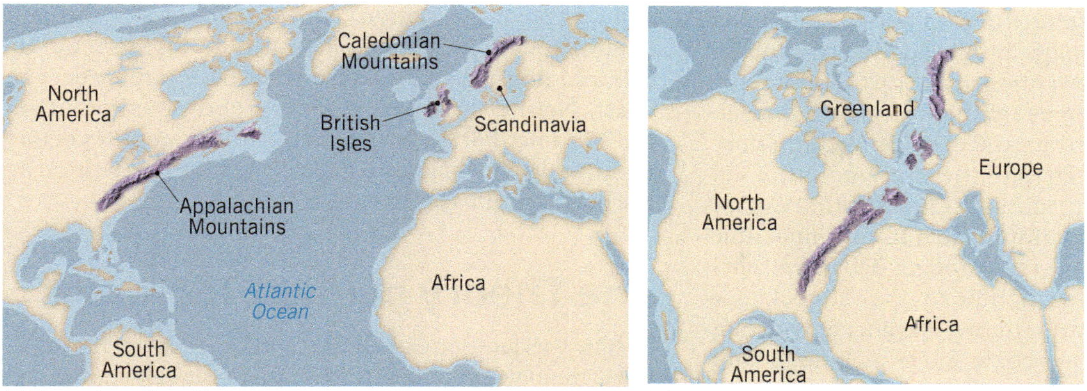

(a) Present configuration (b) Pangaea

▲ **Figure 14-3** Mountain ranges match on both sides of the North Atlantic Ocean. (a) The Appalachian Mountains are of similar age and structure to mountains in the British Isles and Scandinavia. (b) Before Pangaea rifted apart, these mountain belts formed a nearly continuous chain.

petrologic (rock) records on both sides of the Atlantic show many distributions—such as ancient coal deposits—that would be continuous if the ocean did not separate them. Moreover, when models of the continents are moved back into the Pangaean configuration, mountain belts in Scandinavia and the British Isles match up with the Appalachian Mountains in eastern North America (**Figure 14-3**).

Supporting evidence came from paleontology: the fossils of some dinosaur and other reptile species, such as the freshwater swimming reptile *Mesosaurus*, are found on both sides of the southern Atlantic Ocean but nowhere else in the world (**Figure 14-4**). Fossilized plants, such as the fernlike *Glossopteris*, are found in similar-aged rocks of South America, South Africa, Australia, India, and Antarctica—its seeds too large and heavy to have been carried across the expanse of the present-day oceans by wind.

Wegener worked with climatologist Wladimir Köppen (Wegener's father-in-law) to study the past climate patterns of Earth. They examined glacial deposits that indicated that large portions of the southern continents and India were extensively glaciated about 300 million years ago. The pattern of deposits made sense if the continents had been together in Pangaea when this glaciation took place (**Figure 14-5**).

LearningCheck 14-1 **How did the shape of continental coastlines and evidence of past glaciation support Wegener's idea of continental drift? (Answer on p. AK-4)**

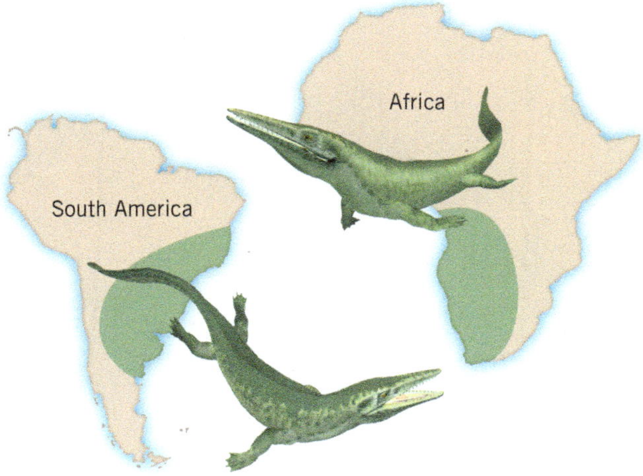

▲ **Figure 14-4** Mesosaurus fossils have been found in southeastern South America and southwestern Africa. These animals were alive at the time that present-day South America and Africa were together as part of a larger landmass in the geologic past.

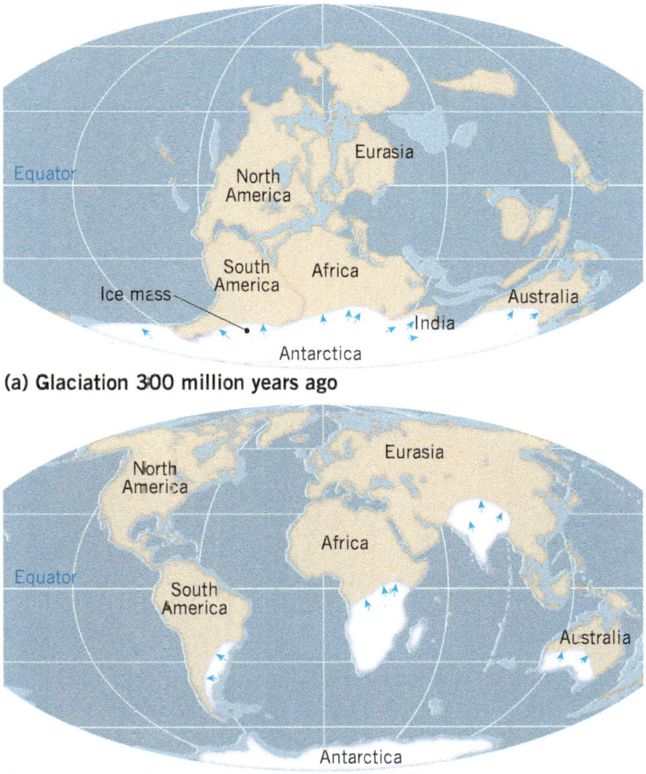

(a) Glaciation 300 million years ago

(b) Same glacial extent in present configuration

▲ **Figure 14-5** (a) About 300 million years ago, extensive areas of Pangaea were glaciated. When the continents are shown in their former positions, (b) today's pattern of glacial deposits make sense.

Rejection of Continental Drift: Wegener's accumulated evidence could be most logically explained by continental drift. His ideas attracted attention in the 1920s—and generated much controversy. Some Southern Hemisphere geologists responded with enthusiasm. The general response, however, was disbelief. Despite the vast amount of evidence Wegener presented, most scientists felt that two difficulties made the theory improbable if not impossible: (1) Earth's crust was believed to be too rigid to permit such large-scale motions—after all, how could solid rock plow through solid rock? (2) Furthermore, Wegener did not offer a suitable mechanism that could move such large landmasses for a long journey. For these reasons, most Earth scientists ignored or even debunked the idea of continental drift for the better part of half a century after Wegener presented his theory.

Although disappointed that his ideas on continental drift were rejected by most scientists, Wegener continued his other scientific work—most notably, in meteorology and polar research, where his contributions are widely acknowledged.[1] In 1930, Wegener led a meteorological expedition to Greenland. After delivering supplies to scientists stationed in the remote research outpost of Eismitte in the middle of the ice cap (see Figure 8-30c for a climograph of this station), on November 1 Wegener and fellow expedition member Rasmus Villumsen set out by skis and dogsled to return to their base camp near the coast. Neither arrived. Wegener's body was found six months later buried in the snow. He died decades before his ideas on continental drift received serious attention by the majority of Earth scientists.

[1]Before Wegener developed his ideas on continental drift, his 1911 textbook *The Thermodynamics of the Atmosphere* had become a standard in German universities. Swedish meteorologist Tor Bergeron openly acknowledged Wegener's contribution to our understanding of the raindrop formation process known today as the Bergeron process (see Chapter 6).

The Theory of Plate Tectonics

Despite the questions about the validity of continental drift, throughout the middle of the twentieth century continuing research revealed more and more about our dynamic planet. Evidence that eventually supported continental drift came from many disciplines.

The Evidence

Among the many gaps in scientific knowledge at the time of Alfred Wegener was an understanding of the dynamics of the ocean floors. By the 1950s, geologists, geophysicists, seismologists, oceanographers, and physicists had accumulated a large body of data about the ocean floor and the underlying crust.

One of the most intriguing early findings came in 1957 when thousands of depth soundings from the oceans were used by American geologist and cartographer Marie Tharp and geologist Bruce Heezen to construct the first detailed map of ocean floor topography. The result was remarkable (**Figure 14-6**): vast abyssal plains were seen dotted with

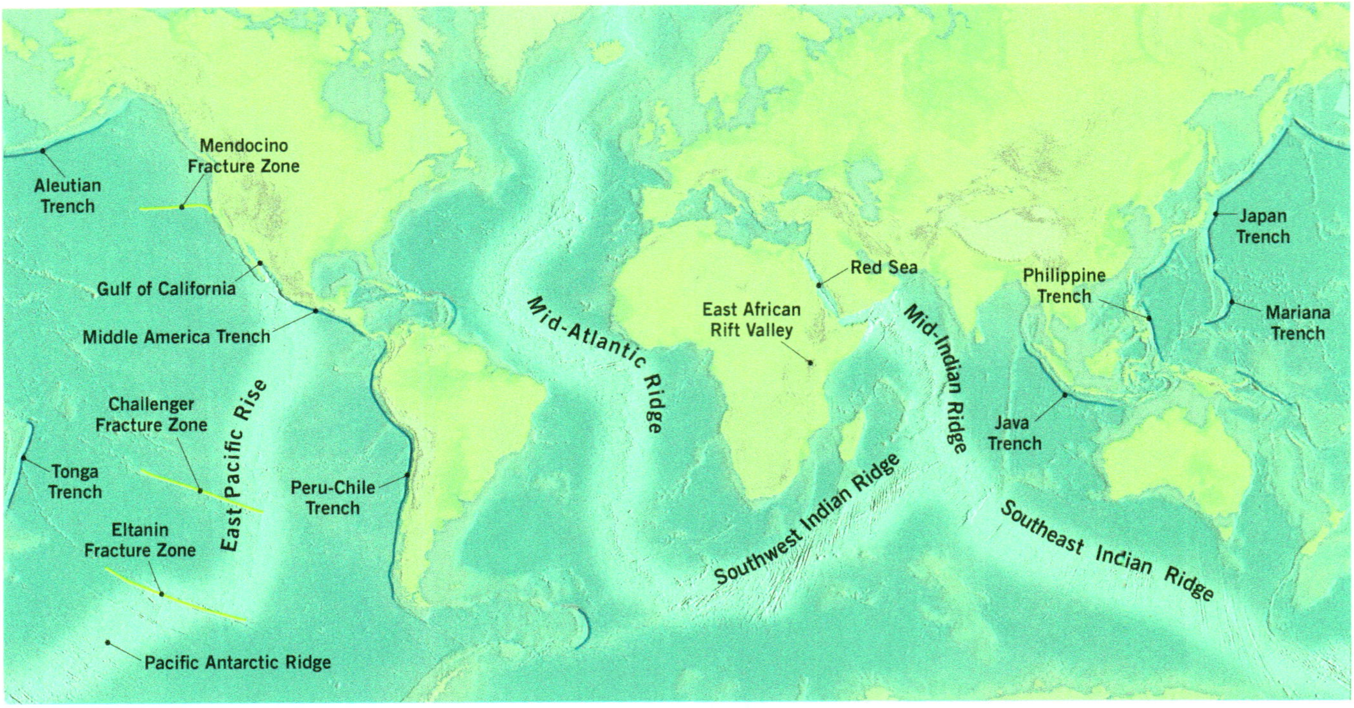

▲ **Figure 14-6** A continuous system of ridges—including the Mid-Atlantic Ridge, the East Pacific Rise, and the Southwest and Southeast Indian Ridges—runs across the floor of the world ocean. This map also shows major oceanic trenches, fracture zones, and rift valleys.

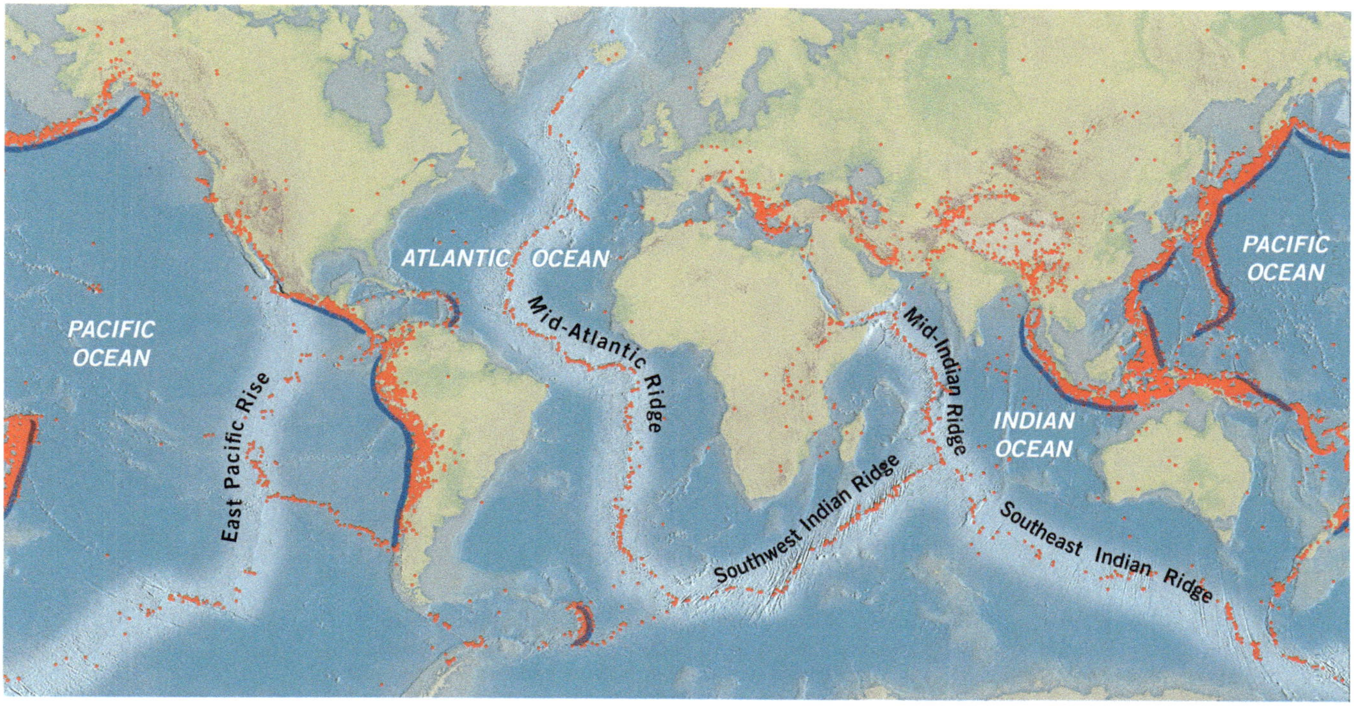

▲ **Figure 14-7** The distribution of epicenters for all earthquakes of at least 5.0 magnitude over a 10-year period. Their relationship to midocean ridges and oceanic trenches is striking.

chains of undersea volcanoes known as *seamounts*. Narrow, deep *oceanic trenches* occurred in many places, often around the margins of the ocean basins.

Perhaps most stunning of all was a continuous ridge system running across the floors of all the oceans for 64,000 kilometers (40,000 miles), wrapping around the globe like the stitching on a baseball. The mid-Atlantic segment of this *midocean ridge system* is especially striking, running exactly halfway between—and matching the shape of—the coastlines on both sides, almost as if a giant seam had opened up in the ocean floor between the continents.

By the 1960s, a world network of seismographs was able to pinpoint the location of every significant earthquake in the world. When earthquake locations were mapped, it was clear that earthquakes do not occur randomly around the world; instead, most earthquakes occur in bands, often coinciding with the pattern of the midocean ridge system and oceanic trenches (Figure 14-7).

Seafloor Spreading

Animation
Seafloor Spreading
https://media.pearsoncmg.com/ph/esm/esm_mcknight_physgeo_13/media/

In the early 1960s, a new theory was proposed, most notably by American oceanographer Harry Hess and geologist Robert S. Dietz, that could explain the significance of the midocean ridges, the oceanic trenches, and the pattern of earthquakes—and might provide a mechanism for Wegener's continental drift. Known as **seafloor spreading**, this theory stated that midocean ridges are formed by currents of magma rising up from the mantle and that volcanic eruptions create new basaltic ocean floor, which then spreads away laterally from the ridge (Figure 14-8). Thus, midocean ridges contain the newest crust formed on the planet. At other places in the ocean basin—at the oceanic trenches—older lithosphere descends into the asthenosphere in a process called **subduction**, where it is ultimately "recycled." The amount of new seafloor created at midocean ridges is compensated for by the amount lost at trenches, which are part of *subduction zones*.

LearningCheck 14-2 **How do midocean ridges and oceanic trenches fit in with the idea of seafloor spreading?**

Verification of Seafloor Spreading: Seafloor spreading was confirmed by two main lines of evidence: paleomagnetism and core sampling of the ocean floor.

When any rock containing iron—such as iron-rich ocean floor basalt—forms, it is magnetized such that the magnetic field within its iron-rich grains become aligned with Earth's magnetic field. This orientation then becomes a permanent record of the polarity of Earth's magnetic field at the time the rock solidified, known as **paleomagnetism**. Over the last 100 million years, for reasons that are not fully understood, the polarity of Earth's magnetic field has reversed more than 130 times—with the north magnetic pole becoming the south magnetic pole.

In 1963, Fred Vine and D. H. Matthews used paleomagnetism to test the theory of seafloor spreading by studying magnetic polarity data from a portion of the midocean ridge system. If the seafloor has spread laterally by the addition of new crust at the oceanic ridges, the pattern of magnetic orientation should be relatively symmetrical—normal polarity,

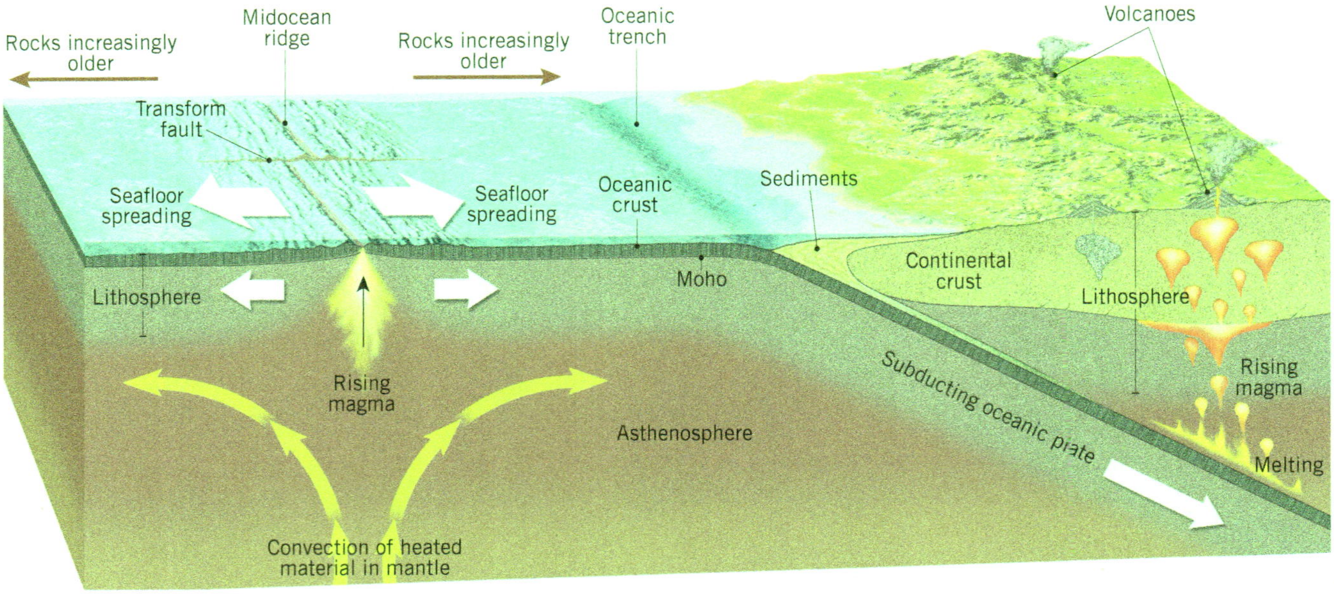

▲ **Figure 14-8** Seafloor spreading. Convection currents bring magma from the asthenosphere up through fissures in the oceanic lithosphere at a midocean ridge. The cooling magma forms new rock at the center of the ridge along the ocean floor, and the two sides of the ridge spread apart. Where oceanic lithosphere converges with less-dense continental lithosphere, the edge of the oceanic plate slides under the continental plate in a process called *subduction*. Magma produced by subduction rises, forming volcanoes and igneous intrusions.

reversed polarity, normal polarity, and so on—on both sides of the ridges (**Figure 14-9**). Such was found to be the case (**Figure 14-10**).

Final confirmation of seafloor spreading was obtained from sediment cores drilled into the ocean floor by the research ship *Glomar Challenger* in the late 1960s. Several thousand **ocean floor cores** of the sediments that slowly settle and accumulate on the sea bottom were analyzed. This work showed that, almost invariably, sediment thickness and the age of fossils in the sediment increase with increasing distance from the midocean ridges, indicating that sediments farthest from the ridges are oldest. At the ridges, ocean floor material is almost all igneous, with little buildup of sediment. Any sediment near the ridges is thin and young because it has been accumulating for only a short time.

Thus, the seafloors are almost like gigantic conveyor belts, moving outward from the midocean ridges toward the

▶ **Figure 14-9** New basaltic ocean floor is magnetized according to Earth's existing magnetic field. As ocean floor spreads away from a ridge (shown from top to bottom), a symmetrical pattern of normal and reversed magnetic polarities develops on both sides of the spreading center.

Animation
Paleomagnetism
https://media.pearsoncmg.com/ph/esm/esm_mcknight_physgeo_13/media/

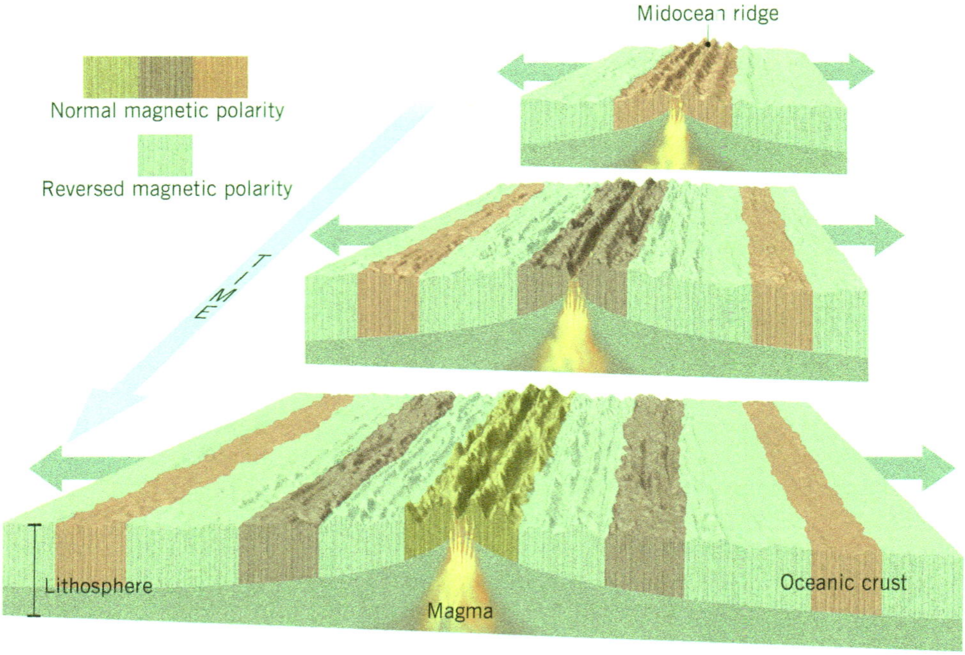

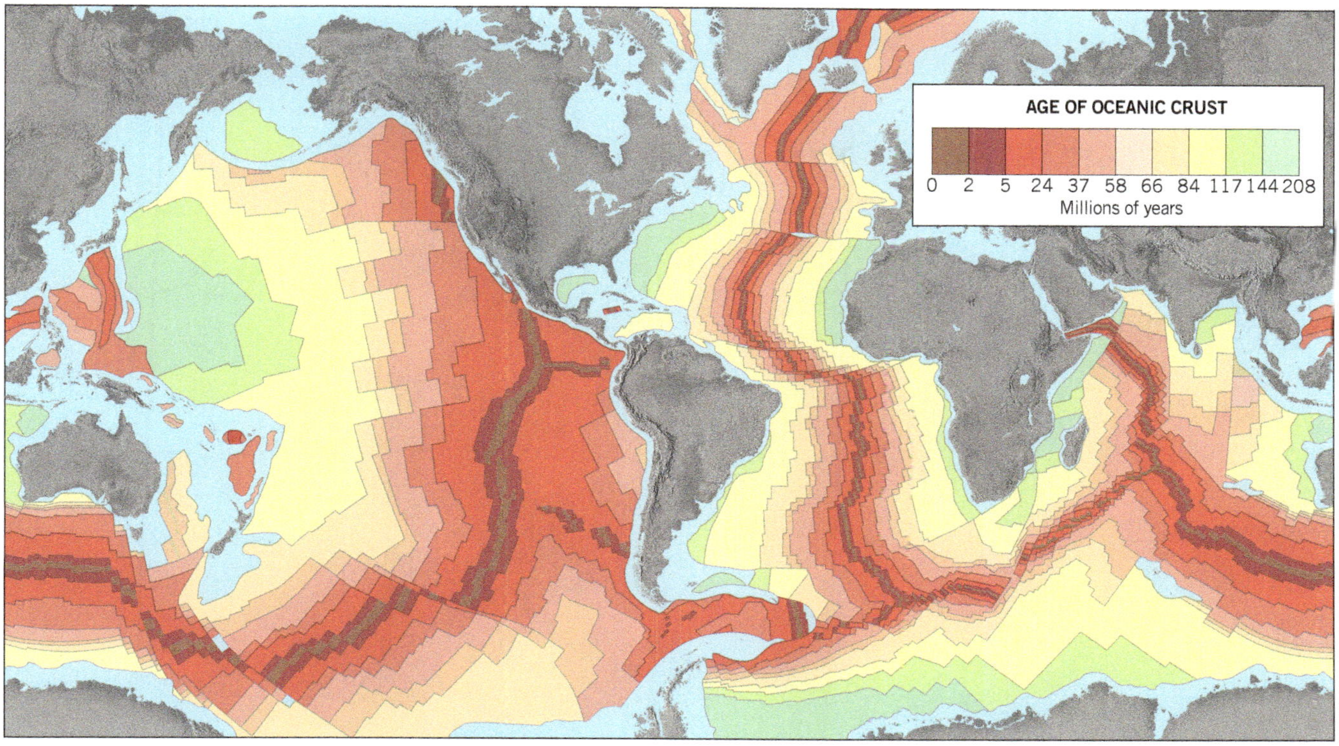

▲ **Figure 14-10** The age of the ocean floors. The pattern of magnetic reversals recorded in the ocean floor as it spreads away from midocean ridges has helped establish the age of oceanic crust. The youngest ocean floor is found at midocean ridges, whereas ocean floor farther from the ridges is the oldest. (*Note*: The length of the age categories varies.)

trenches. Oceanic lithosphere has a relatively short life at Earth's surface. New crust forms at the oceanic ridges and, within 200 million years, is returned to the mantle by subduction.

Because it is less dense than oceanic lithosphere, continental lithosphere cannot be subducted; once it forms, it is virtually permanent. Indeed, the interiors of continents have a "core" or *craton* of very old crystalline rock exposed in places as a *continental shield*. The continuous recycling of oceanic crust means that its average age is only about 100 million years, whereas the average age of continental crust is 20 times that. Indeed, some areas of continental crust have been discovered that are more than 4 billion years old—nearly nine-tenths of the age of Earth!

As it turns out, Alfred Wegener was wrong about one important detail in his theory of continental drift: it is not just the continents that are drifting. The continents are embedded in the thicker lithospheric plates, carried along by the action of seafloor spreading.

LearningCheck 14-3 **Describe the age pattern of the ocean floors. How does this pattern help verify seafloor spreading?**

Plate Tectonic Theory

By 1968, on the basis of these details and a variety of other evidence, the theory of **plate tectonics**, as it had become known, came to be accepted by the scientific community. Plate tectonics provides a framework with which we can understand and relate a wide range of internal processes and topographic patterns around the world.

The lithosphere is a mosaic of rigid plates floating over the underlying plastic asthenosphere. These lithospheric plates, consisting of the crust together with the upper mantle, vary considerably in area; some are almost hemispheric in size, whereas others are much smaller (**Figure 14-11**). Seven major plates, an equal number of intermediate-sized plates, and perhaps a dozen smaller plates are recognized. Many of the smaller plates are remnants of once-larger plates that are now being subducted. These plates are about 65 to 100 kilometers (40 to 60 miles) thick, and most consist of both oceanic and continental crust.

Animation **Convection and Plate Tectonics** https://media.pearsoncmg.com/ph/esm/esm_mcknight_physgeo_13/media/

Mechanism for Plate Tectonics: The driving mechanism for plate tectonics is thought to be convection within Earth's mantle. (We discussed convection in Chapter 4 in the context of atmospheric processes.) A very sluggish thermal convection system appears to be operating within the planet, bringing deep-seated hot, lower-density rock slowly to the surface (see Figure 14-8). Plates may be "pushed" away from midocean ridges to a certain extent, but it appears that much of the motion is a result of the plates being "pulled" along by the subduction of colder, denser oceanic lithosphere down into the asthenosphere.

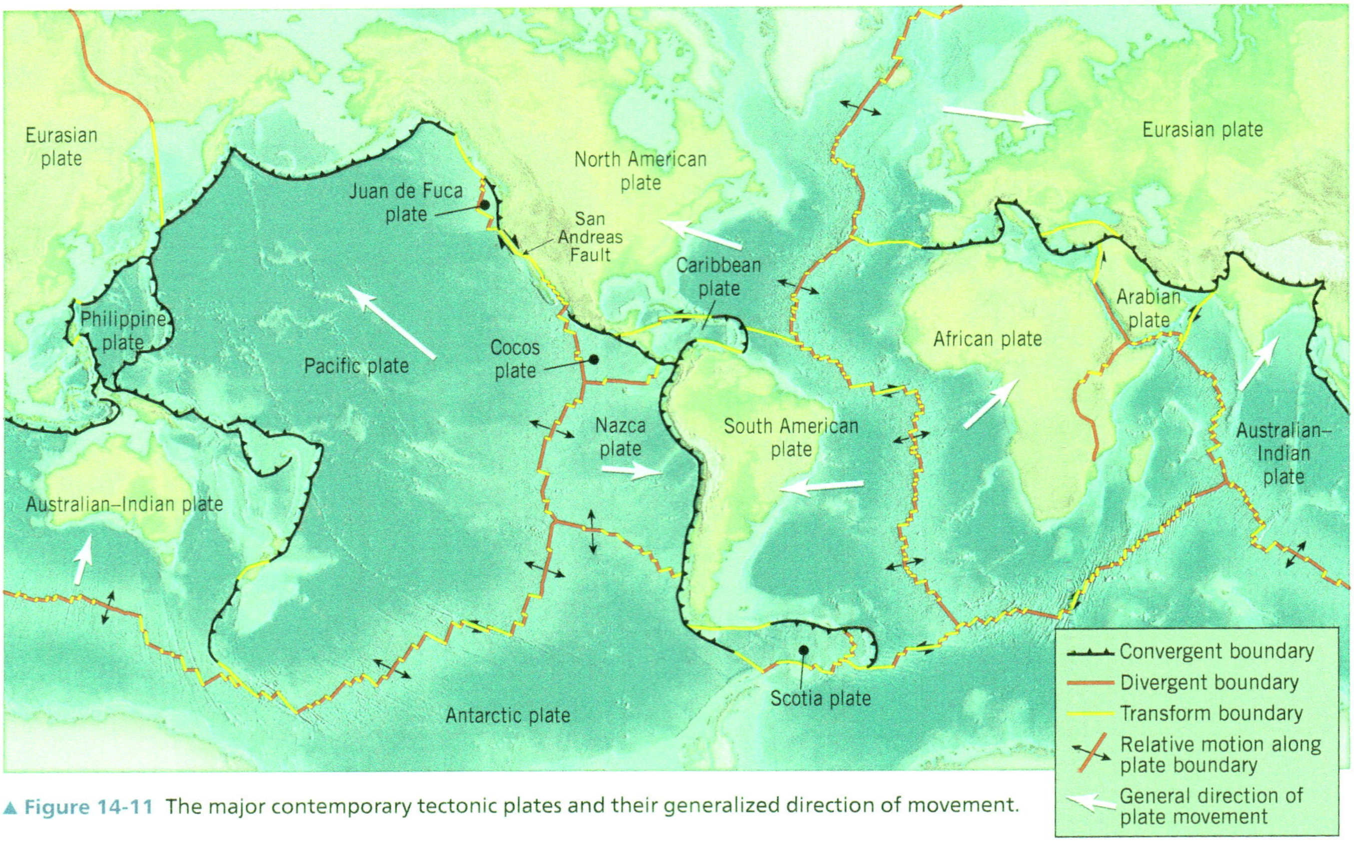

▲ **Figure 14-11** The major contemporary tectonic plates and their generalized direction of movement.

The complete details of thermal convection within the mantle and the ultimate fate of subducted plates remain to be established.

These plates move slowly over the asthenosphere. The rates of seafloor spreading vary from less than 1 centimeter (0.4 inch) per year in parts of the Mid-Atlantic Ridge to as much as 10 centimeters (4 inches) per year in the Pacific-Antarctic Ridge.

LearningCheck 14-4 **What are "plates," and how does the theory of plate tectonics explain their movement?**

Plate Boundaries

Plates are relatively cold and rigid and therefore deformed significantly only at the edges, where one plate meets another. Most of the "action" in plate tectonics takes place along such plate boundaries. Three general types of plate boundaries are possible: two plates may diverge from one another (a divergent boundary), converge toward one another (a convergent boundary), or slide laterally past one another (a transform boundary) (Figure 14-12).

Divergent Boundaries

At a **divergent boundary**, magma from the asthenosphere rises up between two plates. This upward flow of molten material produces a line of volcanic vents that spill out basaltic lava onto the ocean floor, with the plutonic rock gabbro solidifying deeper below.

Midocean Ridges: A divergent boundary is usually represented by a **midocean ridge** (Figure 14-13). The midocean ridges of the world are either active or extinct spreading ridges. Such *spreading centers* are associated with *shallow-focus earthquakes* (meaning that the ruptures that generate the earthquakes are within about 70 kilometers [45 miles] of the surface). They are also noted for volcanic activity and hydrothermal metamorphism—as well as the presence of remarkable marine life-forms thriving in the hostile environment of hydrothermal vents on the ocean floor. Divergent boundaries are "constructive"—material is being added to the crustal surface.

Continental Rift Valleys: Divergent boundaries can also develop within a continent, resulting in a **continental rift valley**, such as the Great East African Rift Valley that extends from Ethiopia southward through Mozambique (Figure 14-14). Iceland, which straddles the Mid-Atlantic Ridge, also has a prominent rift valley.

The Red Sea is also the outcome of spreading taking place within a continent; in this case the spreading has been great enough to form a "proto-ocean" (see Figure 14-6).

LearningCheck 14-5 **How are midocean ridges and continental rift valleys related?**

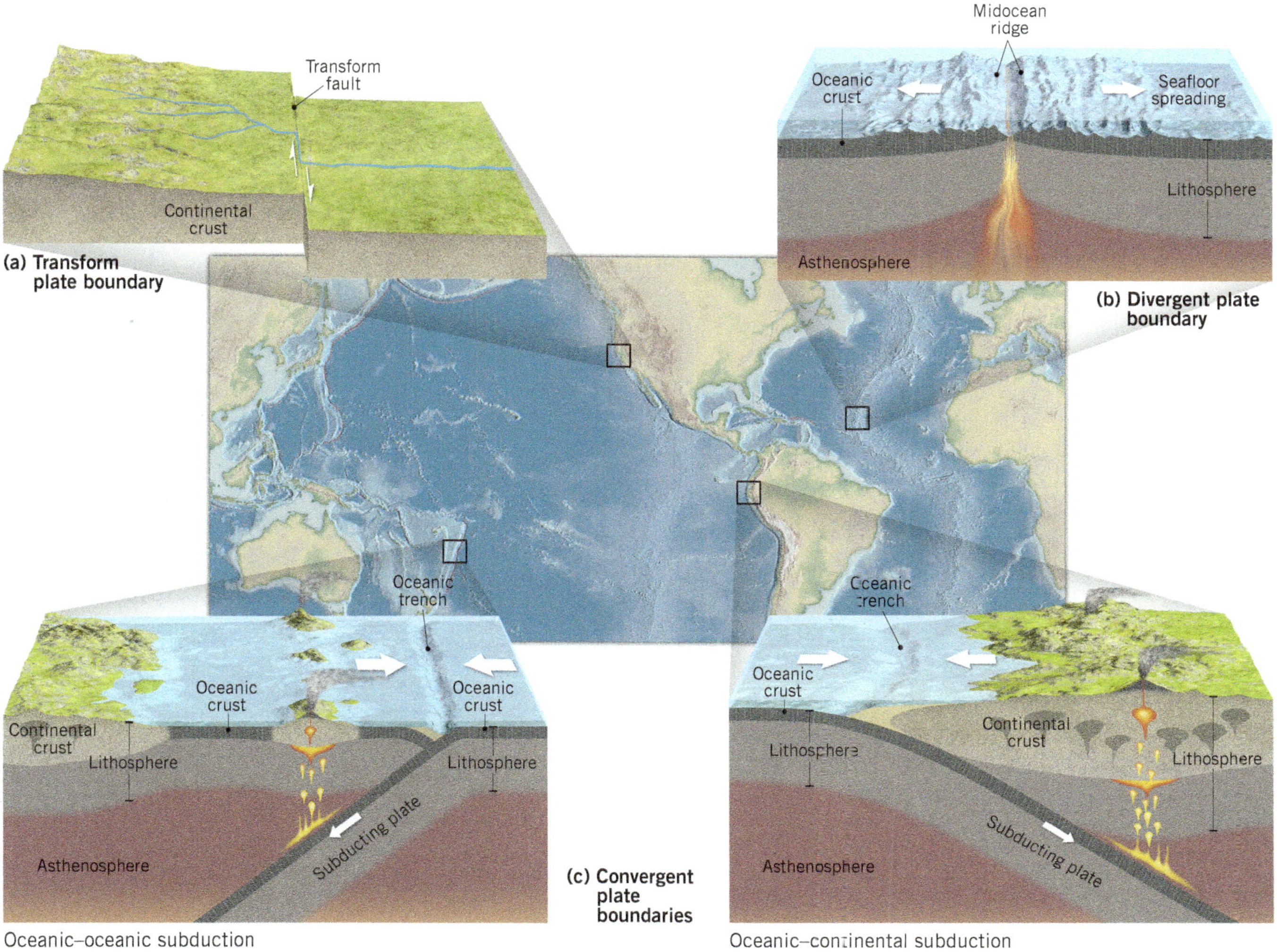

▲ **Figure 14-12** Three kinds of plate boundaries. The edges of lithospheric plates (a) slide past each other along transform boundaries; (b) move apart at divergent boundaries such as midocean ridges or continental rift valleys; and (c) come together at convergent boundaries such as oceanic–oceanic plate subduction zones and oceanic–continental plate subduction zones or continental collision zones.

Animation
Plate Boundaries
https://media.pearsoncmg.com/ph/esm/esm_mcknight_physgeo_13/media/

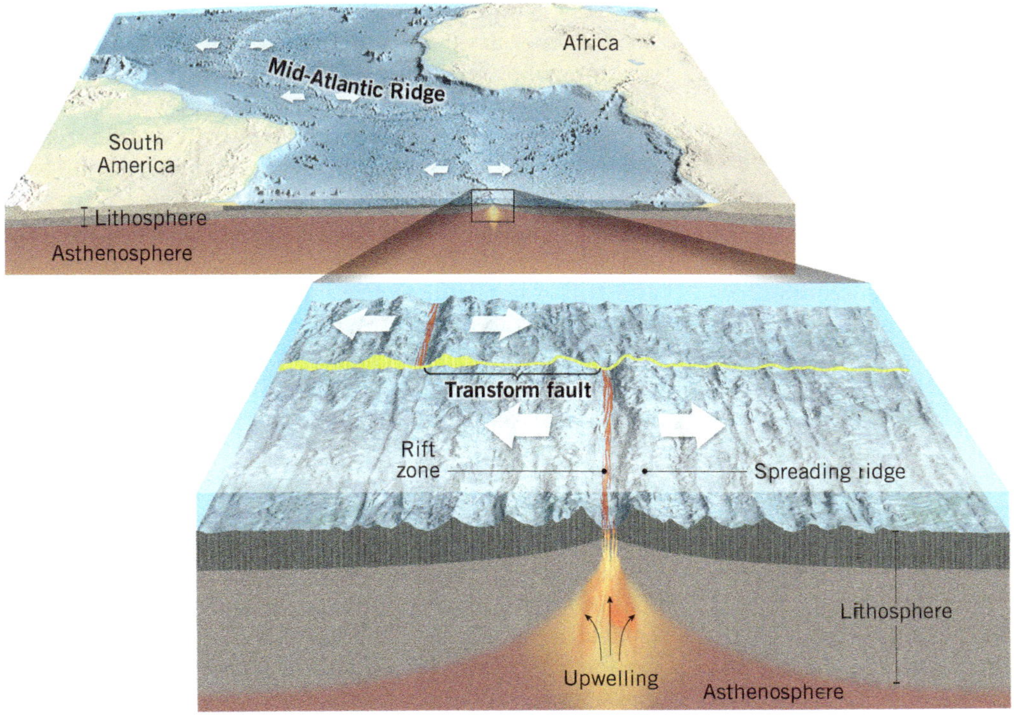

Animation
Divergent Boundaries
https://media.pearsoncmg.com/ph/esm/esm_mcknight_physgeo_13/media/

◄ **Figure 14-13** Midocean ridge spreading center. Seafloor spreading involves the rise of magma from within Earth and the lateral movement of new ocean floor away from the zone of upwelling. This gradual process moves the older material farther from the spreading center as it is replaced by newer material from below. Transform faults are found along the short offsets associated with slight bends in the ridge system.

▲ **Figure 14-14** (a) A continental rift valley develops where divergence takes place within a continent. (b) As spreading proceeds, blocks of crust drop down, forming a rift valley. (c) Ol Doinyo Lengai volcano and the East African Rift Valley in Tanzania.

Convergent Boundaries

Plates collide along a **convergent boundary** and are sometimes called "destructive" boundaries (Figure 14-15). Convergent plate boundaries are responsible for some of the most massive and spectacular of Earth's landforms: major mountain ranges, volcanoes, and oceanic trenches. Indeed, most *orogenesis*, or mountain building, is caused by the compression and volcanism associated with plate convergence.

Three types of convergent boundaries are found: *oceanic–continental convergence*, *oceanic–oceanic convergence*, and *continental–continental convergence*.

Oceanic–Continental Convergence: Because oceanic lithosphere includes dense basaltic crust, it is denser than continental lithosphere, so oceanic lithosphere always underrides continental lithosphere when the two collide (Figure 14-15a). The dense oceanic plate slowly sinks into the asthenosphere by subduction. The subducting slab of lithosphere pulls on the unsubducted portion of the plate. Such "slab pull" is probably the main cause of most plate movement—pulling the rest of the plate in after itself, as it were. Wherever such an oceanic–continental convergent boundary exists, a mountain range forms inland of the continent's edge (such as the Andes in South America and the Cascades in northwestern North America) and a parallel **oceanic trench** develops as the subducting plate pulls the seafloor downward.

Earthquakes take place along the margin of a subducting plate. Shallow-focus earthquakes are common at the trench, but as the subducting plate descends into the asthenosphere, the earthquakes become progressively deeper. Some subduction zones generate earthquakes as deep as 700 kilometers (435 miles) below the surface.

Volcanoes develop from magma generated in the subduction zone. According to recent research, it is unlikely that a subducted plate would completely melt when pushed down into the hot asthenosphere. Oceanic crust is relatively cold when it approaches a subduction zone and would take a long time to melt. It is more likely that beginning at a depth of about 100 kilometers (about 60 miles), water is driven off from the oceanic crust as it subducts, and this water reduces the melting temperature of the mantle rock above, causing it to melt. This magma rises through the overriding plate, producing both extrusive and intrusive igneous rocks. The chain of volcanoes that develops in association with an oceanic–continental plate subduction zone is referred to as a *continental volcanic arc*. As we see later in this chapter, such subduction zone volcanoes frequently erupt explosively.

Metamorphic rocks often develop in association with subduction zones. The margin of a subducting oceanic plate is subjected to increasing pressure, although relatively modest heating, as it begins to descend. This can lead to the formation of high-pressure, low-temperature metamorphic rocks, such as *blueschist*. In addition, the magma generated

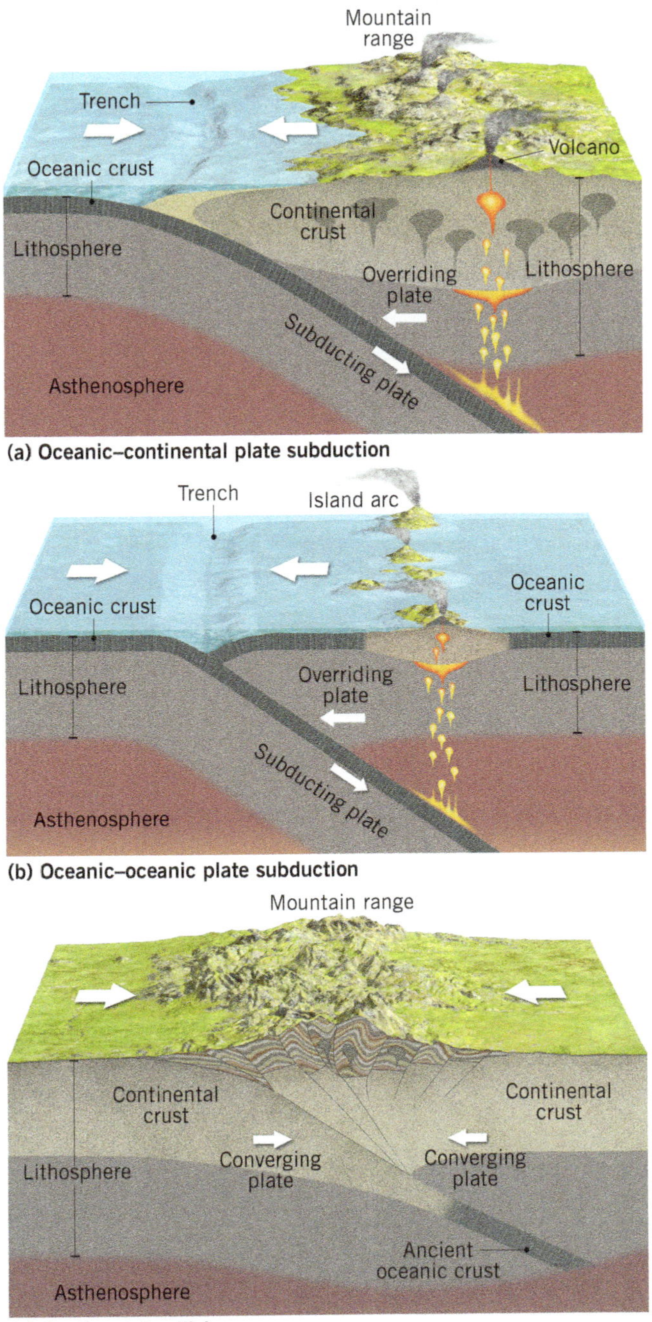

(a) Oceanic–continental plate subduction

(b) Oceanic–oceanic plate subduction

(c) Continental collision

▲ **Figure 14-15** Idealized portrayals of three kinds of convergent plate boundaries. (a) Where an oceanic plate converges with a continental plate, the oceanic plate is subducted, and an oceanic trench and coastal mountains with volcanoes form. (b) Where an oceanic plate subducts beneath another oceanic plate, an oceanic trench and volcanic island arc result. (c) Where two continental plates collide, no subduction takes place, but mountains are thrust upward.

in the subduction zone may cause contact metamorphism as it rises through the overlying continental rocks.

Oceanic–Oceanic Convergence: If the convergent boundary is between two oceanic plates, subduction also takes place (**Figure 14-15b**). As one oceanic plate subducts beneath the other, an oceanic trench forms, shallow-, intermediate-, and deep-focus earthquakes occur (**Figure 14-16**), and volcanoes form on the ocean floor. With time, a **volcanic island arc** (such as the Aleutian Islands and Mariana Islands) develops. Such an arc may eventually become a more mature island arc system (such as Japan and the islands of Sumatra and Java in Indonesia are today).

Continental–Continental Convergence: Where there is a convergent boundary between two continental plates, no subduction takes place because continental crust is too buoyant to subduct. Instead, huge mountain ranges, such as the Alps, are uplifted (**Figure 14-15c**).

The most dramatic example of continental collision resulted in the formation of the Himalayas (**Figure 14-17**). The Himalayas began to form more than 45 million years ago, when the subcontinent of India started to collide with the rest of Eurasia. Under the conditions of continental collision, volcanoes are rare, but shallow-focus earthquakes and regional metamorphism are common (as you will see in Figure 14-48).

LearningCheck 14-6 **Describe and explain the major features associated with oceanic–continental convergence and oceanic–oceanic convergence.**

Transform Boundaries

At a **transform boundary**, two plates slip past one another laterally. This movement occurs along great vertical fractures called *transform faults* (a type of *strike-slip fault* discussed in more detail later in this chapter). Because the plate movement is basically parallel to a transform boundary, these boundaries neither create new crust nor destroy old. Transform faults are associated with a great deal of seismic activity, commonly producing shallow-focus earthquakes.

Most transform faults are found along the midocean ridge system, where they form short offsets in the ridge perpendicular to the spreading center, allowing the midocean ridges to "curve" (see Figure 14-13). Extending away from midocean ridges, inactive "fossil" transform faults remain as *fracture zones* on the ocean floor, such as the Mendocino Fracture Zone off the coast of California (see Figure 14-6).

In some places, transform faults extend for great distances, occasionally through continental lithosphere. For example, the most famous fault system in the United States, the San Andreas Fault in California, is on a transform boundary between the Pacific and North American plates (**Figure 14-18**).

Plate Boundaries over Geologic Time

Animation
Breakup of Pangaea
https://media.pearsoncmg.com/ph/esm/esm_mcknight_physgeo_13/media/

Plate tectonics provides us with a framework for understanding the extensive lithospheric rearrangement that has taken place during the history of Earth. A brief summary of major events in Earth's history starts over a billion years ago and is described in the list that follows.

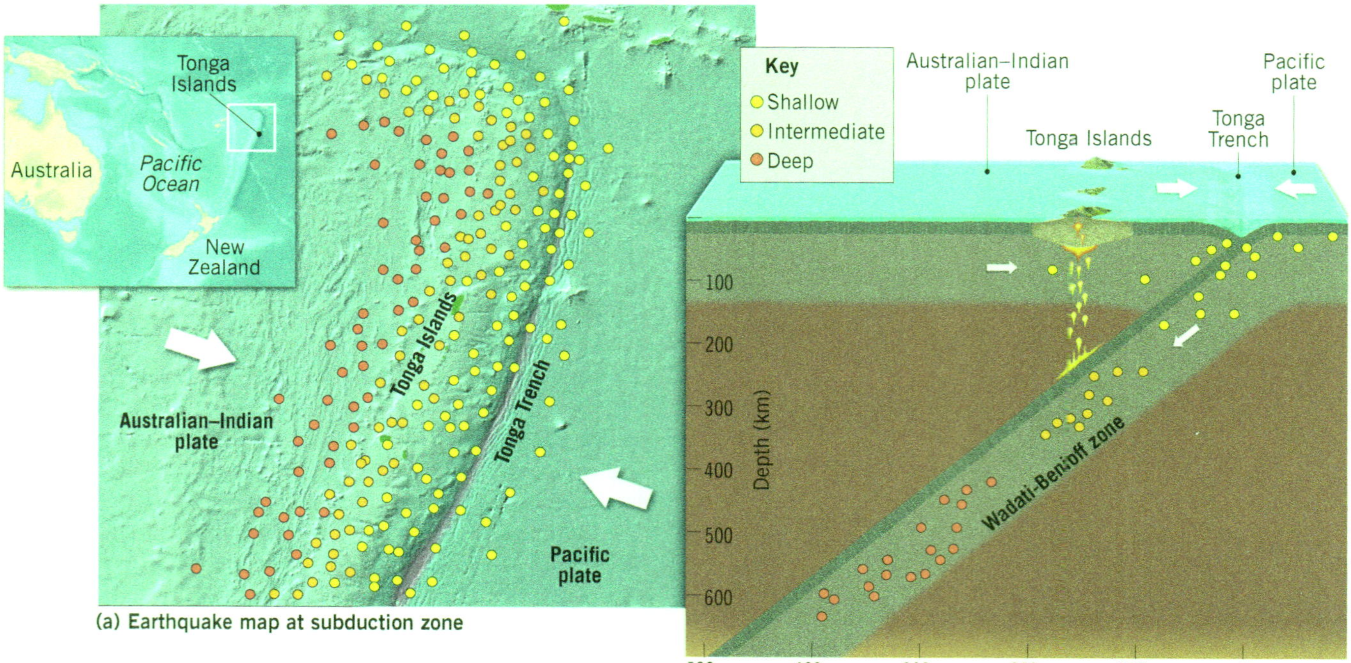

(a) Earthquake map at subduction zone

(b) Side view of subduction zone earthquakes

▲ **Figure 14-16** Earthquake patterns associated with the Tonga Trench subduction zone, shown in (a) map view and (b) side view. Shallow-focus earthquakes occur where the Pacific Plate begins to subduct at the trench. Intermediate- and deep-focus earthquakes occur as the subducting oceanic plate goes deeper into the asthenosphere below. The *Wadati–Benioff zone* is named for seismologists Kiyoo Wadati and Hugo Benioff, who first described these inclined zones of earthquakes.

Animation
Subduction Zones
https://media.pearsoncmg.com/ph/esm/esm_mcknight_physgeo_13/media/

Animation
Collision of India with Eurasia
https://media.pearsoncmg.com/ph/esm/esm_mcknight_physgeo_13/media/

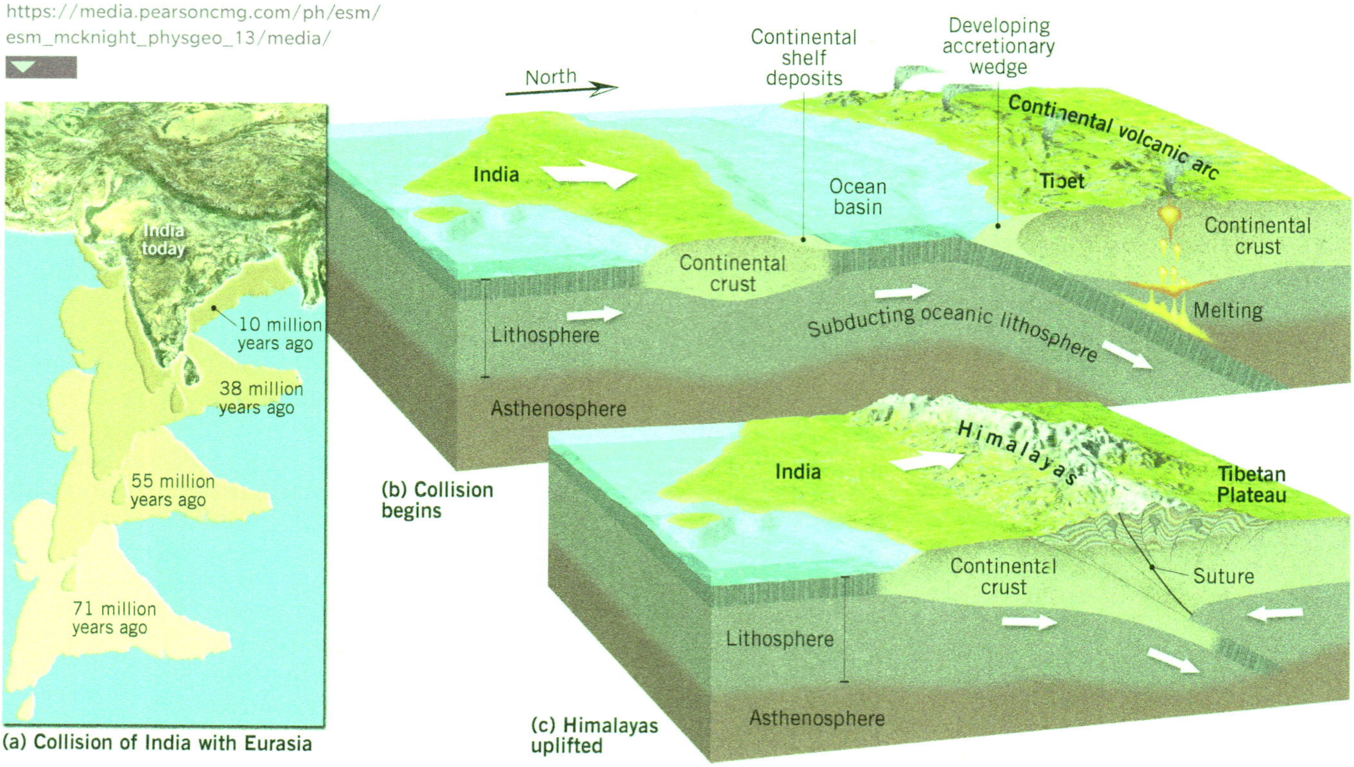

(a) Collision of India with Eurasia

(b) Collision begins

(c) Himalayas uplifted

▲ **Figure 14-17** (a) The collision of the subcontinent of India with Eurasia began about 45 million years ago. This (b) collision and (c) continental "suture" uplifted the Himalayas and the Tibetan Plateau.

Animation
Transform Faults and Boundaries
https://media.pearsoncmg.com/ph/esm/esm_mcknight_physgeo_13/media/

(b) San Andreas Fault in Carrizo Plain

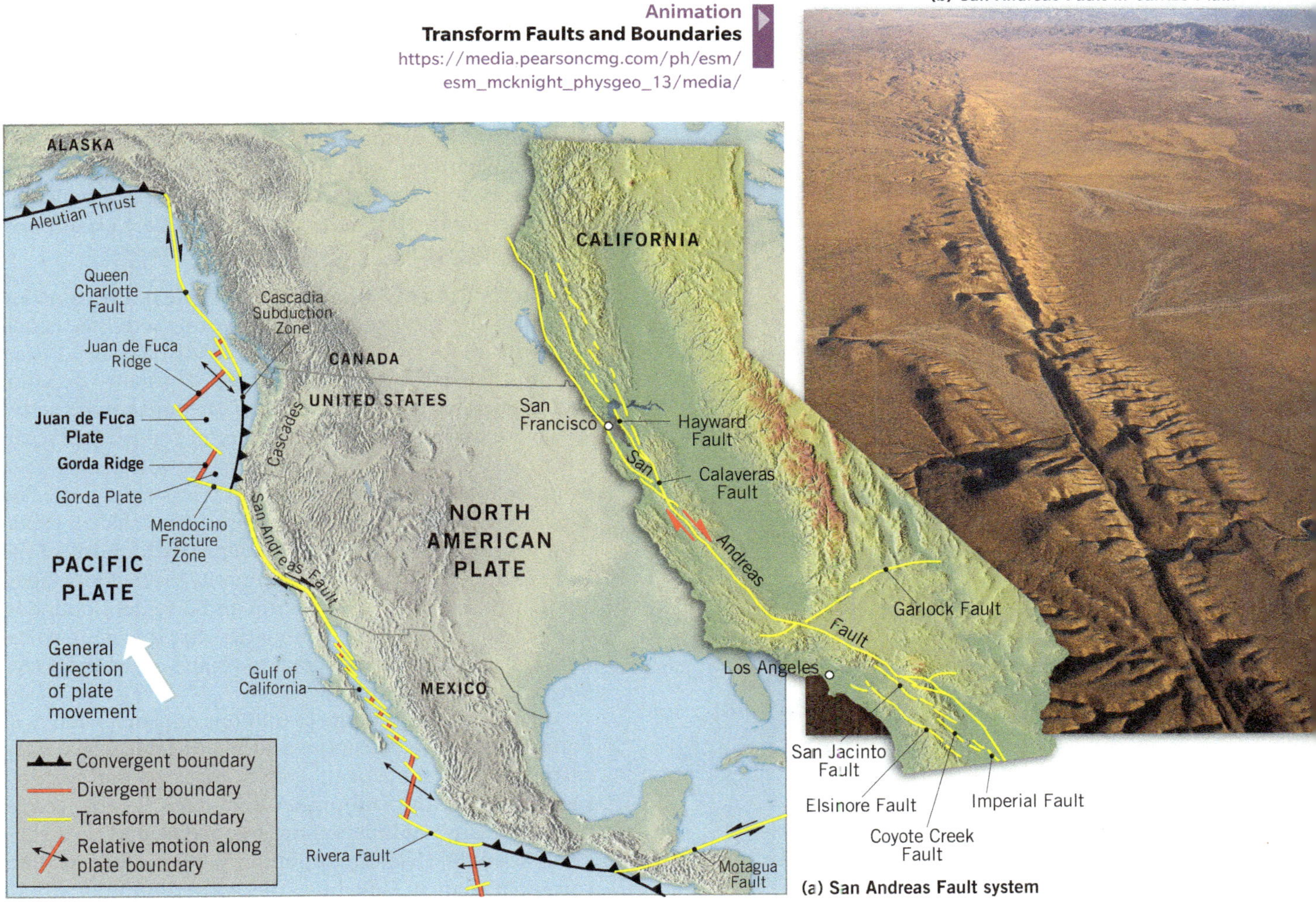

(a) San Andreas Fault system

▲ **Figure 14-18** (a) The San Andreas Fault system of California is the major component of the transform boundary between the Pacific Plate and North American Plate. North of the San Andreas Fault system, several small oceanic plates are subducting in the Cascadia subduction zone in association with the Cascade volcanoes. (b) Aerial view of the fault through the Carrizo Plain in central California.

Mobile Field Trip
San Andreas Fault
https://media.pearsoncmg.com/ph/esm/esm_mcknight_physgeo_13/media/

- Between about 1.2 billion and 750 million years ago—before Pangaea existed—there was an earlier supercontinent called *Rodinia*.
- By about 700 million years ago, Rodinia was rifting apart into continental pieces that would eventually "suture" (fuse) together again—first into a large southern continent called *Gondwana* (which included present-day South America, Africa, India, Australia, and Antarctica) and later into a northern continent called Laurasia (comprising present-day North America and Eurasia). By about 250 million years ago, Gondwana and Laurasia had joined to form Pangaea.
- About 200 million years ago, when Pangaea began to rift apart, there was only one largely uninterrupted ocean (**Figure 14-19**).
- By 100 million years ago, the North Atlantic Ocean began to open, and the South Atlantic began to separate South America from Africa. Antarctica is the only continent that has remained near its original position.
- By 50 million years ago, both the North and South Atlantic Oceans had opened, and South America was a new and isolated continent that was rapidly moving westward. The Andes were growing as South America overrode the Pacific Ocean basin; the Rockies and the ancestral Sierra Nevada had risen in North America.
- About 3 million years ago, South America connected with North America. North America had separated from western Eurasia, Australia has split from Antarctica, and India has collided with Eurasia to thrust up the Himalayas. Today, Africa is splitting along the Great Rift Valley and slowly rotating counterclockwise.

LearningCheck 14-7 **How does the theory of plate tectonics explain the history of the Atlantic Ocean, the Himalayas, and the Andes Mountains?**

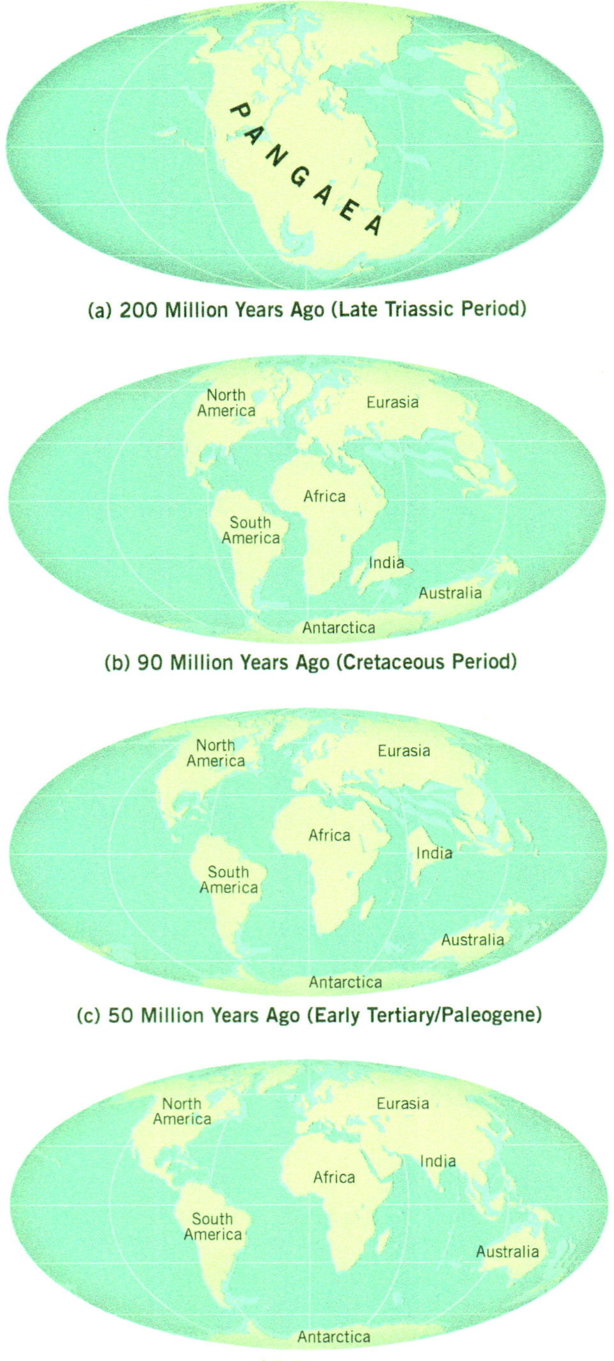

▲ **Figure 14-19** The breakup of Pangaea, beginning about 200 million years ago.

Plate Motion into the Future: If current plate movement continues, 50 million years into the future Australia will straddle the equator as a huge tropical island. Africa may pinch shut the Mediterranean, and East Africa may become a new large island like Madagascar. The Atlantic will widen while the Pacific shrinks. Southern California—perhaps along with much of the rest of the state—will slide past the rest of North America en route to its ultimate destination in the Aleutian Trench in the Gulf of Alaska.

One of the great triumphs of the theory of plate tectonics is that it explains broad topographic patterns. It can account for the formation of many *cordilleras* (groups of mountain ranges), midocean ridges, oceanic trenches, island arcs, and the associated earthquake and volcanic zones. Where these features appear, plates are usually colliding or separating.

The Pacific Ring of Fire: Perhaps nowhere in the world are the consequences of tectonic and volcanic activities associated with plate boundaries more vividly displayed than around the rim of the Pacific Ocean. For many decades, geologists noted the high number of earthquakes and active volcanoes occurring around the rim of the Pacific Ocean basin. About three-quarters of all active volcanoes in the world lie within the Pacific Rim, but it was only in the late 1960s that the theory of plate tectonics provided an explanation for this pattern. Plate boundaries are found all of the way around the Pacific basin—primarily subduction zones, along with segments of transform and divergent boundaries. It is along these plate boundaries that the many volcanoes and earthquakes take place in what is called the *Pacific Ring of Fire* (**Figure 14-20**).

The Pacific Rim is home to millions of people. Active or potentially active volcanoes and major fault systems are within sight of some of the largest metropolitan regions, such as Mexico City, Los Angeles, and Tokyo. Recent decades have had many reminders of the ever-active Ring of Fire: the 1980 Mount St. Helens eruption; the 1985 Nevado del Ruiz volcano tragedy in Colombia; the 1991 eruption of Mount Pinatubo in the Philippines; the 1994 Northridge earthquake in California; the December 2004 Sumatra, Indonesia, earthquake and tsunami that killed more than 227,000 people; and the March 2011 earthquake and tsunami that killed 15,000 people in Japan.

Additions to Plate Tectonic Theory

Each year we learn more about plate tectonics. Two examples of important additions to plate tectonic theory are *hot spots* and *accreted terranes*.

Hot Spots and Mantle Plumes

One augmentation to plate tectonic theory was introduced at the same time as the original model. The basic theory of plate tectonics can explain tectonic and volcanic activities along the margins of plates, but there are places on Earth where magma rising from the mantle comes to the surface at locations that are not near a plate boundary. These locations of volcanic activity in the interior of a plate are referred to as hot spots. As many as 50 hotspots have been identified.

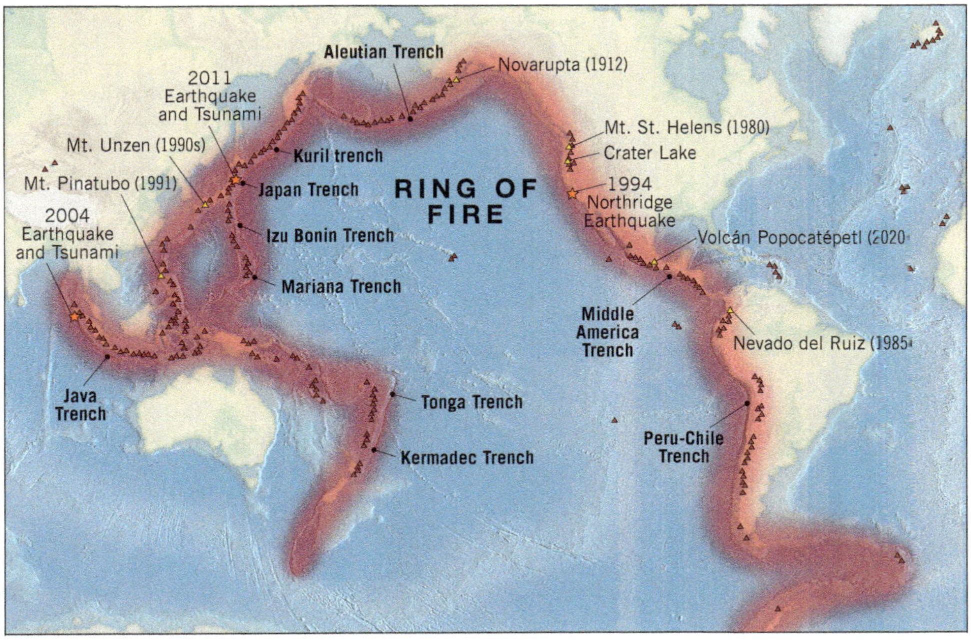

◀ **Figure 14-20** The Pacific Ring of Fire. Triangles indicate the locations of active volcanoes.

Explaining Hot Spots: To explain the existence of hot spots, the mantle plume model was proposed in the early 1960s. In this model, midplate volcanic activity develops over narrow plumes of heated material rising through the mantle, perhaps originating as deep as the core–mantle boundary (**Figure 14-21**). Such mantle plumes are believed to be relatively stationary over long periods of time (as long as tens of millions of years). As the magma rises through the plate above, hot spot volcanoes or hydrothermal (hot water) features form on the surface—often after an initial large outpouring of lava known as *flood basalt* (discussed later in this chapter).

The plate above the hot spot is moving, so the volcanoes or other hot spot features are eventually carried off the plume and become inactive. In turn, new volcanic features develop over the plume, so a straight-line *hot spot trail* is generated. Volcanic islands carried off the hot spot may eventually subside, forming underwater *seamounts* as the oceanic lithosphere cools and becomes denser. Because many hot spots seem to be effectively fixed in position for long periods of time, the hot spot trails they produce can indicate both the direction and speed of plate motion. Seamounts become progressively older in the direction of plate movement.

The Hawaiian Hot Spot: The most dramatic present-day example of a hot spot is associated with the Hawaiian Islands. Although both developed over the same hot spot, the ancient volcanic remnants of Midway Island are now 2430 kilometers (1510 miles) northwest of the presently active volcanoes on the so called "Big Island" of Hawai'i, separated in time by more than 27 million years. The volcanoes of the Hawaiian chain are progressively younger from west to east; as the Pacific Plate drifts northwestward, new volcanoes are produced on an "assembly line" moving over the persistent hot spot (**Figure 14-22**). After

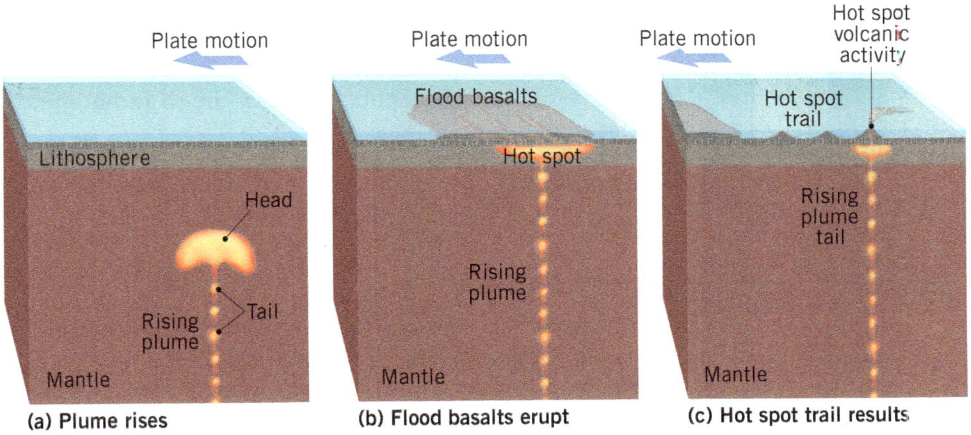

◀ **Figure 14-21** The mantle plume model of hot spot origin. (a) A plume of heated material rises from deep within the mantle. (b) When the large head of the plume reaches the surface, flood basalts spill out. (c) Plate motion carries the flood basalts off the stationary plume, and a new volcano or volcanic island forms. As the moving plate carries each volcano off the hot spot, it becomes extinct, resulting in a straight-line hot spot trail. As volcanic islands move off the hot spot, the plate cools, becomes denser, and subsides. Some islands eventually sink below the surface as seamounts.

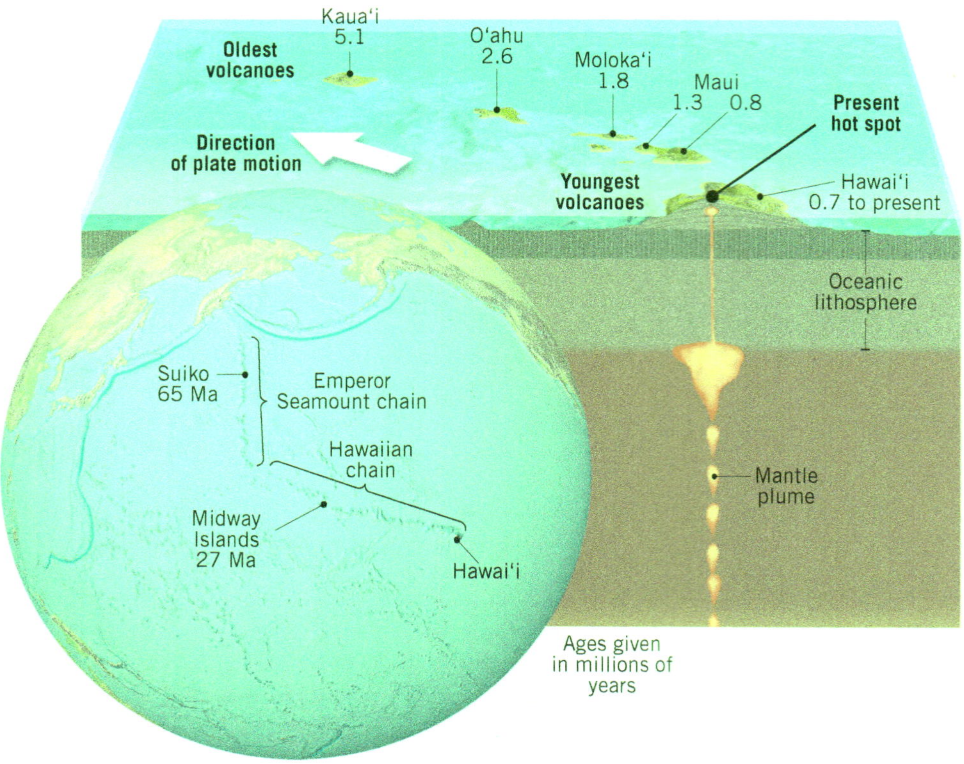

◀ **Figure 14-22** The Hawaiian hot spot. A hot spot has persisted here for many millions of years. As the Pacific Plate moved northwest, a progression of volcanoes was created and then died as their source of magma was shut off. Among the oldest is Midway Island. Later volcanoes developed down the chain. The numbers on the main islands indicate the age of the basalt that formed the volcanoes, in millions of years (Ma) before the present.

Animation
Hot Spot Volcano Tracks
https://media.pearsoncmg.com/ph/esm/esm_mcknight_physgeo_13/media/

plate movement carries the Big Island off the hot spot, the next Hawaiian island rises in its place. In fact, scientists are already studying the undersea volcano Lō'ihi as it builds up on the ocean floor just southeast of the Big Island (Figure 14-23). Other well-known hot spot locations are Yellowstone National Park, Iceland, and the Galápagos Islands.

▲ **Figure 14-23** The only recently active volcanoes in the Hawaiian chain are on the Big Island of Hawai'i—Mauna Loa and Kīlauea have been repeatedly active during historic times. Lō'ihi, an underwater volcano southeast of the Big Island, is the next volcanic island being built.

Recent research indicates that the complete explanation of hot spots may turn out to be more complex than the mantle plume model suggested. *Seismic tomography*—a technique that uses earthquake waves to produce a kind of "ultrasound" of Earth—suggests that the magma source of at least some hot spots is quite shallow, whereas the source for others are mantle plumes originating deep from within the mantle. Furthermore, some researchers cite evidence suggesting that several mantle plumes may have changed location in the geologic past. For example, the Emperor Seamounts—a chain of seamounts to the northwest of Midway Island—are part of the Hawaiian hot spot trail, but they appear to divert quite significantly in direction from the straight line of the rest of the Hawaiian chain (see Figure 14-22). This "bend" in the hot spot trail is due either to a significant change in direction of the Pacific Plate about 43 million years ago or to the migration of the hot spot itself—perhaps both.

As additional information is gathered, a more complete understanding of hot spots, mantle plumes, and midplate volcanic activity will likely emerge.

LearningCheck 14-8 **How does the mantle plume model explain the formation of the Hawaiian Islands?**

Accreted Terranes

Animation
Terrane Formation
https://media.pearsoncmg.com/ph/esm/esm_mcknight_physgeo_13/media/

A more recent discovery has helped explain the often-confusing juxtaposition of different types of rock seen along the margins of some continents. A **terrane** is a small-to-medium mass of lithosphere—bounded on all sides by faults—that may have been carried a long distance by a moving plate, eventually to converge with the edge of another plate. The terrane is too buoyant to be subducted in the collision and instead is fused ("accreted") to the other plate, often being fragmented in the process. In some cases, slices of oceanic lithosphere have accreted in terranes (including the accumulated sediment in what is called the *accretionary wedge* of a subduction zone); in other cases, it appears that entire old island arcs have fused with the margin of a continent (**Figure 14-24**).

Terranes are distinctive geologically because they are composed of rock generally quite different from that of the plate to which they are accreted. It is widely believed that every continent has grown outward by the accumulation of accreted terranes on one or more of its margins. North America is a prominent example (**Figure 14-25**): most of Alaska and much of western Canada and the western United States consist of a mosaic of several dozen accreted terranes, some of them traced to origins south of the equator.

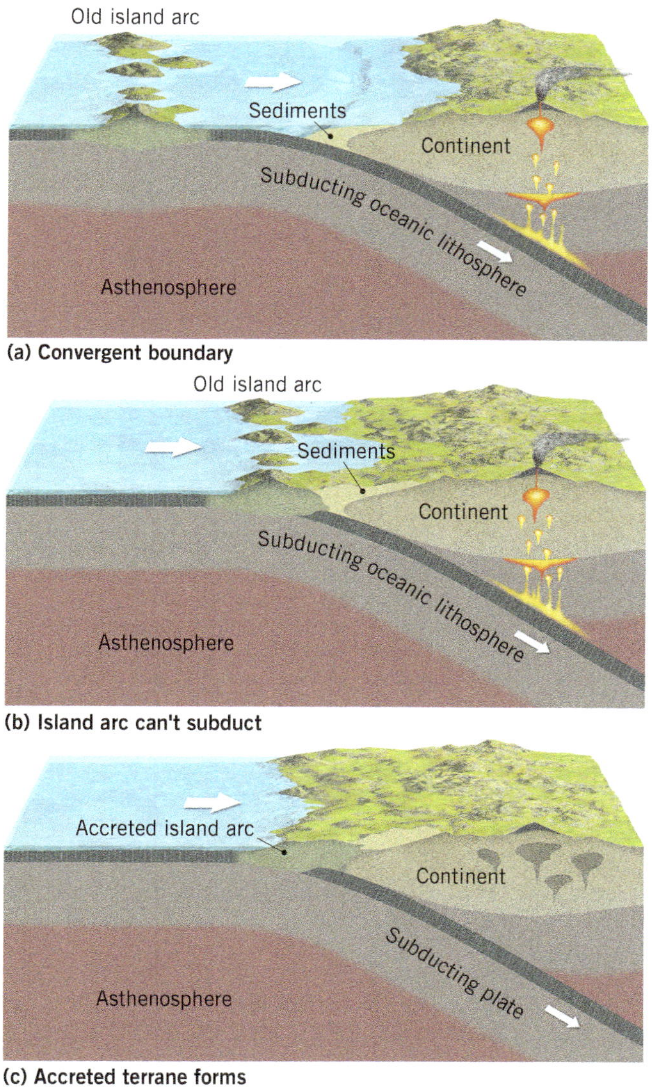

▲ **Figure 14-24** The origin of an accreted terrane in a convergent boundary. (a) A moving oceanic plate carries along an old island arc. (b) The oceanic plate converges with a continental plate. (c) The island arc, too buoyant to subduct, is accreted to the continental plate.

LearningCheck 14-9 How does the idea of a terrane help explain the geologic history of western North America?

Remaining Questions

Plate tectonic theory has advanced our understanding of the internal processes of Earth dramatically. However, a number of questions remain unanswered. For example, several major mountain ranges in North America and Eurasia are in the middle of plates rather than in boundary zones. Although the origin of some midplate ranges, such as the Appalachians in North America and the Ural Mountains in Eurasia, can be traced to continental collisions in the geologic past, other midplate mountain ranges or regions of seismic activity are not yet fully understood. Why are some plates so much larger than others, and what determines the zones of weakness where plate boundaries first develop? What explains tectonic activity, such as earthquakes, in the middle of some continental plates? Furthermore, although convection of heated material within the mantle provides the general mechanism for plate movement, the details of heat flow within Earth and the possible relationships of mantle plumes to these overall patterns are still being worked out.

Our present state of knowledge about plate tectonics, however, provides a firm basis for understanding the patterns of most of the world's major relief features—the size, shape, and distribution of the continents, major mountain ranges, and ocean basins. To understand more localized topographic features, however, we turn to a set of internal processes that are often—but not always—associated with tectonic movement.

Volcanism

Volcanism (or *igneous processes*) is a general term that refers to all phenomena connected with the origin and movement of molten rock. These phenomena include the well-known explosive volcanic eruptions that are among the most spectacular and terrifying events in nature, along with much more quiet events, such as the slow solidification of magma below the surface.

We noted in Chapter 13 the distinction between *volcanic* (extrusive) and *plutonic* (intrusive) igneous rocks; a similar distinction is made between extrusive and intrusive volcanism. When molten magma is expelled onto Earth's surface, the activity is extrusive and is called *volcanism*. When magma solidifies below the surface, *intrusive*, or *plutonic*, *activity* is said to occur and results in intrusive igneous features.

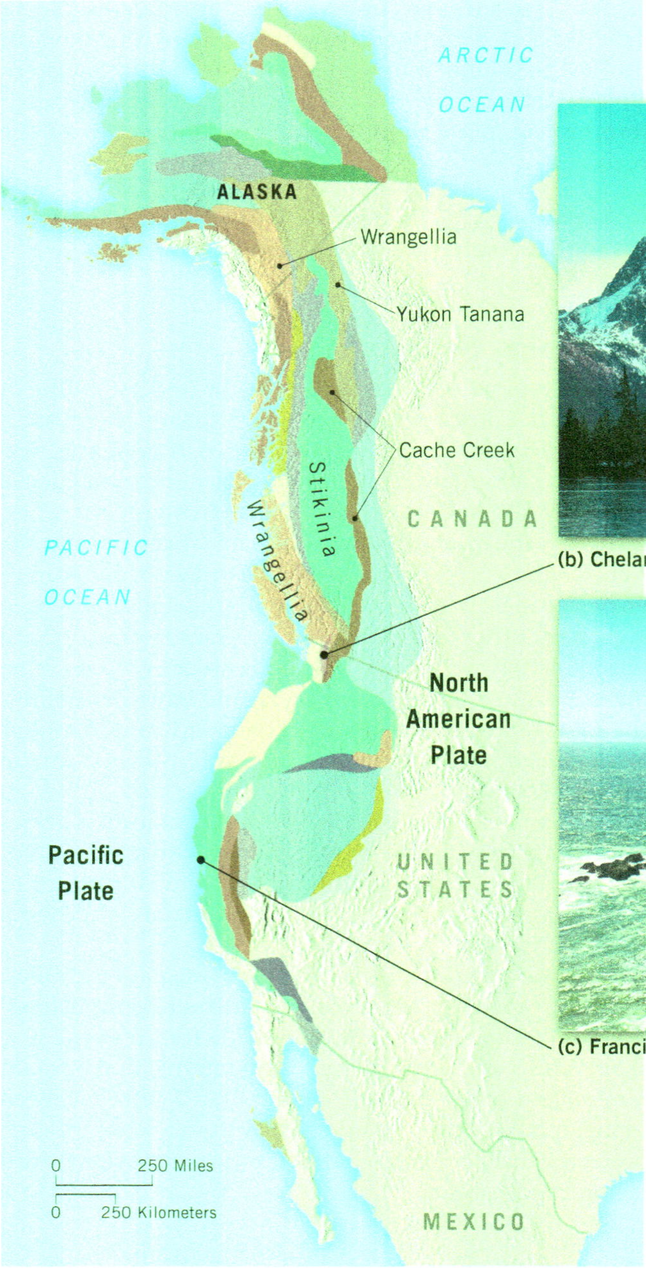

(a) Western North America Terranes

(b) Chelan Mountains Terrane, North Cascades National Park

(c) Franciscan Terrane, Golden Gate National Recreation Area

▲ **Figure 14-25** The western part of North America (a) consists of a complicated mixture of terranes that have been accreted to the North American Plate. (b) Metamorphosed rock in the Chelan Mountains Terrane is exposed in Mount Shuksan in North Cascades National Park, Washington. (c) Pillow basalt formed when lava cooled quickly on the ocean floor is now exposed in the Franciscan Terrane in the Golden Gate National Recreation Area north of San Francisco.

Volcano Distribution

Areas of volcanism are widespread over the world (**Figure 14-26**); they are primarily associated with plate boundaries and hot spots. At divergent boundaries, magma wells up from the interior both by eruption from active volcanoes and by flooding out through fissures in the lithosphere. At convergent boundaries where oceanic lithosphere is subducting, volcanoes form in association with the generation of magma. Hot spots are responsible for volcanic and hydrothermal activity in many places, such as Yellowstone, Hawai'i, and the Galápagos Islands.

The most notable area of volcanism in the world is around the margin of the Pacific Ocean in the Pacific Ring of Fire (see Figures 14-20 and 14-26)—also called the *Andesite Line* because the volcanoes consist primarily of the volcanic rock andesite. About 75 percent of the world's volcanoes, both active and inactive, are associated with the Pacific Rim.

Volcanic Activity: A volcano is considered active if it has erupted at least once within historical times and is considered likely to do so again. About 500 volcanoes in the world are active, and another 1000 are potentially active.

In addition to surface eruptions on continents and islands, a great deal of volcanic activity takes place underwater. Indeed, it is estimated that more than three-fourths of all volcanism is undersea activity, such as at midocean ridge

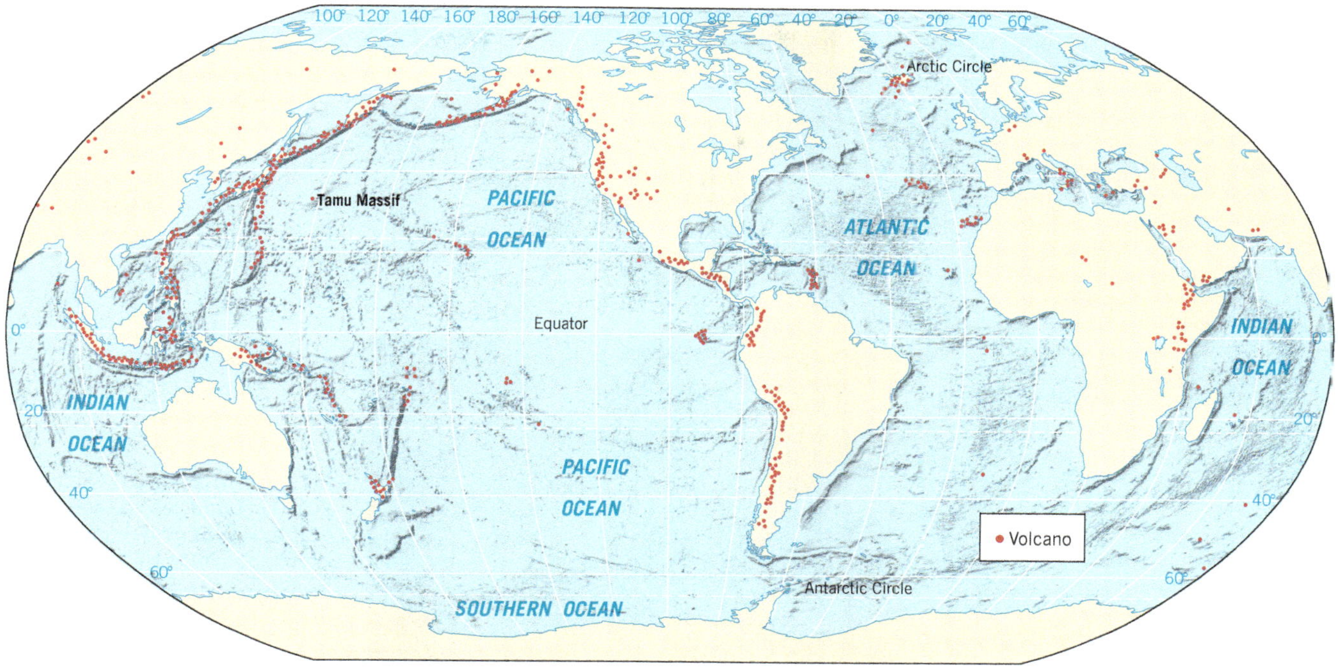

▲ **Figure 14-26** Distribution of volcanoes known to have erupted at some time in the recent geological past. The Pacific Ring of Fire is quite conspicuous.

spreading centers. Earth's largest single volcano in area may be the Tamu Massif, an extinct basaltic volcano on the seafloor about 1600 kilometers (1000 miles) east of Japan in the Shatsky Rise in the North Pacific Ocean.

Both Alaska and Hawai'i have many active volcanoes. Within the conterminous 48 states, only Mount St. Helens in Washington and Lassen Peak in California (which last erupted in 1917)—both in the Cascade Range of western North America—have erupted in the last century. A number of other volcanoes are potentially active—notably, California's Mount Shasta and Long Valley Caldera, Washington's Mount Baker and Mount Rainier, and the Yellowstone Caldera (sometimes referred to as the *Yellowstone Super Volcano*)—see the box *Featured National Park: Yellowstone*, in Chapter 17. There are hundreds of extinct volcanoes, primarily in West Coast states.

Active volcanoes are relatively temporary features of the landscape. Some may have an active life of only a few years, whereas others are sporadically active for thousands of years. New volcanoes occasionally make spectacular entries: in 1943, an eruption in a Mexican corn field built the volcanic cone Parícutin. In 1963, the volcanic island of Surtsey rose out of the sea above a hot spot off the coast of Iceland. In 2011, a new island formed from an eruption in the Red Sea. In 2015, ongoing eruptions connected the two young islands of Hunga Tonga and Hunga Ha'apai in the South Pacific island country of Tonga (**Figure 14-27**).

Despite the destruction they cause, volcanoes provide vital services. Much of the water on Earth today was originally released as water vapor during volcanic eruptions during the early history of our planet. Magma also contains elements such as phosphorus, potassium, calcium, magnesium, and sulfur, which are required for plant growth. When this magma is extruded as lava that hardens into rock (**Figure 14-28**), weathering releases the nutrients into soil in the course of decades or centuries. When the magma is ejected as ash, however, nutrients can be leached into the soil within months. It is no coincidence that Java, one of the most volcanically active parts of the planet, is also one of the world's most fertile areas.

Magma Chemistry and Styles of Eruption

As we saw in the previous chapter, **magma** (molten mineral material below the surface) extruded onto Earth's surface is called **lava**. The ejection of lava into the open air is sometimes volatile and explosive, devastating the area for many kilometers around; in other cases, it is gentle and quiet, affecting the landscape more gradually. All eruptions, however, add new material to Earth's surface.

During an explosive volcanic eruption, solid rock fragments, solidified lava blobs, cinders, and dust—collectively called **pyroclastics** (or *tephra*)—as well as gas and steam, may be hurled upward in extraordinary quantities. In some cases, the volcano literally explodes. The supreme example of such self-destruction within historic times was the final eruption of Krakatau, a volcano that occupied a small island in Indonesia between Sumatra and Java. When it exploded in 1883, the noise was heard 2400 kilometers (1500 miles) away in Australia, and 9 cubic kilometers (2.2 cubic miles) of material blasted into the air. The island disappeared, leaving only open sea where it had been. The *tsunamis* it generated (great seismic sea waves; discussed in Chapter 20) drowned more than 30,000 people, and sunsets in various parts of the world were colored by fine volcanic dust for many months afterward.

◀ **Figure 14-27** In December 2014, an undersea volcano erupted in Tonga between the islands of Hunga Ha'apai and Hunga Tonga. NASA depiction of (a) the islands as they would have appeared in 2014 and (b) the two islands as they are now connected.

The nature of a volcanic eruption is determined largely by the chemistry of the magma that feeds it, especially the relative amount of silica (SiO_2) in the magma. Recall the common magma types: relatively high-silica *felsic* magma (which produces the volcanic rock rhyolite and the plutonic rock granite), intermediate-silica *andesitic* magma (which produces the volcanic rock andesite and the plutonic rock diorite), and relatively low-silica *mafic* magma (which produces the volcanic rock basalt and the plutonic rock gabbro). See Figure 13-5 in Chapter 13 for a review of igneous rocks resulting from different magma chemistries.

▲ **Figure 14-28** Young basalt in Hawai'i Volcanoes National Park. This smooth, ropy lava is called *pāhoehoe*.

Felsic Magmas: In high-silica felsic magmas, long chains consisting of silicate structures can develop even before crystallization of minerals begins, greatly increasing the viscosity (thickness or "stickiness") of the magma. High silica content also usually indicates cooler magma in which some of the heavier minerals have already crystallized and a considerable amount of gas has separated. Some of this gas is trapped in pockets in the magma under great pressure. Gas bubbles can rise only slowly through viscous felsic magma, unlike more fluid lavas. As the magma approaches the surface, the confining pressure is diminished, and the pent-up gases are released explosively, generating an eruption in which large quantities of pyroclastic material are ejected. Any lava flows are likely to be very thick and slow moving.

Mafic Magmas: Low-silica mafic magma is likely hotter and a lot more fluid because of its lower silica content. Dissolved gases can bubble out of very fluid mafic magma much more easily than from viscous felsic magma. The resulting *effusive eruptions* usually yield a great outpouring of lava, without explosions or large quantities of pyroclastics. The highly active volcanoes of Hawai'i erupt in this fashion.

Intermediate Magmas: Volcanoes with intermediate-silica content magmas erupt in a style somewhat between the styles of felsic and mafic magmas: periodically venting fairly fluid andesitic lava flows and periodically having explosive eruptions of pyroclastics. Many of the major volcanoes associated with subduction zones are this type.

LearningCheck 14-10 **Which factor determines whether a volcano's lava is viscous or fluid, and whether an eruption will be explosive or effusive?**

▲ **Figure 14-29** Devils Postpile National Monument east of Yosemite in California. Hexagonal columns are a common result when a fluid basaltic lava flow cools uniformly.

Lava Flows

A lava flow spreads outward approximately parallel with the surface. Although some viscous flows cling to steep slopes, the vast majority eventually solidify in a near-horizontal layer that may resemble the stratification of sedimentary rock, particularly if several flows have accumulated on top of one another.

The surface of a basalt flow may be smooth and ropy *pāhoehoe* lava (see Figure 14-28) if it is very hot and fluid, or it may be rough and irregular *'a'ā* lava if it has cooled slightly and is more viscous (see Figure 13-6a). Very fluid basaltic lava may also flow below the surface though *lava tubes*—tunnels that form when the surface of a flow solidifies while molten lava continues to flow below.

Columnar Basalt: One of the most distinctive of all volcanic landscape features commonly develops from flows of fluid lava such as basalt. When such a lava flow cools uniformly, it contracts and forms a distinctive pattern of vertical joints (cracks in the rock), leaving prominent hexagonal columns known as *columnar basalt*. The Giant's Causeway—or *Clochán an Aifir*—in Northern Ireland and Devils Postpile near Yosemite National Park in California are famous examples of columnar basalt (**Figure 14-29**).

Flood Basalt: Many of the world's most extensive lava flows were not extruded from volcanic peaks but rather poured from fissures associated with hot spots. The lava that flows from these vents is nearly always basaltic and frequently comes out in great volume. As we saw in Figure 14-21, many scientists think that the initial consequence of a large mantle plume reaching the surface can be a huge outpouring of lava.

The term **flood basalt** is applied to the vast accumulations of lava that build up, layer upon layer, sometimes covering tens of thousands of square kilometers to depths of many hundreds of meters. A prominent example of flood basalt in the United States is the Columbia Plateau, which covers 210,000 square kilometers (81,000 square miles) in Washington, Oregon, and Idaho (**Figure 14-30**). Larger outpourings are seen on other continents, most notably the Deccan Traps of India, which cover an area of 520,000 square kilometers (200,000 square miles). (*Trap* comes from the Dutch-Swedish word *trappa* meaning "stairs" in reference to the layers of lava flows.) Globally,

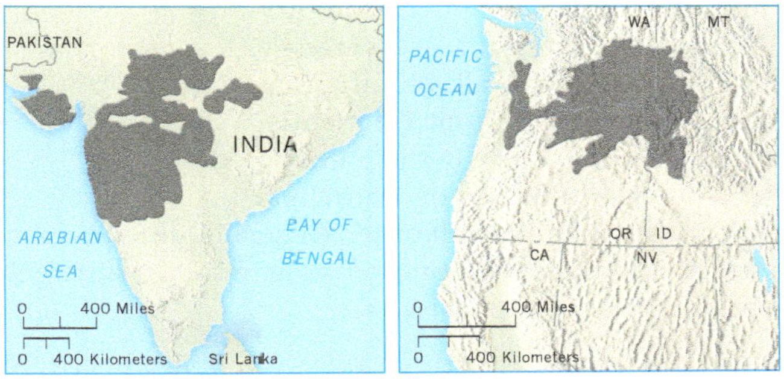

▲ **Figure 14-30** Two extensive outpourings of flood basalt: the Deccan Traps (also called the Deccan Plateau) in India and the Columbia Plateau in the northwestern United States.

TABLE 14-1 Principal Types of Volcanoes

Volcano Type	Shape and Size	Structure	Magma and Eruption Style	Examples
Shield 	Broad, gently sloping mountain; much broader than high; size varies greatly.	Layers of solidified lava flows.	Magma usually basaltic; effusive eruptions of fluid lava.	Hawaiian Islands; Tahiti
Composite (Stratovolcano) 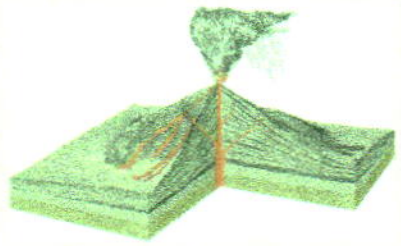	Steep-sided symmetrical cone; over 3700 m (12,000 ft.) high.	Layers of lava flows, pyroclastics, and hardened volcanic mudflow deposits.	Magma usually intermediate in chemistry, often andesitic; long life span; both explosive eruptions of pyroclastics and effusive eruptions of lava.	Mt. Fuji, Japan; Mt. Rainier, Washington; Mt. Shasta, California; Mt. Vesuvius, Italy; Mt. St. Helens, Washington
Lava Dome (Plug Dome) 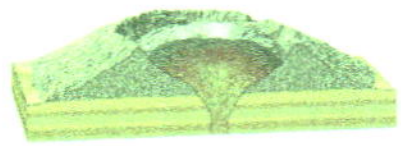	Usually small, typically less than 600 m (2000 ft.) high; sometimes irregular shape.	Solidified lava that was thick and viscous when molten; plug of lava often covered by pyroclastics; common within crater of composite volcano.	Magma usually high in silica, often rhyolitic; dome grows by expansion of viscous lava from within; explosive eruptions common.	Lassen Peak, California; Mono Craters, California
Cinder Cone 	Small, steep-sided cone; up to 500 m (1500 ft.) high.	Loose pyroclastics; may be composed of ash or cinder-size pieces.	Magma varies, often basaltic; short life span; pyroclastics ejected from central vent; occasionally produce lava flows.	Parícutin, Mexico; Sunset Crater, Arizona

more lava has issued effusively from fissures than from the combined outpourings of all volcanoes.

Research indicates that the timing of several major flood basalt eruptions in the geologic past correlate with mass extinctions of plants and animals—perhaps caused by the environmental disruption brought by the massive lava flows and "outgassing" (release of volcanic gases) from the eruptions. For example, a few scientists think that the major extinctions about 65 million years ago that ended the reign of the dinosaurs were as much, or more, a consequence of the flood basalt eruptions of the Deccan Traps than of the asteroid impact that occurred at the same time.

Volcanic Peaks

Animation
Volcanoes
https://media.pearsoncmg.com/ph/esm/esm_mcknight_physgeo_13/media/

Volcanoes vary greatly in size and configuration. Many volcanic peaks take the form of a cone. Nearly all volcanic peaks have a crater at the top. Frequently, smaller subsidiary cones develop around the base or on the side of a principal peak or even in its summit crater. Generally, differences in magma, and therefore eruption style, result in different types of volcanic peaks (Table 14-1).

Shield Volcanoes: Basaltic lava tends to flow quite easily over the surrounding surface, forming broad, low-lying **shield volcanoes**, built up of layer upon layer of solidified lava flows with relatively little pyroclastic material (Figure 14-31a). Some shield volcanoes are massive and very high, but they are never steep sided.

The Hawaiian Islands are composed of numerous shield volcanoes. Produced by the Hawaiian "hot spot," Mauna Loa on the Big Island of Hawai'i is the world's tallest volcano (Figure 14-31b)—it is more than 9 kilometers (6 miles) high from its base on the ocean floor to its summit (Figure 14-32a). Kīlauea, currently the most active of the Hawaiian shield volcanoes, is on the southeast flank of Mauna Loa (see the box, *Featured National Park: Hawai'i Volcanoes*).

Composite Volcanoes: Volcanoes that emit intermediate lavas, such as andesite, tend to erupt explosively and develop into symmetrical, steep-sided volcanoes known as **composite volcanoes** or *stratovolcanoes* (Figure 14-33a). These mountains build up steep sides by having layers of ejected pyroclastics (ash and cinders) from explosive eruptions alternate with lava flows from nonexplosive eruptions. The pyroclastics produce the steep slopes, whereas the solidified lava flows hold the pyroclastics together. Famous examples of composite volcanoes include Mt. Fuji in Japan, Mt. Rainier in Washington (Figure 14-32b), and Volcán Popocatépetl near Mexico City, which is currently active (Figure 14-33b).

LearningCheck 14-11 **Contrast the shape, structure, and formation of composite volcanoes and shield volcanoes.**

(a) Shield volcano

(b) Mauna Loa

▲ **Figure 14-31** (a) Shield volcanoes have gentle slopes and consist of layer after layer of solidified lava flows with little pyroclastic material. A wide, shallow depression at the top of a shield volcano is known as a *summit caldera*. (b) The gentle slopes of the shield volcano of Mauna Loa on the Big Island of Hawai'i. The shield volcano Mauna Kea is in the distance.

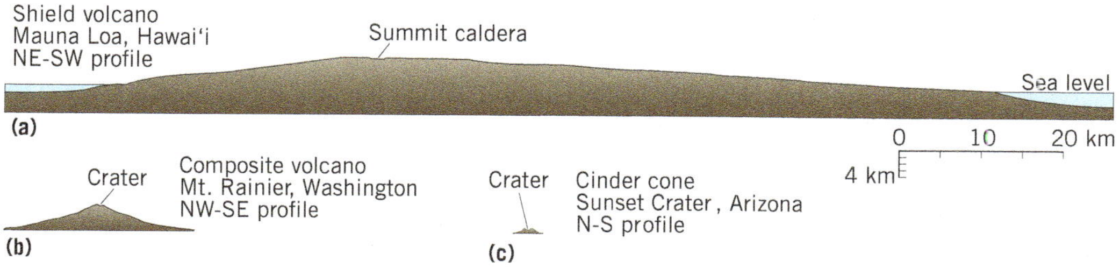

◄ **Figure 14-32** Profiles of volcanoes, found in three national parks, drawn at identical scales. (a) Mauna Loa, Hawai'i: a shield volcano. (b) Mt. Rainier, Washington: a composite volcano. (c) Sunset Crater, Arizona: a cinder cone.

(a) Composite volcano

(b) Volcán Popocatépetl

▲ **Figure 14-33** (a) Composite volcanoes consist of layers of pyroclastics and solidified lava flows. (b) Volcán Popocatépetl, a composite volcano near Mexico City, which is currently in an eruptive period.

Featured National Park

Hawai'i Volcanoes

A Timeline: Kīlauea's 35-Year Eruption

▲ **Figure 14-B** Fissure 8 eruption along Kīlauea's Lower East Rift Zone in the Leilani Estates subdivision on May 22, 2018.

The people of Hawai'i have long thought of *Pele*, the goddess of fire and volcanoes, as the creator of their island home. For the 35 years, those living on the Big Island have watched and experienced this creation firsthand through the eruption of Kīlauea. During this continuous eruption, lava flows from the volcano have destroyed hundreds of houses and closed parts of Hawai'i Volcanoes National Park—while at the same time, Kīlauea added fresh new land to the island's coast.

1983: On January 3, 1983, Kīlauea's eruption begins. Several fissures break open along Kīlauea's East Rift Zone (ERZ; **Figure 14-A**). Fountains of lava build up a 250 meter (800 foot) high "splatter cone" named Pu'u 'Ō'ō. Large areas around the vent are covered with thick, rough *'a'ā* lava that destroys several homes in the nearby Royal Gardens subdivision.

1986: The eruption shifts to a new location along the ERZ, and for the next 5 years the volcano continuously emits fluid, smooth, *pāhoehoe* lava forming a low shield peak called Kupaianaha. Lava flows cut through the coastal town of Kapa'ahu, across Highway 130 in the Park, and to the ocean. By the end of 1990, 100 homes in the nearby community of Kalapana have been buried in lava.

1992: Kupaianaha stops erupting and activity shifts back to Pu'u 'Ō'ō. A new shield builds on its west flank, effusively emitting pāhoehoe lava for the next 15 years. In June 2007, its magma chamber depressurizes, and Pu'u 'Ō'ō collapses, forming an 80 meter (260 foot) deep crater. In July 2007, new fissures develop between Pu'u 'Ō'ō and Kupaianaha, sending flows of 'a'ā lava down toward communities to the northeast.

2008: In March 2008, the summit caldera of Kīlauea also becomes active. An explosive eruption of old lava along the walls of Halema'uma'u—the volcano's deepest summit crater—opens a new vent (informally called the "Overlook" crater). For the next 10 years, this small crater repeatedly fills with lava and then drains.

2011: East Rift Zone eruptions continue. By March 2011, Pu'u 'Ō'ō again fills with lava. That summer, lava breaks though the side of the crater, creating a 3.5 kilometer (2.2 mile) long lava flow. By June 2014, new fissures develop around Pu'u 'Ō'ō, and lava flows 20 kilometers (12 miles) east to the town of Pāhoa.

2018: On April 30, 2018, Pu'u 'Ō'ō again collapses, and by early May, 10 new fissures open in the lower ERZ, pouring lava into the residential subdivision of Leilani Estates (**Figure 14-B**). By the middle of May, 10 more fissures open. In June, lava flows from "Fissure 8" reach the ocean at Kapoho Bay, creating a new lava delta.

In early May 2018, as the ERZ eruption continues, magma drains from beneath the summit of Kīlauea, and the floor of Halema'uma'u—along with much of the surrounding caldera—begins to collapse, forcing the temporary closure of much of the Park. By August, the depth of Halema'uma'u crater has increased from 85 meters (280 feet) to about 600 meters (1970 feet).

The End: By August 2018, lava flows from Fissure 8 and activity at the summit caldera have significantly decreased, and it appears that the 35-year long eruption of Kīlauea has come to an end.

In all, about 150 square kilometers (58 square miles) of the island were covered with new lava flows, more than 600 homes were destroyed, and about 14 kilometers (9 miles) of highway were buried. The current pause in eruption activity is most certainly temporary. Kīlauea remains an active volcano, and volcanologists have every expectation that it will erupt again.

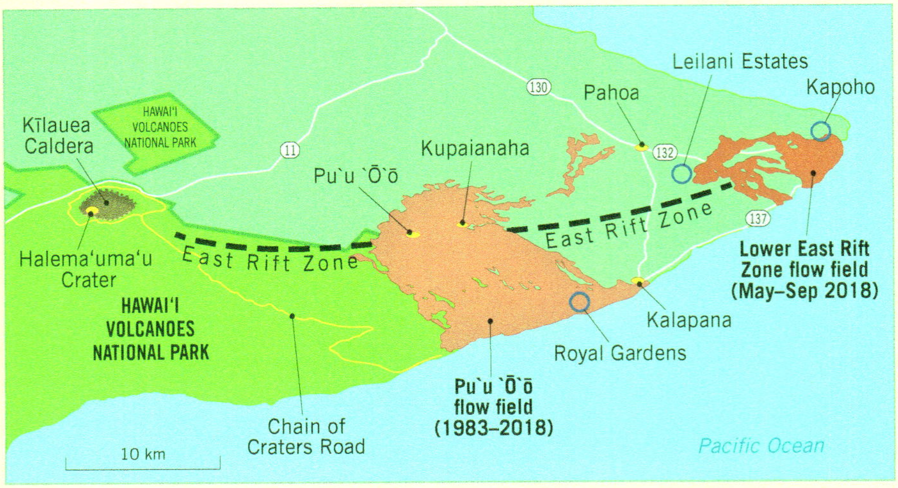

▲ **Figure 14-A** Kīlauea volcano and lava flows from its 1983 to 2018 East Rift Zone eruptions on the Big Island of Hawai'i.

Questions

1. Does the lava flow in Figure 14-B appear to be quite fluid or very viscous? How do you know?
2. Why have the recent eruptions of Kīlauea caused property damage, but not a loss of life as have some volcanic eruptions in other parts of the world?

Mobile Field Trip
Kīlauea volcano
https://media.pearsoncmg.com/ph/esm/esm_mcknight_physgeo_13/media/

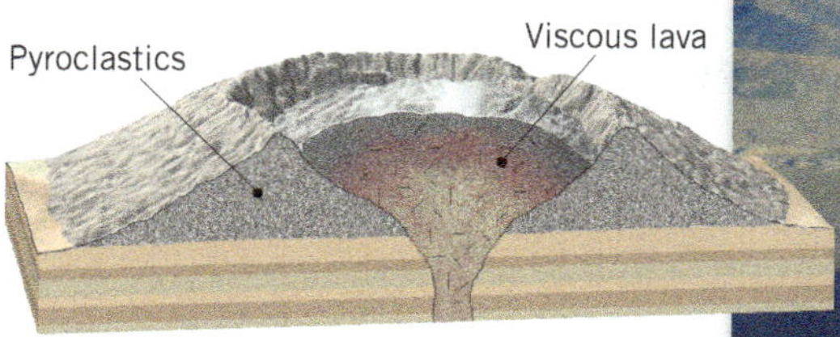

(a) Plug dome

(b) Mono Craters

▶ **Figure 14-34** (a) Lava domes (plug domes) develop when viscous lava (commonly rhyolite) is "squeezed" up into a volcanic vent. The plug of lava may be covered or surrounded by explosively ejected pyroclastics. (b) Crater Mountain is the highest peak in the Mono Craters chain of rhyolitic plug domes east of Yosemite National Park and south of Mono Lake in California. The irregular summits of these volcanoes was formed by the bulging-up of viscous lava.

Lava Domes: Lava domes—also called **plug domes**—have masses of very viscous lava, such as high-silica rhyolite, that are too thick and pasty to flow very far. Instead, lava bulges up from the vent, and the dome grows largely by expansion from below and within (Figure 14-34a). The Mono Craters are a chain of young rhyolitic plug domes just east of the Sierra Nevada and Yosemite National Park in California (Figure 14-34b). Their most recent activity took place just a few hundred years ago.

Lava domes may also develop within the craters of composite volcanoes when viscous lava moves up into the vent. Shortly after the large eruption of Mount St. Helens in 1980, such a lava dome began to develop.

Cinder Cones: Cinder cones are the smallest of the volcanic peaks (Figure 14-32c). Their magma chemistry varies, but basaltic magma is most common. They are cone-shaped peaks built by the unconsolidated pyroclastics that are ejected from the volcanic vent (Figure 14-35a). The size of the ejected particles determines the steepness of the slopes. Tiny particles ("ash") can support slopes as steep as 35 degrees, whereas the larger *ejecta* ("cinders") will produce slopes up to about 25 degrees. Cinder cones are generally less than 450 meters (1500 feet) high and are often associated with other volcanoes (Figure 14-35b). Lava flows are occasionally emitted from the same vent that produces a cinder cone.

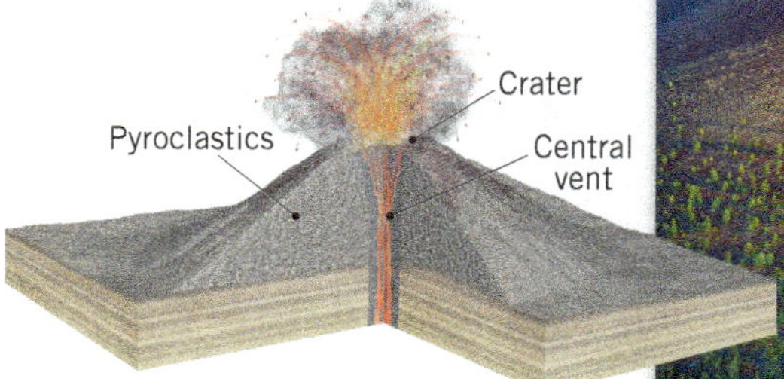

(a) Cinder cone

(b) Sunset Crater

▲ **Figure 14-35** (a) Cinder cones are small volcanoes consisting of pyroclastics. (b) Sunset Crater is a classic example of a cinder cone volcano. Sunset Crater Volcano National Monument in northern Arizona is part of the San Francisco volcanic field near Flagstaff.

Calderas: Uncommon in occurrence but spectacular in result is the formation of a **caldera**, which is produced when a volcano explodes, collapses, or both. An immense basin-shaped depression, generally circular, forms with a diameter many times larger than that of the original volcanic vent or vents. Some calderas are tens of kilometers in diameter. North America's most famous caldera is Oregon's misnamed Crater Lake (see the box, *Featured National Park: Crater Lake*). Other major calderas in North America include California's Long Valley Caldera and Wyoming's Yellowstone Caldera.

Shield volcanoes may develop *summit calderas* in a different way. When large quantities of fluid lava are vented from rift zones along the sides of a volcano, the magma chamber below the summit can empty and collapse, forming a relatively shallow caldera. Both Mauna Loa and Kīlauea on the Big Island of Hawai'i have calderas that formed in this way.

LearningCheck 14-12 **How does a caldera such as the one now occupied by Crater Lake in Oregon form?**

Volcanic Hazards

Millions of people around the world live in close proximity to active or potentially active volcanoes. In the United States alone, more than 50 volcanoes have erupted within the last 200 years, and others could well become active in the near geologic future. Some of these volcanoes are located near population centers in Washington, Oregon, California, Alaska, Hawai'i, or the region around Yellowstone National Park. Future eruptions could expose many people to a wide range of volcanic hazards (Figure 14-36).

Volcanic Gases

A volcano emits large quantities of gas during an eruption. Water vapor makes up the bulk of the gas emitted, but other volcanic gases include carbon dioxide, sulfur dioxide, hydrogen sulfide, and fluorine. The hazards resulting from these gases vary. For example, in 1986 the sudden release of carbon dioxide from a magma chamber beneath Lake Nyos in Cameroon killed nearly 1800 people.

Sulfur dioxide released during an eruption may combine with water in the atmosphere. A mist of sulfuric acid forms and falls to the surface as acid rain, harming vegetation and causing corrosion. High in the atmosphere, these same droplets may reflect incoming solar radiation, altering global weather. As we described in Chapter 8, the large quantities of sulfur dioxide emitted by Mount Pinatubo in the Philippines in 1991 reduced insolation enough to lower global temperatures slightly for more than a year.

Lava Flows

Surprisingly, lava flows rarely cause loss of life. The speed and distance covered by a lava flow depends mostly on its viscosity, which in turn depends on its silica content.

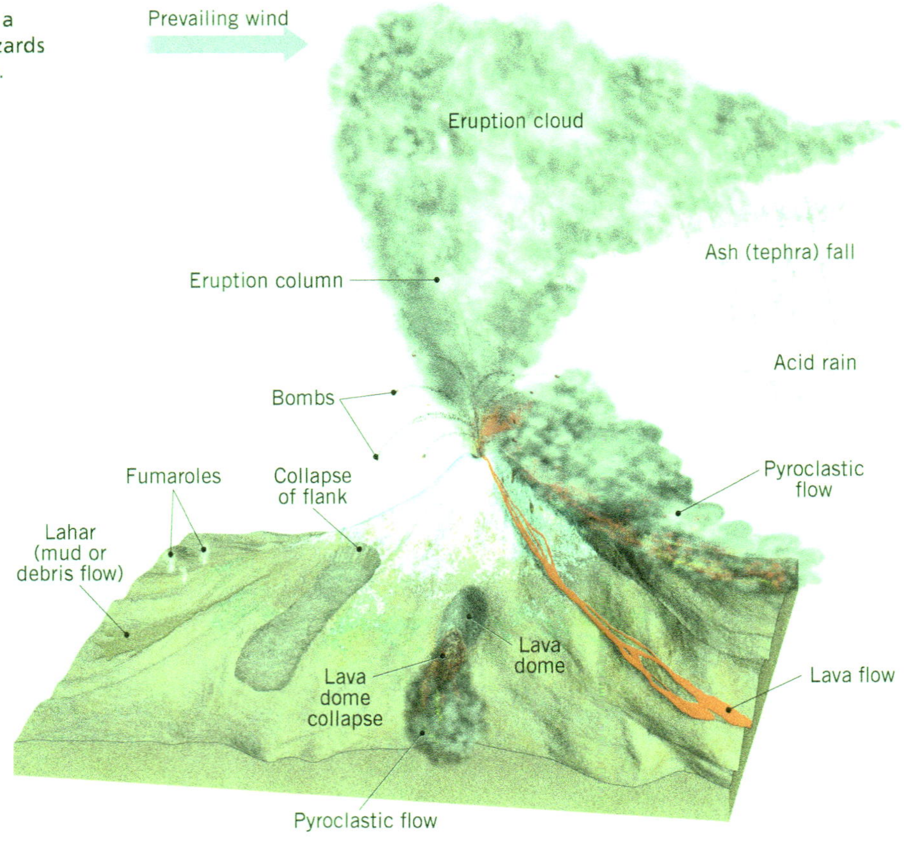

▶ **Figure 14-36** The hazards associated with a typical composite volcano. Some of these hazards may also occur with other kinds of volcanoes.

Featured National Park

Crater Lake

A Spectacular Caldera

▲ **Figure 14-C** Oregon's Crater Lake occupies an immense caldera. Wizard Island, within the lake, is a more recent subsidiary volcanic cone.

Oregon's Crater Lake is the deepest lake in the United States (**Figure 14-C**). Its pristine waters fill the caldera of a collapsed volcano. Before it gave way, the volcano—posthumously called Mount Mazama—was like many of the other tall volcanoes we see in the Cascade Range today.

Mount Mazama was a composite volcano that reached an estimated elevation of 3660 meters (12,000 feet) above sea level. During a major eruption about 7700 years ago, the walls of Mount Mazama weakened and collapsed as enormous volumes of pyroclastics were ejected from the volcano (**Figure 14-D**). The partial emptying of the magma chamber below Mount Mazama may have also contributed to this collapse. The final, cataclysmic eruption removed—by explosion and collapse—the upper 1220 meters (4000 feet) of the peak and produced a caldera whose bottom is 1220 meters (4000 feet) below the crest of the remaining rim.

After the collapse of Mount Mazama, water from rain and snowfall filled 592 meters (1943 feet) of the caldera with water, creating one of the deepest lakes in North America.

A subsidiary volcanic cone subsequently built up from the bottom of the caldera and now breaks the surface of the lake as Wizard Island. Bathymetry of Crater Lake's floor shows evidence of both postcollapse volcanic activity and massive landslides off the caldera walls (**Figure 14-E**).

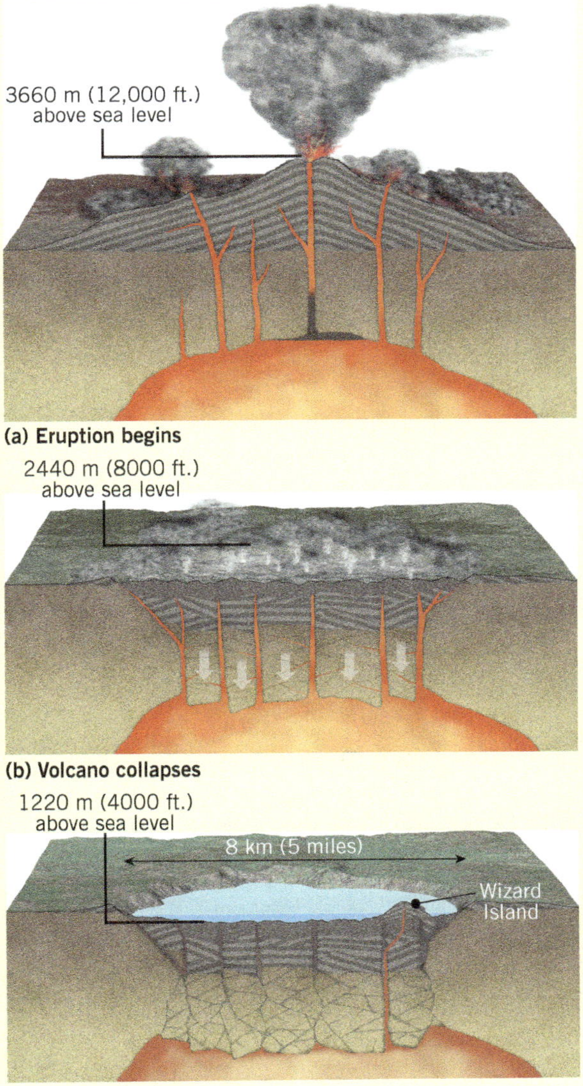

◄ **Figure 14-D** Formation of Crater Lake. (a) Mount Mazama about 7700 years ago. (b) During the cataclysmic eruption, an enormous volume of pyroclastics erupted from the magma chamber and the volcano collapsed, forming a caldera 1220 meters (4000 feet) deep. (c) The caldera partially filled with water, forming a lake; a new fissure formed the volcano known as Wizard Island.

Questions

1. Describe the sequence of events that led to the formation of present-day Crater Lake.
2. How do we know that volcanic activity at Crater Lake did not end with the collapse of the original volcano?

Animation
Formation of Crater Lake
https://media.pearsoncmg.com/ph/esm/esm_mcknight_physgeo_13/media/

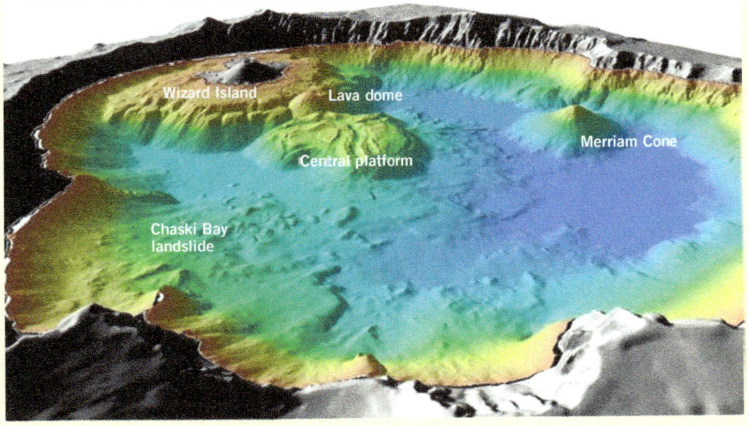

▲ **Figure 14-E** Bathymetry of Crater Lake shows postcollapse volcanic activity and debris from large landslides coming off the caldera walls.

Low-silica basaltic lava associated with shield volcanoes, such as those on Hawai'i, tends to be quite fluid and fast moving. Although most basaltic lava flows move more slowly than a person can walk, some flows travel at speeds of over 25 kilometers (15 miles) per hour and cover distances of several tens of kilometers before solidifying. Because lava tends to flow in predictable paths, they cause few injuries, although many dozens of homes have been destroyed by lava flows from Kīlauea volcano in Hawai'i over the last few decades (see the box *Featured National Park: Hawai'i Volcanoes*).

Higher-silica lava, such as the andesitic lava associated with the many composite volcanoes around the Pacific Rim, tends to be thicker than basaltic lava, usually moving only short distances down the slopes of the volcano. Very viscous rhyolitic lava often does little more than squeeze out of a vent, bulging up to form a lava dome. Although lava flows from intermediate- and high-silica magmas rarely represent a direct danger to people, the explosive nature of these volcanoes does often produce a number of significant hazards.

Eruption Column and Ash Fall

Composite volcanoes and lava domes often erupt explosively. The violent ejection of pyroclastics and gases from a volcano can form an *eruption column* reaching altitudes of 16 kilometers (10 miles) or more. Large fragments of solid rock, called volcanic *bombs*, drop to the ground immediately around the volcano; smaller fragments of volcanic ash and dust form an enormous *eruption cloud* from which great quantities of ash may fall. A heavy covering of ash can damage crops and even cause buildings to collapse.

Pyroclastic Flows

The collapse of a lava dome or the rapid subsidence of an eruption column during an explosive eruption of a volcano can lead to a terrifying high-speed avalanche of searing hot gases, ash, and rock fragments known as **pyroclastic flow**, or *nuée ardente*. A pyroclastic flow can travel down a volcano at speeds of more than 160 kilometers (100 miles) per hour, burning and burying everything in its path (Figure 14-37). Probably the most famous example of a pyroclastic flow took place in 1902 on the Caribbean island of Martinique. An explosive eruption of Mount Pelée sent a massive pyroclastic flow onto the port city of St. Pierre, destroying the town and killing nearly all of its 28,000 inhabitants in a matter of moments.

In the 1990s, the Unzen volcano complex on the Japanese island of Kyushu began a series of eruptions. After a period of escalating seismic activity, a lava dome formed in Fugen-dake, a peak adjacent to the large Mayu-yama dome. By 1993, pyroclastic flows and volcanic mudflows had destroyed 2000 structures, forced the evacuation of as many as 12,000 local residents, and took the lives of 43 people, including French filmmakers and volcanologists Maurice and Katia Krafft.

Volcanic Mudflows (Lahars)

One of the most common hazards associated with composite volcanoes is a **volcanic mudflow**—also known by its

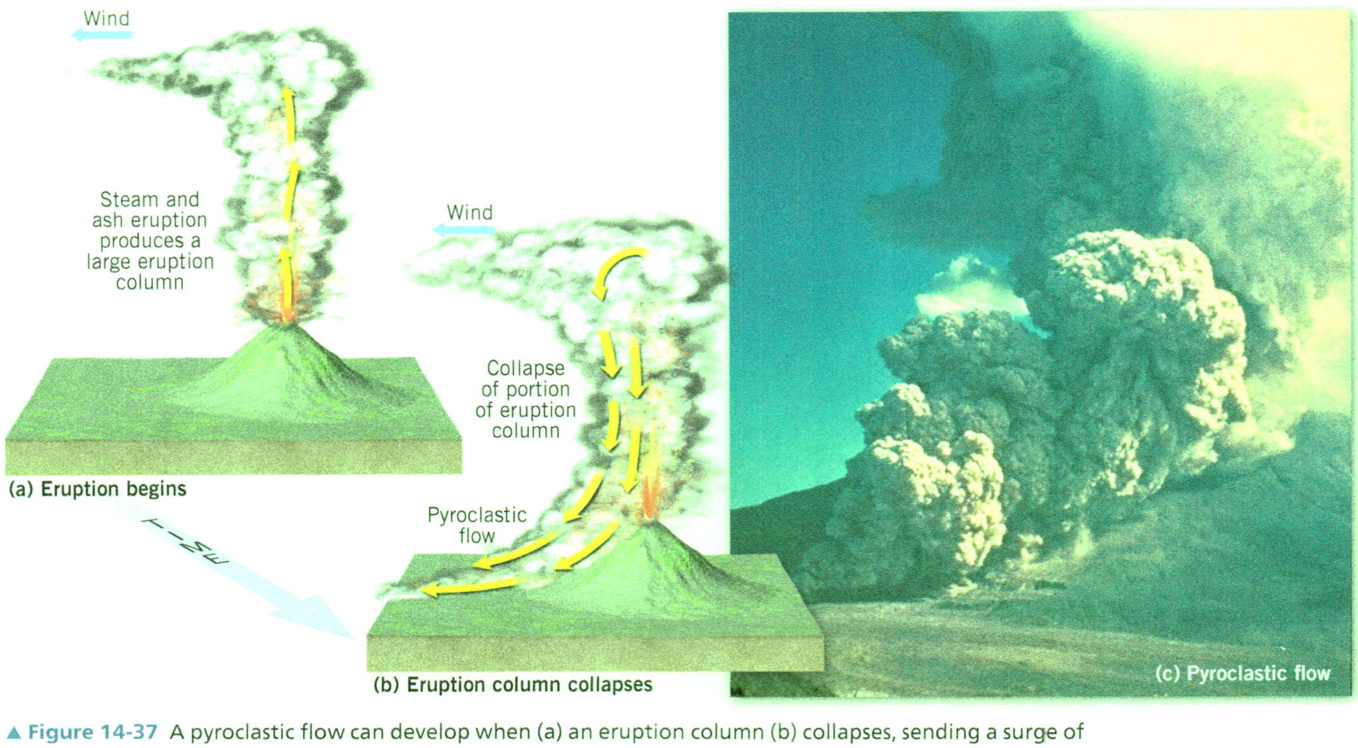

▲ Figure 14-37 A pyroclastic flow can develop when (a) an eruption column (b) collapses, sending a surge of hot gases and pyroclastics down the side of a volcano. (c) Pyroclastic flow heading down the flank of Mount St. Helens during an eruption on August 7, 1980.

▲ **Figure 14-38** Lahar deposits from a 2015 eruption of the Villarrica Volcano fill a river valley near the town of Pucón, Chile.

Indonesian name, **lahar**. A loose mantle of ash and pyroclastic flow deposits on the slopes of a volcano can be mobilized easily by heavy rain or by the melting of snow and glaciers during an eruption. The water mixes with unconsolidated pyroclastics, producing a fast-moving—and sometimes hot—slurry of mud and boulders, flowing with the consistency of wet concrete. Lahars typically flow down stream valleys off the slopes of a volcano, leaving the valley floor buried in thick mud and debris (**Figure 14-38**). They can reach speeds of more than 80 kilometers (50 miles) per hour and travel more than 80 kilometers (50 miles).

One of the most tragic examples of a lahar took place in 1985 when the Nevado del Ruiz volcano in Colombia produced a mudflow that inundated the town of Armero about 50 kilometers (30 miles) away, killing more than 23,000 people.

LearningCheck 14-13 **Explain the differences between a pyroclastic flow and a volcanic mudflow (lahar).**

Monitoring Volcanoes

Volcanoes may lie dormant for hundreds of years between major eruptions, so the volcanic hazards in the surrounding regions are not always evident. In North America, the Cascade Range alone contains more than a dozen potentially active volcanoes (**Figure 14-39**), including seven that have exhibited activity since the late 1700s. Because of rapidly expanding populations around the Pacific Rim, increasing numbers of people are now exposed to the same volcanic hazards that have killed thousands in the past.

In the United States, the U.S. Geological Survey and research universities are looking at historical eruptions, as well as evidence from the geological record, to map out the most likely paths of pyroclastic flows and lahars from volcanoes. The monitoring of active volcanoes includes using sensitive "tiltmeters," which detect swelling of a volcano with magma, to measure slight changes in the slope of a mountain; measuring variations in gas composition and quantity vented from a volcano, which may indicate changes in magma; using remote cameras to document changes to crater features; and monitoring earthquake activity below a volcano—swarms of small earthquakes may indicate the filling of the magma chamber below a volcanic peak. The search for distinctive seismic patterns that accompany the onset of major eruptions is promising.

Eruption of Mount St. Helens: The last major volcanic eruption in the conterminous 48 states occurred in 1980 in Washington. On the morning of May 18, after several months of sporadic eruptions of steam and ash, Mount St. Helens unleashed a devastating eruption—and became one of the most studied volcanoes in history.

When the eruption began, the entire north slope of the mountain broke loose in an enormous landslide. The landslide depressurized the magma within the volcano, triggering a powerful lateral blast that leveled trees more than

▲ **Figure 14-39** Prominent volcanoes of the Cascade Range in the Pacific Northwest.

24 kilometers (15 miles) north of the volcano (see Figure 10-19), destroying an area of more than 500 square kilometers (200 square miles). At the foot of the volcano, Spirit Lake, once one of the most scenic bodies of water in North America, was instantly filled with debris and dead trees.

The lateral blast was followed by a strong vertical explosion of ash and steam; the eruption column rose to an altitude of 19 kilometers (12 miles) in less than 10 minutes. More than 470 billion kilograms (520 million tons) of ash were carried to the east by the prevailing winds; 400 kilometers (250 miles) away, Spokane, Washington, was in darkness at midday. A series of pyroclastic flows rolled down the mountain. Lahars, some created when the eruption melted snow and ice that capped the volcano, poured down nearby river valleys. The largest of the lahars knocked out every bridge on the Toutle River for 48 kilometers (30 miles) and clogged the Columbia River shipping channel to less than half of its normal depth.

The eruption reduced the elevation of Mount St. Helens by 400 meters (1300 feet), removed about 2.8 cubic kilometers (0.67 cubic miles) of rock from the volcano (**Figure 14-40**), spread volcanic ash over an area of 56,000 square kilometers (22,000 square miles), caused more than $1 billion in property and economic losses, and killed 57 people.

More than 40 years after its last major eruption, Mount St. Helens continues to show activity. Whether another major eruption comparable to the one in 1980 will occur anytime soon is unclear, but in the long run, Mount St. Helens is quite likely to erupt vigorously again.

It is sobering to realize that the 1980 eruption of Mount St. Helens was not an extraordinarily large one. The 1991 eruption of Mount Pinatubo in the Philippines was more than 10 times larger, and the 1912 eruption of Novarupta in

▼ **Figure 14-40** (a) Mount St. Helens prior to the 1980 eruption. The peak in the background is Mount Adams. (b) Mount St. Helens after the 1980 eruption. Its post-eruption elevation is 400 meters (1300 feet) less than the pre-eruption elevation.

(a) Before 1980 eruption

(b) After 1980 eruption

Animation
The Eruption of Mount St. Helens
https://media.pearsoncmg.com/ph/esm/esm_mcknight_physgeo_13/media/

Alaska was 30 times larger. It is hoped that the knowledge gained from monitoring Mount St. Helens and other active volcanoes will enable authorities to make informed choices about when and where to evacuate local populations when the next volcano erupts.

Intrusive Igneous Features

Magma that solidifies below Earth's surface produces plutonic igneous rocks. In general, these types of igneous rocks form structures called **igneous intrusions**. Over time, many intrusions are exposed at the surface through the action of the external processes. Usually resistant to erosion, intrusions become conspicuous when they stand higher than the surrounding land.

Plutons

Pluton is a general term used to refer to intrusive igneous bodies of nearly any size. Intrusions come in all shapes, sizes, and compositions. The intrusive process usually disturbs the preexisting rock. The invading magma can assimilate that country rock (the surrounding rock into which magma intrudes) or heat it enough to make it flow out of the way. Adjacent to the area the magma has invaded, the country rock usually undergoes *contact metamorphism* from the heat and pressure of the rising intrusion.

Although igneous intrusions take an almost infinite variety of forms, most can be broadly classified into just a few types (Figure 14-41).

Batholiths: The largest intrusion is the **batholith**, a subterranean igneous body of enormous size (typically with a surface area of at least 100 square kilometers [40 square miles]) and perhaps of unknown depth. A large batholith may comprise dozens of plutons, emplaced gradually as a series of "pulses" of magma collected over millions of years.

Uplifted and exposed batholiths form the core of many major mountain ranges, such as the Sierra Nevada in California and Colorado's Front Range (Figure 14-42). Most of the plutonic bedrock exposed in these ranges consists of granite or granodiorite (a felsic plutonic rock with a mineral composition between that of granite and diorite; see Figure 13-5).

Volcanic Necks: A *volcanic neck* is the remnant of the pipe, or "throat," of an old volcano that filled with solidified lava after its final eruption. A neck is typically a sharp spire rising above the surrounding land, left after less resistant material that made up the volcano's cone eroded away (Figure 14-43).

Laccoliths: A *laccolith* develops when viscous magma is forced between horizontal layers of preexisting rock, bulging up a mushroom-shaped mass that domes the overlying strata. Some laccoliths, such as the Henry, Abajo, and La Sal Mountains in southeastern Utah, are large enough to form the cores of hills or mountains as batholiths do. Devils Tower in Wyoming is thought to be either an exposed laccolith or the neck of an extinct volcano.

Sills: A *sill* is a long, thin intrusive body formed when magma, typically basaltic, is forced between strata. The

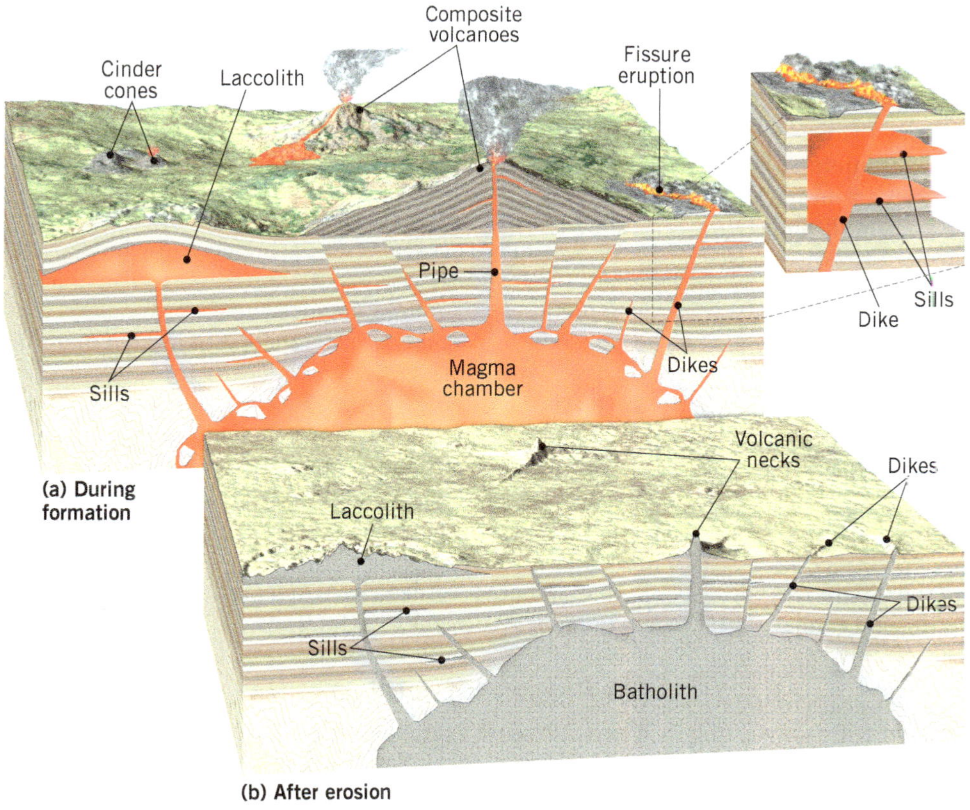

◀ Figure 14-41 The formation of common types of igneous intrusion. (a) Volcanic eruptions and intrusion of magma. (b) After erosion.

Animation
Igneous Features
https://media.pearsoncmg.com/ph/esm/esm_mcknight_physgeo_13/media/

Condor Video
Cinder Cones and Lava Flows
https://media.pearsoncmg.com/ph/esm/esm_mcknight_physgeo_13/media/

▲ **Figure 14-42** The major batholiths of western North America. These extensive intrusions are mostly associated with a former subduction boundary along the western edge of the North American Plate.

▲ **Figure 14-43** Shiprock in northwestern New Mexico is a prime example of a volcanic neck. It rises 490 meters (1600 feet) above the surrounding landscape. A *dike* extends behind Shiprock.

result is often a horizontal igneous sheet between horizontal sedimentary layers.

Dikes: One of the most widespread of intrusive forms is the **dike**, formed by the intrusion of a vertical or nearly vertical sheet of magma into preexisting rock. Dikes are vertical, narrow (a few centimeters to a few meters wide), and usually quite resistant to erosion. In some cases, dikes extend for kilometers or even tens of kilometers. When exposed at the surface by erosion, they commonly form sheer-sided walls that rise above the surrounding terrain. Dikes, such as those around Shiprock in New Mexico (see Figure 14-43) and the Spanish Peaks in Colorado, often occur as radial "walls" extending outward from a volcano like spokes of a wheel.

A special kind of dike complex is found below the layers of basaltic lava on the ocean floor. *Sheeted dikes*—a sequence of vertical dikes one right next to another—develop in a midocean ridge spreading center as basaltic magma is injected into the ever-spreading gap produced by diverging plates.

LearningCheck 14-14 **What are the differences and similarities among a batholith, a laccolith, and a dike?**

Veins: Least prominent among igneous intrusions but widespread in occurrence are thin *veins* of igneous rock that may occur individually or in profusion. They commonly form when hydrothermal fluids are forced into small fractures in the preexisting rocks.

Tectonism: Folding

Rocks may be bent or broken in a variety of ways in response to great pressure exerted in the crust or mantle. *Tectonism* (or *diastrophism*) is a general term that refers to such deformation of Earth's crust.

Sometimes the *stress* (pressure exerted on an object) that results in tectonism is caused by the rise of molten material from below. In some cases, stress is clearly the result of plate movement, but it may be unrelated directly to plate boundaries. Tectonic movements are particularly obvious in sedimentary rocks because nearly all sedimentary strata are initially deposited horizontally or nearly so—if they now are bent, broken, or tilted, we know that they experienced tectonism (Figure 14-44).

In this chapter we describe two types of tectonism: *folding* and *faulting*. We begin with folding.

The Process of Folding

When crustal rocks are subjected to stress, particularly lateral compression, they can be deformed by being bent in a process called **folding**. Our common experience is that rocks are

▲ **Figure 14-44** The Waterpocket Fold in Capitol Reef National Park in Utah showing the approximate extent of this fold before erosion.

Animation
Folding
https://media.pearsoncmg.com/ph/esm/esm_mcknight_physgeo_13/media/

hard and brittle, and we might expect them to break if subjected to stresses. However, when great pressure (and sometimes heat) is applied for long periods—especially deep below the surface—the result is often a slow plastic deformation that can produce folded structures of incredible complexity. Folding can occur in any kind of rock, but it is most recognizable when once-flat sedimentary strata have been deformed.

Folding can take place at almost any scale. Some folds are measured in no more than centimeters, whereas others develop over such broad areas that crest and trough are tens of kilometers apart.

Types of Folds

Folds are generally classified according to the orientation of the bend (**Figure 14-45**). A *monocline* is a one-sided fold—a slope connecting two horizontal or gently inclined strata. A simple symmetrical upfold is an **anticline**, and a simple downfold is a **syncline** (see Figure 14-45b). If an upfold is pushed extensively from one side, it may become an *overturned fold*.

Topographic Features Associated with Folding

The simplest relationship between structure and topography, and one that often occurs in nature, is that the upfolded anticlines produce ridges and the downfolded synclines form valleys. The converse relationship is also possible, however, with valleys developing on the anticlines and ridges on the synclines (see Figure 14-45c). This "inverted" topography is most easily explained by the effects of *tension* (pulling apart) and *compression* (pushing together) on the folded strata. Where a layer is arched over an upfold, tension cracks can form and provide easy footholds for erosional processes to remove materials and incise downward into the underlying strata. Conversely, the compression that acts on the downfolded beds increases their density and therefore their resistance to erosion. Thus, over a long period of time, the upfolds may be eroded faster than the downfolds, producing *anticlinal valleys* and *synclinal ridges*.

All these types of folding, as well as many variations, are found in the Ridge-and-Valley section of the Appalachian Mountains, a world-famous area noted for its remarkably parallel sequence of mountains and valleys developed from folds (**Figure 14-46**). The section extends for about 1600 kilometers (1000 miles) in a northeast–southwest direction across parts of nine states, with a width of 40–120 kilometers (25–75 miles).

LearningCheck 14-15 What are synclines and anticlines, and what kind of topography is associated with them?

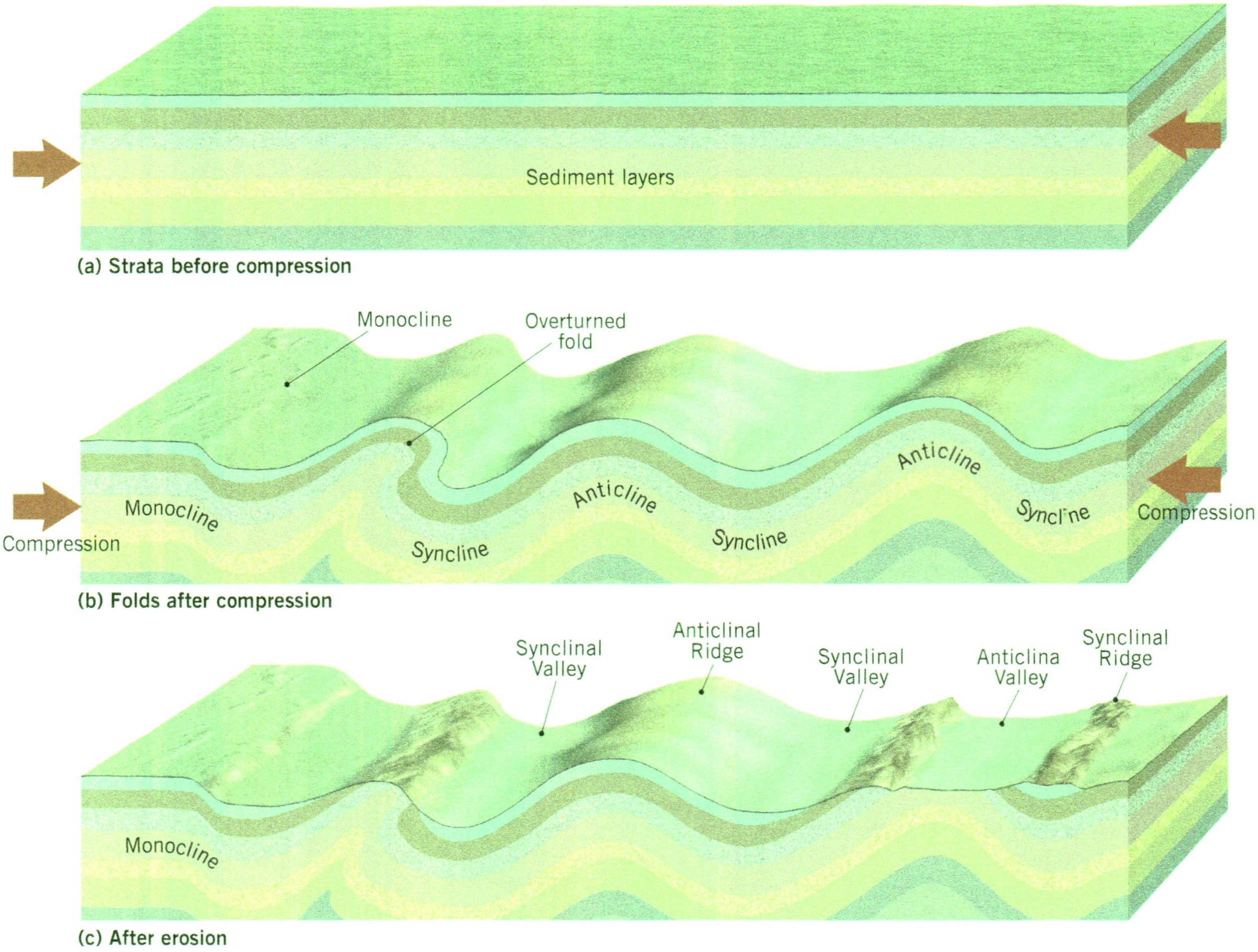

▲ **Figure 14-45** The development of folded structures. (a) Compressive stresses cause sedimentary strata that are initially horizontal to fold. (b) The basic types of folds. Before erosion, anticlines form ridges and synclines form valleys. (c) After erosion, anticlinal valleys and synclinal ridges may remain.

Condor Video
Monoclines
https://media.pearsoncmg.com/ph/esm/esm_mcknight_physgeo_13/media/

Condor Video
Anticlines and Synclines
https://media.pearsoncmg.com/ph/esm/esm_mcknight_physgeo_13/media/

▼ **Figure 14-46** The prominently folded Appalachian Mountains are clearly shown in this shaded relief map of south-central Pennsylvania.

Tectonism: Faulting

Another prominent result of tectonism is the breaking of rock structures. When a rock structure is broken and one side is forcibly displaced relative to the other, the action is called **faulting**. The displacement can be vertical, horizontal, or both. Faulting takes place where the crust is weak, along a fracture zone called a *fault plane* (or *fault zone*). The intersection of that plane with Earth's surface is called a *fault line* (Figure 14-47).

Major faults penetrate many kilometers into Earth's crust. Deep fault zones can serve as conduits to allow both water

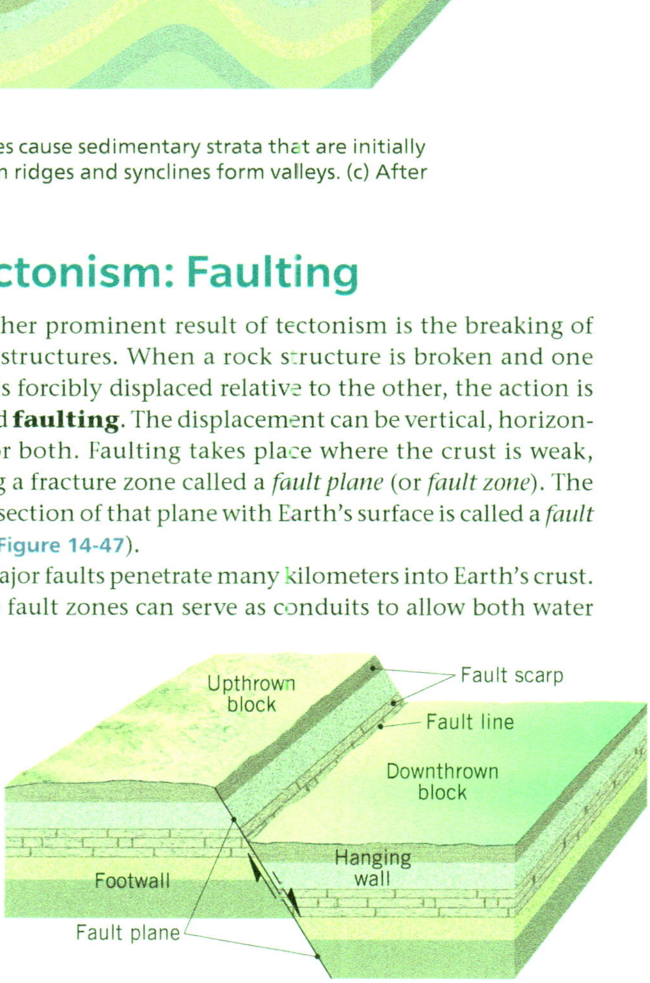

▲ **Figure 14-47** A simple fault structure.

▲ **Figure 14-48** Damage in Kathmandu from the April 2015 Nepal earthquake.

and magma to approach the surface. Frequently springs are found along fault lines, sometimes with hot water flowing out. Volcanic activity is also associated with many fault zones as magma forces its way upward in the zone of weakness.

Movement of crust along a fault zone can be very slow, but it commonly occurs as a sudden rupture. A single rupture may result in a displacement of only a few centimeters, but the movement can be as much as 10 meters (30 feet). Successive ruptures may be years or even centuries apart, but the cumulative displacement over millions of years can amount to hundreds of kilometers horizontally and tens of kilometers vertically. The sudden rupture and displacement along a fault cause most large earthquakes, such as the ones that devastated Haiti in 2010, Nepal in 2015 (Figure 14-48), and Chiapis, Mexico, in 2017.

Types of Faults

Faults can be categorized into four principal types on the basis of direction of stress and angle of movement:

1. **Normal faults:** A normal fault (Figure 14-49a) results from *tension stresses* (pulling apart or extension) in the crust. It produces a very steeply inclined fault zone, where one block slides down the fault plane ("normal" to the pull of gravity). A prominent *fault scarp* usually forms.
2. **Reverse faults:** A **reverse fault** (Figure 14-49b) is produced from *compression stresses* (pushing together), where one block slides up the incline of the fault plane ("reverse" to the pull of gravity), so the fault scarp would be severely oversteepened if erosion did not act to smooth the slope somewhat.
3. **Thrust faults:** Related to reverse faults are **thrust faults** (Figure 14-49c), or *overthrust faults*. Compression forces the upthrown block to override the downthrown block at a relatively low angle, sometimes for many kilometers. Thrust faulting occurs frequently in mountain building (including buried *blind thrust faults*), resulting in unusual geologic relationships, such as older strata being piled atop younger rocks. Both thrust faults and reverse faults are commonly associated with subduction zones or continental collision zones.
4. **Strike-slip faults:** In a **strike-slip fault** (Figure 14-49d), the movement is horizontal (side to side), with the adjacent blocks being displaced laterally relative to each other. Strike-slip faults result from *shear stresses* (stress causing two parallel surfaces to slide past one another). Transform faults, described earlier in this chapter, are one variety of strike-slip fault.

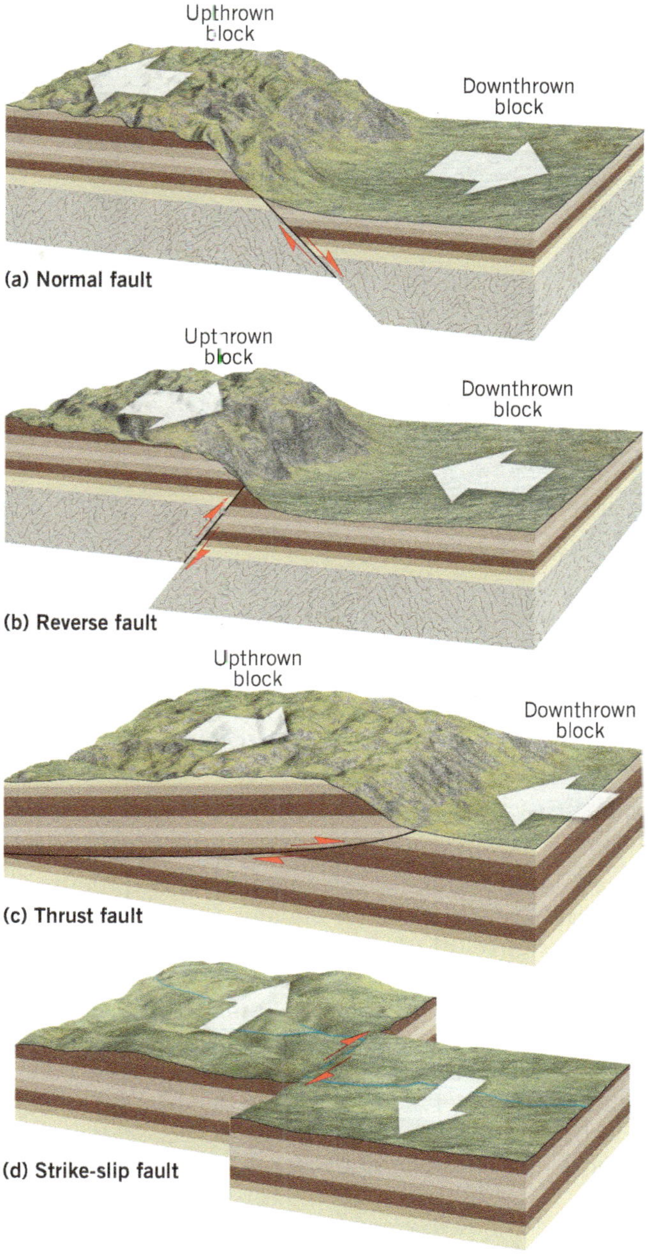

▲ **Figure 14-49** The principal types of faults. The large arrows show the direction of stress. (a) Normal fault from tension stresses; (b) reverse fault and (c) thrust fault from compression stresses; (d) strike-slip fault from shear stresses.

▲ **Figure 14-50** Scientists study a fault scarp near Canyon Lake, Texas. The vertical displacement along the fault plane is nearly 3 meters (10 feet).

Animation
Faulting
https://media.pearsoncmg.com/ph/esm/esm_mcknight_physgeo_13/media/

Condor Video
Faulting versus Joints
https://media.pearsoncmg.com/ph/esm/esm_mcknight_physgeo_13/media/

LearningCheck 14-16 **Contrast the direction of stress and displacement along normal, reverse, and strike-slip faults.**

Fault Scarps: Fault lines are often marked by prominent topographic features. Most obvious perhaps are **fault scarps**—steep cliffs that represent the edge of a vertically displaced block (Figure 14-50). Slight amounts of vertical displacement along strike-slip faults may leave a scarp, but scarps are most commonly associated with normal faulting. Some fault scarps are as much as 3 kilometers (2 miles) high and extend for more than 150 kilometers (100 miles) in virtually a straight line. The abruptness of their rise, steepness of their slope, and linearity of their orientation combine to make some fault scarps extremely spectacular landscape features.

Landforms Associated with Normal Faulting

Under extensional stresses, a surface block of crust may be faulted along one side, leaving the block tilted asymmetrically. The result is a **tilted fault-block mountain** range (also called a **fault-block mountain**), with a steep slope along the fault scarp and a relatively gentle slope on the other side of the block. The classic example is California's Sierra Nevada (Figure 14-51), an immense block nearly 640 kilometers (400 miles) north–south and about

▼ **Figure 14-51** The western slope of the Sierra Nevada is long and gentle, whereas the eastern slope is short and steep. (a) This gentle/steep combination is the result of enormous block faulting on the eastern side. (b) The steep east side of the Sierra Nevada Mountains in California is a prominent fault scarp.

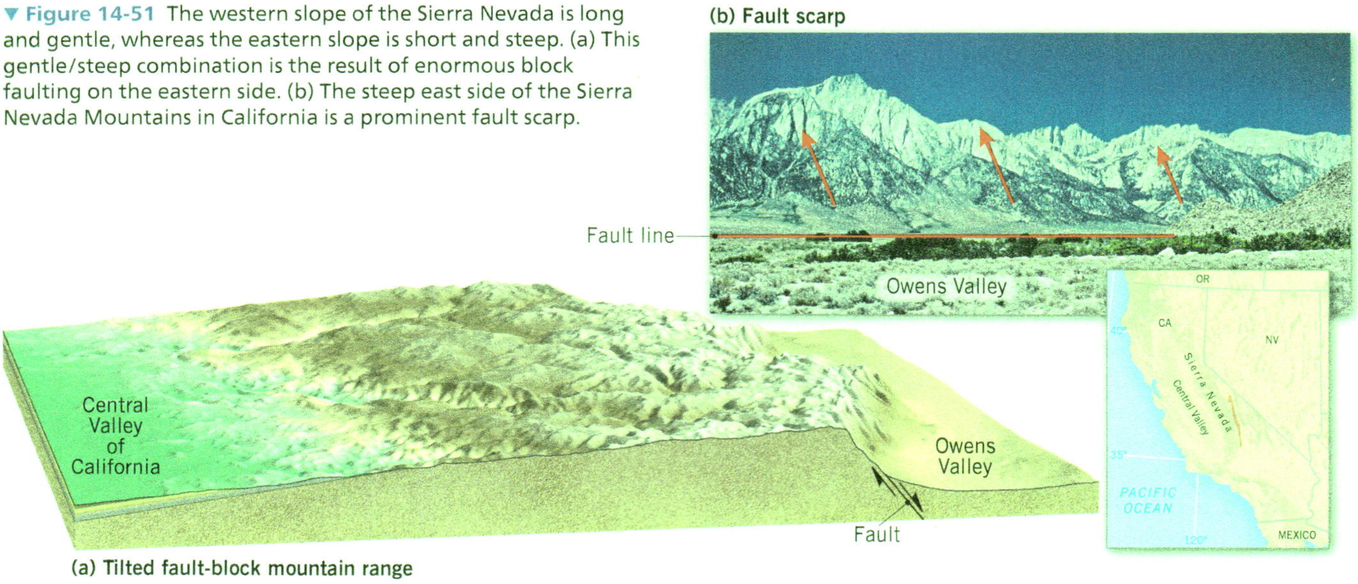

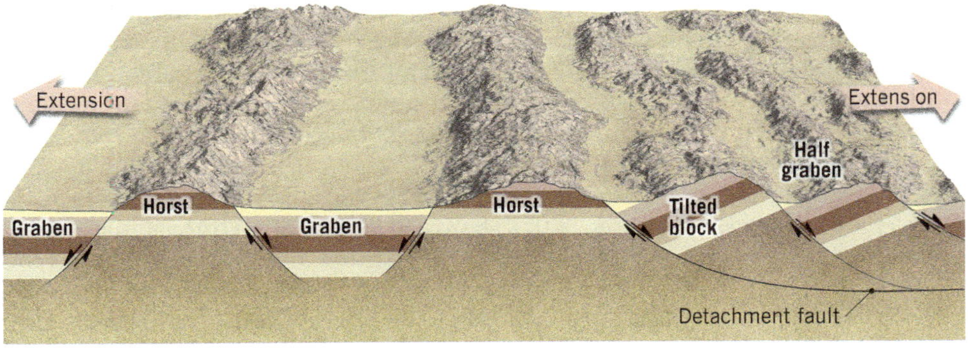

▲ **Figure 14-52** Landforms associated with tensional stresses and normal faulting. Horsts and grabens are bound by faults on both sides. Tilted blocks have been faulted along just one side, forming fault block mountains and "half grabens." Extension may also result in detachment faults, where a steep normal fault ties in with a nearly horizontal fault below.

110 kilometers (70 miles) east–west. The spectacular eastern face is a fault scarp that has a vertical relief of about 3 kilometers (about 2 miles) over a horizontal distance of only about 20 kilometers (12 miles). In contrast, the gentle slope of the western flank of the range covers a horizontal distance of more than 80 kilometers (50 miles). Other processes have modified the shape of the range, but its general configuration was determined by block faulting.

If faulting leaves a block of crust uplifted between two parallel faults, the resulting structure is called a **horst** (Figure 14-52). Frequently, horsts are the result of the land on both sides being downthrown rather than the block itself being uplifted. In either case, the horst may take the form of a plateau or a mountain mass with two steep, straight sides.

The opposite kind of structure is a **graben**: a down-dropped block of land bounded by parallel faults. A graben is a distinctive structural trough with straight, steep-sided fault scarps on both sides (see Figure 14-52).

Tilted fault-block mountains, grabens, and horsts may occur side by side. The basin-and-range region of the western United States (covering most of Nevada and portions of surrounding states; see Figure 18-29) is a vast sequence of tilted fault-block mountains and *half-grabens* (basins formed by blocks tilting down along just one fault), along with a few true grabens and horsts. As the basin-and-range region undergoes sustained extensional faulting, the landscape continues to spread. It develops along normal faults and *detachment faults*, which form where steeply dipping normal faults tie into low-angle (or nearly horizontal) faults below the surface. Detachment faults form along the boundary between brittle crustal rocks above and more ductile (flexible) crustal rocks below.

Rift Valleys: Where a divergent plate boundary develops within a continent, the resulting downfaulted grabens occasionally extend for extraordinary distances as linear valleys enclosed between steep fault scarps. Such lengthy troughs are called rift valleys (see Figure 14-14). They comprise some of Earth's most spectacular linear features, particularly the Great East African Rift Valley.

Landforms Associated with Strike-Slip Faulting

A wide variety of landforms can result from the lateral movement of strike-slip faults (Figure 14-53a). The surface trace of a large strike-slip fault may be marked by a **linear fault trough** or valley, formed by repeated movement and fracturing of rock within the fault zone. Small depressions known as *sags* develop through the settling of rock within

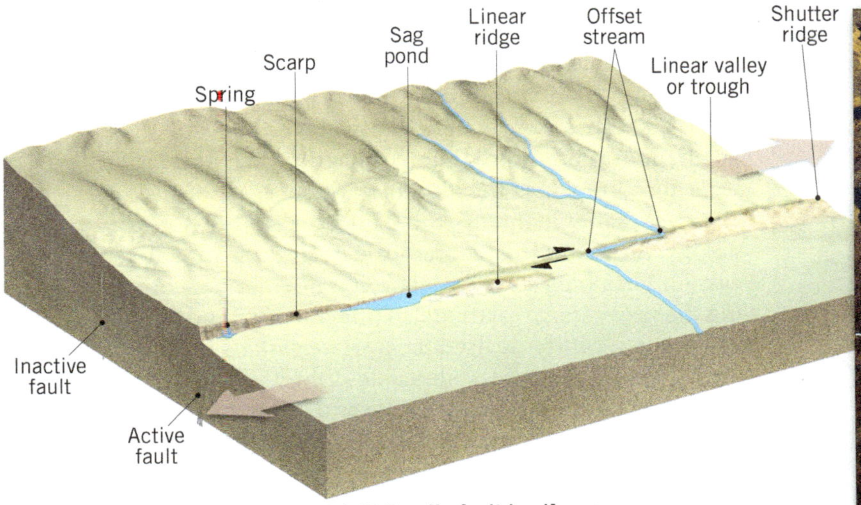

(a) Strike-slip fault landforms

(b) Offset stream

▲ **Figure 14-53** (a) Common landforms produced by strike-slip faulting. (b) Offset stream along the San Andreas Fault in the Temblor Range of California.

the fault zone and may fill with water, forming **sag ponds**. Linear ridges and scarps trending parallel to the trace of the fault may develop from slight vertical displacement or compression that can occur along a strike-slip fault.

Perhaps the most conspicuous landform produced by strike-slip faulting is an **offset stream** (see Figure 14-53b). Streams flowing across the fault are displaced by periodic fault movement or diverted when a *shutter ridge* is faulted in front of a drainage channel, closing off a valley.

LearningCheck 14-17 Explain the formation of grabens and offset streams.

Earthquakes

Usually, but not exclusively, associated with faulting is the abrupt shaking of the crust known as an *earthquake*. As we learned, most earthquakes are associated with plate boundaries (see Figure 14-7).

An **earthquake** is essentially a vibration in Earth produced by shock waves resulting from a sudden displacement along a fault. (Earthquakes may also develop from the movement of magma, sudden ground subsidence, or human activities such as the injection of fracking fluids.) The fault movement allows an abrupt release of energy, usually after a long, slow accumulation of strain. Although a fault rupture can take place at the surface, displacement can also take place at considerable depth—as deep as 700 kilometers (435 miles) beneath the surface at subduction zones.

Seismic Waves

Animation **Seismic Waves** https://media.pearsoncmg.com/ph/esm/esm_mcknight_physgeo_13/media/

The energy released in an earthquake moves through Earth in several kinds of seismic waves that originate at the center of fault motion, called the *focus* of the earthquake (Figure 14-54). These waves travel outward in widening circles, like the ripples produced when you throw a rock into a pond, gradually diminishing in amplitude with increasing distance from the focus. The strongest shocks, and greatest crustal vibration, are often felt on the ground directly above the focus, at the location known as the **epicenter** of the earthquake.

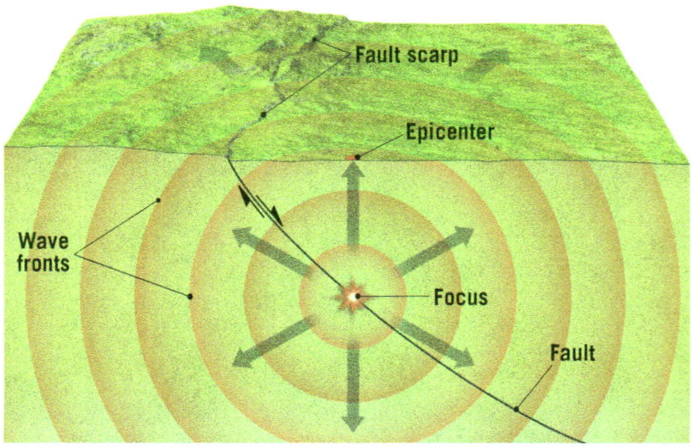

▲ Figure 14-54 Relationship among focus, epicenter, and seismic waves of an earthquake. The seismic waves are indicated by the concentric circles.

The fastest-moving seismic waves, and the first to be felt during an earthquake, are known as *P* or *primary waves*. P waves move through Earth in the same fashion as sound waves, by alternately compressing and relaxing the medium through which they are traveling. The initial jolt of the P waves is followed by the strong side-to-side and up-and-down shearing motion of the slower moving *S* or *secondary waves*. Both P waves and S waves travel through the body of Earth (so they are also known as *body waves*), whereas the motion of a third type of seismic wave is limited to the surface. These *surface waves* typically arrive immediately after the S waves and produce strong side-to-side movement as well as the up-and-down "rolling" motion often experienced during a large earthquake.

Animation **Seismographs** https://media.pearsoncmg.com/ph/esm/esm_mcknight_physgeo_13/media/

At the surface, P waves travel through bedrock at about 6 kilometers (3.7 miles) per second, while S waves travel at about 3.5 kilometers (2.2 miles) per second. Because P waves and S waves travel at different speeds, we can determine the distance to an earthquake's focus. The farther away an earthquake, the greater the time lag between the arrival of P waves and the arrival of S waves. By using a network of *seismographs* (instruments that record earthquakes) and comparing arrival times of the seismic waves, seismologists can pinpoint the focus of an earthquake with great precision.

Earthquake Magnitude

Perhaps the most widely mentioned—and misunderstood—aspect of an earthquake is its magnitude. **Magnitude** describes the relative amount of energy released during an earthquake and is calculated on a logarithmic scale. From one level of magnitude to the next, seismic waves differ in amplitude by a factor of 10—but they differ in energy by a factor of almost 32. The difference between small earthquakes (magnitudes less than 3) and large earthquakes (magnitude 7 and higher) is enormous (Figure 14-55). A magnitude 4 earthquake releases about 32 times more energy than one of magnitude 3; a magnitude 5 releases 1000 times more energy than a magnitude 3; and a magnitude 7 releases 1,000,000 times more energy than a magnitude 3!

The most commonly quoted magnitude is the *Richter scale* (also known as the *local magnitude scale*), devised in 1935 by Charles Richter at the California Institute of Technology. Although the Richter scale magnitude is relatively easy for seismologists to calculate, it is not ideal for comparing the sizes of large earthquakes (magnitude 7 and higher). The more recently devised *moment magnitude* is now the most commonly used scale to describe the size of large quakes, although the Richter scale and other scales are still in use. Because the method of calculation varies among the magnitude scales, slightly different magnitude numbers are often reported for the same earthquake.

Each year, tens of thousands of earthquakes occur around the world. The vast majority of these are too small for people to feel, but perhaps 100 are large enough to cause damage or loss of life (commonly magnitude 6 or higher).

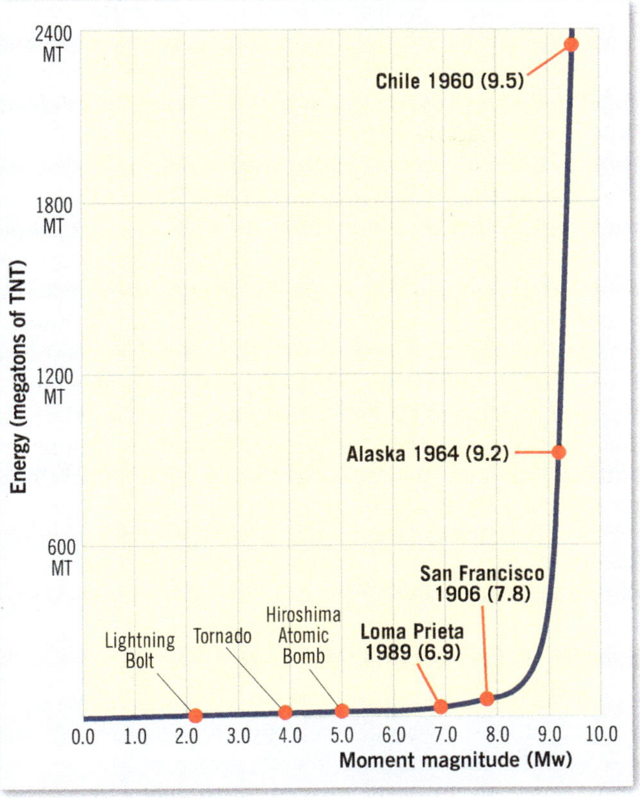

▲ **Figure 14-55** The amount of energy released during an earthquake increases exponentially as magnitude increases. A magnitude 9 earthquake releases 1,000,000 times the energy of a magnitude 5 earthquake. The approximate energy equivalents are given in megatons of TNT.

Great earthquakes—those with magnitude 8 or higher—might occur only once every few years. The two largest earthquakes yet recorded were the 1960 Chile earthquake with a moment magnitude of 9.5, and the 1964 Alaska earthquake with a moment magnitude of 9.2. The great Sumatra earthquake of December 2004 had a magnitude of 9.1; the March 2011 Japan earthquake, a magnitude of 9.1; and the April 2015 Nepal earthquake, a magnitude of 7.8. For comparison, the famous 1906 San Francisco earthquake had a moment magnitude of 7.8; the 1989 Loma Prieta earthquake, a magnitude of 6.9; and the 1994 Northridge earthquake, a magnitude of 6.7.

Shaking Intensity

Even though each earthquake can be assigned a single magnitude number to describe its relative size, every earthquake generates a wide range of local ground-shaking intensities. It is the strength of local ground shaking that directly influences the amount of damage that results from an earthquake. Local variations in the shaking intensity of an earthquake can be quantitatively measured—usually as the peak horizontal ground acceleration expressed as a percentage of gravity—but the most widely used intensity scale was devised in 1902 by Italian Earth scientist Giuseppe Mercalli. Updated by the U.S. Coast and Geodetic Survey, the **modified Mercalli intensity scale** assigns the strength of local shaking to 1 of 12 categories, based on the observed effects and damage (Table 14-2). Seismologists also use this scale to describe the anticipated intensity of ground shaking that may occur during future earthquakes in a region.

LearningCheck 14-18 **How is earthquake magnitude different from earthquake shaking intensity?**

TABLE 14-2 Modified Mercalli Intensity Scale

I	Not felt except by very few people under especially favorable circumstances.
II	Felt only by a few persons at rest, especially on upper floors of buildings.
III	Felt quite noticeably indoors, especially on upper floors of buildings, but many people do not recognize it as an earthquake.
IV	During the day felt indoors by many, outdoors by few; sensation like heavy truck striking building.
V	Felt by nearly everyone, many awakened; disturbances of trees, poles, and other tall objects sometimes noticed.
VI	Felt by all; many frightened and run outdoors; some heavy furniture moved; few instances of fallen plaster or damaged chimneys; damage slight.
VII	Everybody runs outdoors; damage negligible in buildings of good design and construction; slight to moderate in well-built ordinary structures; considerable damage in poorly built or badly designed structures.
VIII	Damage slight in specially designed structures; considerable in ordinary substantial buildings, with partial collapse; great damage in poorly built structures; fall of chimneys, factory stacks, columns, monuments, and other vertical features.
IX	Damage considerable in specially designed structures; buildings shifted off foundations; ground cracked conspicuously.
X	Some well-built wooden structures destroyed; most masonry and frame structures destroyed with foundations; ground badly cracked.
XI	Few, if any, masonry structures remain standing; bridges destroyed; broad fissures in ground.
XII	Damage total; waves seen on ground surfaces; objects thrown upward into the air.

Earthquake Hazards

Most of the damage from an earthquake is a direct or indirect consequence of ground shaking.

Ground Shaking

Generally, the strength of ground shaking decreases with increasing distance from the epicenter of the earthquake, but this pattern can be significantly modified by the local geology. Loose, unstable regolith, sediments, and soil tend to amplify ground shaking—buildings on such "soft" ground will shake much more strongly than those located on bedrock the same distance from the epicenter.

Liquefaction: During the shaking of an earthquake, loose, water-saturated sediments, such as coastal landfill, may undergo **liquefaction**—that is, become fluid. The result is subsidence, fracturing, and horizontal sliding of the surface. Much of the damage in the 1989 Loma Prieta earthquake in California and the Japan earthquakes in Kobe (1995) and Tohoku (2011) was due to amplified ground shaking and liquefaction in artificial fill (Figure 14-56).

Landslides: Landslides are often triggered by earthquakes. During the 1964 Alaska earthquake, for example, dozens of homes in the Turnagain Heights area of Anchorage were destroyed when a 2.6-kilometer (1.6-mile) wide section of the development moved toward the sea in a series of massive landslides. The devastating April 2015 earthquake in Nepal generated dozens of large landslides.

Tsunami

Another kind of hazard associated with earthquakes involves water movements in lakes and oceans. Abrupt crustal movement can set great waves (known as *seiches*) in motion in lakes and reservoirs, causing them to overflow shorelines or dams in the same fashion that water can be sloshed out of a dishpan.

Much more significant, however, are great seismic sea waves called *tsunami* (the Japanese term for these large waves), which are sometimes generated by undersea fault ruptures or landslides. These waves, often occurring sequentially, move quickly across the ocean. They are barely perceptible in deep water, but when they reach shallow coastal waters, they sometimes build up to heights of more than 15 meters (50 feet). With devastating effects, they crash onto the shoreline (where they are often incorrectly called "tidal waves"). Most of the deaths from the 1964 Alaska earthquake were the result of tsunami. As far away as Crescent City, California, 12 people were killed when a 6-meter (19-foot) high tsunami came on shore. In December 2004, about 230,000 people were killed when a tsunami generated by a subduction zone earthquake off the Indonesian island of Sumatra spread through the Indian Ocean. The March 2011 Japan earthquake generated a tsunami as high as 30 meters (100 feet) that killed about 16,000 people. A detailed description of tsunami formation and the 2011 Japan earthquake and tsunami is found in Chapter 20, where we discuss the dynamics of ocean waves.

LearningCheck 14-19 **What causes liquefaction during an earthquake?**

Earthquake Hazard Warnings

In the Pacific Ocean, a tsunami warning system made up of buoys and sensors can detect a tsunami in midocean and provide advanced warning to endangered coastal populations. Currently, however, no technology can predict earthquakes—see the box *Focus: Earthquake Prediction*.

Earthquake Early Warning Systems: Although earthquake prediction remains elusive, the technology exists to detect the impending arrival of the strong ground shaking from a nearby large earthquake. The USGS, in cooperation with the California Governor's Office of Emergency Services and several research universities, is developing and currently testing statewide an Earthquake Early Warning (EEW) system called *ShakeAlert*. Such a system can provide 10–60 seconds advance warning of the arrival of destructive seismic waves from a large earthquake on the U.S. West Coast (Figure 14-57).

▲ **Figure 14-56** Manhole cover popped up along street in Urayasu, Japan, due to liquefaction and subsidence of the street during the 2011 magnitude 9.1 Tohoku earthquake.

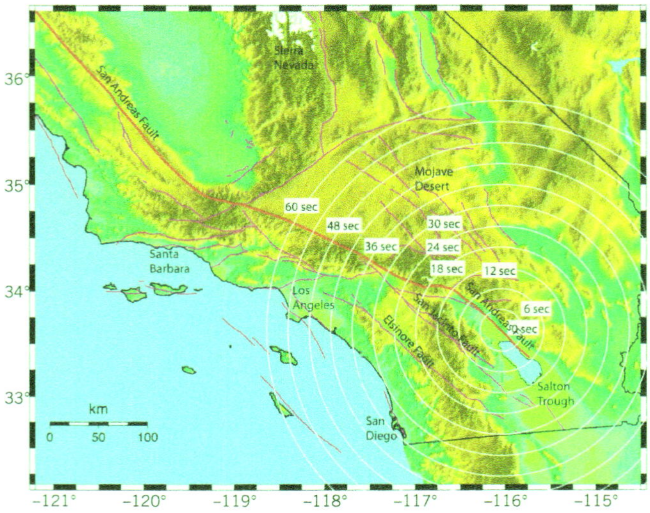

▲ **Figure 14-57** Estimated warning times from an Earthquake Early Warning system in southern California for a major earthquake occurring on the southern San Andreas Fault.

Focus

Earthquake Prediction

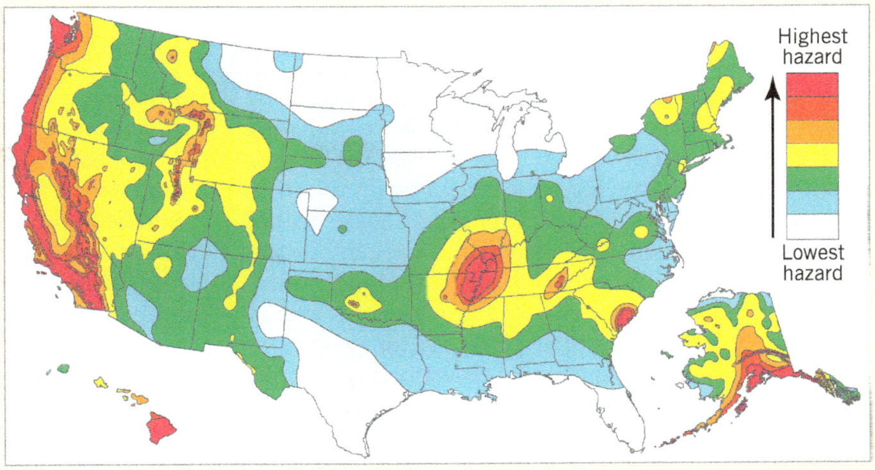

▲ **Figure 14-G** Earthquake hazard map of the United States. Hazards are defined by the probability of a location experiencing a damaging earthquake over a 50-year period beginning in 2018.

Predicting that an earthquake will occur on a specific day or in a specific month depends on the existence of "precursors"—warning signs of an impending earthquake. Possible warning signs that are being studied by scientists around the world include patterns of small earthquakes that sometimes occur before major quakes (*foreshocks*), changes in water-well levels, changes in the amount of dissolved radon gas in groundwater, variations in electrical conductivity of rocks, the bulging of the ground surface, and changes in the behavior of animals.

The Parkfield Project: One of the most elaborate and longest-running earthquake precursor research projects began in 1985 in the tiny California town of Parkfield, which sits atop the San Andreas Fault. Since the mid-1800s, this segment of the fault has ruptured at fairly regular intervals—1857, 1881, 1901, 1922, 1934, 1966, and 2004—each time producing an earthquake of about magnitude 6. In addition to many kinds of instruments designed to "capture" earthquakes as they happen, Parkfield is also the location of the San Andreas Fault Observatory at Depth (SAFOD)—an ambitious drilling project that put instruments 2–3 kilometers (1.2–1.9 miles) deep in the active fault zone, providing seismologists with valuable data on fault-zone rocks and fluids that may influence rupture recurrence (**Figure 14-F**).

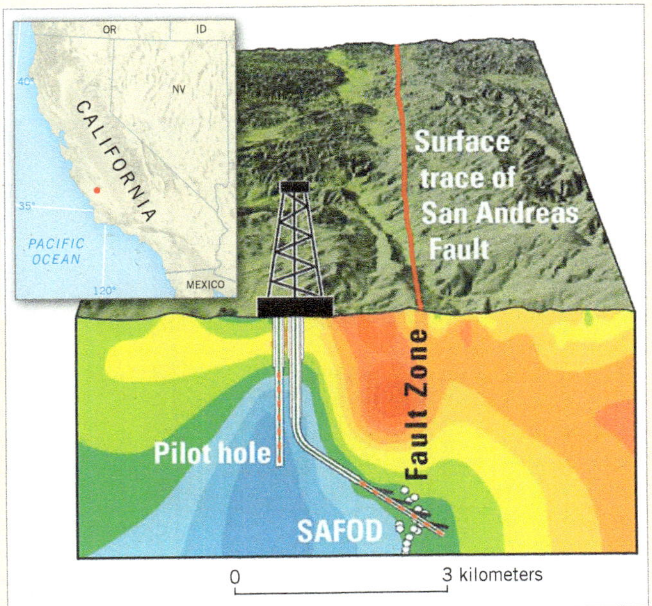

▲ **Figure 14-F** The San Andreas Fault Observatory at Depth (SAFOD) in Parkfield, California. White dots show minor earthquake activity. Subsurface rocks shown in red have low electrical resistivity and may represent fluid-rich areas.

Earthquake Forecasts: Current research has found no reliable method of predicting earthquakes. However, seismologists have made great strides with long-term forecasts. A forecast establishes the probability of an earthquake occurring along a given segment of a fault over a period of several decades. Long-term probabilities are based on the balance of two processes: the loading of strain onto faults and the slip along the faults (which should, over time, release the strain). Such long-term forecasts can help local governments, businesses, and the general public establish priorities for retrofitting structures and for disaster preparation.

The most recent earthquake forecasts for California from the USGS are based on a new model known as the third *Uniform California Earthquake Rupture Forecast* (UCERF3). Developed in 2014, UCERF3 incorporates data on more than 350 fault segments, takes into account the possibility of multifault earthquakes, and includes fault "readiness" estimates. (Faults that ruptured long ago are more ready to rupture than more-recently ruptured faults, which haven't had time to build up as much strain.)

The latest forecast for California has some good news and some bad news: the probability of moderate-size earthquakes has decreased slightly from previous estimates, but the probability of large earthquakes has increased slightly. Over the next 30 years, the probability that California will experience at least one magnitude 6.7 or larger earthquake is greater than 99 percent. In the San Francisco region, the probability is 72 percent; in the Los Angeles region, it is 60 percent. However, Los Angeles has a higher probability than San Francisco of experiencing at least one magnitude 7.5 or larger earthquake over the next three decades.

Seismic Hazards across the United States: The probabilities of earthquakes across the country were updated in 2018 (**Figure 14-G**). The high probabilities in western states are associated with the San Andreas Fault system, the Cascadia subduction zone, and active faulting in the Rockies. High hazards in Alaska are from the Aleutian subduction zone, and those in Hawai'i from ongoing volcanism on the Big Island. The high earthquake hazard in the Midwest is associated with the New Madrid Fault zone, which produced several of the largest earthquakes in American history in 1811 and 1812. Such intraplate earthquakes are not well understood but probably relate to buried faults associated with an ancient "failed rift" within the usually stable interior of the continent.

Questions

1. What is the importance of "precursors" to earthquake prediction?
2. What kinds of information are used to develop long-term earthquake forecasts?

The EEW system is based on existing seismic monitoring networks. The damaging high-amplitude S waves radiating from an earthquake focus travel more slowly than the P waves; the farther the focus, the longer it takes S waves to arrive. When a seismic network detects a strong earthquake, within about 10 seconds the quake's location and size can be computed. The EEW system then sends out an automatic warning—for people to move to a safe location; for trains to slow down and airports to cancel or divert flights; for fire stations to open their doors and for emergency generators to be started; and for medical personnel to stop surgical procedures.

The amount of warning time depends on the distance to the epicenter. Because seismic waves travel so quickly, locations within about 30 kilometers (20 miles) of the epicenter receive effectively no warning. Locations about 65 kilometers (40 miles) get about a 10-second warning, whereas locations 225 kilometers (140 miles) have a 60-second warning.

Complexities of the Internal Processes—Example of the Northern Rockies

Examining each internal process in turn as we have done is a helpful way to systematize knowledge. In nature, however, these processes are interrelated. As an example, let's consider the development of northwestern Montana that encompasses the spectacular scenery of the Rocky Mountains around Glacier National Park.

Most of the rocks in this region formed in the Precambrian Era (more than 800 million years ago), when much of the area consisted of a large, shallow, seawater-filled trough. Silt and sand washed into this Precambrian sea for millennia, and a vast thickness of sedimentary strata built up. Limestones, shales, and sandstones accumulated as six distinct formations, each with a conspicuous color variation known collectively as the Belt Series.

During this lengthy interval, a vast outpouring of flood basalt issued from fissures in the ocean floor. Further sediments were then deposited atop the lava flow. Igneous intrusions were also injected from time to time, including one large sill and a number of dikes. The igneous activity initiated contact metamorphism, changing mainly limestone to marble and sandstone to quartzite.

After a long gap in the geologic record, during which the Rocky Mountain region was mostly above sea level, the land once again sank below the ocean during the Cretaceous Period (145 million to 66 million years ago) and another thick series of sediments was deposited. A period of mountain building called the *Laramide orogeny* followed. In the Glacier National Park area, the rocks were compressed and uplifted, converting the site of the former sea into a mountainous region.

Along with uplift came extreme lateral pressure from the west (Figure 14-58a), convoluting the gently downfolded strata into a prominent anticline. Continuing pressure then overturned the anticline toward the east. This additional strain eventually caused a vast rupture. The entire block was pushed eastward along one of the greatest thrust faults known, the *Lewis Overthrust*. The faulting forced the Precambrian sedimentary rocks above the Cretaceous strata that underlie the plains to the east by as much as 80 kilometers (about 50 miles)—placing older rock layers on top of much younger strata. The terrain thus produced is referred to as "mountains without roots." Chief Mountain is widely known as a rootless mountain because of its conspicuous location as an erosional outlier (called a *klippe*) east of the main range (Figure 14-58b).

LearningCheck 14-20 **Describe how Chief Mountain, Montana, is associated with faulting.**

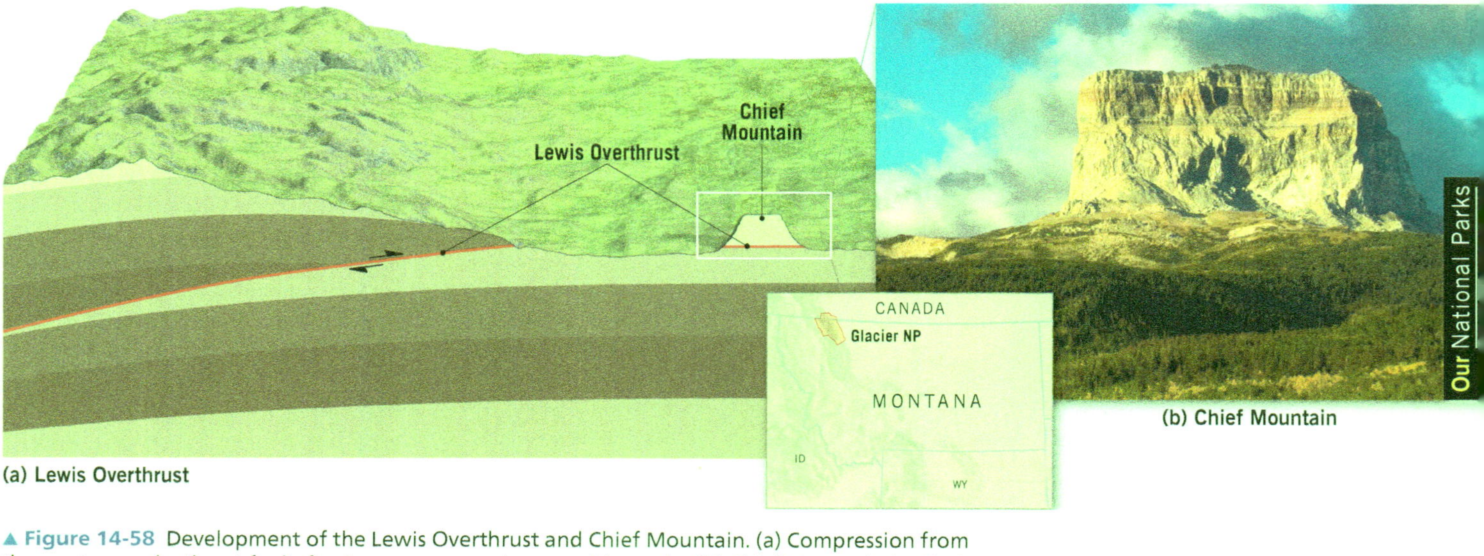

▲ Figure 14-58 Development of the Lewis Overthrust and Chief Mountain. (a) Compression from the west caused a thrust fault, forcing younger rocks over older rocks. (b) Chief Mountain in Glacier National Park, Montana, sits in splendid isolation because of the massive movement associated with the Lewis Overthrust and subsequent erosion.

After studying this chapter, you should be able to answer the following questions. Key terms from each text section are shown in **bold type.** Definitions for key terms are also found in the glossary at the back of the book.

Learning Review 14

Key Terms and Concepts

From Rigid Earth to Plate Tectonics (*p. 406*)

1. Describe and explain at least two lines of evidence that Alfred Wegener used to support his theory of **continental drift** and the existence of the supercontinent of **Pangaea**.
2. Why was Wegener's theory of continental drift rejected for so long?

The Theory of Plate Tectonics (*p. 408*)

3. What lines of evidence confirm that **seafloor spreading** has been taking place? You should be able to explain evidence from both **paleomagnetism** and **ocean floor cores**.
4. Describe and explain the driving mechanism for **plate tectonics**—in other words, explain why and how lithospheric plates move.

Plate Boundaries (*p. 412*)

5. Describe the fundamental differences among **divergent**, **convergent**, and **transform plate boundaries**.
6. Describe and explain the tectonic activity, volcanic activity, and general topographic features associated with the two kinds of divergent plate boundary: **midocean ridges** and **continental rift valleys**. Mention at least one present-day example of each of these kinds of divergent boundaries.
7. Describe and explain **subduction** and the tectonic activity, volcanic activity, and general topographic features associated with the three kinds of convergent plate boundary: *oceanic–continental convergence*, *oceanic–oceanic convergence*, and *continental–continental convergence*. Mention at least one present-day example of each of these kinds of convergent boundaries.
8. Describe the differences between an **oceanic trench** and a midocean ridge.
9. How does the San Andreas Fault system fit in with the theory of plate tectonics?
10. Why are volcanoes and earthquakes concentrated around the margin of the Pacific Ocean—the region referred to as the **Pacific Ring of Fire**?

Additions to Plate Tectonic Theory (*p. 418*)

11. What is a **hot spot**? Name at least one present-day example of a hot spot.
12. How is a *hot spot trail* different from a **volcanic island arc**?
13. How does the **mantle plume** model explain the existence of hot spots?
14. What is a **terrane**, and how does one form?

Volcanism (*p. 421*)

15. What is meant by the term **volcanism**?
16. Define and contrast the following: **magma**, **lava**, and **pyroclastics**.
17. Explain the general differences in silica content and style of volcanic eruption (i.e., effusive lava flows versus explosive eruptions of pyroclastics) associated with basaltic magma, andesitic magma, and rhyolitic magma.
18. What is **flood basalt**?
19. Describe and explain the general formation, shape, and structure of the following kinds of volcanic peaks: **shield volcanoes**, **composite volcanoes**, **lava domes (plug domes)**, and **cinder cones**.
20. Explain the formation of a **caldera** such as Crater Lake in Oregon.

Volcanic Hazards (*p. 430*)

21. Describe and explain the origin and characteristics of **pyroclastic flows** and **volcanic mudflows (lahars)**.

Intrusive Igneous Features (*p. 435*)

22. Define the following terms: **igneous intrusion**, **pluton**, **batholith**, and **dike**.

Tectonism: Folding (*p. 436*)

23. How is **folding** different from **faulting**?
24. What is a **syncline**? What is an **anticline**?

Tectonism: Faulting (*p. 438*)

25. Explain the differences in stress direction and displacement among the four basic kinds of faults: **normal faults**, **reverse faults**, **thrust faults**, and **strike-slip faults**.
26. What is a **fault scarp**?
27. Describe and explain the formation of landforms that result from normal faulting (such as **grabens**, **horsts**, and **tilted fault-block mountains**).
28. Describe and explain the formation of landforms that result from strike-slip faulting (such as **linear fault troughs**, **sag ponds**, and **offset streams**).

Earthquakes (*p. 442*)

29. What is an **earthquake**?
30. What is the difference between the focus of an earthquake and the epicenter of an earthquake?
31. What is the difference between earthquake **magnitude** and earthquake shaking intensity?
32. What does the **modified Mercalli intensity scale** convey?
33. Briefly describe the differences among the *P waves*, *S waves*, and *surface waves* of an earthquake.

Earthquake Hazards (*p. 444*)

34. Under what conditions does **liquefaction** occur?

Complexities of the Internal Processes—Example of the Northern Rockies (*p. 446*)

35. Explain how faulting can leave older rocks resting over younger rocks.

Study Questions

1. What is the general relationship between global earthquake activity and plate boundaries?
2. Why doesn't subduction take place in a continental plate collision zone?
3. Why are deep-focus earthquakes (those originating many hundreds of kilometers below the surface) associated with subduction zones but not with divergent boundaries or transform boundaries?
4. In what ways do hot spots not fit in with the basic theory of plate tectonics and plate boundaries?
5. In what ways have hot spots been used to verify that plate motion is taking place?
6. How is it possible for a syncline to be associated with a topographic ridge and for an anticline to be associated with a topographic valley?
7. Why do areas far away from the epicenter sometimes experience greater damage during an earthquake than areas closer to the epicenter? (Consider factors other than differences in building construction.)

Exercises

1. If the Mid-Atlantic Ridge continues to spread at the same rate into the future (about 2.5 centimeters [1 inch] per year), how long would it take for the Atlantic Ocean to become 1 kilometer wider? ______ years
2. Assuming that the long-term rate of displacement along the San Andreas Fault (about 3.5 cm/year [1.4 inches/year]) remains the same and that the location of the fault remains the same, how long would it take Los Angeles (on the west side of the fault) to move north 620 kilometers (385 miles) to the position of San Francisco (on the east side of the fault)?
3. If a lahar is traveling down a stream valley at a speed of 30 km/hour [18.6 miles/hour], how long will it take to reach a town 10 kilometers (6.2 miles) away?
4. About how much more energy is released during a magnitude 6 earthquake than during a magnitude 5 earthquake?
5. About how much more energy is released during a magnitude 8 earthquake than during a magnitude 5 earthquake?

Environmental **Analysis** RECENT EARTHQUAKES

Data
Recent Quakes
https://media.pearsoncmg.com/ph/esm/esm_mcknight_physgeo_13/media/

As tectonic plates move across Earth, stress builds up. This stress is released in the form of earthquakes, which occur around the world every day. We can use the location and magnitude of earthquakes to learn more about plate boundaries.

Activities

Go to the USGS Earthquake Hazards Program's "Latest Earthquakes" map at http://earthquake.usgs.gov/earthquakes/map. Zoom in to view only the contiguous United States. The earthquakes displayed on the left are only the ones displayed on your current map. Use the + and − to zoom in or out.

1. When and where was the most recent earthquake of magnitude 2.5 or greater? What was its magnitude?

Using the Settings icon, select "1 Day, All Magnitudes U.S." Click the icon again to close the settings panel.

2. How many total earthquakes occurred in the conterminous United States in one day?

3. Zoom out to the world view. How many earthquakes occurred globally in one day?

Using the Settings menu, for the last seven days worldwide, compare the number of earthquakes of all magnitudes with those having magnitudes greater than 4.5.

4. How many earthquakes of all magnitudes occurred over the last seven days? How many with magnitudes greater than 4.5?
5. What can you say about the relationship between the magnitude of earthquakes and their frequency?

Using the Settings icon, compare the locations of earthquakes having magnitudes greater than 4.5 with those greater than 2.5 for the past 30 days. Refer to Figure 14-11 for plate boundary types.

6. Along which type of boundary do most strong earthquakes occur?
7. Why might that type of boundary produce more frequent strong earthquakes than the boundary along the coast of California?
8. Notice that divergent boundaries appear to have the fewest earthquakes. One explanation is that these boundaries produce the fewest earthquakes. What might be another explanation?
9. In the United States, where do frequent earthquakes occur that are not close to a plate boundary, and what might cause them? Why are there many earthquakes on the Big Island of Hawai'i?

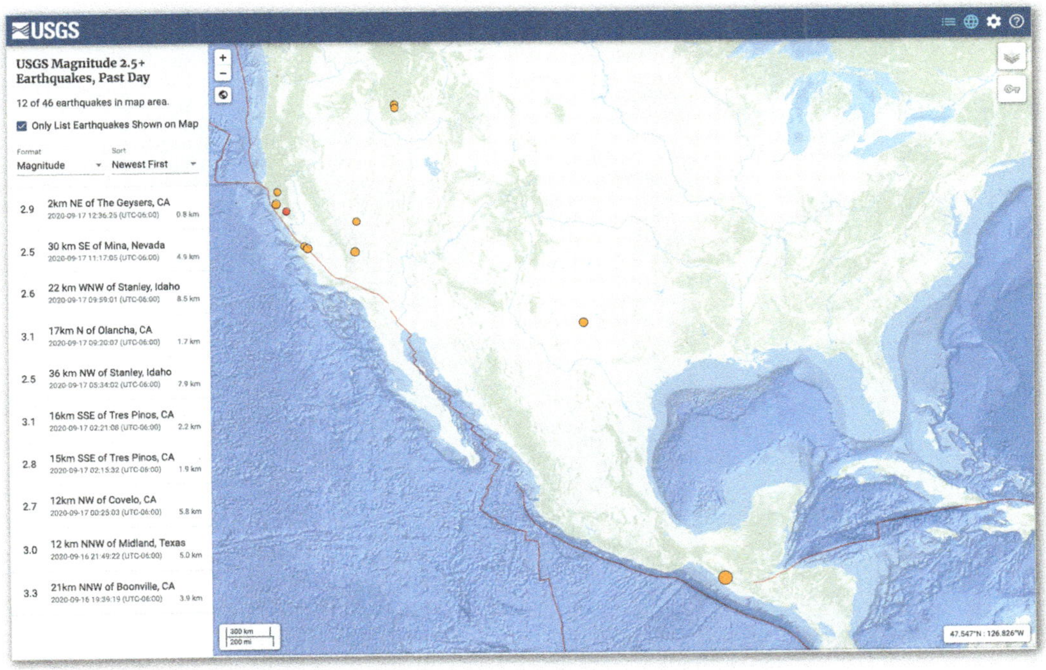

Seeing Geographically in Our National Parks

Look again at the eruption of Kīlauea Volcano on the island of Hawai'i at the beginning of the chapter (p. 404). Which type of volcano is it likely to be? Why do you say this? What kinds of volcanic hazards are associated with such a volcano? What explains the volcanic activity on this Hawaiian island?

Mastering Geography®

Looking for additional review and test prep materials? Visit the Study Area in **Mastering Geography** to enhance your geographic literacy, spatial reasoning skills, and understanding of this chapter's content by accessing a variety of resources, including Map Master interactive maps, geoscience animations, *Mobile Field Trips*, videos, *Project Condor* Quadcopter videos, *In the News* RSS feeds, flashcards, web links, self-study quizzes, and an eText version of *McKnight's Physical Geography*.

Seeing Geographically in Our National Parks

Thunderstorm over Conquistador Aisle in the Grand Canyon, Arizona. What kinds of structural features are most obvious in the rock? Which class of rocks is most clearly exposed? Has bedrock been removed from this landscape? How do you know? Why is it likely that both internal and external processes helped shape the Grand Canyon?

Grand Canyon National Park, AZ

Weathering and Mass Wasting

15

HAVE YOU EVER WONDERED how the Grand Canyon formed? The obvious answer might seem to be that it was carved by the Colorado River. But that's not quite right. The river certainly played a role, but the canyon is actually the outcome of a wide range of actions that we collectively call the *external processes*.

If the internal processes of landscape formation are awe-inspiring, the external processes are inexorable. All the while that lithospheric plates are moving, the crust is folding and faulting, volcanoes are erupting, and intrusions are invading country rock, another suite of natural processes is at work. These are the external processes, often mundane in contrast to those described in the preceding chapter: *weathering* breaks down bedrock, *mass wasting* moves weathered rock downslope, and *erosion* transports that rock away. Yet the cumulative effect of the external processes is awesome; they are capable of wearing down anything the internal processes can erect. No rock is too resistant and no mountain too massive to withstand their unrelenting power. Ultimately, the details of peaks, slopes, valleys, and plains are created by the work of gravity, water, wind, and ice.

Although weathering and mass wasting, the topics of this chapter, are often de-emphasized as processes of landform modification and shaping, the role they play is critical. The formation of the Grand Canyon—one of the most impressive landscape features in the world—is usually attributed only to the erosive powers of the Colorado River. However, weathering and mass wasting have produced most of the structures we see in the Grand Canyon. Besides deepening its channel, the river's main role is to transport away the sediment loosened by weathering and pulled into the stream by gravity and mass wasting.

As you study this chapter, think about these **Key**Questions:

- **How do *weathering*, *mass wasting*, and *erosion* change the landscape together?**
- **Why are *joints* important to weathering processes?**
- **What is *mechanical weathering*?**
- **What is *chemical weathering*?**
- **What are the four main types of mass wasting?**

The Impact of Weathering and Mass Wasting on the Landscape

The most readily observable landscape effect of weathering is the fragmentation of rock—the reduction of large rock units into more numerous and less cohesive smaller units. Mass wasting involves the downslope movement of this weathered rock. Its most obvious mark on the landscape is normally twofold: an open scar may be left on the surface where the movement began, and an accumulation of debris is deposited somewhere downslope from the scar.

Denudation

The overall effect of the disintegration, wearing away, and removal of rock material is generally referred to as **denudation**: the processes that wear down the landscape. Denudation is accomplished by the interaction of three types of activities (Figure 15-1):

1. **Weathering** is the breaking-down of rock into smaller pieces by atmospheric and biotic action.
2. **Mass wasting** involves the relatively short-distance downslope movement of broken rock material under the direct influence of gravity.
3. **Erosion** entails the removal, transportation, and eventual deposition of fragmented rock material over wider areas and sometimes to greater distances than is the case for mass wasting.

LearningCheck 15-1 What is the main effect of weathering on the landscape? Of mass wasting? (Answer on p. AK-5)

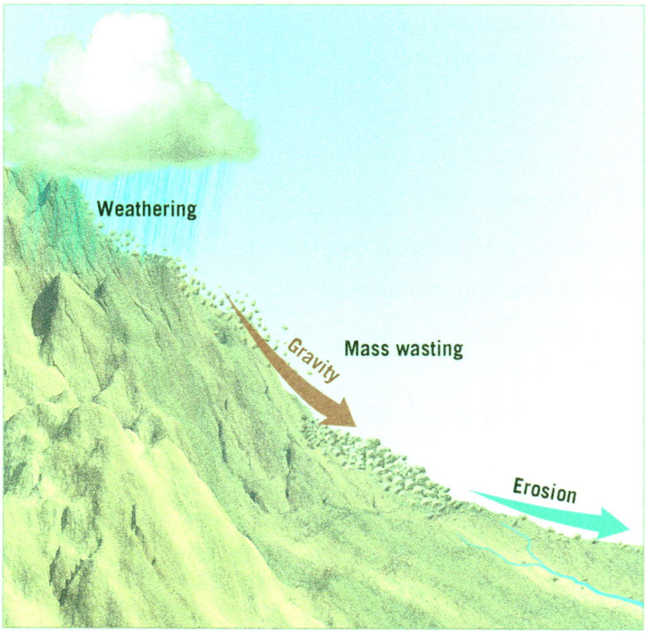

▲ Figure 15-1 Denudation, the lowering of continental surfaces, is accomplished by a combination of weathering, mass wasting, and erosion.

Weathering and Rock Openings

The first step in shaping Earth's surface by external processes is *weathering* or "rock decay," the mechanical disintegration or chemical decomposition that begins to fragment rock into progressively smaller pieces. Weathering occurs wherever the lithosphere and atmosphere meet. It can occur with great subtlety, however—breaking chemical bonds, separating mineral grain from mineral grain, pitting smooth surfaces, and fracturing solid rock. It is the aging process of rock surfaces, the process that must precede the other, more active, forms of denudation.

Significance of Rock Openings: Whenever bedrock is exposed, it weathers. Weathered rock often has a different color or texture from neighboring unexposed bedrock. Most significant from a topographic standpoint, exposed bedrock is likely to be looser than the underlying rock. Blocks or chips may be so loose that they can be detached with little effort. Sometimes pieces are so "rotten" that they can be crumbled by finger pressure. Slightly deeper in the bedrock, firmer, more solid rock is present, although along cracks or crevices weathering may extend as much as several hundred meters beneath the surface. This penetration is made possible by open spaces in the rock bodies and even between the mineral grains. Subsurface weathering is initiated along these openings, which such weathering agents as water, air, and plant roots can penetrate. As time passes, the weathering effects spread from the immediate vicinity of the openings into the denser rock beyond (Figure 15-2).

Openings in the rock surface, whether microscopic or huge, provide avenues along which weathering agents can attack the bedrock and break it apart.

LearningCheck 15-2 Why are rock openings important in weathering processes?

Types of Rock Openings

Broadly speaking, five types of rock openings are common:

1. **Microscopic openings:** Microscopic openings in the rock surface occur in abundance. They may consist of spaces between crystals of igneous or metamorphic rocks, pores between grains of sedimentary rocks, or minute fractures within or alongside mineral grains.
2. **Joints:** The most common structural features of the rocks of the lithosphere are **joints**—cracks that develop as a result of stress—but the rocks do not show noticeable displacement along these breaks. Joints are innumerable in most rock masses, dividing them into blocks of various sizes. Because of their ubiquity, joints are among the most important of all rock openings in facilitating weathering.

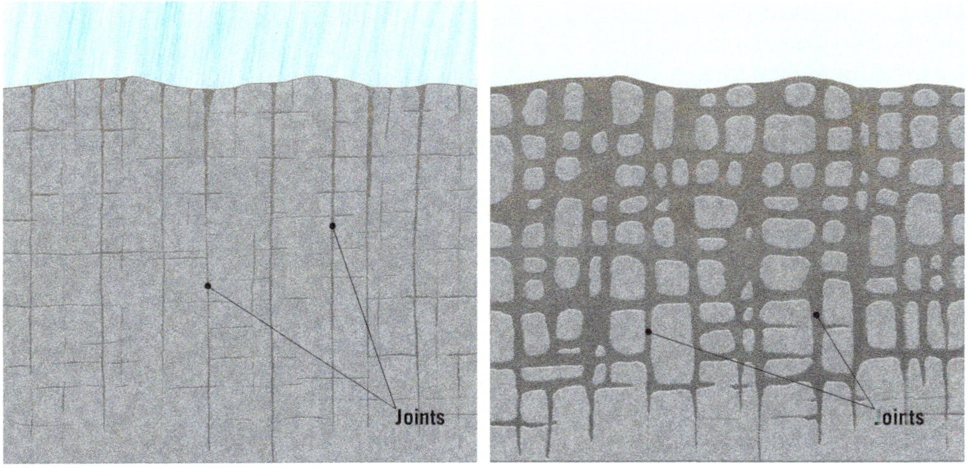

(a) Before weathering (b) After weathering

◀ **Figure 15-2** Weathering below the surface can take place in bedrock containing many openings such as joints.

3. **Faults:** As we saw in Chapter 14, **faults** are breaks in bedrock along which there has been forced displacement of the rock structure (**Figure 15-3**). Faults generally occur individually but may appear as major landscape features, extending for tens or even hundreds of kilometers, whereas joints are typically more numerous and are normally minor structures extending only a few meters. Faults, like joints, allow easy penetration of weathering agents into subsurface rock.
4. **Lava vesicles:** Lava *vesicles* are holes of various sizes, usually small, that develop in cooling lava when gas is unable to escape as the lava solidifies.
5. **Solution cavities:** *Solution cavities* are holes formed in carbonate rocks (particularly limestone) as the soluble minerals are dissolved and carried away by percolating water. Most solution cavities are small, but sometimes huge holes and even massive caverns are created when large amounts of solubles are removed. (Solution processes and topography are discussed further in Chapter 17.)

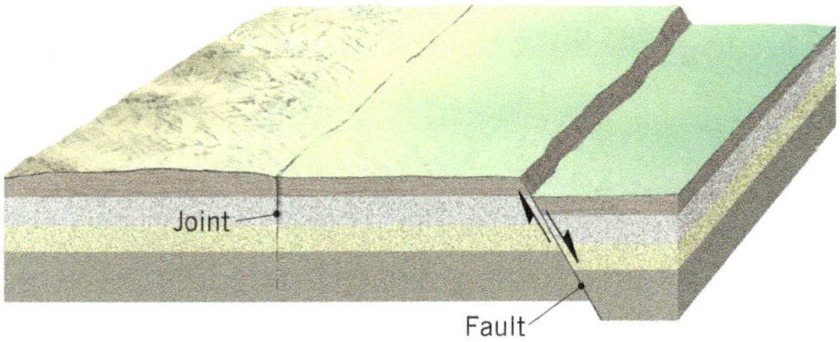

▲ **Figure 15-3** The essential difference between joints and faults is that joints exhibit no displacement along either side of the break.

The Importance of Jointing

Almost all lithospheric bedrock is jointed, resulting sometimes from the contraction of molten material as it cools, sometimes from contraction of sedimentary strata as they dry, and sometimes from tectonic stresses. At the surface, a joint may be quite conspicuous because weathering tends to increase the size of the fracture. Below the surface, however, the visible separation is minimal.

Joints are relatively common in most rock, but they are clearly more abundant in some places than in others. Where numerous, joints are usually arranged in sets, each set part of a series of approximately parallel fractures (**Figure 15-4**).

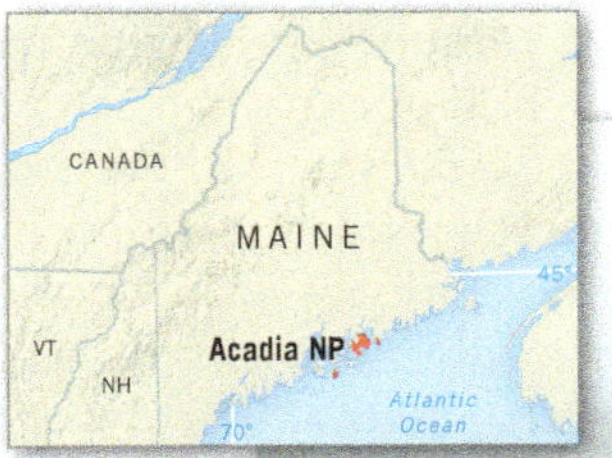

Condor Video
Joints and Faults
https://media.pearsoncmg.com/ph/esm/esm_mcknight_physgeo_13/media/

▶ **Figure 15-4** Jointing in the shoreline rocks of Acadia National Park in Maine.

Frequently, two prominent sets intersect almost at right angles; such a combination constitutes a *joint system*. A well-developed joint system, particularly in sedimentary rock having prominent natural bedding planes, can divide stratified rock into a remarkably regular series of close-fitting blocks. In general, jointing is more regularly patterned, and the resulting blocks are more sharply defined, in fine-grained rocks than in coarse-grained ones.

Large joints or joint sets may extend for long distances and through a considerable thickness of rocks; these are termed **master joints**. Master joints play a role in topographic development by functioning as a plane of weakness, a plane more susceptible to weathering and erosion than the rock around it. Thus, the location of large features of the landscape, such as valleys and cliffs, may be influenced by the position of master joints (see the box, *Snapshots from Our National Parks: Landscapes Shaped by Jointing*).

LearningCheck 15-3 Explain the difference between a joint and a fault.

Weathering Agents

As the term *weathering* suggests, most weathering agents are atmospheric in origin. The atmosphere penetrates readily into all cracks and crevices in bedrock. Oxygen, carbon dioxide, and water vapor are the three atmospheric components of greatest importance in rock weathering.

Temperature changes are a second important weathering agent. Most notable, however, is liquid water, which can penetrate downward effectively into openings in the bedrock. Biotic agents also contribute to weathering, in part through the burrowing activities of animals and the growth of plant roots, but especially through the production of chemical substances that can decompose the rock.

The total effect of these agents is complicated and is influenced by a variety of factors. For analytical purposes, however, it is convenient to recognize three principal categories of weathering: *mechanical*, *chemical*, and *biological*. Although we now consider each of them in turn, keep in mind that they often act together.

Mechanical Weathering

Animation **Mechanical Weathering** https://media.pearsoncmg.com/ph/esm/esm_mcknight_physgeo_13/media/

Mechanical weathering (or *physical weathering*) is the physical disintegration of rock material without any change in its chemical composition. In essence, big rocks are mechanically weathered into little ones by various stresses that cause the rock to fracture into smaller and usually angular fragments. Most mechanical weathering occurs at or very near the surface, but under certain conditions it may occur at considerable depth.

Frost Wedging: Probably the most important agent of mechanical weathering is the freeze–thaw action of water.

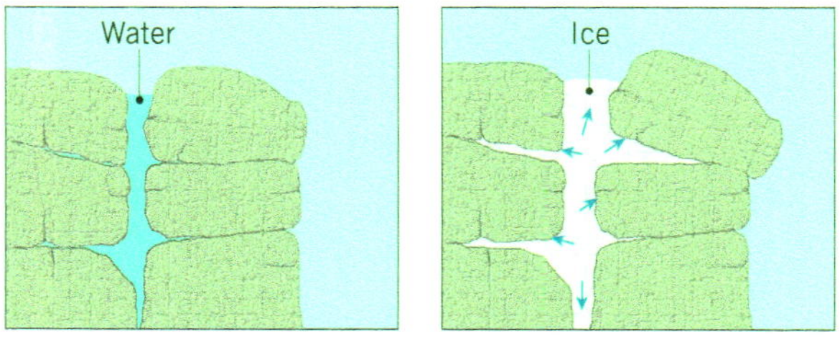

▲ **Figure 15-5** Schematic illustration of frost wedging. When water in a rock crack freezes, the ice expansion exerts a force that can deepen and widen the crack, especially if the process is repeated many times.

As we first saw in Chapter 6, when water freezes, it expands by almost 10 percent. Moreover, the upper surface of the water freezes first, so the principal force of expansion is exerted against the wall of the confining rock rather than upward. This expanding wedge of ice splits the rock (Figure 15-5).

Even the strongest rocks cannot withstand frequent alternation of freezing and thawing. Repetition is the key to understanding the inescapable force of **frost wedging**, or *frost shattering*. If an opening in rock contains water when the temperature falls below 0°C (32°F), ice forms, wedging its way downward, regardless of the opening's size. When the temperature rises above freezing, the ice melts, and the water sinks farther into the slightly enlarged crack. With renewed freezing, the wedging is repeated. It is most noticeable above the treeline of mountainous areas (Figure 15-6), where broken blocks of rock are likely to be seen in abundance everywhere except on slopes that are too steep to allow them to lie without sliding downhill.

Frost wedging in large openings may produce large boulders, whereas that occurring in small openings may granulate the rock into sand and dust particles. Every size gradation in between occurs, too. A common form of breakup in coarse-grained crystalline rocks is a shattering caused by frost wedging between grains. This type of shattering produces gravel or coarse sand in a process termed *granular disintegration*.

Salt Wedging: Related to frost wedging is **salt wedging**, which happens when salts crystallize out of solution as water evaporates. In areas of dry climate, water is often drawn upward in rock openings by capillary action. (*Capillarity* was discussed in Chapter 6.) This water nearly always carries dissolved salts. When the water evaporates, the salts are left behind as tiny crystals. With time, the crystals grow, prying apart the rock grain by grain, much as freezing water does, although less intensely (Figure 15-7).

Salt wedging may also be a weathering factor along ocean coastlines. Above the tideline, seawater from ocean spray gets between mineral grains; after the water evaporates, the growth of salt crystals can slowly pry off mineral grains.

LearningCheck 15-4 Explain the process of frost wedging.

Snapshots From Our National Parks

Zion, Bryce Canyon, and Arches

Landscapes Shaped by Jointing

▲ **Figure 15-A** In Utah's Zion National Park, master joints are widely spaced, allowing for the development of massive blocks and precipitous cliffs.

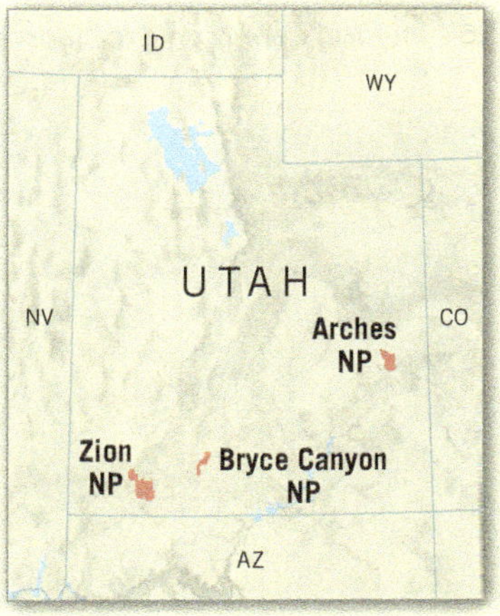

▲ **Figure 15-B** The topography of Bryce Canyon in southern Utah from Inspiration Point. The closely spaced joints and bedding planes contribute to intricate sculpting by weathering and erosion.

Of Utah's many scenic national parks, three of its most popular—Zion, Bryce Canyon, and Arches—owe much of their dramatic landscapes to the same underlying geomorphic factor. Although the parks are quite different in their overall appearance, the spacing and orientation of joints in the bedrock have strongly influenced the shaping of the topography of each.

The rocks exposed in Bryce Canyon and Zion are part of the Grand Staircase—a thick sequence of sedimentary rock layers stretching from Utah to Arizona that were uplifted with the Colorado Plateau, then tilted and eroded, exposing a nearly continuous sequence of geologic history going back 650 million years. Going "down" the staircase from its top layers at Bryce Canyon, we pass through layers that also extend into Zion and finally into the Grand Canyon. The rocks in the bottom (and so oldest) rock layer exposed in Bryce Canyon, are found in the *top* layer of Zion Canyon—and the rocks in Zion's bottom-most layer are exposed in the top layer of the Grand Canyon (see the feature on the Grand Canyon in Chapter 13).

Zion The magnificent sandstone cliffs of Zion Canyon were carved by the Virgin River as the entire region was slowly uplifted. The near vertical walls we see today developed out of massively jointed rock. Major fractures run through the bedrock—the widely spaced breaks in the cliffs here occur along such *master joints* (**Figure 15-A**). The lack of closely spaced joints prevented the easy weathering and erosion of the rock, leaving the steep, uninterrupted walls of Zion Canyon.

Bryce Canyon On the other hand, the mudstones, thinly bedded sandstones, and conglomerates found in Bryce Canyon have been broken by a myriad of closely spaced vertical joints (**Figure 15-B**). Although called a "canyon," the amphitheatre of Bryce Canyon was not carved by streams. Rather, weathering of the rock along these joints has left the hundreds of tall, thin towers or *hoodoos* for which Bryce Canyon is famous.

Arches In Arches National Park, jointing also helped shape a remarkable landscape that includes more than 2000 natural rock arches. Deep deposits of salt and gypsum bulged up an anticline running northwest–southeast through the region, fracturing the red-stained Entrada Sandstone and producing a set of closely spaced vertical joints. As the sandstone weathers along these parallel joints, narrow walls or "fins" of rock remain. Over time, a small "window" may weather completely through such a wall. As the window slowly enlarges, an arch develops (**Figure 15-C**). Even today the arches in the Park continue to change. Periodically rock falls from an arch, and sometimes an arch becomes weak enough to suddenly collapse—as did Wall Arch in August 2008.

▼ **Figure 15-C** Delicate Arch in Utah's Arches National Park formed out of a thin "fin" of sandstone remaining between two parallel vertical joints.

Questions

1. How did jointing influence the development of the widely different appearances of Zion Canyon and Bryce Canyon?
2. Explain the sequence of development of a sandstone arch in Arches National Park.

NPS Video
Geology of Arches
https://media.pearsoncmg.com/ph/esm/esm_mcknight_physgeo_13/media/

◀ **Figure 15-6** Frost wedging is especially pervasive on mountaintops above the treeline, as with these rocks near the summit of Glyder Fach mountain in northwestern Wales.

Temperature Changes: Temperature changes not accompanied by freeze–thaw cycles may also weather rock mechanically, but they do so much more gradually than the processes just described. The fluctuation of temperature from day to night and from summer to winter can cause minute changes in the volume of most minerals, forcing expansion when the rock that contains them is heated and contraction when the rock is cooled, and weakening the coherence of the mineral grains. Many thousands of repetitions are normally required for much weakening or fracturing to occur. This factor is most significant in arid areas and near mountain summits, where direct solar radiation is intense during the day and radiational cooling is prominent at night.

Probably more important but less widespread than simple solar heating and cooling is heating from forest fires or brushfires. The intense heating of a fire can cause a rock to expand and break apart.

▼ **Figure 15-7** Salt wedging. This crystalline rock is being shattered by salt wedging in the desert near the floor of Death Valley in California.

Exfoliation: One of the most striking of all weathering processes is **exfoliation**, in which curved layers peel off bedrock. Curved and concentric sets of joints develop in the bedrock, and parallel shells of rock break away in succession, somewhat analogous to the way layers of an onion separate. The sheets that split off are sometimes only a few centimeters thick; in other cases, however, they may be several meters thick.

The results of exfoliation are conspicuous. If the rock mass is a large one, such as Half Dome or one of the other granitic monoliths overlooking Yosemite Valley, California, its surface consists of imperfect curves punctuated by several partially fractured shells of the surface layers, and the mass is referred to as an **exfoliation dome** (Figure 15-8a). Overall, especially in regions of exposed plutonic bedrock, exfoliation tends to smooth the landscape gently.

The dynamics of exfoliation are not fully understood. The most widely accepted explanation of massive exfoliation is that the rock cracks after an overlying weight has been removed, a process called *unloading* or *pressure release* (Figure 15-8b). The intrusive bedrock may originally have been deeply buried beneath a heavy overburden—perhaps several kilometers deep. When the overlying material is stripped away by erosion, the release of pressure allows expansion in the rock. The outer layers cannot contain the expanding mass, and the expansion can be absorbed only by cracking along the sets of *sheeting joints* (Figure 15-8c).

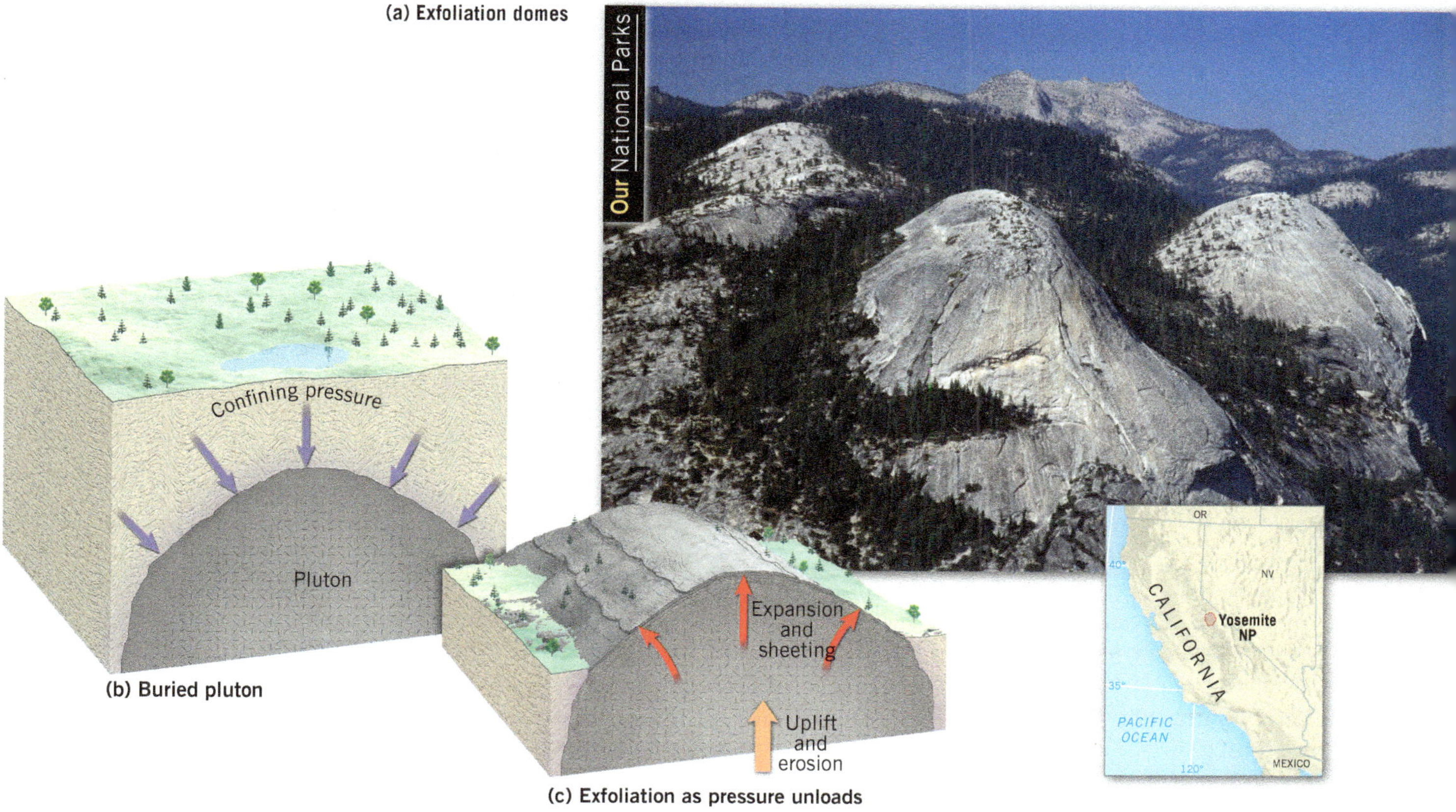

▲ **Figure 15-8** Exfoliation. (a) Two large exfoliation domes in Yosemite National Park, California. (b) When formerly buried rocks such as granite are exposed at the surface, (c) the unloading of confining pressure causes slight expansion of the rock mass. As a consequence, curved layers of rock peel off in the process of exfoliation.

Exfoliation occurs mainly in granite and related intrusive rocks, but under certain circumstances it is seen in sandstone and other sedimentary strata.

Recent research has suggested an alternative explanation of exfoliation in some locations: it takes place when convex exposures of strong rock (such as granite) are subjected to significant sideways compression. As the rock is compressed, vertical tension within the rock mass is enough to cause joints just below the convex surface to "pop open" parallel to the surface. Further research may clarify if this mechanism is more important than simple unloading of pressure.

LearningCheck 15-5 **Explain the formation of exfoliation domes.**

Other Mechanical Weathering Processes: Chemical changes may also contribute to mechanical weathering. Various chemical actions can cause an increase in volume of the affected mineral grains. This swelling sets up strain that weakens the coherence of the rock and causes fractures.

Some biotic activities also contribute to mechanical weathering. Most notable is the penetration of growing plant roots into cracks and crevices, which exerts an expansive force that widens the openings. This factor is especially obvious where trees grow out of joint or fault planes, with their large roots showing amazing tenacity and persistence as wedging devices (**Figure 15-9**). In addition, burrowing animals sometimes cause rock disintegration.

When acting alone, mechanical weathering breaks up rock masses into ever smaller (and often angular) pieces, producing boulders, cobbles, pebbles, sand, silt, and clay-sized particles. As more and more rock surface area is exposed over time, the process proceeds at an accelerating rate (**Figure 15-10**).

Chemical Weathering

Mechanical weathering is frequently accompanied by **chemical weathering**, which is the decomposition of rock by the chemical alteration of its minerals. Almost all minerals are subject to chemical alteration when exposed to atmospheric and biotic agents. Some minerals, such as quartz, are extremely resistant to chemical change, but many others are very susceptible. Very few rocks cannot be significantly affected by chemical weathering because the alteration of even a single significant mineral constituent can lead to the eventual disintegration of an entire rock mass.

One important effect of mechanical weathering is to expose bedrock to the forces of chemical weathering. The greater the surface area exposed, the more effective the

(a) Small-scale wedging by roots

(b) Large-scale wedging by roots

▲ **Figure 15-9** Roots can serve as formidable natural tools to enlarge cracks and crevices in bedrock. In California's Desolation Valley: (a) a small plant growing out of a rock crack; (b) a full-sized lodgepole pine tree growing out of a joint in the granitic bedrock.

chemical weathering. Thus, finer-grained materials, with more exposed surface area, decompose more rapidly than coarser-grained materials of identical composition (see Figure 15-10).

Virtually all chemical weathering requires moisture. Thus, an abundance of water enhances the effectiveness of chemical weathering, and chemical processes operate more rapidly in humid climates than in arid areas. Moreover, chemical reactions proceed more rapidly at higher temperatures. Consequently, chemical weathering is most efficient and conspicuous in warm, moist climates. In cold or dry lands, mechanical weathering tends to dominate.

Some of the chemical reactions that affect rocks are very complex, but others are simple and predictable. The principal reacting agents are oxygen, water, and carbon dioxide, and the most significant processes are *oxidation*, *hydration*, *hydrolysis*, and *carbonation*. These processes often take place more or less simultaneously, largely because they all involve water. Water percolating into the ground acts as a weak acid because it contains dissolved gases and decay products from the local vegetation. The presence of these impurities increases the water's capacity to drive chemical reactions.

LearningCheck 15-6 Describe the main difference between mechanical weathering and chemical weathering.

Oxidation: When the oxygen dissolved in water comes in contact with certain minerals in rock, the minerals undergo **oxidation**. In this process, the oxygen atoms combine with atoms of various metallic elements making up the minerals, and new products form. The new substances are usually more voluminous, softer, and more easily removed than the original compounds.

When iron-bearing minerals react with oxygen (in other words, become oxidized), *iron oxide* is produced:

$$\underset{\text{Iron}}{4Fe} \quad + \quad \underset{\text{Oxygen}}{3O_2} \quad \rightarrow \quad \underset{\text{Iron Oxide (Hematite)}}{2Fe_2O_3}$$

This reaction, probably the most common oxidation in the lithosphere, is called *rusting*. The prevalence of rusty red

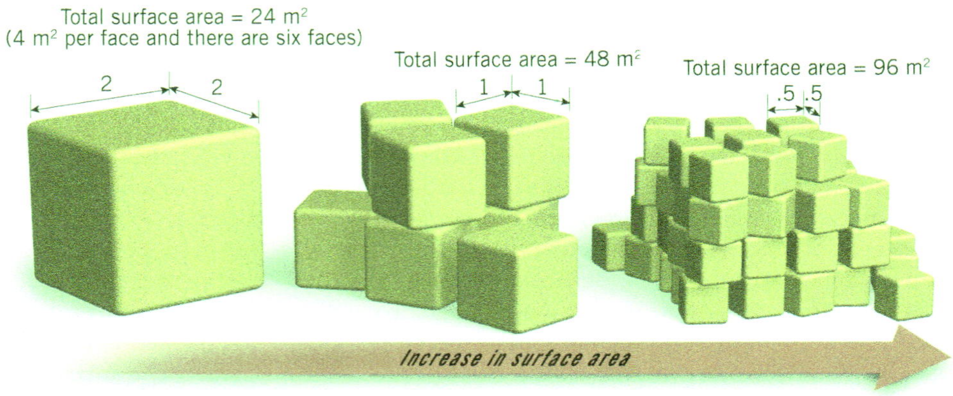

▶ **Figure 15-10** As mechanical weathering fragments rock, the amount of surface area exposed to further weathering increases. Each successive step shown doubles the surface area of the preceding step.

▶ **Figure 15-11** Iron oxide (rust) stains on a sandstone cliff in Capitol Reef National Park in Utah.

stains on the surface of many rocks attests to its widespread occurrence (**Figure 15-11**). Similar effects are produced by the oxidation of aluminum. Because iron and aluminum are very common in Earth's crust, a reddish-brown color is seen in many rocks and soils, particularly in tropical areas, where oxidation is the most notable chemical weathering process. Rusting contributes significantly to weathering because oxides are usually softer and more easily removed than the original iron and aluminum compounds from which the oxides were formed.

Hydration: In **hydration**, water is added to a compound and becomes part of its composition without breaking up the compound. Minerals that have incorporated water into their structure are called *hydrates*. For example, when calcium sulfate ($CaSO_4$) undergoes hydration, it forms gypsum ($CaSO_4 \cdot 2H_2O$). Hydration increases the volume of the mineral, and this expansion can contribute to mechanical disintegration of rock.

Hydrolysis: Conversely, in **hydrolysis**, water is added to a compound and breaks it up. This produces a new compound that is nearly always softer and weaker than the original. Silicate minerals, such as the feldspars in igneous rocks, commonly undergo hydrolysis. Silicate minerals can react with the normally slightly acidic rainwater, forming clay minerals (such as kaolinite) and leaving behind resistant particles of silica (quartz). In tropical areas, where water frequently percolates to considerable depth, hydrolysis often occurs far below the surface.

Carbonation: When carbon dioxide is dissolved in water, *carbonic acid* (H_2CO_3) forms. In **carbonation**, the carbonic acid reacts with carbonate rocks such as limestone, producing the very soluble product calcium bicarbonate. Calcium bicarbonate is readily removed by runoff or percolation and can also be deposited in crystalline form if the water evaporates. We discuss this process known as *dissolution* in greater detail in Chapter 17.

Spheroidal Weathering: Especially due to chemical weathering, very thin rock layers—usually only a few millimeters thick—peel off exposed boulders, gently rounding them over time (**Figure 15-12**). This process is called *spheroidal weathering*. Spheroidal weathering is primarily the outcome of chemical weathering processes such as hydration, which swells minerals and weakens the rock surface.

Chemical weathering proceeds continuously at and beneath Earth's surface. Most chemically weathered rocks are changed physically: they are weakened, and the loose particles produced at the surface are unlike the parent material. Beneath the surface, the rock holds together—but in a chemically altered condition. The eventual products of chemical weathering are most often clays.

▼ **Figure 15-12** Spheroidal weathering. A thin layer of rock is peeling from the surface of this granite boulder in California's Joshua Tree National Park.

LearningCheck 15-7 How are oxidation and carbonation similar? How are they different?

Biological Weathering

Plants and animals occasionally contribute to weathering; such processes that involve living organisms are called **biological weathering**. Most notable is the penetration of growing plant roots into cracks and crevices (see Figure 15-9).

Lichens are primitive organisms that consist of algae and fungi living as a single unit. Typically they live on bare rock, bare soil, or tree bark (Figure 15-13). They draw minerals from the rock by ion exchange, and this leaching can weaken the rock. Moreover, as lichens get alternately wet and dry, expansion and contraction of the symbiotic pair flake off tiny particles of rock.

Burrowing by animals mixes soil effectively and is sometimes a factor in rock disintegration. (For example, see Figure 12-5 in Chapter 12.)

Differential Weathering

As we've just seen, all rock does not weather at the same rate or to the same extent. Some kinds of rock are relatively weak and easily weathered, whereas others are strong and more resistant to weathering. This is the simple but crucial concept in geomorphology of **differential weathering**. As we see in the chapters that follow, differential weathering often leaves a prominent mark in the landscape: exposures of weaker rock are more obviously susceptible to mass wasting and erosion than exposures of stronger rock.

Although the inherent strength of the rock significantly influences differential weathering, the local environment also plays a role. For example, rock that resists weathering in an arid environment might be relatively "weak" in a humid environment. Climate is one of the key variables influencing weathering.

LearningCheck 15-8 Explain the concept of differential weathering.

Climate and Weathering

In general, weathering—particularly chemical weathering—is enhanced by a combination of high temperatures and abundant precipitation. Moisture is usually more important than temperature. For example, in most desert regions (see Chapter 18), due to a general lack of precipitation, mechanical weathering may be more conspicuous than chemical weathering. The climatic regime significantly influences patterns of soil development (see Chapter 12).

The connection between weathering and climate has many variations (Figure 15-14). All else being equal, the depth of active weathering tends to be relatively shallow in regions of tundra and desert but relatively deep in regions of tropical rainforest.

▲ Figure 15-13 Rocks covered with multicolored lichens in the Lake District of England.

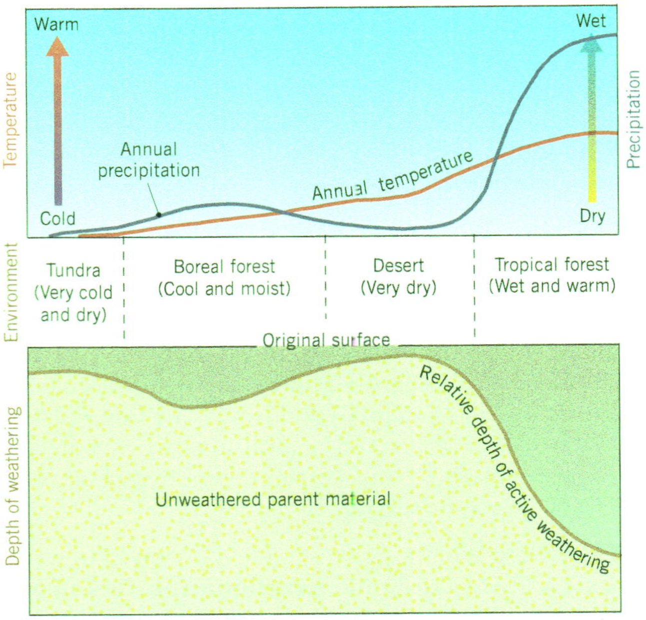

▲ Figure 15-14 Relationship between important climatic elements and depth of weathering The extreme effectiveness of weathering in the humid tropics and the less-effective weathering in the tundra and desert are clearly shown. Humid midlatitude regions are omitted from this diagram because the relationships are more complex there.

Mass Wasting

Animation
Mass Wasting
https://media.pearsoncmg.com/ph/esm/esm_mcknight_physgeo_13/media/

The ultimate destiny of all weathered material is to be carried away by erosion, a topic we cover in the remaining chapters of this book. The remainder of this chapter is concerned with mass wasting (also called *mass movement*), the processes whereby weathered material is moved a relatively short distance downslope under the direct influence of gravity. Although it is sometimes bypassed when erosion acts directly on weathered rock, mass wasting is normally the second step in the three-step denudation process: weathering, mass wasting, and erosion.

Gravity is inescapable; everywhere on the surface it pulls objects toward the center of Earth. Where the land is flat, the influence of gravity on topographic development is minimal. Even on gentle slopes, however, minute effects are likely to be significant in the long run, and on steep slopes the results are often immediate and obvious. Any loosened material is impelled downslope by gravity—in some cases falling abruptly or rolling rapidly, in others flowing or creeping with imperceptible gradualness.

Gigantic boulders respond to the pull of gravity in much the same fashion as particles of dust do, although the larger the object, the more immediate and pronounced the effect. Of particular importance, however, is the implication of "mass" in mass wasting: the accumulations of material moved—fragmented rock, regolith, and soil—are often extremely large and contain enormous amounts of mass. Mass wasting can be a costly or even deadly natural hazard.

Factors Influencing Mass Wasting

A variety of factors determine the characteristics of a mass-wasting event, including the slope and the nature of the material involved.

Angle of Repose: All rock materials, from individual fragments to cohesive layers of soil, lie at rest on a slope if undisturbed unless the slope attains a critical steepness. The steepest angle that can be assumed by loose fragments on a slope without downslope movement is called the **angle of repose** (Figure 15-15). This angle, which varies with the nature and internal cohesion of the material, represents a fine balance between the pull of gravity and the cohesion and friction of the rock material. For example, dry sand has an angle of repose of about 34°, whereas it can be as high as 40° for larger blocks of weathered rock. If additional material accumulates on a debris pile lying on a slope that is near the angle of repose, the newly added material may upset the balance (because the added weight overcomes the friction force that has kept the pile from sliding) and may cause all or part of the material to slide downward.

Water: If water is added to weathered rock through rainfall, snowmelt, or subsurface flow, the rock fragments are more likely to move—particularly if they are small. Water is a "lubricating" medium, and it diminishes friction between particles so that they can slide past one another more readily. Water also adds to the buoyancy and weight of the weathered material, which makes for a lower angle of repose and adds momentum once movement is under way. For this reason, mass wasting is particularly likely during and after heavy rains.

Clay: Another facilitator of mass wasting is *clay*. As noted in Chapter 12, clays readily absorb water. This absorbed water, combined with the fine-grained texture of the material, makes clay very slippery and mobile. Any material resting on clay can often be set in motion by rainfall or an earthquake shock, even on very gentle slopes. Indeed, some clay formations are called *quick clays* because they spontaneously change from a relatively solid mass to a near-liquid condition as the result of a sudden disturbance or shock.

◄ Figure 15-15 The angle of repose of the sand in these dunes in Great Sand Dunes National Park, Colorado, is about 34°.

Permafrost: In subarctic regions and at high latitudes, mass wasting is often initiated by the heaving action of frozen groundwater. The presence of thawed, water-saturated ground in summer overlying permanently frozen subsoil (*permafrost*, discussed in Chapter 9) also contributes to mass wasting in such regions. Some geomorphologists assert that, in the subarctic, mass wasting is the single most important means of transport of weathered material.

LearningCheck 15-9 What is meant by the angle of repose of a rock deposit?

Types of Mass Wasting

Although some types of mass wasting are rapid, others are slow and gradual (Figure 15-16). Here, as a way of highlighting the key characteristics of movement, we group all kinds of mass wasting into just four categories—*fall*, *slide*, *flow*, and *creep*—although in nature the various types often overlap.

Fall

The simplest and most obvious form of mass wasting is **rockfall**, or simply **fall**: the falling of pieces of rock downslope. When loosened by weathering on a very steep slope, a rock fragment may simply be dislodged and fall, roll, or bounce down to the bottom of that slope. This is a very characteristic event in mountainous areas, often as a result of frost wedging.

Pieces of unsorted, angular rock that fall in this fashion are referred to collectively as **talus** or **scree**. (Some geomorphologists use the term *talus* when referring to larger blocks and *scree* when the material is smaller, but the terms are often used interchangeably.) When the fragments accumulate along the base of the slope, the resulting landform is called a *talus slope* or *talus apron*. More characteristically, however, the dislodged rocks collect in sloping, cone-shaped piles called **talus cones** (Figure 15-17a). This cone pattern is commonplace because most steep bedrock slopes and cliffs are seamed by vertical ravines and gullies that funnel the falling rock into piles

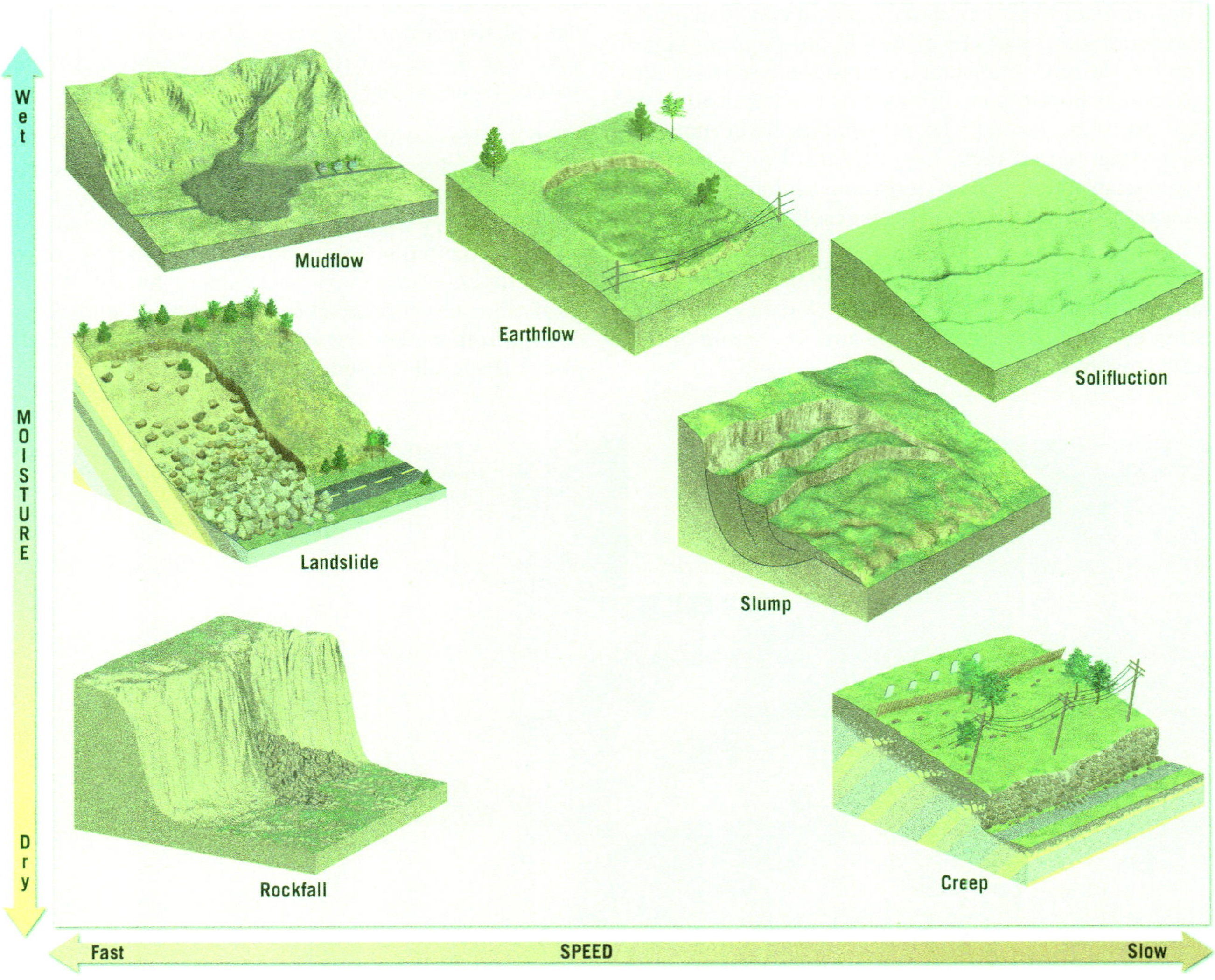

▲ Figure 15-16 Speed and moisture relationships for the various types of mass wasting. For example, *mudflow* entails the rapid movement of very wet material, whereas *rockfall* is the very rapid movement of dry material.

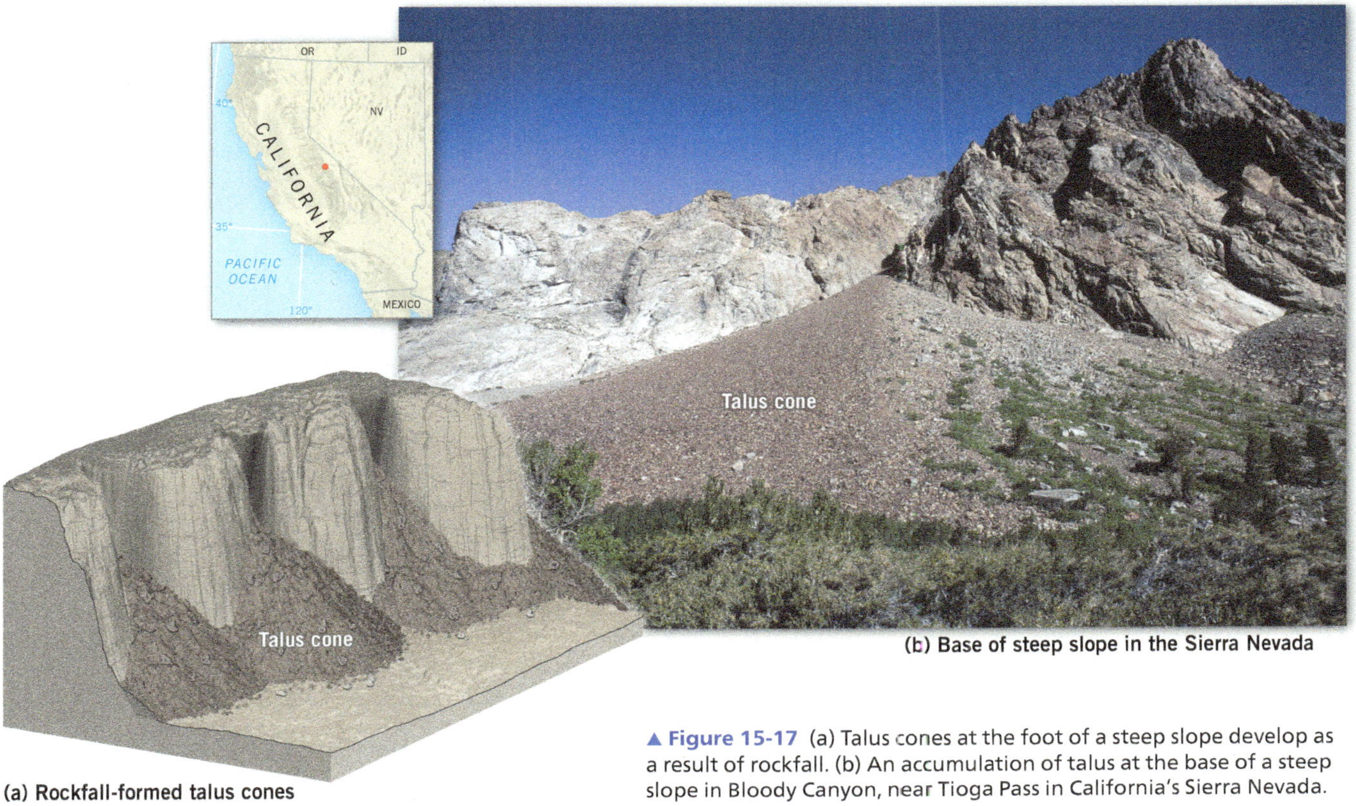

(a) Rockfall-formed talus cones

▲ **Figure 15-17** (a) Talus cones at the foot of a steep slope develop as a result of rockfall. (b) An accumulation of talus at the base of a steep slope in Bloody Canyon, near Tioga Pass in California's Sierra Nevada.

directly beneath the ravines, usually producing a series of talus cones side by side along the base of the slope or cliff. Some falling fragments, especially larger ones with their greater momentum, tumble and roll to the base of the cone. Most of the new talus, however, comes to rest at the upper end of the cone. The cone thereby grows up the mountainside (Figure 15-17b).

The angle of repose for talus is very high, generally about 35° and sometimes as great as 40°. Each new piece that falls onto the talus cone may dislodge material, causing rock to tumble down the cone until it again stabilizes at the angle of repose.

Rock Glaciers: In some mountain areas, great masses of accumulated talus may move slowly downslope under their own weight. As we shall see in Chapter 19, glaciers move in a somewhat similar way, and for this reason these extremely slow-moving masses of talus are called **rock glaciers**. The movement of rock glaciers is primarily the result of gravity, aided by the freeze–thaw cycle, especially if they contain a core of frozen water. Rock glaciers occur primarily in glacial environments; many are found in midlatitude mountains that are relics of periods of glaciation. They are normally found on relatively steep slopes but can extend far downvalley and even out onto an adjacent plain.

In some parts of the world it appears that rockfalls are becoming more common as climate warms (Figure 15-18). Perhaps in part due to the melting of ice deep within talus cones or rock glaciers, talus slopes may become more unstable and susceptible to mass movement—see the box *Global Environmental Change: Are Rockfalls Becoming More Common around the World?*

▲ **Figure 15-18** This rockfall from El Capitan in Yosemite National Park on September 28, 2017, was 10,324 cubic meters in volume, or nearly 28,000 tons.

NPS Video
Rock Fall
https://media.pearsoncmg.com/ph/esm/esm_mcknight_physgeo_13/media/

Learning Check 15-10 How does a talus cone form?

Global Environmental Change

Are Rockfalls Becoming More Common around the World?

Kerry Lyste, Everett Community College

▲ **Figure 15-D** On August 23, 2017, around 3.1 million cubic meters of rock crashed down from Pizzo Cengalo near the Swiss village of Bondo, burying eight hikers.

We could argue that tropical regions are experiencing increased cataclysmic cyclonic events and that polar regions are undergoing severe reductions in the cryosphere. Other effects of global climate change are being seen in temperate regions. For example, a reduction in winter snowfall is resulting in much less spring and summer runoff, whereas increased temperatures in fall and winter seasons are leading to more frequent extreme rainfall events. Although the relationship is unclear, El Niño years as well may be becoming more frequent. Another, perhaps unexpected, consequence of global climate change is an increase in the number of rockfalls around the world.

The Pacific Northwest: Current hydrologic conditions are causing unstable conditions in steep-sloped mountains in the Pacific Northwest of the United States. The rock here is experiencing more prolonged periods of saturation and therefore increased chemical weathering. Due to higher temperatures, these areas may be experiencing less mechanical weathering from frost wedging but more frequent movement of talus from rockfall. One example is the Newhalem rockfall in the North Cascades of Washington state. In 2003, a 750,000-cubic-meter rockfall wiped out a section of highway Route 20 near the town of Newhalem; another rockfall occurred in 2006. These rockfalls were triggered by an increase in groundwater associated with high-rainfall events that weakened weathered rock.

The Alps: Five major rockfall incidents of volume greater than 1 million cubic meters have occurred in the Swiss Alps since 1987 (Figure 15-D) and six events greater than 5000 cubic meters took place in part of the French Alps in 2007 and 2008 alone. In these areas, Alpine permafrost has warmed 0.5–0.8°C (0.9–1.4°F) in the upper parts of mountains for two reasons. First, southern slopes have incurred more summer "shock" from high temperatures (from incoming shortwave radiation); second, northern slopes have encountered gentler thawing from gradual increases in air temperature (associated with outgoing longwave radiation). The result has been the destabilization of large segments of talus slopes due to more active hydrology, exposing joints and crevices in rock slopes and weakening stress points that normally hold exposed faces together (Figure 15-D).

Pakistan: In northern Pakistan, a massive rockfall and landslide buried part of a village and blocked the Hunza River on January 4, 2010. Within months, the blocked river submerged 5 kilometers (3 miles) of the Karakorum Highway. Although the steep incised valleys of this area have long been prone to avalanches, recent increases in spring snowmelt during the rainy season have destabilized the clasts in rock deposits on the steep valley walls. Sudden rock wall failure can result in rapid runout of dry rubble, involving millions of cubic meters of bedrock. Mistaken for glacial outwash in the past, these rockfalls are now being studied to evaluate the risk they pose to the communities located in the river valleys below.

With global temperatures projected to continue to rise over the next century, rockfall events such as these are likely to become even more common.

Questions

1. How does reduced snowfall affect mass wasting activities in the Pacific Northwest?
2. Suggest two ways in which slopes in Switzerland are being destabilized by climate change.
3. How could rockfall events in the Karakorum Mountains be mistaken for glacial outwash?

Slide

In mountainous terrain, landslides carry large masses of rock and soil downslope abruptly and often catastrophically. A **landslide** is any type of *slope failure* involving an instantaneous collapse of a slope and movement along a generally flat sliding plane (Figure 15-19). In other words, the sliding material represents a rigid mass that is suddenly displaced without any fluid flow. Landslides do not require the lubricating effects of water or clay, although the presence of water may contribute to the action; many slides are triggered by rains that add weight to already overloaded slopes. Landslides may be activated by other stimuli as well—notably, by earthquakes (Figure 15-20).

Animation
The Eruption of Mount St. Helens
https://media.pearsoncmg.com/ph/esm/esm_mcknight_physgeo_13/media/

Some slides move only regolith, but many large slides also involve masses of bedrock detached along joint planes and fracture zones. The term *rock avalanche* (or *debris avalanche*) describes slides consisting primarily of rock without large quantities of soil or other fine-grained material.

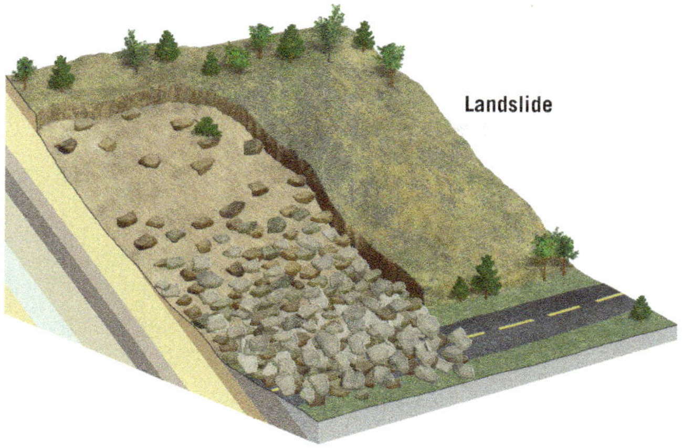

▲ **Figure 15-19** Landslide is a type of slope failure that entails the rapid downslope movement of masses of weathered rock along a flat sliding plane.

Landslide action is not only abrupt but also rapid. Precise measurement of the rate of movement is difficult, but eyewitness accounts report speeds of 160 kilometers (100 miles) per hour. Thunderous noise accompanies the slide, and the blasts of air that the slide creates can strip leaves, twigs, and even branches from nearby trees.

The immediate topographic result of a landslide may be threefold. First, on the hill where the slide originated, there is a deep and extensive *scar*, usually exposing a mixture of bedrock and scattered debris.

Second, because most landslides occur in steep, mountainous terrain, the great mass of material (displaced volume is sometimes measured in cubic kilometers) that roars downslope may choke the valley at the bottom with debris. Its surface consists of a jumble of unsorted material, ranging from immense boulders to fine dust. Moreover, the momentum of the slide may push material several hundred meters up the slope on the other side of the valley.

▲ **Figure 15-20** Park staff gathering GPS data in 2019 of the "Pretty Rocks" landside along the Denali National Park Road, Alaska.

Finally, in the valley bottom where the slide material comes to rest, a natural dam across the width of the valley may form, blocking the valley-bottom stream. That produces a new lake, which becomes larger and larger until it either overtops the dam or cuts a path through it (**Figure 15-21**).

Slump: An extremely common form of mass wasting is the type of slide called a **slump**. Slumping involves slope failure in which the rock or regolith moves downward and at the same time rotates outward along a curved slide plane that has its concave side facing upward (**Figure 15-22**). (For this reason, a slump is sometimes called a *rotational slide*.) The upper portion of the moving material tilts down and back, and the lower portion moves upward and outward. The top of the slump is usually marked by a crescent-shaped scarp face, sometimes with a steplike arrangement of smaller scarps and terraces below. The bottom of the slumping block consists of a bulging lobe of saturated debris that protrudes downslope or into the valley bottom.

LearningCheck 15-11 **Explain how heavy rainfall can trigger a landslide or slump.**

Flow

In another form of mass wasting, a *flow*, a section of a slope becomes unstable with the absorption of water, and so flows downhill. The flow can range from fairly rapid to gradual and sluggish. Usually the center of the mass moves more rapidly than the base and sides, which are retarded by friction.

Many flows are relatively small, often encompassing an area of only a few square meters. More characteristically, however, they cover tens or hundreds of hectares. Normally, they are relatively shallow phenomena, including only soil and other regolith, but under certain conditions a considerable amount of bedrock may be involved.

As with other forms of mass movement, gravity is the impelling force, but water is the catalyst to the movement. Surface materials become unstable with the added weight of water, and their cohesion is diminished by waterlogging, so they begin to move downslope. The presence of clay also promotes flow, as some clay minerals become very slippery when moistened.

Earthflow: The most common kind of flow is **earthflow**, in which a portion of a water-saturated slope moves a limited distance downhill, normally during or immediately after a heavy rain. Where the flow originates, the surface of the slope usually has a distinct scar—either cracks or a prominent oversteepened scarp face (**Figure 15-23**). An earthflow is most conspicuous in its lower portion, where a bulging lobe of material pushes out onto the valley floor.

This type of slope failure is relatively common on hillsides that are not densely vegetated and often results in blocked roads or railways in valley bottoms. Property damage can be extensive, but the rate of movement is usually so sluggish that life is not threatened.

Mobile Field Trip
Landslide!
https://media.pearsoncmg.com/ph/esm/esm_mcknight_physgeo_13/media/

◄ **Figure 15-21** On August 17, 1959, a magnitude 7.3 earthquake near Yellowstone National Park triggered an enormous landslide down into the Madison River Valley. The landslide killed at least 28 people, and formed "Quake Lake" behind the debris.

Slope failures such as slides and earthflows often exhibit repeated movement over a period of years or decades. Once a slope has been destabilized, it may move again when the combination of factors is right—especially after heavy rainfall. One example of such repeated mass-wasting events is described in the box *People & the Environment: The Oso Landslide.*

LearningCheck 15-12 **In what ways are landslides and earthflows similar? In what ways are they different?**

Mudflow: A **mudflow** originates in drainage basins when a heavy rain, often following a long dry spell, produces a cascading runoff too great to be absorbed into the soil. Loose debris is picked up from the hillsides

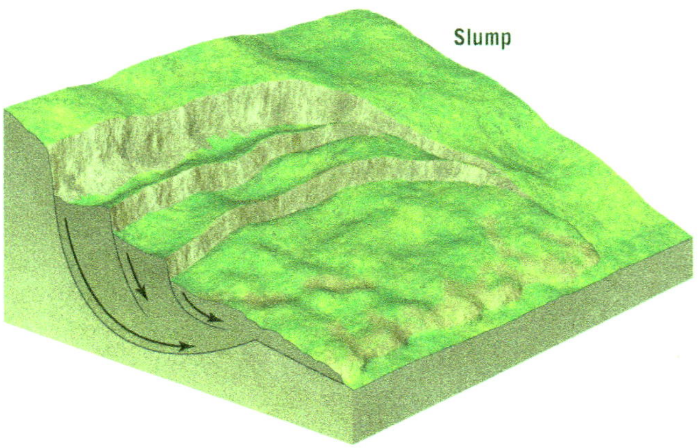

▲ **Figure 15-22** Slump involves movement along a curved sliding plane.

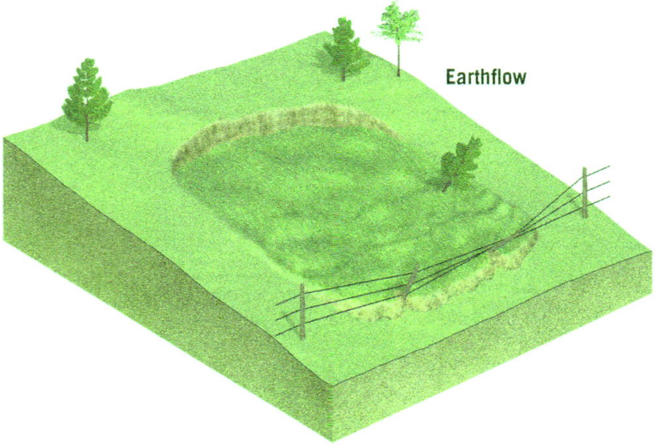

▲ **Figure 15-23** Earthflow takes place on hillsides when wet surface material begins to flow downslope.

People & the Environment

The Oso Landslide

• **Kerry Lyste,** Everett Community College; Pat Stevenson, Natural Resources Department, Stillaguamish Tribe

On March 22, 2014, at 10:37 A.M., a massive slide came down the north bank of the North Fork Stillaguamish River in Washington state (**Figure 15-E**). The slide wiped out the community that lived on Steelhead Lane, burying it in as much as 8 meters (25 feet) of mud and killing 43 people.

Born from Fire and Ice: The underlying cause of this slide is linked to the area's geologic history, both in terms of previous impacts of local volcanic activity and continental glaciation. Approximately 14,000 years ago, during the Pleistocene Epoch, when the Cordilleran ice sheet was receding in the northern area of Washington state, there was a great deal more water and sediment deposition than that region has today. Many river valleys, such as those in the Oso area, were dammed by ice jams. Inland-river lakes formed, capturing silt and clay and giving birth to a deep gray soil that today forms an unstable foundation for the river banks. These events were compounded by isostatic rebound, which lifted much of the base level in this area by 21–61 meters (70–200 feet). The precursor of the North Fork Stillaguamish River received drainage from the Sauk, Suiattle, Skagit, and North Stillaguamish Rivers until 13,000 years ago, when lahars from nearby Glacier Peak sealed off the eastern drainage divide.

Afterward, the river had much less volume and incised the old prehistoric river banks. This new era left a brown soil with a higher organic component on top of the gray soil.

Four Main Factors: What caused the catastrophe in 2014? Slopes have four main limiting factors: gravity correlated to the degree of slope, the amount of water in the soil, loading or destabilizing influences at the top of the slope, and destabilization or lack of support at the bottom of the slope. The Oso area had all four of these.

Not the First Slide; Not the Last: The North Fork Stillaguamish is an active river system that has shifted course a great deal and, as a result, has cut deeply into the slopes that confine its river bed in many places. Soft soil in many of these locations results in extreme mass-wasting events. Mega slides are common to the North Fork Stillaguamish, and the Oso slide was not the largest to have occurred (**Figure 15-F**).

In 2014, the Oso region experienced very heavy rainfall, which severely destabilized this part of the river valley slopes. In addition, clear-cut logging had occurred above the landslide area, allowing increased runoff and erosion. Numerous attempts to stabilize slopes at the bottom of the river proved to be unsuccessful (although most of these were intended to prevent material from entering the river). A slide in 2006 had moved the river 213 meters (700 feet) south. Local residents reported that several creeks around the slide area turned brown from precursor slide events days before.

March 22, 2014: The ensuing slide occurred in two phases. The first phase was the movement of the bottom, remobilizing the 2006 slide and wiping out the Steelhead Haven community in less than a minute. That slide was followed by the second phase, in which the remaining portion of the upper slope collapsed. What made this slide unique and devastating was that this was a *runout slide*, which moved a great distance, jumping the river and wiping out almost everything in its path.

▲ **Figure 15-E** The March 22, 2014, landslide near Oso, Washington.

Questions

1. What makes future slides in the Oso area likely?
2. Using Figure 15-F, describe slides that were larger than the 2014 slide.
3. What could be done to reduce the likelihood of future slides?

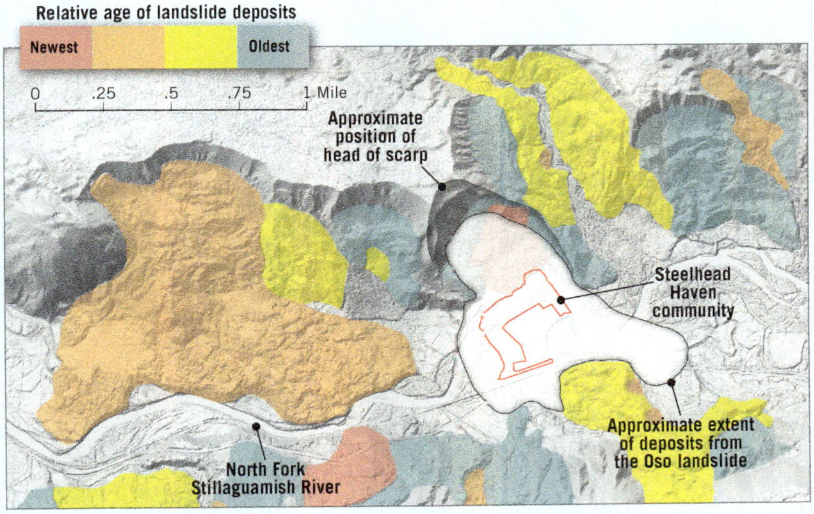

▲ **Figure 15-F** Relative age of Oso area landslides.

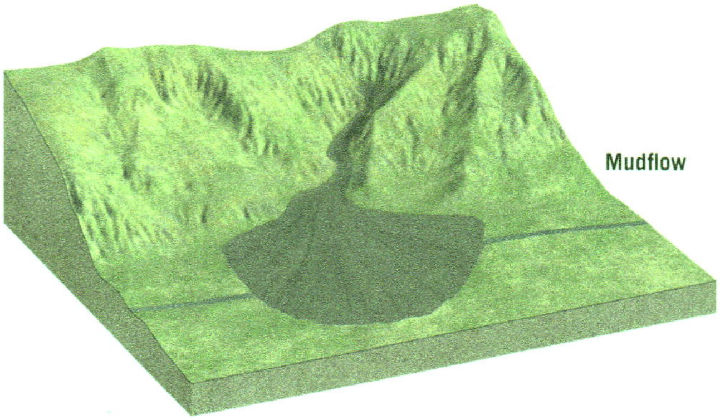

▲ **Figure 15-24** Mudflows entail the rapid movement of very wet material through a canyon or valley. The mud and debris are typically deposited at the mouth of the valley in a fan-shaped deposit.

by the runoff and concentrated in the valley bottoms, where it flows down-valley with the consistency of liquid concrete. The leading edge of the mudflow continues to accumulate load, becoming increasingly stiff, sometimes retarding the flow of the more-liquid upstream portions. When a mudflow reaches the mouth of the valley, the pent-up liquid breaks through with a rush, spreading muddy debris into a wide sheet (**Figure 15-24**). A *volcanic mudflow* (*lahar*), described in Chapter 14, is a special kind of mudflow that develops on the slopes of volcanoes.

Mudflows often pick up and carry large rocks, including huge boulders. When the large rocks are numerous, the term **debris flow** may be used instead of *mudflow*.

An important distinction between earthflow and mudflow is that mudflows (and debris flows) move down established drainage channels, whereas earthflow involves a slope failure and has little or no relationship to the drainage network. Moreover, mudflows are normally much more rapid, with a rate of movement intermediate between the sluggish surge of an earthflow and the rapid flow of a stream of water. Mudflows and debris flows are potentially more dangerous than earthflows to humans because of the more rapid movement, the larger quantity of debris involved, and the fact that the mudflow often discharges abruptly at the foot of a mountain, which is frequently an area of human settlement and agriculture (**Figure 15-25**).

As we will see in Chapter 18, mudflow and debris flow are especially important mass-wasting processes in arid regions. In the mountain areas of deserts, an intense, local thunderstorm can quickly mobilize loose, weathered material. The flow can move rapidly down a desert canyon, depositing the mud and debris at the foot of the mountains.

The media often incorrectly uses the rather ambiguous term *mudslide* to refer to most kinds of mass wasting. As we have seen, there are significant differences among the various kinds of slides and flows.

LearningCheck 15-13 **Describe and explain the differences between an earthflow and a mudflow.**

Creep

The slowest and least perceptible form of mass wasting is **creep**, or **soil creep**. It consists of a very gradual downhill movement of regolith so unobtrusive that it is normally recognized by indirect evidence only. Generally the entire slope is involved. Creep is such a pervasive phenomenon that it occurs all over the world on sloping land. Although most notable on steep, lightly vegetated slopes, it also occurs on gentle slopes that have dense plant cover. Wherever weathered materials are available for movement on land that is not flat, creep is a persistent form of mass wasting.

◀ **Figure 15-25** Following a devastating fire season that destabilized slopes, heavy rains in January 2018 triggered a large debris flow that buried houses in the southern California city of Montecito.

Causes of Creep: Creep is caused by the interaction of various factors, the most significant being alternation of freeze–thaw or wet–dry conditions. When water in the soil freezes, soil particles tend to be displaced outward from the ground surface due to ice expansion (**Figure 15-26**). After thawing, however, the particles settle downward, pulled slightly downslope by gravity. With countless repetitions, this process can result in downhill movement of the entire slope.

Any activity that disturbs regolith on a sloping surface contributes to creep because gravity affects every rearrangement of particles, attracting them downslope. For example, burrowing animals pile most of their excavated material downslope, and subsequent infilling of burrows is mostly by material from the upslope side. As plant roots grow, they also tend to displace particles downslope. Animals that walk on the surface exert a downslope movement as well. Even the shaking of earthquakes or thunder produces disturbances that stimulate creep.

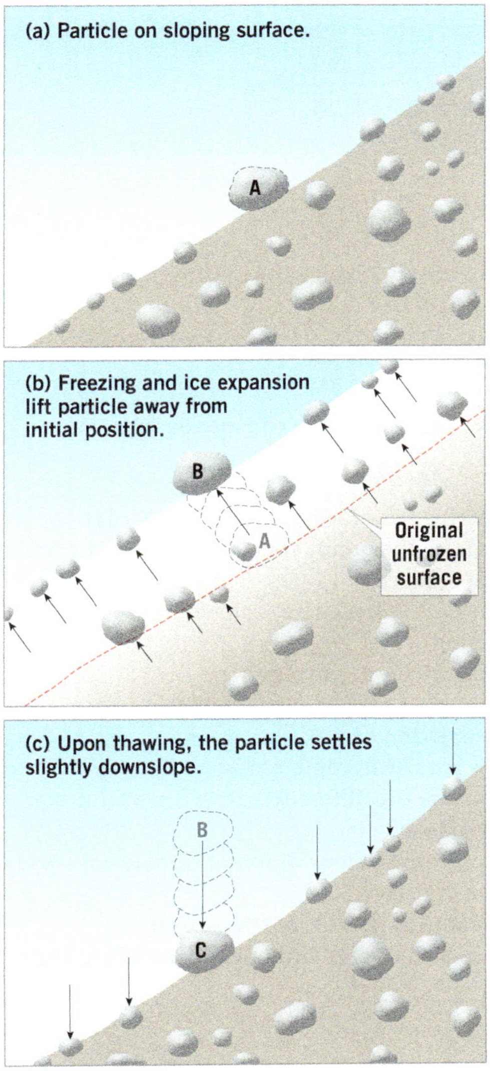

▲ **Figure 15-26** The movement of a typical rock particle in a freeze–thaw situation. Freezing and ice expansion lift the particle perpendicular to the slope (from A to B); upon thawing, the particle settles slightly downslope (from B to C).

▲ **Figure 15-27** Visual evidence of soil creep: displacement and/or bending of fences, utility poles, and retaining walls.

Whenever it occurs, creep is a very slow process, commonly just a fraction of a centimeter per year. Creep operates faster on steep or water-saturated slopes than on gentle or dry or vegetated slopes. Even then, it is much too slow for the eye to perceive, and the results are all but invisible. Creep is usually recognized only when human-built structures are displaced—most commonly when fence posts or utility poles are tilted downhill. Retaining walls may be broken or displaced, and roadbeds may be disturbed (**Figure 15-27**). Unlike the other forms of mass wasting, creep produces few distinctive landforms. Rather, it induces an imperceptible diminishing of slope angles and gradual lowering of hilltops—in other words, a widespread but minute smoothing of the land surface.

Under certain conditions, usually on steep grassy slopes, grazing animals accentuate soil creep through the formation of a network of hillside ridges known as *terracettes*. Over time, the entire hillside may be covered with a maze of terracettes.

LearningCheck 15-14 **What processes are responsible for soil creep?**

Solifluction: A special form of creep that produces a distinctive surface appearance is **solifluction** (meaning "soil flowage"), a process largely restricted to high-latitude and high-elevation tundra landscapes above the treeline (**Figure 15-28**). In summer, the near-surface portion of the ground (called the *active layer*) thaws, but the meltwater cannot percolate deeper because of the permafrost below. Spaces between the soil particles become saturated, and the heavy surface material sags slowly downslope. Movement is erratic and irregular, with lobes overlapping one another in a haphazard, fish-scale pattern. The lobes move only a few centimeters per year but remain very obvious in the landscape, in part because of the scarcity of vegetation. Where solifluction occurs, drainage channels are usually scarce because water flow during the short summer is mostly lateral through the soil rather than across the surface.

▶ **Figure 15-28** Solifluction in the high country of Colorado's Rocky Mountain National Park. The surface has gradually slipped downslope about 60 meters (200 feet). For scale, note the cars on Trail Ridge Road on the far right.

15 Learning Review

After studying this chapter, you should be able to answer the following questions. Key terms from each text section are shown in **bold type.** Definitions for key terms are also found in the glossary at the back of the book.

Key Terms and Concepts

Denudation (*p. 452*)

1. What is meant by **denudation**?
2. Distinguish among **weathering**, **mass wasting**, and **erosion**.

Weathering and Rock Openings (*p. 452*)

3. What roles do rock openings play in weathering processes?
4. What is the difference between a **joint** and a **fault?**
5. What are **master joints**, and how can they influence topography?

Weathering Agents (*p. 454*)

6. What are the general differences between **mechanical weathering** and **chemical weathering**?
7. Explain the mechanics of **frost wedging**.
8. Explain the process of **salt wedging**.
9. Explain the weathering process of **exfoliation** ("unloading") that is responsible for features such as **exfoliation domes**.
10. What is the relationship between **oxidation** and rust?
11. Briefly describe the chemical weathering processes of **hydration**, **hydrolysis**, and **carbonation**.
12. Describe and explain one example of **biological weathering**.
13. What is meant by **differential weathering**?

Mass Wasting (*p. 461*)

14. How does the **angle of repose** affect mass wasting?

Types of Mass Wasting (*p. 462*)

15. Describe the process of **rockfall (fall)**.
16. Describe the origin and general characteristics of **talus (scree)**.
17. What is a **talus cone**, and where does one typically develop?
18. What is a **rock glacier**?
19. What roles may heavy rain play in triggering a **landslide**?
20. How is a **slump** different from other kinds of landslide?
21. What are the differences between a landslide and a **mudflow**?
22. How is **earthflow** different from mudflow and **debris flow**?
23. Explain the process of **soil creep**.
24. In what kinds of environments is **solifluction** common?

Study Questions

1. Explain how it is possible for weathering to take place beneath the surface of bedrock.
2. Why is chemical weathering more effective in humid climates than in arid climates?
3. What is the relationship between gravity and mass wasting?
4. What is the role of clay in mass wasting?
5. In what ways can rainfall expedite mass wasting?

Exercises

1. A large square slab of granite, 1 meter thick and 100 meters to a side, falls off an exfoliation dome. The density of granite is 2.7 grams/cm^3. Calculate the total mass of the rock that fell: ________ kilograms
2. Assume that the rate of movement of the solifluction lobe in Figure 15-28 is 5 centimeters per year. How many years did it take for the lobe to travel 60 meters (200 feet) downslope? ________ years
3. The chemical weathering of rocks such as basalt produces a "rind" of altered minerals. Although the rate of weathering varies by location (especially due to differences in climate), the thickness of the rind offers a rough estimate of how long a rock has been exposed to weathering: the thicker the rind, the longer the exposure. One study found that the thickness of the weathering rind increased by about 0.007 millimeters for every 1000 years of exposure. Assuming a constant rate of weathering, estimate how long the rocks have been exposed to weathering under the following conditions:
   a. Weathering rind is 0.5 mm thick = ________ years of exposure
   b. Weathering rind is 1.5 mm thick = ________ years of exposure

Environmental Analysis LANDSLIDES IN OREGON

Data **Landslides** https://media.pearsoncmg.com/ph/esm/esm_mcknight_physgeo_13/media/

Many regions of Oregon are prone to landslides because of an unfortunate combination of steep slopes, on which gravity readily acts, and abundant rainfall, which further destabilizes the sloped land.

Activities

1. Why does Oregon get so much rainfall? (Refer to Figure 6-37.)

Go to the Statewide Landslide Information Database for Oregon (SLIDO) at gis.dogami.oregon.gov/maps/slido/. Click on the Help link; then read the first few sections about how to navigate the map.

2. According to the "Landslide Inventory" section, what is talus-colluvium? What is a fan?

Return to the SLIDO map. In the "Basemap" menu, choose "Imagery with Labels." In the "Layers" menu, click the arrow beside "Landslide Inventory" and check "Historic Landslide Records."

3. In which portion of Oregon do most landslide events (represented by dots) occur?
4. Click on several events in the eastern half of Oregon. What type of landslide is common there? What does the number and type of events tell you about the land in eastern Oregon?

Type "Cannon Beach" into the "Enter address" field and hit "Enter." Zoom out until you can also see Arch Cape.

5. Notice that historic events follow a path. Click on several events. What type of "Movement class" occurs along this path?
6. "Zoom to" a few of these events. Along what type of feature do these landslides occur? What is the most likely cause of these landslides?

Go to http://oregonstate.edu/ua/ncs, an Oregon State University site. In the "Search" field, type "New landslide maps" and select the link "Map outlines western Oregon landslide risks from a subduction zone . . . ".

7. What future event is likely to induce a significant number of landslides?
8. Why is it important to consider landslide and transportation issues in addition to damage to structures?
9. Where else in the western United States are similar hazards found? Why?

Seeing Geographically in Our National Parks

Look again at the photograph of the Grand Canyon at the beginning of the chapter (p. 450). What evidence of differential weathering is visible? Which kinds of mass-wasting processes have likely helped shape this landscape? Why do you say this? What kinds of mass-wasting events might the intense thunderstorm trigger?

Mastering Geography®

Looking for additional review and test prep materials? Visit the Study Area in **Mastering Geography** to enhance your geographic literacy, spatial reasoning skills, and understanding of this chapter's content by accessing a variety of resources, including Map Master interactive maps, geoscience animations, *Mobile Field Trips*, videos, *Project Condor* Quadcopter videos, *In the News* RSS feeds, flashcards, web links, self-study quizzes, and an eText version of *McKnight's Physical Geography*.

Seeing Geographically in Our National Parks

The Green River in Canyonlands National Park, Utah. Describe the topography of this river valley. What kind of climate is likely found here? Why do you say so? What is the structure of the rocks in this area? Does the river appear to be eroding more prominently on the outside bank of a turn or on the inside bank? How do you know?

Canyonlands National Park, UT

Fluvial Processes

16

HAVE YOU EVER WONDERED how a river changes over time? Or why the Green River here in Canyonlands National Park has such a winding course? You probably know that the amount of water in a river changes from time to time. But it's usually less obvious that nearly all rivers change course over time—sometimes subtly, just shifting the channel a short distance to the side; other times quite dramatically, abandoning an old course and adopting a new one. It is to the work of running water—the *fluvial processes*—that we turn in this chapter.

Of all the external agents of erosion and deposition—running surface water, subsurface water, wind, glaciers, and coastal waves and currents—by far the most important is running water. The flow of water in streams probably contributes more to shaping landforms than all other external processes combined. This is true not because running water is more powerful than other agents—glaciers and pounding waves often apply much more force per unit area—but rather because running water is nearly ubiquitous. At one time or another, running water contributes to landform development almost everywhere on land surfaces, in the wettest, driest, coldest, and hottest places on Earth. (Wind is ubiquitous, too, but compared with running water has much less power as a sculptor of terrain.) So, as we begin our study of erosional and depositional processes in the remaining chapters of this book, it is appropriate for us to start with the work of running water.

As you study this chapter, think about these **Key**Questions:

- **What is the difference between *streamflow* and *overland flow*?**
- **Why are floods so important to the erosional and depositional work of streams?**
- **Why do streams develop sinuous, meandering, or braided channels?**
- **How do stream drainage patterns reflect the underlying structure of a landscape?**
- **How do stream valleys deepen, widen, and lengthen?**
- **What are the typical landforms on a floodplain?**
- **What is *stream rejuvenation*?**

The Impact of Fluvial Processes on the Landscape

Running water is so widespread and effective as an agent of erosion and deposition that its influence on the landscape is usually the dominant process at work. The shapes of most valleys are strongly influenced by the transportation and deposition of sediment. Areas above the valleys are less affected, but even there, flowing water may influence the shape of the land. Running water smooths irregularities, wears down the topography by erosion and transportation of debris, and fills up valleys by deposition.

Streams and Stream Systems

As we learned in Chapter 9, in the study of geomorphology we call any channeled flow of water a **stream**—whether it is a tiny creek or an enormous river. Before we get into the specifics of stream processes and their topographic results, we need to introduce a few basic concepts.

Streamflow and Overland Flow

Fluvial processes are processes that involve running water. They encompass both the unchanneled downslope movement of surface water, called **overland flow**, and the channeled movement of water along a valley bottom, called **streamflow**.

Valleys and Interfluves

All surfaces of the continents can be thought of as consisting of two topographic elements: *valleys* and *interfluves*. A **valley** is that portion of the terrain in which a drainage system is clearly established (Figure 16-1). It includes the valley bottom, which is partially or totally occupied by the channel of a stream, as well as the valley walls that rise above the valley bottom on both sides. The upper limit of a valley is not always apparent, but it can be clearly conceptualized as a lip or rim at the top of the valley walls above which drainage channels are absent or indistinct.

An **interfluve** (from the Latin, *inter*, meaning "between," and *fluvia*, meaning "rivers") is the higher land above the valley walls that separates adjacent valleys. Some interfluves consist of ridgetops or mountain crests, but others are simply broad and flattish divides between drainage systems. Conceptually, all parts of the terrain that are *not* in a valley are part of an interfluve. We can envision that, on an interfluve, water will move downslope through unchanneled overland flow until it reaches the lip of the interfluve. There, as the water drops off the lip of the interfluve into the first small gullies of the valley system, streamflow begins.

However, these simplistic definitions are not always applicable in nature. Swamps and marshes, for example, are typically found in sections of valleys where no drainage system is clearly established, but they may also be found on interfluves.

Drainage Basins

The **drainage basin**, or **watershed**, of a stream is all of the area that contributes overland flow, streamflow, and groundwater to that stream. In other words, the drainage basin consists of a stream's valley bottom, valley sides, and those portions of the surrounding interfluves that drain toward the valley. The drainage basin ends at a **drainage divide**, which is the line of separation between runoff that descends in the direction of one drainage basin and *runoff* that goes toward an adjacent basin

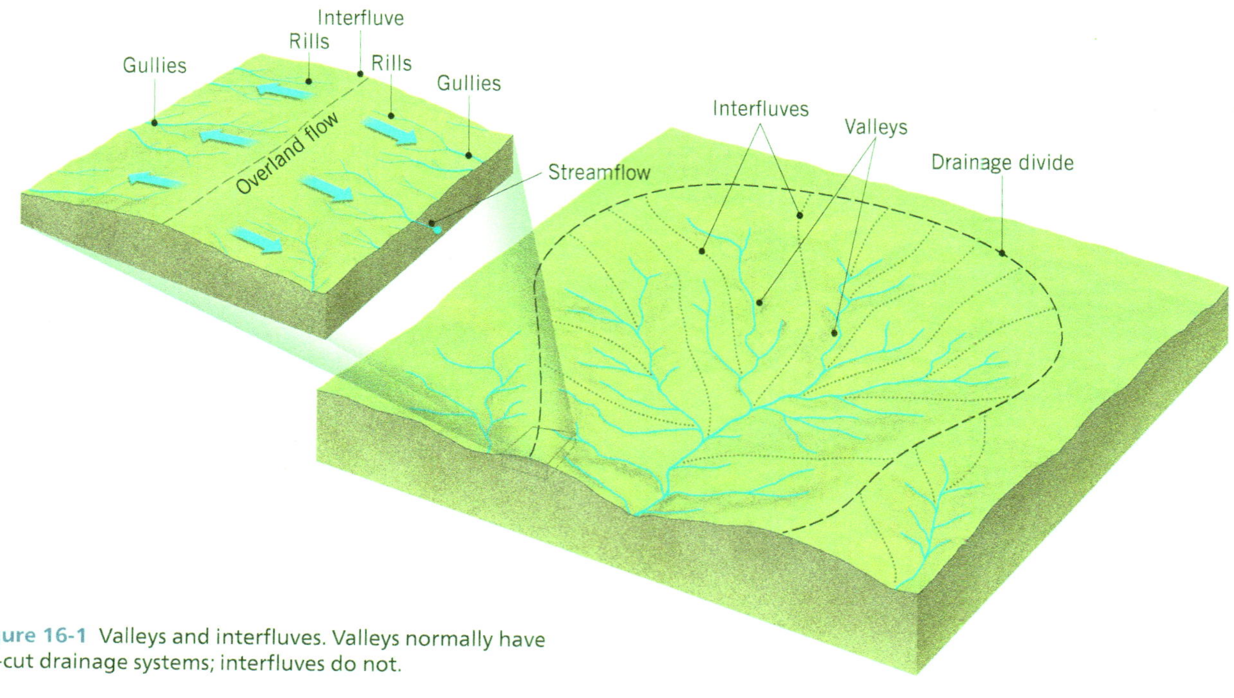

▲ Figure 16-1 Valleys and interfluves. Valleys normally have clear-cut drainage systems; interfluves do not.

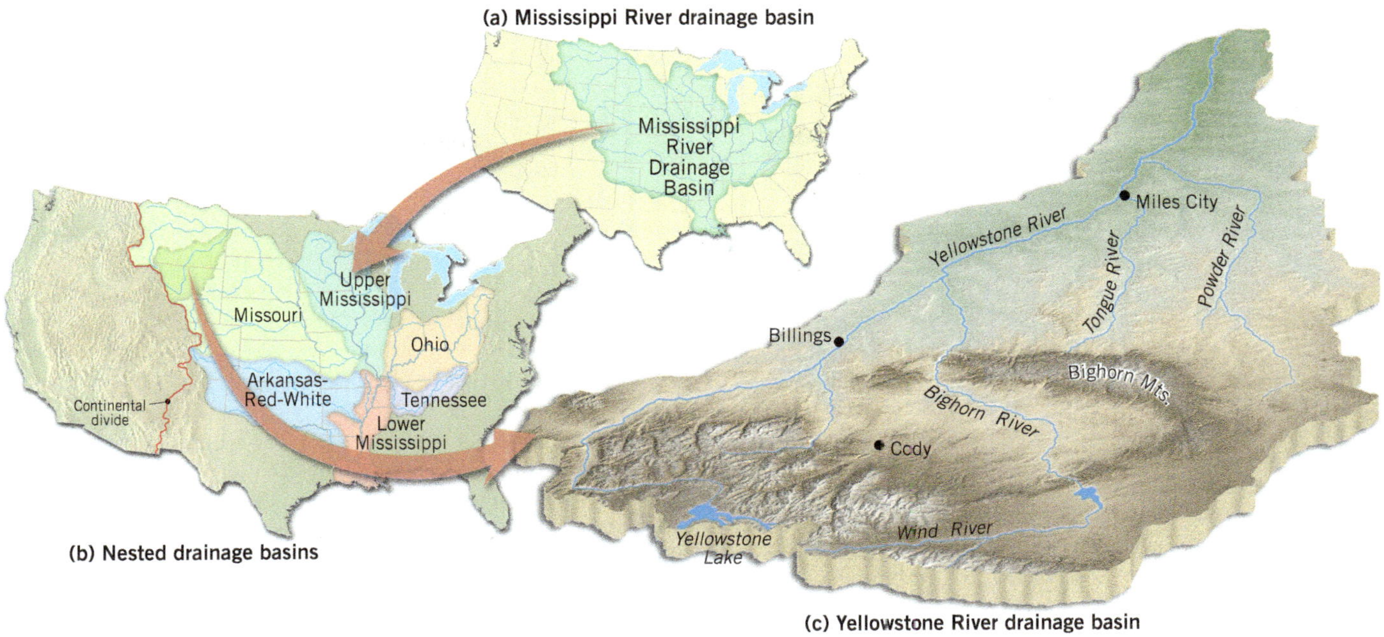

▲ **Figure 16-2** A nested hierarchy of drainage basins. Many sub-basins are found within the overall Mississippi River drainage basin. For example, beginning in the Yellowstone River drainage basin, the Bighorn River flows into the Yellowstone River, which flows into the Missouri River, which flows into the Mississippi River; finally, the Mississippi empties into the Gulf of Mexico. The continental divide separates rivers flowing toward the Pacific Ocean from those flowing toward the Atlantic Ocean or Gulf of Mexico.

(see Figure 16-1). A drainage divide can be a sharp ridge between drainage basins or a more subtle separation of basins in the form of an interfluve.

Every stream has its own drainage basin, but for practical purposes, the term is often reserved for major streams. The drainage basin of a principal river encompasses the smaller drainage basins of all of its tributaries; consequently, larger basins include a hierarchy of smaller tributary basins (**Figure 16-2**).

The *continental divide* of North America is the set of ridges separating the drainage basins of streams that flow eastward or southward (ultimately, to the Atlantic Ocean or Gulf of Mexico) from those that flow westward to the Pacific Ocean (see Figure 16-2) and from those that flow north into the Arctic Ocean.

Dams are often built to capture the runoff of a drainage basin. The impounded water may serve a variety of purposes, including the generation of electrical power. See the box *Energy for the 21st Century: Hydropower*.

LearningCheck 16-1 **Explain the relationship of streamflow to interfluves and valleys. (Answer on p. AK-5)**

Stream Orders

In every drainage basin, small streams come together to form successively larger ones. The concept of **stream order** describes the arrangement and organization of all streams within a watershed (**Figure 16-3**).

A *first-order stream*, the smallest stream, has no tributaries. Where two first-order streams unite, they form a *second-order stream*. At the confluence (joining) of two second-order streams, a *third-order stream* begins, and so on through successively higher orders.

Notice in Figure 16-3 that stream orders are not added to each other. For example, the confluence of a first-order stream and a second-order one does not produce a third-order stream. A third-order stream forms only where two second-order streams meet.

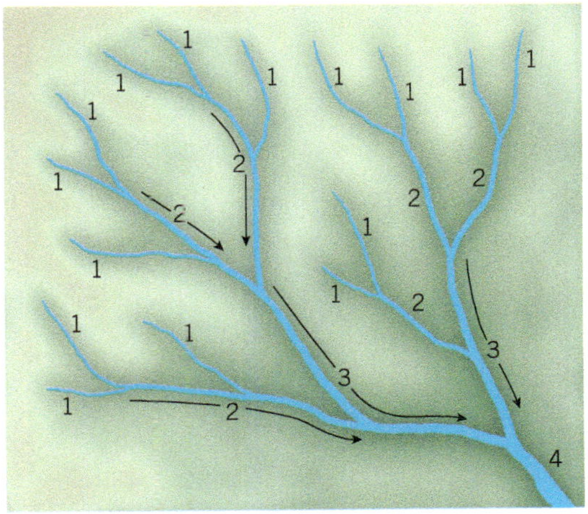

▲ **Figure 16-3** Stream orders. The branching component of a stream and its tributaries can be classified into a hierarchy of segments, ranging from smallest to largest (here, from 1 to 4).

Energy for the 21st Century

Hydropower

• **Nancy Lee Wilkinson**, San Francisco State University

What's the most economical way to produce energy? Because moving water has the power to do work, water wheels powered the mills and looms of the Industrial Revolution. Hydroelectric generation began on a commercial scale in the United States in the early 1880s and now provides as much as 20 percent of the world's electricity. The largest facilities are in China (**Figure 16-A**), Canada, Brazil, the United States, and Russia.

How Hydropower Works: Hydroelectric generators convert the potential energy of water stored behind a dam into kinetic energy (energy of motion). Dams store water in reservoirs. Water exits a reservoir through large pipes (*penstocks*) near the base of the dam, turning turbines several meters across. As a turbine spins giant shafts topped by a generator, electric current is produced (**Figure 16-B**).

How much electricity does a hydropower plant generate? That depends on the volume of water available and the elevation difference between the water supply and the turbine. The best hydropower sites are deep, narrow canyons with solid bedrock where elevation changes rapidly and a tall dam can be built with little material.

Hydropower's Advantages: Once built, dams generate minimal air and water pollution, providing an appealing alternative to coal, gas, or nuclear power plants. They require no fuel and only a small labor force and can be turned on quickly to meet short-term surges in power demand.

Dams built for hydropower production can store water for irrigation or other uses, facilitate navigation, and provide recreational opportunities and flood protection. Revenues from these uses can offset construction and operating costs. Finally, dams are impressive public projects that symbolize progress and permanence and can continue generating electricity for a hundred years or more. For these reasons, hydropower dams have been an attractive tool for economic development.

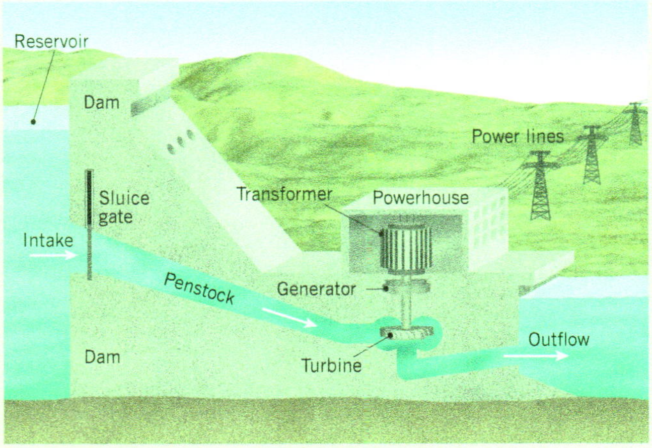

▲ **Figure 16-B** Hydroelectric generation.

Social and Environmental Impacts of Dams: Dams can have substantial social and environmental impacts. They flood forests, fertile fields, and communities, displacing millions of people worldwide. They block the upstream migration of fish that spawn in freshwater, flood upstream habitats, and dry up wetlands that harbor great biodiversity. They block transport of river sediment, increasing riverbank erosion and starving downstream beaches of sand and gravel. Aging or poorly built dams can collapse, posing a threat to downstream communities.

The consequences of dam building became apparent as hydroelectric development boomed in the southeastern and western United States during the twentieth century and dams became a major tool in international development. More than 60 percent of the world's large rivers have been dammed, and many new hydropower projects are under construction. Some of these projects have encountered angry resistance. Meanwhile, the United States has led the way in decommissioning dams that have outlived their usefulness and in pursuing new challenges in the art and science of stream and watershed restoration.

Hydropower's Future: Dam building has declined since a peak in the 1970s, but many massive projects are still under construction, and more are proposed. China completed the world's largest dam at Three Gorges in 2012; international interests continue to support dam construction in Amazonia, Southeast Asia, and Africa.

Hydropower's environmental and social impacts make it difficult to build large dams in the United States today. Off-stream reservoirs that store water but do not disrupt the stream's flow are one way to reduce the impacts of hydropower. Alternatively, "small hydro" facilities can utilize stream flows to generate electricity at a smaller scale, without dams.

Changing precipitation patterns associated with global climate change may call for reassessment of hydroelectricity's role in regional energy production. For example, the recent multiyear drought reduced California's hydropower production by more than half. If midlatitude mountain ranges receive less snowfall in the coming decades, reservoirs will receive less inflow in summer, reducing hydropower's usefulness in meeting seasonal power demands.

▲ **Figure 16-A** Three Gorges Dam in China.

Questions

1. How might climate change reduce hydropower potential? What other energy resources might take its place, and with what environmental consequences?
2. How might the environmental consequences of dams in tropical regions differ from their impacts in temperate regions?

In a well-developed drainage system, we can predict with some certainty that first-order streams and valleys are more numerous than all others combined and that each succeeding higher order is represented by fewer and fewer streams. More importantly, average stream length generally increases with increasing order, average watershed area generally increases with increasing order, and average stream **gradient** (elevation change over a given distance) decreases with increasing order.

Fluvial Erosion and Deposition

All external processes remove fragments of bedrock and regolith (erosion), transport them to another location, and set them down (deposition). Fluvial processes produce one set of landforms by erosion and another quite different set by deposition.

Erosion by Overland Flow

On the interfluve, fluvial erosion begins when rain starts to fall. Unless the impact of rain is absorbed by vegetation or some other protective covering, the collision of raindrops with the ground is strong enough to blast fine soil particles upward and outward, shifting them a few millimeters laterally. On sloping ground, most particles move downhill by this *splash erosion*.

In the first few minutes of a rain, much of the water infiltrates the soil; consequently, there is little runoff. During heavy or continued rain, however, infiltration is reduced, and most of the water moves downslope as overland flow. The water flows across the surface as a thin sheet, transporting material already loosened by splash erosion, in a process termed *sheet erosion*.

As overland flow moves downslope and its volume increases, the resulting turbulence tends to break up the sheet flow into tiny channels called *rills*. This more concentrated flow picks up additional material and grooves the slope with numerous parallel seams through *rill erosion*. If the process continues, the rills begin to coalesce into fewer and larger channels called *gullies*, and *gully erosion* becomes recognizable. As the gullies get larger, they tend to become incorporated into the drainage system of the adjacent valley, and the flow changes from overland flow to streamflow (see Figure 16-1).

Erosion by Streamflow

Once surface flow is channeled, its ability to erode and transport material is greatly enhanced by the increased volume and velocity of water. Erosion is accomplished in part by the direct hydraulic power of the moving water along the bottom and sides of the channel. Channel banks can also be undercut by streamflow, particularly at times of high flow.

The erosive capability of streamflow is also increased by the abrasive "tools" it picks up and carries along with it. All sizes of rock fragments, from silt to boulders, chip and grind the streambed as they travel downstream in the moving water. These rock fragments break off more fragments from the bottom and sides of the channel, and they collide with one another, becoming smaller and rounder from the wear and tear of the frequent collisions.

A certain amount of chemical action also accompanies streamflow. Some chemical weathering processes—particularly solution action and hydrolysis—also help erode the stream channel through *corrosion* (dissolution of soluble rock by water).

The erosive effectiveness of streamflow varies enormously, determined primarily by the speed and turbulence of the flow and by the resistance of the bedrock. Flow speed is governed by the gradient (slope angle) of the streambed (the steeper the gradient, the faster the flow), by the shape of the channel (generally the narrower the channel, the faster the flow), and by the volume of flow (more water normally means higher speed). The degree of turbulence is determined in part by the flow speed (faster flows are more turbulent) and in part by the roughness of the stream channel (an irregular channel surface increases turbulence).

LearningCheck 16-2 **Explain the difference in erosion between streamflow and overland flow.**

Transportation

Any water moving downslope, whether moving as overland flow or as streamflow, can transport rock. The load carried by overland flow is likely to be small in comparison with what a stream can transport. Eventually, most of this material reaches the streams in the valley bottoms, where it is added to the stream-eroded debris and material contributed by mass wasting to constitute the **stream load**.

Animation **Stream Sediment Movement** https://media.pearsoncmg.com/ph/esm/esm_mcknight_physgeo_13/media/

The stream load contains three components (Figure 16-4):

1. Some minerals, mostly salts, are dissolved in the water and carried in solution as the **dissolved load**.
2. Very fine particles of clay and silt are carried in suspension, moving along with the water without ever touching the streambed. These tiny particles,

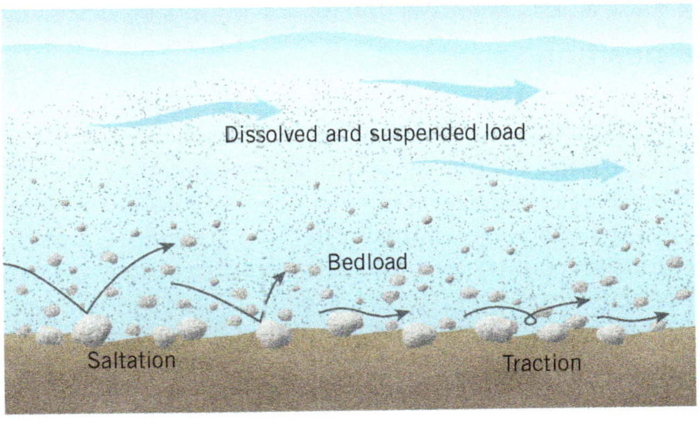

▲ Figure 16-4 A stream moves its load in three ways. The dissolved and suspended loads are carried in the general water flow. The bedload is moved by traction (dragging) and saltation (bouncing).

called the **suspended load**, have a very slow settling speed, even in still water. (Fine clay may require as much as a year to sink 30 meters [100 feet] in perfectly quiet water.)

3. Sand, gravel, and larger rock fragments constitute the **bedload**. The smaller particles are moved along with the general streamflow in a series of jumps or bounces collectively referred to as *saltation*. Coarser pieces are moved by *traction*, which is defined as rolling or sliding along the streambed. The bedload is normally moved intermittently, especially during floods: debris is transported some distance, dropped, and then picked up later and carried farther.

Competence: Geomorphologists employ two concepts—*competence* and *capacity*—to describe the load a stream can transport. **Stream competence** is a measure of the particle size a stream can transport, expressed by the diameter of the largest particles the stream can move. Competence depends mainly on flow speed, with the power of the water generally increasing by the square of its speed. In other words, if the flow speed doubles, the size of the largest movable particle increases fourfold ($2^2 = 4\times$). If the flow speed triples, the force increases ninefold ($3^2 = 9\times$)—as stream speed increases even moderately, the largest moveable particle increases significantly. Thus, a stream that normally can transport only sand grains might easily be able to move large boulders during a flood.

Capacity: **Stream capacity** is a measure of the amount of solid material a stream has the potential to transport, normally expressed as the volume of material passing a given point in the stream channel during a given time interval. Capacity may vary tremendously over time, depending mostly on fluctuations in volume and flow speed but also on the characteristics of the load (particularly the mix of coarse and fine sediments). It is difficult to overemphasize the significance of the greatly expanded capacity of a stream to transport material during floods.

LearningCheck 16-3 **What factor determines the size of rock a stream is able to transport?**

Deposition

Whatever is picked up must eventually be set down, so erosion is inevitably followed by deposition when either flow speed or water volume decreases. Diminished flow is often the result of a change in gradient, but it may also occur when a channel widens or changes direction. Therefore, stream deposits are commonly found at the mouths of canyons, on floodplains, and along the inside bank of river bends. Eventually, however, much of a stream's load is deposited into a body of water, such as a lake or the ocean.

Alluvium: The general term for stream-deposited sediment is **alluvium**. Such deposits may include all sizes of rock debris, but small particles constitute the bulk of the total. Alluvial deposits often exhibit one or more of the following characteristics:

- Alluvium is typically smooth and round due to the battering the rocks receive from each other as they flow downstream. Over the great distances of many stream systems, this battering and abrasion (along with chemical weathering) eventually reduces larger rocks (boulders, cobbles, and pebbles) to smaller sizes (sand and silt).
- Alluvial deposits often display distinct *strata*, or layers, due to episodes of deposition following periodic floods.
- Alluvium is often "sorted"—that is, an alluvial deposit often consists of rocks of just about the same size. Sorting can occur when the speed of water flow diminishes. As stream competence drops, the heaviest (and therefore largest) rocks are deposited first, while smaller rocks are carried away and deposited elsewhere.

LearningCheck 16-4 **Why is alluvium often rounded, stratified, and sorted by size?**

The work of fluvial erosion and deposition continues to shape the landscapes of the Appalachian Mountains in the eastern United States—see the box *Featured National Park: Great Smoky Mountains*.

Perennial and Intermittent Streams

We tend to think of rivers as permanent, but in fact many of the world's streams do not flow year-round. In humid regions, the large rivers and most tributaries are **perennial streams**—that is, permanent. However, in more arid parts of the world, many of the major streams and most tributaries carry water only part of the time. These impermanent flows are called **intermittent streams** or *seasonal streams* if they flow for only part of the year (the wet season) and **ephemeral streams** if they carry water only during and immediately after a rain. The term *intermittent* is sometimes applied to both cases (Figure 16-5).

Even in humid regions, many first-order and second-order streams are intermittent. These are generally short streams with relatively steep gradients and small watersheds. If rain is not frequent or snowmelt not continuously available, these low-order streams run out of water. High-order streams in the same regions are likely to have permanent flow because they have a larger drainage area and because previous rainfall or snowmelt that sank into the ground can emerge in the valleys as groundwater runoff long after the rains have ended.

Floods as Agents of Erosion and Deposition

Our discussion of factors associated with effective stream erosion and deposition leads us to an important concept in fluvial geomorphology: the role of floods as agents of erosion and deposition.

Featured National Park

Great Smoky Mountains

A Landscape of Streams and Ancient Rocks

▲ **Figure 16-D** Rounded cobbles and boulders in Ramsey Prong of Little Pigeon River.

▲ **Figure 16-E** Flooding stream in Great Smoky Mountains National Park.

Great Smoky Mountains National Park in the southern Appalachians of North Carolina and Tennessee encompasses a diverse natural environment of forests, mountains, streams, and waterfalls. Here, in our most-visited national park, ancient rocks, tectonic processes, and the work of fluvial erosion, transportation, and deposition are on striking display.

The Mountains: The Great Smoky Mountains are part of the Blue Ridge, a rugged mountain range east of the folded "Ridge and Valley" section of the Appalachians. This region mostly consists of metamorphosed sedimentary rocks, some as old as 800 million years.

The ancestral Appalachian Mountains began forming about 300 million years ago, when ancient continents collided to assemble the supercontinent of Pangaea. This collision folded and uplifted the original horizontal layers and pushed a huge mass of older rock northwest over relatively younger rocks along the Great Smoky Thrust Fault. Since then, the mountains have been slowly wearing down.

The most resistant rock of the Great Smoky Mountains is now left exposed in the highest peaks, such as Clingmans Dome, and along the *knickpoints* that create the many waterfalls found in the Park (**Figure 16-C**). Interestingly, some of the younger rocks in the Park were uncovered by erosion at the *bottom* of valleys, such as Cades Cove—a "window" through the older rock that makes up the surrounding hills.

The Streams: The streams of the Park are fed by the high rainfall of the region—as much as 220 centimeters (85 inches) on some mountain peaks. The gently rounded boulders and cobbles that line the streambeds (**Figure 16-D**) are reminders that periodic floods are responsible for most change in this landscape. Only the strongest water flows can move these large rocks, tumbling and smoothing them as they advance downstream (**Figure 16-E**).

▼ **Figure 16-C** Waterfall along the Great Smoky Thrust Fault at White Oak Sink. Older metamorphic rocks were pushed over younger limestone along this fault.

▲ **Figure 16-F** Sorted deposit of gravel-size alluvium in Little River.

When the water slows at the end of a flood, a deposit of alluvium is left. As is often the case, these alluvial deposits are typically rounded and sorted by size (**Figure 16-F**). Over time, the enormous quantities of sediment eroded from these mountains reach their final destination: the Atlantic Ocean or the Gulf of Mexico.

Questions

1. Why are the stream boulders shown in Figure 16-D rounded and smooth?
2. Why are the pieces of gravel shown in Figure 16-F all about the same size?

Mobile Field Trip
Streams of the Great Smoky Mountains
https://media.pearsoncmg.com/ph/esm/esm_mcknight_physgeo_13/media/

▲ **Figure 16-5** Most desert streams are either intermittent or ephemeral, carrying water only rarely. This flash flood is in a usually dry side canyon of the San Juan River in southeastern Utah.

A deep valley with a small stream at its bottom is a common sight. By invoking three geomorphic factors, we can understand how such a seemingly modest stream can shape an extensive valley: (1) the extraordinary length of geologic time, (2) the role of mass wasting in moving weathered rock down into valleys where it can be transported away, and (3) the remarkable effectiveness of floods.

We've seen how the vast expanse of geologic time provides the opportunity for countless repetitions of a landform-development process and how the repetitive action of even a tiny flow of water can wear away the strongest rock. We've also discussed the range of processes that can move masses of rock downslope, where later it can be picked up and carried away by a stream. Of perhaps equal importance is variation in the amount of water flowing in a stream.

Stream Discharge: From place to place and from time to time, most streams have variable flow regimes with pronounced fluctuations in **discharge**: the volume of flow per unit time. The discharge of a stream is described as

$$Q = wdv.$$

(here Q is the discharge of a stream, w is the width of the channel, d is the depth of the channel, and v is the stream velocity). The discharge of a stream can change if any of the variables change. For example, if the width and depth of a channel remain constant but stream velocity increases, discharge will increase. Stream discharge is usually described in cubic meters per second (cms or m^3/s) or cubic feet per second (cfs or ft^3/s).

Many of the world's streams carry a relatively modest amount of water most of the year but a relatively large volume during floods—generally defined as times when water overtops a stream's normal banks. In some streams, flood flow occurs for several weeks or even months each year; however, in the vast majority, the duration of high-water flow is much more restricted. In many cases, the "wet season" may be only one or two days a year. Yet the amount of denudation that can be accomplished during flood flow is much greater than during periods of normal discharge.

As stream discharge increases during a flood, the competence and capacity of the stream increase. Recall that as the speed of flow increases, the competence of the stream increases exponentially. Thus, the epic work of most streams—the excavating of great valleys, the forming of vast floodplains—is primarily accomplished during flood flow.

LearningCheck 16-5 **Why are floods so important in fluvial geomorphology?**

Measuring Streamflow: Each stream system has its own characteristic flow regime, influenced by local factors such as the climate and area of drainage. To understand the flow characteristics of a particular stream, data are gathered from a network of *streamgages*. Some 14,000 such stream gaging stations are operating across the United States and Canada, with many thousands more in operation around the world. On an ongoing basis, streamgages log information about the local water height in a stream (the *gage height* or *stage* above a local reference point). They also record flow velocity, from which the discharge of a stream is calculated and plotted over time on a graph called a *hydrograph*.

By correlating stream discharge and gage height to local weather events, we can estimate when floods will arrive or crest at different places along the course of a river. The amount of urbanization in a region significantly influences this relationship. For example, in heavily urban areas, where there tends to be less infiltration of rain because much of the region is covered with impermeable surfaces such as pavement and buildings, a heavy rain tends to produce greater and faster runoff into local streams than in areas with more permeable surfaces (**Figure 16-6**).

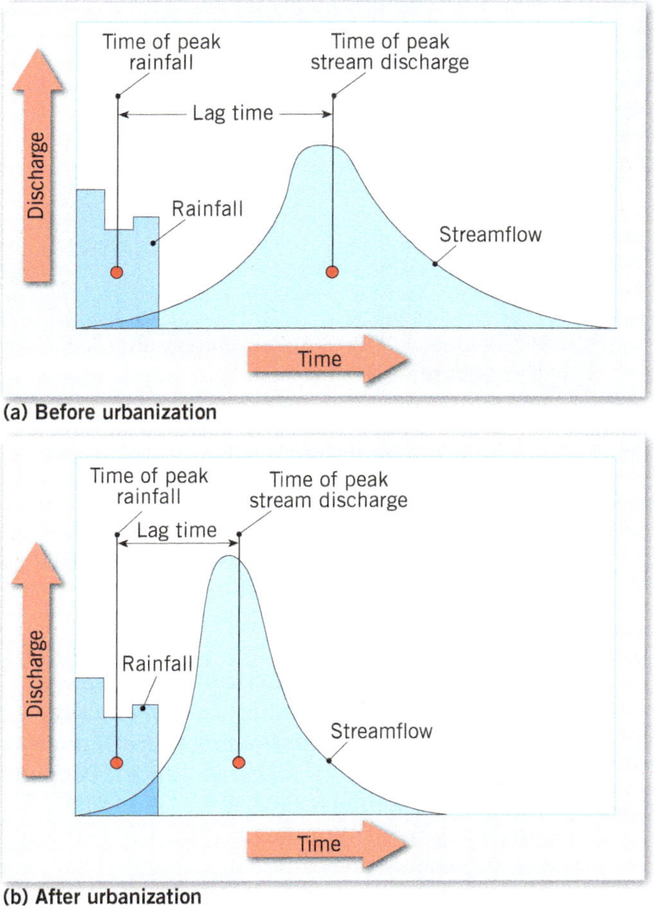

▲ **Figure 16-6** Hypothetical hydrographs show that the lag time between the peak of rainfall and peak stream discharge during a flood is usually reduced in urban areas, where little infiltration of runoff occurs.

▲ **Figure 16-7** Flooding in Davenport, Iowa, in May 2019.

Flood Recurrence Intervals: Streamflow data gathered for a stream over many decades can also provide information about the relative frequency and size of floods. The probability of a given-size flood occurring is usually described as a **recurrence interval** (or *return period*). For example, a "100-year flood" refers to stream discharge that has a 1 in 100 (1 percent) probability of being exceeded in any single year. Such terminology is somewhat misleading, since it may leave the impression that such a flood is expected only once every 100 years, but that is not the case. It is possible to have more than one 100-year flood each century, or no such floods (**Figure 16-7**).

Flood recurrence intervals are calculated on the basis of available streamflow data; the longer the record, the more accurate the probabilities. But changing circumstances—such as greater urbanization or climate change—can alter these probabilities.

Stream Channels

Overland flow is a relatively simple process. It is affected by such factors as rainfall intensity and duration, vegetative cover, surface characteristics, and slope shape, but its general characteristics are straightforward. Streamflow, however, is more complicated.

Channel Flow

Unlike overland flow, streamflow is confined to channels, which gives it a three-dimensional nature. In any channel with even a slight gradient, gravitational pull overcomes friction forces and moves the water downhill. Except under unusual circumstances, however, the water in a stream does not flow exactly straight downhill.

Channel Cross Section: A principal cause of flow irregularity is friction along the bottom and sides of the channel, which causes the water to move most slowly there and fastest in the center of the stream. The amount of friction is determined by the width and depth of the channel. For example, a shallow, wide channel has a greater retarding effect on streamflow than a deep, narrow one (**Figure 16-8**). Where the channel is deep and smooth, a stream may exhibit *laminar flow*, in which all of the water moves smoothly downstream in parallel paths.

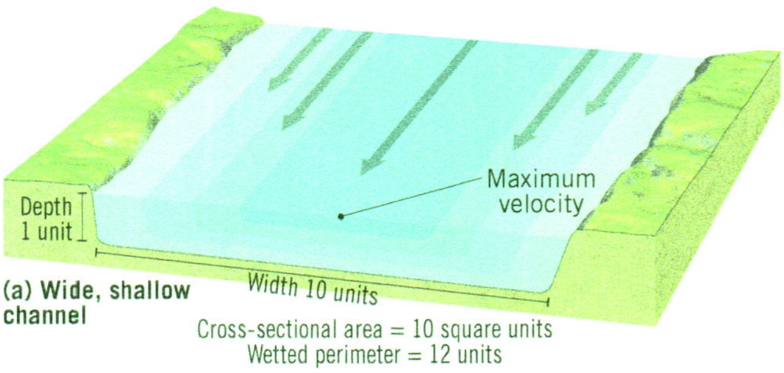

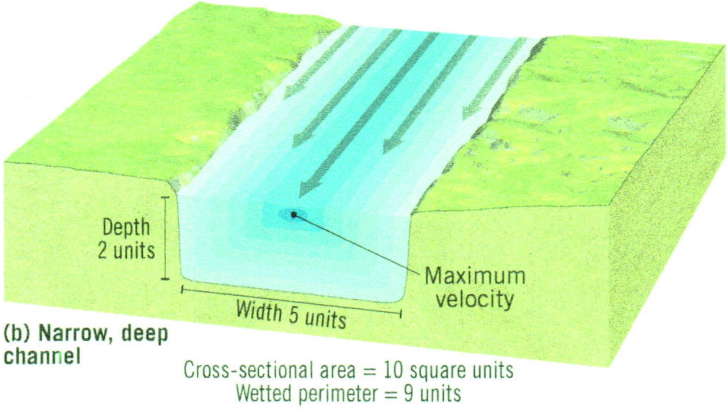

▲ **Figure 16-8** Friction between the stream bank and liquid surfaces cause a stream to flow fastest in the middle and at the surface. (a) A wide, shallow channel slows the water more than (b) a narrow, deep channel.

Turbulence: In turbulent flow, conversely, the general downstream movement is interrupted by irregularities in the direction and speed of the water. Such irregularities produce momentary currents that can move in any direction, including upward. Turbulence in streamflow is caused partly by friction, partly by internal shearing stresses between currents within the flow, and partly by surface irregularities in the channel. Stream speed also contributes to the development of turbulence; faster streams are more turbulent than slow-moving ones.

Eddies and whirlpools are conspicuous results of turbulence, as is the roiling whitewater of rapids. Even streams that appear very placid and smooth on the surface are often turbulent at lower levels.

Turbulent flow creates a great deal of frictional stress. This stress dissipates much of the stream's energy, decreasing the amount of energy available to erode the channel and transport sediment. Yet, turbulence contributes to erosion by creating flow patterns that pry and lift rock materials from the streambed.

Stream Channel Patterns

Irregularities in streamflow are manifested in various ways, but perhaps most conspicuously by variations in channel patterns. If streamflow were smooth and regular, we might expect stream channels to be straight and direct. However, few natural stream channels are straight and uniform for any appreciable distance. Instead, they curve to a greater or lesser extent, sometimes developing remarkable sinuosity.

Stream channel patterns are generally grouped into four categories—straight, sinuous, meandering, and braided.

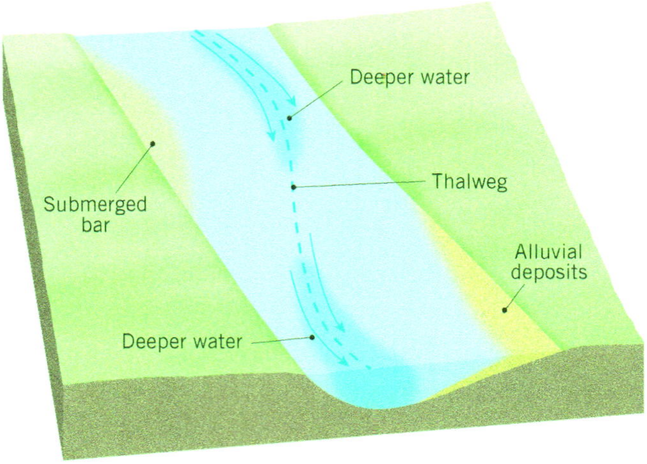

▲ **Figure 16-9** Even in a straight stream channel, the course of the deepest and fastest flow (the *thalweg*) tends to be slightly sinuous.

1. Straight channels are short and uncommon and usually indicative of strong control by the underlying geologic structure. A straight channel does not necessarily mean straight flow, however. A line running in the direction of the water and indicating the deepest parts of the channel, called the *thalweg* (from the German *thal*, meaning "valley," and *weg*, meaning "way"), rarely follows a straight path midway between the stream banks. Rather, it wanders back and forth across the channel (Figure 16-9). Alluvium is likely to be deposited opposite the place where the thalweg approaches one bank. Thus, straight channels are likely to have many of the characteristics of sinuous channels.
2. **Sinuous channels** are much more common than straight ones. They are winding and occur in almost every type of topographic setting (Figure 16-10). Their curvature is usually gentle and irregular. Stream channels are likely to be sinuous even when the stream flows down a steep slope. Where gradients are low (over flatter land), many stream channels tend to develop greater sinuosity and begin to meander.

▲ **Figure 16-10** Sinuous channel pattern of the New River in New River Gorge National River, West Virginia.

▲ **Figure 16-11** The meandering stream channel and floodplain of Alatna River in Gates of the Arctic National Park in Alaska.

▲ **Figure 16-12** A typical braided stream in the valley of the Toklat River in Denali National Park, Alaska.

3. **Meandering channels** exhibit an extraordinarily intricate pattern of smooth curves. The stream follows a serpentine course, twisting and contorting and turning back on itself, forming tightly curved loops and then abandoning them, cutting a new, different, and equally twisting course. Meandering generally occurs when the land is flat and the gradient is low, especially when most of the rock being transported by the stream is a fine-grained suspended load (Figure 16-11). A meander shifts its location almost continuously through erosion on the outside of curves and deposition on the inside. In this fashion, meanders migrate across the floodplain. They also tend to shift down-valley, producing rapid and sometimes abrupt changes in the channel. (Landforms that develop along a meandering stream channel are discussed later in this chapter.)
4. **Braided channels** consist of multiple interwoven and interconnected channels separated by low bars or islands of sand, gravel, and other loose debris (Figure 16-12). Braiding often takes place when a very flat stream channel has a heavy sediment load (such as the coarse-grained bedload supplied by glaciers to a meltwater stream) and in regions with prominent dry seasons and periods of low stream discharge (as is common in arid environments). At any time, the active channels of a braided stream may cover less than one-tenth of the width of the entire channel system, but in a single year most or all of the surface sediments may be reworked by the flow of the laterally shifting channels.

LearningCheck 16-6 **Under what conditions does a meandering channel pattern commonly develop? A braided channel pattern?**

Structural Relationships

Many factors affect stream development, and perhaps the most important is the geologic–topographic structure over or through which the stream flows. Each stream takes the path of least resistance, usually responding directly and conspicuously to structural controls, their courses guided and shaped by the nature and arrangement of the underlying bedrock. Sometimes, however, we must understand the geomorphic history of a region before we can comprehend the location of a drainage channel.

Consequent and Subsequent Streams

In the simplest, most common relationship between underlying structure and channel development, the stream follows the initial slope of the land. Such a *consequent stream* is normally the first to develop on newly uplifted land, and many streams remain consequent throughout their evolutionary development (Figure 16-13). Streams that develop along zones of structural weakness are *subsequent streams*—they may erode their channels along an outcrop of weak bedrock or perhaps follow a fault zone or a master joint. Subsequent streams often trend at right angles to other drainage channels.

Antecedent and Superposed Streams

Some streams seem to "defy" the structure with courses that cut through ridges or other significant structures. One way

▲ Figure 16-13 A consequent stream follows the initial slope of the land; a subsequent stream flows along structural weaknesses.

this can occur is when an established stream is interrupted by an uplift of land that is so slow that the stream maintains its previously established course by downward erosion, carving a deep gorge through hills or mountains. Because such a stream antedates (predates) the uplift, it is called an **antecedent stream** (Figure 16-14).

A *superposed* (or *superimposed*) *stream* existed when the landscape was higher, but this older, higher landscape has since largely or entirely eroded away. The original drainage pattern was incised into an underlying sequence of rocks of quite different structure. The result may be a drainage system that seems to bear no relation to the present surface structure.

Stream Drainage Patterns

In addition to the response of individual streams to the underlying structure, entire stream systems often form conspicuous drainage patterns in the landscape. These patterns develop largely in response to the underlying structure and slope of the land surface. Geologic–topographic structure can often be deduced from a stream drainage pattern, and conversely, drainage patterns can often be predicted from the underlying structure.

Dendritic Pattern: A treelike, branching **dendritic drainage pattern** (Figure 16-15) consists of a random merging of streams ("dendritic" comes from the Greek word *dendron* for "tree"). Its tributaries join larger streams irregularly but always at an angle smaller than 90°. With dendritic

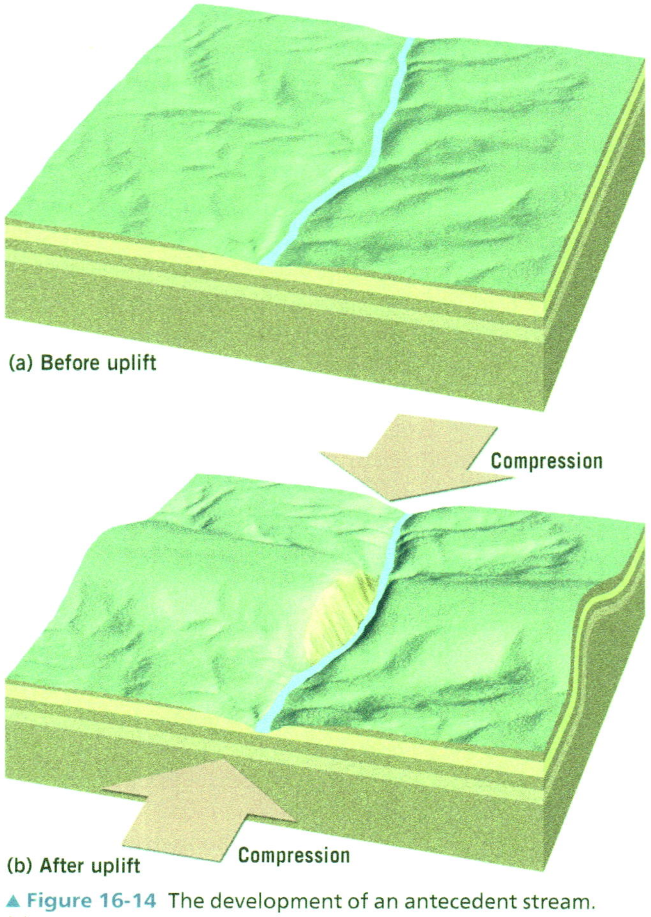

▲ Figure 16-14 The development of an antecedent stream. (a) The stream course is established before uplift begins. (b) The stream maintains its course and erodes through the slowly rising anticlinal ridge.

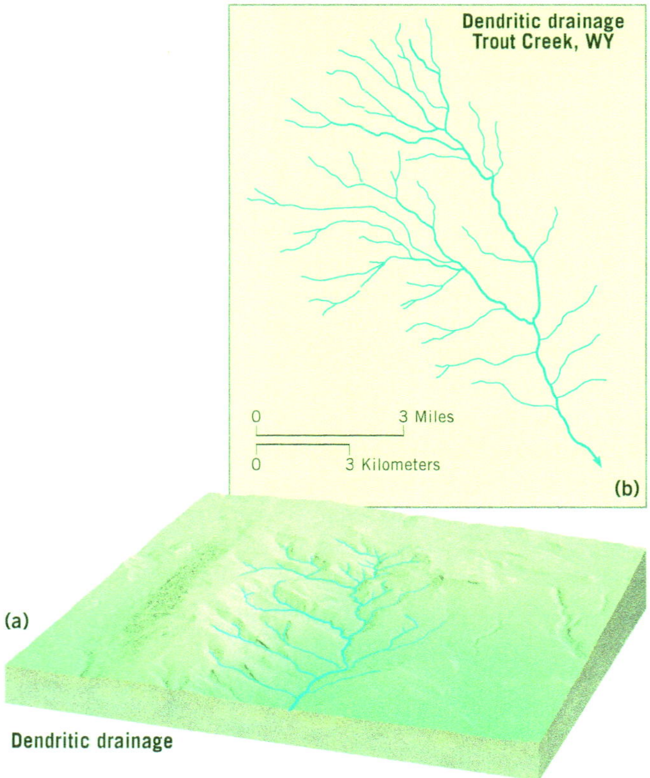

▲ Figure 16-15 (a) Dendritic drainage pattern. (b) Trout Creek, from the *Pat O'Hara Mountain, Wyoming*, topographic quadrangle.

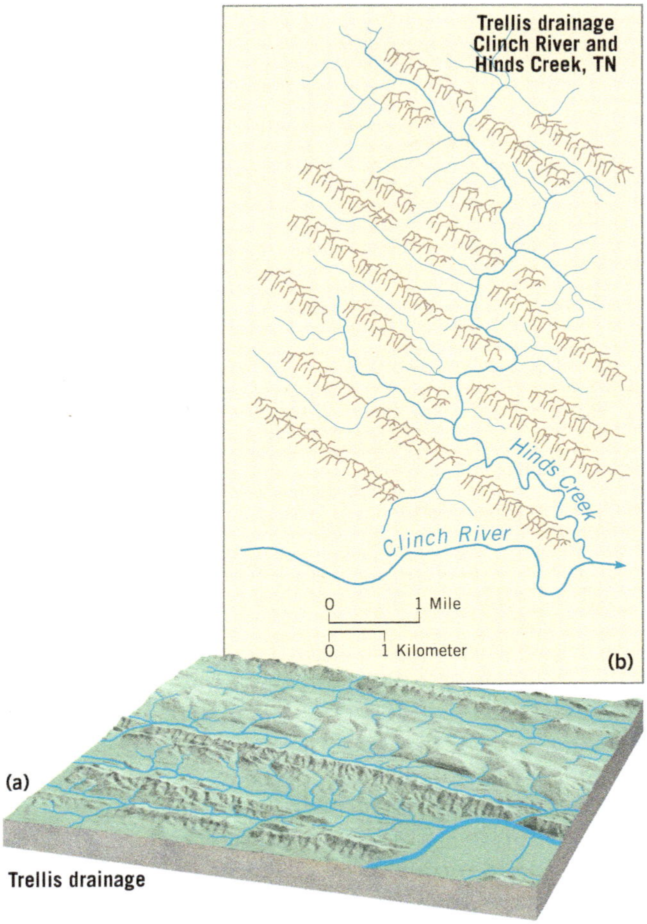

▲ **Figure 16-16** (a) Trellis drainage pattern. (b) Clinch River and Hinds Creek, from the *Norris, Tennessee*, topographic quadrangle.

drainage, the underlying structure does not control the evolution of the drainage pattern: the underlying rocks are more or less equally resistant to erosion. Dendritic patterns are more common than all others combined and are found almost anywhere.

Trellis Pattern: A **trellis drainage pattern** develops as a response to underlying structural control: alternating bands of tilted hard and soft strata. In this pattern, long, parallel streams are linked by short, right-angled segments (**Figure 16-16**).

Two regions of the United States are noted for their trellis drainage patterns: the Ridge-and-Valley section of the Appalachian Mountains and the Ouachita Mountains of western Arkansas and southeastern Oklahoma. In the Ridge-and-Valley section, which extends northeast–southwest for more than 1600 kilometers (1000 miles) from New York to Alabama, the drainage pattern developed in response to tightly folded Paleozoic sedimentary strata forming a series of parallel ridges and valleys (see Figure 14-46). Parallel streams flow in the valleys between the ridges, with short, right-angled connections here and there cutting through the ridges.

The marked contrast between trellis and dendritic patterns is shown dramatically by the principal streams of West Virginia (**Figure 16-17**). The folded structures of the eastern part of the state produce trellis drainage, whereas the nearly horizontal strata of the rest of the state are dendritic.

LearningCheck 16-7 **Contrast the influence of the underlying geologic structure in areas of dendritic drainage to that in areas of trellis drainage.**

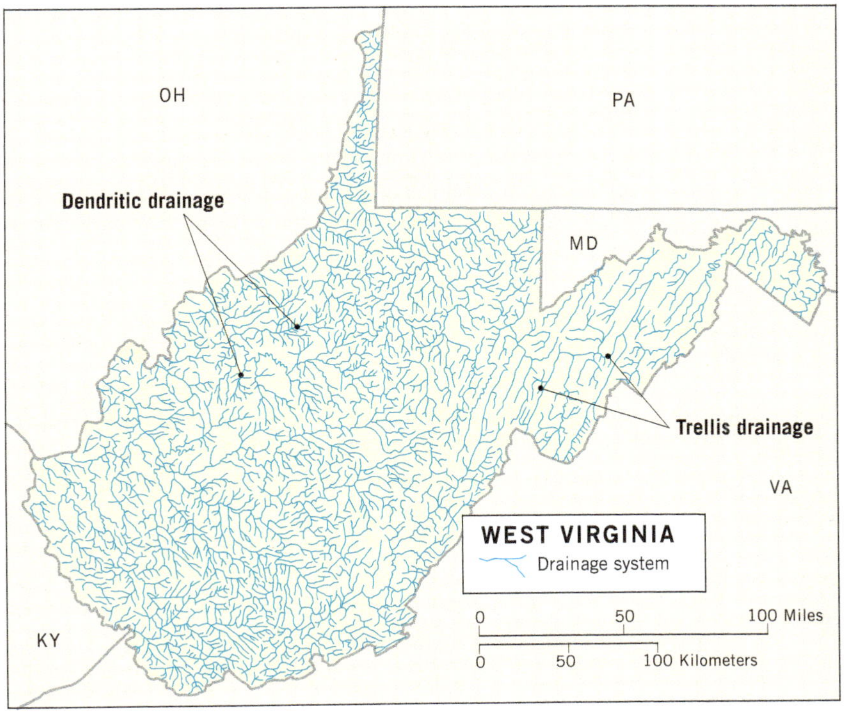

◀ **Figure 16-17** Drainage pattern contrasts in West Virginia. The trellis systems in the eastern part of the state are a response to parallel folding; the dendritic drainages in the western part developed because there are no prominent structural controls.

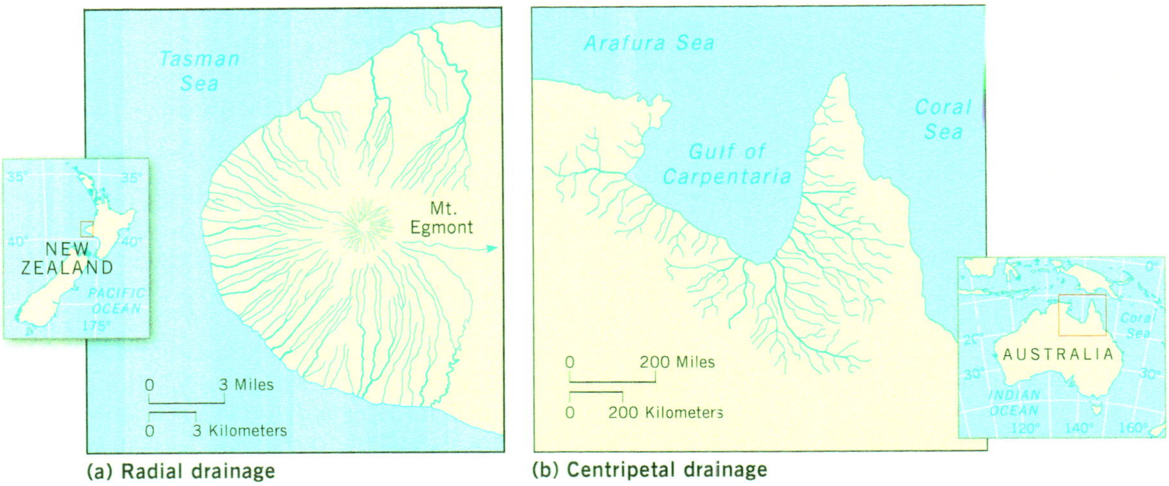

▲ **Figure 16-18** (a) The extraordinary radial drainage pattern of Mount Egmont in New Zealand. (b) The centripetal drainage pattern around the Gulf of Carpentaria in northeastern Australia.

Radial Pattern: A *radial drainage pattern* is usually found when streams descend from some sort of concentric uplift, such as an isolated volcano (**Figure 16-18a**).

Centripetal Pattern: A *centripetal drainage pattern*, essentially the opposite of a radial one, is usually associated with streams converging in a basin (**Figure 16-18b**).

Annular Pattern: More complex is a ring-shaped *annular drainage pattern* ("annular" comes from the Latin word *annularis*, meaning "ring"). This pattern can develop on a dome or in a basin where erosion has exposed alternating concentric bands of tilted hard and soft rock. The principal streams follow curving courses on the softer material, occasionally breaking through the harder layers in short, right-angled segments. At Maverick Spring Dome in Wyoming, ancient crystalline rock was pushed up through a sedimentary overlay and deeply eroded, thus exposing erosion-resistant crystalline rocks in the higher part of the hills, with upturned concentric sedimentary ridges (called *hogbacks*) around the margin (**Figure 16-19**). The streams are mostly incised into the softer sedimentary layers.

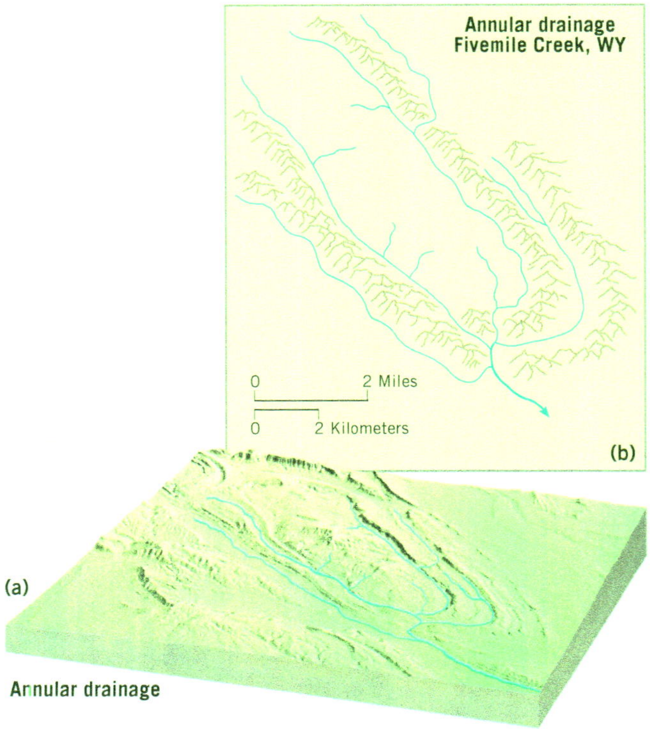

▲ **Figure 16-19** (a) Annular drainage pattern. (b) Fivemile Creek, from the *Maverick Spring Dome, Wyoming*, topographic quadrangle.

The Shaping and Reshaping Of Valleys

Running water shapes terrain partly by overland flow on interfluves but mostly by streamflow in the valleys. So, by focusing on the processes of erosion and deposition through which streams shape and reshape their valleys through time, we can understand the development of fluvial landforms in general.

Valley Deepening

Wherever a stream has a relatively rapid speed or relatively large volume, it expends most of its energy in **downcutting**. This lowering of the streambed involves the hydraulic power of the moving water, the prying and lifting capabilities of turbulent flow, and the abrasive effect of the stream's bedload as it rolls, slides, and bounces along the channel. Downcutting is most prevalent in the upper reaches of a stream, where the gradient is usually steep and the valley narrow. In general, downcutting produces a deep valley with steep sides and often a V-shaped cross section (**Figure 16-20**).

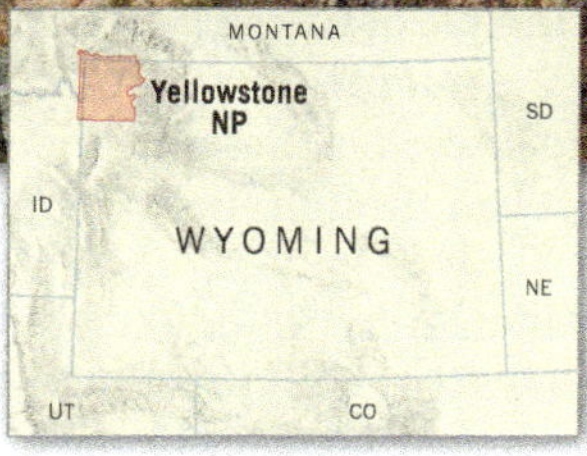

▲ **Figure 16-20** The Yellowstone River occupies a conspicuous V-shaped valley, shaped in part by downcutting.

Base Level: A stream deepens its valley by eroding the channel bed. If only downcutting were involved, the resulting valley would be a narrow, steep-sided gorge. Such gorges sometimes occur, but usually other factors are at work also, and the result is a wider valley. In either case, there is a lower limit to how much downcutting a stream can do; this limit is called the stream's **base level**. Base level is an imaginary surface extending beneath the continents from sea level at the coasts (Figure 16-21). This imaginary surface is not simply a horizontal extension of sea level, however; inland it is gently inclined at a gradient that allows streams to maintain some flow. Sea level, then, is the absolute, or *ultimate base level*, or lower limit of downcutting for most streams.

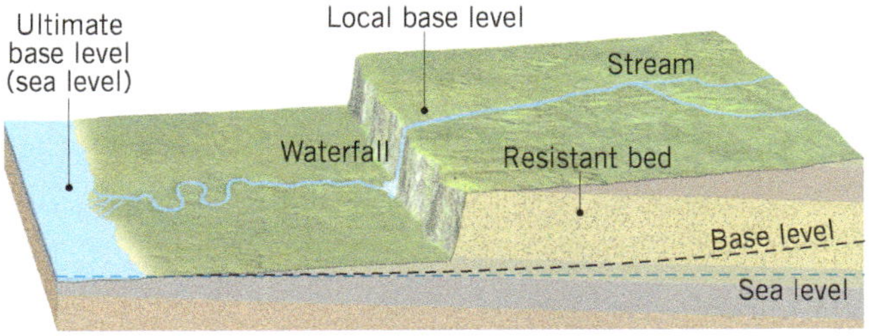

▲ **Figure 16-21** Comparison of sea level, base level, and local base level.

There are also local, or temporary, base levels, which are limits to downcutting resulting from structural or drainage conditions. For example, no tributary can cut deeper than its level of confluence with the higher-order stream it joins, so the level of their junction is a local base level for the tributary. Similarly, a lake normally serves as the temporary base level for all streams that flow into it.

Some valleys have been downfaulted to elevations below sea level (Death Valley in California is one example), producing a temporary base level lower than the ultimate base level. This can occur because the stream does not reach the ocean but terminates in an inland basin or body of water that is itself below sea level.

LearningCheck 16-8 **Under what conditions does a stream typically downcut rapidly?**

Graded Streams: The longitudinal profile of a stream (the down-valley change in elevation from source to mouth) is ultimately restricted by base level, but the profile at any given time—whether smooth, stepped, or a combination of both—depends on a variety of factors. Generally, a stream has the steepest gradient near its top and a gentler gradient toward its mouth. In the long term, a stream has a tendency to approach, or achieve, a state of dynamic equilibrium: the channel slope adjusts, changing the water velocity such that the amount of sediment entering a stretch of a stream is equal to the amount leaving it. A stream in which the gradient has adjusted to the point of allowing just the transportation of its load is called a **graded stream**.

Stream equilibrium can be upset easily. For example, a change in stream discharge or a sudden deposit of material from mass wasting temporarily changes the balance between transportation and deposition. For this reason, most streams do not achieve a graded condition throughout their course all of the time.

Knickpoint Migration: Waterfalls and rapids are often found in valleys where downcutting is prominent. They occur in steeper sections of the channel, and their faster, more turbulent flow intensifies erosion. These irregularities in a channel are collectively termed **knickpoints** (or *nickpoints*). Knickpoints are commonly the result of abrupt changes in bedrock resistance. The more resistant material inhibits downcutting, and as the water plunges over the waterfall or rapids, the increased speed tends to erode the channel above and along the knickpoint and fill the channel immediately downstream. This intensified action eventually erodes away the harder material, so the position of the knickpoint migrates upstream with a successively lower profile until it finally disappears and the channel gradient is smoothed (Figure 16-22). Such **knickpoint migration** is relatively rapid when the bedrock consists of soft sedimentary rocks but slower when the rock is resistant plutonic or metamorphic rock.

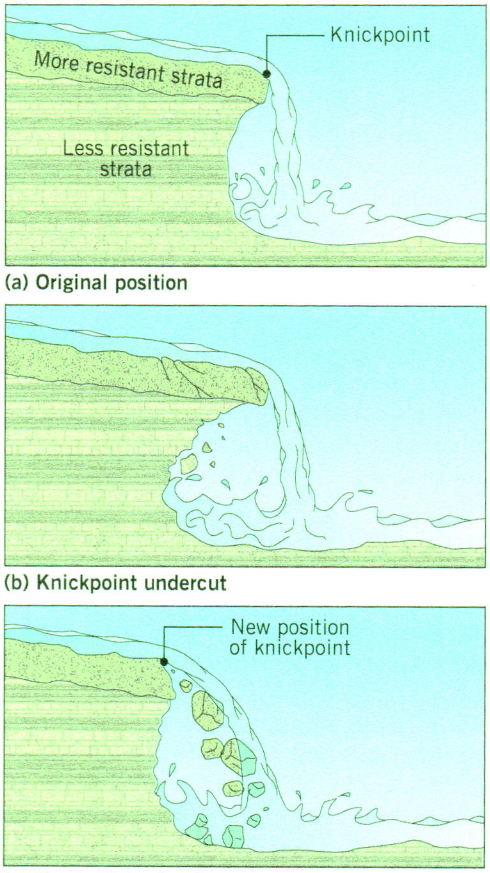

▲ **Figure 16-22** (a) Knickpoint formed where a stream flows over a resistant layer of rock. (b) The water flow undercuts the lip and (c) causes it to collapse. The knickpoint has migrated upstream.

Niagara Falls is a good example of knickpoint migration (**Figure 16-23a**). The Niagara River flows between Lake Erie and Lake Ontario. As the contemporary drainage system became established after the last retreat of Pleistocene ice sheets about 10,000 years ago, Lake Ontario was about 50 meters (about 150 feet) higher than its present level, and the Niagara River had no falls. However, an easterly outlet, the Mohawk Valley, developed for Lake Ontario, and the lake drained down to approximately its present level, exposing a prominent escarpment directly across the course of the river. The Niagara Escarpment is formed by a massive bed of resistant limestone that dips gently toward Lake Erie and is underlain by similarly dipping but softer strata of shale, sandstone, and limestone (**Figure 16-23b**).

As the river pours over the escarpment, the swirling water undermines the hard limestone by erosion of the weaker beds beneath, leaving a lip of resistant rock projecting without support. Through the years, the lip has collapsed, with block after block of limestone tumbling into a gorge below. After each collapse, rapid undermining takes place again, leading to further collapse. In this fashion the falls has gradually retreated upstream, moving southward a distance of about 11 kilometers (7 miles) from its original position along the trend of the escarpment, the retreat being marked by the deep gorge.

This principle of **knickpoint migration** illustrates dramatically the manner in which valley shape often develops first in the lower reaches and then proceeds progressively upstream, even though the water flows downstream!

LearningCheck 16-9 **Why does the position of a waterfall generally shift upstream over time?**

▲ **Figure 16-23** (a) Niagara Falls from the American side in Niagara Falls National Heritage Area. The Niagara River drops about 55 meters (180 feet) over a knickpoint formed by the Niagara Escarpment. (b) Niagara Falls was originally located where the Niagara River crosses the Niagara Escarpment but has retreated upstream to its present location.

Valley Widening

Where a stream gradient is steep and the channel is well above the local base level, downcutting is usually the dominant activity; as a result, valley widening is likely to be slow. Downcutting diminishes with time as the stream gradient is reduced, or wherever the stream course flows down a gentle slope. The stream's energy is then increasingly diverted into a meandering flow pattern. As the main flow of the current swings from one bank to the other, **lateral erosion** begins: erosion on the outside of the curves (the *cut bank*), where the water speed is greatest. Alluvium is deposited on the inside of the curves, forming a *point bar*, where the speed is slowest (**Figure 16-24**). The channel shifts position often, so undercutting is not concentrated in just a few locations. Rather, over a long period of time, most or all parts of the valley sides are undercut. (Landforms associated with such meandering streams are described later in this chapter.)

All the while a valley floor is widening through lateral erosion, mass wasting is usually helping wear back the valley walls—and in some valleys, mass wasting is the dominant process responsible for widening. In addition, similar processes along tributary streams also contribute to the general widening of the main valley.

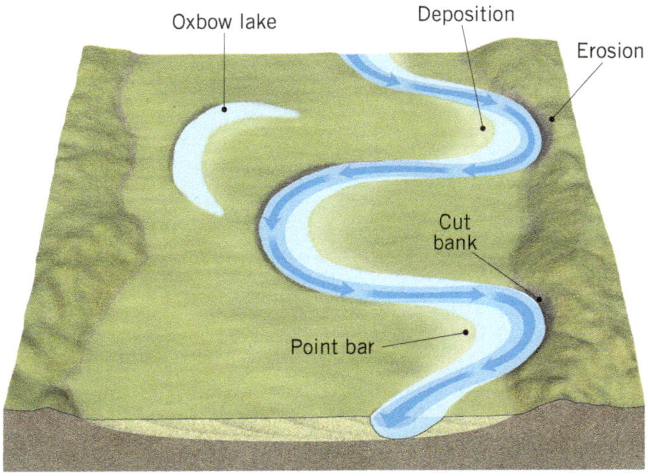

▲ **Figure 16-24** Lateral erosion in a meandering stream. Erosion occurs on the outside of the meander bend, where the water flow is fastest, forming a cut bank. Deposition of alluvium is common on the inside of a bend, forming a point bar. If the stream cuts through the neck of a meander, an oxbow lake forms.

Valley Lengthening

A stream may lengthen its valley in two quite different ways: by *headward erosion* at the upper end or by *delta formation* at the lower end. For our purposes, understanding the fluvial processes that lead to valley lengthening is more important than the actual distances involved, which may be quite modest.

Headward Erosion: No concept is more fundamental to an understanding of fluvial processes than **headward erosion**. It is the basis of rill, gully, and valley formation and extension. The upper perimeter of a valley is the line where the gentle slope of an interfluve changes to the steeper slope of a valley side. Overland flow from the interfluve drops abruptly over this slope break. The fast-moving water tends to undercut the rim of the interfluve, weakening it and often causing a small amount of material to collapse (**Figure 16-25**).

The result is a decrease in interfluve area and a corresponding increase in valley area. As the overland flow of the interfluve becomes part of the streamflow of the valley, there is a minute but distinct extension of rills and gullies into the drainage divide of the interfluve—in other words, a headward extension of the valley. When multiplied by a thousand gullies and a million years, this extension can lengthen a valley by tens of kilometers and expand a drainage basin by hundreds of square kilometers. Thus, the valley lengthens at the expense of the interfluve (**Figure 16-26**).

Stream Capture: Headward erosion is illustrated dramatically when a portion of the drainage basin of one stream is diverted into the basin of another stream by natural processes. This event, called **stream capture** or **stream piracy**, is relatively uncommon in nature, but evidence indicates that it does occur.

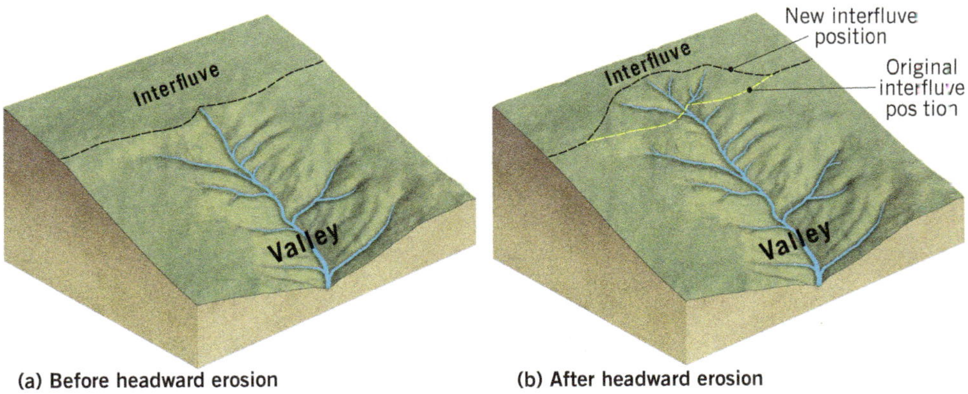

◀ **Figure 16-25** (a) Headward erosion occurs at the upper end of a stream where overland flow pours off the lip of the interfluve into the valley. (b) The channeled streamflow erodes the lip of the interfluve—over time, extending the valley headward at the expense of the interfluve.

▲ **Figure 16-26** Headward erosion of ephemeral streams into flat-lying sedimentary rock in Canyonlands National Park.

As a hypothetical example, let us consider two streams flowing across a coastal plain (**Figure 16-27**). Their valleys are separated by an interfluve, with stream A aligned such that headward erosion is extending its valley in the direction of stream B. Eventually stream A reduces the drainage divide between them and cuts into the channel of stream B, capturing its flow. In the parlance of the geographer, stream A has "captured" stream B. Stream A is called the *captor stream*, the lower part of B is the *beheaded stream*, the upper part of B is the *captured stream*, and the abrupt bend in the stream channel where the capture took place is called the *elbow of capture*.

Stream capture on a grand scale has taken place in West Africa. The headwaters of the Niger River are relatively near the Atlantic Ocean, but it flows inland rather than seaward for nearly 2000 kilometers (1200 miles). It then makes an abrupt turn to the southeast and continues for another 2200 kilometers before emptying into the Atlantic. In the past, the upper reaches of the Niger belonged to a separate river that flowed northeast into a great inland lake in what is now the central Sahara (**Figure 16-28**). This river was captured by the ancestral Niger, producing a great elbow of capture and leaving the beheaded stream to wither and dry up as the climate became more arid.

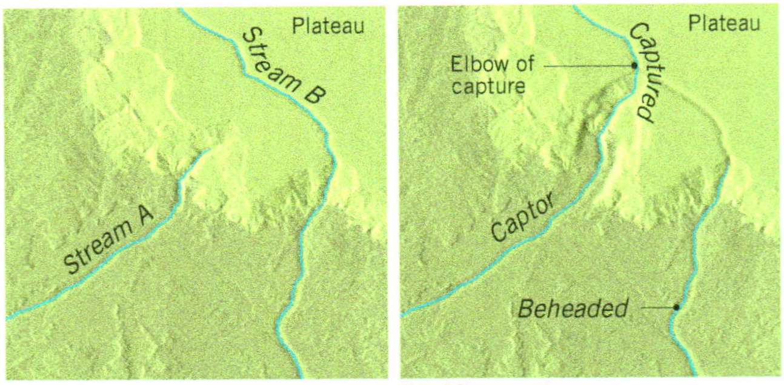

(a) Before capture (b) After capture

▲ **Figure 16-27** A hypothetical stream-capture sequence. The valley of stream A extends by headward erosion until it captures and beheads stream B.

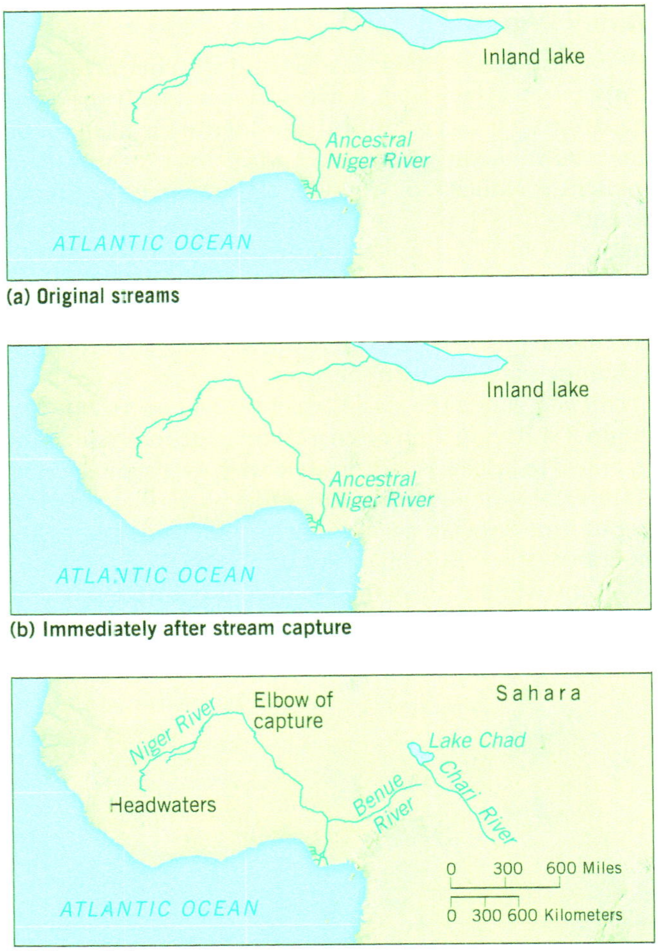

(a) Original streams

(b) Immediately after stream capture

(c) Present day

▲ **Figure 16-28** Stream capture, actual and anticipated in West Africa. (a) The upper Niger River was once part of an unnamed stream that flowed into a large lake in what is now the Sahara. (b) This ancient stream was captured by headward erosion of the ancestral Niger. (c) Through headward erosion, the Benue River may capture the Chari River in a few thousand years.

The map of Africa also provides us with an example of stream capture that may take place in the future. The Chari River of central Africa flows northwesterly into Lake Chad (see Figure 9-22). West of the Chari is an active and powerfully downcutting river in Nigeria: the Benue, a major tributary of the Niger. Some of the tributary headwaters of the Benue originate in a flat, swampy interfluve only a short distance from the floodplain of the Chari. As the valley of the Benue extends by means of headward erosion, the Benue may cut into the Chari, beheading it before our very eyes, so to speak, provided we can wait a few thousand years.

LearningCheck 16-10 **Explain the role of headward erosion in stream capture.**

Delta Formation: A valley can be lengthened not only at its headward end but also at its seaward end—in this case, by deposition. Flowing water slows down whenever it enters the quiet water of a lake or ocean and deposits its load. Most of this debris is dropped right at the mouth of the river in a landform called a **delta**, after a fancied resemblance to the Greek capital letter delta, Δ (Figure 16-29). The classic triangular shape is maintained in some deltas (known as *arcuate*, or arc-shaped, deltas), but it is severely modified in others because of variations in sediment load or the removal of those sediments by ocean waves and currents.

The stream slows down, losing both competence and capacity, and drops much of its load, which partially blocks the channel and forces the stream to seek another path (Figure 16-30). Later this new path is likely to become clogged, and the pattern is repeated. As a result, deltas usually consist of a maze of roughly parallel channels called *distributaries* (or *bayous*, as some are called in the Mississippi Delta) through which the water flows slowly toward the sea. Coarser sediments tend to be deposited immediately where the stream enters the water in sloping *foreset beds*. The foreset beds become covered with thin horizontal *topset beds* during floods. The finest sediments settle in *bottomset beds* on the ocean floor beyond the above-water portion of the delta. Rich alluvial sediments and an abundance of water favor the establishment of vegetation, which provides a base for further expansion of the delta. In this fashion, the stream valley extends downstream.

The largest delta in the world, covering an area of about 105,000 square kilometers (41,000 square miles), is formed by the combined deposits of the Ganges and Brahmaputra Rivers where they empty into the Bay of Bengal in South Asia. Although many other large rivers—such as the Mekong, Nile, and Mississippi Rivers—form large deltas, not all rivers do. At the mouths of some rivers, coastal currents are so strong that no delta forms; the stream sediment is simply swept away and deposited elsewhere along the coast or offshore (Figure 16-31).

▲ Figure 16-29 The Nile River delta, as viewed by a NASA satellite using the Multi-angle Imaging Spectroradiometer's (MISR's) nadir camera.

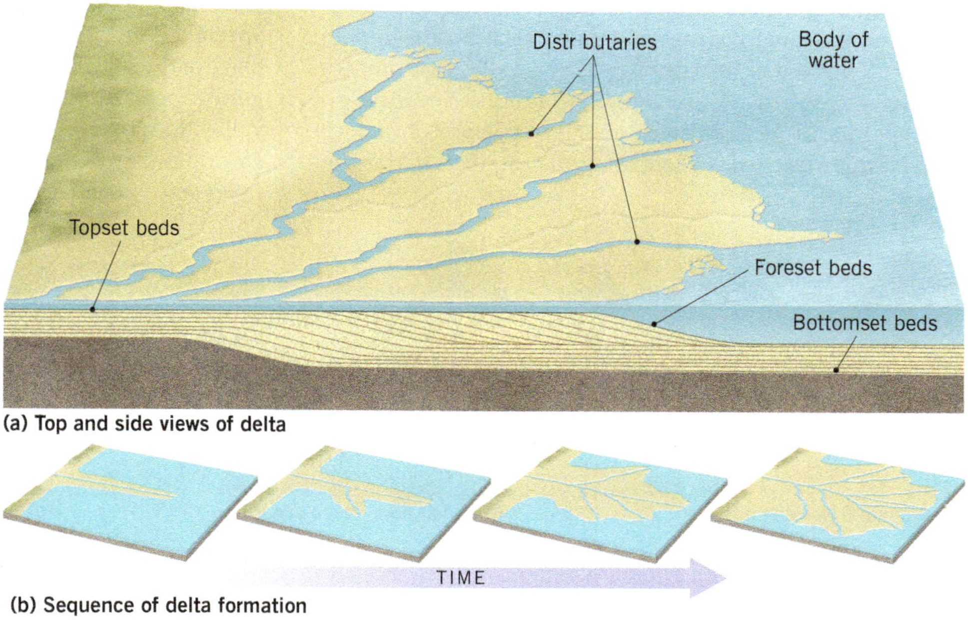

▶ Figure 16-30 Hypothetical sequence of delta formation in a quiet body of water. (a) *Foreset bed* deposits consist of coarser sediments dropped immediately by the stream; *topset beds* are deposited over foreset beds during floods; *bottomset beds* are composed of fine sediments that settle some distance from the mouth of the stream. (b) The sequential process of delta extension.

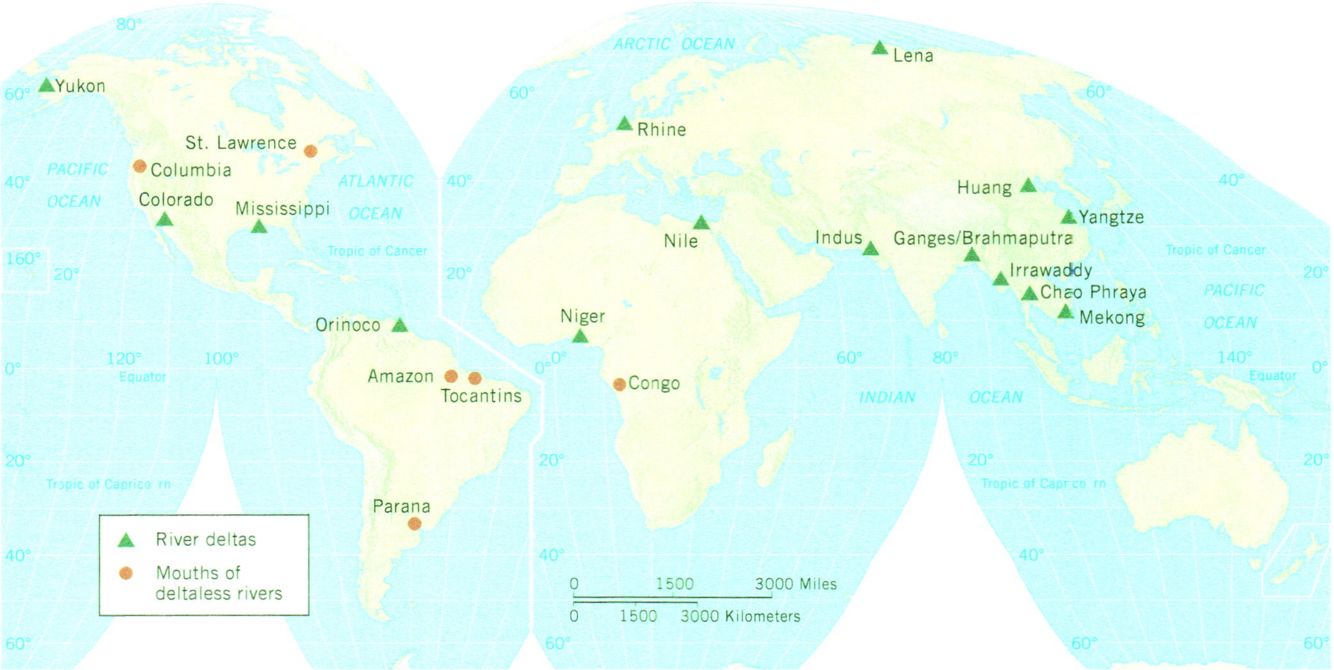

▲ **Figure 16-31** Locations of the world's largest deltas and deltaless rivers. Not all large rivers form deltas. Major rivers such as the Amazon, Congo, and the Paraná have not developed significant deltas.

As we see later in this chapter when we discuss flood control strategies, many of the world's major deltas, such as the Nile and Mississippi, are becoming smaller as a direct consequence of human manipulation of the rivers that feed them, as well as slowly rising sea levels.

Deposition in Valleys

Thus far, we have emphasized the prominence of the removal and transportation of material in the formation and shaping of valleys—but deposition, too, has a role in these processes.

In nearly every stream, sediment is continuously rearranged in response to variations in flow speed and volume. Any time a stream loses power to transport its load, alluvium can be deposited almost anywhere in a valley bottom—on the stream bottom, on the sides, in the center, at the base of knickpoints, in overflow areas, and in a variety of other locations. **Aggradation** is a general term that refers to this process of deposition.

During high-water periods, when flow is fast and voluminous, the stream scours its bed and shifts most or all sediment downstream. During low-water periods, particularly after a flood, the flow is slowed and sediment is more likely to settle to the bottom, which results in filling of the channel. Under some circumstances, alluvium may accumulate on the streambed to such an extent that the level of the stream bottom is raised (Figure 16-32).

LearningCheck 16-11 **Under what conditions will a stream begin to deposit some of its load as alluvium?**

Floodplains

Special attention needs to be given to an important assemblage of fluvial landforms. The most prominent depositional landscape is the **floodplain**: a low-lying, nearly flat alluvial valley floor that is periodically inundated with floodwaters. Floodplains frequently form where a meandering stream flows across a wide, nearly level valley floor.

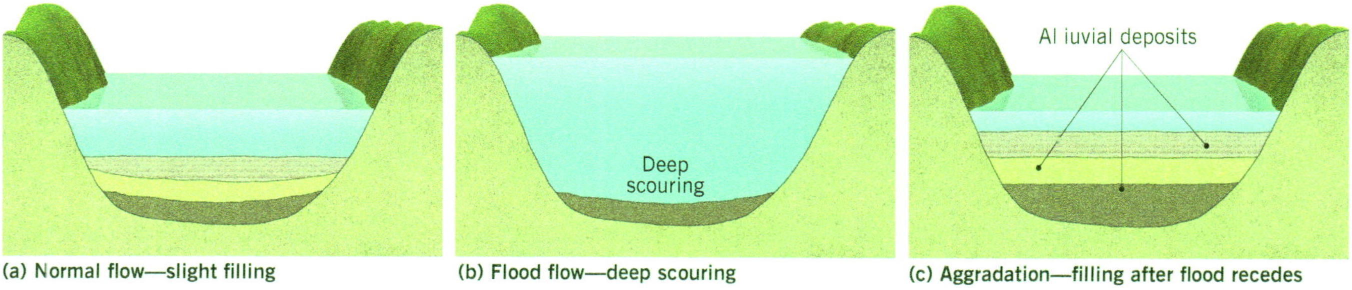

▲ **Figure 16-32** Changing channel depth and shape during a flood: (a) Normal flow with slight filling. (b) Flood flow significantly deepens the channel by scouring. (c) As the flood recedes, considerable filling raises the channel bed again.

Floodplain Landforms

The frequent shifting of stream meanders produces an increasingly broad, flattish valley floor largely or completely covered with alluvium left by periodic floods. At any given time, a stream is likely to occupy only a small portion of the flatland, although during periods of flood flow, the entire floor may be flooded. For this reason, the valley bottom is properly termed a *floodplain*.

The outer edges of a floodplain are usually bounded by an increase in slope, sometimes marking the outer limit of lateral erosion and undercutting where the flat terrain abruptly changes to a line of *bluffs*. Valley widening and floodplain development may extend for great distances. The floodplains of many of the world's largest rivers are so broad that if you stood on the bluffs on one side, you could not see the bluffs on the other.

Cutoff Meanders: The most conspicuous feature of a floodplain is often the meandering channel of its river (Figure 16-33). A meandering river swings back and forth in ever-expanding loops. Eventually, when the radius of a loop reaches about 2.5 times the stream's width, the loop stops growing. Often a meander loop is bypassed as the stream channel shifts by lateral erosion, cuts a new channel across its neck, and starts meandering again, leaving the old meander loop as a **cutoff meander**. The cutoff portion of the channel may remain filled with water for a period of time as an **oxbow lake** (so named because its rounded shape resembles the bow part of yokes used on teams of oxen; oxbow lakes are called *billabongs* in Australia). Oxbow lakes gradually fill with sediment and vegetation, becoming oxbow swamps. Eventually they dry up, leaving **meander scars** (Figure 16-34).

Natural Levees: A floodplain is slightly higher along the banks of the stream channel than elsewhere. As the stream overflows its banks during a flood, friction with the floodplain surface slows the current when it leaves its normal channel. This causes deposition to take place along the margins of the main channel, producing **natural levees** (derived from the Latin *levare*, meaning "to raise") on each side of the stream (Figure 16-35). Natural levees merge outwardly and almost imperceptibly with the less well-drained, lower portions of the floodplain, generally referred to as *backswamps*.

Yazoo Streams: Sometimes a tributary stream entering a floodplain cannot flow directly into the main channel

▲ Figure 16-33 The floodplain of the North Platte River in southern Wyoming. As with other floodplains, the meandering river channel is the most conspicuous feature.

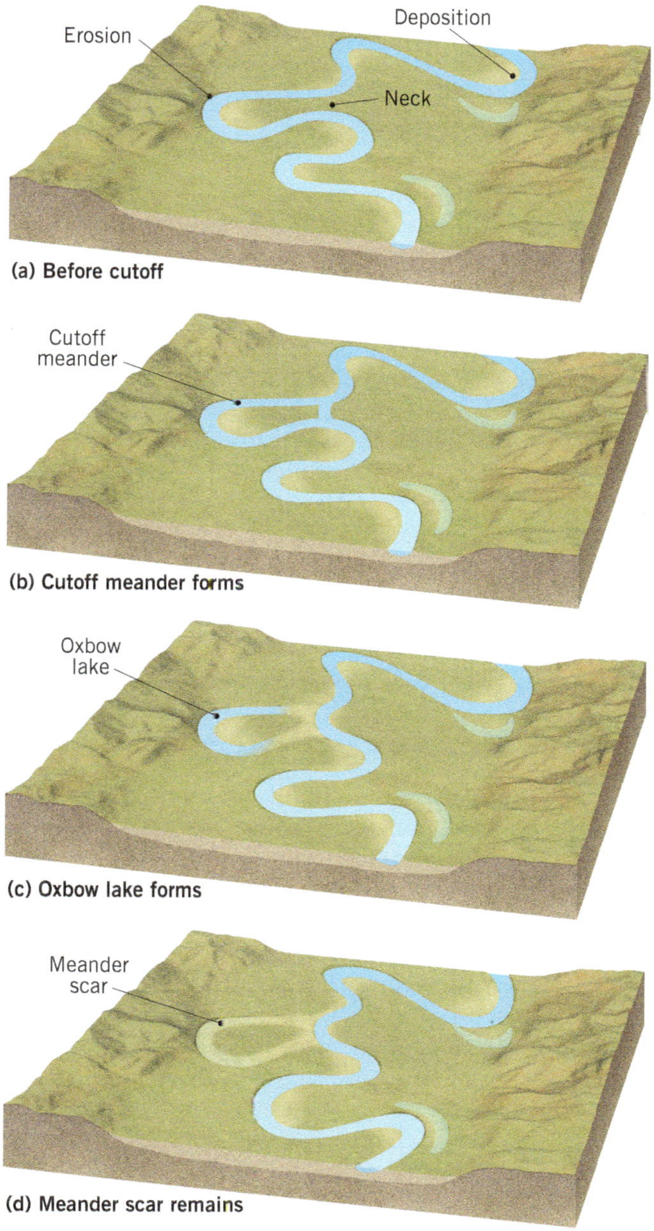

▲ Figure 16-34 The formation of a cutoff meander on a floodplain. As the river cuts across the narrow neck of a meander, the river bend becomes an oxbow lake, which becomes an oxbow swamp, which becomes a meander scar.

Condor Video
Meandering Rivers
https://media.pearsoncmg.com/ph/esm/esm_mcknight_physgeo_13/media/

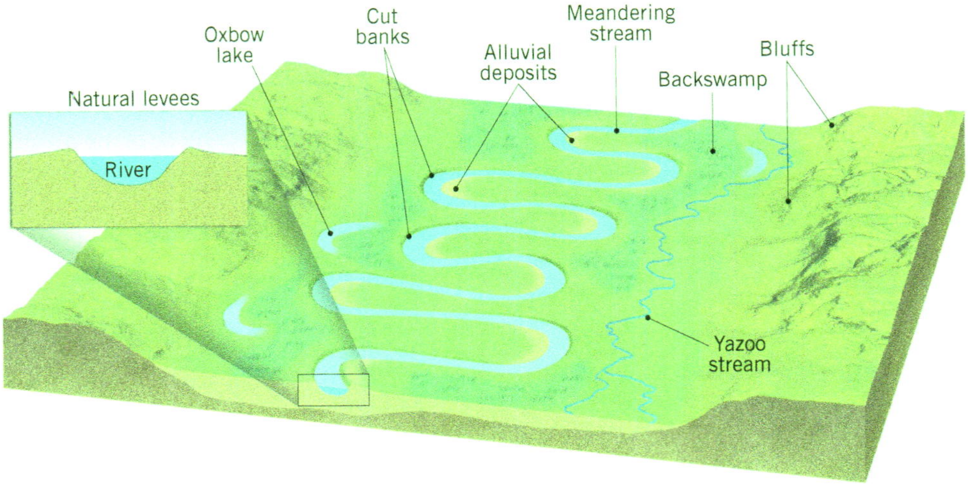

Animation
Oxbow Lake Formation
https://media.pearsoncmg.com/ph/esm/esm_mcknight_physgeo_13/media/

Animation
Floods and Natural Levee Formation
https://media.pearsoncmg.com/ph/esm/esm_mcknight_physgeo_13/media/

◂ **Figure 16-35** Typical landforms in a floodplain. Natural levees build up from alluvium deposited along the sides of a stream during floods.

because it encounters prominent natural levees. The tributary then flows down-valley in the backswamp zone, running parallel to the main stream for some distance before finding an entrance. Such a tributary stream is called a **yazoo stream**, after Mississippi's Yazoo River, which flows parallel to the Mississippi River for about 280 kilometers (175 miles) before joining it.

Although the landforms we have just described are most prominently displayed on the floodplains of large streams, nearly identical processes and landforms may be found anywhere a stream is meandering over a nearly flat surface and down a gentle slope. Thus, examples of cutoff meanders and oxbow lakes might well be seen on the valley floor of a small creek meandering across a flat alpine meadow.

LearningCheck 16-12 How does an oxbow lake form?

Modifying Rivers for Flood Control

Flat land, abundant water, and productive soils attract humans to valley bottoms, so such areas are often places of intensive agriculture, transportation routes, and urban development. However, every stream is subject to at least occasional flooding; nature and humans must coexist in an uneasy juxtaposition on floodplains.

Accordingly, wherever we have settled in large numbers in river valleys, people have gone to extraordinary lengths to mitigate potential flood damage. The principal means for averting disaster are structures such as dams, artificial levees, and overflow floodways. As an example of the remarkable efforts that go into such endeavors, we consider the major river system of North America: the Mississippi.

Flood Control on the Mississippi River

The Mississippi originates in Minnesota and flows more or less directly southward to the Gulf of Mexico below New Orleans (**Figure 16-36**). It is joined by a number of right-bank tributaries along the way, of which the Missouri is by far the most important. There are also many left-bank tributaries, of which the Ohio—with its own tributary, the Tennessee—is the most notable. (Tributary streams are designated as "right bank" or "left bank" from the perspective of an observer looking downstream.)

Dams: All four of these rivers have been thoroughly dammed, largely for flood control but also for other benefits such as hydroelectricity production, navigation stabilization, and recreation. On the Mississippi, the Army Corps of Engineers operates and maintains 29 low dams between St. Louis and the head of navigation at Minneapolis, most equipped with hydroelectricity facilities and each with locks to allow barges and other shallow-draft boats to pass. (The dams maintain a water depth of at least 2.7 meters [9 feet] throughout the navigable course of the river.) The Missouri has fewer but larger dams, six in all, widely scattered from Montana to Nebraska. They are primarily flood-control dams. The Ohio is punctuated by more than three dozen low dams, mainly to keep

▴ **Figure 16-36** Sites of major dams on the Mississippi, Missouri, Ohio, and Tennessee Rivers.

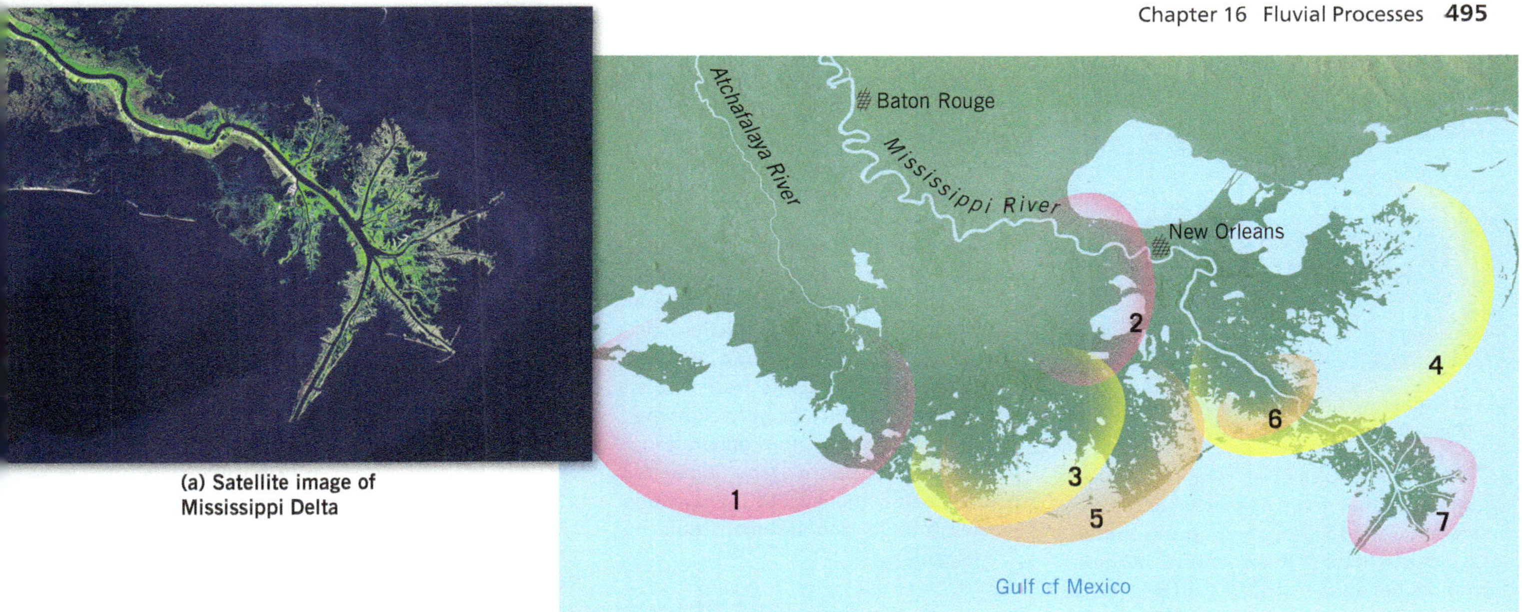

(a) Satellite image of Mississippi Delta

(b) Evolution of the delta

▲ **Figure 16-37** (a) Satellite image of the bird's-foot delta of the Mississippi River. (b) Over the last 5000 years, seven subdeltas were built by the Mississippi River ("1" is the oldest; "7" is the youngest). The present bird's-foot delta ("7") began to form about 600 years ago.

pools deep enough for barge navigation; flood control is secondary. Most thoroughly dammed is the Tennessee, whose nine mainstream dams have reduced it to a series of quiet reservoirs for its entire length, apart from the upper headwaters. These are Tennessee Valley Authority (TVA) dams built during the 1930s for flood control and power generation.

Artificial Levees: These four river valleys contain an extensive series of artificial levees designed to protect the local floodplain and move floodwaters downstream. Levee building has a vicious-circle aspect: anytime a levee is raised in an upstream area, most downstream locales require higher levees to pass the floodwaters on without overflow. These levees are usually the highest parts of the landscape, particularly in the ever-flatter and more extensive floodplains downstream. In addition, although the system of artificial levees may work to contain flooding in most years, if the levees fail, the results can be catastrophic for the population living on the adjacent floodplain. (Many experienced hydrologists and geologists say, somewhat cynically, that there are only two kinds of flood-control levees: "those that have failed and those that will fail.")

Managing the Lower Mississippi: At the lower end of the Mississippi system, in southern Louisiana, river control is extremely complicated, and the results of human efforts are somewhat ambiguous. This region, which receives the full flow of North America's mightiest river system, is exceedingly flat and thus has poor natural drainage. During the past 5000 years, the lower course of the river has shifted several times, producing at least seven subdeltas that are the principal elements of the present *bird's-foot delta* of the Mississippi—a series of narrow, sediment-lined distributary channels (**Figure 16-37**). The main flow of the river during the past 600 years or so has been along its present course, southeast from New Orleans. This portion of the delta was built out into the Gulf of Mexico at a rate of more than 10 kilometers (6 miles) per century in that period.

Normally, the main flow of the river would now be shifting to the shorter and slightly steeper channel of the Atchafalaya River, a prominent distributary of the ancestral delta a few kilometers west of the present main channel. However, enormous flood-control structures—part of the Old River Control Project—were erected upstream from Baton Rouge in an effort (thus far successful) to prevent the river from abandoning its present channel and delta (**Figure 16-38**).

▲ **Figure 16-38** Old River Control Project on the Mississippi River upriver from Baton Rouge, Louisiana. Diversion dams keep most of the Mississippi (upper left) from draining into the Atchafalaya River.

People & the Environment

The Future of the Mississippi River Delta

• **Natalie Peyronnin,** Mississippi River Delta Restoration

The Mississippi River is a tenth-order stream that drains 41 percent of the continental United States and part of southern Canada as it flows 3730 kilometers (2320 miles) from its headwaters to the Gulf of Mexico. One of the largest rivers in the world in discharge and sediment load, the Mississippi River and its distributaries constructed a delta plain of about 25,000 square kilometers (9653 square miles) over 7000 years through overbank flow and crevasses (breaks in the natural levee that allow streamflow into the floodplain). Levees, canals, and other human-made alterations, as well as natural processes, such as subsidence, have led to the loss of nearly 5000 square kilometers (1931 square miles) of deltaic wetlands in the last 80 years. To put it another way, in just 80 years, the delta has lost the land it took the river nearly 1400 years to build.

One major cause of this land loss is the Mississippi River and Tributaries system of levees and floodgates constructed by the U.S. Army Corps of Engineers after the flood of 1927. The resulting land loss was foreseen, as described by E. L. Corthell in an 1897 issue of *National Geographic*:

> No doubt the great benefit to the present and two or three following generations accruing from a complete system of absolutely protection levees, excluding the flood waters entirely from the great areas of the lower delta country, far outweighs the disadvantages to future generations from the subsidence of the Gulf delta lands below the level of the sea and their gradual abandonment due to this cause.

Future of the Delta without Action: Hurricanes such as Hurricane Katrina in 2005 demonstrated that coastal Louisiana communities have become more vulnerable to storm surge flooding with the loss of the area's protective wetlands. If no new actions are taken, Louisiana could lose another 4500 square kilometers (1737 square miles) of land in the next 50 years (Figure 16-G), and annual damages from storm surges could increase from \$2.4 billion to \$23.4 billion by 2060.

Coastal Master Plan: Faced with this dismal future, the Louisiana state government embarked on an extensive comprehensive planning effort that resulted in the 2017 Coastal Master Plan. This \$50 billion, 50-year plan outlines 124 restoration and protection projects that have clear economic, social, and environmental benefits. The master plan unified policymakers, scientists, stakeholders, and the public and was unanimously passed by the Louisiana legislature. Implementation of the master plan projects could result in no net loss of land after 20 years and annual net gain after 30 years, as well as reduce damages from flooding by more than 75 percent.

Sediment Diversions: Reconnecting the Mississippi River to its floodplain is a keystone activity of the 2017 Coastal Master Plan. Through structures called *sediment diversions*, freshwater and sediment from the river would be allowed to flow back into the wetlands, depositing sediment in shallow open water bodies or degraded marshes to build and sustain the coastal ecosystem. Sediment diversions are designed to mimic natural processes—overbank flow and streamflow through crevasses—during flood stages. Because they are so effective, the Master Plan concludes that sediment diversions are essential to maintaining coastal Louisiana. Work is ongoing to design and engineer these projects.

Questions

1. What actions have caused the loss of Louisiana's wetlands?
2. What actions could help build new land in the delta, and how are they designed to mimic natural processes?

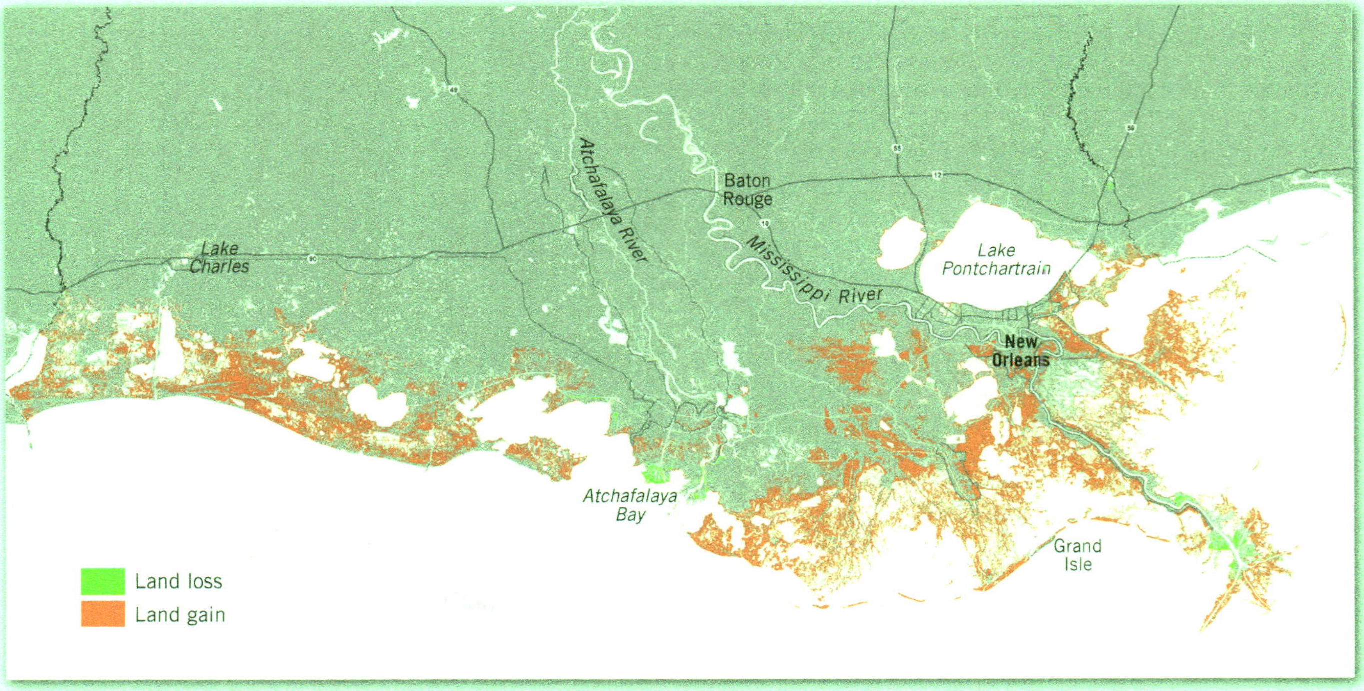

▲ **Figure 16-G** Predicted land changes over the next 50 years without future actions (Louisiana's 2017 Coastal Master Plan).

▲ **Figure 16-39** Marshes in the Mississippi River Delta near Pilottown, Louisiana.

Mobile Field Trip
Mississippi Delta
https://media.pearsoncmg.com/ph/esm/esm_mcknight_physgeo_13/media/

The natural processes operating on the floodplain and delta of the Mississippi River are changing: the delta is sinking, and natural marshlands are retreating (**Figure 16-39**). Part of the reason is human-built flood control and navigation measures—see the box *People & the Environment: The Future of the Mississippi River Delta*.

LearningCheck 16-13 **How has human activity contributed to changes in the Mississippi River delta?**

Living on the Floodplain: Can people live on a floodplain without incurring untenable economic and ecological costs? Part of the answer rests in sensible land-use practices. Local governments in most urban areas adjacent to streams have drafted maps showing the extent of floodplain inundation for a *design flood*, such as the 100-year flood; zoning regulations can then restrict land use so as to reduce the likelihood of costly damage. Diversion or bypass channels can be constructed to divert floodwaters out onto an "artificial" floodplain. These diversion areas can be used profitably for other activities, such as farming, most of the year.

Restoring Floodplains: In some locations, levees are being removed in an effort to give streams back at least part of their original floodplains. This approach restores valuable wetlands habitat and may dampen the surge of a flood downstream by increasing the "storage capacity" of a floodplain upstream. Even streams within large cities are being restored—see the box *Global Environmental Change: Restoring Urban Streams*.

Stream Rejuvenation

From time to time, the elevation of continental surfaces relative to sea-level changes. This change can be caused by a drop in sea level (as occurred around the world during the various ice ages when frozen water accumulated on the land, reducing the amount of water in the oceans), but it is much more commonly the result of tectonic uplift of the land surface. When such uplift occurs, it "rejuvenates" streams in the area by increasing their gradients. The increased gradient causes the streams to flow faster, which provides renewed energy for downcutting. (A dramatic, sustained increase in discharge may also rejuvenate a stream.) Vertical incision, which may have long been dormant, is initiated or intensified by such **stream rejuvenation**.

Stream Terraces: If a stream occupied a broad floodplain prior to rejuvenation, with renewed downcutting the stream carves a new valley floor (**Figure 16-40**). The old floodplain can no longer function as an overflow area and instead becomes an abandoned stretch of flat land overlooking the new

Animation
Stream Terrace Formation
https://media.pearsoncmg.com/ph/esm/esm_mcknight_physgeo_13/media/

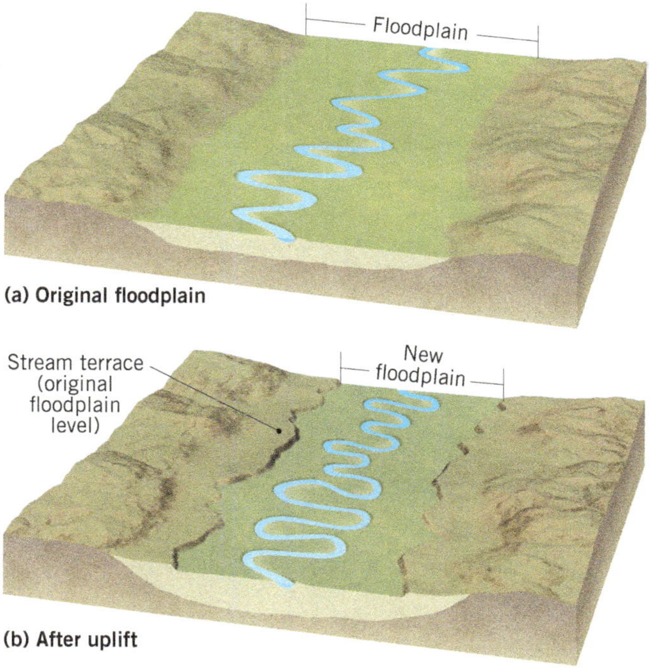

▲ **Figure 16-40** Stream terraces normally represent sequences of uplift and rejuvenation. (a) A river meanders across an alluvial floodplain. (b) After uplift, the stream downcuts and then widens a new floodplain, leaving remnants of the original floodplain: a pair of stream terraces

Global Environmental Change

Restoring Urban Streams Around the World

• **Nancy Lee Wilkinson,** San Francisco State University

(a) Before restoration

(b) After restoration

▲ **Figure 16-I** (a) Dismantling the freeway along the Cheonggyecheon River in Seoul, South Korea. (b) Cheonggyecheon River Park after restoration.

Around the world, rivers once altered for flood control or other purposes are being restored.

The Los Angeles River: The Los Angeles River (**Figure 16-H**) was the original water supply for the city of Los Angeles—as well as the city's first sewer. After devastating floods in 1914, 1934, and 1938, the U.S. Army Corps of Engineers encased the river in a concrete channel 82 kilometers (51 miles) long to hasten transport of floodwaters to the sea. By the 1980s, few Angelenos realized the massive concrete channel housed a river. Friends of the Los Angeles River (FoLAR) was organized in 1986 to protect and restore the river and its riparian habitat, with the goal of creating a swimmable, fishable, boatable river. FoLAR's efforts have been supported by federal and local funding. The U.S. Army Corps of Engineers is now overseeing implementation of a comprehensive ecosystem restoration (for more see http://www.lariver.org).

International Efforts: River and stream restoration is a worldwide phenomenon. For example, in Australia attention has been directed to improving the ecological conditions of rivers and riparian systems damaged by 150 years of European settlement. An estimated 2000–3000 river restoration projects are currently underway across Australia, including riparian management; bank stabilization; in-stream habitat improvement; channel reconfiguration; storm-water management; fish-passage improvements; water-quality management; and aesthetic, recreational, and educational projects (for more see https://arrc.com.au). Similar efforts are underway worldwide, including in Europe (see, for example, www.ecrr.org).

Benefits of River Restoration: The Cheonggyecheon River in Seoul, South Korea, was once little more than a sewer capped by a six-lane freeway. In 2002, the city's mayor pledged to tear down the freeway, restore the river, and create a park 11 kilometers (7 miles) long on the riverbanks (**Figure 16-I**). Upon the project's completion in 2005, environmental improvements were immediately apparent. For example, summer temperatures along the restored river were much cooler and more pleasant. Riparian birds, fish, and wildlife returned. Citizens flocked to the water's edge to play, relax, or exercise, and bus service was improved. Cheonggyecheon Park has become a leading tourist destination and a model for other cities seeking to link river restoration and urban regeneration.

Restoration Science: Although consensus about the importance of river restoration is growing, scientific analysis of project results is needed to support future restoration efforts. Limited access to information makes it difficult to monitor and evaluate past results. The United States, Australia, and Europe have established organizations to develop restoration databases that will enable scientists to analyze the extent, nature, scientific basis, and success of stream restoration projects (e.g., Australia's National Water Commission). The ongoing health of many river systems worldwide could depend on their efforts.

▼ **Figure 16-H** The Los Angeles River looks more like a freeway than a river.

Questions

1. What might be the most challenging aspects of stream restoration in an urban setting?
2. What physical impacts might stream restoration have on urban climate, biogeography, nutrient cycling, and other physical geographic patterns and processes?

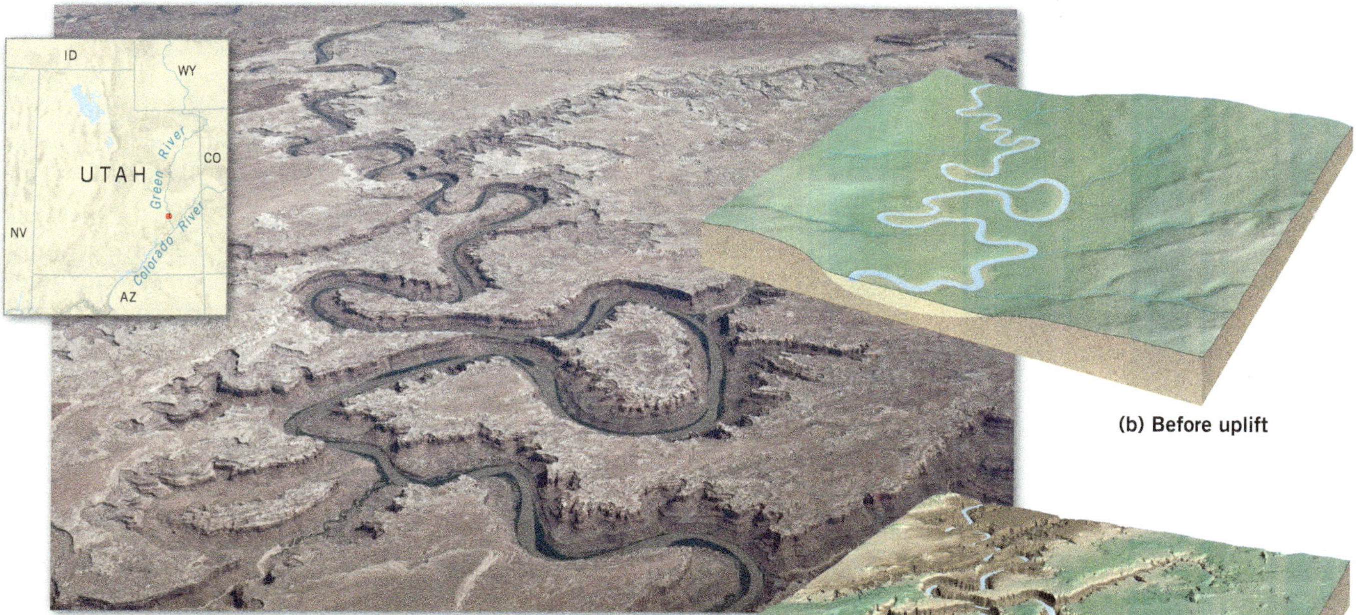

(b) Before uplift

(a) Green River in Utah

Condor Video
River Terraces and Base Level
https://media.pearsoncmg.com/ph/esm/esm_mcknight_physgeo_13/media/

(c) After uplift

▲ **Figure 16-41** (a) Deeply entrenched meanders of the Green River in southeastern Utah. (b) Floodplain meanders may become rejuvenated by uplift. (c) If a stream maintains its meandering pattern during uplift, renewed downcutting can produce entrenched meanders.

valley. This remnant of the previous valley floor is called a **stream terrace**. Terraces often occur in pairs, one on either side of the newly incised stream channel.

Entrenched Meanders: Under certain circumstances, uplift and rejuvenation have a different, more conspicuous topographic effect: **entrenched meanders** (Figure 16-41). These topographic features form when an area containing a meandering stream is uplifted slowly and the stream incises downward while retaining its meandering course—often with the entrenchment extending itself headward upstream. In some cases, such meanders may become entrenched in narrow gorges hundreds of meters deep.

LearningCheck 16-14 How do stream terraces develop?

Theories of Landform Development

The internal and external processes operating on Earth's surface produce an infinite variety of landscapes. Systematizing this vast array into a coherent body of knowledge has been the goal of many geomorphologists. The role of running water is key in comprehensive theories of landform development.

Davis's Geomorphic Cycle

The first, and in many ways most influential, model of landscape development was propounded by William Morris Davis, an American geographer and geomorphologist active in the 1890s and early 1900s. Sometimes called the *cycle of erosion*, Davis's theory is now usually referred to as the *geomorphic cycle*.

Davis believed that we can interpret any landscape by analyzing structure, process, and stage. *Structure* refers to the type and arrangement of the underlying rocks and surface materials, *process* is concerned with the internal and external forces that shape the landforms, and *stage* is the length of time during which the processes have been at work.

Davis metaphorically likened the development of landforms to the life cycle of an organism, recognizing stages of development he called "youth," "maturity," and "old age" (Figure 16-42). He envisioned a continuous sequence of terrain evolution in which a relatively flat surface is uplifted rapidly and little erosion takes place until uplift is complete:

- **Youth:** The initial flat surface is dissected by streams, which begin to incise deep, narrow, steep-sided, V-shaped valleys separated by flattish interfluves. The streams flow rapidly and have courses marked by waterfalls and rapids.

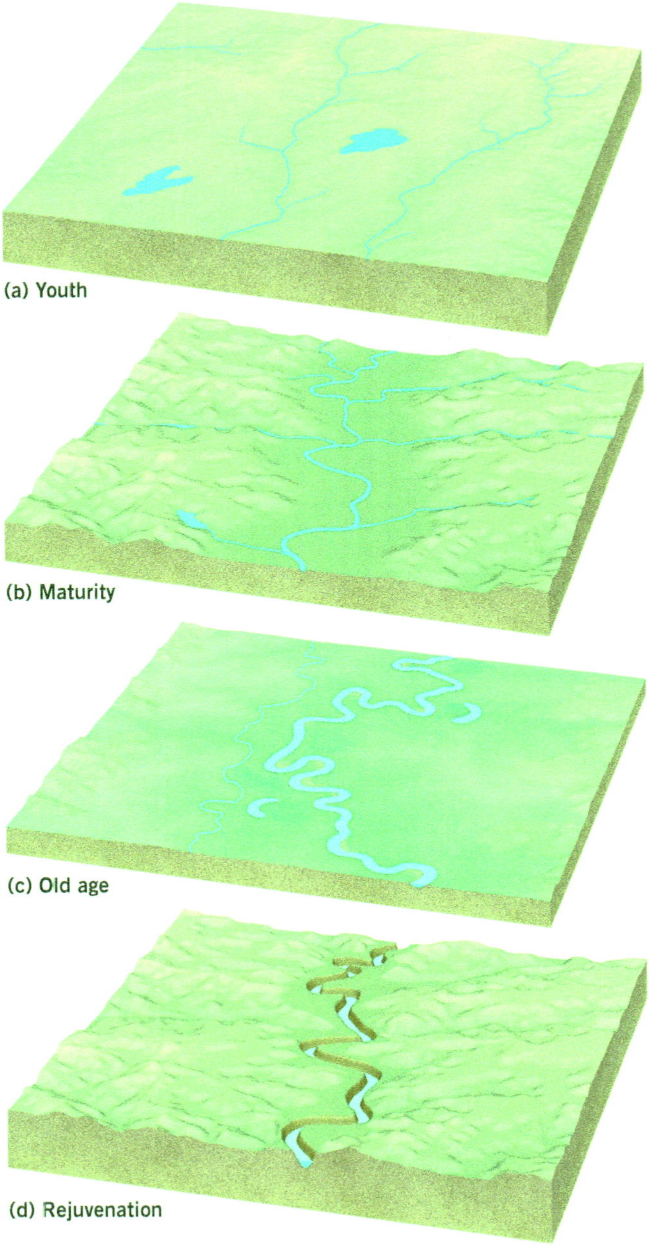

▲ **Figure 16-42** Davis's idealized geomorphic cycle.

- **Maturity:** The streams approach an equilibrium condition, having worn away the falls and rapids and developed smooth profiles. As stream gradient is reduced, streams begin to meander and floodplains form. The drainage system expands, dissecting all of the interfluves.
- **Old Age:** Over time, erosion reduces the entire landscape to near base level. The region is dominated by extensive floodplains over which a few major streams meander broadly and slowly. They leave a flat, featureless landscape with minimal relief, which Davis called a *peneplain* (*paene* is Latin for "almost"; hence, "almost a plain"). He envisioned occasional remnants of resistant rock rising slightly above the peneplain surface; such erosional remnants are called *monadnocks*, after Mount Monadnock in New Hampshire.
- **Rejuvenation:** Davis's theory also took into account rejuvenation, whereby regional uplift could raise the land, initiate a new period of downcutting, and restart the cycle.

Strong dissenters questioned some of Davis's assumptions and conclusions. For example, apparently no intact peneplains exist—only remnants. Furthermore, many geomorphologists didn't accept his idea that little erosion takes place while the initial surface is being uplifted. Finally, many researchers felt that his biological analogy was more misleading than helpful. Although some landscapes appear youthful, mature, or old, there is little evidence that one stage necessarily precedes another, even in a single valley. The terms *youth*, *maturity*, and *old age* are best used as descriptive summaries of regional topography; they should not be taken to imply sequential development.

Penck's Theory of Crustal Change and Slope Development

Walther Penck, a young German geomorphologist and prominent critic of Davis, pointed out in the 1920s that slopes assume various shapes as they erode. Penck stressed that uplift stimulates erosion immediately and that slope form is significantly influenced by the rate of uplift. He argued that slopes tend to maintain a constant angle as they erode, retaining their steepness as they retreat rather than being worn down at a continuously lower slope angle. In this way, the shape of many initial surfaces is retained long after they would have been worn away in Davis's concept. Many of Penck's ideas in his *theory of crustal change and slope development* have been substantiated by subsequent workers.

LearningCheck 16-15 **Explain some of the problems with Davis's geomorphic cycle theory.**

Equilibrium Theory

A third model of landform development became known as *equilibrium theory*. In the last 60 years or so, many geomorphologists studied the physics of landform development, emphasizing the delicate balance between form and process in the landscape. Crustal movement and the resistance of the underlying rock are as significant as differences in process in determining terrain. Equilibrium theory suggests that slopes adjust to geomorphic processes such that a balance of energy exists: the energy provided is just adequate for the work to be done. For example, harder rock develops steeper slopes and higher relief, and softer rock forms gentler slopes and lower relief. The uniformity inherent in both the Davis and Penck theories is thus called into question.

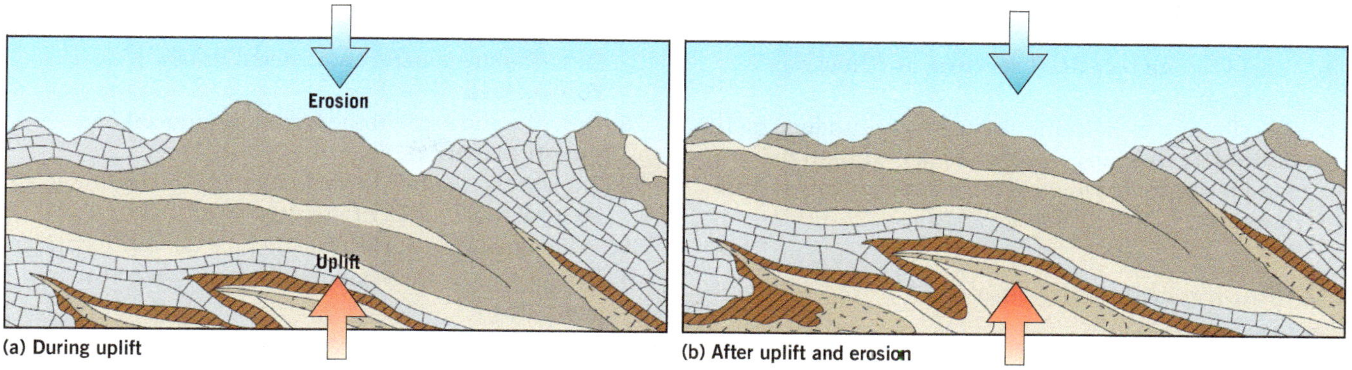

▲ Figure 16-43 The dynamic equilibrium concept. This vertical cross section through an area of the Swiss Alps shows that (a) erosion reduces relief about as rapidly as uplift raises the land. (b) As a result, the elevation and shape of the mountains remains approximately the same because removal and replacement are in balance.

We can apply the equilibrium theory where the land is being simultaneously uplifted tectonically and eroded fluvially, as is happening in the Alps and the Himalayas today (Figure 16-43). If the slopes are in equilibrium, they are being denuded by mass wasting and erosion at the same rate as they are being regenerated by uplift, so the general form of the surface topography remains largely the same over time. A change in either the rate of erosion or the rate of uplift forces the landscape through a period of adjustment until the slopes again reach a gradient at which the rate of erosion equals the rate of uplift.

Equilibrium theory has serious shortcomings in areas that are tectonically stable or have limited streamflow (deserts, for example). It does, however, focus more precisely than the other two models on the relationship between geomorphic processes and surface forms. For this reason, it has dominated fluvial geomorphology since the 1960s.

After studying this chapter, you should be able to answer the following questions. Key terms from each text section are shown in **bold type.** Definitions for key terms are also found in the glossary at the back of the book.

Learning Review 16

Key Terms and Concepts

Streams and Stream Systems (*p. 474*)

1. In geomorphology, what is meant by a **stream**?
2. What are **fluvial processes**?
3. Describe the difference between **streamflow** and **overland flow**.
4. How do an **interfluve** and a **valley** differ?
5. What is a **drainage basin (watershed)**? A **drainage divide**?
6. Using the concept of **stream order**, explain the difference between a first-order stream and a second-order stream, and the general relationship between stream order and stream **gradient**.

Fluvial Erosion and Deposition (*p. 477*)

7. Describe the different components of **stream load**: **dissolved load**, **suspended load**, and **bedload**.
8. What is the difference between stream **competence** and stream **capacity**?
9. How does a stream sort **alluvium** by size?
10. Contrast **perennial streams** with **intermittent streams** and **ephemeral streams**.
11. What is meant by stream **discharge**?
12. What is meant by a flood **recurrence interval** of 50 years—in other words, a 50-year flood?

Stream Channels (*p. 481*)

13. In what way is the pattern of water flow in a straight stream channel likely to be similar to that of a **sinuous channel**?
14. Under what circumstances is a stream likely to have a **meandering channel** pattern?
15. Under what circumstances is a stream likely to have a **braided channel** pattern?

Structural Relationships (*p. 483*)

16. What is an **antecedent stream**, and how might one form?
17. Describe and explain the circumstances under which **dendritic drainage patterns** develop and the conditions under which **trellis drainage patterns** develop.

The Shaping and Reshaping of Valleys (*p. 486*)

18. Explain the process of valley deepening through **downcutting**.
19. What is meant by **base level**? A **graded stream**?
20. What is a **knickpoint**?
21. Describe and explain the process of **knickpoint migration**.
22. Explain how a meandering stream widens its valley through **lateral erosion**.
23. Describe and explain **headward erosion**.
24. Explain **stream capture (stream piracy)**.
25. Describe the formation of a **delta**.
26. Under what conditions does **aggradation** of a stream channel take place?

Floodplains (*p. 492*)

27. Describe the general characteristics of a **floodplain**.
28. Describe and explain the formation process of a **cutoff meander**.
29. Explain the relationships among a cutoff meander, an **oxbow lake**, and a **meander scar**.
30. What are **natural levees**, and how do they form?
31. What explains the presence of a **yazoo stream** on a floodplain?

Stream Rejuvenation (*p. 497*)

32. Explain the circumstances that can lead to **stream rejuvenation**.
33. Describe and explain the formation of **stream terraces**.
34. How is it possible for **entrenched meanders** to form?

Theories of Landform Development (*p. 499*)

35. Describe and explain at least one problem with Davis's geomorphic cycle model of fluvial landform development.
36. How does equilibrium theory differ from earlier theories of topographic development?

Study Questions

1. What factors influence the erosional effectiveness of a stream?
2. What "abrasive" tools does a stream use in erosion?
3. What factor determines the competence of a stream?
4. Why is alluvium often rounded and stratified?
5. Explain how talus from rockfall (Chapter 15) is likely to look different from alluvium.
6. Why are floods so important in the development of fluvial landforms?
7. Describe radial drainage patterns and one kind of location where such a drainage pattern might be found.
8. What prevents a stream from downcutting to sea level throughout its entire course?
9. Why are few depositional features found in most V-shaped stream valleys?
10. Why don't all large rivers form deltas where they enter the ocean?
11. Why might meandering rivers make poor political boundaries?
12. What are some of the negative consequences of using artificial levees for flood control along a major river?

Exercises

1. Assume that the rate of knickpoint migration for Niagara Falls has been constant. If it took 12,000 years for the position of the waterfall to shift 11 kilometers upstream, what is the average annual rate of movement? _____ meters
2. Assume that the rate of knickpoint migration for Niagara Falls has been constant. If the annual rate of knickpoint migration continues into the future, how long will it take before the position of the falls migrates the remaining 27 kilometers to Lake Erie? _____ years
3. Assume that the competence of a stream increases with the square of the speed of flow. If a stream can move a 2-kilogram rock, what is the mass of a rock that stream can move if its speed increases by a factor of 3 during a flood? _____ kilograms
4. Using the map of Trout Creek, Wyoming, in Figure 16-15 b, what is the stream order where the stream runs off the map? _____ order stream

Environmental Analysis RIVER WATER STAGES

Data
Flood Stage
https://media.pearsoncmg.com/ph/esm/esm_mcknight_physgeo_13/media/

River water levels are monitored across the United States to evaluate water resources and determine *flood stage* (the level at which the water surface has risen enough to be a hazard). The National Weather Service also creates forecasts for river depths to predict flooding.

Activities

Go to the National Weather Service's River Observations map at http://water.weather.gov/ahps.

1. How many total river gauges (also written as "gages") are there in the conterminous United States?
2. What is the current level of flooding indicated by the majority of gauges?
3. How many gauges indicate major flooding? Moderate flooding? Minor flooding?

Click on a gauge located near you to view a regional map showing local river gauges. Hovering over a gauge location will display a graph showing river stage data. Forecasted river stage is also graphed for some locations. ("Flood Category Not Defined" stations will not show a flood stage value.)

4. Hover over the button "What is UTC time?" Why is time displayed in Universal Time (UTC) as well as local time ("Site Time")?
5. Hover over the gauge site closest to your location. Where is this gauge located? What is the current river stage? What is the flood stage?
6. Using local time, what are the earliest time and date displayed on the graph? What are the latest time and date?

Click on the "River Forecasts" tab and hover over a gauge site in your vicinity that shows forecasted river stage data.

7. What is the current river stage? How much is this river stage projected to change over the next few days?
8. How far above or below flood stage is the current river stage?

Return to the national map and click on a river gauge on the East Coast.

9. Why do river levels vary in this way along the East Coast?

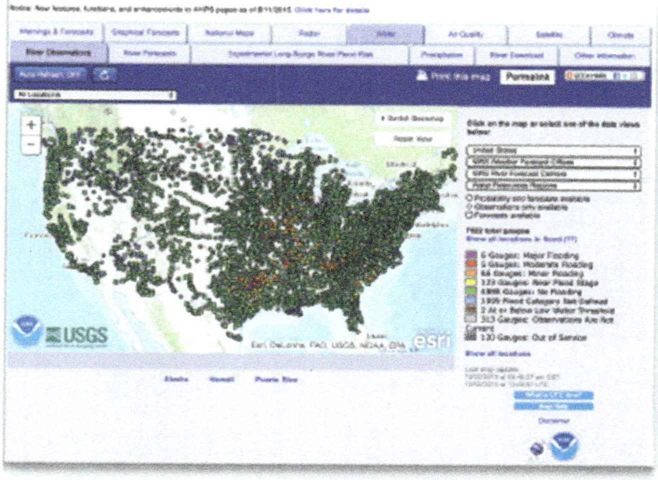

Seeing Geographically in Our National Parks

Look again at the photograph of the Green River at the beginning of the chapter (p. 472). Is this meandering river flowing on a floodplain? What has happened to the river to produce the topography here? How might the course of the river change over time? How are vegetation patterns influenced by alluvium?

Mastering Geography®

Looking for additional review and test prep materials? Visit the Study Area in **Mastering Geography** to enhance your geographic literacy, spatial reasoning skills, and understanding of this chapter's content by accessing a variety of resources, including Map Master interactive maps, geoscience animations, *Mobile Field Trips*, videos, *Project Condor* Quadcopter videos, *In the News* RSS feeds, flashcards, web links, self-study quizzes, and an eText version of *McKnight's Physical Geography*.

Seeing Geographically in Our National Parks

Old Faithful geyser in Yellowstone National Park. Describe the rock you see below the geyser. What suggests that this rock might have been deposited by water? What factors might account for the differences in vegetation below the geyser and in the forest behind? Does Old Faithful appear to be ejecting only steam? Why do you say this?

Yellowstone National Park, WY, ID & MT

Karst and Hydrothermal Processes

17

HAVE YOU EVER WONDERED how Old Faithful geyser works?

Hydrothermal features such as the geysers here in Yellowstone National Park can develop only in areas with a special combination of conditions: an adequate amount of groundwater, a heat source (usually magma near the surface), and a network of passageways and chambers in the bedrock to allow the buildup of heated water and steam. The end result is a landscape that can suddenly come to life in a captivating show of sight and sound—and even of smell!

In studying topographic development, we pay a great deal of attention to the role of water. We have already noted the significance of surface water as a shaper of terrain, and we will soon see that coastal waters produce distinctive landforms around the margins of oceans and lakes. In both cases, the water is moving relatively rapidly as it accomplishes erosion, transportation, and deposition. Water below the surface can also shape terrain, but through a quite different set of processes.

Water under the ground is largely unchanneled and therefore generally diffused, and for the most part it moves very slowly. Consequently, it is almost totally ineffective in terms of hydraulic power and other kinds of mechanical erosion. However, groundwater can leave a distinctive mark on the surface landscape through a variety of solution processes associated with the development of karst topography and hydrothermal features—the topics of this chapter.

As you study this chapter, think about these **Key**Questions:

- **How is bedrock such as limestone affected by water below the surface?**
- **How do limestone caverns develop?**
- **What kinds of surface features develop in areas of limestone?**
- **What causes a hot spring or geyser to develop?**

The Impact of Solution Processes on the Landscape

The mechanical effects of water below the surface have limited influence on topographic development. Through its chemical weathering, however, groundwater becomes an effective shaper of the topographic landscape. (Strictly speaking, *groundwater* refers to water below the *water table* only, but here we use that term to refer to any underground water.) Water is a solvent for certain minerals, dissolving them from rock and then carrying them away in solution and depositing them elsewhere. The aboveground results of this dissolution can be widespread and distinctive.

Groundwater also affects surface topography through the creation of hydrothermal features such as hot springs and geysers, formed when hot water from underground is discharged at the ground surface.

Dissolution and Precipitation

The chemical reactions involving groundwater are relatively simple. Although pure water is a relatively poor solvent for rocks, almost all groundwater is laced with enough chemical impurities to make it a good solvent for the compounds that make up a few common minerals (**Figure 17-1**). Basically, groundwater is a weak solution of **carbonic acid** (H_2CO_3) because it contains dissolved carbon dioxide gas. As we first saw in Chapter 15, the resulting *carbonation* can lead to the dissolution of bedrock.

Dissolution Processes

Dissolution is the removal of bedrock through chemical weathering and the action of water. Dissolution affects many kinds of rocks, but it is particularly effective on carbonate

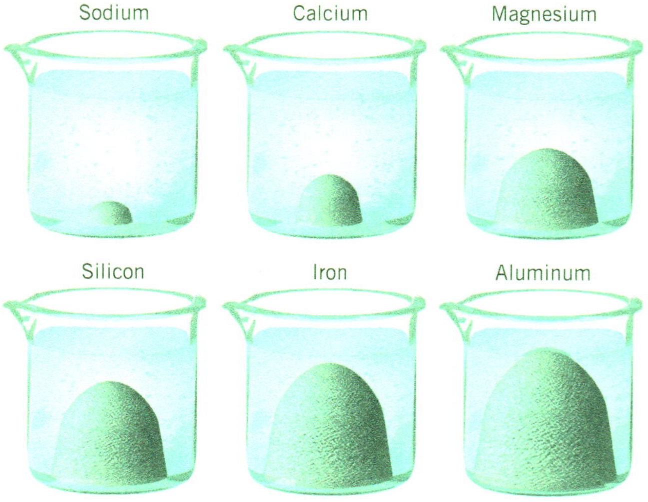

▲ **Figure 17-1** Relative solubility in water of some common rock-forming elements. The "lumps" in each beaker represent the proportion of the element that remains undissolved in a given amount of water. Sodium and calcium dissolve almost completely, for instance, whereas iron and aluminum are essentially insoluble.

sedimentary rocks, especially limestone. Limestone is composed largely of calcium carbonate ($CaCO_3$), which reacts strongly with carbonic acid. That yields calcium bicarbonate, a compound that is very soluble and easily removed by water:

$$\underset{\text{limestone (calcium carbonate)}}{CaCO_3} + H_2O + \underset{\text{carbon dioxide}}{CO_2} \rightarrow \underset{\text{calcium bicarbonate}}{Ca(HCO_3)_2}$$

Other carbonate rocks, such as gypsum and even the metamorphic rock marble, undergo similar reactions. Dolomite is a calcium magnesium carbonate rock that dissolves almost as quickly as limestone and also yields calcium bicarbonate:

$$\underset{\text{dolomite}}{CaMg(CO_3)_2} + \underset{\text{water}}{2H_2O} + \underset{\text{carbon dioxide}}{2CO_2} \rightarrow \underset{\text{calcium bicarbonate}}{Ca(HCO_3)_2} + \underset{\text{magnesium bicarbonate}}{Mg\,(HCO_3)_2}$$

Because limestone and related rocks are largely composed of soluble minerals, great volumes of rock are sometimes dissolved and carried away by water percolating down into carbonate bedrock, leaving conspicuous voids in the rock structure. This action occurs more rapidly and on a wider scale in humid climates, where abundant rain provides plenty of water containing the dissolved carbon dioxide necessary for dissolution. In arid regions, evidence of dissolution is unusual except for relict features dating from a more humid past.

Although reactions with carbonic acid are the most common processes involved in the dissolution of carbonate rocks, recent studies suggest that in some locations sulfuric acid (H_2SO_4) may also be important. For example, it appears that Lechuguilla Cave in New Mexico was at least in part enlarged through dissolution of limestone by sulfuric acid, formed when hydrogen sulfide (H_2S) from deeper petroleum deposits combined with oxygen in the groundwater.

Role of the Bedrock Structure: Bedrock structure is also a factor in dissolution. A profusion of *joints* and *bedding planes* permits groundwater to penetrate the rock readily. That the water is moving also helps because, as a given volume of water becomes saturated with dissolved calcium bicarbonate, it can drain away and be replaced by fresh unsaturated water, which can dissolve more rock. Such drainage is enhanced by some outlet at a lower level, such as a deep subsurface stream.

Most limestone is resistant to mechanical weathering and erosion and often produces rugged topography. Thus, its easy solubility contrasts with its mechanical durability—a vulnerable interior beneath a durable surface.

LearningCheck 17-1 **Why is limestone so susceptible to dissolution? (Answer on p. AK-6)**

Precipitation Processes

Complementing the removal of calcium carbonate is its chemical *precipitation* from solution back into solid form. Mineral-rich water may trickle in along a cave roof or wall. The reduced air pressure in the open cave induces precipitation of the minerals the water is carrying.

Hot springs and geysers nearly always provide an accumulation of precipitated minerals—frequently brilliant white but sometimes orange, green, or some other color due to associated algae. Groundwater, when it comes in contact with magma, becomes heated and sometimes finds its way back to the surface so rapidly through a natural opening that it is still hot when it reaches the open air. Hot water is generally a much better solvent than cold water, so a hot spring or geyser usually contains a significant quantity of dissolved minerals. When exposed to the open air, the hot water precipitates much of its mineral content as its temperature and the pressure on it decrease, and as the dissolved gases that helped keep the minerals in solution dissipate or as algae and other organisms living in it secrete mineral matter. These deposits (such as *travertine*, *tufa*, and *sinter*) contain a variety of calcareous minerals and take the form of mounds, terraces, walls, and peripheral rims.

However, the solubility of carbon dioxide *decreases* as water temperature increases. Thus, cool water often is more potent than hot water as a solvent for calcium carbonate.

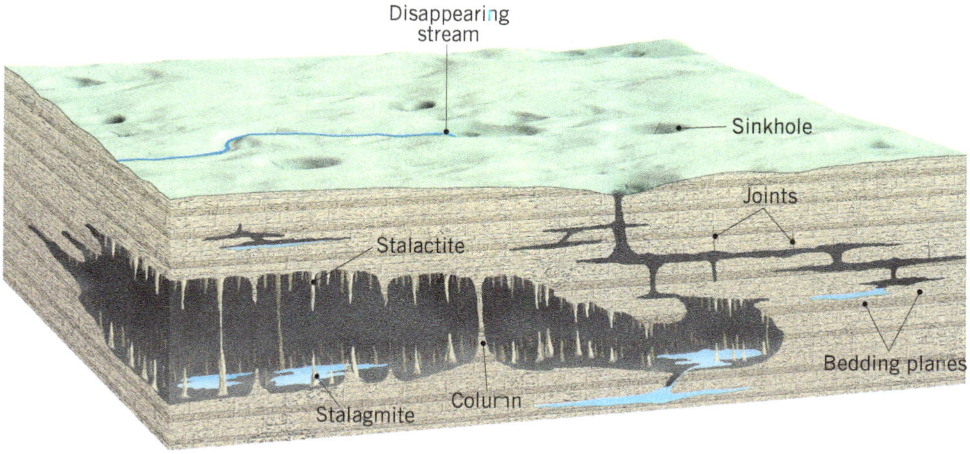

▲ **Figure 17-2** Caverns form by the solution action of underground water as it trickles along bedding planes and joint systems.

LearningCheck 17-2 **How do rocks such as travertine and tufa form?**

Caverns and Related Features

Some of the most spectacular landforms produced by dissolution are not visible at Earth's surface. Dissolution along joints and bedding planes in limestone beneath the surface often creates large open areas called **caverns**. The largest of these openings are usually more expansive horizontally than vertically, indicating a development along bedding planes. In many cases, however, the cavern pattern has a rectangularity that demonstrates a relationship to the joint system (**Figure 17-2**).

Speleothems

Cavern formation has two principal stages. First is the initial excavation, wherein percolating water dissolves the carbonate bedrock and leaves voids. In some cases, these openings are enormous: the "Big Room" in New Mexico's Carlsbad Caverns is about 3.3 hectares (8.2 acres) in area—large enough to accommodate six football fields! A stream may flow along the floor of a large cavern, adding another dimension to erosion and deposition.

This dissolution stage is followed, often after a drop in the water table, by a "decoration stage" in which ceilings, walls, and floors are decorated with a wondrous variety of **speleothems** (from Greek *spelaion*, "cave," and *thema*, "deposit"). These forms are deposited when water leaves behind the compounds (principally carbon dioxide and calcite) it was carrying in solution (**Figure 17-3**). Once out of solution, the carbon dioxide gas diffuses into the cave atmosphere, and calcite is deposited. Speleothems can be used to

▲ **Figure 17-3** Speleothems in Carlsbad Cavern National Park, New Mexico.

Global Environmental Change

Caverns Hold Evidence of Climate Change

• **Chris Groves**, Western Kentucky University

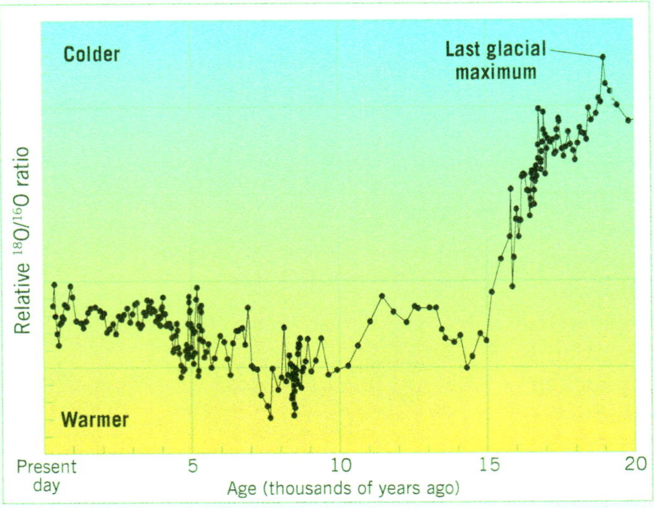

◀ **Figure 17-B** Temperature changes over the last 20,000 years, determined from oxygen isotopes from speleothems in Soreq Cave, Israel.

Stalagmites and other speleothems form when the mineral calcite precipitates as water drips in limestone caverns. The calcite has been dissolved by CO_2-rich waters that flow downward through fractures in the limestone above the cave and is left behind when CO_2 is lost into the cave atmosphere from a drip of water that reaches the cave ceiling.

Speleothems form growth bands (**Figure 17-A**) that can be dated by using uranium isotopes that occur in trace amounts in the calcite crystals. Over time, radioactive uranium decays at a known rate into thorium, so the age of a particular band of the speleothem can be determined from the ratio of these two elements. Within a single speleothem cross section, banding may be annual, with alternating lighter and darker bands called *laminae*, which can reflect seasonal changes in drip water flow and chemistry. However, banding may also record more complex histories of deposition, as the banding pattern can vary with time: deposition can speed up, slow down, or even stop in response to changes in the hydrology and chemistry of the waters feeding the drip. Once researchers establish the timing of the laminae, the chemistry of the calcite in the cave can in turn provide information about the climate of the region where the cave is found. These records often span tens of thousands of years.

▲ **Figure 17-A** Banding within a speleothem.

The Record in the Rock: Speleothems provide multiple indicators of paleoclimate. For example, the calcite contains isotopes of oxygen and carbon, whose relative abundances depend on the conditions in the cave at the time of speleothem deposition. The conditions inside the cave, in turn, reflect environmental conditions outside the cave. Air and water temperatures deep in the cave approximate the mean annual temperature of the region, and speleothem growth rates vary with rainfall amounts. The ratio of the oxygen isotopes ^{18}O and ^{16}O (discussed in Chapter 8) in the calcite is influenced by the temperature of the dripping water in the cave and therefore by the outside temperature. Thus, the $^{18}O/^{16}O$ ratio in a single stalagmite records fluctuations in temperature and rainfall, often spanning thousands of years. For example, an oxygen isotope signal derived from speleothems within Soreq Cave, in central Israel, shows the ending of the last glacial maximum about 17,000 years ago (**Figure 17-B**). Similarly, $^{13}C/^{12}C$ ratios within the calcite can provide information about the vegetation above the cave. ^{13}C and ^{12}C are two stable carbon isotopes with different masses that are preferentially selected by various plant species in different photosynthetic processes. Plants that are common in cooler, wetter climates have characteristic $^{13}C/^{12}C$ signatures that differ from plants that thrive in warmer conditions. Plant types can also be identified by pollen grains trapped within the calcite.

What Is the Value of Speleothem Records? Paleoclimate evidence—whether from glaciers, seafloor sediments, or other sources—records information about the specific locations where samples were collected. Because limestone caves and the speleothems within them occur throughout the world, these records have greatly improved spatial coverage of the world's climate history. Individual cave records have shown changes in temperature and rainfall and have helped explain the atmospheric processes—changing monsoon behavior, for example—that caused these changes. The correlation of many data sources in both space and time reveals patterns of climate change and the mechanisms that create those patterns.

Questions

1. How do variations in the isotopic composition of oxygen within the calcite of speleothems tell us about past climates?
2. How can the bands, or laminae, within speleothems be useful in determining past conditions in Earth's history?

decipher changes in climate—see the box *Global Environmental Change: Caverns Hold Evidence of Climate Change*.

Much of the deposition occurs on the sides of the cavern, but the most striking features are formed on the roof and floor. Where water drips from the roof, a pendant structure—a **stalactite** (from the Greek *stalaktos*, meaning "dripping")—grows slowly downward like an icicle. Where the drip hits the floor, a companion feature—a **stalagmite** (from the Greek *stalagma*, meaning "a drop")—grows upward. Stalactites and stalagmites may extend until they meet, forming a *column* (**Figure 17-4**). In some caverns, long, slender *soda straws* hang down from the ceiling; little more than one water drop wide, these delicate hollow tubes may eventually grow into stalactites.

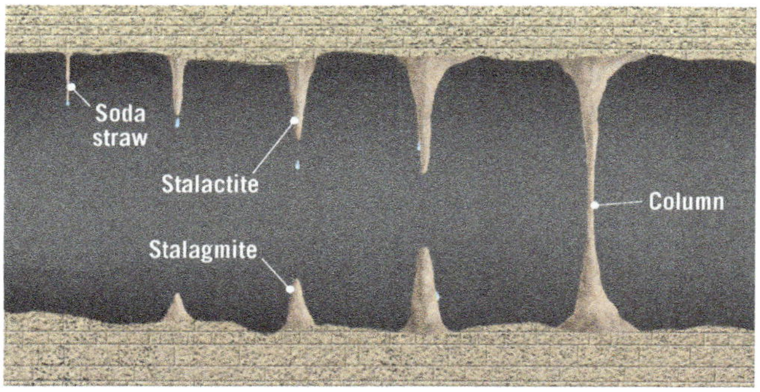

▲ **Figure 17-4** The development of soda straws, stalagmites, stalactites, and columns.

LearningCheck 17-3 **Explain the two stages of cavern development.**

Karst Topography

In many areas where the bedrock is limestone or similarly soluble rock, dissolution has been so widespread and effective that a distinctive landform assemblage has developed at the surface, in addition to whatever caves may exist underground. The term **karst** (a Germanized form of an ancient Slavic word meaning "barren land") is applied to this topography. The name derives from the Kras or Krš Plateau region of Slovenia, a rugged hilly area shaped almost entirely by dissolution in limestone formations (Figure 17-5).

The term *karst* connotes both a set of processes and an assemblage of landforms. It is the catchall name of a cornerstone concept that describes the special landforms that develop on exceptionally soluble rocks, although there is a broad international vocabulary to refer to specific features in specific regions.

Karst Landforms

Typical landforms in karst regions include sinkholes, disrupted surface drainage, and underground drainage networks that have openings formed by dissolution. The openings range in size from enlarged cracks to huge caverns.

Karst landscapes usually evolve where there is massive limestone bedrock. However, karst features may also occur where other highly soluble rocks—dolomite, gypsum, or halite (rock salt)—predominate. Karst landforms are worthy of study not only because of their dramatic appearance but also because of their abundance. It is estimated that about 15 percent of Earth's land area has soluble carbonate rocks at or near the surface (Figure 17-6); in the conterminous United States, this total rises to about 23 percent (Figure 17-7). See the box *Featured National Park: Mammoth Cave—A Landscape of Caverns and Karst.*

Sinkholes: The most common surface features of karst landscapes are **sinkholes** (also called *dolines*), which occur by the hundreds or even thousands. Sinkholes are rounded depressions formed by the dissolution of surface carbonate rocks, typically at joint intersections. Sinkholes erode more

▲ **Figure 17-5** Irregular topography and collapsed caves are prominent in the karst region of Primorska in Slovenia.

rapidly than the surrounding area, forming closed depressions (Figure 17-8a). Their sides generally slope inward at the angle of repose (usually 20° to 30°) of the adjacent material. A sinkhole that results from the collapse of the roof of a subsurface cavern is called a **collapse sinkhole** (or *collapse doline*); these may have vertical walls or even overhanging cliffs (Figure 17-8b).

Sinkholes range in size from shallow depressions a few meters in diameter and a few centimeters deep to major features kilometers in diameter and hundreds of meters deep. The largest are associated with tropical regions, where they develop rapidly and where adjacent holes often intersect.

The bottom of a sinkhole may lead into a subterranean passage, down which water pours during rainstorms. More commonly, however, the subsurface entrance is blocked by rock rubble, soil, or vegetation, and rains form temporary

Featured National Park

Mammoth Cave

A Landscape of Caverns and Karst

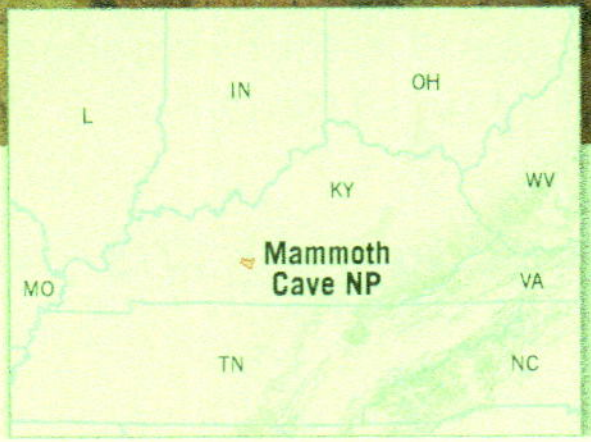

▲ **Figure 17-D** A multitude of stalactites hang from the ceiling of this room in Mammoth Cave. The smooth, gently rounded surfaces are often referred to as *flowstone*.

Caverns are found almost anywhere a massive limestone deposit is at or near the surface. The area around Mammoth Cave National Park in Kentucky is a prime example of this.

Much of this region in Kentucky is underlain by extensive deposits of limestone, such as the 350 million-year-old Girkin Formation, the St. Louis Formation, and the Ste. Genevieve Limestone. South of the cave itself this limestone layer is exposed at the surface to form the "Sinkhole Plain"—a region of classic karst topography with dozens of sinkholes and disappearing streams (**Figure 17-C**).

Mammoth Cave developed within the Mammoth Cave Plateau to the north of the Sinkhole Plain. Here, a layer of resistant sandstone and shale covers the more permeable limestone. All of these rock layers tilt down gently to the north, so groundwater slowly percolates from the Sinkhole Plain along the bedding planes of the limestone within the Mammoth Cave Plateau before finally entering the Green River to the north of Mammoth Cave.

The dissolution of limestone within the Mammoth Cave Plateau has left an extensive network of openings—Mammoth Cave itself has more than 663 kilometers (412 miles) of known passages. As with other large cavern systems, Mammoth Cave includes an elaborate network of galleries and passageways, often very irregular in shape and sometimes including massive openings ("rooms") scattered along the galleries. Redecoration of these openings with speleothems has left one of the most remarkable caverns in the country, with hundreds of examples of stalactites, stalagmites, and flowstone (**Figure 17-D**).

▼ **Figure 17-C** The Pennroyal Plateau or "Sinkhole Plain" south of Mammoth Cave, Kentucky, is dominated by sinkholes.

Questions

1. What is the relationship of the rock of the Sinkhole Plain to that of Mammoth Cave?
2. How does water from the Sinkhole Plain end up in the Green River to the north?

Mobile Field Trip
Mammoth Cave
https://media.pearsoncmg.com/ph/esm/esm_mcknight_physgeo_13/media/

lakes until the water percolates away. Indeed, sinkholes are the karst equivalent of river valleys, in that they are the fundamental unit of both erosion and weathering.

In many karst locations, depressions dominate the landscape. Sinkholes are commonplace in central Florida and parts of the Midwest and South—particularly Kentucky, Illinois, and Missouri. For example, at the University of Missouri in Columbia, both the football stadium and the basketball fieldhouse are built in sinkholes, and parking lots around them occasionally sink.

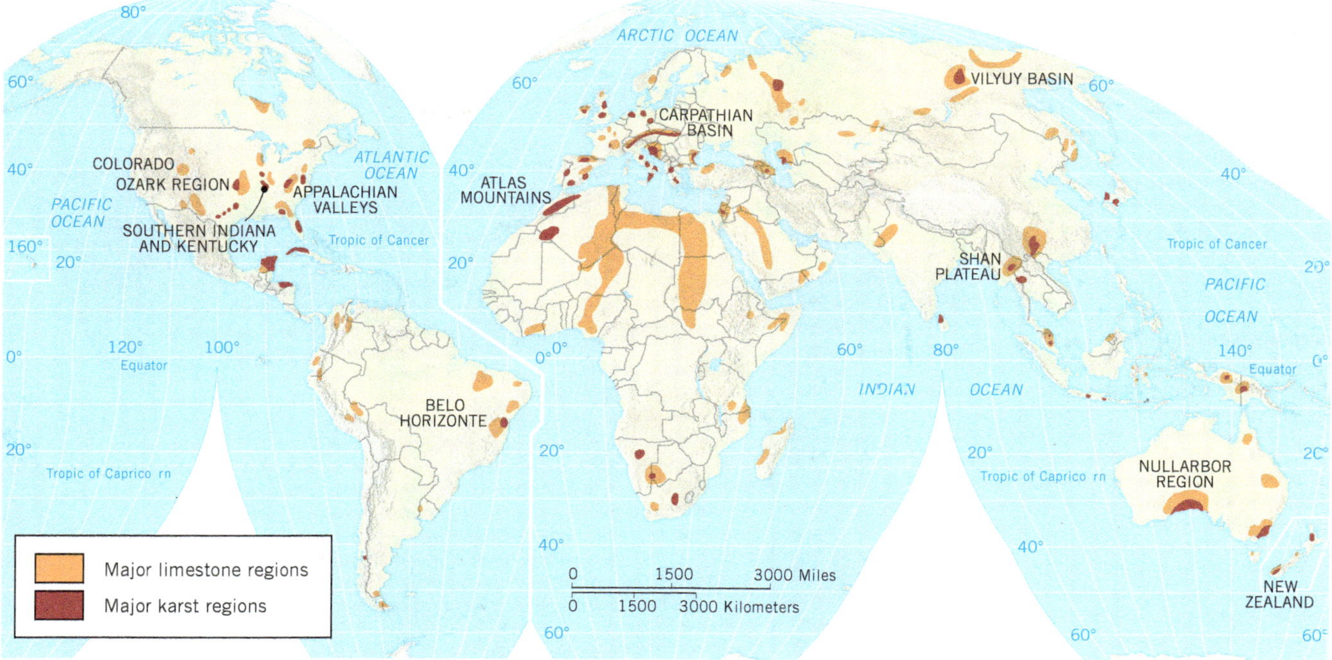

▲ **Figure 17-6** Major limestone and karst regions of the world.

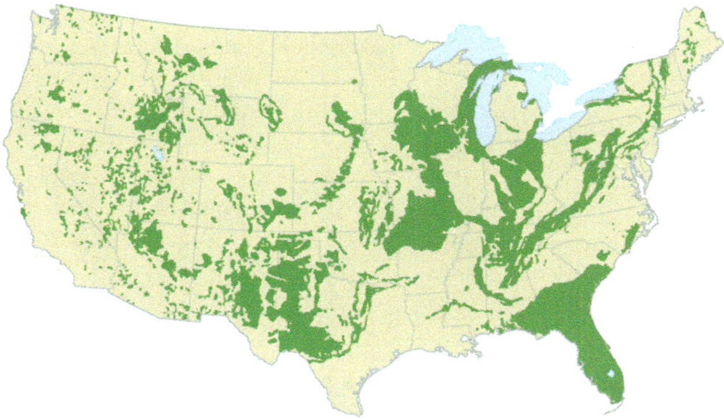

▲ **Figure 17-7** Karst areas (shown in green) of the conterminous United States.

Apart from the ubiquity of sinkholes, karst areas show considerable topographic diversity. Where the relief is slight, as in central Florida, sinkholes are the dominant features. Where the relief is greater, however, cliffs and steep slopes alternate with flat-floored, streamless valleys. Limestone bedrock exposed at the surface tends to be pitted, grooved, etched, and fluted with a great intricacy of erosive detail.

Surface Drainage in Karst Areas: Where sinkholes occur in profusion, they often channel surface runoff into the groundwater circulation, leaving networks of dry valleys as relict surface forms. The Serbo-Croatian term **uvala** refers to such a chain of intersecting sinkholes. In many cases, sinkholes evolve into uvala over time.

In many ways, the most notable feature of karst regions is what is missing: surface drainage. Most rainfall and snowmelt seep downward along joints and bedding planes, enlarging them by dissolution. Surface runoff that does become channeled does not usually go far before it disappears into a sinkhole or joint crack. Such streams are often termed **disappearing streams** (see Figure 17-8a). The water that collects in sinkholes generally percolates downward, but some sinkholes have distinct openings at their bottom, called **swallow holes**, through which surface drainage can pour directly into an underground channel, often to reappear at the surface through another hole some distance away. Where dissolution has been effective for a long time, a complex underground drainage system may have superseded any sort of surface drainage network. An appropriate generalization concerning surface drainage in karst regions is that valleys are relatively scarce and mostly dry.

Tower Karst: Residual karst features, in the form of very steep-sided hills, dominate some parts of the world—especially in the tropics, where high rainfall contributes to dissolution around the base of the hills (**Figure 17-8c**). These formations are sometimes referred to as **tower karst** because of their almost vertical sides and conical or hemispherical shapes. Such towers are sometimes riddled with caves. The tower karst of southeastern China, adjacent parts of northern Vietnam, and southern Thailand is famous for its spectacular scenery (**Figure 17-9**), as are the *mogotes* (haystack hills) of western Cuba and northwestern Puerto Rico.

LearningCheck 17-4 **What is a sinkhole and how does one form?**

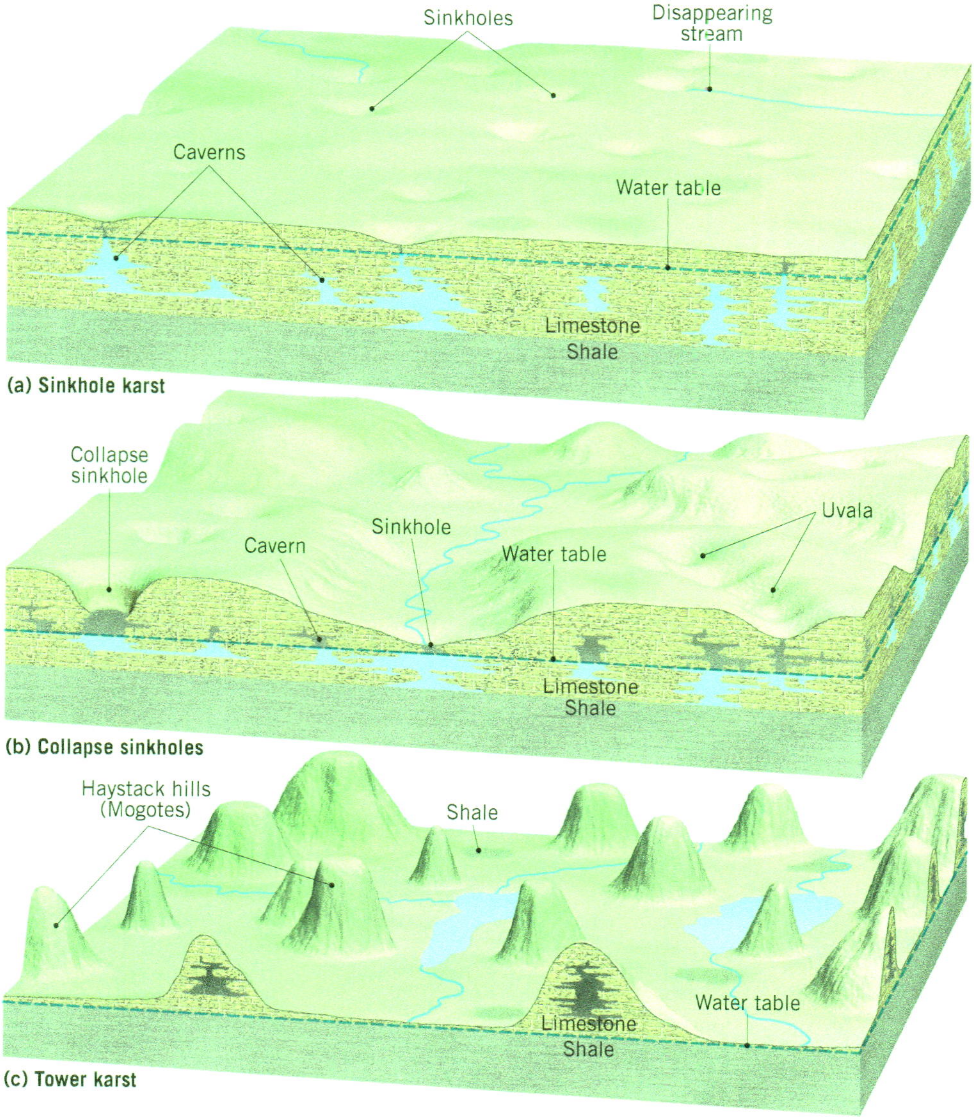

▶ **Figure 17-8** The development of karst topography. (a) Landscape dominated by sinkholes and disappearing streams. Dissolution of bedrock below the water table may leave openings that develop into caverns. (b) Where underground caverns collapse, a collapse sinkhole forms. Streams may disappear into sinkholes through swallow holes. If the water table drops enough to leave open caverns, speleothems may gradually form. (c) Tower karst topography consists of residual towers of limestone (haystack hills, or mogotes).

▶ **Figure 17-9** Spectacular tower karst hills (mogotes) near Yangshuo, Guangxi Province, China.

◀ **Figure 17-10** This sinkhole in Clermont, Florida, formed overnight, damaging three buildings.

Groundwater Extraction and Sinkhole Formation: Human activities can have direct and immediate consequences in some areas of karst topography. For example, most of central Florida is underlain by massive limestone bedrock, which is particularly susceptible to the formation of sinkholes and collapse sinkholes. This process is accelerated when the water table drops. Sinkholes have been forming in Florida for a long time, with several thousand of them having appeared in the twentieth century. Indeed, most of central Florida's scenic lakes began as sinkholes.

The population growth that Florida has experienced in recent decades put a heavy drain on its groundwater supply, causing a drawdown of the water table. As a result, the number and size of Florida sinkholes increased to a disturbing pace—reaching a peak of about one per day in the early 1980s (Figure 17-10).

Hydrothermal Features

Many parts of the world contain small areas where hot water comes to the surface through natural openings. Such outpouring of hot water, often accompanied by steam, is known as **hydrothermal activity** and usually takes the form of either a hot spring or a geyser.

In many places around the world, people have tapped hydrothermal resources as a source of clean energy—see the box *Energy for the 21st Century: Geothermal Energy.*

Hot Springs

The appearance of hot water at Earth's surface usually indicates that the underground water has come in contact with heated rocks or magma and has been forced upward through a fissure by the pressures that develop when water is heated. The usual result at the surface is a **hot spring** (Figure 17-11), with water bubbling out continuously or intermittently. The hot water invariably contains a large amount of dissolved mineral matter, and a considerable proportion of this load precipitates out as soon as the water reaches the surface and both its temperature and pressure decrease.

The deposits around and downslope from hot springs can take many forms. If the opening is on sloping land, terraces usually form. Where the springs emerge onto flat land, there may be cones, domes, or irregular concentric deposits. Because calcium carbonate is so readily soluble in water containing carbonic acid, the deposits of most springs are composed largely of massive (*travertine*) or porous (*tufa* or *sinter*) accumulations of calcium carbonate. Various other minerals, especially silica, may also exist in the deposits but are much less common than calcium compounds.

Many hot springs are decorated with brilliantly colored algae or "extremophile" microbes—forms of life adapted to the extreme heat of this environment. Such

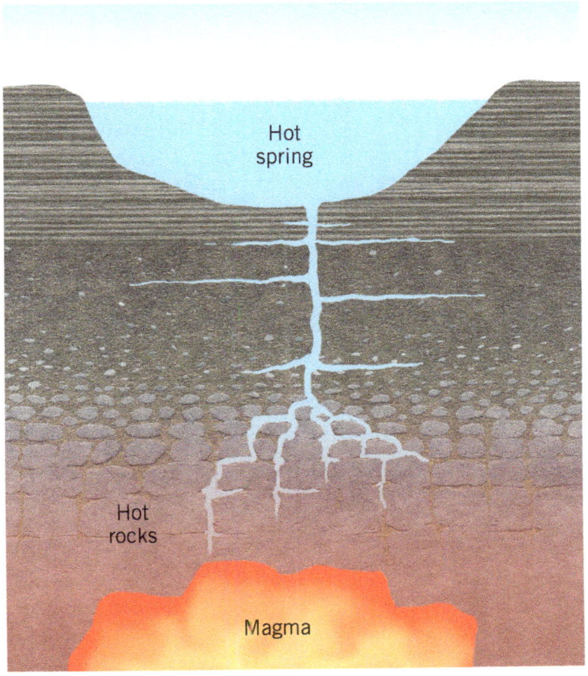

▲ **Figure 17-11** Cross section through a hot spring.

Energy for the 21st Century

Geothermal Energy

• Nancy Lee Wilkinson, San Francisco State University

Geothermal technology harnesses the heat produced in Earth's interior by the radioactive decay of elements such as uranium and potassium, as well as the heat from Earth's cooling core. This heat is easily accessed at "hot spots" and along tectonic plate boundaries, where molten magma rises close to Earth's surface.

Harnessing Geothermal Energy: People have enjoyed bathing in hot springs for hundreds of thousands of years. Ancient Roman engineers went further, developing public baths and circulating water from hot springs to heat buildings—a practice still employed in Iceland and elsewhere. Today's geothermal heat pumps also make direct use of geothermal energy by circulating air or antifreeze through buried piping systems that draw heat from the ground.

Commercial technologies have been developed to generate electricity from geothermal energy. All involve drilling holes into the ground to capture steam:

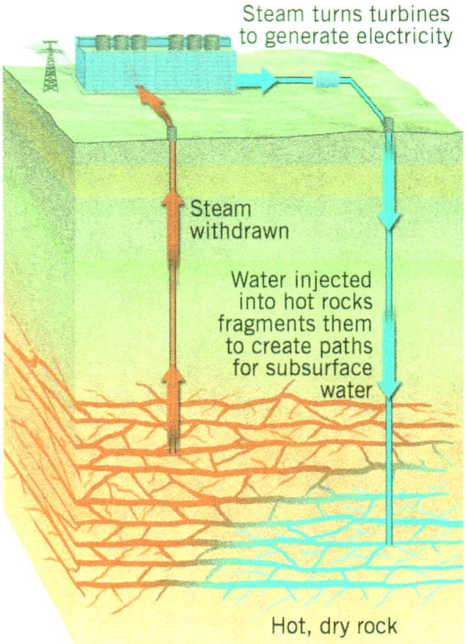

▲ **Figure 17-E** Electrical generation by an Enhanced Geothermal System. Water is injected into hot but dry rock. The steam produced is extracted to power the turbines.

▲ **Figure 17-F** Geothermal power generation facilities at The Geysers in northern California.

- Dry steam power plants extract steam from inside Earth to turn turbines; flash steam power plants spray steam into a tank at much lower pressure, where it vaporizes ("flashes") and drives a turbine. In both cases, the used steam is condensed and returned to the ground to be reheated naturally.
- Binary cycle power plants circulate geothermal water and a secondary fluid (such as isobutene) that has a lower boiling point than water through separate pipes and a heat exchanger. Heat from the geothermal fluid causes the secondary fluid to vaporize, driving a turbine. This technology exploits low-heat geothermal resources.
- Enhanced Geothermal Systems pump cold water into the ground in geothermally active, dry regions to produce steam (Figure 17-E). These systems might make possible a vast expansion of geothermal power plants in arid lands of the western United States and Australia.

Advantages and Disadvantages: Geothermal energy plants produce energy day and night with no mining or transportation of raw materials, no carbon emissions, and very minimal pollution. Although CO_2 is released during extraction of steam, the amount is a small fraction of that generated by coal-fired facilities. Steam plants also release hydrogen sulfide gas, which gives hot springs their characteristic "rotten egg" smell, but geothermal plants emit 97 percent less sulfur than fossil fuel power plants. Geothermal energy plants can induce earthquakes by drilling, water extraction, or water injection. This is of particular concern at Enhanced Geothermal System plants. Because geothermal plants operate in seismically active regions, it can be difficult to distinguish natural seismic activity from induced seismicity. Finally, if steam extraction at a power plant exceeds the natural recharge rate, a reduction of steam pressure can result. Several wells at the Geysers geothermal area in California were depleted in this way.

Geography of Geothermal Technology: Geothermal energy technology is booming. Geothermal plants in 29 countries produce about 15,950 megawatts. The United States leads production, with more than 3700 megawatts of installed capacity, most notably at The Geysers north of San Francisco (Figure 17-F). However, utility-scale plants provide just 0.3 percent of U.S. electricity. The Philippines, Kenya, Indonesia, and El Salvador have smaller generating capacities, but their geothermal plants account for 25 percent to 30 percent of national electricity production. Geothermal power is also important in other volcanically active regions, such as Mexico, New Zealand, and Italy. According to the Geothermal Energy Association, almost 700 new geothermal projects are currently proposed in 76 countries, including large plants in Chile, Turkey, and Ethiopia. Small island nations such as Nevis and St. Lucia are also investing heavily in geothermal power to capitalize on local, nonpolluting energy resources.

Questions

1. What heats the water in geothermal reservoirs? If not captured for use, where does the heat go?
2. Identify three things that are needed to generate electricity from a geothermal resource.
3. Why might small island nations see geothermal plants as an especially good way to generate electricity?

NPS Video
Science in the Park
Yellowstone: Thermofiles
https://media.pearsoncmg.com/ph/esm/esm_mcknight_physgeo_13/media/

▲ **Figure 17-12** The sides and bottoms of hot springs are often brilliantly colored by algal growth. This is Beauty Pool in Yellowstone National Park. The black area at the bottom of the hot spring is the opening to the fissure that brings to the surface hot water at a temperature that is typically about 77°C (170°F).

▲ **Figure 17-13** Researchers collect samples and measure the acidity of Champagne Pool in the Wai-O-Tapu geothermal area on the North Island of New Zealand.

organisms may add to the striking appearance of hot springs as well as contribute mineral secretions to the deposit (**Figure 17-12**). Certain extremophiles are further adapted to the very acidic water of some hot springs (**Figure 17-13**).

Geysers

A special form of intermittent hot spring is the **geyser**. Hot water tends to flow from a geyser only sporadically, and most or all of the flow is a temporary ejection (called an *eruption*) in which hot water and steam spout upward. Then the geyser subsides into apparent inactivity until the next eruption.

Eruption Mechanisms: The process leading to a geyser eruption begins when groundwater seeps into subterranean openings that are connected in a series of narrow caverns and shafts. Heated rocks or magma is close enough to these storage reservoirs to provide a constant source of heat. As the water accumulates in the deeper reservoirs, it is heated to 200°C (400°F) or higher without boiling. Such superheating is possible because of high pressure due to the weight of the water above. The heating of water in the upper chambers, where the pressure is lower, causes the water to expand and surge out onto the surface (**Figure 17-14a**). This surge suddenly reduces the pressure of the superheated water below, allowing it to boil. The resulting flash of steam triggers an eruption, causing an upward surge that sends hot water and steam showering out of the geyser vent (**Figure 17-14b**). This eruption releases the pressure, and when the eruption subsides, groundwater again begins to collect in the reservoirs in preparation for a repetition of the process.

A tremendous supply of heat is essential for geyser activity. Recent studies in Yellowstone Park's Upper Geyser Basin indicate that the heat emanating from that basin is at least 800 times greater than the heat flowing from a nongeyser area of the same size.

LearningCheck 17-5 **Describe the conditions that typically lead up to a geyser eruption.**

Eruption Patterns: Some geysers erupt continuously, indicating that they are really hot springs that have a constant supply of water through which steam escapes. Most geysers are only sporadically active, however, apparently depending on the accumulation of sufficient water to force an eruption. Some eruptions are very brief, whereas others continue for many minutes. Most erupt at variable intervals of a few hours or a few days, but some wait years or even decades between eruptions. The temperature of the erupting water generally is near the boiling point for pure water (100°C or 212°F at sea level). In some geysers, the erupting water column goes up only a few centimeters in the air, whereas in others the column rises to more than 45 meters (150 feet).

The term *geyser* comes from the Icelandic word *geysa* ("to gush" or "to rage"); the Great Geysir in southern Iceland is the namesake origin. The most famous of all geysers is Old Faithful in Yellowstone National Park (see the chapter opening photograph). Its reputation is based partly on the force of its eruptions (the column reaches more than 40 meters or 130 feet high) but primarily on its regularity. Since first timed by scientists more than a century ago, Old Faithful has erupted at intervals of 35 to 120 minutes, day

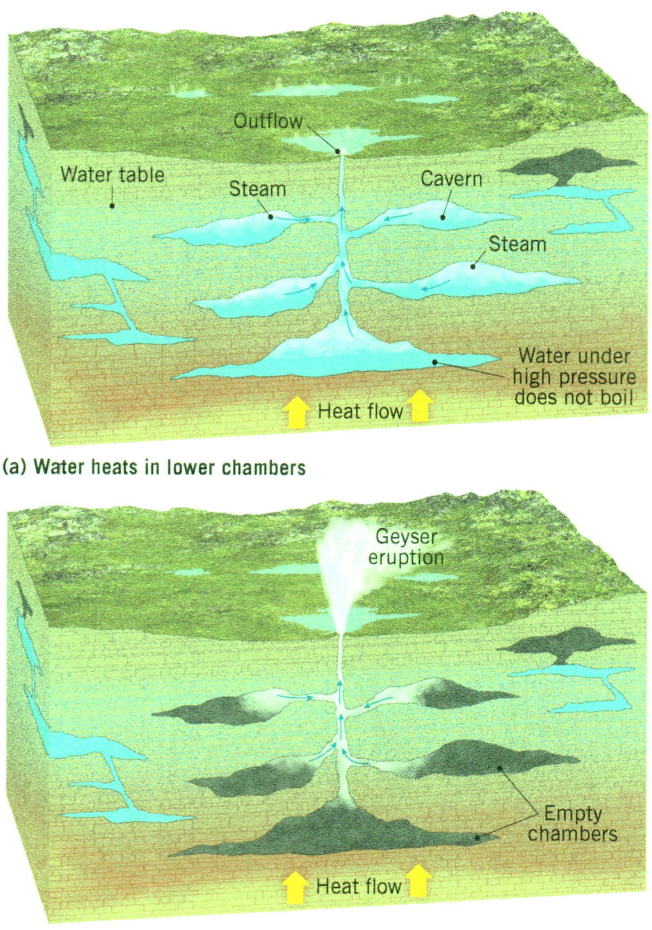

▲ **Figure 17-14** The eruption sequence of a typical geyser. (a) Water in lower chambers heats but does not boil at high pressure. Water in upper chambers heats, turns to steam, expands, and flows out onto the surface. (b) Surface outflow reduces pressure below, allowing the superheated deep water to boil. The sudden formation of steam below causes an eruption. Afterward, the chambers refill with water and the process begins again.

and night, winter and summer, year after year. In the early 1980s, however, several consecutive earthquakes on the Yellowstone plateau apparently upset the geyser's internal plumbing slightly, making Old Faithful less regular. It currently has an average interval of 60 to 110 minutes between eruptions.

Mineral Deposits: The deposits resulting from geyser activity are usually much less notable than those associated with hot springs. Some geysers erupt from open pools of hot water, throwing tremendous sheets of water and steam into the air but usually producing relatively minor depositional features. Other geysers are of the "nozzle" type: they build up a depositional cone and erupt through a small opening in it. Most deposits resulting from geyser activity are simply sheets of precipitated mineral matter spread irregularly over the ground.

Fumaroles

A third hydrothermal feature is the **fumarole**, a surface crack that is directly connected to a deep-seated heat source (Figure 17-15). Very little water drains into the tube of a fumarole. The water that does drain in is instantly converted to steam by the heat, and a cloud of steam is then expelled from the opening—often with a roaring or hissing sound. Thus, a fumarole is marked by steam issuing continuously or sporadically from a surface vent; in essence, a fumarole is simply a hot spring that lacks liquid water. If steam from below heats water and mud in an impermeable surface depression, it can form a boiling *mudpot*, which may be colored by variations in the accumulated minerals.

Hydrothermal features are found in many volcanic areas, being particularly notable in Iceland, New Zealand, Chile, Siberia's Kamchatka Peninsula, and most famously in Yellowstone National Park. See the box *Featured National Park: Yellowstone—A Hydrothermal Wonderland.*

◀ **Figure 17-15** A fumarole is like a geyser except that it erupts no liquid water—only steam. This scene is near Mývatn Lake, Iceland.

Featured National Park

Yellowstone

A Hydrothermal Wonderland

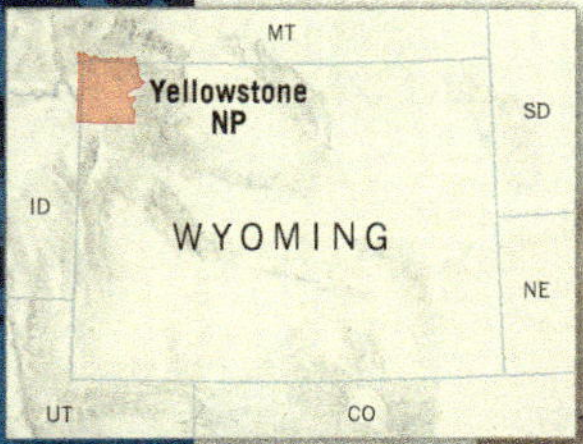

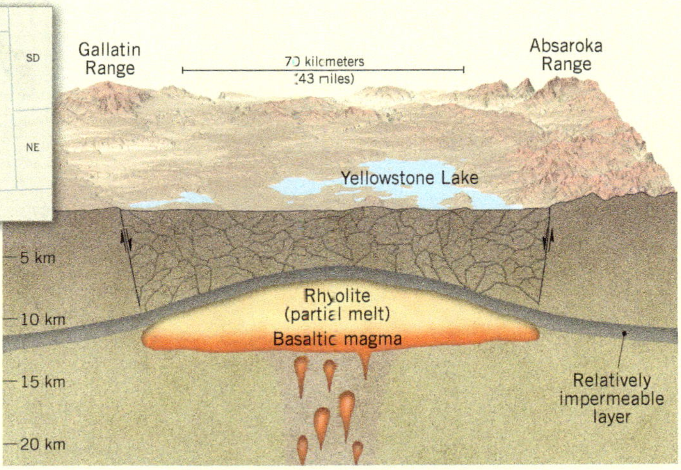

▲ **Figure 17-H** Schematic west–east cross section through the Yellowstone Plateau. Notice the extent of the Yellowstone Caldera and the magma chamber below.

By far the largest concentration of hydrothermal features in the world is found in Yellowstone National Park, located mostly in northwestern Wyoming. Yellowstone contains about half of the world's geysers, and more than 10,000 hydrothermal features (**Figure 17-G**).

Geologic Setting: Yellowstone National Park consists of a broad plateau bordered by extensive mountains (the Absaroka Range) on the east and by more limited highlands (particularly the Gallatin Mountains) on the west. The bedrock surface of the plateau is almost entirely volcanic, although no volcanic cones are in evidence. About 640,000 years ago, a catastrophic volcanic eruption here ejected about 1000 cubic kilometers (240 cubic miles) of pyroclastics—about 1000 times more material than the 1980 Mount St. Helens eruption. The surrounding region was covered with thick deposits of volcanic ash, and a *caldera* 65 kilometers (40 miles) in diameter formed (**Figure 17-H**). An even larger eruption took place about 2.1 million years ago, and geologists think that future eruptions of the Yellowstone "super volcano" are a distinct possibility.

▼ **Figure 17-G** Most geysers erupt from vents or hot pools, but some like Castle Geyser build up prominent depositional "nozzles" through which water is expelled.

Hydrothermal Conditions: The uniqueness of Yellowstone's geologic setting stems from the presence of a large, shallow magma chamber beneath the plateau—thought to be the result of a *hot spot* formed by a mantle plume (see Figure 14-21). Test boreholes and geophysical studies indicate that the top of the magma chamber is perhaps only 8 kilometers (5 miles) below the surface. Between 2004 and 2010, the ground in the caldera rose by 25 centimeters (10 inches), most likely as a result of swelling of the magma chamber below. After that, it subsided, only to rise again. Volcanologists have observed such cycles of ground movement in the past. The shallow magma pool provides the heat source—the most important of the three conditions necessary for the development of hydrothermal features.

The second required condition is an abundance of water that can seep downward and become heated. Yellowstone receives copious summer rain and a deep winter snowpack (averaging more than 380 centimeters [150 inches]).

The third condition for hydrothermal development is a weak or broken ground surface that allows water to move up and down easily. Here, too, Yellowstone fits the bill. Frequent earthquakes, faulting, and volcanic activity have left the ground with many fractures and weak zones, providing easy avenues for vertical water movement.

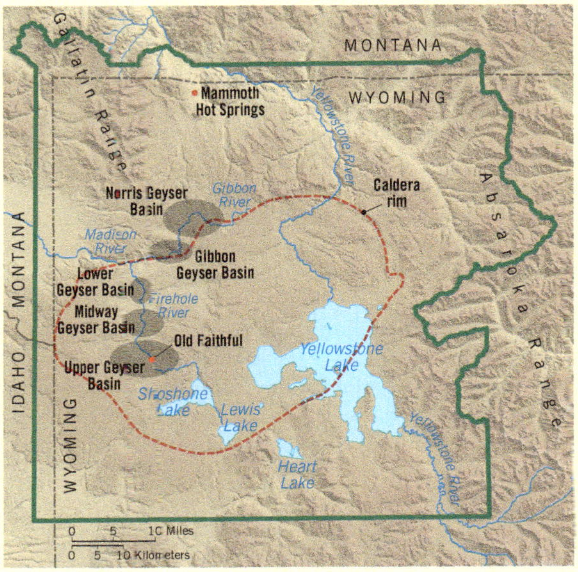

▲ **Figure 17-I** Yellowstone National Park, its major geyser basins, and the approximate outline of the Yellowstone Caldera.

Geyser Basins: The Park contains about 500 geysers, more than 3000 hot springs, and 7000 other thermal features (fumaroles, steam vents, hot-water terraces, and hot mudpots). There are five major geyser basins and many minor ones or small groups of thermal features.

All of the principal geyser basins are in the same watershed on the western side of the Park (**Figure 17-I**). The Gibbon River from the north drains the Norris and Gibbon geyser basins, and the Firehole River from the south drains the Upper, Midway, and Lower geyser basins. The Firehole River derives its name from the great quantity of hot water fed into it from about two-thirds of all hydrothermal features in Yellowstone. The Gibbon and Firehole rivers unite, forming the Madison River, which flows westward into Idaho.

continued on next page

(a) Mammoth Hot Springs

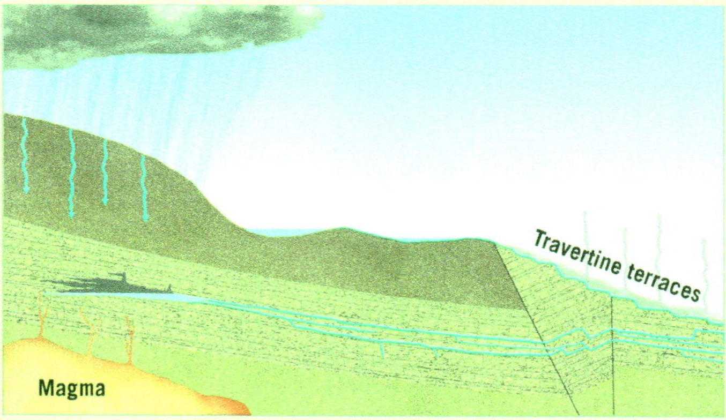

(b) Subsurface structure

▲ **Figure 17-J** Yellowstone's Mammoth Hot Springs. (a) Although the white material looks like snow or ice, it is actually travertine. (b) Water from rainfall and snowmelt percolates down into the underlying limestone, where it is heated and flows downslope. Hot water flows onto the surface and precipitates travertine deposits when exposed to the air.

All of the major geyser basins consist of gently undulating plains or valleys covered mostly with glacial sediments and large expanses of whitish siliceous material called *geyserite*. Each basin contains a few to several dozen geysers and a number of hot springs and fumaroles. These features exhibit an extraordinary range of eruption behaviors.

Mammoth Hot Springs: In the northwestern portion of the Park is the most remarkable aggregation of hot-water terraces in the world: the Mammoth Hot Springs Terraces (**Figure 17-J**). There, groundwater percolates down from surrounding hills into thick layers of limestone. Hot water, carbon dioxide, and other gases rise from the heated magma, mingle with the groundwater, and produce a mild carbonic acid solution that rapidly dissolves great quantities of the limestone. Saturated with dissolved minerals, the temporarily carbonated water seeps downslope until it gushes forth near the base of the hills as the Mammoth Hot Springs. The carbon dioxide escapes into the air, and the calcium carbonate precipitates as massive deposits of travertine in the form of beautiful flat-topped, steep-sided terraces.

Questions

1. Identify three conditions that make Yellowstone ideal for the development of hydrothermal features.
2. How do the terraces of Mammoth Hot Spring form?

NPS Video
Yellowstone in Depth: Geysers
https://media.pearsoncmg.com/ph/esm/esm_mcknight_physgeo_13/media/

17 Learning Review

After studying this chapter, you should be able to answer the following questions. Key terms from each text section are shown in **bold type.** Definitions for key terms are also found in the glossary at the back of the book.

Key Terms and Concepts

Dissolution and Precipitation (*p. 506*)

1. How does **carbonic acid** form?
2. What is meant by **dissolution**?
3. What kinds of rock are most susceptible to solution processes? Why?

Caverns and Related Features (*p. 507*)

4. What is the importance of jointing and bedding planes to the underground structure of **caverns**?
5. Describe and explain the formation of **speleothems** such as **stalactites**, **stalagmites**, and columns.

Karst Topography (*p. 509*)

6. In what kinds of rocks does **karst** topography develop?
7. Explain how a **sinkhole** forms.
8. Describe the formation of a **collapse sinkhole** and an **uvala**.
9. Describe the characteristics of **tower karst**.
10. What is a **swallow hole**? A **disappearing stream**?
11. Why is there a scarcity of surface drainage in karst areas?

Hydrothermal Features (*p. 513*)

12. What is **hydrothermal activity**?
13. What are the differences among a **hot spring**, a **geyser**, and a **fumarole**? What causes these differences?
14. Briefly explain the eruption sequence of a typical geyser.

Study Questions

1. Which is more important for the weathering action of underground water: mechanical or chemical weathering? Why?
2. How does the underground structure of the bedrock influence the dissolution process?
3. How is it possible for percolating groundwater to remove mineral material as well as deposit it?
4. How can groundwater pumping by people lead to sinkhole formation?
5. What three conditions are necessary for hydrothermal features to develop?
6. What is the importance of jointing and bedding planes to the development of hot springs and geysers?
7. Why don't most geysers erupt at regular intervals?
8. The 1912 eruption of Mount Katmai in Alaska buried a nearby river valley beneath a thick layer of volcanic ash. Today the area is called "The Valley of 10,000 Smokes." What do you think this name refers to? Explain.

Exercises

1. If a "soda straw" speleothem in a cavern grows at an average rate of 1.7 mm per year, how long does it take to form a soda straw 6 centimeters long?
2. During the morning of August 14, 2011, Old Faithful Geyser in Yellowstone National Park erupted at the following times (to the nearest minute): 12:07 A.M., 1:42 A.M., 3:05 A.M., 4:41 A.M., 6:07 A.M., 7:37 A.M., 9:08 A.M., and 10:34 A.M. What was the average interval between eruptions on this morning? _____ minutes

Environmental **Analysis** SINKHOLES IN TENNESSEE

Data Sinkholes https://media.pearsoncmg.com/ph/esm/esm_mcknight_physgeo_13/media/

Karst is common in regions that have carbonate bedrock, such as limestone, beneath the surface. Cracks allow rainwater to seep in and dissolve the bedrock, forming chambers underground. Sinkholes develop when these chambers collapse.

Activities

Go to the sinkhole count map of the state of Tennessee at http://tnlandforms.us/heatmap/3x3heat.html. Bright red indicates a high number of sinkholes; pale green indicates few sinkholes. Zoom in and out as needed.

1. Which regions of Tennessee are prone to sinkholes?
2. Identify three towns that have a problem with sinkholes.
3. What is the overall pattern or directional trend of sinkhole distribution?

Click on the "Map" button and select the "Terrain" checkbox.

4. Do sinkholes tend to form along mountains or in valleys? Why?

In a separate window go to http://https://mrdata.usgs.gov/geology/state/map-us.html, the USGS map of U.S. geology. Click on the menu icon and choose "US States" under "Geographic reference." Zoom in to Tennessee and click on the various shaded regions—regions that have sinkholes and those that don't, based on the sinkhole count map—to distinguish the different geologic units.

5. What rock types in Tennessee are prone to sinkholes?
6. What rock types in Tennessee are not prone to sinkholes?
7. Based on the pattern shown in Tennessee, in which portions of other eastern states would you expect to find sinkholes?
8. What information can you find that verifies your predictions?

Seeing Geographically in Our National Parks

Look again at the photograph of Old Faithful at the beginning of the chapter (p. 504). Describe the characteristics of the water ejected from Old Faithful. What kinds of rocks are likely to exist around the base of the geyser? How do they form? What is the role of steam in triggering an eruption of a geyser?

Mastering Geography®

Looking for additional review and test prep materials? Visit the Study Area in **Mastering Geography** to enhance your geographic literacy, spatial reasoning skills, and understanding of this chapter's content by accessing a variety of resources, including Map Master interactive maps, geoscience animations, *Mobile Field Trips*, videos, *Project Condor* Quadcopter videos, *In the News* RSS feeds, flashcards, web links, self-study quizzes, and an eText version of *McKnight's Physical Geography*.

Seeing Geographically in Our National Parks

Looking across Death Valley to the Panamint Range from Zabriskie Point. What suggests that this is a dry environment? Describe the overall topography. Is Manly Beacon (the sharp peak in the right foreground) made of more resistant or less resistant rock than the adjacent badlands? How do you know? What suggests that, at times, this environment has running water?

Death Valley National Park, CA

18

The Topography of Arid Lands

HAVE YOU EVER WONDERED what happens during a flash flood in the desert? Even in the driest of deserts, such as here in Death Valley National Park, a *flash flood*—a sudden, short-lived surge of water flowing down a dry streambed—is not unusual. Fed by a downpour of rain from a thunderstorm perhaps many kilometers away, a bone-dry ephemeral stream can fill with a torrent of muddy water in just a few minutes with little warning. Perhaps surprisingly, such events are among the most important shapers of desert topography. Most of the erosion and deposition in many desert areas takes place during such events. Evidence of flash floods and *debris flows* (rapidly moving flows of mud and boulders) is a common sight in almost any of our desert National Parks.

Arid lands are in many ways distinct from humid ones, but no obvious boundaries separate the two. In this chapter, we focus on the dry lands of the world without attempting to establish precise definitions or borders. We are concerned not with where such borders lie but rather with the processes that shape desert landscapes. Both the processes and the landforms, however, occur more widely than the term *desert* might imply, and some of the landforms we describe may be seen even in humid regions.

It is important to understand that some of today's deserts had quite different climates in the geologic past. Parts of the Sahara Desert, for example, were much wetter just a few thousand years ago than they are today. Thus, in addition to processes operating today, some desert landscapes have been shaped by different processes that were at work in the past.

As you study this chapter, think about these **Key** Questions:

- **What special conditions influence landform development in deserts?**
- **Why are streams the most important mechanism of erosion and deposition in most deserts?**
- **What is the role of wind in shaping desert landforms?**
- **Why are landforms such as playas and alluvial fans so prevalent in the basin-and-range desert region of North America?**

NPS Video
Death Valley Explorer (E1)
https://media.pearsoncmg.com/ph/esm/esm_mcknight_physgeo_13/media/

A Specialized Environment

Desert terrain is usually stark and abrupt, unsoftened by regolith, soil, or vegetation (see Chapter 11). Despite the great difference in appearance between arid lands and humid lands, most of the terrain-forming processes active in humid areas are also at work in desert areas. Deserts, however, have special conditions that significantly affect landform development. As we saw in Chapter 8, in terms of climate, deserts are defined as places where potential evapotranspiration exceeds precipitation. Here, however, we focus broadly on those conditions that influence arid land topography.

Special Conditions in Deserts

A wide variety of factors influence geomorphic development in deserts.

Weathering: Because water is required for nearly all kinds of chemical weathering, in many desert regions mechanical weathering is dominant. Chemical weathering is likely to be absent in only the driest of deserts, however. Mechanical weathering processes such as *salt wedging* are more common in arid regions than in humid ones (see Figure 15-7). This predominance of mechanical weathering results in both a slower rate of total weathering in deserts and the production of more angular pieces of weathered rock.

Soil and Regolith: In deserts, the covering of soil and regolith is thin or absent in most places. This condition exposes the bedrock to weathering and erosion and contributes to the stark, rugged, rocky terrain (Figure 18-1). Soil creep is a relatively minor phenomenon on most desert slopes. This is due partly to the lack of soil but primarily to the lack of the lubricating effects of water. Creep is a smoothing phenomenon in more humid climates, and its lack in deserts accounts in part for the angularity of desert slopes.

Impermeable Surfaces: Much of the desert surface is impermeable to infiltrating water, permitting little moisture to seep into the ground. *Caprocks* (resistant bedrock surfaces) and *hardpans* (hardened and generally water-impermeable subsurface soil layers) of various types are widespread, and what soil has formed is usually thoroughly compacted and often does not readily absorb water. Importantly, such impermeable surfaces lead to high runoff when it rains.

Sand: Some deserts have much more sand than other parts of the world do. This is not to say, however, that deserts are mostly sand covered. Indeed, the notion that all deserts consist of great seas of sand is incorrect. Nevertheless, the relatively large amount of sand in some deserts has three important influences on topographic development: (1) A sandy cover allows water to infiltrate the ground and thereby reduces runoff by way of both streams and overland flow; (2) sand is readily moved by heavy rains; and (3) sand can be transported and redeposited by wind. (The development of desert sand dunes is discussed later in this chapter.)

LearningCheck 18-1 **When it rains in a desert, how is the amount of runoff influenced by impermeable surfaces? (Answer on p. AK-6)**

Rainfall: Although rainfall is limited in desert areas, much of the rain that does fall comes from intense convective thunderstorms—which result in very high and rapid runoff. Floods, although often brief and covering only a limited area, are the rule rather than the exception in deserts. Thus, fluvial erosion and deposition, however sporadic and rare, are remarkably effective and conspicuous.

Fluvial Deposition: Almost all streams in desert areas are *ephemeral*, flowing only during and immediately after a rain. Such streams are effective agents of erosion, shifting enormous amounts of material in a short time. This is mostly short-distance transportation, however. A large volume of unconsolidated debris is moved to a nearby location; as the stream dries up, the debris is dumped on slopes or in valleys, where it is readily available for the next rain. As a consequence, depositional features of alluvium are quite common in desert areas.

Wind: Another fallacy associated with deserts is that their landforms are produced largely by wind action. This is not true, even though high winds are characteristic of most deserts and even though sand and dust particles are easily shifted.

Basins of Interior Drainage: Many desert areas contain watersheds that do not drain ultimately into any ocean. For most continental surfaces, rainfall can flow all the way to the sea. In dry lands, however, drainage networks are frequently underdeveloped, and the terminus of a drainage system is often a basin or valley with no external outlet (Figure 18-2). Any rain that falls in Nevada, for example, except in the extreme southeastern and northeastern corners of the state, has no chance of flowing in a stream to the sea.

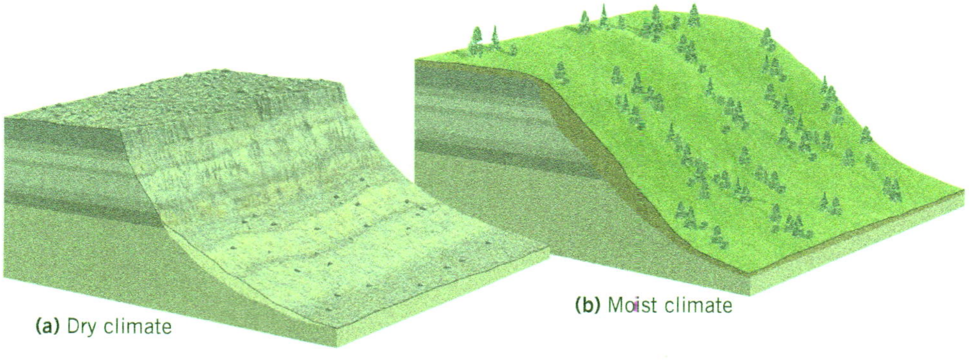

▶ Figure 18-1 Slopes in (a) dry and (b) moist climates. The steep relief is more easily seen in a dry climate, which has very little vegetation cover, unlike that in a humid climate.

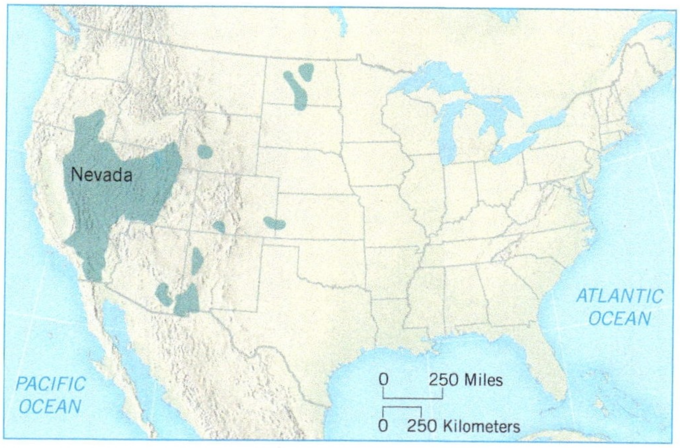

▲ **Figure 18-2** All of the basins of interior drainage (shown in green) in the conterminous United States are in the west. By far the largest area of interior drainage is in the "Great Basin" centered in Nevada and the adjoining states.

Vegetation: All of the previous environmental factors have important effects on topographic development, but perhaps the most obvious feature of dry lands is the lack of a continuous cover of vegetation. The plant cover consists mostly of widely spaced shrubs or sparse grass. They provide little protection from the force of raindrops and inadequately bind the surface material with roots, making it easier for a desert's scarce rainfall to loosen and transport sediment (Figure 18-3). The less vegetation a desert has, the greater the rate of weathering.

LearningCheck 18-2 **How does the sparse cover of soil and vegetation influence weathering and erosion in deserts?**

▼ **Figure 18-3** In most arid regions, vegetation cover is sparse. The creosote bush is the green shrub in this scene in Death Valley, California.

Running Water in Waterless Regions

Probably the most fundamental fact of desert geomorphology is that running water is by far the most important external agent of landform development. The erosional and depositional work of running water influences the shape of the terrain surface almost everywhere outside of areas of extensive sand accumulation. The lightly vegetated ground is defenseless to whatever rainfall may occur, and erosion by rain splash or streamflow is enormously effective. Despite the rarity of precipitation, its intensity and the presence of impermeable surfaces produce abrupt runoff, and great volumes of sediment can be moved in a very short time.

The steep gradients of mountain streams increase the capacity of these streams for carrying large loads, but the flow of mountain streams in arid lands is sporadic. At any given time, large amounts of alluvium sit at rest in the dry streambed of a desert mountain, awaiting the next flow. Loose surface material is thin or absent on the slopes, and bedrock is often clearly exposed, with the more resistant strata standing out as caprocks and cliff faces.

Where slopes are gentle in an arid land, the streams rapidly become choked with sediment as a brief flood subsides. Here stream channels are readily subdivided by braiding, and main channels often break up into distributaries in the basins. Much silt and sand are thus left on the surface for the next flood to move, unless wind moves them first.

Surface Water in the Desert

Surface water in deserts is conspicuous by its absence. These are lands of sandy streams and dusty lakes, in which the presence of surface water is usually episodic and brief.

Exotic Streams: Permanent streams in dry lands are few, far between, and, with rare exceptions, *exotic*—meaning that they are sustained by water that originates outside the desert. The water that feeds **exotic streams** comes from an adjacent wetter area or a higher mountain area in the desert and has enough volume to survive flowing across the dry lands. The Nile River of North Africa is a classic example of an exotic stream (Figure 18-4). Its water comes from the mountains and lakes of central Africa and Ethiopia in sufficient quantity to survive a 3200-kilometer (2000-mile) journey across the desert without benefit of tributaries. The Colorado River is a prominent exotic stream in North America.

In humid regions, a river becomes larger as it flows downstream, nourished by tributaries and groundwater inflow. In dry lands, however, the flow of exotic rivers diminishes downstream because the water seeps into the riverbed, evaporates, or is diverted for irrigation.

Ephemeral Streams: Although some deserts have prominent exotic rivers, the vast majority of all desert streams are **ephemeral streams**, which carry water only seasonally or after a rainstorm. The brief periods during which these streams flow are times of intense erosion, transportation, and deposition, however. Most ephemeral desert streamflow

▲ **Figure 18-4** The preeminent example of an exotic stream, the Nile River, flows for many hundreds of kilometers without being joined by a tributary.

▲ **Figure 18-5** Ephemeral stream channel in the Mojave Desert near Baker, California.

eventually dissipates through seepage and evaporation, although sometimes such a stream reaches the sea, a lake, or an exotic river (Figure 18-5).

The normally dry beds of ephemeral streams typically have flat floors, sandy bottoms, and steep sides. In the United States, they are variously referred to as *arroyos*, *gullies*, *washes*, or *coulees*. In North Africa and Arabia, the name *wadi* is common; in South Africa, *donga*; in India, *nullah*.

LearningCheck 18-3 **Distinguish between an exotic stream and an ephemeral stream in a desert.**

Playas: Although lakes are uncommon in desert areas, dry lake beds are not. We have already noted the prevalence of basins of interior drainage in dry lands; most of them have a lake bed occupying their area of lowest elevation, which functions as the local base level for that basin. These dry lake beds are called **playas** (Figure 18-6); the term **salina** may be used if there is an unusually heavy concentration of salt in the lake-bed sediments. If a playa surface is heavily impregnated with clay, the formation is called a *claypan*. On rare occasions, the intermittent streams may have enough flow to bring water to the playa, forming a temporary *playa lake*.

Playas are among the flattest and most level of all landforms—some are several kilometers across but have a local relief of only a few centimeters. A playa develops such a flat surface when it is periodically covered with water as a playa lake. The suspended silt in the shallow lake eventually settles out, and the water evaporates, leaving a flat layer of dried mud. Through repeated episodes such as this, over the centuries a playa attains its flat surface.

Saline Lakes: A few desert lakes are permanent. The smaller ones are nearly always the product of subsurface structural conditions that provide water from a permanent spring or exotic streams or from streams flowing down from nearby mountains. Many permanent lakes in desert areas are **saline lakes**, in which high rates of evaporation relative to the inflow or basins of interior drainage lead to the accumulation of dissolved salts.

Many of the largest natural desert lakes are remnants of still larger bodies of water that had formed in a wetter climate. Utah's Great Salt Lake is the outstanding example in the United States. In terms of surface area, it is the second largest lake wholly in the United States (after Lake Michigan). It is a mere shadow, however, of the former Lake Bonneville, which formed during the wetter conditions of the Pleistocene Epoch (see Figure 19-6).

LearningCheck 18-4 **What are playas? Why do they form?**

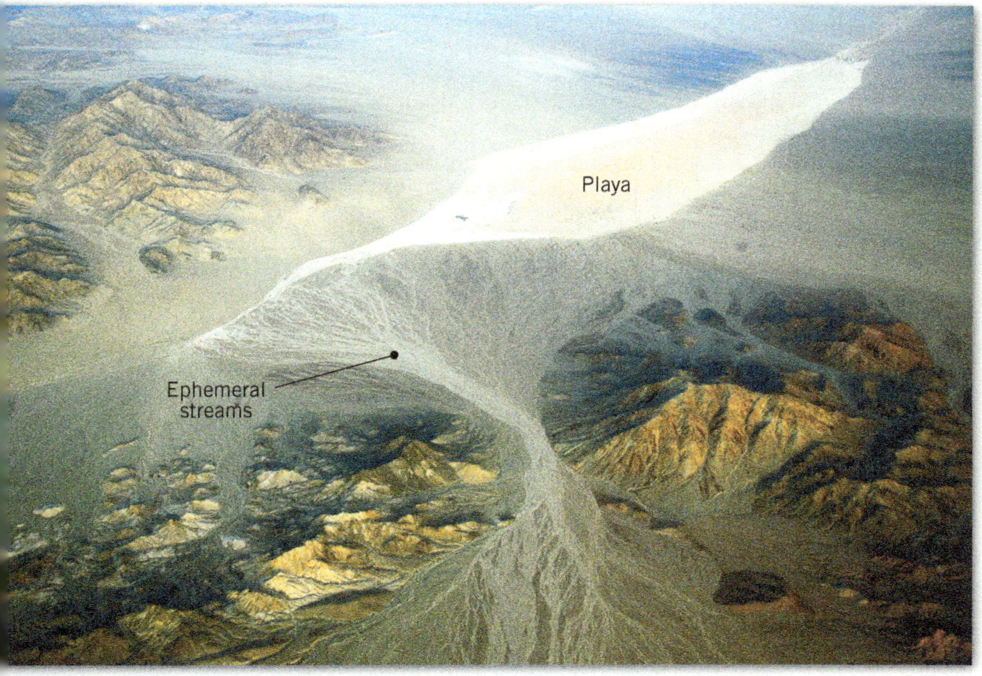

◀ **Figure 18-6** Broadwell Lake in California is a desert playa. Playas are among the flattest of all landforms.

Fluvial Erosion in Arid Lands

Although fluvial erosion takes place in desert areas during only a small portion of each year, it works rapidly and effectively, and the results are obvious. In desert areas of any significant relief, large expanses of exposed bedrock are common because soil and vegetation are lacking. During the rare rains, this bedrock is both mechanically weathered and eroded by running water, leaving steep, rugged, rocky surfaces.

Fluvial erosion in a desert might typically begin with a brief but intense thunderstorm, perhaps over a mountain drainage basin. The localized thunderstorm puts a lot of water on the ground, but because the desert surface is either exposed bedrock or some other kind of relatively impermeable surface, most of the water quickly runs off into a nearby dry ephemeral stream channel. This stream channel quickly fills with water, perhaps developing into a flash flood or a debris flow (described in Chapter 15) that travels for many kilometers from the desert mountains down onto a basin floor. Flash floods and debris flows can move a remarkable amount of material in just a few minutes (**Figure 18-7**). It is through such localized but infrequent events that most change takes place in the desert.

Flash Flood Hazards: Both flash floods and debris flows pose a significant hazard to humans in arid lands. Travelers are cautioned against camping or parking in dry stream washes because such floods can arrive with little local warning. In many desert cities, flood-control channels and basins have been constructed as protection from these infrequent but often catastrophic events. Deep concrete-lined stream channels carrying only a trickle of water or huge empty catchment basins may seem unnecessary to the casual observer—until one of these rare deluges suddenly fills these structures to overflowing!

LearningCheck 18-5 **Explain the significance of flash floods and debris flows in deserts.**

Differential Weathering and Erosion in Deserts: As we saw in Chapter 15, variations in rock type and structure influence the rate of weathering and, in turn, the ease of erosion. Such **differential weathering and erosion** frequently produce differences in the slope and shape of the resulting landform. Because of the generally sparse cover of soil and vegetation in deserts, the mark of differential weathering and erosion is often striking.

In many cases, rocks resistant to weathering and erosion form the cliffs, pinnacles, spires, and other sharp crests, while softer rocks wear away more rapidly, producing gentler slopes. Differential erosion is very common in sedimentary landscapes because there can be significant differences in resistance from one stratum to the next. Such areas often have vertical *escarpments* (steep, clifflike slopes) and abrupt

◀ **Figure 18-7** A flash flood caused by runoff from rainfall in a usually arid area of the lower Park Avenue area of Arches National Park, Utah.

▲ **Figure 18-8** Differential weathering and erosion are conspicuous on the Red Cliffs near Gateway, Colorado. Resistant top layers form an abrupt escarpment; softer layers below weather and erode into gentler slopes.

▲ **Figure 18-9** Uluru (Ayers Rock) is a bornhardt in the desert of central Australia. Kata Tjuta (the Olgas) in the distance is also a bornhardt.

changes in slope angle (Figure 18-8). In areas dominated by igneous or metamorphic bedrock, differential weathering and erosion may be less obvious because there is not much difference in resistance from one part of the bedrock to another.

Residual Surfaces and Features: Scattered throughout arid and semiarid lands are isolated landforms that rise abruptly from the surrounding plains. Such steep-sided mountains or ridges are referred to as **inselbergs** ("island mountains") because they stand out as rocky islands. A distinctive type of inselberg is the *bornhardt*, which is composed of highly resistant rock and has a rounded form (Figure 18-9). Differential weathering and erosion lower the surrounding terrain, leaving the resistant bornhardt standing high (Figure 18-10). Bornhardts are very stable and may persist for tens of millions of years.

Along the lower slopes of desert mountains and hills, a gently inclined bedrock platform, called a **pediment**, may form. This "residual" surface (in other words, a surface left by weathering and the removal of rock rather than by deposition) extends outward from the mountain front. Pediments were long thought to develop as desert mountains are worn down and back by weathering and erosion during previous periods of wetter climate (Figure 18-11a). Recent analysis suggests, however, that some pediments may be the result of deep subsurface weathering. After the overlying weathered material is removed by erosion, a flat or gently sloping bedrock surface remains. Other pediments may have formed as a consequence of *detachment faulting* in areas undergoing tectonic extension (see Figure 14-52), whereby blocks of crust are displaced along nearly horizontal fault planes—the bedrock surface of the pediment representing the fault plane (Figure 18-11b).

Pediments can be found in many deserts and are sometimes the dominant terrain feature. They are not easily recognizable, however, because almost invariably they are covered with a veneer of debris deposited by water and wind.

Fluvial Deposition in Arid Lands

Except in hills and mountains, depositional features are more prominent than erosional ones in a desert landscape. Depositional features consist mostly of talus accumulations

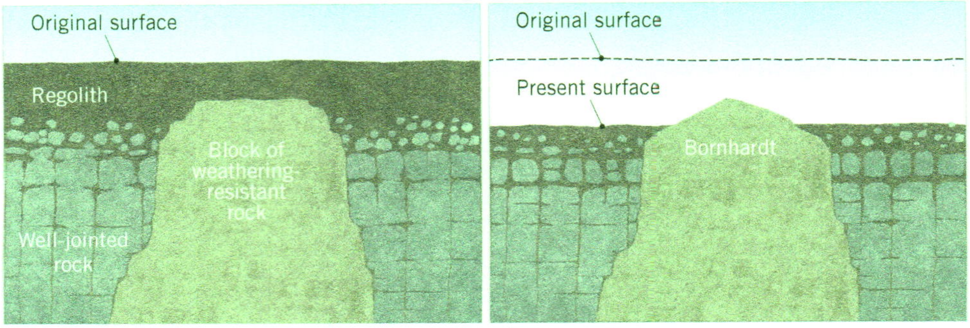

(a) Before erosion (b) After erosion

► **Figure 18-10** The development of a bornhardt. (a) The well-jointed rock is more susceptible to weathering and erosion than the resistant block in the center. (b) The bornhardt is the result of this differential weathering and erosion.

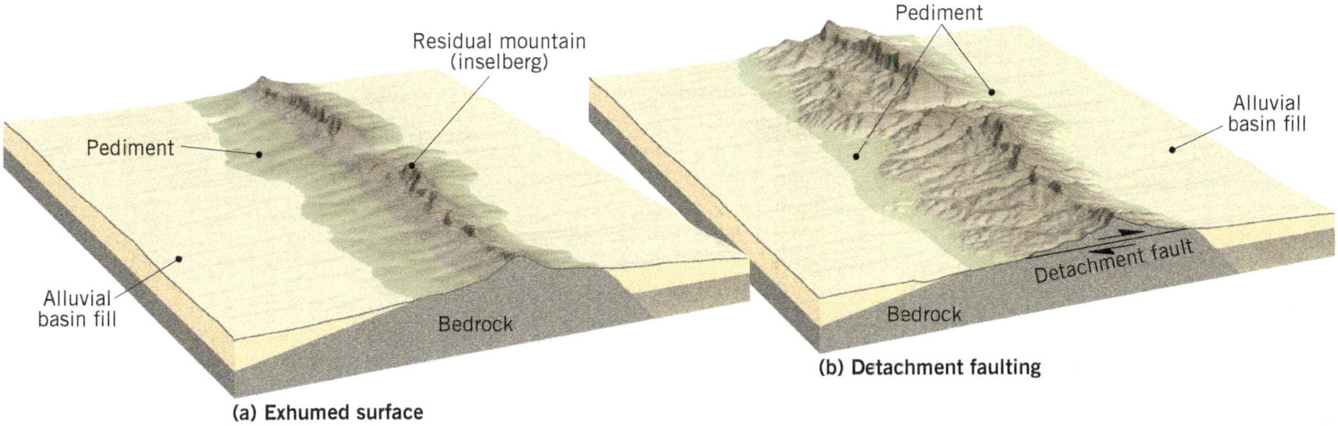

▲ **Figure 18-11** Desert pediments. (a) Some pediments may represent the top of a now-exhumed, weathered bedrock surface. (b) Others may develop as a consequence of detachment faulting and subsequent erosion. The residual desert hills are inselbergs.

at the foot of steep slopes and deposits of alluvium in ephemeral stream channels.

Piedmont is a generic term meaning any zone at the foot of a mountain range. (The term *pediment* comes from the same Latin root but refers to a specific landform.) The **piedmont zone** of a desert mountain range is a prominent area of fluvial deposition. There is normally a pronounced change in slope at the mountain base, with a steep slope giving way abruptly to a gentle one (Figure 18-12). The break in slope greatly reduces the speed of any sheetwash, streamflow, or debris flow that travels into the piedmont zone. The resulting fluvial deposition in the piedmont often reaches depths of hundreds of meters.

LearningCheck 18-6 **Why is the piedmont zone an area of fluvial deposition in most deserts?**

Alluvial Fans: The piedmont zone typically includes one of the most prominent and widespread topographic features in any desert area, the **alluvial fan**. (Although a dominant feature in many deserts, alluvial fans also can develop in humid areas.) As a stream leaves the narrow confines of a mountain gorge and emerges onto the open piedmont zone, its flow slows. The flow breaks into distributary channels that wend their way down the piedmont slope, sometimes cutting shallow new channels in the loose alluvium but frequently depositing more debris atop the old (Figure 18-13). In this fashion, a moderately sloping, fan-shaped landform builds up at the mouth of the canyon. When one part of the fan is built up, the channeled flow shifts to another section and builds that up. The entire fan is eventually covered more or less symmetrically with alluvium. In general, large boulders are dropped near the "apex" (top) of the fan and finer material accumulates around the margins, with a considerable mixing of particle sizes throughout. As deposition continues, the fan extends outward across the piedmont zone and onto the basin floor.

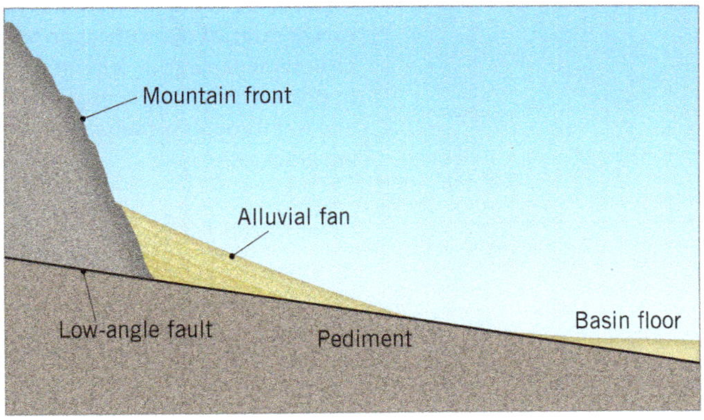

▲ **Figure 18-12** An idealized cross section of a desert piedmont zone. The pediment surface (in this case, associated with a low-angle fault) is often covered with alluvial deposits at the mountain front and along the basin floor.

Deposition in Desert Basins: The flatter portions of desert areas also often exhibit a significant accumulation of alluvium, in part because streamflow is not abundant enough to carry sediments very far from a mountain front—and it has no place to go if it terminates in a basin of interior drainage. Large rock fragments are rarely transported into the middle of a basin; instead, the middles of basins are covered with fine particles of sand, silt, and clay, sometimes to a considerable depth.

LearningCheck 18-7 **What is an alluvial fan, and how does one form?**

Climate Change and Deserts

Recall from Chapter 6 that rainfall in arid regions tends to exhibit high year-to-year variability and that climate projections are increasingly suggesting that rainfall in deserts and other semiarid environments will become even more erratic. Of great concern to policy planners are regions of the world—especially in poor developing countries—where desert areas have been expanding in recent decades in a process called *desertification*. Desertification is brought on by both natural and human-caused reasons—see the box *Global Environmental Change: Desertification*.

Global Environmental Change

Desertification

• **Mike Pease,** Central Washington University

Desertification is the process by which a desert encroaches on adjacent lands. Deserts grow and contract over millennia. The overtaking of agricultural lands by deserts is a major environmental issue in arid and semiarid regions. Nearly every continent has experienced desertification in the last century. From the 1930s Dust Bowl in the western Great Plains of the United States to the current expansion of the Gobi Desert into central China, desertification is among the top 10 global environmental issues. Desertification and accompanying land-use issues can impact global food security and thus global political security. The United Nations Convention to Combat Desertification (UNCCD) was established in 1994 to "mitigate the effect of drought in affected areas." This ad hoc committee funds conferences and scientific task forces to research ways to lessen the effects of water scarcity and land degradation.

Mechanism of Desertification: Typically, regions surrounding deserts are transition zones with semiarid conditions and sparse vegetation. Soils in these regions are vulnerable to degradation, particularly via wind erosion (**Figure 18-A**). With poor land-use management or periods of drought, soils can quickly lose their remaining vegetation. They then become less stable and increasingly susceptible to erosion by water and wind. If these conditions persist for several years, the lands can become deserts.

The Sahel in Sub-Saharan Africa is one such region. This 5900-kilometer-wide east–west band of semiarid shrublands and savanna south of the Sahara Desert is a transitional zone extending through Mali, Niger, Chad, Sudan, South Sudan, and Ethiopia (**Figure 18-B**). Annual precipitation varies from less than 100 millimeters (4 inches) in the north on the border of the Sahara to 500 millimeters (20 inches) in the south; this precipitation is highly seasonal and subject to high interannual variability.

The IPCC reports "the Sahel has experienced the most substantial and sustained decline in rainfall recorded anywhere in the world." Some climate models suggest that drought conditions may intensify as climate anomalies cause changes in seasonal patterns of the ITCZ, thereby intensifying the normal drought cycles. For Sahelian countries, desertification is an issue of national importance because most residents engage in subsistence agriculture or pastoral nomadism. The endemic poverty in the region leaves many ill-equipped to deal with even short-term losses of agricultural productivity. Expansion of the desert into the Sahel affects habitat for wildlife and rangeland for livestock as well as agricultural lands. Similar concerns exist for central Asia, northern China, and northern and eastern Australia. Even in the United States, extended droughts, such as those experienced in California between 2012 and 2015, can intensify soil loss.

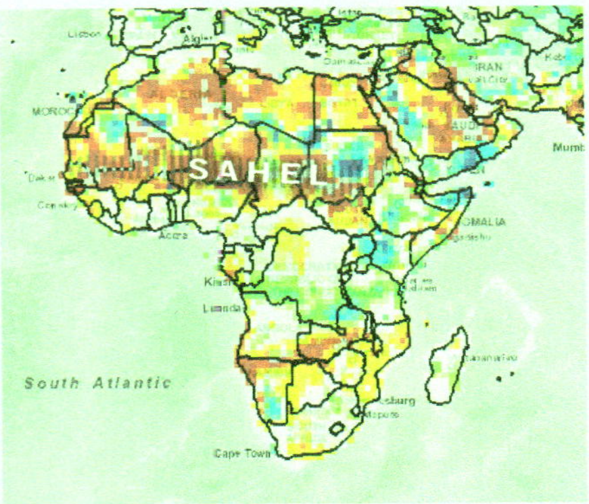

▲ **Figure 18-B** Drought in northern Africa and the Sahel region during 2020. Orange and red regions show extreme drought based on the Global Precipitation Climatology Centre Drought Index. Data from NOAA and the Global Drought Information System.

The Climate–Desertification Nexus: The models presented in the IPCC's *Fifth Assessment Report* in 2014 suggest that over the next 50 years, most places on Earth will experience an increase in severe weather phenomena, including a rise in intensified drought periods in water-scarce areas. This issue will be conflated with an increase in the amount of precipitation that falls during high-intensity storms. Regional climate models in semiarid areas and in regions already vulnerable to drought, such as the southwestern United States and central China, may experience a higher frequency of drought periods.

Agricultural Security: A primary driving force of desertification is the excessive cultivation of erosion-vulnerable soils. This trend is likely to intensify; the UNCCD estimates that by 2030, global food demand will increase 50 percent over 2012 levels. Desertification is leading to the loss of what UNCCD estimates to be 12 million hectares (26.5 million acres) of agricultural land annually. The result is a land-use paradox: increased land conversion for agriculture will be required to meet food demand. Future rates of desertification will be determined by our commitment to protection of native vegetation and soil through sustainable farming practices.

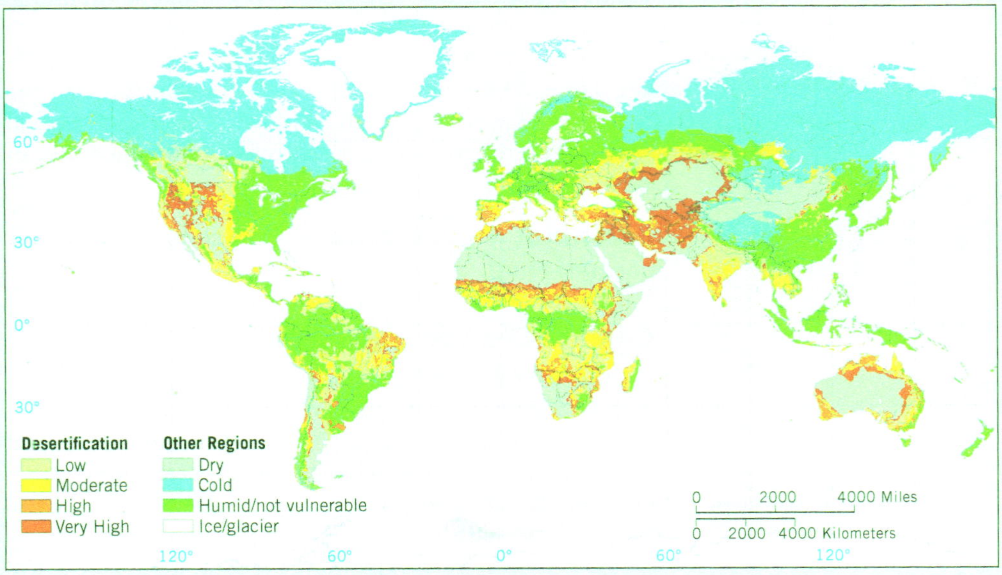

▲ **Figure 18-A** Areas of vulnerability to increased desertification. Notice the near-continuous band of very high conditions in the Sahelian countries in central Africa.

Questions

1. Other than the Sahel, what regions with large population centers are vulnerable to desertification?
2. What are some likely effects of desertification, other than land-use loss?

(a) Alluvial fan formation

(b) Alluvial fan in Death Valley

▲ **Figure 18-13** (a) An alluvial fan develops at the mouth of a canyon. (b) Alluvial fan in Death Valley, California.

The Work of the Wind

The frequent strong winds characteristic of many deserts can create spectacular sand and dust storms and can reshape minor details of the landscape. However, the effect of wind as a sculptor of terrain is very limited, with the important exception of such relatively impermanent features as sand dunes (**Figure 18-14**).

In general, the motion of air passing over the ground is similar to that of water flowing over a streambed. In a thin layer right at the ground surface, wind speed is zero, just as the speed of the water layer touching the banks and bed of a stream is zero, but wind speed increases with distance above the ground. The shear developed between different layers of air moving at different speeds causes turbulence similar to that in a stream of water. Wind turbulence can also be caused by warming from below, which causes the air to expand and rise.

Aeolian processes are those related to wind action. (Aeolus was the Greek god of the winds.) They are most pronounced, widespread, and effective wherever fine-grained, unconsolidated sedimentary material is exposed to the atmosphere, without benefit of vegetation, moisture, or some other form of protection—in other words, in deserts and along sandy beaches. Our primary focus here is wind action in desert regions.

Aeolian Erosion

The erosive effect of wind can be divided into two categories: *deflation* and *abrasion*.

Deflation: When loose particles are blown through the air or along the ground, they are said to be shifted by **deflation**. Except under extraordinary circumstances, wind is not strong enough to move anything more than dust and sand grains, so deflation creates few significant

▼ **Figure 18-14** Wind can be a prominent force in rearranging loose particles. Here, wind blows sand off of a sand dune in White Sands National Park, New Mexico.

▲ **Figure 18-15** A blowout forms where wind deflation removes loose material from the surface, leaving a depression, as here in Nebraska's Sandhills.

landforms. A **blowout**, or *deflation hollow*—a shallow depression from which an abundance of fine material has been deflated—may form (Figure 18-15). Most blowouts are small, but some exceed 1.5 kilometers (1 mile) in diameter.

Abrasion: Aeolian abrasion is analogous to fluvial abrasion, except that the aeolian variety is much less effective. Whereas deflation is accomplished entirely by air currents, abrasion requires "tools" in the form of airborne sand and dust particles. The wind drives these particles against rock and soil surfaces in a form of natural sandblasting. Wind abrasion does not construct or even significantly shape a landform; it merely sculpts those already in existence. The principal results of aeolian abrasion are the pitting, etching, faceting, and polishing of exposed rock surfaces and the further fragmenting of rock fragments. Rocks faceted by such wind "sandblasting" are called **ventifacts** (Figure 18-16).

Larger wind-eroded landforms known as *yardangs* are found in many deserts. Most yardangs consist of poorly consolidated sedimentary rock that has been sandblasted by persistent winds (Figure 18-17). They generally align parallel with the dominant wind direction and typically stand a few meters high, although some are tens of meters high.

Aeolian Transportation

Rock materials are transported by wind in much the same fashion as they are moved by water—but less effectively. The finest particles are carried in suspension as dust. Strong, turbulent winds can lift and carry thousands of tons of suspended dust. Some dust storms extend for hundreds of meters above Earth's surface (see Figure 3-4) and may move material through more than 1600 kilometers (1000 miles) of horizontal distance.

▲ **Figure 18-16** A sand-blasted rock, or ventifact. This piece of basalt is in Death Valley, near Badwater.

Particles larger than dust are moved by wind through *saltation* and *traction*, just as in streamflow (Figure 18-18). Wind is unable to lift particles larger than the size of sand grains, and even these are likely to be carried less than 1 meter above the surface. Indeed, most sand, even when propelled

▲ **Figure 18-17** Yardangs produced by wind abrasion in the Sahara Desert of southwestern Egypt.

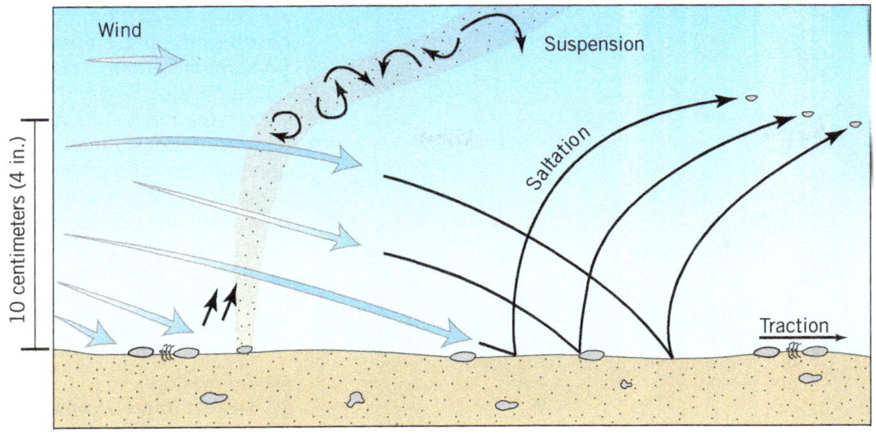

Animation
Wind Transportation of Sediment
https://media.pearsoncmg.com/ph/esm/esm_mcknight_physgeo_13/media/

◀ **Figure 18-18** Wind carries tiny dust particles in suspension; larger particles are moved by saltation or traction.

by a strong wind, leaps along in the low, curved trajectory typical of saltation, striking the ground at a low angle and bouncing onward. Larger particles move by traction, being rolled or pushed along the ground by the wind. It is estimated that three-fourths of the total volume of all wind-moved material in dry lands is shifted by saltation or traction, particularly the former. At the same time, the entire surface layer of sand moves slowly downwind as a result of the frequent impact of the saltating grains; this process is called *creep*. (Do not confuse this with *soil creep* discussed in Chapter 15.)

Because wind can lift particles only so high, a true *sandstorm* is a cloud of horizontally moving sand that extends only a few centimeters or feet above the ground surface. People standing in its path have their legs peppered by sand grains, but their heads are most likely above the sand cloud. The abrasive impact of a sandstorm, while having little erosive effect on the terrain, may be quite significant for the works of humans near ground level. Unprotected wooden poles and posts can be rapidly cut down by the sandblasting, and cars traveling through a sandstorm are likely to suffer etched windshields and chipped paint.

LearningCheck 18-8 **Why is evidence of wind erosion of bedrock usually quite limited?**

Aeolian Deposition

Sand and dust moved by the wind are eventually deposited when the wind dies down. The finer material, which may be carried long distances, is usually laid down as a thin coating of silt and has little or no landform significance. The coarser sand, however, is normally deposited locally. Sometimes it is spread across the landscape as an amorphous sheet called a *sandplain*. The most notable of all aeolian deposits, however, is the **sand dune**, in which loose, windblown sand is piled into a mound or low hill.

Desert Sand Dunes: In some instances, dunefields are composed entirely of unanchored sand. It is mostly uniform grains of quartz (occasionally gypsum; rarely some other minerals) and usually brownish-gray, although sometimes a brilliant white. Unanchored dunes are deformable obstructions to airflow that can move, divide, grow, or shrink. They develop sheltered air pockets on their leeward sides that slow down and baffle the wind, so deposition is promoted there.

Unanchored dunes are normally moved by local winds. Wind erodes the windward slope of the dune, forcing the sand grains up and over the crest. They are then deposited on the steeper leeward side, or **slip face** (Figure 18-19). The slip face of a dune typically maintains an angle of 32° to 34°—the *angle of repose* of dry sand (see Figure 15-15). If the wind prevails from one direction for many days, the dune may migrate downwind without changing shape. Such migration is usually slow, but in some cases dunes can move tens of meters in a year.

Not all dunes are unanchored, however. In another characteristic dunefield arrangement, the dunes are mostly or entirely anchored by vegetation and therefore do not shift with the wind. Dunes provide little nourishment or moisture for plant growth, but desert vegetation is remarkably hardy and persistent and often able to survive in a dune environment. Where vegetation manages to gain a foothold, it may proliferate and anchor the dunes.

Dune patterns are almost infinite in their variety. Several characteristic dune forms are widespread in the world's deserts, their configuration largely a consequence of the

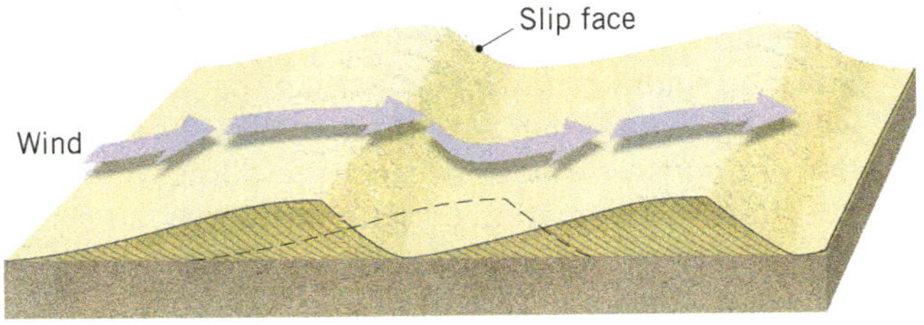

▲ **Figure 18-19** Sand dunes migrate downwind as sand grains move up the gentle windward slope and are deposited on the steep slip face.

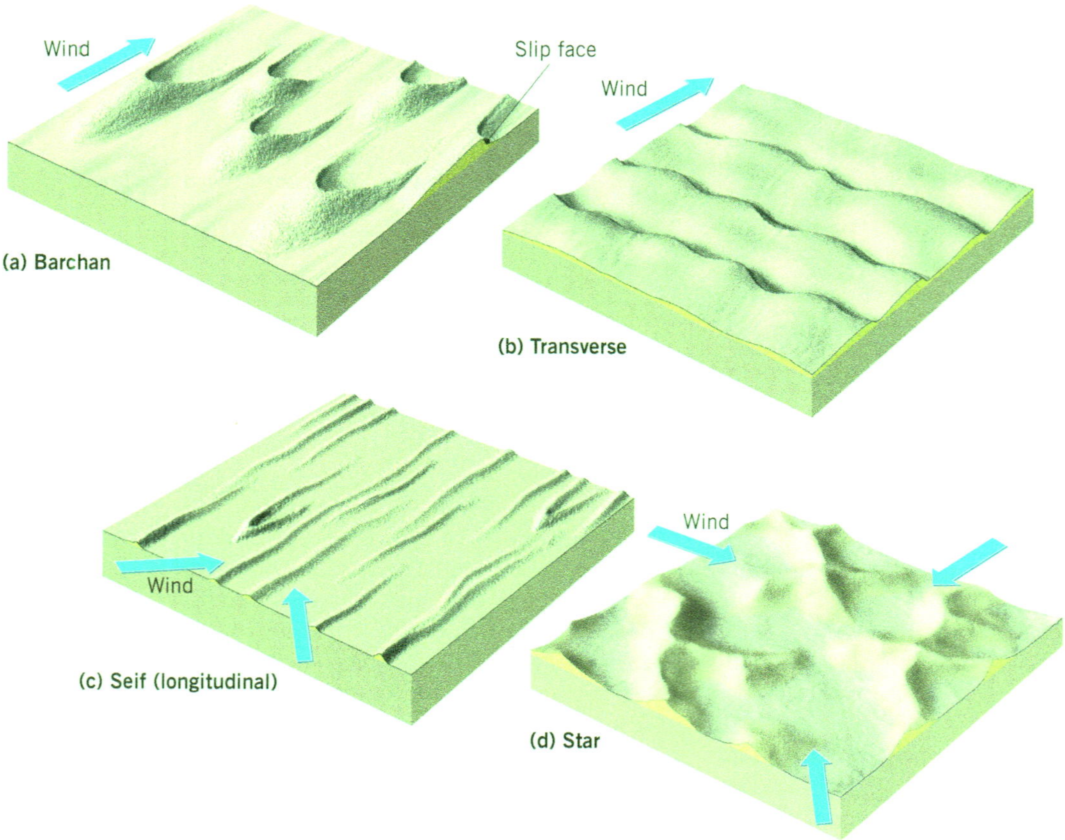

Figure 18-20 Common desert sand dune types: (a) barchans and (b) transverse dunes develop where the wind direction is consistent; (c) seifs or longitudinal dunes generally develop where the wind blows from one direction part of the year and another direction the rest of the year; (d) star dunes develop where the wind direction varies throughout the year.

Animation
Desert Sand Dunes
https://media.pearsoncmg.com/ph/esm/esm_mcknight_physgeo_13/media/

relative amount of sand present and the persistence of the wind direction. We consider four of the most common:

1. Best known of all dune forms is the **barchan**, which usually occurs as an individual dune migrating across a nonsandy surface, although barchans are also found in groups. A barchan is crescent-shaped, with the "horns" of the crescent pointing downwind (Figures 18-20a and 18-21). Sand movement in a barchan is not only over the crest, from windward side to slip face, but also around the edges of the crescent, extending the horns. Barchans form where strong winds blow consistently from one direction. They tend to be the fastest moving of all dune types and are found in all deserts except those of Australia. They are most widespread in the deserts of central Eurasia (Thar and Takla Makan) and in parts of the Sahara.
2. **Transverse dunes**, which are less uniformly crescent-shaped than barchans (Figure 18-20b), occur where normally the entire landscape is sand covered. The sand supply is much greater than that in locations where barchans occur. As with a barchan, the convex side of a transverse dune faces the prevailing wind direction. In a formation of transverse dunes, all of the crests are perpendicular to the wind direction, and the dunes are aligned in parallel waves across the land. They migrate downwind as barchans do; if the sand supply decreases, they are likely to break up into barchans.

Figure 18-21 The characteristic crescent shape of barchan dunes in Namibia's Namib-Naukluft National Park. The dominant wind direction is from left to right.

▲ **Figure 18-22** The parallel linearity of seifs in the Simpson Desert of central Australia.

▲ **Figure 18-23** Star dunes in Namib-Naukluft National Park, Namibia.

3. **Seifs** are a type of linear or *longitudinal dune*. They are long, narrow dunes that usually occur in abundance and in a generally parallel arrangement (**18-20c** and **Figures 18-22**). They are typically a few dozen to a few hundred meters high, a few tens of meters wide, and kilometers or even tens of kilometers long. Their lengthy, parallel orientation apparently represents an intermediate direction between two dominant wind directions—blowing from one direction part of the year and from another direction the rest of the year. Seifs are rare in American deserts but may be the most common dune forms in other parts of the world, such as northern Australia.
4. **Star dunes** are large pyramid-shaped dunes with arms radiating out in three or more directions (**Figure 18-20d**). Star dunes develop in areas where the wind frequently varies in direction (**Figure 18-23**).

LearningCheck 18-9 **Why does the leeward side of a sand dune (the slip face) have a steeper slope than the windward side?**

Fossil Sand Dunes: Some locations have what are called "fossil" sand dunes. For example, in parts of the southwestern United States, vast deposits of sandstone exhibit the characteristic *cross-bedding* of wind-deposited sand, rather than the more typical horizontal strata of sediments accumulated in bodies of water (**Figure 18-24**). Cross-bedding in sand dunes develops when wind-blown sand slides down the slip face of a dune, leaving thin layers inclined relative to the ground surface (see Figure 18-19).

Aeolian Processes in Nondesert Regions

Although our discussion thus far has concerned the work of wind in desert areas, two kinds of wind-produced features are frequently found in nondesert areas as well: *coastal dunes* and *loess*.

▼ **Figure 18-24** Cross-bedded sandstone in "fossil" sand dunes in Zion National Park in Utah.

Coastal Dunes: Winds are active in dune formation along many stretches of ocean and lake coasts, whether the climate is dry or not. On almost all flattish coastlines, ocean waves deposit sand along the beach. A prominent onshore wind can blow some of the sand inland, often forming dunes. In areas where vegetation becomes established on the sand, *parabolic dunes* develop; they look much like barchan dunes but with the "horns" pointing into the wind offshore. Most coastal dune aggregations are small, but they sometimes cover extensive areas. The largest area is probably along the Atlantic coastline of southern France, where dunes extend for 240 kilometers (150 miles) along the shore and reach inland for 3 to 10 kilometers (2 to 6 miles).

Loess: Another form of aeolian deposit not associated with dry lands is **loess**, a wind-deposited silt that is fine grained, calcareous, and usually buff colored. (Pronounced "luhs"—rhyming with "hearse" without the "r"—this German word is derived from the name of a village in Alsace.) Despite its depositional origin, loess lacks horizontal stratification. Perhaps its most distinctive characteristic is its great vertical durability, which results from its fine grain size, high porosity, and vertical, jointlike cleavage planes. The tiny grains have great molecular attraction for one another, making the particles very cohesive. Moreover, the particles are angular, which increases porosity. Thus, loess accepts and holds large amounts of water. Although relatively soft and unconsolidated, loess maintains almost vertical slopes when it is exposed to erosion because of its structural characteristics, as though it were firmly cemented rock (Figure 18-25). Prominent bluffs are often produced as erosional surfaces in loess deposits.

The formational history of loess is complex, although much of the silt was produced in association with Pleistocene glaciation (discussed in Chapter 19). During both glacial and interglacial periods, rivers carried large amounts of debris-laden meltwater from glaciers, producing many broad floodplains. During periods of low water, winds whipped the smaller, dust-sized particles from the floodplains and dropped them in very thick deposits. Some loess also seems to have been generated by deflation of dust from desert areas, especially in central Eurasia.

▲ **Figure 18-25** Loess has a remarkable capability for standing in vertical cliffs, as seen in the Loess Plateau of Shaanxi Province, China.

Most deposits of loess are in the midlatitudes, where some are very extensive, particularly in the United States, Russia, China, and Argentina (Figure 18-26). Indeed, some 10 percent of Earth's land surface is covered with loess; in the conterminous United States, the total approaches 30 percent. Loess deposits provide fertile possibilities for

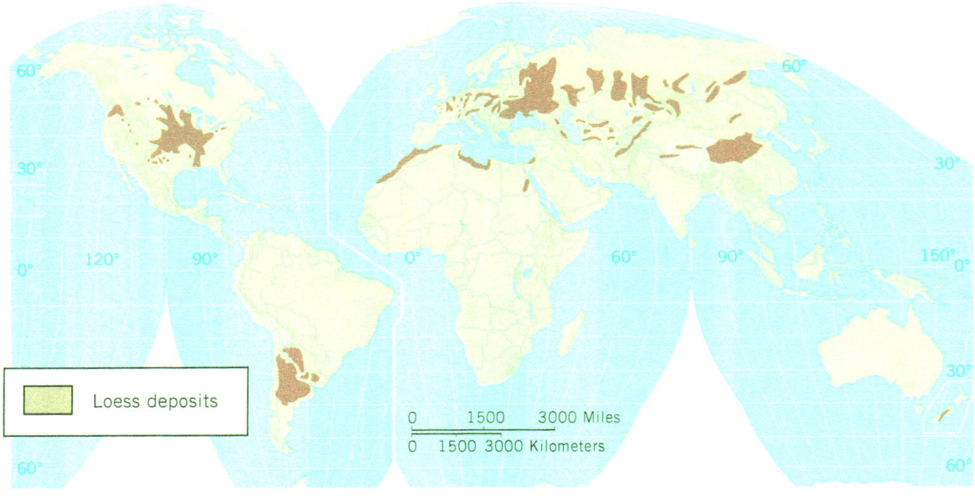

◀ **Figure 18-26** Major loess deposits of the world.

agriculture: they are the parent material for some of the world's most productive soils, especially for growing grain.

The loess areas have been particularly significant in China because of their agricultural productivity. The Yellow River (Hwang Ho) received its name from the vast amount of buff-colored sediment it carries, as did its destination, the Yellow Sea. Also, numerous cave dwellings have been excavated in the Chinese loess because of its remarkable capability for standing in vertical walls. Unfortunately, however, this region is also prone to earthquakes, and the cave homes collapse readily when tremors occur. Some of the world's greatest earthquake disasters in terms of loss of life have taken place there.

Characteristic Desert Landscape Surfaces

Surface features vary considerably from one desert to the next, and it is impossible to describe all of the possible landforms found in arid regions. So instead, for the remainder of this chapter we first describe distinct kinds of landscape surfaces found in many deserts of the world, and then we offer detailed descriptions of two representative desert landform assemblages found in North America.

Over time, the weathering, erosional, and depositional processes at work in deserts produce distinctly different kinds of landscapes. Three common types of landscape surfaces are found only in desert areas: the *erg*, the *reg*, and the *hamada*.

Erg—A Sea of Sand

The most notable desert surface is the **erg** (Arabic for "dune field"), the classic "sea of sand" often associated with the term *desert* (Figure 18-27). An erg is a large area covered with loose sand generally arranged in some sort of dune formation by the wind. The accumulation of the vast amount of sand necessary to produce an erg probably cannot be explained only by the slow weathering processes operating in deserts today. Rather, much of this sand probably accumulated during a period of more humid climate. After forming, these products were carried by streams into an area of accumulation. Then the climate became drier; as a result, wind rather than water became the principal agent of transportation and deposition of the sand.

Several large ergs occur in the Sahara and Arabian deserts, and smaller ergs are found in most other deserts. "Relict" ergs (usually in the form of sand dunes covered with vegetation) are sometimes found in nonarid areas, indicative of a drier climate in the past. The "Sandhills" of western Nebraska are such relict ergs, now stabilized by prairie grasses.

Reg—Stony Deserts

A second type of desert landscape surface is the **reg** (Arabic for "stone"), a tight covering of coarse gravel, pebbles, or boulders from which wind and water have removed all sand and dust. A reg is a stony desert, although the surface

▲ Figure 18-27 An erg landscape of sand dunes in White Sands National Park, New Mexico.

covering of stones may be very thin (in some cases, just one pebble deep). The finer material was removed through surface erosion—perhaps aided by sediment movement below the surface through the action of rainwater percolation—so the surface pebbles often fit closely together, sealing whatever material is below from further erosion. For this reason, a reg is often referred to as **desert pavement** or *desert armor* (Figure 18-28). In Australia, where regs are widespread, they are called *gibber plains*. Because this surface usually takes hundreds to thousands of years to form, the presence of

▼ Figure 18-28 Desert pavement in California's Panamint Valley. Notice the dark brown coating of desert varnish on the larger rocks.

well-developed desert pavement is an indication of a relatively undisturbed surface.

Desert Varnish: A striking feature of some deserts, one particularly but not exclusively associated with regs, is **desert varnish**. This is a dark, shiny coating that forms on the surface of pebbles, stones, and larger outcrops after long exposure to the desert air. (Notice the larger rocks shown in Figure 18-28.) Desert varnish is characterized by a high content of iron and manganese oxides along with wind-delivered clay. The relatively high concentrations of manganese in desert varnish seem to be a consequence of a biochemical process involving bacteria. Desert varnish can be used as a relative dating tool for geomorphologists because the longer a rock surface has been exposed to weathering, the greater the concentration of the oxide coating and thus the darker the color.

Hamada—Barren Bedrock

A third desert landscape surface is the **hamada**, the Arabic term for a barren surface of consolidated material. A hamada surface usually consists of exposed bedrock, but it is sometimes composed of sedimentary material that has been cemented together by salts evaporated from groundwater. In either case, fragments formed by weathering are quickly swept away by the wind, so little loose material remains.

LearningCheck 18-10 **Explain what the presence or absence of desert pavement and desert varnish can reveal about the likelihood of relatively recent erosion or deposition on a desert surface.**

Although the extent of ergs, regs, and hamadas is significant in some desert areas, the majority of the arid land in the world contains only a limited number of these surfaces. For example, only one-third of the Arabian Desert, the sandiest desert of all, is covered with sand, and much of that is not in the form of an erg. Ergs, regs, and hamadas are restricted to plains areas. Regs and hamadas are exceedingly flat, whereas ergs are as high as the sand dunes built by the wind. The boundaries of these landscapes are often sharp because of the abrupt change in friction-layer speed as the wind moves from a sandy surface to a nonsandy surface or vice versa.

Two Representative Desert Landform Assemblages

We now turn to more detailed examples of landform development in arid lands. Two prominent assemblages of landforms are found in the deserts of North America: basin-and-range terrain and mesa-and-scarp terrain. Their patterns of development are repeated time and again over thousands of square kilometers of the American Southwest (Figure 18-29). Although desert landscapes vary greatly around the world, these two landform assemblages are good examples of the outcome of the special conditions and the external desert-landform-shaping processes we describe in this chapter.

▲ **Figure 18-29** The southwestern interior of the United States contains two principal assemblages of landforms: basin-and-range and mesa-and-scarp.

Basin-and-Range Landforms

Most of the southwestern interior of the United States is characterized by basin-and-range topography (see Figure 18-29). This is a land largely without external drainage, with only a few exotic rivers (notably, the Colorado and Rio Grande) flowing through or out of the region. This region of North America has undergone extensive normal faulting—including movement along low-angle detachment faults. That faulting left a landscape consisting of numerous fault-block mountain ranges surrounding a series of interior drainage basins, including many down-dropped *grabens* and down-tilted *half-grabens* (see Figure 14-52). The aridity of the basin-and-range region is mostly due to rain shadows—especially that of the Sierra Nevada range in eastern California.

Basin-and-range terrain has three principal features: ranges, piedmont zones, and basins (Figure 18-30).

The Ranges

If we stand in the basin of any basin-and-range landscape, rugged mountain ranges dominate the horizon in all directions. Although the tectonic origins of these mountains vary (most were tilted by faulting; others formed by folding or volcanism), their surface features have largely been shaped by weathering, mass wasting, and fluvial processes.

Most ridge crests and peaks are sharp, steep cliffs are common, and rocky outcrops protrude at all elevations. The mountain ranges of a basin-and-range formation tend to be long, narrow, and parallel to one another. Most of them are seamed by numerous gullies, gorges, and canyons that rarely

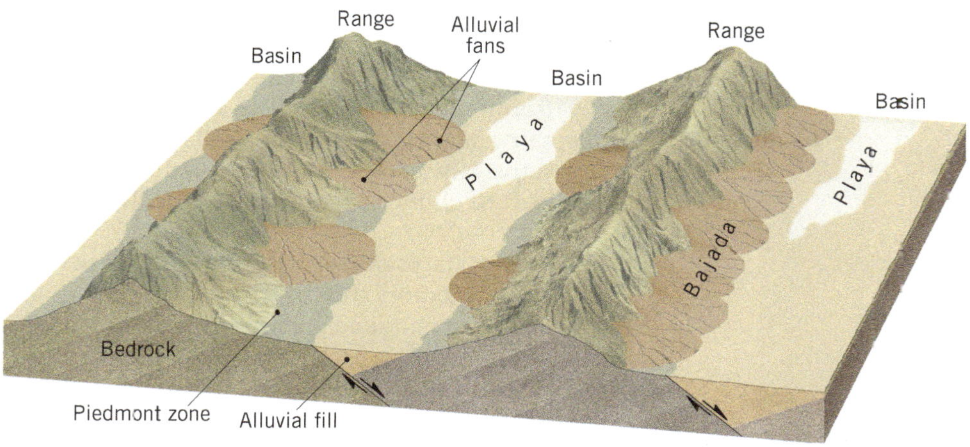

Condor Video
Alluvial Fans
https://media.pearsoncmg.com/ph/esm/esm_mcknight_physgeo_13/media/

◀ **Figure 18-30** A typical basin-and-range desert landscape.

have flowing streams. These dry drainage channels are usually narrow and steep sided and have a V-shaped cross section. Typically, the channel bottoms are filled with sand and other loose debris.

Wineglass canyons are found in some ranges. The "cup" of the wineglass is the open area of dispersed headwater tributaries high in the range, the "stem" is the narrow gorge cut through the mountain front, and the "base" is the alluvial fan that opens out onto the piedmont zone. As we saw earlier, if the range stands in isolation and the alluvial plains and basins roundabout are extensive, the term *inselberg* is applied to the mountain remnant (see Figure 18-11).

Piedmont Zone

At the base of the ranges, the piedmont zone marks the change from the ranges' steep slopes to the near-flatness of the basin floors (see Figure 18-12). Much of the piedmont zone may be underlain by a pediment. Rarely visible, the pediment is normally covered with several meters of unconsolidated sediment because the piedmont zone is an area of fluvial deposition. During the occasional rainfall, flash floods and debris flows come roaring out of the gullies and gorges of the surrounding ranges, heavily laden with sedimentary material. As the sediments burst out of the mouths of the confining canyons onto the piedmont zone, their speed and load capacity drop abruptly, and deposition results.

Alluvial Fans and Bajadas: Particularly characteristic of the basin-and-range region are alluvial fans, occasionally developing to enormous sizes (**Figure 18-31**). As alluvial fans become larger, neighboring ones often overlap. Continued growth and more complete overlap may eventually result in a continuous alluvial surface all across the piedmont zone, making it difficult to distinguish individual fans. This feature is known as a **bajada** (see Figure 18-30). Near the mountain front, a bajada surface is undulating, with convex sections near the canyon mouths and concave sections in the overlap areas between the canyons.

LearningCheck 18-11 **Why are alluvial fans and bajadas so common in the basin-and-range desert?**

The Basins

Beyond the mountain front is the flattish floor of the basin, which very gently slopes from all sides toward some low point. A playa is usually found in this low point (see Figure 18-6). Drainage channels across the basin floor are often shallow and ill defined, frequently disappearing before reaching the low point. This low point thus functions as the drainage terminal for all overland and streamflow from the near sides of the surrounding ranges, but

▼ **Figure 18-31** Alluvial fans at the mouths of canyons at Badwater Basin in Death Valley National Park, California.

▲ Figure 18-32 The Death Valley "salt pan" consists mostly of ordinary table salt (NaCl). The polygonal ridges form as salt crystals grow through the evaporation of water from a salty "slush" just below the surface.

only sometimes does much water reach it. Most is lost by evaporation and seepage long before.

Salt accumulations are commonplace on the playa surrounding the low point of a desert basin because water-soluble minerals wash out of the surrounding watershed. Water is usually abundant enough to allow flow across the outer rim of the basin floor and into the playa. After the water evaporates or seeps away, salts become increasingly concentrated; the playa is then more properly called a *salina*. The presence of the salt usually gives a brilliant whitish color to the surface. Many different salts can be involved, and their accumulations may be large enough to support mining. In some desert basins, such as in California's Death Valley, evaporation has been sufficient for large accumulations of salts to develop, leaving a *salt pan* or *salt flat* (Figure 18-32).

In the rare occasions when water does flow into a playa, the formation becomes a playa lake. Such lakes are usually very shallow and normally persist for only a few days or weeks. Unlike shallow freshwater lakes, which have muddy water, saline lakes are marked by clear water and a salty froth around the edges. Why? In a process called *flocculation*, saltwater introduces cations that neutralize the often-negative charge of silt and clay particles, allowing the particles to clump together and settle, leaving the water clear.

The basin floor is covered with very fine-grained material because the contributory streams are too weak to transport large particles. Silt and sand predominate and sometimes accumulate to remarkable depths. Indeed, the normal denudation processes in basin-and-range country tend to raise the floor of the basins. Debris from the surrounding ranges has nowhere to go but the basin of interior drainage. Thus, as the mountains are wearing down, the basin is gradually filling up. The fine material of basin floors is very susceptible to wind, so small concentrations of sand dunes are often found in some corner of the basin.

Death Valley in California is a remarkable basin-and range landscape—see the box *Featured National Park: Death Valley—A Laboratory of Basin-and-Range Terrain.*

LearningCheck 18-12 **Why are playas found in almost all basins of the basin-and-range region?**

Mesa-and-Scarp Terrain

The other major landform assemblage of the American Southwest is mesa-and-scarp terrain (see Figure 18-29). It is most prominent in Four Corners country—the place where Colorado, Utah, Arizona, and New Mexico come together in the regionally uplifted Colorado Plateau. **Mesa** is Spanish for "table" and implies a flat-topped surface. *Scarp* is short for "escarpment" and refers to the steep, more or less vertical cliffs (Figure 18-33).

◄ Figure 18-33 The stair-step pattern of mesa-and-scarp terrain is shown on an imposing scale in Arizona's Grand Canyon National Park.

Featured National Park

Death Valley

A Laboratory of Basin-and-Range Terrain

▲ **Figure 18-D** An extensive bajada formed at the foot of the Panamint Range.

California's Death Valley is a vast topographic museum of basin-and-range terrain. Located in east-central California, the "valley" is actually a down-dropped graben 225 kilometers (140 miles) long and 6–26 kilometers (4–16 miles) wide (**Figure 18-C**).

Death Valley is not a classic graben. The basin floor tilts down more along its eastern side than its western side. It was created at least in part as a "pull-apart" basin, formed where land drops down between two parallel strike-slip faults. The downfaulting has been so pronounced that much of the valley floor is below sea level, reaching a depth of 86 meters below sea level (–282 feet).

The lengthy, tilted fault-block mountain ranges that border the valley on both sides are classic desert mountains: rugged, rocky, and generally barren. The Panamint Range on the west is the most prominent (**Figure 18-D**); the high point, at Telescope Peak (3368 meters [11,049 feet] above sea level), is only 29 kilometers (18 miles) due west from the low point. The Amargosa Range on the east is a bit lower overall. The canyons that seam the ranges are deep, narrow, V-shaped gorges, including many "wineglass" canyons. The series of high mountain ranges to the west—most notably, the Sierra Nevada range—generate the rain shadow that makes Death Valley so dry.

Alluvial Fans: The piedmont zone at the foot of the Panamints and Amargosas is largely covered with alluvium in one of the most extensive fan complexes imaginable (see Figures 18-13 and 18-31). Every canyon mouth is the apex of a fan or fan-shaped debris flow deposit. Most of the fans at the foot of the Panamint Range are thoroughly coalesced into a conspicuous bajada that averages about 8 kilometers (5 miles) wide. The Amargosa fans are much smaller, primarily because they are supplied by smaller drainage basins and because faulting has tilted down the eastern side basin floor, so the fans nestle close to the base of the range.

Basin Floor: In Death Valley, as much as 900 meters (3000 feet) of young alluvium rests atop another 1800 meters (6000 feet) of Tertiary sediment. The surface of the valley floor has little relief and slopes gently toward the low point near Badwater, a permanent saltwater pond. Several extensive crusty-white salt pans (see Figure 18-32), along with several areas of sand dunes (one covering 36 square kilometers [14 square miles]), occupy the basin floor.

During the most recent Ice Age, Death Valley was occupied by Lake Manly, more than 140 kilometers (90 miles) long and 180 meters (600 feet) deep. It was fed by three rivers that flowed into the valley from the west, carrying meltwater from Sierra Nevada glaciers. As the climate became drier and warmer, the lake disappeared, but traces of its various shoreline levels can still be seen on the lower slopes. Much of the salt in the valley accumulated from the evaporating waters of Lake Manly and from more recent lakes. The accumulation of borate minerals, such as borax, in parts of the salt pan attracted mining operations by the 1880s. The "twenty-mule teams" used to transport borax by cart remain iconic symbols of Death Valley.

Questions

1. What factors explain the extensive alluvial fans in Death Valley?
2. Why has so much salt accumulated in the bottom of Death Valley?

Mobile Field Trip
Desert Geomorphology
https://media.pearsoncmg.com/ph/esm/esm_mcknight_physgeo_13/media/

NPS Video
Death Valley Explorer (E2)
https://media.pearsoncmg.com/ph/esm/esm_mcknight_physgeo_13/media/

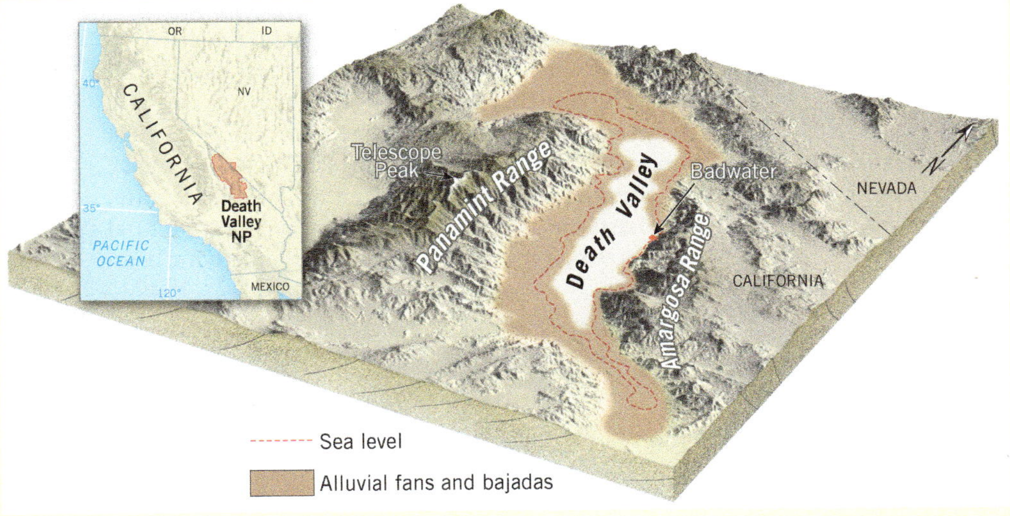

▲ **Figure 18-C** The mountains and basins of Death Valley are a consequence of extensional faulting. Alluvial fans and bajadas are found all along the mountain fronts.

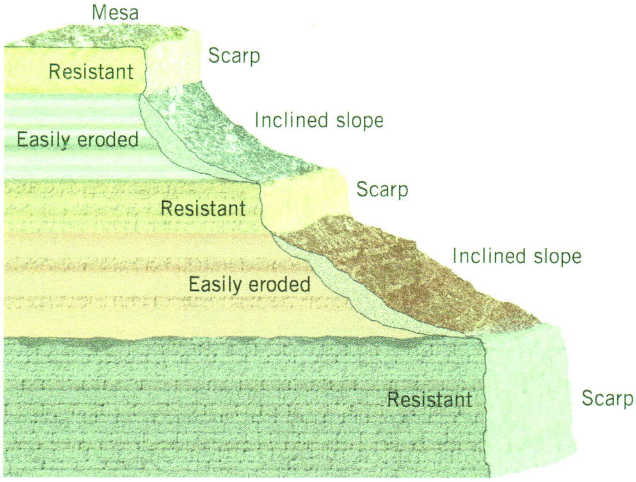

▲ **Figure 18-34** Cross section of a mesa-and-scarp formation. Differential weathering and erosion show up prominently: The resistant strata weather and erode into mesas or scarps, whereas the more easily eroded strata yield gentler inclined slopes.

▲ **Figure 18-36** The spectacular starkness of arid-land topography is demonstrated dramatically in the view of the Mitten Buttes in Arizona's Monument Valley.

Structure of Mesa-and-Scarp Landforms

Mesa-and-scarp terrain is normally associated with horizontal sedimentary strata—shown off prominently in this arid region because it lacks a continuous cover of trees and other vegetation. Such strata invariably offer different degrees of resistance to weathering and erosion, so abrupt changes in slope angle are characteristic of this terrain. The most resistant layers, typically limestone or sandstone, form an extensive caprock, which becomes the mesa; at the eroded edge of the caprock, that hard layer then protects underlying strata and produces an escarpment. Thus, the resistant layers are responsible for both mesa and scarp elements of this terrain type.

Often mesa-and-scarp topography has a broad, irregular stair-step pattern (Figure 18-34). An extensive, flat platform (the mesa) in the topmost resistant layer of a sedimentary accumulation terminates in an escarpment (the scarp) that extends downward to the bottom of the resistant layer(s). From here, another slope, steep but not as steep as the escarpment, continues down through softer strata. This inclined slope extends downward as far as the next resistant layer, which forms either another escarpment or another mesa ending in an escarpment.

The top platforms are properly referred to as **plateaus** if they are bounded on one or more sides by a prominent escarpment. If a scarp edge is relatively inconspicuous or absent, the platform is called a *stripped plain*.

Erosion of Escarpment Edge

The escarpment edge is worn back by weathering, mass wasting, and fluvial erosion. The cliffs retreat, maintaining their perpendicular faces, as they are undermined by the more rapid removal of the less resistant strata (commonly shale) beneath the caprock. Much of the undermining is accomplished by *sapping*, a process in which groundwater seeps and trickles out of the scarp face, eroding fine particles and weakening the cohesion of the face. Blocks of the caprock break off, usually along vertical joint lines. Throughout this process, the harder rocks are the cliff-formers, and the less-resistant beds develop more gently inclined slopes. Talus often accumulates at the base of the slope.

Although the term *mesa* is applied to many flat surfaces in dry environments, it properly refers to a flat-topped, steep-sided hill with a limited summit area. It is a remnant of a formerly more extensive surface, most of which has been worn away (Figure 18-35). It may stand in splendid isolation, but more commonly it occurs as an outlying mass not very far from the retreating escarpment face to which it was once connected.

A related but smaller topographic feature is the **butte**, an erosional remnant that has a very small surface area and cliffs that rise conspicuously above their surroundings. Most buttes are formed by the mass wasting of mesas (Figure 18-36). With further denudation, a still smaller residual feature, called a **pinnacle** or *pillar*, may be all that is left—a final spire of resistant caprock protecting weaker underlying beds.

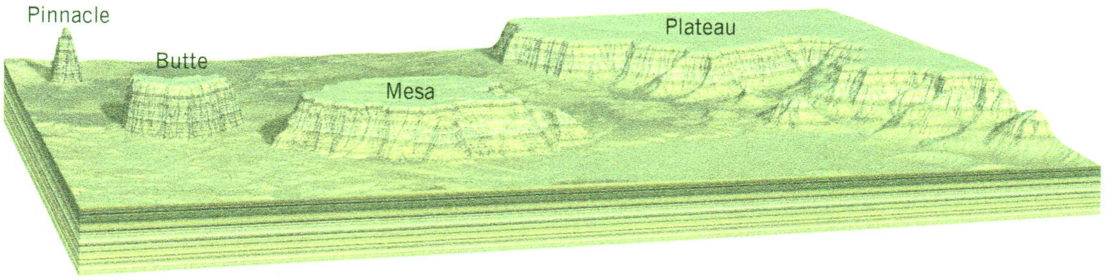

▶ **Figure 18-35** Typical development of residual landforms in horizontal sedimentary strata with a hard caprock. Over time, larger features erode into smaller features.

Buttes, mesas, and pinnacles are typically near some retreating escarpment face (see Figure 18-35).

LearningCheck 18-13 **Explain the role of resistant caprock in the formation of mesa-and-scarp terrain.**

Arches and Natural Bridges

Mesa-and-scarp terrain is also famous for numerous minor erosional features, most produced by a combination of weathering, mass wasting, and fluvial erosion. These features are not confined to arid regions, but the mesa-and-scarp region has many examples. An *arch* can form when the lower portions of a narrow "fin" of sedimentary rock—associated with prominent, closely spaced vertical joints—weaken and collapse, leaving more-resistant rock above (see Figure 15-C). A *natural bridge* can form when the rock over which water flows changes from an erosion-resistant type to a less-resistant type (Figure 18-37). A natural bridge frequently forms where an entrenched meander wears away the rock in a narrow neck between meander loops (see Figure 16-45).

▲ **Figure 18-37** Rainbow Bridge in Rainbow Bridge National Monument, Utah, was formed when an entrenched meander cut away the narrow neck between two river loops.

Pedestals and pillars, sometimes larger at the top than at the bottom, rise abruptly above their surroundings, their caps resistant material but their narrow bases continuously weathered by rainwater trickling down the surface. This water dissolves the cementing material that holds the sand grains together, and the loosened grains easily blow or wash away.

Mesa-and-scarp terrain is notable for its vivid colors. The sedimentary outcrops and sandy debris of these regions are often resplendent in various shades of red, brown, yellow, and gray, due mostly to iron compounds.

Badlands

One of the most striking topographic features of arid and semiarid regions is the intricately dissected and barren terrain known as **badlands**. In areas underlain by horizontal strata of shale and other clay formations that are poorly consolidated, overland flow after the occasional rains is an extremely effective erosive agent. Innumerable tiny rills that develop over the surface evolve into a maze of short but steep-sloped gullies and gorges, scattered with a great many ridges, ledges, and other erosional remnants. Erosion is too rapid to permit soil to form or plants to grow, so badlands are barren, lifeless wastelands of almost impassable terrain (Figure 18-38). They are found in scattered locations in every western state, the most famous areas being in Bryce Canyon (southern Utah), Badlands (western South Dakota), and Theodore Roosevelt (western North Dakota) National Parks.

◀ **Figure 18-38** Badlands are characterized by innumerable ravines and gullies dissecting the land and forming a maze of low but very steep slopes. This scene is from Badlands National Park, South Dakota.

18 Learning Review

After studying this chapter, you should be able to answer the following questions. Key terms from each text section are shown in **bold type.** Definitions for key terms are also found in the glossary at the back of the book.

Key Terms and Concepts

A Specialized Environment *(p. 522)*

1. List several ways in which topographic development in arid lands is different from that in humid regions.
2. What is an impermeable surface and how does such a surface influence the runoff from rainfall in a desert?
3. What is a basin of interior drainage?

Running Water in Waterless Regions *(p. 523)*

4. What is the difference between an **ephemeral stream** and an **exotic stream** in a desert?
5. What is a **playa**, and why does one form?
6. What is the difference between a playa and a **salina**?
7. Why is a desert lake in a basin of interior drainage likely to be a **saline lake**?
8. Explain **differential weathering and erosion**.
9. Describe the formation of an **inselberg**.
10. What is the difference between **pediment** and **piedmont zone**?
11. What is an **alluvial fan**, and how does one form?

The Work of the Wind *(p. 529)*

12. Contrast the **aeolian processes** of **deflation** and abrasion.
13. How does a **blowout** form?
14. What is a **ventifact** and how does one form?
15. Describe and explain the general cross section of an unanchored desert **sand dune**. Be sure to contrast the windward side with the **slip face** of the dune.
16. Explain the shape and movement of **barchan** sand dunes.
17. How are **transverse dunes** different from **seifs**?
18. Under what circumstances do **star dunes** form?
19. Most **loess** is not found in arid regions; why is it discussed in this chapter?

Characteristic Desert Landscape Surfaces *(p. 535)*

20. Distinguish among **ergs**, **regs**, and **hamadas**.
21. Describe and explain the formation of **desert pavement** and **desert varnish**.

Basin-and-Range Landforms *(p. 536)*

22. How is a **bajada** different from an alluvial fan?

Mesa-and-Scarp Terrain *(p. 538)*

23. How does a **plateau** differ from a **mesa**?
24. Describe and explain the processes involved that can change a mesa into a **butte** or **pinnacle**.
25. What is distinctive about **badlands** terrain? How did it become that way?

Study Questions

1. Although deserts have very little rainfall, running water is still the most important process of erosion and deposition in arid environments. Describe and explain at least two special conditions in deserts that tend to make fluvial erosion more likely when it does rain.
2. Why are playas so flat and level?
3. Why are depositional features of alluvium so prominent in many desert regions?
4. Overall, how important is wind in the erosion of desert landforms?
5. Why are some playas in the basin-and-range desert so salty?
6. How does an alluvial fan differ from a delta?
7. Why are few deep stream channels cutting across basin floors in the basin-and-range desert?

Exercises

1. Assume that the dunes in a field of barchan sand dunes are moving at an average long-term rate of 20 meters (about 65 feet) a year. How long does it take a dune to move a distance of 500 meters? _____ years
2. If one of the dunes of Exercise 1 is 10 meters wide, how long does it take that dune to cross an abandoned 100-meter-wide airport runway completely? _____ years
3. Today, the lowest part of Death Valley is near Badwater, an elevation of 86 meters (282 feet) below sea level. One of the abandoned shorelines, cut by waves when the basin was filled with water as a lake during the Pleistocene, is found at an elevation of 80 meters (262 feet) above sea level. Estimate the depth of this lake stand during the Pleistocene: _____ meters
4. What factors might complicate calculating the true depth of the lake that left the shorelines cited in Exercise 3?

Environmental Analysis DROUGHT MONITORING

Data
Drought Monitoring
https://media.pearsoncmg.com/ph/esm/esm_mcknight_physgeo_13/media/

Drought is a prolonged period of abnormally low precipitation. The degree of dryness compared with normal precipitation and the length of dry spells are used to classify droughts. For a drought to be declared, the dry period must be much longer in arid places (such as Arizona) than in wetter places (such as Georgia).

Activities

Go to http://droughtmonitor.unl.edu, the U.S. Drought Monitor map.

1. Which regions have the most intense drought?
2. What is the percentage of drought-covered area and intensity in your state (or the nearest state that has drought conditions)? Is this a long-term or short-term drought?

Click on "Maps" and choose "Compare Two Weeks."

3. How has the drought situation changed in your state?

Click on "Current Map" and scroll down to "This Week's Drought Summary" for your region.

4. Briefly describe the factors that influenced the drought change from last week to this week.

Click "Maps" and select "Comparison Slider." The "Right" date indicates the current drought map. The "Left" date is the comparison map. Move the slider bar to compare the maps.

5. How does the current drought coverage and intensity in your state compare with coverage and intensity one month ago? Six months ago? One year ago?

Go to the Climate Prediction Center's Three-Month Outlooks at https://www.cpc.ncep.noaa.gov/products/predictions/long_range/seasonal.php?lead=1 Click on the maps to see larger versions.

6. Precipitation is the strongest control of drought. How will precipitation in your state change next month? Over the next three months?
7. How will drought conditions in your state change over the next few months? Why?
8. Based on the predicted precipitation map for the next three months, will the regions with the most intense drought improve, worsen, or not change? Why?

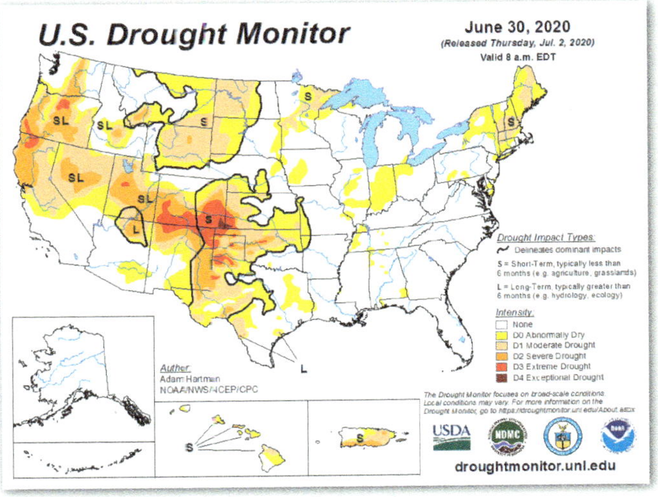

Seeing Geographically in Our National Parks

Look again at the photograph of Death Valley at the beginning of the chapter (p. 520). What kind of gently sloping desert landform is seen at the foot of the Panamint Mountains on the far side of the basin? Why is it riddled with stream channels? How did it form, and why is it located here? What kind of landform lies along the bottom of the flat basin floor shown as an area of light tan and white? What is it made of, and how did it form?

Mastering Geography®

Looking for additional review and test prep materials? Visit the Study Area in **Mastering Geography** to enhance your geographic literacy, spatial reasoning skills, and understanding of this chapter's content by accessing a variety of resources, including Map Master interactive maps, geoscience animations, *Mobile Field Trips*, videos, *Project Condor* Quadcopter videos, *In the News* RSS feeds, flashcards, web links, self-study quizzes, and an eText version of *McKnight's Physical Geography*.

Seeing Geographically in Our National Parks

Nabesna Glacier in Wrangell-St. Elias National Park in Alaska. What suggests that the speed of ice movement may not be the same everywhere on the glacier? What might explain the lines of dark material in the middle and along the sides of the glacier? Do you think that all of this material came from the immediate valley walls here? Why?

Wrangell-St. Elias National Park, AK

Glacial Modification of Terrain

19

HAVE YOU EVER WONDERED how we know that glaciers once covered large areas of the continents? If you stand outside on a hot summer's day in Iowa or Illinois, it's hard to imagine that just a few thousand years ago you would have been covered with ice! However, much of the evidence of these past glaciers is quite easy to recognize—as is the evidence that this modern-day glacier in Wrangell-St. Elias National Park in Alaska has gotten smaller over the last century.

An unknown number of ice ages have occurred during Earth's long history. With one outstanding exception, however, most of this recognizable evidence of past glacial periods has been eliminated by more recent geomorphic events. Thus, when we refer to the *Ice Age*, capitalized, we usually mean this most recent glacial period. That is the hallmark of the geologic epoch known as the *Pleistocene*, a period that began about 2.6 million years ago and ended less than 12,000 years ago.

In this chapter, we are concerned with Pleistocene events both because they significantly modified pre-Pleistocene topography and because their effects are so thoroughly imprinted on many parts of the continents. Glaciers are still at work today, but their importance as shapers of terrain is much less now than just a few thousand years ago simply because so much less glacial ice exists today. However, glaciers are among the most sensitive indicators of contemporary climate change.

For us to understand the geomorphic effects of glaciers, this chapter largely focuses on processes—in this case, the erosional and depositional work of glacial ice and meltwater. The landforms resulting from continental glaciation conform quite closely to the clearly defined limits of continental ice sheets during the Pleistocene, and landforms from mountain glaciation can be found in almost all high mountain areas—even in the tropics.

As you study this chapter, think about these **Key**Questions:

- **How has the extent of glaciers changed since the Pleistocene epoch?**
- **How do glaciers form and move?**
- **How do glaciers erode, transport, and deposit rock?**
- **What landforms do the erosion, deposition, and meltwater of continental ice sheets and mountain glaciers produce?**

The Impact of Glaciers on the Landscape

Wherever glaciers have developed, they have had a significant impact on the landscape. Moving ice grinds away almost anything in its path: it carries away most soil and polishes, scrapes, gouges, plucks, and abrades bedrock. Moreover, the rock that is picked up is eventually deposited in a new location, further changing the shape of the terrain. Perhaps 7 percent of all contemporary erosion and transportation of rock debris on the continents is accomplished by glaciers. In short, they dramatically transformed the preglacial topography.

Types of Glaciers

A **glacier** is more than a mass of ice high up in a mountain valley or on a polar plain. As we see shortly, ice must be moving or flowing to be considered a glacier. Although glacial ice behaves in similar fashion wherever it accumulates, its pattern of movement and its effect on topography vary considerably depending on its quantity and on the environment. These variations are best understood by first considering the different types of glaciers: *mountain glaciers* and *continental ice sheets*.

Mountain Glaciers

In a few high-mountain areas today, ice accumulates in an unconfined sheet that may cover a few hundred or few thousand square kilometers. It overlies all the underlying topography except perhaps for some protruding pinnacles called *nunataks*. Such **highland icefields** are notable in parts of the high country of western Canada and southern Alaska and on various Arctic islands (particularly Iceland). Their outlets are often "tongues" of ice that travel down mountain valleys and are called **valley glaciers** (Figure 19-1). If the leading edge of a valley glacier reaches a flat area and escapes from the confines of its valley walls, it is called a *piedmont glacier* (Figure 19-2).

Sometimes the term **alpine glacier** is used to describe glaciers that develop individually, high in the mountains rather than as part of a broad icefield, usually at the heads of valleys. Very small alpine glaciers confined to the basins where they originate are called **cirque glaciers** (and the basins are called *cirques*, as we see later in this chapter). Normally, however, alpine glaciers flow out of their originating basins and continue down-valley as long, narrow valley glaciers.

Continental Ice Sheets

Glaciers that form in nonmountainous areas of the continents are called **continental ice sheets**. During the Pleistocene, these vast blankets of ice completely inundated the underlying terrain to depths of hundreds or even thousands of meters. Because of their immense size, ice sheets were significant agents of glaciation across large expanses of some continents. Only two exist today, in Antarctica and Greenland (Figure 19-3).

The ice in an ice sheet accumulates to great depths in the interior of the sheet but is much thinner at the outer edges. Around the margin of the sheet, some long tongues of ice, called *outlet glaciers*, extend between rimming hills to the sea. In other places, the ice reaches the ocean along a massive front, where it sometimes projects out over the sea as an *ice shelf* (Figure 19-4). As we saw in Chapter 9, great

▲ Figure 19-1 The Kennicott Glacier flows down from Mount Blackburn in southern Alaska.

Mobile Field Trip
Glaciers of Alaska
https://media.pearsoncmg.com/ph/esm/esm_mcknight_physgeo_13/media/

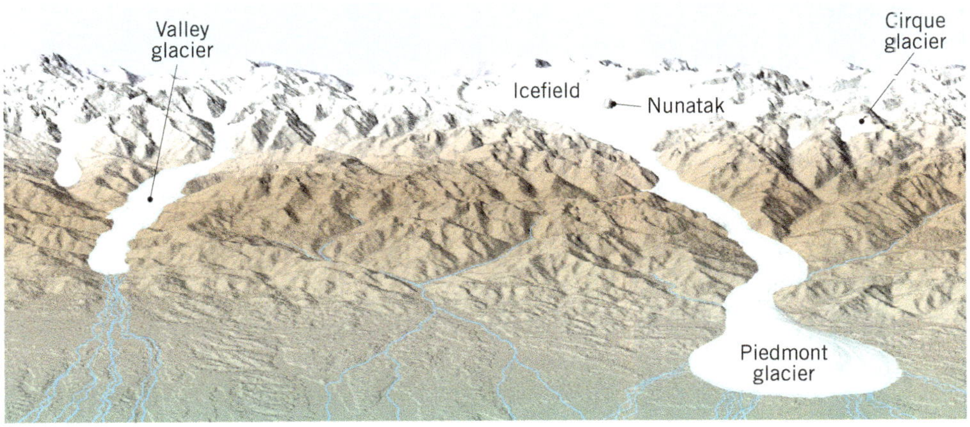

◀ **Figure 19-2** Types of mountain glaciers. Valley and piedmont glaciers can originate in a highland icefield or as alpine glaciers that overflow their basin (cirque) and flow down-valley. Nunataks are pinnacles rising above the ice of a highland icefield.

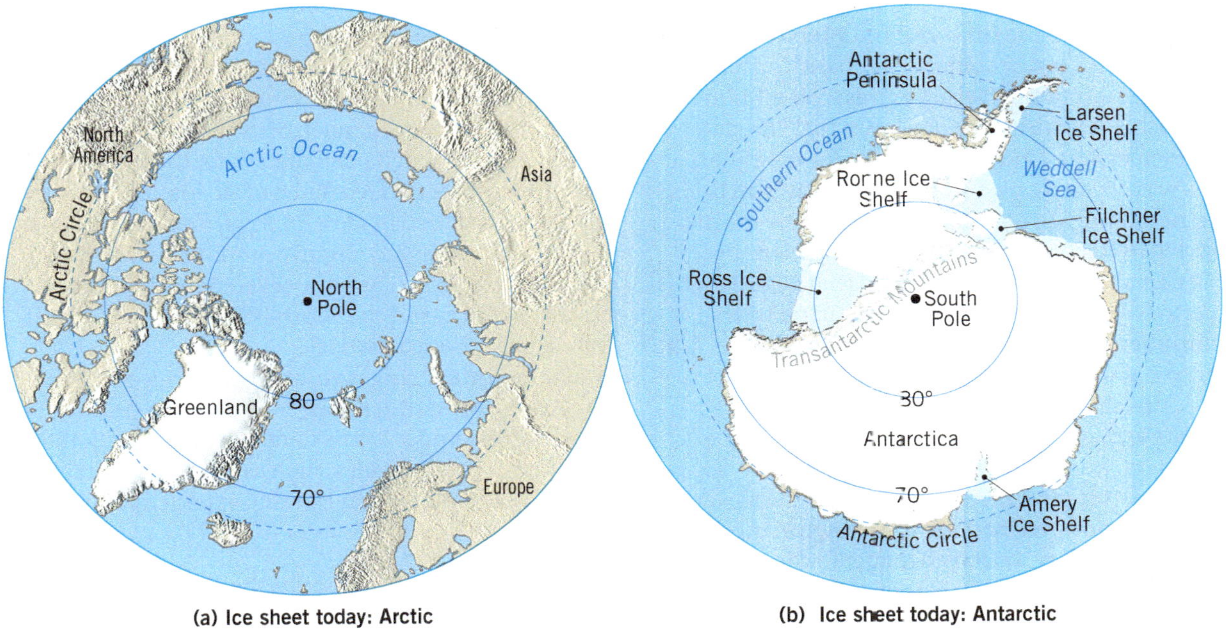

▲ **Figure 19-3** Generalized extent of continental ice sheets in the (a) Arctic and (b) Antarctic. Over the last few decades, both ice sheets have retreated somewhat around the edges and have thinned appreciably.

NASA Animation
Arctic Sea Ice Minimum
https://media.pearsoncmg.com/ph/esm/esm_mcknight_physgeo_13/media/

NASA Video
16 Years of Ice Loss
https://media.pearsoncmg.com/ph/esm/esm_mcknight_physgeo_13/media/

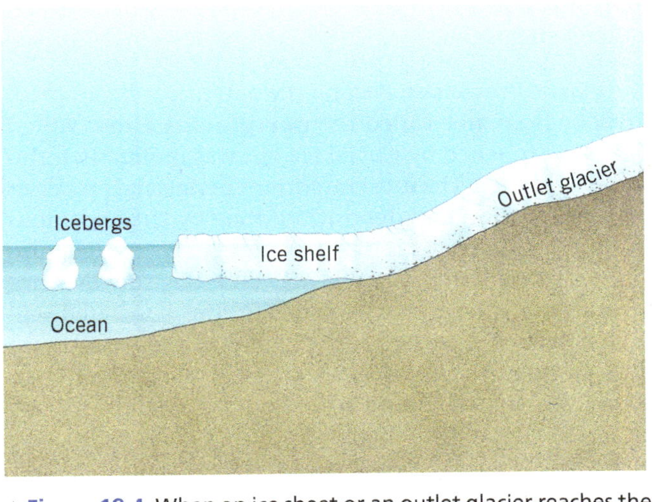

▲ **Figure 19-4** When an ice sheet or an outlet glacier reaches the ocean, some of the ice may extend out over the water as an ice shelf. Icebergs form when the hanging ice breaks off and floats away, a process called *calving*.

chunks of ice frequently break off (from both ice shelves and the ends of outlet glaciers) and fall into the sea; this process is called *calving*. The ice masses float away as *icebergs*.

LearningCheck 19-1 **Compare mountain glaciers and continental ice sheets. (Answer on p. AK-6)**

Glaciations Past and Present

The amount of glacial ice on Earth's surface has varied remarkably over the last few million years, with periods of ice accumulation and advance alternating with times of ice retreat. A great deal of evidence was left behind by the moving and melting ice; nevertheless, the record is incomplete and often approximate. As is to be expected, the more recent events are best documented; the further we delve into the past, the murkier the evidence becomes.

Pleistocene Glaciation

The dominant environmental characteristic of the Pleistocene was the cooling of high-latitude and high-elevation areas, so a vast amount of ice accumulated in many places. However, the epoch was by no means universally icy. During several lengthy periods, most or all of the ice melted, only to be followed by intervals of more ice accumulation. In broad terms, the Pleistocene consisted of an alternation of *glacial* periods (times of ice accumulation) and *interglacial* periods (times of ice retreat). Current evidence suggests that perhaps 20 glacial episodes took place during the Pleistocene.

Dating the Ice Age: The precise timing of the **Pleistocene epoch** (see Figure 13-20) is still subject to debate. Current estimates define the start of the Pleistocene as 2.58 million years ago, but geochronologists now recognize that the glaciations began even earlier. Evidence suggests, for example, that Antarctica was covered by an ice cap similar in size to its current extent as long ago as 10 million years. The most recent findings tell us that by the start of the Pleistocene, the "amplitude" of climate fluctuations—from glacial period to interglacial period—had increased and some parts of the Northern Hemisphere were covered by glaciers.

New evidence has also changed the date of the close of the Pleistocene. Although the end of the Pleistocene is now set at 11,700 years ago, evidence suggests that some of the Pleistocene glaciers were still retreating as recently as 9000 years ago. Even this most recent estimate for the close date is a bit uncertain, however, because the Ice Age may not yet have ended—a possibility we consider later in the chapter. For now, let us say just that, to the best of our knowledge, the Pleistocene epoch occupied almost all of the most recent two and a half million years of Earth's history.

The end of the Pleistocene epoch coincided with the conclusion of what is known in North America as the "Wisconsin" glacial stage (known as the "Würm" in the Alps), approximately 11,700 years ago. The period since then is identified as the *Holocene Epoch*. Conceptually, then, the Holocene is either a postglacial epoch or the latest in a series of interglacial interludes.

Extent of Pleistocene Glaciations: At its maximum Pleistocene extent, ice covered one-third of the total land area of Earth—nearly 47,000,000 square kilometers (18 million square miles) (Figure 19-5a). Ice thickness varied and can be estimated only roughly, but we do know that in some areas it reached a depth of several thousand meters. During the Pleistocene:

- The greatest total area of ice-covered land was in North America. The *Laurentide* ice sheet, which covered most of Canada and a considerable portion of the northeastern United States, was the most extensive Pleistocene ice mass; its area was slightly larger than that of the glacier covering Antarctica today. It extended southward into the United States to approximately the present location of Long Island, the Ohio River, and the Missouri River.
- Most of western Canada and much of Alaska were covered by an interconnecting network of smaller ice sheets. For reasons we do not fully understand, however, a small area in northwestern Canada as well as extensive portions of northern and western Alaska were never glaciated during the Pleistocene. Moreover, a small area (62,000 square kilometers [24,000 square miles]) in southwestern Wisconsin and parts of three adjoining states was also left uncovered (Figure 19-5b). This area, referred to as the *Driftless Area*, was apparently never completely surrounded by ice; rather, ice encroached first on one side during one glacial advance and then on the other side during a different advance.
- More than half of Europe was overlain by ice (Figure 19-5c). Asia was less extensively covered, presumably because in much of its subarctic portion there was not enough precipitation for the ice to last. Nevertheless, ice covered much of Siberia, and extensive glaciation occurred in most Eurasian mountain ranges.
- Antarctic ice was only slightly more extensive than it is today. A large ice complex covered southernmost South America, and the South Island of New Zealand was largely covered with ice.
- Major mountain ranges around the world—such as the Andes, Alps, Himalayas, Sierra Nevada, and Rocky Mountains—underwent extensive glaciation. Some mountains in the tropics, such as those in central Africa, New Guinea, and Hawai'i, experienced limited glaciation.

LearningCheck 19-2 **Describe the extent of glaciation during the peak of the Pleistocene glaciations.**

Indirect Effects of Pleistocene Glaciations

The accumulation of ice and the movement and melting of the resulting glaciers had an enormous effect on topography and drainage, a topic we discuss in detail shortly. In addition, however, Pleistocene glaciations had several indirect effects.

Periglacial Processes: Beyond the outermost extent of ice advance is an area called the **periglacial zone**, which was never touched by glacial ice but was indirectly influenced by the ice. The most important periglacial processes were the erosion and deposition done by the enormous amounts of meltwater released as the glaciers melted. Also important were frost weathering caused by the low temperatures in the periglacial zone and the associated *solifluction* of frozen subsoil (see Chapter 15). It is estimated that periglacial conditions extended over more than 20 percent of Earth's land area. (Periglacial landforms are discussed later in this chapter.)

Sea-Level Changes: When continents were covered with ice, less water was available to drain from the continents into the oceans. That condition resulted in a worldwide lowering of sea level during every episode of glacial advance;

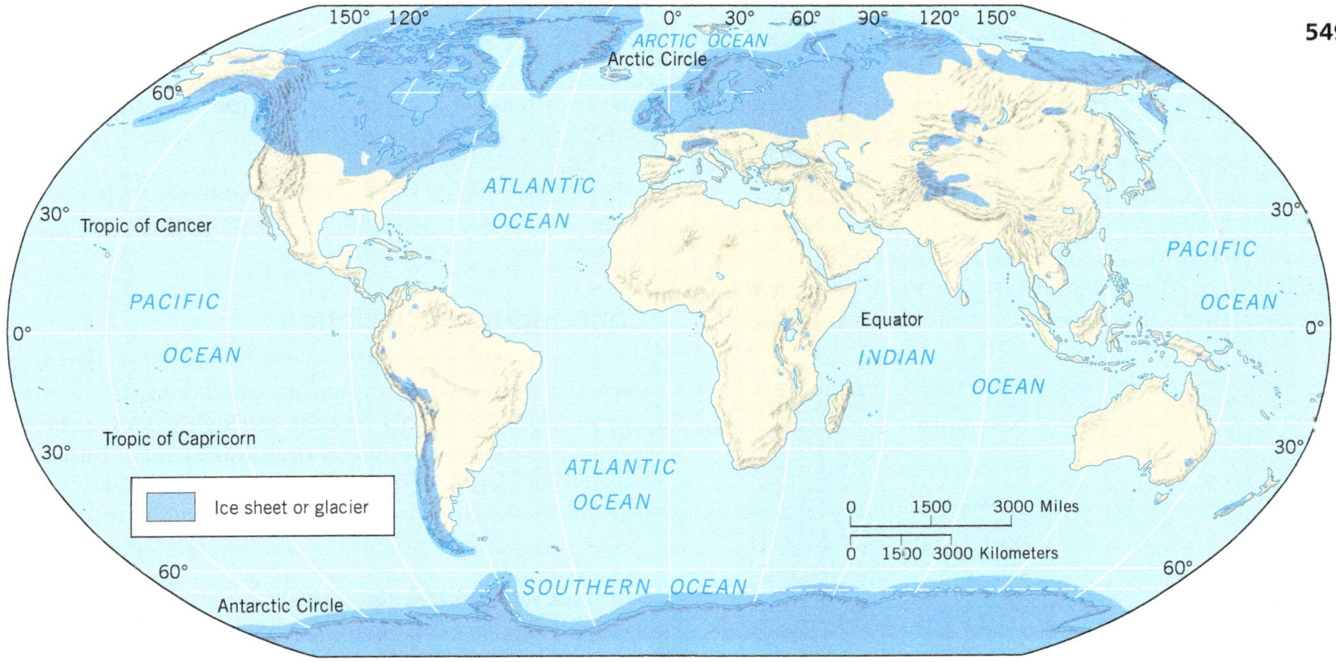

(a) Glacial extent in Pleistocene: global

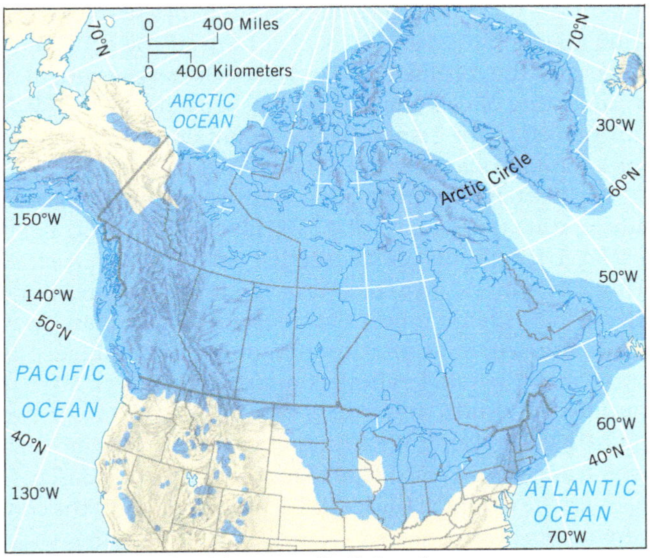

(b) Glacial extent in Pleistocene: North America

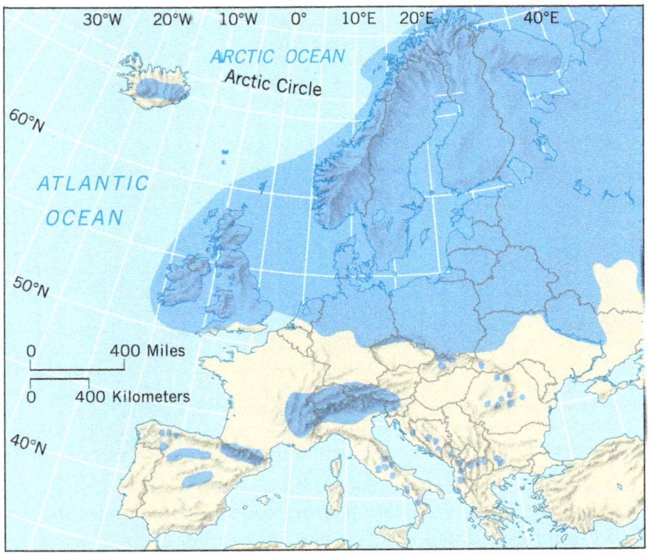

(c) Glacial extent in Pleistocene: western Eurasia

▲ **Figure 19-5** The maximum extent of Pleistocene glaciation: (a) worldwide; (b) in North America; (c) in western Eurasia.

Animation
End of the Last Ice Age
https://media.pearsoncmg.com/ph/esm/esm_mcknight_physgeo_13/media/

when the glaciers retreated, sea level would again rise as meltwater returned to the ocean. At the peak of the Pleistocene glaciations, global sea level was at least 120 meters (395 feet) lower than it is today. These fluctuations in the amount of ocean water caused a significant difference in drainage patterns and topographic development on seashores and coastal plains. (The influence of Pleistocene sea-level changes on coastal topography is discussed in Chapter 20.) During the Pleistocene glaciations, the Bering Strait between present-day Alaska and Russia was a dry land bridge, allowing the migration of both animals and humans (see Figure 11-21).

LearningCheck 19-3 **Why did sea level fluctuate during the Pleistocene?**

Crustal Depression: The enormous weight of accumulated ice on the continents caused portions of Earth's crust to sink at least 400 meters (1300 feet). After the ice melted, the crust slowly began to rebound, and this *isostatic adjustment* has not yet been completed. Some portions of Canada and northern Europe are still rising as much as 10 centimeters (4 inches) per decade. (Isostasy is discussed in detail in Chapter 13; see Figure 13-16.)

Animation
Isostasy
https://media.pearsoncmg.com/ph/esm/esm_mcknight_physgeo_13/media/

Pluvial (Increased Rain) Developments: During the Pleistocene glaciations, there was, on almost all areas of the continents, a considerable increase in the amount of moisture available. This increase was caused by a combination of meltwater runoff, increased precipitation, and decreased evaporation. A prominent result of these **pluvial effects** was the creation of many lakes in areas where none had previously existed. Most of these **Pleistocene lakes** have subsequently drained or significantly reduced in size, but they left lasting imprints on the landscape, especially in the western part of the United States (Figure 19-6). The Great Salt Lake in Utah is a tiny remnant of a much larger Pleistocene lake known as Lake Bonneville, and today's Bonneville Salt Flats were once the floor of this enormous lake.

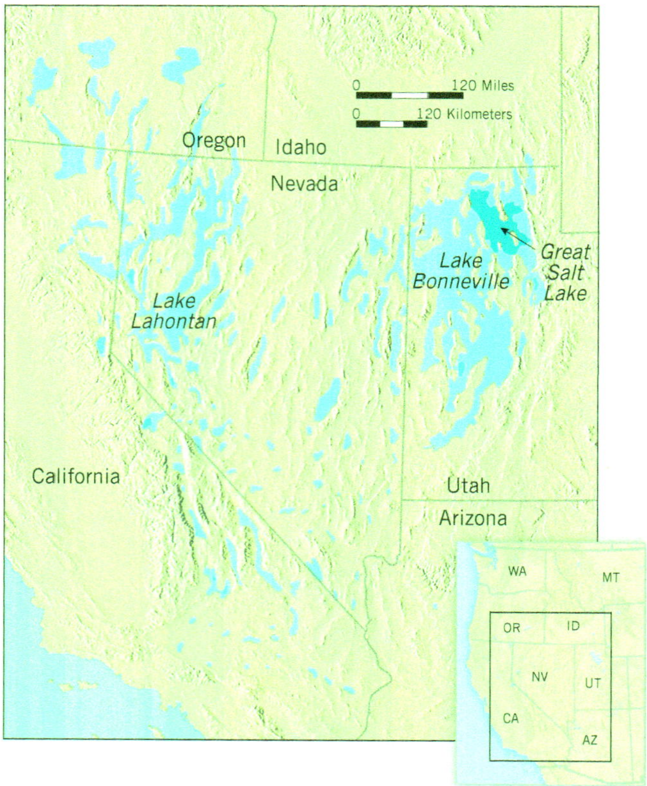

▲ Figure 19-6 Pleistocene lakes of the western United States. Today's Great Salt Lake (Utah) is shown outlined in blue inside the boundaries of the ancestral Lake Bonneville.

LearningCheck 19-4 **What explains the presence of large lakes, such as Lake Bonneville, during the Pleistocene?**

Contemporary Glaciation

In marked contrast to Pleistocene glaciation, the extent of ice covering the continental surfaces today is very limited (Figure 19-7). About 10 percent of Earth's land surface—15 million square kilometers (6 million square miles)—is covered with ice today, but more than 96 percent of that total is in Antarctica and Greenland. More than two-thirds of all freshwater is at this moment frozen into glacial ice.

Antarctic Ice Sheet: Antarctic ice is by far the most extensive ice sheet on Earth (see Figure 19-3b). About 98 percent of its surface is covered with glacial ice, representing almost 90 percent of the world's land—ice total. This ice is more than 4000 meters (13,000 feet) thick in some places and about 2160 meters (7100 feet) thick over most of the continent. Physically, the continent and its ice sheets can be thought of as consisting of two unequal sections separated by the wide upland belt of the Transantarctic Mountains, which extend for some 3500 kilometers (2200 miles; Figure 19-8).

West Antarctica, the smaller of the two sections, is generally mountainous, much of it 2500 meters (8200 feet) above sea level. It contains several interior valleys that are ice free. This "Dry Valleys" area consists of about 3900 square kilometers (1500 square miles) that, because winds blast away snow and keep precipitation out, does not build ice. The three major parallel valleys contain several large lakes, a number of ponds, and a river that flows for one or two months each year. If West Antarctica were to lose its ice, it would appear as a considerable number of scattered islands because the weight of the ice has depressed much of the continent here below sea level.

▶ Figure 19-7 World distribution of glacial ice today. The area covered by mountain glaciers is exaggerated to show approximate locations; the Greenland and Antarctic Ice Sheets have thinned in recent decades, especially around the edges.

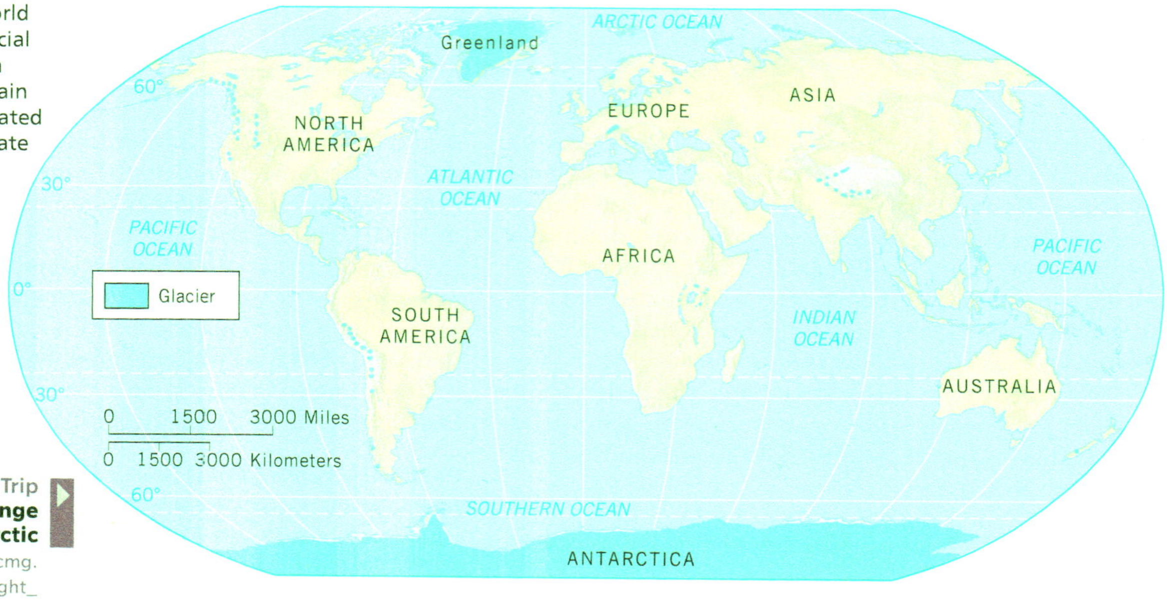

Mobile Field Trip
Climate Change in the Arctic
https://media.pearsoncmg.com/ph/esm/esm_mcknight_physgeo_13/media/

▲ **Figure 19-8** The Transantarctic Mountains separate the ice sheets of West and East Antarctica.

East Antarctica contains about nine times the volume of ice of West Antarctica. Recent radar mapping studies have confirmed the presence of the Gamburtsev Subglacial Mountains—a mountain range comparable to the Alps in size but buried under at least 600 meters (2000 feet) of ice.

Several hundred subglacial lakes containing liquid freshwater exist below the Antarctic ice sheet (**Figure 19-9**). The largest, Lake Vostok, is 3700 meters (12,100 feet) below the ice sheet in East Antarctica and has a surface area of 12,500 square kilometers (4800 square miles). It is more than 500 meters (1700 feet) deep in places. In 2013, scientists drilled down through 800 meters (2600 feet) of ice into Lake Whillans, under the West Antarctica ice sheet. Water samples contained thousands of species of microbes. These species have likely inhabited the lake—without the benefit of solar energy—for at least 120,000 years.

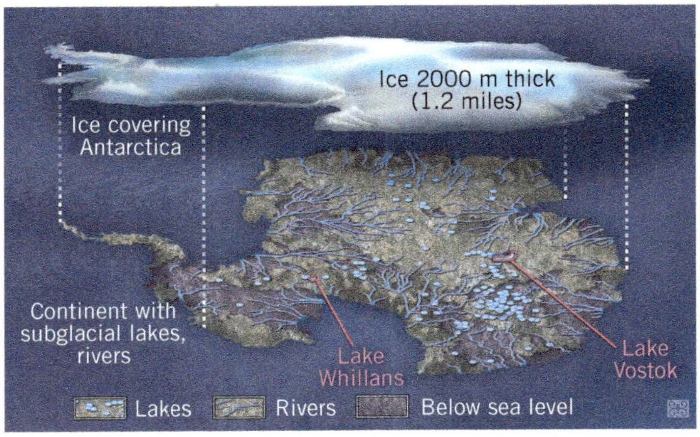

▲ **Figure 19-9** The location of subglacial lakes in Antarctica. The largest is Lake Vostok.

Greenland Ice Sheet: Greenland ice is much less extensive than that of Antarctica but still impressive: 1,700,000 square kilometers (656,000 square miles) in area (see Figure 19-3a). Elsewhere, only relatively small ice masses are on certain islands in the Canadian Arctic, Iceland, and some of the islands north of Europe.

Mountain Glaciers: Other than the two major ice caps, the remainder of the world's present-day glaciers are concentrated in high mountain areas. In the conterminous United States, most glaciers are in the Pacific Northwest; more than half of these are in the North Cascade Mountains of Washington (**Figure 19-10**). Alaska has 75,000 square kilometers (29,000 square miles) of glacial ice, amounting to about 5 percent of the state's total area. The largest Alaskan glacier is the Bering Glacier, near Cordova; it covers 5175 square kilometers (2000 square miles) and is nearly twice the size of Rhode Island.

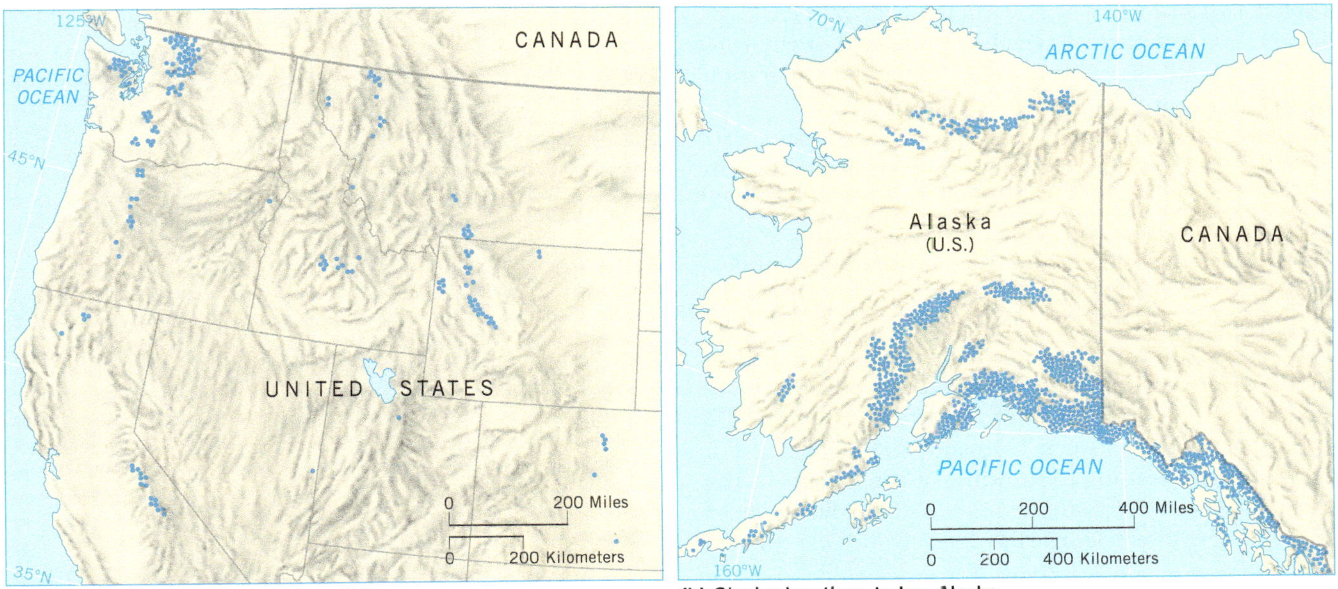

(a) Glacier locations today: western U.S.

(b) Glacier locations today: Alaska

▲ **Figure 19-10** The locations—but not extent—of contemporary glaciers and perennial snowfields in (a) the conterminous western United States and (b) Alaska. The dots considerably exaggerate the size of each glacier.

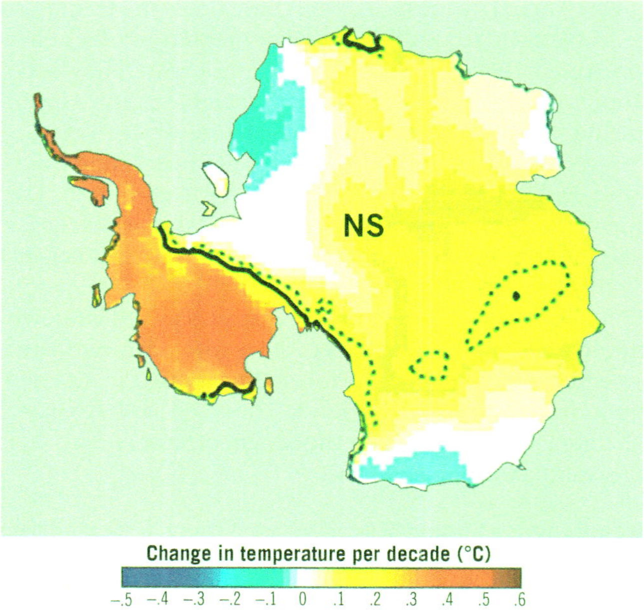

▲ Figure 19-11 Rate of temperature change per decade in Antarctica between 1958 and 2012. The thick black line marking off West Antarctica and the Antarctic Peninsula shows areas where the data confidence level is high; within the dotted lines, the data confidence level is medium; in the region marked NS (most of central and eastern Antarctica) the data confidence levels are not statistically significant.

Climate Change and Contemporary Glaciation: As we saw in earlier chapters, global climate change is significantly influencing the extent of contemporary glaciation. The retreat of the Arctic sea ice pack and the loss of mass of Greenland's ice sheets is indicative of the higher temperatures experienced in the high latitudes of the Northern Hemisphere over the last 50 years (for example, see Figure 8-28). As we see later in this chapter (*Featured National Park: Glacier—The Shrinking Glaciers of Glacier National Park*), glaciers are sensitive indicators of environmental change—reflecting variations in both temperature and precipitation.

Perhaps nowhere today is the attention of scientists focused more keenly on contemporary ice sheets than in Antarctica. Not only does the Antarctic ice sheet provide invaluable information about Earth's past climate (see Chapter 8), but also changes in the ice sheet today point to a contemporary climate shift. Antarctica has been warming at a rate of about 0.12°C (0.22°F) per decade over the last 50 years—much faster than scientists thought even a few years ago. Furthermore, West Antarctica has been warming much faster than the continent as a whole, about 0.33°C (0.59°F) per decade between 1958 and 2012 (Figure 19-11). This warming is leading to thinning of the ice sheets, the breakup of Antarctic ice shelves, and to higher flow rates of outlet glaciers as well. See the box *People & the Environment: Disintegration of Antarctic Ice Shelves.*

LearningCheck 19-5 How is climate change affecting glaciers today?

Glacier Formation and Movement

Glaciers are *open systems* with inputs and outputs of both material and energy (see Figure 1-5). They are associated with several climate feedback loops linked to contemporary climate change. For example, as glaciers retreat due to higher temperatures, surface albedo decreases; in turn, the decreased reflectance results in greater solar energy absorption, which can lead to still higher temperatures.

A glacier may begin to develop when there is a net year-to-year accumulation of snow. That is, over a period of years, the amount of snow that falls in a winter is greater than the amount that melts the following summer. The snow that falls the next winter weighs down on the old snow and turns it to ice. After many years of such accumulation, the ice mass begins to move under the pull of gravity—and a glacier forms. Glaciers depend on just the right combination of temperature and moisture to survive. The persistence of any glacier depends on the balance between **accumulation** (addition of ice by incorporation of snow) and **ablation** (wastage of ice through melting and sublimation).

Changing Snow to Ice

Snow is not simply frozen liquid water. Rather, it crystallizes directly from water vapor in the atmosphere and floats to Earth as lacy, hexagonal crystals that are only about one-tenth as dense as liquid water. Sooner or later (within a few hours if the temperature is near freezing, but only over a period of years in very cold situations), crystalline snow is compressed by overlying snow into granular form. In the process, its density approximately doubles. With more time and further compression, the granules are packed more closely and begin to coalesce, the density increasing steadily until it is about half the density of water (Figure 19-12). This material is called **névé** or *firn*. As time passes, the weight of

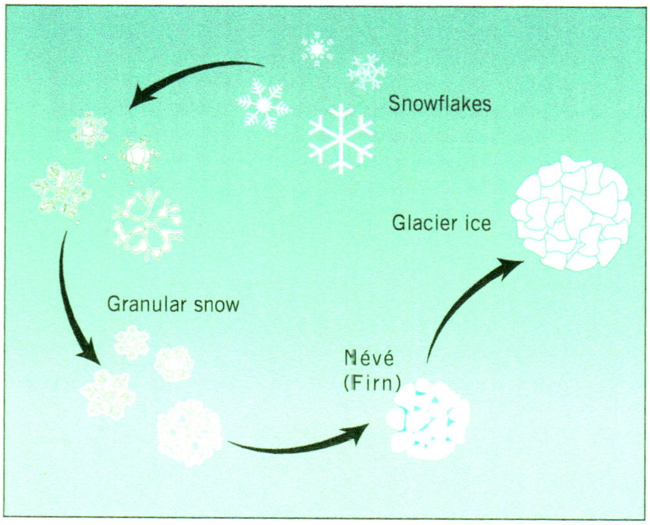

▲ Figure 19-12 Compression and coalescence change snow to ice, following a sequence from snowflake to granular snow to névé to glacial ice.

People & the Environment

Disintegration of Antarctic Ice Shelves

▲ **Figure 19-B** Images of the Larsen-B Ice Shelf of Antarctica, taken by NASA's Terra satellite. (a) On January 31, 2002, blue "melt ponds" of liquid water are seen on top of the ice shelf. (b) By March 7, 2002, the ice shelf had collapsed. The blue area shows the shattered floating ice from the shelf.

Despite its remoteness, Antarctica exerts a prominent influence on the world's environment: conditions in Antarctica affect global sea level, ocean temperature, ocean nutrient content, and patterns of atmospheric circulation. At the same time, the global environment exerts its own influence on the conditions in Antarctica. Over the last half-century, average temperatures on the Antarctic Peninsula rose at least 2.8°C (5.0°F)—a much greater increase than the worldwide average. On February 6, 2020, the highest temperature ever recorded in Antarctica was reached on the northern Antarctic Peninsula: 18.3°C (64.9°F).

Although the complete melting of the Antarctic ice cap is unlikely, the long-term equilibrium between accumulation and wastage of the ice is changing in response to higher temperatures. Some of the most dramatic changes are occurring in Antarctica's ice shelves.

Antarctic Ice Shelves: The Antarctic ice sheets flow outward from the interior of the continent in nearly all directions toward the sea (**Figure 19-A**), so icebergs are being calved (broken off) into the sea more or less continuously around the perimeter of the continent. Some of these icebergs originate through outlet glaciers, but many break off from ice shelves. The Ross Ice Shelf in West Antarctica is the largest (480,000 square kilometers, or 183,330 square miles). A number of smaller ice shelves are on the Antarctic Peninsula, such as the Larsen Ice Shelves.

Over the last few decades, large sections of the ice shelves along the Antarctic Peninsula have disintegrated—more than 8000 square kilometers (3100 square miles) of ice shelf have disappeared since the 1990s. In 1995 the Larsen-A Ice Shelf collapsed and disappeared. In 2002, much of the Larsen-B Ice Shelf suddenly collapsed (**Figure 19-B**). The Wilkins Ice Shelf began to disintegrate in February 2008; in January 2010, a piece of floating ice the size of Rhode Island broke off the Ronne-Filchner Ice Shelf; and in July 2017, an iceberg the size of Delaware broke off the Larsen-C Ice Shelf.

Warmer Currents Melting Ice Shelves: More important than the ice area loss in Antarctica is the total ice mass loss. A 2019 study found that the ice mass loss in Antarctic accelerated 280 percent over the last 40 years.

Many West Antarctica ice shelves are losing mass primarily because changes in wind and ocean currents are directing warmer water beneath the ice shelves. The ice is melting from below—a process known as *basal melt*. Higher air temperatures also appear to be contributing significantly to the thinning of the Larsen-C Ice Shelf.

Although the breakup of ice shelves in themselves does not raise global sea level (for the same reason that a floating ice cube does not raise the water level in a glass as it melts), changing ice shelves can trigger a change in the flow of land-based ice off the continent—and this can contribute to rising sea levels (a concern because 100 million people around the world live within 1 meter of sea level).

To a certain extent, an intact ice shelf holds back the flow of continental ice into the ocean. Once an ice shelf is gone, the continental ice may flow into the ocean faster. Since the breakup of the Larsen-B shelf, the flow rate of glaciers feeding it accelerated by as much as six times their previous rate.

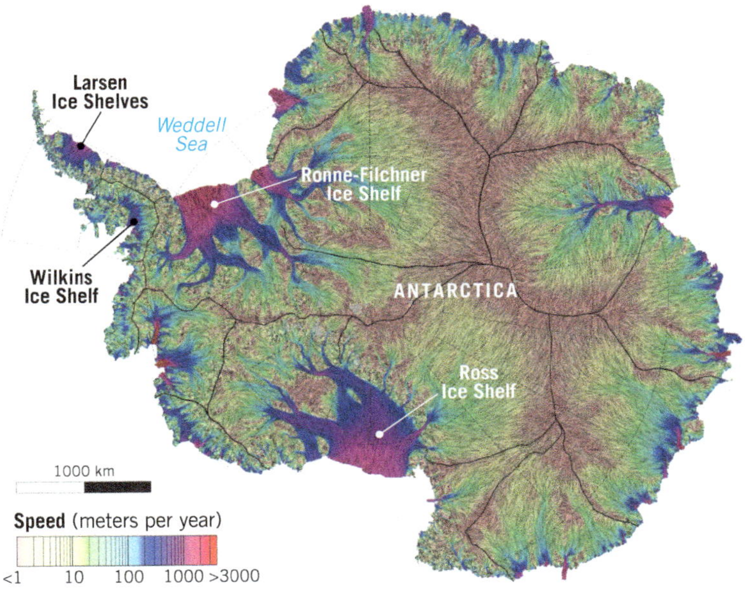

▲ **Figure 19-A** Ice flow speed in Antarctica. Data gathered between 2013 and 2018 show the role of ice shelves as outlets for Antarctica's continental ice sheets. As the ice shelves continue to shrink, the flow rates of ice off the continent are expected to increase.

Questions

1. What processes are responsible for the current loss of ice mass in Antarctic ice shelves?
2. How does the presence of an ice shelf influence the rate of ice flow off Antarctica into the ocean?

NASA Videos
Operation IceBridge
https://media.pearsoncmg.com/ph/esm/esm_mcknight_physgeo_13/media/

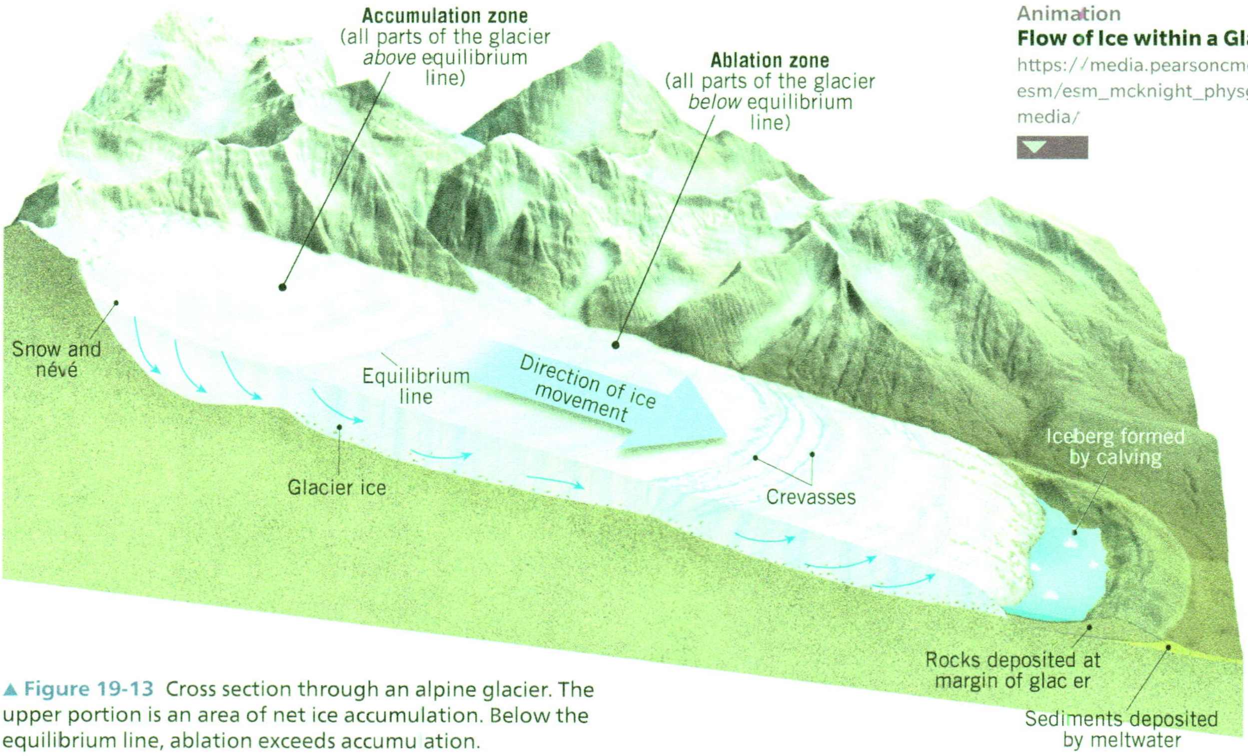

Animation
Flow of Ice within a Glacier
https://media.pearsoncmg.com/ph/esm/esm_mcknight_physgeo_13/media/

▲ **Figure 19-13** Cross section through an alpine glacier. The upper portion is an area of net ice accumulation. Below the equilibrium line, ablation exceeds accumulation.

the overlying snow gradually squeezes out the air trapped in pore spaces among the whitish névé crystals. The density approaches 90 percent of that of liquid water, and the material takes on the bluish tinge of glacial ice. (Dense glacial ice absorbs most wavelengths of visible light but reflects and scatters blue light.) This ice continues to change, although very slowly, with more air being forced out, the density increasing slightly, and the crystals increasing in size.

Every glacier can be divided into two portions on the basis of the balance between accumulation and ablation (Figure 19-13). The upper portion is called the **accumulation zone** because here the amount of new ice from snowfall added each year exceeds the amount lost by melting and sublimation. The lower portion is called the **ablation zone** because here the amount of new ice added each year is less than the amount lost. Separating the two zones is a theoretical **equilibrium line**, where accumulation exactly balances ablation.

LearningCheck 19-6 **Explain what conditions are necessary for a glacier to form. Describe the difference between the zones of accumulation and ablation.**

Glacier Movement

Despite the fact that glaciers are often likened to rivers of ice, there is little similarity between liquid flow and glacial movement.

We usually think of ice as being a brittle substance that breaks rather than bends. This is generally true for surface ice, as indicated by the cracks and *crevasses* that often appear at the surface of a glacier. However, ice under confining pressure, as below the surface of a glacier, behaves quite differently: it deforms rather than breaks. Moreover, partial melting, due to the stresses within and the pressure at the bottom of the glacier, aids movement because the meltwater sinks to the bottom of the glacier and becomes a slippery layer on which the glacier can slide.

Plastic Flow of Ice: When a mass of ice attains a thickness of about 50 meters (165 feet)—less on steep slopes—the **plastic flow of ice** begins in response to the overlying weight. The entire mass does not move uniformly. Rather, ice oozes outward from around the edge of an ice sheet or down-valley from the toe (the end) of an alpine glacier.

Basal Slip: In a second kind of glacier movement, **basal slip** at the bottom of the glacier, the entire mass slides over its bed on a lubricating film of water. The glacier more or less molds itself to the shape of the terrain over which it is riding.

Rates of Movement: Glaciers usually move very slowly; indeed, the adjective "glacial" is synonymous with "exceedingly slow." The movement of many glaciers can be measured in a few centimeters per day, although an advance of several meters per day is not unusual. Extreme examples of nearly 30 meters (100 feet) in a 24-hour period have been recorded. Also, the flow is often erratic, with irregular pulsations and surges over a short span of time.

As we might expect, all parts of a glacier do not move at the same rate. The fastest moving ice is that at and near the surface, with speed generally decreasing with depth. If the glacier is confined—the way a valley glacier is, for example—the center of the surface ice moves faster than the sides, which is similar to streamflow patterns.

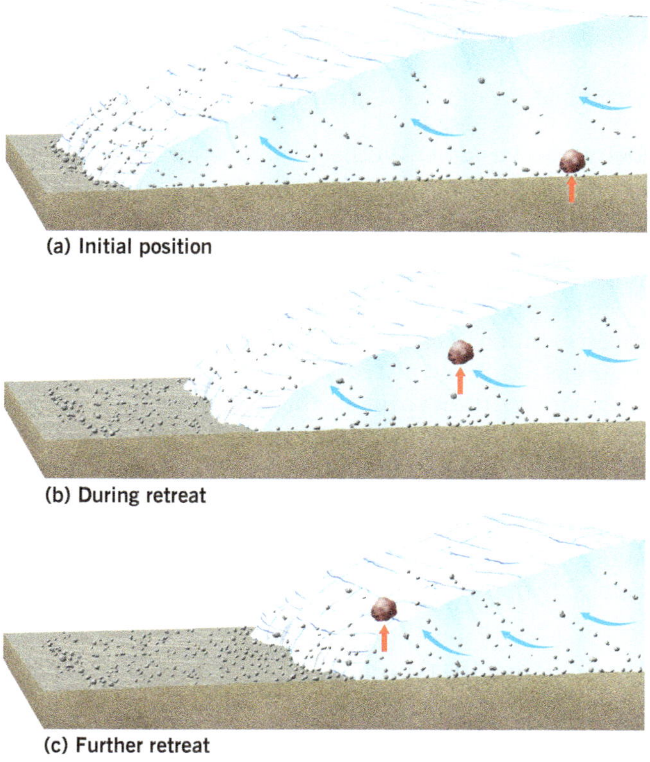

▲ **Figure 19-14** A flowing glacier is not necessarily an advancing glacier. In this sequence, the front of the glacier is retreating, but the ice continues to flow forward, as shown by the boulder marked with a red arrow.

Glacier Flow versus Glacier Advance

In discussing glacier movement, we must distinguish between glacier *flow* and glacier *advance*. As long as a glacier exists, the ice in it is flowing, either downhill or laterally outward. The outer edge of the ice is not necessarily advancing, however. The ice in a glacier always moves forward, but the outer margin of the glacier may or may not be advancing, depending on the balance between accumulation and ablation (**Figure 19-14**). Even in a retreating glacier (one whose outer margin is retracting toward its point of origin due to heavy ablation), the ice is flowing forward.

During wetter or cooler periods when the accumulation of ice is large, a glacier can flow farther before it finally wastes away, so the outer margin of the glacier advances. During warmer or drier periods when the rate of ablation is increased, the glacier continues to flow but wastes away sooner, so the end or terminus of the glacier retreats.

LearningCheck 19-7 **Why can a retreating glacier continue to erode and transport rock?**

The Effects of Glaciers

As glaciers move across a landscape, they can reshape the topography through the erosion, transportation, and deposition of rock.

Animation
Glacial Processes
https://media.pearsoncmg.com/ph/esm/esm_mcknight_physgeo_13/media/

Erosion by Glaciers

The amount of erosion caused by a glacier is roughly proportional to the thickness of the ice and its rate of flow. The depth of the erosion is limited in part by the structure and texture of the bedrock and in part by the relief of the terrain.

Glacial Plucking: The direct erosive power of moving ice is greater than that of flowing water but not remarkably so. As the slowly moving ice scrapes against bedrock, friction between rock and ice causes the lowermost ice to melt, and the layer of water created reduces the pressure on the rock. This water can refreeze around rocky protrusions, however, and the refrozen ice can exert a significant force as it is pushed by the ice behind it. Probably the most significant erosive work of glacial ice is accomplished by this **glacial plucking**. Rock fragments beneath the ice are grasped as meltwater refreezes in bedrock joints and fractures, where *frost wedging* further loosens the rock. As the ice moves along, these particles are plucked out and dragged along. This action is particularly effective on leeward slopes (slopes facing away from the direction of ice movement) and in well-jointed bedrock.

Glacial Abrasion: Glaciers also erode by abrasion, in which the bedrock is worn down by rock debris being dragged along in the moving ice. Abrasion generally produces minor features, such as polished surfaces when the bedrock is of highly resistant material and striations (fine parallel indentations) and grooves (deeper indentations) in less-resistant bedrock (**Figure 19-15**). Whereas plucking tends to roughen the underlying surface, abrasion tends to polish it and to dig striations and grooves.

Subglacial Meltwater Erosion: A third process can also contribute to glacial erosion: not only can meltwater

▼ **Figure 19-15** Glacial polish on granite in Yosemite National Park. The glaciers flowed from upper left to lower right.

▲ **Figure 19-16** Grooves and striations near Cusco, Peru, smoothed by meltwater erosion below a glacier.

streams flowing below the glacier transport rock, they can also smooth grooves and channels in the bedrock below a glacier (Figure 19-16).

In plains areas, the topography produced by glacial erosion may be inconspicuous. Prominences are smoothed, and small hollows may be excavated, but the general appearance of the terrain changes little. In hilly areas, however, the effects of glacial erosion are much more noticeable. Mountains and ridges are sharpened; valleys are deepened, steepened, and made more linear; and the entire landscape becomes more angular and rugged.

LearningCheck 19-8 **Explain the process of glacial plucking.**

Transportation by Glaciers

Glaciers are extremely competent, as well as indiscriminate, in their ability to transport rock debris. They can move blocks of rock the size of houses up to hundreds of kilometers. Most of a glacier's load, however, is a heterogeneous collection of particles of all sizes, including finely ground rock material known as **glacial flour**.

Most of the material transported by continental glaciers is plucked or abraded from the underlying surface and carried along the base of the ice. Thus, a narrow zone at the bottom of the glacier is likely to be heavily packed with rock debris frozen into it, with most of the rest of the glacial ice relatively free of rock fragments.

With mountain glaciers, in addition to rock transported within the ice, some of the material is also transported on top of the ice, deposited by rockfall or other forms of mass wasting from the surrounding slopes. Mass wasting supplies more rock debris to some mountain glaciers than is added by glacial erosion.

A glacier transports its load outward or down-valley at a variable speed. The rate of flow usually increases in summer and slows in winter but also depends on variations in ice accumulation and in the gradient of the underlying slopes.

Meltwater Streams: Another important aspect of transportation by glaciers is the role of flowing water on, in, and under the ice. During the warmer months, streams of meltwater normally flow along with the moving ice. Such streams may run along the surface of the glacier until they reach cracks or crevasses—including steep drainage shafts in the ice called *moulins*—into which they can plunge (Figure 19-17). They then continue their flow as a subglacial stream within the ice or along the interface between the glacier and bedrock. Such streams transport rock debris, particularly smaller particles and glacial flour, from the ice surface to the middle or bottom of the glacier. Furthermore, where subglacial streams encounter finely ground rock along the bottom of a glacier, the lubricating effect can accelerate the flow of ice.

Deposition by Glaciers

Probably the major role of glaciers in landscape modification is to pick up rock from one area and take it to a distant region, where it is left in a fragmented and vastly changed form. This is clearly displayed in North America, where an extensive portion of central Canada has been glacially scoured of its soil, regolith, and much of its surface bedrock. That scouring left a relatively barren, rocky, gently undulating surface dotted with many lakes. Much of the removed material was transported southward and deposited in the midwestern United States, producing an extensive plains area of remarkably fertile soil. Thus, the legacy of Pleistocene ice sheets for the U.S. Midwest was the evolution of one of the largest areas of productive soils ever known, at the expense of central Canada, which was left impoverished of soil by those same glaciers. (However, many valuable

▲ **Figure 19-17** A researcher examines a meltwater stream flowing into a moulin on the surface of the Greenland Ice Sheet.

Canadian mineral deposits were exposed when the glaciers removed soil and regolith.)

The general term for all material moved by glaciers is **drift**, a misnomer coined in the eighteenth century when it was believed that the vast debris deposits of the Northern Hemisphere were leftovers from biblical floods. Nonetheless, the term is still used to refer to any material deposited by glaciers or their meltwater.

Direct Deposition by Glacial Ice: Rock debris deposited directly by moving or melting ice, with no meltwater redeposition involved, is given the more specific name **till** (Figure 19-18). Direct deposition by ice occurs when melting takes place around the margin of an ice sheet or near the lower end of a mountain glacier, but it is also accomplished whenever debris is dropped on the ground beneath the ice, especially in the ablation zone. In either case, the result is an unsorted and unstratified agglomeration of fragmented rock. Most of the fragments are angular because they have been held in position while carried in the ice (so they had little opportunity to become rounded by frequent impact as pebbles in a stream would) or were deposited on top of the ice by mass wasting.

Sometimes outsized boulders are deposited by a retreating glacier. Such enormous fragments, which may be very different from the local bedrock, are called **glacial erratics** (Figure 19-19).

LearningCheck 19-9 **How and why does a deposit of glacial till look different from a deposit of alluvium?**

Secondary Deposition by Meltwater: Glacial stream runoff has several peculiarities—peak flows in midsummer, distinct day-and-night differences in volume, large silt content, and occasional floods—that set meltwater streams apart from other kinds of streams. Much of the debris carried by glaciers is eventually deposited or redeposited by meltwater. This may be accomplished by subglacial streams issuing directly from the ice and carrying sedimentary material washed from the glacier. Much meltwater deposition, however, involves debris that was originally deposited by ice and subsequently picked up and redeposited by the meltwater well beyond the outer margin of the ice. Such **glaciofluvial deposition** occurs around the margins of all glaciers, as well as far out in some periglacial zones.

▼ Figure 19-18 Unsorted glacial till in a roadcut along Tioga Road in Yosemite National Park, California.

▲ Figure 19-19 Bubble Rock in Acadia National Park, Maine, is a glacial erratic.

Continental Ice Sheets

Apart from the oceans and continents, continental ice sheets are the most extensive features ever to appear on the face of the planet. Their actions during the Pleistocene significantly reshaped both the terrain and the drainage of nearly one-fifth of the total surface area of the continents.

Development and Flow

Pleistocene ice sheets, with the exception of the one covering Antarctica, did not originate in the polar regions. Rather, they developed in subpolar and midlatitude locations and then spread outward in all directions, including poleward. Several (perhaps several dozen) centers of original ice accumulation have been identified. The accumulated snow/névé/ice eventually produced such a heavy weight that the ice began to flow outward from each center of accumulation.

The initial flow was channeled by the preexisting terrain along valleys and other low-lying areas, but in time the ice developed to such depths that it overrode almost all preglacial topography. In many places, it submerged even the highest points under thousands of meters of ice. Eventually the various ice sheets coalesced into only one, two, or three massive sheets on each continent. These vast ice sheets flowed and ebbed as the climate changed, always modifying the landscape with their enormous erosive power and the great masses of debris they deposited. The result was nothing less than a total reshaping of the land surface and a complete rearrangement of the drainage pattern.

Erosion by Ice Sheets

Except in mountainous areas of great initial relief, the principal topography resulting from the erosion caused by an ice sheet is a gently undulating surface. The most conspicuous features are valley bottoms gouged and deepened by the moving ice. Such troughs are deepest where the preglacial valleys were oriented parallel to the direction of ice movement, particularly in areas of softer bedrock. A prime example of such development is the Finger Lakes District of central New York, where a set of parallel stream valleys was accentuated by glaciation into a group of long, narrow, deep lakes (**Figure 19-20**). Even where the preglacial valley was not oriented parallel to the direction of ice flow, however, glacial erosion normally produced a large number of shallow basins or troughs that became lakes after the ice disappeared. Indeed, the postglacial landscape in areas of ice-sheet erosion is notable for its abundance of lakes.

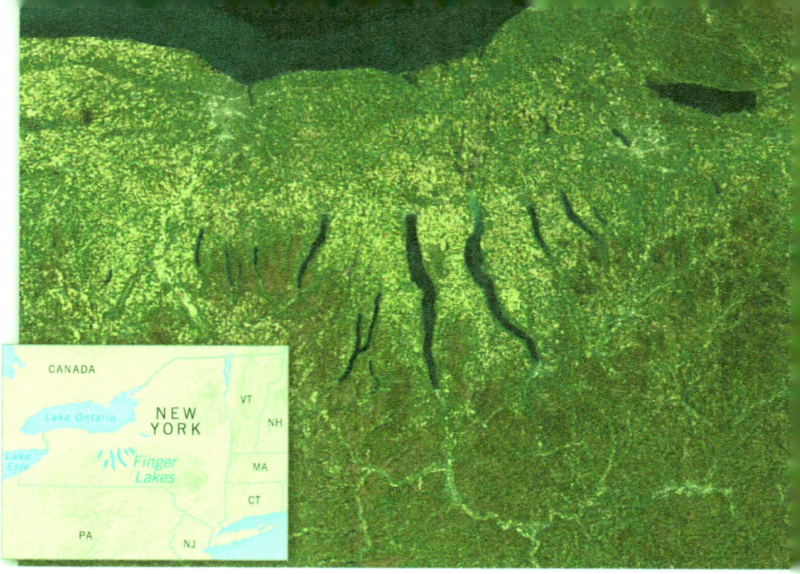

▲ **Figure 19-20** The Finger Lakes of upstate New York occupy glacial valleys that were accentuated by ice sheets moving from north–northwest to south–southeast. The large body of water at upper left is Lake Ontario. This true-color photograph was taken with the MODIS instrument aboard NASA's Terra satellite.

Roche Moutonnée: Hills are generally sheared off and rounded by the moving ice. A characteristic shape produced by both continental ice sheets and mountain glaciers is the **roche moutonnée**, which is often produced when a bedrock hill is overridden by moving ice (**Figure 19-21**).[1] The *stoss side* (facing in the direction from which the ice came) of a roche moutonnée is smoothly rounded and streamlined by grinding abrasion as the ice rides up the slope. The lee side (facing away from the direction from which the ice came) is shaped largely by plucking, which produces a steeper and more irregular slope.

[1]The origin of the French term *roche moutonnée* is unclear. It is often translated as "sheep's back," but some authorities believe it is based on a fancied resemblance to wavy wigs that were fashionable in France in the late 1700s and were known as *moutonnées* because they were pomaded with mutton tallow.

The postglacial landscape produced by ice sheets is one of relatively low relief but not absolute flatness. The principal terrain elements are ice-scoured rocky knobs and scooped-out depressions. Soil and weathered materials are largely absent, with bare rock and lakes dominating the surface. Stream patterns are erratic and inadequately developed because the preglacial drainage system was "deranged" by ice erosion. Once eroded by the passing ice sheet, however, most of this landscape was subjected to further modification by glacial deposition. Thus, the starkness of the erosional landscape is modified by depositional debris.

LearningCheck 19-10 **Explain the shape and formation of a roche moutonnée.**

▼ **Figure 19-21** (a) Lembert Dome in Yosemite National Park is a roche moutonnée. The glaciers moved from right to left (east to west) across this granite dome. (b) When a roche moutonnée forms, the glacier rides over a resistant bedrock surface, smoothing the stoss side by abrasion and steepening the lee side by plucking. An asymmetrical hill is left once the ice melts.

(a) Roche moutonnée

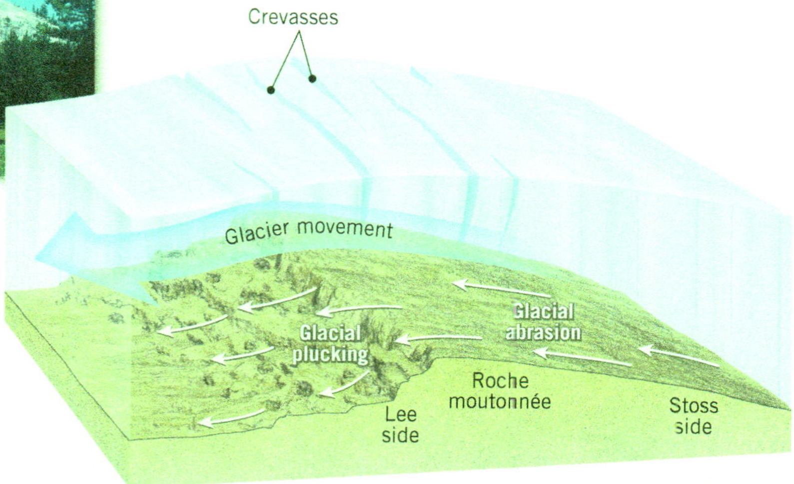

(b) Formation of roche moutonnée

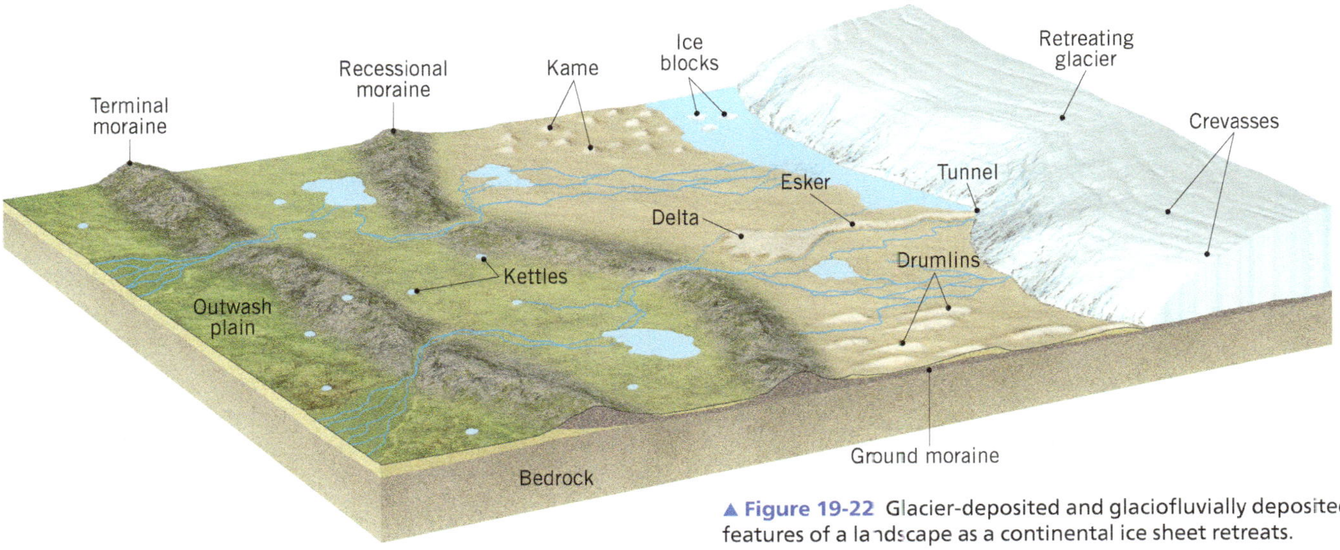

▲ **Figure 19-22** Glacier-deposited and glaciofluvially deposited features of a landscape as a continental ice sheet retreats.

Deposition by Ice Sheets

In some cases, the till transported by ice sheets is deposited heterogeneously and extensively; a veneer of unsorted debris is simply laid down over the preexisting terrain. In some places, till is deposited to a depth of several hundred meters, completely obliterating the shape of the preglacial landscape. Deposition tends to be uneven, producing an irregularly undulating surface of broad, low rises and shallow depressions. Such a surface is referred to as a *till plain*.

Moraines: In many instances, glacial deposits are laid down in more defined patterns, creating characteristic and identifiable landforms. **Moraine** is a general term for glacier-deposited landforms composed entirely or largely of till. Moraines are usually much longer than they are wide. Some moraines are distinct ridges, such as the *end moraine* that forms along the active edge of a glacier; other moraines are much more irregular in shape. Their relief is not great, varying from a few meters to a few hundred meters. When originally formed, moraines tend to have relatively smooth and gentle slopes. They become more uneven over time, as the blocks of stagnant ice included within the till eventually melt, leading to the collapse of the surface of the moraine.

Three types of moraines are commonly associated with deposition from continental ice sheets, but they are left by mountain glaciers as well (Figure 19-22). A **terminal moraine** is a ridge of till that marks the maximum advance of a glacier. It can vary in size from a conspicuous ridge tens of meters high and several kilometers wide to a low, discontinuous wall of debris (Figure 19-23). A terminal moraine forms as an end moraine when a glacier reaches its equilibrium point—it is wasting at the same rate that it is being

◄ **Figure 19-23** The maximum advance of Exit Glacier in Kenai Fjords National Park, Alaska, is marked by a prominent terminal moraine.

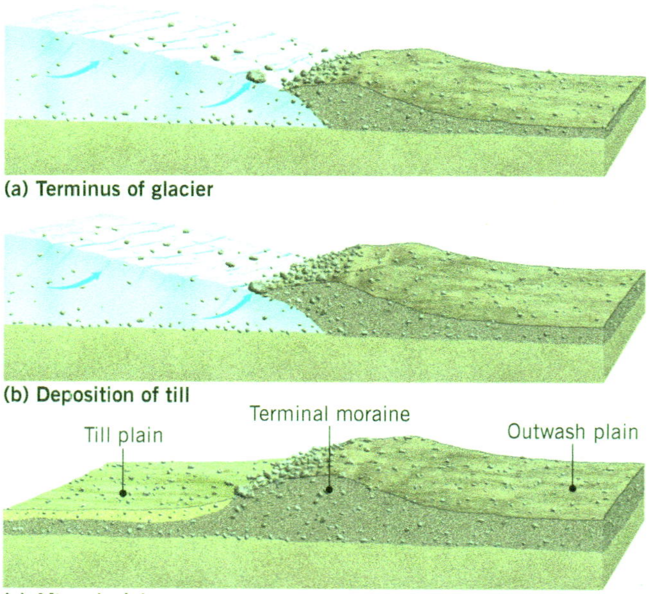

▲ **Figure 19-24** Moraine growth at the terminus of a glacier. (a) Rock may be carried within and atop the ice, emerging at the end of the glacier, (b) where it is deposited in a moraine. (c) After the ice has melted, a terminal moraine remains, with an outwash plain beyond.

nourished. Although the end of the glacier is not advancing, the interior continues to flow forward, delivering a supply of till. As the ice melts around the margin, the till is deposited, and the moraine grows (**Figure 19-24**).

Behind the terminal moraine, **recessional moraines** may develop as the glacier recedes. These are ridges that mark positions where the ice front was temporarily stabilized during the final retreat of the glacier. Both terminal and recessional moraines normally occur in the form of concave arcs that bulge outward in the direction of ice movement, indicating that the ice sheets advanced not along an even line but rather as a connecting series of great tongues of ice, each with a curved front (**Figure 19-25**).

The third type of moraine is the **ground moraine**, formed when large quantities of till are laid down from underneath the glacier rather than from its edge. A ground moraine usually means gently rolling plains across the landscape. It may be shallow or deep and often consists of low knolls and shallow *kettles*.

Kettles: Depressions known as **kettles** form when large blocks of ice left by a retreating glacier become surrounded or even covered by glacial drift. After the ice block melts, the morainal surface collapses, leaving an irregular depression (**Figure 19-26**). Today, many kettles remain filled with water as lakes—the water in the kettles is not the same water that was left by the melting ice; the ice simply left the depression that forms the basin of the lake. Walden Pond in Massachusetts is such a kettle lake.

Drumlins: Another prominent feature deposited by ice sheets is a low, elongated hill called a **drumlin** (from *druim*, an old Irish word for "ridge"). Drumlins are much smaller than moraines but composed of similarly unsorted till. The long axis of the drumlin is parallel to the direction of ice movement (**Figure 19-27**). The end of the drumlin facing the direction from which the ice came is blunt and slightly steeper than the opposite end. Thus, the configuration is the reverse of that of a roche moutonnée.

The origin of drumlins is complex, but most of them are apparently the result of ice readvance into an area of previous glacial deposition. In other words, they are depositional features subsequently shaped by erosion. Drumlins usually

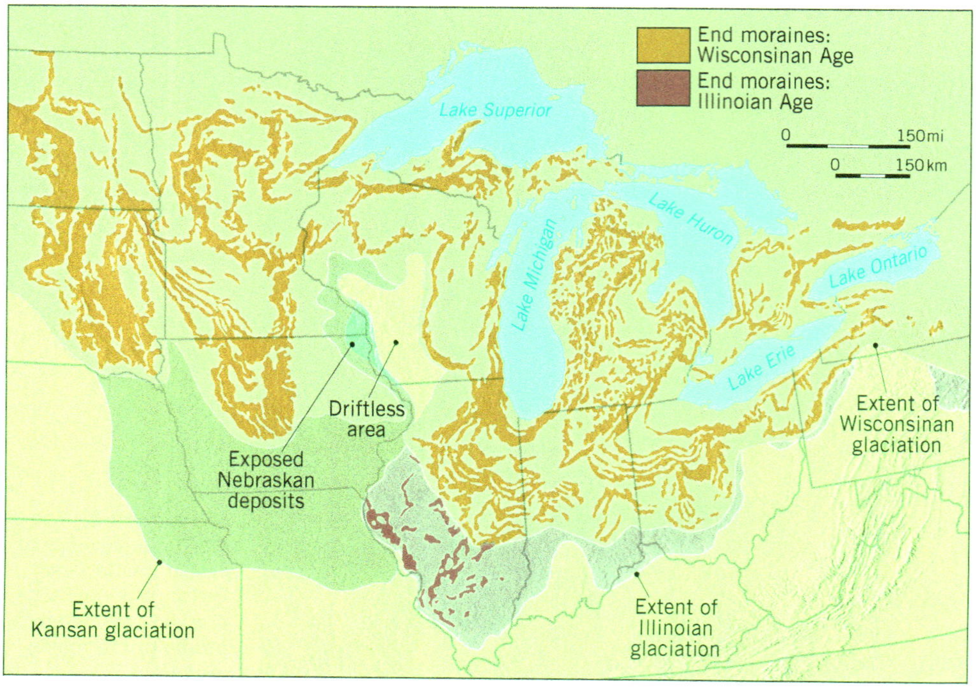

◀ **Figure 19-25** Terminal and recessional moraines (collectively called *end moraines*) near the Great Lakes in the United States. Moraines left by the more recent Wisconsin-stage glaciations are more prominent than those left by the earlier Illinoian-stage glaciations. Evidence of the older Kansan-stage glaciations is found to the southwest, but moraines left by the oldest, Nebraskan-stage glaciations were largely covered here by the three more recent glaciations. The advances and retreats of these ice sheets formed the Great Lakes.

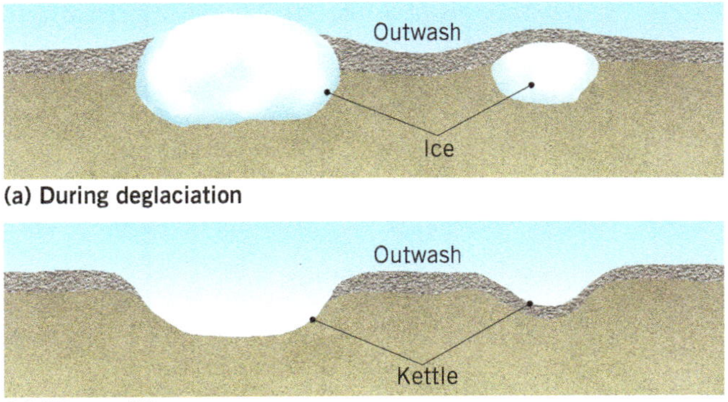

▲ **Figure 19-26** The formation of kettles. (a) During deglaciation, isolated masses of ice are often mixed in with the glacial debris and outwash and melt slowly because of the insulation provided by the surrounding debris. (b) When the ice eventually melts, sizable depressions known as kettles may pit the surface of the outwash.

occur in groups, sometimes numbering in the hundreds, with all drumlins in a group oriented parallel to each other. The greatest concentrations of drumlins in the United States are found in central New York and eastern Wisconsin.

LearningCheck 19-11 **Contrast the formation of a terminal moraine and a recessional moraine.**

Glaciofluvial Features

The deposition or redeposition of debris by ice-sheet meltwater produces certain features both where the sheet covered the ground and in the periglacial zone. These features are composed of **stratified drift**, deposits that resemble alluvium in that they may appear both layered and somewhat sorted because they were carried along by the meltwater. Glaciofluvial features are composed largely or entirely of gravel, sand, and silt because meltwater is rarely capable of moving larger material.

Outwash Plains: The most extensive glaciofluvial features are **outwash plains**, which are smooth, flat alluvial aprons deposited beyond recessional or terminal moraines by streams issuing from the ice (see Figure 19-22). Streams heavily loaded with reworked till, or with debris washed directly from the melting ice, form a braided pattern of channels across the area beyond the glacial front (see Figure 16-12). As they flow away from the ice, these braided streams, choked with debris, rapidly lose their speed and deposit their load. Such outwash deposits sometimes cover many hundreds of square kilometers. They are occasionally pitted by kettles, which often become ponds or small lakes. Beyond the outwash plain, a lengthy deposit of glaciofluvial alluvium may be confined to a valley bottom; such a deposit is termed a **valley train**.

Eskers: Less common but more conspicuous than outwash plains are long, sinuous ridges of stratified drift called **eskers** (from *eiscir*, another Irish word for "ridge"). These landforms are composed largely of glaciofluvial gravel and are thought to have originated when streams flowing through tunnels in the interior of the ice sheet became choked with debris. As the ice stagnates and melts, these subglacial streams deposit much of their load in the tunnel. Eskers are the debris that are exposed once the ice melts. They are usually a few dozen meters high, a few dozen meters wide, and up to 160 kilometers (100 miles) long (**Figure 19-28**).

Kames: Small, steep mounds or conical hills of stratified drift are found sporadically in areas of ice-sheet deposition. These *kames* appear to be of diverse origin, but they are clearly associated with meltwater deposition in stagnant ice. (*Kame* derives from *comb*, an old Scottish word referring to an elongated steep ridge.) They are mounds of poorly sorted sand

▲ **Figure 19-27** Drumlin near Dunmore, Ireland. The glacier flowed from right to left.

▲ **Figure 19-28** The sinuous ridge of an esker near Whitefish Lake, Northwest Territories, Canada.

▲ Figure 19-29 A series of kames in the Leaf Hills of Minnesota.

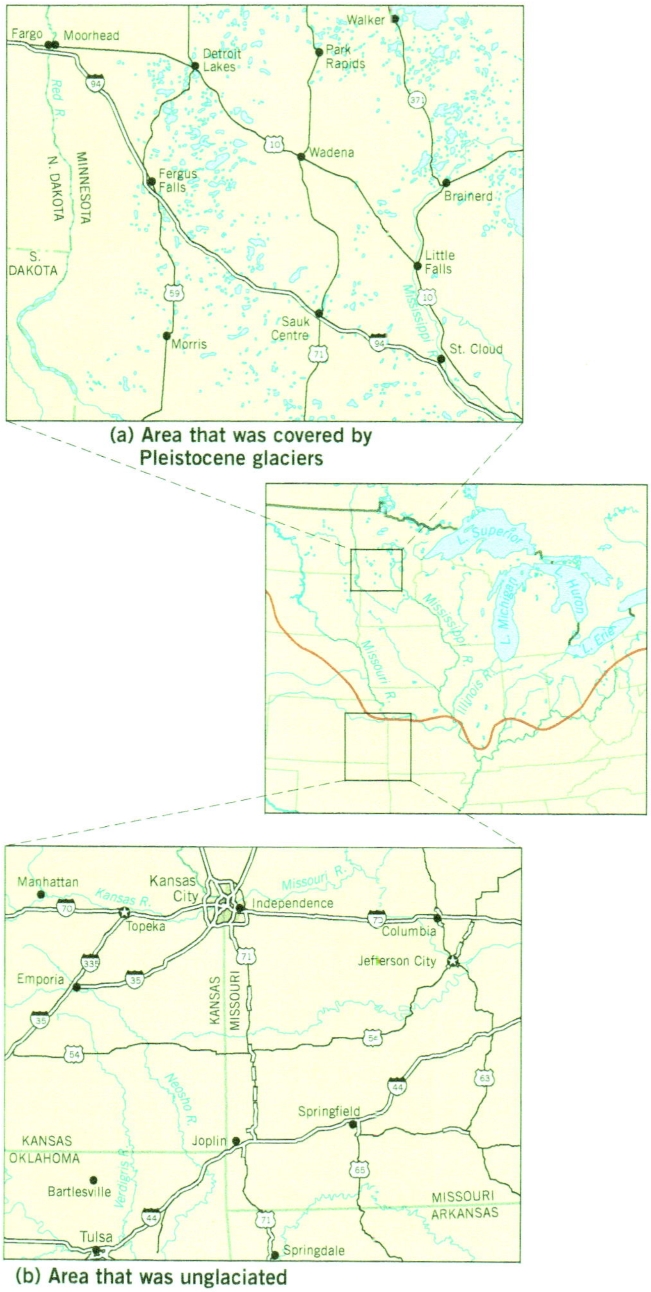

▲ Figure 19-30 (a) There is an abundance of natural lakes in the region north of the Ohio and Missouri rivers, a result of glacial action. (b) South of these rivers, however, there was no glaciation; consequently, few natural lakes exist there

and gravel that probably formed within glacial crevasses or between the glacier and the land surface (Figure 19-29). Many seem to have been built as steep fans or deltas against the edge of the ice that later collapsed partially when the ice melted. Morainal surfaces containing a number of mounds and depressions are called *kame-and-kettle topography*.

Lakes: Lakes are very common in areas that were glaciated during the Pleistocene. The old stream systems were obliterated or deranged by the ice sheets, and water remains ponded in the many erosional basins and kettles and behind morainal dams. One has only to compare the regions north and south of the Pleistocene end moraines to recognize this fact (Figure 19-30). Most of Europe and the northern part of Asia demonstrate a similar correlation between past glaciation and present-day lakes.

The Great Lakes of North America formed as a result of both glacial erosion and deposition during the Pleistocene. Their geomorphic history is complex, the result of repeated advances and retreats of ice sheets, in which earlier patterns of lakes and drainage were altered by subsequent advances and retreats. Erosion deepened the preexisting stream valleys, and deposition during glacial retreats left drift and large moraines around the margins of the present-day lakes (see Figure 19-25).

LearningCheck 19-12 **Explain the formation and characteristics of a glacial outwash plain.**

Mountain Glaciers

Most of the world's high-mountain regions experienced extensive Pleistocene glaciation. Many mountain glaciers exist today, although most are getting smaller as a consequence of a warming climate—see the box *Featured National Park: Glacier—The Shrinking Glaciers of Glacier National Park.*

Mountain glaciers do not normally reshape the terrain as completely as continental ice sheets. The effect of glacial action, particularly erosion, on mountainous topography is to create slopes that are often steeper and relief that is greater than the preglacial slope and relief (Figure 19-31). In contrast, continental ice-sheet action tends to smooth and round the terrain.

Development and Flow of Mountain Glaciers

A highland icefield can extend broadly across the high country, coating all but the uppermost mountain peaks, with outlet glaciers extending down adjacent drainage channels (Figure 19-32). Individual alpine glaciers usually form in sheltered depressions near the heads of stream valleys (often far below the peaks). Glaciers from either source flow downslope, usually along a preexisting stream valley. A system of merging glaciers typically develops, with a *trunk glacier* in the main valley joined by tributary glaciers from smaller valleys.

Featured National Park

Glacier

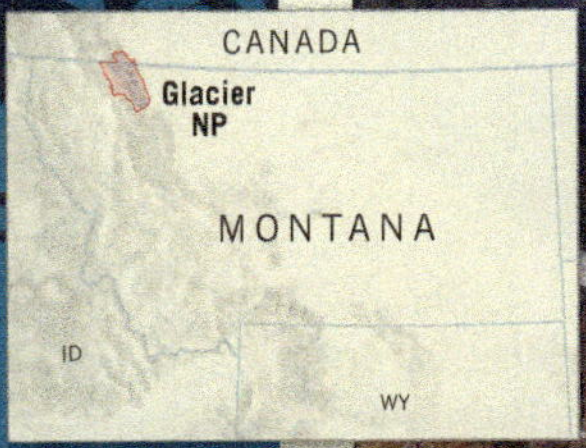

The Shrinking Glaciers of Glacier National Park

▲ **Figure 19-E** Meltwater from Grinnell, Gem, and Salamander Glaciers feeds the teal lakes of Grinnell Valley in the Many Glacier area of the Park.

Glaciers are sensitive indicators of environmental change. The size of a glacier is a delicate balance between the accumulation and ablation of ice. A relatively small increase in summer temperature or decrease in winter precipitation for a number of years may lead to the retreat of a glacier. But the relationships are sometimes counterintuitive: a glacier could advance if slightly higher temperatures lead to an increase in winter precipitation.

In places such as Glacier National Park in Montana, higher temperatures over the last century have led to a dramatic reduction in the size of its glaciers.

Glaciers in the Park: Glacier National Park had 35 named and active glaciers in the mid-1960s—down from about 146 glaciers at the peak of the Little Ice Age in the mid-1800s. By 2015 that number had shrunk to only 26 because of the warming climate. And not only did the numbers shrink, but so did the glaciers. On average, glaciers in the Park have lost 39 percent of their area between 1967 and 2015, with some losing as much as 85 percent (**Figure 19-C**).

Repeat Photography: Some of the most striking evidence of the retreat of glaciers comes from old photographs. Beginning in 1997, the U.S. Geological Survey began comparing photographs of glaciers in the Park taken in the early twentieth century with photographs taken from the exact same location more recently. To ensure that the true extent of glacial ice was shown, both in the past and today, photographs of glaciers were taken in late August or early September, after the last winter's snow covering the ice had melted away but before the first snow of autumn arrived (**Figure 19-D**). Similar repeat photography projects around the world have shown comparable loss of glaciers over the last century.

Will the Glaciers Completely Disappear? Glaciers in the Park are expected to continue shrinking in the coming decades if temperature patterns don't reverse. Small remnants of a few glaciers may persist in the Park through the end of this century or even beyond, but uncertainty surrounds those projections. In 2019, a global study of the glaciers in UNESCO World Heritage sites, such as Glacier National Park (part of the joint Waterton–Glacier International Peace Park of the United States and Canada), projected an ice mass loss of 33 percent to 60 percent by the year 2100 compared with 2017. Unless we can stop or at least slow the current trends, it is indeed possible that Glacier National Park may lose all of its namesakes.

▲ **Figure 19-C** The Jackson Glacier in Glacier National Park in 1911 and 2009.

▲ **Figure 19-D** The Shepard Glacier in Glacier National Park in 1911 and 2005.

Questions

1. How could a time-series photograph comparison, such as in Figure 19-D, be used to estimate the volume of ice lost as a glacier retreats?
2. Other than higher summer temperatures causing increased melting, what other factors could influence the rate of retreat of a glacier?

NPS Video
Glacial Change at Kenai Fjords National Park
https://media.pearsoncmg.com/ph/esm/esm_mcknight_physgeo_13/media/

▲ **Figure 19-31** The Teton Range of Wyoming has been sculpted in part by the action of mountain glaciers. The Teton Glacier is a small cirque glacier.

▲ **Figure 19-32** Valley glaciers extend down from the highland icefield on top of Mt. Rainier, the highest volcano in the Cascade Range in Washington.

Erosion by Mountain Glaciers

The erosion and transportation of rock debris by both highland icefields and alpine glaciers can reshape the topography of mountains in dramatic fashion. These processes remodel the peaks and ridges and also transform the valleys that lead down from the high country (Figure 19-33).

Cirques: The basic landform feature in glaciated mountains is the **cirque**, a broad amphitheater hollowed out at the uppermost head (top) of a glacial valley (Figure 19-34). It has very steep, often perpendicular, head and side walls and a floor that is flat, gently sloping, or deepened enough to form a basin. A cirque marks the place where an alpine glacier originated. It is the first landform feature produced by alpine

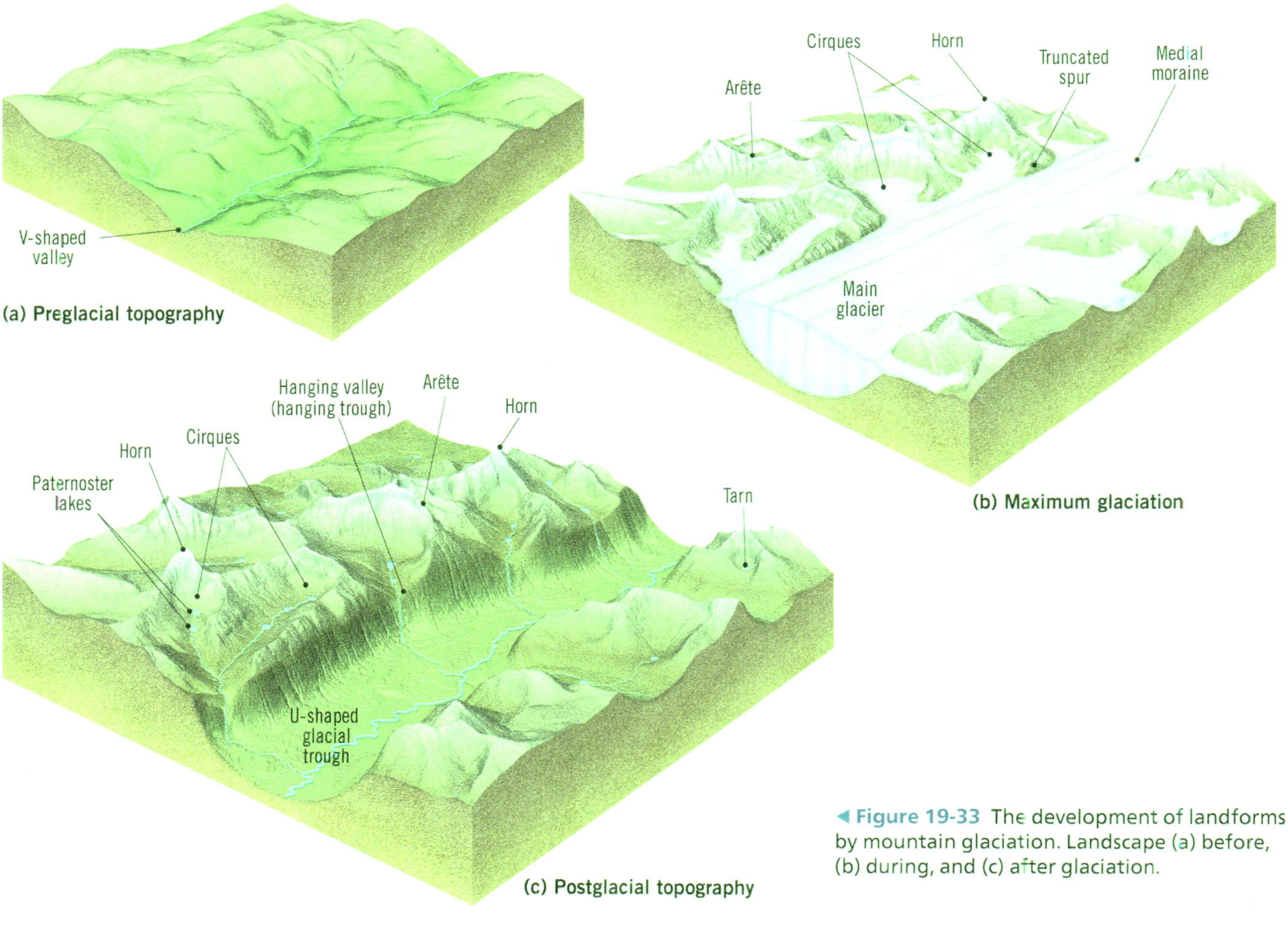

◀ **Figure 19-33** The development of landforms by mountain glaciation. Landscape (a) before, (b) during, and (c) after glaciation.

glaciation, essentially being quarried out of the mountainside by plucking, mass wasting, and frost wedging.

By the middle of summer, a large crevasse known as a *bergschrund* opens at the top of the glacier, exposing part of the headwall to frost wedging. The shattered rock from the headwall is eventually incorporated into the glacier. As the glacier grows and begins to extend down-valley out of its cirque, quarried fragments from the cirque are carried away with the flowing ice (Figure 19-35). Cirques range from a few hectares to a few square kilometers in extent. Many large cirques apparently develop from repeated episodes of glaciation.

When the glacial ice in a cirque has melted away, enough of a depression to hold water may be left behind. Such a cirque lake is called a **tarn**.

Arêtes and Cols: A cirque grows steeper, in part through frost wedging, as its glacier plucks rock from the head and side walls. Where cirques are close together, the upland interfluve between neighboring cirques is reduced to little more than a steep rock wall. Where several cirques have been cut back into an interfluve from opposite sides of a divide, a narrow, jagged, serrated spine of rock may be all that is left of the ridge crest. This ridge is called an **arête** (French for "fishbone"; derived from the Latin *arista*, meaning "spine"). If two adjacent cirques on opposite sides of a divide cut back enough to remove part of the arête between them, they form a **col** (*collum* is Latin for "neck"), a sharp-edged pass or saddle through the ridge (Figure 19-36). (One of the most popular climbing routes to the summit of Mount Everest is along the South Col, a narrow ridge between Everest and Lhotse, its neighboring peak.)

Horns: An even more prominent feature of glaciated highland summits is a **horn**, a steep-sided, pyramid-shaped mountain peak formed by expansive quarrying of the head walls where three or more cirques intersect (Figure 19-37).

▲ **Figure 19-34** Each of three small glaciers begins in a cirque below a steep peak in Nahanni National Park, Northwest Territories, Canada.

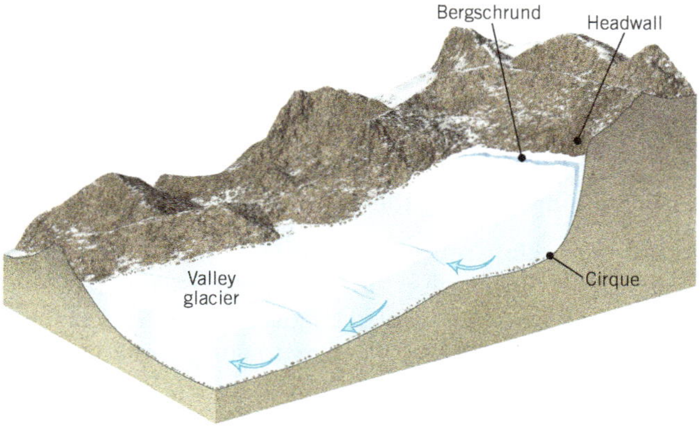

▲ **Figure 19-35** The development of a cirque at the head of a valley glacier.

▲ **Figure 19-36** A glaciated section of Rocky Mountain National Park, in the Front Range in north-central Colorado. A col is seen between the two sharp peaks.

▲ **Figure 19-37** Flinsch Peak in Glacier National Park, Montana, is a prominent glacial horn.

The name is derived from Switzerland's Matterhorn, the most famous example of such a glaciated spire.

LearningCheck 19-13 What is a cirque, and how does one form?

Glacial Troughs: Some alpine glaciers do little more than erode cirques, presumably because of insufficient accumulation of ice to force a down-valley movement. As a result, the valleys below are unmodified by glacial erosion. Most alpine glaciers, however, flow down preexisting valleys and reshape them (see Figure 19-33b).

When a glacier moves down a mountain valley, it often does so with greater erosive effectiveness than a stream. The glacier is thicker and capable of eroding through both abrasion and plucking. The lower layers of ice can even flow uphill for some distance if blocked by resistant rock on the valley floor, permitting rock fragments to be dragged out of depressions in the valley floor.

An alpine glacier can deepen, steepen, and widen its valley. Abrasion and plucking take place not only on the valley floor but also along the sides. The cross-sectional profile is often changed from a stream-cut "V" shape to an ice-eroded "U" shape flared at the top (see Figure 19-33c). Moreover, the general course of the valley may be straightened somewhat because the ice does not meander like a stream. Rather, it grinds away the protruding spurs that separate side canyons, creating what are called *truncated spurs*, and thereby replacing the sinuous course of the stream with a somewhat straighter, U-shaped **glacial trough** (Figure 19-38). Not all glacial valleys are U-shaped, however; the resistance of the bedrock, jointing patterns, and processes such as exfoliation and rockfall also influence the final shape of a glacial trough.

▲ Figure 19-38 Gunsight Lake in Glacier National Park, Montana, lies in a conspicuous U-shaped glacial trough.

Glacial Steps: As we might expect, glacial erosion along the floor of a glacial trough does not always produce a smooth surface. Because of differential erosion, resistant rock on the valley floor is gouged less deeply than weaker or more fractured rock. As a result, the down-valley profile of a glacial trough is often marked by an irregular series of rock steps or benches, separated by steep (although usually short) cliffs on the down-valley side. Such landforms are known as **glacial steps** (Figure 19-39).

A postglacial stream that flows out of a cirque and down a glacial trough usually has a relatively straight course but typically includes rapids and waterfalls—particularly down the cliffs that mark the series of glacial steps. Small lakes may remain in shallow depressions on the benches of the glacial steps, forming a sequence called **paternoster lakes**, named after a fancied resemblance to beads on a rosary.

As we discuss in Chapter 20, some of the most spectacular glacial troughs occur along coastlines where valleys have been partly drowned by the sea, creating *fjords*.

Hanging Glacial Troughs: Just as large streams may have tributaries, valley glaciers also may be fed by smaller tributary glaciers. When occupied by glaciers and thus covered by a relatively level field of ice, main and tributary valleys may appear equally deep. When the ice melts, however, the valleys are of different depths because the erosive effectiveness of a glacier is determined largely by its volume of ice: tributary glaciers cannot erode as deeply as the larger valley glaciers. The mouths of the tributary valleys are characteristically perched high along the sides of the major troughs, forming **hanging valleys**—or, more properly, **hanging troughs** (see Figure 19-33c). Typically, streams that drain the tributary valleys plunge over waterfalls to the floor of the main trough. Several of the world-famous falls of Yosemite National Park are of this type. See the box *Featured National Park: Yosemite—Our Most Famous Glacial Valley.*

LearningCheck 19-14 In what ways is a glacier likely to alter the former stream valley through which it flows?

▼ Figure 19-39 A sequence of glacial steps, shown in a longitudinal cross section of a glaciated valley in hilly or mountainous terrain.

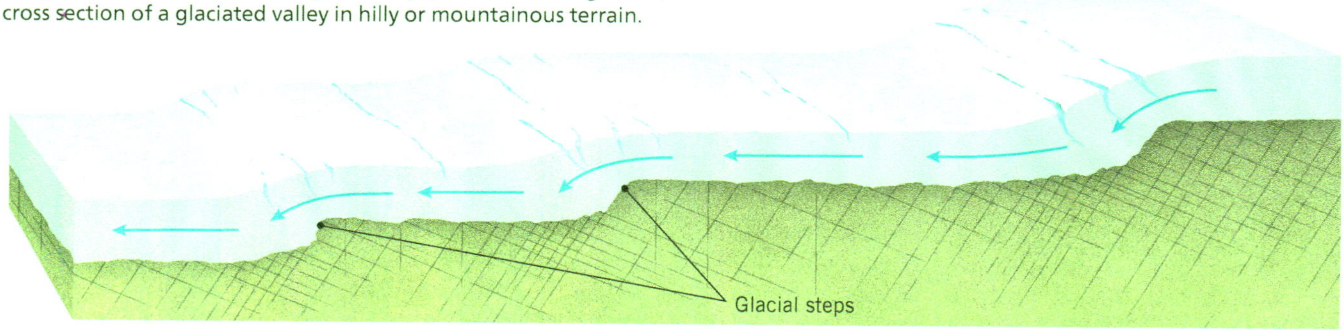

Featured National Park

Yosemite

Our Most Famous Glacial Valley

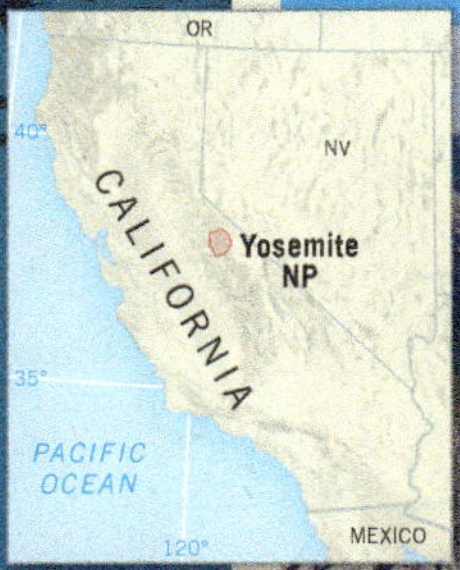

▲ **Figure 19-G** Half Dome and Little Yosemite Valley with Vernal Fall and Nevada Fall.

Known for its steep granite walls, high waterfalls, and iconic landforms such as Half Dome, Yosemite Valley in California is perhaps the most famous glacial valley in the world. Although nearly all visitors know that glaciers flowed through Yosemite, helping sculpt this breath-taking landscape, it is less well known that Yosemite is by no means a "classic" or typical glacial valley—its walls are steeper and its floor is flatter than most glacial troughs (**Figure 19-F**).

Valley Walls: The near vertical walls of the valley are a consequence of its resistant, massively jointed granite. Streams originally carved the valley, and during the Pleistocene, glaciers deepened the valley somewhat. However, some geomorphologists think that glaciers didn't widen the valley a great deal. The granite was so strong and the joints so widely spaced—especially where the valley narrows between El Capitan and the Cathedral Rocks—that glacial erosion simply wasn't very effective. The gentle slopes of "The Rockslides" are from the accumulation of talus from rockfall in weaker rock.

Valley Floor: The nearly flat valley floor is due to deposition of glacial outwash and alluvium after the glaciers receded. The valley is filled with 600 meters (2000 feet) of sediment in some places, forming the current floodplain of the Merced River.

Waterfalls: The Park's most famous waterfalls, Yosemite Falls and Bridalveil Fall, drop from hanging valleys left by smaller tributary glaciers. Up in Little Yosemite Valley, Vernal Fall and Nevada Fall plunge off glacial steps where the glaciers encountered resistant bedrock on the valley bottom (**Figure 19-G**).

Half Dome: Half Dome was not cut in two by glaciers as many tourists believe. Although glaciers flowed past Half Dome as they moved down Tenaya Canyon toward the main Yosemite Valley, ice probably never reached more than 210 meters (700 feet) below the top of the dome—even at the peak of the Pleistocene glaciations, the top of Half Dome remained above the ice. The vertical face of Half Dome is due to a prominent vertical joint running through in the granite here. The rounded back side of Half Dome is the result of exfoliation—the same process that left granite domes throughout the Sierra Nevada (see Figure 15-8).

Questions

1. What explains the steepness of Yosemite Valley's walls?
2. Why is the floor of Yosemite Valley nearly flat and not U-shaped?

Mobile Field Trip
Yosemite
https://media.pearsoncmg.com/ph/esm/esm_mcknight_physgeo_13/media/

NPS Video
Experience Your Yosemite
https://media.pearsoncmg.com/ph/esm/esm_mcknight_physgeo_13/media/

▲ **Figure 19-F** Yosemite Valley was glaciated, but in some ways it is not a typical glacial valley.

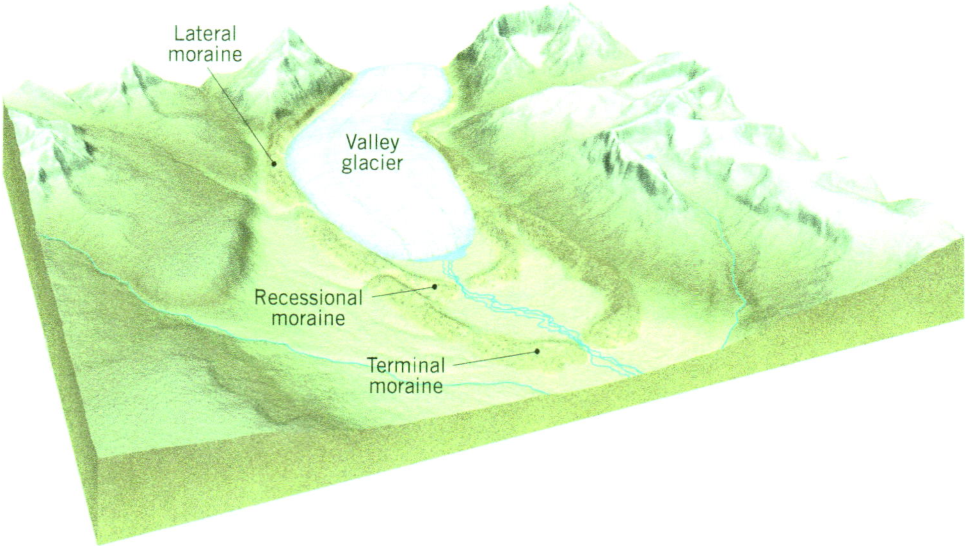

▶ Figure 19-40 Common types of moraines in mountainous areas.

Deposition by Mountain Glaciers

Depositional features are less significant in areas of mountain glaciation than in areas where continental ice sheets have been at work. The high country is often devoid of till; only in the middle and lower courses of glacial valleys does much deposition occur.

The principal depositional landforms associated with mountain glaciation are moraines. Terminal and recessional moraines form just as they do with ice sheets. Moraines resulting from mountain glaciation are much smaller and less conspicuous, however, because they are restricted to glacial troughs (Figure 19-40).

Lateral Moraines: The largest depositional features produced by mountain glaciation are often **lateral moraines**: well-defined ridges of till built up along the sides of valley glaciers (Figure 19-41). The debris is partly material deposited by the glacier and partly rock that falls or is washed down the valley walls (Figure 19-42).

Where a tributary glacier joins a main valley glacier, their lateral moraines (and debris carried on top of the ice along the sides of the glaciers) unite at the intersection. They commonly continue together down the middle of the combined glacier as a dark band of rocky debris known as a **medial moraine** (Figure 19-43). Medial moraines are sometimes

◀ Figure 19-41 The Athabasca valley glacier issuing from a highland icefield in the Canadian Rockies. The distance from the highest icefall to the toe is about 5 kilometers (3 miles). The glacier has conspicuous lateral moraines along both sides.

▲ **Figure 19-42** Emerald Bay on the California side of Lake Tahoe is nearly closed off from the rest of the lake by the arms of two lateral moraines left by Pleistocene glaciers.

found in groups of three or four running together, indicating that several glaciers have joined and produced a striped effect of black (moraine) and white (ice) bands that extend down the valley.

The debris left by meltwater below mountain glaciers is similar to that bordering ice sheets because similar outwash is produced.

LearningCheck 19-15 **What is a lateral moraine, and how does one form?**

The Periglacial Environment

The term *periglacial* means "on the perimeter of glaciation." More than 20 percent of the world's land area is presently periglacial, but most of this was covered by ice on one or more occasions during the Pleistocene epoch.

Periglacial lands are found either in high latitudes or at high elevations. Almost all are in the Northern Hemisphere. Southern Hemisphere continents do not extend far enough into the high latitudes to be significantly affected by contemporary glaciation (Africa, South America, Australia) or are mostly ice covered (Antarctica).

Nonglacial land-shaping processes function in periglacial areas, but in addition the pervasive cold imparts some distinctive characteristics. The most notable is *permafrost* (discussed in Chapter 9). Whether continuous or discontinuous, permafrost occurs over most of Alaska and more than half of Canada and Russia. There are also extensive high-altitude areas of permafrost in Asia, Scandinavia, and the western United States. In some cases, the frozen ground extends to extraordinary depths; a thickness of 1000 meters (3000 feet) has been found in Canada's Northwest Territories and 1500 meters (4500 feet) in north-central Siberia.

Patterned Ground

The most eye-catching periglacial terrain is **patterned ground**, the generic name applied to various geometric patterns that repeatedly appear over large areas in the

◀ **Figure 19-43** Prominent medial moraines on the Barnard Glacier in Wrangell–St. Elias National Park, Alaska.

▲ **Figure 19-44** Ice-wedge polygons near Prudhoe Bay, Alaska, formed by the slow growth of ice wedges in cracks below the surface of the tundra.

▲ **Figure 19-45** The channeled scablands of central Washington below Pothole Reservoir were scoured by a series of enormous floods during the Pleistocene.

Arctic (**Figure 19-44**). The patterns vary; some, related to the freeze–thaw cycle that slowly disrupts uniform surfaces of soil and regolith, form rough polygons. In other cases, ice wedges—formed over many centuries by ice expansion in deep cracks in the tundra ground—develop large polygonal patterns.

Proglacial Lakes

Another sometimes conspicuous development in periglacial regions is a **proglacial lake** (here, *pro* means "marginal to" or "in advance of"). Where ice flows across a land surface, the natural drainage is impeded or blocked, and meltwater from the ice can become dammed by the ice front, forming a proglacial lake. Such an event sometimes occurs in alpine glaciation but is much more common along the margin of continental ice sheets, particularly when the ice stagnates. Most proglacial lakes are small and quite temporary because subsequent ice movements cause drainage changes and because normal fluvial processes, accelerated by the growing accumulation of meltwater in the lakes, cut spillways or channels to drain the impounded waters. Some proglacial lakes, however, are large and relatively long lived. Such major lakes are characterized by considerable fluctuations in size due to the changing location of the receding or advancing ice front. Several huge proglacial lakes were impounded along the margins of the ice sheets as they advanced and retreated during the Pleistocene Epoch in North America, Europe, and Siberia.

Washington's Channeled Scablands: Among the most dramatic consequences of lakes once impounded by continental glaciers is a landscape found in eastern Washington near Spokane. A series of great floods during the Pleistocene scoured eastern Washington's "channeled scablands" when enormous volumes of water were periodically let loose from an ice sheet–dammed lake near present-day Missoula, Montana. Dozens of times during the Pleistocene, a lobe of the continental ice sheets blocked the Clark Fork River, forming Pleistocene Lake Missoula, which at times was more than 600 meters (2000 feet) deep. When the ice dam failed, a wall of water surged across the landscape, scouring deep channels and leaving enormous ripple marks before the water eventually drained into the Columbia River (**Figure 19-45**).

LearningCheck 19-16 **How do proglacial lakes form?**

Causes of the Pleistocene Glaciations

Ice ages are fascinating not only because of the landscape changes they bring about but also because of the questions they raise. What initiates massive accumulations of ice on the continental surfaces, stimulates their advances and retreats, and finally causes them to disappear? Scientists have pondered these questions for decades. Any complete theory of the causes of the Pleistocene glaciations must be able to account for four main glacial characteristics:

1. The accumulation of ice masses more or less simultaneously at various latitudes in both hemispheres but without uniformity (e.g., much less in Siberia and Alaska than in similar latitudes in Canada and Scandinavia)
2. The apparently concurrent development of pluvial conditions in dry land areas

3. Multiple cycles of ice advance and retreat, including both minor fluctuations over decades and centuries and major glaciations and deglaciations over tens of thousands of years
4. Eventual total (or near total) deglaciation

We know that glaciers grow when there is a net accumulation of snow over a period of time and that glaciers waste away when summer melting exceeds winter snowfall. Beyond that simplistic statement, however, in some cases it is not always clear whether a colder climate would be more conducive to glaciation than a warmer but wetter one! Although colder conditions would inhibit summer wastage and thus enhance the longevity of the winter accumulation, cold air cannot contain much water vapor. Hence, warmer winters would favor increased snowfall, whereas cooler summers are needed for decreased melting. Even a theory that accommodated significantly increased snowfall or significantly decreased melting—or a combination of both—would have to take into account the timing of glacial advances and retreats.

In recent years, oxygen isotope analysis has dramatically enhanced our understanding of the sequence of glacial and interglacial episodes during the Pleistocene. As we saw in Chapter 8, the ratio of $^{18}O/^{16}O$ derived from carbonate sediments on the ocean floor and from ice cores in glaciers can be used as a proxy for temperature. The climate record that emerged from these studies shows a sequence of major glacial and interglacial episodes, along with a great many minor advances—some of which ended quite abruptly (**Figure 19-46**).

Climate Factors and the Pleistocene

A lot is known about many of the factors associated with the Pleistocene glaciations—but certainly not *all* of the factors.

Milankovitch Cycles: As discussed in Chapter 8, cyclical variations in Earth–Sun relations, referred to as *Milankovitch cycles*, played a part. The combination of slight variations in the inclination of Earth's axis and the eccentricity of Earth's orbit, as well as the changing orientation of Earth's axis relative to the stars (known as the "precession of the equinoxes"), correlates fairly well with some—but not all—of the major glacial advances and retreats during the Pleistocene.

Animation Orbital Variations and Climate Change
https://media.pearsoncmg.com/ph/esm/esm_mcknight_physgeo_13/media/

Other Climate Factors: Other factors that have been postulated as triggers or components of glacial episodes include variations in the energy output of the Sun, variations in the level of carbon dioxide and other greenhouse gases in the atmosphere, the changing positions of continents and the configuration of ocean basins and ocean circulation patterns, atmospheric circulation changes due to increased elevation of continental masses after a period of tectonic uplift, and reductions in insolation reaching the surface due to particulates released during massive volcanic eruptions.

We make no attempt here to offer details of these many hypotheses, primarily because none of them alone or in combination yet fully explains the climate changes during the Pleistocene. The search for a complete explanation continues.

LearningCheck 19-17 **What factors other than Milankovitch cycles might have contributed to Pleistocene glaciations?**

Are We Still in an Ice Age?

An intriguing question about the Pleistocene remains: has Earth's recent ice age actually ended? Are we now living in a postglacial period or in an interglacial period? Is the cycle of glaciation and deglaciation that marked the Pleistocene really over, or have the glaciers merely receded temporarily? Based on the climate patterns of the last two and a half million years, it is possible that Earth will enter another period of glaciation within several tens of thousands of years. However, the possibility also exists that the human-enhanced greenhouse effect and global warming will "postpone" the onset of another glacial period.

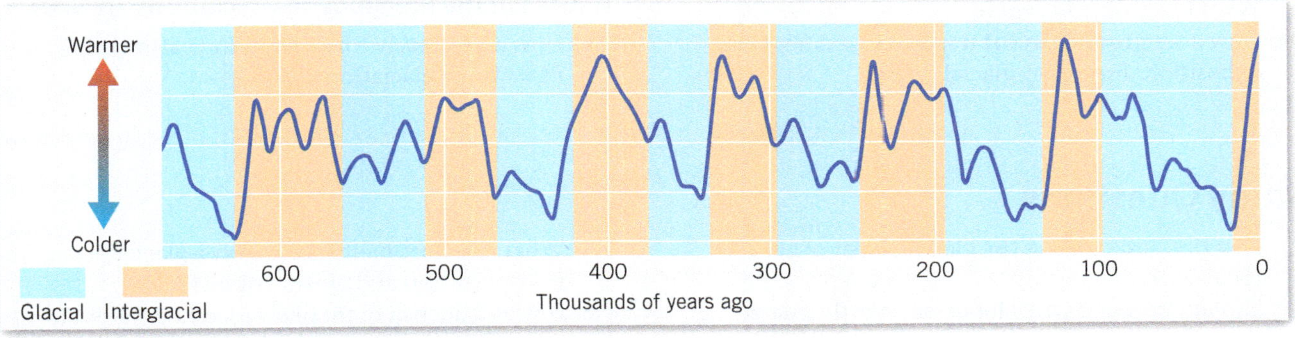

▲ **Figure 19-46** Global temperature fluctuations from 650,000 years ago to the present.

19 Learning Review

After studying this chapter, you should be able to answer the following questions. Key terms from each text section are shown in **bold type.** Definitions for key terms are also found in the glossary at the back of the book.

Key Terms and Concepts

Types of Glaciers (p. 546)

1. Describe and contrast the different kinds of **glaciers**: **continental ice sheets**, **highland icefields**, **alpine glaciers**, **valley glaciers**, and **cirque glaciers**.

Glaciations Past and Present (p. 547)

2. Why is the **Pleistocene epoch** so important to physical geography, whereas other ice ages are not?
3. Briefly describe the worldwide extent of ice cover during the peak of the Pleistocene glaciations.
4. What is a **periglacial zone**?
5. What is the relationship of large continental ice sheets to the crustal depression associated with isostatic adjustment (isostasy)?
6. What were the **pluvial effects** of the Pleistocene?
7. Describe and explain the formation of large **Pleistocene lakes** in western North America.

Glacier Formation and Movement (p. 552)

8. Describe and contrast the processes of glacial ice **accumulation** and **ablation**.
9. Describe the transformation of snow to **névé (firn)** to glacial ice.
10. What is the relationship of the **equilibrium line** of a glacier to its **accumulation zone** and **ablation zone**?
11. Discuss the different components of glacial movement: **plastic flow of ice** and **basal slip**.

The Effects of Glaciers (p. 555)

12. Contrast the erosional processes of **glacial plucking** and glacial abrasion.
13. What is **glacial flour**? Glacial **drift**?
14. Describe the characteristics of glacial **till**.
15. What is a **glacial erratic**?
16. How does **glaciofluvial deposition** differ from deposition directly by glacial ice?

Continental Ice Sheets (p. 557)

17. Describe and explain the formation of a **roche moutonnée**.
18. A **moraine** is made from what kind of glacial material?
19. Explain the formation of **terminal moraines** and **recessional moraines**.
20. Describe the general appearance of **ground moraine**.
21. Describe and explain the formation of a **kettle**.
22. Describe and explain the formation and orientation of a **drumlin**.
23. What is **stratified drift**?
24. Describe the formation of a glacial **outwash plain** and a **valley train**.
25. How does an **esker** form?

Mountain Glaciers (p. 562)

26. Describe and explain the formation of **cirques**, **horns**, **arêtes**, and **cols**.
27. Where in a glaciated landscape is a **tarn** found?
28. Contrast the general cross-sectional shape of a typical stream valley in the high mountains with that of a typical **glacial trough** (glacial valley).
29. Describe the general down-valley profile of a glacial trough and the formation of **glacial steps**.
30. What are **paternoster lakes**, and where do they form?
31. Explain the formation of a **hanging valley (hanging trough)**.
32. What is a **lateral moraine**?
33. How does a **medial moraine** form?

The Periglacial Environment (p. 569)

34. Briefly describe **patterned ground** in periglacial areas.
35. What is a **proglacial lake**?

Causes of the Pleistocene Glaciations (p. 570)

36. How do *Milankovitch cycles* fit the pattern of Pleistocene glaciations?

Study Questions

1. Describe and contrast the global extent of glaciers today with that during the Pleistocene.
2. Explain how and why global sea level fluctuated during the Pleistocene.
3. Explain how the balance between ice accumulation and ablation influences the advance or retreat of a glacier.
4. Why can a glacier continue to erode and transport rocks even while it is retreating?
5. What role does meltwater below a glacier play in the transportation and erosion of rock?
6. How is a deposit of till likely to look different from a deposit of alluvium?
7. What accounts for the difference in the quantity and quality of soil found in central Canada and the Great Plains near the United States–Canada border?

8. Why are there so many lakes in areas that were glaciated by continental ice sheets during the Pleistocene?
9. Why are most mountainous areas that have experienced glaciation quite rugged?
10. What is the relationship of cirques to the formation of arêtes and horns?
11. Why might the lateral moraines left by a Pleistocene mountain glacier be more prominent today than the terminal or recessional moraines?
12. Why is it difficult to know whether the Pleistocene glaciations are over?

Exercises

1. Assume that on the surface in the center of a mountain glacier, the long-term average speed of ice movement is 1 meter per day. How long will it take a rock that has fallen onto the glacier to travel 3 kilometers to the glacier's terminus? _____ days
2. The Grinnell Glacier in Glacier National Park had a surface area of 1,206,000 square meters in 1966. Its area was 615,454 square meters in 2005. The 2005 surface area is what percentage of the 1966 surface area? _____ percent
3. The Helheim glacier in Greenland enters the ocean in a 6.3-kilometer-wide fjord. The terminus of the glacier retreated 7.5 kilometers between May 2001 and June 2005. How many square kilometers of glacier surface area in the fjord were lost between 2001 and 2005? _____ square kilometers

Environmental **Analysis** GLACIAL FLOW PATTERNS

Data
Glacial Flow
https://media.pearsoncmg.com/ph/esm/esm_mcknight_physgeo_13/media/

Glacial ice flows much like water—only very slowly. The flow patterns are determined by snow accumulation and the local terrain.

Activities

Go to NASA's Earth Observatory at http://earthobservatory.nasa.gov, *search for "Columbia Glacier, Alaska" and select the "World of Change: Columbia Glacier, Alaska" link. As you step forward in time, notice that sometimes the (blue) ice has flow streaks and sometimes it does not. The flowing part is the glacier; the nonflowing part is chunks of calved icebergs that have rafted together.*

1. In 1986, what is the distance between the beginning of the medial moraine (where the glaciers first meet) and the underwater terminal moraine? Between the top of the medial moraine and the terminus?
2. What role does the underwater terminal moraine play in ice accumulation downstream of the terminus?
3. Why is there no ice accumulation downstream of the terminus in some years?
4. How far has the terminus retreated in 1996? 2006? 2014?

Go to http://cdn.antarcticglaciers.org. Select "Antarctica," then "Antarctic datasets." Open the "BEDMAP 2" map in a new window to see a compilation of bedrock topography data for Antarctica.

5. Where are the highest elevations located? The lowest elevations?

Return to the "Antarctic datasets" page and open the "Ice velocity" map in a new window.

6. What controls the direction in which the ice streams flow?
7. Where are the fastest ice velocities located? Why?

Return to "Antarctic datasets" page and open the "Landsat Image Mosaic of Antarctica" map in a new window.

8. Name the ice sheets with the fastest ice velocities.

Seeing Geographically in Our National Parks

Look again at the photograph of the Nabesna Glacier at the beginning of the chapter (p. 544). What kind of glacial landform is developing along the sides of the glacier in the foreground? What comprises that landform? What are the dark lines running down the middle of the glacier? How do they form? What indicates that this glacier has retreated in recent decades?

Mastering Geography®

Looking for additional review and test prep materials? Visit the Study Area in **Mastering Geography** to enhance your geographic literacy, spatial reasoning skills, and understanding of this chapter's content by accessing a variety of resources, including Map Master interactive maps, geoscience animations, *Mobile Field Trips*, videos, *Project Condor* Quadcopter videos, *In the News* RSS feeds, flashcards, web links, self-study quizzes, and an eText version of *McKnight's Physical Geography*.

Seeing Geographically in Our National Parks

The coastline of Acadia National Park in Maine. Does it appear to be high tide or low tide? Why do you say this? Is the rock in the cliffs here relatively resistant or easily eroded by waves? What do you see that supports your answer? Does the sediment on the small beach consist of very fine sand or larger gravel-size particles? What might explain this?

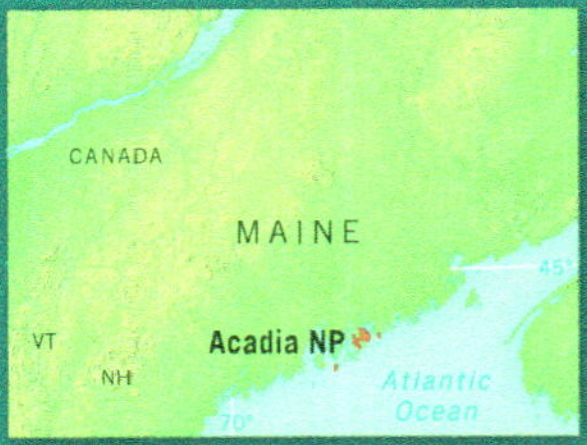

Acadia National Park, ME

Coastal Processes and Terrain

20

HAVE YOU EVER WONDERED how a beach forms? Although you might think that the sediment on this small beach in Acadia National Park in Maine was eroded from the rocky cliffs nearby, it's rarely that simple. Much of the sand on many beaches has come from far away. And unlike many of the landscapes we've studied so far, where change usually happens slowly or infrequently, beaches can change quickly. Beaches can be changing even when they appear to be staying the same!

Coastlines are among the most varied of all landscapes. This diversity is a result of the great variety of local geology, climate, and geomorphic processes at work. In addition to virtually all of the internal and external processes we've discussed so far, coastlines are shaped by waves and local ocean currents. The result is that coastal terrain is often quite different from the landscape just a short distance inland.

The coastal environment often shows changes in appearance and process from day to day or even hour to hour. Part of this dynamism is a consequence of shorelines acting as an interface of the lithosphere, hydrosphere, and atmosphere—and quite often the cryosphere and biosphere as well. As we saw in the chapters on weather and climate, bodies of water—especially the oceans—are components of a number of key Earth systems. We have seen the ocean's role of regulating the temperature of Earth, but in addition to that, some of the energy received from the Sun ultimately powers wind that creates ocean currents and waves that, as we see in this chapter, can in turn shape coastal landforms.

As you study this chapter, think about these **Key**Questions:

- **How are waves generated, and how do waves affect shorelines?**
- **How do changes in sea level, coastal deposition, and coastal transport influence shorelines?**
- **Why can beaches and spits change size and shape quickly?**
- **What coastal features are seen with shorelines of submergence and emergence?**
- **How do coral features such as fringing reefs, barrier reefs, and atolls develop?**

The Impact of Waves and Currents on the Landscape

Coastal processes affect only a tiny fraction of the total area of Earth's surface, but they create a landscape that is almost completely different from any other on the planet. Along shorelines, waves are agents of erosion, often forming rocky cliffs and headlands. Coastal currents are agents of transportation and deposition, frequently leaving dynamic but impermanent landscape features such as beaches at the interface between land and water.

Coastal Processes

The coastlines of the world's oceans and lakes extend for hundreds of thousands of kilometers. Every conceivable variety of structure, relief, and topography can be found somewhere along these coasts. Coastlines are a dynamic and highly energetic environment, primarily because of the restless motions of the waters (Figure 20-1).

The Role of Wind in Coastal Processes

We saw in Chapter 18 that wind can shape landforms on the continents, especially through the deposition of sand. Along coastlines, the wind has an even greater influence on topography: wind blowing over the surface of a body of water generates waves and ocean currents.

Wind is not the only force causing water to move. Oceanic coastlines also experience daily tidal fluctuations that often move enormous quantities of water. Tectonic events, particularly earthquakes, contribute to water motion, as can volcanic activity. More fundamental are long-term variations in sea or lake level caused by tectonic forces and *eustatic sea-level change* (sea-level change due to an increase or decrease in the amount of water in the world ocean or a change in the volume of an ocean basin). However, in terms of geomorphic effects, wind is the most important cause of waves and currents.

▼ **Figure 20-1** Hydrosphere, lithosphere, and atmosphere meet at coastlines, interfaces of ceaseless movement and energy transfer. Pounding waves may not erode evenly. Sea stacks and cliffs are results of differential erosion at Rialto Beach, Olympic National Park, Washington.

Coastlines of Oceans and Lakes

The processes that shape the topography of oceanic coastlines are similar to the processes acting on lakeshores, with three important exceptions:

1. Along lakeshores, the range of tides is so small that they are insignificant to landform development.
2. The causes of sea-level fluctuations are quite different from the causes of lake-level fluctuations.
3. Coral reefs are built in tropical and subtropical oceans only, not in lakes.

With these exceptions, the topographic forms produced on seacoasts and lakeshores are generally similar. Even so, the larger the body of water, the greater the effects of coastal processes. Thus, topographic features along seacoasts are normally larger, more conspicuous, and more distinctive than those along lakeshores, and the focus of this chapter is largely ocean shorelines.

LearningCheck 20-1 **In what ways are ocean shoreline-shaping processes different from those shaping lakeshores? (Answer on p. AK-7)**

Many processes contribute to the shaping of coastal features. In addition to the internal and external processes discussed in earlier chapters, a number of processes largely confined to shorelines are at work. By far the most important is the work of waves.

Waves

Waves entail the transfer of energy through a cyclical rising and falling motion in a substance. Our interest here is in water waves: undulations in the surface layers of a body of water.

Wave Motion

Animation
Wave Motion
https://media.pearsoncmg.com/ph/esm/esm_mcknight_physgeo_13/media/

Although water waves appear to move horizontally, this appearance is somewhat misleading. In open water, the form of a wave (and therefore its energy) moves along the surface of the water, although the water itself moves forward only very slightly. This motion changes in shallow water, where waves crest and break.

Most water waves are wind generated, set in motion largely by the friction of air blowing across the water. This transfer of energy from wind to water initiates wave motion. Some water waves (called *forced waves*) are generated directly by wind stress on the water surface; they can develop to considerable size if the wind is strong, but these waves usually last for only a limited time and do not travel far. Water waves become **swells** when they travel beyond the location where they were generated by wind and, in so doing, can travel enormous distances. A small number of all water waves are generated by something other than the wind, such as a tidal surge, volcanic activity, or undersea tectonic movement (discussed later in this chapter).

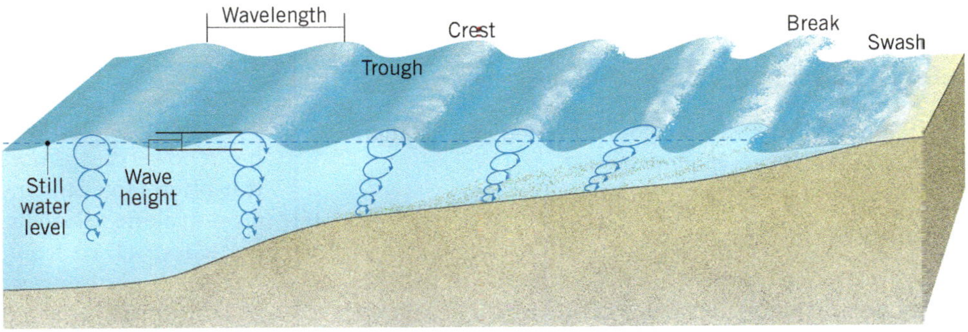

▶ **Figure 20-2** In deep water, passage of a wave involves almost circular movements. Water movement (blue orbits) diminishes rapidly with increasing depth (see the decreasing diameters downward). As the wave moves into shallow water, the orbits become more elliptical, the wavelength shortens, and the wave steepens. Eventually, it "breaks" and dissipates its remaining energy as it washes up onto the beach.

Waves of Oscillation: As a wave passes a given point on the water surface, the water at that location makes a small circular or oscillatory movement, with very little forward motion. ("Oscillate" means to move back and forth again and again.) Such a wave is called a **wave of oscillation**. As the wave passes, the water moves upward, producing a *wave crest* (Figure 20-2). Crest formation is followed by a sinking of the surface that creates a *wave trough*. The horizontal distance from crest to crest or from trough to trough is called the **wavelength**. The vertical distance from crest to trough, equivalent to the diameter of the circular orbit of the surface water as the wave form passes, is called the **wave height**. The height of a wave depends on wind speed, wind duration, water depth, and *fetch* (the area of open water over which the wind blows). For this reason, much higher waves are generated in large lakes (such as the Great Lakes) than in small ponds.

The passage of a wave of oscillation normally moves water only very slightly in the direction of travel. Thus, when a wave passes through, an object floating on the surface simply bobs up and down without advancing (unless the wind pushes it). The influence of wave movement diminishes rapidly with depth; even very high waves stir the subsurface water to a depth of only a few tens of meters.

Waves of Translation: Waves often travel great distances across deep water with relatively little change in speed or shape. As they roll into shallow water, however, a metamorphosis occurs. When the water depth becomes equal to about half the wavelength, the wave motion begins to be affected by frictional drag on the sea bottom. The waves of oscillation

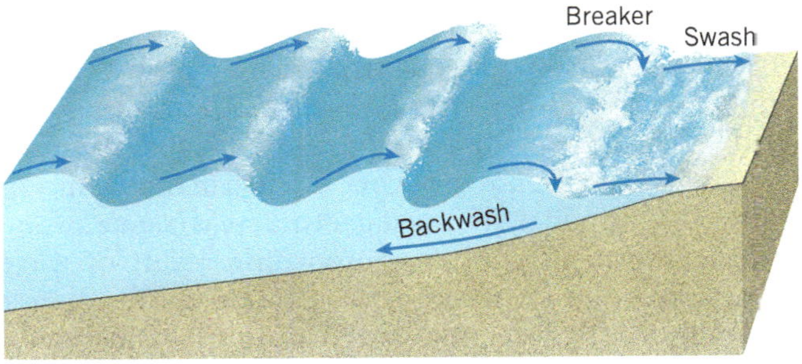

▲ **Figure 20-3** A breaking wave. In shallow water, the ocean bottom impedes oscillation, causing the wave to steepen until it collapses and tumbles forward as a breaker. The surging water then rushes up the beach as swash and then drains off the beach below the waves as backwash.

then rapidly become **waves of translation**. The result is significant horizontal movement of the surface water. Friction retards the progress of the waves, so they are slowed and bunch together, marking a decrease in wavelength, while their height increases. As the wave moves into still shallower water, frictional drag becomes even greater. As the wave becomes higher and steeper, it tilts forward. Soon and abruptly the wave *breaks* (Figure 20-3), collapsing into whitewater surf or plunging forward as a breaker or, if the height is small, perhaps surging up the beach without cresting.

The breaking wave rushes toward shore or up the beach as **swash**. This surge can carry sand and rock particles onto the beach or can pound onto rocky headlands and coastal cliffs with considerable force (Figure 20-4). The momentum of the surging swash is soon overcome by friction and gravity.

◀ **Figure 20-4** The continuous pounding of waves can erode even the most resistant shoreline rocks. This scene is in Pictured Rocks National Lakeshore, Lake Superior, Michigan

A return flow, called **backwash**, then drains much of the water seaward again, carrying loose material with it, usually to meet the oncoming swash of the next wave.

LearningCheck 20-2 **How do waves of oscillation change as they reach shallow water?**

Wave Refraction

Waves often change direction as they approach the shore, a phenomenon known as **wave refraction**. It occurs when a line of waves does not approach exactly parallel to the shore, or where the coastline is uneven or water depth is irregular in the near-shore zone. The portion of a wave that reaches shallow water first is slowed down. As a result, the wave line bends (*refracts*) as it pivots toward the obstructing area (Figure 20-5a), finally breaking roughly parallel to shore. Thus, wave energy tends to be concentrated in the vicinity of an obstruction and diminished in other areas (Figure 20-5b).

The most conspicuous geomorphic result of wave refraction is the focusing of wave action on headlands (Figure 20-6). They are subjected to the direct onslaught of pounding waves, whereas an adjacent bay experiences much gentler, low-energy wave action. Other things (such as the resistance of the bedrock) being equal, the differential effect of wave refraction tends to smooth the coastal outline by wearing back the headlands and increasing sediment accumulation in the bays.

Wave Erosion

The most notable erosion along coastlines is accomplished by wave action. The incessant pounding of even small waves is a potent force in wearing away the shore, and the enormous power of storm waves almost defies comprehension—it is often large storm waves that accomplish most of the erosion along a shoreline. Waves break with abrupt and dramatic impact, with spray moving as fast as 115 kilometers (70 miles) per hour; small jets from breaking waves have been measured at more than twice that speed. This speed, coupled with the sheer mass of the water involved in such hydraulic pounding, is responsible for much coastal erosion, which is made much more effective by the abrasive rock particles the waves carry.

Along a rocky shoreline, wave erosion has another dimension: air is forced into cracks in the rock as the wave hits the shore. The resulting compression is abruptly released as the water recedes, allowing instant expansion of the air. This pneumatic action is often very effective in loosening rock particles of various sizes.

Chemical action also plays a part in the denudation of rocks and cliffs because most rocks are to some extent soluble in seawater. In another form of weathering action, salts from seawater crystallize in the crevices and pores of onshore rocks and cliffs, and this deposition further weakens and breaks up the rock (as in *salt wedging*, discussed in Chapter 15).

(a) Wave refraction

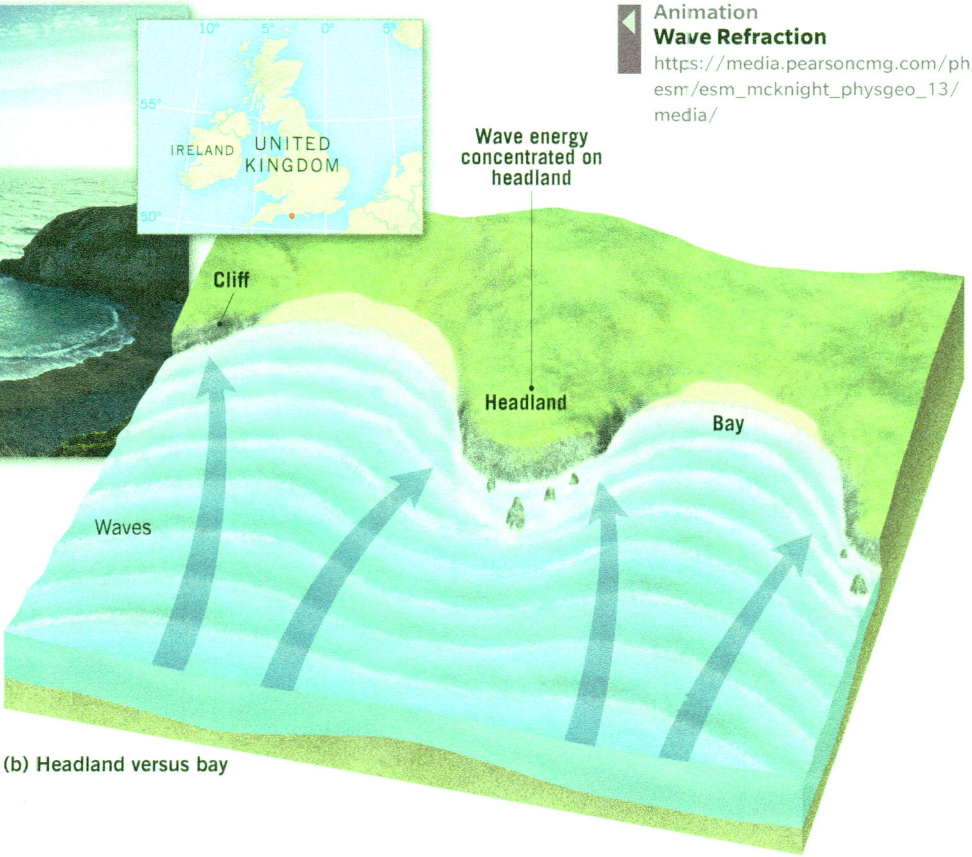

(b) Headland versus bay

Animation
Wave Refraction
https://media.pearsoncmg.com/ph/esm/esm_mcknight_physgeo_13/media/

▲ **Figure 20-5** Wave refraction. (a) Waves roll in obliquely to Man O'War Cove, in St. Oswald's Bay along the Dorset coast of southern England, conforming to the curved shape of the shoreline. (b) The waves reach the headland first and "wrap" around it, breaking nearly parallel to the coastline as a result of wave refraction. Thus, wave energy is concentrated on the headlands and diminished in the bays.

(a) Sea arches

(b) After collapse

▲ **Figure 20-6** The incessant pounding of waves on this soft-rock headland on the southern coast of the state of Victoria, Australia, produced a double arch that eventually eroded into a single arch. The view is in Port Campbell National Park (a) in 1985, (b) in 1992.

On shorelines made up of cliffs, the most effective erosion takes place at or slightly above sea level, so a notch is cut in the base of the cliff. The cliff face then retreats as the slope above the undercutting collapses (**Figure 20-7**). Wave action breaks up the resulting debris, eventually carrying most of it seaward.

As we see later in this chapter, where a shoreline is composed of sand or other unconsolidated material, currents and tides may also cause rapid erosion. Storms greatly accelerate the erosion of sandy shores; a violent storm can remove an entire beach in just a few hours, cutting it right down to bedrock.

(a) Waves undercut cliff

(b) Cliff retreats

(c) Wave-cut cliffs

▲ **Figure 20-7** (a) Waves pounding an exposed rocky shoreline erode the rock most effectively at water level. As a result, a notch may be cut in the face of the headland. (b) The presence of the notch undermines the higher portion of the headland, causing it to collapse and the cliff face to retreat. (c) A tectonically active coastline has helped produce the sharp cliffs that are wearing back through wave erosion in this scene along the north coast of California.

Whether they are awesome storm waves or mild swells, the peculiar contradiction of water waves is that they normally pass harmlessly under such fragile things as boats or swimmers in open water but can devastate even the hardest rocks of a shoreline. In other words, a wave of oscillation is a relatively gentle phenomenon, but a wave of translation can be a powerful force of destruction.

LearningCheck 20-3 **Why are storm waves so important in coastal erosion?**

Tsunami

Animation **Tsunami** https://media.pearsoncmg.com/ph/esm/esm_mcknight_physgeo_13/media/

Occasionally, a sudden disruption of the ocean floor triggers major oceanic wave systems. These waves are called **tsunami** (from the Japanese *tsu* for "harbor" and *nami* for "wave") or *seismic sea waves* (improperly called "tidal waves").

Tsunami Formation: Most tsunami are a consequence of abrupt movement along an ocean floor fault—especially from the vertical displacement caused by reverse or thrust faulting along a subduction zone. Tsunami may also result from underwater volcanic eruptions and major underwater and coastal landslides.

The great destructive power unleashed by some tsunami comes from the way in which the ocean is disrupted when such a wave forms. Recall that with wind-generated waves, only the surface of the ocean exhibits significant movement; the orbital movement of the water extends down to a depth of only about one-half the wavelength of the wave (rarely more than a few tens of meters). When fault rupture on the ocean floor generates a tsunami, however, the entire water column—from the ocean floor to the surface—is disrupted, displacing an enormous volume of water (Figure 20-8).

Although tsunami can travel at speeds exceeding 700 kilometers (435 miles) per hour, in the open ocean they are usually inconspicuous because they are low and have very long wavelengths. (In the open sea, a tsunami might have a wave height of only 0.5 meter [1.5 feet] with a wavelength of perhaps 200 kilometers [125 miles].) When a tsunami reaches shallow water, however, it changes considerably. As a tsunami approaches a coast, it slows—as do all waves—causing the wavelength to decrease and the wave height to increase.

Effects of Tsunami: When they strike a shoreline, tsunami rarely form towering breaking waves. Instead, most tsunami arrive as a very rapidly advancing surge of water up to 40 meters (130 feet) high. Unlike many large, wind-generated waves, however, immediately behind the wave crest of a tsunami is an enormous volume of water that can surge great distances inland before receding. In many cases, before a tsunami arrives, the water withdraws from the coast, appearing like a very sudden, very low tide; this happens when the trough of the tsunami arrives. Unfortunately, people sometimes venture out on the freshly exposed subtidal areas to collect shellfish or stranded fish, only to be caught minutes later by the rapid surge when the tsunami crest comes onshore. Frequently, a series of surges and withdrawals ensues, with the largest surge not necessarily the first to arrive.

The 2004 Sumatra–Andaman Earthquake and Tsunami: On December 26, 2004, one of the greatest natural disasters in recent history was triggered after a magnitude 9.1 earthquake shook the northern coast of Sumatra in Indonesia. A section of the interplate thrust fault (or "megathrust"), 1200 kilometers (750 miles) long and formed where the Australian–Indian Plate subducts beneath the Burma Plate, ruptured, uplifting the ocean floor by as much as 4.9 meters (16 feet). The sharp movement of the ocean floor generated a tsunami that spread in all directions.

About 28 minutes after the earthquake struck, a wave 24 meters (80 feet) high rushed onshore at the city of Banda Aceh on the northern tip of Sumatra, Indonesia, just 160 kilometers (100 miles) from the epicenter of the earthquake (Figure 20-9). The tsunami spread across the Indian Ocean, striking Sri Lanka, the Maldives, and the coast of Somalia in

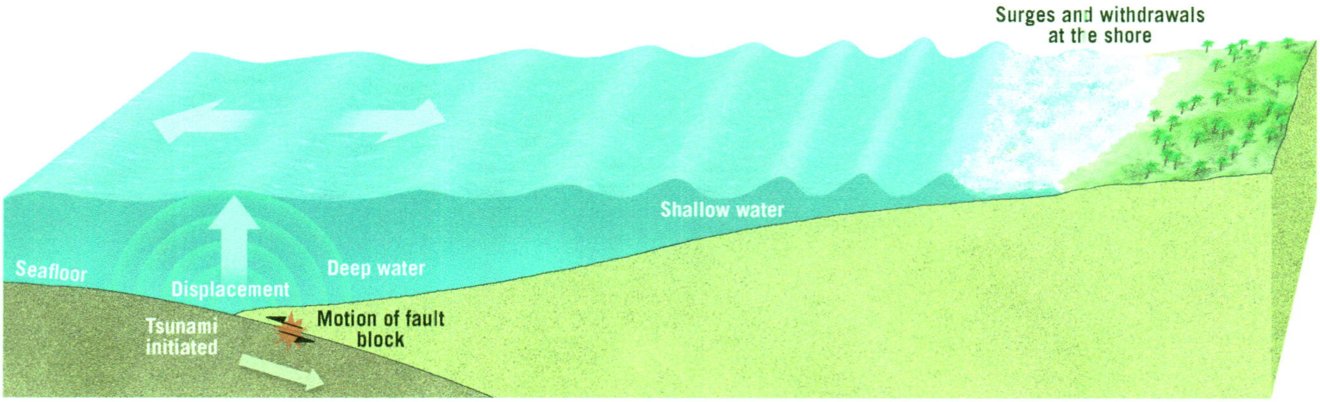

▲ Figure 20-8 Tsunami formation. A vertical disruption of the ocean floor, such as from faulting, displaces the entire water column from ocean floor to surface. In the open ocean, the tsunami may be almost indistinguishable because of its great wavelength. Once it reaches shallow water, wave height increases. The tsunami may come onshore as a series of surges and withdrawals.

▲ **Figure 20-9** Damage from the December 26, 2004, tsunami in Banda Aceh, Indonesia.

▲ **Figure 20-10** Tsunami coming onshore in Natori, Miyagi Prefecture, Japan, on March 11, 2011.

northeast Africa. Estimates suggest that nearly 228,000 people died and many tens of thousands more were seriously injured. In a few locations, entire villages were quite literally washed away.

Tsunami Warnings: Because most tsunami originate from sudden fault displacement in a subduction zone, the resulting earthquakes are readily detectable by seismographs. For decades, the Pacific Tsunami Warning Center in Hawai'i has used seismographic and other data to detect tsunami heading for coastlines around the Pacific basin. With such information, a tsunami warning can usually be issued long enough in advance to allow time for evacuation of the impact area. However, as the Sumatra tsunami disaster of 2004 revealed, if local warning systems are not in place, evacuation orders may not reach coastal populations in time.

Furthermore, as we saw in Sumatra in 2004 and in Japan in March 2011, if a large tsunami-generating earthquake originates close to a populated coastline, the waves can arrive so quickly that coastal residents may have only a few minutes to evacuate (Figure 20-10).

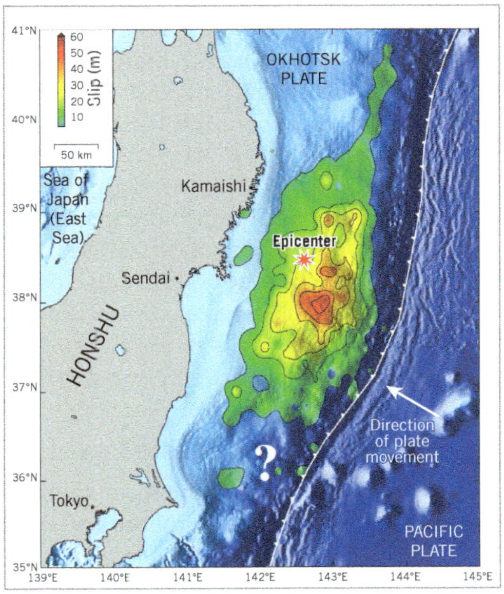

▲ **Figure 20-11** The 2011 Japan earthquake fault rupture zone. The maximum movement along the fault plane was more than 50 meters (164 feet), shown in red.

The 2011 Japan Earthquake and Tsunami: On March 11, 2011, a magnitude 9.0 earthquake struck the northeast coast of Japan, along a thrust fault where the Pacific Plate subducts beneath the Okhotsk "microplate" (Figure 20-11). The severe ground shaking lasted more than 3 minutes. In that time, a 300-kilometer-long by 150-kilometer-wide (185-mile by 90-mile) segment of the subduction zone slipped as much as 50 meters (164 feet). Afterward, the northeastern coast of Japan had jumped about 2.4 meters (8 feet) to the east, and parts of the coastline in Miyagi Prefecture had subsided by more than 1 meter (3.3 feet).

The result of this Great Tohoku Earthquake was devastation and loss of life almost unimaginable in a country that is as well prepared for large earthquakes as any in the world. Nearly 16,000 people lost their lives. Initially, more than 130,000 people were left homeless; telecommunications, transportation, and water supplies were widely disrupted. Utilities were damaged, most dramatically when several nuclear reactors were damaged near Fukushima, Japan.

Although the earthquake caused extensive damage, the tsunami that followed was most deadly and destructive. The ocean floor above the fault rupture was abruptly uplifted, displacing an enormous volume of water. Tsunami warnings were issued within 10 minutes of the earthquake, but the coastal populations closest to the epicenter had almost no time to evacuate. The height of the tsunami was typically about 10 meters (33 feet), but in some confined harbors the height was as much as 40 meters (130 feet). In low-lying coastal plains, such as parts of the Sendai region north of Tokyo, the surge of water advanced as much as 10 kilometers (6 miles) inland, leveling buildings, ruining roads, and depositing tons of debris.

LearningCheck 20-4 **Why are tsunami often so much more destructive than even very large storm waves?**

Important Shoreline-Shaping Processes

In addition to wind-generated waves and tsunami, a variety of other processes also modify coastlines in ways that range from gradual and subtle to sudden and spectacular.

Tides

As we learned in Chapter 9, the waters of the world ocean oscillate in a regular and predictable pattern called *tides*, resulting from the gravitational influence of the Sun and Moon (see Figure 9-8). The tides rise and fall in a cycle that takes about 12 hours, producing two high tides and two low tides a day on most (but not all) seacoasts.

Despite the enormous amount of water moved by tides and despite the frequency of this movement, the topographic effects are surprisingly small. Tides are significant agents of erosion only in narrow bays, around the margin of shallow seas, and in passages between islands, where they produce currents strong enough to scour the bottom and erode cliffs and shorelines (**Figure 20-12**). The movement of water through tides is, however, a promising source of power for generating electricity—see the box *Energy for the 21st Century: Tidal Power*.

Changes in Sea Level and Lake Level

Sea-level changes can result from the local uplift or sinking of a landmass (tectonic cause) or from **eustatic sea-level change**—an increase or decrease in the amount of water in the oceans. During Earth's recent history, there have been many changes in sea level, sometimes worldwide and sometimes only around one or a few continents or islands. The eustatic changes of greatest magnitude and most extensive effect are those associated with seawater volume before, during, and after the Pleistocene glaciations. As we saw in Chapter 19, at the peak of the Pleistocene glaciations, sea level around the world was at least 120 meters (395 feet) lower than today.

As a result of both tectonic and eustatic sea-level changes, many present-day ocean coastlines have been submerged, with a portion of a previous landscape now underwater. Others show emergent characteristics, in which shoreline topography of the past is now well above the contemporary sea level (see Figure 20-7c). We consider the topographic consequences of these circumstances later in this chapter.

Most water-level changes in lakes are less extensive and less notable than those along ocean shorelines. These changes are usually the result of the total or partial drainage of a lake. Their principal topographic expression is exposed ancestral beach lines and wave-cut cliffs above present lake levels.

Global Warming and Sea-Level Change: In Chapters 4 and 8, we discussed the consequences of global climate change—especially what is commonly called "global warming." We noted that as global climate warms, a rise in sea level is associated with the thermal expansion of ocean water as well as the increase in volume from the melting of alpine glaciers and continental ice sheets and ice caps. (The melting of sea ice doesn't add to sea-level rise.) If worldwide temperature continues to warm, we can anticipate an ongoing period of deglaciation, with the ice sheets of Antarctica and Greenland slowly melting. Such a situation would cause a global eustatic rise in sea level that would inundate many islands and coastal plains and would expose coastal populations to greater risks from storm waves, such as those generated by hurricanes.

Animation
Tides
https://media.pearsoncmg.com/ph/esm/esm_mcknight_physgeo_13/media/

◀ **Figure 20-12** A large tidal range can influence the shaping of coastal landforms. These gigantic pedestal rocks on the edge of the Bay of Fundy in New Brunswick, Canada, were carved by waves in this region, which has the world's greatest tidal range: up to 15 meters. For scale, the spruce trees on top of the rocks are about 9 meters (30 feet) tall.

Energy for the 21st Century

Tidal Power

• **Jennifer Rahn,** Samford University

Before large-scale commercial electrical power generation, many civilizations used the force of moving water to power machines such as textile mills and lumber mills. Just as wind turns the blades of a windmill, recent technological advances utilize the inflow and outflow of tidal water during *flood tides* and *ebb tides* (see Chapter 9) to turn a turbine to generate electricity.

Tides are more predictable than solar power or wind energy—in most places, tides change twice a day like clockwork. The higher the *tidal range*, the greater the potential for tidal energy generation. Seawater is 840 times denser than air, so a 5-knot (5.7-mph) tidal current has more kinetic energy than a 191-knot (220-mph) wind. The ideal locations for tidal power are areas with high flow volumes or high tidal ranges (**Figure 20-A**).

Tidal Power Installations: The first tidal power plant was the Rance Tidal Power Station built in 1966 in Brittany, France (**Figure 20-B**), and the first one in North America was installed in the 1980s in an inlet of the Bay of Fundy in eastern Canada. One of the world's largest *tidal stream* power installations was completed in 2008 in Strangford Lough in Northern Ireland, although it has since been decommissioned. The first pilot tidal power installation in the United States became operational in September 2012 off eastern Maine. Currently, the world's largest tidal power installation is the Sihwa Lake Tidal Power Station in South Korea—it uses 10 turbines in a seawall as a *tidal barrage* to generate power during the incoming tide. Tidal power facilities have also been built in China, Russia, and Australia.

▲ **Figure 20-B** Tidal power station built across an estuary of the Rance River in Brittany, France.

Limitations of Tidal Power: Tidal power is not widely used, in part because few places in the world have high-flow tidal regimes. Tidal-power projects are also extremely expensive because massive structures must be built in difficult saltwater environments. However, this technology has many long-term advantages. Tidal power is a renewable energy resource with a virtually unlimited supply and no greenhouse gas emissions.

Estimates suggest that tidal power could meet 10 percent of the United States' and perhaps 50 percent of the United Kingdom's electricity demand within the next few decades.

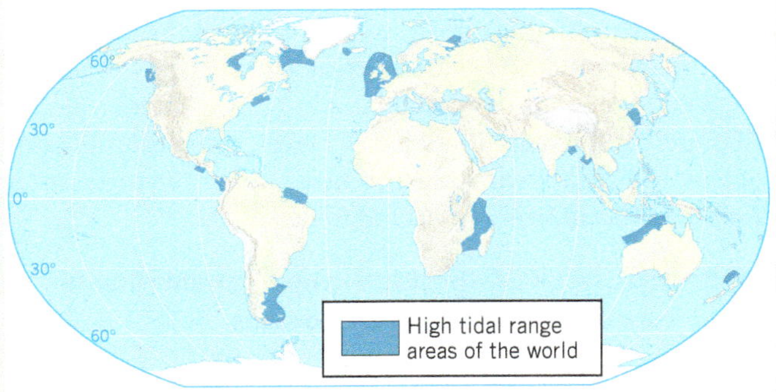

▲ **Figure 20-A** Areas with tidal ranges exceeding 4 meters (13 feet).

Technical Obstacles: The technology needed to turn the tides into electricity efficiently and economically is still being developed. Several new models for generating tidal power have been developed in recent years and will begin to operate commercially. Other problems include damage to the local aquatic environment, mostly with small fish getting caught in the rotors. Many people are hesitant to adopt large-scale tidal power plants because of the unknown harm they might cause the environment. As we are now seeing with dammed rivers, when humans alter the flow of energy and the surrounding environment, often there are unintended consequences. The growth of tidal power generation will depend on the public and political desire for governments to make the large investment needed for this type of alternative energy.

Questions

1. Water is much more dense than air. What does this mean for the size and shape of tidal turbine rotors?
2. Using Figure 20-A, list favorable locations (other than those mentioned) where tidal energy plants could be installed.
3. What challenges limit the rapid adoption of tidal power?

Should the ice caps of Antarctica and Greenland melt completely (a result not anticipated by most climate scientists, even over the next century or so), global sea level would rise by about 70 meters (230 feet). However, even a slight increase in global sea level could be devastating to populations now living in low-lying coastal areas (**Figure 20-13**). Given the rising levels of greenhouse gases and the associated temperature rise due to global warming, in 2013 the *Fifth Assessment Report* of the Intergovernmental Panel on Climate Change (IPCC) projects a *likely* sea-level rise of 0.26–0.55 meter (10.2–21.7 inches)

Global Environmental Change

Impact of Sea-Level Rise on Islands

As sea level rises due to melting glaciers, ice sheets, and ice caps, reef islands are at great risk because their maximum elevation is not much above the ocean surface. Because these islands were formed entirely from sediment captured by coral reefs (discussed later in this chapter), they are prone to erosion that will likely accompany higher sea levels. Ten populated islands in the west-central Pacific Ocean may disappear altogether if eustatic sea-level change is as large as predicted for the next century. New findings suggest, however, that reef islands are more adaptable than previously thought.

Dynamic Islands: Geomorphologists studying reef islands to determine the likely impact of sea-level rise are discovering that reef islands are very dynamic. Reef islands grow, shrink, and even move at the whim of the ocean. As sea level rises, one side of a reef island erodes while the other side collects more sediment, causing the island to shift position. Hurricanes and higher tides also deposit sediment, building the island surface instead of washing it away. Therefore, many of these islands may adapt to rising sea levels instead of being submerged.

Populated islands, however, have buildings—which are not as mobile as the island (**Figure 20-C**). These islands, where humans have disrupted the natural sedimentation process, will be particularly hard hit by sea-level rise. Island nations such as Tuvalu (population: ~12,000), the Republic of the Marshall Islands (population: ~59,000), and the Republic of Kiribati (population: ~115,000) face a range of potential impacts from global climate change and sea-level rise, including eroding coastlines, flooding, and intrusion of saltwater into the soil and drinking water.

▲ **Figure 20-C** Kwajelein Atoll, part of the Republic of the Marshall Islands, which are composed entirely of coral reefs.

Questions

1. How do reef islands adapt to eustatic changes in the ocean?
2. Why will populated islands be hard hit by sea-level rise?
3. What are the impacts of sea-level rise on populated reef islands?

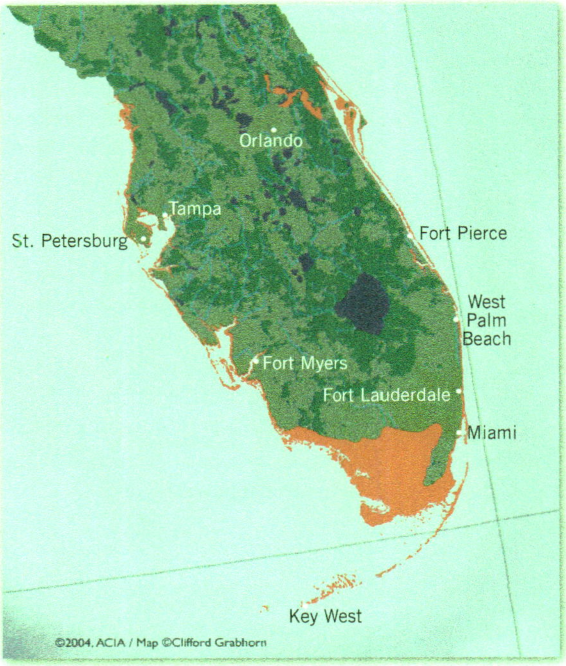

▲ **Figure 20-13** Shown in red, the estimated coastal areas of Florida that would be flooded by a 1-meter rise in sea level.

by the end of this century under their lowest-emission scenario, and a rise of 0.52–0.98 meter (20.4–38.6 inches) under their highest-emission scenario (see Chapter 8). Some scientists consider the IPCC projections to be too conservative. Even the smallest projected sea-level rise would cause shorelines around the world to retreat, eliminating thousands of square kilometers of coastal land in North America alone. With such an increase in global sea level, some island countries would disappear—see the box *Global Environmental Change: Impact of Sea-Level Rise on Islands*.

LearningCheck 20-5 **Explain how current climate change may affect the coastlines of oceans.**

Ice Push

The shores of bodies of water that freeze over in winter are sometimes significantly affected by *ice push*, the result of the contraction and expansion of ice along a water's edge. As more and more water turns to ice and expands in volume (recall the discussion of *frost wedging* in Chapter 15), near-shore ice is shoved onto the land, pushing against the shore more or less in the fashion of a small glacial advance.

Ice push is most significant in the Arctic and Antarctic, but it can be responsible for erosion along the shores of the Great Lakes and around smaller high-latitude or high-elevation lakes.

Organic Secretions

Several aquatic animals and plants produce solid masses of rocklike material by secreting calcium carbonate. By far the most significant of these organisms is the coral *polyp*, a tiny animal that builds a hard external skeleton of calcium carbonate. Coral polyps are of many species, and they cluster together in social colonies of uncounted billions of individuals. Under favorable conditions (clear, shallow, salty warm water), the coral can accumulate into enormous masses, forming reefs, platforms, and atolls, all commonplace features in tropical and subtropical oceans. (Coral reef structures are discussed later in this chapter.)

Stream Outflow

The source of most sediment deposited in shoreline beaches and other depositional features is the outflow from streams, although in some locations all or part of the sediment may come directly from the erosion of coastal rocks. As we saw in Chapter 16, the sediment carried by streams may be deposited as alluvium in a delta. Even in such cases, at least some of the sediment carried by streams into the ocean is further transported and then deposited elsewhere by coastal waters (**Figure 20-14**).

LearningCheck 20-6 **What is the source of most coastal sediment?**

▼ **Figure 20-14** The sediment plume of the Connecticut River where it enters Long Island Sound near Old Lyme, Connecticut.

Coastal Sediment Transport

Many kinds of currents flow in the oceans and lakes of the world, but nearly all transportation of sediment along coastlines is accomplished by wave action and local currents.

Longshore Currents: Coastal topography is affected most by **longshore currents** (or *littoral currents*), in which the water and sediment move roughly parallel to the shoreline. (Think of *longshore* as a contraction for "along the shore.") Longshore currents develop just offshore and are set up by the action of the waves striking the coast at a slight angle (**Figure 20-15a**). Because most waves are generated by wind, the direction of a longshore current typically reflects the local wind direction. Longshore currents are prominent transporters of sand and other sediment along many shorelines.

Beach Drifting: Another significant mechanism of coastal sediment transport involves the short-distance shifting of sand directly onshore by breaking waves and the retreating water from the beach. This movement takes the form of **beach drifting** along a coastline, a zigzag movement of sediment that results in a general downwind displacement parallel to the coast (**Figure 20-15b**). Nearly all waves approach the coast obliquely rather than at a right angle, and therefore the sand carried onshore by the breaking wave moves up the beach at an oblique angle. Some of the water soaks into the beach, but much of it returns seaward directly downslope, which is normally at a right angle to the shoreline. This return flow takes some of the sand with it, much of which is picked up by the next surging wave and carried shoreward again along an oblique path. This infinitely repetitious pattern of movement shifts the debris farther and farther along the coastline. Because wind is the driving force for wave motion, the strength, direction, and duration of the wind are the principal determinants of beach drifting.

Some sediment transport along shorelines is accomplished directly by the wind. Wherever waves have carried or hurled sand and finer-grained particles to positions above the water level, these particles can be picked up by a breeze and moved overland. This type of movement frequently results in dune formation and sometimes moves sand a considerable distance inland. (See Chapter 18 for a detailed discussion of sand dunes.)

LearningCheck 20-7 **Explain how longshore currents and beach drifting transport coastal sediment.**

Coastal Depositional Landforms

Although waves and local currents can accomplish considerable erosion and transportation, in many cases the most conspicuous topographic features of a shoreline are formed by the deposition of sediment, especially sand-size sediment. Just as with streamflow on the surface of a continent, coastal deposition occurs wherever the energy of moving water is diminished.

(a) Waves create longshore current

(b) Beach drifting and longshore current

Mobile Field Trip
Gulf Coast
https://media.pearsoncmg.com/ph/esm/esm_mcknight_physgeo_13/media/

Animation
Coastal Sediment Transport
https://media.pearsoncmg.com/ph/esm/esm_mcknight_physgec_13/media/

▲ **Figure 20-15** (a) Waves striking shore at an angle set up longshore currents and beach drifting along the Fort Morgan Peninsula, Alabama. Sediment would be transported from the top of the photograph toward the bottom. (b) Longshore currents develop just offshore and move sediment parallel to the coastline. Beach drifting involves a zigzag movement of sand along the coast. Sand is brought obliquely onto the beach by the wave and returned seaward by backwash. Because most waves develop in response to the wind, longshore currents and beach drifting typically move sediment in a general downwind direction along a shore.

Sediment Budget of Depositional Landforms

Deposits along coastlines tend to be more ephemeral than those farther inland. This is due primarily to the relatively small particle size of coastal deposits (sand and gravel) and to the fact that the sand tends not to be stabilized by vegetation cover. Most coastal deposits are under constant onslaught by waves, which can rapidly wash away portions of the sediment. Consequently, the **sediment budget** must be in balance if the deposit is to persist: the removal of sand must be offset by the addition of sand. Most coastal deposits have a continuing sediment flux, with sediment arriving in some places and departing in others. During storms, the balance is often upset. As a result, coastal deposits are significantly reshaped or totally removed, only to be replaced when calmer conditions prevail.

Beaches

The most widespread marine depositional feature is the **beach**, which is an exposed deposit of loose sediment adjacent to a body of water. Although the sediment can range in size from fine sand to large cobbles, it is usually relatively homogeneous on a given section of beach. Beaches composed of sand are normally broad and slope gently seaward, whereas those formed of larger particles (gravel, cobbles) generally slope more steeply. (Silt and clay get carried away in suspension, so they rarely form beaches.)

Beach Profiles: Beaches occupy the transition zone between land and water, sometimes extending well above the normal sea level into elevations reached by only the highest storm waves. In a typical beach profile (Figure 20-16), the beach on the seaward side generally extends down to the level of the lowest tides. It can often be found at still lower levels, where it merges with muddy bottom deposits.

The *backshore* is the upper part of the beach, landward of the high-water line. It is usually dry, being covered by waves during severe storms only. It contains one or more *berms*, which are flattish, wave-deposited sediment platforms. The *foreshore* is the zone that is regularly covered and uncovered by the rise and fall of tides. The *nearshore* extends seaward from the low-tide mark to where the low-tide breakers begin to form. The nearshore is not exposed to the atmosphere, but it is the place where waves break and surf action is greatest. The *offshore* zone is permanently underwater and deep enough that wave action rarely influences the bottom. Some beaches extend for dozens of kilometers along straight coastlines, particularly if the relief of the land is slight and the bedrock unresistant. Along irregular shorelines, beach development may be restricted largely or entirely to bays, with the bays frequently alternating with rocky headlands.

Beach shape may change greatly from day to day and even from hour to hour—anytime the sediment budget of the beach changes. Normally beaches are built up slowly during quiet weather and removed rapidly during storms.

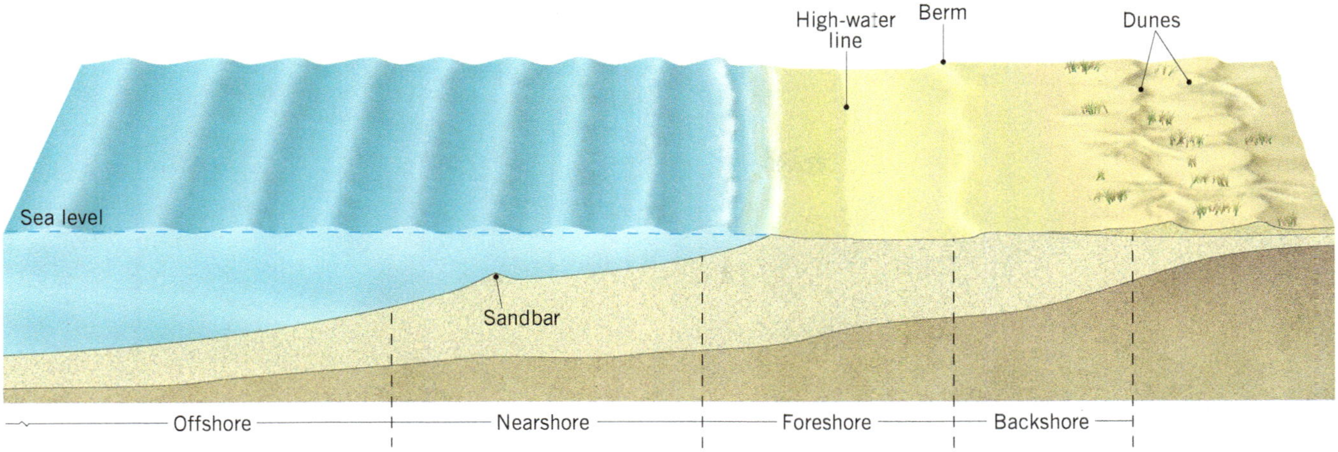

▲ **Figure 20-16** An idealized beach profile.

Video
Summertime/Wintertime Beach Conditions
https://media.pearsoncmg.com/ph/esm/esm_mcknight_physgeo_13/media/

Most midlatitude beaches are longer and wider in summer and greatly worn away by the storminess of winter.

LearningCheck 20-8 **What is the sediment budget of a beach, and how can it change?**

Spits

At the mouth of a bay, sediment transported by longshore currents moves into deeper water. There the flow speed slows, and the sediment is deposited. The growing bank of land guides the current farther into the deep water, where still more material is dropped. Any such linear deposit attached to the land at one end and extending into open water in a down-current direction is called a **spit** (Figure 20-17).

Although most spits are straight, sandy peninsulas projecting out into a bay or other coastal indentation, variations in local currents, winds, and waves often give them other configurations. If the spit becomes extended clear across the mouth of a bay and connects with land on the other side, it becomes a **baymouth bar**, transforming the bay into a *lagoon*. If instead water movements in the bay cause the deposits to curve toward the mainland, a *hook* forms at the outer end of the spit (see Figure 20-F in the box on Cape Cod National Seashore).

A less common but even more distinctive development is a **tombolo**: a depositional feature that connects a near-shore island with the mainland (Figure 20-18). Some tombolos form as a type of spit where waves deposit sediment on the landward side of the island. Other tombolos develop where a bedrock structure at, or just below, sea level connects the island with the mainland and traps sand.

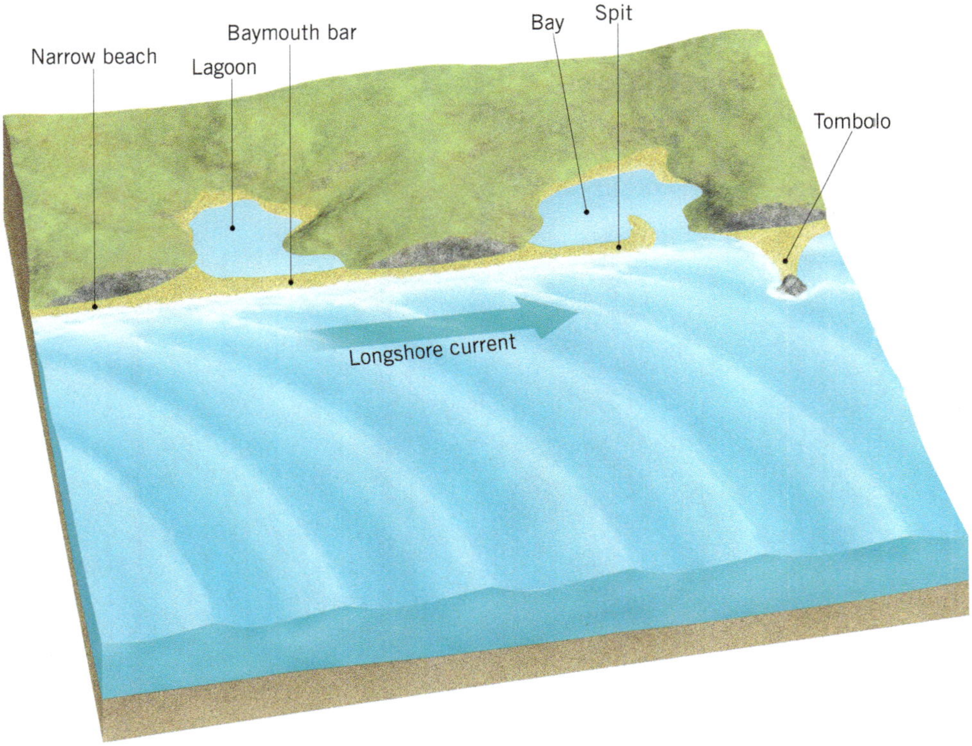

◀ **Figure 20-17** Common depositional landforms along a coastline include spits, baymouth bars, and tombolos. Notice the orientation of the spit to the direction of the longshore current.

Video
Movement of Sand in Beach Compartment
https://media.pearsoncmg.com/ph/esm/esm_mcknight_physgeo_13/media/

▲ **Figure 20-18** This tombolo connects the nearshore island of Point Stephens to the mainland in Tomaree National Park, New South Wales, Australia.

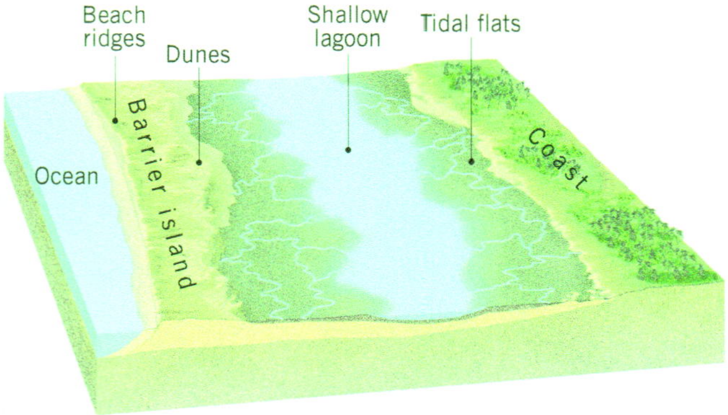

▲ **Figure 20-19** A typical relationship between ocean, barrier island, and lagoon.

Barrier Islands

Another prominent coastal deposition is the **barrier island**: a long, narrow island built up in shallow offshore waters. Sometimes a barrier island is only a few hundred meters from shore, but often they are several kilometers away. Barrier islands are always oriented approximately parallel to the shore (**Figure 20-19**). They are believed to result from the deposition of sediment where large waves (particularly storm waves) begin to break in the shallow waters of continental shelves. Larger barrier islands may have more complicated histories linked to the lowered sea level during the Pleistocene.

Barrier islands often become the dominant element of a coastal terrain. Although they usually rise at most only a few meters above sea level and are typically only a few hundred meters wide, they may extend many kilometers in length. Most of the Atlantic and Gulf of Mexico coastline of the United States, for instance, is paralleled by lengthy barrier islands, several more than 50 kilometers (30 miles) long (**Figure 20-20**).

(a) Gulf Coast barrier islands

◄ **Figure 20-20** (a) The longest of the barrier islands off the Gulf Coast of the United States is (b) Padre Island, Texas.

(b) Padre Island

▲ **Figure 20-21** Crew using a Surface Elevation Table to determine if sedimentation in salt marsh is keeping pace with sea level rise in Fire Island National Seashore, New York.

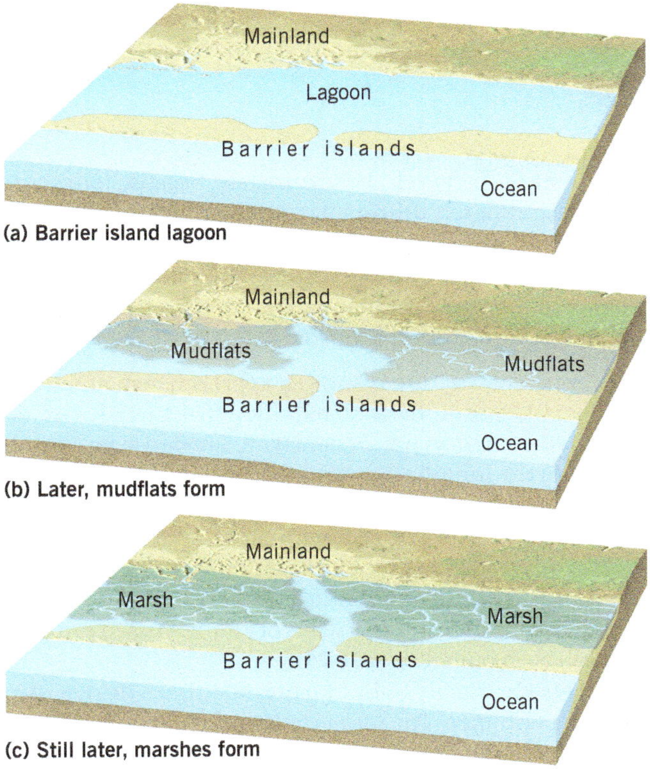

▲ **Figure 20-22** (a) Barrier islands are separated from the mainland by a lagoon. With the passage of time, the lagoon often becomes choked with sediment and is converted to (b) mudflats and, eventually, (c) marshes.

▲ **Figure 20-23** St. George Island, Florida, was overtopped by the storm surge from Hurricane Michael in 2018. (a) April 2017, before Hurricane Michael. (b) October 13, 2018, after Michael had passed. Sand eroded by the storm surge covered the access road (shown with arrow) to St. George Island State Park.

Barrier Island Lagoons: An extensive barrier island isolates the water between itself and the mainland, forming a body of quiet saltwater or brackish water (a mixture of saltwater and freshwater) called a **lagoon**. Over time, a lagoon becomes increasingly filled with water-deposited sediment from coastal streams, wind-deposited sand from the barrier island, and tidal deposits if the lagoon has an opening to the sea. All three of these sources contribute to the buildup of mudflats on the edges of the lagoon (Figure 20-21). Unless tidal inlets across the barrier island permit vigorous tides or currents to carry lagoon debris seaward, the ultimate destiny of most lagoons is to be slowly transformed—first to mudflats and then to coastal marshes (Figure 20-22).

Animation
Movement of a Barrier Island
https://media.pearsoncmg.com/ph/esm/esm_mcknight_physgeo_13/media/

Infilling by sediment is not the only factor that contributes to a lagoon's disappearance. After a barrier island becomes a certain size, it often begins to migrate slowly shoreward as waves wear away its seaward shore and sediments accumulate, building up its landward shore. Eventually, if the pattern is not interrupted, such as by sea-level changes, the island and the mainland shore will merge.

Because most barrier islands rise just a few meters above sea level and are largely built up of coastal sediment, they can be quite susceptible to damage from large storms. A number of inhabited barrier islands along the Gulf Coast and East Coast of the United States experienced severe erosion during hurricanes such as Lili in 2002, Katrina in 2005, Ike in 2008, Sandy in 2012, and Michael in 2018 (Figure 20-23). With even the slight rise in global sea level anticipated this century by the IPCC, barrier islands will become even more vulnerable to storm waves.

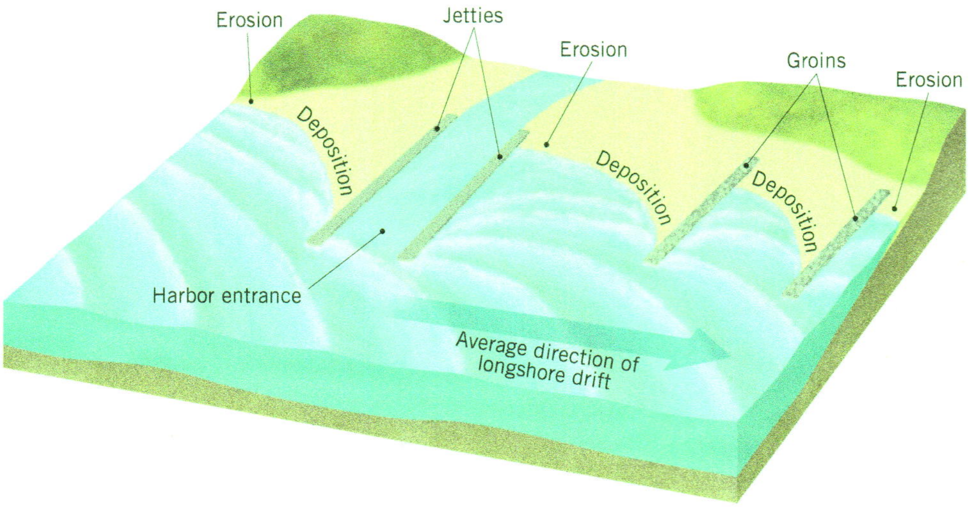

◀ **Figure 20-24** Jetties and groins along a shoreline trap sediment on their upcurrent side, while erosion tends to remove sediment on their downcurrent side.

Animation
Coastal Stabilization Structures
https://media.pearsoncmg.com/ph/esm/esm_mcknight_physgeo_13/media/

Human Alteration of Coastal Sediment Budgets

Over the last century, human activity has disrupted the sediment budgets of beaches along many shorelines of the world. This has been especially true in many coastal areas of North America. For example, dams built along rivers for flood control or hydroelectric power generation effectively act as sediment traps. With less sediment reaching the mouths of rivers, less sediment is available for transport by longshore currents and beach drifting along the shoreline, so the down-current beaches begin to shrink. In addition, artificial structures built to increase or stabilize one community's beaches may reduce the amount of sediment transported farther down a shoreline, thus causing down-current beaches to be reduced in size.

Beach Nourishment: Local communities have taken a number of approaches to solving the problem of shrinking beaches. One direct, but relatively expensive, approach is to "nourish" a beach by dumping tons of sand just slightly up-current of the beach. Unfortunately, because longshore currents and beach drifting eventually transport the sand away, such nourishment must be undertaken repeatedly to maintain the beach at a desired size. Many of the most famous beaches in the world are maintained through nourishment, including Hawai'i's iconic Waikīkī Beach in Honolulu.

Stabilization Structures: Another approach to maintaining beaches is the use of "hard" stabilization structures. For example, a **groin** is a short wall or dam built outward from a beach to impede the longshore current and force sand deposition on the up-current side of the structure (Figure 20-24). Although groins do trap sediment on their up-current side, erosion tends to take place on their down-current side. To reduce this erosion, another groin can be built just down-current. In some locations, a series of groins, known as a *groin field*, has been built (Figure 20-25).

Jetties are usually built in pairs on either side of a river or harbor entrance. The idea is to confine the flow of water to a narrow zone, thereby keeping the sand in motion and inhibiting its deposition in the navigation channel. Jetties

◀ **Figure 20-25** Groin field along Norderney Island in northern Germany.

tend to interfere with longshore currents in the same way as groins, trapping sand on the up-current side while causing erosion on the down-current side.

After undertaking such expensive projects as building groin fields and carrying out regular beach nourishment, some communities have found that it is a losing proposition—the beaches continue to shrink. No clear solution seems to be available.

LearningCheck 20-9 **How do groins and jetties typically affect the beach around them?**

Shorelines of Submergence and Emergence

One of the most conspicuous influences on coastal topography, especially oceanic coastlines, is a change in the height of the water in relation to the land elevation: a relative rise in sea level leading to a *shoreline of submergence*, or a relative rise in the land leading to a *shoreline of emergence*. As we've seen, such changes can occur when the amount of ocean water increases or decreases or when the land rises tectonically or sinks relative to the ocean.

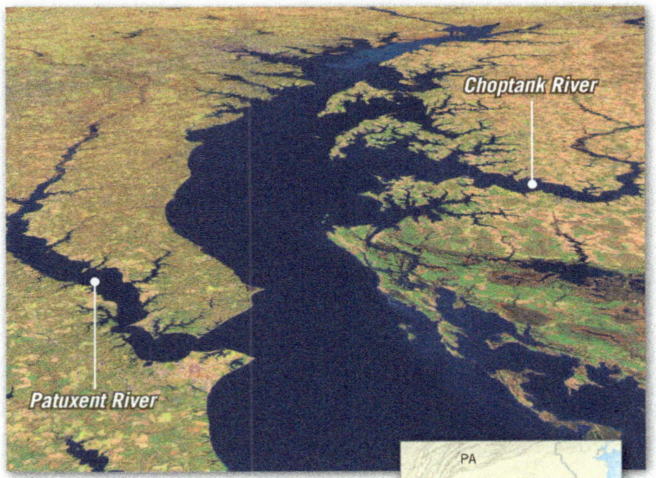

▲ **Figure 20-26** Landsat image of Chesapeake Bay. The bay includes a series of flooded river mouths.

Coastal Submergence

In the recent geological past, sea level has fluctuated sharply. For example, during a warmer climatic interval in an interglacial period 125,000 years ago, sea level was about 6 meters (20 feet) higher than it is today. During the last glacial peak (about 20,000 years ago), sea level is estimated to have been about 120 meters (395 feet) lower than it is at present.

Almost all oceanic coastlines show evidence of submergence during the last 15,000 years or so, a result of the melting of the Pleistocene ice sheets. As water from melting glaciers returned to the oceans, rising sea level caused widespread submergence of coastal zones. Furthermore, as contemporary global warming leads to a slight increase in sea level, a slow but gradual expansion of flooded shorelines is expected to continue during this century.

Ria Shorelines: The most prominent result of submergence is the drowning of previous river valleys, such as North America's Chesapeake Bay, which produces *estuaries*, or long fingers of seawater projecting inland (Figure 20-26). A coast along which numerous estuaries project is called a **ria shoreline**. A ria (from the Spanish *ría*, meaning "river") is a long, narrow inlet of a river that gradually decreases in depth from mouth to head. If a hilly or mountainous coastal area is submerged, numerous offshore islands may indicate the previous location of hilltops and ridge crests.

Cape Cod in Massachusetts is formed mostly of glacial moraines left by Pleistocene ice sheets. When sea level rose at the end of the last glacial period, some of these deposits were surrounded by the rising sea, leaving the coastal islands of Martha's Vineyard and Nantucket—see the box *Featured National Park: Cape Cod National Seashore—A Landscape Shaped by Glaciers and Waves*.

Fjorded Coasts: Spectacular coastlines often occur where high-relief coastal terrain has undergone extensive glaciation. Troughs once occupied by valley glaciers or by continental ice sheets may be so deep that their bottoms are presently far below sea level; as sea level rose at the end of the Pleistocene, the troughs filled with seawater. In some localities these deep, sheer-walled coastal indentations—called **fjords**—are so numerous that they create an extraordinarily irregular coastline, often with long, narrow fingers of saltwater reaching more than 160 kilometers (100 miles) inland. The most extensive and spectacular fjorded coasts are in Norway, western Canada, Alaska, southern Chile, the South Island of New Zealand, Greenland, and Antarctica (Figure 20-27).

LearningCheck 20-10 **Why do so many coastlines around the world show signs of submergence?**

Coastal Emergence

Evidence of previously higher sea levels is sometimes related to ice melting during past interglacial ages but is more often associated with tectonic uplift. The clearest topographic results of coastal emergence are shoreline features raised well above the present water level. Often the emerged portion of a continental shelf appears as a broad, flat coastal plain, with erosion wearing back the land at a cliff dropping down to the sea.

Wave-Cut Cliffs and Platforms: Wave-cut cliffs, sea stacks, and wave-cut platforms comprise one of the most common coastal landform complexes (see Figure 20-7). As we discussed earlier, as waves erode away at a rocky headland, steep wave-cut cliffs form; these cliffs receive the greatest pounding at their base, where the power of the waves is concentrated.

Featured National Park

Cape Cod National Seashore

A Landscape Shaped by Glaciers and Waves

Cape Cod National Seashore in Massachusetts is known for its long sandy beaches, historic lighthouses, beautiful coastal marshes, and sweeping ocean vistas. This landscape, however, traces its earliest origins not to the sea, but to glaciers.

Pleistocene Glaciation: There is no bedrock in Cape Cod. The entire peninsula and the associated islands of Nantucket and Martha's Vineyard are made of sand, gravel, and boulders deposited by glaciers as recently as 21,000 years ago. At the height of the last glacial advance of the Pleistocene, sea level was lower, and the area of present-day Cape Cod was a low coastal plain. As the Laurentide Ice Sheet reached its greatest extent, moraines were deposited along the margin of the ice—Cape Cod began as these deposits of till. When the glaciers receded, outwash from meltwater further deposited and redeposited glacial drift, leaving features such as kettles dotting the landscape (**Figure 20-D**).

Waves Reshape Cape Cod: At the end of the Pleistocene, as the ice sheets melted, sea level rose, leaving Cape Cod nearly surrounded by the ocean. Waves then started to reshape the landscape. Along what is now Marconi Beach and the Outer Beaches to the north, waves began cutting back the deposits of till, leaving a high coastal bluff (**Figure 20-E**). Along this eastern edge of the peninsula, northward-moving longshore currents and beach drifting transported sand along the shore, eventually depositing some of it into the narrow, curved spit at Provincetown (**Figure 20-F**).

▼ **Figure 20-D** Great Pond in Cape Cod formed in a glacial kettle.

▲ **Figure 20-E** Bluffs along the eastern edge of Cape Cod are made of glacial till, now being worn back by waves.

▲ **Figure 20-F** This spit near Provincetown, Cape Cod, is formed of sand deposited by beach drifting and longshore currents.

Still Changing: Cape Cod is still evolving. Waves continue to wear away at the bluffs. Sand continues to be transported along the shore. Currently, more dry land is lost each year to the sea than is added.

These changes to the landscape can be observed at even the human timescale. In the early 1800s, low-lying Billingsgate Island at the mouth of Wellfleet Harbor on Cape Cod Bay was the location of 30 homes and a lighthouse. A storm in 1855 cut the island in two; a seawall was built in the late 1800s, but waves continued to erode the island, and by the 1940s only a shoal was left to be exposed at low tide.

Such loss of shoreline to natural processes has been ongoing for centuries. However, sea-level rise as a result of current climate change will accelerate the process. The fate of Billingsgate Island may be a preview of the not-so-distant future for much of Cape Cod.

Questions

1. What is the origin of the rock material now found in Cape Cod?
2. Over time, what is the eventual fate of the bluffs along Marconi Beach?

Mobile Field Trip
Cape Cod: Sculpted by Ice & Storm
https://media.pearsoncmg.com/ph/esm/esm_mcknight_physgeo_13/media/

◄ **Figure 20-27** Western Brook Pond fjord in Newfoundland's Gros Morne National Park, Canada.

A combination of hydraulic pounding, abrasion, pneumatic push, and chemical solution at the cliff base cuts a notch at the high-water level (**Figure 20-28a**). As the notch enlarges, the overhang sporadically collapses; the cliff recedes as the ocean advances. Where wave action cuts through the bottom of a cliff-topped headland, a *sea arch* may form (see Figure 20-6), while *sea stacks* develop where wave erosion leaves towers of rock isolated just offshore from the coastal cliff (see Figure 20-1).

Seaward of the cliff face, the pounding and abrasion of the waves create a broad erosional surface called a **wave-cut platform** (or *wave-cut bench*) usually slightly below water level. The combination of wave-cut cliff and wave-cut platform produces a profile that resembles a letter "L," with the steep vertical cliff descending to a notched base and a nearly flat, gently sloping platform extending seaward.

Most of the debris eroded from cliff and platform is removed by the swirling waters. The larger fragments are battered into smaller and smaller pieces until they are transportable. Some of the sand and gravel produced in this fashion may be washed into an adjacent bay and become, at least temporarily, a part of the beach. Much of the sediment, however, is shifted directly seaward, where it may be deposited. As time passes and the cliff wears away by weathering and erosion, these deposits may eventually cover the wave-cut platform, resulting in a beach that extends to the base of the cliff.

Marine Terraces: When a wave-cut platform is uplifted along a tectonically rising coast, a **marine terrace** forms (**Figure 20-28b**). It appears that fluctuations of sea level during the Pleistocene played a part in the formation of at least some marine terraces. When sea level drops during a glacial period, the wave-cut platform is left well above sea level; gradual tectonic uplift during the period of low sea level leaves the terrace high enough to be preserved after sea level rises during the subsequent interglacial period.

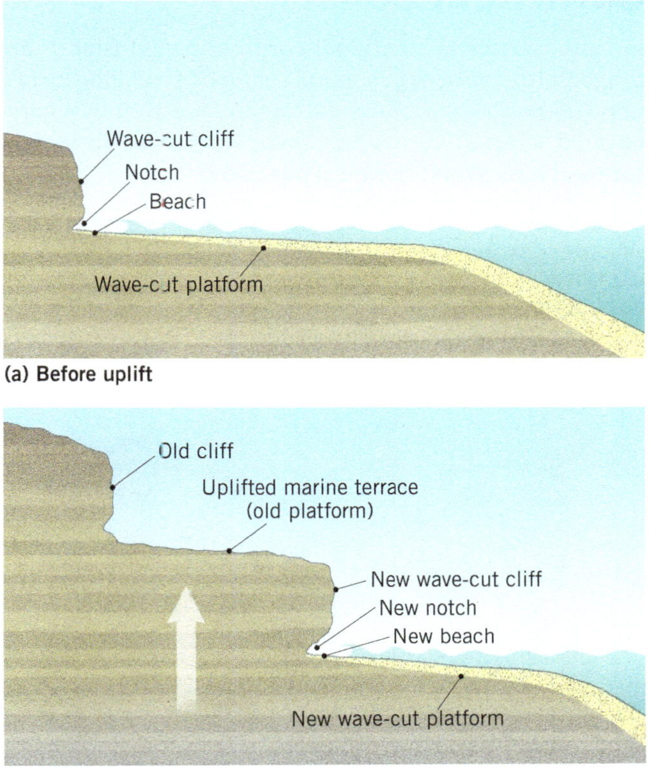

▲ **Figure 20-28** (a) A wave-cut platform develops where a coastal cliff is worn back by wave erosion. (b) A marine terrace develops when a wave-cut platform is tectonically uplifted above sea level.

◀ Figure 20-29 Uplifted marine terraces, well above the present shoreline. This scene is on the north coast of California near Fort Ross.

Along some shorelines, a series of marine terraces is present, reflecting several episodes of terrace formation (Figure 20-29).

LearningCheck 20-11 **What can you infer about a coastline that has a series of marine terraces?**

Coral Reef Coasts

In tropical oceans, nearly all continents and islands are fringed with *coral reefs* or some other type of coralline formation (Figure 20-30). Coralline structures are built by a complicated series of events that involve animals, algae, and various physical and chemical processes.

Coral Polyps

The critical element in the development of coral reefs is a group of anthozoan animals (members of the class *Anthozoa* that are closely related to jellyfish and sea anemones) called *stony corals*. These tiny creatures (most only a few millimeters long) live in colonies of countless individuals, attaching themselves to one another both with living tissue and with their external skeletons (Figure 20-31). Most massive corals grow 0.5–2 centimeters (0.2–0.8 inches) per year, with branching corals growing faster. Because of this slow rate of growth, it can take up to 10,000 years for a coral reef to become established. Some types of reefs existing today began forming 30 million years ago.

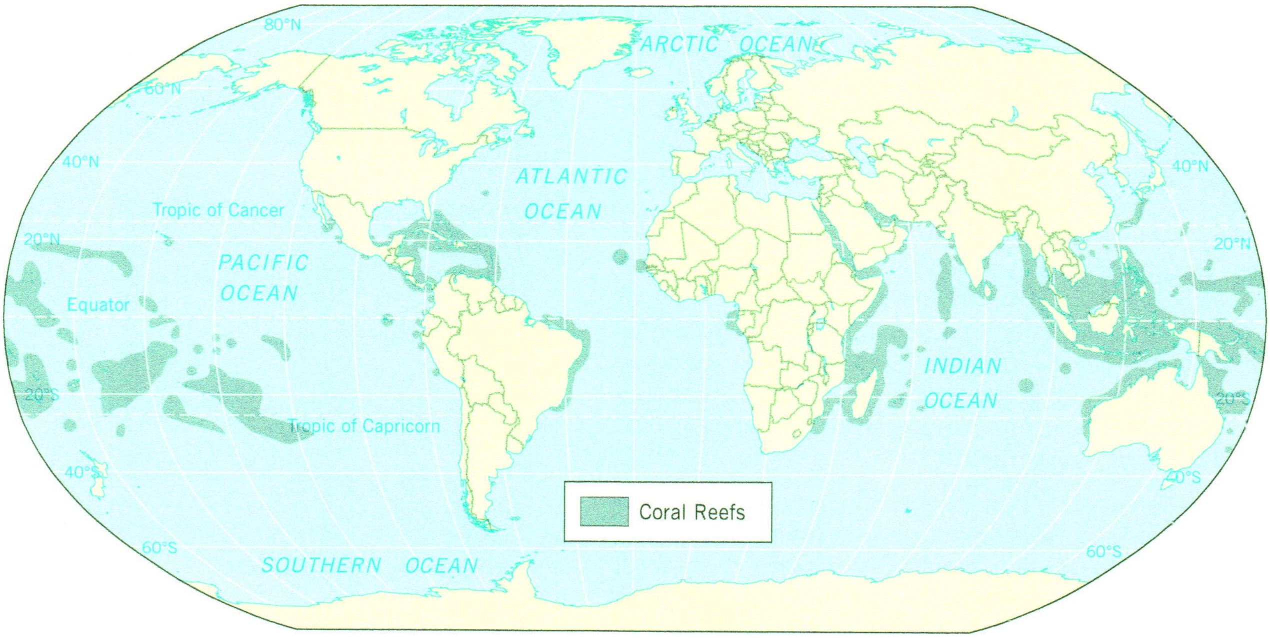

▲ Figure 20-30 Distribution of coral reefs and other coralline structures in the oceans.

▲ **Figure 20-31** Coral species have an extraordinary variety of sizes, shapes, and forms. This close-up reef scene in the Bahamas shows a "Christmas tree" worm atop a colony of coral.

Each individual *coral polyp* extracts calcium carbonate from seawater and secretes a limy skeleton around the lower half of its body. Most polyps withdraw into their skeletal cups during the day and extend their arm-like feeding structures at night. At the top of the body is a mouth surrounded by rings of tentacles, which gives them a blossom-like appearance that for centuries caused biologists to believe that they were plants rather than animals. They feed on minute animal and plant plankton. Although the coral polyp is an animal, reef-building hard corals are hosts to symbiotic algae that provide additional food for the coral polyp through photosynthesis.

The ubiquity of coral reefs in shallow tropical waters is a tribute to the remarkable productiveness of the polyps because they are not very hardy creatures. They cannot survive in water that is very cool, very fresh, or very dirty. Moreover, they require considerable sunlight; most cannot live more than a few tens of meters below the surface of the ocean. (Recently explored *deep-* or *cold-water corals*, found at depths of more than 2000 meters [6600 feet]), are a fascinating exception.)

Many coral reefs around the world show signs of degradation—from both natural and human-generated causes. For example, see the box *Focus: Imperiled Coral Reefs*.

Coral Reefs

Coral polyps can build coralline formations almost anywhere in shallow tropical waters where a coastline provides a stable foundation. Coral reefs in the shallows off the coasts of Florida, for instance, are built on such stable bases. The famous Great Barrier Reef off the northeastern coast of Australia is an immense shallow-water platform of bedrock largely, but not entirely, covered with coral. Its enormously complex structure includes many individual reefs, irregular coral masses, and a number of islands (Figure 20-32).

One favored location for coral reefs is around a volcanic island in tropical waters. As the volcano forms and then subsides, a sequence of different kinds of reefs may grow upward: *fringing reefs*, *barrier reefs*, and *atolls*.

Fringing Reefs: When the volcano first forms, as, for example, over the Hawaiian hot spot described in Chapter 14, coral accumulates on the part of the mountain flank just below sea level because polyps live in these shallow waters. The result is a reef built right onto the volcano (Figure 20-33a). Such an attached reef is called a **fringing reef** (Figure 20-34). The plate, as it moves off the hot spot, cools and becomes denser, so the volcano begins to sink.

Barrier Reefs: As new layers are laid down over old, the coral builds upward around the volcano as a cylinder of irregular height. At the same time, the volcano is subsiding and pulling the original reef base downward. When the coral has been built up enough and the volcano has sunk

Animation
Seamounts & Coral Reefs
https://media.pearsoncmg.com/ph/esm/esm_mcknight_physgeo_13/media/

◀ **Figure 20-32** A view of Australia's Great Barrier Reef.

Focus

Imperiled Coral Reefs

• **Kristine DeLong**, Louisiana State University

▶ **Figure 20-G** (a) Healthy coral and (b) bleached coral in American Samoa.

(a) December 2014 **(b) February 2015**

Coral reefs are some of the most biologically diverse ecosystems in subtropical and tropical oceans. However, coral reefs worldwide are in decline because of multiple stressors. In the Florida Keys and Caribbean, coral reefs were once large and complex, yet today live coral coverage is greatly reduced in these reefs. The largest documented change in the Atlantic reefs occurred after the 1970s, when the dominant branching corals experienced population declines of greater than 90 percent. Two of these coral species, *Acropora palmata* and *Acropora cervicornis*, were listed in 2008 as critically endangered. In 2014, an additional 20 coral species were listed as threatened. Reefs in the eastern tropical Pacific have suffered similar declines; the largest and most robust coral reefs occur in the Indo-Pacific region, where declines are more recent, especially after the 1998 El Niño.

Sources of Coral Stress: Above-normal water temperatures can lead to *coral bleaching*, an event in which corals expel the symbiotic algae that provide these animals with nutrients (**Figure 20-G**). Corals can recover if bleaching lasts for a few days, but drastic temperature changes that last for weeks can result in coral death. Such events are occurring with increased frequency. Bleaching events take place every summer—but the longest lasted from 2014 to 2017. Bleaching events are followed by disease epidemics that result in further coral mortality (**Figure 20-H**).

Increasing CO_2 in the atmosphere leads to CO_2 dissolving in the ocean, which changes the ocean's pH. Increasing seawater acidity reduces growth rates in organisms that have calcified skeletons, such as corals, clams, snails, crustaceans, and some plankton. Future projections of continued CO_2 emissions and ocean acidification suggest increased coral stress, resulting in more bleaching and coral disease and widespread coral mortality by 2100.

Rising sea levels increase the water depth of coral reefs and reduce the amount of light that reaches the corals, which they need to support the symbiotic algae that provide the coral with food. Coral reefs can respond to slow rates of sea-level rise, as they have during past glacial–interglacial cycles. Coral reefs may not be able to keep up with the predicted rates of sea-level rise caused by climate change by the end of the twenty-first century.

Possible Solutions: The *Coral Reef Watch* project (http://coralreefwatch.noaa.gov/satellite/index.php) records surface ocean temperatures via remote sensing to monitor potential thermal stress to corals (**Figure 20-I**). Researchers are actively investigating the impact of climate change on corals and their capacity to adapt. Some studies report declining coral growth rates, whereas others find some species are growing faster but with less dense skeletons. The concern is that high CO_2 levels may lead to the demise of corals, whereas some researchers suggest that corals may be able to adapt or live in some locations. Removing direct human-caused stressors (such as boat and diver damage), reducing runoff and sedimentation from land, and removing human-sourced debris on reefs may give corals a chance to survive and be healthy.

Questions

1. How could El Niño events impact coral reefs?
2. Why is protecting coral reefs important?
3. Would ocean acidification cause coral reefs to dissolve completely? Explain.

▲ **Figure 20-H** A coral with a black band disease.

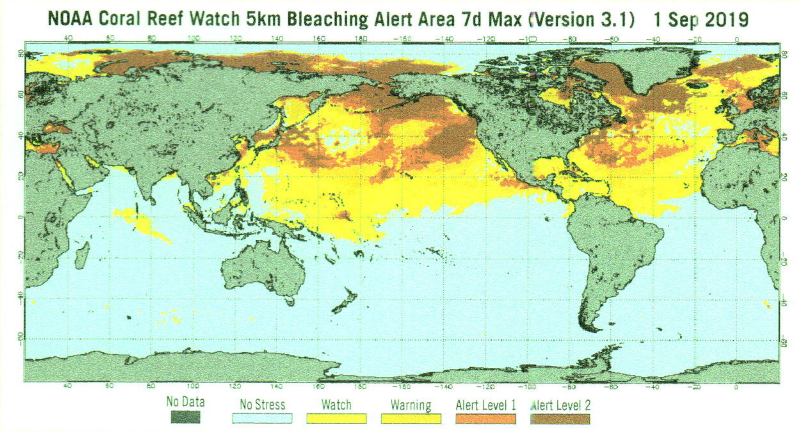

▲ **Figure 20-I** Global map of coral thermal stress levels in September 2019, based on satellite data of sea-surface temperatures. The darkest red shows the highest potential for coral bleaching.

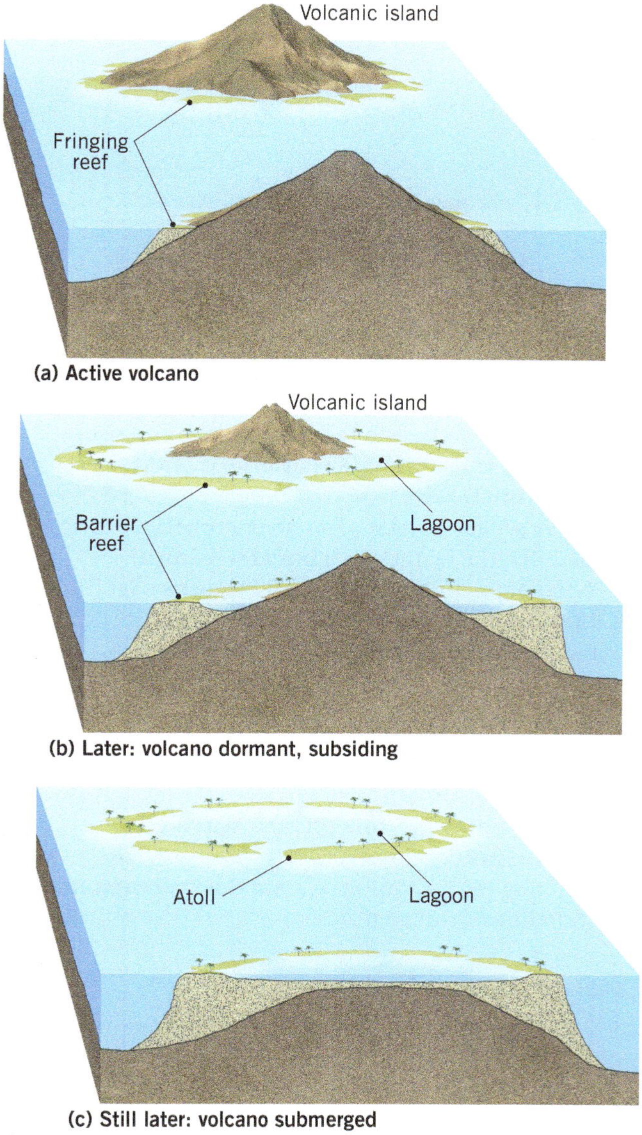

▲ **Figure 20-33** Coral reef formation around a sinking volcano. (a) Secretions from coral polyps living along the shallow-water flanks of a newly formed volcano accumulate into a fringing reef attached to the volcanic island as it rises above a tropical ocean. (b) As the volcano becomes dormant and begins to subside, the coral continues to grow upward over the original base, essentially a cylinder surrounding the island. Such a reef separated from its mainland by a lagoon is a barrier reef. (c) Once the volcano is completely submerged, the coral surrounding a landless lagoon is called an atoll reef.

▲ **Figure 20-34** The fringing reef on Monuriki Island, Fiji.

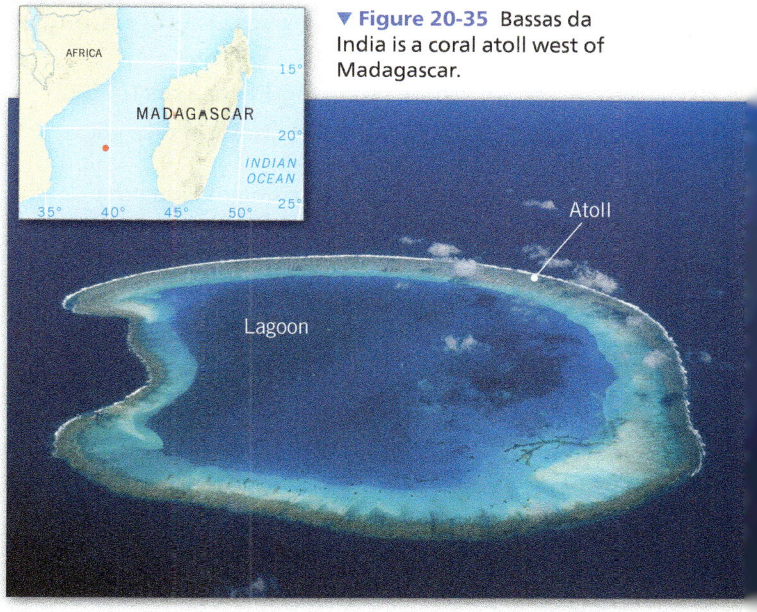

▼ **Figure 20-35** Bassas da India is a coral atoll west of Madagascar.

enough, the result at the water surface is a coral ring separated by a lagoon from the part of the volcano still above water (Figure 20-33b). Called a **barrier reef**, this ring of coral may be a broken circle because of the varying thickness of the coral, appearing to "float" around a central volcanic peak. (The ring is actually attached to the flanks of the sinking mountain far below the water surface.) The surface of a barrier reef is usually right at sea level, with some portions projecting upward into the air.

Atolls: Coral polyps continue to live in the upper, shallow-water portions of a barrier reef, so the reef continues to grow upward. Once the top of the volcano sinks below the water surface, the reef surrounding a now landless lagoon is called an **atoll** (Figure 20-33c). The term *atoll* implies a ring-shaped structure (Figure 20-35). The ring is rarely unbroken, however; rather, it consists of a string of closely spaced coral islets separated by narrow channels of water. Each islet is called a *motu*. Because coral cannot live above sea level, much of the coral debris that makes up the above-water portion of an atoll was deposited by storm waves.

LearningCheck 20-12 **How can an atoll develop from a fringing reef or a barrier reef?**

20 Learning Review

After studying this chapter, you should be able to answer the following questions. Key terms from each text section are shown in **bold type.** Definitions for key terms are also found in the glossary at the back of the book.

Key Terms and Concepts

Coastal Processes (*p. 576*)

1. What are some of the ways that landform development along ocean shorelines is different from that along lakeshores?

Waves (*p. 576*)

2. How are most ocean waves generated?
3. What are **swells** in the ocean?
4. How is a **wave of oscillation** different from a **wave of translation**?
5. Contrast ocean waves in deep water with waves in shallow water. Especially note how **wavelength** and **wave height** change as a wave comes onshore.
6. Contrast **swash** and **backwash** on a beach.
7. Describe and explain the process of **wave refraction**.
8. Explain the formation and characteristics of a **tsunami**.

Important Shoreline-Shaping Processes (*p. 582*)

9. Why did **eustatic sea-level changes** take place during the Pleistocene?
10. What is the source of most sediment along the shorelines of the continents?
11. Describe **longshore currents** and their formation.
12. Describe the process of **beach drifting**.

Coastal Depositional Landforms (*p. 585*)

13. Explain the concept of the **sediment budget** of a coastal depositional landform such as a beach.
14. What causes a **beach** to change shape and size?
15. Describe and contrast a coastal **spit** and a **baymouth bar**.
16. Under what circumstances does a **tombolo** form?
17. Describe the features of a **barrier island**.
18. What happens to most coastal **lagoons** over time?
19. How do **groins** and **jetties** typically affect the beaches around them?

Shorelines of Submergence and Emergence (*p. 591*)

20. Explain the formation of **ria shorelines** and **fjords**.
21. Explain how wave-cut cliffs and **wave-cut platforms** develop.
22. Describe the formation of a **marine terrace** along a shoreline of emergence.

Coral Reef Coasts (*p. 594*)

23. Explain how a coral **fringing reef** forms and how it can subsequently become first a **barrier reef** and then an **atoll**.

Study Questions

1. What factors influence the erosional power of waves striking a coastline?
2. How does air serve as a tool of erosion in wave action?
3. What will likely happen to a down-current beach when a major river flowing into the ocean is dammed? Why?
4. What is beach nourishment, and is it a good investment for a coastal community? Why or why not?
5. Why are shorelines of submergence so common today?
6. How might global warming and other environmental changes caused by human activities affect coral reefs?

Exercises

1. The steepness of a wave is described with the ratio of the wave height to wavelength (in other words, wave steepness = wave height ÷ wavelength). When the steepness is greater than 1/7, the wave breaks.
   a. What is the maximum wave height for a wavelength of 8 meters? ______ meters
   b. What is the maximum wave height for a wavelength of 21 meters? ______ meters
   c. If a breaking wave is 2 meters high, what is the wavelength? ______ meters
2. Most waves will break when they reach a water depth that is about 1.3 times the wave height. If a breaking wave has a height of 1.5 meters, how deep is the water below? ______ meters
3. Most waves will break when they reach a water depth that is about 1.3 times the wave height. If a breaking wave has a height of 7 meters, how deep is the water below? ______ meters

Environmental **Analysis** SEA-LEVEL RISE

Data
Sea-Level Rise
https://media.pearsoncmg.com/ph/esm/esm_mcknight_physgeo_13/media/

Climate projections indicate that sea level will rise globally as ice caps and glaciers melt. This change will impact coastal regions, with some areas being more vulnerable than others.

Activities

Go to the Sea Level Rise and Coastal Flooding Impacts site at https://coast.noaa.gov/slr. Read the Disclaimer and then click "Get Started."

1. What is the purpose of this data viewer?

Zoom in to a state along the East Coast. Use the "Sea Level Rise" slider to raise the sea level.

2. What is the effect of rising sea level? What percentage of the coastline is affected?

Return sea level to its current value. Click on the "Vulnerability" icon and zoom in to a city with high (in red) "Social Vulnerability." Toggle between "High Tide Flooding" and "Vulnerability" to determine the geography of the area.

3. Identify the city. Describe the geographic features that make this city vulnerable.

Toggle between "Sea Level Rise" and "High Tide Flooding."

4. How will high tide flooding change as sea level rises in this city?

Click on "Local Scenarios" and then click on the question mark. Complete the "Local Scenarios Tutorial."

5. Estimate the year when Key Biscayne, Florida, will experience widespread flooding using a high sea-level rise scenario. Why is this information useful?
6. You are on a committee that is planning to avoid widespread flooding in Key Biscayne, Florida. What water level leads to widespread flooding in 50 years?
7. Which sea-level rise scenario should the committee use to plan for the next 50 years?

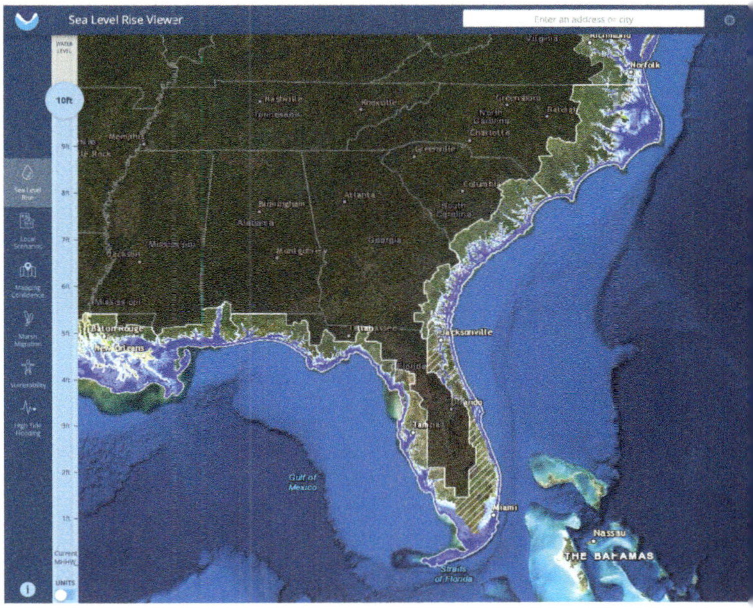

Seeing Geographically in Our National Parks

Look again at the photograph of the coast of Acadia National Park at the beginning of the chapter (p. 574). Based on the direction that the waves are approaching this shoreline, in which direction should the longshore current transport the sediment from the beach? Explain where and how a sea stack could form along this section of the coast.

Mastering Geography®

Looking for additional review and test prep materials? Visit the Study Area in **Mastering Geography** to enhance your geographic literacy, spatial reasoning skills, and understanding of this chapter's content by accessing a variety of resources, including Map Master interactive maps, geoscience animations, *Mobile Field Trips*, videos, *Project Condor* Quadcopter videos, *In the News* RSS feeds, flashcards, web links, self-study quizzes, and an eText version of *McKnight's Physical Geography*.

Learning Check Answers

Chapter 1

1-1: *Physical geography* primarily focuses on patterns in the natural environment and on human interaction with the environment; *human geography* primarily focuses on patterns of human activity and culture. **1-2:** Many patterns and processes in the natural environment can be influenced by economic activity—such as the consequences of the extraction of resources or the burning of fossil fuels; economic activity in one part of the world can influence the environment in another part of the world. **1-3:** In conversational English, a *theory* often means a guess or a hunch; in scientific terms, a theory is a well-tested and widely accepted explanation. **1-4:** The *lithosphere* is the solid part of Earth; the *atmosphere*, the gases surrounding Earth; the *hydrosphere*, the waters of Earth; the *cryosphere*, the frozen water of Earth; and the *biosphere*, the living organisms on Earth. **1-5:** A system is in equilibrium when the inputs are balanced by the outputs, and the conditions within a system remain the same over time. **1-6:** With a *positive feedback loop*, change in one direction tends to reinforce change in that direction; with a *negative feedback loop*, change tends to bring the system back toward equilibrium. **1-7:** Terrestrial planets are relatively small and are composed primarily of mineral material; Jovian planets are larger, less dense, and composed primarily of gases, liquids, and ices. **1-8:** Earth's highest point is Mount Everest (8848 meters; 29,029 feet), and its lowest point is the bottom of the Mariana Trench (−11,033 meters; −36,198 feet)—a difference of 19,881 meters (65,227 feet). **1-9:** The largest circle that can be drawn on a sphere, dividing it into two hemispheres; the equator is a great circle. **1-10:** All parallels are parallel to each other—they never cross or touch. **1-11:** North America is west of the prime meridian, so it is described by west longitude; China is east of the prime meridian, so it is described by east longitude. **1-12:** Earth rotates, or "spins," on its axis; Earth revolves, or "orbits," around the Sun. **1-13:** No; the North Pole leans most directly toward the Sun on the June solstice and leans most directly away from the Sun on the December solstice. **1-14:** 23.5° N. **1-15:** Not at all—the equator has virtually 12 hours of daylight every day of the year. **1-16:** On the March equinox and September equinox. **1-17:** 6 months. **1-18:** It becomes one hour later. **1-19:** It becomes the previous day.

Chapter 2

2-1: Because it is impossible to flatten a sphere without distortion. **2-2:** 10,000 centimeters. **2-3:** A small-scale map shows a large area but with little detail; a large-scale map shows a small area in great detail. **2-4:** A system that mathematically transfers the graticule and features of Earth onto the flat surface of a map. **2-5:** An *equivalent (equal area) map* shows correct area (size) relationships over the entire map; a *conformal map* maintains correct angular (shape) relationships over the entire map. **2-6:** No; a *Mercator projection* shows correct shapes but severely distorts apparent area in the high latitudes, making it unsuitable for studying area distribution patterns. **2-7:** A *Goode's interrupted projection* is equivalent, making it suitable for studying area distribution patterns; in addition, Goode's maintains reasonable shape relationships for the continents (although it isn't conformal). **2-8:** A line of equal value on a map; for example, temperature patterns can be mapped with *isotherms* and topography mapped with *elevation contour* lines. **2-9:** An oblique view of the landscape is mathematically constructed from elevation data, using shaded relief, as if the Sun were illuminating the topography from the northwest. **2-10:** A GPS unit receives timing signals from at least three satellites; by determining the slight difference in arrival time of the signals, the GPS unit calculates the distance to each satellite and, from that information, triangulates a location. **2-11:** Near infrared images measure short infrared radiation, typically resulting in a "false-color" image that can, for example, be used to detect differences in living vegetation; thermal infrared imagery detects differences in emitted longwave radiation, so, in effect, it measures differences in temperature. **2-12:** Remote sensing that detects radiation in several bands of radiation simultaneously, such as visible light, near infrared, middle infrared, and thermal infrared. **2-13:** Geographic information systems (GIS) integrate databases and maps, allowing for sophisticated overlay analysis; the Global Positioning System (GPS) determines location precisely.

Chapter 3

3-1: Density of the atmosphere decreases with increasing altitude. **3-2:** Nitrogen (N_2); virtually no effect on weather processes; ozone absorbs a large portion of incoming ultraviolet (UV) radiation from the Sun. **3-3:** Temperature generally decreases from the surface up to the *tropopause* because the surface is the source of warmth for the troposphere; in the *stratosphere*, temperature increases with increasing altitude due to the absorption of ultraviolet radiation by the ozone layer. **3-4:** The *homosphere* has a homogeneous composition, and the *heterosphere* has gas layers by weight. **3-5:** *Weather* refers to the short-run conditions in the atmosphere; *climate* is the aggregate of weather conditions over at least 30 years. **3-6:** The *controls of weather and climate* influence or act on the *elements of weather and climate*, producing variations in atmospheric conditions; latitude is the most fundamental control of weather and climate, primarily because of latitudinal differences in the receipt of solar energy. **3-7:** As a result of Earth's rotation, the path of any free-moving object is deflected to the right in the Northern Hemisphere and to the left in the Southern Hemisphere. **3-8:** CFCs (chlorofluorocarbons) and other ozone-depleting chemicals in the ozone layer undergo a photochemical reaction that destroys ozone. **3-9:** *Smog* forms from a photochemical reaction of primary pollutants such as nitrogen dioxide; smog can damage human tissue in the eyes, lungs, and nose.

Chapter 4

4-1: *Temperature* is a description of the average kinetic energy of the molecules in a substance; *heat* is the energy that transfers from one object to another because of a difference in temperature. **4-2:** *Shortwave radiation* is emitted by the Sun and includes ultraviolet, visible, and short (near) infrared wavelengths of electromagnetic radiation; *longwave radiation* is emitted by Earth's surface and atmosphere and consists of thermal infrared (far infrared) wavelengths. **4-3:** *Radiation* is the flow (or emission) of electromagnetic energy; *absorption* entails the assimilation of electromagnetic energy by an object; *reflection* occurs when an object repels the electromagnetic waves that strike it. Black asphalt has low *albedo* because that surface is a poor reflector. **4-4:** Shortwave radiation from the Sun transmits through the atmosphere and is absorbed by Earth's surface; the surface is warmed, so it emits longwave radiation that is absorbed by *greenhouse gases* (such as water vapor and carbon dioxide), thus delaying the eventual loss of this energy to space. **4-5:** *Conduction* involves the transfer of energy through molecule-to-molecule collision; *convection* involves the transfer of energy through the vertical circulation of a fluid (such as air). **4-6:** Rising air expands and cools adiabatically; descending air compresses and warms adiabatically. **4-7:** Through conduction and convection, through the emission of longwave radiation and the greenhouse effect, and through the transfer of latent heat that is absorbed during the evaporation of water and released during condensation. **4-8:** When the Sun is high in the sky, the energy from a beam of sunlight passes through less atmosphere and is concentrated onto a smaller surface area than when the Sun is lower in the sky. **4-9:** Because of its high *specific heat*, water must absorb about five times more energy than land to exhibit a comparable temperature increase. **4-10:** The interior location, because water only slowly warms in summer, whereas land warms quickly. **4-11:** The ocean current flowing along the west coast of a continent is cool; the current flowing along the east coast is warm. **4-12:** An increase in temperature with increasing altitude—the opposite, or inverse, of the usual situation. A temperature inversion inhibits the upward movement and mixing of air—a condition that can allow air pollutants to accumulate. **4-13:** From the equator toward the poles, temperature generally decreases because low latitudes receive more total insolation over the year than high latitudes do. **4-14:** The tropics and coastal areas have small annual temperatures ranges; high latitudes in the middle of continents experience the greatest temperature ranges. **4-15:** Increasing greenhouse gas emissions, especially carbon dioxide released through the burning of fossil fuels since the mid-1700s, have increased the greenhouse effect in the atmosphere.

Chapter 5

5-1: The density of the atmosphere decreases with altitude, so pressure decreases. **5-2:** Descending air is most likely associated with high atmospheric pressure at the surface; rising air is more likely to be associated with low pressure. **5-3:** In the lowest 1000 meters of the atmosphere, friction slows the wind, so the amount of Coriolis effect deflection decreases. **5-4:** Wind converges in a counterclockwise direction in a surface *cyclone* in the Northern Hemisphere; wind diverges clockwise from a Northern Hemisphere *anticyclone*. **5-5:** Air rises in a *cyclone* and descends toward the surface in an *anticyclone*. **5-6:** Warm air rises near the equator; by an altitude of about 15 kilometers, the air spreads northward and southward and descends toward the surface at about 30° N and S, where it moves back toward the equator; there it is uplifted again in the *Hadley cells*. **5-7:** *Subtropical highs* are generally located over the ocean, just off the west coasts of continents at about 25° to 30° N and S; weather is generally sunny and dry. **5-8:** The ITCZ is generally near the equator; weather is characterized by clouds, thunderstorms, and rain. **5-9:** The *trade winds* and *westerlies* diverge from the subtropical highs. **5-10:** High-speed currents of air in the upper troposphere; in each hemisphere, one *jet stream* is generally located just poleward of the subtropical highs and one just equatorward of the polar front. **5-11:** The position of the ITCZ shifts with the seasons, following the location of greatest insolation and surface warming. **5-12:** During the summer *monsoon*, wind blows onshore, off the ocean; during the winter monsoon, wind blows offshore. **5-13:** Warming of land during the day creates relatively lower pressure over land, so the wind blows off the water toward land in a *sea breeze*. **5-14:** During an *El Niño* event, the trades weaken or reverse direction. **5-15:** A connection between the weather or the ocean in one part of the world with that in another part; for example, during a strong El Niño, the North Atlantic Ocean usually experiences fewer hurricanes than normal.

Chapter 6

6-1: The attraction between the negative (oxygen) side of one water molecule and the positive (hydrogen) side of another. **6-2:** Its volume increases. **6-3:** About five times as much energy is needed to increase the temperature of water as a comparable mass of soil or rock, slowing water's rate of warming in summer. **6-4:** In order for ice to melt, it must absorb energy (the *latent heat* of melting) to break bonds and allow a phase change; this energy does not increase the temperature of the water. **6-5:** The rate of evaporation drops off sharply. **6-6:** Ten grams of water as vapor is in each kilogram of air. **6-7:** $5\text{ g} \div 20\text{ g} = {}^1/_4$, or 25 percent *relative humidity*. **6-8:** As temperature decreases, water vapor capacity decreases, so relative humidity increases. **6-9:** The air must reach 100 percent relative humidity (saturation) and surfaces, such as *condensation nuclei*, must be available. **6-10:** Above the *lifting condensation level*, rising air is expanding and cooling adiabatically, but latent heat released during condensation counteracts some of this cooling. **6-11:** *Cirriform* (wispy clouds), *stratiform* (widespread layered clouds), and *cumuliform* (puffy clouds). **6-12:** A parcel of air will be unstable (will rise without being forced) if it is warmer, thus less dense, than the surrounding air. **6-13:** Highly unstable air is associated with strong updrafts in cumulonimbus clouds; such updrafts are necessary to lift liquid water drops high up into a cloud, where they freeze into hail. **6-14:** When air is forced to rise on the windward side of a mountain, adiabatic cooling often leads to cloud formation and precipitation; air on the leeward side of the mountain (in the *rain shadow*) has lost moisture rising up the windward side. If this air descends down the leeward side, it will warm adiabatically, reducing its relative humidity even more. **6-15:** Warm, moist air rising in the ITCZ leads to high rainfall. **6-16:** The descending air of the subtropical highs leads to dry conditions. **6-17:** High *precipitation variability* from year to year. **6-18:** Sulfur dioxide (SO_2) from power plants and nitrogen oxides (NO_x) from vehicle exhaust are common sources of the chemicals that lead to acid rain.

Chapter 7

7-1: *Air masses* form over uniform surfaces, such as the ocean or tundra, when wind stops for a few days, allowing the surface to impart its characteristics to the air above. In the band of the westerlies (the midlatitudes), the atmosphere is rarely quiet long enough for an air mass to form. **7-2:** A *maritime polar* air mass is moist and cool; a continental tropical air mass is hot and dry. **7-3:** A *cold front* forms when a cold air mass advances into the area of a warm air mass; a cold front tends to have a steeper slope, move faster, and produce more precipitation than a *warm front*. **7-4:** The convergence of air into the low of a *midlatitude cyclone* brings together two unlike air masses. **7-5:** Precipitation develops where air is rising and cooling adiabatically within a midlatitude cyclone—typically along the cold front, the warm front, and in the center of the low. **7-6:** An *occluded front* forms when the faster-moving cold front catches up with the warm front, lifting all of the warm air mass off the surface. **7-7:** A *trough* is found along the surface position of a well-developed cold front; as the front approaches, the trough is getting closer, so pressure falls; as the front moves away, pressure rises because the trough is moving away. **7-8:** Descending air cannot make clouds and rain, and the diverging wind pattern does not bring together air masses (so no fronts form). **7-9:** Warm, moist air cools as it converges and rises in a *hurricane*; condensation releases latent heat that increases instability, strengthening the updrafts and low pressure of the storm. **7-10:** Warm ocean currents are found along the east coasts of continents; the cool currents along the west coast quickly weaken a hurricane. **7-11:** The strong winds of a hurricane generate high waves (and the very low pressure allows sea level to rise slightly): when the storm reaches a coastline, this bulge of ocean water is pushed onshore. **7-12:** Initially, updrafts dominate, and the cumulonimbus cloud grows; when it is mature, falling precipitation initiates downdrafts that eventually lead to storm dissipation. **7-13:** Slower winds near the surface, with faster winds above, generate air spinning along a horizontal axis;

strong thunderstorm updrafts lift this spinning air into a rotating *mesocyclone*, from which a tornado may form.

Chapter 8

8-1: The *Köppen climate classification system* is based on a location's average monthly temperature and precipitation. **8-2:** A *climograph* plots a location's average monthly temperature with a solid line, and average monthly precipitation with bars coming up from the bottom of the chart. **8-3:** Areas of Af climate are generally under the influence of the ITCZ all year, whereas areas of Aw climate come under the influence of the ITCZ when it migrates into the area during summer. **8-4:** Most deserts are caused by the atmospheric stability associated with cool ocean currents and the presence of the subtropical high all year, are in the rain shadows of mountains, or are deep in a continent, far from sources of moisture. **8-5:** The subtropical highs are most persistent over the cool water along the west coasts of continents. **8-6:** BWh climates have hot summers and mild winters; BWk climates are found in the midlatitudes, so they have warm summers but cold winters. **8-7:** Areas of both Cs and Cfa climate receive rain in winter from midlatitude cyclones traveling in the westerlies. In summer, Cfa climates receive rain from thunderstorms and tropical cyclones; however, along the west coasts, where Cs climates are found, the subtropical highs migrate poleward in summer, bringing stable conditions and dry weather. **8-8:** The prevailing westerlies bring cold, continental air masses to the eastern side of the continent. **8-9:** The very cold air has almost no water vapor capacity, so little precipitation is possible. **8-10:** As elevation increases, temperature generally decreases—roughly similar to what happens to temperature with increasing latitude. **8-11:** Water molecules containing the lighter oxygen isotope ^{16}O evaporate more easily than those with the heavier ^{18}O; therefore, when climate is cold, more water molecules with ^{16}O than ^{18}O will evaporate and can fall as snow. By comparing the ratio of ^{18}O to ^{16}O in glacial ice, we can infer whether that ice layer accumulated during a colder period or a warmer period. **8-12:** Tiny air bubbles frozen in the glacial ice retain the atmospheric gas mixture present when the snow fell. **8-13:** One example: high concentrations of volcanic ash in the atmosphere can block incoming solar radiation, lowering global temperatures for a period of time. **8-14:** Long-term but predictable variations in Earth–Sun relations taking place over periods of thousands of years; these cycles combine to influence the onset and end of glacial periods. **8-15:** One example: higher temperatures lead to reduced ice and snow cover; the resulting lower albedo increases absorption of solar radiation by the surface, leading to higher temperatures. **8-16:** Differences from the long-term average temperature; over the last century, both land and ocean temperatures have increased. **8-17:** The IPCC projects that, in addition to a continued increase in average global temperature, sea level will rise slightly, polar sea ice will diminish, and the frequency of extreme weather events will increase.

Chapter 9

9-1: No; more water falls on the continents as precipitation than is added to the air above through evapotranspiration—the "extra" precipitation falling on the continents comes from water evaporated from the oceans. **9-2:** Salinity is generally highest in subtropical areas, where evaporation is high. Salinity is generally lowest near the mouths of major rivers; near the equator, where rainfall is high; and in polar areas, where evaporation is low and meltwater from glaciers enters the ocean. **9-3:** The highest high tides (*spring tides*) occur around the time of the new moon and full moon, when the Sun, Earth, and Moon are in a line; the lowest high tides (*neap tides*) occur when the Moon and Sun are at right angles to each other. **9-4:** It is associated with *thermohaline circulation*, in which cold, dense, relatively salty ocean water sinks in high-latitude ocean areas; this cold, dense water slowly flows along the bottom of the ocean, eventually to rise into warmer, shallow currents. **9-5:** Most ice is locked in continental ice sheets in Antarctica and Greenland. **9-6:** With thawing *permafrost*, buildings and roads may be damaged by subsidence, coastal areas may erode more quickly, and greenhouse gases may be released from decomposing, newly thawed organic matter. **9-7:** The Aral Sea has shrunk significantly, mainly because dams and irrigation projects diverted much of the water that once flowed into the lake. **9-8:** A layer of permeable rock into which groundwater can infiltrate. **9-9:** When a well pumps groundwater faster than it recharges, the *water table* of an *aquifer* drops around the well, forming an inverted cone pattern, sometimes leaving nearby shallow wells dry. **9-10:** Because withdrawal has exceeded recharge, the water table of the Ogallala Aquifer has dropped significantly (more than 45 meters [150 feet]) in some places.

Chapter 10

10-1: *Biodiversity* describes the number of different kinds of organisms in a location. **10-2:** *Photosynthesis* is the pathway through which most solar energy enters the biosphere. **10-3:** Photosynthesis converts CO_2 into *carbohydrates*; sediments derived from this organic material can be stored in carbonate rocks or coal, later to be released into the atmosphere as CO_2 through rock weathering or the burning of fossil fuels. **10-4:** Nitrogen becomes usable to plants after it has been "fixed" into *nitrates*, primarily by nitrogen-fixing bacteria in soil. **10-5:** *Producers* are plants that fix carbon and store solar energy through photosynthesis; *primary consumers* are plant-eating animals; *secondary consumers* are animals that eat other animals. **10-6:** Typically only about 10 percent of the energy stored in the *biomass* of one organism can be passed on and stored by an organism that consumes it. **10-7:** Because acacias evolved before the continents began to rift apart millions of years ago, they are found widely in the Southern Hemisphere; the range of cattle egrets increased recently, following the expansion of cattle ranching around the world. **10-8:** Over time, the lake will develop marsh vegetation around its edges; as sediment continues to fill the lake, it will eventually become a meadow, and finally a forest. **10-9:** *Climax vegetation* is a stable association of plants that tends to change little over time; such plants can be thought of as representing the "optimal" vegetation for that environment. **10-10:** All plants require light for photosynthesis, so in dense groves, trees are likely to be tall and slender in order to gain access to light. Wind especially has a drying effect on plants, favoring growth on protected sides of branches. **10-11:** Wildfire improves the health of some ecosystems by clearing away old vegetation; the seedpods or cones of some trees are opened by the heat of a fire. **10-12:** Nutrients cycle very quickly through the tropical rainforest; in effect, most of the nutrients are held in the living vegetation.

Chapter 11

11-1: A *biome* is a large, recognizable assemblage of plants and animals in a functional interaction with the environment. **11-2:** *Annual* plants die off during the winter or dry season, leave seeds, and then germinate when the conditions are favorable; *perennial* plants live through weather and environmental fluctuations on an ongoing basis. **11-3:** *Evergreen trees* are fully leaved at all times; *deciduous trees* lose their leaves each year during the cold or dry season. **11-4:** *Xerophytic adaptations* allow plants to survive dry conditions; *hygrophytic adaptations* allow plants to survive in wet, terrestrial environments. **11-5:** *Vertical zonation* refers to distinct patterns of climate and vegetation resulting from increasing elevation along mountain slopes. A location's upper *treeline* is generally determined by a limiting low summer temperature and to a lesser extent the availability of moisture. **11-6:** Mammals and birds are *endothermic*; unlike *ectotherms*, they can maintain a constant body temperature, even in harshly cold environments. **11-7:** *Hibernation* during the winter allows animals to survive a harsh winter without the need to find food. **11-8:** Many desert animals are nocturnal, sparing them the need to be out during the hottest part of the day; some desert animals can delay reproduction until conditions are favorable. **11-9:** *Mutualism* is a mutually beneficial relationship between organisms; *parasitism* entails one organism obtaining nourishment from another, weakening the host organism. **11-10:** The Australian continent was isolated from other continents quite early in geologic terms, so species here adapted and evolved with little influence from the outside. **11-11:** The *tropical rainforest* biome is characterized by a great many species of plants and animals—although in any given area, there might be only a few individuals of each species. **11-12:** The *tropical savanna* biome is associated with a marked rainfall seasonality—often the tropical savanna (Aw) climate, caused by the seasonal shift of the ITCZ. **11-13:** The *mediterranean woodland and shrub* biome is most closely associated with the winter wet/summer dry mediterranean (Cs) climate. **11-14:** The *boreal forest* biome has much lower biodiversity than the tropical rainforest biome—typically just a few species of trees dominate the boreal forest. **11-15:** Tropical rainforest loss results in the extinction of many species, and once the forest cover is removed, the soil quickly loses fertility and becomes susceptible to rapid erosion from the high rainfall in these areas. **11-16:** An *exotic species* is a nonnative plant or animal that becomes established in a new environment—introduced sometimes unintentionally by human activities; an *invasive species* is an exotic plant or animal that rapidly outcompetes a native species.

Chapter 12

12-1: Soil formation generally begins with the weathering of bedrock. **12-2:** Although the characteristics of the parent rock is important in the early stages of soil development, climate—especially temperature and moisture—is the most important soil-forming factor in the long run. **12-3:** Earthworms aerate soil and facilitate drainage; the organic material in earthworm *casts* increases the fertility of most soils. **12-4:** Many plant nutrients are held to clay particles; *humus* is a source of organic material that can increase soil fertility. **12-5:** *Capillary water*—held on the surface of soil particles by surface tension—is the most important source of moisture for plants because it remains available to plant roots after *gravitational water* has drained away. **12-6:** *Porosity* refers to the amount of open space between soil particles (influencing the amount of water a soil can hold); *permeability* refers to the interconnectedness of the pore spaces (influencing how easily water can move through a soil). **12-7:** *Cation exchange capacity* is the ability of a soil to attract and exchange cations, along with the nutrients associated with them. **12-8:** From top to bottom, *solum* includes a thin *O horizon* of organic material at the surface; an *A horizon* ("topsoil"), with a mixture of mineral material and humus; an *E horizon*, where *eluviation* removes clay, iron, and aluminum; and a *B horizon*, where *illuviation* deposits clay, iron, and aluminum from above. **12-9:** *Laterization* takes place in warm, wet climates and entails the rapid decomposition of organic material and leaching-away of most minerals (including silica)—leaving relatively infertile, brick-red soils with high concentrations of iron and aluminum oxides; *podzolization* takes place in the cool, moist boreal forests of subarctic regions—leaving acidic, sandy, leached soils of low fertility. **12-10:** *Salinization* takes place in areas of high evaporation; as water is drawn up toward the surface, it evaporates and leaves salts deposited in the soil. **12-11:** *Soil Taxonomy* is based on observable characteristics of the soil, not on the local environment or processes of formation. **12-12:** *Entisols* are typically very immature; they show little soil profile development. **12-13:** *Gelisols* are found in the Arctic and subarctic, where permafrost is present. **12-14:** *Aridisols* are generally found in areas of desert climate. **12-15:** *Vertisols* are often good agricultural soils (although sometimes difficult to plow because of their high clay content); *Mollisols* are grassland soils that are highly productive; *Alfisols* are typically fertile soils and second only to Mollisols in agricultural productivity. **12-16:** Both *Spodosols* and *Oxisols* are highly leached, so they are generally quite infertile.

Chapter 13

13-1: The *crust* is Earth's thin outer shell; the *mantle*, below the crust, makes up the bulk of Earth's volume; the solid, dense *inner core* is surrounded by the molten *outer core*. **13-2:** The *lithosphere* (the "plates") consists of the crust and upper, rigid mantle; the *asthenosphere* is the dense, warm, plastic (deformable) layer of the mantle over which the lithosphere moves. **13-3:** *Silicates* are the most abundant family of minerals; these minerals contain silica (SiO_2), composed of the elements silicon and oxygen, often in combination with other elements. **13-4:** *Magma* is molten rock below the surface; *lava* is molten rock at the surface. *Felsic* rocks contain large portions of light-colored silicate minerals (such as quartz and feldspar); *mafic* rocks include large portions of dark-colored minerals (such as olivine and pyroxene); fine-grained igneous rocks form from the rapid cooling of lava, whereas coarse-grained igneous rocks develop from the slow cooling of magma below the surface. **13-5:** *Granite* forms from the slow cooling of felsic magma below the surface (so it has a coarse-grained texture with both light- and dark-colored minerals); *basalt* forms from the rapid cooling of mafic lava (so it has a fine-grained texture of only dark-colored minerals). **13-6:** *Detrital sedimentary rocks*, such as sandstone, form from the accumulated fragments of weathered rock; *chemical sedimentary rocks*, such as limestone, develop from the precipitation of dissolved compounds (such as calcium carbonate); *organic sedimentary rocks*, such as bituminous coal, consist of compacted dead plant material. **13-7:** *Foliation* is the prominent alignment or orientation in some metamorphic rocks; it develops when directed stress is applied to rocks during metamorphism, leaving the minerals oriented with a distinctly platy or banded texture. **13-8:** Because oceanic lithosphere includes basaltic crust, it is denser than continental lithosphere. The dominant crustal rocks of the continents, such as granite, are less dense than basalt, the dominant crustal rock of the ocean floors. **13-9:** *Structure* refers to the nature and orientation of the rocks in a landform. **13-10:** *Internal processes* originate within Earth and include the lithospheric rearrangement associated with plate tectonics, as well as volcanism, folding, and faulting—processes that can increase the relief of the landscape; *external processes*—weathering, mass wasting, and erosion—operate on, or just below, Earth's surface and can wear down the landscape and diminish relief. **13-11:** In order for the incrementally slow internal and external processes to operate and make a significant difference in the landscape, very long periods of time (by human standards) are required. **13-12:** An individual landform is best studied at the local scale (as in Figure 13-21a); continental structure is best studied from a more distant perspective (as in Figure 13-21e).

Chapter 14

14-1: Continental coastlines (especially across the Atlantic Ocean) match up like the pieces of a jigsaw puzzle; patterns of ancient glaciation also align if the continents are put back into their positions in Pangaea, when this glaciation took place. **14-2:** Ocean floor is created and spreads away in both directions from a *midocean ridge*; ocean floor is "recycled" through the process of *subduction* at *oceanic trenches*. **14-3:** The ocean floor is youngest at midocean ridges and becomes progressively older in both directions away from the ridge—showing that the ocean floor is indeed spreading away from the ridges. **14-4:** Plates are large slabs of lithosphere (the crust and upper mantle) that move over the dense, warm asthenosphere below and are put in motion by the slow convection of heated rock within the mantle. **14-5:** Both *midocean ridges* and *continental rift valleys* are types of *divergent plate boundaries*, where two plates spread away from each other. **14-6:** Both are subduction zones, associated with shallow- and deep-focus earthquakes; oceanic–continental convergence produces a deep oceanic trench and a chain of volcanoes just inland from the coastline; oceanic–oceanic convergence produces a deep oceanic trench and a *volcanic island arc* parallel to the trench. **14-7:** The Atlantic Ocean is the result of the divergence of the continents away from the Mid-Atlantic Ridge; the Himalayas are being uplifted by the continental collision of India with Eurasia; the volcanoes of the Andes are a result of the subduction of the Nazca Plate beneath the South American Plate. **14-8:** The chain of islands is forming as the Pacific Plate moves over a stationary plume of magma rising from the mantle. **14-9:** The western edge of North America has grown through the accretion of material—especially lithosphere that was too buoyant to subduct. **14-10:** The relative amount of silica in the lava; high-silica (felsic) lava tends to be viscous and often explosive, whereas lower-silica (mafic) lava tends to be fluid and less explosive during an eruption. **14-11:** *Composite volcanoes* are steep-sided cones composed of alternating layers of *pyroclastics* from explosive eruptions and *lava flows* from more effusive eruptions; *shield volcanoes* are wide, gently sloped volcanoes consisting of layer after layer of solidified lava flows. **14-12:** A *caldera* develops when a volcano expels an enormous quality of pyroclastic material during an explosive eruption; the weakened volcano collapses, leaving a wide, low depression. **14-13:** A *pyroclastic flow* is a dense, rapidly moving avalanche of pyroclastics and searing-hot gases; a *volcanic mudflow* is a fast-moving mixture of water and pyroclastics flowing down a volcano with the consistency of wet concrete. **14-14:** All are *igneous intrusions* formed by magma solidifying below the surface; a *batholith* is the largest kind of intrusion; a *laccolith* develops when magma is forced between horizontal layers of strata, bulging up the overlying rock; a *dike* forms when a sheet of magma is injected into a vertical fracture. **14-15:** A *syncline* is a simple structural downfold, and an *anticline* is a simple upfold; sometimes a syncline is associated with a topographic valley and an anticline with a topographic ridge. **14-16:** *Normal faults* are caused by tension (extension) and result in vertical displacement along the fault plane; *reverse faults* are caused by compression and also result in vertical displacement; *strike-slip faults* are associated with shearing, in which one side of the fault moves sideways past the other. **14-17:** A *graben* develops when a block of crust drops down between two parallel faults (most commonly normal faults); an *offset stream* develops when the course of a stream is displaced by lateral movement along a strike-slip fault. **14-18:** *Magnitude* describes the relative "size" or amount of energy released in an earthquake; *earthquake intensity* describes how violently the local ground shakes during an earthquake. **14-19:** *Liquefaction* occurs when water-saturated sediments lose strength and liquefy while the ground shakes during an earthquake. **14-20:** Chief Mountain is older rock that has been moved over younger rock along a *thrust fault*.

Chapter 15

15-1: *Weathering* fragments the bedrock into smaller pieces that can be transported; *mass wasting* moves this weathered rock a relatively short distance down slope. **15-2:** Rock openings allow moisture and atmospheric gases to penetrate the bedrock, increasing the surface area of the rock exposed to weathering processes. **15-3:** Both are breaks in the rock structure. Along a *fault*, there has been displacement; along a *joint*, there has not. **15-4:** Liquid water seeps into a joint or crack in the rock; when the water freezes, the ice expands and pries open the crack. **15-5:** When formerly buried rock (such as granite) is exposed at the surface, the unloading of pressure allows the outer part of the rock to expand, and layer after layer of rock breaks off along curved *sheeting joints* in the process of *exfoliation*, much like the skin of an onion. **15-6:** In *mechanical weathering*, rocks are physically broken apart through stress; in chemical *weathering*, chemical reactions decompose the minerals in a rock. **15-7:** Both are chemical weathering processes that require water. *Oxidation* involves the combination of metallic elements and oxygen, producing softer oxides (such as rust); *carbonation* involves a weak acid (*carbonic acid*) that is especially effective in weathering carbonate rocks, such as limestone. **15-8:** Some rock types are strong and resistant to weathering, whereas other types are weak and easily weathered. **15-9:** The steepest angle that can be maintained by loose material before it begins to move downslope. **15-10:** *Talus cones* develop at the foot of steep slopes from the accumulation of material deposited by *rockfalls*. **15-11:** Rain can soak in and add weight to weathered rock on a slope, making it more susceptible to sliding or slumping. **15-12:** Both landslides and earthflows are slope failures that occur on hillsides. *Earthflows* involve more water and the flow of material down a slope; *landslides* involve material moving down a sliding plane and can be completely dry. **15-13:** Both involve quite a bit of water. *Earthflows* are slope failures that occur on hillsides; *mudflows* entail more water and flow down valleys. **15-14:** The freeze–thaw cycle aids in slowly moving soil particles downslope (as ice forms, the soil expands slightly outward; with melting, soil particles drop slightly downslope), as does the slight swelling and contraction of soil through simple wetting and drying.

Chapter 16

16-1: Channeled *streamflow* occurs in *valleys*, whereas unchanneled *overland flow* occurs on *interfluves*. **16-2:** Channeled *streamflow* is generally more effective at erosion and transportation of rock than unchanneled *overland flow*. **16-3:** The speed of flow determines the largest size of rock that a stream can move (known as stream *competence*). **16-4:** *Alluvium* is often rounded because as rocks are tumbled along in a stream, rough edges are worn off; stratification is often visible because alluvium is deposited after a periodic flood; sorting may occur because as stream speed drops, rocks of the same weight are deposited at the same time. **16-5:** Because stream *competence* and *capacity* increase significantly during flood periods, when a stream has its fastest flow and highest *discharge*, most streams accomplish most of their erosion (and eventual deposition of alluvium) during floods. **16-6:** *Meandering channel* patterns typically develop when a stream flows down a gentle slope (especially streams with large *suspended loads*); *braided channel* patterns typically develop when a stream has a large *bed load* (especially larger-size rocks) but doesn't have enough flow to keep this material moving. **16-7:** *Dendritic* drainage patterns tend to develop where the underlying structure does not significantly influence erosion; *trellis* drainage patterns develop in response to structural control, especially in regions of parallel ridges and valleys. **16-8:** *Downcutting* typically occurs when a stream flows down a steep slope (or has especially high speed and large volume). **16-9:** Undercutting and erosion of the lip of the waterfall lead to collapse of the *knickpoint*, so the location of the waterfall gradually shifts upstream as it is worn away. **16-10:** *Headward erosion* may allow one stream to extend its valley into the course of another stream and capture the other stream's water. **16-11:** Deposition of alluvium will occur any time and any place a stream loses power, such as when flow speed diminishes after a flood, on the inside bank of a meander, or where the stream enters a quiet body of water. **16-12:** An *oxbow lake* develops when one bend of a meander turn cuts into an adjacent bend, forming a *cutoff meander*—leaving an old portion of the stream channel filled with water but isolated from the rest of the stream. **16-13:** Dams have trapped sediment, reducing the amount of alluvium that would have been deposited in the delta; artificial levees and other structures have prevented the river from naturally changing course and thus from depositing material in different parts of the delta. **16-14:** *Stream terraces* develop when a stream flowing across a *floodplain* is rejuvenated (gains downcutting ability due to regional uplift or higher discharge); the stream downcuts into its old floodplain, leaving remnants of its original floodplain as terraces on both sides of the new stream level. **16-15:** Davis's theory assumed that little erosion occurred during the initial uplift of a landscape (very unlikely); no *peneplains* are found anywhere.

Chapter 17

17-1: *Dissolution* (chemical weathering) exposes limestone (and other carbonate rocks) to *carbonic acid*—and most surface water and underground water are slightly acidic because they contain at least some dissolved carbon dioxide. **17-2:** Travertine and tufa are carbonate rocks that form from the precipitation of calcium carbonate when carbon dioxide leaves solution, such as inside the openings of a cave or around a surface opening of a hot spring. **17-3:** The first stage of *cavern* formation entails the removal of carbonate rocks by percolating underground water; the second ("decoration") stage involves the precipitation of calcium carbonate *speleothems* into the voids left by the excavation stage. **17-4:** A *sinkhole* is a depression that develops from the dissolution of rock, often at the intersection of a joint. **17-5:** Pressure builds up in an underground opening where steam and superheated water (liquid water well above the boiling temperature were it at surface pressure) accumulate; finally, pressure is great enough to expel the water and steam in a surge out of the *geyser*. **17-6:** Yellowstone has magma and hot rocks close to the surface, lots of water that seeps downward and becomes heated, and highly fractured rocks that allow water to move up and down easily.

Chapter 18

18-1: Because impermeable surfaces don't readily absorb water, whenever it rains, most of the water runs off into the nearest dry stream channel. **18-2:** With little soil and vegetation covering it, the bedrock in deserts is directly exposed to weathering and erosional processes. **18-3:** An *exotic stream* gets its water from a humid area, but flows across a desert; an *ephemeral stream* is dry most of the time (perhaps for many years), but fills with water whenever there is a local rainstorm. **18-4:** A nearly level, dry lake bed composed of fine-grained, dried mud; a *playa* forms where desert basins have interior drainage. Ephemeral streamflow will occasionally fill the bottom of the basin with water (forming a playa lake); once the water evaporates and the silt settles out, the playa remains. **18-5:** Most fluvial transportation of material in deserts occurs through infrequent events such as flash floods and debris flows—in effect, in a desert the streams are either dry or are flooding. **18-6:** In regions with little rainfall, a flash flood or debris flow out of a mountain canyon transports material until it reaches the gentle slopes of the *piedmont zone*; as flow speed decreases, deposition of alluvium occurs at the foot of the mountain. **18-7:** A gently sloping, fan-shaped accumulation of flash flood and debris flow deposits (alluvium) that builds up at the mouth of a canyon, where flow speed decreases. **18-8:** Under all but extremely rare circumstances, wind is not capable of exerting enough force to move any material larger than sand, so erosion of bedrock is limited to abrasion from "sandblasting." **18-9:** Wind blows sand grains up the gentle windward side of a dune (through the bouncing action of *saltation*); at the top of the dune crest, sand drops down the leeside *slip face* of the dune, typically remaining at the *angle of repose* for dry sand (about 33°). **18-10:** Both *desert pavement* and *desert varnish* require many centuries to develop but can be disturbed or covered by erosion or deposition; so, we can infer that a surface with well-developed pavement and/or varnish has been unaffected by significant erosion or deposition for many years. **18-11:** *Alluvial fans* and *bajadas* are common in the basin-and-range region because most of the fault-produced mountains here have abrupt slope changes in the *piedmont zone* (so fluvial deposition will occur here). In addition, in this desert environment there isn't enough streamflow to move this alluvium much beyond the piedmont zone. **18-12:** Basins of interior drainage are found throughout the basin-and-range region; ephemeral streamflow drains to the lowest part of a basin, temporarily forming a playa lake and eventually leaving a dry playa. **18-13:** The caprock of resistant horizontal sedimentary rock maintains the steep scarp below; the scarp retreats as this resistant caprock is undercut by weathering and erosion of the weaker rock below (which forms a gently sloping surface).

Chapter 19

19-1: Mountain *glaciers* are found in high-elevation mountain areas and are much smaller in area than *continental ice sheets*, which are continuous covers of glacial ice in nonmountainous areas. **19-2:** During the *Pleistocene*, glaciers covered about one-third of the continents, including all of Antarctica, much of the mid- and high-latitude portions of North America and Eurasia, and high mountain areas. **19-3:** Sea level dropped during glacial advances because water evaporated from the oceans was locked up as ice on the continents; during interglacial periods, this water flowed back into the oceans. **19-4:** *Pleistocene lakes* developed in basins away from the ice sheets due to increased rainfall and lower evaporation. **19-5:** Warmer global temperatures are causing the retreat or thinning of most mountain glaciers and continental ice sheets today. **19-6:** Snow left at the end of summer is covered by new winter snow; as this continues for some years, the old snow is increasingly compacted and recrystallized into glacial ice. Once this ice begins to move, a glacier has formed. In the *zone of accumulation*, addition of snow and ice is greater than ablation (loss) of ice; in the *zone of ablation*, loss of ice exceeds accumulation. **19-7:** Glaciers are always flowing or moving downslope. Even when a glacier is getting smaller because *ablation* is greater than *accumulation* of ice, the glacier continues to move (and transport rock). **19-8:** Meltwater seeps down into cracks below a glacier and refreezes; *frost wedging* shatters the rock below. When the glacier flows, this shattered rock is plucked out and incorporated into the glacier. **19-9:** Because it is melted out of a glacier at its terminus or edges, glacial *till* is typically unsorted, unstratified, and often angular; *alluvium* is moved by water, so it is often sorted by size, deposited in distinct strata, and rounded by tumbling in a stream. **19-10:** A *roche moutonnée* has a gentle *stoss side* smoothed by glacial abrasion on the way up; the steep, irregular lee side is due to glacial plucking as the glacier pulls away. **19-11:** A *terminal moraine* is the ridge of till that marks the maximum advance of a glacier; a *recessional moraine* forms as the glacier retreats and stagnates for a period of time, leaving another ridge of till. **19-12:** Meltwater streams from a glacier carry glacial *drift* out beyond the area directly glaciated, leaving deposits of *stratified drift*, typically in a gently undulating *outwash plain* of low relief. **19-13:** A *cirque* is a bowl-shaped glacial valley head, eroded out of bedrock primarily through glacial plucking and frost wedging of the headwall. **19-14:** Glaciers may slightly straighten a stream valley, and they typically deepen and steepen the valley into a U-shaped *glacial trough*; irregular *glacial steps* are often left in the down-valley profile. **19-15:** A *lateral moraine* is a ridge of till that accumulates along the sides of a valley glacier, especially near the mouth of the glacial valley. **19-16:** *Proglacial lakes* develop in front of a glacier where ice blocks the natural drainage or meltwater from the glacier. **19-17:** Variations in greenhouse gases, particulates released during volcanic eruptions, and changes to ocean circulation patterns.

Chapter 20

20-1: Along ocean coastlines, tides are significant, and coral reefs can develop in tropical areas; the causes of sea-level changes are different from those that lead to lake-level changes. **20-2:** As a *wave of oscillation* reaches shallow water, it is slowed by friction with the bottom. As the wave slows, its wavelength decreases and wave height increases until the wave falls forward and breaks, forming a *wave of translation*. **20-3:** Storm waves are larger than normal waves; when they break on a shoreline, their higher speed and greater volume of water are capable of rapid erosion. **20-4:** Because the entire water column, from the ocean floor to the surface, is displaced when a *tsunami* forms, the volume of water that comes onshore with even a modest-size tsunami is much greater than that of a comparable-height storm wave. **20-5:** With a warming climate, the melting of glaciers is increasing global sea level slightly, leading to the gradual flooding of low-lying coastal areas. **20-6:** Most coastal sediment is deposited into the ocean through the outflow of streams. **20-7:** *Beach drifting* moves sand down a beach in a zigzag fashion—waves push sand up on a beach at an angle, but this sand drains straight back down off the beach. *Longshore currents* are set up by the action of waves striking a coastline at an angle—water offshore transports sediment down the coast, parallel to shore. **20-8:** The size of a beach is the result of the balance between the amount of sediment coming into that beach and the amount of sediment being carried away; if for a period of time more sand is carried away than is deposited, a beach will shrink. **20-9:** The upcurrent side of a *groin* or *jetty* tends to trap sand, expanding the width of the beach; however, more sand is eroded on the downcurrent side of these structures, so the beach tends to become narrower there. **20-10:** At the end of the Pleistocene glaciations about 10,000 years ago, water flowed back into the ocean, raising sea level and submerging most coastlines. **20-11:** The presence of *marine terraces* usually indicates that the coastline has been rising tectonically, lifting former *wave-cut platforms* above sea level. **20-12:** A *fringing reef* develops along the shore of a tropical volcanic island (such as those formed by a *hot spot*); the plate cools and becomes denser as it carries the island off the hot spot, so the island slowly begins to sink. Coral continues to build up toward the surface, but as the volcanic island sinks, the ring of coral may become a *barrier reef* or even an *atoll* if the island sinks below sea level.

Appendix I

The International System of Units (SI)

With the major exception of the United States, the system of weights and measures used worldwide both in scientific work and in everyday life is the International System, usually abbreviated SI from its French name, *Système International*. The system has seven base units and supplementary units for angles (**Table I-1**). The beauty of the system lies in its reliance on multiples of the number 10, with the prefixes shown in **Table I-2** being used to cover a magnitude range from the astronomically large to the infinitesimally small. **Table I-3** on the next page lists the most frequently used conversion factors.

TABLE I-1 SI Units

Quantity	Unit	Symbol
Base Units		
Length	Meter	m
Mass	Kilogram	kg
Time	Second	s
Electric current	Ampere	A
Temperature	Kelvin	K
Amount of substance	Mole	mol
Luminous intensity	Candela	Cd
Supplementary Units		
Plane angle	Radian	rad
Solid angle	Steradian	Sr

TABLE I-2 Common Multiples and SI Prefixes

Multiple	Value	Prefix	Symbol
1,000,000,000,000	10^{12}	tera	T
1,000,000,000	10^{9}	giga	G
1,000,000	10^{6}	mega	M
1000	10^{3}	kilo	k
100	10^{2}	hecto	h
10	10^{1}	deka	da
0.1	10^{-1}	deci	d
0.01	10^{-2}	centi	c
0.001	10^{-3}	milli	m
0.000001	10^{-6}	micro	μ
0.000000001	10^{-9}	nano	n
0.000000000001	10^{-12}	pico	p

TABLE I-3 SI–English Conversion Units

Multiply	By	To Get
Length		
Inches	2.540	Centimeters
Feet	0.3048	Meters
Yards	0.9144	Meters
Miles	1.6093	Kilometers
Millimeters	0.039	Inches
Centimeters	0.3937	Inches
Meters	3.2808	Feet
Kilometers	0.6214	Miles

Multiply	By	To Get
Area		
Square inches	6.452	Square centimeters
Square feet	0.0929	Square meters
Square yards	0.8361	Square meters
Square miles	2.590	Square kilometers
Acres	0.4047	Hectares
Square centimeters	0.155	Square inches
Square meters	10.764	Square feet
Square meters	1.196	Square yards
Square kilometers	0.3861	Square miles
Hectares	2.471	Acres

Multiply	By	To Get
Temperature		
(°F − 32°) ÷ 1.8 = °C		
(°C × 1.8) + 32° = °F		

Multiply	By	To Get
Volume		
Cubic inches	16.387	Cubic centimeters
Cubic feet	0.028	Cubic meters
Cubic yards	0.7646	Cubic meters
Fluid ounces	29.57	Milliliters
Pints	0.47	Liters
Quarts	0.946	Liters
Gallons	3.785	Liters
Cubic centimeters	0.061	Cubic inches
Cubic meters	35.3	Cubic feet
Cubic meters	1.3079	Cubic yards
Milliliters	0.034	Fluid ounces
Liters	1.0567	Quarts
Liters	0.264	Gallons

Multiply	By	To Get
Mass (Weight)		
Ounces	28.3495	Grams
Pounds	0.4536	Kilograms
Tons (2000 lb.)	907.18	Kilograms
Tons (2000 lb.)	0.90718	Tonnes
Grams	0.03527	Ounces
Kilograms	2.2046	Pounds
Kilograms	0.0011	Tons (2000 lb.)
Tonnes	1.1023	Tons (2000 lb.)

Appendix II

U.S. Geological Survey Topographic Maps

The U.S. Geological Survey (USGS) is one of the world's largest mapping agencies and is primarily responsible for the National Mapping Program of the United States. The USGS produces a broad assortment of maps, but its topographic "quadrangles" are among the most widely used by geographers. The quadrangles come in various sizes and scales (from 1:24,000 to 1:1,000,000), but all are rectangles bordered by parallels and meridians rather than by political boundaries. For many years, the quadrangles were produced by surveys on the ground. Today, however, they are created from aerial photographs, satellite imagery, and digital elevation databases.

Topographic maps convey both human-built features—such as roads and buildings—and natural features—such as rivers, glaciers, and areas of forest cover. Topographic maps use *elevation contour lines* to depict the topography of a landscape (**Figure II-1**). With these two-dimensional maps, we can envision the three-dimensional topography, making "topo" maps especially useful for studying landform patterns.

The USGS has produced more than 55,000 different printed topographic quadrangles and about 285,000 digital orthorectified aerial images. In 2001, the USGS began an ambitious multiyear plan to develop *The National Map*, a seamless, continuously maintained, nationally consistent set of online, public-domain, geographic-based information for the United States. The National Map incorporates topographic maps, aerial and satellite imagery, and a wide range of geospatial map layers. It is available online at http://nationalmap.gov. The latest large-scale topographic maps from the USGS are part of the *US Topo* series, downloadable 1:24,000 scale quadrangles with data layers that include contour lines and aerial imagery.

Elevation Contour Lines

USGS topographic maps use elevation contour lines to portray the shape, slope, elevation, and relief (the difference in elevation between the highest and lowest locations) of a landscape (**Figure II-2**). Use the following guidelines when you interpret contour lines:

1. A contour line connects points of equal elevation.
2. The difference in elevation between two contour lines is called the *contour interval*.
3. Usually every fourth or fifth contour line is a darker *index contour*.
4. Elevations on one side of a contour line are higher than those on the other side.
5. Contour lines never touch or cross one another (except at some cliffs).
6. Contour lines have no beginning or end—every line closes on itself, either on or off the map.
7. Uniformly spaced contours indicate a uniform slope.
8. If spaced far apart, contour lines indicate a relatively gentle slope; if spaced close together, they represent a steep slope.
9. A contour line bends upstream in a "V" shape when it crosses a stream valley, gully, or "draw" (the "V" points uphill).
10. Along a spur or ridge running down a hillside, a contour line forms a "V" pointing downhill.
11. A contour line that closes within the limits of the map represents a hill or rise. The land inside that contour is higher than the elevation of the closed contour itself.
12. A depression may be represented by a closed contour that is hachured on the side leading into the depression. The elevation of such a *depression contour* is the same as that of the adjacent lower regular contour (unless otherwise marked).

Topographic Map Symbols

Topographic quadrangles also use symbols and colors to portray a variety of other features. Standard colors are used to distinguish various kinds of map features:

- **Brown:** contour lines and other topographic features
- **Blue:** hydrographic (water) features
- **Black:** features constructed or designated by humans, such as buildings, roads, boundary lines, and names
- **Green:** areas of vegetation, such as woodlands, forests, orchards, and vineyards
- **Red:** important roads and lines of the public land survey system
- **Gray or red tint:** urban areas
- **Purple:** features added from aerial photos during map revision

The principal standard symbols used on USGS topographic maps are shown in **Figure II-3**. Notice that in a few cases, more than one symbol is used to show the same kind of feature.

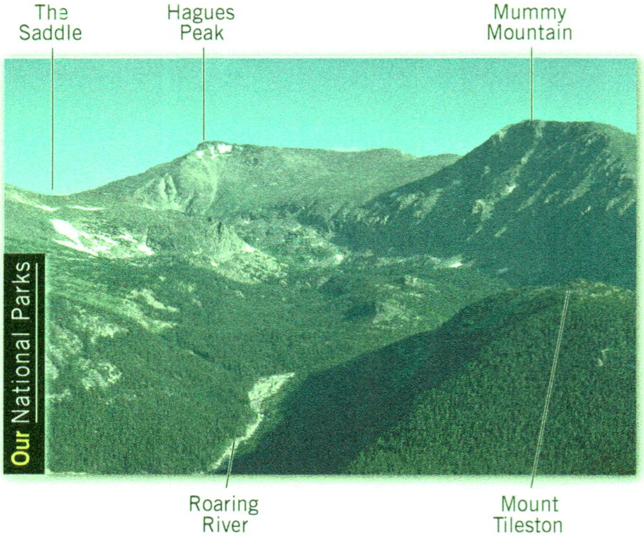

(a) Photograph of terrain in Colorado.

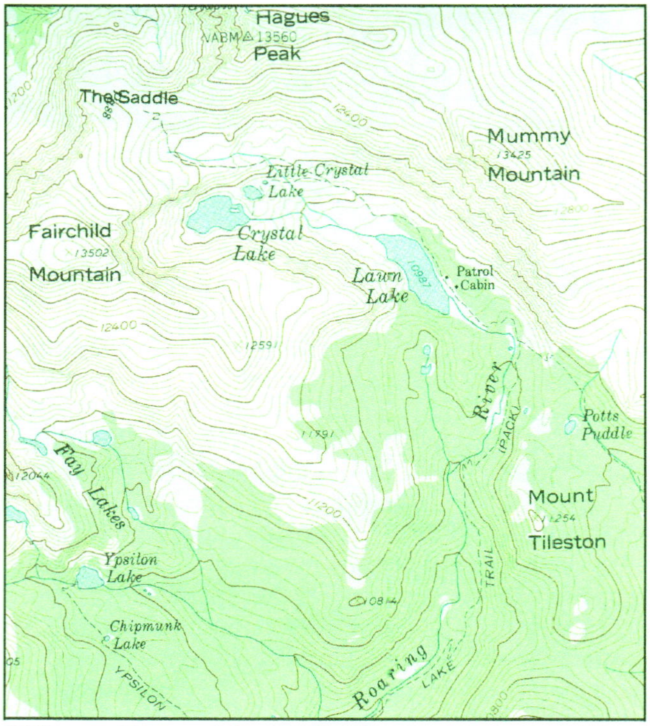

(b) Topographic map of the same terrain.

▲ **Figure II-1** Portrayal of terrain by means of elevation contour lines. (a) Here is a photo of Hagues Peak and Mummy Mountain in Colorado's Rocky Mountain National Park, along with (b) a standard topographic map of the same area. (Original map scale 1:125,000; contour interval is 80 feet [27 meters].)

(a) Drawing of a landscape.

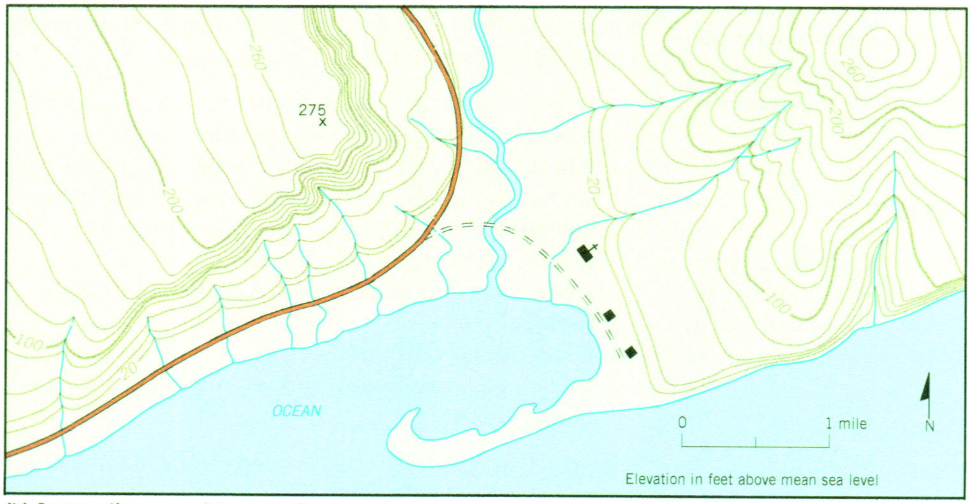

(b) Contour line map of the same landscape.

▲ **Figure II-2** (a) A fictitious landscape and (b) a contour line map of the same area. Where the contours are close together, the slope is steep; where the lines are far apart, the slope is gentle.

BATHYMETRIC FEATURES	
Area exposed at mean low tide; sounding datum line***	
Channel***	
Sunken rock***	+

BOUNDARIES	
National	
State or territorial	
County or equivalent	
Civil township or equivalent	
Incorporated city or equivalent	
Federally administered park, reservation, or monument (external)	
Federally administered park, reservation, or monument (internal)	
State forest, park, reservation, or monument and large county park	
Forest Service administrative area*	
Forest Service ranger district*	
National Forest System land status, Forest Service lands*	
National Forest System land status, non-Forest System land*	
Small park (county or city)	

BUILDINGS AND RELATED FEATURES	
Building	
School; house of worship	
Athletic field	
Built-up area	
Forest headquarters*	
Ranger district office*	
Guard station or work center*	
Racetrack or raceway	
Airport, paved landing strip, runway, taxiway, or apron	
Unpaved landing strip	
Well (other than water), windmill or wind generator	
Tanks	
Covered reservoir	
Gaging station	
Located or landmark object (feature as labeled)	
Boat ramp or boat access*	
Roadside park or rest area	
Picnic area	
Campground	
Winter recreation area*	
Cemetery	Cem

*USGS-USDA Forest Service Single-Edition

**Provisional-Edition maps only.

***Topographic Bathymetric maps only.

COASTAL FEATURES	
Foreshore flat	Mud
Coral or rock reef	Reef
Rock, bare or awash; dangerous to navigation	
Group of rocks, bare or awash	
Exposed wreck	
Depth curve; sounding	18 23
Breakwater, pier, jetty, or wharf	
Seawall	
Oil or gas well; platform	

CONTOURS	
Topographic	
Index	6000
Approximate or indefinite	
Intermediate	
Approximate or indefinite	
Supplementary	
Depression	
Cut	
Fill	
Continental divide	
Bathymetric	
Index***	
Intermediate***	
Index primary***	
Primary***	
Supplementary***	

CONTROL DATA AND MONUMENTS	
Principal point**	3-20
U.S. mineral or location monument	USMM 438
River mileage marker	Mile 69
Boundary monument	
Third-order or better elevation, with tablet	BM 9134 BM 277
Third-order or better elevation, recoverable mark, no tablet	5628
With number and elevation	67 4567
Horizontal control	
Third-order or better, permanent mark	Neace Neace
With third-order or better elevation	BM 52 Pike BM393
With checked spot elevation	1012
Coincident with found section corner	Cactus Cactus
Unmonumented**	

▲ **Figure II-3** Topographic map symbols used on USGS quadrangles.

(Continued)

CONTROL DATA AND MONUMENTS — *continued*

Vertical control

Third-order or better elevation, with tablet	BM × 5280
Third-order or better elevation, recoverable mark, no tablet	× 528
Bench mark coincident with found section corner	BM 5280
Spot elevation	×7523

GLACIERS AND PERMANENT SNOWFIELDS

- Contours and limits
- Formlines
- Glacial advance
- Glacial retreat

LAND SURVEYS

Public land survey system

Range or Township line	
Location approximate	
Location doubtful	
Protracted	
Protracted (AK 1:63,360-scale)	
Range or Township labels	R1E T2N R3W T4S
Section line	
Location approximate	
Location doubtful	
Protracted	
Protracted (AK 1:63,360-scale)	
Section numbers	1 - 36 1 - 36
Found section corner	
Found closing corner	
Witness corner	WC
Meander corner	MC
Weak corner*	

Other land surveys

- Range or Township line
- Section line
- Land grant, mining claim, donation land claim, or tract
- Land grant, homestead, mineral, or other special survey monument
- Fence or field lines

MARINE SHORELINES

- Shoreline
- Apparent (edge of vegetation)***
- Indefinite or unsurveyed

MINES AND CAVES

Quarry or open pit mine	⚒
Gravel, sand, clay, or borrow pit	×
Mine tunnel or cave entrance	≺
Mine shaft	
Prospect	x
Tailings	Tailings
Mine dump	
Former disposal site or mine	

PROJECTION AND GRIDS

Neatline	3915′ 9037′30″
Graticule tick	55′
Graticule intersection	
Datum shift tick	

State plane coordinate systems

Primary zone tick	640 000 FEET
Secondary zone tick	247 500 METERS
Tertiary zone tick	260 000 FEET
Quaternary zone tick	98 500 METERS
Quintary zone tick	320 000 FEET

Universal transverse mercator grid

UTM grid (full grid)	273
UTM grid ticks*	269

RAILROADS AND RELATED FEATURES

- Standard gauge railroad, single track
- Standard gauge railroad, multiple track
- Narrow gauge railroad, single track
- Narrow gauge railroad, multiple track
- Railroad siding
- Railroad in highway
- Railroad in road
- Railroad in light duty road*
- Railroad underpass; overpass
- Railroad bridge; drawbridge
- Railroad tunnel
- Railroad yard
- Railroad turntable; roundhouse

RIVERS, LAKES, AND CANALS

- Perennial stream
- Perennial river
- Intermittent stream
- Intermittent river
- Disappearing stream
- Falls, small
- Falls, large
- Rapids, small
- Rapids, large
- Masonry dam
- Dam with lock
- Dam carrying load

▲ **Figure II-3** (*Continued*)

RIVERS, LAKES, AND CANALS — *continued*

Perennial lake/pond	
Intermittent lake/pond	
Dry lake/pond	Dry Lake
Narrow wash	
Wide wash	Wash
Canal, flume, or aqueduct with lock	
Elevated aqueduct, flume, or mud pot	
Aqueduct tunnel	
Water well, geyser, fumarole, or mud pot	
Spring or creep	

ROADS AND RELATED FEATURES

Please note: Roads on Provisional-edition maps are not classified as primary, secondary, or light duty. These roads are all classified as improved roads and are symbolized the same as light duty roads.

Primary highway	
Secondary highway	
Light duty road	
Light duty road, paved*	
Light duty road, gravel*	
Light duty road, dirt*	
Light duty road, unspecified*	
Unimproved road	
Unimproved road*	
4WD road	
4WD road*	
Trail	
Highway or road with median strip	
Highway or road under construction	Under Const
Highway or road underpass; overpass	
Highway or road bridge; drawbridge	
Highway or road tunnel	
Road block, berm, or barrier*	
Gate on road*	
Trailhead*	TH

*USGS-USDA Forest Service Single-Edition
**Provisional-Edition maps only.
***Topographic Bathymetric maps only.

SUBMERGED AREAS AND BOGS

Marsh or swamp	
Submerged marsh or swamp	
Wooded marsh or swamp	
Submerged wooded marsh or swamp	
Land subject to inundation	Max Pool 431

SURFACE FEATURES

Levee	Levee
Sand or mud	Sand
Disturbed surface	
Gravel beach or glacial moraine	Gravel
Tailings pond	Tailings Pond

TRANSMISSION LINES AND PIPELINES

Power transmission line; pole, tower	
Telephone line	Telephone
Aboveground pipeline	
Underground pipeline	Pipeline

VEGETATION

Woodland	
Shrubland	
Orchard	
Vineyard	
Mangrove	Mangrove

▲ **Figure II-3** (*Continued*)

Appendix III

Meteorological Tables

Determining Relative Humidity

A *psychrometer* is an instrument used for measuring relative humidity. It consists of two thermometers mounted side by side. One is an ordinary thermometer (called a *dry bulb*), which simply measures air temperature. The other thermometer (called a *wet bulb*) has its bulb encased in a covering of muslin or gauze that is saturated with distilled water prior to use. The two thermometers are then thoroughly ventilated either by being whirled around (this instrument has a handle around which the thermometers can be whirled and is referred to as a *sling psychrometer*) or by fanning air past them. This ventilation encourages evaporation of water from the covering of the wet bulb at a rate that is directly related to the humidity of the surrounding air. Evaporation is a cooling process, causing the temperature of the wet bulb to drop. More evaporation occurs, and therefore more cooling, in dry air than in moist air. The difference between the resulting wet-bulb and dry-bulb temperatures (called the *depression of the wet bulb*) is an expression of the relative saturation of the surrounding air. A large difference indicates low relative humidity; a small difference means that the air is near saturation. If the air is completely saturated, no net evaporation will take place; thus the two thermometers will have identical readings.

To determine relative humidity, use a sling psychrometer to measure the wet-bulb and dry-bulb temperatures. Then use these two values to read the relative humidity from **Table III-1** (for degrees Celsius) or **Table III-2** (for degrees Fahrenheit).

TABLE III-1 Relative Humidity Psychrometer Tables (°C)

Air Temp (°C)	Depression of Wet-Bulb Thermometer (°C)																					
	1	*2*	*3*	*4*	*5*	*6*	*7*	*8*	*9*	*10*	*11*	*12*	*13*	*14*	*15*	*16*	*17*	*18*	*19*	*20*	*21*	*22*
−4	77	54	32	11																		
−2	79	58	37	20	1																	
0	81	63	45	28	11																	
2	83	67	51	36	20	6										Relative Humidity (%)						
4	85	70	56	42	27	14																
6	86	72	59	46	35	22	10	0														
8	87	74	62	51	39	28	17	6														
10	88	76	65	54	43	33	24	13	4													
12	88	78	67	57	48	38	28	19	10	2												
14	89	79	69	60	50	41	33	25	16	8	1											
16	90	80	71	62	54	45	37	29	21	14	7	1										
18	91	81	72	64	56	48	40	33	26	19	12	6	0									
20	91	82	74	66	58	51	44	36	30	23	17	11	5									
22	92	83	75	68	60	53	46	40	33	27	21	15	10	4	0							
24	92	84	76	69	62	55	49	42	36	30	25	20	14	9	4	0						
26	92	85	77	70	64	57	51	45	39	34	28	23	18	13	9	5						
28	93	86	78	71	65	59	53	45	42	36	31	26	21	17	12	8	4					
30	93	86	79	72	66	61	55	49	44	39	34	29	25	20	16	12	8	4				
32	93	86	80	73	68	62	56	51	46	41	36	32	27	22	19	14	11	8	4			
34	93	86	81	74	69	63	58	52	48	43	38	34	30	26	22	18	14	11	8	5		
36	94	87	81	75	69	64	59	54	50	44	40	36	32	28	24	21	17	13	10	7	4	
38	94	87	82	76	70	66	60	55	51	46	42	38	34	30	26	23	20	16	13	10	7	5

TABLE III-2 Relative Humidity Psychrometer Tables (°F)

Depression of Wet-Bulb Thermometer (°F) — Relative Humidity (%)

Air temp (°F)	1	2	3	4	5	6	7	8	9	10	11	12	13	14	15	16	17	18	19	20	21	22	23	24	25	26	27	28	29	30
0	67	33	1																											
5	73	46	20																											
10	78	56	34	13																										
15	82	64	46	29	11																									
20	85	70	55	40	26	12																								
25	87	74	62	49	37	25	13	1																						
30	89	78	67	56	46	36	26	16	6																					
35	91	81	72	63	54	45	36	27	19	10	2																			
40	92	83	75	68	60	52	45	37	29	22	15	7																		
45	93	86	78	71	64	57	51	44	38	31	25	18	12	6																
50	93	87	74	67	61	55	49	43	38	32	27	21	16	10	5															
55	94	88	82	76	70	65	59	54	49	43	38	33	28	23	19	11	9	5												
60	94	89	83	78	73	68	63	58	53	48	43	39	34	30	26	21	17	13	9	5	1									
65	95	90	85	80	75	70	66	61	56	52	48	44	39	35	31	27	24	20	16	12	9	5	2							
70	95	90	86	81	77	72	68	64	59	55	51	48	44	40	36	33	29	25	22	19	15	12	9	6	3					
75	96	91	86	82	78	74	70	66	62	58	54	51	47	44	40	37	34	30	27	24	21	18	15	12	9	7	4	1		
80	96	91	87	83	79	75	72	68	64	61	57	54	50	47	44	41	38	35	32	29	26	23	20	18	15	12	10	7	5	3
85	96	92	88	84	81	77	73	70	66	63	59	57	53	50	47	44	41	38	36	33	30	27	25	22	20	17	15	13	10	8
90	96	92	89	85	81	78	74	71	68	65	61	58	55	52	49	47	44	41	39	36	34	31	29	26	24	22	19	17	15	13
95	96	93	89	86	82	79	76	73	69	66	63	61	58	55	52	50	47	44	42	39	37	34	32	30	28	25	23	21	19	17
100	96	93	89	86	83	80	77	73	70	68	65	62	59	56	54	51	49	46	44	41	39	37	35	33	30	28	26	24	22	21
105	97	93	90	87	84	81	78	75	72	69	66	64	61	58	56	53	51	49	46	44	42	40	38	36	34	32	30	28	26	24

For example, if the dry-bulb temperature is 20°C and the wet-bulb temperature is 14°C, the depression of the wet bulb is 6°C In Table III-1, find 20°C in the "Air Temperature" column; move across to the 6° column under "Depression of Wet-Bulb Thermometer," and at the point of intersection, read the relative humidity, 51 percent.

The Beaufort Scale of Wind Speed

Early in the nineteenth century, Admiral Francis Beaufort of the British Navy developed a scale of wind speed widely used in the English-speaking world. It has been modified through the years, but the essentials have not changed. The scale is shown in **Table III-3**.

Wind Chill

Wind chill is the popular name used to describe what cold weather feels like at various combinations of low temperature and high wind. On a cold, windless day, your body heat is conducted sluggishly to a thin layer of atmospheric molecules near the skin. These heated molecules diffuse away slowly and are replaced by other, cooler molecules. Your body thus comes into contact with a relatively small number of cool molecules, and your body heat dissipates slowly.

With an increase in wind speed, however, body heat dissipates much more rapidly, as the protective layer of warmer molecules is speedily removed and supplanted by a continually renewing supply of cold air molecules against the skin. Up to a certain speed, the greater the wind velocity, the greater the cooling effect on the body. Wind chill

TABLE III-3 Beaufort Scale

Beaufort Force	Speed: Kilometers per Hour	Speed: Miles per Hour	Speed: Knots	Description
0	<1	<1	<1	Calm
1	1–5	1–3	1–3	Light air
2	6–11	4–7	4–6	Light breeze
3	12–19	8–12	7–10	Gentle breeze
4	20–29	13–18	11–16	Moderate breeze
5	30–38	19–24	17–21	Fresh breeze
6	39–49	25–31	22–27	Strong breeze
7	50–61	32–38	28–33	Near gale
8	62–74	39–46	34–40	Gale
9	75–87	47–54	41–47	Strong gale
10	88–101	55–63	48–55	Storm
11	102–116	64–72	56–63	Violent storm
12	117–132	73–82	64–71	Hurricane
13	133–148	83–92	72–80	Hurricane
14	149–166	93–103	81–89	Hurricane
15	167–183	104–114	90–99	Hurricane
16	184–201	115–125	100–108	Hurricane
17	202–219	126–136	109–118	Hurricane

affects only organisms that generate heat; inanimate objects have no such heat to lose, so their temperatures are not affected by wind movement.

The term *wind chill* was apparently first used in 1939 by Paul Siple, a geographer and polar explorer. Meteorologists from the National Weather Service of the United States and from the Meteorological Services of Canada have refined the calculations several times, most recently in 2002. The revised wind-chill index currently in use accounts for the wind effects at face level (the "old" system relied on observed winds 10 meters [33 feet] above the ground) and is a better calculation for body heat loss.

The updated index also includes a new frostbite chart, which shows how long skin can safely be exposed to the air, given varying temperatures and wind (Tables III-4 and III-5).

The Heat Index

Sensible temperatures may be significantly influenced by humidity, which can make the weather seem either colder or warmer than it actually is—although humidity is more likely to impinge on our lives in hot weather. Quite simply, high humidity makes hot weather seem hotter.

The National Weather Service has developed a *heat index* that combines temperature and relative humidity to produce an "apparent temperature" that quantifies how hot the air feels to your skin. A sample heat index chart is given in Table III-6.

To this index has been added a general heat stress index to indicate heat-related dangers at various apparent temperatures, as shown in Table III-7.

TABLE III-4 Wind Chill (°C)

Wind Speed (kilometers per hour)	Actual Air Temperature (°C)									
Calm	5	0	-5	-10	-15	-20	-25	-30	-35	-40
5	4	-2	-7	-13	-19	-24	-30	-36	-41	-47
10	3	-3	-9	-15	-21	-27	-33	-39	-45	-51
15	2	-4	-11	-17	-23	-29	-35	-41	-48	-54
20	1	-5	-12	-18	-24	-31	-37	-43	-49	-56
25	1	-6	-12	-19	-25	-32	-38	-45	-51	-57
30	0	-7	-13	-20	-26	-33	-39	-46	-52	-59
35	0	-7	-14	-20	-27	-33	-40	-47	-53	-60
40	-1	-7	-14	-21	-27	-34	-41	-48	-54	-61
45	-1	-8	-15	-21	-28	-35	-42	-48	-55	-62
50	-1	-8	-15	-22	-29	-35	-42	-49	-56	-63
55	-2	-9	-15	-22	-29	-36	-43	-50	-57	-63
60	-2	-9	-16	-23	-30	-37	-43	-50	-57	-64
65	-2	-9	-16	-23	-30	-37	-44	-51	-58	-65
70	-2	-9	-16	-23	-30	-37	-44	-51	-59	-66
75	-3	-10	-17	-24	-31	-38	-45	-52	-59	-66
80	-3	-10	-17	-24	-31	-38	-45	-52	-60	-67

Frostbite within 30 minutes | Frostbite within 10 minutes | Frostbite within 5 minutes

TABLE III-5 Wind Chill (°F)

Wind Speed (miles per hour)	Actual Air Temperature (°F)																
Calm	40	35	30	25	20	15	10	5	0	-5	-10	-15	-20	-25	-30	-35	-40
5	36	31	25	19	13	7	1	-5	-11	-16	-22	-28	-34	-40	-46	-52	-57
10	34	27	21	15	9	3	-4	-10	-16	-22	-28	-35	-41	-47	-53	-59	-66
15	32	25	19	13	6	0	-7	-13	-19	-26	-32	-39	-45	-51	-58	-64	-71
20	30	24	17	11	4	-2	-9	-15	-22	-29	-35	-42	-48	-55	-61	-68	-74
25	29	23	16	9	3	-4	-11	-17	-24	-31	-37	-44	-51	-58	-64	-71	-78
30	28	22	15	8	1	-5	-12	-19	-26	-33	-39	-46	-53	-60	-67	-73	-80
35	28	21	14	7	0	-7	-14	-21	-27	-34	-41	-48	-55	-62	-69	-76	-82
40	27	20	13	6	-1	-8	-15	-22	-29	-36	-43	-50	-57	-64	-71	-78	-84
45	26	19	12	5	-2	-9	-16	-23	-30	-37	-44	-51	-58	-65	-72	-79	-86
50	26	19	12	4	-3	-10	-17	-24	-31	-38	-45	-52	-60	-67	-74	-81	-88
55	25	18	11	4	-3	-11	-18	-25	-32	-39	-45	-54	-61	-68	-75	-82	-89
60	25	17	10	3	-4	-11	-19	-26	-33	-40	-48	-55	-62	-69	-76	-84	-91

Frostbite within 30 minutes | Frostbite within 10 minutes | Frostbite within 5 minutes

TABLE III-6 Heat Index (Apparent Temperature)

Temperature	Relative Humidity 10%	20%	30%	40%	50%	60%	70%	80%	90%
46°C 115°F	43°C 110°F	49°C 121°F	57°C 134°F	66°C 152°F	78°C 173°F	*	*	*	*
43°C 110°F	40°C 105°F	44°C 112°F	50°C 122°F	58°C 136°F	67°C 152°F	77°C 171°F	*	*	*
41°C 105°F	37°C 99°F	40°C 104°F	44°C 112°F	49°C 121°F	57°C 134°F	65°C 149°F	74°C 166°F	*	*
38°C 100°F	34°C 94°F	36°C 97°F	39°C 102°F	43°C 109°F	48°C 118°F	54°C 129°F	62°C 143°F	70°C 158°F	*
35°C 95°F	32°C 89°F	33°C 91°F	34°C 94°F	37°C 99°F	41°C 105°F	45°C 113°F	51°C 123°F	57°C 134°F	*
32°C 90°F	29°C 85°F	30°C 86°F	31°C 88°F	33°C 91°F	35°C 95°F	38°C 100°F	41°C 106°F	45°C 113°F	50°C 122°F
29°C 85°F	27°C 81°F	28°C 82°F	28°C 83°F	29°C 84°F	30°C 86°F	32°C 89°F	34°C 93°F	36°C 97°F	39°C 102°F
27°C 80°F	26°C 78°F	26°C 79°F	26°C 79°F	27°C 80°F	27°C 81°F	28°C 82°F	28°C 83°F	29°C 84°F	30°C 86°F

*Temperature–relative humidity conditions rarely observed in the atmosphere.

TABLE III-7 General Heat Stress Index

Danger Category	Heat Index	Heat Syndrome
IV. Extreme Danger	Above 52°C (125°F)	Heat/sunstroke highly likely with continued exposure.
III. Danger	39–51°C (103–124°F)	Sunstroke, heat cramps, or heat exhaustion likely; heatstroke possible with prolonged exposure and/or physical activity.
II. Extreme Caution	32–39°C (90–102°F)	Sunstroke, heat cramps, or heat exhaustion possible with prolonged exposure and/or physical activity.
I. Caution	27–32°C (80–89°F)	Fatigue possible with prolonged exposure and/or physical activity.

Appendix IV

The Weather Station Model

Weather data are recorded at regular intervals for a great many locations on Earth, each location being called a *weather station*. These data are then plotted on weather maps according to a standard format and code. The format for a standard *station model* is shown in **Figure IV-1**, along with an explanation of the code. **Figure IV-2** shows the same model but with the codes replaced by sample data, and **Figures IV-3** and **IV-4** list some of the codes and symbols used by meteorologists. **Tables IV-1**, **IV-2**, and **IV-3** give additional codes and symbols, and **Figure IV-5** is a sample weather map.

The wind symbol (Table IV-3) gives two pieces of information. (1) Wind direction is indicated by an arrow shaft entering the station circle from the direction in which the wind is blowing, as at the one-o'clock position in Figure IV-2. (2) Wind speed is indicated by the number of "feathers" and half-feathers protruding from the shaft. Each half-feather represents a 5-knot increase in speed (a knot is one nautical mile per hour, which is the same as 1.15 statute miles per hour or 1.85 kilometers per hour); each full feather represents a 10-knot increase. A triangular pennant represents a 50-knot increase.

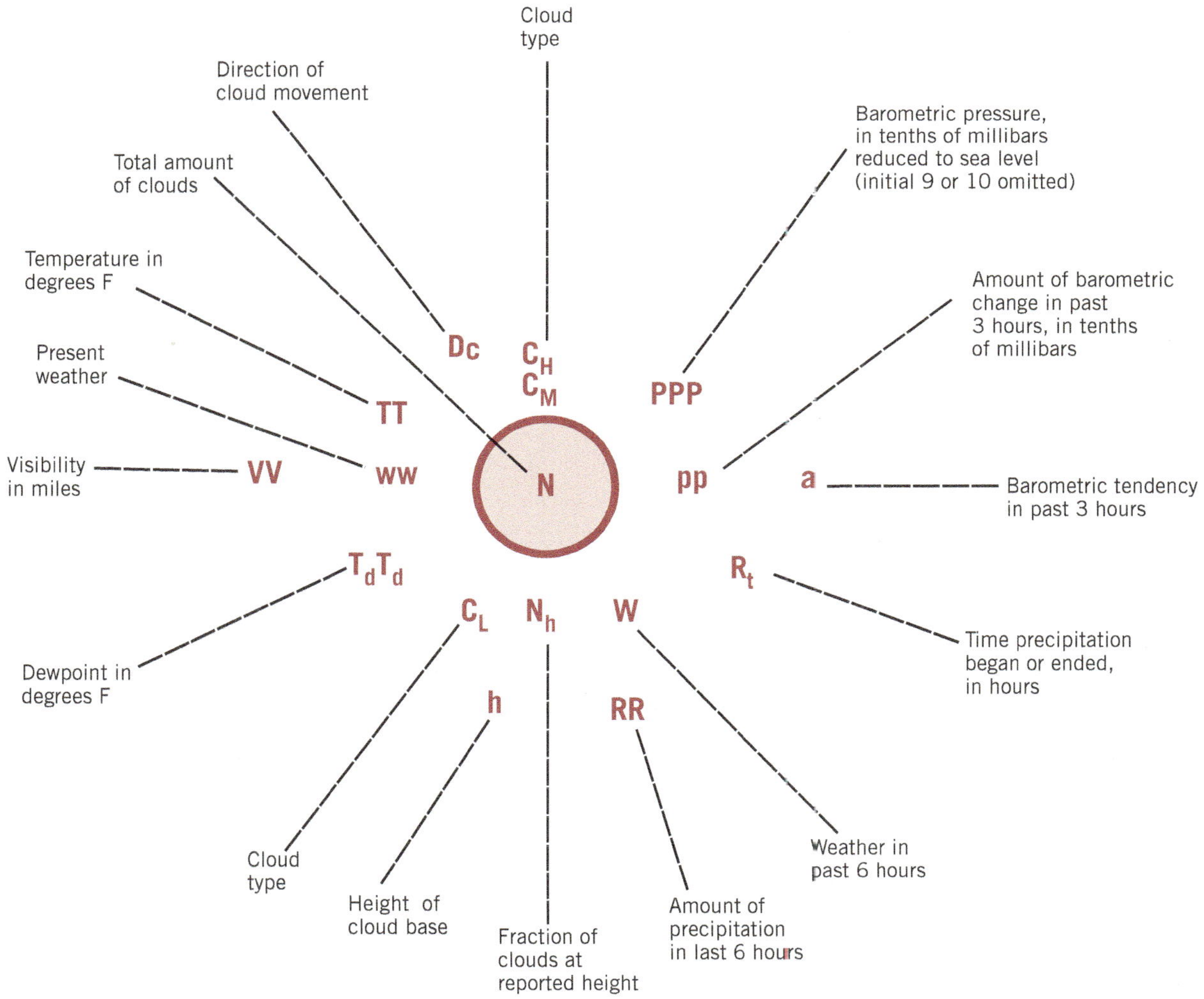

▲ **Figure IV-1** A standard weather station model. (No symbol for wind speed and direction is shown here because there is no one assigned place for this symbol; instead, its position on the model depends on wind direction.)

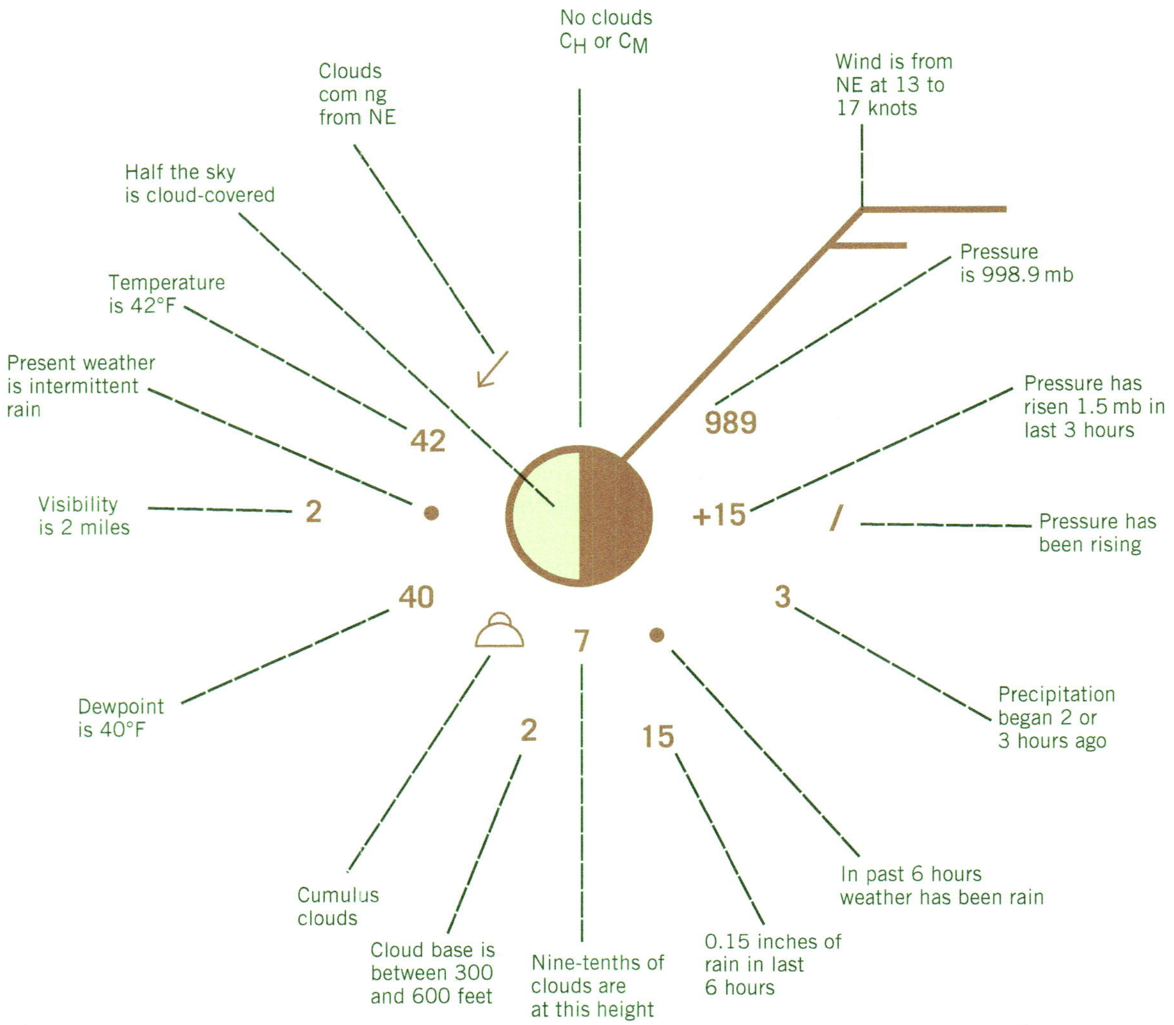

▲ **Figure IV-2** A standard weather station model with the codes replaced by sample data.

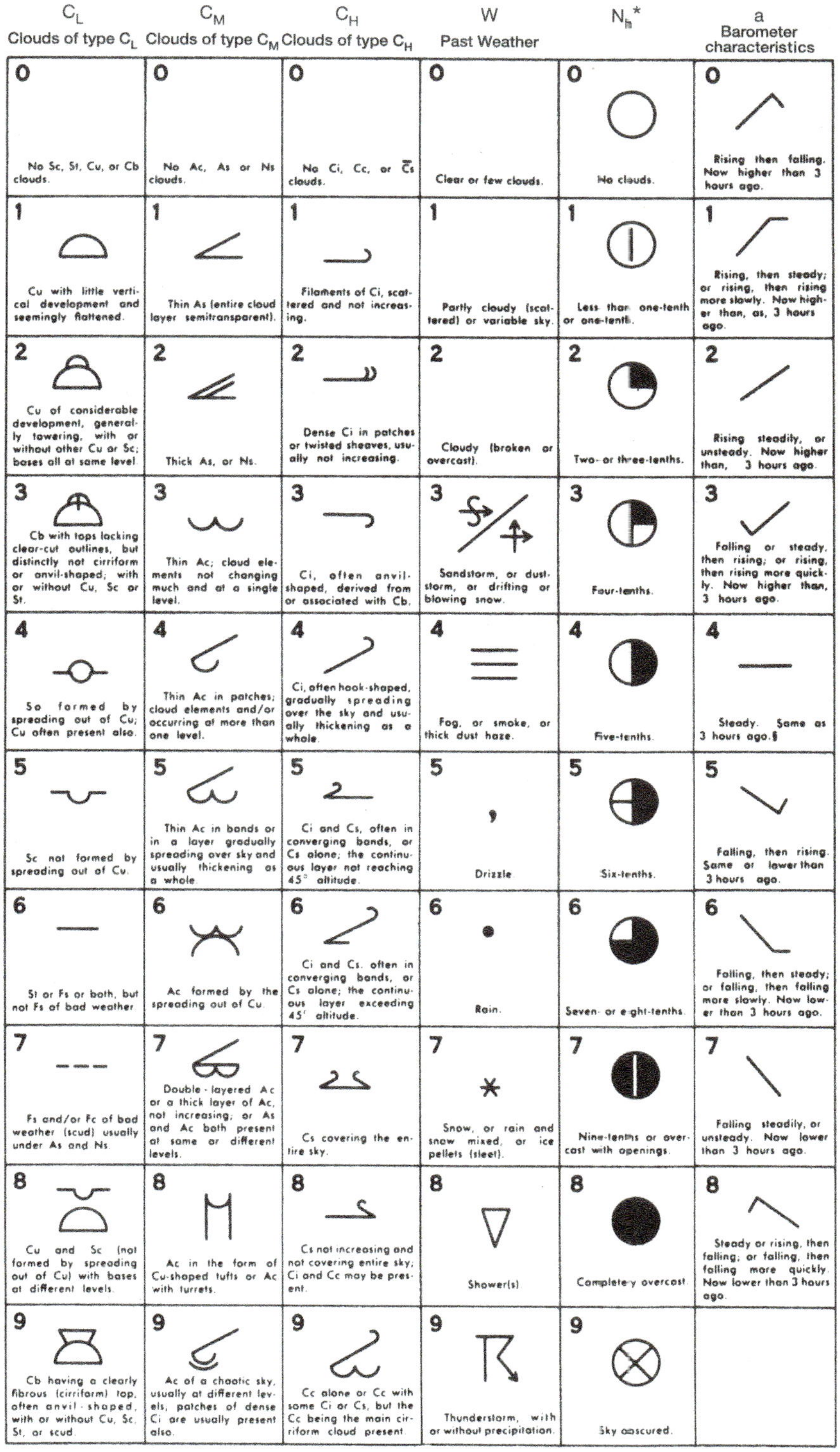

C_L Clouds of type C_L	C_M Clouds of type C_M	C_H Clouds of type C_H	W Past Weather	N_h*	a Barometer characteristics
0 No Sc, St, Cu, or Cb clouds.	0 No Ac, As or Ns clouds.	0 No Ci, Cc, or Cs clouds.	0 Clear or few clouds.	0 No clouds.	0 Rising then falling. Now higher than 3 hours ago.
1 Cu with little vertical development and seemingly flattened.	1 Thin As (entire cloud layer semitransparent).	1 Filaments of Ci, scattered and not increasing.	1 Partly cloudy (scattered) or variable sky.	1 Less than one-tenth or one-tenth.	1 Rising, then steady; or rising, then rising more slowly. Now higher than, or, 3 hours ago.
2 Cu of considerable development, generally towering, with or without other Cu or Sc; bases all at same level	2 Thick As, or Ns.	2 Dense Ci in patches or twisted sheaves, usually not increasing.	2 Cloudy (broken or overcast).	2 Two- or three-tenths.	2 Rising steadily, or unsteady. Now higher than, 3 hours ago.
3 Cb with tops lacking clear-cut outlines, but distinctly not cirriform or anvil-shaped; with or without Cu, Sc or St.	3 Thin Ac; cloud elements not changing much and at a single level.	3 Ci, often anvil-shaped, derived from or associated with Cb.	3 Sandstorm, or dust-storm, or drifting or blowing snow.	3 Four-tenths.	3 Falling or steady, then rising; or rising, then rising more quickly. Now higher than, 3 hours ago.
4 Sc formed by spreading out of Cu; Cu often present also.	4 Thin Ac in patches; cloud elements and/or occurring at more than one level.	4 Ci, often hook-shaped, gradually spreading over the sky and usually thickening as a whole.	4 Fog, or smoke, or thick dust haze.	4 Five-tenths.	4 Steady. Same as 3 hours ago.§
5 Sc not formed by spreading out of Cu.	5 Thin Ac in bands or in a layer gradually spreading over sky and usually thickening as a whole	5 Ci and Cs, often in converging bands, or Cs alone; the continuous layer not reaching 45° altitude.	5 Drizzle	5 Six-tenths.	5 Falling, then rising. Same or lower than 3 hours ago.
6 St or Fs or both, but not Fs of bad weather	6 Ac formed by the spreading out of Cu.	6 Ci and Cs often in converging bands, or Cs alone; the continuous layer exceeding 45° altitude.	6 Rain.	6 Seven- or eight-tenths.	6 Falling, then steady; or falling, then falling more slowly. Now lower than 3 hours ago.
7 Fs and/or Fc of bad weather (scud) usually under As and Ns.	7 Double-layered Ac or a thick layer of Ac, not increasing; or As and Ac both present at same or different levels.	7 Cs covering the entire sky.	7 Snow, or rain and snow mixed, or ice pellets (sleet).	7 Nine-tenths or overcast with openings.	7 Falling steadily, or unsteady. Now lower than 3 hours ago.
8 Cu and Sc (not formed by spreading out of Cu) with bases at different levels.	8 Ac in the form of Cu-shaped tufts or Ac with turrets.	8 Cs not increasing and not covering entire sky; Ci and Cc may be present.	8 Shower(s)	8 Completely overcast	8 Steady or rising, then falling; or falling, then falling more quickly. Now lower than 3 hours ago.
9 Cb having a clearly fibrous (cirriform) top, often anvil-shaped, with or without Cu, Sc, St, or scud.	9 Ac of a chaotic sky, usually at different levels; patches of dense Ci are usually present also.	9 Cc alone or Cc with some Ci or Cs, but the Cc being the main cirriform cloud present.	9 Thunderstorm, with or without precipitation.	9 Sky obscured.	

*Fraction representing how much of the total cloud cover is at the reported base height.

▲ **Figure IV-3** Standard symbols used to indicate cloud conditions, past weather, and barometer characteristics. The numbers in the upper-left-hand corner of the cells are used in a standard model, and the icons are used on weather maps.

W W
Present weather

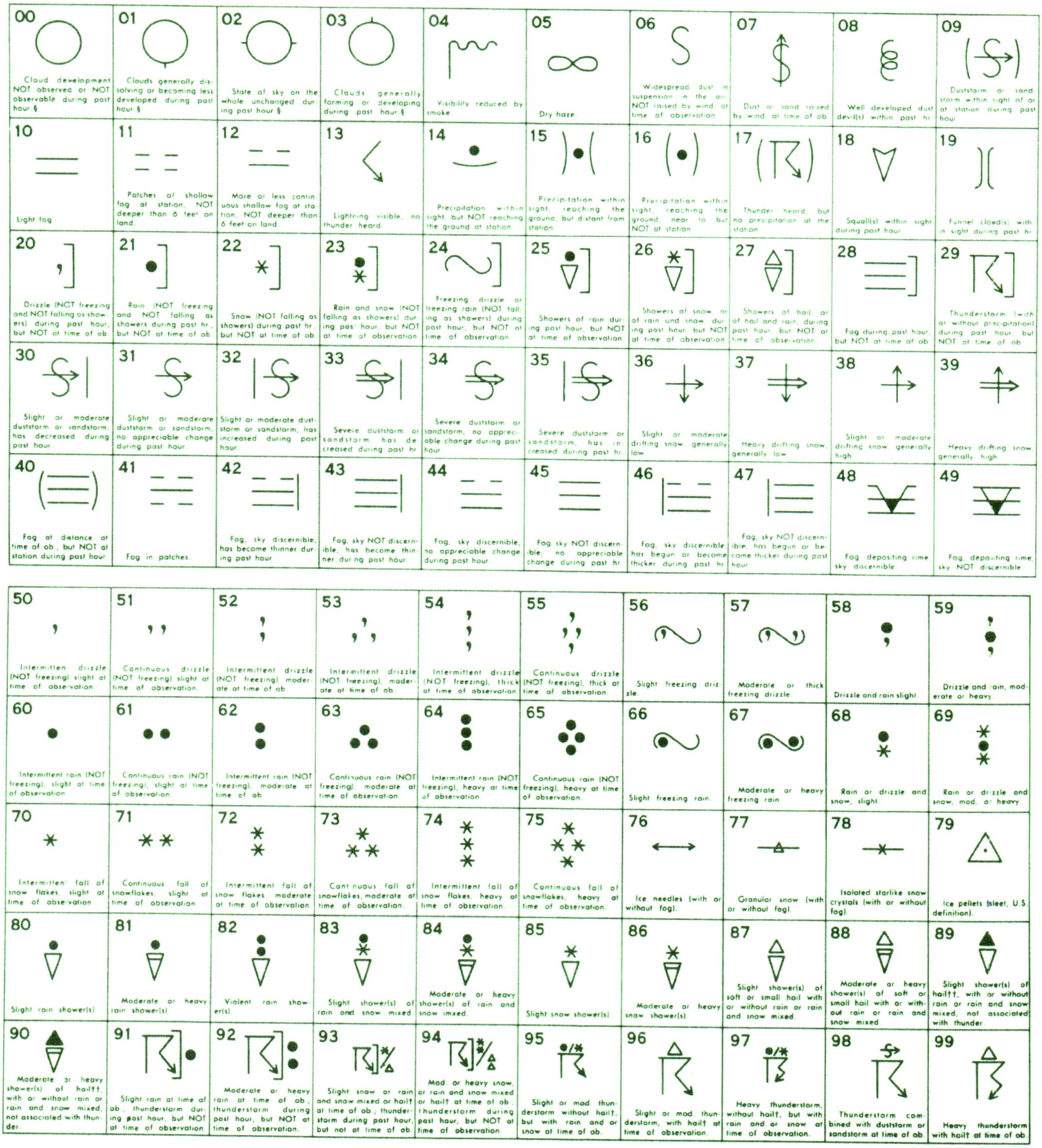

§ The symbol is not plotted for "ww" when "00" is reported. When "01, 02, or 03" is reported for "ww," the symbol is plotted on the station circle. Symbols are not plotted for "a" when "3 or 8" is reported.

† Refers to "hail" only.

††Refers to "soft hail," "small hail," and "hail."

▲ **Figure IV-4** Standard weather map symbols used to indicate present weather.

TABLE IV-1 Standard Cloud Height Codes

	Approximate Cloud Height	
h (height of cloud base)	Meters	Feet
0	0–49	0–149
1	50–99	150–299
2	100–199	300–599
3	200–299	600–999
4	300–599	1000–1999
5	600–999	2000–3499
6	1000–1499	3500–4999
7	1500–1999	5000–4699
8	2000–2499	6500–7999
9	> 2500 or no clouds	>8000 or no clouds

TABLE IV-2 Standard Precipitation Codes

R_t Code	Time of Precipitation
0	No precipitation
1	Less than 1 hour ago
2	1 to 2 hours ago
3	2 to 3 hours ago
4	3 to 4 hours ago
5	4 to 5 hours ago
6	5 to 6 hours ago
7	6 to 12 hours ago
8	More than 12 hours ago
9	Unknown

TABLE IV-3 Wind Speed/Direction Symbols

Symbol	Wind Speed (knots)
◎	Calm
	1–2
	3–7
	8–12
	13–17
	18–22
	23–27
	28–32
	33–37
	38–42
	43–47
	48–52
	53–57
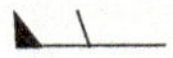	58–62
	63–67
	68–72
	73–77

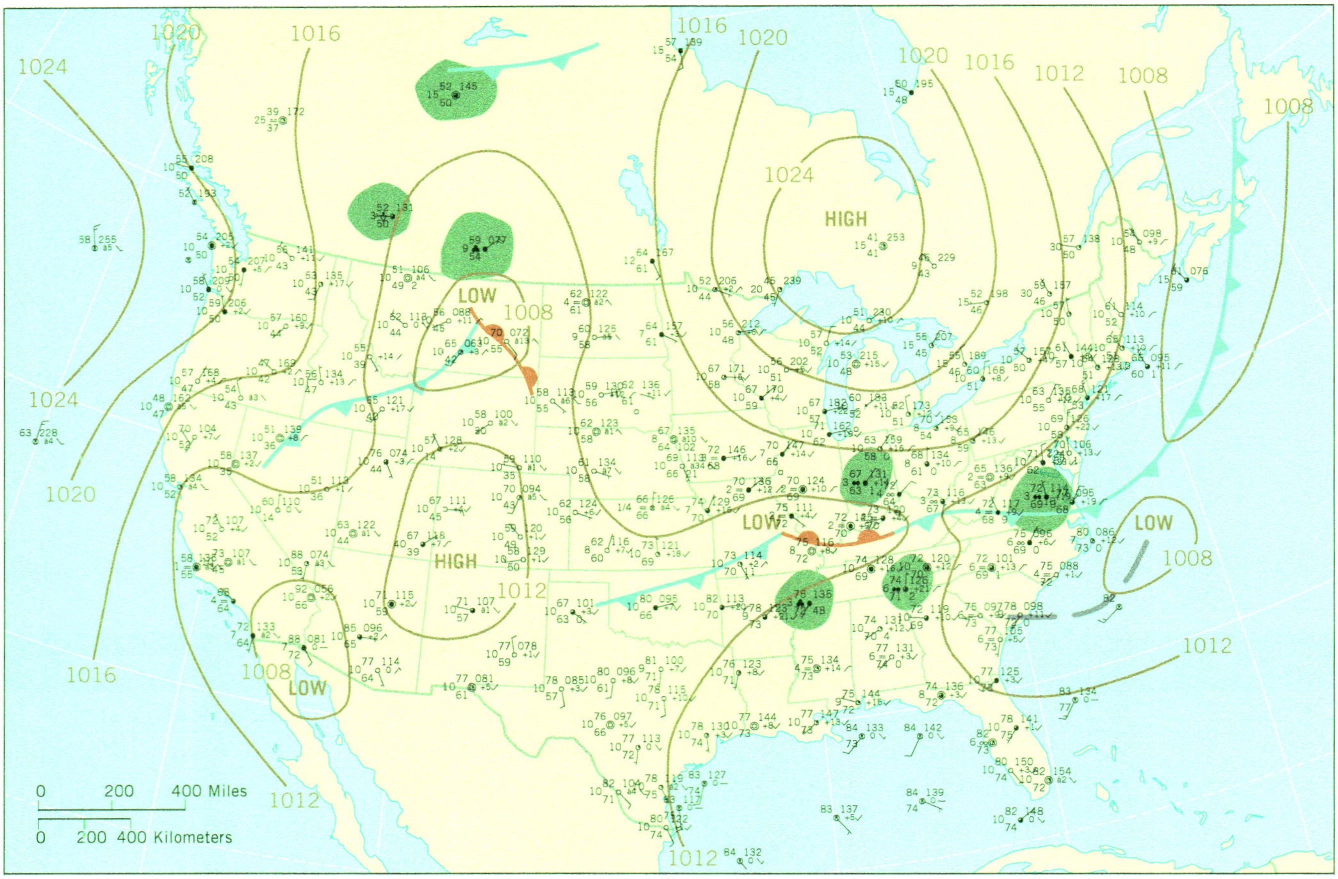

▲ **Figure IV-5** A sample weather map.

Appendix V

Köppen Climate Classification

Table V-1 provides the definitions of the code letters used in the modified Köppen climate classification system. The exact classification of a climate using the Köppen system may require additional calculations, so the definitions of each letter given here may be approximate. For more detailed directions for classifying climates with the Köppen system, see the *Physical Geography Laboratory Manual* for McKnight's Physical Geography: A Landscape Appreciation, 13th edition, by Darrel Hess.

TABLE V-1 Code Letters of the Modified Köppen Classification System

Letters 1st	2nd	3rd	Description	Definitions
A			Low-latitude humid climates	Average temperature of each month above 18°C (64°F)
	f		No dry season [German: *feucht* ("moist")]	Average rainfall of each month at least 6 cm (2.4 in.)
	m		Monsoon; short dry season compensated by heavy rains in other months	1 to 3 months with average rainfall less than 6 cm (2.4 in.)
	w		Winter dry season (low-Sun season)	3 to 6 months with average rainfall less than 6 cm (2.4 in.)
B			Dry climates	Evaporation exceeds precipitation
	W		Desert [German: *wüste* ("desert")]	Average annual precipitation generally less than 38 cm (15 in.) in low latitudes; 25 cm (10 in.) in midlatitudes
	S		Steppe (semiarid)	Average annual precipitation generally between 38 cm (15 in.) and 76 cm (30 in.) in low latitudes; between about 25 cm (10 in.) and 64 cm (25 in.) in midlatitudes; without pronounced seasonal concentration
		h	Low-latitude (subtropical) dry climate [German: *heiss* ("hot")]	Average annual temperature more than 18°C (64°F)
		k	Midlatitude dry climate [German: *kalt* ("cold")]	Average annual temperature less than 18°C (64°F)
C			Mild midlatitude climates	Average temperature of coldest month between 18°C (64°F) and −3°C (27°F); average temperature of warmest month above 10°C (50°F)
	s		Summer dry season	Wettest winter month has at least 3× precipitation of driest summer month
	w		Winter dry season	Wettest summer month has at least 10× precipitation of driest winter month
	f		No dry season [German: *feucht* ("moist")]	Does not fit either s or w above
		a	Hot summers	Average temperature of warmest month more than 22°C (72°F)
		b	Warm summers	Average temperature of warmest month below 22°C (72°F); at least 4 months with average temperature above 10°C (50°F)
		c	Cool summers	Average temperature of warmest month below 22°C (72°F); less than 4 months with average temperature above 10°C (50°F); coldest month above −38°C (−36°F)

(Continued)

TABLE V-1 Code Letters of the Modified Köppen Classification System (*continued*)

Letters				
1st	**2nd**	**3rd**	**Description**	**Definitions**
D			Humid midlatitude climates with severe winters (2nd and 3rd letters same as in C climates)	Warmest month above 10°C (50°F); coldest month below −3°C (27°F)
		d	Very cold winters	Average temperature of coldest month less than −38°C (−36°F)
E			Polar climates; no true summer	No month with average temperature more than 10°C (50°F)
	T		Tundra climates	At least one month with average temperature more than 0°C (32°F) but less than 10°C (50°F)
	F		Ice cap climates ("frost")	No month with average temperature more than 0°C (32°F)
H			Highland climates	Significant climatic changes within short horizontal distances due to altitudinal variations

Appendix VI

Biological Taxonomy and Nomenclature

Taxonomy: *Taxonomy* is the science of classification. As a term, it was originally applied to the classification of plants and animals, although its meaning has been broadened to encompass any sort of systematic classification. Our concern here is with biological taxonomy only.

People have attempted to devise meaningful classifications of plants and animals for thousands of years. One of the most useful of the early classification systems was designed by Aristotle 2300 years ago, and this system was in general use for nearly 20 centuries. In the late 1700s, Aristotelian classification was finally replaced by a much more comprehensive system developed by the Swedish naturalist Carolus Linnaeus. Linnaeus made use of ideas from other biologists, but the system is largely his own work.

The original Linnaean system of classification was based on *observable characteristics* of organisms—primarily their anatomy, structures, and details of reproduction. However, contemporary biology tends to focus on *phylogenies*—in other words, analyzing organisms in terms of *evolutionary relationships,* and may reject or redefine relationships accepted in traditional taxonomies.

The intellectual beauty of a good biological taxonomy is twofold: (1) every organism that has ever existed can be fitted into the scheme in a logical and orderly manner because the system is capable of indefinite expansion, and (2) the various categories in the system provide a conceptual framework for understanding the relationships among different organisms or groups of organisms.

The fact that this approach has lasted so long is not to say that it is perfect. For example, Linnaeus believed that species are unchanging entities, and his original system had no provision for the variations that occur as species evolve. The concept of subspecies was a major modification of the system that was introduced subsequently to accommodate observed conditions of evolution. Nor is the system even completely objective. Whereas its usefulness is understood by scientists throughout the world, there is often room for differing interpretations of the relevant data. Consequently, some details of contemporary biological taxonomy are disputed and even controversial. Despite points of debate, the system of taxonomy described below provides a magnificent framework for biological classification.

Levels of Taxonomic Classification: Taxonomies based on Linnaean principles are hierarchical—organisms are placed into ranked levels. Eight main taxonomic levels recognized today, in order from most inclusive to least, are (1) domain, (2) kingdom, (3) phylum, (4) class, (5) order, (6) family, (7) genus, and (8) species.

Domain is the broadest (and most recently recognized) level of classification and describes the largest number of organisms. Three domains are widely accepted: the Bacteria, the Archaea (unicellular organisms lacking a defined nucleus, i.e., prokaryotes), and the Eukarya (single-celled or multicellular organisms consisting of cells with a defined nuclei).

Kingdom is the next broadest category of organisms. In the past, the system recognized only two kingdoms: one encompassing all plants and the other including all animals. Increasingly, however, as microscopes became available, taxonomists encountered difficulty accommodating the many varieties of one-celled and other microscopic organisms into such a two-kingdom system. The category underwent multiple refinements.

In the 1960s, five kingdoms were recognized: *Monera, Protista, Fungi, Plantae,* and *Animalia*. Since then, the accepted number of kingdoms has varied—ranging from six to eight—with no scheme achieving universal acceptance. In one recent scheme, six kingdoms are recognized:

1. *Bacteria* (or *Eubacteria*), the "true" bacteria. Single-celled organisms lacking a nucleus; prokaryotes.
2. *Archaea* (or *Archaebacteria*), also prokaryotes. Recognized more recently than bacteria, archaea live in harsh environments such as underwater hydrothermal vents, hot sulfur springs, and hypersaline water. Some may also live in the open ocean and can be either autotrophs or heterotrophs.
3. *Protozoa* (or *Protista*) consist of unicellular organisms and some unspecialized multicellular algae, most of which were formerly classified as plants.
4. *Fungi* were also previously classified as plants, but we now recognize that they differ in origin, direction of evolution, and primary nutrition from plants and therefore deserve separation.
5. *Plantae* include the multicellular green plants and algae with specialized tissues.
6. *Animalia* are multicellular organisms that include the familiar vertebrate and invertebrate organisms like fish, amphibians, reptiles, insects, and mammals.

Phylum is a third major level of classification. Of the approximately three dozen phyla within the animal kingdom, one, *Chordata*, includes all animals with backbones, which means almost every animal more than a few centimeters in length (**Figure VI-1**). In the plant kingdom, the term *division* is often used in place of phylum. Most large plants belong to the division *Tracheophyta*, or vascular plants, which have efficient internal systems for transporting water and sugars and a complex differentiation of organs into leaves, stems, and roots.

A fourth principal level is *class*. Among the several dozen animal classes, the most familiar are *Mammalia* (mammals), *Aves* (birds), *Reptilia* (reptiles), *Amphibia* (amphibians), *Insecta* (insects), and two classes that encompass

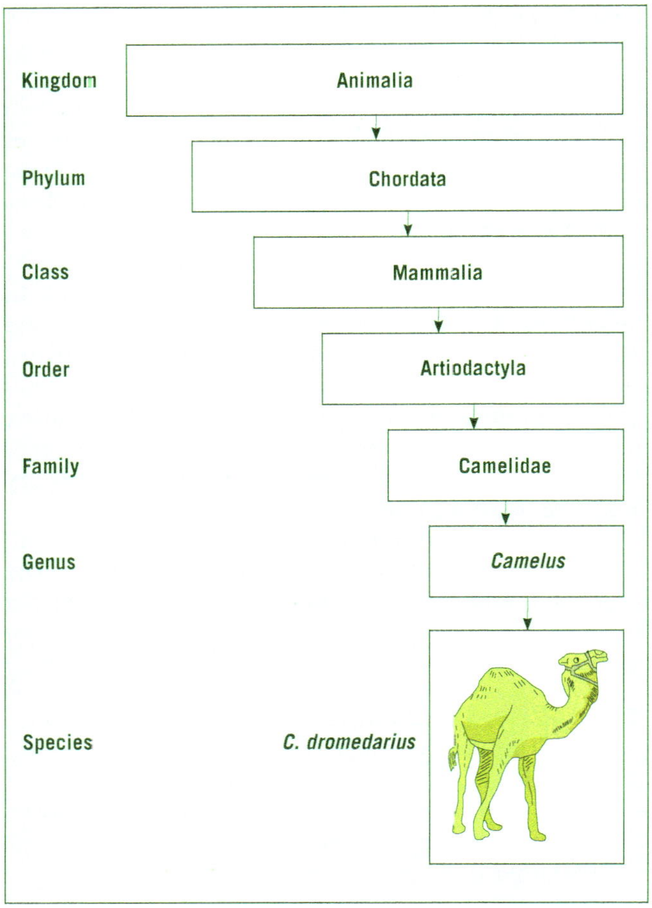

▲ **Figure VI-1** The taxonomy of an animal, as illustrated by the Arabian camel, or dromedary.

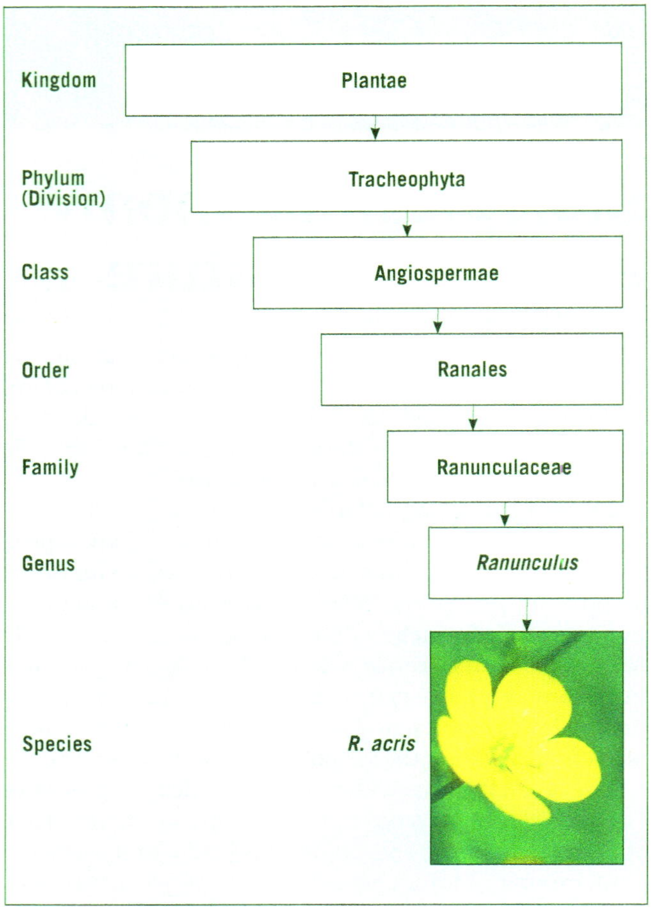

▲ **Figure VI-2** The taxonomy of a plant, as illustrated by the meadow buttercup.

fishes. There are somewhat fewer classes of plants, of which the most well-known might be *Angiospermae*, the flowering plants (**Figure VI-2**).

A fifth level of classification is called *order*, and three lower levels are *family*, *genus*, and *species*. As in all levels of the hierarchy, each succeeding lower level contains organisms that are increasingly alike. Species is the basic unit in traditional systems of classification. In theory, only members of the same species are capable of breeding with one another. In practice, however, interbreeding is possible among just a few species (always within the same genus), although the offspring of such interspecific breeding are nearly always infertile (i.e., incapable of reproducing). In some cases, species are further subdivided into subspecies, also called *varieties* or *cultivars*.

The relative diversity of living and extinct species is worthy of note. To date, over 1.8 million species of living organisms have been scientifically identified, with more being discovered every year. From 3 to 10 million additional species, mostly microscopic in size, may not yet be identified. An estimated 500 million species have become extinct over Earth's history. Thus, more than 95 percent of all evolutionary lines have already disappeared.

Scientific Nomenclature (Binomial Names): Nomenclature is not the same thing as classification, although the two activities are related. Classification orders organisms into groups based on similarities or differences. Nomenclature simply assigns names in agreement with a chosen classification system.

The *binomial* naming of organisms is highly systematized. Each type of living thing can have a unique name with two parts. The first part, in which the first letter is capitalized, designates the *genus*, or group; the second part, which is not capitalized, indicates the *species*, or specific kind of organism. The combination of genus and species is referred to as the *binomial* ("two-terms") or *scientific name*; it is always in Latin, although many of the words have Greek derivations.

Each type of organism, then, can have a scientific name that distinguishes it from all other organisms. Although vernacular names may be variable, or even indefinite, the scientific name is unvarying. Thus, in different parts of the Western Hemisphere, the large native cat may be called a mountain lion, cougar, puma, panther, painter, or leon, but its scientific name would always be *Felis concolor*.

Appendix VII

The Soil Taxonomy

The *Soil Taxonomy*, described in Chapter 12, utilizes a nomenclature "invented" for the purpose. This nomenclature consists of "synthetic" names, which means that syllables from existing words are rearranged and combined to produce new words for the names of the various soil types. The beauty of the system is that each newly coined name is highly descriptive of the soil it represents.

The awkwardness of the nomenclature is threefold:

1. Most of the terms are new and have never before appeared in print or in conversation. Thus, they look strange, and the words do not easily roll off the tongue. It is almost a new language.
2. Many of the words are difficult to write, and great care must be taken in the spelling of seemingly bizarre combinations of letters.
3. Many of the syllables sound so much alike that differences in pronunciation are often slight, although enunciating these slight differences is essential if the new nomenclature is to serve its purpose.

Nevertheless, the Soil Taxonomy has a sound theoretical base, and once you are familiar with the vocabulary, every syllable of every word gives important information about a soil. Almost all the syllables are derived from Greek or Latin roots, in contrast to the English and Russian terms used in previous systems. In some cases, the appropriateness of the classical derivatives may be open to question, but the uniformity and logic of the terminology extend throughout the system.

The Hierarchy

The top level in the Soil Taxonomy hierarchy is *soil order*. The names of orders are made up of three or four syllables, the last of which is always *sol* (from the Latin word *solum*, "soil"). The next-to-last syllable consists of a single linking vowel, either *i* or *o*. The syllable (or two) that begins the word contains the formative element of the name and gives information regarding some distinctive characteristic of the order. For example, in the order Entisol, *ent* comes from the word "recent." Thus, each of the 12 soil order names contains a distinctive syllable that (1) identifies the formative element of that soil order and (2) appears in all the names in the next three levels of the hierarchy for that soil order.

Suborder is the second level in the hierarchy. All suborder names contain two syllables: the first indicates some distinctive characteristic of the suborder; the second identifies the order to which the suborder belongs. For example, *aqu* is derived from the Latin word for "water," and *Aquent* is the suborder of wet soils of the order Entisols. The names of the four dozen suborders are constructed from the two dozen root elements shown in Table VII-1.

The third level of the hierarchy contains *great groups*, the names of which are constructed by grafting one or more syllables to the beginning of a suborder name. Hence, a *Cryaquent* is a cold soil that is a member of the Aquent suborder of the Entisol order (*cry* comes from the Greek word for "coldness"). The formative prefixes for the great group names are derived from about 50 root words, samples of which are presented in Table VII-2.

The next level is called *subgroup*, and more than 1000 subgroups are recognized in the United States. Each subgroup name consists of two words: the first derived from a formative element higher up in the hierarchy (with a few exceptions), the second the same as that of the relevant great group. Thus, a *Sphagnic Cryaquent* is of the order Entisol, the suborder Aquent, and the great group *Cryaquent*, and it contains sphagnum moss (derived from the Greek word *sphagnos*, "bog").

The *family* is the penultimate level in the hierarchy. It is not given a proper name but is simply described by one or more lowercase adjectives, such as "a skeletal, mixed, acidic family."

Finally, the lowest level of the hierarchy is the *series*, named for geographic location. Table VII-3 gives a typical naming sequence.

Summary of Soil Suborders

Alfisols

Alfisols have five suborders. *Aqualfs* have characteristics associated with wetness. *Boralfs* are associated with cold boreal forests. *Udalfs* are brownish or reddish soils of moist midlatitude regions. *Ustalfs* are similar to Udalfs in color but subtropical in location and usually have a hard surface layer in the dry season. *Xeralfs* are found in mediterranean climates and are characterized by a thick, hard surface horizon in the dry season.

Andisols

Seven suborders are recognized in the Andisols, five of which are distinguished by moisture content. *Aquands* have abundant moisture, often with poor drainage. *Torrands* are associated with a hot, dry regime. *Udands* are found in humid climates; *Ustands* are in dry climates with hot summers. *Xerands* have a pronounced annual dry season. In addition, *Cryands* are found in cold climates, and *Vitrands* are distinguished by the presence of glass.

Aridisols

Two suborders are generally recognized on the basis of degree of weathering. *Argids* have a distinctive subsurface horizon with clay accumulation, whereas *Orthids* do not.

TABLE VII-1 Name Derivations of Soil Suborders

Root	Derivation	Connotation	Example of Suborder Name
alb	Latin *albus*, "white"	Presence of bleached eluvial horizon	Alboll
and	Japanese *ando*, a volcanic soil	Derived from pyroclastic material	Andept
aqu	Latin *aqua*, "water"	Associated with wetness	Aquent
ar	Latin *arare*, "to plow"	Horizons are mixed	Arent
arg	Latin *argilla*, "white clay"	Presence of a horizon containing illuvial clay	Argid
bor	Greek *boreas*, "northern"	Associated with cool conditions	Boroll
ferr	Latin *ferrum*, "iron"	Presence of iron	Ferrod
fibr	Latin *fibra*, "fiber"	Presence of undecomposed organic matter	Fibrist
fluv	Latin *fluvius*, "river"	Associated with floodplains	Fluvent
fol	Latin *folia*, "leaf"	Mass of leaves	Folist
hem	Greek *hemi*, "half"	Intermediate stage of decomposition	Hemist
hum	Latin *humus*, "earth"	Presence of organic matter	Humult
ochr	Greek *ochros*, "pale"	Presence of a light-colored surface horizon	Ochrept
orth	Greek *orthos*, "true"	Most common or typical group	Orthent
plag	German *plaggen*, "sod"	Presence of a human-induced surface horizon	Plaggept
psamm	Greek *psammos*, "sand"	Sandy texture	Psamment
rend	Polish *rendzino*, a type of soil	Significant calcareous content	Rendoll
sapr	Greek *sapros*, "rotten"	Most decomposed stage	Saprist
torr	Latin *torridus*, "hot and dry"	Usually dry	Torrox
trop	Greek *tropikos*, "of the solstice"	Continuously warm	Tropert
ud	Latin *udud*, "humid"	Of humid climates	Udoll
umbr	Latin *umbro*, "shade"	Presence of a dark surface horizon	Umbrept
ust	Latin *ustus*, "burnt"	Of dry climates	Ustert
xer	Greek *xeros*, "dry"	Annual dry season	Xeralf

TABLE VII-2 Name Derivations for Great Groups

Root	Derivation	Connotation	Example of Great Group Name
calc	Latin *calcis*, "lime"	Presence of calcic horizon	Calciorthid
ferr	Latin *ferrum*, "iron"	Presence of iron	Ferrudalf
natr	Latin *natrium*, "sodium"	Presence of a natric horizon	Natraboll
pale	Greek *paleos*, "old"	An old development	Paleargid
plinth	Greek *plinthos*, "brick"	Presence of plinthite	Plenthoxeralf
quartz	The German name	High quartz content	Quartzipsamment
verm	Latin *vermes*, "worm"	Notable presence of worms	Vermudoll

TABLE VII-3 A Typical Soil Taxonomy Naming Sequence

Order	Entisol
Suborder	Aquent
Great group	Cryaquent
Subgroup	Sphagnic Cryaquent
Family	Skeletal, mixed, acidic, Sphagnic Cryaquent
Series	Aberdeen

Entisols

There are five suborders of Entisols. *Aquents* occupy wet environments where the soil is more or less continuously saturated with water; they may be found in any temperature regime. *Arents* lack horizons because of human interference, particularly that involving large agricultural or engineering machinery. *Fluvents* form on recent water-deposited sediments that have satisfactory drainage. *Orthents* develop on recent erosional surfaces. *Psamments* occur in sandy situations, where the sand is either shifting or stabilized by vegetation.

Gelisols

The three suborders of Gelisols—*Histels*, *Orthels*, and *Turbels*—are distinguished largely on the basis of quantity and distribution of organic material.

Histosols

The four suborders of Histosols—*Fibrists*, *Folists*, *Hemists*, and *Saprists*—are differentiated on the basis of degree of plant-material decomposition.

Inceptisols

The six suborders of Inceptisols—*Andepts*, *Aquepts*, *Ochrepts*, *Plaggepts*, *Tropepts*, and *Umbrepts*—have relatively complicated distinguishing characteristics.

Mollisols

The seven suborders of Mollisols—*Albolls*, *Aquolls*, *Borolls*, *Rendolls*, *Udolls*, *Ustolls*, and *Xerolls*—are distinguished largely on the basis of relative wetness/dryness.

Oxisols

The five suborders of Oxisols—*Aquox*, *Humox*, *Orthox*, *Torrox*, and *Ustox*—are distinguished from one another primarily by what effect varying amounts and seasonality of rainfall have on the profile.

Spodosols

Of the four suborders of Spodosols, most widespread are the *Orthods*, which represent the typical Spodosols. *Aquods*, *Ferrods*, and *Humods* are differentiated on the basis of the amount of iron in the spodic horizon.

Ultisols

Five suborders of Ultisols—*Aquults*, *Humults*, *Udults*, *Ustults*, and *Xerults*—are recognized. The distinction among them is largely on the basis of temperature and moisture conditions and on how these parameters influence the epipedon.

Vertisols

The four principal suborders of Vertisols are distinguished largely on the frequency of "cracking," which is a function of climate. *Torrerts* are found in arid regions, and the cracks in these soils remain open most of the time. *Uderts* are found in humid areas, and in these soils cracking is irregular. *Usterts* are associated with monsoon climates and have a relatively complicated cracking pattern. *Xererts* occur in mediterranean climates and have cracks that open and close regularly once each year.

Glossary

A

ablation Wastage of glacial ice through melting and sublimation.

ablation zone The lower portion of a glacier, where there is a net annual loss of ice due to melting and sublimation.

absolute humidity The measure of the actual water vapor content of air, expressed as the mass of water vapor in a given volume of air, usually as grams of water per cubic meter of air.

absorption The ability of an object to assimilate energy from electromagnetic waves that strike it.

accumulation (glacial ice accumulation) Addition of ice into a glacier by incorporation of snow.

accumulation zone The upper portion of a glacier, where there is a greater annual accumulation of ice than there is wastage.

acid rain Precipitation with a pH less than 5.6. It may involve dry deposition without moisture.

adiabatic cooling Cooling by expansion, such as in rising air.

adiabatic warming Warming by compression, such as in descending air.

adret slope A slope oriented such that the Sun's rays arrive at a relatively high angle. It tends to be relatively warm and dry.

advection Horizontal transfer of energy, such as through the movement of wind across Earth's surface.

aeolian processes Processes related to wind action that are most pronounced, widespread, and effective in dry lands.

aerosols Solid or liquid particles suspended in the atmosphere; particulates.

aggradation The process in which a stream bed is raised as a result of the deposition of sediment.

A horizon The upper soil layer, in which humus and other organic materials are mixed with mineral particles.

air mass An extensive body of air that has relatively uniform properties in the horizontal dimension and moves as an entity.

albedo The reflectivity of a surface. The fraction of total solar radiation that is reflected back, unchanged, into space.

Alfisol A widely distributed soil order distinguished by a subsurface clay horizon and a medium-to-generous supply of plant nutrients and water.

alluvial fan A fan-shaped depositional feature of alluvium laid down by a stream issuing from a mountain canyon.

alluvium Any stream-deposited sedimentary material.

alpine glacier Individual glacier that develops near a mountain crest line and normally moves down-valley for some distance.

Andisol The soil order derived from volcanic ash.

angiosperms Plants that have seeds encased in some sort of protective body, such as a fruit, a nut, or a seedpod.

angle of incidence The angle at which the Sun's rays strike Earth's surface.

angle of repose The steepest angle at which loose fragments on a slope can be assumed to remain without downslope movement.

annual plants (annuals) Plants that perish during times of environmental stress but leave behind a reservoir of seeds to germinate during the next favorable period.

Antarctic Circle The parallel of 66.5° south latitude.

antecedent stream A stream that predates the existence of the hill or mountain through which it flows.

anticline A simple symmetrical upfold in the rock structure.

anticyclone A high-pressure center.

antitrade winds Tropical upper-atmosphere westerly winds at the top of the Hadley cells that blow toward the northeast in the Northern Hemisphere and toward the southeast in the Southern Hemisphere.

aphelion The point in Earth's elliptical orbit at which Earth is farthest from the Sun (about 152,100,000 kilometers [94,500,000 miles] away).

aquiclude An impermeable rock layer that is so dense as to exclude water.

aquifer A permeable subsurface rock layer that can store, transmit, and supply water.

Arctic Circle The parallel of 66.5° north latitude.

arête A narrow, jagged, serrated spine of rock; the remainder of a ridge crest after several glacial cirques have been cut back into an interfluve from opposite sides of a divide.

Aridisol The soil order occupying dry environments that do not have enough water to remove soluble minerals from the soil; typified by a thin profile that is sandy and lacking in organic matter.

artesian well The free flow that results when a well is drilled from the surface down into an aquifer and the confining pressure is sufficient to force the water to the surface without artificial pumping.

asthenosphere The plastic layer of the upper mantle that underlies the lithosphere. Its rock is dense but very hot and therefore weak and easily deformed.

atmosphere The gaseous envelope surrounding Earth.

atmospheric pressure The force exerted by the atmosphere on a surface.

atoll A coral reef in the general shape of a ring or partial ring that encloses a lagoon.

average annual temperature range The difference in temperature between the average temperature of the hottest and coldest months for a location.

average lapse rate The average rate of temperature decrease with height in the troposphere—about 6.5°C per 1000 meters (3.6°F per 1000 feet).

B

backwash Water moving seaward after the momentum of the wave swash is overcome by gravity and friction.

badlands Intricately rilled and barren terrain of arid and semiarid regions, characterized by a multiplicity of short, steep slopes.

bajada A continual alluvial surface that extends across the piedmont zone, slanting from the range toward the basin, in which it is difficult to distinguish among individual alluvial fans.

barchan dune A crescent-shaped sand dune with cusps of the crescent pointing downwind.

barometer An instrument used to measure atmospheric pressure.

barrier island A narrow offshore island composed of sediment; generally oriented parallel to shore.

barrier reef A prominent ridge of coral that roughly parallels the coastline but lies offshore, with a shallow lagoon between the reefs and the coast.

basal slip The sliding of the bottom of a glacier over its bed on a lubricating film of water.

basalt Fine-grained, dark (usually black) volcanic rock; forms from mafic lava (which has relatively low silica content).

base level An imaginary surface extending underneath the continents from sea level at the coasts and indicating the lowest level to which land can erode.

batholith The largest and most amorphous of igneous intrusions.

baymouth bar A spit that extends entirely across the mouth of a bay, transforming the bay into a lagoon.

beach An exposed deposit of loose sediment, normally composed of sand and/or gravel, and occupying the coastal transition zone between land and water.

beach drifting The zigzag movement of sediment caused by waves washing particles onto a beach at a slight angle; the net result is the movement of sediment along the coast in a general downwind direction.

bedding plane A flat surface separating one sedimentary layer from the next.

bedload Sand, gravel, and larger rock fragments moving in a stream by saltation and traction.

B horizon The mineral soil horizon located beneath the A horizon; subsoil.

biodiversity The number of different kinds of organisms present in a location.

biogeochemical cycles Interconnected series of cycles and pathways that entail the flow of energy and chemical elements and compounds (such as water and various nutrients) through the biosphere.

biogeography The study of the distribution patterns of plants and animals, and how these patterns change over time.

biological weathering Rock weathering processes involving the action of plants or animals.

biomass The total mass (or weight) of all living organisms in an ecosystem or per unit area.

biome A large, recognizable assemblage of plants and animals in functional interaction with its environment.

biosphere The living organisms of Earth.

biota The total complex of plant and animal life.

blowout (deflation hollow) A shallow depression from which an abundance of fine material has been deflated by wind.

boreal forest (taiga) An extensive needleleaf forest in the subarctic regions of North America and Eurasia.

braided channel pattern (braided stream) A stream that consists of a multiplicity of interwoven and interconnected shallow channels separated by low islands of sand, gravel, and other loose debris.

broadleaf trees Trees that have flat and expansive leaves.

butte An erosional remnant of very small surface area and clifflike sides that rises conspicuously above the surroundings.

C

calcification One of the dominant pedogenic regimes in areas where the principal soil moisture movement is upward because of a moisture deficit; characterized by a concentration of calcium carbonate ($CaCO_3$) in the B horizon, forming a hardpan.

caldera A large, steep-sided, roughly circular depression resulting from the explosion and/or collapse of a large volcano.

capacity (stream capacity) The maximum load that a stream can transport under given conditions.

capacity (water vapor capacity) The maximum amount of water vapor that can be present in the air at a given temperature.

capillarity The action by which water can climb upward in restricted confinement as a result of its high surface tension, and thus the ability of its molecules to stick closely together.

carbonation A process in which carbon dioxide in water reacts with carbonate rocks, producing a very soluble product (calcium bicarbonate) that can readily be removed by runoff or percolation and that can be deposited in crystalline form if the water evaporates.

carbon cycle The change from carbon dioxide to living matter and back to carbon dioxide.

carbon dioxide CO_2; a minor gas in the atmosphere; one of the greenhouse gases; a by-product of combustion and respiration.

carbonic acid H_2CO_3; a mild acid formed when carbon dioxide dissolves in water.

cation An atom or group of atoms with a positive electrical charge.

cation exchange capacity (CEC) The capability of soil to attract and exchange cations.

cavern A large opening or cave, especially in limestone; often decorated with speleothems.

chemical weathering The chemical decomposition of rock by the alteration of rock-forming minerals.

chinook A localized downslope wind of relatively dry and warm air, which is further warmed adiabatically as it moves down the leeward slope of the Rocky Mountains.

chlorofluorocarbons (CFCs) Synthetic chemicals commonly used as refrigerants and in aerosol spray cans; destroy ozone in the upper atmosphere.

C horizon The lower soil layer composed of weathered parent material that has not been significantly affected by translocation or leaching.

cinder cone A small, common volcano that is composed primarily of pyroclastic material blasted out from a vent in small but intense explosions; structure is usually a conical hill of loose material.

circle of illumination The edge of the sunlit hemisphere that is a great circle separating Earth into a light half and a dark half.

cirque A broad amphitheater hollowed out at the head of a glacial valley by glacial erosion and frost wedging.

cirque glacier A small glacier confined to its cirque and not moving down-valley.

cirrus cloud A high cirriform cloud of feathery appearance.

clay A combination of very small inorganic particles produced by chemical alteration of silicate minerals.

climate An aggregate of day-to-day weather conditions and weather extremes over a long period of time, usually at least 30 years.

climax vegetation A stable plant association of relatively constant composition that develops at the end of a long succession of changes.

climograph (climatic diagram) A chart showing the average monthly temperature and precipitation for a weather station.

cloud A visible accumulation of tiny liquid water droplets or ice crystals suspended in the atmosphere.

col A pass or saddle through a ridge, produced when two adjacent glacial cirques on opposite sides of a divide are cut back enough to remove part of the arête between them.

cold front The leading edge of a cool air mass actively displacing warm air.

collapse sinkhole A sinkhole produced by the collapse of the roof of a subsurface cavern; a collapse doline.

colloids Organic and inorganic microscopic particles of soil that represent the chemically active portion of particles in the soil.

competence (stream competence) The size of the largest particle that can be transported by a stream.

composite volcano A volcano with the classic symmetrical, cone-shaped peak, produced by a mixture of lava outpouring and pyroclastic explosion; also called *stratovolcano*.

compromise map projection A map projection that is neither conformal nor equivalent but a balance of those, or other, map properties.

condensation The process by which water vapor is converted to liquid water; a warming process because latent heat is released.

condensation nuclei Tiny atmospheric particles of dust, bacteria, smoke, and salt that serve as collection centers for water molecules.

conduction The movement of energy from one molecule to another without changing the relative positions of the molecules. It enables the transfer of heat between different parts of a stationary body.

cone of depression An inverted cone where the water table has sunk in the immediate vicinity of a well as the result of the removal of a considerable amount of groundwater.

confined aquifer An aquifer between two impermeable layers (aquicludes).

conformal map projection A projection that maintains proper angular relationships over the entire map; over limited areas, it shows the correct shapes of features.

conic projection A map projection in which one or more cones is set tangent to, or intersecting, a portion of the globe and the geographic grid is projected onto the cone(s).

conifer See *gymnosperm*.
consumer An animal that consumes plants or other animals; a heterotroph.
contact metamorphism Metamorphism of surrounding rocks by contact with magma.
continental drift The theory proposing that the present continents were originally connected as one or two large landmasses that broke up and drifted apart over the last several hundred million years.
continental ice sheet A large ice sheet covering a portion of a continental area.
continental rift valley A fault-produced valley resulting from spreading or rifting of continent.
controls of weather and climate The most important influences acting on the elements of weather and climate.
convection Energy transfer through the vertical circulation and movement of fluids, such as air, due to density differences.
convection cell A closed pattern of convective circulation.
convective lifting Air lifting with showery precipitation resulting from convection.
convergent [plate] boundary A location where two lithospheric plates collide.
convergent lifting Air lifting as a result of wind convergence.
Coordinated Universal Time (UTC) The world time standard reference; previously known as *Greenwich Mean Time (GMT)*.
Coriolis effect (Coriolis force) The apparent deflection of free-moving objects to the right in the Northern Hemisphere and to the left in the Southern Hemisphere, in response to Earth's rotation.
creep (soil creep) The slowest and least perceptible form of mass wasting, consisting of a very gradual downhill movement of soil and regolith.
crust The outermost solid layer of Earth.
cryosphere The subsphere of the hydrosphere that encompasses water frozen as snow or ice.
cumulonimbus cloud A tall cumulus cloud associated with rain, thunderstorms, and other kinds of severe weather, such as tornadoes and hurricanes.
cumulus cloud A puffy, white cloud that forms from rising columns of air.
cutoff meander A portion of an old meandering stream course left isolated from the present stream channel because the narrow meander neck has been cut through by stream erosion.
cyclogenesis The process of formation or "birth" of cyclones.
cyclone A low-pressure center.
cylindrical projection A family of maps derived from the concept of projection onto a paper cylinder that is tangential to, or intersects with, a globe.

D

daylight-saving time The shifting of clocks forward one hour.
debris flow The streamlike flow of dense, muddy water heavily laden with sediments of various sizes; a mudflow containing large boulders.
December solstice The day of the year when the vertical rays of the Sun strike the Tropic of Capricorn; on or about December 21; winter solstice in the Northern Hemisphere.
deciduous tree A tree that experiences an annual period in which all leaves die and usually fall from the tree, due either to a cold season or a dry season.
declination of the Sun The latitude receiving the vertical rays of the Sun.
decomposers Mainly microscopic organisms such as bacteria that decompose dead plant and animal matter.
deflation The shifting of loose particles by wind blowing them into the air or rolling them along the ground.
delta A landform composed of alluvium at the mouth of a river, produced by the sudden reduction of a stream's velocity and the resulting deposition of the stream's load.
dendritic drainage pattern A treelike, branching pattern that consists of a random merging of streams, with tributaries joining larger streams irregularly at acute angles; generally develops where the underlying structure does not significantly control the drainage pattern.
dendrochronology The study of past events and past climate through the analysis of tree rings.
denitrification The conversion of nitrates into free nitrogen in the air.
denudation The total effect of all actions (weathering, mass wasting, and erosion) that lower the surface of the continents.
deposition The process by which water vapor is converted directly to ice.
desert The climate, landscape, or biome associated with extremely arid conditions.
desert pavement A hard and relatively impermeable desert surface of tightly packed small rocks.
desert varnish A dark, shiny coating of iron and manganese oxides that forms on rock surfaces exposed to desert air for a long time.
dew The condensation of beads of water on relatively cold surfaces.
dew point temperature (dew point) The critical air temperature at which water vapor saturation is reached.
differential weathering and erosion The process whereby different rocks or parts of the same rock weather and/or erode at different rates.
digital elevation model (DEM) A computer-generated, shaded-relief image of a landscape derived from a database of precise elevation measurements.
dike A vertical or nearly vertical sheet of magma that is thrust upward into preexisting rock.
disappearing stream A stream that abruptly disappears from the surface as it flows into an underground cavity; common in karst regions.
discharge The volume of flow of a stream per unit of time.
dissolution The removal of bedrock through chemical action of water; includes removal of subsurface rock by the action of groundwater.
dissolved load The minerals, largely salts, that are dissolved in water and carried invisibly in solution.
divergent [plate] boundary A location where two lithospheric plates spread apart.
doldrums The belt of calm air associated with the region between the trade winds of the Northern and Southern hemispheres, generally near the equator; the region of the intertropical convergence zone (ITCZ).
downcutting The action of a stream eroding a deeper channel; occurs when the stream is flowing swiftly and/or down a steep slope.
drainage basin An area that contributes overland flow and groundwater to a specific stream; also called *watershed* or *catchment*.
drainage divide The line of separation between runoff that descends into two different drainage basins.
drift (glacial drift) All material carried and deposited by glaciers.
drumlin A low, elongated hill formed by ice-sheet deposition and erosion; the long axis is aligned parallel with the direction of ice movement. The blunt, steeper end faces the direction from which the ice came.
dry adiabatic rate (dry adiabatic lapse rate) The rate at which a parcel of unsaturated air cools as it rises (10°C per 1000 meters [5.5°F per 1000 feet]).
dynamic high A high-pressure cell associated with prominently descending air.
dynamic low A low-pressure cell associated with prominently rising air.

E

earthflow A mass wasting process in which a portion of a water-saturated slope moves a short distance downhill.

earthquake Vibrations generated by abrupt movement of Earth's crust.

easterly wave A long but weak migratory low-pressure trough in the tropics.

ebb tide A periodic falling of sea level during a tidal cycle.

ecosystem The totality of interactions among organisms and the environment in a given area.

ecotone The transition zone between biotic communities, in which the typical species of one community intermingle with those of another.

edaphic factors Soil-related factors.

E horizon The light-colored, eluvial layer that usually occurs between the A and B horizons.

electromagnetic radiation The flow of energy in the form of electromagnetic waves; radiant energy.

electromagnetic spectrum Electromagnetic radiation arranged according to wavelength.

elements of weather and climate The basic ingredients of weather and climate: temperature, pressure, wind, and moisture.

elevation contour line (contour line) A line on a map joining points of equal elevation.

El Niño The periodic atmospheric and oceanic phenomenon of the tropical Pacific Ocean that typically involves the weakening or reversal of the trade winds and the warming of surface water off the west coast of South America.

eluviation The process by which gravitational water picks up fine particles of soil from the upper layers and carries them downward.

emission See *radiation*.

endemic Relating to an organism found only in a particular area.

endotherm A warm-blooded animal.

energy The ability to do work; anything that has the ability to change the state or condition of matter.

Enhanced Fujita Scale A classification scale of tornado strength, with EF-0 being the weakest tornadoes and EF-5 being the most powerful.

ENSO (El Niño–Southern Oscillation) A linked atmospheric and oceanic phenomenon of pressure and water temperature. The Southern Oscillation is a periodic seesaw of atmospheric pressure in the tropical southern Pacific Ocean basin. Also see *El Niño*.

Entisol The least developed of all soil orders, with little mineral alteration and no pedogenic horizons.

entrenched meanders A winding, sinuous stream valley with abrupt sides; the possible outcome of rejuvenation of a meandering stream.

environmental lapse rate The observed vertical temperature gradient of the troposphere.

ephemeral stream A stream that carries water only during the "wet season" or during and immediately after rains.

epicenter The location on the surface directly above the center of fault rupture during an earthquake.

equal area projection See *equivalent map projection*.

equator The parallel of 0° latitude.

equilibrium line A theoretical line separating the ablation zone and accumulation zone of a glacier; along it, accumulation exactly balances ablation.

equivalent map projection A projection that maintains constant area (size) relationships over the entire map; also called *equal area projection*.

erg A large area covered with loose sand, generally arranged in some sort of dune formation by the wind; "sea of sand."

erosion The detachment, removal, and transportation of fragmented rock material.

esker A long, sinuous ridge of stratified glacial drift composed largely of glaciofluvial gravel and formed by the choking of subglacial streams during a time of glacial stagnation.

eustatic sea-level change A change in sea level due to an increase or decrease in the amount of water in the world ocean, or a change in the volume of an ocean basin; also called *eustasy*.

eutrophication An oversupply of nutrients to lakes or streams that can lead to a plant or algae bloom.

evaporation The process by which liquid water is converted to gaseous water vapor; a cooling process because latent heat is stored.

evapotranspiration The transfer of moisture to the atmosphere by transpiration from plants and evaporation from soil and plants.

evergreen tree A tree or shrub that sheds its leaves on a sporadic or successive basis but at any given time appears to be fully leaved.

exfoliation A weathering process in which curved layers peel off bedrock in sheets; commonly occurs in granite and related intrusive rocks after overlying rock has been removed, allowing the rock mass to expand slightly; also called *unloading*.

exfoliation dome A large rock mass with a surface configuration that consists of imperfect curves punctuated by several partially fractured shells of the surface layers; the result of exfoliation.

exotic species (exotics) Organisms that are introduced into "new" habitats in which they did not naturally occur.

exotic stream A stream that flows into a dry region, bringing its water from somewhere else.

external [geomorphic] processes Destructive processes that serve to denude, or wear down, the landscape; includes weathering, mass wasting, and erosion.

extrusive igneous rock Igneous rock formed on the surface of Earth; also called *volcanic rock*.

eye (eye of tropical cyclone) The nonstormy center of a tropical cyclone; a singular area of calmness, with a diameter of 16–40 kilometers (10–25 miles) in the maelstrom that whirls around it.

F

fall A mass wasting process in which pieces of weathered rock fragments fall to the bottom of a cliff or steep slope; also called *rockfall*.

fault A fracture or zone of fracture where the rock structure is forcefully broken and one side is displaced relative to the other. The movement can be horizontal or vertical, or both.

fault-block mountain (tilted fault-block mountain) A mountain formed where a surface block is faulted and relatively upthrown on one side without any faulting or uplift on the other side. The block is tilted asymmetrically, producing a steep slope along the fault scarp and a relatively gentle slope on the other side of the block.

faulting The production of a fault.

fault scarp A cliff formed by faulting.

fauna Animals.

field capacity The maximum amount of water that soil can retain after gravitational water has drained away.

fjord A glacial trough that has been partly drowned by the sea.

flood basalt A large-scale outpouring of basaltic lava that may cover an extensive area of Earth's surface.

floodplain A flattish valley floor covered with stream-deposited sediments (alluvium), subject to periodic or episodic inundation by overflow from the stream.

flood tide The movement of ocean water toward the coast in a tidal cycle; from the ocean's lowest surface level, the water rises gradually for about 6 hours and 13 minutes.

flora Plants.

fluvial processes Processes involving the work of running water on the surface of Earth.

foehn See *chinook*. The word *foehn* is used predominantly in Europe.

fog A cloud whose base is at or very near ground level.

folding The bending of crustal rocks by compression and/or uplift.
food chain The sequence of predation: organisms feed on one another, with organisms at one level providing food for organisms at the next level, and so on. Energy is thus transferred through the ecosystem.
food pyramid A conceptualization of energy transfer through the ecosystem—from large numbers of energy-storing organisms through succeedingly smaller numbers of consumers, as the organisms at one level are eaten by the organisms at the next higher level. Also see *food chain*.
forest An assemblage of trees growing so closely together that their individual leaf canopies generally overlap.
fossil fuels Naturally occurring fuels (such as coal, petroleum, and natural gas) that form over geologic time from organic materials.
fractional scale (fractional map scale) The ratio of distance measured on a map and the actual distance that represents on Earth's surface, expressed as a ratio or fraction; assumes that the same units of measure are used on the map and on Earth's surface.
freezing The process by which liquid water is converted to ice.
friction layer The zone of the atmosphere, between Earth's surface and an altitude of about 1000 meters (3300 feet), where most frictional resistance to air flow occurs.
fringing reef A coral reef built out laterally from the shore, forming a broad bench that is only slightly below sea level; often the tops of individual coral "heads" are exposed to the open air at low tide.
front A sharp zone of discontinuity between unlike air masses.
frontal lifting The forced lifting of air along a front.
frost wedging The fragmentation of rock due to the expansion of water that freezes into ice within rock openings.
fumarole A hydrothermal feature consisting of a surface crack that is directly connected with a deep-seated heat source. When the little water that drains into this tube is instantly converted to steam by heat and gases, a cloud of steam is expelled from the opening.
funnel cloud A funnel-shaped cloud extending down from a cumulonimbus cloud; a tornado forms when the funnel cloud touches the surface.

G

Gelisol The soil order that develops in areas of permafrost.
geographic information systems (GIS) Computerized systems for the capture, storage, retrieval, analysis, and display of spatial (geographic) data.
geography The study of the spatial distribution and interconnections of phenomena; the study of distribution patterns in the landscape.
geomorphology The study of the characteristics, origin, and development of landforms.
geostrophic wind A wind that moves parallel to isobars as a result of the balance between the pressure gradient force and the Coriolis effect.
geyser A form of intermittent hot spring with water issuing only sporadically as a temporary ejection; hot water and steam are spouted upward for some distance.
glacial erratic An outsize boulder in glacial till; it may be very different from the local bedrock.
glacial flour Rock material that has been ground to the texture of very fine talcum powder by glacial action.
glacial plucking The action whereby rock fragments beneath glacial ice are loosened and grasped by the freezing of meltwater in joints and fractures, and then pried out and dragged along in the glacier's general flow; also called *glacial quarrying*.
glacial steps A series of level or gently sloping bedrock benches alternating with steep drops in the down-valley profile of a glacial trough.
glacial trough A valley reshaped by an alpine glacier, usually U-shaped.
glacier A long-lasting mass of ice that is moving or flowing from its source region.
glaciofluvial deposition The action whereby rock debris that is carried by glaciers is eventually deposited or redeposited by glacial meltwater.
gleization The dominant pedogenic regime in areas where the soil is saturated with water most of the time due to poor drainage.
global conveyer-belt circulation The slowly moving circulation of deep ocean water that forms a continuous loop from the North Atlantic to the Southern Ocean around Antarctica, into the Indian and Pacific Oceans, and back into the North Atlantic.
Global Navigation Satellite System (GNSS) A satellite-based system for determining accurate positions on or near Earth's surface; includes the American GPS.
Global Positioning System (GPS) The American satellite-based system for determining accurate positions on or near Earth's surface.
global warming The common name given to the recent warming of Earth's climate due to human-released greenhouse gases.
graben A block of land bounded by parallel faults; it has been downthrown, producing a distinctive structural valley with a straight, steep-sided fault scarp on both sides.
graded stream A stream in which the gradient has adjusted to the point of allowing just the transportation of its load.
gradient The elevation change of a stream over a given distance.
granite The most common and well-known plutonic (intrusive) rock; coarse-grained rock consisting of both dark- and light-colored minerals; forms from felsic magma (which has relatively high silica content).
graphic scale (graphic map scale) The use of a line marked off in graduated distances as a map scale.
grassland A plant association dominated by grasses and forbs.
great circle The circle formed by the intersection of Earth's surface with any plane that passes through Earth's center.
greenhouse effect The warming in the lower troposphere caused by the differential transmissivity of radiation through the greenhouse gases in the atmosphere; the atmosphere easily transmits incoming shortwave radiation from the Sun but inhibits the transmission of outgoing longwave radiation from the surface.
greenhouse gases Gases that can transmit incoming shortwave radiation from the Sun but absorb outgoing longwave terrestrial radiation; the most important natural greenhouse gases are water vapor and carbon dioxide.
Greenwich Mean Time (GMT) Time in the Greenwich time zone; more commonly called *Coordinated Universal Time (UTC)*.
groin A short wall built perpendicularly from the beach into the shore zone to interrupt the longshore current and trap sand.
ground moraine A moraine consisting of glacial till deposited widely over a land surface beneath an ice sheet.
groundwater Water found underground in the zone of saturation.
gymnosperms Seed-reproducing plants that carry their seeds in cones; "naked seeds."

H

Hadley cells Two complete vertical convective circulation cells between the equator, where warm air rises in the ITCZ, and 25° to 30° of latitude, where much of the air subsides into the subtropical highs.
hail Rounded or irregular pellets or lumps of ice produced in cumulonimbus clouds as a result of active turbulence and vertical air currents. Small ice particles grow by collecting moisture from supercooled cloud droplets.

hamada A barren desert surface of consolidated material that usually consists of exposed bedrock but is sometimes composed of sedimentary material that has been cemented together by salts evaporated from groundwater.

hanging valley (hanging trough) A tributary glacial trough, the bottom of which is considerably higher than the bottom of the principal trough that it joins.

headward erosion Erosion that cuts into the interfluve at the upper end of a gully or valley.

heat Energy that transfers from one object or substance to another because of a difference in temperature; also called *thermal energy*.

high [pressure cell] An area of relatively high atmospheric pressure.

highland climate A high-mountain climate where altitude is the dominant control; designated H in the Köppen system.

highland ice field A largely unconfined ice sheet in high-altitude areas.

Histosol The soil order characterized by organic, rather than mineral, soils; invariably saturated with water all or most of the time.

horizon (soil horizon) A more or less distinctly recognizable layer of soil, distinguished from other soil layers by differing characteristics and forming a vertical zonation of soil.

horn A steep-sided, pyramidal rock pinnacle formed by expansive glacial plucking and frost wedging of the headwalls where three or more cirques intersect.

horse latitudes Areas in the subtropical highs that are characterized by warm sunshine and an absence of wind.

horst A relatively uplifted block of land between two parallel faults.

hot spot An area of volcanic activity within the interior of a lithospheric plate; associated with magma rising up from the mantle below.

hot spring Hot water at Earth's surface that has been forced upward through fissures or cracks by the pressures that develop when underground water comes in contact with heated rocks or magma beneath the surface.

human geography The study of the human and/or cultural elements of geography.

humid continental climate A severe midlatitude climate characterized by hot summers, cold winters, and precipitation throughout the year.

humid subtropical climate A mild midlatitude climate characterized by hot summers and precipitation throughout the year.

humus A dark-colored, gelatinous, chemically stable fraction of organic matter on or in soil.

hurricane A tropical cyclone with wind speeds of 119 kilometers per hour (74 mph; 64 knots) or greater affecting North or Central America.

hydration A chemical process in which water is added to a compound, becoming part of its composition without breaking up the compound; the resulting increase in volume can weaken a mineral.

hydrogen bond An attraction between water molecules; the negatively charged, oxygen side of one water molecule is attracted to the positively charged, hydrogen side of another water molecule.

hydrologic cycle A series of storage areas interconnected by various transfer processes, in which there is a ceaseless interchange of moisture in terms of its geographical location and its physical state.

hydrolysis A chemical union of water with another substance, producing a new compound that is nearly always softer and weaker than the original.

hydrosphere The total water realm of Earth, including the oceans, surface waters of the lands, groundwater, and water held in the atmosphere.

hydrothermal activity The outpouring or ejection of hot water, often accompanied by steam, which usually takes the form of a hot spring or a geyser.

hydrothermal metamorphism Metamorphism associated with hot, mineral-rich solutions circulating around preexisting rock.

hygrophytic adaptations Terrestrial plants adapted to very wet environments.

I

iceberg A great chunk of floating ice that breaks off an ice shelf or the end of an outlet glacier.

ice cap climate A polar climate characterized by temperatures below freezing throughout the year.

ice floe A mass of ice that breaks off from larger ice bodies (ice sheets, glaciers, ice packs, or ice shelves) and floats independently in the sea; generally refers to large, flattish, tabular masses.

ice pack The extensive and cohesive mass of floating ice that is found in the Arctic and Southern Oceans.

ice shelf A massive portion of an ice sheet that projects out over the sea.

igneous intrusion A feature formed by the emplacement and cooling of magma below the surface.

igneous rock Rock formed by solidification of molten magma.

illuviation The process by which fine particles of soil from the upper layers are deposited at a lower level.

Inceptisol An immature order of soils that has relatively faint characteristics; not yet prominent enough to produce diagnostic horizons.

inclination [of Earth's axis] The tilt of Earth's rotational axis relative to its orbital plane (the plane of the ecliptic).

infrared [radiation] Electromagnetic radiation in the wavelength range of about 0.7 to 1000 micrometers; wavelengths just longer than those of visible light.

inner core The solid, dense, innermost portion of Earth, believed to consist largely of iron and nickel.

inselberg "Island mountain"; an isolated summit rising abruptly from a low-relief surface.

insolation Incoming solar radiation.

interfluve The higher land or ridge that separates adjacent valleys; drained by overland flow.

intermittent stream A stream that carries water only during the "wet season" or during and immediately after rains.

internal [geomorphic] processes Geomorphic processes originating below the surface; include volcanism, folding, and faulting.

International Date Line The line marking a time difference of an entire day from one side of the line to the other; it falls on the 180th meridian, except where it deviates to avoid separating an island group.

International System of measurement (S.I.) The metric system of measurement.

intertropical convergence zone (ITCZ) The region near or on the equator where the northeast trade winds and southeast trade winds converge; associated with rising air of the Hadley cells and frequent thunderstorms.

intrusive igneous rock Igneous rock formed below ground from the cooling and solidification of magma; also called *plutonic rock*.

invasive species An introduced species (exotic species) that proliferates in its new environment at the expense of native species.

invertebrates Animals without backbones.

island arc See *volcanic island arc*.

isobar A line (on a map) joining points of equal atmospheric pressure.

isohyet A line (on a map) joining points of equal numerical value of precipitation.

isoline A line (on a map) connecting points that have the same quality or intensity of a given phenomenon.
isostasy The maintenance of the hydrostatic equilibrium of Earth's crust; the sinking of the crust as weight is applied and the rising of crust as weight is removed.
isotherm A line (on a map) joining points of equal temperature.
ITCZ See *intertropical convergence zone.*

J

jet stream A rapidly moving current of wind in the upper troposphere; jet streams can be thought of as the high-speed "cores" of the high-altitude westerly wind flow that frequently meander in a north–south direction over the midlatitudes.
jetty A wall built into the ocean at the entrance of a river or harbor to protect against sediment deposition, storm waves, and currents.
joints Cracks that develop in bedrock due to stress, but with no appreciable movement parallel to the walls of the joint.
June solstice The day of the year when the vertical rays of the Sun strike the Tropic of Cancer; on or about June 21; the summer solstice in the Northern Hemisphere.

K

karst Topography developed as a consequence of subsurface solution.
katabatic wind A wind that originates in cold upland areas and cascades toward lower elevations under the influence of gravity.
kettle An irregular depression in a morainal surface created when blocks of stagnant ice eventually melt.
kinetic energy The energy of movement.
knickpoint A sharp irregularity (such as a waterfall, rapid, or cascade) in a stream-channel profile; also called *nickpoint.*
knickpoint migration The upstream shift in location of a knickpoint due to erosion.
Köppen climate classification system The most widely used climatic classification of the world; devised by Wladimir Köppen.

L

lagoon A body of quiet salt or brackish water in an area between a barrier island or a barrier reef and the mainland.
lahar A volcanic mudflow; a fast-moving muddy flow of volcanic ash and rock fragments.
lake A body of water surrounded by land.
land breeze A local wind blowing from land to water, usually at night.
landform An individual topographic feature of any size; the term *landforms* refers to topography.
landslide An abrupt and often catastrophic event in which a large mass of rock and/or soil slides downslope in only a few seconds or minutes; an instantaneous collapse of a slope.
La Niña The atmospheric and oceanic phenomenon associated with cooler than usual water off the west coast of South America; sometimes simplistically described as the opposite of El Niño.
large-scale map A map with a scale that is a relatively large representative fraction and therefore portrays only a small portion of Earth's surface, but in considerable detail.
latent heat Energy stored or released when a substance changes state. Evaporation is a cooling process because latent heat is stored, and condensation is a warming process because latent heat is released.
latent heat of condensation Heat released when water vapor condenses back to liquid form.
latent heat of evaporation Energy stored when liquid water evaporates to water vapor.
lateral erosion Erosion that occurs when the principal current of a stream swings laterally from one bank to the other, eroding where the velocity is greatest (on the outside bank) and depositing alluvium where it is least (on the inside bank).
lateral moraine A well-defined ridge of unsorted debris (till) built up along the sides of a valley glacier, parallel to the valley walls.
laterization The dominant pedogenic regime in areas that have relatively high temperatures throughout the year and are characterized by rapid weathering of parent material, dissolution of nearly all minerals, and the speedy decomposition of organic matter.
latitude Location described as an angle measured north or south of the equator.
lava Molten magma that is extruded onto the surface of Earth, where it cools and solidifies.
lava dome (plug dome) A dome or bulge formed by the pushing-up of viscous magma in a volcanic vent.
leaching The process in which dissolved nutrients are transported down in solution and deposited deeper in soil.
lidar The use of reflected laser light to measure distances and produce three-dimensional depictions of surface features.
lifting condensation level (LCL) The altitude at which rising air cools sufficiently to reach 100 percent relative humidity at the dew point temperature and condensation begins.
lightning A luminous electric discharge in the atmosphere, caused by the separation of positive and negative charges associated with cumulonimbus clouds.
limiting factor A variable that is important, or most important, in determining the survival of an organism.
linear fault trough A straight-line valley that marks the surface position of a fault, especially a strike-slip fault; formed by the erosion or settling of crushed rock along the trace of a fault.
liquefaction A phenomenon observed during an earthquake when water-saturated soil or sediments become soft or even fluid when ground shaking is strong.
lithosphere The part of the crust and upper rigid mantle that comprises tectonic plates; also refers to the entire solid Earth (one of the Earth "spheres").
litter The collection of dead plant parts that accumulate at the surface of soil.
loam A soil texture in which none of the three principal soil separates—sand, silt, and clay—dominates the other two.
loess A fine-grained, wind-deposited silt that lacks horizontal stratification; its most distinctive characteristic is its ability to stand in vertical cliffs.
longitude Location described as an angle measured (in degrees, minutes, and seconds) east or west from the prime meridian on Earth's surface.
longshore current A current in which water moves roughly parallel to the shoreline in a generally downwind direction; also called *littoral current.*
longwave radiation Wavelengths of thermal infrared radiation emitted by Earth and the atmosphere; also called *terrestrial radiation.*
low [pressure cell] An area of relatively low atmospheric pressure.
loxodrome (rhumb line) A true compass heading; a line of constant compass direction.

M

magma Molten material below Earth's surface.
magnitude [of an earthquake] The relative amount of energy released during an earthquake. Several magnitude scales are in current use, such as the *moment magnitude* and the *Richter scale.*
mantle The portion of Earth beneath the crust and surrounding the core.

mantle plume A column of mantle magma that rises to, or almost to, Earth's surface; associated with many hot spots but not directly with most lithospheric plate boundaries.

map A flat representation of Earth at a reduced scale, showing only selected detail.

map projection A systematic representation of all or part of the three-dimensional Earth surface on a two-dimensional flat surface.

map scale The relationship between the distance measured on a map and the actual distance on Earth's surface.

March equinox One of two days of the year when the Sun's vertical rays strike the equator; every location on Earth has equal day and night; occurs on or about March 20 each year.

marine terrace A platform formed by marine erosion that has been uplifted above sea level.

marine west coast climate A mild midlatitude climate characterized by mild temperatures and precipitation throughout the year.

marsh A flattish surface area that is submerged in water at least part of the time but is shallow enough to permit the growth of water-tolerant plants, primarily grasses and sedges.

mass wasting The short-distance downslope movement of weathered rock under the direct influence of gravity; also called *mass movement.*

master joints Major joints that run for great distances through a bedrock structure.

meandering channel pattern (meandering stream channel) A highly twisting or looped stream channel pattern.

meander scar A dry former stream channel meander through which the stream no longer flows.

mechanical weathering The physical disintegration of rock material without any change in its chemical composition; also called *physical weathering.*

medial moraine A dark band of rocky debris down the middle of a glacier; created by the union of the lateral moraines of two adjacent glaciers.

mediterranean climate A mild midlatitude climate characterized by dry summers and wet winters.

mediterranean woodland and shrub A woodland and shrub plant association found in regions of mediterranean climate.

melting The process by which ice is converted to liquid water.

Mercator projection A cylindrical projection that is mathematically adjusted to attain complete conformality and has a rapidly increasing scale with increasing latitude; straight lines on a Mercator projection are lines of constant compass heading (loxodromes).

meridian An imaginary line of longitude extending from pole to pole, crossing all parallels at right angles, and aligned in true north–south directions.

mesa A flat-topped, steep-sided hill with a limited summit area.

mesocyclone The cyclonic circulation of air within a severe thunderstorm; a diameter of about 10 kilometers (6 miles).

metamorphic rock Rock that has been drastically changed by massive forces of heat, pressure, and/or hydrothermal fluids that worked on it from within Earth.

midlatitude anticyclone An extensive migratory high-pressure cell of the midlatitudes that moves generally with the westerlies.

midlatitude cyclone A large migratory low-pressure system that occurs within the midlatitudes and moves generally with the westerlies; also called *extratropical cyclone* or *wave cyclone.*

midlatitude deciduous forest A broadleaf forest plant assemblage composed of mostly deciduous trees.

midlatitude desert climate A desert climate characterized by warm summers but cold winters.

midlatitude grassland A grassland plant assemblage in semiarid regions of the midlatitudes; regionally called *steppe, prairie, pampa,* or *veldt.*

midocean ridge A lengthy system of deep-sea mountain ranges, generally located at some distance from any continent; formed by divergent plate boundaries on the ocean floor.

Milankovitch cycles The long-term astronomical cycles involving Earth's inclination, precession, and eccentricity of orbit; in combination, believed to be at least partially responsible for major periods of glaciation and deglaciation. Named for Milutin Milankovitch, the astronomer who studied these cycles.

millibar A measure of pressure, consisting of one-thousandth of a bar.

mineral A naturally formed solid inorganic substance that has a specified chemical composition and crystal structure.

modified Mercalli intensity scale A qualitative scale, from I to XII, used to describe the relative strength of ground shaking during an earthquake.

Mohorovičić discontinuity The boundary between Earth's crust and mantle; also called *Moho.*

Mollisol The soil order characterized by the presence of a mollic epipedon—a mineral surface horizon that is dark, thick, contains abundant humus and base nutrients, and remains soft when it dries out.

monsoon A seasonal reversal of winds; a general onshore movement in summer and a general offshore flow in winter, with a very distinctive seasonal precipitation regime.

moraine The largest and generally most conspicuous landform feature produced by glacial deposition of till; consists of irregular rolling topography that rises somewhat above the level of the surrounding terrain.

mountain breeze A downslope breeze from a mountain due to the chilling of air on its slopes at night.

mudflow A rapid, downslope movement of a dense mixture of weathered rock and water through or within a valley.

multispectral [remote sensing] A remote sensing instrument that collects multiple digital images simultaneously in different electromagnetic wavelength bands.

N

natural levee An embankment of slightly higher ground fringing a stream channel in a floodplain; formed by deposition during floods.

neap tides The lower-than-normal tidal variations that occur twice a month as the result of the alignment of the Sun and Moon at a right angle to one another.

needleleaf trees Trees adorned with thin slivers of tough, leathery, waxy needles rather than typical leaves.

net primary productivity The net photosynthesis of a plant community over a period of one year, usually measured in the amount of fixed carbon per unit area (kilograms of carbon per square meter per year).

névé Snow granules that have become packed and begin to coalesce due to compression, achieving a density about half as great as that of water; also called *firn.*

nitrogen cycle An endless series of processes in which nitrogen moves through the environment.

nitrogen fixation The conversion of gaseous nitrogen into forms that can be used by plant life.

normal fault The result of tension (extension) producing a steeply inclined fault plane, with the block of land on one side being upthrown in relation to the block on the other side, which is downthrown.

North Pole The latitude of 90° N.

O

occluded front A complex front formed when a cold front overtakes a warm front, lifting all of the warm air mass off the ground.

occlusion The process of a cold front overtaking a warm front, forming an occluded front.
ocean floor cores Rock and sediment samples removed from the ocean floor.
oceanic trench A deep linear depression in the ocean floor where subduction is taking place.
offset stream A stream course displaced by lateral movement along a fault.
O horizon The immediate surface layer of a soil profile, consisting mostly of organic material.
orographic lifting Uplift that occurs when air is forced to rise over topographic barriers.
outcrop A surface exposure of bedrock.
outer core The liquid (molten) shell beneath the mantle that encloses Earth's inner core.
outwash plain An extensive glaciofluvial feature that is a relatively smooth, flattish alluvial apron deposited beyond recessional or terminal moraines by streams issuing from ice.
overland flow The general movement of unchanneled surface water down the slope of the land surface.
oxbow lake A cutoff meander that initially holds water.
oxidation The chemical union of oxygen atoms with atoms from various metallic elements, forming new products that are usually more voluminous, softer, and more easily eroded than the original compounds.
Oxisol The most thoroughly weathered and leached of all soils; this soil order invariably displays a high degree of mineral alteration and profile development.
oxygen cycle The movement of oxygen by various processes through the environment.
oxygen isotope analysis Using the ratio of ^{16}O ("oxygen 16") and ^{18}O ("oxygen 18") isotopes in compounds such as water and calcium carbonate to infer temperature and other conditions in the past.
ozone A gas composed of molecules consisting of three atoms of oxygen, O_3.
ozone layer The layer in the atmosphere at an altitude of 16–40 kilometers (10–25 miles), where the concentration of ozone is greatest; it absorbs much of the incoming ultraviolet solar radiation.

P

Pacific ring of fire The name given to the rim of the Pacific Ocean basin, the site of widespread volcanic and seismic activity; associated with lithospheric plate boundaries.
paleoclimatology The study of past climates.
paleomagnetism Past magnetic orientation.
Pangaea The massive supercontinent that Alfred Wegener first postulated to have existed about 200 million years ago; it broke apart into several large sections that have continuously moved away from one another and that now comprise the present continents.
parallel A line connecting all points of equal latitude; such a line is parallel to all other parallels.
parallelism [of Earth's axis] See *polarity*.
parent material The source of the weathered fragments of rock from which soil is made; solid bedrock or loose sediments that have been transported from elsewhere by the action of water, wind, or ice.
particulate Composed of distinct tiny particles or droplets suspended in the atmosphere; also called *aerosol*.
paternoster lakes A sequence of small lakes found in the shallow excavated depressions or steps within a glacial trough.
patterned ground Polygonal patterns in the ground that develop in areas of seasonally frozen soil and permafrost.
ped A clump into which individual soil particles tend to aggregate; it determines the structure of the soil.
pediment A gently inclined bedrock platform that extends outward from a mountain front, usually in an arid region.
pedogenic regimes Soil-forming regimes that can be thought of as environmental settings in which certain physical/chemical/biological processes prevail.
perennial plants (perennials) Plants that can live more than a single year despite seasonal environmental variations.
perennial stream A permanent stream that contains water year-round.
periglacial zone An area of indefinite size beyond the outermost extent of ice advance that was indirectly influenced by glaciation.
perihelion The point in its orbit where Earth is nearest to the Sun (about 147,100,000 kilometers [91,400,000 miles]).
permafrost Permanent ground ice or permanently frozen subsoil.
permeability The characteristic of soil or rock by which water can move through interconnected pore spaces.
photochemical smog A form of secondary air pollution caused by the reaction of nitrogen compounds and hydrocarbons to ultraviolet radiation in strong sunlight.
photoperiodism The response of an organism to the length of exposure to light in a 24-hour period.
photosynthesis The basic process whereby plants produce stored chemical energy from water and carbon dioxide; it is activated by sunlight.
physical geography The study of the physical or environmental elements of geography.
piedmont zone The zone at the "foot of the mountains."
piezometric surface The elevation to which groundwater will rise under natural confining pressure in a well.
pinnacle An erosional remnant in the form of a steep-sided spire that has a resistant caprock; normally found in an arid or semiarid environment; also called *speleothem column*.
planar projection (plane projection) A family of maps derived by the perspective extension of the geographic grid from a globe to a plane that is tangent to the globe at some point.
plane of the ecliptic The imaginary plane that passes through the Sun and through Earth at every position in its orbit around the Sun; Earth's orbital plane.
plant respiration The oxidation of carbohydrates in plants, releasing water, carbon dioxide, and stored energy (heat).
plant succession The process whereby one type of vegetation is replaced naturally by another over time.
plastic flow of [glacial] ice The slow, nonbrittle movement of ice under pressure.
plateau A flattish erosional platform bounded on at least one side by a prominent escarpment.
plate tectonics A coherent theory of massive lithospheric rearrangement based on the movement of continent-sized plates.
playa A dry lake bed in a basin of interior drainage.
Pleistocene Epoch The epoch of the Cenozoic Era between the Pliocene and the Holocene; from about 2.6 million to about 11,700 years ago.
Pleistocene lakes Large freshwater lakes that formed in basins of interior drainage because of higher rainfall and/or lower evaporation during the Pleistocene.
plucking See *glacial plucking*.
plug dome A volcano dome or bulge formed by the pushing-up of viscous magma in a volcanic vent; also called *lava dome*.
pluton A large, intrusive igneous body.
plutonic rock An igneous rock formed below ground from the cooling and solidification of magma; also called *intrusive rock*.
pluvial (pluvial effects) Pertaining to rain; often used in connection with a past rainy period.
podzolization The dominant pedogenic regime in areas where winters are long and cold; characterized by slow chemical weathering of soils and rapid mechanical weathering from frost action, resulting in soils that are shallow, acidic, and have a fairly distinctive profile.

polar easterlies A global wind system that occupies most of the area between the *polar highs* and about 60° of latitude. The winds move generally from east to west and are typically cold and dry.

polar front The contact between unlike air masses in the subpolar low-pressure zone at about 60° N and S.

polar high A high-pressure cell situated over either polar region.

polarity [of Earth's rotation axis] A characteristic of Earth's axis wherein it always points toward Polaris (the North Star) at every position in Earth's orbit around the Sun; also called *parallelism*.

porosity The amount of pore space between soil particles and between peds; it is a measure of soil's capacity to hold water and air.

precipitation Drops of liquid or solid water falling from clouds.

precipitation variability The expected departure from an area's average annual precipitation in any given year.

pressure gradient The difference in atmospheric pressure over some horizontal distance.

primary consumer An animal that eats plants; the first stage in a food pyramid or food chain.

primary pollutants Contaminants released directly into the air.

prime meridian The meridian passing through the Royal Observatory at Greenwich, England, from which longitude is measured.

producers Organisms that produce their own food through photosynthesis; plants.

proglacial lake A lake formed when ice flows across or against the general slope of the land and the natural drainage is impeded or blocked, such that meltwater becomes impounded against the ice front.

pseudocylindrical projection (elliptical projection) A family of map projections in which the entire world is displayed in an oval shape.

pyroclastic flow A high-speed avalanche of hot gases, ash, and rock fragments emitted from a volcano during an explosive eruption; also called *nuée ardente*.

pyroclastics (pyroclastic material) Solid rock fragments thrown into the air by volcanic explosions.

R

radar Radio detection and ranging.

radiant energy See *electromagnetic radiation*.

radiation The process in which electromagnetic energy is emitted from a body; the flow of energy in the form of electromagnetic waves.

rain The most common and widespread form of precipitation, consisting of drops of liquid water.

rain shadow An area of low rainfall on the leeward side of a mountain range or topographic barrier.

recessional moraine A glacial deposit of till formed during a pause in the retreat of the ice margin.

recurrence interval [of a flood] The probability of a given-size flood occurring in a year; also called *return period*.

reflection The ability of an object to repel waves without altering the object or the waves.

reg A desert surface of coarse material from which all sand and dust have been removed by wind and water erosion; also called *desert pavement* or *desert armor*.

regional metamorphism The widespread subsurface metamorphism of rock as a result of prolonged exposure to heat and high pressure, such as in areas of plate collision or subduction.

regolith A layer of broken and partly decomposed rock particles that covers bedrock.

relative humidity The ratio of the amount of water vapor in the air (the water vapor content) to the maximum amount that could be there if the air were saturated (the capacity); expressed as a percentage.

relief The difference in elevation between the highest and lowest points in an area; the vertical variation from mountaintop to valley bottom.

remote sensing The measurement or acquisition of information by a recording device that is not in physical contact with the object under study; commonly includes cameras and satellites.

reverse fault A fault produced from compression, with the upthrown block rising steeply above the downthrown block.

revolution [around the Sun] Earth's orbital movement around the Sun.

R horizon The consolidated bedrock at the base of a soil profile.

ria shoreline An embayed coast with numerous estuaries; formed by the flooding of stream valleys by the sea.

ridge [of atmospheric pressure] A linear or elongated area of relatively high atmospheric pressure.

rift valley See *continental rift valley*.

riparian vegetation Streamside growth, particularly prominent in relatively dry regions, where stream courses may be lined with trees, although no other trees are to be found in the landscape.

roche moutonnée A characteristic glacial landform produced when a bedrock hill or knob is overridden by moving ice. The stoss side is smoothly rounded, but the lee side is steeper and more irregular.

rock Solid material composed of aggregated mineral material.

rock cycle The long-term "recycling" of mineral material from one kind of rock to another.

rockfall (fall) A mass wasting process in which weathered rock drops to the foot of a cliff or steep slope.

rock glacier An accumulated talus mass that moves slowly but distinctly downslope under its own weight.

Rossby wave A very large north–south undulation of the upper-air westerlies and jet stream.

rotation [of Earth] The spinning of Earth around its imaginary north–south axis.

runoff The flow of water from land to oceans by overland flow, streamflow, and groundwater flow.

S

Saffir-Simpson Hurricane Scale The classification system of hurricane strength with category 1 the weakest and category 5 the strongest.

sag pond A pond caused by the collection of water from springs and/or runoff into sunken ground, resulting from the crushing of rock in an area of fault movement.

salina A dry lake bed that contains an unusually heavy concentration of salt in the lake-bed sediment.

saline lake A salt lake; commonly caused by interior stream drainage in an arid environment.

salinity A measure of the concentration of dissolved salts.

salinization One of the dominant pedogenic regimes in areas where principal soil moisture movement is upward because of a moisture deficit.

salt wedging Rock disintegration caused by the crystallization of salts from evaporating water.

sand dune A mound, ridge, or low hill of loose, windblown sand.

Santa Ana winds Dry, usually warm, and often very strong winds blowing offshore in southern California.

saturated adiabatic rate (saturated adiabatic lapse rate) The diminished rate of cooling, averaging about 6°C per 1000 meters (3.3°F per 1000 feet) of rising air above the lifting condensation level; a result of the latent heat of condensation counteracting some of the adiabatic cooling of rising air.

saturation [with water vapor] The condition in which air contains the maximum amount of water vapor at a given temperature.

scattering The deflection of light waves in random directions by gas molecules and particulates in the atmosphere; shorter wavelengths of visible light are more easily scattered than longer wavelengths.

scree Pieces of weathered rock, especially small pieces, that fall directly downslope; also called *talus*.

sea breeze A wind that blows from the sea toward the land, usually during the day.

seafloor spreading The pulling-apart of lithospheric plates, permitting the rise of deep-seated magma to Earth's surface at midocean ridges.

secondary consumer An animal that eats other animals, as the second and further stages in a food pyramid or food chain.

secondary pollutant Pollutants formed in the atmosphere as a consequence of chemical reactions or other processes; for example, see *photochemical smog*.

sediment Small particles of rock debris or organic material deposited by water, wind, or ice.

sedimentary rock Rock formed of sediment that is consolidated by the combination of pressure and cementation.

sediment budget [of a beach] The balance between the sediment deposited on a beach and the sediment transported away from a beach.

seif (longitudinal) dune A long, narrow desert dune that usually occurs in multiplicity and in parallel arrangement.

sensible temperature The relative apparent temperature that is sensed by a person's body.

separates The size groups within the standard classification of soil particle sizes.

September equinox One of two days of the year when the Sun's vertical rays strike the equator; every location on Earth has equal day and night; occurs on or about September 22 each year.

shield volcanoes Volcanoes built up in a lengthy outpouring of very fluid basaltic lava; they are broad mountains with gentle slopes.

shortwave radiation Wavelengths of radiation emitted by the Sun; especially ultraviolet, visible, and short infrared radiation.

shrubland A plant association dominated by relatively short woody plants.

silicates A category of minerals composed of silicon and oxygen combined with another element or elements.

sinkhole (doline) A small, rounded depression that is formed by the dissolution of surface limestone, typically at joint intersections.

sinuous channel pattern (sinuous stream channel) A gently curving or winding stream channel pattern.

slip face [of sand dune] The steeper leeward side of a sand dune.

slump A slope collapse slide with rotation along a curved sliding plane.

small-scale map A map with a scale that is a relatively small representative fraction and therefore portrays a large portion of Earth's surface in limited detail.

snow Solid precipitation in the form of ice crystals, small pellets, or flakes, which forms by the direct conversion of water vapor into ice.

soil An infinitely varying mixture of weathered mineral particles, decaying organic matter, living organisms, gases, and liquid solutions; it is that part of the outer "skin" of Earth occupied by plant roots.

soil order The highest (most general) level of soil classification in the Soil Taxonomy.

soil profile A vertical cross section from Earth's surface down through the soil layers into the parent material beneath.

Soil Taxonomy The system of soil classification currently in use in the United States. It is genetic in nature and focuses on the existing properties of the soil rather than on environment, genesis, or the properties it would possess under virgin conditions.

soil-water balance The relationship among gain, loss, and storage of soil water.

soil-water budget An accounting that demonstrates the variation of the soil-water balance over a period of time.

solar altitude The angle of the Sun above the horizon.

solifluction A special form of soil creep in tundra areas; associated with summer thawing of the near-surface portion of permafrost, causing the wet, heavy surface material to sag slowly downslope.

solum The true soil that includes only the top four horizons: O, the organic surface layer; A, the topsoil; E, the eluvial layer; and B, the subsoil.

sonar Sound navigation and ranging—an active remote sensing system that permits underwater imaging.

Southern Oscillation The periodic "seesaw" of high and low atmospheric pressures between northern Australia and Tahiti; first recognized by Gilbert Walker in the early twentieth century.

South Pole The latitude of 90° S.

specific heat capacity (specific heat) The amount of energy required to raise the temperature of 1 gram of a substance by 1°C.

specific humidity A direct measure of water-vapor content expressed as the mass of water vapor in a given mass of air (in grams of vapor/kilograms of air).

speleothem A feature formed by precipitated deposits of minerals on the wall, floor, or roof of a cave.

spit A linear deposit of marine sediment that is attached to the land at one or both ends.

Spodosol The soil order characterized by the occurrence of a spodic subsurface horizon, which is an illuvial layer where organic matter and aluminum accumulate, and which has a dark, sometimes reddish, color.

spring tide A time of maximum tide that occurs as a result of the alignment of Sun, Moon, and Earth.

stable [air] Air that rises only if forced.

stalactite A pendant structure hanging downward from a cavern's roof.

stalagmite A projecting structure growing upward from a cavern's floor.

star dune A pyramid-shaped sand dune with arms radiating out in three or more directions.

stationary front The common boundary between two air masses in a situation in which neither air mass displaces the other.

storm surge A surge of wind-driven water as much as 8 meters (25 feet) above normal tide level, which occurs when a hurricane advances onto a shoreline.

storm warning A weather advisory issued when a severe thunderstorm or tornado has been observed in an area; people should seek safety immediately.

storm watch A weather advisory issued when conditions are favorable for strong thunderstorms or tornadoes.

strata Distinct layers of sediment or layers in sedimentary rock.

stratified drift Drift that was sorted as it was carried along by flowing glacial meltwater.

stratosphere The atmospheric layer directly above the troposphere.

stratus clouds Layered, horizontal clouds, often below altitudes of 2 kilometers (6500 feet); they sometimes occur as individual clouds but more often appear as a general overcast.

stream A channeled flow of water, regardless of size.

stream capacity The maximum load that a stream can transport under given conditions.

stream capture (stream piracy) The diversion of a portion of the flow of one stream into that of another by natural processes.

stream competence The size of the largest particle that a stream can transport.

streamflow The channeled movement of water along a valley bottom.

stream load The solid matter carried by a stream.

stream order The concept that describes the hierarchy of a drainage network.

stream piracy See *stream capture*.

stream rejuvenation An increase in a stream's downcutting ability, usually through regional tectonic uplift.

stream terrace A remnant of a previous valley floodplain of a rejuvenated stream.
strike-slip fault A fault produced by shearing, with adjacent blocks being displaced laterally with respect to one another; the movement is mostly or entirely horizontal.
subarctic climate A severe midlatitude climate found in high-latitude continental interiors, characterized by very cold winters and an extreme annual temperature range.
subduction The descent of the edge of an oceanic lithospheric plate under the edge of an adjoining plate.
sublimation The process by which ice is converted directly to water vapor.
subpolar low A zone of low pressure that is situated at about 50° to 60° of latitude in both Northern and Southern Hemispheres; also called *polar front*.
subtropical desert climate A hot desert climate; generally found in subtropical latitudes, especially on the western sides of continents.
subtropical gyres The closed-loop pattern of surface ocean currents around the margins of the major ocean basins; the flow is clockwise in the Northern Hemisphere and counterclockwise in the Southern Hemisphere.
subtropical high (STH) An area of large, semipermanent, high-pressure cells centered at about 30° N and S over the oceans; it has an average diameter of 3200 kilometers (2000 miles) and is usually elongated east–west.
supercooled water Water that persists in liquid form at temperatures below freezing.
supersaturated [air] Air in which the relative humidity is greater than 100 percent but condensation is not taking place.
surface tension The tendency of liquid water molecules to stick together because of electrical polarity; a thin "skin" of molecules forms on the water surface, causing it to "bead."
suspended load The very fine particles of clay and silt that are in suspension and move along with the flow of water without ever touching the streambed.
swallow hole The distinct opening at the bottom of some sinkholes through which surface drainage can pour directly into an underground channel.
swamp A flattish surface area that is submerged in water at least part of the time but is shallow enough to permit the growth of water-tolerant plants—predominantly trees.
swash The cascading forward motion of a breaking wave that rushes up the beach.
swell An ocean wave, usually produced by stormy conditions, that can travel enormous distances away from the source of the disturbance.
symbiosis A mutually beneficial relationship between two different types of organisms.
syncline A simple downfold in the rock structure.

T

talus Pieces of weathered rock, of various sizes, that fall directly downslope; also called *scree*.
talus cone A sloping, cone-shaped heap of dislodged talus.
tarn A small lake in the shallow excavated depression of a glacial cirque.
teleconnection The coupling or relationship of weather and/or oceanic events in one part of the world with those in another.
temperature A description of the average kinetic energy of the molecules in a substance; the more vigorously the molecules "jiggle" (and therefore the greater the internal kinetic energy), the higher the temperature of a substance; in popular terms, a measure of the degree of hotness or coldness of a substance.
temperature inversion A situation in which temperature increases as altitude increases, so the normal condition is inverted.
terminal moraine A glacial deposit of till that builds up at the outermost extent of ice advance.
terrane A mass of lithosphere, bounded on all sides by faults, that has become accreted to a lithospheric plate margin with different lithologic characteristics from those of the terrane; often composed of lithosphere that is too buoyant to subduct; also called *accreted terrane*.
terrestrial radiation Longwave radiation emitted by Earth's surface or atmosphere.
thermal high A high-pressure cell associated with cold surface conditions.
thermal infrared radiation (thermal IR) The middle and far infrared part of the electromagnetic spectrum.
thermal low A low-pressure cell associated with warm surface conditions.
thermohaline circulation The slow circulation of deep ocean water because of differences in water density that arise from differences in salinity and temperature.
thrust fault A fault created when compression forces the upthrown block to override the downthrown block at a relatively low angle; also called *overthrust fault*.
thunder The sound that results from the shock wave produced by the instantaneous expansion of air that is abruptly heated by a lightning bolt.
thunderstorm A relatively violent convective storm accompanied by thunder and lightning.
tidal bore A wall of seawater, several centimeters to more than a meter in height, that rushes up a river as the result of enormous tidal inflow.
tidal range The vertical difference in elevation between high and low tide.
tides The rise and fall of coastal water levels; caused by the alternate increasing and decreasing gravitational pull of the Moon and the Sun on varying parts of Earth's surface.
till Rock debris that is deposited directly by moving or melting ice, with no meltwater flow or redeposition involved.
tilted fault-block mountain See *fault-block mountain*.
time zone A region on Earth (generally a north–south band defined by longitude) within which the agreed-upon local time is the same.
tombolo A spit formed by sand deposition that connects an island to the mainland.
topography The surface configuration of Earth.
tornado A localized cyclonic low-pressure cell surrounded by a whirling cylinder of violent wind; characterized by a funnel cloud extending below a cumulonimbus cloud.
tower karst The tall, steep-sided hills in an area of karst topography.
trade winds The major easterly wind system of the tropics, issuing from the equatorward sides of the subtropical highs and diverging toward the west and toward the equator.
transform [plate] boundary A location where two lithospheric plates slip past one another laterally.
transmission The ability of a medium to allow electromagnetic waves to pass through it.
transverse dune A crescent-shaped sand dune ridge that has convex sides facing the prevailing direction of wind and occurs where the supply of sand is great. The crest is perpendicular to the wind direction and aligned in parallel waves across the land.
treeline The elevation above (or below) which trees do not grow.
trellis drainage pattern A drainage pattern that is usually developed on alternating bands of hard and soft strata, with long parallel subsequent streams linked by short, right-angled segments and joined by short tributaries.
trophic level The position of an organism in a food chain; animals in the same trophic level tend to consume the same kinds of food.
tropical cyclone A storm most significantly affecting the tropics and subtropics; an intense low-pressure center that is essentially circular in shape; when wind speed reaches 119 kilometers per hour (74 mph; 64 knots); called *hurricane* in North America and the Caribbean.

tropical deciduous forest A tropical forest found in regions with a pronounced dry period, during which many of the trees shed their leaves.

tropical depression An incipient tropical cyclone with winds not exceeding 33 knots (by international agreement).

tropical monsoon climate A tropical humid climate with a pronounced winter dry season and a very wet summer rainy season; associated with a monsoon wind pattern.

tropical rainforest A distinctive assemblage of tropical vegetation that is dominated by a great variety of tall, high-crowned trees; also called *selva.*

tropical savanna A tropical grassland dominated by tall grasses.

tropical savanna climate A tropical humid climate with a dry winter season and a moderately wet summer season; associated with the seasonal migration of the ITCZ.

tropical scrub A widespread tropical assemblage of low trees and bushes.

tropical storm An incipient tropical cyclone with winds between 34 and 63 knots (by international agreement).

tropical wet climate A tropical humid climate that is wet all year; usually under the influence of the ITCZ all year.

Tropic of Cancer The parallel of 23.5° N latitude, which marks the northernmost location reached by the vertical rays of the Sun in the annual cycle of Earth's revolution.

Tropic of Capricorn The parallel of 23.5° S latitude, which marks the southernmost location reached by the vertical rays of the Sun in the annual cycle of Earth's revolution.

troposphere The lowest thermal layer of the atmosphere, in which temperature decreases with height; the layer of the atmosphere in contact with Earth's surface.

trough [of atmospheric pressure] A linear or elongated band of relatively low atmospheric pressure.

tsunami An oceanic wave of very long wavelength generated by a submarine earthquake, landslide, or volcanic eruption; also called *seismic sea wave.*

tundra A complex mix of very low-growing plants, including grasses, forbs, dwarf shrubs, mosses, and lichens, but no trees; it occurs only in the perennially cold climates of high latitudes or high altitudes.

tundra climate A polar climate in which no month of the year has an average temperature above 10°C (50°F).

U

ubac slope A slope oriented such that sunlight strikes it at a low angle and hence is much less effective in heating and evaporating than on the adret slope, thus producing more luxuriant vegetation of a richer diversity.

Ultisol A soil order similar to Alfisols but more thoroughly weathered and more completely leached of bases.

ultraviolet (UV) radiation Electromagnetic radiation in the wavelength range of 0.1–0.4 micrometers.

uniformitarianism The concept that the "present is the key to the past" in geomorphic processes; the processes now operating also operated in the past.

unstable [air] Air that rises without being forced.

upwelling This rising of cold, deep ocean water to the surface where wind patterns deflect surface water away from the coast; especially common along the west coasts of continents in the subtropics and midlatitudes.

urban heat island (UHI) effect The situation in which higher temperatures are measured in an urban area compared with the surrounding rural area.

UTC See *Coordinated Universal Time.*

uvala A compound sinkhole (doline) or chain of intersecting sinkholes.

V

valley That portion of the total terrain in which a stream drainage system is clearly established.

valley breeze An upslope breeze up a mountain due to heating of air on its slopes during the day.

valley glacier A long, narrow feature resembling a river of ice that spills out of its originating basins and flows down-valley.

valley train A lengthy deposit of glaciofluvial alluvium confined to a valley bottom beyond the outwash plain.

vapor pressure The total pressure exerted by water vapor in the atmosphere.

ventifact A rock that has been sandblasted by the wind.

verbal map scale The scale of a map as stated in words; also called *word scale.*

vertebrates Animals that have a backbone that protects their spinal cord: fishes, amphibians, reptiles, birds, and mammals.

vertical zonation The horizontal layering of different plant associations on a mountainside or hillside.

Vertisol A soil order comprising a specialized type of soil that contains a large quantity of clay and has an exceptional capacity for absorbing water.

visible light Waves in the electromagnetic spectrum in the narrow band between about 0.4 and 0.7 micrometers in length; wavelengths of electromagnetic radiation to which the human eye is sensitive.

volcanic island arc A chain of volcanic islands associated with an oceanic plate–oceanic plate subduction zone; also called *island arc.*

volcanic mudflow A fast-moving, muddy flow of volcanic ash and rock fragments; also called *lahar.*

volcanic rock Igneous rock formed on the surface of Earth; also called *extrusive rock.*

volcanism The movement of magma from Earth's interior to or near the surface.

W

Walker Circulation A general circuit of air flow in the southern tropical Pacific Ocean. Warm air rises in the western side of the basin (in updrafts of the ITCZ); flows aloft to the east, where it descends into the subtropical high off the west coast of South America; and flows back to the west in the surface trade winds. Named for Gilbert Walker, the meteorologist who first described this circumstance.

warm front The leading edge of an advancing warm air mass.

watershed See *drainage basin.*

waterspout A funnel cloud in contact with the ocean or a large lake; similar to a weak tornado over water.

water table The top of the saturated zone within the ground.

water vapor Water in the form of a gas.

water vapor capacity The maximum amount of water vapor that air can contain at a given temperature.

wave-cut platform A gently sloping, wave-eroded bedrock platform that develops just below sea level; common where a coastal cliff is being worn back by wave action; also called *wave-cut bench.*

wave height The vertical distance from a wave's crest to its trough.

wavelength The horizontal distance from the crest of one wave to the crest of another or from the trough of one wave to the trough of another.

wave of oscillation The motion of wave in which the individual particles of the medium (such as water) make a circular orbit as the wave form passes through.

wave of translation The horizontal motion produced when a wave reaches shallow water and finally "breaks" on the shore.

wave refraction The change in a wave's directional trend as it approaches a shoreline; results in ocean waves generally breaking parallel to the shoreline.

weather The short-term atmospheric conditions for a given time and a specific area.

weathering The physical and chemical disintegration of rock that is exposed to the atmosphere.

westerlies The great wind system of the midlatitudes that flows basically from west to east around the world in the latitudinal zone between about 30° and 60° both north and south of the equator.

wetland A landscape characterized by shallow, standing water all or most of the year, with vegetation rising above the water level.

wilting point The point at which plants are no longer able to extract moisture from the soil because the capillary water is all used up or evaporated.

wind Horizontal air movement.

wind shear (vertical wind shear) A significant change in wind direction or speed in the vertical dimension.

woodland A tree-dominated plant association in which the trees are spaced more widely apart than those of forests and do not have interlacing canopies.

X

xerophytic adaptations The characteristics that allow certain types of plants to withstand protracted dry conditions.

Y

yazoo stream A tributary unable to enter the main stream because of natural levees along the main stream.

Z

zone of aeration (vadose zone) The topmost hydrologic zone within the ground; contains a fluctuating amount of moisture (soil water) in the pore spaces of the soil (or soil and rock).

zone of saturation (phreatic zone) The second hydrologic zone below the ground surface; its uppermost boundary is the water table. The pore spaces and cracks in the bedrock and the regolith of this zone are fully saturated.

zoogeographic regions The major realms of the world's land areas, differentiated by characteristic fauna.

Credits

Photo Credits

Chapter 1

Opening Photo Terry Virts/NASA; p.3: Michael Collier; p.5: Jan Bryant/Alamy Stock Photo; p.6: National Park Service; p.6: Jim Peaco/NPS; p.7: Peter Unger/Stone/Getty Images; p.9: John Warden/The Image Bank/Getty Images; p.15: Felix Stensson/Alamy Stock Photo; p.16: Tom L. McKnight; p.23: Alex Hubenov/Shutterstock; p.26: NASA; p.26: Terry Virts/NASA

Chapter 2

Opening Photo NASA; p.32: kzww/Shutterstock; p.33: National Park Service; p.33: National Park Service; p.33: National Park Service; p.33: USGS; p.33: Frank D. Granshaw; Andrew G. Fountain; p.44: NASA; p.44: NASA; p.44: NASA; p.44: NASA; p.46: Pearson Education; p.46: NASA; p.47: Jesse Allen/NASA; p.48: NASA; p.49: NASA; p.49: NOAA; p.49: National Park Service; p.49: Patrick Myers/NPS; p.51: NASA; p.51: NOAH SEELAM/AFP/Getty Images; p.52: NOAA; p.52: NASA; p.53: NASA; p.55: Google Earth

Chapter 3

Opening Photo Simon Crumpton/Alamy Stock Photo; p.58: NASA; p.60: US Marines Photo/Alamy Stock Photo; p.63: Grant Dixon/Lonely Planet Images/Getty Images; p.63: James Smith/Alamy Stock Photo; p.63: National Park Service; p.63: Nathan Kostegian/National Park Service; p.65: Kent Miller/NPS; p.67: Tom L. McKnight; p.68: ESA/Science Source; p.71: NASA; p.72: National Oceanic and Atmospheric Administration; p.73: NASA; p.74: Tao Zhang/Getty Images News/Getty Images; p.75: JEWEL SAMAD/AFP/Getty Images; p.75: Yawar Nazir/Getty Images

Chapter 4

Opening Photo National Park Service; p.80: Tom Walker/Alamy Stock Photo; p.80: Gaston Piccinetti/Shutterstock; p.83: Dennis Schroeder/National Renewable Energy Laboratory/U.S Department of Energy; p.85: Paul Mayall/imageBROKER/Agefotostock; p.86: PhotoviewPlus/Moment/Getty Images; p.88: Darrel Hess; p.93: NULL/Design Pics Inc/Alamy Stock Photo; p.98: Cooperative Institute for Meteorological Satellite Studies (CIMSS), University of Wisconsin-Madison; p.100: Alexander Chaikin/Shutterstock; p.104: NASA; p.104: NASA; p.104: NASA; p.106: NOAA; p.106: Rolf Peterson; p.106: Kaitlyn Knick/National Park Service; p.109: NASA

Chapter 5

Opening Photo Menno Boermans/Getty images; p.112: Darrel Hess; p.118: Pearson Education Inc.; p.123: Jackie Frost/NPS Photo; p.124: NOAA Geostationary Satellite Server; p.128: Christina Felschen/Alamy Stock Photo; p.128: Sheralee Stoll/Alamy Stock Photo; p.133: NASA; p.133: National Park Service; p.133: Anela Kopshever/NPS; p.136: NOAA; p.138: NOAA; p.141: NOAA

Chapter 6

Opening Photo Bill Lea/Dembinsky Photo Associates/Alamy Stock Photo; p.146: AlasdairJames/Getty Images; p.148: digi_guru/Getty Images; p.148: prajit boonpa/123RF Images; p.153: John D Sirlin/Shutterstock; p.153: kzww/Shutterstock; p.157: Homer Sykes/Alamy Stock Photo; p.160: JONATHAN EASTLAND/Ajax News & Feature Service/Alamy Stock Photo; p.160: Roberta Olenick/All Canada Photos/Alamy Stock Photo; p.160: zebra0209/Shutterstock; p.160: Pavel_Klimenko/Shutterstock; p.160: Michelle Marsan/Shutterstock; p.160: JuliusKielaitis/Shutterstock; p.161: Pearson Education Inc.; p.162: Dominique Braud/Dembinsky Photo Associates/Alamy Stock Photo; p.162: NOAA; p.162: Jim Peaco/National Park Service; p.162: NPS Photo/Alamy Stock Photo; p 162: Alex VanDerStuyf/NPS photo; p.164: National Park Service; p.165: University Corporation for Atmospheric Research; p.166: NOAA; p.166: NOAA; p.166: NOAA; p.167: Ryan McGinnis/Alamy Stock Photo; p.169: Michael Collier; p.174: USGS; p.174: National Park Service; p.174: National Park Service

Chapter 7

Opening Photo Cavan Images/Alamy Stock Photo; p.184: National Park Service; p.184: USGS; p.184: GIPhotoStock Z/Alamy Stock Photo; p.186: NASA Goddard Space Flight Center; p.190: NASA; p.191: NOAA; p.191: NOAA; p.196: NASA; p.196: NASA; p.196: DANIEL MARTINEZ/UPI/Alamy Stock Photo; p.196: Juan Carlos Munoz/Nature Picture Library/Alamy Stock Photo; p.197: NOAA; p.197: FEMA/Alamy Stock Photo; p.198: USGS; p.200: Universal Images Group/SuperStock; p.201: John Sirlin/Alamy Stock Photo; p.201: J. B. Forbes/MCT/Newscom; p.205: Douglas Pulsipher/Alamy Stock Photo; p.205: NOAA; p.205: NOAA; p.208: NOAA

Chapter 8

Opening Photo National Park Service; p.217: Juan Carlos Muñoz/Agefotostock; p.218: Nature Picture Library/Alamy Stock Photo; p.219: SANJAY KANOJIA/AFP/Getty Images; p.222: Peter Groenendijk/robertharding/Alamy Stock Photo; p.224: National Park Service; p.226: Hudson Fleece/Alamy Stock Photo; p.227: Jeff Lepore/Alamy Stock Photo; p.228: U.S Forest Service; p.230: Don Breneman/Alamy Stock Photo; p.231: National Park Service; p.233: Steven Kazlowski/Nature Picture Library/Alamy Stock Photo; p.233: NASA Earth Observatory; p.234: Thomas Sbampato/imageBROKER/Alamy Stock Photo; p.235: H. Mark Weidman Photography/Alamy Stock Photo; p.237: Ecuadorpostales/Shutterstock; p.238: NOAA; p.238: National Park Service; p.238: National Park Service; p.238: U.S Forest Service; p.240: Matauw/Shutterstock; p.241: AB Forces News Collection/Alamy Stock Photo; p.243: Carlos Gutierrez/Reuters; p.252: Ken Hawkins/Alamy Stock Photo

Chapter 9

Opening Photo Pat & Chuck Blackley/Alamy Stock Photo; p.259: Universal Education/Universal Images Group/Getty Images; p.260: National Park Service; p.262: Andrew Syred/Science Source; p.265: Georg Gerster/Science Source; p.266: Andrew Payne/Alamy Stock Photo; p.268: Paul Souders/Stone/Getty Images; p.270: Ashley Cooper/Alamy Stock Photo; p.271: Sasha Buzko/Shutterstock; p.272: Deposit Photos/Glow Images; p.272: NASA; p.272: NASA; p.272: NASA; p.272: NASA Earth Observatory; p.273: Bryan Mullennix/Tetra Images/Alamy Stock Photo; p.273: Pete Mcbride/National Geographic Image Collection/Getty Images; p.274: National Park Service; p.278: Tom L. McKnight; p.281: Jim Wark/Agstockusa/AGE Fotostock p.Adrees Atif/REUTERS/Alamy Stock Photo

Chapter 10

Opening Photo John Barger/Alamy Stock Photo; p.286: Roine Magnusson/AGE Fotostock; p.286: National Park Service; p.287: USGS; p.291: Banks Photos/Getty Images; p.291: Luiz Claudio Marigo/Nature Picture Library; p.297: Eric Baccega/AGE Fotostock; p.298: Martin Moxter/imageBROKER/Alamy Stock Photo; p.299: Bill Bachman/Alamy Stock Photo; p.299: Tom L. McKnight; p.300: Stephen R Ausmus/USDA; p.301: National Park Service; p.302: National Park Service; p.303: USGS; p.304: NASA; p.304: NASA; p.304: NASA; p.304: National Park Service; p.305: National Park Service; p.306: Michael Collier; p.306: Tom L. McKnight; p.307: Don Johnston_WU/Alamy Stock Photo; p.307: Tom L. McKnight; p.307: Tom L. McKnight; p.308: Justin Sullivan/Getty Images; p.309: Darrel Hess

Chapter 11

Opening Photo Newman Mark/Prisma by Dukas Presseagentur GmbH/Alamy Stock Photo; p.316: National Park Service; p.317: Victoria Stauffenberg/National Park Service; p.317: Scott Sharaga/NPS; p.318: Anton Foltin/Shhutterstock; p.319: Shirley Kilpatrick/Alamy Stock Photo; p.320: Michael Collier; p.321: Tom L. McKnight; p.321: TOM WILSON/NPS; p.322: National Park Service; p.323: Stephanie Jackson/Australian wildlife collection/Alamy Stock Photo; p.323: Dave Watts/Alamy Stock Photo; p.323: H Lansdown/Alamy Stock Photo; p.324: Gary Hartley/NPS; p.324: Douglas Herr/Alamy Stock Photo; p.324: National Park Service; p.325: Barrett & MacKay/All Canada Photos/Alamy Stock Photo; p.325: Joe Austin Photography/Alamy Stock Photo; p.325: Rick & Nora Bowers/Alamy Stock Photo; p.325: John Serafin; p.326: Tom L. McKnight; p.326: Tom L. McKnight; p.330: David Shaw/Alamy Stock Photo; p.330: Michael Collier; p.331: Nick Turner/Nature Picture Library; p.332: TEOCARAMEL/Moment/Getty Images; p.332: Tom L. McKnight; p.333: National Park Service; p.334: Robb Hannawacker/NPS; p.335: Darrel Hess; p.335: Jack Goldfarb/Getty Images; p.335: Tom L. McKnight; p.336: David R. Frazier Photolibrary, Inc./Alamy Stock Photo; p.337: Jochen Schlenker/Photolibrary/Getty Images; p.337: Tom L. McKnight; p.338: Tom L. McKnight; p.338: Thomas Sbampato/imageBROKER/Alamy Stock Photo; p.339: CSP_deberarr/AGE Fotostock; p.339: Tom L. McKnight; p.340: Nigel Cattlin/Alamy Stock Photo; p.341: NASA; p.341: NASA; p.341: NASA; p.341: NASA; p.341: Nigel Smith/SuperStock/Alamy Stock Photo; p.343: National Park Service; p.343: Tom L. McKnight; p.344: arnet117/Shutterstock; p.344: Mark Conlin/Alamy Stock Photo; p.346: NASA; p.347: Michael Collier

Chapter 12

Opening Photo George Ostertag/agefotostock/Alamy Stock Photo; p.352: Ivan Kuzmin/Alamy Stock Photo; p.353: Darrel Hess; p.354: National Park Service; p.358: National Park Service; p.362: National Park Service; p.364: Eduardo Pucheta/Alamy Stock Photo; p.365: Randall J. Schaetzl; p.365: Randall J. Schaetzl; p.365: Randall J. Schaetzl; p.368: USDA; p.368: USDA; p.369: Loyal A. Quandt/USDA; p.369: USDA; p.370: Randall J. Schaetzl; p.371: Randall J. Schaetzl; p.371: USDA; p.371: Loyal A. Quandt/USDA; p.372: USDA; p.373: USDA; p.373: USDA; p.374: USDA; p.736: USDA; p.377: Michael Collier

Chapter 13

Opening Photo TravelCollection/Image Professionals GmbH/Alamy Stock Photo; p.381: Mark A Schneider/Dembinsky Photo Associates/Alamy Stock Photo; p.381: Dennis Tasa; p.381: Albert Russ/Shutterstock; p.384: Dennis Tasa; p.384: Dennis Tasa; p.384: Dennis Tasa; p.384: Dennis Tasa; p.384: Dennis Tasa; p.384: Tyler Boyes/Shutterstock; p.385: Darrel Hess; p.385: Michael Collier; p.386: Janice and Nolan Braud/Alamy Stock Photo; p.386: Clement Philippe/Arterra Picture Library/Alamy Stock Photo; p.387: Alan Spencer/Powered by Light/Alamy Stock Photo; p.387: Siim Sepp/Alamy Stock Photo; p.387: Tom L. McKnight; p.388: Dennis Tasa; p.388: Paola Moschitto-Assenmacher/EyeEm/Getty Images; p.389: Michael Collier; p.389: Dennis Tasa; p.389: Tom L. McKnight; p.391: Lynn Bystrom/123RF GB LIMITED; p.391: Kevin Schafer/The Image Bank/Getty Images; p.391: Doug Martin/Science Source; p.391: Tyler Boyes/Shutterstock; p.394: Dennis MacDonald/Agefotostock; p.394: USGS; p.398: Dennis Tasa; p.398: National Park Service; p.401: USGS

Chapter 14

Opening Photo USGS; p.412: Ulrich Doering/Alamy Stock Photo; p.415: Michael Collier; p.418: Stocktrek Images, Inc./Alamy Stock Photo; p.420: Michael Marfell/123RF; p.420: Rileysmithphotography.com/Shutterstock; p.422: Darrel Hess; p.422: NASA; p.422: NASA; p.423: John Elk/The Image Bank/Getty Images; p.425: Violeta Schmidt/Reuters Pictures; p.425: USGS; p.426: USGS/Handout/Anadolu Agency/Getty Images; p.427: Michael Collier; p.427: Michael Collier; p.429: Greg Vaughn/Alamy Stock Photo; p.429: Dipper Historic/Alamy Stock Photo; p.430: Peter W. Lipman/USGS; p.431: Cristobal Saavedra/Reuters Pictures; p.432: Pearson Education; p.432: Pearson Education; p.435: Niebrugge Images/Alamy Stock Photo; p.436: USGS; p.437: Ishara S. Kodikara/AFP/Getty Images; p.438: Corbin17/Alamy Stock Photo; p.438: Darrel Hess; p.439: Michael Collier; p.442: Koki Nagahama/Getty Images; p.443: Michael Collier; p.443: USGS; p.444: David Cobb/Alamy Stock Photo; p.447: USGS

Chapter 15

Opening Photo Johnny Adolphson/Shutterstock; p.451: Francisco Blanco/123RF; p.453: Darrel Hess; p.453: Pearson Education; p.453: Tom L. McKnight; p.454: Jason Friend/Loop Images/Alamy Stock Photo; p.454: Darrel Hess; p.455: Michael Runkel/SuperStock; p.456: Darrel Hess; p.456: Darrel Hess; p.457: Tetra Images/AGE Fotostock; p.457: Charlie Ott/Science Source; p.458: Martin Bond/Science Source; p.459: Richard Maschmeyer/Alamy Stock Photo; p.461: Darrel Hess; p.461: Hennette Olsboe Foreyen/NPS; p.462: Bruno SOURZAC/AFP/Getty Images; p.463: Bruno SOURZAC/AFP/Getty Images; p.464: Larry Mayer/The Billings Gazette/AP Photo; p.464: USGS; p.465: Michael Collier; p.466: ZUMA Press Inc/Alamy Stock Photo; p.468: Pearson Education

Chapter 16

Opening Photo Michael Weber/imageBROKER/Alamy Stock Photo; p.474: Imaginechina Limited/Alamy Stock Photo; p.477: Danita Delimont/Alamy Stock Photo; p.477: Pat & Chuck Blackley/Alamy Stock Photo; p.477: Michael Collier; p.477: USGS; p.478: Leon Werdinger/Alamy Stock Photo; p.479: ZUMA Press, Inc./Alamy Stock Photo; p.480: National Park Service; p.481: Science History Images/Alamy Stock Photo; p.481: National Park Service; p.485: National Park Service; p.486: Nathan Blaney/Photographer's Choice RF/Getty Images; p.488: National Park Service; p.489: NASA EOS Earth Observing System; p.491: Jim Havey/Alamy Stock Photo; p.493: US Army Corps of Engineers; p.493: USGS; p.494: Costal Protection and Restoration Authority; p.495: Michael Collier; p.496: Francisco Anzola; p.496: steinphoto/Getty Images; p.496: Erik Moller; p.497: Tom L. McKnight; p.501: NOAA

Chapter 17

Opening Photo Corbin17/Alamy Stock Photo; p.507: Peter Jones/NPS photo; p.508: Ted Kinsman/Science Source; p.509: David Robertson/Alamy Stock Photo; p.510: Michael Collier; p.510: Michael Collier; p.512: GRANT ROONEY PREMIUM/Alamy Stock Photo; p.513: John Raoux/AP Photo; p.514: Tom Tracy Photography/Alamy Stock Photo; p.515: Darrel Hess; p.515: Thierry Berrod, Mona Lisa Production/Science Source; p.516: Iñaki Caperochipi/Agefotostock; p.517: Tom L. McKnight; p.518: Tom L. McKnight

Chapter 18

Opening Photo Steve Whiston - Fallen Log Photography/Moment/Getty Images; p.523: Darrel Hess; p.524: Darrel Hess; p.525: Michael Collier; p.525: National Park Service; p.526: Sergio Pitamitz/Robert Harding World Imagery; p.526: Tom L. McKnight; p.528: National Centers for Environmental Information; p.529: Michael Collier; p.529: National Park Service; p.530: Darrel Hess; p.530: Robert and Jean Pollock/Science Source; p.530: rickyd/Shutterstock; p.532: Frantisek Staud/Alamy Stock Photo; p.533: Tom L. McKnight; p.533: Tom L. McKnight; p.533: Hauke Dressler/Image Professionals GmbH/Alamy Stock Photo; p.534: Keren Su/China Span/Alamy Stock Photo; p.535: Darrel Hess; p.535: National Park Service; p.537: Michael Collier; p.538: Darrel Hess; p.538: A Hartl/Blickwinkel/Age Fotostock; p.539: Dennis Tasa; p.540: Tom L. McKnight; p.541: National Park Service; p.541: National Park Service; p.543: The National Drought Mitigation Center

Chapter 19

Opening Photo National Park Service; p.546: Michael Collier; p.551: NASA; p.553: NASA; p.553: NASA; p.553: Mouginot, J., E. Rignot, and B. Scheuchl. 2019. Continent-wide, interferometric SAR phase-mapping of Antarctic ice velocity, Geophysical Research Letters. 46. 9710-9718. https://doi.org/10.1029/2019GL083826; p.555: Darrel Hess; p.556: Toño Labra/Agefotostock; p.556: Robbie Shone/Alamy Stock Photo; p.557: GeoStills/Alamy Stock Photo; p.557: Michael Collier; p.558: Darrel Hess; p.558: NASA; p.559: Michael Collier; p.561: Grambo Photography/All Canada Photos/ SuperStock; p.561: Paul Heinrich/Alamy Stock Photo; p.562: Gary A Nelson/Dembinsky Photo Associates/Alamy Stock Photo; p.563: National Park Service; p.563: Tim Rains/NPS photo; p.563: USGS; p.563: USGS; p.563: USGS; p.564: Darrel Hess; p.564: Michael Collier; p.565: Peter Mather/Design Pics Inc/Alamy Stock Photo; p.565: Dan Oldenburg/Alamy Stock Photo; p.565: Tom L. McKnight; p.566: National Park Service; p.567: Eric Murphy/Alamy Stock Photo; p.567: NielsVK/Alamy Stock Photo; p.568: blickwinkel/Baesemann/Alamy Stock Photo; p.569: John Schwieder/Alamy Stock Photo; p.569: Darrel Hess; p.570: Steven J. Kazlowski/Alamy Stock Photo; p.570: Michael Collier

Chapter 20

Opening Photo Harry Collin/Alamy Stock Photo; p.576: Craig Ellenwood/Alamy Stock Photo; p.577: Craig Sterken/Alamy Stock Photo; p.578: Derek Croucher/Alamy Stock Photo; p.579: Darrel Hess; p.579: Tom L. McKnight; p.579: Tom L. McKnight; p.581: Choo Youn Kong/AFP/Getty Images; p.581: Kyodo News/Getty Images; p.582: Tom L. McKnight; p.583: Hemis/Alamy Stock Photo; p.584: Travel Pix/Alamy Stock Photo; p.585: NASA; p.586: Michael Collier; p.588: David Wall/Alamy Stock Photo; p.588: Michael Collier; p.589: USGS; p.589: National Park Service; p.589: Erika Rocks/NPS photo; p.590: Karl Johaentges/Image Professionals GmbH/Alamy Stock Photo; p.591: NASA; p.592: dbimages/Alamy Stock Photo; p.592: Michael Collier; p.592: Michael Collier; p.593: Russ Heinl/All Canada Photos/SuperStock; p.594: Darrel Hess; p.595: Gonzalo Azumendi/Age Fotostock; p.595: Darrel Hess; p.596: Kristine DeLong/Pearson Education Inc.; p.596: THE OCEAN AGENCY/XL CATLIN SEAVIEW SURVEY; p.597: David Wall/Alamy Stock Photo; p.597: A & J Visage/Alamy Stock Photo; p.599: NOAA

Cover Page Eleanor Scriven/Getty Images

Text Credits

Chapter 1

p.2: Organic Act of 1916, NPS, U.S. Department of Interior; p.23: After Robert J. List, Smithsonian Meteorological Tables, 6th rev.ed Washington, D.C.: Smithsonian Institution, 1963, Table 171.

Chapter 3

p.169: U.S. Global Change Research Program

Chapter 5

p.138: http://research.jisao.washington.edu/pdo/, University of Washington

Chapter 7

p.199: Climate Change 2013, The Physical Science Basis, IPCC; p.206: NOAA; p.205: National Weather Service

Chapter 8

p.247: Synthesis Report Climate Change 2014, Synthesis Report, Fifth Assessment Report, Intergovernmental Panel on Climate Change, 2020.; p.247: American Geophysical Union, AGU Position Statement: Human-Induced Climate Change Requires Urgent Action, adopted by AGU December 2003, revised and reaffirmed August 2013 and November 2019.; p.248: www.globalchange.gov/climate-change; p.248: Climate Change 2013, The Physical Science Basis, IPCC; p.248: Flannery, Tim, The Weather Makers: How Man Is Changing the Climate and What It Means for Life on Earth (New York: Atlantic Monthly Press, 2005), p. 246.

Chapter 9

p.262: NASA (https://salinity.oceansciences.org/smap-salinity.htm); p.269: USGS; p.279: NASA; p.280: NCA 2018/NOAA

Chapter 11

p.343: Climate Change 2013, Impacts, Adaptation, and Vulnerability, IPCC; p.342: Food and Agriculture Organization of the United Nations, 2020, Global Forest Resources Assessment 2020 – Key findings. Rome, http://www.fao.org/documents/card/en/c/ca9825en. Reproduced with permission.

Chapter 18

p.528: Cited in Hulme, M., & Kelly, M. (1993). Exploring the links between desertification and climate change. Environment: Science and Policy for Sustainable Development, 35(6), 4-45. P.528 Global Drought Information System, https://www.drought.gov/gdm/current-conditions

Chapter 19

p.547: NASA; p.553: Mouginot, J., E. Rignot, and B. Scheuchl. 2019. Continent-wide, interferometric SAR phase-mapping of Antarctic ice velocity, Geophysical Research Letters. 46. 9710-9718. https://doi.org/10.1029/2019GL083826

Index

Page numbers with f indicate figures or photos, t indicate tables, and A indicate appendices at back of book.

A

D

H

I

N

O

P

T

U

V

World – Political

Central America and the Caribbean